Proceedings of the 20th International Symposium on Advancement of Construction Management and Real Estate

Yuzhe Wu · Sheng Zheng · Jiaojiao Luo
Wei Wang · Zhibin Mo · Liping Shan
Editors

Proceedings of the 20th International Symposium on Advancement of Construction Management and Real Estate

Volume I

Springer

Editors
Yuzhe Wu
Department of Land Management
Zhejiang University
Hangzhou, Zhejiang
China

Wei Wang
Department of Land Management
Zhejiang University
Hangzhou, Zhejiang
China

Sheng Zheng
Department of Land Management
Zhejiang University
Hangzhou, Zhejiang
China

Zhibin Mo
Department of Land Management
Zhejiang University
Hangzhou, Zhejiang
China

Jiaojiao Luo
Department of Land Management
Zhejiang University
Hangzhou, Zhejiang
China

Liping Shan
Department of Land Management
Zhejiang University
Hangzhou, Zhejiang
China

ISBN 978-981-10-0854-2 ISBN 978-981-10-0855-9 (eBook)
DOI 10.1007/978-981-10-0855-9

Library of Congress Control Number: 2016935985

Printed on acid-free paper

This Springer imprint is published by Springer Nature
The registered company is Springer Science+Business Media Singapore Pte Ltd.

Acknowledgements

The symposium gratefully acknowledges the support of organizations below:

Conference Organizer

The Chinese Research Institute of Construction Management (CRIOCM),
Zhejiang University

Conference Executive Organizer

School of Public Affairs, Zhejiang University
College of Civil Engineering and Architecture, Zhejiang University
Land Academy for National Development, Zhejiang University
Institute of Land Science and Real Estate, Zhejiang University
Institute of Construction Management, Zhejiang University
Institute of Urban and Rural Planning Theories and Technologies,
Zhejiang University

Conference Co-organizers

Enjoy-town Investment Management (Hangzhou) Co., Ltd.
Department of Real Estate and Construction, The University of Hong Kong
Department of Building and Real Estate, The Hong Kong Polytechnic University
International Research Center for Sustainable Built Environment,
Chongqing University
School of Urban-rural Planning and Management, Zhejiang University of Finance
and Economics

Academic Committee of Civil Engineering Management in Zhejiang Province
Journal of Urban Management
International Journal of Construction Management

Advisory Committee (In Alphabetical Order by Last Name)

Organizing Committee

Chairman

Co-chairman

Executive Chairman

Secretary-General

Executive Secretary

Members (In Alphabetical Order by Last Name)

Prof. Haoying Han, Zhejiang University
Dr. Lifeng Han, Macau University of Science and Technology
Dr. Jane J.L. Hao, Ryerson University, Canada
Dr. Dezhi Li, Southeast University
Prof. Xinhai Lu, Huazhong University of Science and Technology
Dr. Wilson W.S. Lu, The University of Hong Kong
Dr. Yi Peng, Zhejiang University of Finance and Economics
Dr. Chengshuang Sun, Harbin Institute of Technology
Dr. Vivian W.Y. Tam, University of Western Sydney, Australia
Prof. Hui Wang, Zhejiang University
Prof. Zhigang Wu, Nanjing Normal University
Dr. Bo Xia, Queensland University of Technology, Australia
Dr. Michael C.Y. Yam, The Hong Kong Polytechnic University
Prof. Kunhui Ye, Chongqing University
Dr. Mingxuan Yu, Renmin University of China
Dr. Hongping Yuan, Southwest Jiaotong University
Prof. Wenze Yue, Zhejiang University
Prof. Hong Zhang, Tsinghua University
Dr. Xiaoling Zhang, City University of Hong Kong
Prof. Zhenyu Zhao, North China Electric Power University
Prof. Yimin Zhu, Louisiana State University, USA
Dr. Jian Zuo, University of South Australia, Australia

International Academic Committee

Prof. S.P. Low, National University of Singapore, Singapore
Prof. Ping Lv, Renmin University of China
Prof. Mona Shah, National Institute of Construction M&R, India
Prof. Geoffrey Shen, The Hong Kong Polytechnic University
Prof. Jonathan Shi, Louisiana State University, USA
Prof. Qian Shi, Tongji University
Prof. Dalu Tan, Sichuan University
Prof. Jiayuan Wang, Shenzhen University
Prof. Shouqing Wang, Tsinghua University
Prof. Xueqing Wang, Tianjin University
Prof. Xiaolong Xue, Harbin Institute of Technology
Prof. Donglang Yang, Xi'an Jiaotong University
Prof. Saixing Zeng, Shanghai Jiao Tong University
Prof. Qing Zhu, Southwest Jiaotong University
Prof. George Zillante, The University of Adelaide, Australia
Prof. Patrick X.W. Zou, Swinburne University of Technology, Australia

Conference Secretariat

Dr. Sheng Zheng, Zhejiang University
Ms. Lu Song, Zhejiang University
Ms. Jiaojiao Luo, Zhejiang University
Ms. Liping Shan, Zhejiang University
Ms. Wei Wang, Zhejiang University
Mr. Zhibin Mo, Zhejiang University

Contents

Chapter 1
Granger Causality Test of Science & Technology Innovation Expense and New Urbanization—A Case Study of Hangzhou City

Yuezhen Lyu, Yuzhe Wu and Ye Lin

Abstract Based on the data of Hangzhou in 2005–2013, a comprehensive evaluation system of new urbanization from the demographic aspect, economic aspect and social aspect is built to measure the level of new urbanization development. Meanwhile, the dynamic relationship of science and technology innovation expense and the level of new urbanization is tested by co-integration test and Granger causality test. Results show that the comprehensive level of the new urbanization of Hangzhou is improved continuously from 2005 to 2013. There is a cointegrated relation for a long equilibrium between the S&T innovation expense and the level of new urbanization development. The indicators of S&T innovation expense, which includes the R&D expenditure and the number of R&D activities personnel, have a Granger impact on the level of new urbanization development. And the number of R&D activities personnel has more significant impact on the level of new urbanization development.

Keywords Science and technology innovation expense · New urbanization · Co-integration test · Granger causality test · Hangzhou

1.1 Introduction

Urbanization is a transformation process from tradition to modern and from countryside to city, aiming to shift to the people-oriented and sustainable development way [1]. Report at 18th Party put forward clearly to implement the innovation-driven development strategy, and stressed to focus on the promotion the

Y. Lyu · Y. Lin
Hangzhou Science and Technology Information Institute, Hangzhou 310001, China
e-mail: lyz@hznet.com.cn

Y. Wu (✉)
Department of Land Management, Zhejiang University, Hangzhou 310001, China
e-mail: wuyuzhe@zju.edu.cn

Y. Wu et al. (eds.), *Proceedings of the 20th International Symposium on Advancement of Construction Management and Real Estate*,
DOI 10.1007/978-981-10-0855-9_1

new urbanization, make an effort to solve the major structural problems which constrained economic sustainable and healthy development.

Existed research shows that S&T innovation is a powerful force to promote the new urbanization development [2]. Shanghai, Wuhan, Nanjing, Hangzhou and other cities have a try on information infrastructure construction, information technology application, wisdom medical services, etc. to promote the construction of innovation city, wisdom city and knowledge city.

Under the new situation of the speed-up development of new urbanization in China, the driven effort of Science and technology innovation to the development of new urbanization will be more prominent. By gathering elements to promote the division of labor, reduce the cost of information exchange, etc., the process of urbanization accelerates the S&T innovation of city, and becomes to be a key factor in promoting the S&T innovation [3]. So, S&T innovation have interacted with the urbanization, and scholars at home and abroad have some studies on the relationship between them. Such as, William Arthur Lewis proposed Dual Sector model of agriculture and industry, and considered the birth of the industrial sector caused by S&T innovation would accept all surplus labor, to promote coordinated development. Hollis B. Chenery also proposed there was an interaction between urbanization and industrialization. With the balanced development of the national economy, urbanization is the inevitable result of the interaction of various manifestations. S&T innovations have promoted the transfer of surplus rural labor to cities and towns. K.J. Button thought there was a necessary connection among the urbanization, industrialization and S&T innovation. He noted industrialization and the progress of S&T were the basis of the urbanization, and played an important role in promoting the development of urbanization. Joseph Eugene Stiglitz, economist of American, thought the important factors which affected the process of world and changed the face of world in the 21st century were the development of high-tech industry and the process of social urbanization in the United States. In the domestic, Hu and Wang [4] proposed the concept of coordinated development of industrialization, urbanization and S&T innovation, and from the angle of industrialization and urbanization to study the value of S&T innovation. Zhen [5] argued that S&T innovation played an important role in the process of urbanization, and there was a positive relationship between them. Gan [1] studied the collaborative relationship between S&T innovation and urbanization, and thought only the depth of integration of S&T innovation and urbanization could improve the level of urbanization. Zheng [6] had made an empirical analysis on the correlation between the S&T innovation and the new urbanization in the example of Jiangsu province.

Overall, there are more theoretical literatures and qualitative analyses on the study of the relationship between the S&T innovation and the development of new urbanization, and relatively less quantitative researches. Moreover, most studies are based on the assumption of stable data. However, with the rapid development of new urbanization in China, most of the time series data have showed non-stationary characteristics.

The method of Granger causality test is an effective way to study the relationship between two economic variables. It has be used widely in the studies on the relationship between economic variables [7]. Therefore, based on Granger causality test of non-stationary variables, a case study of S&T innovation and the development of new urbanization have be conducted by Granger causality test in the example of Hangzhou city. The S&T innovation expense is inseparable from S&T innovation, and it is an important material basis for S&T innovation. In order to quantify the impact of S&T innovation on the new urbanization, the paper, taking the example of Hangzhou city, has studied the relationship between S&T innovation and the new urbanization from the perspective of S&T innovation expense through the co-integration test and Granger causality analysis, to seek the sensitivity factor in S&T innovation expense which has an influence on the development of new urbanization.

1.2 Research Methods and Data Sources

1.2.1 Research Methods

Using the method proposed by Clive W.J. Granger (1969), called Granger causality test, the data of S&T innovation expense and the level of new urbanization development has been tested in stata12. Granger causality test was used to analyze the relationship between two variables which has priority in time, to explain the probability of one variable being the reason of another variable in statistics. If there is a significant Granger causality between the two stationary series, there must be a causal relationship between them in economic sense. On the contrary, it may not hold. When stationary series exist truly significant causal relationship, it can be tested out clearly by Granger causality method. On the contrary, even if there is a causal relationship between stationary series but not significantly, it may not be tested out by Granger causality method. In this paper, the test result showed that S&T innovation expense and the new urbanization existed causation by Granger causality test.

1.2.2 Data Sources

In this paper, all data are derived from the "Hangzhou Statistical Yearbook" and the reports of "Hangzhou Innovation Index" in each year.

Indicators of S&T Innovation Expense

The S&T innovation expense includes R&D expenditure and S&T personnel investment. Taking into account the availability of data, this paper selects the data of Hangzhou from 2005 to 2013. The percentage of R&D expenditure in GDP (%)

indicates the strength of R&D expenditure, and the number of R&D activities personnel (10,000 years) indicates the S&T personnel investment.

Evaluation System of Urbanization

The urbanization of population is usually used to measure the level of urbanization, namely the percentage of non-agricultural population in total population, which mainly reflects the distribution of population in urban and rural space, can't fully reflect the level of new urbanization. Therefore, with the existing literature [3], we build the evaluation system of new urbanization development from the demographic, economic and social aspects, then determine the weight of each indicator combining the Delphi method and entropy method, and the results show in Table 1.1.

Eventually we use integrated weight grade method to obtain the composite index of new urbanization, which is most likely a comprehensive, scientific evaluation of the level of new urbanization development.

1.2.3 Co-integration Test

When performing time series analysis, time series are required to be stable, and the original data must be difference processed normally to become stable,which will

Table 1.1 The evaluating system and weight of new urbanization development

Rule layer	Weight	Indicator layer	Weight
Population urbanization	0.1541	Non-agricultural population accounts for the proportion of total population %	0.3240
		The urban population of the city accounts for the proportion of total population %	0.1010
		The proportion of tertiary industry employment %	0.2756
		The registered urban unemployment rate %	0.2994
Economy urbanization	0.6913	Per capita GDP (Yuan)	0.2021
		Tertiary industry output value accounts for the proportion of GDP %	0.2084
		Total fixed asset investment (billion yuan)	0.3651
		Total retail sales of consumer goods(billion yuan)	0.2244
Society urbanization	0.1546	The number of sick beds per ten thousand people (a/ten thousand)	0.1261
		Information application level %	0.2250
		Greening coverage in built up area %	0.2074
		Urban air pollution index	0.2219
		Urban and rural residents medical insurance coverage %	0.2196

lose the long-term information of total quantity. Therefore, the variables must be verified by co-integration test to know whether there is a long-term equilibrium relationship between them. Its residual formula is as follows.

$$e_t = y_t - a - bx_1 \tag{1.1}$$

The e_t is the residual, and if the residual error is stationary, the variable sequences present co-integration. Co-integration relationship does not exist and vice versa.

1.2.4 Granger Causality Analysis

For the two time series variables, if under the conditions of containing the past information of variables X and Y, the prediction effect of variable Y is better than the prediction of variable Y that made just with the past information of variable Y, namely the variable X is helpfull to explain the change of variable Y in the future. Variable X is considered the Granger causality of variable Y [8]. The p-order lag equation of Yt is as follows.

$$Y_t = b + \sum_{i=1}^{r} a_i X_{t-i} + \sum_{j=1}^{r} \beta_i Y_{t-i} + e_t \tag{1.2}$$

$$X_t = b + \sum_{i=1}^{p} a_i X_{t-i} + \sum_{j=1}^{p} \beta_i Y_{t-i} + e_t \tag{1.3}$$

In the equation, b is a constant term. α_i, β_i are the regression coefficients. i is the lag order. p is the maximum order lag. e_t is the residual. The first formula (1.2) is used to test whether variable X_t is the Granger causality of variable Y_t. The second formula (1.3) is used to test whether variable Y_t is the Granger causality of variable X_t.

1.3 Empirical Analysis: A Case Study of Hangzhou

The level of new urbanization development is set to U. The percentage of R&D expenditure in GDP is set to R_1, the number of R&D activities personnel is set to R_2. According to the data of Hangzhou from 2005 to 2013, the comprehensive index of new urbanization development is calculated through the evaluation system of new urbanization development level. The results show in Table 1.2, which reflect the level of new urbanization development in Hangzhou increasing

Table 1.2 The level of new urbanization and the S&T innovation expense from 2005 to 2013 in Hangzhou

Year	Level of new urbanization development (U)	Percentage of R&D expenditure in GDP (R_1)	Number of R&D activities personnel (R_2)
2005	0.090	2.29	2.32
2006	0.171	2.42	2.81
2007	0.233	2.56	3.08
2008	0.308	2.56	4.15
2009	0.396	2.7	6.17
2010	0.520	2.8	6.71
2011	0.616	2.88	7.41
2012	0.747	2.92	7.83
2013	0.904	2.98	8.18

continuously, with an average annual growth rate of 33.46 %. We can know that the new urbanization have developed rapidly from 2005 to 2013.

1.3.1 Stationary Test (ADF Root Test)

Before performing the co-integration test of the level of new urbanization development (U), the percentage of R&D expenditure in GDP (R1) and the number of R&D activities personnel (R2), we carry on ADF root test of their stationaries to determine whether these time series stationary. The results of the ADF root test are showed in Table 1.3. From the table, we can know that the variable U, the variable R1 and the variable R2 are non-stationary at the 10 % significance level. The first-order difference sequence of the variable R1 is stationary time series at the 5 % significance level. The second-order difference sequence of the variable U is stationary time series at the 5 % significance level, and the second-order difference sequence of the variable R2 is stationary time series at the 10 % significance level.

Table 1.3 Stationary test results of time sequence of indicators data selected

Variable	ADF test statistic t	1 % critical value	5 % critical value	10 % critical value
U	0.693	−4.380	−3.600	−3.240
R_1	−2.272	−4.380	−3.600	−3.240
R_2	−1.363	−4.380	−3.600	−3.240
ΔU	−3.237	−4.380	−3.600	−3.240
ΔR_1	−3.067	−3.750	−3.000	−2.630
ΔR_2	−1.931	−3.750	−3.000	−2.630
$\Delta^2 U$	−3.508	−3.750	−3.000	−2.630
$\Delta^2 R_2$	−2.680	−3.750	−3.000	−2.630

Thus, time series variables of the percentage of R&D expenditure in GDP are integrated of order one, and the time series variables of the level of new urbanization development and the number of R&D activities personnel are both integrated of order two.

1.3.2 Co-integration Test (EG-ADF Test)

The first-order difference sequence of variable U is set to be dependent variable. The Variable R1 and the first-order difference sequence of variable R2 are set to be independent variables. Using ordinary least squares to estimate the residual sequence, we carry on the residual sequence by ADF test. If the residuals have unilaterally root process, there is no equilibrium relationship between variables, and vice versa for stationary series, namely there is a equilibrium relationship between variables. The results of residuals examined by ADF test show in Table 1.4. As showed in the table, in the 1 % threshold level, it rejects the original hypothesis of e as the unilateral roots, namely e is a stationary process. So there is a co-integration relationship among the three time sequence variables of the level of new urbanization development, the percentage of R&D expenditure in GDP and the number of R&D activities personnel.

1.3.3 Granger Causality Test

Because the time sequence variables of the percentage of R&D expenditure in GDP are integrated of order one, and the time sequence variables of the level of new urbanization development and the number of R&D activities personnel are both integrated of order two. So we can use Granger causality test to examine their relationship. The results of the test among them are showed in Tables 1.5, 1.6 and 1.7.

Table 1.4 Co-integration test results of non-stationary time sequence

Variable	T-statistic	1 % critical value	5 % critical value	10 % critical value
e	−3.159	−2.660	−1.950	−1.600

Table 1.5 Granger causality test results of U and R1

Lag period	Null hypothesis	F-test value	Significance level
1	R_1 does not Granger cause U	0.09	0.81
	U does not Granger cause R_1	0.13	0.75
2	R_1 does not Granger cause U	0.81	0.53
	U does not Granger cause R_1	1.27	0.38

Table 1.6 Granger causality test results of U and R2

Lag period	Null hypothesis	F-test value	Significance level
1	R_2 does not Granger cause U	2.47	0.26
	U does not Granger cause R_2	0.84	0.46
2	R_2 does not Granger cause U	5.52	0.14
	U does not Granger cause R_2	2.06	0.29

Table 1.7 Granger test results of R1 and R2

Lag period	Null hypothesis	F-test value	Significance level
1	R_2 does not Granger cause R_1	1.13	0.40
	R_1 does not Granger cause R_2	3.94	0.14
2	R_2 does not Granger cause R_1	0.02	0.89
	R_1 does not Granger cause R_2	3.27	0.17

Table 1.5 shows the results of Granger causality test of U and R1. In the lag period 1 and period 2, the probabilities of accepting the null hypothesis that variable R1 is not the Granger cause of variable U are 0.81 and 0.53 separately. In the lag period 2, the probability of rejecting the null hypothesis that variable U is not the Granger cause of variable R1 is 0.62, namely under the 62 % confidence level, variable U is the Granger cause of variable R1. It indicates that increasing the level of new urbanization development will bring R&D expense changed.

The development of new urbanization has promoted the S&T innovation. It is the center of new urbanization to enhance the connotation of urban culture, public services in order to become being a high quality and suitable living place. So it is urgent need to promote the S&T innovation, and the financial input will increase correspondingly, which is consistent with the test results.

Table 1.6 shows the results of Granger causality test of U and R2. In the lag period 1 and period 2, variable R2 is the Granger cause of variable U under the 74 % and 86 % confidence level separately. In the lag period 1 and period 2, variable U is the Granger cause of variable R2 under the 54 % and 71 % confidence level separately. It shows that the number of R&D activities personnel has a more direct impact on the new urbanization, while the new urbanization has a lag effect, and also has an impact on the number of R&D activities personnel.

Comparing the test results of Table 1.5 and Table 1.6, the number of R&D activities personnel has more significant impact on the new urbanization than the percentage of R&D expenditure in GDP. The reason may be that the R&D activities personnel investment is a long-term elements of the new urbanization. High-tech talents are initiative and high value-added resources, who undertake the task of the S&T innovation in the process of new urbanization. Therefore, for the policy makers of new urbanization development, it is necessary to increase the investment of S&T innovation personnel to enhance the new urbanization development. Specifically, it should play more attention on attracting and retaining talent, improving the mechanism of training and incentive for them in order to guide them into the development of new urbanization.

Table 1.7 shows the results of Granger causality test of R1 and R2. In the lag period 1 and period 2, the probabilities of accepting the null hypothesis that variable R2 is not the Granger cause of variable R1 are 0.4 and 0.89 separately. While in the lag period 1 and period 2, variable R1 is the Granger cause of variable R2 under the 86 % and 83 % confidence level separately. It shows that the increasing of R&D expenditure has brought the significant changes in the number of R&D activities personnel, while the change of the number of R&D activities personnel has little effect on the R&D expenditure.

1.4 Conclusion and Discussion

1.4.1 Conclusion

Taking Hangzhou for example, we build the evaluation system of new urbanization development, and use the co-integration test and Granger causality test to analyze the index of new urbanization and the S&T innovation expense dynamically. Then we give the following conclusions.

1. The results of new urbanization development of Hangzhou: from 2005 to 2013, the comprehensive index of new urbanization of Hangzhou is growing, with the average annual growth rate of 33.46 %. It indicates that the level of new urbanization of Hangzhou has improved constantly with the fast development speed. And the population urbanization, economic urbanization and social urbanization are the three main factors of promoting the development of new urbanization.
2. The results of the co-integration test of urbanization and S&T innovation investment: the stationary test result shows that the time series of the index of new urbanization, the percentage of R&D expenditure in GDP and the number of R&D activities personnel have unilaterally root process, namely they are non-stationary sequences. The first-order difference sequence of the percentage of R&D expenditure in GDP is stationary sequence. And the second-order difference sequences of the level of new urbanization development and the number of R&D activities personnel are stationary time series. The co-integration test result shows that there is a long-run equilibrium relationship among the three time sequences variables of the level of new urbanization development, the percentage of R&D expenditure in GDP and the number of R&D activities personnel. That is, the development of urbanization and the S&T innovation expense are highly correlated statistically.
3. The results of Granger causality test of the new urbanization and S&T innovation expense: the development of new urbanization is the Granger cause of the percentage of R&D expenditure in GDP and the Number of R&D activities

personnel. While the number of R&D activities personnel is the Granger cause of new urbanization, and the percentage of R&D expenditure in GDP is not the Granger cause of new urbanization. The percentage of R&D expenditure in GDP and the Number of R&D activities personnel are the single relationship of Granger causality.

1.4.2 Discussion

In this paper, we conduct the co-integration test and Granger causality test on the relationship between S&T innovation expense and the new urbanization development. Then we have the results that there is a long-term equilibrium relationship between the S&T innovation expense and the new urbanization development, and the new urbanization development is the Granger cause of the S&T innovation expense, while the number of R&D activities personnel has more significant effect on the new urbanization development. This conclusions are consistent with the case of Hangzhou city. However, in the process of quantitative analysis, the results affected by the sample length and lag period, so the results have yet to be further explored whether it is fit for other regions. And we will study on the relationship and quantitative effect between the other indicators of S&T innovation and new urbanization development on the further.

References

1. Gan DL (2014) The research about collaborative development strategies between technological innovation and the new urbanization. Sci Technol Prog Policy 31(6):42–45
2. Meng ZQ (2013) Study on the development path of new urbanization under innovation-driven strategy. Theory Study 7:27–29
3. Yuan B, Liu FC (2014) Collaborative development between technological innovation capability and the level of urbanization. J Dalian Univ Technol (Soc Sci) 35(2):56–61
4. Hu ZY, Wang R (2012) Research of technological innovation and joint development. Sci Manage Res 30(6):5–8
5. Zhen XF (2013) Process of science and technology innovation promoting new rural urbanization. J Northwest A&F Univ (Soc Sci Ed) 5:33–37
6. Zheng LX (2014) Correlation between science & technological innovation investment and new urbanization development. Rural Econ Sci Technol 25(12):176–178
7. Okunev J, Wilson P, Zurbruegg R (2000) The causal relationship between real estate and stock markets. J Real Estate Finance Econ 21(3):251–261
8. Liu YJ (2008) Dynamic econometric analysis of the relationship between urbanization and ecological environment in Jiangxi Province. Resour Sci 30(6):829–836

Chapter 2
Study on Functional Mechanism and Effectiveness of Land Ticket System in the Process of New-Type Urbanization

Jiangtao Fang

Abstract Most districts of China have already taken measurements in system innovation and exploration upon rural land management system ever since the middle and later periods of 1990s, which caused wide public concern. Since the twenty-first century, with the support and recognition of the authority, local governments have been constantly deepening the reformation on rural land management system, establishing long-term mechanism to promote industry nurturing agriculture and cities supporting countryside as long-term guiding principles, which are positive and effective. Among them, the land ticket system reformation in Chongqing is most worthy of focus. Land tickets are the quotas acreage of land, which were reclaimed from rural construction land, such as farmers' homesteads, available for urban construction according to the urban and rural construction planning. Since the land ticket system has been put into effect from 2008 in Chongqing, the layout of urban and rural construction land has been optimized, the relationship between the population urbanization and the land urbanization has been better coordinated, the rural economic has developed fast, and the new-type urbanization construction has been carried out smoothly.

Keywords Land ticket system · New-type urbanization · Functional mechanism · Effectiveness

2.1 Introduction

Most districts of China have already taken measurements in system innovation and exploration upon rural land management system ever since the middle and later periods of 1990s, which caused wide public concern. Since the twenty-first century,

---

J. Fang (✉)
Economics College of Sichuan University, Chengdu, China
e-mail: 676043596@qq.com

Y. Wu et al. (eds.), *Proceedings of the 20th International Symposium on Advancement of Construction Management and Real Estate*,
DOI 10.1007/978-981-10-0855-9_2

with the support and recognition of the authority, local governments have been constantly deepening the reformation on rural land management system, establishing long-term mechanism to promote industry nurturing agriculture and cities supporting countryside as long-term guiding principles, which are positive and effective. Among them, the land ticket system reformation in Chongqing is most worthy of focus. Land tickets are the quotas acreage of land, which were reclaimed from rural construction land, such as farmers' homesteads, available for urban construction according to the urban and rural construction planning.[1] Since the land ticket system has been put into effect from 2008 in Chongqing, the layout of urban and rural construction land has been optimized, the relationship between the population urbanization and the land urbanization has been better coordinated, the rural economic has developed fast, and the new-type urbanization construction has been carried out smoothly. But how all these happen, how to guide the land ticket system in a healthy mode to promote the new-type urbanization. Obviously, it is a subject worthy of study. The key to answer these questions is to disentangle the functional mechanism and effectiveness of land ticket system in the process of new-type urbanization.

Since the "National Planning on New-type Urbanization (2014–2020)" was issued, new-type urbanization became a research focus. Wu [1] discusses the path of New-type urbanization of Chongqing. Zhang [2] argues that to ensure the China's new-type urbanization sustainable development, we must implement a national urbanization strategy top-level design, construct lead to metropolitan, medium and small cities and towns simultaneously, having a poor urban area development system ordering pattern. Guo Jun (2014) [3] argues that the urbanization of the kernel is the human urbanization, which requires the government as a positive, while focusing on the allocation of resources to play a decisive role in the market, so that farmers into the city to work, live, science promote New-type urbanization. However, few studies have been done to explore the relationship between the land ticket system and new-type urbanization. As a participant in the reform of the land ticket system, the author tried to deploy the study about the relationship. Firstly, this study sets up a theory framework of the land ticket system influence upon the new-type urbanization; then, discusses the functional mechanism of the land ticket system impacting on the new-type urbanization; after that, evaluates the effectiveness of the ticket system; at length, puts forward some policy adjustment measures and suggestions in the land ticket system innovation, expecting to play a better role in accelerating the new-type urbanization in China.

[1]2015 "Chongqing Land Ticket Regulations" (Working Paper).

2.2 The Functional Mechanism of the Land Ticket System Affecting New-Type Urbanization

The concept of new-type urbanization has been put forward for more than 10 years, but the clear and unified definition is still been lack. It has been approved by many scholars that the key of new-type urbanization is the coordination development of population urbanization and land urbanization, more economical utilization and more optimized layout of urban and rural construction land, and also considering the construction and development of rural economy and society. The land ticket system fits the goal and requirement of the new-type urbanization.

The operation of the land ticket system could be distributed into three stages, which are reclamation of the rural construction land, trading and using of land ticket, income distribution of land ticket. Based on the concept of land ticket, the operation of the land ticket system and the theories of new-type urbanization, the theoretical framework of the land ticket system affecting new-type urbanization has been constructed, as shown in Fig. 2.1. It shows that the impact that the land ticket system influence new-type urbanization could be systematic characterized through a certain approach and a functional role.

Specifically, the approaches that the land ticket system influences new-type urbanization can be summarized as the following four aspects.

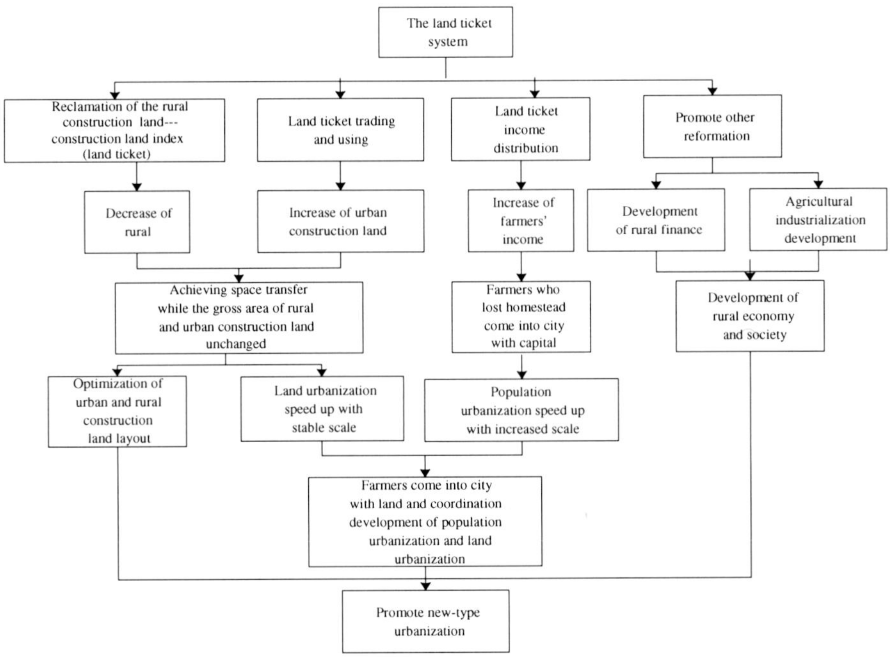

Fig. 2.1 The theoretical framework of the land ticket system affecting new-type urbanization

2.2.1 The Land Ticket System Affecting Land Urbanization

At present, Chinese urban construction land supply is tight, which can't meet the land demand in the process of industrialization and new-type urbanization. In contrast, large area of the rural collective construction land is serious idled and wasted. Under the premise of not increasing the total scale of urban and rural construction land, revitalizing the storage construction land and guiding the rural collective construction land decrease and urban construction land reasonably increase are the target of policy formulation. In this context, the land ticket system has been proposed. The main steps of the land ticket includes rural construction land reclamation and cultivated land formation, public trading for the qualified construction land index (land ticket), land ticket utilization in city's urban and rural planned construction districts. Newly increased commercial construction land must use land ticket according to the regulations of Chongqing. Real estate developers must use the land ticket when they choose land that in the scope of urban construction land planning but hasn't completed the land acquisition. The land ticket will be used to apply for the government to complete the land acquisition procedures and to handle procedures of agricultural conversion. Then the land department will organize the "bid, auction and listing" to present that the increase of urban construction land. In other words, in the target of following urban construction land planning, the accelerated transformation from agricultural land to construction land in urban planning area has been achieved. Thus the land urbanization speeds up.

2.2.2 The Land Ticket System Affecting Population Urbanization

In China, the process of land urbanization is faster than population urbanization as a result of urban-rural binary management structure, including household registration system and land acquisition system. Taking Chongqing as an example, both the land urbanization index and population urbanization index keep in a continuous increasing trend since Chongqing administrated under the Central Government. The growth rate of population urbanization index was obviously slower than that of land urbanization index since 2001, which is not conducive to the coordinated development of human and land. To solve this problem, in August 2010, Chongqing launched the reform of the household registration system for migrant workers to promote the agglomeration of urban population and industries and to speed up the process of urbanization. After the farmers migrate to the city, large area of rural construction land such as homestead is idled and wasted, which brings waste of resources and also affects the farmers' psychology of the migration. While the land ticket system provides the channel for the migration farmers to dispose the rural land property such as homestead and removes their worries and provide much capital for the migration farmers. To a certain extent, the land ticket system promotes the reform of household registration system, which leading the total scale of the urban population increases and population urbanization process accelerates.

2.2.3 The Land Ticket System Affecting the Layout of Urban and Rural Construction Land

According to the land survey in 2012, urban and rural construction land of Chongqing is 7,770,000 acres, among which urban construction land is 2,412,000 acres and rural residential land (village land) is 5,358,000 acres. Based on the overall land use planning of Chongqing approved by state council, the whole city can increase 567,000 acres of urban and rural construction land during 2013–2020, among which urban construction land increases 982,500 acres and rural construction land decreases 415,500 acres. Thus Chongqing continues to promote land management system innovation and develop reclamation of rural construction. Establishing the land ticket system to encourage and guide the increase of the urban construction land while rural construction land decreases. It can be said that the land ticket system explores the space replacement mode of urban and rural construction land of "Increase and decrease, without total amount change". The structural adjustment and layout optimization of urban and rural construction land have been realized.

2.2.4 The Land Ticket System Affecting the Development of Rural Economic Society

Urbanization in China is the direction and result of rural economic society development. The land ticket system affects the rural economic society development from at least three aspects. Firstly, increase farmer's income. The land ticket system specified that 85 % of net income of the reclamation rural collective construction land, such as homestead, belongs to farmers and 15 % belongs to rural economic organization after the application of land ticket. Secondly, promote financial development in rural area. The successful operation of the land ticket system provides a reference for the evaluation of rural housing mortgage loan, which not only improves rural land and housing evaluation criteria, but also makes rural assets have the ability of mortgage, thus financial development in rural area be promoted. Thirdly, promote the development of agricultural industrialization. Land ticket system promotes the process of population urbanization. Both per capita share of rural resources and intensive agricultural production level be improved after the migration of people from rural to urban area, which is benefit for the realization of rural scale management and sustainable utilization of land resources. As a result, the rural economic society development is promoted.

2.3 Summary of the Land Ticket System's Influence on New-Type Urbanization

The land ticket system has been implemented for 6 years, leading Chongqing's urban and rural construction land layout be more optimized, population urbanization and land urbanization in a more coordinated condition, rural economic society develops rapidly, as well as new-type urbanization in a high speed promotion.

2.3.1 The Land Ticket System Promoting Urban and Rural Construction Land Layout More Optimized

The land ticket system breaks the strict timing constraints for land development of national annual land use plan. Under the premise of the fixed total scale of construction land, the operation of the land ticket system can provide urban construction land quota and create development space for Chongqing's new-type urbanization. By the end of 2014, the reclamation rural land in Chongqing had accumulated to 166,800 acres, among which 151,900 acres complete land ticket trading and 100,400 acres land tickets has been used, as shown in Fig. 2.2. The urban construction land quota increased by 159,000 acres and the increased scale of urban construction land was 100,400 acres. As a result, urban and rural construction land layout is more optimized and reasonable.

According to the Chongqing's Land Consolidation Potential Analysis (2011–2020), Chongqing's rural land reclamation potential is about 820,000 acres from 2011 to 2020, which can form 770,000 acres construction land quota.

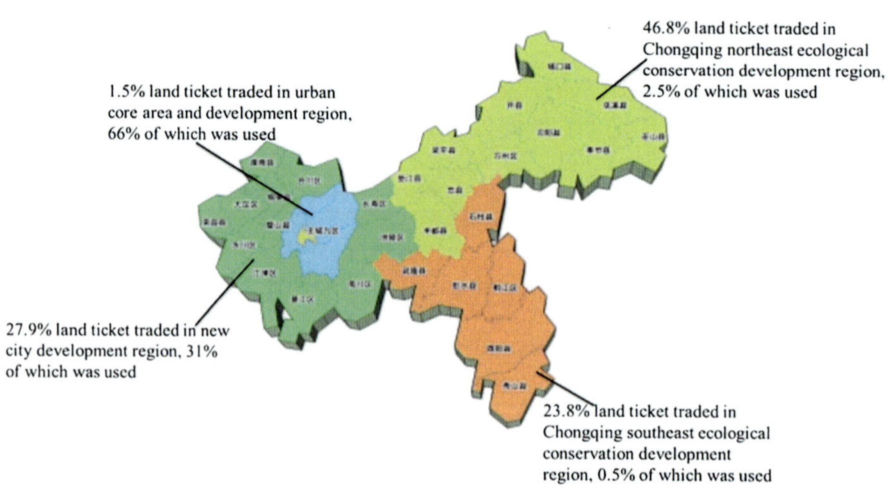

Fig. 2.2 Trading and using of land ticket

Until now, about 100,000 acres land tickets have been used, there is still about 600,000 acres construction land quota can be used (exclude national planning quota). Thus Chongqing's new-type urbanization has land guarantee.

What's more, those who own land ticket can freely chose the land according to their own willingness and then apply to the government for land acquisition. The operation process introduces market willingness into land planning and changes the government-leading mode in the planning implementation, which reduces the situation of land being granted but not being acquired and land being acquired but not being supplied. It makes the implementation timing and sequence of the planning more reasonable and urban development fast and orderly.

In addition, the land ticket system leads rural construction land to decrease orderly through market mechanism and tends to more intensification and marketization. Farmers are going to live in new residential villages and in a more concentrated living condition. The living condition of farmers has been improved in a large degree and the gap between urban and rural public service has been narrowed, which promotes the integrating of urban and rural space.

2.3.2 The Land Ticket System Promoting the Coordination of Population Urbanization and Land Urbanization

In China, the process of land urbanization is faster than population urbanization as a result of urban-rural binary management structure, including household registration system and land acquisition system. Taking Chongqing as an example, both the land urbanization index and population urbanization index keep in a continuous increasing trend since Chongqing administrated under the Central Government. The growth rate of population urbanization index was obviously slower than that of land urbanization index since 2001, which is not conducive to the coordinated development of human and land. To solve this problem, in August 2010, Chongqing launched the reform of the household registration system for migrant workers to promote the agglomeration of urban population and industries and to speed up the process of urbanization. After the farmers migrate to the city, large area of rural construction land such as homestead is idled and wasted, which brings waste of resources and also affects the farmers' psychology of the migration and lagging the process of population urbanization. At this point, rural land disposal policy in the reform of the household registration system has become the key.

To promote the reform in a linkage situation, the land ticket system has been proposed. The land ticket system support farmers obtain cash compensation after they transfer the rural construction land such as homestead, through which to solve the farmer's dilemma for the land that "not will to abandon also not will to use". The land ticket system strengthens the willingness and confidence for farmer to transfer their land and then move to city. It is clearly defined that the household registration transferred farmers have their own rights to decide to reserve, circulate or exit their homestead, among which the farmers who want to exit their homestead and the affiliated facilities could obtain property through the land ticket system. The regulation supply great support to the household registration transferred farmers own the property to participate the society security and to rent or buy houses in the city, which can attract more farmers to transfer their household registration and then live in the city. The land ticket system promotes the reform of the household registration system and achieves that farmers go to city with land during the process of urbanization, which further coordinate the land use between rural and urban area.

In Sep 2013, Chongqing proposed the development strategy that "(1) Industry follows the function orientation; (2) Population follows the industry; (3) Construction land follows the population and industry". The land ticket system is the best practice that leading construction land to follow the population and industry. As mentioned above, among the 151,900 acres trading land ticket in Chongqing, 70 % comes from Chongqing northeast ecological conservation development region and southeast ecological conservation development region. The transferred population of the two regions is 1.935 million and decreased the construction space by 109,400 acres, which is benefit for agricultural space aggregation and ecological function recovery. 97 % of the 100,400 acres used land tickets are in urban core area and development region, which has vigorously promoted the industrial development and population aggregation.

2.3.3 The Land Ticket System Promoting the Economic and Social Development of Rural Area

The land ticket system not only improves the overall welfare of rural land use, but also promotes rural economic and social development, and even leveraged the rural finance development. Firstly, during the process of implementing the land ticket system, the reclamation farmers can not only get 85 % of the land ticket funds (net of costs), but also can continue to cultivate the arable land and get planting income. Meanwhile, reclaimed farmland connects to the surrounding land, which is beneficial to large-scale agricultural production. Chongqing Rural Land Exchange Institute organized to investigate reclaimed farmland utilization of rural construction land in Ma Wu Town, Fuling District, Chongqing City (totally 2716 households in 117 clubs, 18 villages) in October 2014. The results showed that the interests of reclaimed farmland was gratifying, a total of 2382 households chose to cultivate the reclaimed farmland, except 124 households moved to the town or go to the city to earn money, 91.9 % of farmers planted grain crops and 8.1 % planted cash crops. The cash crops have shaped in a large scale and stepped into the trend of industrialization. Secondly, rural collective economic organizations can get 15 % of benefit of Land ticket from the reclaimed homestead. For public welfare land be reclaimed into land ticket, the rural collective economic organizations can get the total benefit of land ticket. Thus the strength of collective economic was significantly improved, also the benefit of the villagers can be guaranteed. Taking Mawu Tow in Fuling District of Chongqing City as an example, in the 18 villages, there were 11 villages whose collective economic income were zero prior to the implementation of the land ticket system, and the remaining 7 villages whose yearly per capital income were lower than 10,000 yuan. The reclamation of construction land brought a total of 61.61 million yuan income for 18 collective economic organizations in Mawu Town, with 3.42 million yuan per village till October 2014. The land ticket system greatly promoted the local economic and social development. Thirdly, the land ticket system leveraged the development of rural finance. The use right of rural land such as rural buildings, homestead and affiliated facility land can't be served as mortgaged goods for a loan from the bank because of restriction of laws, regulations and policies for a long time, which limited the development of rural economy and society. After the implementation of the land ticket system, the unique ticket price formation mechanism provided a visible reference value for rural housing mortgage, which not only reduced the operational risk of financial institutions, but also promoted financial development in countryside. The mortgage finance of Chongqing rural residents housing reached to 13.22 billion yuan by the end of 2014.[2] Meanwhile, Chongqing Rural Commercial Bank and Agricultural Bank of China Chongqing Branch introduced a pledged loan for rural construction land reclamation project to meet the financing needs and secured by the expected return of land ticket. By the end of 2014, Chongqing handled 13.5 billion yuan pledged loan for construction land reclamation projects.[3]

2.4 Conclusions

The land ticket system affecting the new urbanization mainly characterized by means of rural construction land use changing and spatial displacement. The effect mainly realized through four ways, which are land urbanization, population urbanization, urban and rural construction land spatial layout, the rural economy society. Therefore, it is necessary for the land ticket system to continue plays the positive role and then to speed up the process of new urbanization under the premise that advancing the new-type urbanization has become a consensus. However, there are still some problems need to be addressed in the existing the land ticket system: Firstly, lacking of support of the upper law. Secondly, low degree of market. Thirdly, the long-term development of the land ticket system production area is affected.

It requires to deep the reform of the land ticket system, accelerate the step of rural land management system innovation and boost new-type urbanization. Firstly, adapt the mature experience to laws and regulations timely. Land ticket system has made innovation and exploration in the realization form of rural construction land usufructuary right and the share of rural land ownership and the using right, so it is

[2]Source: Chongqing Finance Office.

[3]Source: Chongqing Rural Land Exchange Institute.

necessary to be strengthened and summarized to form relevant law to provide support for the reform of the land ticket system. Secondly, expand the scope of the pilot reasonably. It should be considered to expand the scope of the pilot reform of the Chongqing land ticket system under the law bottleneck which limited the circulation of rural homestead and find the price of rural construction land through the land ticket system in other areas to increase farmers' income. Thirdly, coordinate the promotion of reform. Rural land reform not only involves a wide range, but also it has far-reaching significance, so it is necessary to promote synergy cooperating with other reforms. Therefore, it should pay attention to the common development of the balance of rural construction and urban construction, the rural homestead management, the innovation of land use planning program, and the reform of the household registration system.

References

1. Land ticket transaction: optimal combination of modes within the system of land and industry (2010). Contemp Finance Econ, 05
2. Wu J (2010) Path of Chongqing to promote the new-type urbanization. Southwest University doctoral thesis
3. Zhang H (2013) China's new-type urbanization theory and practice of innovation. Sociolo Stud, 3

Chapter 3
On Urban Land Development Boundary Oriented by the New Type Urbanization Policy: A Case of Yiwu City

Liping Shan and Yuzhe Wu

Abstract The paper summarizes the traditional method to delimit the urban land development boundary which aims to make a good balance of urban sprawl and ecology protection in a "industrial" model, while under the new type urbanization policy, a "urbanization" mode seems to be more appropriate with China's development. In the paper, on the basis of traditional methods to delimit urban land development boundary under the background of the new urbanization policy, we try to explore a delimitation method which is more closer to the "urbanization" mode. We take the downtown of Yiwu for an example and obtain results of the urban land development boundary of Yiwu in the year of 2020 in both of the two modes, one of them is in the "industrial" model and guided with the "sector theory", and the other is in the "urbanization" mode and guided with the "multiple nuclei theory", and then we compare the thoughts and the results. The paper chooses the second method, namely the method which is under the new urbanization background and is close to the "urbanization" mode as the better. At last, we propose the requirements and methods to apply the method, following with some recommendations and opinions.

Keywords Urban land development boundary · The new urbanization policy · Sector theory · Multi-core theory

3.1 Introduction

Since the new century, the urbanization process of China has got into a rapid developing period, result in the proportion of urban population increased from 36.22 % in 2000 to 53.73 % in 2013, with the 17.51 % increase, and the urbanization

L. Shan · Y. Wu (✉)
Department of Land Management, Zhejiang University, Hangzhou, China
e-mail: wuyuzhe@zju.edu.cn

L. Shan
e-mail: shanliping@zju.edu.cn

Y. Wu et al. (eds.), *Proceedings of the 20th International Symposium on Advancement of Construction Management and Real Estate*,
DOI 10.1007/978-981-10-0855-9_3

showed an astonishing average annual growth of 1.35 %.The rapid advance of China's urbanization has played an indispensable role in China's economic development and improvement of people's welfare. At the same time, it should be seriously seen that since the year of 2000, it has a huge change in the statistical caliber of China's urban population with the year of 1999. For example, the definition of the population of permanent residents has been modified from the original "temporarily stay more than one year in the city" to "temporarily stay over 6 months". It means that the number of urban residents population will significantly increase in statistics, with the development of economy, the number of "stay more than 6 months" population has become really bigger and those people has made the bigger contribution to the development of urbanization. However, in comparison of the contribution, the "temporarily stay" population has not been incorporated into the city's social security system with the difficulties in housing burden, children education, health care, pension, etc.

It is in this context that China issued the national new urbanization planning (2014–2020) at the year of 2014. The core of the planning was the urbanization for population and it emphasised on orderly promoting the urbanization of the agricultural transfer population, people-oriented and fair share; Demanded constantly enhance the quality connotation and level of urbanization development, with the center to enhance the connotation of urban culture, public service, to make the cities and towns become high quality living place; Emphasis on optimal layout and promote the coordinated development of small and medium cities...County in China is a very important administrative level, "county governed well, so will be the country", the county administrative units take an significant role in the national power structure and the development strategy, not only have relatively independent administrative, judicial and financial rights, but also are capable to the exercise the rights of land transfer. Therefore, it is of great practical significance to study how new urbanization planning implemented in county.

China is a country with large population and lack land resource, arable land should be protected under the background of food security, however, with the urbanization process, arable land is always occupied. Therefore, combining the reality that China's urban construction land scale extents continuously, to border city construction land expansion boundary is also necessary for limiting urban land. And on the other hand, the urban construction land radio between life and production use is not harmonious, the increase speed of living land is far less than which of the production land, which affects the improvement of the life quality and social security question of new urban population and is not conducive to improvement of the urban development quality and connotation. Therefore, under the background of the new urbanization, it put forward new requirements for urban construction land expansion boundary.

3.2 Literature Review

Urban Growth Boundary (UGB) was originally promoted for American urban sprawl-one of the urban space development problems. In the 1950s, there had been differences in the urban development management for Salem between the Marion and Polk counties in the US state Oregon Salem, after which they negotiated to delimit UGB of urban and rural land [1]. Sybert [2] defined that UGB was a designated line around the city to limit the urban to develop outside. In 1997, American planning association (APA) put forward that urban growth boundary should be delimited to promote the compact and continuous space development, so that with the development the cities were able to obtain more effective public service, meanwhile protect agricultural land, public open space and environmentally sensitive areas; The urban growth boundary areas should include the developed and the developing land, and more importantly, the land should have enough land use intensity which can guarantee the cities to obtain effective growth in the future [3]. After then, Howard, Benton and Paul defined the city and suburb boundaries respectively from the city and nature point of view, pointed out that the city boundary was the independent continuous border around cities which used to limit urban growth, and the suburban boundary protected suburbs from the violation. The Two lines may be separate or overlapping [4].

In the late 1980s, China began to use the "green belt" strategy of the UK for reference in the planning circles, and thought to established boundaries, to limit urban sprawl [5]. Zhang [6] first introduced the urban growth boundary into China, and domestic scholars have the similar understanding of the urban growth boundary. Huang [5] pointed out that on the one hand, firstly, the urban growth boundary was not a time boundary during the planning period, secondly, it was not an eternal boundary which contained city sprawl, it was the final reasonable scale named the "rigid boundary" which the city can achieve with the inside and outside factors (i.e., vision scale); On the other hand, urban growth boundary was a kind of "elastic" boundary from the perspective of urban construction land, it was to distinguish the construction and non-construction land. Lv and Xu defined the urban growth boundary was the isoline of urban function, ecological function and agricultural zones to balance growth and constraints, demand and supply, power and resistance, and suggested to divide prospective and permanent urban growth boundary [7].

UGB is implemented in various ways, most related to the development within the boundaries in 10–25 years [8]. As a means of application, the aim of UGB is to control urban sprawl rather than eliminate it, and also to protect the urban open space, curb urban sprawl [2], adapt to cities' forward population growth and the supply of public service facilities, improve the land use efficiency, meet the needs of housing and employment, protect farmland, land compatible character and settle with solving the problems, such as the society, economy and environment [9]. In Oregon, the law prescribes that land use planners must consider some factors like housing affordability and transportation infrastructure when delimiting UGB, which combines UGB with people's living and travel demand. And at present, UGB has already been applied in the United States, Canada, UK, Albania, Australia, South Korea, China and other countries. In China, urban planners guide the urban space development mainly by designating four zones—"banning construction zone, limiting construction zone, constructable zone and built up zone" as well as the "blue line", "green line", "purple line", "red line" and other means. Because China's methods of delimiting urban planning zones for urban overall planning are most followed the former Soviet union, they have much path dependence and personal judgment makes big difference, and also lack theory [10]. And the means like four zones play a limited role in space growth guiding, which generally control the urban expansion from ecology, environment and resource constraints. However, when curbing the urban development, they cannot balance the land demand for city development [11].

Since delineating UGB may cause a series of negative effects like growth house prices growth, we need to focus on the housing affordability. Some scholars have further study about the relationship between the urban containment means and the housing prices and supply. Fischel [12] suggested that the city containment opponents argued that strict land use control caused rising housing prices. Glaeser et al. [13] thought that the reason for the high housing prices was that the housing supply was inelastic, and the local land use restrictions mainly caused inelastic housing supply. Casey and Arthur [14] examined the impact of urban containment policies to the housing prices and concluded that the land sample with more stringent restrictions would have higher land and housing prices. On this basis, there are some studies about UGB connected with housing supply, some scholars have found that UGB had a significant impact on the land prices [15, 16] and housing prices [17], however, other scholars believed that there was no effect [18]. Many scholars have demonstrated that urban growth boundaries would lead to rising housing prices, and coupled with transportation cost people's housing affordability would decrease. The more scientific policies and measures are to increase the supply of affordable housing to low-income people, which is able to help promote a diversified community, weaken wealth concentration phenomenon and promote the social capital. However, it may cause the rising prices and reduce the effective supply of all types of residences in the market. Arthur [19] analyzed the literature on the relationship between urban growth management and affordable housing and found that the traditional land-use planning and growth management methods would make house price rise, if rising prices were inevitable, what to determine the quality of land planning was the residence choices.

3.3 Methods

The paper is to explore the delimitation logic and methods of urban land development boundary oriented by the new type urbanization and study the results. Yiwu city's urban development has followed the industrialization development model, which expanding outward in the monocentric model, market and urban space are overlapped greater and so do the employment and residences. However, under the background of the new type urbanization, the coordination and common development between all kinds of functions will become the focus of urban development, and also pay attention to the employment-living balance. This paper chooses the downtown of Yiwu City as the study area, and delineates the expansion boundary of the Yiwu City downtown in the year of 2020 based on the land use of Yiwu in 2013.

Traditional delineating methods of urban land development boundary is in the industrialized development model, and the main features of urban sprawl in the "industrialization" model are: mainly based on the industrial and industrial land, supplemented with urban function land use, on the basis of ensuring the industry land layout, then delineating urban land development boundary, urban sprawl is oriented by the expansion of industrial land. People in cities are thought as a labor force, and then the land demand is considered, it is right to make the fan theory as the guidance in this delineation thought, cities planning follows the original development model which expanding outward, and increase the corresponding land according to the population growth. As for the forecast of construction land quantity, employment forecast is the main content, it is necessary to provide the construction land which the employment in various industrial sectors needed and satisfy the land demand of supporting industries development. And on the spatial distribution, it claims to expand outward linearly on the basis of the development trend. Delineating methods in this idea is considered to expand outward unilaterally, have significant mechanical properties, and are likely to cause traffic stress and increase the travel time of residents.

Under the new type urbanization, people are considered as residents and the combination of the employment and housing should be concerned about. It delineates the function zones according to the regional status of the areas and the planning distribution of residences, which reflecting the demand of the population's living in the urban land development boundary. Guidance for multi-core theory provides a new idea, that is, providing a good living environment for the residents is to fill the interior space of cities rather than just expand epitaxial to satisfy the rising urban population simply, using the sub-centers' functions to meet some of the needs from residents and industry. The cities' current features of the function zones and the analysis of the land use status is important in this idea. As for the forecast of construction land quantity, the idea increases the living forecast module, corresponding the employment forecast module, the main purpose of the living forecast module is to provide a variety of housing options for regional residents, satisfy the demand of residents' residential land and the needs of infrastructure land supply. To ensure the balance between employment and housing is an important criterion to evaluate the prediction. And then get into the space layout, in addition to the necessary steps like the environmental sensitivity assessment and planning convergence, the idea combines the development of the non-developed land and the exploration of functional areas with the development of the external land space, the incremental land with the stock, and then delineates the space distribution (Fig. 3.1).

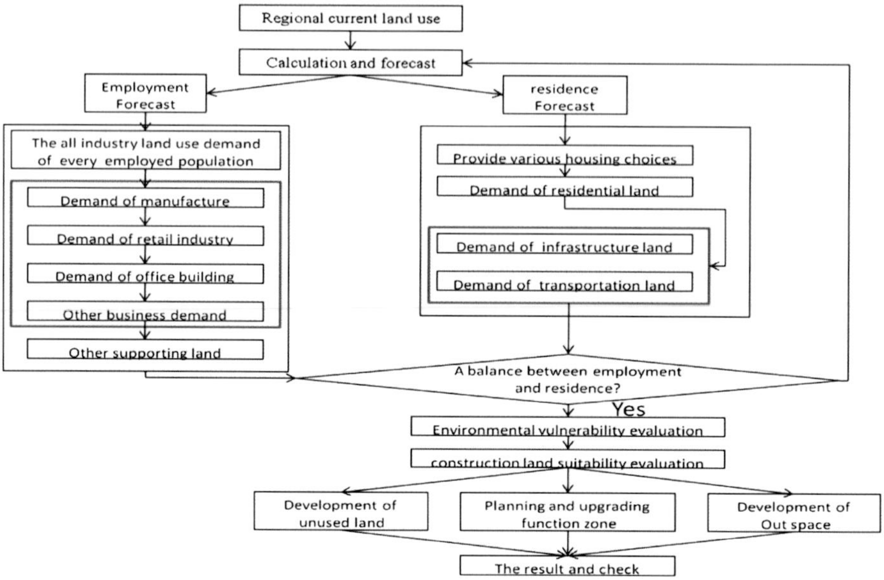

Fig. 3.1 The progress map of urban land development boundary under the new type of urbanization

3.4 Data and Application

3.4.1 Analysis of Study Area

In recent years, Yiwu has the accelerate urbanization, according to the household population statistics, the urbanization rate of Yiwu City in 2006 was 68.50 %, while it rose to 70.12 % in 2013. With the urbanization, Yiwu has seen the urban land expanding rapidly. Yiwu's urban and rural construction land in 2006 was 19,049.21 ha, which rose to 23,005.54 ha in 2013, with an average annual growth rate of 2.6 %. And the average annual growth rate of urban construction land was 7.24 %. The downtown of Yiwu city has borne the main functions of the city such as the financial commerce, city living and municipal and administrative office and so on, thus Yiwu downtown has obvious concentrate population, the radio of the permanent residents population to the overall population of the downtown has increased in general, and the number of permanent residents has been increasing, which is shown in Table 3.1.

However, the land can be developed in the downtown is limited. Within the region of 266 km^2, the remaining land can be developed is only 206 km^2, except 60 km^2 of the mountains, water and other land can not be developed. In recent years, the construction land of downtown has been expanding continuously, from 2006 to 2013, the total amount of downtown's construction land increased from

Table 3.1 The changes of permanent resident population

Year	2003	2004	2005	2006	2007	2008	2009	2010	2011	2012
The residential population in the central city (10,000 people)	90.3	88.9	95.8	109.2	114.7	118.3	126.2	126.1	130.3	134.5
The residential population of Yiwu (10,000 people)	128.9	129.8	137.3	151.1	159.8	165.4	178.7	180.6	185.1	191
Ratio (%)	70.05	68.49	69.77	72.27	71.78	71.52	70.62	69.82	70.39	70.42

Source Statistical yearbook of Yiwu

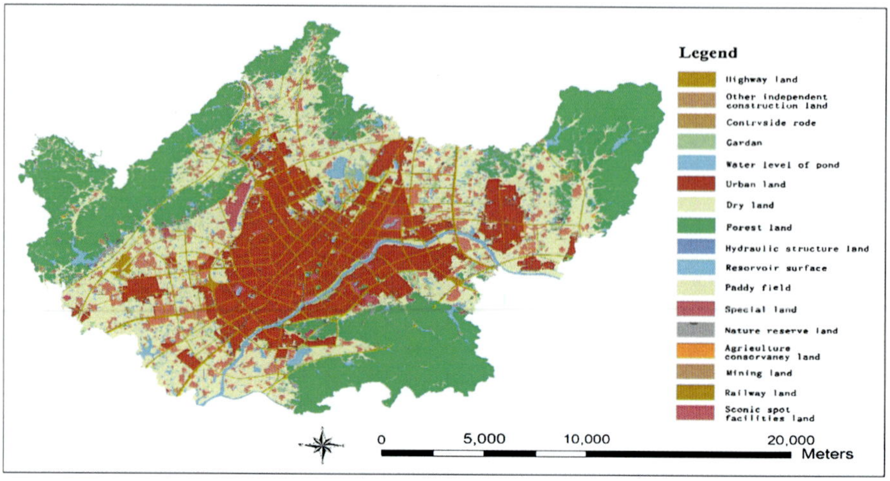

Fig. 3.2 The current land use of Yiwu city's central city

Table 3.2 The construction land ratio of central city of Yiwu

Urban construction land	Ratio (%)
Residential land	32.89
Land for public management and service facilities	6.11
Commercial and service facilities land	7.73
Industrial land	19.10
Logistics and storage land	4.50
Land for roads and traffic facilities	20.23
Land for public facilities	1.04
Land for green space and square	8.40
Total	100.00

Source Overall planning of Yiwu city (2013–2030)

12,229.58 to 15,245.96 ha. With the outward expansion of downtown, the construction land has been extending, and in Fig. 3.2, the construction land located around the downtown scattered.

As can be seen in Table 3.2, the structure of urban construction land of the downtown is unreasonable. And on the basis of the resident population of 1.345 million, the per capita housing area is 23.39 m^2 for the city of Yiwu, which is unable to meet the housing demand of the residents. Although it has a high proportion of residential land, the residential land named "four and a half" accounts for more than 70 % of the residential land, resulting in a waste of land. On the other hand, due to the high degree of function mixing, it will impact the housing quality, and make it difficult to form a living community. The proportion of business services land is far less than industrial land, industrial upgrading is needed; ecological function can not play well with the low green land radio.

3.4.2 The Results of the Urban Land Development Boundary

Figure 3.3 shows the urban land development boundary of Yiwu City's downtown in 2020 and the present land use status, the black line is the urban land development boundary result under the traditional idea, while the red line shows the results under the new type urbanization. Both of them are similar in general, they are almost delineated along the current construction land scope and expanding in all directions. The current land use has restrictions and influence on the direction and scope of urban construction land expansion boundary, so we can see the two urban land development boundary under the different approaches are similar in the direction of boundary's extension, which are overlapping in some directions.

Under the traditional approach, the urban land development boundary is delineated on the basis of urban industrialization, therefore, the sector theory is more suitable, the city has only one center, other land expands around this center with the stable land use and quantities, urban land development boundary expands outward mechanically, a certain type of land use increases means other to increase too, and, in this approach, the dominant type of land is always industrial use. Figure 3.4 shows the status and the expansion trend of downtown's urban construction land, commercial and services land, industrial land and residential land showing the sector distribution, the Fig. 3.4 shows that the land formed as sector and circle with the Xiuhu Square being the the center and different radius or concentric. We found that the status quo of land layout of the Yiwu city's downtown is just according with the sector theory, and urban sprawl on this basis should be radial direction along various land use (away from the land can't be used).

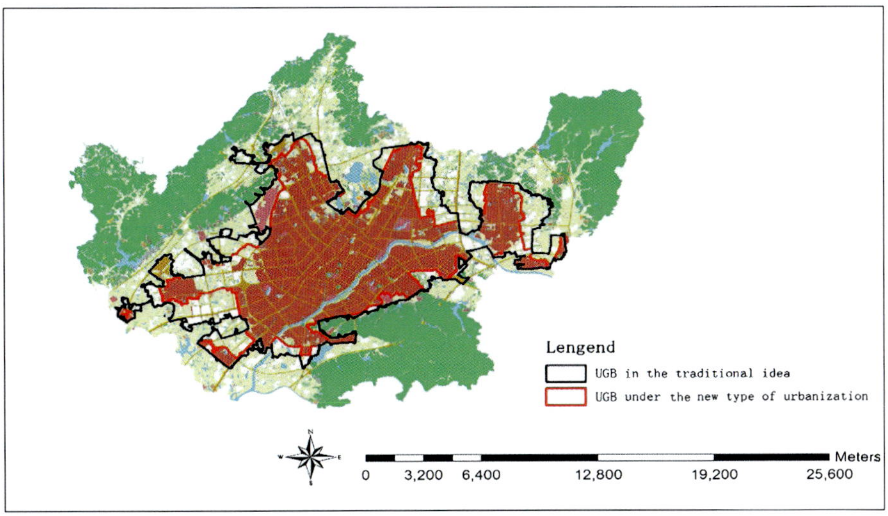

Fig. 3.3 The comparative map of urban expansion boundaries

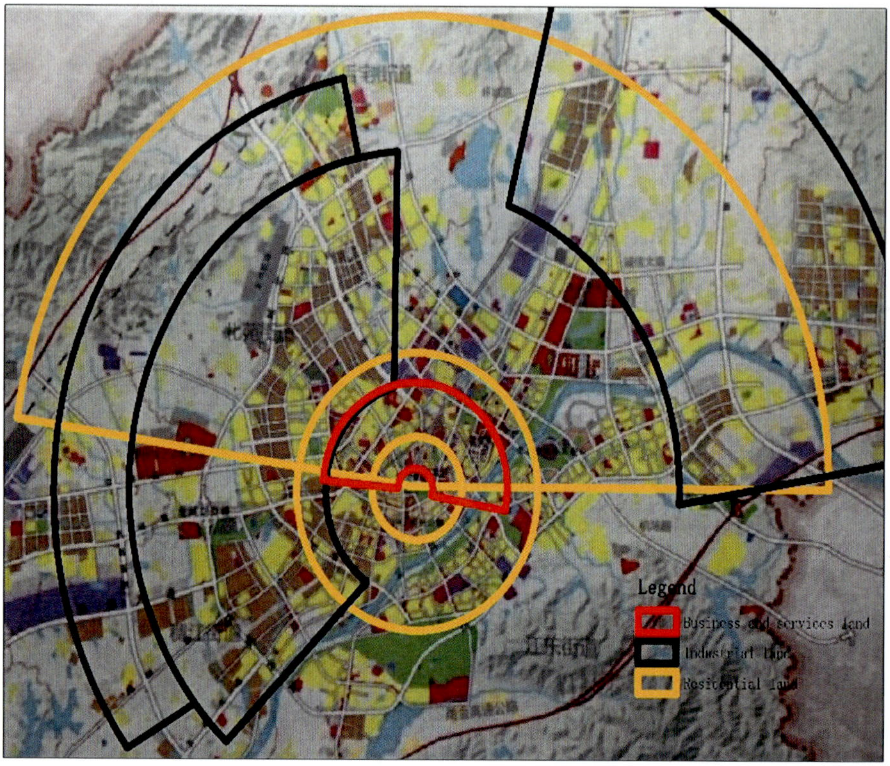

Fig. 3.4 The construction land trend map in the traditional idea

As for the type of urbanization, multi-core theory is the main basis, the city has only one center, however, there are some sub-centers which provide the corresponding level and quantity of service, the land use shows several similar land concentric circles and is adjusted according to the main function the sub-centers. Taking the current situation and the future direction of development of the downtown into account, in the form of "one core-four zone", the various internal and functions are developed and have balance with living. Firstly, the four residential branches with technological innovation, commerce, entertainment and logistics are surrounding the coreround-Xiuhu center. Under the background of the new urbanization, the downtown of Yiwu don't have blocks with clear functional zone, but have functional blocks in the form of the main center (core circle), in which the land adjustment of the living and employment is taken into the blocks, implementing the effective land use and development along transit lines (or planning traffic lines), shown in Fig. 3.5.

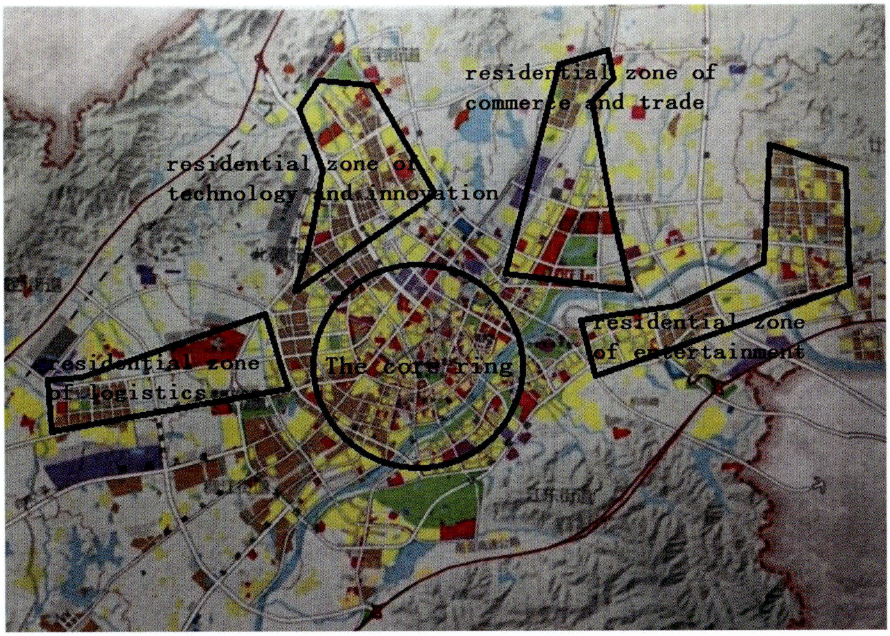

Fig. 3.5 The core zone map under the new type of urbanization

3.5 Comparison

3.5.1 Comparison in Construction Land Scale and Structure

The paper has a comparison between the results of traditional ideas and which is under the new type of urbanization, and finds that the urban construction land scope within the expansion boundary under the new type of urbanization is smaller than that under the traditional idea. As for the incremental urban construction land, the difference is 2211.69 ha, if the new urban construction land is classified as urban and rural construction land, the results are shown in Table 3.3. Similarly, the current construction land within the boundaries exhibits the same difference, the expansion

Table 3.3 The comparison of construction land between the different delimitation methods

The delimitation method of urban land development boundary	The incremental urban-rural construction land scale within the boundary (ha)	The construction land scale within the boundary (ha)
In the traditional idea	3549.3	11,943.37
Under the new type of urbanization	1337.61	10,498.1
Difference	2211.69	1445.26

boundary in traditional ideas is smaller. In the year of 2013, the total construction land of Yiwu city's downtown is 15,245.96 ha, and with the new urban construction land and other construction land within the boundaries, the construction land scope under the traditional idea of the downtown of the Yiwu city in the year 2020 is going to reach the ceiling. Maybe such rapid expansion of construction land will bring economic growth, but overly aggressive expansion is not conducive to the sustainable development of the city and the environment around (Table 3.3).

As for the urban construction land structure of the downtown, in the traditional idea, the urban land development boundary shows linear expansion mainly, on the basis of employment forecast, ensure various types of land for industrial development, and then give the respective support of residential land and other facilities land, therefore, the proportion of the land inside the urban construction land follows the current proportion, which makes the downtown's urban construction land to have the problems, including: the high proportion of industrial land, the low proportion of residential land, green space and place land, however, for the absolute terms, the result in the traditional idea has larger residential land area than which under the new type of urbanization. Under the new type of urbanization, the delimitation of urban land development boundary mainly takes the balanced of employment and living into consideration, and combines the employment land forecast with residential land. Therefore, for the proportion, the proportion of residential land increases accordingly, which will increase housing choice of residents; and considering the needs of industrial upgrading, the proportion of industrial land reduces, the proportion of the service and commercial land increases; at the same time, on the basis of the entire central city's ecological environment, there has some increases in the proportion of green space and land square, Table 3.4.

Table 3.4 The comparison of construction land ratio between different delimitation methods

Urban construction land	The land ratio in the traditional idea (%)	The land ratio under the new type of urbanization (%)
Residential land	32.89	35.89
Land for public management and service facilities	6.11	9.11
Commercial and service facilities land	7.73	10.73
Industrial land	19.10	5.60
Logistics and storage land	4.50	3.00
Land for roads and traffic facilities	20.23	21.73
Land for public facilities	1.04	1.04
Land for green space and square	8.40	12.90
Total	100.00	100.00

3.5.2 Comparison in Expansion Space Morphology

Despite the space morphology of urban land development boundary is largely restricted by terrain and current land use, it has great differences from various urban expansion boundaries in different ideas. Overall, the urban land development boundary has wider spatial scale of the traditional idea of the space under the new type of urbanization than which in the traditional idea. In traditional idea, the boundary radiates in all directions and is in the typical sector model, however under the new type urbanization, the city has smaller radiation and tends to be more clustered.

From the specific direction, in the traditional idea, the boundary extends considering the layout of industrial land and logistics warehousing land in the north, northeast and west of the city. However, under the new type of urbanization, taking the inner function combination and the cluster effect into account, the expansion in the directions of the North, Northeast and West will not be reflected, while taking into account the current construction land use, there may be some crushing boundary in the west and east.

3.5.3 Comparison in the Economics, Society and Ecology Effect

Different urban construction land scope and spatial morphology of the expansion boundaries will have a different impact on the downtown of Yiwu city. From an economic point of view, although urban land development boundary in industrialization mode will ensure the recent needs of industrial development, but from a long-term point of view, Yiwu's development mainly relies on the low-end manufacturing, business and services do not account for a high ratio, if the original model of development has been followed, the urban development must be in a backward state, however, Yiwu City has been aggressively pursuing the restructuring in recent years, and under the trend of new type of urbanization the concern with the "people" is also putting forward a greater challenge to Yiwu's development model. Under the new type of urbanization, the demarcation of urban land development boundary pay attention to the meaning of "people", and in the urbanization model, it requires to reduce the proportion of industrial land, increase which of the commercial and residential land, which carried out urbanization plan from land use.

As for the social effects, in the industrialization mode, people are put into the lower position.

On the basis of the industry development, there is less residential land distribution with the confused living quality. One of the most important issue at this stage of residential land in Yiwu City is what are called "Four and a half", they mixed the functions of apartments, shops, and warehousing and so on and have poor living conditions, result in the waste of land. So in the urbanization mode, the balance between residential land and employment land is put into the most important position, and it request to provide residents with appropriate housing choices, a good living and employment environment. In general, the former is “industry more important than people”, while the latter is “industry equal to people”, which is the truly urbanization. In addition, the urban development the traditional idea has clear distinction between internal functions, and is likely to cause social stratification; under the new type of urbanization, it will pay more attention to mixing function, and promoting social integration.

As for the ecological effects, the intention to delimit urban land development boundary was originally to curb excessive expansion of urban land use in order to protect the environment, and was the result for the balance of the demand of urban sprawl and ecological protection. From this point of view, the space of the urban land development boundary under the new type of urbanization is more likely to protect the environment and ecology, and in the intuitive point of view, the urban land expansion range is smaller, at the same time do less damage to the land structure; As for the internal structure of urban construction land, there is higher green space proportion, which benefit the ecological environment of the city, the upgrading of industry and the decreasing of low-end manufacturing and other industrial land will contribute to sustainable development.

3.6 Solution

3.6.1 Conclusion of the Comparison

Give a comparison conclusion and you will have a more comprehensive understanding of the two kinds of urban land development boundary delimitation, the results are shown in Table 3.5.

The urban land development boundary under the traditional idea is in a mode of industrialization for urban development, which has the development more in line with the expansion of the sector theory. It is a more extensive way for urban development and expansion, and is easy to cause urban sprawl and “pie” phenomenon that affecting the traffic pressure from city center outward, creating traffic jams and so on, and also will have the restrictions of urban and rural construction land indicators in China. On the other hand, the city forms significant functional partitioning from the center outward, which, from a social point of view, is easier to form social stratification, and do bad to integration and development of the city.

Under the new urbanization background the paper takes urbanization mode into account, it considers to expand with a balanced between the living and employment, focuses on the coordinated development of urban center and sub-centers, at the same time within the center focuses on a land balance. It is compared to the former that under the idea not only the transport links between primary and secondary centers

Table 3.5 The summary of the comparison between the two methods

	In the traditional idea	Under the new type urbanization
Concept	The expansion in the industrialization mode	The expansion in the urbanization mode, focus on 'people's demand'
Focus	Ensure the growth of industrial land in order to keep the economic development, on the basis of the employment	Provide with various choices of housing, pay attention to the industry upgrading and the balance between employment and living
Theory	The sector theory	The multi-core theory
Construction land scale	Both the incremental and total construction land scale is larger	Both the incremental and total construction land scale is smaller
Spatial layout	Expand in all directions, larger urban land space	Smaller land space, the expansion in the West and North is decreased
Advantages	Convenient calculation and simple the space determining, expand roughly in all directions	Conform to the requirements of the new type urbanization, industrial upgrading will promote sustainable development, the focus on people is the trend of The Times
Disadvantages	Do not conform to the requirements of the urban and rural construction land index; industrialization has do not conform to the requirements of the development of Yiwu city, there is an urgent need to industry upgrading, it is not wise to blindly ensure the demand of the industrial land and reduce residential land and other land use types; large space expansion increases traffic pressure; the obvious functional partition causes the social stratification	It requires to make sure the balance between employment and living when calculating and determining space, higher request for spatial analysis and understanding of Yiwu's needs; ease traffic pressure; mixed district function which benefits to the merge

but also the traffic layout within the centers is taken into account, therefore it may reduce traffic congestion. Most importantly, the centers have mixed functions, housing, employment, recreation, medical, etc. have the land layout of different levels, which will help strengthen the integration within the city.

3.6.2 Some Solutions to Follow

Delimiting city's construction land expansion border under the background of new urbanization is in conformity with the law of city development. Summarized in this paper is something with general significance in research ideas about city's construction land expansion border demarcation. Details are as follow:

1. Keynote-balance in employment and living needs under the background of new urbanization

What people need in city shall be taken into consideration when delimiting city's construction land expansion border under the background of new urbanization, especially employment and living needs. Balance in employment and living needs should be considered throughout the delimitation of city's construction land expansion boundary, from scale forecast to spatial arrangement. Employment is the major concern under the traditional ideas which may pay more attention to the layout of industrial land, thus leads in poor quality, living confusion and low efficiency in land use. On the contrary, the keynote that keep balance in employment and living needs will pay more attention to employment and living, and the layout about industrial land, residential land and other land more reasonable, which can promote city's healthy development.

2. Function-Primary and secondary city center layout under multiple core

City construction land expansion boundary delimitation under the guidance of multiple core theory is a good interpretation of urban space form. After city's positioning, on the basis of rational analysis of existing urban functional areas, to determine the city center, primary and secondary as well as the center, radiation area, service population, strengthen the service function of every center. When predicting the construction land, the first thing is making population projections and analysis for primary and secondary centers, then making predictions for each type of land. We should pay attention to the mixed function and strengthen land coordination.

3. Scale-the coordination between the scale and structure

Urban construction land scale forecast is the most important link before delimiting the expansion boundary because a reasonable construction land scale and structure is the basis of spatial layout. The proportion of residential land, industrial land, the land for business services and other infrastructure should be coordinated. However, many cities in China at present stage have a higher proportion of industrial land and a relatively low percentage of residential land, land for business services and other infrastructure, which is not conducive to industrial restricting and the development of urbanization.

4. Pattern-urban space filling under economical and intensive land use

City development under multi-core mode is a filling of urban space, which can alleviate the blind expansion at a certain extent and focus on the internal function coordination. When delimiting the expansion boundary, we need to analyze the internal space filling in the area and the possibility of urban renewal, rather than blindly spread outward. At the same time, we should pay attention to economical and intensive land use.

3.7 Conclusion

It seems be inconceivable to change the train of thought on urban land development boundary delimitation. During the past several decades, Yiwu City has formed the future of small commodities market, exists in which is the problems such as extensive use of land, mixed use of live and warehousing land and the quantities of floating populations. Under the mutual influence of these problems, the urban area of Yiwu has a huge demand for the future construction land. In the context of the new urbanization, however, the expanded boundaries and new construction lands of the urban area of Yiwu are much smaller than ever before. In that matched constructions are needed to promote a stable and healthy development under the space limitation of urban land development boundary.

This study aimed at putting forward a different thinking on urban land development boundary, which may exert some effects on the delimitation of our urban land development boundary. In the process of the study rise some problems as well as inevitable flaws, so we need a more standardize research methods to obtain better results. Firstly, because of difficulties on data acquisition this paper only choose center city of Yiwu as the research area and cannot represent the country's development situation and level. Secondly, the research findings are on the base of assessment analysis of Yiwu's social-economic development status, but there shall be some deviation due to the knowledge limitations of the researches. Yet this paper is not for an accurate delimitation result but for an urbanization-oriented method of urban land development boundary delimitation method which has a better adaption to the current new urbanization and a good influence on urban land development boundary. Finally the whole paper has not formed a scientific and reasonable delimitation method on the urban construction land expansion boundary, we need a further study.

Acknowledgements Authors would like to thank to the nationally-supported projects which has helped fund, National Natural Science Foundation of China (Grant No. 71373231). Thanks also to all the teachers and friends for the research.

References

1. Feng K et al (2008) The theory discussion and application of urban growth boundary. Econ Geogr 3:425–429
2. Sybert R (1991) Urban growth boundaries. Governor's Office of Planning and Research (California) and Overnor's Interagency Council on Growth Management
3. Ding CR (2012) The theory models of urban growth boundary. Forum Planners 3(28):5–11
4. Duany A, Elizabeth PZ (1998) Time-saver standard for urban design. In: Lexicon of the new urbanism, pp 5–11
5. Huang MH, Cou CH, Qu W (2012) The combination of "rigidity" and "flexibility"-thinking of urban growth boundary. Forum Planners 3(28):12–16
6. Zhang J (2002) American urban growth management. Foreign Urban Plan 2:37–40

7. Lv B, Lv QZ (2010) The technology and institutional problems for application of urban growth boundary in China. In: Planning innovation, proceedings of 2010 annual meeting of China's urban planning, pp 1–14
8. Calthorpe P, Fulton W (2001) The regional city: planning for the end of sprawl. Island Press, Washington, DC, p 304
9. Zhang RP, Zhou CS (2010) The research progress and review of American urban growth boundary. Extraterritorial Plan 11(26):89–96
10. Du WP et al (2009) The scope delimiting study in the less developed regions—using Woyang county Anhui province as an example. J Shandong Normal Univ (Nat Acad) 24(3):65–68
11. Zhu CS et al (2010) The main discussion of the urban non-construction land planning. Plan Des 3:32–37
12. Fischel W (1989) Do growth controls matter? A review of empirical evidence on the effectiveness and efficiency of local government land use regulation. Lincoln Institute of Land Policy
13. Glaeser EL, Gyourko J, Sakes RE (2006) Why is Manhattan so expensive? Housing price and regulations [EB/OL], 21 Oct 2006. http://www.nber.orgpapers/w10124
14. Casey JD, Arthur CN (2002) Urban containment policies and housing prices: an international comparison with implications for future research. Land Use Policy 19:1–12
15. Knaap GJ (1985) The price effects of urban growth boundaries in metropolitan Portland, Oregon. Land Econ 61(1):26–35
16. Grout CA, Jaeger WK, Plantinga AJ (2011) Land-use regulations and property values in Portland, Oregon: a regression discontinuity design approach. Regul Sci Urban Econ 41: 98–107
17. Phillips J, Goodstein E (2000) Growth management and housing prices: the case of Portland, Oregon. Contemp Econ Policy 18(3):334–344
18. Pendell R (1999) Do land-use controls cause sprawl? Environ Plan B Plan Des, 555–571
19. Arthur CN (2002) The link between growth management and housing affordability: the academic evidence. A discussion paper prepared for the brookings institution center on urban and metropolitan policy

Chapter 4
The Study on Urban-Rural Land Transfer System Reform in the Process of New Urbanization

Afan Sun and Suiquan Yang

Abstract The new urbanization relies on the reorganization of the key elements of production, including land, labor and capital. The existing urban-rural dual land system in China created the single model of land urbanization. In addition, the unfair allotment of the land income created many social issues related to stabilization, harmony and environment. Therefore, taking the opportunity of the marketization of rural collective profitable constructive land, constructing urban-rural land transfer system is important to solve the urban and rural land partition and promote the development of new urbanization, which including land development rights aimed at land use controlling, assisting the transforming function of the local government, and through the market regulation to realize the reasonable disposition of rural land resources.

Keywords New urbanization · Urban-rural dual land system · Urban-rural land transfer system

4.1 Introduction and Background

Urbanization is an inevitable trend along with the development of industrialization and modernization. Since China's reform and opening up, urbanization has made significant achievements with rapid economic development. According to statistics, between 1978 and 2014, the country's urbanization rate increased from 17.9 to 54.77 %. However, the traditional urbanization is a one-side growth comprehensively dominated by the government which attaches great importance to the matter and ignores the human element. It's also a single-direction assemble of main factors from a countryside to an urban, such as capital, labor and technology, which embodies in

A. Sun (✉)
College of Economics, Sichuan University, Chengdu, China
e-mail: 80243334@qq.com

S. Yang
College of Law, Sichuan University, Chengdu, China

Y. Wu et al. (eds.), *Proceedings of the 20th International Symposium on Advancement of Construction Management and Real Estate*,
DOI 10.1007/978-981-10-0855-9_4

the gradual expansion from the cities to the suburbs and the villages. The result shows that land urbanization is faster than the population urbanization, which refers that many farmers virtually were excluded in fringes of urbanization (Tables 4.1 and 4.2). They can't enjoy the fruit of development, but also bear the burden of development cost, causing interest division among various social groups. All these run contrary to essential requirement of new urbanization.

Scholars introduced the disadvantages of traditional urbanization from different angles, and put forward the reform measures to new-type urbanization. Some experts

Table 4.1 The rate of population increase and construction land increase in urban (2004–2014)

	Land sales		Urban population (ten thousand)	The growth rate than the last year (%)	The construction land area (km^2)	The growth rate than the last year (%)
	Amount (billion yuan)	The weight of land sale income in local governments' fiscal revenue (%)				
2004	6412.0	54.83	54,283	–	30,781.28	–
2005	5884.0	39.53	56,212	3.55	29,636.83	−3.72
2006	8077.6	53.02	58,288	3.69	31,765.70	7.18
2007	12,216.7	51.83	60,633	4.02	36,351.65	14.44
2008	10,259.5	35.81	62,403	2.92	39,140.46	7.67
2009	17,179.5	52.69	64,512	3.38	38,726.92	−1.06
2010	27,464.5	67.62	66,978	3.82	39,758.42	2.66
2011	32,126.1	61.14	69,079	3.14	41,860.61	5.29
2012			71,182	3.04	45,750.67	9.29
2013			73,111	2.71	47,108.5	2.97

Source China Statistical Yearbook and China Land and Resources Almanac

Table 4.2 Change of urban construction land and urban population in China in 1990–2000

	The east area	The central area	The west area	The average area
The per construction land area in 1990/m^2	94.4	90.6	87.4	92.3
The per construction land area in 2000/m^2	104.0	83.6	86.2	99.5
The expansion of urban construction land/%	44.0	17.8	33.1	39.8
Urban population growth	29.5	27.6	35.0	29.7
Growth of per capita land	10.2	−7.7	−1.4	7.8

Source The expansion of urban construction land and its occupation to cultivated in China in the 1990s [1]

held an opinion that the traditional urbanization mainly take the campaign-style governance with the essence of policy governance, which led to go after political achievement and rely on administrative means, meanwhile lack of strategic planning and law-guaranty in urbanization. To realize the institutional governance of new-type urbanization, it should through make and perform the urbanization's strategic programming, strengthen the reform of the related field system, such as land, household registration, employment, social security, taxation and finance, and perfect regulatory system [2]. Some experts held an opinion from a legal point view that the "development mode" of urbanization is highly correlated with factors including local government competition, land financial dependence and vague institutional arrangement. The new-type urbanization should transfer from the single center completely dominated to the polycentric limited guided by government, from policy totalitarian to legislation-centred doctrine, from the "material-oriented concept" to the "human-oriented concept" [3]. Meanwhile, it should take a series of measures that can strengthen market forces, harden power constraints, convert dynamic mechanism, clarify institution rules and reconstruct the encouragement system of local government [4]. Other experts held an opinion from economics that the core value of urbanization is to improve people's well-being during urbanization, encourages migrants' assimilation and citizenry and people themselves are the main body and the city-builders for a vibrant urban development, and make efforts to avoid the social exclusion of those domestic migrants working and living in cities [5]. Therefore, it need to correctly balance the relationship between land urbanization and population urbanization, accelerate the land system reform [6], and manage the relationship with the key supported industry and basic public services cover all agriculture transferring people, so that gradually eliminate the differences between town and country [7]. But none of them has introduced a measure to solve the problem of urban-rural dual land system in new-type urbanization. We will analyze the relation between the existing land system and urbanization, and find land system of urban and rural integration should be constructed so as to put forward concrete proposals for realizing the new-urbanization. Our paper will be organized as follows. Section 4.2 analyzes the relations of elements in new-type urbanization. Section 4.3 describes the urban-rural dual land system and its influence. Section 4.4 puts some suggestion. The final section contains the conclusion.

4.2 Essential Requirement and Relations of Three Elements in New-Type Urbanization

New-type urbanization is a new concept put forward at the 18th National Congress of the Communist Party of China (CPC), which emphasizes realizing the harmony, integration and harmonized development of industrialization, IT application, urbanization and agricultural modernization. This requires that holding the scientific development outlook to promote human's comprehensive development, and then to achieve the full-scale, harmonious and sustainable development of economy and society without sacrificing the benefits of rural regions and peasants. First, it should set up the people-oriented awareness and solve the problem of farmers where they go. While defined a clear status, farmers as the main element of urbanization should become social subject, profit gainer and business subject, so as to fairly share social development fruits. Second, it must be on the basis of the right of the estate to solve the problem of land conversion. The land is the material carrier of urbanization, which is favorable for the good circulation of urbanization by identifying the urban-rural land transfer system and improving the high efficient and intensified use and rational disposition of land. Last, based on the money and problem-solving where the money will come from. To solve this problem, it should change traditional practice of land financial and expand the origin and ways of local government revenue to provide necessary guarantees for sustainable urbanization. As the basic elements of urbanization, it is interdependent and mutual influence among farmers, land and funds (Chart. 4.1). For one thing, it requires a lot of land for construction in urbanization, which promotes the large scale circulation of rural land. For another thing, it also releases large numbers of farmers in the process of land conversion, which needs to be transferred to the town and provide matched public services and infrastructure. At the same time, land conversion offers aggressive fund to urbanization and helps finally to realize population urbanization. Thus it can be seen that the land is the core of the urbanization, which is also a certificate that every country's history development. On the one hand, the land benefits distribution system of rural land transference related to the survival and development of farmers; on the other hand, the transfer system of rural land related to the integration with urban-rural factor market and the way of rural land system reforms. It's also the key to realize urbanization from land to person.

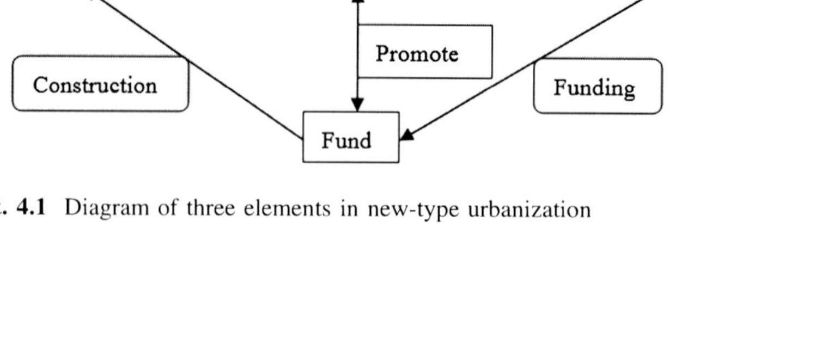

Chart. 4.1 Diagram of three elements in new-type urbanization

4.3 The Existing Urban-Rural Dual Land System and Its Influence

The obvious policy biased toward urban make a lot of factors of production and capital flow into the city in China's development, while fastened industrialization and urbanization, which widen the gap between urban and rural areas. Land, as the mother of wealth, the effect is more apparent of wealth paste in the existing urban-rural dual land system. At present, the urban-rural dual land system is mainly embodied as:

4.3.1 Difference in Nature

The current land in our country is divided into state-owned land and collective land. That is to say, land in urban or in villages and suburb belonged to the state prescribed by law is owned by the state. Land in rural and suburban areas is owned by peasant collective, except for those portions of land which belong to the State as provided for by law, including house sites and privately farmed plots of cropland and hilly land.

4.3.2 Difference in Subject

Land in the cities is owned by the state, local governments of various levels often exercise the right of land seller. According to the article 3 of "The People's Republic of China urban state-owned land use right transfer and transfer of the provisional regulations", state-owned land use rights can be obtained by no unit or person including foreign units and individuals, who has the right to use of State-owned land complying with the law. For collective land, the article 43 of "The Land Administrative Laws" provided that all units and individuals that need land for construction shall, in accordance with law, apply for the use of stated-owned land, with the exception of the collective economic organizations and farmers of such organizations that have lawfully obtained approval of using the land owned by the farmer collective of the organizations to build township or town enterprises or to build houses for villagers and the units and individuals that have lawfully obtained approval of using the land owned by a farmer collective to build public utilities or public welfare undertakings for a township (town) or village. Thus, users of collective land limit to the collective organization farmer, rural township enterprises and township-village collective economic organizations, and the enterprise which has right to realize the circulation of right to use collective land for merges and bankruptcies.

4.3.3 *Difference in Form and Scope of Land Conversion*

The form of urban land conversion includes compensatory transfer and allotment. The compensatory transfer includes tender, auction, agreement and alienating, exchanging, using as equity contribution, endowing or mortgaging the right to use land. Its scope does not have any limitation. The usage of rural collective land mainly includes homestead, township and village enterprises, public utilities and public welfare undertakings for a township or village. Under current law, after the application is approved, the land use right of first transfer can be obtained by free allocation, circulated again is not allowed in principle. While for bankruptcy, consolidation or other reasons, enterprise which obtained constructed-land lawfully and consisted with land use planning can transfer its land-use right.

These differences built the urban-rural dual land system. It is duality that hinders the free flow of construction land between urban and rural. On the one hand, local government has double monopoly. It is the only part for need of the rural construction land and the only supplier of the urban construction land. It also earns huge profits from expropriating rural collective land in a low price and remising the right to use of state-owned land at a good price, forming the land finance of local government. On the other hand, relatively low amounts of land compensation make farmers lose the existing land-use right, along with the assurance system of the future development. It also led to many social problems that farmers are not satisfied with the compensation for land expropriation, such as government employed the policy force in the expropriation of land, and produced the conflict with peasants and collective appealing.

4.4 Construction Ideas of Urban-Rural Land Transfer System

Where consistent with planning and use control, sale, lease and shares of rural collective profitable construction land is admission and enters the market with the same right and price as state land, which is put forward in the Third Plenary Session of the 18th CPC Central Committee. The legal system of local government taking and auctioning land is replaced by the marketization of rural collective constructive land. The policy is seen as a breakthrough for Chinese new turn of land reform. Take this as an opportunity, constructing urban-rural land transfer system, it can meet requirement of new urbanization and structural reform of finance and taxation, realizing the transformation from land urbanization to person urbanization and the transition from land finance to tax revenues. Then, it can kill two birds with one stone to ensure the local economic and social sustainable development.

4.4.1 Taking the Opportunity of Marketization of Rural Profitable Collective Constructive Land

The rural profitable collective constructive land is a new concept which put forward in the Third Plenary Session of the 17th CPC Central Committee in 2008. It claimed that the integrated construction land market should be established between urban and suburban regions, and the right of land use should be transferred in the unified tangible market through public standards and equal rights and interests with state-owned land under planning can be shared. At present, the transfer policy of land is increasingly clear. Under the premise of consisting in planning and use control, it allowed the rural collective profitable construction land to sale, lease, share, and access to market in same right and price with state-owned land. But what is the rural profitable collective constructive land? How to transfer? The details of scope and method are not clear. Taking the opportunity of the marketization of rural collective profitable constructive land, it can standardize the land transfer mechanism between town and country, clear circulation ways, procedure and coverage, so as to gradual adjust land policy and guide the flow of social resources and capital to the countryside, which contribute to promote the new rural development and reconstruction and countryside urbanized.

4.4.2 Taking Right for Land Development as the Core

In common law system, land development right is a kind of property right can be separated from ownership of land and solely disposed. Its basic meaning is the right of developing land for changing the nature of use. From private land ownership, the land development right was determined to the land owner in United States, and could trade through transfer and acquisition. While the land development right is owned by the state in England, and land owners need to comply with the national strict control on land usage. In case of change of use, it is required to purchase the development right or expropriate the land for development by government. The land development right hasn't formal setting in China. It's appeared with the form of other right or in the fuzzy state when it should play a role. From the current transfer system of rural collective land in China, the farmer's land development right can embody the right in imposition and confiscation and development and use. In fact, the imposition at a low price and sell in a high price of local governments didn't adequately protect the right. Therefore, constructing urban-rural land transfer system, it should respect the farmer's land right of development, both to protect farmer's land property right and distribute the huge increment interest of land resources rationally and effectively.

4.4.3 Controlling According to the Land Use

The land usage must be changed in urbanization. According to the function to confirm the right of land and realize the same price with the same usage between rural collective land and state-owned land, which is the key of constructing urban-rural land market. At macro level, it should as far as possible to ensure the use-planning certainty of urban and rural unified function, and improve constraint mechanism through strengthening authority restraint mechanism. Make consistent of the planning and use control as the line of collective land entering market. It requires that when the specific planning is set up, the following precautions should always be taken. Firstly, it should compliance with the national regulations that strict restriction shall be imposed on turning farm into constructing-land, and kept the total area of the cultivated land away from reduction. Secondly, it should pay attention to the coordination and supporting among various types of planning. When determining the land planning, not only ensure the coordination of urban planning and land use planning among the different city, town and village, but also realize coordination among the plan of town development, eco-environmental protection planning, and the safety planning of grain production and so on. Lastly, it needs to adhere to the use control principle, and refine and conscientiously follow the general plan and annual plan for land use of countryside. Starting with the coordination of ecological function, economic function and social function of the rural collective land, it should deal well with relation among the rural collective land, economic progress, environmental protection, countryside and town construction, and ecological system of traditional dwellings in coordination, eliminating the disorder in the development and construction of land.

4.4.4 Fully Play the Basic Role of Allocating Resources on the Market and Strengthening Institutional Construction

The existing land profits monopoly of the city government must be influenced by the transfer system construction of urban-rural land. It should deal with the relationship between the government and the market, as well as give full play to the basic role of market forces in allocating under the macroeconomic guidance and regulation of the government, which is the assurance for the smoothly transfer of collective land in the new-type urbanization. Firstly, the determining mode for plans of land utilization gradually shifts from the traditional generated by orders of the upper government, to the usage that can reflect the market oriented demand. At the same time, the local government should be changed the identity from direct participant in market into the executives and supervisor of land use planning. Secondly, the determining mechanism and market mechanism of commercial value of the new integrated urban-rural land use right for construction should be established. Under defining the ownership of rural community land and its subject, refine the subject and condition, scope and period, procedure and pattern of rural profitable collective constructive land conversion. Then, gradually establish and strengthen dynamic management system such as cadastral survey, land information management, contract management etc., and price assess mechanism including land rating and pricing, evaluation of assets etc., intermediary service company system and multi-factor supervising system should be constructed, including investment and financing of land, legal advice, principal-agent. This can decrease transaction costs. Thirdly, it should set up fairly income distribution mechanism of collective construction land, which taking account of the state, the collective and individual, and fully safeguarding peasants' land rights and interests to improve personal income.

4.4.5 Perfect Laws and Regulations, and Reasonable Arrange the Profit of Circulation of Collective Land

Providing the Legal Protection for Collective Land Entering Market

First of all, to provide basic protection, it should modify the confliction with permission of the market exchange of the collective land in China rural area in "The Land Administrative Laws". The article 8 of "land in urban areas of citied belongs to the state" should be deleted, just to provide legalizing foundation for occupying the collective land without changing the nature of the collective land in the process of urbanization. A second point is to modify the article 43, "All units and individuals that need land for construction shall, in accordance with law, apply for the use of Stated-owned land". It can be replaced by "All units and individuals that need land for construction shall, in addition to the public interest, obtain the use of Stated-owned land or collective land through fair trade of market in accordance with land plans". To suit for free flow of profitable collective land, the article 63 should be deleted. In addition, it should add to some articles in the Property Law, such as "The State implements equal protections for peasants' collective land ownership and state ownership", "Land owned by the State and land collectively owned by peasants exercise equal land enjoying equal rights and prices." Furthermore, the article 151 in Chap. 12 should be deleted, and definite the scope and pattern and permission of the right of collective construction land.

Appreciation Income Belongs to the Public of Collective Land in Firstly Circulation

The profit of the collective land firstly-transfer in market is relevant to its geographic location and facilities. Besides the improvement of the village collective, what's more, the externalities in investment play a decisive role, which caused by government implementing the function such as urban planning, social services and infrastructure construction. Based on "who invests, who gains", it's reasonable that the rise of price belongs to the public by taxation. It can expand the scope of land value increment tax to the rural collective constructive land marketization, and modify the article 2 of the land VAT provisional regulations of the People's Republic of China, changing state-owned land to state or collectiveness owned land. Meanwhile, the compensation standard of article 47 in "The Land Administrative Laws" should be improved, and change compensation according to the original uses of land to compensation depending on the market price.

Gradually Establish a New Integrated Urban-Rural Land Tax System

Besides the land Value-Added Tax, it should revise the urban land used tax to the land use tax, and sprawls the revenue scope from cities, counties, administrative towns, and industrial and mining areas to rural areas, including rural construction land. Land idling fees should be uniformly applied. In the future, the housing property tax and the land use tax can be amalgamated, and adopted across the country. At the same time, "city real estate management law" also should be revised, deleting the "city" and extending to the whole country. Learning and introducing the specifications and procedures for circulation of land sales, land evaluation, land auction and bidding auctions in the system of land, and control measures of land conversion to the system of rural collective construction land. Finally, local government and village collective economic organization should respect the law of sources arrangement and cost-benefit rule, which in the marketalization system of collective profitable constructive land. Meanwhile, through reforming the land tax gradually reduce the land transferring fee, until replace it.

4.5 Conclusions

In conclusion, urbanization is a complicated system including the transformation of many aspects, such as population shifts, space expansion, social transformation and government role change. In the process, for one thing, people's model of production, life style and living pattern is transforming from agriculture to industry and commerce, and from countryside to urban areas; for another, economic industry model, social organization system, cultural values, and public administration and services are also changing. It embodied the transformation of the whole social patterns from traditional countryside to modern urban society, from agricultural civilization to industrial and urban civilization. However, the existing urban-rural dual land system in China, to a certain extent, both affects the rational flow and normal function of three elements in urbanization, and hinders the sustainable development of urbanization. It's difficult to realize the normal transformation of the whole society. In fact, seizing the opportunity of the marketization of rural collective profitable constructive land, and constructing the urban-rural land transfer system is important. It can ensure successfully implement the new urbanization and fairly collocate the profit of circulation of right to use collective land, getting rid of the disadvantage of land finance. Moreover, it need perfect legal system of land taxation and land management. History has proven that land system related to the configuration of other production factors, and land system reform related to several aspects of the system of society. Therefore, the construction of urban-rural integration land system can not easy to solve. More research needs to do in every sense.

References

1. Tan M, Li X, Lv C (2004) The expansion of urban construction land and its occupation to cultivated in China in the 1990s. J Sci China (Ser D) Earth Sci 34(12):1157–1165
2. Wang D, Zhang Y (2013) From campaign-style governance to institutional governance: the select of governance model of new-type urbanization. J Explor Contention 11:47–50
3. Zhu J (2015) The transfer of governance model of new-type urbanization and implementation mechanism under the rule of law. J Tianjin Adm Inst 17(1):90–96
4. Li Y, Yang L (2015) The development dilemma and transition path of the 'development mode' of urbanization in China. J Chin Public Adm 6:53–57
5. Ren Y (2014) People-centered urbanization: essence of the new urbanization. J FuDan J (Soc Sci) 4:134–139
6. Xu J (2014) People-centered urbanization is the core of urbanization. J Tribune Study 2:26–29
7. Wang Y (2013) Core problems of new-type urbanization and land system reform. J Res Econ Manage 12:48–52

Chapter 5
Research on Pension Model of "Reverse Mortgage"—A Case Study of Chongqing

Jing Bai and Yongfei Pan

Abstract Today, the problem of aging of society has become increasingly prominent in Chinese mainland. Problems such as inadequate pensions, safeguard imperfect pension agencies and so on are also need to be resolved. Pension model existed in Chinese mainland is mainly based on traditional family pension, which is supplemented by social old-age pension and the community. In this paper, Chongqing was the main object of study. Expert interviews, questionnaires, visits to nursing homes and other ways to conduct site visits, and combined with the government work report and Statistical Yearbook of the relevant data, detailed analysis of the Reverse Mortgage pension model a. Aiming to survey the willingness of all ages elderly Select "reverse mortgage" mode, who hold the property rights and analyze the obstacles in the implementation process of the model that will be encountered. The paper from "reverse mortgage" pension model the importance of the implementation in the Chongqing region, feasibility and market acceptance and other detailed analysis of proposed measures "reverse mortgage" policy successfully implemented in Chongqing. Reference was provided for relevant government departments.

Keywords Chongqing · Reverse mortgage · Pension model

5.1 Introduction

Over the past decade, China's over 65 years older by the total population increased from 9.07 to 9.68 %, the elderly dependency ratio rose to 13.1 from 12.7, China is about to enter the aging is an irreversible state has implemented. Ageing with many

J. Bai (✉)
Department of Chongqing Jiaotong University, Chongqing, China
e-mail: 504984337@qq.com

Y. Pan
School of Chongqing Jiaotong University, Chongqing, China

Y. Wu et al. (eds.), *Proceedings of the 20th International Symposium on Advancement of Construction Management and Real Estate*,
DOI 10.1007/978-981-10-0855-9_5

different Western countries, China is facing "old before getting rich" and the family "one child more than the old" double pressure [1]. In China, researchers are focusing more on the west, rural and other underdeveloped regions pension problem, but implementation of China's family planning policy has been more thoroughly implemented in the city, and thus the formation of the "forty-two" a family pension structure, this pension structure for supporting the elderly, children and other good presents greater challenges from the economy and many other issues, but also makes this pension structure, there are many uncertainties. Thus, the city pension problem with increasing the proportion of China's urban population is gradually aging is valued by experts and scholars from all walks of life. Currently more scholars compare endorsed "reverse mortgage" pension model, can be used as an important complement to the city pension model implemented in Chinese cities [2].

"Reverse mortgage" pension model is also called "to the Housing Endowment" mode or "reverse mortgage" pension model, refers to the elderly will have their own mortgage to financial institutions (such as banks), which can be regular way to get mortgage funds give financial institutions. Currently, this pension model achieved good results in the United States and other countries. China's real estate market originated in the late 1990s, and in the subsequent two decades has been in a state of rapid development, which is what makes many urban residents have their own set or sets of real estate, which will be to "reverse mortgage" pension model offers the possibility.

"Reverse mortgage" pension model as an important complement to solve the current shortage of China's current pension policy, in our upcoming aging society can play an important role. And now China's traditional pension model is different, "reverse mortgage" pension model can rely on a number of financial institutions to the Housing Endowment, to solve the shortage of pension funds prevalent in existing pension mode. Chongqing is one of China's four municipalities, is the economic center of China southwest area, its vast, undulating terrain, the terrain is more complicated than in the mountains and hills, this significant geographical features, resulting in the main urban area small, large population-density characteristics, which were much less would restrict features to build a large community nursing homes, nursing homes or government. Chongqing City housing policy probably originated in 1998, according to the exercises had been the past 20 years, at the same time due to the implementation of family planning policy, then those who buy real estate most of the one-child families, leading to a lot of families are now a couple to simultaneously support four elderly, weight gain in the intangibles of their burden. With the arrival of aging, but also the need for economic practicality of the pension system in order to address the shortfall in the pension problem now widespread. "Reverse mortgage" pension model plays an important complementary role in resolving the pension problem on the aspect shortage of funds, however, compared with the current Chongqing traditional pension model, it is the specific advantage that it faces in the actual operation of the process What the problem is, and how specific embodiment, it is now the local government and academics need to be resolved.

5.2 Applications in China and Abroad of "Reverse Mortgage" Pension Model

5.2.1 Applications in Foreign of "Reverse Mortgage" Pension Model

The United States for the study of "reverse mortgage" pension model in various countries around the world are started relatively early state, which originated in 1961. But until 1987 the Federal Housing and Urban Development Department in accordance with the "National Housing Act," introduced housing conversion mortgages, the US government is the real start to accept and implement the "Reverse Mortgage" pension system (HECM). The HECM in American society is just a temporary pilot project, which was guaranteed by the Federal Housing Administration, the then American society, a lot of financial risk and profitability can not really predict the business, leading to "reverse mortgage" pension model Development struggling. The US Congress in 1998 to determine the status of HECM program is permanent, and the Housing Authority of the United States "reverse mortgage" business expansion that is carried out to provide the insurance, these changes for the United States "reverse mortgage" market has brought rapid development. As in 2005, the United States issued HECM loans grew by 77 % to 43,131 pens, than the sum of ten years after the launch of this model in volume. But the rapid development of the US This "reverse mortgage" business is entirely the result of irrational exuberance of the real estate market, the financial crisis in 2008 severely curbed the development of "reverse mortgage" pension model in the United States, and in later years, drastically reduced the HECM contracted amount. So, as the birthplace of the financial crisis, the US subprime mortgage crisis has also indirectly led to the US reverse mortgage market by the gloom and doom.

From the "reverse mortgage" pension model in the United States development path, we can see that price HECM volume and housing are closely related, and the United States too much of the "reverse mortgage" to the market mechanism, not a good to the government's macroeconomic adjustment mechanism, once the shock when the financial markets, it will inevitably bring disaster to "reverse mortgage" pension system. However, the US financial markets and our financial markets there is a difference in nature, and the United States to "reverse mortgage" pension model over the market, which in our country is not desirable. We need to combine our specific conditions, starting from the actual situation of each region, to analyze how to better implement the "reverse mortgage" pension system.

5.2.2 Applications in China of "Reverse Mortgage" Pension Model

Since the beginning of the implementation of the family planning policy of eight years, when the only child gradually began to form their own family, and gradually formed the current "4-2-1" family system [3], in the middle part of the one-child couples will bear great pressure, so the traditional family pension has been unable to meet current pension problems. According to the relevant experience of foreign countries, China in the early twenty-first century, there have been Nanjing, Shanghai and other cities began the trial "reverse mortgage" pension model. Nanjing after the outcome of future attempts as expected, the Shanghai trial period of six years in just six of the application is approved. Reverse mortgages in 2012, CITIC Bank in Beijing trial, but also few people interested, mainly because of the high capital threshold other reasons [4].

5.3 Analysis of the Existing Pension Model of Chongqing

5.3.1 Family Pension

Family style pension is a pension model family unit, this pension model is now most families take a pension model. In China always carry out "100 Yoshitaka first" ideology, so for elderly parents and family support is an obligation of every young man and every family should fulfill. And with the gradual arrival of China's aging society. This kind of family as the main carrier of the pension mode, differentiation of social responsibility in large part, to reduce the burden on society for pension funding requirements. However, due to the law of lifestyle habits and the elderly and young people have a lot of differences, which will be family members of two generations or even generate big contradiction between the generations, leading to family disharmony. Moreover, since many of them are now married newlyweds double single, so young people should simultaneously support four elderly, which would lead to a fiscal crisis household spending. So although this pension model is now a universal pension model adopted by society [5], but it still exists great drawbacks.

5.3.2 Social Pension

Social pension formula mainly refers to nursing homes and other institutions in the community opened, some of the elderly people in the community come together to carry out a pension model of centralized services. In Chongqing City opened a number of public and private nursing homes, such as the elderly hospital Banan

District of Chongqing Municipality, etc., by the author of this article visited a number of pension institutions found that Chongqing, private pension institutions than public pension agency facilities, as well as access to services standards are relatively high number, many private pension agency already has, medical care, rehabilitation and hospice services, of course, its price is relatively high. And the different generations living under one roof style pension model, this way pension model is mainly through social channels, as well as some of the safeguards provided by the Government to offer [6]. Pen with through visits found that the advantages of this pension model is that it can provide centralized pension service, but can also provide some special services for special elderly, this centralized conducive to the effective and optimal allocation of resources, and will not produce waste of social resources, while a centralized pension can make pension institutions more focused on professional services for the elderly, and the pension institutions, will be a companion to each other between the elderly and the elderly. But it also has some drawbacks, such as charges that can provide a better service private pension institutions generally high, so the elderly and their families unbearable, followed many elderly people do not want to live in nursing homes, because in here They will not see their children a long time and loved ones, the last is the lack of beds in many of the relatively high cost of pension institutions, as well as to provide a particular service pension agency.

5.3.3 Community-Based Pension

Community-based pension through government support, community participation and market operation, and gradually establish a family pension as the core, community service-based, relying on professional services, to provide life care home for the elderly, health care, mental comfort, culture and entertainment as the Service main content. But community care is not a simple family pension, Social Security Model while there are differences, he was mainly the aged is to be introduced into service in the community, the implementation of home retirement communities. He focused absorbed the family pension and old-age social pension are two ways a bit and operability, the best combination of family support and institutional care focused on the community. But also for the huge problem of aging of our society will soon face a new kind of multi-proposed pension model. Chongqing municipal civil affairs, news bureau received, as of November 2014, the city put into operation various community care services center 826, a total investment of $ 182 million to complete construction. As a pension model in its infancy, there must be some drawbacks: First, its close government support funds, funds constituted over a single, negative effect on their long-term development. Second, the management of standardization needs to be further improved, due to its infancy stage, many measures is still in the pilot period, the construction, service model and business model is not perfect, to be further improved.

In this paper, the existing pension model in Chongqing conducted field research, and the available information and data were statistically and finishing, for each pension model put forward their own views, and points out some problems of its existence, but by Comparative study of the three modes, easy to see they are a common problem, is the shortage of funds, in other words, to be in the city among the elderly in the old age may have a more or house to live, but There is not enough money for retirement, which should cause widespread concern in society.

5.4 "Reverse Mortgage" Mode Feasibility Study in Chongqing

Research team selected in Chongqing Chongqing more representative of the South Bank area, Yuzhong District, Shapingba District, Jiangbei District, Banan District five areas to five regional as the main object, and in these five areas Business Center area were visited and questionnaires survey research. The questionnaire design is mainly composed of three parts: The first part is mainly used to distinguish between different subjects, including residence, gender, age and education, etc., through this part, I mainly want to get their different age structure crowd, as well as different cultural backgrounds reverse mortgage for the implementation of views on perceptual pension model; the second part systematically analyzes the composition of employment, family members interviewed staff, the number of property owners, and may have after retirement a series of questions and number of funds at their disposal, from the interviewees answer to these questions can be drawn, they fell for the implementation of a rational understanding of the mortgage pension model, which can draw some more reliable information; Part III mainly through the research staff of the respondents were "reverse mortgage" pension model after a simple explanation, and choose their wishes this pension model. In this research, the research team shared the five days, using one to one field filled out a question from the scene to explain the "reverse mortgage" pension model, and the site collected questionnaires were collected in the field. persons interviewed some of the problems down the mortgage for the proposed pension model.

5.4.1 How to Make a Proper Assessment of the Property

The research team found that during the investigation process, when a lot of people understand their own house can exchange pension funds, so as to provide financial security for their future retirement, they have to accept the will, but they do not know how to own property valuation [7]. Our research also found that in the survey, respondents who own property, their own property location, age and so there is a relatively big difference. Whether on the property to make a correct and fair valuation, to a certain extent, restricted the people for the "reverse mortgage" pension model of acceptance.

5.4.2 *How to Pay Pension*

The research team conducting the research process found that many people are more concerned about the future when they have mortgage to financial institutions, they were going to get a pension will be paid in what manner. How to pensions, in what way issuance, payment of how fixed point in time and for the loss of independence and the loss of self-care for the elderly and children are not around the elderly, who are going to receive their pensions, these are "reverse mortgage" pension mode in the implementation process need to be resolved.

5.4.3 *"Reverse Mortgage" Period Expires, How to Deal with Real Estate*

The research team conducting the research process found that many people are concerned about how their property will be disposed of in a hundred years after his own, if you want the property left to their own children what to do, and if she wanted to give up what to do ? This is also the implementation of the "reverse mortgage" pension model mortgage process need to think about.

5.5 How to Implement the "Reverse Mortgage" Pension Model in the Chongqing

In this research, we found that high levels of consumption in Chongqing and the generally low income levels of the public funds, and Chongqing population age structure is gradually entered the aging, the implementation of "reverse mortgage" pension model as the existing pension system an important supplement, can solve the existing pension system in Chongqing mainly to insufficient funds many problems. But how to implement the "Reverse Mortgage" pension system in Chongqing, this paper presents the following recommendations.

5.5.1 *Establish a Complete Set of Evaluation System*

Chongqing real estate industry is a relatively early start of the city, the construction of the main city there for decades, and the maturity of the urban area of Chongqing regional development there are great differences. For this difference, according to the Chongqing area real estate prices now as a basis for further evaluation on the property age, size, area, terrain and so on. The government can introduce a set of standards specific property assessment, specific assessment can be carried out by the private sector, but the results of the assessment by the government to check and prevent black-box operation, deceive homeowners behavior. This can be formed in the community, led by financial institutions, to assess the company implemented three-dimensional evaluation system of government oversight, can better implement the real estate assessment.

5.5.2 *Diversification of Funds Paid Out, Release Time*

Since everyone needs money paid out for different financial institutions can target different needs of the population, setting diversified way pensions. For example, for a person to take the way of family support, they need the cash discretionary, to take the social pension manner of people, due to the relatively small their participation in social activities, they are required to pay more to the community nursing home pensions. Take a flexible pensions mechanisms, modalities and time granted the freedom to decide according to the different groups [8]. This will allow a greater level of "reverse mortgage" pension model applicants to obtain more control over pension funds.

5.5.3 *Effective Treatments Property*

Somebody have used the "reverse mortgage" pension model, when the elderly death, they can take a variety of ways to deal with their estate. First, the family of the elderly can be on the property at a certain time, within the purview have a priority right of redemption. Secondly, if in the middle, "reverse mortgage" applicant for a waiver for the "reverse mortgage" pension purposes, so that they have been issued to repay the pension, and after the payment of liquidated damages can get back to some real estate. Finally, if the old man in "reverse mortgage" pension model implementation process, unexpected death, the remaining funds under the inheritance by relatives, until complete payment of all pension funds so far. Through the implementation of these measures points to be maximum to avoid disputes between the applicant and the financial institution "reverse mortgage" pension model in the implementation process.

References

1. Zhao LZ, Xia YX, Ms ZR, Qiu Y (2014) Problems and countermeasures of housing endowment in China's urban area. Urban Dev Stud 11(21):16–19
2. Cui XY, Li Y, Yu T (2013) Reverse mortgage: China's aging social research to the housing endowment mode. Mod Manage Sci 11:70–72
3. Qiu F (2012) "To the housing endowment": new mode of pension. China Real Estate Finance 2012(9):27–30

4. Zhang JW, Han Q (2014) Macro-strategies studies on reverse mortgage development. Urban Dev Stud 21(6):73–79
5. Wang X (2014) Out "to the housing endowment" dilemma of countermeasures. Mod Econ Res 2014(2):20–22
6. Bao LM (2011) The analysis of the reverse mortgage pricing. Res Econ Manage 6:124–128
7. Liu SQ (2013) On the obstacles and solutions about China's reverse mortgage. J Fujian Jiangxia Univ 3(5):124–128
8. Han XL, Kan XX (2010) The reasons and the suggestions on the policy of housing reverse mortgage in rural areas. J Shanxi Agric Univ 9(6):662–664

Chapter 6
Private Capital Rate of Return on Investment in PPP Projects Based on the Perspective of Government Regulation

Hui Gao, Yacheng Xiang and Shiqing Xie

Abstract PPP projects are usually more difficult to implement than other traditional procurement models because of their complexity. Earlier research works on several PPP projects in China showed that a number of problems exist in the project returns and an objective, reliable and practical return assessment model for PPP projects is essential to the successful implementation of PPP projects. The related research on many aspects of PPP projects is numerous; however, actual empirical research studies in this research area are rather limited. The decision-making process, based on the established return allocation principles, such as some floating points based on bank interest rates, requires qualitative judgment and experiential knowledge of construction experts, which is partial, subjective, and implicit in actual application. This paper aims to develop a return on investment evaluation model for determining an equitable return allocation between the government and the private sector based on the perspective of the government regulation. The rate of return within a reasonable range was deduced based on making marginal conversion price equal the rate of return from trading. This novel approach can not only assist the government to transform rate of return on investment into a more usable and systematic quantitative-based analysis, substituting only from the perspective of market cost control in the past, but also help PPP participants make a decision whether to participate in this project.

Keyword PPP rate of return regulation model

H. Gao · Y. Xiang (✉) · S. Xie
College of Economic and Trade Management, Zhejiang University of Technology, Hangzhou, China
e-mail: daxiangxyc@163.com

Y. Wu et al. (eds.), *Proceedings of the 20th International Symposium on Advancement of Construction Management and Real Estate*,
DOI 10.1007/978-981-10-0855-9_6

6.1 Introduction

As an effective way of delivering public infrastructure or services, the Public-private partnership (PPP) form of procurement have been studied and adopted in the last decades in China. More and more governments have acted as a project sponsor and attracted private capital due to public budget constraints and the massive demand for new or upgraded infrastructure. Involving the private sector in infrastructure development is expected to not only have longer-term benefits, such as "value-for-money" [1], but also shorter-term benefits, such as a reduction in cost and time taken to deliver infrastructure services, higher quality service delivery, lower administrative costs, and the transfer of risks to the private sector [2]. In recent years, there have been an increasing market of PPP for the development and operation of infrastructure projects in China. With the fast pace of market-oriented transformation in the economy of China and the severe need for all kinds of infrastructure, a delicate balance has to be sought among private sector capacity, government regulatory function, and public satisfaction [3].

Governments foster private sector involvement in public investment projects in order to maximize the social welfare utility, satisfying the requirements of the private sector's interests at the same time. On account of the determination of the current project return on investment, to some extent, lacking of pertinence and scientificity, the government, as a representative of the consumer, need approval a reasonable level of investment returns. Therefore, determining a reasonable return on investment, which can not only satisfying investor returns, but also meeting the requirements of the relevant government departments, becomes an extremely important issue of the marketization of infrastructure financing. The aim of this paper is to develop an equitable return allocation proportion for the delivery of PPP projects in China.

This article is structured as follows. After this introduction, a literature review looks at the main aspects associated with influencing the private sector's rate of return. The third section presents the summary of return resulting from the revision of the literature and the novel methodology used. Section 6.4 comprises the innovation points of the research and deficiencies respectively.

6.2 Literature Review

Infrastructure construction projects are intrinsic long-term, capital-intensive and difficult to assess, having sometimes monopoly characteristics which impose huge costs to society while providing positive externalities [4]. Therefore, governments have responsibility to provide adequate and efficient infrastructure and public services even when provided through a market-based solution. The characteristics inherent to these projects, the economic constraints of public funds available for infrastructure, as well as the lack of transparency, of expertise and procedures to prevent corruption and, ultimately, the difficulty in obtaining credit led governments to invite a new player into this area-the private sector [5].

The term 'public–private partnership' has been in general use since 1990s, yet there is no widely unified single definition or model of a PPP [6]. Broadly with substantial risk shared by the private party, it is a long term partnership between government and private sector for provisioning of public services. Variations exist in different countries according to their earlier practices in the split of responsibilities between the State and private companies regarding provision of services of public interest [7].

Establishing a consensual definition of PPPs is difficult due mainly to the fact that they are applied to a broad range of models. For the objective of this research, we adopt the definition suggested by Ministry of Finance of the People's Republic of China which states that "PPP is a long-term relations of cooperation established in the field of infrastructure and public service. In this pattern, social capital undertakes most of the work, such as design, construction, operation, maintenance, infrastructure, getting a reasonable return through user-pay and some necessary pay by government, while the government departments responsible for price and quality supervision of the infrastructure and public service, to ensure the public interest maximization".

Research into PPPs has mainly focused on tendering, procurement, risk allocation, and the financing of PPPs, however relatively little research has been carried out on the extent to which benefits of private sector are realized. Here are some key factors affecting the rate of return on investment.

6.2.1 Risk Allocation

A variety of risks and uncertainties are inherent and difficult to deal with by reason of the large construction cost and the long concession period commonly associated with PPP projects [8]. Reliable risk identification and allocation of risks between the government and private sectors are essential to the success of PPP projects [9]. During the implementation phase, managers are faced with a variety of challenges and delivery difficulties. These challenges and difficulties often come to managers as events, stemming from the project's socio-physical context [10]. Some categories of risks are identified, such as (i) political risk containing government corruption, government's breach of contract, and poor government decision making; (ii) financial risk containing foreign exchange fluctuation, interest rate fluctuation, and inflation; (iii) legal effectiveness issues; (iv) operation risk containing operation cost over-spending, tariff/toll change, and expense payment risk; (v) market demand change; and (vi) environment risk. The process of risk allocation is complex and diverse, and a equitable mechanism of risk allocation is imperative [11].

How the project risks being shared should be taken adequate care for, and then governments would seek to combine the advantages of private investors' capability and public's capability of taking these risks. Governments procuring a PPP project would state its preference, while private investors would assess their capability, and then put forward a bidding price. Risk-sharing scheme is one of the most important items of the contract negotiations. A general principle is that each risk should be allocated to the participant which one would best able to manage it and at the least cost [12]. As a matter of fact, an optimal risk allocation scheme is not to pass all risks to the private sector, but to look for a pattern minimizing both the total management and construction costs of the governments and private sectors. Capitalizing on the Chinese government's increased PPP experience in the last decade, they have made a large amount of efforts in order to improve the investment environment, including moving towards the adoption of international contractual practices and working out an equitable risk-sharing scheme. All these work are used to prevent from the happening of the risk, and guarantee the delivery of projects and the supply of public services, in the meantime, private could gain reasonable return by operating projects.

6.2.2 Concession Period for PPPs

Several critical aspects related to a PPP project need to be managed if the government would like to fully gain a variety of benefit provided by PPP, among these the determination of the concession period, considered one of the most important issues in the PPP contracts. Under the concession-based PPP, the private sector has the responsibility to provide funding the scheme, while their capital investment will be recovered through the operation revenue over the concession period [13]. Therefore, calculating a reasonable investment return of private capital over the concession period, as a very important aspect that influences success of the PPP project, is necessary, particularly so as the concessionaires may attempt to increase their toll/tariff while the revenue fall short of their expected. Nevertheless, to make the decision of the concession period, the government would conduct the traditional net present value (NPV) and payback period analyses on account of the difficulties in estimating the long-term uncertainties and wider-risk profiles at the tendering stage.

In the common international practice, the government usually presets the concession period to a fixed length, requests the concessionaire to bid for other project aspects, and guarantees the concessionaire a certain level of internal rate of return [14]. However, generally, this practice does not lead to an efficient selection of concessionaires and it also results in the frequent failure or renegotiation of concession contracts [15].

Private sector finances under the guarantee of the government and achieve the construction of the project, in return, the private sector has the right to operate project in order to obtain operating revenue in the concession period. Therefore, determining an appropriate concession period is important to the success of a PPP project. Being protected by an assured minimum 'revenue stream' guaranteed by the government, the concessionaire is entitled to raise the toll/tariff in case their actual profit falls short of the anticipated return. Projects with a shorter concession period could hence lead to a higher toll/tariff regime, and from the perspective of public service consumers, this is obviously not desirable. On the contrary, from the government's perspective, granting an excessively lengthy concession period could mean a loss in public interest and welfare especially when the equipment of project would reach the peak of its economic life towards the end of the concession period. Therefore, for the government, identifying an optimum concession period is necessary which is not only long enough to guarantee an attractive financial return for the private capital but yet soon enough for the public service facilities to be transformed to the government for subsequent operation [13].

To calculate the concession period, most of the models proposed in the literature utilize the net present value (NPV) of the project cash flow as standard [16]. Some of these determine the concession period considering the maximization of the benefits of private capital. For example, it is proposed that the least-present-value of revenue (LPVR) method to determine the concession period of toll roads [17]. Other models adopt a win–win approach, which means that the concession period is determined in order to maximize benefits of both the government and the concessionaire [18]. The length of concession period of project usually depends on two variables, construction period and operation period. For the sake of Regulation o private capital return, the government ought to establish an appropriate concession period.

6.2.3 Price Regulation

Price regulation is an integral part of the infrastructure under the public-private partnerships. The infrastructure, having characteristics that non-exclusive, non-competitive, natural monopoly, positive externalities and so on, cannot be formed a reasonable price by the market, so it is essential that the government implement price regulation of infrastructure. The regulation of PPP project price could be usually in two ways: the price cap regulation and return regulation. Price cap regulation belongs to incentive regulation, as the most common way in practice and easily leading to cause the renegotiation problems, being an effective way of stimulate private enterprise to improve the way of operation and enhance enterprise's market competitiveness; Due to financial capital cost of return regulation is lower, by contrast, it is considered as a guarantee of government revenue. In practice, government often adopt a hybrid pattern combining the advantage of two regulation ways, not merely bringing incentive effect to the enterprises, but also having government compensation based on the return regulation. Therefore, the two forms of regulation have convergence and complementation to each other to a certain extent. As a strong demand for investment, return regulation is an appropriate way for developing countries. Price set according to the rate of return on investment is a common infrastructure price regulation model with high frequency and a long history, being derived based on making the private enterprise investment to be paid back with a reasonable return. The difficulties of investment rate of return regulation model are focus on how to determine the return on investment and rate base, in other words, how the regulatory sector to determine a return on capital for the natural monopoly manufacturer in a closing to "competitive" and "fair" market, making the same amount of capital could get the average market return under the condition of similar risk and similar technique.

6.3 The Derivation of Return

6.3.1 The Existing Method of Revenue Distribution

The approach brings additional resources to fill the fiscal gap, assists in transfer of technology, and imparts efficiency in project construction and operation through the involvement of private sector. The procurement of PPP projects evolves through project development, construction, and operation. The private' investment in the project through development, construction, and operation of the project is recouped with the returns in the form of either user's fee or grant [19]. As a form of risk sharing in essence, certain guarantees often provided to investors by governments in PPP project. This allows governments to effectively attract domestic and foreign capital. However, the appropriate level of guarantee should not only satisfies the investors, but also remains within the government's affordable range. The following two ways could be taken by government to allocate the profits: (1) output sharing mode; (2) fixed pay mode. Government usually adopts a hybrid mode which is combined with the advantages of the two ways. Methods of profit distribution are mainly as follows: (1) Shapley value method; (2) Minimum Cost—Remaining Savings; (3) Colony barycenter model; (4) Nash negotiation model.

The PPP projects adopt the idea of mutual cooperation, from the perspective of reasonable cooperation, restraining the behavior of project participants through certain agreement, to make project participants always focus on the project benefit maximization goal. At this time, we can apply cooperative game theory to solve the issue of collaboration. How to get a satisfactory profit distribution plan for every project participants, Nash negotiation model of the cooperative game can usually be used to solve the problem of the distribution.

1. Revenue distribution under complete information
   Every partner, from their own interests, analyzes and argues the problems such as revenue distribution and risk sharing until getting an agreement in cooperation negotiations. Under the condition of complete information, the government and other external investors, knowing each other's investment and the degree of risk-taking, therefore, could make predictions about next bargaining strategy of the other participant. From this perspective, the process of revenue distribution between the government and other external investors is essentially a process of bargaining, so the bargaining model based on complete information can be used

to calculate the optimal solution of the revenue distribution, which is the optimal allocation scheme.

2. Revenue distribution under incomplete information
Although they have a common interest objective, out of self-interest,they always choose the appropriate strategy to maximize their own interests, which could be understood as 'individual rationality', so the two sides will hide some important information to each other in the negotiations. We can apply the incomplete information game theory to solve revenue distribution issues in this case. This process is divided into two stages: the first stage is cooperation between two participants. Through their common assessment of the project, two participants determine investment and revenue distribution plan between the two sides; In the case of equilibrium interest allocation scheme has been proposed in the first stage, the second stage is that both sides attempt to maximize the profitability of their own by their own investment level, respectively.

Reviewing related research literature of PPP, we can come to the conclusion that, in the field of infrastructure construction, especially the project areas of public-private partnership, usually, revenue distribution is merely based on risks, overemphasizing the crucial role of risk sharing on revenue distribution, ignoring the other factors, such as action strategies for both sides, the innovative cost of two participant and so on. This paper aims to develop a new return evaluation model for determining an equitable return allocation between the government and the private sector based on the perspective of the government regulation, through the regulatory model, providing a novel approach to derive the return of PPP protect.

6.3.2 *Rate of Return Deduced by Regulation Model*

On behalf of the consumer, the government negotiates with project company total output (Q) and payment amount (R). The government has the utility function: $U(Q, x)$, where x is an alternative goods, ω is the government's original endowment. As a representative of the consumers, the indirect utility of the government can be derived by the function: $V(Q,R) = U(Q, \omega - R)$. The cost and profits of the project company could be expressed as $C(Q)$ and $\Pi(Q,R) = R - C(Q)$. The utility function of the government is concave and increasing, $U_x \geq 0$, while marginal cost is not decreasing, $C''(Q) \geq 0$.

We assume that $\bar{V}$, $\bar{\Pi}$ is defined as the consumers' and the project company's the opportunity cost of entering the regulated market, respectively. $V = P(\Pi)$ is expressed as valid collection or contract curve.

Nash Bargaining Solution (NBS) $f^N(\cdot)$ is defined as:

$$f^N(S, \bar{V}, \bar{\Pi}) = \arg \max_{(V,\Pi) \in S} (V - \bar{V})(\Pi - \bar{\Pi})$$

Nash solution is a result in the effective collection (p). Nash equilibrium (V^*, Π^*) can solve the problem of profit Π maximization, $\max_{\Pi}[P(\Pi) - \bar{V}](\Pi - \bar{\Pi})$, derived by the function: $P'(\Pi) = \frac{-(V^* - \bar{V})}{\Pi^* - \bar{\Pi}}$, where $V^* = P(\Pi^*)$, making marginal conversion price equals the rate of return from trading.

Equilibrium output depends on the rate of return, $Q = Q(s)$, and effective boundary can be expressed as:

$$V = P(\Pi) = U\{Q(s), \omega - R[Q(s)]\}$$

where $s = s(\Pi) = \Pi^{-1}(\Pi)$. Applying envelope condition, marginal conversion price could be inferred using:

$$P'(\Pi) = -U_x[Q, \omega - R(Q)]QP'(Q)\frac{\partial Q}{\partial s}\frac{ds}{d\Pi}$$

where $Q = Q(s)$ and $s = s(\Pi)$, $\frac{ds}{d\Pi} = \frac{1}{\xi K}$, the above function could be simplified as:

$$P'(\Pi) = U_Q[Q, \omega - R(Q)]\frac{\partial Q}{\partial s}\frac{1}{\eta_D \xi K}$$

where $\frac{1}{\eta_D} = -\frac{Qp'(Q)}{p(Q)}$, ξ is shadow price, K is capital stock. The hypothesis is that relationship ($P = p(Q)$) between price and production output is known. Due to the Nash Bargaining Solution can be used to calculate the rate of return, this paper make the marginal conversion price equals the rate of return from trading:

$$-U_Q[Q, \omega - R(Q)]\frac{\partial Q}{\partial s}\frac{1}{\eta_D \xi K} = \frac{V^* - \bar{V}}{\Pi^* - \bar{\Pi}}$$

where $V^* = U[Q, \omega - R(Q)]$, $\Pi^* = R(Q) - C(Q)$, and then if we could get the price's relationship with the production output of public services ($P = p(Q)$) through some way, such as principal-agent theory, a reasonable rate of return (s^*) could be derived by above function.

6.4 Conclusions

This research has adopted a novel approach to develop a practical rate of return assessment model for determining an equitable return allocation between the government and the private sector based on the perspective of the government regulation. The approach provide a new idea which not only assist the government to transform rate of return on investment into a more usable and systematic quantitative-based analysis, but also help PPP participants make a decision whether to participate in this project. This article has the following innovation points: (1) This article has researched the return on investment of private capita when it participates in the construction of public utilities from the perspective of government regulation, attempting to make quantitative analysis about the reasonable profit margins of public utilities, instead of only from the perspective of market cost control in the past. (2) This article has established a return on investment evaluation model for determining an equitable return allocation between the government and the private sector by regulation theory, and the innovation of research method may be provide more scientific basis to measure project reasonable return on investment for the government. The deficiency of this article is that we do not work out the price's relationship with the production of public services, as well as not make the utility function of the government materialize, and only provide a new train of thought to derive the return of private capital. Further research will be carried out later.

Acknowledgements This research is supported by Natural science foundation project of Zhejiang Province, the research on risk sharing and return on investment while private capital participates in the infrastructure construction: based on the viewpoint of regulating the market behavior of private capital (Y15G030097).

References

1. Grimsey D, Lewis MK (2004) Public private partnerships: the world wide revolution in infrastructure provision and project finance. Edward Elgar, Cheltenham
2. Flyvbjerg B, Bruzelius N, Rothengatter W (2003) Megaprojects and risk: an anatomy of ambition. Cambridge University Press, Cambridge
3. Ke Y, Wang SQ (2010) Preferred risk allocation in China's public–private partnership(PPP) projects. Int J Project Manage 28:482–492
4. D Grimsey, M Lewis (2002) Evaluating the risks of public private partnerships for infrastructure projects. Int J Proj Manage 20(2):107e118
5. Martinsb J, Marquesb RC, Cruz CO (2014) Maximizing the value for money of PPP arrangements through flexibility: an application to airports. J Air Transp Manage 39:72–80
6. Bernardino J, Hrebicek Zdenek, Marques Carlos (2010) Applying social marginal cost pricing in rail PPPs: present state, drawbacks and ways forward. Res Transp Econ 30:59–73
7. Viegas JM (2010) Questioning the need for full amortization in PPP contracts for transport Infrastructure. Res Transp Econ 30:139–144
8. Song J, Song D, Zhangb X, Sun Y (2013) Risk identification for PPP waste-to-energy incineration projects in China. Energy Policy 61:953–962
9. Jin X, Doloi H (2008) Interpreting risk allocation mechanism in public-private partnership projects: an empirical study in a transaction cost economic perspective. Constr Manage Econ 26(7):707–721
10. Love PED, Holt GD, Shen LY, Li H, Irani Z (2002) Using systems dynamics to better understand change and rework in construction project management systems. Int J Proj Manage 20:425–436
11. Jin XH (2010) A neurofuzzy decision support system for efficient risk allocation in public-private partnership infrastructure projects. J Comput Civ Eng 24(6):525–538

12. Cooper DF, Grey S, Raymond G, Walker P (2005) Project risk management guidelines: managing risk in large projects and complex procurements. Wiley, England
13. Nga ST, Xiea J, Cheunga YK, Jefferiesb M (2007) A simulation model for optimizing the concession period of public–private partnerships schemes. Int J Proj Manage 25:791–798
14. Zhang XQ (2009) Win–win concession period determination methodology. J Constr Eng Manage 135(6):550–558
15. Gustavo N, Rus G (2004) Flexible-term contracts for road franchising. Transp Res A Policy Pract 38(3):163–179
16. Carbonara N, Costantino N, Pellegrino R (2014) Concession period for PPPs: a win–win model for a fair risk sharing. Int J Proj Manage 32:1223–1232
17. Engel EMRA, Fischer RD, Galetovic A (2001) Least-present-value of revenue auctions and highway franchising. J Polit Econ 109(5):993–1020
18. Shen LY, Li H, Li QM (2002) Alternative concession model for build operate transfer contract projects. J Constr Eng Manage 128(4):326–330
19. Singh LB, Kalidindi SN (2006) Traffic revenue risk management through annuity model of PPP road projects in India. Int J Proj Manage 24:605–613

Chapter 7
An Idea on Housing Vouchers Policy Making: Based on New-Type Urbanization

Jiaojiao Luo, Yuzhe Wu and Xiaoling Zhang

Abstract Urbanization and industrialization have achieved great promotion since market economy was adopted last century in China. City, the most important carrier to support urbanization, of course, realizes quality and quantity optimization on production factors such as population, land and capital. Whereas, housing inequality becomes serious due to urban-rural dual structure during urbanization. People pay more attention to floating population urban housing problems than before. Based on new-type urbanization, the paper firstly analyzes indemnificatory housing's difficulty in construction as well as its implement and combines the reality of commercial residential building in stock, utilization to justify the rationality of housing vouchers policy. Finally, for the sake of easing urban housing problems, the paper puts up an idea and constructs floating population urban housing purchase system via housing vouchers, where containing housing vouchers financial channels, subsidy objects and cash approaches.

Keywords Housing vouchers · Floating population · Indemnificatory housing · Capitation grant · Housing inequality

---

J. Luo · Y. Wu (✉)
Department of Land Management, Zhejiang University, Hangzhou, China
e-mail: wuyuzhe@zju.edu.cn

J. Luo
e-mail: luojiaojiao@zju.edu.cn

X. Zhang
Department of Public Policy, City University of Hong Kong, Kowloon Tong
Hong Kong
e-mail: xiaoling.zhang@cityu.edu.hk

Y. Wu et al. (eds.), *Proceedings of the 20th International Symposium on Advancement of Construction Management and Real Estate*,
DOI 10.1007/978-981-10-0855-9_7

7.1 Introduction

Ever since China stepped into market economy, the State Council officially announced stopping housing allotment in kind and establish housing allocation monetization system. This abolition of the policy on the provision of welfare housing in 1998 is a vital milestone in Chinese urban housing reform, which resulted in the market-oriented urban housing provision system [1].

Welfare housing distribution system generally refers to the unit builds or purchases houses and distribute gratuitously to internal employee who just need paying low rent has recorded the special housing system after the founding of the PRC in a planned economic environment [2]. After the liberation, China achieved dramatic growth in population followed with a continuous rising demand for housing. Whereas, during that period, namely the late 1970s, domestic economy was still in a relatively depressed state which made welfare housing distribution hard to gain a virtuous circle. In practical terms, at the time of planning system, housing construction rely on the state's financial and business investment entirely. The house's lives is no more than one hundred years, in contrast to centuries for investors to have the money back due to low-rent welfare housing system, housing construction, definitely, is a loss of investment [3]. Country's financial supply cannot meet the need in expansion of housing demand. Hence, in order to ease country's financial burden as well as to dig out a new "growth points" for national economy, the state promoted the housing allocation monetization reform in 1998. But, unfortunately, housing market strides into the other extreme. In the face of today's housing price in China, even in second and third tier cities, more than 85 % residents have no purchasing power. To say the least, if some resident buy a house, this action will certain produce enormous crowding out effect on household consumption [4]. Approximately from 2004, the central adopted some economic policy measures trying to utilize macro-control to standard housing market order and balance total supply with total demand, supply structure with demand structure. Yet, these policies do not seem to play an ideal role, some scholars [5] think macro-control is not designed with home ownership philosophy, but rather to safeguard housing as a new economic point. In that case, if you are rich, you can easily access to buy a house, if poor, you cannot afford a house.

Urban housing system reform ought to be established as a multi-level urban housing supply system, namely: low-income families live in low-rent house provided by government or institutions; middle-income families purchase affordable houses; high-income families buy or rent commodity houses. The main types of indemnificatory housing in our country contains affordable houses, low-rent houses, public rental houses, capped-price houses, and so on. All this houses are under the unified planning by government and restricted in construction standards and selling (or renting) price. Of course, authorities have stipulated low-income families and migrant workers who have worked for the cities a few years to be the housing security object.

According to the State Statistics Bureau data, in 2014, China's urban permanent resident population reaches to 749.16 million accounted for 54.77 % of the total population, an increase of 18.05 million than last year; rural population is 618.66 million, reducing 10.95 million one year. In addition, there exists 253 million floating population, aggrandizing 800 million over the previous year. CAS ever predicted that urbanization rate in 2050 in China would be the rate around 77–81 %, meaning if relevant policies of population migration have not changed significantly, floating population in that year could up to 350 million [6]. As the majority of the world population will migrate from rural to city in the coming decades, the track of global poverty is also moving to the cities, a process what we called the urbanization of poverty [7]. In other words, along with urbanization progress, more and more rural-to-urban migrant turn into floating population in the future. Without doubt, floating population is an important safeguard object of indemnificatory housing, if only relying on indemnificatory housing construction to deal with migrants' housing problem is bound to put enormous financial stress on government. Can official attract enough money to ensure continuing promotion of indemnificatory housing policy? Maybe cannot.

New-type urbanization emphasizes healthy urbanization development by promoting human-centered principle as well as accelerates urban fringe groups' integration. What's more, new-type urbanization proposed constructing indemnificatory houses to work out floating population housing problem. However, the paper mentioned above, since the future will greet large number of floating population, government are lacking in funds. Then, we may consider how about introducing "housing vouchers" into housing system? Follows attempts to analyze such trial under the new-type urbanization background, including setup foundation, reasons and concrete paths.

Section 7.2 reviews and links some relative literatures to illustrate analytical perspectives of this paper, including housing inequality during urbanization and housing vouchers. Section 7.3 introduces recent development on affordable construction as well as commercial housing in China to imply the rationality of housing vouchers implement. Section 7.4 presents how to design housing vouchers in detail from the perspective of financial channels, subsidy object, et al. and Sect. 7.5 concludes.

7.2 Analytical Perspectives

7.2.1 Housing Inequality

Housing inequality is a worldwide problem, which means it is not just happened in developing countries. For instance, in the United States, as the homeownership rate reached its highest ever level of 69 % in 2004, the distribution of homeownership remain uneven along racial and ethnic lines. Less than half of black and Hispanic households and approximately 60 % of Asian households owned a home, compared to more than 70 % of white households [8]. When we shift our attention to other countries, like Britain, France, scholars often assess housing inequality on racial or ethnic identity [9, 10]. Scholars from developed countries almost all relate housing inequality to ethnic identity. Although there is no racial discrimination in China, urban-rural dual registration construction, in some ways, is similar to ethnic identity. Due to dual registration construction, famers who flow into city are not equal to what we called "citizens", but migrant workers instead. They create wealth for the city, yet they don't acquire urban residence certificate (chengshi hukou) and theirs income and life quality generally remain in a not so good level.

Dating from the end of last century, the housing reform generates housing inequality to be the most prominent social injustice due to its own flaws in new system [11]. The housing reform seems to reward the privileged groups who have already benefited under welfare-oriented distribution system since migrants continues to live in crowed housing and are unlikely to become homeowners [12]. Although more and more migrant households have been settling in urban areas, they are still alienated from housing reform policy [13]. Here we carve up housing inequality into two segments, one is excessive-priced housing, and the other is the defective public housing system.

Excessive-price housing turns into the specialist feature in housing market when towards to China. Zhang [14] points out modernization process can bring in middle class, a colossal development stabilizer, on the contrary excessive price damages social class structure through falling middle class into a role called home mortgage slave (fangnu). Li [15] emphasizes sustainable growing housing price has magnified wealth inequality many times, which will certainly clamp down housing tenure. Currently, two types of people win a place as customers, ones don't have large income but with purchasing desire; the others already possess one or more sets of houses but still take an active participant in housing market.

Goulden [16] comments the lack of affordable, decent housing is increasingly seen to be a major socioeconomic crisis when he analyze housing inequality in Syria. Academics have raised numbers of problems happening during indemnificatory housing system implementation. For instance, Yeung and Howes [17] insist the housing provident fund, as an important role of affordable housing development in China, inevitably imposes a financial burden on enterprises. One day must the source of funding in enterprises will diminish and finally exhausted; Long [18] considers limited land sources, unreasonable supply mode, low government enthusiasm matter indemnificatory housing construction. Given indemnificatory housing couldn't bring benefits for government, instead, officials need to invest more capital in housing conservation after closeout, thus it can be said funds have become a bottleneck restricting indemnificatory housing policy.

7.2.2 Housing Vouchers

Housing vouchers was founded in 1974, and has been largest direct housing assistance program currently in the US. Housing vouchers is an income supplement earmarked to meet housing costs [19]. One of its main goals is to provide low-income families with housing security service while utilize private market saving government expenditure efficiently.

In the eyes of most officials, vouchers have replaced supply-side subsidies and rent control as the most effective method to guarantee low or moderate-income families for affordable housing [20]. Housing vouchers have been implemented for some time now in the United States, Britain, the Netherlands [21, 22]. In China, researches on housing vouchers are not so much if compared with abroad: some scholars build housing vouchers system from the perspective of public rental housing. For example, Ma and Hao [23], Ma et al. [24] combine public rental housing institution with "pothook of city construction land increase and rural residential land decrease" policy to design migrant workers housing vouchers mechanism innovatively. But they ignore the problem how to integration the mechanism with existing indemnificatory housing; other scholars [25, 26] illustrate indemnificatory implementation troubles while fail analysis housing vouchers system construction in depth. In this paper, we will do some research on dilemmas during carrying out primary indemnificatory housing construction based on new-type urbanization through analyzing the stock, utilization of commodity housing to justify housing vouchers' rationality. Finally, for the sake of easing urban housing problems, the paper puts up an idea and constructs floating population urban housing purchase system via housing vouchers, where containing housing vouchers financial channels, subsidy objects and cash approaches.

7.3 Implementation Environment Analysis

7.3.1 Present Situation of Affordable Housing

Early in 1990s, affordable housing construction sprang up everywhere while the term "indemnificatory housing" was first proposed in 2004. Overall, low-rent housing, public rental housing and orientation placement housing occupy less share among indemnificatory housing. In other words, indemnificatory housing always functions in the shape of affordable housing. Hence, the paper will analyze current affordable housing construction situation thorough to explore practical problems existing along with indemnificatory housing policy implementation.

Annual variation tendency of proportion, namely the value equals to affordable housing investment divided by total housing investment, is shown in Fig. 7.1 on a national level together with investment amount growth rate in different types of housing. Evidence reflects the value of affordable housing investment proportion

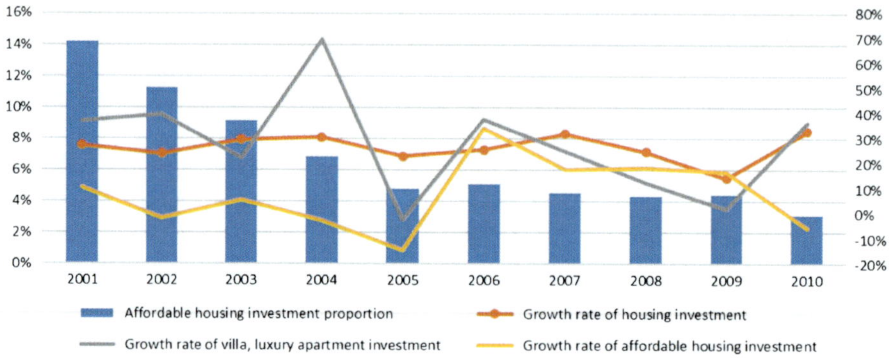

Fig. 7.1 Investment proportion of affordable housing and various types of housing growth rate. *Data source* China statistical yearbook

holds a decreasing trend, from 14 % in 2001 to 3 % in 2010. Specifically, prior period during 2001 to 2005 occurs great proportion drop, from 14 to 5 % mainly due to government negative attitudes towards to affordable housing construction for its endless drain on capital and inadequate absorption on revenue. Later in 2005–2010, affordable housing investment accounts for stable share in total housing investment, only decreasing 2 % points. At the same time, growing rates of affordable housing investment amount declines while villa, luxury apartments' increment rates are stable with a slight rise. Especially after 2008 financial crisis, China comes on slack fiscal policy like abolishing commercial bank credit constraint, which lead to more funds from developers are input into villas, luxury building construction. In addition, given the evidence that growth rate of affordable housing investment and its proportion on total investment reach to the lowest level since 2006, we can say affordable housing is snubbed.

On a macro level, affordable housing sales area continue the downward trend with contrast to ascending trend of villas, luxury apartments (see Fig. 7.2). Whereas, due to financial crisis in 2008, people hold pessimism attitudes towards real economy resulting in scale fluctuation which shows up as sales areas decease on villa, luxury apartments and increase on affordable houses. Later, housing

Fig. 7.2 Affordable housing and villa, luxury apartment sales area in 2001–2010. *Data source* China statistical yearbook

market gets warmed as the consequence of national macro-control. As the result, local's low passion on affordable housing construction leading to its sales reduction.

According to the data collected from the sixth national census, how did homeownership obtain the houses and what career affordable housing homeownerships are can be describe as follows:

1. Main sources to possess a house among homeownership are purchasing commodity houses, self-built houses and renting other houses, the rates are 27, 16 and 29 % respectively. People who living in indemnificatory houses are small, only 7 %, including 4 % of purchasing affordable houses and 3 % leasing low-rent houses.
2. Affordable housing homeownerships are divided into seven types in accordance with occupation, namely people working for state organs or enterprises, professional & technical personnel, officers, business services personnel, agriculture, forestry, husbandry and fishing personnel, transport equipment operators and others. A lot of low-income people gather in agriculture, equipment operators and business services fields, although theirs' total rate surpasses 50 %, the proportion needs improved if take the consideration of indemnificatory housing's original intention that is building for middle and low income groups.

7.3.2 Gegenwart Condition of Housing Vouchers

Different from brick subsidy of indemnificatory housing, housing vouchers introduces "capitation grant" innovatively. Based on qualification examination, middle and low income people will be subdivided into several groups following the principle of income levels, age structure, and number of family members, etc. to receive established subsidies. Then, these groups go to purchase (or rent houses) and pay developers (or owners) with housing vouchers as part of purchase fund (or charter money). Governments have guaranteed that developers or owners can exchange for cash with housing vouchers. In this way, a sufficient amount of commodity housing (new houses or resold apartments both are available) is certainly a momentous prerequisite for housing vouchers implementation. As for China, how is the status quo of its commodity housing, can it be a cornerstone of housing vouchers policy?

In fact, commodity houses referring to the houses that is legally permissible and can be traded freely in the market in our country appears at a time around 1980s, earlier than indemnificatory houses. Since part of commodity housing data is not available, the paper roughly snoops current development of commodity housing with the acquirable data.

Commodity housing completed sizes, sales areas and average sales price all show a good rise tendency since new century except 2008 due to financial crisis (Fig. 7.3). In the wake of urbanization, fast economic development make people hold optimism about the future of market which can explain the steady growth of

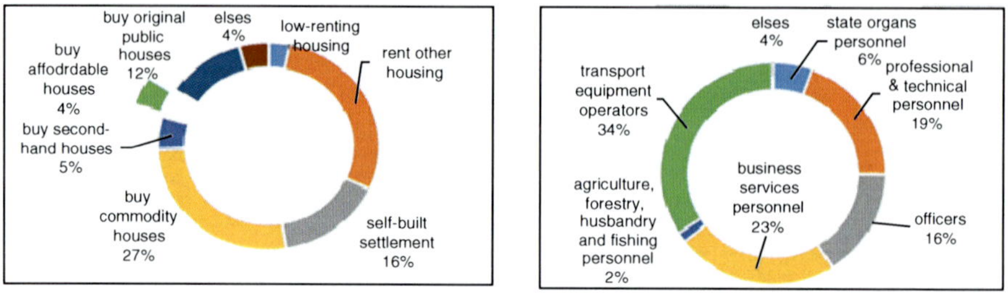

Fig. 7.3 Owner career and housing source situation. *Data source* Sixth nationwide population census

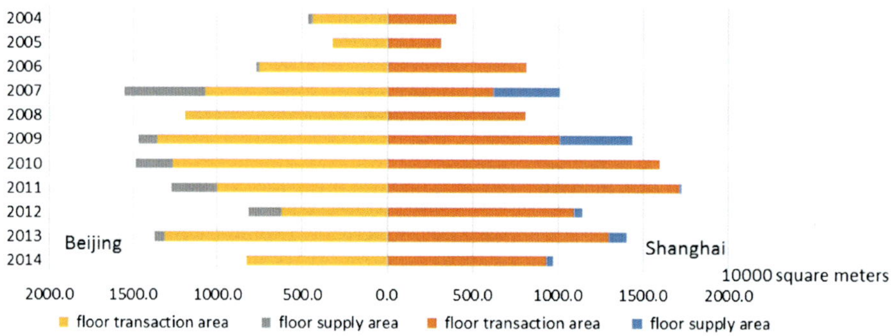

Fig. 7.4 Floor supply area and floor transaction area in Beijing, Shanghai. *Data source* CRIC (http://www.cricchina.com/)

investment intensity in housing market. Commodity housing sales areas first exceed its completed sizes in 2005 and reaches to 900 million square meters, adding up to commodity housing stock begins to be consumed promptly. New-type urbanization stresses more on coordinative development of urban-rural area as well as "people oriented" concept, such as allowing more migrants to become citizens. Large-scale population migration, from poor city to large cities, especially megalopolis area has been a significant social characteristic in China during urbanization. So, it is necessary to analyze urban housing stock together with supply and demand situation for the sake of determining whether reasonable or not to implement housing vouchers. This part tempts to interpret gegenwart condition of housing vouchers from the perspective of real estate construction floor areas, transaction area (Fig. 7.4) in two first-tier cities: Beijing and Shanghai, and commodity housing vacancy rate (Fig. 7.6) in all first-tier cities.

Supply and transaction areas in Beijing and Shanghai still receive growth in turnover fluctuations during 2004–2014. Beijing advocates reduction development for people's livelihood guarantee in 2010. Government cuts down construction land deliver quantity which leads to housing supply decrease in the second real estate

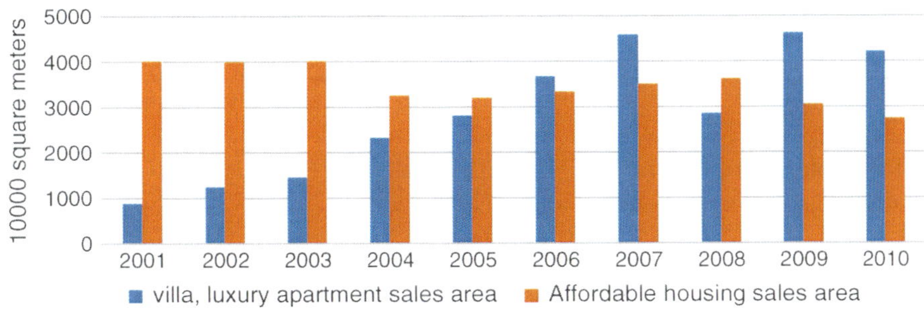

Fig. 7.5 Overview on commodity housing. *Data source* China statistical yearbook

market. Thus commodity housing supply starts gliding until 2013 when meets land market momentum. Developers expect that a large number of low price owner-occupied commodity housing will flock in markets in early 2014 which signifies remission state of demand exceeding supply. Given such mentality, we can see from Fig. 7.5, housing supply receives large rise in 2010 but declines in 2014. Of course, commodity housing in Shanghai can be interpreted in similarly way. As for commodity housing transaction areas in two cities, there is no big gap when contrast with supply floor areas. Overall, commodity housing supply areas in Shanghai are around 10 million square meters while Beijing varies from 7.5 million to 12 million square meters. If make a careful comparison, commodity floor areas are a little more than transaction areas in gross over the years.

Vacant housing belongs to a waste of resources in some extent for it is likely to reduce market vitality. When facing a huge volume of commodity housing in China, vacant housing is a problem that concerns. From what in Fig. 7.6, we can obviously find that China's commodity housing vacancy rate is up to 25.6 %, followed by affordable housing which is 23.3 %. Combining commodity housing

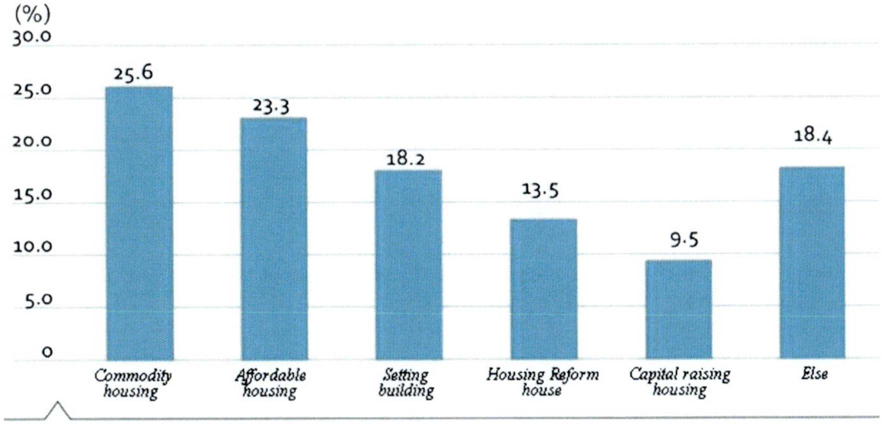

Fig. 7.6 Urban vacancy rates in 2014. *Data source* China household finance survey center

supply with vacancy rate, the paper concludes that domestic commodity housing stock could be favorable condition for housing vouchers policy implementation.

7.3.3 Housing Vouchers Versus Affordable Housing

Capitation Grant Versus Brick Subsidy

Indemnificatory housing policy generally has two types: capitation grant and brick subsidy. Previous years, indemnificatory housing policy always adopts "brick subsidy" method, namely constructing affordable housing, low-rental housing and etc. and then determine protect objects. The government provide various subsidies forms for developers in order to cut down housing price and pass these benefits to low income families indirectly, like exempting developer from land leasing fees, urban infrastructure or other administrative fees. On the contrary, whether brick subsidy are able to reconcile efficiency and fairness remains questionable. Evidence shows (see Fig. 7.3) almost half of the affordable housing possessed by people working for state organs or enterprises, professional & technical personnel and officers where low income groups proportion is small. We may infer maybe there is some transaction or rent-seeking behaviors between government and enterprises so that part of affordable housing drifts away from the scope of suitable main body. Housing vouchers are kind of "capitation grant", which delineates security object first and give housing funds later, vouchers could play a role in the payment behavior in the event of housing purchase. City new low-income residents, if meet certain criteria (such as age), could be incorporated into housing security groups. Different with brick subsidy, recipients receive housing vouchers directly whose subsidy denomination is affected by income level, household population and or so. Since housing subsidy is delivered as the form of vouchers instead of cash, people couldn't use these subsidy in other field except housing market. In addition, brick subsidy prevents recipients choosing houses at liberty while housing vouchers gives full respect for people's right to choose alternative housing because they can select freely-flow commodity housing instead of some affordable housing specified by the government or the developers.

One Market Versus Two Markets

From economic perspective, housing vouchers only involves one trading market while indemnificatory housing is concerned in two markets: first of all, indemnificatory housing, simple operation but lack maintain money, is the coexist production generated from planned economy and market economy; for another, housing vouchers can help absorb vacant commodity houses, including new houses and secondary houses, which indeed active secondary housing market effectively. However, one thing that we cannot ignore is housing vouchers draw people into housing market and give them housing subsidies may lead to market price fluctuate which in turn will give rise to policy failure. Though housing vouchers have a certain economic risk, affordable housing funding problems are more severe. As is known to all, the most prominent feature of affordable housing is its immediate policy effect, such as it can provide a lot of housing to meet requirement in a short time [27]. Given the reality in China's abundant housing supply, allow housing vouchers to be the main type of indemnificatory housing to guarantee low-income people with alternative houses instead of constructing indemnificatory housing on a large scale.

7.4 Housing Vouchers System

"Brick subsidy" indemnificatory housing procedure is simple if compared with housing vouchers. In other words, how to implement housing vouchers policy is a little more complex (Fig. 7.7):

1. Floating population can submit an application to the government to become recipients if they tally with some established standards.
2. Government reviews applicant qualification, and then decides to hand out subsides or not.
3. Recipients go to purchase (or rent houses) and pay developers (or owners) with housing vouchers as part of purchase fund (or charter money).
4. Developers can pay government with housing vouchers as part of land leasing fees, and owners can exchange for cash from government.
5. Purchased houses which can be listed circulation become common property of recipients and government

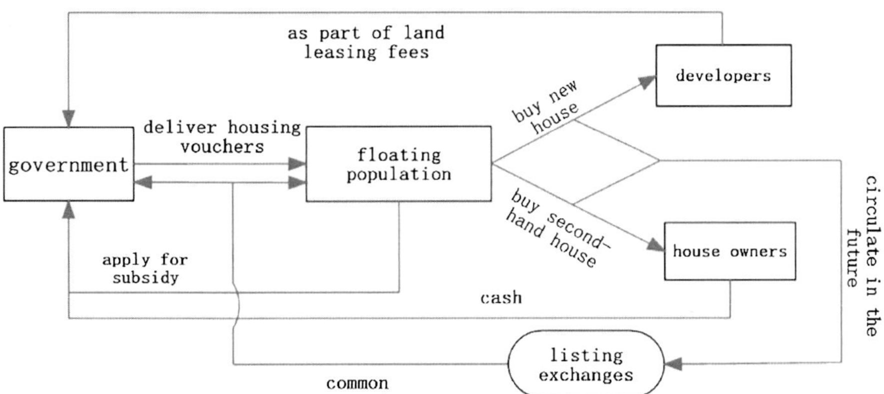

Fig. 7.7 Process map of housing vouchers policy

All above procedures relates to four aspects: housing vouchers' financial channels, subsidy objects, cash approaches and property allocation.

7.4.1 Financial Channels

Housing vouchers belongs to a kind of government subsidy essentially, so we need discuss something about financial channels. Differ from long cycle and large costing of indemnificatory housing construction, housing vouchers' expenditures are relatively lower. There are three main housing vouchers financial channels: part of land leasing fee, government's public budget as well as community donations. Officials could set up a special policy-oriented financial institution or establish partnerships with other relevant social management department to unify management operation for funds.

7.4.2 Subsidy Objects

Subsidy policies are required to set up a standard to determine who the subsidy objects are, how much should they be endowed. Policy failure always comes from rent-seeking caused by fuzzy rules. Based on new-type urbanization, floating people, or we say new low-income people living in cities become dominating subsidy objects. Housing vouchers promise to support people's housing demand the mount applies conform to some standards (such as age, income levels, property status). Of course, apply results sure to be publicized.

7.4.3 Cash Approaches

With the purpose of effective working capital, the flow of housing vouchers among government, floating population and developers must be an iterative process. Supposed that floating population get the housing vouchers from government and pay them to developers or owners, how developers or owners can do to return these vouchers back to government "financial warehouse"? We call this return behavior as housing vouchers cash. The paper suggests government should allow developer to disburse part of land leasing fee with housing vouchers; by the way, owners can utilize housing vouchers gained from sale to exchange cash for next housing purchase.

7.4.4 Property Allocation

Housing vouchers policy doesn't mean that government give recipients purchase money gratuitous. What does this mean? Both government and housing purchasers are the property owner and have the right to dispose housing according to theirs' own investment ratio. In order to avoid recipients embezzle state property through black-box operation, either one who possesses housing once sells the house, the other has the preemptive right. Certainly, government should encourage recipients to conduct housing repo and become independent house property owner. Only in this way can government obtain subsidies rolling cumulative effect.

7.5 Conclusion

At present, government relies on affordable housing construction to dealing with migrants' housing. As we discussed above, brick subsidy indemnificatory housing urges for much financial support, and at the same time, deprive person of free choice opportunity. Affordable housing construction sprang up early in 1990s, the indexes of these project, such as investment, sales area have been in a state of low growth when compare with other sorts of housing since recent years. Still, numbers of low-income people in community did not enjoy the benefits brought by indemnificatory housing construction. From another point of view, commodity construction booms along with passionate economic momentum. Large quantities of commodity housing comes on the scene, together with formidable skyrocket in sales event. For all this, unfortunately, some investigation shows that a considerable volume of commodity housing is vacant, a form of resource waste. Thus, in a manner of speak, the existing housing stock looks promising to meet numbers of migrants housing problem through "housing vouchers".

The most prominent feature of affordable housing is its immediate policy effect, applicable to acute shortage of housing market for it can provide a lot of housing to meet requirement in a short time. Given the reality in China's abundant housing supply, allowing housing vouchers to be the main type of indemnificatory housing to guarantee low-income people with alternative houses instead of constructing indemnificatory housing can avoid power rent-seeking. At the end of paper, we set up running process for housing vouchers policy and discuss on financial channels, subsidy objects, cash approaches and property allocation with the hope to provide certain reference significance for indemnificatory housing policy.

Acknowledgements The authors appreciate the kind financial support from the National Natural Science Foundation in China (71373231).

References

1. Chen J, Guo F, Wu Y (2011) One decade of urban housing reform in China: urban housing price dynamics and the role of migration and urbanization, 1995–2005. Habitat Int 35(1):1–8
2. Chou Y (2012) Introduction to occupational welfare. Shandong people's publishing house, Jinan, p 251 (Compile)
3. Liang Y (1998) Attention on monetization in distributing dwelling houses '98 a grand thought of China housing reform. China Prices Press, Beijing, pp 27–28
4. Yi X (2014) Excessive-priced housing becomes a domestic demand growth obstacle: over 90 % wage-earners cannot afford it. Economic Information Dail 2014(11):14
5. Zhang Q (2009) Public policy and social justice. Jilin People's Press, Changchun, pp 180–182
6. Wei J, Wei H (2010) Annual report on urban development of China. Social Sciences Academic Press, Beijing (Compile)
7. Jiang L (2006) Living conditions of the floating population in urban China. Hous Stud 21 (5):719–744
8. DeSilva S, Elmelech Y (2012) Housing inequality in the United States: explaining the white-minority disparities in homeownership. Hous Stud 27(1):1–26
9. Ratcliffe P (1999) Housing inequality and 'race': some critical reflections on the concept of'social exclusion'. Ethnic Racial Stud 22(1):1–22
10. Verdugo G (2011) Public housing and residential segregation of immigrants in France, 1968–1999
11. Chen Y (2010) What causes humble abode. Xiangyin 7:007
12. Huang Y, Jiang L (2009) Housing inequality in transitional Beijing. Int J Urban Reg Res 33 (4):936–956
13. Hiroshi SATO (2006) Housing inequality and housing poverty in urban China in the late 1990s. China Econ Rev 17(1):37–50
14. Zhang Z (2010) High housing price: an economic issue or a political issue-a political economic analysis. J Northwest Normal Univ (Soc Sci) 47(2):86–90
15. Li SM (2007) Transition to homeownership: implications for wealth redistribution. China's Emerging Cities: The Making of New Urbanism, pp 143–160
16. Goulden R (2011) Housing, inequality, and economic change in Syria. Br J Middle East Stud 38(2):187–202
17. Yeung SCW, Howes R (2006) The role of the housing provident fund in financing affordable housing development in China. Habitat Int 30(2):343–356
18. Long K (2013) Effective strategies to increase indemnificatory housing land supply. Econ Rev 10:40–42
19. Sard B (2001) Housing vouchers should be a major component of future housing policy for the lowest income families. Cityscape, pp 89–110
20. Priemus H, Kemp PA, Varady DP (2005) Housing vouchers in the United States, Great Britain, and the Netherlands: current issues and future perspectives. Hous Policy Debate 16(3–4):575–609
21. Peterson GE (2000) Housing vouchers: the US experience. In: Eugene SC et al. (Hrsg.) Vouchers and the provision of public services. Washington. S, pp 139–175
22. Priemus H (2003) Social housing management: concerns about effectiveness and efficiency in the Netherlands. J Hous Built Environ 18(3):269–279
23. Ma Z, Hao M (2014) Migrant workers' housing vouchers system: based on overall urban-rural development. Jiangsu Agric Sci 1:386–389
24. Ma M, Nie Xi, Xiaolin C et al (2014) Study on migrant workers' housing Vouchers in urban and rural reform experimental zone. J Chongqing Inst Technol 10:52–58

25. Liao Y, Li C, Pang X (2011) Primary investigation on indemnificatory housing fusion. Commercial Times 20:109–110
26. Hu X (2012) New research on Wuhan indemnificatory housing: based on housing vouchers. Jin Ri Hu Bei 7:39
27. Zheng Y, Li J (2009) Analyzing the subsidy pattern of the affordable housing in our country. J Party Sch Fuzhou 4:67–71

Chapter 8
Research on Evaluation of Public Engineering Core Value System Based on Stakeholder Theory

Jianjun She, Hu Cheng, Xin Yuan and Hongtao Wu

Abstract With the deepening of the urbanization process in China, the construction of public engineering also develop rapidly, public engineering has non-competitive and non-exclusive characteristics, the core of its decision-making is value thinking. Based on the philosophy of value and stakeholder theory, this paper firstly defines the core stakeholders of public works, using VBS method to analyze the value demands of key stakeholders, building core value system of public engineering, and using structure entropy method to determine the index weight, finally the paper study the core value system evaluation by Planguage method. The objective of the paper is to provide reference for the investment decisions theoretical and empirical studies of public engineering.

Keywords Public engineering · Stakeholder · Core values · Value evaluation

8.1 Introduction

With the rapid development of economy, the investment of public engineering is increasing. On the one hand, the implementation of public engineering meet the needs of the public, and promote the development of social economy. On the other hand, public engineering has caused environmental pollution, ecological damage,

J.J. She (✉) · H. Cheng
College of Civil Engineering, Southeast University, Nanjing, China
e-mail: shejj1016@126.com

H. Cheng
e-mail: 101001849@seu.edu.cn

J.J. She · X. Yuan
College of Civil Engineering, Nanjing Tech University, Nanjing, China
e-mail: 1142088865@qq.com

H.T. Wu
The First Branch of China Construction Third Engineering Bureau, Wuhan, China
e-mail: 772517523@qq.com

Y. Wu et al. (eds.), *Proceedings of the 20th International Symposium on Advancement of Construction Management and Real Estate*,
DOI 10.1007/978-981-10-0855-9_8

poor investment performance and a series of problems can't be ignored. Therefore, in order to make the construction of public engineering to get better development, evaluating the value of public works has very important significance. At present, domestic and foreign pay more attention to the engineering value itself and engineering ethics value, but the evaluation criteria, the research on the value of the public engineering are still in the nascent stage [1]. From the core stakeholder perspective, this paper use the VBS method to analyze the value requirement of the core stakeholders, and on this basis, building the core value system of public engineering, and evaluating the core value system by Planguage method. The final results of this paper is to improve the satisfaction degree of key stakeholders by realizing of the interests of stakeholders, and make the value of public engineering tends to be maximized.

8.2 Public Engineering Core Value Theory

8.2.1 Connotation of Public Engineering

Public engineering are the construction project built by the state as an investment in order to meet the main needs of urban development, public demand, the target population survival needs, security needs and other purposes [2]. Public engineering have the following characteristics:
(1) Public engineering is to meet the public needs, the provision and production of public goods (including services) of the project, it is usually fixed asset investment; (2) The ultimate owner of a public engineering is the government, and the government has decision-making and leading rights in the supply of public goods; (3) Public engineering pursue of public interest, with public goods and public welfare objectives.

According to the project whether there is a mechanism for charging, Public engineering can be subdivided as pure operating public engineering, quasi-operational public engineering and non-operational public engineering, this paper studies the quasi-operational public engineering, which have not only social but also economic benefits, such as: gas, subway, light rail, water, roads and so on.

8.2.2 The Core Value of Public Engineering

On the definition of "value", there are different arguments and debates in the academic, such as "Labor", "production cost" theory, the article defines the meaning of Value is that the usefulness of things, usefulness of the project to stakeholders constitutes the value of each stakeholder.

Engineering value is a kind of relationship between engineering activities and their results that meet human needs [3]. Compared with the general value, engineering value is a special kind of value, which is created and realized in the field of engineering activities. Based on the multi interests of public engineering, this paper study the value of the project and discuss the value of the interests of the main body. In view of the engineering value is the product of the engineering practice, the engineering practice in the field of different, engineering purpose that meet the needs of the main body will produce different engineering value, such as the economic value of the project, the political value of the project, the project's social value and military value engineering [4].

The core value of public engineering can be understood as "the demand of the core stakeholders of public project with the optimal allocation of resources under the premise of equity" [5], Project core value come from the analysis of the value of the public project stakeholders and a single abstraction core value should have the following characteristics: easy to understand, power, credibility, dynamic, stability, subjectivity [6].

8.3 Definition of Core Stakeholders in Public Engineering Investment Decision Stage

8.3.1 Identification of Stakeholders in Public Engineering

The stakeholders of public engineering refers to some individuals, groups and organizations, which have some interest in the project, and which will be affected by the project decision-making, implementation and operational activities, but also can affect the achievements of project objectives in varying degrees [7]. This paper uses the method of literature review to collect the interests of stakeholders involved in the public engineering investment decision-making stage, there are mainly the following 7, namely the government, investors, financial companies, environmental protection departments, supervision companies, design companies, the public [8].

8.3.2 Selection of Core Stakeholders in Public Engineering

According to Michel dynamic Changes of stakeholder theory, the core stakeholders of public project should have two attributes [9]. Attribute 1: have a significant influence in project and assume the corresponding risk. Attribute 2: able to reflect fairness at the core of social property. According to this standard definition, it is a normative analysis of the stakeholders of the public engineering investment decision-making stage, the results are shown in Table 8.1.

Table 8.1 Stakeholder normative analysis

Stakeholder	Attribute 1	Attribute 2	Stakeholder	Attribute 1	Attribute 2
Financial companies	✓	–	Environmental protection departments	✓	✓
Investors	✓	✓	Supervision companies	✓	–
The public	✓	✓	Design companies	✓	✓
Government	✓	✓	–	–	–

From the Table 8.1, the government, investors, the public, environmental protection departments and design companies meet the attribute 1 and 2, so the 5 stakeholders are defined as the core stakeholders of public engineering investment decision-making stage [10].

8.4 Construction of the Core Value of Public Engineering Investment Decision Stage

8.4.1 Value Demand Analysis of Stakeholders in the Investment Decision Stage

The 5 core stakeholders of public project investment decision-making stage have been identified. As to the stakeholder theory and the research method of VBS, the value needs of the core stakeholders are analyzed to extract the value elements of the core stakeholders [11]. Secondly, according to the previous literature integration, the various value elements are classified into the core value of public engineering criteria, the results of the Table 8.2.

According to the above results, the 21 value indexes are summarized and the value standard is formed by the index, On the basic of 12 value standards, the questionnaire survey was conducted by the 5 levels of Li Ke Liang's table method, and 1–5 points corresponding to the extent of the impact from low to high, and the reliability analysis of the data of the questionnaire survey was conducted. From Table 8.3, we can see that KMO test sample precision coefficient of $0.792 > 0.6$, with the probability of Bartlett spherical test gives a value of 0.000, less than the significance level 0.05, it is proved that 12 variables are not independent, and factor analysis can be done, and it is proved that the investigation is effective and reliable. Using SPSS statistical software to calculate, 13 value criteria indicators further refined to the value of the dimension. Using factor analysis, we can draw a small number of common factors representative, the results are shown in Table 8.4.

Table 8.2 Value elements and value criteria of core stakeholders in the investment decision-making stage

Use stage	Core value standard	Value element	Core stakeholder
Investment decision-making stage	X1 Project brings reasonable economic benefits to various stakeholders	1.1 Promoting regional economic development	1 Government
		1.2 Increasing local revenue	1 Government
		4.3 Raising wages, social welfare	4 The public
	X2 The project is completed according to regulations, with a high return on investment	2.1 Higher ROI	2 Investors
		3.1 Maximizing design costs	3 Design units
	X3 Project focusing on environmental protection	5.1 Low carbon environmental protection system	5 Environmental protection departments
	X4 Ecological environment meets the requirements of the Sustainable	4.1 Improving local air quality and reducing environmental noise	4 The public
	X5 Promoting local employment	1.3 Increasing the employment rate	1 Government
		4.2 Increasing employment opportunities	4 The public
	X6 Reasonable cost reduction	2.2 Reasonable cost reduction	2 Investors
		3.4 Low-carbon design, reducing waste	3 Design units
	X7 Project makes enterprises get a good reputation	2.5 Ensuring project quality and building brands	2 Investors
		3.4 Creating design brand, improving customer satisfaction	3 Design unit
	X8 Establishing a good working relationship with all parties	2.4 Establishing long-term cooperation with all the parties	2 Investors
	X9 Promoting technological innovation	1.4 Promoting technological progress, scientific and Technological Development	1 Government
		3.2 Design optimization and technological progress	3 Design units

(continued)

Table 8.2 (continued)

Use stage	Core value standard	Value element	Core stakeholder
	X10 Strengthening to promote local culture	1.5 Influencing cultural education of the residents	1 Government
		3.3 Design meets the surrounding environment, history and culture	3 Design units
		4.4 Cultural heritage	4 The public
	X11 Protection of historical monuments	5.2 Reducing loss of monuments	5 Environmental protection departments
	X12 Promoting sustainable industrial development	2.3 Training highly qualified personnel	2 Investors
		1.1 Rich industrial structure	1 Government
		3.2 Design optimization and technological progress	3 Design units

Table 8.3 MKO measure and Bartlett test table

KMO and bartlett test		
Sampling sufficient Kaiser-Meyer-Olkin metric		0.792
Spherical degree test of Bartlett	Approximate chi square	911.152
	Df	15
	Sig.	0.000

Due to the accumulated variance contribution rate, the cumulative variance contribution rate of the first 5 factors is 82.1 %, which has given a full summary of the majority of factors, so the first 5 factors are appropriate. This can play a role in reducing the dimension and simplifying the data. In order to more clearly reflect the relationship between the variables of factors, using maximization of variance orthogonal rotation (Varimax) to give the rotation factor loading matrix, as shown in Table 8.5.

According to the factor load matrix, the 5 kinds of factors are summarized, and the factor variables are named, economic value, social value, environmental value, technological value and cultural value, the specific classification is shown in Table 8.6. Therefore, the value index system of the core stakeholders of public works is finished.

Table 8.4 Total variance explained

Ingredients	Initial eigenvalues			Extracting square sum and loading			Rotating square sum and loading		
	Total	Variance (%)	Cumulative (%)	Total	Variance (%)	Cumulative (%)	Total	Variance (%)	Cumulative (%)
1	3.519	29.328	29.328	3.519	29.328	29.328	2.860	23.833	23.833
2	1.984	16.532	45.861	1.984	16.532	45.861	2.139	17.825	41.658
3	1.627	13.556	59.416	1.627	13.556	59.416	1.722	14.351	56.009
4	1.420	11.834	71.250	1.420	11.834	71.250	1.643	13.688	69.697
5	1.302	10.851	82.101	1.302	10.851	82.101	1.488	12.404	82.101
6	0.667	5.559	87.660						
7	0.560	4.669	92.329						
8	0.369	3.072	95.401						
9	0.258	2.151	97.552						
10	0.179	1.493	99.044						
11	0.076	0.637	99.681						
12	0.038	0.319	100.000						

Extraction Method: Principal Component Analysis

Table 8.5 Rotation factor loading matrix table

	Ingredients				
	1	2	3	4	5
X1	0.085	0	−0.334	0.701	0.395
X2	−0.267	0.184	−0.306	0.795	−0.190
X3	−0.229	0.068	0.698	0.230	0.387
X4	−0.217	0.004	0.888	−0.041	−0.187
X5	0.948	−0.019	−0.072	−0.119	0.056
X6	−0.095	0.249	−0.218	−0.669	0.356
X7	0.957	−0.069	−0.166	−0.001	−0.041
X8	0.882	−0.181	−0.186	0.031	−0.044
X9	0.094	0.814	0.006	0.017	0.432
X10	−0.017	−0.028	0.020	−0.084	0.873
X11	−0.218	0.232	0.126	0.159	0.787
X12	0.161	0.838	−0.332	0.105	0.147

Extraction Method: Principal Component
Rotation Method: a Kaiser standardized orthogonal rotation method

Table 8.6 Public factor explanation

Public factor number	Value standard variable	Name public factor
F1	X5, X7, X8	Social value
F2	X1, X2, X6	Economic value
F3	X3, X4	Environmental value
F4	X9, X12	Technological value
F5	X10, X11	Cultural value

8.4.2 Core Value Index System of Public Engineering Investment Decision Stage

The value demands of core stakeholders in the investment decision stage are analyzed by using VBS (value breakdown structure) method firstly. On the basis of which, the 12 value standards of core stakeholders are extracted, then the 12 value standards can be classified as 5 value dimensions through factor analysis method. Finally, the core value system of public engineering investment decision stage is built, shown in Fig. 8.1.

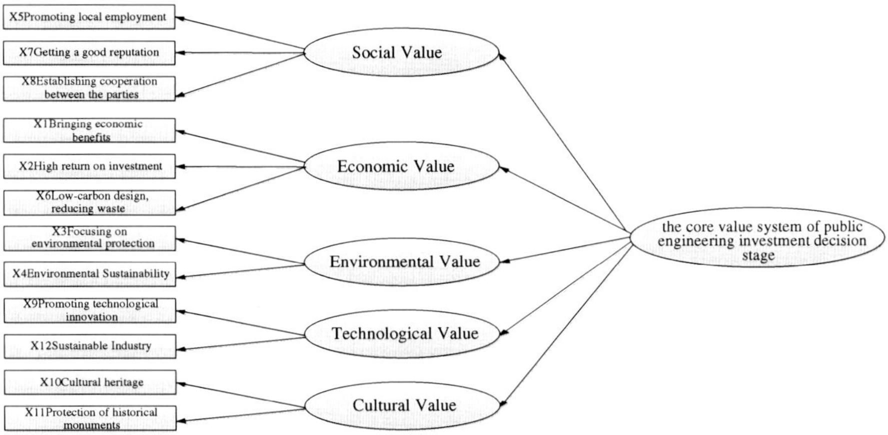

Fig. 8.1 Core value system of public engineering investment decision stage

8.5 Evaluation Method Based on Structure Entropy Weight and Planguage Method

Structure Entropy method is used in this paper to determine the core values of public engineering evaluate the weight, the method of the Delphi expert opinion survey method and fuzzy analysis method combined to form a "typical sort", and then calculate the entropy, and analyze "Blind degree" [12]. It is a kind of qualitative analysis and quantitative analysis method for determining the weights of the index weights.

8.5.1 Index Weight Determination

This paper use the Delphi method and extract experts in the relevant conditions as the research object, according to the importance of the typical sort of indicators, respectively, the most important, more important, important and unimportant. If that is the most important index, the value is 1 point, the more important value is 2 point, and so on, allowing several indicators that are equally important and "blind" analyzing the investigators survey results, denoted by Q_j.

$$Q_j = \left| \left\{ \begin{array}{c} [\max(b_{1j} + b_{2j} + \ldots + b_{kj}) - b_j] + \\ [\min(b_{1j} + b_{2j} + \ldots + b_{kj}) - b_j] \end{array} \right\} / 2 \right| \quad (8.1)$$

$$b_{ij} = \mathrm{u}(a_{ij}) = -\frac{\ln(m - a_{ij})}{\ln(m - 1)} \quad (8.2)$$

$$b_j = (b_{1j} + b_{2j} + \ldots + b_{kj})/\mathrm{k} \quad (8.3)$$

Its b_{ij} is the corresponding membership function value of a_{ij}, b_i is the average recognition of the experts. a_{ij} is expressed as a i expert for the evaluation of the u_j of the j index, m is used as the conversion parameter, and the m = j + 2. Experts on the overall awareness of the indicators, denoted by x_j, then:

$$x_j = \mathrm{b_j}(1 - \mathrm{Q_j})x_j > 0; a_j = x_j / \sum\nolimits_{i=1}^{n} x_i \quad (8.4)$$

L = {a_1, a_2, …, a_n} is called the index set U = {u_1, u_2, …, u_n} of the weight vector. Repeat the above steps, the i of typical survey indicators u_j weight value y_{ij}. Entropy using a structured method to determine the credibility of the five investigations affecting the right of people weight x_j, the normalization process:

$$b_j = \sum_{i=1}^{m} x_{iy_{ij}} / \sum_{j=1}^{n} \sum_{i=1}^{m} x_i y_{ij} \quad (8.5)$$

W = {b_1, b_2, …, b_n} is the evaluation index set U = {u_1, u_2, …, u_n} of the final weight vector, the final value evaluation index weight determined.

8.5.2 Planguage Evaluation Method

Planguage Overview

When the value comprehensive evaluation of Public engineering, we need to consider the project in many indicators of economic, social, environmental, etc., but these tend to be similar to the performance indicators, multiple qualitative indicators of reliability, security and so on, it is difficult for a reasonable measurement. This paper used the Planguage evaluation method, the main advantages lie in qualitative analysis. Because the key words of Planguage contain all the important dimensions, the possibility of missing information is reduced. This paper mainly uses the following four key words, as in Table 8.7.

Table 8.7 Planguage keywords

Keywords	Standard meaning
METER	Standard measurement process or method
MUST	to reach the required minimum level in order to avoid failure
PLAN	Planning to achieve the level of success
WISH	The desired results through effective means

Quantitative Evaluation of the Core Value of Public Engineering and Ratings Explained

The above indicators can be quantified by using Planguage, for Planguage quantitative indicators, compare the actual value of the project with the three values of MUST, PLAN and WISH. If the actual value of the project and the description of the MUST, then get 1 point, to reach 3 points in PLAN, WISH was 5 points, the situation between the two levels, according to the actual situation score S, do not meet the MUST index deducted 5 points, so the score range: $1 \le S \le 5$ or S = −5.

Final Score Calculation and Judging Rating Reference Value

The basis score for the project is 5 points, which is calculated according to the following formula:

$$\begin{aligned}\text{Core value total evaluation score} &= \text{Reference points } 5 + \text{a} \\ &= 5 + \sum_{i=1}^{5} b_i \sum_{j=1}^{n} c_k d_k\end{aligned} \tag{8.6}$$

where: b_i represents the weight of the criteria layer; c_k represents the weight of a judgment index; d_k represents the measured value of a certain judgment.

The core value of public engineering investment decision-making stage is divided into 5 levels:

(1) the core value of public engineering can be achieved completely (green): the total score of the project is 8–10; (2) the core value of public engineering can be achieved (blue): the total score of the project is 6–8; (3) part of the core value of public engineering can be achieved (yellow): the total score of the project is 4–6; (4) the core value of public engineering is basically not realized (red): the total score of the project is 2–4; (5) the core value of public engineering is completely not achieved (black): the total score of the project is 0–2 (Table 8.8) [5].

8.6 Case Study—A Case Study of Nanjing Metro

Nanjing subway is the Urban Rail Transit which services in Nanjing city and Nanjing metropolitan area within the urban. As of May 2015, Nanjing subway has 6 lines and 121 stations, Line length of 225.4 km, average daily traffic of more than 2.2 million visitors. To the end of 2014, Nanjing Subway Line 3 had been completed and began operations in April 2015.

The completion of the line 3 will become the second north-south traffic backbone line through both sides of the Yangtze River in Nanjing. It will have a vital effect to achieve complementary advantages for various transportation hub along

Table 8.8 Public Engineering core value "colored map" score table

Judgment stage		Public Engineering did not realize the core values (0–2)	Public Engineering did not achieve the basic core values (2–4)	Part of the core values of public Engineering can be achieved(4–6)	Public Engineering fundamental core values can be achieved (6–8)	Public Engineering core values are fully able to achieved (8–10)
Full life cycle	Investment decision stage					
	Construction stage					
	Use stage					

Table 8.9 Core value weights and the final score of Nanjing metro

Level indicators	Weights	Secondary indicators	Weights	Overall weights	Index score
F1	0.245	X5	0.301	0.074	4
		X7	0.347	0.085	4
		X8	0.352	0.086	3
F2	0.232	X1	0.362	0.084	5
		X2	0.347	0.081	4
		X6	0.291	0.068	3
F3	0.198	X3	0.491	0.097	4
		X4	0.509	0.101	4
F4	0.176	X9	0.457	0.080	4
		X12	0.543	0.096	4
F5	0.149	X10	0.523	0.078	3
		X11	0.477	0.071	5

and the sharing of resources between the main city and new urban. It has very important significance that how to evaluate the core value of the subway project.

8.6.1 Quantitative Analysis of the Core Value of Nanjing Metro

Based on the structure entropy method to determine the weight, and the use of expert scoring method, through expert scoring tables to the experts consulted and 3 feedback and adjustment, and ultimately determine the Nanjing Metro Line 3 of each index score, as shown in Table 8.9.

8.6.2 The Final Score of Nanjing Metro Core Values

$$\begin{aligned}\text{The final score of Nanjing Metro Core Values} &= 5 + \sum\nolimits_{i=1}^{5} b_i \sum\nolimits_{j=1}^{n} c_k d_k \\ &= 5 + (0.074 * 4 + 0.085 * 4 + 0.086 * 3) \\ &\quad + (0.084 * 5 + 0.081 * 4 + 0.068 * 3) \\ &\quad + (0.097 * 4 + 0.101 * 4) + (0.08 * 4 + 0.096 * 4) \\ &\quad + (0.078 * 3 + 0.071 * 5) = 8.669\end{aligned}$$

In summary, the project total score between 8 and 10 points, Nanjing engineering core values achieved completely.

8.7 Conclusion

This article analyzes the stakeholders of public engineering, series problems of the core values index system and evaluation methods. By using the existing value philosophy and stakeholder theory, the use of SPSS data processing analysis method and Planguage evaluation method, we get the core value system of public engineering and give the evaluation criteria of the core value of public engineering. This paper is only a preliminary reference, there are still many problems need to be improved through repeated practice, the purpose of the article is to give some constructive reference for the evaluation of public engineering value.

Acknowledgements The paper is supported by philosophy and social science research project funds in Jiangsu colleges and universities, NO. 2013SJD720003.

References

1. Kolltveit BJ, Karlsen JT, Gronhaug K (2004) Exploiting opportunities in uncertainty during the early project phase. J Manage Eng 20(4)
2. Hu F (2012) Performance evaluation of large public works projects. Hunan University
3. Sun ML, Li B (2015) Analysis of project value. Proj Manage Technol 65–69
4. Morns P, Jamieson A (2004) Translating corporate strategy into project strategy: realizing corporate strategy through project management. Newtown Square. Project Management Institute, Pennsylvania
5. Wang J (2008) The successful standard of large engineering projects. Central South University
6. Miller R, Lessard D (2001) The strategic management of large engineering projects: shaping risks, institutions and governance. MIT Press, Cambridge
7. Yin Y, Hu J (2006) Research on the success criteria of public projects based on the core value of stakeholder analysis. China Soft Sci 149–155
8. Bai L (2009) Based on the full life cycle of water conservancy project stakeholder classification management of. Construction economy, pp 98–100
9. Li Y (2010) Performance evaluation index system of public project based on stakeholder theory. Tianjin University of Technology
10. Mwita JI (2000) Performance management model. J Public Sect Manage 40–51
11. Chen B (2007) The value of the study on the value of the multidimensional perspective. J Hubei Univ Econ (Humanit Soc Sci Ed) 15–16
12. Wang QR (2003) The value of urban green space and its assessment

Chapter 9
How Do Top Construction Companies Diversify in the International Construction Market?

Meng Ye, Weisheng Lu, Kunhui Ye and Roger Flanagan

Abstract Globalization has created an international market that allows construction companies transcend traditional national boundaries and conduct business overseas. With new opportunities being brought to contractors, competition also grows exponentially in this market. Diversification is frequently adopted by these contractors as a strategy by the contractors for either growth, or risk management, or both in this competitive environment. However, the pattern of diversification has not been well measured, mapped, and analyzed. The aim of this research is to develop a Diversify Index (DI) and examine international contractors' diversification pattern. The data is from the Top 225 International Contractors' reports ranked by ENR (Engineering News-Record) from 1995 to 2014. Distributions of the DI were explored first and then case studies were used to investigate the specific diversification strategies adopted by the top international contractors. It is discovered that a downward trend of DI with subsequent changes in rankings expresses that larger contractors increasingly adopt various diversification strategies in international competition. The results provide valuable sights on the relationship between the competitive success and their diversification strategies as well as the tendency of diversification strategies adopted by top international contractors in different regions.

Keywords International construction · Diversification · Market competition · Competitiveness

M. Ye (✉) · W. Lu
Department of Real Estate and Construction, Faculty of Architecture,
University of Hong Kong, Pokfulam, Hong Kong, China
e-mail: megan828@hku.hk

K. Ye
Faculty of Construction Management and Real Estate,
Chongqing University, 400045 Chongqing, China

R. Flanagan
School of Construction Management and Engineering,
University of Reading, Reading RG6 6AW, UK

Y. Wu et al. (eds.), *Proceedings of the 20th International Symposium on Advancement of Construction Management and Real Estate*,
DOI 10.1007/978-981-10-0855-9_9

9.1 Introduction

Globalization attracts lots of attention from last century on, which brings a larger market, richer resources as well as advanced technology, fast transportation, convenient communications and effective knowledge transfer. All of these have helped lower traditional barriers and encourage firms from different countries to compete and enlarge their markets. With the globalization of the world economy, today's construction business is fast becoming an internationally interdependent marketplace (Lu et al. [1]). Many large construction firms operate not only in their home country but also across country board to pursue the long-run profits, balance development, make use of resources, increase assets turnover, and most importantly, meet the market demands. Construction is a demand-oriented industry; changes of demands in one marketplace affect the profits of construction firms directly. Compared to the developed countries, less developed countries or regions, such as some Asian countries, Middle East and African area, have more demands in construction. High-speed development of economy and the progress of urbanization raise the demands of buildings and basic infrastructural services, such as electricity reticulation, roads and other means of transport, port facilities, and water supply facilities and so on. According to S.N. Subrahmanyan ([2], p. 7), senior executive vice president of India's Larsen & Toubro Ltd., India and China will continue to be the most important construction markets, driven primarily by the irreversible forces of urbanization and infrastructure growth. Such unbalanced development of region economy drives internationalization of construction firms, making the global more internationalized than ever. According to Engineering-News Records (ENR) (2014), the ENR Top 250 International Contractors (TIC 250) had \$543.97 billion in contracting revenue in 2013 from projects outside their home countries, up 6.4 % from \$511.05 billion in 2012, and up 224.78 % from \$167.5 billion in 2004.

With new opportunities being brought to international contractors, ever-growing competition also exists. Faced with an increasingly competitive international construction market, international contractors need to enhance their competitive advantages for surviving and sustainable development. Diversification become one of the solutions for fierce competition [3]. Companies diversify to compensate for technological obsolescence, to distribute risk, to utilize excess productive capacity, to reinvest earnings, to obtain top management, and so forth [3]. Since then, many studies have focused on this topic, but the results vary due to different theories. On the basis of resource-based theory and economy of scope, diversification can be driven by employing the firm's intangible assets to enter into different industries (Arasti et al. [4]), and it is regarded as one way to make full use of the surplus resources or core ability and decrease transaction costs [5]. Prahalad and Hamel [6] advocate that diversified corporations should not be seen as a portfolio of discrete businesses, but as a collection of competitively important competencies that could be used in different products and markets. In this view, diversification is conducive to the growth of the corporations and related diversification is better than unrelated diversification in the case of effective use of resources. Organizational management theory provides another view into diversification. Gary [7] presents the higher initial slack strategy, which shows a policy in which management embarks on a diversification move only when there is at least 10 % slack in the organization. Additional organizational slack enables management to maintain the balance between shared resources and total workload demands with the extra buffer of excess resources in place before the diversification. Agency theory suggests that diversification can benefit managers and hurt shareholders [8]. Chandler [9] believes that diversification would increase the complexity of organizations so that new management problems would be brought. Based on risk management theory, Berger and Ofek [10] argue that one of the potential benefits of diversification arises from combining business with imperfectly correlated earnings streams since debt capacity can be increased by increasing interest tax shields. Chang and Thomas [11] present their view that diversified firms would avoid specific risks. In this view, unrelated diversification performs better than related diversification.

The prolific theories provide different conjectures about diversification. Meanwhile, diversification should vary from one industry to another too. In the construction industry, the output is not specific transferable goods with short production cycle, which is so different from those in the manufacturing industry. Diversification in the construction industry, especially to the international markets, needs more labor, materials, machinery, and expertise to the host markets rather than just exporting commodities from the home countries. Kurien [12] found that some construction contractors prefer to seek one particular project to enter a new market. However, how construction companies diversify in the international construction market is not entirely clear.

The primary aim of this study is to provide the empirical findings on the overall diversification pattern for international contractors. To achieve the aim, several research questions, each representing a specific objective, should be identified. It is mentioned by [13] that firms diversify for proactive and defensive reasons. Poorly performing firms would have high motivation to search for alternate business sectors even if they are risker while highly performing firms may use diversification to improve their performance by reducing their market risks. But it is known that large firms would be rich in resources, which is beneficial for diversifying to other sectors. As international contractors are ranked based on the international revenue, one question could be raised that Q1: Would the diversification levels vary with the rankings following a specific rule? Theories related to diversification tell us different views on it, which give us a signal that diversification could not be chosen as nostrums. "The more, the better" is not the rule for any diversified firm. There may exist one proper diversification level for the international contractors. There comes the second question that Q2: What is the proper diversification level for the international contractors? Along with the rapid process of economic globalization, competition in the international market becomes fiercer. Whether diversification becomes one of the strategies accepted by the contractors or not is a big question. So, we can raise the third question, Q3: Would the diversification level be improved as time goes on?

9.2 Methods

9.2.1 Sample

Engineering News-Record (ENR) has published annual data about contractors and designers in the U.S.A and some other international markets every year since 1958. The Top 225 International Contractors (TIC225) List (has been changed to Top 250 International Contractors [TIC 250] since 2013 due to increase of the contractors) is selected as the sample given its comprehensiveness and reliability [1] in describing the presence of top contractors in the international market. International revenues, gross revenues, new contract awards, and previous year's rank of the top 225 or 250 international contractors are presented in the lists. Meanwhile, the lists also provide the percentage of international revenues in different construction business sectors for ranked firms. If the sum of the percentage is less than 100 %, it is explained that other kinds of businesses contribute to the international revenues. Nine construction business sectors are classified by ENR: general building (GB), manufacturing (MNF), power (PW), water supply (WS), sewerage/solid waste (SSW), industrial process/petroleum (IPP), hazardous waste (HW), and telecommunications (TC). Two of them, water supply and telecommunications, have been added to the lists since 2001.

9.2.2 Data

The data used in the current study cover 20 years from 1995 to 2014. Top 225 International Contractors Lists in 20 years are collected in an excel profile. What's more, top 100 international contractors ranked in 2014 are selected to track the history, by which, one can identify how the contractors' businesses have changed through diversification over the data period. For tracking, an excel profile was developed to reorganize separate annual data sets for each of contractors. As of result, about 36 out of 100 contractors have lasted in the ENR list through the entire 20 years. The other 64 contractors have had shorter durations. The average length of appearance of the 100 firms is 14.3 years.

9.2.3 Measure of Diversification

Various measures of firm diversification have been developed over the past decades. The current study uses entropy measure as shown in Eq. [9.1] to calculate the levels of diversification, which is called diversify index (DI). This measure was first applied to the assessment of diversification by [14] and it has several mathematical advantages: the measure is additive (it can be divided into additive elements that contribute to the total) and sensitive (it accounts better for low levels of diversification compared to other measures) [15, 16].

$$\text{Entropy} = \mathrm{K}\sum_{i=1}^{N} pi \times \ln\frac{1}{pi} \tag{9.1}$$

In Eq. [9.1], pi is the revenue percentage of the firm's international revenues in the ith construction business sector, N is number of market sectors plus 1 (most sums of business percentage are less than 100 %, one more sector is assumed for the balance; therefore, N = 8 before the year 2000, and N = 10 after the year 2000), K is the coefficient of standardization, one dividing by the logarithmic value of N, which is the possible maximum value, so that all measures are scaled onto the interval [0.0, 1.0]. In information theory, entropy is a measure of unpredictability of information content. The entropy of the message is its amount of uncertainty; it increases when the message is closer to random, and decreases when it is regular. So, when the firm diversity index (DI) is closer to 1, the level of diversification is higher while there is no diversification when the DI is 0.

9.3 Analysis, Results and Discussion

9.3.1 Results Based on the Top 225 International Contractors List

Entropy measure was used to calculate diversify index (DI) of the top 225 international contractors from 1995 to 2014 and an average DI could be calculated for each ranking. Therefore, ignoring the specific contractors, we can get a table including the averages of DI and rankings. Figure 9.1 shows us the relationship between the averages of DI and rankings.

The linear trend line in Fig. 9.1 shows us a downward trend of DI with subsequent changes in rankings. We divide the rankings into 4 groups, which are [1–56], [57–112], [113–168], [169–225] and use the software SPSS to do the analysis of variance. *P*-value for test of homoscedasticity is 0.904 showing the equal variance of four groups while *p*-value for ANOVA analysis is 0, showing the significant difference among the means of four groups. Figure 9.2 shows the distributions of the four groups, in which the average DIs for firms ranked 1 to 56 are between 0.3 and 0.55, while the firms ranked 169–225 only have the average DIs below 0.3.

According to the analysis above, the first question (Q1) could be answered. Diversification levels measured by diversify index (DI) show a downward trend with subsequent changes in rankings. Since rankings in top 225 international contractors list are based on the international revenues, which on one hand represent

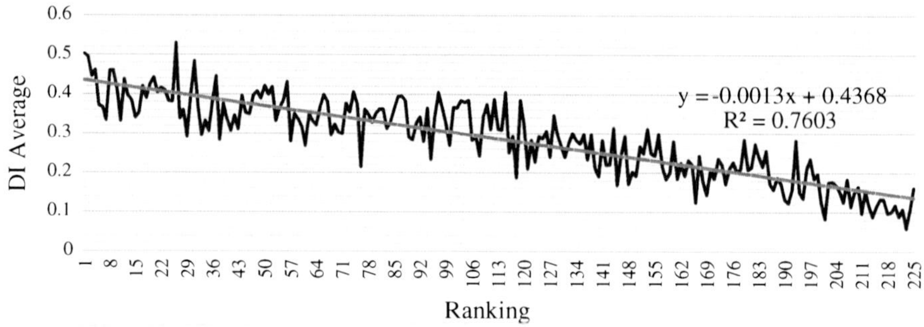

Fig. 9.1 The relationship between averages of diversity index (DI) and rankings

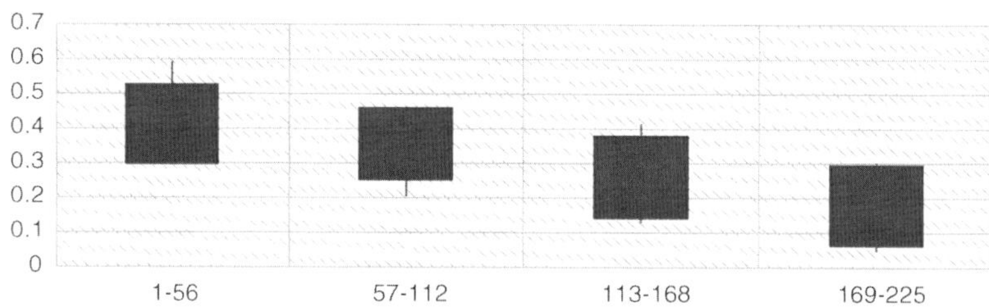

Fig. 9.2 The relationship between averages of diversification index (DI) and rankings in groups

the size of the firms and on the other hand stand for performances, the results can approximately show that large contractors may be more diversified than smaller contractors.

9.3.2 *Results Based on the Top 100 International Contractors Ranked in 2014*

As mentioned above, top 100 international contractors ranked in 2014 are selected to track the history and deal with the problems accordingly. Separate annual data sets for each of contractors from 1995 to 2014 are reorganized. The answer to Question 1 has been mentioned and results would be tested based on the top 100 international contractors. The other research questions mentioned are answered.

Q1: Would the diversification levels vary with the rankings following a specific rule?
Each contractor has its ranking and diversification index (DI) each year accordingly. Averages of DI and averages of rankings could be calculated for each contractor to do the regression analysis. The scatter diagram and linear trend line for the relationship between averages of DIs and averages of rankings are shown in

Fig. 9.3 (in which, the little dots represent the selected contractors). The negative correlation relationship could be reflected in the figure, which is to say, contractors with high rankings would be more likely diversified than those with lower rankings, although the results seem to be less obvious. For large contractors, they may have more resources, including intangible assets and tangible assets. On one hand, large contractors hold more mechanical equipment and labors to do work in different business sectors in the international market, and they may be less worried about the cash flows and hold the capital to diversify; on the other hand, large contractors with good performance may keep in touch with the local government in the overseas market, providing good opportunities to perform not only in one business sector.

Q2: What is the proper diversification level for the international contractors?

Frequencies of contractors in different diversification levels are calculated for each year, which are shown in Fig. 9.4. The patterns represented by the distribution of the contractors are similar over 20 years from 1995 to 2014. In Fig. 9.5, the maximum and minimum frequencies show the range within which the largest variations have occurred during the data period. Both figures show the highest

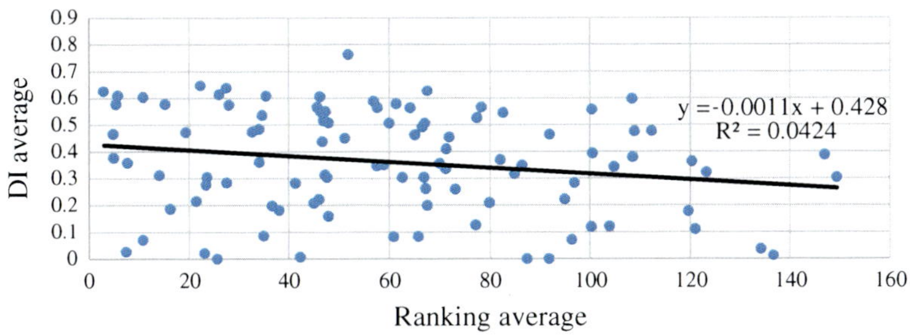

Fig. 9.3 The relationship between averages of diversification index (DI) and averages of rankings for the top 100 contractors ranked in 2014

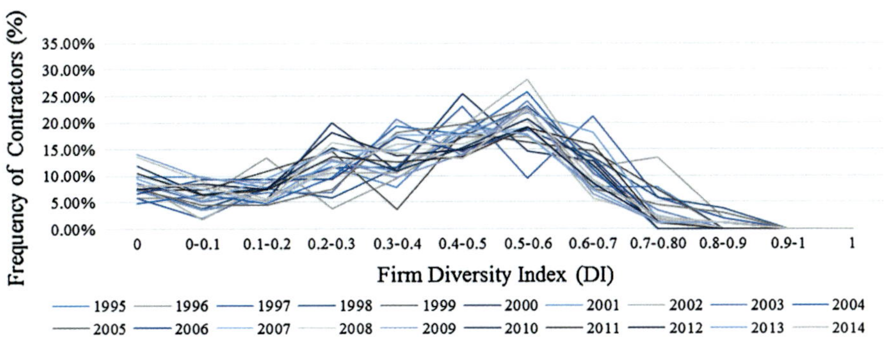

Fig. 9.4 Frequency of contractors in different levels of DI for each year

frequency in the level between 0.5 and 0.6. The frequency of specialized contractors range from 5 to 15 % with DI chose to zero. And on average, about 9 % of contractors were fully specialized. Most of the contractors (above 85 %) performed their business in multiple sectors. However, there are few contractors with DI = 1.0, the maximum level of diversification, which can be attained with a uniform distribution of revenues over all sectors.

Diversification levels with DI from 0.5 to 0.6 contribute to the highest proportion almost in every single year. On average, almost 20 % of the contractors has the diversification level with DI between 0.5 and 0.6, which is the most popular diversification level. Among the top 100 international contractors ranked in 2014, about 36 of 100 contractors have lasted in the ENR list through the entire 20 years. The average length of appearance of the 100 contractors is 14.3 years, which show a well sustainable ability in the international competitive environment. The most popular diversification level seems to stand for the proper diversification level, with DI ranging from 0.5 to 0.6.

Q3: Would the diversification level be improved as time goes on?

Figure 9.6 shows the total number of contractors among the top 100 international contractors ranked in 2014 on a yearly basis. Those contractors without specific data of business sectors were excluded. Therefore, we get the contractors from 51 in 1995 to 98 in 2014. The averages of diversity index (DI) for each year could be

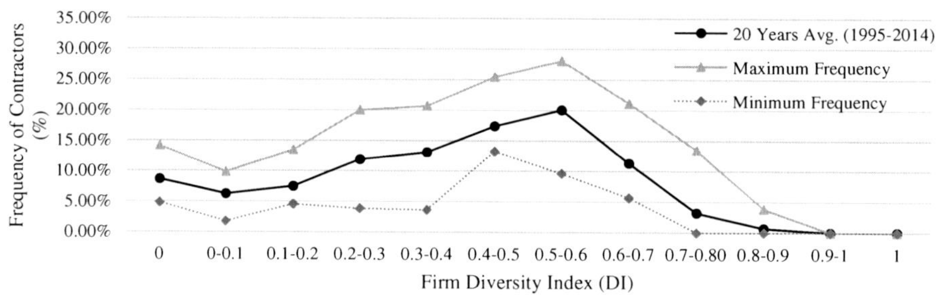

Fig. 9.5 Overall frequency of contractors in different levels of DI

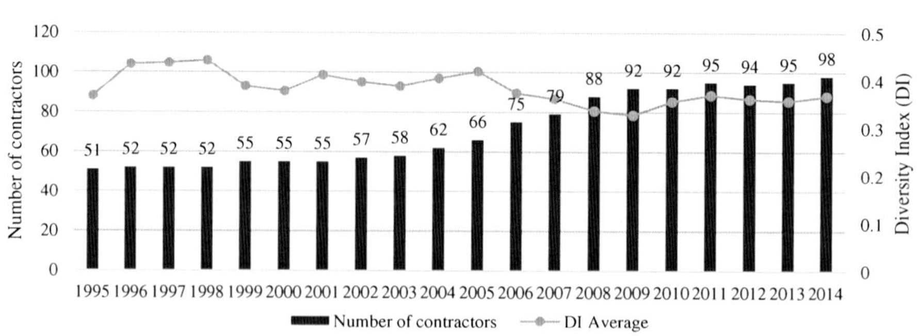

Fig. 9.6 Number of international contractors and averages of DI for each year

calculated, showing a flat level, which hold the level between 0.35 and 0.45. Although competition in the international construction market may become fiercer with the progress of globalization and internationalization, the average diversity index remains the almost the same even a little downward level. It seems that diversification is not the first choice for competing in the international construction market. Contractors should balance all the strategies to compete with other contractors rather than just diversifying to other business sectors.

9.4 Conclusions

The study has analyzed the pattern of international contractors' market diversification using the data of Top 225 International Contractors ranked by Engineering-News Record, which cover the past 20 years from 1995 to 2014. Diversify index (DI) measured by entropy is used to reflect the diversification level of the international contractors. A downward trend of DI with subsequent changes in rankings (225 rankings in total) expresses that larger contractors increasingly adopt more diversification strategies in the international competition.

For more detailed analysis, longitudinal data of the ENR top 100 international contractors ranked in 2014 were reorganized by tracking individual contractors' annual DI and rankings over the data period. The negative correlation relationship between the average DI and average rankings shows that contractors with high rankings would be more likely diversified than those with lower rankings. This confirmed the resource-based view, which implies the competitive advantage of a firm lying primarily in the application of a bundle of valuable tangible or intangible resources at the firm's disposal. On average, almost 20 % of the contractors has the most popular diversification level with DI between 0.5 and 0.6, which indicates the proper diversification level for successfully surviving in the international competitive construction market. Moreover, the average DI in individual years show a flat level, even though the competition in the international construction market becomes fiercer.

The research provides an overall pattern of diversification, through which issues of business diversification for the international contractors can be investigated. Our findings suggest that large firms with rich resources may do well in diversification. The rule for diversification is not "the more, the better", and it may not even be the best strategy to compete. The international contractors could follow the suggested proper diversification level but decide their own diversification level accordingly.

Acknowledgements The work is supported by the Hong Kong Research Grants Council (RGC) General Research Fund (GRF) (Project No.: HKU 749312B).

References

1. Lu W (2014) Reliability of engineering news-record international construction data. Constru Manage Econ 1–15. doi:10.1080/01446193.2014.919403
2. Reina P, Tulacz GJ (2013) The top 250 international contractors. Eng News Rec 271(8):1–19
3. Ansoff HI (1957) Strategies for diversification. Harvard Bus Rev 35(5):113–124
4. Arasti MR, Khaleghi M, Noori J (2010) The linkage of technology strategy and overall strategy of multi business diversified groups: literature review and theoritical framework (pp 1–12)
5. Jewell C, Flanagan R, Lu WS (2014) The dilemma of scope and scale for construction professional service firms. Constru Managem Econ 32(5):473–486
6. Prahalad C, Hamel G (1990) The core competence of the corporation. Harvard Bus Rev 68 (5):79–91
7. Gary MS (2005) Implementation strategy and performance outcomes in related diversification. Strateg Manage J 26(7):643–664. doi:10.1002/smj.468
8. Mackey TB (2006) Essays on corporate diversification and firm value. Ph.D. thesis, The Ohio State University
9. Chandler AD (1969) Strategy and structure: chapters in the history of the industrial enterprise: Massachusetts Institute of Technologie
10. Berger PG, Ofek E (1995) Diversifications effect on firm value. J Financ Econ 37(1):39–65. doi:10.1016/0304-405x(94)00798-6
11. Chang Y, Thomas H (1989) The impact of diversification strategy on risk-return performance. Strateg Manage J 10(3):271–284
12. Kurien SA (2004) Business development strategies used by general contracting construction companies in Texas for market diversification. M.S. thesis, Department of Construction Science, Texas A & M University, College Station, Tex
13. Reed R, Luffman GA (1986) Diversification: The growing confusion. Strateg Manage J 7 (1):29–35
14. Shannon CE (1948) A mathematical theory of communication. Bell Syst Tech J 27(3)
15. Choi J, Russell JS (2005) Long-term entropy and profitability change of united states public construction firms. J Manage Eng 21(1):17–26. doi:10.1061/(ASCE)0742-597X(2005) 21:1 (17)
16. Jacquemin AP, Berry CH (1979) Entropy measure of diversification and corporate growth. J Ind Econ 27(4):359–369. doi:10.2307/2097958

Chapter 10
Assessment of Ahmedabad (India) and Shanghai (China) on Smart City Parameters Applying the Boyd Cohen Smart City Wheel

Mona N. Shah, Shekhar Nagargoje and Chiranjay Shah

Abstract The Indian Government launched a new urban initiative '100 Smart Cities by the year 2022', under The Smart City Vision Plan, 2014, with a budget allocation of USD 1.2 billion for 2014–15 (PRNewswire 2015). The initiative revolves around the concept of new urbanisation or urbanism. Currently, no city in India or Asia, fully qualifies to be called a smart city, mainly because many parameters do not match the prevailing standards, either European or American. In China 311 smart cities are initiated for development by Chinese government and European companies in the 12th Five Year Plan (2011–2015), (EU China, 2014). Some works like, The Boyd Cohen Smart City Wheel (BCSCW) are a frequently cited tool to help cities, communities, and companies to become smart, innovative and a green economy. It covers six Dimensions viz. Environment, Mobility, Government, Economy, People, and Living divided into 18 'Working Areas', factorised into 24 'Indicators' and further factorised into 64 base level parameters (Cohen in The smartest cities in the world 2015, [1]). This paper makes an assessment of two cities, Ahmedabad and Shanghai using an extensive web based search, and assesses the degree of existing smartness of the cities based upon the BCSCW. Using weighted averages a composite city index for 27 matched parameters between Ahmedabad and Shanghai, has been calculated which enables us to view Ahmedabad and Shanghai's performance against that of a global smart city benchmark of Toronto City. With this, the research aims at providing an idea of the current state of smartness of Ahmedabad and Shanghai vis a vis Toronto City.

Keywords Smart cities · Indicators · Parameters · Assessment · Boyd cohen

M.N. Shah (✉) · S. Nagargoje
National Institute of Construction Management and Research (NICMAR) Pune, Pune, India
e-mail: mnshah@nicmar.ac.in

C. Shah
CEPT University, Ahmedabad, India

Y. Wu et al. (eds.), *Proceedings of the 20th International Symposium on Advancement of Construction Management and Real Estate*,
DOI 10.1007/978-981-10-0855-9_10

10.1 Background and Introduction

Cities are thick concentrations of human habitats that are stylised, fragile and dependent on systems and subsystems to sustain them. Cities have been a place for social and economic exchanges due to countless human interactions that take place within its boundaries daily. To accommodate this socio-economic growth, city limits expand and are termed as 'sprawls'. For planners 'sprawls' have negative connotations, which grew in the first place due to the centrifugal forces of low density growth. Sprawls tend to convert open space and environmentally sensitive lands into urban use, reduce the choices in housing and neighbourhoods, lead to traffic congestion and air pollution caused by automobiles, costly expansion of surface roads, new sewers, water systems, decaying old neighbourhoods and segregated land uses that reduce the need for travel. According to [2] traditional development of the industrial, electrical, or early information technology revolution happened with the intensive use of natural resources. Refer Fig. 10.1. This shortage has led to emphasis being laid on 'sustainable development', 'smart growth', 'intensive development' proposed by the governments and academia all over the world. This is done on all fronts but requires a good infusion of Information and Communication Technology (ICT) to achieve operational efficiencies in city management.

To counter uncontrolled sprawls, the concept of 'smart growth' has emerged [1, 3–5]. It is to be noted that there is no standardised definition of what constitutes a "smart city". It is described as a city that pushes the quality of resource management and service provision to the limit possible at the time [6]. Following this concept, smart city projects are part of city's renewal and modernisation.

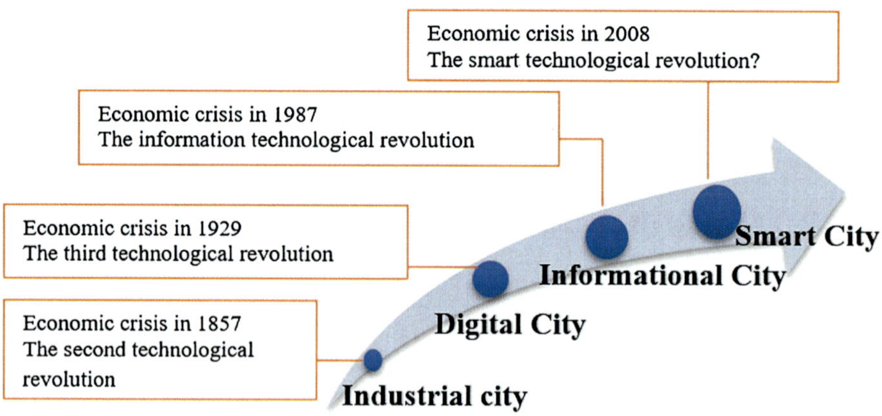

Fig. 10.1 The background of smart city [2]. *Source* The path towards smart cities in China: from the case of Shanghai Expo 2010, http://www.corp.at/archive/CORP2014_132.pdf

Another version is that it is "a space for coexistence among people who, based on the available technologies, can thrive and develop, while taking into account economic, social and environmental sustainability", [7]. Cavada [8], maintain that definitions hitherto available revolve around business strategy or resilience but smart refers to services, security and growth of stakeholders (people, academics, government and commercial) in an urban environment. Lee et al. [9], introduced the layered concept of the basic smart city architecture that integrates aspects of (1) integration of city infrastructures (2) governance (3) intelligence and sustainability (4) urban openness (5) urban innovativeness, and (6) service innovation as main aspects. They listed the six domains mentioned above and eighteen perspectives that fall within the domains. Alternatively some scholars have identified six key dimensions a little differently as that of [9] as follows: natural resource preservation, community development, housing, economic development, transportation choice, and planning for smart growth [10, 11].

10.2 Growth and Environment

Sustainable city planning addresses widespread issues in a single comprehensive manner. In the USA, according to George and Gerard [12], cities in the US like Austin in Texas, has developed a local plan to manage growth known as the 'Smart Growth Initiative' [12]. The initiative covers a variety of programmes viz. undertake renewal in the urban core with intensified land use, preservation of environmentally sensitive areas, and address larger regional strategies for future growth. Smart growth at local levels include: (a) In-fill development projects on vacant or underutilized parcels of land and (b) Planning "Transit-oriented development" (TOD), in which cities use subway and commuter rail stations as hubs for mixed-use developments within a half-mile radius, including housing, retail, and day care [12]. Other growth initiatives that can be dubbed in the 'smart' category include permitting systems that help retain or develop environmentally sensitive areas, land pooling and banking, and the use of priority funding areas (PFAs) that direct development toward existing community centers (Smart Growth Network 2002). In case of smart growth, policy choices are shaped by the difference in responsiveness of locally elected officials to interest group demands [13]. Vienna considered as a topmost smart city in the world has the highest scores due to its public transport network, extensive social housing activities and social services with high penetration and affordability for everyone, and also others factors [14]. In another example of Seoul under Mayor Myung Bak Lee from 2001 to 2006, has transformed the city into an easy commuting, working and living hub through new infrastructure construction while providing greater opportunities to business to function efficiently [15]. Other concerns associated with sprawl focus on the environment: the extensive land consumption associated with sprawl has been criticized for habitat loss and fragmentation, wetland destruction, and degradation of air and water quality.

10.3 Mobility

The most crucial aspect is the mobility choices the city already has, or would aim at creating to become efficient. Mobility options begin with primarily planning and forecasting the needs of a growing population and businesses and then incorporated into the city design (horizontal or vertical), and the planning of its transportation routes and mediums. Smart choices of mobility veer around the creation of new development patterns and implementing transportation improvements in order to maximise opportunities for multi-modal transportation. It has to create a balanced mix of land uses and a transportation system that provides mobility for pedestrians, bicyclists, transit and automobiles. In Britain efforts are on to improve conventional engines and increase the use of bio-fuels, and introduce ultra-low carbon vehicles (ULCV) and creation of 'plugged in places', in London, Milton Keynes and the North-East of England [16].

The Navigant Research (2014) report examines the key smart urban mobility infrastructure and services being offered in smart cities related to car sharing and rideshare services; public electric vehicle charging equipment and services; smart parking systems; congestion charging schemes; and advanced intelligent transportation systems (ITS) and other innovations in transportation infrastructure. The Report forecasts that global smart city technology revenue will grow from $8.8 billion annually in 2014 to $27.5 billion in 2023. In the case of Seoul, South Korea, their mobility choice marked a shift from surface roads that enhanced 'automobility' in favour of enhanced public amenities and quality of urban living through mass transportation systems. 74 lineal km of road lanes were built to accommodate a BRT network. Many regular bus routes were reconfigured to better feed into the city's extensive subway system, and an integrated fare and transfer system between buses and the metro rail network established [17]. Bus-operating speeds increased from an average of 11 to more than 21 kph. According to a study by Seoul Development Institute [18], bus reliability improved, because of regular traffic flows; dedicated lanes reduced accidents by 27 % one year and ridership improved by 60 % for BRT services; introduction of real-time traffic information system and in-vehicle navigation aids was introduced to guide traffic flows and alert motorists; one-way arterial couplets and curbside parking was substantially curtailed to help expedite traffic flows; and an innovative license plate scheme that required the motorists to leave their cars at home, once every ten days helped ease the mobility issues of Seoul. Good information systems help enable the overall city's system to function to its optimum, with the least spillages of stressed and scarce resources. Agile systems evolve smartly, based on memory and experience.

10.4 Government Role and Initiatives

To make cities enabled, empowered and smart, changes in traditional governance are very important. The traditional "closed and top down" approach will have to be replaced by a more "open model" for the management of its stakeholders. City officials need to encourage innovative and inclusive, open and transparent governance systems. Use of open and inclusive networks, and data infrastructure, citizen engagement, and integrated management structures enable the same. In the new millennium environmental concern of citizens at the local level have led them to increase their support for environmental and conservation goals. All these initiatives require balancing by national and sub-national governments in terms of economy and environment, public and private institutional arrangements that affect the resource flows, and the socio-technical infrastructure systems through which they flow [19]. For e.g. Milwaukee's RiverWalk Project (USA), launched in 1994 transformed a heavily industrialized and isolated riverfront area, into a clean recreational spot with a partnership between the city and downtown property owners. This locality has fuelled a housing boom, spawned a number of new restaurants, shops, and green space, and in the process created a broader constituency for cleaning up the Milwaukee River [20]. Cities like Denver and Dallas adopted the transit-oriented, mixed-use communities' model. Dallas has laid 44 miles of light rail systems, with a commuting potential of 60, 000 riders a day that has increased the retail, residential, and office development around rail stations. In case of Denver, a light rail system connects suburbs from all directions to the downtown, and a free shuttle bus to the downtown spine. A land reuse converted an obsolete shopping mall into a transit-oriented, mixed use community where apartments are constructed over office and retail space, within walking distance of the Englewood light rail stop.

The Government of India smart cities strategy is based upon 3 primary programmes namely retrofitting, redevelopment, and greenfield development (www.Smartcities.gov.in 2015). Ninety-eight cities have been earmarked for conversion into smart cities in the first phase. Ahmedabad city is listed in it. In the case of China a white paper drawn up to identify a Europe—China smart city cooperation states that in 2013, a total of 311 cities in China have drawn up smart city development plans, including all cities above the sub-provincial-level, 89 % cities at the prefectural-level and above, and 47 % cities at the county-level and above [6]. Shanghai Pudong New Area is included in it. In Shanghai in the year 2015, the internet penetration rate has risen to 68 % from 45 % compared to 2010. In 2010, Shanghai's mobile phone penetration rate was 110 %, indicating multiple phone ownership (Chinese Economic Review 2015).

10.5 City Economy

Cities are agglomerations of population and economic activity. European researchers, as per the Espon 1.1.1 project criteria, state that a 'smart' economy has characteristics of innovativeness, entrepreneurship, economic image and trademarks, productivity, flexibility of labour markets, international embeddedness and ability to transform [21]. According to Rose [22], a city may deplete critical resource within its own boundaries or its hinterlands, lose its comparative advantage in cross-border trade, or suffer severe social ills. It may also be subjected to external shocks from natural and man-made disasters. American states are promoting smart economy concept to revive existing cities through initiatives such as improving strategic public subsidies for projects that create new, wage job opportunities [23]. So also governments of Shanghai, a global city, and Ahmedabad a leading city attracts far lower level of international investment compared to Shanghai.

10.6 People Strengths

The characteristics of people in 'smart' include their level of qualification, affinity to lifelong learning, composition of social and ethnic plurality of the city, flexibility, creativity, cosmopolitanism/open mindedness, and propensity to participate in public life [21]. The type and composition of the city's populace strengthens the city's ability to attract investment and to make it a thriving hub of knowledge and culture. In the day to day workings of city life, people are empowered with information and communication technologies to add choices and efficiency to their lives.

10.7 Quality of Life

Citizens in cities that are considered smart are sure to make better use of the natural resources and environment conscious. The city management takes effort to provide better options to the citizens in commuting and recreating, and enjoying a better quality of life compared to other city residents that lack them [24]. These cities score higher on providing cultural opportunities, public safety, health and housing, educational facilities to all segments of the population, touristic attractiveness and social cohesion. Automobile dependency is low hence citizens do not indulge much in sedentary lifestyles. Due to mass transport facilities, the statistics on road accidents is lesser.

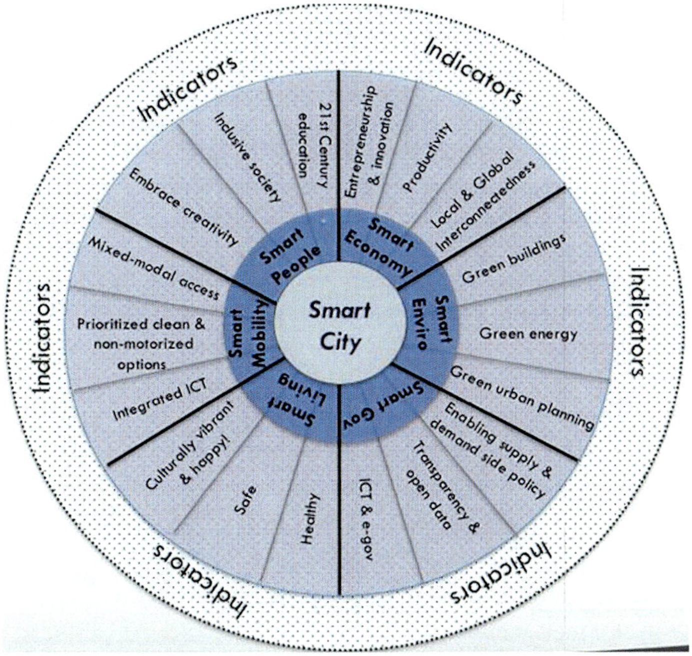

Fig. 10.2 The Boyd Cohen smart cities wheel. *Source* www.ubmfuturecities.com

10.8 Boyd Cohen Smart Cities Wheel

Boyd Cohen is a researcher in the urban and climate strategy area. He has evolved the smart cities framework from his early work in the in the green building and green neighbourhood development communities. According to him the smart city concept is yet in its earliest stages and therefore difficult to standardise. A city that wants to be tagged smart is on "a journey towards being smart and sustainable" [25]. He has introduced a tool named "Smart Cities Wheel" to facilitate the assessment of the cities against 64 select parameters. The 64 parameters themselves are derived from the 100 parameters contained in the ISO 37120 Standard.[1] The Wheel is to be seen as a baseline derived from the ISO 37120. Standard and real scenarios in the cities may either exceed or be below the baseline standard. Refer Fig. 10.2

[1]ISO 37120:2014 is applicable to any city, municipality or local government that undertakes to measure its performance in a comparable and verifiable manner, irrespective of size and location. (http://www.iso.org/iso/catalogue_detail?csnumber=62436).

10.9 Interpretation of the Wheel

No score of the individual parameter is standardised or is in-built into the Wheel, as it is considered a process and the current status of each city may vary. Thus the scores obtained on the individual parameters are not to be considered a final definitive standard. To obtain the current status, however a highest rating can be given to the benchmark unit and the score of other units can be obtained relative to the benchmark rating. Thus, the final scores are to be interpreted only at a point in time and as merely directional in nature. The quantification of smartness will help decision-makers to strategise further in launching sustainability improvements of the cities and regions.

10.10 Toronto City

The reason for choice of Toronto for benchmarking Ahmedabad and Shanghai is that it is ranked No. 2 in the world as a Smart City as per the Boyd Cohen Smart Cities' Index. Also, when searched in the public domain, the Toronto city analysis on Smart Wheel dimensions, indicators, and parameters were available but not of the first ranking city-Vienna [26]. Moreover energy use in Canadian cities is low, at 36000 MJ. It is a low energy city with better transit and increased walking and cycling options [27].

10.11 Methodology

The study is based on secondary research using an extensive web based search to ascertain the evolution of the smart cities movement. The selection of sampled cities was based on non-probabilistic convenient sampling technique. The two cities are not comparable in scale for factors such as area in sq.kms, density of population, existing scale of urban infrastructure etc. They have however a similar characteristic of having embarked on the smart city initiative as found from data in the public sources. Ahmedabad has additionally launched the GIFT Smart City Project, while Shanghai Pudong New Area is also declared as a smart city (GIFT 2011; Chinese Economic Review 2011). Earlier efforts of the researchers to obtain data from all research sources from China, for cities like Guangzhou in China, yielded no results. However data about Shanghai was relatively easily available in the public domain. In case of the BCSCW web based search was carried out for each and every parameter contained in the Boyd Cohen Smart Cities Wheel. Overall, the data sources were drawn from government reports, government websites, World Bank Reports and UN Habitat Reports commissioned research reports, select news articles and public announcements. Refer Table 10.1, for the complete list of sources

Table 10.1 Ahmedabad and Shanghai's parametric assessment matrix

Dimension	Working area	Indicator	Description	Toronto	Ahmedabad	Shanghai
				Benchmark standards	score	score
Environment	Smart buildings	Energy	Total residential energy use per capita (in kWh/year) (ISO 37120: 7.1)	1830	0.071	0.074
		Carbon footprint	Greenhouse gas emission measured in tones per capita (ISO 37120: 8.3)	7.38	0.068	0.056
		Air quality	Fine Particular matter 2.5 concentration (μg/m3) (ISO 37120: 8.1)	8.42	0.008	0.010
		Waste generation	% of city's solid waste that is recycled (ISO 37120: 16.2)	100 %	0.013	0.076
			Total collected municipal solid waste city per capita (in kg) (ISO 37120: 16.3)	300	0.061	0.083
		Water consumption	Total water consumption per capita (liters/day) (ISO 37120: 21.5)	205.6	0.061	0.047
	Sustainable urban planning	Climate resilience planning	Does your city have a public climate resilience strategy/plan in place? (Y/N) If yes provide link	Y	0.167	0.167
		Density	Population weighted density (average densities of the separate census tracts that make up a metro)	4372	0.061	0.197
		Green space per capita	Green areas per 100,000 (in hectares) (ISO 37120: 19.1)	445	0.006	0.195

(continued)

Table 10.1 (continued)

Dimension	Working area	Indicator	Description	Toronto	Ahmedabad	Shanghai
				Benchmark standards	score	score
			Smart Environment Score		**0.516**	**0.904**
Mobility	Efficient transport	Clean-energy transport	Kilometers of bicycle paths and lanes per 100,000 (ISO 37120: 18.7)	18.3	0.007	0.023
	Multi-modal access	Public Transport	Annual # of public transport trips per capita (ISO 37120: 18.3)	201	0.299	0.323
	Technology infrastructure	Access to real-time information	Presence of demand-based pricing (e.g. congestion pricing, variably priced toll lanes, variably priced parking spaces). Y/N	Y	0.111	0.111
			# of public transit services that offer real time information to the public: 1 point for each transit category up to 5 total points (bus, regional train, metro, rapid transit system (e.g. BRT, tram), and sharing modes (e.g. bike sharing, car sharing)	5	0.089	0.111
			Availability of multi-modal transit app with at least 3 services integrated (Y/N)	Y	0.000	0.111
			Smart Mobility Score		**0.505**	**0.680**
Government	Online services	Online procedures	% of government services that can be accessed by citizens via web or mobile phone	100 %	0.25	0.25
		Electronic benefits payments	Existence of electronic benefit payments (e.g. social security) to citizens (Y/N)	Y	0.25	0.25
	Infrastructure	Sensor coverage	# of infrastructure components with installed sensors 1 point for each: traffic, public transit demand, parking, air quality, waste, H_2O, public lighting	7	0.11	0.18
		Integrated health + safety operations	# of services integrated in a singular operations center leveraging real-time data. 1 point for each: ambulance, emergency/disaster response, fire, police, weather, transit, air quality	7	0.25	0.25

(continued)

Table 10.1 (continued)

Dimension	Working area	Indicator	Description	Toronto	Ahmedabad	Shanghai
				Benchmark standards	score	score
			Smart Governance Score		**0.857**	**0.929**
Economy	Entrepreneurship & innovation	R + D	% GDP invested in R&D in private sector	1.69 %	0.001	0.007
		Innovation	Innovation cities index score	54	0.123	0.167
	Productivity	GRP per capita	Gross Regional Product per capita (in US$, except in EU, in Euros)	55102	0.007	0.041
			Smart Economy Score		**0.132**	**0.215**
People	Education	Secondary education	% of students completing secondary education (ISO 37120: 6.3)	100 %	0.250	0.333
		Urban living lab	# of officially registered ENOLL living labs	1	0.000	0.333
		Creative industry jobs	Percentage of labor force (LF) engaged in creative industries	6 % [28]	0.100	0.026
			Smart People Score		**0.350**	**0.693**
Living	Culture and well-being	Investment in culture	Gini coefficient of inequality	4	0.040	0.036
	Health	Life expectancy	# of technologies in use to assist with crime prevention, 1 point for each of the following: live streaming video cameras, taxi apps, predictive crime software technologies	7	0.167	0.250
			Average life expectancy (ISO 37120: 12.1)	82.8	0.268	0.332
			Smart Living Score		**0.474**	**0.618**
			Smart City Index		**2.834**	**4.038**

for the individual parameters. In the process of data collection under BCSCW dimensions and indicators, data for 36 parameters for Shanghai were found and 27 for Ahmedabad. To standardise the data sets of the two cities, 9 parameters were removed from Shanghai, and analysis was carried out for a set of 27 matching indicators for comparison of these cities against the benchmarked Toronto City. Therefore for comparative analysis, Toronto was given a score of 1 for each dimension of the BCSCW, smart environment, smart mobility, smart governance, smart economy, smart living and smart people. Thus the total score for Toronto is 6 for the 6 dimensions. A relative index for Ahmedabad and Shanghai was calculated for each parameter in the smart dimensions. Individual scores were calculated for all matched 27 indicators based on their performance against the benchmark. After computing the scores, the indices were multiplied with the weights and aggregated into a weighted average output. The weights were assigned as 100 % for each smart city indicator and equally distributed for each individual parameter under these dimensions. For e.g. in the Mobility Dimension, there are 4 indicators, (clean energy, public transport, smart cards, and access to real-time information). Each of the indicators is assigned an equal distribution of 25 %. Likewise it was carried out for remaining 5 dimensions and indicators. The measured information was converted and aggregated as weighted averages to a composite city index and its parametric categories. Finally, summation of these index values was made to arrive at the composite smart city index for Ahmedabad and Shanghai, refer appendix 1.

10.11.1 Formulas of Weighted Averages and Index

(a) Formula for Weighted Average

$$\begin{aligned} \text{Weighted Avg x} &= \text{w1x1} + \text{w2x2...wnxn} \\ \text{w} &= \text{relative weight } (\%) \\ \text{x} &= \text{value} \end{aligned}$$

The average values of the parameters is derived in the indicator

(b) Formula for Index

$$\text{Weighted Avg x Index} = \text{w1x1} + \text{w2x2...wnxn}$$

where

w1, w2 … wn are relative weights assigned to each respective indicator

x1, x2 … xn are the values derived from benchmark of Toronto i.e. dividing values of Ahmedabad and Shanghai respectively with that of Toronto

10.12 Analysis

In the dimension of Smart Environment, in the analysis of environmental parameters Shanghai scores 0.90 and Ahmedabad scores 0.51 as compared to the benchmark score of 1 for Toronto. In the analysis it is observed that both cities have a 'climate resilience strategy' plan in place however Ahmedabad's plan is relatively recent. In 'greenhouse gas emission' measured in tonnes per capita, Shanghai scores less compared to Ahmedabad. Ahmedabad scores are lower in 'green areas per 100,000 people' and in 'percentage of city's solid waste that is recycled', which carry the scores of 0.006 and 0.013 respectively compared to Shanghai's score of 0.195 and 0.076 respectively.

In Smart Mobility dimension the overall score of Shanghai in all parameters stands at 0.680 and Ahmedabad at 0.505. The analysis indicates that Shanghai has more 'bicycle lanes per 100000 people' with scores of 0.023 compared to 0.007 for Ahmedabad. In case of 'public transport trips per capita', both cities are almost at par with a score of 0.32 for Shanghai and 0.29 for Ahmedabad and closer to the benchmark city of Toronto City. Presence of 'demand-based pricing' is observed in both the cities. As opposed to Ahmedabad, Shanghai has the 'availability of multi-modal transit apps with at least 3 services integrated'.

In the dimension of Smart Governance, the set of parameters chosen for analysis for both Ahmedabad and Shanghai indicates that, both the cities stand almost at par in urban governance with a score of 0.857 for Ahmedabad and a score of 0.929 for Shanghai. Out of four parameters analysed namely 'online procedures', 'electronic benefits systems', 'sensor coverage' and integrated health and safety operations', on three parameters both cities have same scores, except in the case of 'sensor based coverage of infrastructure components', where Shanghai scores 0.18 compared to Ahmedabad's score of 0.11. The data on 'integrated health and safety operations' indicates that, both cities are absolutely at par with Toronto's benchmark score and therefore have received equal scores.

Under the Smart Economy dimension, in the indicator of 'entrepreneurship & innovation', Toronto has a score of 54, Shanghai's score is 50 and Ahmedabad's score is 37 out of 100. Therefore the score of 100 was taken as the benchmark to calculate weighted average scores. In case of 'gross regional product per capita', Ahmedabad and Shanghai, both have lower scores compared to the benchmark score of 1 for Toronto, with Shanghai's score of 0.041 and Ahmedabad's score of 0.007.

In case of the Smart People the overall score of Shanghai on this dimension is almost twice compared to Ahmedabad, as Shanghai scores 0.693 and Ahmedabad scores 0.35 compared to Toronto's score of 1. Ahmedabad scores are lower because it has received no scores in 'urban living labs' such as membership of ENOLL (European Network of Living Labs) but Ahmedabad's scores are higher 'in percentage of labour force' (LF) engaged in creative industries, its more than the benchmark city of Toronto, the score of 100 was taken as benchmark to calculate weighted average scores for Ahmedabad and Shanghai. Score of Ahmedabad for 'percentage of labour force (LF) engaged in creative industries' is 0.1 and Shanghai is 0.02.

Under the Smart Living, it is observed that overall dimensional score of Ahmedabad is 0.474 and Shanghai has a score of 0.61. In 'Gini's coefficient of inequality' both Ahmedabad and Shanghai have poor scores, compared to Toronto. Toronto's coefficient is 4 and Ahmedabad's score is 33.6 whereas Shanghai's score is 37. In 'number of technologies in use to assist with crime prevention', Shanghai scores higher with a score of 0.250 and Ahmedabad's score is 0.167. It is observed that Shanghai has a higher 'life expectancy' and is at par with benchmark of Toronto with average life expectancy as 82.41 years, whereas Ahmedabad's average life expectancy is 66.5 years.

10.13 Conclusion

Both governments in Asia have embarked upon the smart city initiatives in recognition of the economic strengths that they generate and share in the contribution at the national level. The smart cities concept is a recent trend in 'new urbanism' that attempts to harmonise the demands of humans without disturbing the natural and ecological systems. It is also seen that there is yet to evolve to standardised definition of a smart city. Due to the relatively recent nature the literature that is reviewed describes individual elements that constitute 'smartness' at the human and natural impact levels. The literature itself seems to have evolved from broader concerns of the use of natural resources and economic development pressures. A relatively simple yet comprehensive explanation is available in the works of Boyd Cohen. The BCSCW is a useful tool to measure the degree of smartness of a city. In this study of Ahmedabad and Shanghai it is revealed that when compared with the benchmarked indicators of Toronto, the overall city index of Ahmedabad stands at 2.83 where as Shanghai's index is 4.08. The reasons for Shanghai's smart city index being higher than Ahmedabad may be attributed to Shanghai's earlier start in sustainable planning and development and as a globally recognised financial and commercial leader. However, with the pace of urbanisation in India and pressure on urban infrastructure, Ahmedabad's score of 2.83 out of ideal score of 6 for six smart city dimensions indicates that Ahmedabad is almost halfway in the journey of an ideal smart city. Shanghai outperforms in many departments such as environmental planning, quality of urban life and people facilitation with smarter and efficient technology. Ahmedabad in some areas is closer to Shanghai when it comes to urban mobility plans, urban governance and urban economy. Overall both Asian cities are progressing on improving their respective 'smartness'. Much needs to be done on the ground to ensure that all 64 parameters of the Boyd Cohen Smart City Wheel are adequately covered, to be able to come closer to the European and/or American standards.

10.14 Limitations of the Study

The study is a first tentative attempt to compare the performance of Asian cities with that of an already top ranked benchmarked smart city of Toronto. Out of 64 parameters, data of only 27 parameters (42 %) could be used for comparison in both cities. Therefore in future, an in depth search to uncover more data would be necessary to arrive at better conclusions. Moreover the research is completely sourced from the secondary data sources in case of Shanghai. On ground surveys would have helped to obtain better parametric results.

References

1. Burchell R, Lowenstein G et al (2002) Costs of sprawl 2000 (transit cooperative research program report 74). National Academy Press, Washington, D.C
2. Zhou Y (2014) The path towards smart cities in China: From the case of Shanghai Expo 2010. In: Proceedings of REAL CORP 2014, Tagungsband, 21–23 May 2014, Vienna, Austria. http://www.corp.at. Accessed 25 Aug 2015
3. Burchell RW, Listokin D, Galley CC (2000) Smart growth: more than a ghost of urban policy past, less than a bold new horizon. Hous Policy Debate 821–879
4. Downs A (2001) An approach to analyzing the impacts of "smart growth" upon economic development. Econ Dev Rev 17(4):9–17
5. Downs A (2001) What does "smart growth" really mean? Planning 67(4):20–25
6. Yanrong K et al (2014) Comparative study of smart cities in Europe and China"-White paper —EU-China smart and green city cooperation. Ministry of Industry and Information Technology (MIIT) DG CNECT, EU, Commission with China Academy of Telecommunications Research (CATR)
7. Branchi PE, Fernández-Valdivielso C, Matias IR (2014) Analysis matrix for smart cities. Future Internet 6:61–75. http://www.mdpi.com/1999-5903/6/1/61. Accessed 10 Aug 2015
8. Cavada M, Christopher DFR, Hunt DVL (2014) Smart cities: contradicting definitions and unclear measures. World Sustainability Forum 2014–Conference Proceedings Papers. http://sciforum.net/conference/wsf-4/paper/2454/download/pdf. Accessed 17 Aug 2015
9. Lee JH, Hancock MG (2012) Towards a framework for smart cities: A comparison of Seoul, San Francisco and Amsterdam. Stanford programmes on regions for innovation and entrepreneurship. Copyright 2012 ISi lab, GSI, Yonsei University. All rights reserved, Prepared by Prof. J. H. Lee
10. Hawkins CV (2011) Smart growth policy choice: a resource dependency and local governance explanation. Policy Stud J 39(4):679–707
11. Ye L, Mandpe S, Meyer BP (2005) "What is 'smart growth'—Really"? J Plan Lit 19(3):301–315
12. George A, Gerard D (2000) Smart growth and transportation: opportunities and challenges for Austin. Institute of Transportation Engineers (ITE). Journal 70(11):30–34. (Copyright Institute of Transportation Engineers Nov 2000)
13. Feiock RC (2004) A Quasi-market framework for local economic development competition. J Urban Aff 24:123–142
14. The Smart City Wien Initiative Report (2014) Owner and publisher: Vienna City Administration, © July 2014
15. Cervero R (2000) Urban reclamation and regeneration in Seoul, Republic of Korea. In: Lehmann S (ed) Low carbon cities: transforming urban systems. Routledge, Taylor and Francis Group, New York, pp 224–234

16. Hodson M, Marvin S (eds) (2015) Retrofitting Cities: Priorities, Governance and Experimentation, Routledge: Abingdon
17. Pucher J et al (2005) Public transport reforms in seoul: innovation motivated by funding crisis. J Publ Transp 8(5):41–62
18. Seoul Development Institute (2005) Toward better public transport. Seoul Development Institute, Seoul
19. Hodson M, Marvin S (2014) Low carbon nation: making new market opportunities. In: Graham S, McFarlane C (eds) Infrastructural lives: urban infrastructure in context. Routledge, Taylor and Francis Group, New York, pp 219–243
20. Geller LA (2003) Smart growth: a prescription for livable cities. Am J Publ Health 93 (9):1410–1415. (Copyright American Public Health Association Sep 2003)
21. Giffinger F, Kramer M, Pichler M (2012) European smart model. Universities of Vienna, Universities of Delft, Universities of Lubijana. http://www.smart-cities.eu/model_1.html. Accessed 18 Aug 2015
22. Rose R (2014) Economic resilience and its contribution to the sustainability of cities. In: Paolo G, Gaetano M, Asprone D (eds) Resilience and sustainability in relation to natural disasters: a challenge for future cities. Springer, Italy
23. Poloncarz CM (2013) Initiatives for a smart economy. Erie county economic development strategy. Printed in house by the Erie county division of information and support services. http://www2.erie.gov/environment/sites/www2.erie.gov.environment/files/uploads/pdfs/SmartEconomy%20for%20Web3.pdf. Accessed 24 Aug 2015
24. Shapiro JM (2006) Smart Cities: quality of life, productivity, and the growth effects of human capital. The review of economics and statistics, 8(2):324–335 (© 2006 by the President and Fellows of Harvard College and the Massachusetts Institute of Technology)
25. Cohen B (2015) The smartest cities in the world 2015. http://www.fastcoexist.com/3038818/the-smartest-cities-in-the-world-2015-methodology. Accessed 24 Aug 2015
26. Cohen B (2012) The top 10 smart cities on the planet. http://www.fastcoexist.com/1679127/the-top-10-smart-cities-on-the-planet. Accessed 24 Aug 2015
27. Kenworthy J (2006) The eco-city: ten key transport and planning dimension for sustainable city development. Environ Urbanisation 18(1):67–85
28. Silver D (2011) From the ground up: growing toronto's cultural sector. Martin Prosperity Institute. http://www.toronto.ca/legdocs/mmis/2011/ed/bgrd/backgroundfile-41204.pdf. Accessed 20 Aug 15

Bibliography

29. Batagan L (2015) Smart cities and sustainability models. Informatica Economica 15(3):80–87. (Copyright American Public Health Association Oct 2010)
30. Bright J (2015) How big data is breathing new life into the smart cities concept. http://blogs.oii.ox.ac.uk/policy/how-big-data-is-breathing-new-life-into-the-smart-cities-concept/. Accessed 17 Aug 2015
31. Cohen JR (2002) The vanishing automobile and other myths: How smart growth will harm American cities, American planning association. J Am Plan Assoc 68(3):322–323 (Taylor and Francis Inc, Chicago)
32. Costa F (2005) Smart growth/comment: an ambitious movement and its prospects for success. J Am Plan Assoc 71(4):367–380
33. Goel R, Tiwari G (2014) Promoting low carbon transport in India: case study of metro rails in India. UNEP report. http://www.unep.org/transport/lowcarbon/PDFs/CaseStudy_MetroRails.pdf. Accessed 10 Aug 2015
34. http://www.prnewswire.com/news-releases/upcoming-smart-cities-in-india-2015—budget-allocation-towards-the-smart-cities-program-stood-at-around-usd12-billion-for-2014-15-300052519.html. Accessed 10 Aug 2015

35. http://giftgujarat.in/masterplan/vision.aspx. Accessed 11 Aug 2015
36. Hwang K, Byun M, Nah T (2005) Project Cheong Gye Cheon. Nanam Publishing House, Seoul
37. Kitchin R (2014) The real-time city? Big data and smart urbanism. Geo J 79 (1):1–14 (Springer Netherlands)
38. Lehmann S (2015) Low carbon cities: transforming urban systems. Routledge, Taylor and Francis Group, New York
39. List of metro systems. https://en.wikipedia.org/wiki/List_of_metro_systems. Accessed 31 Aug 2015
40. Ministry of Urban Development (2015) Strategy. http://smartcities.gov.in/writereaddata/Strategy.pdf. Accessed 18Aug 2015
41. Paolo G, Gaetano M, Asprone D (2014) Resilience and sustainability in relation to natural disasters: a challenge for future cities. Springer, Italy
42. Shanghai to become 'smart city' by (2015) China economic review—Daily briefings (25 Jan 2011). Copyright SinoMedia 25 Jan 2011. http://search.proquest.com/docview/867358631/D297C88B0ADA42D8PQ/4?accountid=34791. Accessed 19 Aug 2015
43. Sustainable development of communities - Indicators for city services and quality of life. ISO 37120:2014. (2014). http://www.iso.org/iso/catalogue_detail?csnumber=62436. Accessed 20 Aug 2015
44. Tachizawa EM, Alvarez-Gil MJ, Montes-Sancho M (2015) How smart cities will change the supply chain management. Int J Oper 20(3):237–248
45. Yuhong WPE, Xiaomei D, Marcucci JD, Yuen L (2013) Sustainable development planning of protected areas near cities: Case study in China © 2013 American Society of Civil Engineers. http://www.dnaindia.com/money/report-100-smart-city-projects-gets-cabinet-nod-2081677. Accessed 24 Aug 2015

Chapter 11
Research on the Preference Correction Model Based on Expectation in Demolition Disputes

Zhongfu Qin, Liqing Meng and Xianrong Wei

Abstract In the academic field, it's weak to explain the reality based on the traditional game theory. People have focused on the soft strategy, non-rational, mental game, etc. In the conflict analysis, each of the players has a certain preference for each situation, and the measure of this preferences usually use utility function. The composition of utility is analyzed under the framework of the rational people and the non-rational people, and the utility function is constructed respectively. Among them, the psychological utility is introduced into the utility function model. On the other hand, a number of factors influence the psychological utility. The emotion, expectation and other factors are introduced into the establishment of utility function based on the theoretical framework of the mental game theory in this paper. So that the people's preferences can be measured by the utility function. We construct a preference correction model base on expectation, and use the actual case of the demolition disputes to examine.

Keywords Utility function · Expectation · Emotion · Rational · Non-rational

11.1 Introduction

The game theory is the applied theory of how to make rational investment decisions on the basis of careful calculation by the payoff function or utility function [1]. In conflict analysis, the preference of the individual judgment for each situation is an important step to obtain a balanced situation. At present, the preference judgment is

Z. Qin (✉) · L. Meng · X. Wei
College of Civil Engineering and Architecture Zhejiang University,
Hangzhou, China
e-mail: qinzhongfu@zju.edu.cn

L. Meng
e-mail: 1575840551@qq.com

X. Wei
e-mail: 956331238@qq.com

Y. Wu et al. (eds.), *Proceedings of the 20th International Symposium on Advancement of Construction Management and Real Estate*,
DOI 10.1007/978-981-10-0855-9_11

mainly according to the qualitative analysis; some scholars make quantitative analysis by building a preference correction model (utility function), such as Qin Zhongfu, Wu [2] who research on modeling of preference correction based on non-self-interested behaviors in demolition disputes to study the effect of emotion on preference. But this kind of research is not very much. Therefore, to discuss the constitution of the utility function is meaningful.

We study the construction of the utility function in this paper. The model will be represented by a mathematical formula. The establishment of the utility function is assumed as follows: the players in the game can judge the utility level under the different time and different factors, and always seek to the utility maximization in a certain time point.

11.2 Utility Function of Rational People

In the conflict analysis, it is generally considered that the player is rational and maximize the pursuit of their own interests. At the same time, players are social and economic people who pursue the maximization of economic benefits. Economic interests is to meet their needs through the acquisition of money in reality. Under the framework of traditional game theory, the players' utility (U) is the best interest that the individual can get. Formula is as follows.

$$\mathrm{U}_i = \mathrm{U}_i(T) = \mathrm{T}_i \tag{11.1}$$

Material benefit (Ti) of player (i): material gains are converted into monetary performance (expressed in money terms). The greater the material benefit, the greater the utility. The Player would prefer greater utility value situation.

11.3 Utility Function of Non-rational People

In demolition disputes, the formula (11.1) indicates the utility of the bureau is equal to the wealth. But this does not conform to the reality. Research shows that experiment measured the participants conduct do not meet the pursuit of self-interest maximizing behavior assumptions [3]. Currently empirical research in psychology had been concluded: psychological interests are the fundamental of the economic interests of the people [4]. And the traditional game theory ignores the characteristics of non-rational people, and the pursuit of psychological interests is an important part for the economic interests. The concept of non-rational people is an important amendment to the concept of rational people. Jiang Shuguang, Wei Qian also put forward, from the point of view of the rational people, the player's utility is reflected in the pursuit of the maximization of individual interests and the most direct performance is the amount of money. From the point of view of the non-rational people, the player's utility also includes his psychological feeling, namely, "the psychological utility". There are already many theories to study the composition of the "psychological utility", such as warm glow [5], guilt aversion [6], emotion [7] etc. These studies are in the study of the effect of this kind of psychological utility.

Therefore, the player's utility is not only determined by the material gains, but also by the mental state of the player (psychological utility). Psychological utility is the psychological feeling of the player which is influenced by many factors, such as personal expectations, self-image, social or other people's respect, fair preference, guilt, emotion, etc. Material utility here is the income of the people in the material wealth. Therefore, the construction of the utility function is as follows (i refers to people, different players have different utility values): Total utility (U_i) = Material utility (T_i) + Psychological utility (H_i)

$$U_i = T_i + H_i \tag{11.2}$$

When the player is absolutely rational, that is, the pursuit of maximum personal interests, the formula is Ui = Ti. This is an extremely ideal situation. In reality, it may appear that the player does not pay attention to material compensation, but put the psychological experience in the first place. At this time Ui = Hi. This is also an extremely ideal situation.

Therefore, we need to introduce two parameters α and β which represents the material utility intensity coefficient ($0 \leq \alpha \leq 1$) and the psychological utility intensity coefficient ($0 \leq \beta \leq 1$) respectively. In order to reflect the strength of the player's material or psychological utility in the game, and to solve the problem of the sensitivity of the material and the psychological feeling. Formula is as follows:

$$U_i = \alpha T_i + \beta H_i \tag{11.3}$$

When the psychological utility can be explained by a single factor, that is, the psychological utility is mainly driven by the one hand, such as expectation. Then the utility function is the formula (11.3). When the psychological utility is explained by multi factor explanations, that is, the psychological utility is the combined effect by a variety of reasons, based on each of the reasons has the different effects for different people. Then psychological utility is not the same. The effect of each factor is shown by the influence coefficient. The utility formula is as follows.

$$U_i = U(T_i, H_i), U_i = \alpha T_i + \beta [P_{i1} H_{i1} + P_{i2} H_{i2} + \cdots + P_{in} H_{in}] \tag{11.4}$$

Among them, $0 \leq P_1, P_2, P_1 \ldots .. P_n \leq 1$ H_{in}: the nth mental influence factor by player i; P_{in}: The influence strength coefficient of the nth mental influence factor by player i.

We study the effect of the expected factors on the utility. Namely, the formula (11.3) is the basic frame of the utility function in this paper.

11.4 Utility Function in Demolition Disputes

Demolition disputes arise because of the interests of the parties are not met in engineering demolition. The dispute points in the demolition disputes tend to be concentrated in the negotiation of relocation compensation (material compensation can translate into monetary compensation). Then the utility function is constructed based on various factors in the demolition disputes under the framework of formula 11.3 above.

11.4.1 Utility Under the Influence of Individual Expectation

Participants often have a psychological expectation standards for compensation. Agreement can't be reached when the real compensation is not up to the standard. The participants' psychological expectations is reached or not having a significant impact on the personal psychological feelings. Demolition party and the demolished party is not coordinated on demolition compensation price expectations, which caused a demolition disputes. So, here to discuss the individual expectation impact on total utility.

The players subjective psychological expectation (subjective psychological probability measure), is showed the player's expectation value in demolition disputes. Reference the model of expectations hypothesis, the formula is as follows.

$$U_i = \alpha T_{is} + \beta(T_{is} - T_{iq}) \tag{11.5}$$

T_{is} : the actual amount of compensation in demolition disputes of player i; T_{iq} : The expected compensation in demolition disputes of player i.

When the player's psychological expected value is higher than the actual compensation value in demolition disputes, the psychological utility $(T_{is} - T_{iq})$ is negative. Then the total utility value will be reduced. This is consistent with the actual situation in demolition disputes. That is obtained when the demolished party compensation is smaller than the psychological expected, disappointed, dissatisfied and other emotions will generate. Then the utility value will reduce for the demolished party. It is the same with the demolition party.

11.4.2 Introduction of Loss Aversion Coefficient

People are loss averse in real life which similar to risk aversion. The impact of the loss to the people is greater than the impact of income under the same loss and income. That is, on the basis of expected earnings, more than the expected return and less than the expected return to people's psychological impact is different. Less than the expected earnings will make people feel more pain. Kahneman and

Tversky put forward the famous prospect theory. Research on decision-making behavior of investors from the angle of cognitive psychology. They pointed out that people have a characteristic loss aversion that people are more sensitive to losses than profits [8]. We use the generalized piecewise function to describe the loss aversion in this paper [9]. Combined with the expected impact of the individual utility (function 5), formula is as follows:

$$U_i = \begin{cases} \alpha T_{is} + \beta(T_{is} - T_{iq}), T_{is} \geq T_{iq} \\ \alpha T_{is} + \beta\gamma(T_{is} - T_{iq}), T_{is} < T_{iq} \end{cases} \quad (11.6)$$

γ is the loss aversion coefficient. It's more than 1. And the higher the value is, the higher the degree of the loss aversion. This formula is general.

11.4.3 Introduction of Emotion Coefficient

Emotions appear in the forefront of people's psychological life, about people's mental state, driven people's actions. Elster said that many emotions have important economic significance. Therefore, it is necessary to introduce emotional factors into the construction of utility functions. In this paper, from the perspective of expected, it is believed that different emotions affect personal expectation value. And then affect the total utility value. Different people have different personal expectation value set, and personal expectation value set will change under different emotions. For examples, in compassion, dedication emotions, personal expectation value is expected to weaken appropriately; in jealousy, revenge emotions, personal expectation value will increase. In the empty mood, personal expectation value is more rational. In addition, the article does not discuss the impact of mixed emotions. Because whether a single individual is able to experience mixed emotions is still controversial in the academia [10]. Set the expectation value below under the emotion impact.

According to the above formula (11.6), draw lessons from the Qin Zhongfu and Wu Jiali's modeling of preference correction, we introduce coefficient ω :Compassion, dedication mental strength: $0<\omega<1$. Compassion and dedication are also understood to be positive here. Jealousy, revenge mental strength: $\omega > 1$. The jealousy, revenge can also be understood as negative emotions. Formula is as follows.

$$\mathrm{U} = \begin{cases} \alpha T_s + \beta(T_s - \omega T_q), T_s \geq T_q \\ \alpha T_s + \beta\gamma(T_s - \omega T_q), T_s < T_q \end{cases} \quad (11.7)$$

From the value we can see, under compassion, dedication positive emotional impact, $0<\omega<1$, it makes the people's expectation value cut down in the numerical value, thereby improving the utility value. We can also understand that positive emotions can make one side to help the other side. Positive emotions will lead to a promise.

Under jealousy, revenge negative emotional impact, $\omega > 1$, it makes the people's expectation value Increase in the numerical value, thereby reducing the utility value. In addition, negative emotions may encourage people to make irrational behavior, namely player may do acts of aggression the other side. When $\omega = 1$, it refers to player i selects rational decision-making to player j under empty mood [11], without emotional impact.

Therefore, this paper constructs the preference correction model based on expectation in demolition disputes.

11.5 Case Analysis

This paper take "most cattle nail" in Chongqing as a case to analysis (details can be seen in Sohu News). According to boundary standard from the demolition and demolished party's bid respectively, the case is divided into three stages. In the first stage, in September 2006, demolition party preferred to offer 3.5 million yuan, 5.15 million yuan was expected by demolished party, and the talks broke down. The second stage, in February 2007, demolition party preferred to offer 4.2 million yuan, the demolished party insisted 5.15 million yuan, the talks broke down. The third stage, in April 2007, based on an external force: the court, demolition party and demolished party bargained on 3.6 million yuan. This paper simplify the case. This article focuses on the second stage. Because the expectation, emotion have an important impact on this stage of the game.

In this paper, the players is the demolition party i (Chi-run home company) and the demolished party j (Yang Wu and Wu Ping). Optional strategy of the game: demolition party's optional strategy: raise the price of compensation, reduce the compensation price, no measures are adopted, awaiting demolished party to agree the compensation price; demolished party's optional strategy: claim to increase the compensation price, reduce the compensation price, and agree with each other's compensation price.

Study excluded the infeasible situations from logical and preference. Viable situation is as the following Table 11.1. ("1" indicates that choose this strategy, "0" means that don't choose this strategy, player i selects (0, 0) shows that no measures are adopted, awaiting demolition party to agree the compensation price; player j selects (0, 0) indicates player j agree with each other's compensation price.)

Table 11.1 The feasible situation

Players	Strategy	The feasible situation								
i	Increase	0	1	0	0	1	0	0	1	0
	Reduce	0	0	1	0	0	1	0	0	1
j	Increase	0	0	0	1	1	1	0	0	0
	Reduce	0	0	0	0	0	0	1	1	1
Decimal number		0	1	2	4	5	6	8	9	10

Preference ordering principle: a preference for a situation in which a person may select based on the outcome of a situation's gain, his emotional mood, and his understanding of the preferences of other people and so on. Preference ordering principle is as follows: in the framework of a rational person, preference depends on the size of earnings; in the framework of the non-rational person, preferences depends on income, emotion as well as his understanding of the other players' preferences [11].

Compensation price provided by the demolition party has a larger gap with demolished party's expectation, and the demolition party did not agree to raise prices, the first stage of the game talks broke down. Then enter the second stage of the game, as the negotiations proceed, demolition party concerned the overall situation, took concessions behavior, preferred to raise prices (from the original 3.5 million yuan to 4.2 million yuan). This is a signal for the demolished party: the price of compensation was not reasonable, it was low, as long as you insisted, and they also could further improve the price. The demolished party was in a state of non-rational now, the factors of expectations and mood changed his strategy or preference. He may propose a higher price or make non-rational decisions, so that result in the breakdown of negotiations.

When under the assumption of the rational people, we could easily get a balanced situation: the situation 6 that the demolition party reduced the price and the demolished party raised prices, so the negotiations broke down.

When the situation is ordered under the assumption of non-rational people, correction model based on expectation in demolition disputes is introduced. First, we need to calculate the utility value of the player.

1. Utility value calculation of play i

   Firstly, calculate the demolition party i's utility value. The demolition party is relatively rational because of the company's identity. Set up the material utility intensity coefficient $\alpha = 1$, set up the psychological utility intensity coefficient $\beta = 0.5$. When player has no loss aversion, set up $\gamma = 1$. When the player has no emotional impact, set up $\omega = 1$. We use the preference correction model based on expectation to calculate the utility value under different situations. In each case, the numerical values of the parameters and the values of the variables are shown in the below table.

   Utility value can be calculated from the above Table 11.2 and utility value sort from big to small order as follows: 10, 2, 8, 9, 6, 0, 1, 4, 5.

Table 11.2 The demolition party's utility value calculation table

Situation	0	1	2	4	5	6	8	9	10
T_{is}	420	420	420	420	420	420	420	420	420
T_{iq}	420	430	360	420	430	360	420	430	360
ω	1	1	0.9	1.1	1.1	1.1	0.9	0.9	0.8
γ	1	1.1	1	1.1	1.2	1.1	1	1	1
$\mathbf{U_i}$	420	414.5	468	396.9	388.2	433.2	441	436.5	486

2. Utility value calculation of play j
   Then calculate the demolished party j's utility value. In the game of second stage, the demolished party Wu Ping said: "In fact, I wish to resolve it immediately. But I do not sign the agreement absolutely because of self-esteem hurt. I have nothing to lose, the most fancy is my face and self-esteem." From the words of the demolition party revealed that his psychological utility intensity greatly enhanced. At the same time, the material utility is also focused on, but it has weakened. So set up the material utility intensity coefficient $\alpha = 0.95$, set up the psychological utility intensity coefficient $\beta = 1$. From Wu Ping's words we can know that he is satisfied with the 4.2 million yuan of the compensation price. So he would not raise prices too much if he adopted the strategy of raising the price, set up to raise to 5.2 million yuan. Of course, because the influence of factors of self-esteem and the face, he would not reduce the prices too much if he adopted the strategy of reducing the price, set up to raise to 5.0 million yuan. We use the preference correction model based on expectation to calculate the utility value under different situations. In each case, the numerical values of the parameters and the values of the variables are shown in Table 11.3.

Utility value can be calculated from the above Table 11.3 and utility value sort from big to small order as follows: 5, 1, 4, 9, 8, 0, 6, 10, 2. From Table 11.2, Table 11.3 to obtain a balanced situation in Table 11.4.

According to the above table shows, there are three equilibrium situations in the second stage of game under the assumption of non-rational people, the situation 6, 8, 9 respectively. The situation 6 is the only equilibrium situation under the

Table 11.3 The demolished party's utility value calculation table

Situation	0	1	2	4	5	6	8	9	10
T_{js}	420	420	420	420	420	420	420	420	420
T_{jq}	515	515	515	520	520	520	500	500	500
ω	1	0.9	1.1	0.9	0.8	1	1	0.95	1.1
γ	1.1	1.1	1.1	1.1	1.1	1.1	1.2	1.2	1.2
$\mathbf{U_i}$	294.5	351.15	237.85	346.2	403.4	289	303	333	243

Table 11.4 Feasible situations under the assumption of non-rational people

Global balance	×	×	E	E	E	×	×	×	×
Priority vector (i)	r	r	s	s	r	u	u	u	u
	10	2	8	9	6	0	1	4	5
			10	10		2	2	6	6
				8			0		4
Priority vector (j)	r	s	r	s	s	s	r	u	u
	5	1	4	9	8	0	6	10	2
		5		5	4	4		6	6
				1		8			10

assumption of rational people equilibrium situation. The equilibrium situation under the assumption of non-rational people is more realistic. We can see if the demolished party take to choose the situation 8 or 9, choose to reduce price, he could obtain the maximum value (4.2 million yuan). But because of the Influence of his high expectations and self-esteem, make him not choose compromise behavior and failure to obtain the maximum monetary value at last. We know from the whole process of the outcome of the game, the demolished got 3.6 million at last, a difference compensation of 0.6 million yuan. Finally, the demolished party lose 0.6 million yuan. The preference correction model based on expectation is consistent with the actual. So we conclude that psychological utility is important in the game.

11.6 Conclusion

In the demolition disputes, it will generate psychological utility under the impact of the factors of expectation, emotion, etc. Those will affect the total utility. Then affect the preferences, and affect the results of conflict analysis at last. In this paper, we introduce the psychological utility in the framework of the non- rational people. And the utility model is established by combining the material utility and psychological utility (formula 11.3). In reality, the fluctuation of the psychological feeling can affect people's judgment on the total utility. Then it affects people's behavior. In the other hand, this paper mainly studies the model of total utility in the demolition disputes. The construction of the model introduce the factors of the emotion, expectation and loss aversion based on utility function model above(-formula 11.3). Then the preference correction model based on expectation in demolition disputes is constructed (formula 11.7). The model proved to be effective. But it has to be said that the preference correction model based on expectation in demolition disputes have certain defects. It is a difficult point in the parameter selection, and it needs to be improved in the following research.

References

1. Shi X (2000) Game theory. University Press, Shanghai
2. Wu J (2013) Research on modeling of preference correction based on non-self-interested behaviors in demolition disputes. Zhejiang University
3. Jiang S, Wei Q (2013) Beliefs and psychological theory: theory, demonstration and application. Econ Res 06:141–154
4. Peng K (2009) Psychological game of economic man: contribution and challenge of social psychology to economics. J Renmin Univ China 03:61–69
5. Andreoni J (1990) Impure altruism and donations to public goods: a theory of warm glow giving. Econ J 100(401):464–477
6. Battigalli P, Dufwenberg M (2007) Guilt in Games. American economic review papers and proceedings 97(2):170–176
7. Elster J (1998) Emotions and economic theory. J Econ Lit 36(1):47–74

8. Kahneman D, Tversky A (1979) Prospect theory: an analysis of decision under risk. Econometrica 47(2):263–291
9. Lin Z, Cai C, Xu B (2010) Supply chain revenue sharing contract under loss aversion. J Manage Sci China 08:33–41
10. Liu H, Hu Z, Peng D (2008) Theory and research on the relationship between positive and negative emotions. Prog Psychol Sci 16(2):295–303
11. Takehiro I, Shingo T, Bunpei N (2000) Credibility of information in 'soft' games with interperception of emotions. Appl Math Comput 115:23–41

Chapter 12
Research on the Impact of Crisis Events on Urban Development—A Case Study of Kunming Railway Station Terrorist Attack

Yue Li and Yuzhe Wu

Abstract In recent years, China is in an important period in the new-type urbanization construction process, but crisis events happened frequently, doing great harm to the urban infrastructure and hindering the urban development. It will be of great importance to reduce the negative impact of crisis events and promote the urban development. A violent incidents happened in Kunming railway station on the evening of March 1 2014, causing a massive impact to the urban development of Kunming. Therefore, this paper will take Kunming railway station terrorist attack for an example, getting data though literature review and media reports to study how the crisis events produce an effect on the urban development. At last, I also give some suggestions to reverse the bad situation.

Keywords Crisis events · Urban development · Kunming terrorist attack

12.1 Introduction

With the trend of economic globalization, the expansion of urban area and the increase of population, urban crisis events caused by natural disasters, social security issues occur frequently, doing great damage to people's life and urban development. On the other hand, accompany with agglomeration of human and capital, city now is much more complex and fragile in front of the crisis than before. Once a crisis occurs in City, often accompany with chain reaction and amplification effect, which requires us to pay more attention on the crisis events in city.

Y. Li · Y. Wu (✉)
Department of Land Management, Zhejiang University,
Hangzhou, China
e-mail: wuyuzhe@zju.edu.cn

Y. Li
e-mail: liyue0628@zju.edu.cn

Y. Wu et al. (eds.), *Proceedings of the 20th International Symposium on Advancement of Construction Management and Real Estate*,
DOI 10.1007/978-981-10-0855-9_12

A series of violent crisis occurred in recent years, including the Indian Ocean Tsunami, SARS, Global Financial Crisis and 911 Terrorist Attack, make people have a clearly awareness of various social and natural crises all around us. However, due to the crisis has always been regarded as unavoidable and irretrievable, people often neglect to manage the urban crisis, resulting in the city in a passive position and suffering great loss.

At 21:20 on March 1, 2014, a terrorist attack sponsored by national separatists, occurred in Kunming railway station, Yunnan province. It shows that this crisis caused severe damage to Kunming—total 29 dead and 143 injured as of March 2, which shocked people all around the world. As soon as the crisis occurred, Chinese government and international organizations such as the United Nations respond quickly and actively to arrest assailants and give first-aid to the wounded. Kunming railway station terrorist attack do harm to urban infrastructure, urban culture and economic of Kunming in different degrees.

The implementation of the new-type urbanization requires a higher level of urban public service. It is necessary to take urban crisis seriously and figure out how it affects urban development. Most domestic and foreign academic research on crisis management focused on crisis itself or how to prevent a crisis, few studies on the impact of crisis events on urban development. This will take Kunming terrorist attack for example to analysis the impact of crisis events on social order, image and tourism of a city.

12.2 Crisis and Urban Crisis

12.2.1 Crisis Events

The first to study is the urban crisis event before the impact of crisis events on urban development. Crisis management is a theory of universality and versatility, and urban crisis is a crisis theory focuses on the impact on the city. Therefore, the crisis as a starting point, analysis the common of crisis theory and select the individuality of urban crisis.

"Crisis" in the dictionary has three meanings: first, the "latent danger"; the second is "a life-or-death moment"; the third is "the root causes dangerous". Different scholars in different periods have different definitions on the "crisis" from different focuses. Foreign scholars define "crisis" mainly as: (1) Rosenthal believes that crisis is an event which has to make decision under time pressure and high uncertainty, doing great harm to the basic framework of values and rules of conduct. (2) Barton [1] thinks that crisis is an event which may has negative and uncertain effects, resulting high damage to members, products, capital, service and reputation of organization. (3) Robrt [2] said that crisis is situation, often happening unexpected, the fundamental objective of its decision maker is threatened and time is limited to for decision maker to change the decision. (4) Otto [3] believes that crisis is an event which lead an organization into a controversy and endanger its profitability, growth, survival in the future. Domestic scholars define "crisis" mostly based on studies of foreign scholars, from crisis itself or effects. (1) Xue [4] believes that crisis is a decision situation under serious threat and pressure, but decision maker has to make critical decisions in limited and tight time; (2) Lin [5] said that crisis is an unexpected major event which would endanger the property and the security of people's life.

Meanwhile, since there is a wide range of crisis with complex contents, a reasonable classification on the crisis will contribute to further study on the theory of crisis. Research on the crisis classification of China and the foreign country have a different focus. Xue Lan made a comprehensive and representative classification in his book *Crisis Management*. Crisis events are classified as natural crisis (natural phenomena, disasters and accidents) and man-made crisis (terrorist attack, criminal behavior, destructive event) according to the type of motivation, as international crisis, domestic crisis and organization crisis according to the effect of spatial and temporal scales, as political crisis, economic crisis, social crisis and value crisis according to the main causes, as peaceful conflict and violent conflict according to the measure.

12.2.2 Urban Crisis

Urban crisis is a natural or social event caused by uncontrollable or uncontrolled factors, which do a lot of damage to life, property and material wealth in the urban system. Since city is a concentration of population, wealth and industry, city has a high degree of aggregation and networking, challenging the urban resources, environment, infrastructure and urban management, hiding explosion of crisis at the same time of urban development. Meanwhile, urban crisis evolves quickly, making the cause, process and influences uncertain.

Domestic and foreign scholars have do a lot of researches from different focus. Based on research of urban crisis caused by natural disasters, recent years west scholars study urban crisis focus on terrorism, social unrest and economic development from a sociological point of view, emphasizing the improvement of emergency rescue mechanism and the capability of regional cooperation in front of crisis events. Domestic studies focus on natural disasters, few refer to social injustice, ethnic conflicts, social psychological imbalance and other social problems caused by urban crisis.

Urban crisis event has the following three forms: (1) public mass panic; (2) the diversification of the disaster's form; (3) race or class conflict. Urban crisis event has the following five characteristics: (1) urban crisis event occurred in high frequency and various field, and more and more crisis events take city as a starting point and end point, spreading to the world; (2) non-traditional security issues, especially man-made crisis, become a major threat to the security of the modern city; (3) unexpected disaster events can easily evolve into social crisis events; (4) the resonance of crisis events; (5) the international impact of urban crisis events.

12.3 Analysis of Kunming Terrorist Attack

12.3.1 Case Background

At 21:20 on March 1, 2014, several criminals in uniform armed with knives burst into the Kunming railway station square and ticket lobby, attacking anyone they met, resulting a number of casualties. More than 10 police cars rushed to the scene to arrest suspects. Then a SWAT team arrived on the spot and killed four criminals, arrested one. It shows that this crisis caused severe damage to Kunming—total 29 dead and 143 injured as of March 2, which shocked people all around the world. The night of the murder, riot police cordoned off the Kunming railway station, trying to identify other criminals from people in the station. Police, SWAT, firefighters, ambulance and other multi-force arrived at the scene to help deal with the crisis.

As soon as the crisis occurred, the Party Central Committee and Yunnan Provincial Government respond quickly and actively to arrest assailants and give first-aid to the wounded. Germany, France, Japan, the United States, Russia and other countries expressed their condolences to Chinese victims and condemned the heinous crime.

12.3.2 Impact on Urban Development

Kunming railway station terrorist attack is an urban crisis caused by social problem and has far-reaching negative effects on the development of Kunming. The impact of crisis on urban development appears in the following three aspects.

Impact on Social Order

Kunming terrorist attack criminals armed with knives attacked and killed innocent people, bringing a huge threat to life and property security of public. Due to the crisis occurred in a crowded transport, urban infrastructure suffered damage as well.

After the occurrence of Kunming terrorist attack, people' daily life in Kunming is influenced in some ways. People under a great mental stress and negative emotions, go to public places as little as possible. A social panic spread around the whole city, destroyed the original harmony social order and impeded the normal development of Kunming.

Impact on Urban Image

Research on urban image is an important part of urban development, including urban history and culture, urban development planning, urban economic and tourism industry. Urban image refers to the impression and feeling of a city. Essential elements of urban image consist of all factors that can make people have a long memory of a city, such as buildings, roads, transport facilities, store, tourist attractions and living facilities. And some factors contribute to form distinctive urban image, such as residents' behavior, administrative style, culture atmosphere, local conditions and customs.

The urban image of Kunming had always been a famous historical and cultural city, the weather of which is like spring all the year round. But the image was broken by the Kunming terrorist attack. After this crisis, a number of media and network platform reported Kunming by labels like "unsafe", "chaos", doing great harm to the urban image of Kunming.

Impact on Tourism

Kunming terrorist attack has damage to the economic development of Kunming, especially in the field of tourism. With the weather gradually warming in March, many people choose to visit Kunming, but change their mind after this crisis. According to several media interviews of staff in travel agency, the number of people asking about tourism business in Kunming was fall significantly after this crisis. Statistics show that in the first half of 2014, there were 537,300 tourists have been to Kunming, dropped as much as 4 %.

Tourism industry has always been a pillar industry of Kunming. However, as a tourism city, Kunming has a delay and even recession in tourism. There are many reasons such as increasing pressure of economic, consumers' weary state, but the effect of Kunming railway station terrorist attack can not be ignored.

12.4 Conclusion

This paper taking the Kunming terrorist attack as an example, studied the impact of crisis events on urban development and found that crisis events will break the social order, impair urban image and hinder the development of urban tourism.

According to the relevant literature review on the crisis and urban development as well as the empirical study of Kunming terrorist attack, this will give four recommendations that Kunming city should take to promote city development: (1) openness and transparency of government information; (2) quick response mechanism; (3) complete legal system; (4) reshape urban image.

Acknowledgements We would like to thank project supported by the National Natural Science Foundation of China (Grant No.71373231) for financing this study.

References

1. Barton R (1996) Measuring the Effects on tourism of violence and of promotion following violent acts. In: Pizam A, Mansfield Y (eds) Tourism, crime and international security issues, Wiley, New York, pp 159–174
2. Robrt H (2001) Crisis management. China Citic Press p 52
3. Otto L (2001) Crisis management. Wu-Nan Book Inc, p 34
4. Xue L, Zhang Q, Zhong KB (2003) Prevention and reconstruction: research on chinese crisis management in transition from SARS. Macroecon. Reform Dev 4:5–20
5. Lin HC Public Relation Planning. Fudan University Press, p 13

Chapter 13
The Status Quo, Problems and Legal Construction of China's Collective Construction Land Circulation

Yi Zhang and Hong Zhang

Abstract Since the late 20th century, there appeared the collective construction land circulation phenomenon in many places in China. After entering the new century, with the rapid economic and social development taking industrialization and urbanization as the leading role, the scale of collective construction land circulation is increasing, showing some new features and new trends. However, the relevant legal construction is in the need of urgent improvement. On the basis of redefining the concepts and connotation of collective construction land and its circulation, the paper thoroughly investigated the status quo of China's collective construction land circulation and existing problems, and proposed a legal construction framework for China's collective construction land circulation.

Keywords Land reform · Collective construction land · Land circulation · Legal construction

13.1 Introduction

With the rapid advance of urbanization in China, the nationwide spontaneous and non-spontaneous collective construction land circulation has been an increasingly common phenomenon, and the circulation mode is also increasingly diverse. However, the central level has not made changes to the relevant laws and regulations till now, which becomes a huge obstacle factors against collective construction land. "Under the premise of conforming to the planning and use control, allowing the sale, leasing and shareholding of rural collective operating construction land, and implementing the same market treatment and the same price and right as the state-owned land" proposed in the Third Plenary Session of the 18th CPC Central Committee was a larger relaxation of collective construction land circulation that was opened at the national level in recent years. But in reality, the

Y. Zhang (✉) · H. Zhang
School of Civil Engineering, Tsinghua University, Beijing, China
e-mail: zhangyiyh@tsinghua.edu.cn

Y. Wu et al. (eds.), *Proceedings of the 20th International Symposium on Advancement of Construction Management and Real Estate*,
DOI 10.1007/978-981-10-0855-9_13

proportion of China's collective construction land in rural areas has been great, with its land types and sources being very complex. Therefore, it is necessary to redefine the concepts and connotation of China's collective construction land and its circulation, re-examine the status quo and problems of collective construction land circulation, and explore specific measures on perfecting collective construction land system from the law, regulation and policy level respectively.

13.2 The Status Quo of Collective Construction Land and Its Circulation

According to Article 43 of the "Land Management Law of PRC", only three kinds of construction land including rural collective ownership township (town) enterprise land, rural residential land and township (town) village public facilities and public welfare construction land can become collective construction land under the legal approval. According to whether for profit, construction land is divided into collective profitable construction land and collective non-profitable construction land by academia. Currently, there is no concept of "collective profitable construction land" in China's laws, which first appeared in the Decision of the Third Plenary Session of the 18th CPC Central Committee. Collective non-operating construction land includes rural residential land and township (town) village public facilities and public welfare construction land; other construction land beside that is collective operating construction land.

According to current laws, there is only one case for collective construction land circulation, that is, the legal transfer of land use rights will happen due to the bankruptcy or merger of township (town) enterprises using collective construction land.[1] On the concept and connotation of collective construction land circulation, there is no clear legal stipulation in China's laws, but it is generally agreed that collective construction land circulation means the behavior that under the premise that the nature of ownership is unchanged, all types of rural collective economic organizations and other rights holders[2] let collective construction land use rights flow between different economic entities legally by ways of sale, leasing, transfer, mortgage, shareholding, joint operation, etc.

The Third Plenary Session of the 17th CPC Central Committee proposed "the gradual establishment of a integrated urban-rural construction land market". However, the measure has not been better implemented due to the unadjustment of supporting laws and regulations. The Third Plenary Session of the 18th CPC

[1]See Article 63 of "Land Management Law of PRC".

[2]Township (town) villages, village groups, collective organizations at all levels and the villagers, and state-owned enterprises, private enterprises, individual industrial and commercial households, urban residents.

Central Committee proposed that allowing the sale of rural collective operating construction land,[3] which not only provides policy support for exploring the direct leasing mode or indirect property leasing mode of collective operating construction land in coastal developed areas such as the Pearl River Delta and Yangtze River Delta, but also contains the idea of establishing two stages of collective construction land market.

13.2.1 The Collective Construction Land Circulation Has Been Widespread All Over the Country Including the Pilot Areas

After 1990, with the rural social and economic development, especially the restructure of township and village enterprises, there appeared many spontaneous phenomenons of collective construction land circulation in the coastal developed areas and the outskirts of major cities. In order to standardize the development and gain experience for the modification of the law, the Ministry of Land and Resources conducted collective construction land circulation pilot worked consecutively in nine cities including Nanjing, Suzhou, Huzhou, Nanhai and so on in 2000 [1]. In recent years, with the assets property of rural collective construction land is increasingly prominent, the spontaneous rural collective construction land use right circulation has become a common phenomenon. A lot of collective construction land circulation has become an indisputable fact. According to statistics, the rural collective construction land acquired by the way of circulation has been more than 50 % of the total collective construction land in the Pearl River Delta.

13.2.2 Throughout the Country, Collective Construction Land Circulation Is of Obvious Local Features, Diverse Circulation Forms and Multiple Bodies

Collective construction land circulation is closely related to the local level and stage of economic development. Rural collective construction land in different regions and different locations is somewhat different in the subject of implementation, object of implementation, transfer mode, transfer price, the power of construction land use rights, the distribution of interest and so on. For example, "Land Lease"

---

[3]*CPC Central Committee Decision on Major Issues Concerning Comprehensively Deepening Reforms* (adopted at the close of the Third Plenary Session of the 18th CPC Central Committee on November 12, 2013).

mode in Nanhai of Guangdong, "Annual Rent" mode in Yixing of Jiangsu, "Land Shareholding Cooperatives" mode in Suzhou of Jiangsu, "Land Bid Invitation, Auction and Listing" mode in Shayang of Hubei, "Circulation Package" mode in Pi County of Sichuan, etc. [2].

Adapting to the diversification of investment subjects in economic development, participating subjects of collective construction land circulation also show a trend of diversification. The outflow side of collective construction land use rights includes not only government agencies such as township governments and other parties, but also autonomous village organizations such as village committees and village groups, rural collective economic organizations such as township enterprises, and land users such as individual farmers, etc. The inflow side includes not only members of local collective economy organizations, but also that of other collective economy organizations, all types of businesses and other individuals.

13.2.3 Collective Construction Land Use Right Circulation Administrative Measures Have Been Developed in Many Areas

With more and more exploration and pilots on nationwide collective construction land circulation, in order to regulate these circulation activities existing in practices, some provinces and cities have consecutively developed local "collective construction land use right circulation administrative measures" and other normative documents. For example, Anhui Provincial People's Government issued the "Anhui Province Pilot Scheme on Compensated Use of Collective Construction Land and Circulation of Land Use Right"[4] on October 23, 2002; in June 2005, Guangdong Provincial Government issued Decree 100 entitled "Guangdong Province Collective Construction Land Use Right Circulation Administrative Measures"; in August 2008, Hebei Province People's Government promulgated the "Hebei Province Collective Construction Land Use Right Circulation Administrative Measures (Trial)".[5] All these local laws or regulations conditionally permitted compensated and time-limited collective construction land use right circulation.

---

[4]See the "*Notice of Anhui Provincial People's Government on the Issuance of "Anhui Province Pilot Scheme on Compensated Use of Collective Construction Land and Circulation of Land Use Right*"" (Anhui Province Govrement [2002] 60).

[5]See the Hebei Provincial People's Government Order No. 11 (2008).

13.2.4 Collective Construction Land Circulation Is Positively Associated with Local Economic Development Level and Geographical Location Condition

The level of activity and the scale of collective construction land circulation are closely related to local economic development level and geographical location condition. Research found that the collective construction land circulation in economically developed regions is generally showing a wide range, large scale and many forms. However, in economically underdeveloped areas, the collective construction land circulation is lightly scattered and small-scale with few forms. From the location condition, circulation activities involving collective construction land circulation areas are mainly concentrated in suburban or urban fringe areas, and are relatively less in exurban areas and rural areas.

13.2.5 New Deployment and Top-Level Design on Collective Construction Land Circulation After the Third Plenary Session of the 18th CPC Central Committee

In 2013, the "CPC Central Committee Decision on Major Issues Concerning Comprehensively Deepening Reforms" once again stressed to "establish a integrated urban-rural construction land market", which pointed a clear direction for the future transfer of collective construction land market. On December 2, 2014, the seventh meeting of Central Leading Group for Comprehensively Deepening Reforms reviewed and approved the "Opinion on Rural Land Acquisition, Collective Profitable Construction Land Market, and Residential Land System Reform Pilot Work". In the end of February 2015, the thirteenth meeting of the 12th NPC Standing Committee reviewed and approved the motion submitted for deliberation by the state council on "Decision on Authorizing the State Council to Temporarily Adjust Related Laws and Regulations in 33 Pilot Counties (Cities, Districts) Administrative Areas Such as Daxing District of Beijing (Draft)".

It is not difficult to find that the explicit collective construction land circulation market will be deployed by the central government and be in closed operation within the pilot areas in the coming period. It can be expected that, with the gradually accumulated, refined and combed experience in the pilot areas and the modified and improved laws and regulations, a large number of implicit collective construction land circulations in China will be gradually explicitated and incorporated into the legal, standardized and orderly track in the future.

13.3 Existing Problems in China's Rural Collective Construction Land Circulation

Collective construction land circulation has played a significant role for the prosperity of the rural economy and the increase of farmers' income. However, there are still some problems in the process of transfer, focusing mainly on the following aspects.

13.3.1 Legal and Institutional Imperfection in Rural Collective Construction Land Circulation

Contradictions among and inside some laws and regulations have brought not so small obstacles for collective construction land circulation and related provisions. For example, the contradiction between "The Constitution" and "The Land Management Law" not only leads to the failure of implementing rural collective land ownership,[6] but also limits the transfer of collective construction land[7]; Contradictions inside the "The Land Management Law" lead to conflicts and inconsistencies among different articles of the same law[8]; Contradictions between "The Land Management Law" and "The Guaranty Law" and the inconsistency of the two systems including restrictions on land use right and free disposal of rural housing result in larger collective land certificate mortgage and guarantee risk.[9]

13.3.2 The Subject of Collective Construction Land Circulation Is not Clear, and Relevant Specifications Need to Be Detailed Urgently

Which level does the owner ship of rural land belong to? How to determine boundaries for collective administration organizations, autonomous organizations and economic organizations? What is the nature of collective economic organizations? Many problems are stated unclear in China's laws, resulting in that the subject of rural collective land ownership and the status and boundary of it are not clear enough. The subject of collective construction land therefrom is also extremely confusing. The outflow side includes not only government agencies such as township governments, but also autonomous village organizations such as village

[6]See Article 10 of "*Constitution*" and Article 43 of "*Land Management Law*".

[7]See Article 10 of "*Constitution*" and Article 63 of "*Land Management Law*".

[8]See Article 2 and Article 63 of "*Land Management Law*"; Article 63 and Article 62.

[9]See Article 37 of "*The Guaranty Law*".

committees and village groups, rural collective economic organizations such as township enterprises, and land users such as individual farmers, etc. The inflow side includes not only members of local collective economy organizations, but also that of other collective economy organizations, various types of enterprises and other social organizations and individuals.

13.3.3 Circulation Forms Are Diversified, and the Scope and Conditions of Circulation Are not Uniform

In the real exploration, there are multiple forms including selling, transfer, leasing, mortgage guarantee, and so on. However, for the reason that current laws and regulations do not define clearly the relationship between subjects of rural collective construction land circulation, the relationship between the owner and the use right holder as well as the relationship among different use right holders are not clear enough, causing that various forms of definition and contonation, conditions for implementation and operational procedures of collective construction land circulation, are blurred. In addition, currently the scope of collective construction land circulation is not uniform, distinguished either by the range of urban planning area in some places, or by whether or not it is newly acquired construction land in some places, or by if they meet the provisions of overall land use planning in other places. At the same time, it is also in the state of chaos that whether the origin of collective construction land as the transaction object is legitimately approved. There are both legitimately approved land and illegal land changing from rural land to urban land without approval.

13.3.4 Non-market and Unreasonable Circulation Prices

Due to a large number of collective construction land in hidden market, in reality there exists not only the phenomenon of arbitrary circulation pricing, but also the phenomenon of "packaged" circulation for land and other rural collective assets. This will not only lead to the undervaluation of collective land, but also cause a huge loss of collective land assets. Meanwhile, low-cost land use also contributes to the extensive land use and land wastage of land use enterprises, being unfavorable for intensive land use. In Nanhai District of Foshan City, field survey found that most industrial plants on rural collective land are with one to two floors, and the building floor area ratio is far less than that of the neighboring state-owned construction land.

13.3.5 Period and Purpose of Collective Construction Land Use Are not Uniform Enough After Circulation

For the period of collective construction land use after circulation, there are different specifications nationwidely. Some places stipulate a one-year contract, some places require a contract no more than five years, some places refer to the use period of state-owned construction land, and still some places do not set the longest period in years. Similarly, for the purpose of land use after circulation, specifications are different in different regions, some places prohibit profitable purposes, some places take the planning requirement as the bottom line, some places restrict commercial housing development, and some places state no particular restriction.

13.3.6 The Attribute of Financial Assets of Collective Construction Land Can not Play Effectively

In addition to a small number of exceptions specified in national laws, collective construction land circulation in market is prohibited in China, not to mention the mortgage and guarantee, which not only seriously hampers the inventory of a large number of collective construction land assets, but also has an adverse impact on the use of collective construction land which has experienced a lot of transfers. Survey found that enterprises worry about the collective construction land after circulation, most banks do not easily accept mortgage loan applications in the form of collective construction land, and problems on poor loan financing channel of rural collective construction land are prevalent throughout the country. In Guangdong Province, although administrative measures stipulating that collective construction land use rights can be used as mortgage guarantee were already issued in 2005, the effects of policy implementation are unsatisfactory because of a conflict between it and the national upper law.

13.3.7 Circulation Income Distribution Lacks Specification, Farmers' and Collective Interests Are Violated to Varying Degrees

Due to the lack of support on collective construction land circulation from laws and regulations at the national level, income distribution relationship among the four aspects including the central government, local governments, the collective and farmers has not been rationalized. There still lacks a theoretical basis and unified specification on how to allocate the collective construction land circulation income among village collectives, township and county-level governments and municipal governments and how to make internal distribution among members after collective economic organizations acquire income.

In the actual process of circulation, due to the lack of market pricing mechanism in invisible market, thus existing the phenomenon that land circulation income is dominated and possessed by a few of towns and village cadres and farmers' rights are seriously violated and causing many potential conflicts.

13.3.8 Prone to Legal Disputes and Poor Relief Channels

By examining the various cases involving collective construction land use right circulation in the judicial practice, it can be found that, the undefined subject of collective land ownership and its representative, together with various reasons including the fact that the state monopolies state-owned land market and collective land use rights are subject to strict legal restrictions and so on, not only lets it be prone to collective construction land disputes, but also provides gray handling space for grassroots courts on whether to accept this type of cases because of the existing blur. Even if this kind of cases are accepted, there still are difficult problems on trials and settlements, causing that all parties of rights entitled by collective construction land use right circulation can not get timely relief.

13.4 Legal Framework of Collective Construction Land Circulation

Collective construction land has a tremendous role in promoting local economic development and speeding the urbanization. It has been an objective requirement to establish and perfect China's land market system with respect to collective construction land circulation in the new situation. Only through strengthening the institutional innovation continually, taking a combination of approaches including "sparsing" and "blocking", and improving relevant laws and regulations continuously to guide and standardize the collective construction land circulation, can it be the correct orientation in theory and practice.

13.4.1 Identify the Subject of Collective Land Ownership, Perfect the Collective Construction Land Ownership

Clear property relations and well-defined subjects of collective construction land are a prerequisite for the orderly circulation. In consideration of the various actual situations in rural China, we can actually put the subject of collective construction land ownership into three levels (kinds) of organizations: the first is the villagers' group, the second is the village committee and the corresponding collective economic organizations, and the third is the township collective economic organizations or township government, in order to prevent the emergence of unclear or equivocated public property ownership [3]. All types of organizations should be legally equal, independent of each other, and not subject to each other.

Moreover, in view of the rural land shareholding reform that has been implementing for many years in China's coastal cities and areas surrounding major cities, there have been established a new type of rural collective economic organizations-land stock cooperatives in many rural areas. It is suggested to transfer the collective land ownership to land stock cooperatives and grant them with legal entity qualifications.

To perfect the collective construction land ownership, the state-owned monopolying land market structure must be broken and the unified construction land market must be established. To do this, we must first of all, through the amendment to Article 10 Paragraphs 1 and 4 of "The Constitution", reaffirm and clarify the equal status of collective and state-owned land use rights. Secondly, on the basis of the revised "The Constitution", further modify Article 43 of "The Land Management Law" and state clearly that units and individuals in the need of construction land can either apply for state-owned land, or apply for collective land.

13.4.2 Set the Conditions, Scope and Duration of Collective Construction Land Circulation Reasonably

Collective construction land circulation shall meet the following conditions: (1) it must belong to the rural; (2) it must have the appropriate collective land ownership certificate; (3) it must comply with local land use planning, annual land-use plans, town planning, village planning, etc.

The scope of collective construction land circulation: (1) The geographical location scope shall be the collective construction land in villages and towns (including the construction land stock approved by the law and the construction land increment approved for the conversion of agricultural land). A part of collective construction land stock having no approval previously can be put into circulation as long as it complies with the existing land use plan and can go through the land-use formalities. (2) As to the use scope, appropriate relaxation can be made according to actual needs to allow the establishment of all kinds of enterprises in the collective construction land.

About the duration of collective construction land circulation, collective construction land can refer to the provisions of the state-owned land use right circulation in terms of maximum transfer period. Specified period in years is negotiated and determined by land owners and land users. At the same time, a combination of land leasehold system and land rent system is encouraged in the implementation of collective construction land circulation, with circulation period being longer or shorter as needed.

13.4.3 Ways of Rural Collective Construction Land Circulation

By learning from beneficial attempts in the exploration of local collective construction land circulation, it is suggested to state clearly a variety of ways of collective construction land circulation, and modify Article 63 of "The Land Management Law" as "collective land use rights shall not be sold, transferred or leased for the purpose of non-agricultural construction. However, under certain circumstances of being in line with the general land use planning and being legally acquired, the rural collective construction land use right can be shifted by ways of transfer, exchange, leasing, mortgage, shareholding based on land pricing, joint funded associates, inheritance, remote replacement and indicators circulation, etc."

Transfer, namely the way of circulation that land owners transfer land use rights to others and others pay land transaction fees to compensate landowners. In specific practices, either one-time transfer can be implemented by referring to the way of state-owned land use right transfer, with its maximum period in years not exceeding the state-owned land use right transfer period, or multiple transfers with a yearly period can be implemented based on the enterprise type of collective construction land use right transferee. Specific transfer ways can be divided into agreement transfer and non-agreement transfer by refering to the state-owned land use right transfer. Wherein, non-agreement transfer includes mainly bid, auction, listing and other market-oriented approaches.

Remote replacement relies mainly on the policy of linked change of urban-rural construction land issued by the Ministry of Land and Resources. Index trading is commonly known as land ticket transaction. Wherein, "land ticket" means the index linked to construction land, especially the construction land index increased in urban area according to the policy of linked change of urban-rural construction land after the land reclamation of rural collective construction land such as land for township and village enterprises, homestead, rural public facilities and public welfare facilities [4]. Land ticket transaction is currently the major mode of rural land transactions in Chongqing.

13.4.4 Establish the Collective Construction Land Price Determination Mechanism

Whether from the aspect of reinforcing government regulation and regulating current collective construction land circulation in many places, or from the aspect of avoiding the loss of collective land assets, the collective construction land circulation must have a reasonable and legitimate price determination mechanism, so as to prevent or avoid artificially determining factors in the collective construction land circulation. Given that currently there has not been established the appropriate valuation and assessment system with respect to the current rural collective land, it is suggested that the following three pricing system for collective construction land can be taken: the minimum price system, the grading valuation system, and the price reporting and public bidding system.

13.4.5 Form a Consumated, Transparent and Open Construction Land Income Distribution System

Collective construction land circulation can release the value of collective construction land, and in the current context of land acquisition mode without major changes, is in its essence a way that rural collectives and local governments share the land value-added income. On the basis of protecting farmers' rights to know and participate the collective construction land circulation income distribution, it should be fully considered the interests of all parts including local governments, investment developers, village collectives and villagers, together with taking into account the interests of the subject of circulation implementation, so that rights, responsibilities and interests are unified and a transparent, open and consumated collective construction land income distribution system is established. Field survey found that the way of circulation plays a decisive role on the amount of circulation income, so the construction of collective construction land circulation income distribution system has to take into account the way of circulation.

13.4.6 Activate the Financial Attributes of Collective Construction Land, and Gradually Realize the Combination of It with Financial Market

On the basis of the gradually appearing market value of the collective construction land, the government will establish and improve the collective construction land use right market price evaluation system [5], optimize the collective construction land mortgage financing guarantee procedures and other safeguards, and consider how to exert the role of land mortgage financing and promote rural land share cooperation, and so on.[10] Meanwhile, we should focus more on the combination of economic value of land and capital market, in order to maximize the overall effectiveness of elements matching [6].

[10]Refer to the 175th page of the [3].

13.4.7 Formulate Unified Rural Collective Construction Land Circulation Administrative Measures

Currently around the country, administrative measures for collective construction land circulation and use are chaotic, diversified and in need of harmonization and standardization. Not only it is needed to amend related laws and regulations, but also it is urgent to formulate a set of administrative measures that should be nationally unified and suitable for everywhere. Administrative measures should state clearly and normatively the property right of collective construction land, the subject and object of circulation, the scope and conditions of circulation, the way and duration of circulation, the manner of income distribution, the mortgage financing guarantee, and so on, so that it can strengthen the guidance and administration on collective construction land and promoting the orderly circulation of collective construction land.

References

1. Tang J et al (2014) Rural land policy reform experiment under the strategy of new urbanization. China Social Sciences Press, p 58
2. Zheng LZ (2013) China land policy blue book. China Social Sciences Press, pp 161–176
3. Huang SY (2013) Legal research on rural collective construction land circulation. Master thesis, Graduate School of Chinese Academy of Social Sciences, pp 30–31
4. Qiu JQ et al (2010) Market allocation of urban and rural construction land change-linked index. Urban Probl 180(7):65–66
5. Tang J et al (2013) Exploring the path of releasing collective construction land value. China Land and Resources News, 28 Nov 2013
6. Ye JP (2012) Land use dilemma in the rural-urban continuum of china: path choice and mechanism design. China Economic Publishing House, p 220

Chapter 14
Gray Correlation Degree Study on Changsha-Zhuzhou-Xiangtan Urban Construction Land Structure and Economic Development

Xiying Hu and Ting Tan

Abstract Reasonable construction land structure can provide the material basis and bearing space for economic development, economic development can promote the rational allocation of construction land resources effectively. This paper uses information entropy model to analyze the evolution features of Changsha-Zhuzhou-Xiangtan (CZT) construction land from 2007 to 2013, and uses gray correlation model to explore the relationship between construction land structure changes and economic development. The results show that the change trend of equilibrium degree was consistent with that of the information entropy, but the dominance trend is the opposite. In addition, the study shows that the gray correlation degree of economic development and information entropy are within the range of 0.5–1. The analysis reveals that the most important factors affecting the change of construction land structure are urban resident consumption level, total retail sales of social consumer goods and per capita GDP. On this basis, firstly, it is recommended that the government should accelerate to make experiments on "multiple gauge unification". Secondly, CZT should break the administrative division barriers and land market segmentation in CZT land market and reduce administrative intervention to achieve land market operation. Thirdly, the government should adjust and optimize the structure and spatial structure of urban land according to the city's different property and function orientation. What's more, it is suggested that the government need vigorously guarantee the land for economic development demands and carry out differential land supply policy for construction land. At last, the government should promote green cycle low carbon production mode and green consumption lifestyle for "Two-Oriented Society" construction.

Keywords Changsha-Zhuzhou-Xiangtan · Construction land structure · Information entropy · Gray correlation analysis

X. Hu (✉) · T. Tan
School of Urban Construction, Jiangxi Normal University, Changsha-Zhuzhou-Xiangtan, China
e-mail: huxy8503@163.com

T. Tan
e-mail: 1591355083@qq.com

Y. Wu et al. (eds.), *Proceedings of the 20th International Symposium on Advancement of Construction Management and Real Estate*,
DOI 10.1007/978-981-10-0855-9_14

14.1 Introduction

With the continuous economy development of Changsha Zhuzhou and Xiangtan Cities Group, the national strategic comes forefront like the central rise, the Urban Agglomeration in the Middle Reaches of the Yangtze River, Yangtze River Economic Zone and The Belt and Road Initiatives and so on. On one hand, the contradiction between urban construction land demand's rigidly grow and land supply's rigid constraint becomes more and more intense. On the other hand, as the China's urbanization and industrialization development steps are speeding up, the urban agglomeration has become a research focus. Compared with a single city (town), the urban agglomeration's construction land-use system is more enormous and complex because its structure status is the result of many cities of economic links and industrial interaction. So the economy development on the promotion of urban construction land expansion plays an increasingly important role, the pushing effect is undisputed.

Therefore, promoting the development of upgrading meanwhile promoting construction land intensive use become an objective requirement and inevitable choice when we want to find a balance between construction urban land expansion and economic development new normal state and taking the lead to realize transformation innovation development requirements. Based on this, this paper introduces the study of CZT construction land structure's change to find its evolution law and driving mechanism, and then uses gray correlation method to analyze the correlation degree of the construction land use structure change and economic development indicators. Finally, we get the direct dynamic factor of CZT construction land structure's change. According to the analysis results, this paper puts forward some suggestions on harmonious development of CZT construction land use and economic development. What more, it has great practical significance because it can provide reference for constructing Two-Oriented Society, the Rising of Central China and other urban agglomeration in Central China's resources elements integration, efficient allocation of city resources and regional leap forward development.

14.2 The General Situation of the Study Region

CZT Urban Agglomeration is located in the northeast of Hunan province, which is consist of three cities of Changsha, Zhuzhou and Xiangtan. It currently exist three prefecture-level cities, four county-level cities and 183 towns, total area of 2.8 million km^2. Changsha, Zhuzhou and Xiangtan three cities show "triangle-shaped" distribution along the Xiangjiang River. Changsha, Zhuzhou and Xiangtan only 40 km away, Zhuzhou is apart from Xiangtan only 20 km, the structure is very compact. At the end of 2014, the total population of CZT has reached 1408.55 million people, GDP has run up to 1.155591 trillion yuan. Its per capita GDP is 82,456.18 yuan and per capita disposable income achieves 33,124 yuan, which is higher than the whole province 15,502 yuan. What's more, urbanization rate has increased to 66 %, higher than the whole province 19.72 %. CZT has achieved 42.7 % of the province's GDP by 13.3 % of the whole province land and 19.2 % of the whole province population. Currently, CZT is in front of national strategy like Two-oriented Society, the Rising of Central China, Yangtze River Economic Zone, the Urban Agglomeration in the Middle Reaches of the Yangtze River and so on, which provides an important strategic fulcrum and a major strategic opportunity to CZT regional development and expands larger space to deepen and promote the development of regional economy in CZT urban agglomeration.

14.3 Data Sources, Data Processing and Study Methods

14.3.1 Data Sources and Data Processing

Urban construction land data used in this paper comes mainly from 2007–2013 years of Hunan Statistical Yearbook and land use change survey statistical data. Data of economic development is mainly from the Hunan Statistical Yearbook and each city's jahrlich national economic and social development bulletin of CZT.

Since pre-and post 2007 land classification standards adopted are inconsistent, this paper takes 2007 as the start year of the study and selects 2007–2013 seven related years data as study period in order to make the study area construction land data sequential and consistency. Meanwhile, the paper refers primarily new edition "Land Use Status Classification" and the classification in "city size and construction land" of "urban construction and environmental protection" in "Hunan Statistical Yearbook" and integrates the old and new version of code for classification of urban land use and planning standards of development land (GB50137-2011 and GBJ1378-90) to make some adjustments for CZT urban construction land classification and process the data base on the classification. This article divides construction land into nine land categories including residential lands, public facilities land, industrial land, storage land, roads and transport facilities land, municipal utilities land, green land and special land.

14.3.2 Study Methods

1. Information Entropy

Entropy was put forward by American mathematician Shannon in 1948 based on information theory, which is used to reflect a measure of system inside configuration information deficiency extent or uncertainty of random events. Information entropy can also be used to reflect the order degree of regional land use system. In general, the higher the entropy value, the lower the order degree, and vice versa.

The study constructs the construction land information entropy (H) [1] based on Shannon entropy equation, the formula is as follows:

$$A = \sum_{i=1}^{N} A_i \tag{14.1}$$

$$P_i = \frac{\mathrm{A_i}}{A} \tag{14.2}$$

$$H = -\sum_{i}^{N} P_i LnP_i \tag{14.3}$$

Here, A represents the total area of construction land; $\mathrm{A_i}$ represents construction land area of each functional class; N indicates the kind of construction land. $\mathrm{P_i}$ is the percentage of all types of construction land; H represents construction land information entropy. In practice, the information entropy cannot reflect the impact of the functional number N, concept of equilibrium degree and dominance are commonly introduced. The equilibrium degree formula based on information entropy function is as follows:

$$J = \frac{H}{H_m} = -\sum_{i}^{N} \frac{P_i LnP_i}{LnN} \tag{14.4}$$

Here, J represents equilibrium degree, H means construction land information entropy and $\mathrm{H_m}$ is the maximum of entropy value. The greater the J value, the stronger the equilibrium of the construction land system, when J = 1, the type of land achieves the idealistic balance state; when J = 0, the type of land is in the most uneven state. Based on the concept of equilibrium degree, the dominance formula is as follows:

$$I = 1 - J \tag{14.5}$$

Here, I means dominance degree, which reflects the degree of one or more types of construction land to dominate construction land in the region. Its meaning is opposite to the equilibrium degree.

2. Gray Correlation Analysis Method

Gray correlation analysis is a quantitative description method of system development's change situation based on gray system theory. Gray correlation analysis method is also applicable to the situation whether the sample data is enough or regular. What's more, it has smaller computational complexity. Therefore, using the gray correlation analysis method is a good choice to portray the association of urban construction land and economic development [2].

Gray correlation analysis thinking is:

1. Determine analysis sequence. This paper uses urban construction land structure entropy sequence group (x_i) as the reference sequence and regards economic development sequence groups (y_i) as comparison sequence.
2. Non-dimensional treatment. Because the selected various indicators of dimension are different, it cannot be used for statistical analysis, so it needs initial value processing. The calculation formula of initialization is as follows:

$$x_i(k) = \frac{x_i'(k)}{x_1'}(k) \quad y_j(k) = y\frac{j'(k)}{y_1'}(k) \quad k = 1, 2, \ldots, n \tag{14.6}$$

3. Calculate sequence difference $z(k)$. Calculating the absolute value of sequence difference between sequence groups (y_j) and the sequence group (x_i) and then finding the minimum and maximum values. The calculation formula is:

$$z(k) = |xi(k) - yj(k)| \tag{14.7}$$

4. Calculate correlation coefficient. Calculating the correlation coefficient of sequence groups (y_i) and the reference sequence group (x_i) at different times. The formula is:

$$r_{ij}(k) = \frac{z(k)_{\min} + \eta z(k)_{\max}}{z(k) + \eta z(k)_{\max}} \tag{14.8}$$

   Here, η is correlation coefficient and its effect is to increase prominence of the different between the correlation coefficient. $\eta \in (0,1)$, usually taking value 0.5.
5. Calculate correlation degree r_{ij}. Correlation degree's calculation formula is:

$$r_{ij} = \frac{1}{N}\sum_{ij=1}^{N} r_{ij}(k) \tag{14.9}$$

   In order to convenient for quantitative analysis, gray correlation degree can be divided into three categories in turn according to strong medium weak: $0 < r_{ij} \leq 0.35$ is weak association; $0.35 < r_{ij} \leq 0.70$ is medium association; $0.70 < r_{ij} \leq 1$ is strong association.

14.4 CZT Construction Land Structure Present Situation and Information Entropy Evolution Analysis

Urban construction land can reflect the changes of the size and layout of construction land in different periods. This paper uses CZT urban construction land data from 2007 to 2013 to analysis its structure present situation and calculate

Table 14.1 Urban construction land information entropy, equilibrium degree and dominance in CZT, 2007–2013

Type/year	2007	2008	2009	2010	2011	2012	2013
Construction land	356.55	386.07	423.04	468.25	978.88	466.86	475.68
Residential land	97.09	114.02	141.40	165.93	496.30	182.52	184.4
Public facilities land	91.16	91.86	98.46	79.98	172.91	79.63	65.02
Industrial land	68.68	65.01	69.38	84.84	85.23	79.89	68.5
Storage land	7.22	7.98	10.24	12.34	87.12	13.50	11.39
Roads and transport facilities land	52.89	62.97	57.69	68.34	33.32	29.18	65.87
Municipal utilities land	7.19	9.65	9.01	12.18	57.44	47.48	7.58
Green land	22.04	24.20	25.49	32.22	13.05	34.66	44.41
Special land	10.28	10.38	11.37	12.42	33.51	0.00	28.51
Information entropy	1.7352	1.7408	1.7123	1.7311	1.5330	1.6722	1.7376
Equilibrium degree	0.8345	0.8371	0.8234	0.8325	0.7372	0.8042	0.8356
Dominance	0.1655	0.1629	0.1766	0.1675	0.2628	0.1958	0.1644

Unit Hm^2/Nat/%/%
Source Hunan Statistical Yearbook

information entropy, equilibrium degree and dominance based on information entropy formula (14.1)–(14.5), the result is shown in Table 14.1.

As shown in Table 14.1, CZT construction land information entropy showed a decrease first and then an increase from 2007 to 2013. In the whole process, information entropy showed three distinct stages: The first stage, from 2007 to 2010 urban construction land use structure information entropy in the period of decrease with fluctuation, but the decrease amplitude was very small. The second stage, from 2010 to 2011, urban construction land information entropy was in the period of decline rapidly, which showed a greater change of construction land structure and order degree of structure had increased. The third stage, from 2011 to 2013, the urban construction land information entropy was in rapid growth period.

From the above analysis, urban construction structure information entropy of CZT urban agglomeration can reflect the change rules of the structure for construction land. Within the study period, the change trend of equilibrium degree was consistent with that of the information entropy. Compared with 2007, the information entropy and equilibrium degree has a slightly increased in 2013, but the dominance trend is the opposite. It showed that order of CZT construction land has declined, equilibrium degree has risen, a single or several construction types which dominate this region's construction land type has enhanced.

14.5 Gray Correlation Analysis of CZT Construction Land Change and Economic Development

Economic development and urban construction land has inseparable relationship. On the one hand, expansion of economic scale will lead to constantly increasing gross demands of urban construction land while the industrial structure will lead to changes in the demand structure of urban construction land. What's more, regional economic development stage and level or the main direction of economic relation and other factors will play a guiding role in the forming of the spatial structure of urban land. On the other hand, rationalization of urban land use structure is the key to improve economic benefit. Land is the first source of productivity, making the best use of local conditions and land can save land and then reduce the cost of various departments, increase total production and profits and result in better economic benefit.

This paper uses gray correlation analysis method, collects statistical analysis data and uses related research and actual situation of CZT construction land and economic development for reference, then selecting 13 index show in four aspects of economic growth, industrial structure, people's living standards and population development. At last, this paper builds economic development indicator system which affects the change of CZT construction land structure. CZT economic development indicators system is as shown in Table 14.2.

Table 14.2 CZT economic development indicator system

Target layer	Criterion layer	Basic index layer
Economic development (Y)	Economic growth	GDP y_1 (billion yuan)
		Per capita GDP y_2(yuan)
		Local financial revenue y_3 (million yuan)
		Investment in fixed assets of the whole society y_4 (billion yuan)
	Industrial structure	Added value of primary industry y_5 (million yuan)
		Added value of secondary industry y_6 (million yuan)
		Added value of tertiary industry y_7 (million yuan)
	People's living standards	The amount of social consumable retail y_8 (billion yuan)
		Urban residents per capita disposable income y_9 (yuan)
		Urban residents consumption level y_{10} (billion yuan)
	Population development	Urbanization rate y_{11} (%)
		Urban population y_{12} (million people)
		Population density y_{13} (person/km^2)

Table 14.3 The weight of each influence factor

Influence factors	Weight
GDP (y_1)	0.100
Per capita GDP (y_2)	0.087
Local financial revenue (y_3)	0.135
The whole society fixed assets investment (y_4)	0.138
Added value of primary industry (y_5)	0.030
Added value of secondary industry (y_6)	0.130
Added value of tertiary industry (y_7)	0.165
The amount of social consumable retail (y_8)	0.084
Urban residents per capita disposable income (y_9)	0.042
Urban residents consumption level (y_{10})	0.072
Urbanization rate (y_{11})	0.004
Urban population (y_{12})	0.007
Population density (y_{13})	0.008

Considering subjectivity of weight assignment, this paper uses entropy method to determine the weight value of each indicator of basic index layer (In Table 14.3). Finally the article determines eight indicators, the indicator are respectively added value of tertiary industry, investment in fixed assets of the whole society, local

Table 14.4 CZT construction land structure information entropy and economic development correlation degree

Index	Correlation degree	Ranking	Strength grade
y1	0.773130	4	Strong correlation (0.7–1)
y2	0.787073	3	Strong correlation (0.7–1)
y3	0.771033	5	Strong correlation (0.7–1)
y4	0.711816	7	Strong correlation (0.7–1)
y6	0.724704	6	Strong correlation (0.7–1)
y7	0.519874	8	Medium correlation (0.35–0.7)
y8	0.804319	2	Strong correlation (0.7–1)
y10	0.829763	1	Strong correlation (0.7–1)

financial revenue, added value of secondary industry, GDP, per capita GDP, the amount of social consumable retail and urban residents consumption level.

Finally, this paper calculates the correlation degree results of CZT construction land structure information entropy and economic development based on the gray correlation model formula (14.6)–(14.9), the calculated result is shown in Table 14.4.

Table 14.4 shows that the ranking of CZT information entropy value is urban residents' consumption level(y10) > the amount of social consumable retail (y8) > per capita GDP (y2) > GDP (y1) > local financial revenue (y3) > added value of secondary industry (y6) > investment in fixed assets of the whole society (y4) > added value of tertiary industry (y7). Among them, urban residents' consumption level is strongest, the weakest is the added value of tertiary industry. As to the strength grade, there is medium correlation between CZT construction land structure information entropy and added value of tertiary industry but strongest correlation with other seven indices.

The results show that the improving of urban residents' consumption level, increasing of the amount of social consumable retail, growth of GDP and other economic factors are the main reason for CZT construction land structure information entropy change. Furthermore, with the rapid growth of the industrialization and urbanization, the economy will continue to have a profound impact on the change of urban construction land structure.

14.6 Conclusions and Suggestions

According to the main economic influence factors of CZT urban construction land change and the status and future trends of economic development in CZT, combining principles of man-nature harmonization development and the outlook of scientific development, then making some suggestions to provide some reference for harmonious development of CZT construction land and economic social development.

Firstly, the government should accelerate to make experiments on "Multiple Gauge Unification" of CZT overall planning, urban planning and land use planning and strengthen the government's ability to control space, in order to promote economic and land-use harmonious development.

Secondly, CZT should strengthen regional coordinate awareness, break the administrative division barriers and land market segmentation in CZT land market to achieve the integration of CZT urban land market. At the same time reducing administrative intervention, to achieve land market operation and improve the economic benefit of the land.

Thirdly, the government should adjust and optimize the structure and spatial structure of urban land which depends on the city's different property and function orientation. For example, strictly control of the Changsha urban construction land inefficiencies expansion due to obtain land financial revenue. Properly increase the proportion of roads and transport facilities land and green land of traffic hub city Zhuzhou. Prohibiting high energy consumption, low cost industrial warehouse land's over expansion and improving the proportion of tertiary industrial land, municipal facilities land and green land.

Fourthly, the government need vigorously guarantee the economic development demand for land. Opening green pathway for land permission and implement differential land supply policy for construction land and vigorously guarantee great infrastructures, key industrial projects, important livelihood engineering and other important and reasonable economic development land demand.

Finally, the government need promote green cycle low carbon production mode and green consumption lifestyle. Only achieving production and lifestyle changes, contradiction between economic growth and resources can be relieved fundamentally. Furthermore, reducing waste and pollution of land resource and breaking through the "bottlenecks" that land issue as a constraint to CZT "Two-Oriented Society" construction.

References

1. Xie T, Liu A, Gao X (2015) Construction land structure temporal and spatial variation and driving forces analysis based on information entropy and gray correlation analysis of Chengdu city. J Agric Modern Res 36(1):118–125
2. Kang J, Zhang Y, Sun G (2015) Land use structure information entropy and gray correlation of social-economic development analysis of Kashgar city. J Anhui Agric Sci 3:301–303

Chapter 15
Industrial Land Intensive Utilization Evaluation by Means of Improved Fuzzy Analytic Hierarchy Process (FAHP): A Case Study on Typical Enterprises of Chengdu Economic and Technological Development Zone

Xiantao Zhang and Fangming Zhu

Abstract As the core unit of the development zone land use, enterprise land intensive utilization degree directly determines the condition of development zone land intensive utilization. Therefore, it is of necessity to make an in-depth research on enterprises land utilization. Taking 19 typical enterprises of Chengdu Economic and Technological Development Zone, the thesis built an evaluation index system for enterprise land intensive utilization, which comprises of 4 sub-objects including: land investment density, land economic efficiency, land utilization structure and land utilization density, along with 11 indexes. Indexes are weighted by using Fuzzy Analytic Hierarchy Process (FAHP) so as to calculate land intensive degree for each enterprise. The research findings indicated that it is relatively higher for the general level of typical enterprises land utilization of Chengdu Economic and Technological Development Zone, but industries differentiate clearly in land utilization, and land intensive use degree also varies in foreign and domestic enterprises. In the end, the thesis suggested lots of measures for innovating land supply model, strictly approving enterprise access to land use, improving supervision and inspection regulations and laws as well as accelerating enterprise technology innovation.

Keywords FAHP · Typical enterprises · Industrial land intensive utilization · Chengdu Economic and Technological Development Zone

X. Zhang (✉) · F. Zhu
School of Economics, Sichuan University, Chengdu, China
e-mail: etao_0426@163.com

Y. Wu et al. (eds.), *Proceedings of the 20th International Symposium on Advancement of Construction Management and Real Estate*,
DOI 10.1007/978-981-10-0855-9_15

15.1 Introduction

Since reform and open-up in 1978, various kinds of economic development zones (EDZ for short) in China, have become gradually intensive investment areas, newly-emerged industry gathering zones and intensively-developed new cities. Moreover, EDZ has been playing key important roles in China's open economy development, in particular, in improving investment environment, promoting industrial structure adjustment [1]. Land, especially, the industrial land, is a development carrier and an essential producing factor to guarantee EDZ economy sustainable development. As the minimum land use units and core subjects of EDZ, enterprises land utilization degree directly determines land use tendency of EDZ.

The academic research focusing on enterprises land utilization of EDZ recently keeps rising. At present, academic research results mainly fall onto hereinafter aspects: (1) To take industrial land as research object, set up research model and use different evaluation approaches to make quantitative evaluation on different sectors of industrial land, and, on a basis of quantitative evaluation results, along with the present facts of EDZ, put forward ways and measures in terms of industrial adjustment and intensive land use [2, 3]. (2) To take advantage of micro-panel data of EDZ enterprises, introduce sustainable development concept into enterprise space level, and propose theory assumption and build theoretical model to evaluate land use efficiency and capability of land intensive use of EDZ enterprise [4, 5]. (3) To take industrial land into production function account, and calculate marginal contribution of land level to economic output through production function model [6, 7]. (4) To analyze enterprises land factor output elasticity differentiation by comparing land bidding, auction and listing system with other ways to make a conclusion that land grand system reform can accelerate land intensive use [8].

In general, in contrast to the whole EDZ land evaluation research at medium level, the number of EDZ typical enterprise industrial land evaluation research is still less, and evaluation approaches also are RAGA-AHP [9], Bayes discriminant analysis [10] as well as GIS space model [5]. Based on this, the paper takes typical enterprises of EDZ as a research view, FAHP as weight determination to build a comprehensive evaluation index system to evaluate the level of EDZ typical enterprises land intensive utilization, with a purpose of providing EDZ administrative departments useful and helpful suggestions on improving land intensive use.

15.2 Index Building

15.2.1 Index Selection

Index selection mainly takes hereinafter aspects into account: (1) Index can reflect enterprises land use structure, land use intensity and input-output on land; Therefore index should be selected out in aspects of land use and land utilization efficiency.

Table 15.1 Index system of typical enterprises land intensive utilization

No.	Objects	Sub-objects	Index	Index kind
1	Status of land use	Land use structure	Workshop and related facility land use rate (%)	Positive
2			Office and life service facilities land use rate (%)	Negative
3			Open storage and outdoor operation land rate (%)	Negative
4			Greening rate (%)	Appropriate
5		Land use intensity	Plot ratio (dimensionless)	Positive
6			Building factor (%)	Positive
7			Idle land rate (%)	Positive
8	Land use efficiency	Land input intensity	Fixed asset investment intensity (ten thousand yuan/ha)	Positive
9			The intensity of labor input (people/ha)	Positive
10		Land economic efficiency	The intensity of land output (ten thousand Yuan/ha)	Positive
11			Tax output intensity (ten thousand Yuan/ha)	Positive

(2) Index data, combined with relevant policies and regulations of EDZ, must be simple and accessible. (3) Index can be measured in enterprise view, also can fully reflect the real problems which enterprises of EDZ encounter in their development. Based on the Demand of Economical and Intensive Use of Land Regulation (Decree No. 60 of the Ministry of land and resources), and having relevant ways of evaluation index as reference regarded, the paper set up index system. Index system is listed in Table 15.1.

15.2.2 Ideal Value Determination

(1) Objective Performance Criteria. With reference to the Related Standards from Investment Intensity Control Indicators and of Chengdu Industrial Project Construction Land, the paper set up plot ratio, building factor, and ideal greening value. (2) Estimation method for development tendency. Combined with the actual needs of enterprise development, open storage and outdoor operation land rate, and idle land optimal rate were determined. (3) Expertise consultation method. Having consulted experts of Sichuan University and Chengdu Economic and Technological Development Zone, the optimal value of fixed asset investment intensity, labor input intensity, land output intensity as well as tax output intensity were determined respectively. (4) Experience learning method. With reference to advanced experience, the optimal value for workshop and related facility land use rate, as well as office and life service facilities land use rate were determined individually.

15.3 Research Object

15.3.1 Data Sources

Data were collected from typical enterprises of Chengdu Economic and Technological Development Zone (CDETDZ is its abbreviation). CDETDZ is a modern manufacturing base featuring in four-wheel vehicle (mechanical engineering) and key components as its industry development orientation, which was approved by Sichuan province and Chengdu city. The paper selected top 3 enterprises in total revenue of leading industry, or total output revenue as typical ones. In the meantime, the paper also picked up companies with total revenue or output revenue ranking top from the non-dominant industry as typical enterprises. Typical enterprises affiliated with industries and their registered categories are shown as Table 15.2.

Table 15.2 Statistic table of typical enterprises affiliated with industries and categories

No.	Industries	Corporate registered categories
Enterprise 1	Auto components and parts manufacturing	Joint-venture
Enterprise 2	Four-wheel automotive manufacturing	Joint-venture
Enterprise 3	Auto components and parts manufacturing	Joint-venture
Enterprise 4	Auto components and parts manufacturing	Joint-venture
Enterprise 5	Auto components and parts manufacturing	Joint-venture
Enterprise 6	Auto components and parts manufacturing	Domestic
Enterprise 7	Four-wheel automotive manufacturing	Joint-venture
Enterprise 8	Petroleum extraction equipment manufacturing	Domestic
Enterprise 9	Four-wheel automotive manufacturing	Domestic
Enterprise 10	Metal seal manufacturing	Domestic
Enterprise 11	Railway locomotive vehicle accessories manufacturing	Domestic
Enterprise 12	Construction machinery manufacturing	Wholly foreign-owned
Enterprise 13	Auto components and parts manufacturing	Domestic
Enterprise 14	Construction engineering contract, design, consultation, construction equipment manufacturing installation	Domestic
Enterprise 15	Engineering machinery manufacturing	Wholly foreign-owned
Enterprise 16	Four-wheel automotive manufacturing	Domestic
Enterprise 17	Auto components and parts manufacturing	Domestic
Enterprise 18	Auto components and parts manufacturing	Domestic
Enterprise 19	Four-wheel automotive manufacturing	Domestic

15.3.2 *Weight Determined by Improved Fuzzy Analytic Hierarchy Process*

Fuzzy Analytic Hierarchy Process (FAHP Stands for Short)

Analytic Hierarchy Process [11] (AHP hereinafter for short) is a method which is more widely applied in multi-targets and multi-criteria system design scheme selection. However, as a result of human judgment one-sidedness, two-two comparison results does not possess the objective consistency, and is lacking of taking human judgment fuzziness into account. Due to the disadvantages, the AHP method is expanded to fuzzy environment, and AHP and fuzzy comprehensive evaluation are combined, then the fuzzy hierarchy analysis comes out [12].

Construct the Fuzzy Complementary Matrix

Assuming there are factors: A_1, A_2, …, A_n, and then the preferential relation matrix can be demonstrated like this: $A = \{a_{ij}, i = 1, 2, \ldots, n; j = 1, 2, \ldots, m\}$. Among them, factor a_{ij} means when factor A_i compares with factor A_j, the membership degree of fuzzy relations which factor A_i and factor A_j can be given a quantitative description by method of nine marks [11]. Just as Table 15.3 shows, the preferential relation matrix $A = (a_{ij})n \times m$ is a fuzzy complementary matrix.

Weight Determined by Improved FAHP

Presuming fuzzy complementary consistency matrix $R(R = (r_{ij})n \times n)$, $R(R = (r_{ij}) n \times n)$ is fuzzy complementary consistency matrix, $\forall$ i, j, k, If R meets formula (15.1)

$$r_{ij} = r_{ik} - r_{jk} + 0.5 \tag{15.1}$$

Table 15.3 Nine marks method of 0.1–0.9 and its meaning

Mark	Definition	Description
0.5	Equally important	Two factors compared, both are equally important
0.6	Moderately important	Two factors compared, one factor is more important than another slightly
0.7	Obviously important	The two factors compared, one factor is more important than another obvious
0.8	Much more important	The two factors compared, one factor is more important than another
0.9	Extremely important	The two factors compared, one factor is extremely more important than another
0.1, 0.2, 0.3, 0.4	Comparison in inversed sequence	The two factors compared, If factor $A_i/A_j=a_{ij}$, and then A_j/A_i, $a_{ji}= 1 - a_{ij}$

The formula above provides a general formula about fuzzy complementary consistency matrix, but it is computationally expensive. After referring to some scholars research method [11, 13, 14], the improved FAHP method are selected to determine each index weight, which enjoys a good advantage of improving the efficiency of data processing, under the premise that the accuracy of the calculation results is guaranteed.

Judgment matrix is calculated according to the row sum: $r_i = \sum_{k=1}^{n} a_{ik}$, and transformed according to the type (15.2), and then fuzzy consistency matrix R = (r_{ij}) n × m is obtained.

$$r_{ij} = \frac{r_i - r_j}{2(n-1)} + 0.5 \tag{15.2}$$

The ordering vector of fuzzy consistent judgment matrix A: ω = $(\omega_1, \omega_2, \ldots, \omega_n)$ T can be calculated by normalizing rank aggregation method. It matches [15]:

$$W_i = \frac{\sum_{j=1}^{n} a_{ij} + \frac{n}{2} - 1}{n(n-1)} \quad i = 1, 2, \ldots, n \tag{15.3}$$

To make weight calculated by above type (15.3) certain, the fuzzy complementary judgment matrix consistency check is needed [12, 14, 15]. Presuming that P = (p_{ij}) m × m和□Q = (q_{ij})m × m are all fuzzy judgment matrix [16, 17], like this:

Table 15.4 Weight and overall ranking levels

No.	Objects	Factors	Weight	Index	Weight value	Overall ranking levels
1	Land use efficiency	Land investment intensity	0.1897	Fixed asset investment intensity	0.5305	0.1006
2				The intensity of labor input	0.4695	0.0890
3		Land economy efficiency	0.3103	The intensity of land output	0.3190	0.0990
4				Tax output intensity	0.6810	0.2113
5	Status of land use	Land use structure	0.2701	Workshop and related facility land use rate	0.3103	0.0838
6				Office and life service facilities land use rate	0.2701	0.0730
7				Open storage and outdoor operation land rate	0.2299	0.0621
8				Greening rate	0.1897	0.0512
9		Land use intensity	0.2299	Plot ratio	0.4238	0.0974
10				Building factor	0.3333	0.0766
11				Idle land rate	0.2428	0.0558

$$E(P,Q) = \frac{1}{m^2}\sum_{i=1}^{m}\sum_{j=1}^{m}|p_{ij} + q_{ij} - 1| \quad (15.4)$$

Last, weight value and overall ranking level [16] is shown as Table 15.4.

15.4 Typical Enterprises Evaluation and the Result Analysis

15.4.1 Intensive Degree Score

The index of CDETDZ typical enterprises land intensive utilization is available and the intensity value results are calculated as Table 15.5.

15.4.2 Results Analysis

As results are shown from chart 2 and chart 4, the land use characteristics of CDETDZ typical enterprises can be summarized as below:

1. The land intensive use of CDETDZ typical enterprises ranks higher level. The highest score is 94.70, while the lowest 48.27. 15 enterprises among 19 ones enjoy score more than 60. The main reason is that CDETDZ lies in suburbs of Chengdu city, and is an important automobile industry development base of Chengdu city, and the location and investment incentive advantages attract high-quality projects to gather here. Because much attention was paid to reasonable land use, investment intensity and land output efficiency of enterprises are both higher, and the typical enterprises land intensive use possesses higher general level.
2. There is loose relevance between intensive enterprise land level and industry. Auto and parts enterprises get both highest and lowest score. The phenomenon is basically related to auto industry development, such as the enterprise 2, which is a joint-venture four-wheel company, ranks top at sales volume and production capacity of domestic auto industry, and the land use status and utilization efficiency are both high. However, as enterprise 19 is a four-wheel automotive manufacturer dealing in business in some regions, because of its narrow production line, the enterprise domestic sales volume is low, and land use status and utilization efficiency are also low.
3. Joint-ventures exceed domestic enterprises in land intensive utilization. Joint-ventures rank top 5, followed by domestic enterprises. Joint-ventures showed a good level of land intensive use than domestic ones did. What caused this mainly is that joint-ventures of CDETDZ possess some advantages concerning land input-output, and land use etc. Therefore, joint-ventures exceed

Table 15.5 Typical enterprises land use intensity value statistic table

No.	Fixed asset investment intensity	The intensity of labor input	The intensity of land output	Tax output intensity	Workshop and related facility land use rate	Office and life service facilities land use rate	Open storage and outdoor operation land rate	Greening rate	Plot ratio	Building factor	Idle land rate	Intensity value
Enterprise1	100.00	100.00	100.00	100.00	80.87	73.76	100.00	65.31	100.00	100.00	100.00	94.70
Enterprise2	100.00	100.00	100.00	100.00	67.85	93.68	91.43	98.97	71.67	100.00	100.00	93.50
Enterprise3	100.00	100.00	98.11	100.00	60.03	45.68	100.00	73.68	86.67	97.03	100.00	89.62
Enterprise4	53.87	100.00	80.25	100.00	100.00	28.57	99.00	95.00	100.00	76.90	100.00	86.10
Enterprise5	89.13	100.00	100.00	100.00	70.19	2.45	87.25	46.89	52.50	100.00	100.00	81.15
Enterprise6	100.00	100.00	19.34	100.00	87.50	42.86	99.00	35.00	66.67	100.00	97.98	80.04
Enterprise7	61.62	98.51	100.00	100.00	35.16	71.74	85.05	7.69	59.17	71.67	89.67	76.13
Enterprise8	29.11	72.95	15.62	100.00	73.08	85.04	93.89	65.97	71.67	100.00	100.00	73.87
Enterprise9	56.43	91.28	52.83	100.00	40.84	20.02	74.96	2.03	58.33	100.00	100.00	68.74
Enterprise10	14.60	100.00	25.13	100.00	38.56	100.00	89.83	1.69	60.83	96.61	86.44	68.34
Enterprise11	23.43	100.00	31.63	100.00	13.26	100.00	100.00	40.00	36.62	71.97	97.73	66.73
Enterprise12	31.75	75.39	41.12	100.00	26.48	42.15	45.17	79.75	41.67	100.00	100.00	64.60
Enterprise13	100.00	100.00	43.18	50.60	57.32	9.30	98.06	38.27	43.33	90.23	100.00	64.19
Enterprise14	25.10	100.00	20.55	54.23	36.01	76.48	87.04	38.27	75.83	100.00	79.42	60.37
Enterprise15	75.21	53.69	36.07	49.58	52.38	64.07	84.34	2.39	62.38	100.00	100.00	60.16
Enterprise16	43.88	100.00	11.43	27.55	63.31	16.02	93.09	4.39	100.00	100.00	100.00	55.74
Enterprise17	52.51	52.55	9.54	13.97	61.19	77.51	89.68	33.17	100.00	100.00	93.84	54.56
Enterprise18	49.84	100.00	42.61	26.76	50.00	28.57	98.86	64.29	41.67	76.90	76.86	53.74
Enterprise19	38.16	41.32	32.34	2.52	62.85	25.07	93.27	97.98	72.50	84.32	100.00	48.27

domestic enterprises in land intensive utilization due to good investment input and land use status.

4. Index value varies. In aspect of index value, labor input intensity, open storage and outdoor operation land rate, building factors and idle land rate achieved overall high scoring, but the score of land output intensity reached low. The main reason is that land of EDZ typical enterprises were used to build workshop, the land space for open storage and outdoor operation is much less, and so was the idle land. A large amount of labor employment can be solved after operation. Hence, index value related to this hit high scoring, but some typical enterprises had less land output which lowered the overall scoring of typical enterprise land output intensity index.
5. Lower input-out level made some typical enterprises lower scoring. As shown from Chart 4, the enterprises, which land use scores is less than 60, ranked low in the fixed asset investment intensity, land output intensity, as well as tax output intensity. These 3 indexes lowed down the overall scoring of typical enterprises. Moreover, other indexes of typical enterprises of CDETDZ still enjoyed improvement space.

15.5 Conclusion and Suggestions

Through the evaluation, we can find that typical enterprises of CDETDZ possessed overall high ranking level, but auto industry enterprises presented two levels of differentiation. The non-auto industry enterprises ranked at medium level and the relevance between enterprise land intensive use and industry is an uncertainty. Joint-venture enterprises exceeded domestic ones in land intensive use, and the influential factors of domestic enterprises are land output, tax output, as well as office and life service facilities land use rate. Improvement concerning these aspects will upgrade the level of enterprises land intensive use. Therefore, the paper provides suggestions hereinafter:

1. To strengthen enterprise land use access for examination and approval. When introducing enterprises, CDETDZ needs formulating standards in terms of project investment intensity, output per unit area, plot ratio, revenue per unit area, and setting up strict laws and regulations in examination and approval. As basic producing elements for enterprises, land must be used in a reasonable and efficient way to make maximum of land use efficiency.
2. Low efficiency of land use must be prevented. Enterprise inefficiently-used land indicates that stock land of enterprises is messy Layout, extensively used, low output, unreasonably used. Although inefficiently-used land is not idle land, it faulted in its low level of investment intensity, building density, output value per acre, and tax per acre as well as land use status. In line with this, a supervision and assessment system must be improved to promote land efficient use of enterprises.

3. Spurring enterprises to improve land use intensity. Enterprises are encouraged to build multi-story buildings. In accordance with land use planning and under non-changed land use premise, land redevelopment should be supported and multi-story workshops need building. Land use ratio and project investment intensity are to be improved in a three-dimensional development mode to ensure the limited land resources can be used in a reasonable and efficient way.
4. Enterprise technology innovation must be promoted. Improving enterprise technology innovation ability and guiding enterprises to make timely adjustment on product structure to meet new market demand. The low-end, low added-value and uncompetitive products need to be weeded out. The products which enjoy a promising market should be upgraded to increase function and the development of high technology content, large capacity, high added value and market competitive products can improve the efficiency of the land output.
5. Creating synergies, and reducing duplication of investment. A detailed guidance is needed for enterprises in the process of operation. Much more attention must be paid to the relevant industries and enterprises internal collaboration in industries. Big companies must work hard on technology innovation and brand construction. Small enterprises need strengthening internal management and quality control of products, and focusing on small and exquisite, small but specialized, small but strong products and paying services for large enterprises. Enterprises work together to form a complete industrial chain to reduce duplication of investment, and to intensify using land in purpose of enhancing the overall efficiency of the land.

References

1. Long KS, Qin J, Chen LG (2014) Causes and governance solutions of idle land in development zone: a case of new & high-tech industry development zone of City "A" in North China. China Popul Resour Environ 24(1):126–131
2. Wang M, Qu FT (2004) Indicator system for land use intensity assessment of industrial enterprises in Kunshan development zone. China Land Sci 18(6):22–27
3. Zhao XF, Huang XJ, Yan CQ, Li H, Zhang XY (2011) Evaluation of industrial land intensive use with RAGA-AHP-A case study of development areas in Jiangsu province. China Land Sci 20(11):1315–1320
4. Wu JW, Yuan F, Lv WG, Mu YF (2013) Evaluation and space layout adjustment of manufacturing land for Taizhou economic development zone. Resour Sci 35(5):918–927
5. Fan YC, Nie YM, Chen HY et al (2014) The land intensive use potential evaluation of industrial enterprises of development zone based on GIS spatial model. J Anhui Agric 42(6):1836–1839
6. Zhang ZL, Li YL (2007) Research on the relationship of land expansion and economy growth of development zone: taking the national economic and technological development zone as a case. China Land Sci 21(6):4–9
7. Lin J, Zhang P, Liu SY, Xiao D (2010) Research on the differential incomes of industrial land based on the production function.—A case study of the typical enterprises in state-level development zones. Urban Stud 17(6):80–85

8. Wang KQ, Xiong ZX, Gao W (2013) Research on the relationship between the conveyance model of industrial land use rights and the output elasticity of land in the development zones. China Land Sci 27(8):4–9
9. Zhao XF, Huang XJ, Li H, Yang CQ, Zhang XY (2011) Evaluation of industrial trades land intensive use with RAGA-AHP: A case study of Jiangsu province. J Nat Resour 26(8):1270–1277
10. Chen Y, Chen YR, Ma WB (2012) Evaluation of industrial land's intensive use and analysis of potential mining with bayes discrimination: a case study of typical enterprises in Hubei province. Resour Sci 34(3):433–441
11. Kang QR, Tan JX, Zhang WZ (2010) Application of improved FAHP in schemes optimization about landslide treatment. J Chongqing Univ 33(9):98–103
12. Mandani EH (1994) Twenty years of fuzzy control: experiences gained and lessons learned. Fuzzy logic technology and applications. In: Marks R (Ed). IEEE Technical Activities Board, pp 19–24
13. Xu ZS (2001) Algorithm for priority of fuzzy complementary judgment matrix. J Syst Eng 16(4):311–314
14. Xu ZS (2002) Research on compatibility and consistency of fuzzy complementary judgment matrices. J PLA Univ Sci Technol 3(2):94–97
15. Zhu SL, Zhou P, Han Y, Yang HC (2004) Research on risk analysis based on fuzzy AHP method. Comput Integr Manuf Syst 10(8):980–984
16. Ministry of Land and Resources (2014) Development zone land intensive utilization evaluation procedures (2014) (trial)
17. Yuan YL (2013) The research on the construction engineering project risk management based on fuzzy analytic hierarchy process

Chapter 16
Evaluation of Land Carrying Capacity of Island Cities: Based on the Modified Ecological Footprint Model

Jiawei Chen, Hong Zhang and Peng Zhou

Abstract According to the characteristic of island cities, this paper constructed a modified ecological footprint model which specialized for island cities. Our goal was to provide an effective method to evaluate the rational utilization of land resource of island cities. Taking Zhoushan city as the example, this paper used this modified ecological footprint model to estimate the ecological footprint and the ecological capacity of Zhoushan city, and analyzed its land use condition. The results of this paper are: (1) according to rich marine resources and the scarcity of freshwater in the island cities, we construct a modified ecological footprint model which specialized for island cities; (2) we use the modified ecological footprint model, from the quality, nutrition and heat perspective respectively to estimate the land use sustainable index (LUSI), and the result is 0.39, 0.44 and 0.49, which means Zhoushan city is highly sustainable developing.

Keywords Land carrying capacity · Island cities · Modified ecological footprint model · Zhoushan city

16.1 Introduction

Land carrying capacity means the sustained and stable population size supported by a country or region's own land resource. This population size is according to the foreseeable technology, the development level of economy and society, and the conformable standard of material life. Compared to the inland cities, island cities have fewer land area and fresh water resource, but abundant marine resource. Evaluating the land carrying capacity of island cities is beneficial to understand their level of sustainable development, and to provide advice to their further development.

J. Chen · H. Zhang (✉) · P. Zhou
Hang Lung Center for Real Estate/Center for Urbanization and Industrial Development, Tsinghua University, Tsinghua, China
e-mail: zhannie@tsinghua.edu.cn

Y. Wu et al. (eds.), *Proceedings of the 20th International Symposium on Advancement of Construction Management and Real Estate*,
DOI 10.1007/978-981-10-0855-9_16

Ecological footprint model is put forward in 1990s. It is a quantification method used to judge whether human activities exceed the carrying capacity of region ecosystem [1]. Not only because the results of this model is simple and comparable, but also it can reveal the relationship between natural capital and economic development, it is widely used by researchers [2]. In the foreign study, van Vuuren [3] used ecological footprint model to measure the change of ecological footprint in regions all around the world. And McDonald [4] used this model to analyze the status of development of New Zealand. In the domestic study, Xu Zhongmin used ecological footprint model to analyze ecological footprint and ecological carrying capacity of different provinces in China in 1999 and finds that the growth of economy in most provinces is at the expense of environment. It is the first time this model used by the researchers in our country. After that, more and more researchers used this model to evaluate the condition of land utilization in our country and find that our country is in a status of ecological deficit [6–9]. Except the analysis on the whole country, some researchers also use it to evaluate the situation in some provinces [10–12]. In the study of the land carrying capacity of island cities, Han [13] measure the ecological footprint and ecological carrying capacity of Changshan archipelago. And Li [14] also used this model to evaluate the land carrying capacity of Changdao County.

Based on the problems above, this paper analyze the shortcoming of the traditional ecological footprint model when it is used to evaluate the island cities by considering the characteristic of island cities. Then improve the traditional model. And finally take Zhoushan City as example to evaluate its ecological footprint and carrying capacity, which can help us to judge the status of its sustainable development.

16.2 Research Method

16.2.1 Shortcoming and Improvement of Traditional Ecological Footprint Model

Compared to the inland cities, island ones have fewer land area, scarcer fresh water and more abundant marine resource. When we use the traditional model to calculate the land carrying capacity of island cities, there are four shortcomings as follows:

1. The traditional model just consider the supply of aquatic product of water area, and neglect the fresh water supplied by it.
2. The traditional model don't consider the particularity of marine resource.
3. The traditional model ignore pollutant which has a strong impact on that area such as sulfur dioxide and dust.

According to the shortcoming above, this paper improve the traditional model in three aspects as follows:

1. This paper not only evaluate the land carrying capacity from the aspect of quality, but also from the aspect of nutrition and heat, which make our analysis more comprehensive.
2. According to characteristic that fresh water is scarce in island cities, this paper takes fresh water into account.
3. This paper don't consider the carbon dioxide. We consider the sulfur dioxide and dust, which make the evaluation of island cities has more reference meaning.

16.2.2 Steps of Ecological Footprint Model for Island Cities

Calculating the Equivalence Factors

The equivalence factors are the ratio of specific kind of land's unit area productivity and all kinds of land's average unit area productivity. Its formula is as follows:

$$e_i = P_i/\overline{P} \tag{16.1}$$

$$P_i = \left(\sum_{j=1}^{n} Q_{ij} \times w_{ij}\right) \Big/ \mathrm{S}_i \tag{16.2}$$

$$\overline{P} = \sum_{i=1}^{n} \left(\sum_{j=1}^{n} Q_{ij} \times w_{ij}\right) \Big/ \sum_{i=1}^{n} \mathrm{S}_i \tag{16.3}$$

e_i is the equivalence factor of the land of type i. P_i is the unit area productivity of the land of type i; $\overline{P}$ is the average unit area productivity of all kinds of land; Q_{ij} is the quality of food of type j from the land of type i; w_{ij} is the weighting coefficient of food; S_i is the acreage of land of type i.

Calculating the Equivalence Factor of Fresh Water

The formula is as follows:

$$D = \frac{M_g \times w_j}{M_s} \tag{16.4}$$

$$e_s = \frac{P_s}{P_g} \times D \times e_g \tag{16.5}$$

In the formula, D is the convert coefficient of fresh water; M_g is the per capita consumption of crops in our country; M_s is the per capita consumption of fresh water; P_s is the fresh water supply of unit area in our country; P_g is the production of unit plough in our country.

Calculating the Yield Factors

Yield factors means the ratio of a specific type of land's production in an area and the average production of this type of land. It can be calculate by the formula as follow:

$$p_{ik} = P_{ik}/\overline{\mathrm{P}}_i \quad (16.6)$$

p_{ik} is the yield factor of land of type i in area k; P_{ik} is the unit area production of land of type i in area k; $\overline{\mathrm{P}}_i$ is the average unit area production of land of type i.

Calculating the Acreage of Different Types of Land We Need

The formulas to calculate the acreage of different types of land are as follows:

$$s_{ij} = (Q_{ij} \times w_{ij})/P_i \quad (16.7)$$

$$S_i = \sum_{j}^{n} s_{ij} \quad (16.8)$$

s_{ij} is the acreage of land of type i we need which used to product food j; Q_{ij} is the quality of food j which produced by land of type i; w_{ij} is the weighting coefficient of products; S_i is the total acreage of land of type i we need.

Calculating the Acreage of Land Which Used to Absorb Different Kinds of Pollutant

The formulas used to calculate the acreage of land which used to absorb different kinds of pollutant are as follows:

$$T_{ij} = G_{ij}/\mathrm{A}_{ij} \quad (16.9)$$

$$T_i = \max\{\mathrm{T}_{i1}, \mathrm{T}_{i2}, \mathrm{T}_{i3}, \ldots, \mathrm{T}_{in}\}, \quad \mathrm{n} = 1, 2, 3, \ldots, \mathrm{j} \quad (16.10)$$

T_{ij} is the acreage of land of type i we need which used to absorb the pollutant j; G_{ij} is the sum of pollutant j which absorb by land of type i; A_{ij} is the average speed of pollutant j absorbed by land of type i; T_i is the total acreage of land of type i we need to absorb the pollutant.

Calculating the Defect Area of Different Kinds of Land Which Used to Absorb the Pollutant

When the lands which used to absorb the pollutant is not enough, the ones used to produce commercial crops will take the responsibility to absorb the pollutant, so that decrease their productivity. Therefore, we need to calculate the defect area, which can make our evaluation more convinced.

Calculating the Ecological Footprint of Island Cities

$$PF = N \times pf = \sum_i (\mathrm{C}_i/\mathrm{P}_i) \times e_i \tag{16.11}$$

PF is the ecological footprint in this area; *pf* is the per capita ecological footprint; *N* is the population size; C_i is the consumption of product i; e_i is the equivalence factor of land of type i.

Calculating the Ecological Carrying Capacity of Island Cities

$$PP = N \times pp = (1 - 12\,\%) \times \sum_i S_i' \times p_i \times e_i \tag{16.12}$$

PP is the ecological carrying capacity; *pp* is the per capita ecological carrying capacity; *N* is the population size; S' is the acreage of land of type i which has deduced defect area. 12 % is the acreage of land to maintain the ecological diversity which can't be used by human beings.

Calculating the Land Use Sustainable Index (LUSI)

According to the analysis above we construct the land use sustainable index LUSI, its calculating formula is as follow:

$$LUSI = \frac{PF}{PP} \tag{16.13}$$

In the formula, PF is the ecological footprint of this area; PP is the ecological carrying capacity of this area.

The standard is shown in Table 16.1.

16.3 Analysis on the Results

16.3.1 Calculating the Equivalence Factors

After referring to the Chinese statistic yearbook, this paper collect the acreage of plough, woodland, meadow and water area and their products from 2002 to 2011. The acreage of water area is substituted by the accumulative acreage of sea area with affirming rights [15]. And then according to the nutrition and heat in each kilograms of different kinds of products, use formula (16.1), (16.2) and (16.3), we obtain the equivalence factors in the aspects of quality, nutrition and heat.

According to the data from Chinese statistic yearbook, the results are shown in Table 16.2.

16.3.2 Calculating the Yield Factors

According to Statistic Yearbook of Zhoushan, the yield factors of all kinds of land in three aspects are shown in Table 16.3.

Table 16.1 The standard of land sustainable use based on LUSI

The status of land use	Criterion (*LUSI*)	The level of land sustainable use
Ecological surplus	$0 < LUSI \leq 0.5$	High level sustainable
	$0.5 < LUSI \leq 0.9$	Low level sustainable
Ecological balance	$0.9 < LUSI \leq 1.1$	Balance
Ecological deficit	$1.1 < LUSI \leq 1.5$	Low level unsustainable
	$1.5 < LUSI$	High level unsustainable

Table 16.2 Equivalence factor of all kinds of land

	Plough	Woodland	Meadow	Water area	Construction land	Fresh water resource
Weight	1.05	1.23	0.48	0.57	1.05	1.35
Nutrition	1.16	0.63	0.25	0.43	1.16	1.50
Heat	1.15	0.61	0.33	0.49	1.15	1.48

Table 16.3 Yield factors of all kinds of land

	Plough	Woodland	Meadow	Water area	Construction land	Fresh water resource
Weight	1.24	1.25	0.86	2.11	1.24	0.05
Nutrition	0.69	0.98	0.76	2.34	0.69	0.05
Heat	0.71	0.97	0.80	2.28	0.71	0.05

16.3.3 Calculating the Demand of All Kinds of Lands

Zhoushan's population can be classified as urban population, rural population and population of fishing villages. After referring to the Statistic Yearbook of Zhoushan, we can obtain the food consumption of urban population.

Without considering the needed lands which used to absorb the pollutant, the results are shown as Table 16.4.

16.3.4 Calculating the Land Area Needed to Absorb the Pollutant

According to the Statistic Yearbook of Zhoushan, we can find that the primary pollutant in Zhoushan city is sulfur dioxide, dust and smoke.

Even though both woodland and meadow absorb sulfur dioxide and dust, but their efficient has difference, woodland's efficient that to absorb dust and smoke is much higher than meadow, and meadow's efficient that to absorb sulfur dioxide is much higher than woodland. Therefore, we can have the assumption that woodland primarily absorb dust and smoke, and meadow primarily absorb sulfur dioxide.

Based on our analysis and the data, we can learn that Zhoushan needed 7876.405 ha woodland to absorb dust and smoke, and 36,466.9 ha meadow to absorb sulfur dioxide.

Table 16.4 All kinds of lands in demand without considering to absorb the pollutant (units: ha)

	Plough	Woodland	Meadow	Water land	Construction area	Fresh water resource
Weight	36,753.96	4713.48	3838.70	6307.71	417.73	6388.18
Nutrition	41,698.80	3704.69	5860.04	6745.98	417.73	6388.18
Heat	34,071.24	3741.05	5546.39	6796.04	417.73	6388.18

16.3.5 The Area of Deficient Land Which Used to Absorb Pollutant

At present, Zhoushan has 51,823.8 ha woodland, 6992.4 ha meadow. And the area of woodland which used to absorb dust and smoke is just 7876.405 ha, so woodland is not deficient in Zhoushan. But Zhoushan's let out much sulfur dioxide in 2012, it need 36,466.9 ha meadow to absorb sulfur dioxide, so the deficient area of meadow is 29,474.46 ha. In other words, the sulfur dioxide let out by Zhoushan can't be absorbed by Zhoushan itself, so we assume that the sulfur dioxide has to be absorbed by meadow of other areas in Zhejiang. Zhoushan has to compensate these area, so we have to take these deficient area into account as Zhoushan's demand of meadow. On the other hand, because meadow in Zhoushan is not enough to absorb the sulfur dioxide, so after deducting the meadow which used to absorb the pollutant, there's no meadow supply in Zhoushan to product food.

16.3.6 Calculating the Ecological Footprint and the Ecological Carrying Capacity of Zhoushan

According to the results above, with formula (16.11), we can calculate the ecological footprint and the ecological carrying capacity of Zhoushan. The results are shown in Table 16.5.

16.3.7 The Evaluation of Land Sustainable Use

Based on the results above, we can calculate the land use sustainable index (LUSI), the result is shown in Table 16.6.

Table 16.5 The ecological footprint and the ecological carrying capacity of Zhoushan in 2012 (ha)

Index	Weight	Nutrition	Heat
Ecological footprint	72,702.35	63,790.83	75,350.64
Ecological carrying capacity	187,001.5	143,572.8	155,097.3

Table 16.6 Land use sustainable index of Zhoushan

	Weight	Nutrition	Heat
LUSI	0.39	0.44	0.49
The situation of land use	Highly sustainable	Highly sustainable	Highly sustainable

We can learn from Table 16.6 that the land use sustainable index is 0.39 in the aspect of weight, 0.44 in the one of nutrition and 0.49 in the one of heat. According to the evaluation standard in Table 16.1, from the three aspects, Zhoushan is all in a condition of ecological remainder and its land use is highly sustainable. Among all the aspects, the degree of ecological remainder is highest in the aspect of weight, in the aspect of nutrition the degree is lower and it is lowest in the aspect of heat. It is because that compared to other inland city, products of Zhoushan is primarily fish, and its crops from plough are relatively less. Meanwhile, unit mass of crops contain more nutrition and heat than unit mass of fish, therefore, in the aspect of nutrition and heat, Zhoushan's degree of ecological remainder is lower. However, the condition of land use is highly sustainable in Zhoushan.

16.4 Conclusion and Discussion

According to the characteristic of island cities, this paper improves the traditional ecological footprint model, and then analyze from the aspect of weight, nutrition and heat. In addition, we take the scarce fresh water resource of island city and its primarily industrial pollutant such as sulfur dioxide, dust and smoke into account. We conduct a modified ecological footprint model which specialized for island cities and use it to analyze Zhoushan, an island city of our country.

Our paper finds that, in the aspect of weight, nutrition and heat, Zhoushan's ecological carrying capacity is all higher than its ecological footprint, it is in a condition of ecological remainder. Thereinto, the LUSI is lowest in the aspect of weight, it is lower in the one of nutrition, and it is lowest in the aspect of heat. Even though LUSI is different in the three aspect, but they are all lower than 0.5, the land use condition is highly sustainable. In other words, in the current economy, technology and population condition, Zhoushan can sustainably develop without destroying the ecological environment.

References

1. Liu Z, Zheng Y (2011) Evaluation of water ecological carrying capacity based on ecological footprint theory: a case study of Huzhou. Resour Sci 33(6):1083–1088
2. Zhang Y, Zhang H, Fengsheng LI et al (2013) Dynamic analysis of the water resource ecological footprint of Guangxi based on an improved mode. Resour Sci 35(8):1601–1610
3. van Vuuren DP, Bouwman LF (2005) Exploring past and future changes in the ecological footprint for world regions. Ecol Econ 52(1):43–62
4. McDonald GW, Patterson MG (2004) Ecological footprints and interdependencies of New Zealand regions. Ecol Econ 50(1):49–67
5. Xu Z, Cheng D (2002) Calculation and analysis on ecological footprints of china. Acta Pedol Sin (03):441–445
6. Wu L (2005) Dynamic evaluation of China sustainable development based on the ecological footprint index. J China Agric Univ

7. Yuan XH (2005) Dynamic analysis of sustainable development in China based on the ecological footprint. China Popul Resour Environ 15(3):38–42
8. Chen M, Zhang LJ, Wang RS et al (2005) Dynamics of ecological footprint of China from 1978 to 2003. Resour Sci 27(6):132–139
9. Liu Y (Fudan University Shanghai, Department CG et al) (2004) Time series of ecological footprint in China between 1962 ~ 2001: calculation and assessment of development sustainability. Acta Ecologica Sinica
10. Zhao W (Liu JS, Science CAO, et al) (2007) Scenario prediction of regional ecological footprint: a case of Jilin Province. Resour Sci 29(1):165–171
11. Yue-Qing XU (2007) Evaluation of ecological carrying capacity based on ecological footprint model in Beijing. Resour Sci
12. Jian-Sheng WU, Ping LI, Zhang YQ (2008) Sustainable development capacity evaluation based on urban ecological footprint: a case study of Shenzhen. Resour Sci 30(6):850–856
13. Han ZL, Guo JK, Liu K (2008) Research on the ecological footprint of the Island region and it's sustainable development. Ecol Econ
14. Li L, Li J (2011) Evaluation research of ecological resource utilization of island based on ecological footprint model: take Changdao county as an example. Ocean Dev Manage 09:93–99
15. Liu R (2014) Regional carrying capacity research based on the improved ecological footprint model. Tsinghua University, Beijing

Chapter 17
Analysis on Chongqing Securitized Land Exchange System from the Perspective of Land Finance

Zhibin Mo and Yuzhe Wu

Abstract Through a rapid economic development of more than 30 years, China has created tremendous wealth in many fields. However, the contradiction caused by urban-rural dual structure, in particular, by the stringent land use regulation in rural area, has been still prominent. Industrialization and urbanization, before the industrial structural upgrading complete, are still based on the expansion of construction land at the cost of farmlands encroached, which has posed a huge challenge to environmental protection and food safety. In 2008, Chongqing government innovatively designed the securitized land exchange system based on the transfer of development right (TDR) in order to balance the land preservation and social development. However, some problems have gradually emerged in the implementation, which homogenized the securitized land exchange system and land finance system. This paper aims to reveal the inter relations between the securitized land exchange system and land finance by analyzing the initiative driving force, the benefits for local government and the negative influence from either vertical or horizontal perspectives. The vertical analysis primarily regards to the government economic man, including the fiscal difficulty in local government under the tax-sharing system, and the lack of supervision in the securitized land exchange system itself. From the horizontal prospective, all participants, including farmers, local government and developers, whose benefits distribution and intentions in the securitized land exchange system have been discussed against the land finance system.

Keywords Securitized land exchange system · Land finance · Homogenization · Path dependence

Z. Mo · Y. Wu (✉)
Department of Land Management, Zhejiang University, Zhejiang, China
e-mail: wuyuzhe@zju.edu.cn

Z. Mo
e-mail: 260877702@qq.com

Y. Wu et al. (eds.), *Proceedings of the 20th International Symposium on Advancement of Construction Management and Real Estate*,
DOI 10.1007/978-981-10-0855-9_17

17.1 Introduction

The fiscal reform in 1999 seriously got the financial power of local governments weakened, but the duties of them were never reduced. Simulating by huge financial gap, local governments were eager to find a new way to set-off it, which, was the origin of land financing. Furthermore, Kong [1], Chen [2] discovered in their research that GDP-oriented performance evaluation mechanism for governments was either one of the reasons prompting land finance.

Admittedly, land finance was efficient as a way financing the urbanization. In 2001, the land conveyance fee was 0.130 trillions CNY, which reached the amount of 2.710 trillions CNY in 2010, accounting for 32.65 % of the total financial revenue in that year [3], which meant land finance had played an important role in Chinese finance. However, the problems of land finance emerged gradually, such as continually increased housing price, decreased farmland areas, and the violation on farmers' property rights.

The securitized land exchange system in Chongqing had made remarkable contribution for integrating the dual-layer land market existing in Chinese urban and rural areas. However, as perfectly designed this new system was, in operational level, problems emerged inevitably, and researches showed that land finance should claim the responsibility for those problems.

Zhang [4] pointed out in their research that securitized land exchange system was originated from two pairs of contradiction, contradiction between tremendous demands of construction lands and the protection requirements of farmlands, and contradiction between the short of construction land in urban area and the low-efficiently use of construction land in rural area. In fact, from 1997 to 2010, there was a 22 % decrease of rural permanent residents in Chongqing, but the rural homesteads increased 46 % at the same time Wei and Jin [5].

To sum up, experts ascribed the born of land securitized exchange system to a dilemma that society suffers the short of construction lands in urban with the waste of construction lands in rural area, but if we take securitized land exchange system and land finance into consideration integrally, there would be a reasonable hypothesis supported by objective researches later in this paper. Irrational fiscal system and government performance evaluation system leaded to the implement of land finance, but the expropriation quotas are getting increasingly strict in nowadays china, so there must be a new way to support the constant lands supply to keep land finance alive. With all these facts and motives, land securitized exchange system born for its enormous potentiality of quota-generating ability.

17.2 The Differences Between the Two Systems

17.2.1 *Difference Reflected in Ways Expanding Urban Area*

Land finance and land securitized exchange system leaded to a same result, large area of rural farmlands have been converting into urban construction lands, but there are still noticeable differences. In land finance, local government uses the expropriation quotas to "buy" rural collective-owned lands at administrative price which is much lower than market price, then selling those lands by market-oriented transaction, by this way, the local government's fiscal revenue get reinforced, but the biggest obstacle is the limitation of farmlands. According to the statistics in China Land Yearbook (2010), in 1999, China had 12.966 million ha cultivated land, but there was a decrease of 7.933 million ha during the later 10 years.

In principle, the newly-generated construction land quotas, namely, the land securities, or Dipiao in Chinese, generated by land securitized exchange system could only be used when the quality of the reclaimed lands reach a certain standard and examined by local government. Though local government played either athlete or referee in this mode without any supervising mechanism, land securitized land exchange system at least declared the requisition-compensation balance in terms of law, which get it essentially different from land finance.

17.2.2 *Difference About the Promotion of Farmers' Benefits*

According to Land Administration Law, the rural collective-owned lands are highly restricted in conversion, which formed a binary structure in lands [6]. In that structure, rural collective-owned lands could only be expropriated but traded in normal market, but the compensation standard and range are not much rational, especially, the potential land value-added profits would just be grabbed by local governments and real estate developers.

By reclaiming the unutilized construction lands in rural area, then converting them into newly-added construction land quotas, land securitized exchange system improved the farmers' land-related profits. According to the statistics cited from land resource management department of Chongqing, the newly-added construction land conveyance fee and land reclaiming fee were about 30–40 CNY/m^2, compared with the 1.333 millions CNY/hm^2 of land security transaction price, the land value increased nearly 3 times [7].

17.2.3 Difference About the Promotion of Pubic Confidence in Government

In land finance, land-lost farmers didn't have bargaining right when facing land expropriation. According to the article 25 of *Land Administrative Law of the PRC*, if land-lost farmers are not satisfied with the compensation, prefectural government or above are responsible for giving conciliation, if the conciliation failed, it is legal for the local government who authorized this expropriation to give the final judgment, but the expropriation would not be interrupted during the judicial process [2]. As a result, farmers cannot reject or bargain with government, which leaded to menace, coercive demolition or other illegal behaviors. Trapped in such a dilemma, land-lost farmers had to resort extreme actions, like self-burning, attacking government building, which seriously damaged the social stability. On the contrary, participating into land securitized exchange system was totally conformed to the willing of rural land-owners, all roles that local government played were just regulator, supplier of needed institution and public server.

17.3 The Succession Between the Two Systems

17.3.1 The Succession of Origin Between the Two Systems

System of tax distribution, or, the fiscal reform in 1999, prompted the emergence of land finance. System of tax distribution in China directly reduced the fiscal revenue that belonged to local government. In 1993, the local fiscal revenue accounted for 78 % of the total, but there was a drastically decline of 34 % in the next year, and have not exceeded 50 % until today. Meanwhile, the pressure of fiscal expenditure remained unchanged of 70 % [8], so it seemed perfect for local government when land finance appeared.

Before discussing the connection between land finance and land securitized exchange system, the introduction of 'rational economic man of government' seems necessary. It is undeniable that government is a complicated politic organization composed of natural persons who have all the features of rational economic man, they are profits-oriented based on Maslow's Hierarchy of Needs theory [8], and they have either the motivation or the strength to take advantage of their public power for private interests [9].

Based on the causes of land finance and the economical man of government, the cause of land securitized exchange system could be discussed in the following from a new perspective. So, the first question is, from the perspective of economic man of government, what does local government really care in land securitized exchange system? The answer is the extra newly-added construction land quota, with which the land finance could be fueled again. In addition, since the promulgation of *Interim Procedures of the Management for Rural Land Exchange*, the newly-added construction land quotas could only be used for commercial industries, which is proper for local government to sustain land finance.

There are four interests groups involved in the securitized land quotas exchange, including farmers who reclaimed their construction lands, local government, farmers who live in receiving zones, and real estate developers [10]. According to the following analysis on the distribution of the income generated by land securities trade, reclaimed farmers just obtained 13 % of the total. With this inappropriate benefits distribution mechanism, local government acquired extra quotas with low cost, which then be used to expropriate suburb farmlands at lower price, trapped into another land finance circulation.

17.3.2 Succession of the Source of Benefits Between the Two Systems

In land finance, the land conveyance fee is the essential part of local government revenue, the sum of which from 2001 to 2003 was 0.91 trillions CNY, accounting for 35 % of the total. 6 years later, there had been a growth of 0.6 trillions CNY in land conveyance fee. Meanwhile, according to the *Land Administration Law*, in expropriating land, compensation should be made according to the original purposes of the land expropriated, but according to the distribution system of the added value generated by the use change of agricultural land in Jiangsu province, farmers could only obtain 5–10 %.

In land securitized exchange system, the benefits that farmers obtained from giving up their lands is determined by market, but the substantial relation between local government and farmers had nothing changed but the role transitioned from expropriator and land-lost farmer to agent and consignor due to the regulation that the reclaimed land must be packed and then auctioned by local government. Local government charged commission in the name of organizing and administrating the land securities transaction for farmers instead of got benefits directly from land conveyance fee, but the commission seems inappropriate too. According to the *Provisional Regulations of Chongqing Rural Collective-owned Construction Land Reclamation Project*, the price of land security was divided into basic price and added-value price [11], the basic price is based on the cost of reclamation, the added-value price means the remained part when the total minus the basic price. As to the distribution of added-value price, after deducting relevant costs, reclaimed farmers could be compensated 96 thousands CNY/mu at least, if it has surplus, the suplus would be distributed between reclaimed farmers and collective economic organization by 85:15. However, in some counties, many farmers required local governments pay the compensation in advance for the concern that they cannot get the compensation over a long period. In order to ensure the land security transaction going well, local governments promised that farmers could get an advance payment of 96 thousands CNY/mu, the surplus would be paid as soon as the whole transaction is finished. However, many county governments misunderstood the advancement intentionally or not. They thought the advance payment is the total compensation, and the surplus had no business with farmers anymore.

As to the distribution of basic price, there is a calculation based on the distribution system introduced in this paper. It hypothesized that the starting price in the auction of land securities was 160 thousands CNY, and there was one reclaimed farmer who participated in the auction with one mu land. According to the *Implementing Regulations of the Compensation and Arrangement for Land Expropriation in Jiangjin District Chongqing City*, there would be a 10 % of the starting price deducted by land administration bureau.

The further distribution would be demonstrated in the following: reclamation construction fee accounted for 10 thousands CNY, the compensation for housing and attachments on ground accounted for 6 thousands CNY, the subsidy for reclaimed farmer to buy new house was 585 CNY, the compensation of homesteads and other ancillary facilities accounted for 7000 thousands CNY with a 20 % deducted by collective economic organization. To be concluded, the sum that farmers and collective economic organization obtained was 19185 CNY, which was just 13 % of the total. In conclusion, local government obtained land conveyance fee through expropriating farmland and then 'selling it', the land securities were substantially the new quotas of urban construction land, whose meaning for local government was quiet right the lands, lands that could be expropriated and sold. Consequently, the inefficient implemented land securitized exchange system was no more than an institutional arrangement for tapping land source to support larger scale "expropriate-sell cycle" of land finance.

17.3.3 Succession of Negative Effects Between Two Systems

Problems of the Quality and Quantity of Reclaimed Lands

By land finance, obtaining extra-budgetary revenue induces local government to expropriate farmlands in large scale. As a result, the farmlands inevitably and dramatically decreased. According to existing statistics, the grain output in 1999 were 508.386 million tons, the seeded area were 113.161 million ha of the same year, so the yield per unit area was 4.5 tons and the per capita grain production was 0.404 tons. By 2010, the grain output reached 546.477 million tons with a decline of 3285 ha of seeded area, so the yield per unit area was 4.973 tons and the per capita of grain production was 0.406 tons, based on the population of 1.341 billions in 2010. Research showed that over the past ten years, there was a distinctly positive relation between land finance and the decrease of farmlands [4].

According the Interim Procedures, the reclamation in land securitized exchange system could be operated as follows: farmers send over the unutilized construction lands to government, the latter reclaim lands and check the quality of them [10]. So, it is obvious that local government plays both roles of referee and player. Driven by the thought of rational economic man, local government might intentionally lower the acceptance standard for obtaining newly-added construction land quotas as soon as possible. Statistics showed that in Shiti, a village of Chongqing, large area of reclaimed lands was not utilized because of the poor quality. 'There are too many rubbles left in the land to be cultivated' said a villager [12].

No matter adopting land finance or land securitized exchange system, the revenue source of local government was always based on the sale of urban construction land use right. By the former, cultivated land decreased directly, by the latter, cultivated land decreased indirectly for the lack of supervision on quality-checking process of reclaimed land.

The Problem of Land-Lost Farmers Existing in Both of Systems

The problem of land-lost farmers resulted from land finance mainly includes two aspects: the decline of farmers' income and the marginalization of rural migrants, of which the essential is still the inappropriate distribution of land added-value. In China, only when farmers' lands expropriated by governments, can the lands be available in market. It is worth noting that the increased value due to the change of use was mostly shared by government and real estate developers. Especially in some under-developed area, what farmers obtained was just a one-off monetary compensation. However, with lower education and skills, lost-land farmers usually got trapped in investment failure and that one-off monetary compensation was easily to be spent up in soon.

By the end of 2010, there had had 17 transactions of land securities finished with 1981.333 ha newly-generated construction land quotas involved. However, the completion of one transaction of land securities means a corresponding area of farmland located in the suburbs is going to be expropriated. According to statistics, by the end of 2030, there would be 3.633 million ha farmland occupied by industrial purpose, which will generate almost 110 million land-lost farmers in China. More and more newly-added construction quotas will be generated with the development of land securitized exchange system, but before the improvement of land expropriation compensation and resettlement, large-scale land security transaction will inevitably lead to the increase of farmers who do not have land, job and social security.

The Succession of Raising Housing Price Between Two Systems

According to earlier researches, the high land conveyance fee could be an essential part of housing price. In order to attract investment, local government usually relied on single strategy that convert the state-owned land to industrial investors with lower price or even for free. As a result, local government had to raise business land price for covering the loss, they stocked lands or even adopted 'land hungry marketing' strategy, as a result, inadequate supply leaded to the continuously rise of land price. However, real estate developers will ultimately shift the cost to customers [13]. According to relevant statistics, with every one percentage point raise in the total revenue of local government, there would be a 0.9 % raise of land price [3], and in fact, with the doubled land-transferring fee, there were 35 cities in China whose housing price had increased 1.2 times during the period from 2001 to 2008.

In land securitized exchange system, only if land securities holders prevails other bidders who even if do not have land securities, can they attain the land use right, or they have to transfer their land securities to the winner in auction with original price. Thus, securities holders have to bear an opportunity cost. In fact, from 2008 to 2011, there had been 23 transactions of land securities, and the unit price of land security had risen from 1.224 millions CNY to 3.786 millions CNY. Constantly climbing land security price reflected the increasing demand for land, but there is another regulation which restricted the supply of land security, that the supply of land security cannot excess 10 % of the total quotas that year. The relatively short supply of land security directly prompted the price, which even exceeded the sum of land conveyance fee. For instance, the sum of land-transferring fee and land reclaiming fare in 2011 was about 1.099 millions CNY/hm^2, but the lowest price of land securities in the first 6 transactions were 1.206 millions CNY/hm^2, there were about an excess of 30 %. In addition, taking a comparison between Figs. 17.1 and 17.2, we could find that there was a rise of 3000 CNY/m^2 in housing price in Chongqing from 2008 to 2011. Meanwhile, the average price of land security had also risen from 1.224 millions CNY/hm^2 to 2.475 millions CNY/hm^2, an apparent positive correlation of which gradually emerged. To be concluded, from the perspective of securities holders, they have to assume the opportunity cost when failing in auction, or they have to pay more cost which is even higher than what they needed to pay for land use right directly in land finance, with any way of them, the extra cost, like what mentioned earlier in this paper, would not be shouldered by developers alone, which would be shifted to the rocketing housing price and ultimately be eaten by customers.

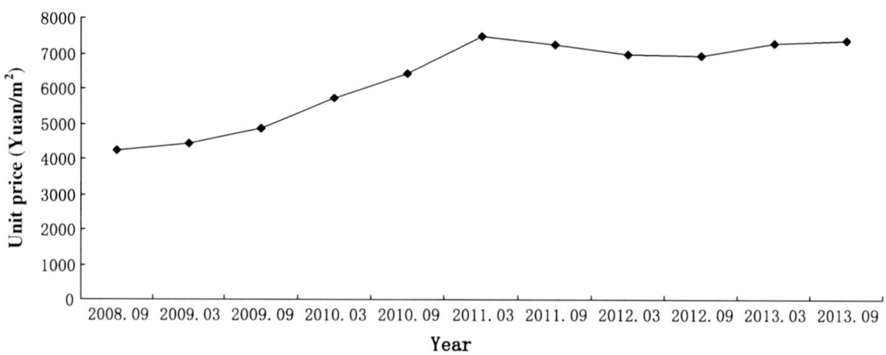

Fig. 17.1 Housing price from 2008 to 2013 in Chongqing

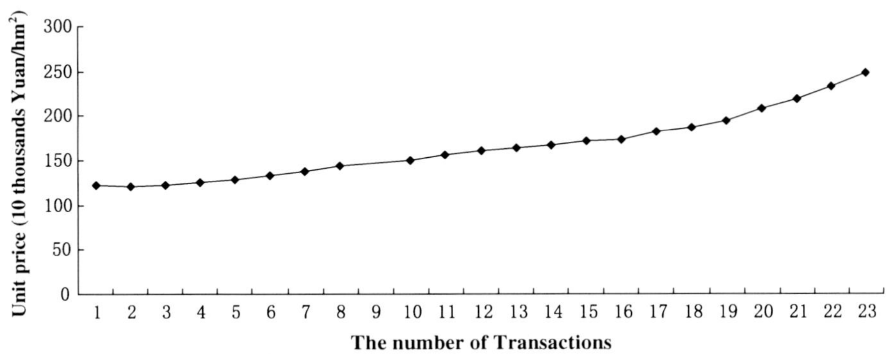

Fig. 17.2 The average price of land securities in each transaction

17.4 Conclusion

It has to be admitted that land securitized exchange system is a really helpful institutional design especially in promoting land value in remote rural areas, optimizing distribution of land resource, improving integral development of urban and rural areas. However, land finance has been implemented for many years in China, whose influence has been deeply engraved into the following attempts of land conversions, land securitized exchange system could not be exceptional. As Zhoucheng said that the manifestation of justice could be different in different societies and periods, but an institution should be designed to conform to the requirements of justice under a particular time-space condition, which, of course, should be one of the inducements to drive institutional change [14]. However, it has been found that land securitized exchange system have much in common with land finance, including the causes, the operational mode, and the negative effects. Although the development prospect of land securitized exchange system is much more optimistic than land finance, the ignorance of the existing problems could direct it onto the road which has been done before by land finance. Therefore, in order to optimize and improve this innovative institution, more attention needs to be paid on the design of fiscal system, the functional orientation of local government, in particular, the distribution of interests.

Acknowledgements The authors appreciate the kind financial support from the National Natural Science Foundation in China (Grant No. 71373231)

References

1. Kong SG (2007) Land finance: the rational behavior of government in increasing interests and the deficiency of relevant institutions. Learn and Practice (5)
2. Chen ZY (2011) The changes of fiscal system, land finance and economic development. Trade Economics (12)
3. Lin CS, Zhu YB (2012) The research on the current operation of securitized land exchange system and its problems. Anhui Agric Sci (5):3062–3064
4. Zhang ZX (2012) Research on land securitized exchange system. Fujian Normal University
5. Wei XY (2012) The performance of land securitized exchange system and reform path. China Land (6):56–58
6. Chi FC (2012) The research on securitized land exchange system based on the need of grain in the future. Chinese Agric Mechanization (6):205–209
7. Fu HT, Duan YM (2012) A new exploration of farmland transfer: research on securitized land exchange system. Agric Econ (10):24–26
8. Zhou LH (2012) The research on land finance from the perspective of the fiscal risk. Shandong University
9. Ellickson RC, Vivki LB (2003) Land use control: cases and materials. Aspen Publishing Inc, pp 191–192
10. Zhang JC (2011) The research on securitized land exchange system based on the integral development of urban and rural areas in Chongqing. Kunming University
11. Xin B, Yu SL (2010) The discussion on land finance and land appreciation. Nowadays Econ (6):36–38
12. Du ZH (2010) The reform of land conveyance system: mode, problem and suggestion. Nowadays Econ Res (2):26–34
13. Lichtenberg E, Ding C (2008) Assessing farmland protection policy in China. Land Use Policy (25):59–68
14. Zhou C (2007) Collections of Zhoucheng. Renmin University of China Press

Chapter 18
Effects of Urban Construction Land Spatial Form on PM2.5 Dispersion

Chuyu Xia, Yanmei Ye, Shaoliang Zhang and Jingming Liu

Abstract Fine particulate matter (PM2.5) is the major air pollutant in China, leading serious threats to human health. Urban construction land spatial form controls the overall aerodynamic roughness of a city, and effects PM2.5 dispersion through wind speed, resulting in PM2.5 concentration uneven spatial distribution. We establish urban construction land use form-PM2.5 dispersion modeling based on Garratt Formula and FLOWSTAR model theoretically, then we use the measured data of the Xuzhou in the winter to verify the model and prove our hypothesis. The PM2.5 concentration that we acquire in Xuzhou is decreasing from the center of the city and the figure of PM2.5 concentration fits the model. This thesis intends to demonstrate that urban construction land spatial form is an important factor effecting the PM2.5 dispersion, and explores the reasonable model to explain it, providing recommendations for urban planning, especially in urban renewal and urban expansion.

Keywords PM2.5 dispersion · Urban construction land spatial form · Model · Urban planning

---

C. Xia · Y. Ye (✉)
Department of Land Science and Real Estate, Zhejiang University, Zhejiang, China
e-mail: yeyanmei@zju.edu.cn

C. Xia
e-mail: xiachuyu1992@126.com

S. Zhang
School of Environment Science and Spatial Informatics, China University of Mining and Technology, Jiangsu, China

J. Liu
Department of Land Resources Administration, China University of Geosciences, Hubei, China

Y. Wu et al. (eds.), *Proceedings of the 20th International Symposium on Advancement of Construction Management and Real Estate*,
DOI 10.1007/978-981-10-0855-9_18

18.1 Introduction

Due to rapid urban sprawl since the early 1980s, fine particulate matter (PM2.5) has been the major air pollutant in mostly China cities [1, 2]. For example, in Beijing, the annual average PM2.5 concentrations were about 70–100 μg/m^3 [3], which were about two to three times higher than the WHO Level 1 Interim Target of 35 μg/m^3 [4]. It has been accepted that PM2.5 concentrations are higher in many regions of China than in other countries from data from model simulation and satellite imagery [5]. Atmospheric visibility and health problems such as lung disease, increased cancer risk, and increased risk of cell mutations have been proved to seriously relate with PM2.5 pollution [6]. Land use change resulted from urban development has been proved to aggravate PM2.5 pollution for playing an increasing role in PM2.5 dispersion because it can influence microscopic meteorological conditions [7]. Han [8] has proved that there is a significant positive correlation between urbanization indicators and PM2.5 pollution, and the difference between mean PM2.5 concentrations of urban area and non-urban area has exceed more than 10 μg/m^3. The obvious characteristic of urban expansion is construction land growth. Therefore, it is seriously significant to study the relationship between PM2.5 pollution and urban construction land space form, for China government is aiming to control air pollution to achieve the goals of the recently released long-term plan.

Several empirical and modeling studies between air pollutant dispersion and land use in urban areas have been developed for many cases. For example, a LTEM (land use-transportation-emissions model) for Charlotte, North Carolina has been performed by Morton et al. [9]. Hang et al. [10] researched how different building heights on street influences air pollutant dispersion and Yuan [11] discussed how to improve air quality in high-density cities by understanding the relationship between air pollutant dispersion and urban morphologies. However, how rapid urban construction land form changed from urban growth impacts PM2.5 dispersion is unclear, especially its mechanism of action. Some studies stay on qualitative investigation level, little researches its relationship for quantitative model. Moreover, the authors did not find any studies concerning modelling urban construction land form with PM2.5 pollution.

There are many ways to describe urban construction land use form, and Spacemate performed by Meta Berghauser Pont widespread. The Combination of building density, floor area ratio, average building height and open space ratio [12] effectively avoids the limitations of single building indicators, and distinguish and present different urban construction land use form. The spatial division of urban construction land leads to spatially-variable roughness, which is defined to be a function of the orientation, spatial density, and height of obstacles to the wind and indicates how air flow interacts with the urban structure [13]. And it creates horizontal variations in the local mean flow and turbulence to impact air pollutant dispersion [14, 15].

The main objective of this work is to prove that urban construction land stereoscopic form resulted of urban expansion is an important factor effecting the PM2.5 dispersion and we will establish model to explain how urban construction land stereoscopic form impact PM2.5 dispersion. With growing population and urban expansion constantly, urban planning proposals for 'urban breathe' is needed to optimize the benefits and can effectively alleviate the pollution to some extent.

18.2 Modeling Method

The modeling method of this study is based on Garratt Formula [16] and FLOWSTAR model [17, 18] to develop relationship between urban construction land use form and air flow rate quantitatively through Intermediate variable named surface roughness. Then we combine it with the atmospheric box model of environmental quality to establish model for urban construction land use form-PM2.5 dispersion under ideal conditions.

18.2.1 Method for Urban Construction Land Use Form—Surface Roughness Modeling

The rough element shape of underlying surface is very complex and in reality. But if we suppose that the urban surface roughness is just consist of buildings for ideal situation, and the urban surface roughness is completely determined by geometrical properties of buildings. Roughness degree varies greatly from the dense, compact and often high-rised areas to the outskirts, especially those of older European cities [19]. Surface roughness degree is relevant with height and density of each roughness element [20, 21], so under ideal situation urban surface roughness varies from part of urban areas determined by building height and building density in that area.

Paeschke firstly recognized rough element height determined the geometric characteristics of rough element, and he presented the ratio of the roughness element height to roughness is a constant ($h_0/z_0 = f$) in 1937. The initial researchers considered the constant *f* about 7.5, but in 1968, Chamberlain through wider tests found that though average value of constant f was around 8, and the dispersion degree of constants from different tests vary greatly from 4.2 to 16.6. Later, Lettau said roughness may be related with surface geometrical properties except roughness element height. Garratt considered z_0/h_0 a single peak function. If density of the rough element is moderate, and z_0/h_0 can achieve the maximum. When the roughness element is sparse, the density increases with the increasing of the roughness element, for it increases the drag force of the atmosphere. But when the roughness element is much denser, with the increase of the density it adds difficulty for the air flow to enter the gap of the rough element. We call it barrier effect which reduce ability of the rough element to the absorb momentum. The appropriate density value is not fixed, and it will be in accordance with other geometrical properties of the rough element, especially the shape of the rough element.

$$z_0/h_0 = \begin{cases} \frac{f_{1m}}{\lambda_0}\lambda, & \lambda < \lambda_0 \\ \frac{f_{1m}}{(1-\lambda_0)}(1-\lambda), & \lambda \geq \lambda_0 \end{cases} \tag{18.1}$$

where h_0 is average height buildings in an area, z_0 is surface roughness, f_{1m} is peak value, λ_0 is moderate density value, and λ is building density.

For peak and moderate density values is unclear and has not much test data for standard, so we can get 0.5 for moderate density value according to visual processing and 0.2 for peak value according to relevant experimental peak [22]. Then we can get General Garratt Formula.

$$z_0/h_0 = \begin{cases} 0.4\lambda, & \lambda < 0.5 \\ 0.4(1-\lambda), & \lambda \geq 0.5 \end{cases} \tag{18.2}$$

From General Garratt Formula, we can know that for urban construction density is generally less than 0.5, so urban surface roughness is determined by $\lambda * h_0$, under ideal situation.

18.2.2 Method for Urban Construction Land Use Form-PM2.5 Dispersion Modeling

The FLOWSTAR model calculates the spatial variation of flow field and turbulence parameters which drive the dispersion. The local value of z_0 represents the mixing close to the surface, and is related to the building height and building density. The model is formulated as:

$$u(z) = \frac{u_*}{k}\left[\ln\frac{z-d}{z_0} + \psi(z, z_0, L, d)\right] \tag{18.3}$$

where u* is friction velocity, k is the von Kármán constant (always 0.4), d is zero-plane displacement height which is two-thirds of height of buddings, z_0 is surface roughness. *L* is the Monin-Obukhov length, and we consider it a constant value. From this formulation, we find mean wind speed at height z(m) is determined by surface roughness z_0, stability through the function and the friction velocity u*. It is difficult to determine PM2.5 source, and there are many disputes in the academic circle. Moreover, urban is a complex system and urban elements Interact with each other.

Assuming unit of non-point pollution is equal with one point pollution source from upper gas flow, so we have a virtual point source. It means adding an initial diffusion coefficient to point source formula of the Gaussian Plume Equation, which is:

$$C(x,y,z) = \frac{q}{\pi\bar{u}(\sigma_y + \sigma_{y0})(\sigma_z + \sigma_{z0})} \exp\left\{ -\frac{1}{2}\left[\frac{y^2}{(\sigma_y + \sigma_{y0})^2} + \frac{h^2}{(\sigma_z + \sigma_{z0})^2} \right] \right\} \tag{18.4}$$

where C is the concentration at point, σ_{y0} and σ_{z0} is the standard deviation of Gaussian distribution function in direction y(m) and z(m) respectively, $\bar{u}$ is wind speed, which is depends on the geostrophic wind and surface roughness. h is the effective plume release height (m), which for road vehicles is taken as 1 m.

Based on the theories above, our hypothesis is that built-up urban construction land use is unreasonable that maldistribution of roughness hinder PM2.5 dispersion. To examine our hypothesis, we update point source formula of the Gaussian Plume Equation as follows:

$$C(x,y,z) = \frac{q}{2\pi\sigma_y\sigma_{\bar{u}}} \exp\left(\frac{y^2}{2\sigma_y^2}\right)\left[\exp\left(-\frac{1}{2}\frac{(h-z)^2}{\sigma_z^2}\right) + \exp\left(-\frac{1}{2}\frac{(h+z)^2}{\sigma_z^2}\right)\right] \tag{18.5}$$

From Box Model of Atmospheric Environment, we know:

$$\frac{dp}{dt} lbh = \bar{u}bh(pb - p) + lbQ - kplbh \tag{18.6}$$

where l, b, H means long, width, height of the box, H is also called the atmosphere mixed layer height, u is average wind speed, Q is intensity of unite source, K is pollution attenuation coefficient, p and pb is pollutant concentrations and pollutant concentrations in real time, f is related to earth rotation speed and latitude. Ignoring the pollutant decay rate, when atmospheric pollutants uniform distribution for a long time, we can get equilibrium concentration:

$$p_b = pb + \frac{Ql}{uH} \tag{18.7}$$

Our modeling idea is that get the parameter formula between the equilibrium concentration and mean wind speed through the wind speed of 10 m and average concentration. Then the parameter formula of the wind speed and the roughness of the ground are obtained by using the modified wind profile function. Moreover, there are some hypotheses for the model: (1) assuming that each monitoring point is the center of each box, and L of each box is the same (we choose average distance between each monitoring point as L). (2) Atmospheric pollutant can reach at an equilibrium concentration only in each box, and it is impossible to mix completely in the global scope u.

18.3 Model Evaluation Set-up, Evaluation

18.3.1 Study Area

As a test case, we use a modelling scenario for Xuzhou, China. Xuzhou City is located in the northwest of Jiangsu province, lying between 33°43′ to 34°58′N, 116°33′ to 118°40′E. There is the warm temperate semi-humid monsoonal climate in the area, and its annual average rainfall is between 800–930 mm, and its annual average temperature is 13.8 °C. Xuzhou, the major city in the Huaihai Economic Zone, currently accelerates the revitalization of the old industrial base in order to enhance position of Huaihai Economic Zone further. In this background, urban construction land expansion has accelerated, resulting 41.3–398 km^2 from 1978 to 2014 [23, 24]. PM2.5 pollution in Xuzhou is very serious, and the air pollution days whose primary pollutant is PM2.5 has exceeded half one year in 2013.

18.3.2 PM2.5 Measurements

Due to the data acquisition limitation, we have acquired only six monitoring stations data which are nearly 4 m from ground and their distribution is as follows (Fig. 18.1): A (Huai tower) in the central, B (Copper Mt. veterinary hospital) in the north, C (Taoyuan Road) and D (Agricultural Academy) in the northeast, E (New zone) in the south-east, F (Copper Mt. Environmental Protection Bureau) in the southernmost, G (The Yellow River Village) in the west. These fixed monitoring points are basically uniform cover in Main city zone of Xuzhou. In order to increase samples numbers, we choose b, c, d, e, f, g points half away from B, C, D, E, F and

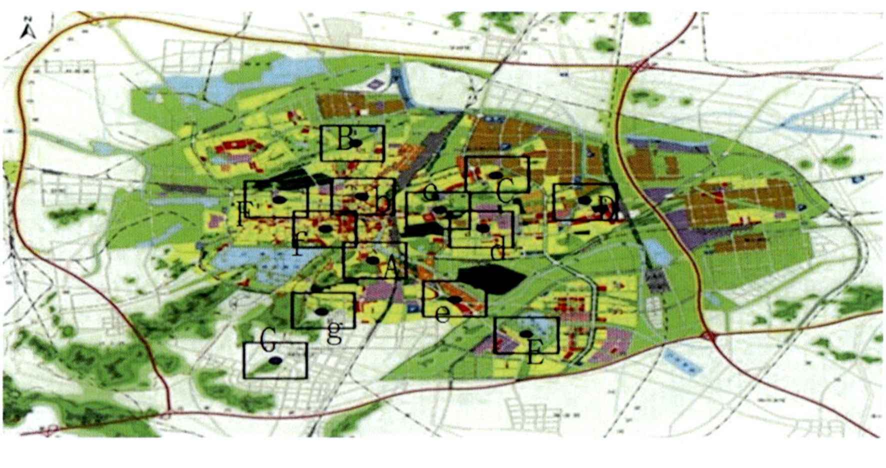

Fig. 18.1 Monitoring point distribution

Table 18.1 Spatial form of urban construction land and the distribution of PM2.5 concentration

	A	B	C	D	E	F	G
Average building density (%)	65	5	45	38	33	25	57
Average building height (m)	8	10	14	19	25	30	8.5
Average roughness (m)	1.28	2.00	2.52	2.89	3.3	3.6	1.46
Mean PM2.5 concentration (μg/m^3)	287	326	240	221	174	150	177
	b	c	d	e	f	g	
Average building density (%)	55	43	35	27	33	41	
Average building height (m)	11.5	12.7	15	23	17	13.2	
Average roughness (m)	2.07	2.18	2.1	2.48	2.8	2.16	
Mean PM2.5 concentration (μg/m^3)	240	227	256	187	210	235	

G to be monitoring points using handheld monitoring instrument (HINAWAY-CW-HAT200) to acquire PM2.5 concentration.

Firstly, we revise readings of handheld monitoring instrument compared with readings of fixed point at the same height, then record data 4 times an hour to keep the same reading frequency as fix monitoring points from 7 am to 7 pm for a week from January 17th to January 23rd, 2014. Finally, we correct all the concentrations to the height of the hand-held instrument about 2 m in which height people daily walk exposed to air. So when we put each monitoring point as the center of each box, wide equal with long (half of distance between the two points), we can get 11 box research areas. The following data is acquired from 7 am to 7 pm for a week from January 17th to January 23rd, 2014.

Table 18.1 considers urban sprawl leads to difference of roughness value from central part to around, which is an important factor of PM2.5 dispersion. And the mean PM2.5 concentration of a week in central part is nearly two times more than the minimum of mean PM2.5 concentration in the outside parts.

18.4 Results and Discussion

18.4.1 PM2.5 Concentration and Mean Wind Speed

This part of the study focuses on Box type atmospheric environmental quality model outcomes, aiming to validate the relationship between PM2.5 concentration and mean wind speed which varies from urban construction land form. Figure 18.2 illustrates mean PM2.5 concentration and average wind speed monitored daily at the height of 2 m from January 17th to January 23rd, 2014.

In order to determine whether the curve fitting the trend or not, we calculate R^2 indicator (Eq. 18.1) for all monitoring.

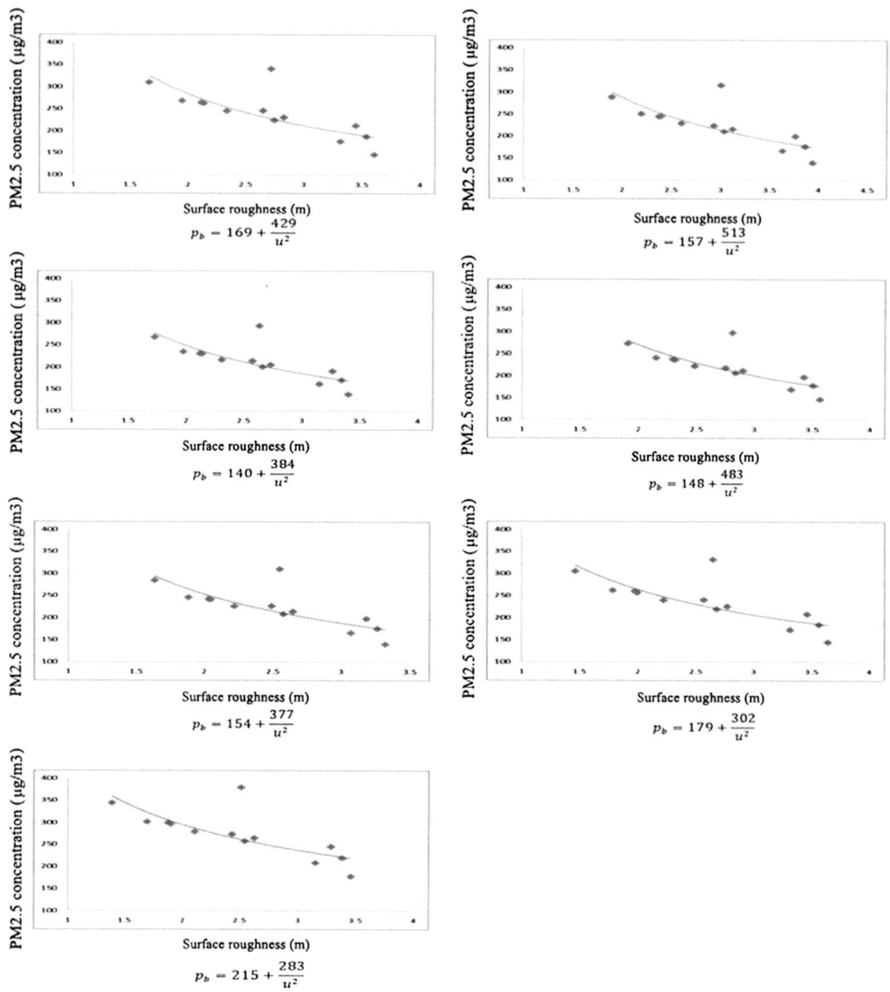

Fig. 18.2 Parameter model of PM2.5 concentration and mean wind speed

$$R^2 = \frac{ESS}{TSS} = 1 - \frac{RSS}{TSS}$$

where RSS means regression sum of squares, TSS means total sum of squares, ESS means regression sum of squares. R^2 of the curves are 0.419, 0.427, 0.474, 0.443, 0.428, 0.403 and 0.381, respectively. According to results, the parameter model can be well explained the relationship between PM2.5 concentration and mean wind speed from urban construction land form.

18.4.2 Wind Speed and Urban Construction Land Form

We use mean wind speed at height of 2 m, average building density and average building height of a box. Figure 18.3 shows that the mean wind speed rises with the increasing of roughness values. The goodness of fit of curves is 0.361, 0.385, 0.447, 0.395, 0.412, 0.402 and 0.412, respectively. Deviation is from the distribution of wind speed for turbulence is developed between streets and building, but the wind speed is considered as uniform in our hypothesis.

18.4.3 PM2.5 Concentration and Urban Construction Land Form

From Fig. 18.4, we consider there is an obvious increase for PM2.5 concentration from central urban area to outside part. The trend of PM2.5 concentration increase is the same as the development direction of urban expansion of Xuzhou. Due to different street plan, building plan and traffic along each street, the effects seen are not uniform, but the patterns are the same.

Figure 18.4 the correlation between roughness value and PM2.5 concentration is −0.717, which mean it can explain mostly spatial differentiation of PM2.5 concentration. However, the figure of C and E deviate the trend line seriously for the PM2.5 concentration of C exceed the trend dramatically and the other is lower from the trend line. To analyze the location of C and E further, we find pint B is located in industrial park and point C is in Dragon Lake Lakeside. It is easily to explain abnormal of point B that industrial Park emissions more atmospheric pollutants.

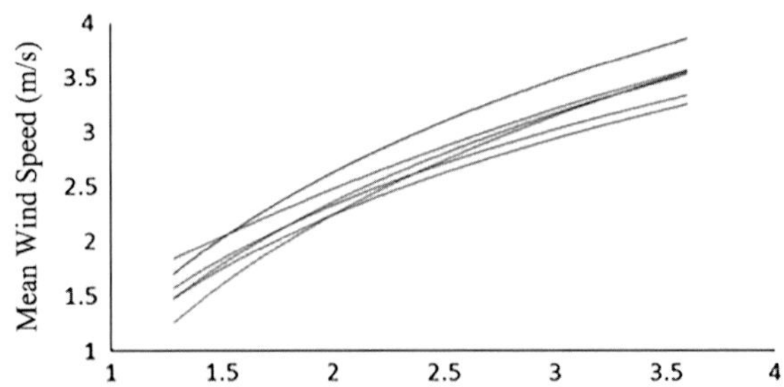

$$u_2 = 1.12\ln\frac{(d-2)}{z_0} + 0.019 z_0 d + 2.056 \qquad u_2 = 1.19\ln\frac{(d-2)}{z_0} + 0.02 z_0 d + 2.311$$

$$u_2 = 0.96\ln\frac{(d-2)}{z_0} + 0.16 z_0 d + 2.068 \qquad u_2 = 0.95\ln\frac{(d-2)}{z_0} + 0.16 z_0 d + 2.247$$

$$u_2 = 0.96\ln\frac{(d-2)}{z_0} + 0.16 z_0 d + 1.977 \qquad u_2 = 1.25\ln\frac{(d-2)}{z_0} + 0.021 z_0 d + 1.903$$

$$u_2 = 1.18\ln\frac{(d-2)}{z_0} + 0.020 z_0 d + 1.808$$

Fig. 18.3 Parameter model of wind speed and urban construction land form

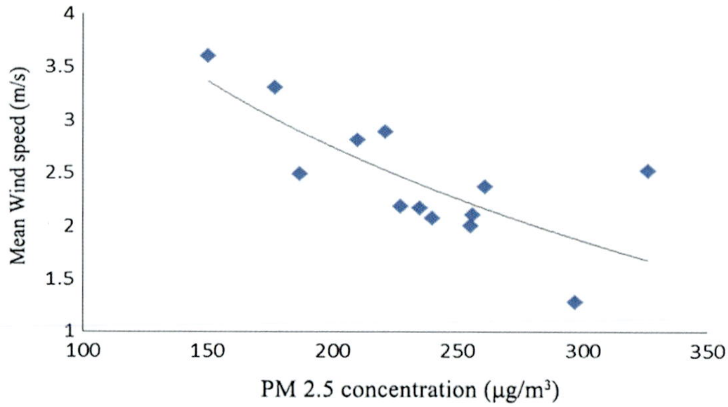

Fig. 18.4 Parameter model of PM2.5 concentration and urban construction land form

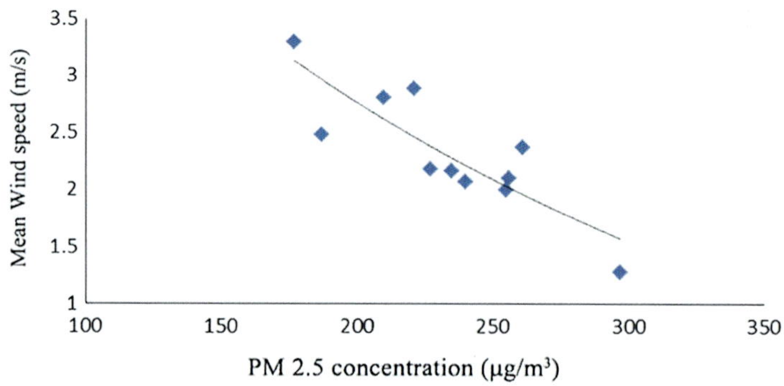

Fig. 18.5 Parameter model of corrected PM2.5 concentration and urban construction land form

Although fine particles can be removed via dry deposition by trees [25], the function of absorption is very limited which can reduce PM2.5 pollution lower than 0.1 $\mu g/m^3$ in American according to Nowak [26]. But presence of water body land use can absorb PM2.5 largely [27]. This is the main reason for anomaly of point C.

After eliminating the abnormal values, the whole trend line fits better.

Figure 18.5 removing B and C, the correlation between roughness value and PM2.5 concentration is −0.858. Due the limitations of the experimental conditions, we have only monitored for 7 days, and someone query it be a case of doubt. So we acquire air quality report of December, and it supports our hypothesis for comparing the degree of PM2.5 pollution of the fixed 7 monitoring points. The acquire air quality report of December once again has proved that the PM2.5 concentration of central part of urban area is higher than around. The days over Severe pollution levels in point A are 2 times higher than in other areas, and basically at the level of serious pollution.

Due to the demolition costs and historical monuments protection, the speed of urban renewal of central part is slow. As a result, unreasonable building density and building height make the urban compacted that seriously block the pollutant dispersion and increase the build-up of pollutants between buildings. Moreover, wind tunnel testing considers windspeeds and turbulence levels between the buildings, this is another reason for reduce of ventilation corridor in the central part of urban. Because of buildings account for large proportion of urban areas, it causes a very complex flow field. Equations (18.1) and (18.2) consider that for our case turbulent mixing effect is the major factors and it controls horizontal ventilation effect. From central urban area to outside part, the average building height increases but the average building density decreases, resulting the decrease of roughness value. Log–wind profile (Eq. 18.3) considers decreased roughness improve horizontal ventilation effect. However, in neutral conditions, the plume spread is relevant with the travel time from the source of pollution (Eq. 18.5). For Eq. (18.5), with σ_z is determined by the value of z_0, using the partial differential of C can examine the effects of parameter values. For example, C changes with even a small change of mean wind speed:

$$\frac{dC}{d\bar{u}} = \frac{\delta C}{\delta \bar{u}} \Delta \bar{u} = \left(-\frac{\Delta \bar{u}}{\bar{u}}\right) C$$

This is a well-known dependence of pollutant dispersion on wind speed, and we can pull the relationship between σ_z and C:

$$C(x,y,z) = \frac{q}{2\pi\sigma_y\sigma_{\bar{u}}} \exp\left(\frac{y^2}{2\sigma_y^2}\right) \left[\exp\left(-\frac{1}{2}\frac{(h-z)^2}{\sigma_z^2}\right) + \exp\left(-\frac{1}{2}\frac{(h+z)^2}{\sigma_z^2}\right)\right]$$

$$\frac{dC}{d\sigma_z} = \frac{\delta C}{\delta \sigma_z}\Delta\sigma_z = \frac{\Delta\sigma_z}{\sigma_z} C \left[\frac{\frac{(h-z)^2}{\sigma_z^2} e^{-\frac{(z-h)^2}{2\sigma_z}} + \frac{(h+z)^2}{\sigma_z^2} e^{-\frac{(z+h)^2}{2\sigma_z}}}{e^{-\frac{(z-h)^2}{2\sigma_z}} + e^{-\frac{(z+h)^2}{2\sigma_z}}} - 1\right] = \frac{\Delta\sigma_z}{\sigma_z} C(A-1)$$

Ignoring the first term, it can be determined by wind speed. As a result, when A is 1, we can get critical height zcri. When z < zcri, the increasing σ_z will reduce C, otherwise it will cause increase of C. Experiments and models by Barnes considers zcri is much larger than z in urban areas. When the average wind speed is <u>, if we ignore transverse direction mixing effect the vertical plume spread at a distance X from pollutant source is: $\frac{1}{c} \sim \langle u \rangle \sigma_z \sim u_* X$. If geostrophic windspeed is unchanged, the increased roughness will decrease pollution concentration as well as surface concentrations of surface emissions.

18.5 Conclusions

On the basis of previous research, the paper discusses whether and how urban construction land space form influences PM2.5 dispersion. Firstly we elucidate its mechanism of action theoretically, then we use he Xuzhou instance data to demonstrate it.

We establish the model of PM2.5 concentration and mean wind speed through Air quality box model firstly, and the average fitting degree of parameter curves by measured data is nearly 0.43. Then we develop the model of PM2.5 and urban construction land space by the ground roughness, and the average R^2 of the curves using measured data is nearly 0.4. Regarding urban construction land use form-PM2.5 dispersion modeling, the fitting degree of each curve of a week is still relatively high. The main deviation of the model is from uneven distribution of pollution sources and large water bodies which can absorb much Atmospheric pollutant. For urban dimensions and its administrative structure are very complex, the data collection for the model was particularly. However, building a model reflecting the relationship between urban expansion and PM2.5 pollution based on empirical factors is proved possible from our study, which makes excellent sense.

We expect our results from study can be viewed when undertaking urban redevelopment, which are most relevant to those post-industrial.

18.6 Limitations and Future Study

This is the first attempt to research PM2.5 pollution and one of aftermath of urban development with limited number of routine monitoring sites. However, we should recognize a few limitations of our research. Firstly, we have just 11 observations, which is much fewer than the 40–80 recommended by [28]. Secondly, we ignore uncharacterized PM2.5 sources and meteorological data, which may be an important role in the spatiotemporal pattern of air pollutants. But it does not affect the accuracy of our study, because land-related factors are proved to be more important role in PM2.5 pollution, which could consequently change meteorological conditions. Our study just use mean value of PM2.5 for a week, and meteorological data and dynamic variables should be used to explain more in the further study.

Due to the needs of urban development, it is hard to reduce overall emissions which can improve air quality effectively, so increase the depositional sink for pollutants or improve the ventilation of city neighborhoods and streets is currently a hot research. So we should study further about how to build the urban ventilation flow corridor on the basis of the existing urban construction land space form using difference of the air pressure to form a remedy for the PM2.5 dispersion.

Acknowledgements We thank Institute of Land Science and Real Estate, Zhejiang University and School of Environment Science and Spatial Informatics, China University of Mining and Technology.

References

1. Cheng S, Lang J, Zhou Y et al (2013) A new monitoring-simulation-source apportionment approach for investigating the vehicular emission contribution to the PM 2.5 pollution in Beijing, China. Atmos Environ 79:308–316
2. Huang W, Tan J, Kan H et al (2009) Visibility, air quality and daily mortality in Shanghai, China. Sci Total Environ 407(10):3295–3300
3. Lang JL, Cheng SY, Li JB et al (2013) A monitoring and modeling study to investigate regional transport and characteristics of PM2. 5 pollution. Aerosol Air Qual Res 13(3):943–956
4. Wang Y, Zhuang G, Sun Y et al (2006) The variation of characteristics and formation mechanisms of aerosols in dust, haze, and clear days in Beijing[J]. Atmos Environ 40 (34):6579–6591
5. Donkelaar AV, Villeneuve PJ (2010) Global estimates of ambient fine particulate matter concentrations from satellite-based aerosol optical depth: development and application. Environ Health Perspect 118(6):847–855
6. Dockery DW, Pope CA, Xu X et al (1993) An association between air pollution and mortality in six US cities. N Engl J Med 329(24):1753–1759
7. Mahrt L (1999) Stratified atmospheric boundary layers. Bound-Layer Meteorol 90(3):375–396
8. Han L, Zhou W, Li W et al (2014) Impact of urbanization level on urban air quality: a case of fine particles (PM2.5) in Chinese cities. Environ Pollut 194(7):163–170
9. Morton BJ, Rodríguez DA, Song Y et al (2007) Using TRANUS to construct a land use-transportation-emissions model of Charlotte, North Carolina. In: Proceedings of the transportation land use, planning, and air quality conference
10. Hang J, Li Y, Sandberg M et al (2012) The influence of building height variability on pollutant dispersion and pedestrian ventilation in idealized high-rise urban areas. Build Environ 56:346–360
11. Yuan C, Ng E, Norford LK (2014) Improving air quality in high-density cities by understanding the relationship between air pollutant dispersion and urban morphologies. Build Environ 71:245–258
12. Berghauser Pont M, Haupt P (2010) Spacematrix: space, density and urban form. Nai Publishers, Rotterdam
13. Mahrt L (1999) Stratified atmospheric boundary layers. Bound-Layer Meteorol 90(3):375–396
14. Bottema M (1997) Urban roughness modelling in relation to pollutant dispersion. Atmos Environ 31(18):3059–3075
15. Salizzoni P, Marro M, Soulhac L et al (2011) Turbulent transfer between street canyons and the overlying atmospheric boundary layer. Bound-Layer Meteorol 141(3):393–414
16. Garratt JR (1992) The atmospheric boundary layer, Cambridge atmospheric and space science series, vol 416. Cambridge University Press, Cambridge, p 444
17. Carruthers DJ, Hunt JCR, Weng WS (1988) A computational model of stratified turbulent airflow over hills–FLOWSTAR I. In: Proceedings of ENVIROSOFT: computer techniques in environmental studies, Springer-Verlag, pp 481–492
18. Belcher SE, Hunt JCR (1998) Turbulent flow over hills and waves. Annu Rev Fluid Mech 30 (1):507–538
19. Grimmond CSB, Oke TR (1999) Aerodynamic properties of urban areas derived from analysis of surface form. J Appl Meteorol 38(9):1262–1292
20. Bottema M (1997) Urban roughness modelling in relation to pollutant dispersion. Atmos Environ 31(18):3059–3075
21. Kastner-Klein P, Rotach MW (2004) Mean flow and turbulence characteristics in an urban roughness sublayer. Bound-Layer Meteorol 111(1):55–84
22. Hu ZB, Yu BF (2008). Review on methods calculating aerodynamic parameters over urban underlying surface. J Meteorol Environ

23. Xin C, Tang et al (2014) Analysis and empirical research on the mechanism of urban scale expansion—a case study of typical regions. Science and technology research (11)
24. Yan Y, Xue L (2014) The study of expansion of urban land in XuZhou City based on GIS
25. Nowak DJ, Crane DE, Stevens JC (2006) Air pollution removal by urban trees and shrubs in the United States. Urban Forest Urban Green 4(3):115–123
26. Yong W, Yanping L, Jiangbo L et al (2015) The effect of PM2.5/PM10 variation based on based on precipitable water vapor and wind speed. J Catastrophology 30(1):5–7
27. Carruthers DJ, Holroyd RJ, Hunt JCR et al (1994) UK-ADMS: a new approach to modelling dispersion in the earth's atmospheric boundary layer. J Wind Eng Ind Aerodyn 52:139–153
28. Hoek G, Beelen R, de Hoogh K et al (2008) A review of land-use regression models to assess spatial variation of outdoor air pollution. Atmos Environ 42(33):7561–7578

Chapter 19
Curbing Urban Sprawl?—The Evolvement and Influences of the Greenbelt Policy in Seoul

Haoying Han and Haifeng Xu

Abstract Greenbelt policy, as the important measure for managing the urban development in Seoul, plays a vital important role in restraining the rapid population increase and the unlimited urban sprawl of Seoul. Based on the elaboration of the evolution process of the Greenbelt policy's "Proposal—Development—Adjustment", this paper points out that the influences of the Greenbelt policy on the comprehensive benefits brought by urban development is a dynamic process. During the proposal and development stage of 1970 to the late 1980, the Greenbelt policy has been of significance in controlling the rapid population increase and holding back urban sprawl, whose positive benefits to the urban development in Seoul exceed the negative effects. However, along with the fast economic development and the improved living standard, the demands for housing and transportation have been constantly increasing by 1990. In this condition, the previous situation was reversed. Despite the adjustment taken by the government in 1999, the adjusting measures are more like the reply to the residents' compensation claims from the government, but not taking the impacts of the Greenbelt policy on Seoul into consideration. What's more, new problems are turning up in carrying out the adjusting measures. This paper puts forward that the Greenbelt policy need further adjusting and be flexible, combining the land protection and urban development harmoniously in line with the idea of "Smart Growth".

Keywords Seoul greenbelt policy · Policy evolvement · Influencing process · Marginal adjustment

H. Han
Department of Urban and Regional Planning, College of Civil Engineering and Architecture, Zhejiang University, Hangzhou, China

H. Xu (✉)
Department of Land Management, College of Public Affairs, Zhejiang University, Hangzhou, China
e-mail: zju-xhf@zju.edu.cn

Y. Wu et al. (eds.), *Proceedings of the 20th International Symposium on Advancement of Construction Management and Real Estate*,
DOI 10.1007/978-981-10-0855-9_19

19.1 Introduction

19.1.1 Definition of Greenbelt

The Greenbelt is a perpetual open area surrounding the city and the built up area, including the woods, farm land and park etc. It has the functions of controlling the urban sprawl, protecting the farm land, controlling flood and storing water, and acting as the recreation place for citizens etc. The Greenbelt scope is determined by the government, and activities within the area are strictly limited in order to control the urban size and promote the sustainability of the city.

19.1.2 Origin of the Greenbelt Concept

The Greenbelt theory was originated from the idea of "Garden Cities", which was put forward in the British sociologist Ebenezer Howard's book Tomorrow–A Peaceful Path to Real Reform in 1898. In the book, Howard imagined a garden city designed for health, lives and industry, whose size was big enough to provide multiple social lives, but within the certain limit. The garden city should be surrounded by perpetual farm land, and factories and companies should at the urban fringe. The city size should be controlled strictly with each garden city has a population of no more than 30 thousand. A new city should be built once the size would exceed a certain scale in order to avoid over concentration and crowding. The "Garden City" of Howard is the embryonic theory of the contemporary theory of "Greenbelt" and "New Town" [1].

The first practitioner of the Greenbelt idea is London in Britain. The green area in London is 484.173 km^2, taking a proportion of the UK area of 3.7 %. The Greenbelt in London plays a vital important role in controlling the urban sprawl and offering a good recreation site for the local citizens [2].

19.2 Establishment and Development of the Greenbelt Policy in Seoul

19.2.1 Reasons for the Establishment of the Greenbelt in Seoul

In 1960s, the Korean government issued the first economic five-year plan, and carried out the export-oriented economic policy. As a result, a large number of Korean people began swarming into the capital city. In 1960, the population in Seoul (Hansung in that time) was just 2.4455 million, but the number was doubled to 5.4332 million in 1970, and the size of Seoul has been expanding accordingly from 268.35 km^2 in 1960 to 613.04 km^2 in 1970. Having been aware of the challenges to the area and even the whole country brought by the expanding of Seoul, the Korean government learned from the Greenbelt policy of London to make the Seoul Greenbelt policy according to its own situation and promoted it to other cities in the country. Korean has built Greenbelts in 14 cities including Seoul across the country from 1971 to 1987 [3].

19.2.2 Development of Greenbelt in Seoul

The Greenbelt in Seoul was first called "Limited Development Zone", which was introduced in the Town Planning Act issued in 1971. This idea was shaped in the Comprehensive National Development Plan (1972–1981) issued in 1973. The Korean government began building Greenbelt in Seoul and the other 13 cities from 1971 to 1987, among which 7 are metropolitan cities and 7 are small and medium-sized cities [4].

The Greenbelt development in Seoul can be divided into four stages (1971–1976). In the first stage, a girdle with a width of 1–9 km was built in the area of 15 km far from the Seoul center. The gross area in Seoul in this stage was

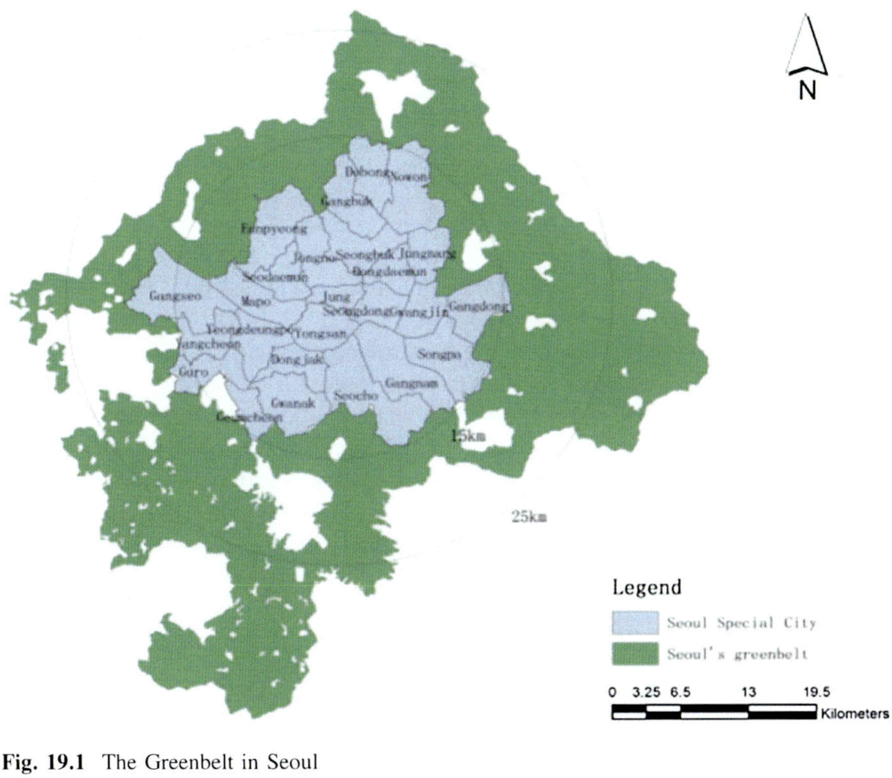

Fig. 19.1 The Greenbelt in Seoul

463.8 km^2. Since land speculations were taking place in Anyang and other districts after the Greenbelt was confirmed in the first stage, the central region as well as scenic places in Anyang were divided into the Greenbelt in the second stage, and the area was increased by 86.6 km^2. The third stage made the most increase with the width increased to 15–35 km. In this stage, the farm land on the east of Seoul and the Hanjiang River water reserve were covered into the Greenbelt, and the area was increased by 768.6 km^2. In the last stage, the Anshan New Town that is near Inchon, Anyang and Suwon was assigned to the Greenbelt, and the Seoul Greenbelt was finally formed. The Greenbelt boundary is 40 km far away from Seoul center at the farthest point, and the gross areas is 1566.8 km^2, taking a 13.3 % of the Seoul Metropolitan Area [5] (Fig. 19.1).

19.3 Adjustment of the Greenbelt Policy in Seoul

19.3.1 Background

The Korean government strictly prohibits any developing activities within the Greenbelt; therefore, old and poor as they are, the houses within this area can not be repaired or rebuilt. What's more, in the policy-making process, government neither negotiated the boundaries with the public, nor issued clear standards in determining the Greenbelt. The Greenbelt boundaries were confirmed more from the perspective of political demands and the purpose of controlling population increase by the government, which is of highly arbitrary. For example, some villages were divided into two sections by the Greenbelt. As a result, the developments of the two sections that were in and out of the Greenbelt were totally different [5]. People enclosed in the Greenbelt took the Greenbelt policy as depriving their properties, so they asked the government to compensate their losses by all means. However, these claims were suppressed by the military government.

In 1990s, the process of democratization in Korea kicked off. The residents whose land development rights within the Greenbelt are limited seized the opportunity of President Election in 1997, and collectively asked the authorities to include the Seoul Greenbelt matter into the topic for discussion in President Election. In December 1998, according to the constitutional court, residents whose land development rights within the Greenbelt are deprived were compensated. Therefore, the voice of adjusting the Greenbelt policy is growing louder in Korea, and these have become the trigger of the policy adjustment [6].

19.3.2 Measures to Adjust the Greenbelt Policy in Seoul

In order to response to the public voice, Korean Ministry of Construction and Ministry of Communication issued the comprehensive measures of adjusting the Seoul Greenbelt policy in July 1999. Main ideas are as follows [6].

1. The Greenbelt boundaries in Seoul will be partly adjusted according to the result of environmental impact assessment. The irrational areas, such as residential areas with more than 20 households (300 houses) can be excluded from the Greenbelt in priority for the convenience of living. Meanwhile, government will repair or rebuild the houses for free for people who remain staying in the Greenbelt.
2. Ranking the Greenbelt in Seoul into A, B, C, D and E five grades according to the five factors—altitude, slope, phytocoenosium, agricultural suitability, forestry suitability and water quality. Areas whose D and E proportions amount over 60 % will be excluded.
3. In order to avoid the result of connecting cities together after the removal, the Korean government limited the excluded Greenbelt to a minimum unit of 0.1 km^2. What's more, the potential excluded area should be at least in a distance of 2 km from the Greenbelt boundaries.

19.3.3 Current Situations of the Greenbelt Policy in Seoul After Adjustment

The initial purpose of Greenbelt policy was to control the urban sprawl in Seoul with the rigid boundaries, and kept the urban size and population size within appropriate scale. The Seoul Greenbelt policy has been strongly carried out by successive governors since its formulation. The adjustment in 1999 was on one hand, the forced measures in responding to the public voices; on the other hand, the long-term Greenbelt limitation did cause many problems to Seoul city. Whether the Greenbelt in Seoul is still the barrier in controlling the urban sprawl has been the heated topic of studies both in Korea and at abroad.

19.4 Influences of Greenbelt on the Urban Development in Seoul

19.4.1 Literature Review

In Korea, the Greenbelt policy of Seoul is heatedly discussed. On one side, it is important in controlling the sprawl and the rapid population increase; on the other side, as the result of the authorization era of the Park Chung-hee government, it violated the legal properties of the residents in the Greenbelt, and led to a series of problems to the urban development in Seoul. In this case, the Greenbelt policy of Seoul has always been the hot issue among Korean scholars.

The studies carried out by Korean experts are mainly from the positive and negative influences on urban development in Seoul. In summary, the positive influences are as fellows: (1) Greenbelt provides good ecological environment for Seoul, and the parks, green land, and recreation facilities are good recreation sites for citizens [7]. (2) Many other scholars think that the Greenbelt in Seoul has successfully control the urban sprawl, and since once these open areas were explored, it would be difficult to recover, so the Greenbelt must be kept to regularize people in using land resources [6, 8].

On the other hand, the opponents hold the arguments: (1) Greenbelt in Seoul is "the symbol of authority". In determining the boundaries, no communications with the public were made, and the Greenbelt in Seoul has serious violated the property rights of the residents in it [3]. (2) The Greenbelt has limited the land supply of the inner city, skyrocketing the housing price and land price of Seoul, and causing the traffic jams in the inner city [9]. (3) The Greenbelt did not control the urban sprawl, and many random "dormitory towns" appeared at the Greenbelt fringe, significantly increasing the commuting costs [10].

19.4.2 Influences of Greenbelt on Urban Development in Seoul

Studies on the Greenbelt in Seoul by Korean scholars mainly focus on the influences of one aspect or several aspects. Since the urban development is dynamic, so is the influences of the Greenbelt on the urban development. Therefore, simple studies on the negative influences or positive influences are biased. The influences can be divided into the following stages. In the early stage of the policy between 1970 and 1980, the Greenbelt played an indispensible role in controlling urban expansion and rapid population increase. However, with the city development and improved living standards, the demands of land are growing while the Greenbelt is restricting the land supply in the city. These factors have caused problems such as flying housing price, traffic jams and increased commuting costs, etc.

1970-Late 1980: The Positive Benefits Surpassed the Negative Influences

This period from 1970 to late 1980 was the proposing and development stage of the Greenbelt policy. The influences of Greenbelt on Seoul mainly included the control of rapid population increase, the limitation of the city boundary, and the grass land and facilities within the belt acting as the recreation sites. Although housing price and land price rose in this period, the negative influences still gave way to the positive benefits.

Greenbelt Decelerated the Rapid Population Increase and Controlled the Urban Expansion of Seoul

According to the data in Table 19.1, in 1960, the city's population was 2.4454 million, but the number has reached 5.4332 million in 1970, which was doubled compared with that in 1960. But after the setting up of the Greenbelt in the 1970s, growth started to slow down: during 1960–1970, the city population growth rate was 8.31 % per year; 1970–1975, it fell to 4.86 %; in the 1990s, the city's population began to appear negative growth (Table 19.1). On the other hand, although the population of Seoul metropolitan circle is increasing year by year, but the proportion of the city's population accounting for the whole metropolitan area is gradually declining, suggesting that Greenbelt promoted people to move to areas with more plentiful land but cheaper prices such as Gyeonggi province, Incheon and Suwon by controlling the city land supply, and suppressed the rest of the population to move in. Greenbelt not only inhibits the rapid growth of population, but also controls the random urban sprawl.

Greenbelt Improved Urban Environment and Provided Recreation Spaces in Seoul

According to the city green space data calculation by the Korea Statistical Office, the city green space per capita is lower than 1.5 m^2 [7]. As an international

Table 19.1 Demographic changes in Seoul City and Seoul Metropolitan Area (SMA), 1960–2010

Year	Seoul's population	SMA's population	Seoul's population accounts for the proportion of SMA's population (%)	Average annual growth rate (%)
1960	2,445,474	–	–	–
1970	5,433,198	8,730,148	62	8.31
1975	6,889,502	10,928,634	63	4.86
1980	8,364,379	13,298,241	63	3.96
1985	9,639,110	15,820,156	61	2.88
1990	10,612,577	18,586,128	57	1.94
1995	10,595,943	20,189,146	52	−0.03
2000	10,373,234	21,354,490	49	−0.42
2005	10,297,004	22,766,850	45	−0.15
2010	10,575,447	23,836,272	44	0.54

Note The above data has synthesized the Korea National Statistical Office's demographic data, and the column of average annual growth rate means the population's annual growth rate during previous ten years metropolis, the Seoul's data is far lower than that of London, Paris and other cities. Seoul, on the other hand, is a densely-populated city with more than 10 million people. The low per capita density and per capita green area can lower people's quality of life, and pose a threat to the sustainable development of the city.

There is a large amount of farmland, forest land, water area and national park facilities in Seoul Greenbelt, among which the land for park facilities is 84.68 km^2 and water area is 85.17 km^2. The total area is far more than the sum of the parks and water area in Seoul center. The park facilities in the Greenbelt can alleviate the deficiency of the city park, and provide citizens with leisure place for play. At the same time, as the mountainous country, the mountains and rivers in the Greenbelt is a good option for outing for. Waters and other land within the green belt can not only provide water for the city, but also provide recreational fishing and other water sports for residents to engage in. The huge forest in the Greenbelt area also has a huge impact on the improvement of the Seoul city environment.

Although this period is the period of rapid economic development, the car ownership in Seoul city grew from the 60,442 in 1970 to 1,193,633 in 1990 [7]. Despite the fast growth rate, the per household car ownership was less than 0.5 in 1990, therefore, the city's traffic pressure was relatively low. As for the urban property prices, in 1986–1990, the city apartment purchase price was only 40 % of 2011 [7], which was acceptable for most people and the demands of improvement housing were not very strong. In this sense, the influences on housing and transportation were in a secondary position compared with the positive benefits on Seoul city development. During this period, the support for Greenbelt policy in Korean surpassed the residents' opposition to it, and the boundary of the Greenbelt almost remained unchanged for more than 20 years [5].

Social Benefits of Greenbelt in Seoul Has Been Declining Since Early 1990s

Along with the economic development and the improved living standards, demands for housing and transportation were increasing gradually. However, due to the Greenbelt, Seoul is unable to provide more land for urban development. To solve the problem of urban residents' housing, the Korean government launched the "New Town Policy" at the fringe of Seoul. Since these new towns were far away from downtown, citizens had to spend a lot of commuting costs in coming back and forth. So since the 1990s, and the social comprehensive benefits of the city brought by Greenbelt began to decline, and the Korean domestic calls for adjusting the Greenbelt policy were getting more and more strong.

The Greenbelt Has Been Increasing the Land Prices and Housing Prices in Seoul Downtown Since Early 1990s

Because the Greenbelt limited the land supply of the city, and 153 km^2 areas were located in Seoul downtown Greenbelt, the really available land area was very limited in Seoul. In 2010, the city's population was 10.5457 million, and the average density of each square km was 17,473 people, which made the city become one of the overpopulated cities in the world. The huge population and scarce land resources made Seoul facing serious housing problem [7].

Because the Korean government did not make sufficient estimation of the development capacity of the city in the future at the beginning of the Greenbelt, it delimited the scope of the city Greenbelt in a hasty, and banned all development activities within the Greenbelt. These factors caused the city land price and housing price stay in a high level, coupled with land speculations, which added fuel to the fire of housing price and land price growth of the city. Apartment rental and sale prices rose apparently in Seoul after 1999 (Fig. 19.2), of which the apartment price of Seoul in 2010 increased by 169.7 % compared with 2000.

Greenbelt Causes More Traffic Jams in Seoul Downtown

In 1990–2010, the city car ownership increased from 1,193,633 to 2,981,400 with an increase of 1,797,767 more than the total car amount in 1970–1990. The private car ownership in 2010 was 2,434,230, accounting for 81.65 % of the total car amount in Seoul, which mean that every household in Seoul got a car. Huge automobile users are severe challenges for any city, especially for Seoul that carries out strict Greenbelt policy.

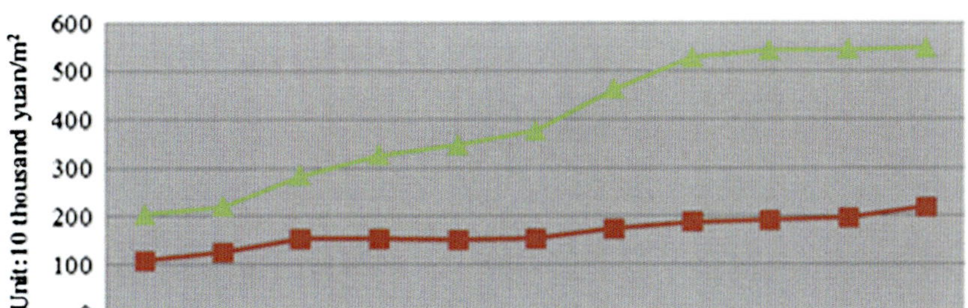

	2000	2001	2002	2003	2004	2005	2006	2007	2008	2009	2010
Aartments' sale price	109.09	126.21	153.33	154.09	151.91	153.88	173.58	187.73	190.45	196.06	218.58
Aartments' rental price	203.00	218.90	284.20	325.60	347.40	375.90	462.10	527.50	542.00	543.60	547.40

Unit:Year

Fig. 19.2 Annual apartment sale prices and rental prices in Seoul, 2000–2010 [11]

Because Greenbelt restricts the urban land supply, the city could not come up with a lot of land for traffic facilities and road construction despite the shortage of housing and construction land [12]. At the same time, the city, unlike Beijing that implements strict restrictions policy throughout the city, only carries out restrictions in some areas, so the problem of traffic congestion has become a huge challenge affecting the development of Seoul city and residents' lives. In order to ease traffic congestion, the city also introduces a large number of policies, and vigorously builds subway and other public transportations, such as the 9 subway lines built successively, but faced with a large population, subway and rail transportations are also insufficient. According to the statistics of Seoul's 1999–2003 annual crowding degree, the average was over 100, indicating that the city's subway transportation capacity has reached a critical level [13].

Greenbelt Increases the Commuting Costs Between Seoul Downtown and the New Towns

In the late 1980s, in order to disperse the population in Seoul, and to cope with rising housing prices and shortage, the government started to set up new towns around Seoul. South Korea's new town development policy is divided into two stages. The first stage began in 1989, in which the Korean government designated PenTang, Issan, Pyeong, Zhongdong and Yamamoto as new towns [14].

Given that housing and land prices are cheaper in new towns, many people choose to work in the downtown, and live in new towns. According to statistics, in 1990, population over 12 years old who commute to the other cities and provinces was about 355,923, accounting for 3.35 % of the population of Seoul. By 2010, the data has reached 637,685, accounting for 6.51 % of the population of Seoul. The commuting population has nearly doubled in 20 years, and the new towns in fact became the dormitory town of Seoul. Huge commuting population means the sharp increase in commuting costs, and in city like Seoul which has a high car ownership, vehicle emissions brought by commuting population have serious impacts on the urban environment [7].

Commuting costs increase appears to be the result of the high urban housing and land prices, but the underlined reasons is the Greenbelt's limit to the land supply in Seoul city. If the Korean government lifts part of urban Greenbelt and expands a certain scope of the urban boundary to peripheral areas, the excluded areas can be provided as the housing for urban residents [15]. Once the housing prices in these areas can be acceptable for citizens, they do not have to go to new towns, which will greatly reduce the urban commuting costs.

19.5 Solutions and Suggestions

Seoul Greenbelt policy needs further adjusting. New Greenbelt policy adjustment should meet the demand for land under the premise of controlling the urban sprawl and limiting the rapid population growth. New problems appeared in the process of excluding Greenbelt should be responded. Specific adjustment measures include the following aspects.

19.5.1 Continue to Relax the Greenbelt Areas in Order to Meet the Demands of Urban Development

The Greenbelt boundaries need further adjusting. Its reasonable adjusting range includes lifting the Greenbelt in downtown, moving the urban boundaries towards the Greenbelt under the premise of ensuring that Seoul would not unit together as a whole with the surrounding cities and the rapid growth of urban population can be controlled. At the same time, to protect the lands for waters, historical sites and facilities within the Greenbelt as much as possible. Based on the GIS data of city land type, 1 km moving out will release about 147 km^2 lands, and added with the released parts from the Greenbelt, these land will have big impacts on curbing the city housing prices and land prices, and ease traffic pressure. What's more, many urban residents can choose to buy houses in the removed areas so that they do not need to travel between the urban areas and new towns, which will save a lot of commuting costs.

19.5.2 Establish Policies to Control Land Prices in Newly Relaxed Zones of the Greenbelt

Relevant legislations about the Greenbelt should be made by the Korean government to manage the prices of those reserved areas and released areas. Given the wide price gap between the lands within and outside the Greenbelt, the released area is inclined to be the object of land speculators. The government needs to crack down on land speculations in the excluded regions. For excluded area with a population of more than 20 households, the original residents should have the priorities to use it for housing improvements. For those who buy the area of land to build private apartment or plants, the government should collect a certain amount of taxes [4].

19.5.3 Introduce Public Participation and Decentralize Management Authorities to Local Administrations

The making of Seoul Greenbelt policy is more of political guidance, without sufficient communications with people, so it lacks public opinion foundations. Since the management right of the Greenbelt is excessively concentrated to central government, as a result, the Greenbelt is seen as the product of "authoritarian" and people question the legitimacy of the Greenbelt [16]. Meanwhile, local governments lack autonomous management to the Greenbelt, so they are performing passively in the Greenbelt management. These factors lead to the constant controversy about the Greenbelt policy. To solve these problems, the Korean government should widely solicit public opinions regarding the Greenbelt policy, and release to the public the plan of land using in retained and excluded Greenbelt area in order to accept the supervision of the people. At the same time, in the management, the central government should play the role of supervisor and coordinator, and take the rights more devolved to local government for the purpose of improving the enthusiasm of local government.

19.5.4 Flexibly Coping with Future Urban Development

The development of the city is a dynamic change, if the boundary of the Greenbelt remains the same, Greenbelt will become the factors hindering the development of the city with the development of the city. Urban dynamic development requires the Greenbelt policy be flexible to meet the needs of land in urban development. The adjustment to Greenbelt boundary here is not unlimited; otherwise Greenbelt would lost its value. So the concept of "Smart Growth" is needed to be introduced, namely the urban development is not against land protection, and management of urban growth should not only confined to the city itself, but also fully consider the land development and urban growth as well as the planning of municipal infrastructure from the perspective of region, and unify urban development and land conservation harmoniously [8].

According to this concept, Greenbelt boundary should be adjusted flexibly based on the location of Seoul in the region and even the whole country. In 2006, the Korean government proposed to change the current Seoul metropolitan spatial structure from the old Seoul-centered metropolitan structure into "multi-core" type in the revision of the "Third Renovation Plan of Seoul Metropolitan Circle (2006–2020)", forming 10 the independent central city circles in Incheon and Gyeonggi province to improve the self-sufficiency by cultivating and renovating the center cities. In traffic system, the radioactive traffic system centering with Seoul will be changed into circular shape traffic system, connecting the central cities and other regions without passing Seoul. At the same time, parts of the Korean government were started to move to Sejong, about 120 km south of Seoul, in 2012, planning to make Sejong as the administrative center to distribute the population and ease the land pressure in Seoul [14].

These policies suggest that the function of Seoul to the Seoul metropolitan area and the whole country is changing. The implementation of these policies will have significant influences on the population increase of the city, the spatial structure and urban development, so the Seoul Greenbelt policy should maintain flexibility to cope with its changing development orientation. This is not just about the improvement of urban competitiveness, also related to achievement of the goal of "Regional Balanced Development" of the Korean government [13].

References

1. Yang XP (2010) London's Greenbelt's policy and its revelation for Chinese cities' Greenbelt construction. Urban Plan Int 25(1):100–106
2. Yang XP (2008) Seoul's Greenbelt policy and New Town's policy. Urban Plan Int 25(1):100–106
3. CPRE and Natural England (2010) Greenbelts: a greener future. https://www.gov.uk/government/organisations/natural-england
4. Christine Bae CH Korea's Greenbelts: impacts and options for change. Pac Rim Law Policy J 7(3):479–500
5. David NB, Youn YC (2004) Seoul's Greenbelt: an experiment in urban containment. Policies for managing urban growth and landscape change: a key to conservation in the 21st century
6. Jin YH (2001) Greenbelt zone regulations are relaxed across the country. Space Environ 16:1–4
7. Korean National Statistic Office. Statistical Database. http://kosis.kr/eng/
8. Liu HL (2005) From disorder sprawl to smart growth—the elaborate of concept of American "Urban Growth Boundary". Urban Problems. 3:67–72
9. Lee CM, Peter L (1998) Dynamics of the Greenbelt amenity effect on the land market—the case of Seoul's Greenbelt. Real Estate Econ 26(1):107–129
10. Jun MJ (2012) The effects of Seoul's Greenbelt on the spatial distribution of population and employment, and on the real estate market. Ann Reg Sci 49:619–642
11. Seoul City Hall. Seoul's various fields statistic database. Seoul City Hall's Chinese Website. http://chinese.seoul.go.kr
12. Cho M (1997) Congestion effects of spatial growth restrictions: a model and empirical analysis. Real Estate Econ 25(3):409–436
13. Piao GX (2011) Innovation research on Seoul metropolitan area's controlling polices. Chinese Academy of Social Science, Beijing
14. Meng YG (2014) Seoul metropolitan's expansion and the development of medium and small cities around it. Chin Soc Sci Net, 21 Mar 2014 [6 Apr 2015]. http://www.cssn.cn/zt/zt_xkzt/zt_jjxzt/jjxzt_czh/czh_gjjj/201403/t20140321_1038804.tml
15. Jun MJ, Hur JW (2001) Commuting costs of "leap–frog" new town development in Seoul. Cities 18(3):151–158
16. Christine Bae CH, Jim MJ (2002) Counterfactual Planning: What if there had no Greenbelt in Seoul. J Plan Educ Res 4(22):374–384
17. Lee CM (1999) An intertemporal efficiency test of a Greenbelt: assessing the economic impacts of Seoul's Greenbelt. J Plan Educ Res 19:41–52

Chapter 20
The Impact of Urbanization on Farmland: Is Division of Right to Contracted Management of Land a Reasonable Way Out?

Xiaobin Zhang and Yanmei Ye

Abstract China is undergoing unprecedentedly rapid urbanization. The migration of squatters from rural to urban areas not only puts pressure on the supply of urban land and public facilities, but also has profound impacts on the use and management of rural land property. The most obvious and important representation of this situation is the large scale of farmland transfer. In this situation, the real manager of contracted land and the original contractor will no longer be the same. As a response to that, scholars put up with the idea to divide right to contracted management of land (Hereinafter referred to as RCML) into contract right and management right, which is further verified by central government in Central Document NO. 1 of 2014 and 2015. The perspective of jurisprudence and institutional changes will prove that, however, division of RCML won't be a reasonable reform path. For one thing, the so called division of RCML completely lacks jurisprudence foundation, embodying the confusion of the characters of different law systems from the macro level, and violation of right creation theory from the micro level, meanwhile it confuses members' right and property right. For another, the division of RCML will confront with large institutional changes cost, which makes it infeasible. This paper thinks that a feasible reform path should be proposed based on our own theory of law system, and it must take farmers, transferees and collective all into account to construct a more comprehensive plan.

Keywords Farmland transfer · Division of RCML · Jurisprudence · Institutional change

---

X. Zhang · Y. Ye (✉)
School of Public Affairs, Zhejiang University, Hangzhou, China
e-mail: yeyanmei@zju.edu.cn

Y. Wu et al. (eds.), *Proceedings of the 20th International Symposium on Advancement of Construction Management and Real Estate*,
DOI 10.1007/978-981-10-0855-9_20

20.1 Introduction

China has been undergoing rapidly urbanization since Reform and Opening, and a more sharply growing of urban population since 1990s, characterized by the average annual growth of 17.88 million. It makes the urbanization rate of China reach 53.73 % in 2013 from 26.41 % of 1990 (Fig. 20.1).

This boom of urban population brings much pressure to city governments on urban construction land supplement and other basic public services of course, while also has as much impacts on rural land, if not more, mainly in two aspects: farmland transfer and land expropriation.

Because of the continuous transfer of agricultural labor force, farmland left behind naturally confront with being transferred to real farmer or agricultural companies. The statistic shows that 380 million hectare farmland has been transferred in 2014, accounting for 28.8 % of all contracted farmland [1], which indicates the separation of contractor and manager of farmland has been a very common state in practice. But farmland transfer will create some vaguely anxiety on farmer of losing land permanently because of the fuzziness of China's rural land property right. Moreover, this anxiety will evolve when taking land expropriation into consideration, which hinders the farmland transfer process of China eventually since it leads to huge disputation between transferors and transferees on distribution of benefits.

A certain kind of reform on farmland property right named division of RCML, which means to divide RCML into contract right and management right, has been proposed as a respond. This proposal has drawn scholars' eyes and even wined so strong support from central government that its implementation has been emphasized for two consecutive years in Central Document No. 1.

Despite the popularity of division of RCML and the fact that it almost becomes an established policy in China, its true nature and function remain vague and lack of comprehensive discussion. Therefore, this paper aims to reveal the essence of

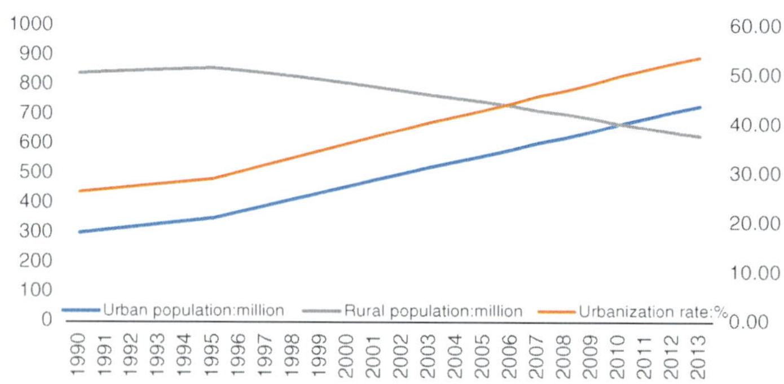

Fig. 20.1 Urbanization level changes of China from 1990 to 2013. *Data source* China Statistical Yearbook 2014

division of RCML and its suitability to China's urbanization process. Moreover, some suggestions will be put forward to for improvement or replacement of division of RCML.

20.2 Literature Review

20.2.1 Main Points of Division of RCML

The Connotation of Division of RCML

Some scholars think there has always been three rights or even more on farmland not only in practice but also in law because the ownership is a bundle of rights that can be separated from each other. The transfer of farmland should never involve in collective ownership and land contract right, but only land management right [2–4]. Another point of view is that the large scale of farmland transfer has proclaimed the separation of land contract right and land management right, what need to do is to confirm it in law, and to define the specific content of each right [1, 5].

There exists three kind of views about how to specific divide RCML. The mainstream is that RCML is somehow simply divided into contract right and management right. The Contract right, which is a members' right shared among collective members only [5, 6], contains the ability of possession, disposal, inheritance and exit. The management right, on the contrary, has no links with owner's identity and thus can be transferred, mortgaged, shared and trusted [7]. Some scholars argue that division of the RCML is one kind of land transfer that allows transferors to keep some ability of RCML when land has been transferred. In line with this comment, not all kinds of transfer but only mortgage, share hold and land trust should use division of the RCML [8]. For the last view, land management right is a created usufruct on RCML, namely a usufruct on usufruct. It contains some but not all ability of RCML to possess, use and seek interests from farmland [9].

The Function of Division of RCML

The main function of division of RCML is summarized as "to stabilize land contract right and to liberalize land management right", which has been confirmed as the direction of rural land reform. The stabilization means one hand to try to extend the contract period to permanent [10], on the other hand, no matter what kind of land transfer happens, farmers should always have some remain rights in their land [1]. This mainly shown in two aspects: when their land confront with being expropriated, they have right to get just compensation; and if contract of farmland is not permanent, they have right to contract farmland again when the next round of contract happens. The liberalization of land management right also has double meanings: to stimulate the enthusiasm of farmers to transfer their land by stabilizing the relationship between farmers and their land as mentioned above, and to break the restriction on mortgaging RCML in law to exert the financing function of land by creating this new right [9]. There is some other benefits of division of RCML such as solving the complex nature of RCML [6], and rationalizing the distribution of right to possess, use, seek interests and disposal among different subjects of rights.

20.2.2 Criticism of Division of the RCML

The criticism of division of the RCML almost all comes from jurists, who mainly argue that right division violates a basic principle in civil law countries, numerus clausus: that the type and content of property rights must be limited in line with provision of property law. Only collective ownership and RCML can be counted as right in our law, to contract and manage farmland are just ability of RCML [11, 12].

In addition, Chen [13] argues that creating a new usufruct on a similar usufruct complicate legal relations artificially. He comments this is degeneration of legislation technology in presence of distinction of property right and contract. Some others think that the proposition of division of the RCML is a result of following the experience of SOEs reform, while ignoring the essential difference between each other [12].

20.2.3 Limitations of Existing Research

As we can see, supporters of division of the RCML indeed rarely conduct law analysis about it, particularly rare from our own law system's perspective. It is a little paradox considering it is a change in farmland property law. However, criticisms from only law aspect are not enough to illustrate the unreasonable and unfitness of it, especially when numerus clausus doctrine, the main argument of critics of division of RCML, is suffering criticism itself because it hinders the creation of new right [14, 15]. Clearly, we need an analysis from more aspects, and a new way of reform should be proposed if division of RCML is not a good reform path.

20.3 Jurisprudence Analysis of Division of the RCML

20.3.1 Confusion of Different Characters of Law System

Although there is some scholars argue common law also has similar restrictions on property rights only without name called numerus clausus [16, 17], most of others, however, deem there is no such restrictions in common law [18, 19]. Undeniably, restrictions on the type and content of property right in common law are more loose than it in civil law, and ownership in common law has no absolute nature which takes precedence over other rights [20]. Ownership is just a combination of rights to possess, use, seek proceeds and dispose, so called "a bundle of rights" [21]. Some of these rights can be treated separately [22], and rights separation is a kind of rights transfer. That is exactly what supporters of division of RCML argue.

However, what is feasible in common law system can be infeasible in civil law system. There are two kinds of rights on property in civil law: creditor's right and property right. The former can be created and defined by parties to the contract almost arbitrarily, but the latter is prior over the former and has more strictly protection by law. Therefore, the type and content of property right must only be clearly define by law, or it will make it difficult to determine who has priority on objects and finally make it costly of transaction. And rights transfer, must be the transfer of an established property right as a whole, or to set creditor's right on objects, but definitely not separation or division of property right. Therefore, the view deems collective ownership as a bundle of rights in law, and rights transfer is a kind of rights division based on a theory from common law, cannot fit in with law theory in our law system.

20.3.2 Violation of Basic Theory of Right Creation

Despite the strict restrictions on type and content of rights in civil law system, surely right creation are allowed according to social changes. But the division of RCML into contract right and management right violet the basic theory in our law system. As we all know, usufructs created are actually restrictions on ownership, when users of land get usufructs from land owner, what left in land owner's hand is still ownership, only with restrictions. That's why collective ownership is still collective after the establishment of RCML rather than be divided into something else. Similar principle goes with the division of RCML: when land management is set on farmland and be transferred to others, what left in farmers' hand is still RCML, not contract right (Fig. 20.2). Contract right doesn't have jurisprudence basis to exist.

20.3.3 Confusion Members' Rights and Property Right

People think, when RCML is divided, farmers can transfer the right at the same time with some right remains, so that farmers don't lost land permanently. This remain right, which they called contract right but should still be RCML just with restrictions on it, is believed to have the ability to grantee farmers qualifications of next round of land contract and more importantly, just compensation when land expropriation happens.

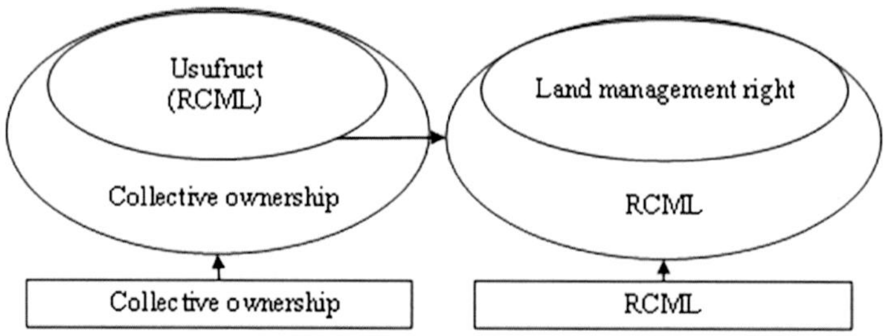

Fig. 20.2 Mode of division of collective ownership and RCML

However, determining who has qualifications to contract land is not the content of RCML but the content of members' right system, similar case shows in determining who has qualifications to own land is not the content of ownership, but the content of ownership system. Even if farmer transfer their land to others, they has qualifications to continue to contract land naturally due to identities of collective members after the expiration of this round of contract, not because they still own some parts of RCML.

As for keeping remain part of RMCL to get just compensation of land expropriation, that's really an argument, but as we will illustrate below, this kind of institutional design can only invite transaction costs, while through members' right system, farmers can still get land expropriation compensation.

20.4 Institutional Change Analysis of Division of the RCML

As we know now, the feasible way to divide RCML is to create a new right called land management right on farmland under law analysis. However, the following analysis will show that this kind of institutional change can hardly protect farmers' right and promote farmland transfer.

20.4.1 Dummy of Collective Ownership

Weakening collective ownership of land has been recognized as the direction of farmland reform to protect farmers' right by many scholars due to the unclearness of "collective" [23]. Division of RCML and extending the contract period of farmland to eternity are both ways to realize it. The reason to create a new but similar usufruct, land management right, on RCML is to substitute the right structure of double usufruct for the previous structure of ownership and usufruct. That will RCML obtain characters of ownership [24]. Plus the extension of farmland contract period, collective ownership will be totally dummied.

Saying that dummy of collective ownership violets a fundamental national policy of China which is public ownership of land may be too political, but it indeed won't adapt to China's unprecedentedly urbanization process. There are almost 12 million farmers who become urban residents and settle down in urban area every year in China. If farmland contract become permanent, and farmers almost "own" their farmland, then there will be hundreds of million people who live and work in unban area, but own farmland in rural area in next 10 years. This is a situation not allowed in many countries of private ownership [25] since it further incur disaster on food security and undermine social equity. Some people argue that they can hand their land to collective when they become urban residents, but that makes so called permanent contract for farmers turn into period uncertain contract, and it will be an obstacle of urbanization.

20.4.2 Incurring of Transaction Costs

Making mortgage of land allowed by law is one important reason of creation of land management right. Meanwhile, because land management just accounts for a part, not all, of RCML, when farmers cannot repay their loan after mortgage, remain part will ensure that farmers won't lost their land thoroughly.

That is true and good from farmers' perspective, but also a point of view totally ignore mortgagees. Over subdivision of rights on farmland will undermine integrity of profitability of farmland, which is a character that mortgagees weigh much on [26], and eventually hinder mortgage, let alone that some other researches show that many financial institutions won't take farmland as pawn due to the low income before division of RCML [27]. In addition, there will be four rights on farmland if mortgage happens, and the complexity of right structure will make it very difficult to evaluated farmland [28], further hampers the process of mortgage.

20.4.3 Huge Costs of Modifications of Law

It is obviously that creation of new property right will cause many costs to change property law.

Firstly, the content of RCML and land management are extremely similar, the only difference is that initial acquisition of RCML are linked with the identities of collective members. Then how to distinguish them in law is quite a problem. Furthermore, there are four rights on land in China's usufruct system: RCML, right to use land for construction, right to use house sites and easement rights, each aims at different type of land or has very different content. Introduction of land management right, a right that extremely similar with RCML, with undermine this system, therefore, need quite an explanation.

Secondly, there are other ways to contract rural land in law except for by farmer households, which are to contract wasteland through bidding, auction and public consultation. Despite different ways to contract land, the right people get is the same RCML. But after the creation of land management right, it can hardly explain what right it should be: it has definitely the same content of land management right, but it seems awkward to call it land management right because people do contract land from collective, just like farmer households.

And of course, these difficulty to distinguish different rights will finally make it costly of land registration and certification.

20.5 Suggestions for Farmland Reform

The analysis above implies that a more adaptive reform path to China's reality should lie in the existing farmland property right structure, and it must be proposed in a more comprehensive way, not only aiming at farmers. There should be an improved farmland transfer institution between transferors and transferees to promote transfer of farmland. To protect farmers, there should be a more clear members' rights system and a stable property right. And a moderate strengthening of collective will make it better for China to adapt to its urbanization (Fig. 20.3).

20.5.1 *Improvement of Farmland Transfer System*

A Clear Distinction of Different Type of Farmland Transfer

As mentioned above, there are two kinds of rights on property: property right and creditor's right. Therefore, all kinds of ways to transfer farmland in practice should be summarized into property right type and creditor's right type in law for reducing transaction costs and promoting land transfer.

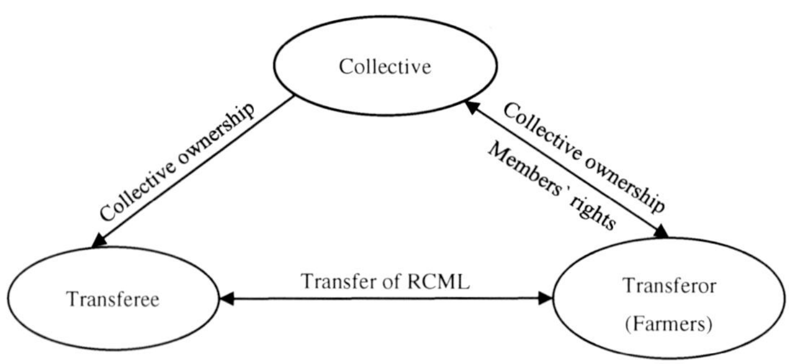

Fig. 20.3 A reform path on farmland

Table 20.1 Different type of farmland transfer

Farmland transfer type	Property right type	Creditor's right type
Specific way to transfer farmland	Transfer, exchange, mortagage	Subcontract, lease, share, trusted

Farmland transfer in property type means that famers give up whole RCML to transferees according to property right law, and farmers will lost all direct connections with their land but remain some indirect connections because they are members of collective and collective always has ownership to all farmland. Transfer in creditor's right type means that famers encumber farmland to others rather than give up RCML to them, and when the contract expire, they get back their land. Theoretically, they can get their land back any time they want as long as they pay satisfactory compensation for breaking contracts. Table 20.1 shows different ways to transfer farmland in practice. Only mortgage needs explanation: mortgage is not 'transfer' in strict sense, but when mortgagors can't repay their debts, farmland will confront with being transferred in property type.

Note that land registration almost is the only proof that can distinguish land transfer in property right type from transfer in creditor's right type. Therefore, RCML registration must be confirm as a symbol of establishment and transfer of RCML in future. This is also beneficial to the uniformity of urban and rural property rights system.

Institutionalization of Land Transfer Restrictions

To restrict farmland transfer by law is common in many developed countries such as Germany [25], France [29] and Norway [30]. China surely should also establish restrictive regulations on farmland transfer, but a simple expression of "approved by collective" seems arbitrary, the restriction on farmland transfer must be more institutional, mainly on following aspects: (1) The type and content of farmland transfer being prohibited must be clearly define by law or regulations; (2) The restrictions on transfer in property type should be more strict than it on transfer in creditor's right; (3) It must be justified when a certain kind transfer is prohibited and remedies to farmers must be established.

20.5.2 *Moderate Strengthening of Collective Ownership*

Improvement of Farmland Contract Period Institution

It has been explain in Sect. 4.1 that extension of farmland contract period indiscriminately can hardly do good to China's urbanization process. We believe there should be a discriminate treatment on farmland when this round of contract expires.

The core idea of it should be to centralize farmland to collective members. Therefore, following measures can be taken based on the identities of RCML owners when contract expires: (1) if owners are still members of collective, they can carry on the contract of it; (2) if owners become urban residents, they must hand over their RCML; (3) the real managers of farmland can keep managing farmland no matter how RCML is transferred, a principle like 'Sales do not break lease' which is common in residential leasing. The last suggestion is just a bold vision which needs further discussed.

The Distribution of Benefits in Land Expropriation

Many thinks that when transfer in property right type happens, farmers will lost their land permanently and when land is expropriated, all compensation will be taken by transferee, often companies, and farmers will get nothing of it.

This is a kind of view that dummies collective ownership. As we all know, transferees of RCML only get RCML of remain term, neither ownership nor a permanent usufruct, when land expropriation happens, what they get are compensations for RCML with limited period, and the collective should also get compensation because of ownership of land. And farmers, who are members of collective, can to some extent enjoy to share the compensation, and that, we think, should be resettlement compensation.

20.5.3 Improvement of Members' Rights System

The improvement of members' rights system is essential to protect farmers' rights under the circumstance that collective ownership has been moderately strengthening.

Members' rights system contains lots of content which beyond the scope of this article, what related is that how to make famers better enjoy the benefits lie in collective ownership. Except for that every collective member has right to contract farmland freely, we think that there are three points to confirm: (1) there must be a clear definition of collective members especially after the differences between urban and rural are removed by household registration reform; (2) every collective member should have right to contract land when next round of land contract begins whether or not they transfer their land in this round; (3) every collective member should get resettlement compensation whether or not they transfer their land when land expropriation happens.

Improvement of members' rights system also relies on a clear distinction of members' rights and property right: to contract farmland freely is a members' right, but once they get RCML, it is not linked with their identities, which means, even if farmers become unban residents they still own RCML before its expiration. Maybe they can be force to hand farmland because of land abandonment, but they cannot be force to do it simply because they are not collective members.

20.6 Conclusion

Rapid urbanization process of China stimulates the transfer of farmland, which can increase farmers' income and improve the degree of large-scale agricultural production. However, the fuzziness of farmland property right and frequent land expropriation make farmers hesitate to transfer their land in fear of losing land forever and getting no expropriation compensation. The division of RCML hence are proposed to solve this tangled situation, to protect farmers' right and to promote farmland transfer.

However, a law analysis in detail shows us that land contract right doesn't has jurisprudence basis to exist. And create a new usufruct called land management right on RCML is feasible under law analysis, but it will cause huge institutional change costs and transaction costs which violets the original purpose of it. So a less costly reform path should lie in the existing right structure on farmland.

We believe that a reform path that can protect farmers, promote farmland transfer and more importantly to adapt to rapid urbanization should consider comprehensively about farmers, collective and transferees, not farmers only. We propose that a clear distinction of different type of farmland transfer would promote land transfer, but an institutional restrictions on farmland transfer should always be there at the same time. The strengthening of collective ownership will actually stabilize RCML to some extent, and improved members' right system will protect farmers against an over strong collective. We just put forward some simple suggestions on all three parts of reform path, and they all should be discussed in further studies.

References

1. Zhu DL, Wang J, Lin RR (2014) Outlook for China's rural collective land system reform: reviews from "Roundtable Forum for Land Policy and Law 2014". China Land Sci 28(9): 89–94
2. Fan HC (2007) Problems and countermeasures of the transfer of agricultural land in hilly areas of Sichuan. Econ Geogr 27(2):317–322
3. Huang ZH, Wang P (2008) The transfer of rural land: status, problems, countermeasures and the influence on Modern Agricultural Development. J Zhejiang Univ (Humanit Soc Sci) 38 (2):38–47
4. Zhang HY (2002) Some comments on China's agricultural land adjustment and use right transfer. Manage World 5:76–87
5. Gao YC (2013) Separation and Reconstruction: A novel theory of land contract management rights transfer. Econ Perspective 2(5):137–139

6. Ye XQ (2014) From separation of ownership to separation of contract and management right: The past and future of farmland property rights. Chin Cadres Tribune 6:7–12
7. Pan J (2014) "The separation of the three rights" on rural land: the content of rights and risk prevention. Acad J Zhongzhou 11:67–73
8. Pan J (2015) The Way of realizing the Division of the right to Contract and Manage Farmland. J Nanjing Agric Univ (Soc Sci Ed) 15(4):98–105
9. Cai LD, Jiang N (2015) The legal structure of the separation of contract right and management right. Chin J Law 3:31–46
10. Zhu GX (2014) Land contract and management rights: the subject, duration and succession. Jilin University Journal Social Sciences Edition 54(4):28–37
11. Gao SP (2014) The legal logic of farmland property rights structure under novel agricultural management system. Chin J Law 4:76–91
12. Shen HW (2015) Legal perspective of the separation of land contract and management right on rural land property rights. China Land Sci 29(3):39–44
13. Chen XJ (2014) The thinking and framework of the transformation of China's rural land legal system. Chin J Law 4:4–25
14. Liang SS (2003) The numerus clausus doctrine: between liberty and compulsory. Chin J Law 3:43–57
15. Yang YX (2002) The numerus clausus doctrine. Study Comp law 1:34–45
16. Hansmann H, Kraakman R (2002) Property, contract, and verification: The numerus clausus problem and the divisibility of right. J Leg Stud 31(2):373–420
17. Merrill TW, Smith HE (2000) Optimal standardization in the law of property: the numerus clausus principle. Yale Law J 5:1–70
18. Danner JC (1997) TDRs-great idea but questionable value. Appraisal J 65:133–142
19. Van ES (2003) A numerus quasi-clausus of property rights as a constitutive element of a future European property law? Electron J Comp Law 7:1–12
20. Ran H (2006) On the concept of property rights in Anglo-American property law and its institutional function. Sci Law: Northwest Univ Polit Sci Law 24(5):33–40
21. Alchian AA, Demsetz H (1973) The property right paradigm. J Econ Hist 33(01):16–27
22. Renard V (2007) Property rights and the 'Transfer of Development Rights' questions of efficiency and equity. Town Plan Rev 78(1):41–60
23. Wang JT (2012) Study on the dilemma of the system of right to the contracted management of land and its solution: on the secondary property-rightization of right to the contracted management of land. Chongqing: Southwest university of polity and law
24. Han J (1999) From separation of ownership to separation of contract and management right. Econ Res 75:30–30
25. Bauer SN (2004) German property law (Upper volume). Law Press, Beijing
26. Zhong TY, Huang XJ, Kong P (2005) Land property rights and farmer households' willingness of farm land leasing. China Land Sci 19(1):49–55
27. Zhong FN, Ji YQ (2009) Land rights, non-agricultural employment opportunities and farmer's Agricultural Investment. Econ Res J 12:43–51
28. Guo ZX, Wang XS, Qu FT (2014) The mechanism design for LCUR-Based Mortgage loan with the property regulation: an analysis based on the Institutional Circumstance and the Management Structure. Manage World (9):48–57
29. Li LN (2014) French agricultural land transfer and its fiscal policy. World Agric 2:88–91
30. OECD (2008) Agricultural support, farm land values and sectoral adjustment. OECD, France

Chapter 21
Analysis of Influential Factors on the WTP of Roof Agriculture's Non-use Value

Qi Yin, Wenjia Zeng and Xiaoqian Liu

Abstract Based on the questionnaire survey of roof agriculture's on-use Value (NUV) in Wenjiang District, Chengdu, China, this paper used contingent valuation method (CVM), to evaluate the influential factors on the willingness to pay (WTP) of roof agriculture through PCA (principal component analysis) and multivariate LOGIT Model. The conclusion is: of all influential factors, annual income, educational level and determination of roof ownership serve as the major factors that influence the respondents' WTP for roof agriculture.

Keywords Roof agriculture · Contingent valuation method (CVM) · Willingness to pay (WTP) · Influential factors

21.1 Introduction

Since the reform and opening-up of China, a vast area of arable land has been occupied in the ongoing process of urbanization, putting land issue in the spotlight. In consideration of the sharp conflict between land supply and demand, people begin to focus on the free urban space, in which urban roof agriculture is one of the important and real approaches [7]. Roof agriculture can harness the urban idle space and relieve the stress on land resources. In addition, roof agriculture can beautify and purify the environment, lower the indoor temperature, lessen UHI (urban heat island) effect, create the energy conservation and low-carbon life, and refresh the urban appearance [10]. Since the mid-20th century, roof greening campaign has

Q. Yin (✉) · W. Zeng · X. Liu
College of Management, Sichuan Agricultural University,
Chengdu 611130, China
e-mail: yinqi@sicau.edu.cn

W. Zeng
e-mail: 349909193@qq.com

X. Liu
e-mail: 526457145@qq.com

Y. Wu et al. (eds.), *Proceedings of the 20th International Symposium on Advancement of Construction Management and Real Estate*,
DOI 10.1007/978-981-10-0855-9_21

been immensely launched by Britain, US, Germany, Japan and other countries [3]. For a decade, roof greening construction has also developed in China, thanks to its urban economic growth [4].

However, in China, roof agriculture has not been rapidly developed with scale effect, and most researches simply focus on its feasibility and discussion on existing problems [1, 4, 8]. Therefore, this paper will probe into the influential factors in respect to the Non-use Value (NUV) of roof agriculture.

21.2 Method and Data

21.2.1 CVM

The basic method employed in this research is Contingent valuation method (CVM). CVM is the standard method, which is recently used at home or abroad to evaluate the public WTP or compensate willingness for environmental resources, acquiring entertainment, selection, existence value and other NUV [9]. Generally, a portion of families or individuals are randomly sampled, a set of hypothetical questions are asked in the form of questionnaire survey, and the consumers' preference for environmental resources and other public goods and service are revealed through the simulated market [12], deducing the respondents' WTP for a certain of environmental improvement benefit or compensate willingness for environmental quality loss. Pay (compensate) willingness of the respondents are calculated and sampling is extended to the whole survey region, and all economic benefits or losses resulting from the planned project will be obtained by the averaged pay(compensate) willingness. Based on this, the feasibility of planned project will be examined through the cost-benefit analysis [6]. The acquisition of WTP stands in the center of researching the value of environmental resources.

21.2.2 Data

General Information of Survey

This questionnaire survey was conducted on the NUV roof agriculture in Wenjiang District, Chengdu; a total of 330 questionnaires were delivered to 24 communities located in Wenjiang District. Of which, 325 are returned with the return rate of 98.48 %; after screening out the useless questionnaires, ultimately 317 valid questionnaires were obtained, with the response rate of 97.53 %.

For the convenience of statistics and analysis, variables and assigned values were given to the questions and answers contained in questionnaire accordingly. To be exact, assigned values are integers, the adjacent properties and equal interval are kept, but equal interval doesn't necessarily represent that the difference between adjacent properties adjacent properties coincides to be identical. For the variables except name, bigger assigned value means higher grade.

Design of Questionnaire

This questionnaire is roughly divided into three parts: Part I is the survey of general information about respondent, including the sex, age, occupation, educational level, annual income and floor at which he or she resides. Part II is the survey of roof agriculture, including the knowledge about roof agriculture, agreement or disagreement, and the benefits of roof agriculture. For example, "roof agriculture can improve UHI effect", "reduce the pollution to urban environment" or alike, each respondents are asked to choose their answers from these benefits. Part III is the survey of WTP, including the survey of WTP and the reason for countering against WTP, in which, the core question is: if a certain amount of funds are required to implement roof agriculture, how much you are willing to pay in a year? (A) RMB 0; (B) RMB 1–10; (C) RMB 10–100; (D) RMB 100–200; (E) RMB 200–500; (F) >RMB 500.

21.3 Results and Analysis

21.3.1 Basic Statistical Results

According to the statistical result of the survey, the positive pay ratio amounts to 89.91 %, this is close to the similar researches. This survey was conducted in Wenjiang, Chengdu city, where most of respondents are community residents, and consequently the questions are relatively true and reliable, as they are the interested parties to roof agriculture. The main statistical results as listed in Table 21.1.

Table 21.1 is the statistical description of each variable of personal attributes of the respondent who is paying, as indicated in the questionnaire survey. As shown in Table 21.1, in questionnaires with non-zero pay, the degree of male's pay willing is slightly higher than that of female's, but there is little influence on WTP. Pay ratio in 18–30 age group is not so high, but WTP is relatively high, the degree of WTP gradually decreases with the age, along with the gradual decrease of pay ratio. WTP of migrant worker is the lowest, and pay ratio is too low, with the degree of WTP averaged only 2.235. WTP of worker, salesman and others is just the second, but it is not very high. WTP and pay ratio for the education level in "Middle school or below" are relatively lower, which may be attributed to more emphasis on social and environmental benefits with higher educational level; WTP and pay ratio are almost on the rise with higher educational level. As to household average income, it exerts a big influence on WTP, more annual income means higher WTP; for the annual income above RMB 10,000, the average WTP is

Table 21.1 Statistical description of the main variables

Classification		Percentage (%)	Pay ratio	Degree of WTP
Sex	Male	50.63	82.5	2.906
	Female	49.37	82.05	2.828
Age	18–30	37.85	86.67	3.041
	30–45	29.97	81.05	2.989
	45–60	22.08	78.57	2.643
	>60	10.1	65.63	2.344
Occupation	Farmer	5.36	64.71	2.235
	Worker	12.93	73.17	2.658
	Student	9.46	96.67	3.433
	Private or public employee	25.55	85.19	3.2
	Teacher	8.2	88.46	3.308
	Self-employed individual	11.67	81.08	2.54
	retiree	9.46	80	2.667
	Salesman	12.62	73.17	2.475
	Others	4.73	80	2.733
Educational level	Middle school or below	21.14	62.69	2.104
	High school or technical school	31.23	80.8	2.747
	Junior college or bachelor degree	44.48	90.07	3.298
	Master degree or above	3.15	80	3.2
Annual income	0–5000	50.47	78.13	2.837
	5000–10,000	36.59	84.48	2.853
	10,000–20,000	9.87	83.87	3.194
	>20,000	3.16	80	2.6

Degree of WTP: the value of this variable is: WTP = 0, then 0; WTP = RMB 1–100, then 1; WTP = RMB 100–500, then 2; WTP>RMB 500, then 3

significantly higher than that in any other range, but the degree of WTP in above RMB 20,000 is only 2.6, which might be related to the quantity of questionnaires: Because of less persons, 2 respondents refuse to pay, lowering the degree of WTP, but it is not indicated to conflict with the result given above.

In Part II survey, the percentage of different WTP is found to be varied, as indicated in Fig. 21.1.

As shown in Fig. 21.1, WTP doesn't satisfy the normal distribution, which coincides with the conclusions drawn by other researchers. The highest percentage of WTP is found in RMB 10–100, and the number of persons in RMB 0–10, 100–200 is approximately the same. The number of persons in >RMB 500 is the least.

In this survey, the percentages of people refuse to pay amounts to 18.93 %, the reasons for refusing to pay are shown in Fig. 21.2.

Figure 21.2 shows that, 34.05 % of people think that roof agriculture will affect waterproof load-bearing, and worry about the security issues. The other 22.7 % of

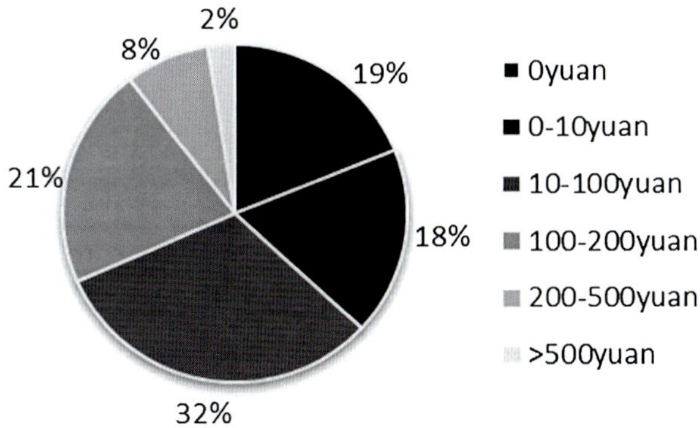

Fig. 21.1 Percentage of different WTP

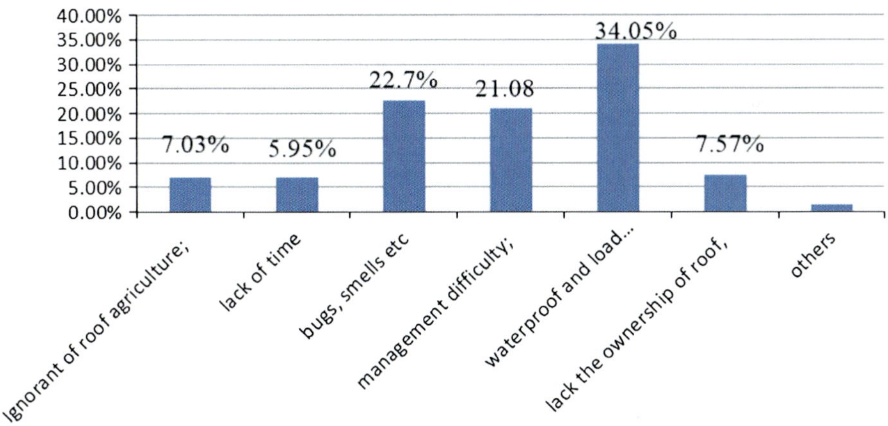

Fig. 21.2 Distribution of reasons refuse to pay

people think that the roof agriculture can cause environmental problems such as bugs, bad smell. 21.08 % of people think that it can't be managed very well, while the other 7.03 % of people don't understand roof agriculture, so they refused to pay. 5.95 % of people think that they don't have enough time, so they can't participate in roof agriculture and are not willing to pay. The remaining 7.57 % of people believe that the top-level households have the rights to dominate roof and other residents don't have the participation rights, so they aren't willing to pay.

Because this survey set many variables affecting the residents' willingness to pay, so further processing of various influencing factors is necessary. Roof agriculture brings about many benefits and the correlation between each problem is bigger, so this research used the principal component analysis method. The multiple variables will be linear combined again into four unrelated comprehensive index. Principal component analysis of factor's loading matrix is shown in Table 21.2.

Table 21.2 The factor loading matrix of principal component method

	Components			
	1	2	3	4
Safe fruits and vegetables (A1)	0.293	0.26	0.724	−0.188
Increase of income (A2)	0.175	−0.02	0.721	0.144
Labor exercise (A3)	0.334	0.67	−0.184	0.305
Agricultural leisure (A4)	0.345	0.619	−0.716	−0.139
Regulating temperature (A5)	0.514	−0.439	−0.19	0.344
Reduction of UHI effect (A6)	0.521	−0.273	0.084	−0.468
Air quality improvement (A7)	0.554	−0.062	−0.053	0.324
Noise reduction (A8)	0.559	−0.036	0.06	−0.405
Pollution reduction (A9)	0.619	−0.079	−0.002	−0.192
Live the top floor or not (A10)	0.364	0.097	0.156	0.744
Knowledge of roof ownership (A11)	0.277	−0.148	0	0.501

Table 21.3 The coefficient matrix of component score

	Components			
	1	2	3	4
Safe fruits and vegetables (A1)	0.122	0.213	0.532	−0.173
Increase of income (A2)	0.073	−0.015	0.519	0.133
Labor exercise (A3)	0.139	0.493	−0.151	0.281
Agricultural Leisure (A4)	0.144	0.258	−0.586	−0.127
Regulating temperature (A5)	0.214	−0.322	−0.156	0.317
Reduction of UHI effect (A6)	0.217	−0.201	0.069	−0.43
Air quality improvement (A7)	0.231	−0.046	−0.044	0.298
Noise reduction (A8)	0.233	−0.027	0.049	−0.373
Pollution reduction (A9)	0.258	−0.109	0	0.461
Live the top floor or not (A10)	0.152	0.072	0.143	0.609
Knowledge of roof ownership (A11)	0.203	−0.109	0	0.461

The principal component score coefficient is shown in Table 21.3.

It can be seen from Table 21.2 that the four principal components can be summed up as: the environmental benefits of roof agriculture, social benefit, economic benefit, roof consciousness. Roof agriculture's environmental effects include some aspects such as regulating temperature, improve the urban heat island effect, improve air quality, reduce noise, reducing environmental pollution. The social benefits include that people have the leisure of agriculture and participate in productive labor exercise. Economic benefits include that city residents can eat the safe vegetables and fruits, increase the economic income, etc. The top roof consciousness include whether live top floor and awareness of roof ownership.

21.3.2 Model of Influencing Factors of WTP

In order to get the relationship between various influencing factors of WTP, according to the above theoretical analysis and speculation, and describe analysis results by combing the data, we established a willingness orderly classification Logit model of roof agriculture about Wenjiang District. Compare Multiple classification Logit model with binary Logit model, the value of the dependent variable is more than one level. Due to varies choices of WTP, it is integrated into four options as following:

A: Y = 0, refuse to pay	Willingness to pay is 0
B: Y = 1, low pay	Willingness to pay is 0–100
C: Y = 2, middle pay	Willingness to pay is 100–500
D: Y = 3, high pay	Willingness to pay is above 500

Establish an orderly more classification Logit back to the model of the measurement equations,

$$\begin{aligned}\text{Logit}\frac{P0}{1-P0} &= -\alpha 0+\beta 1A1+\beta 2A2+\beta 3A3+\beta 4A4\\ &\quad +\beta 5A5+\beta 6A6+\beta 7A7+\beta 8A8+\varepsilon\end{aligned}$$

$$\begin{aligned}\text{Logit}\frac{P0+P1}{1-P0-P1} &= -\alpha 1+\beta 1A1+\beta 2A2+\beta 3A3+\beta+4A4\\ &\quad +\beta 5A5+\beta 6A6+\beta 7A7+\beta 8A8+\varepsilon\end{aligned}$$

$$\begin{aligned}\text{Logit}\frac{P0+P1+P2}{1-P0-P1-P2} &= -\alpha 2+\beta 1A1+\beta 2A2+\beta 3A3+\beta+4A4\\ &\quad +\beta 5A5+\beta 6A6+\beta 7A7+\beta 8A8+\varepsilon\end{aligned}$$

Pi = P(Y = i), the range of values is (0, 1); A1–A8 indicates sex age, educate level, household income, rooftop agriculture environmental benefits, roofs agricultural social benefits, economic efficiency, roof consciousness, αi is constant, i = 0, 1, 2; βi for the equation regression coefficient, i = 0,1 … 8, ε for a random item.

21.3.3 Empirical Results and Analysis

The regression results are shown in Table 21.4.

It can be seen from Table 21.4, under 0.05 level, the yearly income regression coefficient is biggest, 0.448, the education level coefficient is 0.321, It is indicated that the yearly income and the education level have the biggest effect on the

Table 21.4 The Logit model regression results for the impact of each variable on WTP

	Coefficient estimation	Standard error	wald	df	Significance	95 % confidence interval Lower limit	Upper limit
Y = 0	−0.757	0.493	2.36	1	0.124	−1.723	0.209
Y = 1	1.974	0.499	15.69	1	0	0.997	2.952
Y = 2	3.68	0.518	50.5	1	0	2.665	4.695
Sex	0.144	0.179	0.645	1	0.422	−0.208	0.496
Age	−0.195	0.066	8.731	1	0.003	−0.324	−0.07
Education level	0.321	0.1	10.25	1	0.001	0.125	0.518
Annual income	0.448	0.062	51.69	1	0	0.326	0.571
Environmental benefit	0.003	0.003	0.892	1	0.345	−0.003	0.008
Social benefit	0.166	0.181	0.833	1	0.361	−0.19	0.521
Roof consciousness	0.434	0.105	17.2	1	0	0.229	0.639
Economic benefit	−0.016	0.102	0.024	1	0.877	−0.215	0.183

willingness to pay. In addition, the regression coefficient of roof consciousness is 0.434; this suggests the roof ownership concerns and awareness also greatly affect the WTP.

21.4 Conclusions and Suggestions

Though analysis and discussion, the conclusion is: people of different income, different education level and different understandings of roof agriculture show obvious different WTP toward the non-use value of roof agriculture.

This provides important scientific evidence and theory basis for the government in the development of roof agriculture. The suggestions are: first, speed up economic development, increase people's income and the ability to pay. Second, enlarge propaganda dynamics, enhances the familiar and the cognition degree inhabitant to the roof agriculture, let more people understand the roof agriculture and its advantages. Third, the government should start with investment, gradually plan more business models, deal with management issues, and work hard to prepare for extension of the better agricultural development.

References

1. Dai X-g, Huang H, Yang S-w (2011) Investigation and research on roof agriculture in Wuhan: a Shilipu Community of Hanyang District as Example. J Jianghan Univ (Nat Sci Ed) 39 (2):101–103
2. Erner PK (2006) Regulation is needed for China. Roof greening. Landscape Arch 4:39–45
3. Huang J-q (1994) Design and construction roof garden. China Forestry Press, Beijing
4. Huang X-z (2010) Analysis on the development of roof agriculture. Mod Agric Sci Technol (9):316–317
5. Jiao Y, Ao C (2008) Research development of contingent value method in ecological environments evaluation. J Northeast Agric Univ 39(5):131–136
6. Jiao Y (2008) Evaluating the non-use value of sanjiang wetland based on contigent valuation method. Northeast Agricultural University
7. Li B-j (2009) Success and thinking of planting rice on the roof. Bull Agric Sci Technol 11:123–124
8. Li B-j, Sun C-b, Qi X-j (2012) Preliminary study on roof agriculture. Acta Agriculturae Zhejiangensis 24(3):449–454
9. Li J-h, Guo Y-h (2002) Principal component evaluation——A multivariate evaluate method expanded from principal component analysis. J Ind Eng Eng Manage (l):39–43
10. Oberndorfer E, Lundholm J, Betal B (2007) Green roofs as urban cosystems: ecological structures, functions and services. Bio Sci 57:823–833
11. Portney PR (1994) The contingent valuation debate: why economists should care. J Econ Perspect 8(4):3–15
12. Zhang Z-Q, Xu Z-M, Cheng G-D (2001) Valuation of ecosystem services and natural capital. Acta Ecol Sin 21(11):1918–1926

Chapter 22
Study on the Difference of Urban Land Expansion Intensity and Its Influencing Factors—Taking Shaanxi Province as an Example

Donglang Yang, Tianzhe Li and Zhiyong Hou

Abstract This study is based on the data of 10 prefecture level city of 2002–2013 in Shaanxi Province, using the ratio of urban built-up area and urban area as the index of urban land expansion. The research uses decomposition method of Theil index and Econometrics Fixed Effects Model to discuss land expansion intensity difference of prefecture level city and their influencing factors in Shaanxi Province. The results show that the overall level of urban land expansion intensity in Shaanxi Province is low and increases slowly. The differences of land expansion intensity in different cities are significant, and the regional differences in the area of Shan Bei, Guan Zhong, Shan Nan is increasing year by year. The urban land expansion of Guan Zhong is far ahead of Shan Bei and Shan Nan. Regional population density, the development of second industry and the third industry are the significant influencing factors of urban land expansion intensity in Shaanxi Province, but the influence factors are different in the different regions. At the regional level, government expenditure and urban population density are significant influencing factors in the area of Shan Bei and Shan Nan, and the development of second industry and urban population density are significant influencing factors in Guan Zhong.

Keywords Urban land · Expansion intensity difference · Theil index · Influencing factors

D. Yang
Department of Real Estate Institute, Xi'an Jiao Tong University, Xi'an, China
e-mail: yangdl@mail.xjtu.edu.cn

T. Li (✉) · Z. Hou
School of Public Policy and Management, Xi'an Jiao Tong University, Xi'an, China
e-mail: 502923609@qq.com

Y. Wu et al. (eds.), *Proceedings of the 20th International Symposium on Advancement of Construction Management and Real Estate*,
DOI 10.1007/978-981-10-0855-9_22

22.1 Introduction

For a long time, the land expansion is regarded as an important part of the urbanization level, and it is the main index to reflect the urbanization of land in some regions. In foreign countries, the research develops and form the school of the history form, regional economics, social behavior, political economics and the other research methods and theories after 1920s [1, 2]. Since 1990s, with the phenomenon of urban sprawl, smart growth, compact development and other theoretical ideas have emerged [3, 4]. The domestic research started later, the scholars carried on the preliminary exploration to the domestic urban land expansion since 1990s. Since twenty-first Century, the methods of morphology, system, ecology, and GIS, RS have been applied to the analysis of urban land expansion. The main contents of the research include the expansion measurement [5], the dynamic system [6], the efficiency measure [7], the land development model [8], the urban expansion [9, 10] and the economic and ecological effects of urban land expansion [11], but it lacked the research on the differences of land expansion in different cities [12]. Therefore, it is important to explore the differences of urban land expansion intensity and its influencing factors to promote regional coordinated development.

22.2 Theil Index of Difference and Temporal Decomposition

22.2.1 *Land Expansion Intensity Index Selection*

Urban land expansion can be reflected by the Built-up area. The built-up area is the area that city have actually constructed and have basic facilities. This study selects the urban land expansion intensity (the ratio of urban built-up area and urban area) index to analyze the differences of the land area in order to overcome the differences in the base area of land and ensure that the different inspection units is comparable [13].

22.2.2 *Theil Index and Its Decomposition*

The Theil index is mainly used to measure the regional differences. Before the application of Theil index decomposition, this study divided Yanan and Yulin into a group of Shan Bei, Xian, Baoji, Xianyang, Weinan, Tongchuan into the Guanzhong group, Hanzhong, Ankang and Shangluo into the Shan Nan group according to the scale of urban population, economic development level and geographical position of the cities in Shaanxi Province.

The formula for calculating the Theil index of urban land expansion intensity is:

$$Theil = \sum_{i=1}^{k}\sum_{j=1}^{n}\frac{x_{ij}}{x}\ln\frac{x_{ij}/x}{1/N} \tag{22.1}$$

In the Formula, Theil represents Theil index; i represents three regions of Shan Bei, Guan Zhong, Shan Nan; j represents the observation of the city respectively; x_{ij} represents the urban land expansion intensity of i-region's j-observation; x represents the sum of urban land expansion intensity of overall cities; N is the number of the cities.

The Theil index of urban land expansion intensity can be decomposed into the differences between regions and regional. The decomposition formula is:

$$Theil = T_w + T_p = \sum_{i=1}^{k}\left(\frac{x_i}{x}\right)\cdot T_i + T_p \tag{22.2}$$

Among them:

$$T_i = \sum_{j=1}^{n}\frac{x_{ij}}{x_i}\ln\frac{x_{ij}/x_i}{1/N_i} \tag{22.3}$$

$$T_p = \sum_{i=1}^{k}\frac{x_i}{x}\ln\frac{x_i/x}{N_i/N} \tag{22.4}$$

In the formula, T_w represents the region difference; T_p represents the regional differences; T_i represents the differences within the i region, and the x_i represents the sum of urban land expansion intensity within the i region, and N_i is the number of cities within the i region.

22.2.3 Calculation and Analysis of Land Expansion Intensity

Based on the ratio of urban built-up area and urban area, we can get the calculation results of urban land expansion intensity and they have been listed in Table 22.1. It can be found that the urban land expansion intensity in Shaanxi province from 2002 to 2013 has been a steady rise in the annual 0.2–0.4 percentage points. The index rises slowly and the overall level of urban land expansion is not high. As of 2013, the province's land expansion intensity is still no more than 5 %. The land expansion intensity in three region of Shaanxi province is not balanced, and the land expansion intensity of highest region is 20.83 times of the lowest.

Table 22.1 Urban land expansion intensity in Shaanxi Province during 2002–2013

Year	Shan Bei			Guan Zhong						Shan Nan				Average value (year)
	Yulin	Yanan	Average value	Xian	Baoji	Xianyang	Weinan	Tongchuan	Average value	Hanzhong	Ankang	Shangluo	Average value	
2002	0.004	0.006	0.005	0.052	0.012	0.089	0.025	0.008	0.037	0.050	0.007	0.009	0.022	0.026
2003	0.004	0.004	0.004	0.057	0.013	0.089	0.029	0.015	0.041	0.050	0.007	0.009	0.022	0.028
2004	0.004	0.004	0.004	0.062	0.013	0.095	0.029	0.015	0.043	0.050	0.007	0.009	0.022	0.029
2005	0.004	0.007	0.006	0.064	0.017	0.102	0.026	0.015	0.045	0.056	0.007	0.009	0.024	0.031
2006	0.005	0.007	0.006	0.073	0.018	0.11	0.027	0.015	0.049	0.056	0.007	0.009	0.024	0.033
2007	0.005	0.007	0.006	0.075	0.018	0.118	0.029	0.015	0.051	0.049	0.007	0.009	0.022	0.033
2008	0.006	0.007	0.007	0.076	0.033	0.120	0.031	0.015	0.055	0.049	0.008	0.009	0.022	0.035
2009	0.007	0.007	0.007	0.079	0.033	0.146	0.032	0.017	0.061	0.056	0.008	0.009	0.024	0.039
2010	0.007	0.010	0.009	0.091	0.033	0.154	0.036	0.016	0.066	0.059	0.008	0.009	0.025	0.042
2011	0.007	0.010	0.009	0.096	0.027	0.157	0.042	0.017	0.068	0.061	0.009	0.005	0.025	0.043
2012	0.007	0.010	0.009	0.105	0.027	0.157	0.052	0.018	0.072	0.061	0.01	0.008	0.026	0.046
2013	0.010	0.010	0.010	0.118	0.027	0.167	0.058	0.018	0.078	0.061	0.011	0.009	0.027	0.049
Average value (area)	0.006	0.007	0.007	0.079	0.023	0.125	0.035	0.015	0.055	0.055	0.008	0.009	0.024	0.036

According to the data in Table 22.1, it can be found that the land expansion intensity in Shan Bei region has a weak trend of rising in the whole, Guan Zhong region has been rising steadily since 2002, with the annual growth rate of 5.5 %, and Shan Nan region show the trend of waving firstly and then start to rise. The land expansion intensity of Shan Nan presents the fluctuation state of "rising-falling-rising" between 2002 and 2009. In every year during 2002–2009, the average ranking rate of urban land expansion intensity is: Guan Zhong, Shan Bei, Shan Nan.

22.2.4 Analysis of Theil Index of Land Expansion Intensity Difference

Based on the Theil index formula, calculating Theil index of urban land expansion intensity of 10 prefecture level cities in Shaanxi province and decomposition results by different region, and making the Table 22.2 and Fig. 22.1. During 2002–2013, the overall land expansion intensity Theil index of ten cities shows a fluctuated upward trend after the turning point 2008, which shows that urban land expansion intensity difference has a trend to expand.

Among them, urban land expansion intensity Theil index has been fluctuating during 2002–2007 years, the Theil index in 2007 is higher than 2002 1.79 percentage points. In 2008, urban land expansion intensity Theil index plunged to 0.4377, basically unchanged from the 2003 level; From 2008 to 2011, the land expansion intensity Theil index continued to rise, but the data of 2012 and 2013 slipped.

In the comparison of three regions, urban land expansion intensity differences can be ranked as follows: Shan Nan, Guan Zhong, Shan Bei. In addition to 2011, the group differences of Theil index in Shan Nan region suddenly raised to 0.4956, the data of other years basically hovering at 0.4000; In addition to the group differences in 2002, the data 0.3185 is more than 0.3000, the rest year still float below the level of 0.3000; The intensity difference of land expansion in Shan Bei is nearly close to 0.000, because the two cities Yulin, Yanan's land expansion is not obvious and almost same. Overall, the three major regional internal differences are basically in a stable state, which indicates that the degree of different cities' urban land expansion within the Shan Bei, Guan Zhong, Shan Nan groups are close to the same, and they are all in a state of synchronous growth.

In the view of the regional differences, the Theil index rise from 0.1416 to 0.2031 during 2002 to 2013. Its average annual growth is 0.56 percentage points. There is a great gap between three major regional urban land expansion process, the rate of expansion of land in Guanzhong area is significantly faster than Shan Bei and Shan Nan area. The expansion process has the accelerated trend, the gap may will be widening in the future.

Table 22.2 Theil index and regional decomposition of urban land expansion in Shaanxi Province

Year	Overall Theil index	Differences of Theil index decomposition				Differences of Theil index decomposition					
		Regional differences (Tp)		Regional internal differences (Tw)		Shan Bei (NORTH)		Guan Zhong (CENTRE)		Shan Nan (SOUTH)	
		Differences value	Contribution rate (%)	Differences value	Contribution rate (%)	Differences value	Contribution rate (%)	Differences value	Contribution rate (%)	Differences value	Contribution rate (%)
2002	0.4639	0.1416	30.52	0.3223	69.48	0.0201	0.17	0.3185	48.75	0.3786	20.56
2003	0.4390	0.1694	38.59	0.2696	61.41	0.0000	0.00	0.2448	40.87	0.3786	20.55
2004	0.4598	0.1778	38.67	0.2820	61.33	0.0000	0.00	0.2627	42.45	0.3786	18.87
2005	0.4500	0.1564	34.76	0.2936	65.24	0.0377	0.30	0.2666	43.23	0.4166	21.71
2006	0.4648	0.1641	35.31	0.3007	64.69	0.0140	0.11	0.2805	44.85	0.4166	19.74
2007	0.4818	0.1843	38.25	0.2975	61.75	0.0140	0.10	0.2919	46.53	0.3718	15.11
2008	0.4377	0.1914	43.73	0.2464	56.29	0.0030	0.02	0.2330	41.35	0.3500	14.91
2009	0.4793	0.1950	40.68	0.2843	59.32	0.0000	0.00	0.2710	44.06	0.3949	15.26
2010	0.4835	0.1905	39.40	0.2930	60.60	0.0157	0.13	0.2798	45.15	0.4124	15.33
2011	0.5088	0.1975	38.82	0.3113	61.18	0.0157	0.12	0.2853	44.11	0.4956	16.95
2012	0.4813	0.2023	42.03	0.2790	57.97	0.0157	0.12	0.2637	43.22	0.4054	14.63
2013	0.4806	0.2031	42.26	0.2775	57.74	0.0000	0.00	0.2726	45.00	0.3698	12.74

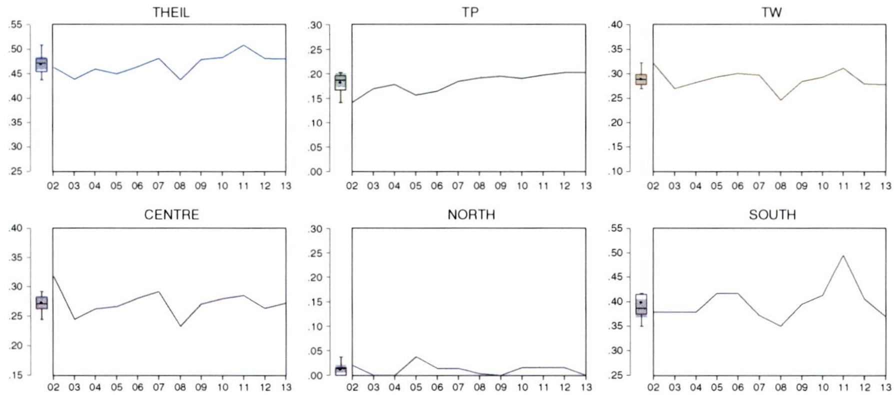

Fig. 22.1 The trend chart of Theil index and regional decomposition of urban land expansion in Shaanxi Province during 2002–2013. *Note* In the figure, THEIL refers to overall Theil index; TP and TW respectively refer to the regional differences and regional internal differences; CENTRE, NORTH, SOUTH refer to the regional internal differences in Guanzhong, Shan Ber and Shan Nan

Through using the Theil index decomposition method to analyze the contribution of regional differences and region differences to the overall differences, we can obtain the contribution share of the three regions and regional differences. From Table 22.2, we can find that the change of differences in urban land expansion intensity has always been influenced by regional differences and regional differences since 2002, and the contribution of regional differences to the overall difference is always less than the contribution rate of region differences. The differences of urban land expansion intensity is mainly due to the differences of the regional differences. The differences between the three regions are the decisive influence on the overall difference.

22.3 Analysis of Difference Based on Fixed Effect Model FGLS

22.3.1 Model Setting, Variable Selection and Data Sources

On the basis of the existing research and the consideration of the parallel data, this study chooses government financial scale, regional industrial structure, urban population density and regional opening degree as the influencing factors of the urban land expansion intensity (LUR) to construct the semi log panel data model. These factors are the key factors that affect the urban land expansion, they can not be replaced by each other, and there is no obvious antagonism. The choice of a semi logarithmic model can avoid the dispute over whether constant term contributes to dependent variable through the appropriate transformation of the constant term.

Therefore, setting the measurement model in the form of:

$$\ln LUR_{it} = \beta_0 + \beta_1 GOV_{it} + \beta_2 SIN_{it} + \beta_3 TIN_{it} + \beta_4 \ln POP_{it} + \beta_5 OPE_{it} + \varepsilon_{it} \quad (22.5)$$

In the influencing factors, the government's fiscal expenditure scale (GOV) is measured by the ratio of local fiscal expenditure and GDP in the municipal district. The regional industrial structure is measured by two index: the ratio of the second industry and GDP(SIN), the third industry and GDP(TIN). The urban population density (POP) is measured by the ratio of the population and the area of the municipal district. The opening degree (OPE) is measured by the ratio of the regional annual import and export (convert by the annual average exchange rate) and GDP [14].

22.3.2 Analysis of Model Estimation Results on Provincial Level

The study uses Stata 11.0 software and the panel data of prefecture level city in Shaanxi Province from "statistical yearbook of Shaanxi Province" and "China City Statistical Yearbook" as the sample to estimate. In the using of Hausman test, the results show that "Prob > chi2" is less than 5 % significance level, therefore the study choose to establish the fixed effect model. In order to avoid the distortion caused by the traditional model OLS and the effect of the individual variance, the study set the FGLS model to estimate [15]. In the independent variable, the population density of the city(POP) is the absolute index, so we take the logarithm of it to reduce the variance. In the formula, t and i represent different cities and different years respectively, i = 1, 2, 3, …, 10, t = 2002, 2003, 2004, …, 2013, ε is a random perturbation.

In the view of statistics, only the significant variables will have a decisive impact on the land expansion intensity, so the study requires that each of the explanatory variables of the land expansion intensity must be significant(at 5 % significant level). In the results of FGLS regression, the variables of TIN, ln POP and SIN are all passing the test, thus after excluding the non-significant variables the established regression model is:

$$\ln LUR_{it} = \beta_0 + \beta_1 SIN_{it} + \beta_2 TIN_{it} + \beta_3 \ln POP_{it} + \varepsilon_{it} \quad (22.6)$$

The result of STATA's calculation is:

$$\ln LUR_{it} = -3.607 + 3.505SIN_{it} + 2.740TIN_{it} + 0.968 \ln POP_{it} + \varepsilon_{it} \quad (22.7)$$

Second Industry's Influence on Land Expansion Intensity

The Table 22.3 shows that the second industrial variable coefficient is positive, and it is significant at 1 % level, which indicates that the development of the second industry has a significant positive effect on urban land expansion intensity. This is because the second industry is the core industry to support the economic development of Shaanxi Province, and its growth is directly related to the province's manufacturing industry and the development of the construction industry. With the development of second industry, the enterprise will have more funds for the industrial and mining storage, residential construction land, leading to the rapid expansion of urban construction land.

Table 22.3 Estimation results in Shaanxi Province

Estimation model		OLS (5)	FGLS (5)	OLS (4)	FGLS (3)
C	Coef.	−2.874	−4.012	−1.655	−3.607
	Prob.	(0.000***)	(0.000***)	(0.053*)	(0.000***)
	Std. Err	0.731	0.547	0.847	0.502
GOV	Coef.	−0.057	0.113	–	–
	Prob.	(0.347)	(0.124)	–	–
	Std. Err	0.060	0.074	–	–
SIN	Coef.	2.859	3.972	2.669	3.505
	Prob.	(0.000***)	(0.000***)	(0.000***)	(0.000***)
	Std. Err	0.726	0.528	0.714	0.463
TIN	Coef.	2.797	3.296	2.810	2.740
	Prob.	(0.000***)	(0.000***)	(0.000)	(0.000***)
	Std. Err	0.782	0.678	0.752	0.606
lnPOP	Coef.	1.074	0.994	1.429	0.968
	Prob.	(0.000***)	(0.000***)	(0.000***)	(0.000***)
	Std. Err	0.071	0.030	0.142	0.026
OPE	Coef.	−6.988	−2.636	−7.323	–
	Prob.	(0.015**)	(0.322)	(0.011**)	–
	Std. Err	2.861	2.662	2.843	–
R^2		0.919	0.932	0.921	0.948
F		146.018	162.764	187.780	203.379

Note Coef. refers to the regression coefficient; Prob. refers to *P* value; Err Std. refers to the standard error; The values in parentheses refers to the number of regression variables; ***, **, * respectively represent that the estimated coefficients are significant at 1, 5 and 10 % level

Three Industry's Influence on Land Expansion Intensity

Table 22.3's estimation shows that the third industry variable coefficient are positive. From the model, it can be found that when the third industry's total value increased by 1 %, the urban land expansion intensity will increase by 2.704 % without considering other factors. This means that the development of the three industry makes the demand of urban land in the circulation and service department increase leading to the growth of urban land expansion intensity. At the same time with the continuous improvement of three production, some major cities in Shaanxi Province moved the industries with low equipment and low added value from the city center to the surrounding suburbs. The transfer of population and a series of ancillary service facilities leads to the built-up area expands to the suburbs, and the land expansion intensity increases.

Urban Population Density's Influence on Land Expansion Intensity

Table 22.3 shows that the estimated coefficient of urban population density is positive, which shows that the urban population density has a significant positive effect on the land expansion intensity. This means that with the improvement of urbanization, more and more people focus on the city, leading to the expansion of its construction land area and inevitably make the urban land expansion intensity continue to grow. With the development of agricultural science and technology, the improvement of agricultural production efficiency will liberate more surplus labor force, which makes people transfer to the nearby prefecture level cities and the trend of population urbanization continues to accelerate.

22.3.3 Analysis of Model Estimation Results on Regional Level

In order to further investigate the difference of regional land expansion influencing factors, the study use regression analysis based on the panel data of Shan Bei, Guanzhong and Shan Nan as a sample to make the Table 22.4. The results show that the influencing factors of land expansion intensity are different in different regions.

In Shan Bei, the second and three industrial development and openness of the city's influence on land expansion intensity is not significant. The government expenditure, urban population density and land expansion intensity are significantly positively correlated, which shows the two factors are important factors that influence the land expansion intensity in Shan Bei. In Guanzhong region, the role of government financial expenditure is not significant. In the model, the second industry development and urban population density is significantly positively

Table 22.4 Estimation results in the Northern, Central and Southern Shaanxi Province

Estimation model		Shan Bei (NORTH)		Guan Zhong (CENTRE)		Shan Nan (SOUTH)	
		FGLS (5)	FGLS (2)	FGLS (5)	FGLS (2)	FGLS (5)	FGLS (2)
C	Coef.	−5.113	−2.300	−2.633	−1.429	0.098	−0.035
	Prob.	0.042**	0.002***	0.003***	0.000***	0.926	0.834
	Std. Err	2.515	0.742	–	–	1.061	–
GOV	Coef.	−0.281	−0.376	0.094	–	0.205	0.211
	Prob.	0.003***	0.000***	0.606	–	0.004***	0.017**
	Std. Err	0.093	0.102	0.182	–	0.072	–
SIN	Coef.	1.980	–	3.157	2.092	1.551	–
	Prob.	0.343	–	0.000***	0.000***	0.141	–
	Std. Err	2.086	–	0.871	–	1.054	–
TIN	Coef.	2.987	–	1.226	–	−1.574	–
	Prob.	0.139	–	0.249	–	0.130	–
	Std. Err	2.018	–	1.063	–	1.040	–
lnPOP	Coef.	0.403	0.527	1.064	1.097	1.299	1.304
	Prob.	0.016**	0.001***	0.000***	0.000***	0.000***	0.000***
	Std. Err	0.168	0.161	0.043	–	0.064	–
OPE	Coef.	−13.631	–	2.479	–	−3.973	–
	Prob.	0.768	–	0.516	–	0.185	–
	Std. Err	46.173	–	3.815	–	2.998	
R^2		0.910	0.960	0.980	0.980	0.940	0.950
F		130.980	182.670	157.830	223.070	151.790	204.680

Note The symbolic interpretation is similar to Table 22.3

correlated. It shows that the impact of the government investment in Guanzhong region is weak and second industry and urban population change is the main reason for the impact of the regional land expansion intensity. In the Shan Nan region, the research conclusion is similar to Shan Bei. Due to the slow development of the two and three industry, the industrial structure has a weak impact on the land expansion intensity, while the government expenditure and the urban population density's impacts are strong.

22.4 Conclusions and Discussion

The study takes the prefecture level cities in Shaanxi Province as the research sample and 2002 to 2013 as the research period, analysing the temporal and spatial differences of urban land expansion intensity and its influencing factors. The conclusion can be listed as follows:

1. The overall level of land expansion intensity in the three regions is relatively low. Compared to the eastern coastal cities, the land expansion intensity of major cities in Shaanxi province is in the middle and lower reaches. In the spatial distribution, the clustering feature of urban land expansion intensity is only reflected in the Guanzhong region.
2. The differences of land expansion intensity in the three major regions is very obvious. In the view of comparisons between the three regions, the contribution rate of the Theil index among the groups of three major regions is still rising from 2002 to 2013, and the gap between the three regions is widening. Guanzhong area's land expansion is far ahead of Shan Bei and Shan Nan area. At the same time, the differences of urban land expansion intensity between the cities of different administrative levels and sizes in Shan Bei, Guanzhong and Shan Nan areas are not same.
3. The regional population density, the development of the second industry and the third industry are important influencing factors which affect the urban land expansion intensity in Shaanxi Province, but its influence factors will also be different because of the regional differences. At the regional level, the government financial expenditure and the urban population density are the important influencing factors that affect the Shan Bei area. The development of second industry and the urban population density are important influencing factors affecting the Guanzhong region. Shan Nan is similar to Shan Bei.

References

1. Shenghe L, Jianmin Z (2001) Western theory and method in urban land-use research. Foreign Urban Plan 1:17–19
2. Shenghe L (2002) Spatial patterns and dynamic mechanisms of urban land use growth. Prog Geogr 21(1):43–50
3. Arendt R (1999) Growing greener: putting conversation into local plans and ordinance. Island Press, Washington, DC, pp 15–45
4. Jenks M, Burgess R (2003) Compact cities: sustainable urban forms for developing countries. J Hous Built Environ 18(4):287–391
5. Qinping L, Yongchun Y, Dongxia Fu (2014) The study of urban space expansion in China based on DMSP-OLS lighting data in 1992–2010. Geogr Sci 34(2):29–136
6. Yali Z, Youzhao L (2013) Research on urban land intensity difference and influence factors. Resour Sci 35(2):380–387
7. Jiating W, Qingfeng C (2013) Measurement of spatial expansion efficiency of 35 large and medium cities in China (10):8–14
8. Zhigang W, Suhong Z (2006) Comparative analysis of urban land development model in the area of rapid urbanization in China. China Land Sci 20(1):33
9. Tao L, Xinhu L, Guoqing Z (2010) Analysis of Urban spatial expansion characteristics and its influencing factors in Xiamen Island. J Geog Sci 65(6):725–726
10. Yue Z, Zhengshi Z (2002) And Chen Hong. The study of multiplier effect and growth model of land development of Beijing metropolitan area. Geogr Res 21(1):90–96

11. Zhang H (2013) Analysis of land use/land cover change, population shift, and their effects on spatiomporal patterns of urban heat islands in metropolitan Shanghai, China.Land Use Policy 35(11):33–39
12. Xiaoling Z, Wu Y, Shen L (2001) An evaluation framework for the sustainability of urban land use: a study of capital cities and municipalities in China. Habitat International, 2001, 35 (1):141—149
13. Yu W, Zhang Peng. The Study of Theil index and Sharplry values of the differences of the intensity of urban land expansion and its causes in Shandong Province. Geogr Geogr Inf Sci 31, 54 (2):59–2015
14. Yan Mu, Zhonghao Q (2012) Analysis of the differences of Chinese farmland non agriculture marketization level area. China Land Sci 26(5):37–43
15. Yali Z, Youzhao L (2013) Research on the differences of urban land development intensity and its influencing factors—Based on the 222 cities' urban panel data. Resour Sci 35, 380 (5): 387–2013

Chapter 23
Spatial Form Analysis of Existing Industrial Buildings Renovation Based on Space Syntax

Lin Wang, Xiao Li, Zewen Zhang, Shirong Li and Bo Xu

Abstract The spatial transformation is an important part of existing industrial buildings renovation, but the traditional ways of spatial transformation focused on theoretical analysis, and lacked quantitative research. The Space Syntax theory is introduced into the existing industrial building renovation in this paper. A space model of Chongqing Industrial Museum & Creative Park was built with convex space using the connectivity, depth, integration to measure the permeability, convenience and distribution degree of the space, to analyze its spatial form quantitatively and explore the effect of spatial arrangement on function. Taking the functional orientation into consideration, analyze whether the transformation plan is reasonable, put forward reasonable transformation strategies, and provide reference for future industrial building renovation.

Keywords Space syntax · Industrial buildings · Renovation

23.1 Introduction

In recent years, with the successful implementation of the Beijing 798 Art Center, Zhongshan Qijiang Park and other industrial building renovation projects, there is growing concern about industrial building renovation, the traditional practice, demolition and reconstruction has gradually been abandoned. Since the 1980s, China began to practice and explore in this area, the economic value of many existing industrial buildings has been significantly improved after the renovation. Thirty years of practice has proved that the transformation and reuse of existing industrial buildings has much ecological, social, cultural and economic value [1], which corresponds with the current principle of sustainable development in China.

L. Wang (✉) · X. Li · Z. Zhang · S. Li · B. Xu
School of Construction Management and Real Estate, Chongqing University, Chongqing, China
e-mail: 1548401183@qq.com

Y. Wu et al. (eds.), *Proceedings of the 20th International Symposium on Advancement of Construction Management and Real Estate*,
DOI 10.1007/978-981-10-0855-9_23

However, because of the short of practice and experience in this field, there are still many difficulties in practice.

One of the important parts of an existing industrial building's renovation is space transformation, and domestic scholars have done a lot of research in this area. Jianguo and Junjiang [2] proposed that the main way of space renovation is to reconstruct and expend the original architectural space (including horizontal expansion and vertical expansion).

In consideration of spatial characteristics of the old industrial buildings, Jiani [3] summed five ways of space reconstruction: planting, spacing, connecting, crossing, combining, from the perspective of space transformation and design. From the perspective of structure aesthetics, Shi et al. [4] proposed several major strategies of remodeling the interior space of the old industrial buildings: dividing space, merging space, nesting space, extending space, and connecting space. Current research in the space transformation of existing industrial building are concentrated on the transformation techniques and manners, but lack of quantitative evaluation for its rationality.

23.2 Space Syntax Overview

23.2.1 Concepts and Applications of Space Syntax

The theory of Space Syntax dated back to the 1970s, and was originated by Bill Hillier and Julienne Hanson from the Bartlett, University of London. Space Syntax is best described as a research program that investigates the relationship between human societies and space from the perspective of a general theory of the structure of inhabited space in all its diverse forms: buildings, settlements, cities, or even landscapes [5].

Different from many other space theories, Space Syntax regards the space as a separate element of research, and as the starting point for further analysis of the relationship between it and the architecture, society, cognition and some other areas [6]. Space Syntax theory does not emphasize the concept of distance, shape in Euclidean geometry, and focuses on the expression of structural system composed of the connection between nodes [7]. It describes a relationship represented by topological relationship, and focuses on the accessibility and relevance of space.

The appearance of Space Syntax theory not only has laid a theoretical foundation for spatial analysis, but also has brought the application of spatial shape analysis in urban planning and architectural design to the new level. Nowadays, the Space Syntax theory has been widely applied to urban planning, architectural design, and many other aspects related to spatial analysis, such as urban form analysis, analysis of urban functions, protection and rehabilitation of urban space, and the analysis of residential space, the museum (Museum of Art) space, work space, and some other architectural spaces [8].

23.2.2 The Basic Law and Morphology Variables of Spatial Syntax

The premise of Space Syntax analysis is that space systems are converted to expression system which can be recognized by the software. Space Syntax theory often uses the following basic laws of space abstract: axis or line, convex space, visible horizon and pixels. When the buildings or building groups within the city are intensive, the axis law can be adopted; When the free space of city is nonlinear layout, use the convex space law, or sight segmentation law [9]. In analyzing issues related to urban planning, the axis law is usually used. But when it comes to issues related to architecture design, the convex space law, or sight segmentation law is usually used. The line law and the pixels law are not used very often. Space Syntax simplifies the space with two-dimensional spatial geometric elements point, line, surface, etc., the linkages between different spaces are abstracted to the connection diagram. And than, according to the basic principles of graph theory, conduct topological analysis for the spatial accessibility of feature points and the axis. Finally, export a series of spatial form variables: connectivity, control value, depth, integration, intelligibility [10]. These morphological variables quantitatively describe the characteristics of the entire space structure and permeability, accessibility, gathering and other characteristics of a single space or space node (Fig. 23.1).

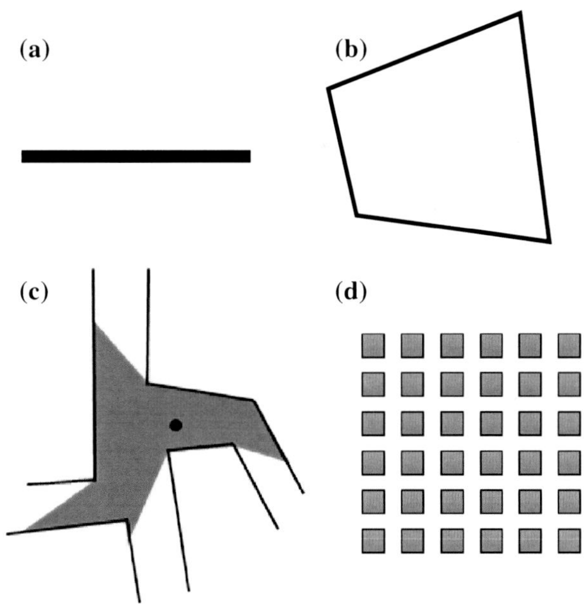

Fig. 23.1 The schematic diagram of space abstract law. **a** Axis or line. **b** Convex space. **c** Visible horizon. **d** Pixels

23.3 Research Laws Overview

23.3.1 Abstract Mode Selection

Industrial buildings often have large interior space, and the space has good plasticity. The common approach of the transformation is to re-partition the internal space, and than make functional partitioning for small divided space. As for single industrial building with relatively small spatial scales, we also need to redefine the original function in order to match the overall function transformation. In order to meet the needs of industrial production and reduction of design, construction cost, the geometry shape of industrial buildings are usually regular. Therefore, whether space repartition for a large plant or functional change in the original space on a small scale, the involving space form are often regular. Such space morphological characteristics provide a great convenience for the application of convex space law. Because most of the space itself is convex space which meets the requirements for analysis, so this can reduce the workload of turning space into two-dimensional graphics in a large scale.

Taking into account the characteristics of the existing industrial building in the space transformation, we use convex space law to describe the spatial form of Chongqing Iron & Steel. Inside a convex polygon area, the connection between any two points can not intersect with the boundary, which means that any two people in the polygon area can be in the eye.

23.3.2 The Selection of Spatial Form Variables

In this paper, three main spatial form variables (connectivity, depth, integration) are chosen to analyze the spatial form of Chongqing Iron & Steel rehabilitation programs.

Connectivity: The number of the sum of other nodes connected with a node in the system [11]. In real space systems, the higher the connectivity of a space is, the better the spatial permeability is.

Depth: Depth represents the shortest topology distances between two nodes in the system, The total depth indicates the minimum topology number required for a node to reach all other nodes in the system, usually used to express the degree of convenience of the node [11].

Integration: The distribution degree of a node in the system, which usually used to express the distribution degree of human and traffic, and it can be divided into global integration and local integration during the analysis [11]. Global integration represents the tightness of the linkage between the node and all other nodes of the entire system, which is usually used to describe the distribution extent of the traffic; While local integration represents the tightness of the linkage between the node and other nodes in several steps, which describes the distribution extent of the flow.

Local integration is usually a value of three topology unit called "radius-3 integration". Integration has removed the interference of the number of system nodes, so that it can be used to compare the distribution extent of space systems with different sizes.

23.4 Introduction of Chongqing Iron & Steel Transformation Plan

The predecessor of Chongqing Iron & Steel was Hanyang iron works founded by Zhang Zhidong, Huguang Governor of Qing Dynasty on September, 1890. It was moved to Dadukou of Chongqing in March 1938.In June, 1995, it was restructured as Chongqing Iron & Steel Group Co. Ltd, which is now the largest state-owned enterprises of Chongqing. Chongqing Iron & Steel has made great contribution to the economic development of Dadukou. On the other hand, it has brought serious environmental pollution. In order to achieve the adjustment of industrial structure and sustainable urban development, the factory has been moved to Changshou. Original area of Chongqing Iron & Steel will be transformed into Chongqing Industrial Museum and Creative Industry Park.

Figure 23.2 is a schematic diagram of the transformation plan. The overall planning of the area includes the introduction of two important axis that follow the perspective axis of preeminent industrial relics such as the gas tank and the chimneys. The use of two mutually perpendicular and distinctive axes can create the atmosphere of culture and commerce. Major industrial relics the entire reconstruction project reserved include large-span steel plant, fifty thousand cubic tank, plant during the time of the Republic of China, main electrical room and chimney. Affiliated industrial

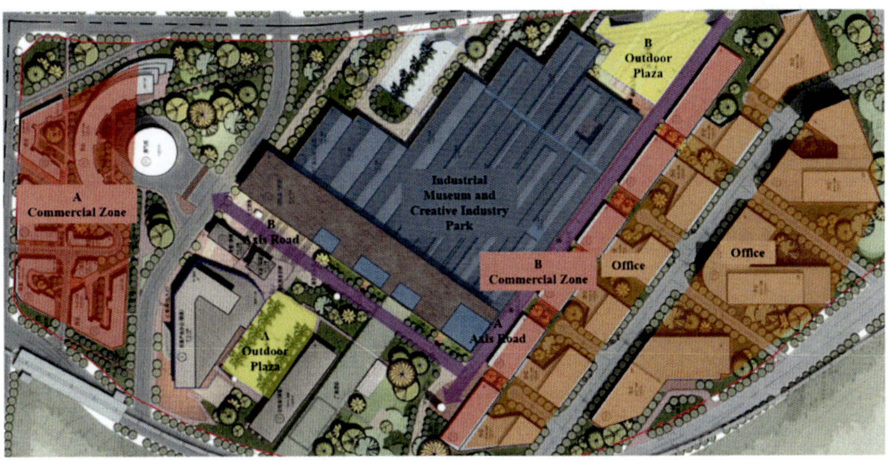

Fig. 23.2 The schematic diagram of transformation plan

cultural landscape buildings include workers canteen, bathroom, waiting stations, etc. The project has multiple open space, including rail square, chairman trails, heritage park, the new stadium and plant exhibition district, old factory courtyard garden and so on, which can meet the need of various types of performance, industrial products release and sales of consumer goods and works of art.

Among them, Chongqing Industrial Museum and Creative Industry Park is consist of the old factory buildings of the plant and the newly built "Rust Stripes" next to chairman trails. The first three floors of the newly built "Rust Stripes" still retains the original structure of the plant. The "Rust Stripes" and the longest factory space develop a mutually engaging spatial relationship. The old factory buildings and adjacent part, which is at the heart of the entire reconstruction project, is the main project of the entire renovation project.

23.5 Space Syntax Analysis of Chongqing Iron & Steel

This paper makes use of space syntax analyzing the status of Industrial Museum and Creative Industry Park in the entire Chongqing Iron & Steel area and the rationality of the arrangement of the whole space. In the study of the entire spatial configuration, the focus is on the spatial relationship with the surrounding architecture. Therefore, some irregular space is described as convex space for processing. The entire space analysis is conducted with the software Depthmap, which uses color to indicate the size of the value. High-value segment is reddish color, and low-value segment is bluish color.

Adjacent to the above two axes, the old factories, which are the core part of the whole area, are the key point of the entire transformation plan. This part has to play an important role of attracting customers. In other words, the transformation result of Industrial Museum and Creative Industry Park determines the economic benefits of the whole project.

23.5.1 Local Integration Analysis

Local integration studies the system how close a convex space and all other convex spaces below three steps are, three steps here refers to the topological distance. The results showed that the level of integration of the trunk road in office area is the highest, namely the red part of Fig. 23.3, which reaches 3.75. The road connects the two sides of the office area, and has relatively close ties with the surrounding space. The second are the two axes ways near the museum and industrial park, its integrations are respectively 2.75 and 2.16. The road on the left of the museum in Fig. 23.3 has established ties between the museum, creative industry park area and the street commerce buildings, so that the two can work together to gather popularity

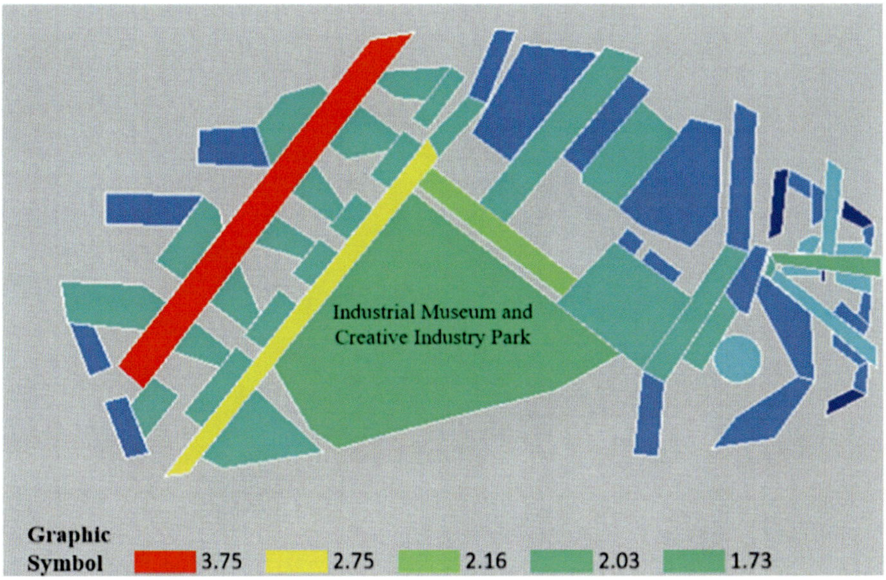

Fig. 23.3 Convex space local integration analysis

and share consumers flow. In contrast, the local integration of the museum area is only 1.73, indicating the lack of good contact with the surrounding space.

23.5.2 Connectivity Analysis

By analyzing the connectivity of museum and creative industry park area, it can be found that the connection of this part is only 3, as showed in Fig. 23.4, which shows that museum lack connects with surrounding area, and the permeability is not satisfying. This has limited the interaction of the region with the surrounding space to a certain extent. While the connectivity values of neighboring street, road are 5 and 10, which once again shows that the space permeability of the axis road is better, and the axis road has played an important role for the entire space system connection.

23.5.3 Depth Analysis

Depth is essential for measuring spatial accessibility. Lower depth values mean lower topological distance required to reach this space, in other words, it's relatively more convenient to go to this space. Global depth indicates the total

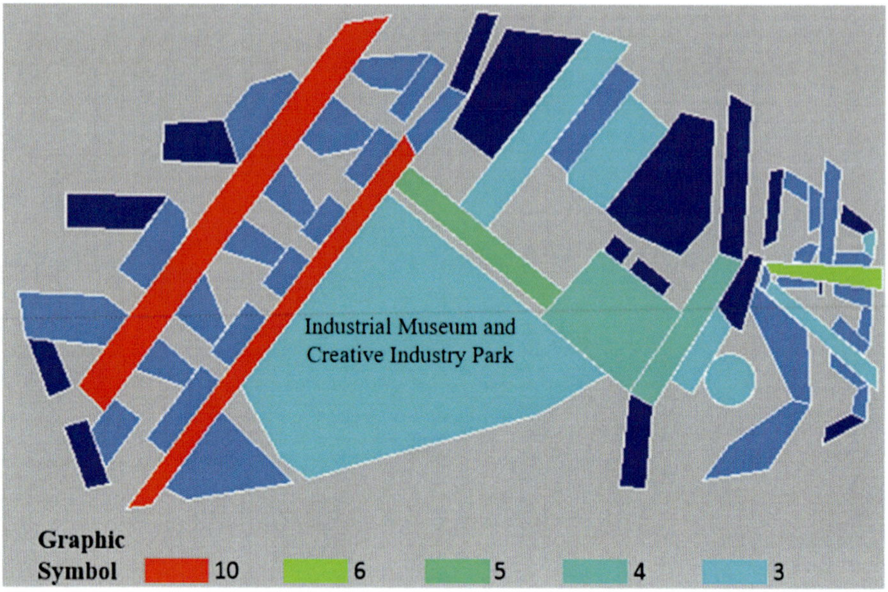

Fig. 23.4 Convex space connectivity analysis

topological distance required to reach a space from each space in the system. It describes how convenient it is to reach some space from various locations in the space system.

Figure 23.5 shows that the global depth value of museum and creative industry park is 144, and the global depth values of the two main roads adjacent to it are respectively 121 and 129. The global depth value of the portion is at the middle level in the entire space system. The smallest global depth value comes from the blue area of Fig. 23.5, ranging from 40 to 60. This shows that the number of spaces required to reach this part is not very big, and this part has better accessibility. At the same time, this part was planned for the commerce, so that it can reduce the difficulty for consumers to reach and explore this part, and this will do certain benefit for commerce.

23.5.4 Space Attribute Analysis of Outdoor Plazas

The connectivity value of two outdoor plazas are respectively 2 and 3, the local integration are 1.37 and 1.49, and the global depth are 161 and 166. As a public space, their various spatial variables are inferior to other public space, and the accessibility, permeability and concentricity are relatively poor, so that it is difficult to achieve its public space functions.

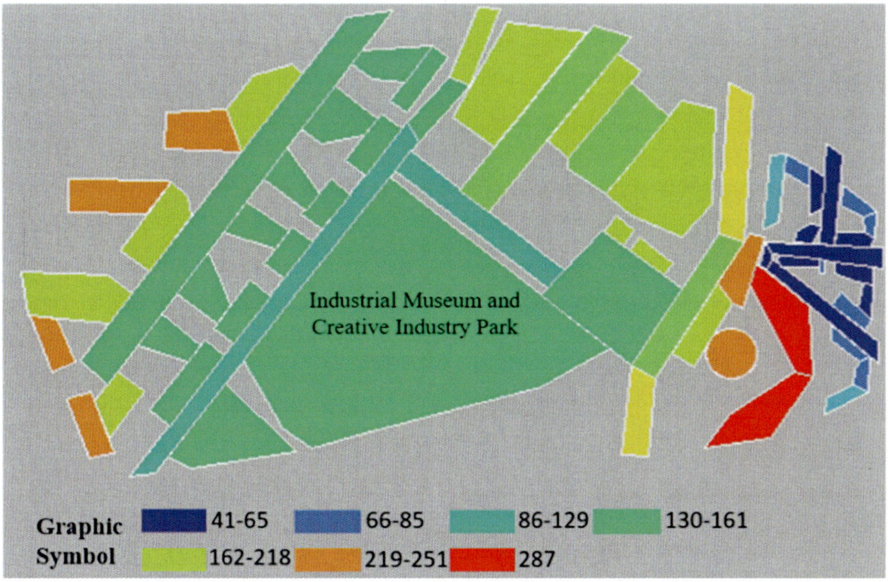

Fig. 23.5 Convex space depth analysis

Outdoor plazas is at the relatively marginal position, and doesn't have very close contact with the entire space system. Because of the poor concentricity of the two outdoor plazas, it is difficult for them to absorb the crowd who want to take a rest. The lack of connection with surrounding spaces also makes it impossible to improve the internal linkages across space systems.

23.5.5 Summary

In summary, the entire office region has good spatial agglomeration and permeability after the transformation of Chongqing Iron & Steel, while the A side of the commercial portion has good spatial accessibility. But the separation of the two commercial zone B and A, leading to the difficult for the commerce to play combined effect. What's more, the A commerce part is in a remote corner of the entire region, and its accessibility and concentricity are not that good, so maybe the operation result is not optimistic. The two main axes adjacent to Industrial Museum and Creative Industry Park, namely the two main roads, have the most balanced performance in the spatial concentricity, permeability and accessibility, and are in the core area of Chongqing Iron & Steel. The two outdoor plazas can't be effective public spaces due to restrictions of location and road links. Although the museum and creative industry park is at the central position of the entire region, but due to the impact of space layout, its accessibility, permeability and concentricity are far from reaching the level of the core position, and the spatial role in Chongqing Iron & Steel has yet to be improved.

23.6 Transformation Strategy Discussion of Chongqing Iron & Steel

The overall transformation program of Chongqing Iron & Steel takes full advantage of existing resources, and conducts protective transformation under the premise of preservation of the original architectural features. The two cleverly planned, mutually perpendicular distinctive axes make it possible to achieve the harmonious co-existence of cultural industries and the commerce atmosphere. The results produced by the space syntax analysis of spatial configurations of Chongqing Iron & Steel transformation plan shows that the spatial distribution of the rehabilitation programs has yet to be improved.

In order to make the overall spatial layout more scientific and reasonable, so that spatial attributes and features definitions can be more matching, the following adjustments can be made to its spatial layout.The commerce zone A and office space exchange their function, so that the office area is relatively independent and will not be influenced by the surrounding noisy commercial environment. At the same time, the two commerce zone can be polymerized to produce a cluster effect. Besides, such changes make the commerce section occupy the position with highest agglomeration and permeability to maximize the benefits and value of this part. After the adjustment, commercial streets and museums area are just across the street, both of them can share consumers flow and build up popularity together. Industrial Museum and Creative Industry Park should strengthen the links between themselves and the two axes roads, increase street entrances, make use of the space geographical advantages of the roads on both sides to enhance their concentricity, permeability and accessibility. Outdoor plazas have to strengthen its transport links with the surrounding spaces, and guide into the dispersing crowd to assume the function of rest.

Aknowledgements We gratefully acknowledge the financial support from Chongqing Graduate Student Research Innovation Project (CYS14036).

References

1. Yanbing Y (2005) Analysis of values of renewal and reuse of old industrial buildings. Ind Constr 06:32–34
2. Jianguo W, Junqiang R (2001) Conservation and adaptive reuse of historical industrial buildings and sites. Time Archit 04:10–13

3. Jiani L (2009) Reconstruction mode exploration of old industrial architecture space. Shanxi Archit 07:31–32
4. Kehui S, Bingjie X, Xuesong H (2013) Studying in reusing of old industrial buildings' space in perspective of structure aesthetics. World Archit 04:112–115
5. Bafna S (2003) Space syntax: a brief introduction to its logic and analytical techniques. Environ Behav 35(1):17–29
6. Duan W (2005) Introduction of related theories to space syntax. World Archit 11:10–15
7. Zhang Yu, Jiaguo Wang (2004) Further discussion about "space syntax". Architecture 03:33–44
8. Jingwen Wang (2010) The research status quo and development direction of space syntax. Huazhong Archit 06:5–7
9. Zhang H, Wang X, Yu R (2006) Space syntax and its research progress. Geospatial Inf 04: 37–39
10. Jiang B, Huang B, Lu F (2002) Spatial analysis and geoscience visualization under GIS environment. Higher Education Press, Beijing
11. Hillier B (2008) Space is the machine: a configurational theory of architecture

Chapter 24
The Social Benefit Evaluation of Shantytown Reconstruction Projects in Xi'an

Donglang Yang, Zhongfei Cui and Yiqi Li

Abstract In order to effectively evaluate the social benefit of shantytown reconstruction projects in Xi'an, through the expert interview, the empirical investigation, and combined with the literature research, the social benefit evaluation system of shantytown reconstruction projects in Xi'an is constructed. Select three typical shantytown reconstruction projects in Xi'an, collect data from the field survey, evaluate the social benefit, and finally get the results of the social benefit evaluation of shantytown reconstruction projects in Xi'an. The results of the study show that the main achievements of Xi'an shantytowns reconstruction projects are in the following aspects: improving the living conditions of residents, promoting the development of education, optimizing the living environment of residents, solving the problem of gas supply and heating problems, completing the disaster prevention facilities, improving the living standards of residents and so on. But there are also the following problems: the large gap between the reform projects; pay attention to the housing placement, and neglect of the residents living; pay attention to the hardware construction and despise software construction; pay attention to the hardware construction, and despise software construction; placement funds are not in place. To solve these problems, the author puts forward policy suggestions in the paper.

Keywords Shantytowns · Social benefit · Analytic hierarchy process · Fuzzy comprehensive evaluation

D. Yang (✉)
Department of Real Estate Institute of Xi'an Jiaotong University,
Xi'an Jiaotong University, Xi'an, China
e-mail: 570362978@qq.com

Z. Cui · Y. Li
School of Xi'an Jiaotong University, Xi'an, China

Y. Wu et al. (eds.), *Proceedings of the 20th International Symposium on Advancement of Construction Management and Real Estate*,
DOI 10.1007/978-981-10-0855-9_24

24.1 Introduction

Shantytowns refers to the places in the city where gathered houses with simple structure, poor performance of disaster, overcrowding, poor function and poor living conditions [1]. With the development of economy, shantytowns increasingly become an obstacle for urban development. Over the past 10 years, shantytown reconstruction projects have made great achievements, reconstruction area of nearly eight million square meters, involving more than 30 million inhabitants, which played a significant role in the development of residents and city construction. But the changes of the people's living quality and satisfaction have not been paid much attention. Shantytowns reconstruction projects can truly bring how much benefit especially social benefit, these are difficult to measure. We study social benefit of shantytown reconstruction projects in Xi'an.

The main content of this research is divided into three parts. The first part is to build the social evaluation index system of shantytowns reconstruction in Xi'an. The main purpose of this part is to build the hierarchy and index evaluation system, and determine the index weight. The second part is to evaluate the social benefit of shantytown reconstruction projects in Xi'an, on the basis of the first part of the results, by means of a questionnaire to collect information, score the social benefit evaluation of shantytown reconstruction in Xi'an which included residents development benefits, regional development benefits and social harmonious development benefit. The third part is the discussion of the results of the evaluation and the countermeasures.

24.2 The Social Benefit Evaluation System

24.2.1 Index System Construction

The first step is to establish the social benefit evaluation index system of the shantytowns reconstruction. In this paper, we establish the system by reading literature, visiting the experts, and combined with the reality of shantytowns reconstruction projects in Xi'an. Ultimately determined these three dimensions, microscopic—the development of residents, meso—regional development, macro—social harmonious development. These indicators are further divided into six aspects including material life, spiritual life, infrastructure, economic development, social security and social order. On this basis, re-refining the 25 indicators, constituting the index system.

The second step is to measure the weight of each index. Using the 25 indicators being made in the first step to get the questionnaire and completed by many experts and relevant staff from the real estate, architecture, social security and other fields. Input the data and using the method of analytical hierarchy process to analysis the data, finally determine the weight of all aspects.

24.2.2 *The Method of Model Construction*

The method of model construction is mainly the analytic hierarchy process [2, 3]. Analytic Hierarchy Process is a multi-objective decision analysis method that changes qualitative analysis into quantitative analysis. The main idea is to decompose the problem into several levels of indicator elements, and then compare the weights of the same levels, establish judgment matrix, calculation the maximum eigenvalue and corresponding feature vector in the judgment matrix, finally draw the weights of each index and each index layer for analysis and judgment.

24.2.3 *Index System and Index Weight*

Using the method of analytic hierarchy process, combined with the weight of experts, finally get the weight of the indicators as shown in Table 24.1.

24.3 Evaluate the Social Benefit of Shantytown Reconstruction Projects in Xi'an

24.3.1 *Data Collection and Analysis*

First, identify data sources and collect data. Selected Taiheju, Dongyaofang, Renjia zhuang these three typical shantytowns project as a survey point for data collection. According to the social benefit evaluation index system, designing the questionnaire, payment and collection. A total of 106 questionnaires were distributed, 98 valid questionnaires, the effective rate was 92.45 %.

Second, using fuzzy comprehensive evaluation method to analyze the collected data. In the process of analyzing data, we use excel software and spss software for data processing, and finally determine the score.

24.3.2 *Analytical Method*

The method of social benefit evaluation is fuzzy comprehensive evaluation method.

The main steps are as follows: [4]

1. Determining factor set F and assessment (language) set E. Factor set F is a collection of evaluation projects or targets, usually have $U = \{U_i\}$, $i = 1, 2, \ldots, n$. In this paper, the factors corresponding to the index level indicators.

Table 24.1 The social benefit evaluation index system of the shantytowns reconstruction

Target layer (A)	Effect layer (B)	Secondary effect layer (C)	Index layer (D)	Index weight
The social benefit evaluation of the shantytowns reconstruction (A) 1.0000	Residents' development (B1) 0.5969	Changes in material life (C1) 0.4175	Changes in family income (D1)	0.2721
			Changes in the level of household consumption (D2)	0.0935
			Changes in living conditions (D3)	0.0520
		Spiritual life changes (C2) 0.1793	Education conditions change (D4)	0.0821
			Changes in leisure activities (D5)	0.0347
			Changes in Social circle (D6)	0.0094
			Life satisfaction after transformation (D7)	0.0152
			Cultural activities and participation (D8)	0.0271
			Political life and participation (D9)	0.0109
	Regional development (B2) 0.2687	Environmental change (C3) 0. 1420	Changes in pollution control (D10)	0.0730
			Changes in green environment (D11)	0.0421
			Changes in health conditions (D12)	0.0269
		Infrastructure construction (C4) 0.0986	Changes in the construction of water, power, gas and heating supply (D13)	0.0337
			Changes of traffic convenience (D14)	0.0175
			Changes in medical conditions (D15)	0.0196
			Disaster prevention facilities (D16)	0.0158
			Changes in pension facilities and service (D17)	0.0120
		Economic development (C5) 0.0282	Employment changes (D18)	0.0201
			Regional economic change (D19)	0.0081
	Social harmonious development (B3) 0.1344	Social security (C6) 0.0536	Social security coverage (D20)	0.0172
			The comprehensive of social security projects (D21)	0.0198
			The convenience of social security services and facilities (D22)	0.0166
		Changes in social order (C7) 0.0809	Changes in social security (D23)	0.0359
			Changes in household registration management (D24)	0.0210
			Changes in the degree of harmony (D25)	0.0250

Table 24.2 Evaluation level score

Evaluation level	Excellent	Good	General	Poor	Very poor
Score	5	4	3	2	1

Evaluation set or comment set E is the set of evaluation level, generally have $V = \{V_i\}$, i = 1,2,…,n.
In this paper, the evaluation set:

$$V = (V_1, V_2, V_3, V_4, V_5)$$

V_1 = excellent; V_2 = Good; V_3 = General; V_4 = Poor; V_5 = Very poor.
The score of each grade is shown in Table 24.2.

2. Determine the membership matrix. Membership degree r_{ij} means the possibility that leach evaluation subject in U_i make a comment on the certain evaluation object in V_i.
   Membership matrix: $R = (R_1, R_2, \ldots, R_n)^T = (r_{ij})$
   In practice, in order to calculate simply, often use U_i on the frequency of V_i to replace the degree of membership.
3. Determine the weight vector W. According to the above results, the weight vector of each index is shown in Table 23.1.
4. According to some algorithm to calculate a comprehensive membership vector S and the composite score μ.

24.3.3 Data Analysis Results

The total score of the social benefit evaluation of the shantytowns reconstruction projects in Xi'an is shown in Table 24.3:

From the Table 24.3, the total score of the social benefit evaluation of the shantytowns reconstruction projects in Xi'an is 3.4706, between the general and good, bias in general. This indicates that Xi'an shantytowns reconstruction projects have made some social benefits, but not obvious.

Using the same method, estimates the benefit evaluation results of effects layer: The evaluation results of the residents development are 3.4709, between the general and good, bias in general, Xi'an shantytowns reconstruction projects have made some social benefits, but not obvious; The evaluation results of the regional development are 3.6444, between the general and good, bias in good, play a benign role; The evaluation results of the social harmonious development are 3.1216, between the general and good, bias in general, Xi'an shantytowns reconstruction projects have played a benign role, but not obvious. In the three aspects, the social coordinated development score is the lowest.

Table 24.3 Shantytown reconstruction projects' social benefit evaluation score table in Xi'an

Index	Weight	Very poor	Poor	General	Good	Excellent
		1	2	3	4	5
Changes in family income(D1)	0.2721	0.000	0.071	0.429	0.398	0.102
Changes in the level of household consumption (D2)	0.0935	0.000	0.000	0.418	0.561	0.020
Changes in living conditions (D3)	0.0520	0.020	0.000	0.143	0.520	0.316
Education conditions change (D4)	0.0821	0.112	0.143	0.480	0.235	0.031
Changes in leisure activities (D5)	0.0347	0.061	0.173	0.408	0.184	0.173
Changes in social circle (D6)	0.0094	0.000	0.245	0.378	0.276	0.102
Life satisfaction after transformation (D7)	0.0152	0.031	0.041	0.255	0.592	0.082
Cultural activities and participation (D8)	0.0271	0.031	0.031	0.490	0.337	0.112
Political life and participation (D9)	0.0109	0.224	0.061	0.643	0.071	0.000
Changes in pollution control (D10)	0.0730	0.020	0.020	0.449	0.429	0.082
Changes in green environment (D11)	0.0421	0.041	0.041	0.204	0.531	0.184
Changes in health conditions (D12)	0.0269	0.010	0.153	0.204	0.418	0.214
Changes in the construction of water, power, gas and heating supply (D13)	0.0337	0.000	0.000	0.082	0.653	0.265
Changes of traffic convenience (D14)	0.0175	0.000	0.061	0.122	0.469	0.347
Changes in medical conditions (D15)	0.0196	0.133	0.184	0.378	0.235	0.071
Disaster prevention facilities (D16)	0.0158	0.102	0.122	0.143	0.459	0.173
Changes in pension facilities and service (D17)	0.0120	0.082	0.204	0.551	0.163	0.000
Employment changes (D18)	0.0201	0.000	0.000	0.381	0.476	0.143
Regional economic change (D19)	0.0081	0.000	0.000	0.524	0.333	0.143
Social security coverage (D20)	0.0172	0.000	0.000	0.898	0.102	0.000
The comprehensive of social security projects (D21)	0.0198	0.000	0.000	0.898	0.102	0.000
The convenience of social security services and facilities (D22)	0.0166	0.102	0.204	0.388	0.214	0.092
Changes in social security (D23)	0.0359	0.133	0.194	0.265	0.255	0.153
Changes in household registration management (D24)	0.0210	0.061	0.214	0.429	0.235	0.061
Changes in the degree of harmony (D25)	0.0250	0.051	0.163	0.347	0.398	0.041
Comprehensive membership degree		0.0338	0.0768	0.3913	0.3875	0.1118
Comprehensive score		3.4706				

24.4 The Main Conclusions and Suggestions

24.4.1 The Main Conclusions

Xi'an shantytowns reconstruction projects' comprehensive evaluation of social benefits is in general, reconstruction projects have achieved some social benefits, but not obvious. In particular, the following conclusions can be summed up.

Evaluation on the residents development is general, the effect of the shantytowns reconstruction projects on residents development is not significant. Among the indicators, changes in household income and household consumption accounting for the highest weight, both the score are about 3.5, achieved some results, but not obvious. Changes in living conditions and life satisfaction improved significantly, but the proportion is low. Achieved grades in education conditions, leisure activities, social circles, cultural activities, political activities are all general. From this we can see that the reconstruction projects does not pay attention to the improvement of residents' mental life. This is an important reason for improving the development of the people ineffectively.

Evaluation on the regional development is good, the effect of the shantytowns reconstruction projects on regional development is significant. Specific to the various indicators, it is found that the effect of the improvement is mainly concentrated in pollution control, green, sanitation, construction of water, power, gas and heating supply, transportation, disaster prevention facilities, employment and economic and other aspects, but the improvements in medical conditions, pension facilities is minimal. As most of the indicators are good, the overall level of regional development is higher. From this result can also be found that the reconstruction projects made enormous improvements mainly in hardware facilities construction and economic development.

Evaluation on the social harmonious development is general, the effect of the shantytowns reconstruction projects on social harmonious development is not significant. Specific to each index, all the indicators included in this aspect (social security coverage, the comprehensive of social security projects, the convenience of social security services and facilities, social security changes, changes in household management, residents harmonious degree change, etc.) are general. Due to the social security work in Xi'an has been fully launched before the renovation, so it is understandable that the impact is not big. However, the residents made the lower evaluation in terms of security, household management, harmonious degree and other aspects, reflecting the shantytowns do not pay attention to software construction. Undoubtedly, from this perspective, shantytowns project is a kind of failure. Shantytowns project as a welfare program, not only should pay attention to the hardware facilities, but also should pay attention to software facilities. In this regard, shantytowns reconstruction projects in Xi'an need to improve.

24.4.2 Suggestions

For the basic characteristics of the shantytowns reconstruction projects in Xi'an, recommendations for improvement are as follows:

1. In the perspective of the residents development, we should keep improving the residents' living standards in the material. At the same time, we should also improve the level of the residents' spiritual life. In education, we should strengthen the construction of education, improve the school's teaching staff, strengthen vocational training, improve the level of education; In leisure activities, we should guide and provide a platform for the leisure activities of the residents; In social circle, we should hold more community recreational activities to strengthen the ties between the residents; In cultural activities, we should build cultural square, publicity projection, the center for the elderly and other cultural facilities, provide a platform for the residents' leisure and recreation, hold cultural activities to enrich the amateur life; In political activities, we should build a platform to facilitate the people to participate and discuss politics, strengthen political propaganda, improve the residents' enthusiasm to participate in political life, cultivate people's awareness of political participation and improve the political quality of the people.
2. Regional development perspective, we should inheritance and keep improving the advanced model that focus on the environment, design and economic development during the process of reconstruction. The main areas of improvement are as follows: ① medical conditions, we should improve the community health services, facilitate the residents to see a doctor; ② pension, we should improve the community pension services and support the elderly residents.
3. Social harmonious development is the aspects needed to focus on. In social security system, the overall level of Xi'an is good, but in shantytowns reconstruction projects, we should pay attention to social security services and facilities, for example, universal the pharmacy in which medicare card is available, build the banks where the residents can conveniently receive social security payments. In the social order, the government should enhance the degree of attention, increase the intensity of resource investment. In social security, we should reasonably configure the police resources, reasonable layout the police, ensure that there are no dead ends. At the same time we should carry out legal education, make legal propaganda to improve people's awareness of the law. Household registration management should update timely to convenient the people handling the relevant business, relocation and domicile' immigration and emigration must be synchronized to avoid unnecessary trouble. Also we should establish the appropriate management services within the community, such as residents committee in order to facilitate the residents. In residents harmonious degree. On the one hand we should combine with recreational activities, promoting their exchange, on the other hand, neighborhood and other

services organizations' management should pay attention to the contradiction between the residents and actively adjust.

4. Macro, we should pay attention to the planning and rational allocation of resources. Most of the problems in the shantytowns reconstruction projects are caused by the bug and vulnerability in planning. To improve the level of the shantytowns reconstruction, from the root, is to improve the level of planning and the efficiency of resource allocation. Specifically, asking for the professional design team to assess the feasibility and benefits of the transformation projects, planning the content strictly and strictly in accordance with the plan to transform. In the process of allocation of resources, we should focus on the rational use of land, health care, police, capital and other resources, improve the resource utilization, reduce expenditure. Meanwhile, we should thinking from the overall consideration, in accordance with the characteristics of Xi'an to integrate allocate resources.

References

1. Naisheng L (2000) Thinking of urban shantytowns prevention. Urban Dev Res 01:32–34
2. Jinyu G, Zhongbin Z, Qingyun S (2008) Research and application of analytic hierarchy process. Chin J Saf Sci 05:148–153
3. Xue D, Jiaming L, Haojian C, Junyang Chen, Junfeng Z (2012) Analytic hierarchy process analysis and its application research. Pract Cogn Math 07:93–100
4. Shiyong L (2004) Engineering fuzzy mathematics and its application. Harbin Institute of Technology press, Harbin, pp 5–7

Chapter 25
Collective Land and Industrial Development in Shenzhen: Current Status and Future Prospects

Yani Lai and Yaning Li

Abstract Industrial development contributes significantly to China's economic growth in the past three decades. As one of the most fundamental institutional arrangements, the urban-rural dual land system plays an important role in China's industrialization process and hence economic growth. Property rights theory suggests that property rights have significant effects on economic performance. In the context of industrial development in China, it implies that different land property rights structure may have different economic effects on industrial development. This study, which is based on available literature, statistical data, a range of unpublished primary sources and the author's own experience, examined the economic implications of collective land property rights on industrial development. Emphasis is put on efficiency of industrial land use. Shenzhen is selected as the case study. An analysis of available data suggests that majority of whole industrial output of Shenzhen comes from industrial development on collective land in recent years. Incomplete collective land property rights structure causes the inefficiency of industrial land use in terms of lower Industrial output value and lower development ratio when compared to complete state-owned urban land system. Two main reasons why collective land leads to inefficiency of industrial land use are identified as follows: collective land has no access to legal land market and credit market; collective land is not included in the land management system, thus cannot be guided and controlled effectively by urban development strategies and policies. The study concludes with an assessment of the recent land policy reforms and a discussion of some likely future changes.

Keywords Collective land · Industrial development · Property rights · Urbanization · China

Y. Lai (✉) · Y. Li
School of Shenzhen University, Shenzhen, China
e-mail: lai.yani@connect.polyu.hk

Y. Wu et al. (eds.), *Proceedings of the 20th International Symposium on Advancement of Construction Management and Real Estate*,
DOI 10.1007/978-981-10-0855-9_25

25.1 Introduction

China experienced rapid and significant economic growth in the past three decades, to which industrial sector makes the most important contribution. According to the survey from state statistics bureau, the proportion of industrial added value account to gross domestic production (GDP) has been increased from 20.5 % in 1979 to 50.1 % in 2007. As one of the most important institutional arrangement, urban-rural dual land system plays a significant role in the process of industrialization and hence economic growth in China.

Two types of land including state and collective ownership coexist in the current land administration system. For state owned land, separating of land ownership and land use right allows transfer of land use rights, and arouses the emergence of urban land market. In another word, the rights of exclusivity, to capture benefits, to transfer all or some of the rights are guaranteed for the land users in the land market. What could be acquired through urban land market are full rights over land during the leasehold period. On the other hand, collective land cannot be sold, transferred or leased for non-agricultural construction including industrial development according to current state land policy.

Classic property rights theory suggests that property rights have significant effects on economic performance. In the more specific context of industrial development over past three decades in China, it implies that different land property rights structure may have different effects on industrial development. Based on the hypothesis, this study examined the economic implications of rural collective land ownership with concern to industrial land use efficiency. Shenzhen is selected for case study.

There has been rich literature contributing China's industrial development performance to different types of ownership. Perkins [1] has compared the productivity performance of China's state owned and non-state-owned enterprises by measuring total factor productivity. Jefferson et al. [2] has examined differences in marginal factor productivity across ownership types including state, foreign-linked, shareholding, and private enterprises in China. Wang and Jin [3] has explores the differences in pollution control performance of industries with different types of ownership in China—State owned (SOE), collectively or community owned (COE), privately owned (POE), foreign directly invested (FDI) companies as well as joint ventures.

However, there is little consideration of property rights structure on land for the different performance of industrial development in China in existing literature. As mentioned before, China's industrialization is based on different arrangements of land property rights. According to previous influential studies by well known scholars, land property rights matter for the performance of land-related investment (Galiani and Schargrodsky [4], Do and Iyer [5], Field and Torero [6], Brasselle et al. [7], Besley [8]). Thus, this study aims to fill this gap by investigating industrial development performance from property rights perspective. Emphasis will be put industrial land use efficiency on collective owned land. Shenzhen is selected for case study.

The Shenzhen special economic zone was established in 1980, to which the state-owned urban land use system was almost thoroughly applied in the four districts from 1992. In the same year, two adjacent new districts were incorporated to Shenzhen city, where most lands were collectively owned. Since then, urban and rural dual land system has come into being in this city. The typical urban-rural land system in Shenzhen city provides an appropriate case for this study.

25.2 Industrial Development: Background and Characteristics

The Shenzhen special economic zone (SEZ) was established in 1980, with the area including current four districts-Luohu, Futian, Nanshan, Yantian District-in which land is rarely developed at that time. In 1992, Shenzhen government carried out urbanization plan for SEZ. According to the plan, all land in SEZ is converted to state owned land. In the same year, two adjacent new districts including Bao'an District and Longgang District were incorporated to Shenzhen city, where most lands were collectively owned. Since then, urban and rural dual land system has come into being in this city. We can roughly say that industrial land in SEZ is state owned land while the industrial land outside of SEZ is collective owned land.

Industrial land increased rapidly during past three decades. Until now, all the land suitable in Shenzhen for industrial use has been developed already. From analysis above, we can see that policy strategy and district location contribute most to the spatial distribution of industrial land in Shenzhen. Specifically, a huge number of industrial districts in Guannei are gradually converted to redi in early times now industry areas now mainly locate in Guanwai districts which are close to traffic arteries. The characteristics of industrial land evolution from 1980 s to now could be summarized in Table 25.1.

Table 25.1 Evolution of industrial land use area in Shenzhen

Year	The whole city (km^2)	SEZ (km^2)	Area outside of SEZ		
			Total (km^2)	Bao'an (km^2)	Longgang (km^2)
1984	–	2.546	–	–	–
1987	–	6.607	–	–	–
1994	82.38	14.12	68.26	37.66	30.60
2001	131.19	15.07	116.12	68.39	47.73
2003	208.33	20.11	188.22	108.99	79.23

25.3 Evaluating Industrial Land Use Efficiency

Land use efficiency may have different meanings and measurements when serving different purposes in different types of studies. Generally speaking, land use efficiency refers to economic efficiency of land use, which can be generated by relationship between land use area and economic output. In a broad sense, land use efficiency also refers to social and environmental efficiency, for example, employed population created per unit of land, energy cost per unit of land etc.

25.3.1 Industrial Output Value

The urban land productivity in Shenzhen is higher than other mainland metropolitan cities in the past decades. However, when compared to overseas cities like Hong Kong, Singapore and Tokyo, the land use efficiency of Shenzhen is still backward. From available date on land use efficiency in 2004, we can see that land use efficiency of Shenzhen is much higher than other mainland metropolitan cities including Guangzhou, Beijing, Shanghai, Tianjin, Suzhou in the respect of GDP per unit of built-up area, yet much lower than overseas cities including Hong Kong, Singapore, Tokyo. Referring to industrial added value per unit of land, Shenzhen and Shanghai are almost equal to each other yet much lower than Hong Kong, Singapore, Tokyo (Table 25.2).

According to the result, industrial land use efficiency of state land with complete property rights is much higher than collective land with incomplete property rights. To be more specific, in respect of industrial output value per unit of land, Bao'an District and Longgang District are 13.8 and 11.7 Billion yuan/km^2 respectively in 2003, while special economic zone is 155.0 Billion yuan/km^2 in the same year,

Table 25.2 Comparison of construction/industrial land productivity in different cities in year 2004

City	Year	GDP per unit of built-up area (Billion yuan/Km^2)	Industrial added value per unit of land (Billion yuan/Km^2)
Shenzhen	2004	5.1	9.32
Suzhou	2004	1.43	–
Guangzhou	2004	2.16	–
Tianjin	2004	1.17	–
Beijing	2004	1.67	–
Shanghai	2004	3.51	9.5
Hong Kong	2003	53.91	30.51
Singapore	2003	24.41	74.48
Tokyo	2002	101.59	57.08

Table 25.3 Industrial land use efficiency in Shenzhen in year 2003

Industrial land use efficiency (year 2003)	The whole city	SEZ	Bao'an district	Longgang district
Industrial output value per unit of land (Billion yuan/km^2)	27.0	155.0	13.8	11.7
Industrial added value per unit of land (Billion yuan/km^2)	7.4	42.9	3.0	4.0

which is roughly ten times higher than the area outside of special economic zone. Referring to industrial added value per unit of land, Bao'an District and Longgang District are 3.0 and 4.0 Billion yuan/km^2 respectively in 2003, while special economic zone is 42.9 Billion yuan/km^2 in the same year (Table 25.3). This result suggests that state industrial land with complete property rights has far better economic performance than collective land with incomplete property rights in terms of industrial output value.

To summarize, state land with complete property rights has higher industrial productivity than collective land with incomplete property rights in terms of industrial output value. In the specific context of Shenzhen, it applies that land in the special economic zone has better performance on industrial output value than the land outside of special economic zone. We have found solid evidence to support this argument as mentioned above.

25.3.2 Development Ratio

Primary data used in this section comes from a survey conducted by the Trade and Industry Bureau of Shenzhen Municipality in 2005. This survey investigates the number of industrial parks, industrial land area and industrial building floor area of each district in Shenzhen city. Based on these data, we can calculate land development ratio in each district and make comparison analysis of these districts which stands for different property rights structure on land (Table 25.4).

Table 25.4 Industrial floor area ratio in Shenzhen in year 2005

District	Number of industrial parks	Area (km^2)	Floor area (Ten thousand m^2)	Floor area ratio
Luohu	29	0.61	109	1.79
Futian	9	2.42	377	1.56
Nanshan	50	1.5	333	2.2
Yantian	2	0.14	34	2.4
Bao'an	379	61.05	6039	0.99
Longgang	157	13.47	1352	1.0
Total	626	79.19	8244	1.04

The result shows that development ratio of land in special economic zone is much higher than land outside of special economic zone. Specifically, land development ratio in Luohu, Futian, Nanshan, Yantian District is 1.79, 1.56, 2.2, 2.4 respectively, while land development ratio in Bao'an and Longgang District is 0.99 and 1.0. Relevant detailed information is shown as following table. This result corresponds well to the argument in the earlier section.

25.4 Mechanism Analysis

25.4.1 Incomplete Property Rights Structure of Collective Land

As mentioned before, urban and rural dual land system is the institutional foundation for land conversion and industrial development in China. Under current dual land system, urban land is owned by the state, while rural land is owned by collectives. As far as urban state land is concerned, separating of land ownership and land use right from Year 1988 allows transfer of land use rights, and arouses the emergence of urban land market. Full rights could be acquired through urban land market. On the other hand, collective land is forbidden to be sold, transferred or leased for nonagricultural use. In another word, if collective land is converted to urban use, it has only exclusive rights and income rights. Collective land can't be transacted through market mechanism, which means it doesn't have transfer rights and collateral rights. Through black land market, which is illegal, collectives may have part of land transfer rights (Table 25.5).

In the meanwhile, industry is prosperously developed on collectively owned land. As mentioned before, collective land doesn't have transfer rights and collateral rights. Collective land users are not allowed to mortgage the land,the mode of "investment with low technology + labors with low capability + collectively owned land with low price" does not compose a necessary condition for higher and better use of land. Low efficiency of land use is observed on most of collective land of every village in Shenzhen. The chain of "low price land—attract investment—industry development and rent collect—new land development—new land rented" goes on, and pushes the villages to develop land at a low level to ensure the revue for all peasants.

Table 25.5 Property rights of state land and collective land in China

Land ownership	Exclusive rights	Income rights	Transfer rights	Collateral rights
State land	Yes	Yes	Yes	Yes
Collective land	Yes	Yes	No	No

25.4.2 Collective Land Is Out of Urban Management System

From a governance perspective, collective industrial land is totally out of urban management system. This situation greatly contributes to low efficiency of collective land use in Shenzhen. To understand the lack of governance on collective industrial land use better, we have divided this problem into several main aspects (Fig. 25.1).

Urban planning and control for collective land in Shenzhen does not function at all in the real practice. Construction on collective land is out the control of urban planning system. Taken Dongfang-Tantou area for example, this area is located in Bao'an District, with most of its land collectively owned. Many plans have been worked out for this area before Year 2005. Here we have chose 2005 plan as a standard of comparison to real situation. According to 2005 plan, Dongfang-Tantou area has an excellent transportation network and a well-designed road system. Certain area of land in the northern part of this district is planned as a park to satisfy the recreation and leisure needs of local people. However, the real situation of built environment in this area in Year 2008 is totally different with what have been planned in Year 2005. There was obvious lack of roads and transportation facilities in Dongfang-Tantou area in 2008. Existing road system was problematic because of too limited width, poor connection and road condition. Industrial buildings in this area were also in severe disorder and thus result in low development ratio as well. No green space was created in this district either. All these aspects demonstrate the failure of urban planning system on collective industrial land.

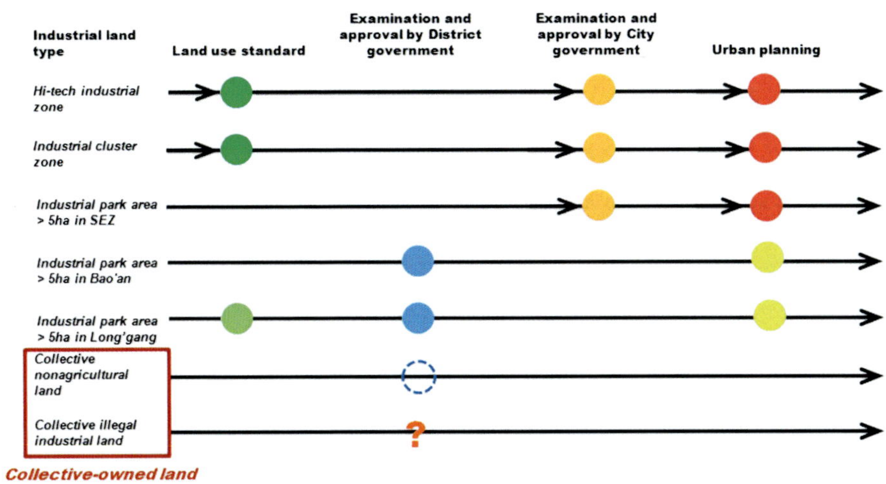

Fig. 25.1 Industrial land management system in Shenzhen

25.5 Land Policy Reforms: Assessment and Prospects

25.5.1 Assessment of Recent Land Policy Reform

Comprehensive urbanization and land conversion of Bao'an and Longgang district in 2004 as the most important institutional change of recent land reforms in Shenzhen city, has significant influence on many aspects in the city's economic transition. Four main policy documents were issued to implement the comprehensive urbanization plan including: (a) Land Management regulations for urbanization of Bao'an and Longgang; (b) Compensation standards of land conversion from collective to state land; (c) Implementation Procedures of land conversion from collective to state land; Regulations of land banking for urbanization of Bao'an and Longgang.

For villages, the property rights structure of their land is substantially changed after 2004. According to Land Management System of Shenzhen municipality, the land is categorized into four groups, Legal Land, Land without complete procedures, Nonagricultural Land and Illegal Land. Legal land has full property rights including exclusive rights, income rights, transfer rights and collateral rights. Land without complete procedures and Nonagricultural Land have exclusive rights and income rights. They can also have transfer rights and collateral rights after completing relevant procedures. Illegal land has only exclusive rights and income rights, but the security of these two kinds of rights is lower compared to pre 2004 because of the regulations of illegal land and construction carried out by city government. Table 25.6 shows land tenure change pre and post 2004 reform in Shenzhen.

The main significant consequences of comprehensive urbanization and land conversion in 2004 can be summarized as follows:

Table 25.6 Land tenure pre and post 2004 reform in Shenzhen (land tenure in this table is classified according to land management system in Shenzhen municipality)

Year	Real user	Ownership	Land tenure	Exclusive rights	Income rights	Transfer rights	Collateral rights
Pre 2004	VSC	Collective	Vague	Yes		No	
Post 2004	VSC	State	Legal land	Yes		Yes	
	VSC	State	Land without complete procedures	Yes		Yes, but need to complete relevant procedures	
	VSC	State	Nonagricultural land				
	VSC	State	Illegal land	Yes		No	

(a) The expropriated land is successfully incorporated to land market and urban land management system, and provides the necessary foundation for future urban development and economic growth.
(b) Rural collectives and individuals are officially entitled to urban economic development opportunity for the first time in China's urbanization process through delineation of nonagricultural land.
(c) Nonagricultural land needs to complete procedures to enter land market. Until today, still some of nonagricultural land isn't converted to urban land with complete property rights, thus is partly out of urban land market and land management system.
(d) Part of land remains illegal and belongs to collectives, which is totally out of land market land management system.

25.5.2 Conclusion: Future Prospects

Collective land in Shenzhen suffers inefficiency of industrial land use in terms of (i) lower Industrial output value and (ii) lower development ratio when compared to state-owned urban land system. The inefficiency is attributed to two main reasons: (i) collective land with incomplete property rights has no access to legal land market and credit market, and (ii) collective land is not included in the land management system, thus cannot be guided and controlled effectively by urban development strategies and policies.

To improve land use efficiency in Shenzhen, the government has carried out a series of gradual land reforms which have produced very important positive effects on the economic development. However, we still need to go through a long and arduous journey to proceed to a more economic effective society. There is still a lot of room to improve land use efficiency and thus more effective industrial development and economic growth. There are two possible directions for future institutional change:

(a) New efforts need to be made to improve the transferability and management of nonagricultural land in Guanwai area.
(b) Effective mechanism and specific policies need to be established to transfer the illegal land to legal land which could have full property rights and be governed by land management system.

References

1. CPF (1996) Productivity performance and priorities for the reform of china's state-owned enterprises. J Dev Stud 32(3):414-444
2. Jefferson GHRT (2000) Ownership, productivity change, and financial performance in chinese industry. J Comp Econ 28(4):786–813
3. Wang HJY (2007) Industrial ownership and environmental performance: evidence from china. Environ Res Econ 36(3):255–273
4. Galiani SASE (2010) Property rights for the poor: effects of land titling. J Public Econ 94 (9):700–729
5. Do QTAI (2008) Land titling and rural transition in Vietnam. Econ Dev Cult Change 56 (3):531–579
6. Field EATM (2006) Do property titles increase credit access among the urban poor? Evidence from a nationwide titling program. Department of Economics, Harvard University, Cambridge, MA
7. Brasselle ASGF (2002) Land tenure security and investment incentives: puzzling evidence from Burkina Faso. J Dev Econ 67(2):373–418
8. Besley T (1995) Property rights and investment incentives: theory and evidence from Ghana. J Polit Econ 103(5):903–937

Chapter 26
A Study of Using Collaborative Briefing Approach to Empower the Briefing Process for Large Scale Infrastructure Projects

Jacky K.H. Chung, Stephen S.Y. Lau and Cynthia H.Y. Hou

Abstract Briefing is the first step in design process that a client either formally or informally informs others of his or her needs, aspirations and desires of a project. A previous study introduced the concept framework named "Collaborative Briefing Approach" (CBA) which was designed to empower the traditionally mobilised briefing team to work collaboratively with a large group of multi-disciplinary stakeholders as an integrated briefing team in the form of a virtual organisation through a shared digital workspace created on a computer network. This paper describes a proposed study designed to investigate the extent to which the use of CBA approach can improve the effectiveness and efficiency as well as the quality of outputs of the briefing process. To achieve the above objectives, a controlled experimental study with 2 × 2 factorial measures will be adopted, under which independent variables (level of support, and type of task) are expected to be controlled and thereby the consequent changes in the dependent variables (process and outcomes) can be observed. Moreover, action research with real-life construction projects will be conducted to examine the real effect and impact on a number of key issues including the levels of task difficulty and user satisfaction in the research process. The research will be one of the pioneer research studies that introduce a shared digital workspace to enable all members to work together remotely and asynchronously in the briefing process. It will demonstrate a new knowledge on how to apply CBA effectively in briefing with a view to improving the process and output quality of construction briefing in a practical manner. Such knowledge will

J.K.H. Chung (✉)
Department of Building, School of Design and Environment,
National University of Singapore, Singapore, Singapore
e-mail: jackychung@nus.edu.sg

S.S.Y. Lau
Department of Architecture, School of Design and Environment,
National University of Singapore, Singapore, Singapore

C.H.Y. Hou
School of Hotel and Tourism Management, The Chinese
University of Hong Kong, Sha Tin, Hong Kong

Y. Wu et al. (eds.), *Proceedings of the 20th International Symposium on Advancement of Construction Management and Real Estate*,
DOI 10.1007/978-981-10-0855-9_26

help to justify the real benefits of the approach and thus, determine a "benefit to cost ratio" so as to identify the most suitable project type for using CBA approach.

Keywords Construction briefing · Stakeholders · Collaboration · Collaborative briefing approach

26.1 Deficiency in Construction Briefing

Briefing is the first step in design process that a client either formally or informally informs others of his or her needs, aspirations and desires of a project [17, 27, 42, 43]. Briefing can be considered to be synonymous with 'Architectural Programming' used in North America [35] and 'Scope Management' used in Australia [47]. These terms essentially describe the same activity and they are interchangeable. Brief, which is the main product of briefing, is a document defining at any point in time the relevant needs, aims, and resources of the clients and users, the context of the project and any project requirements [13, 30].

Briefing is recognised as one of the most important processes in project management. At the pre-design stage, briefing investigates the nature of design problem by helping clients to define, translate, communicate and present their needs and wants into a set of written project requirements in form of specific technical characteristics, functional performance criteria and quality standards. These requirements, which are the root of briefing, act as a basis for approaching designers [6, 7, 12, 42, 43]. At the design stage, these requirements provide guidelines on examining the developed design options so as to determine the optimal one, according to the defined design problem. At the post-design stage, these requirements help clients to review the selected design options during the construction and operation phases [35]. As explained above, significant resources are committed in briefing and therefore, it is crucial to get the brief right at the beginning so as to ensure the effective delivery of the project in time and within budget [36, 38, 40]. As a result, clients are strongly recommended to define and examine their needs in terms of project requirements, before and during briefing [17, 28, 29].

Despite the importance described above, briefing has been reported as persistent problem area in construction and a large number of problems are reported [3, 5, 7, 10, 17, 19–21, 24, 26, 28, 35, 41–43, 46, 48, 49, 51]. For example, there is a tendency for clients to leap to design solutions because of commercial pressure; and to start developing such design solutions without full examination of their needs [35, 48]. In one example, the project brief for a multi-million pounds project was confined to three pages only [3]. As a result, their projects will suffer from poor briefing and they would criticise their completed buildings as not being what they expected [8, 17, 44, 45]. For the above reasons, many major reports come to the same conclusion that there is a strong demand to improve the briefing practice in targeting to enhance project performance [17, 28, 29]. Clients will benefit from investing time and resources in developing a good brief to reduce the likelihood of changes and finally, avoid subsequent delays and cost overruns in the project [17, 29].

In conclusion, briefing is still reported to pose continuing problems in the construction industry although a substantial amount of research activities have been undertaken to address briefing problems (e.g.: [3, 8, 9, 17, 22, 25, 28, 31, 32, 36, 38, 41–43, 52, 56]).

26.2 Construction Briefing and Collaboration Technology

Literature shows that a number of research studies have introduced Information and Communication Technology (ICT) to support construction briefing and some briefing packages including Brief Development Manager from Loughborough University [44–46]; BriefMaker from University of Strathclyde [54, 55]; ClientPro from Loughborough University [33, 34]; CoBrITe from Loughborough University and Salford University [11, 50]; DePlan from Loughborough University [4, 14]; DIVERCITY from Salford University [1, 2]; Team Thinking Tools from University of Cambridge and University of Loughborough [37] etc. are developed. Nevertheless, most of these packages are limited to provide information and analytical support, which applies computerised analytical models, modelling techniques or artificial intelligence techniques to improve the capabilities to capture, organise, analyse, and present collected data and very few of them focus on team collaboration. Hence, it is believed that there is a large potential for collaboration technology, which is a branch of ICT, in supporting the briefing process.

A research study titled "Improving megaproject briefing through enhanced collaboration with ICT" was conducted in 2010 to explore the feasibility of applying collaboration technology to improve the briefing output by capturing stakeholder values through the enhanced collaboration between clients and stakeholders in the briefing process [15, 16]. Under the said study, a new briefing approach named 'Collaborative Briefing Approach' has been developed based on Stakeholder Theory, which is a theory of organisational management and business ethics addressing morals and values in managing an organisation. This theory is an important topic in management literature and there are about a dozen books and more than 100 articles with primary emphasis on the theory over the last two decades [18].

Stakeholder theory was originated from stakeholder concept, which divided major stakeholders into four groups including shareholders, employees, customers, and the general public in General Electric Company in 1929 [39]. This stakeholder concept was fully articulated by Freeman, who drew on various literatures including corporate planning, systems theory, and corporate social responsibility to develop the concept into a theory [23]. The theory identifies and models the groups which are stakeholders of a corporation, and both describes and recommends methods by which management can give due regard to the interests of those groups [18]. It suggests that a modern organisation has relationships with many constituent groups

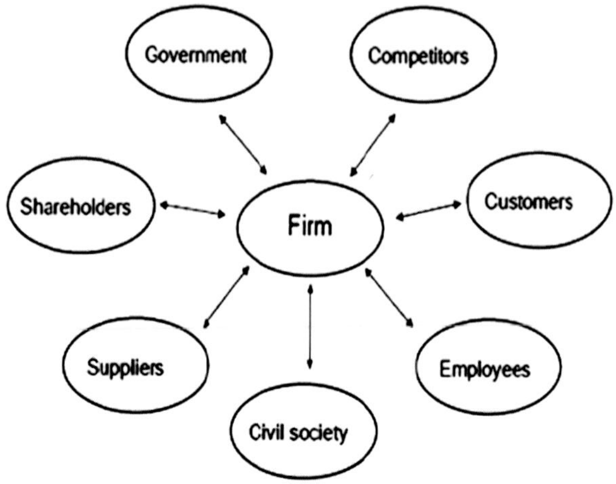

Fig. 26.1 A stakeholder model

but traditional management theories cannot address the "shifts" in business environment induced by the impacts from internal stakeholders and external stakeholders. Consequently, the theory argues that organisations should take into account all of those groups and individuals that can affect, or are affected by, the accomplishment of the business enterprise instead of limited to shareholders only [23]. A basic stakeholder model is shown in Fig. 26.1.

The above model indicates the relationships between various groups and individuals in a single analytical framework. Each oval representing a group of stakeholders. The central oval, which represents the organisation, is surrounded by a number of other ovals with bi-directional arrows toward and from the central oval [23].

26.3 Collaborative Briefing Approach

'Collaborative Briefing Approach' (CBA) is designed to empower the traditionally mobilised briefing team to work collaboratively with a large group of multi-disciplinary stakeholders as an integrated briefing team in the form of a virtual organisation through a shared digital workspace created on a computer network [15, 16]. It suggests that stakeholders own a substantial amount of important information and tacit knowledge, which are kept hidden, without sharing across projects due to the 'wall syndrome' [34]. In line with the idea of the normative aspect, clients should involve their inputs so as to add value to briefing [29]. Nevertheless, stakeholder engagement in briefing is subject to constraints and thus, stakeholder involvement is very limited in the industry [53]. The CBA approach introduces a shared digital workspace that serves as a collaborative platform and enables all members to work together remotely and asynchronously so as to achieve

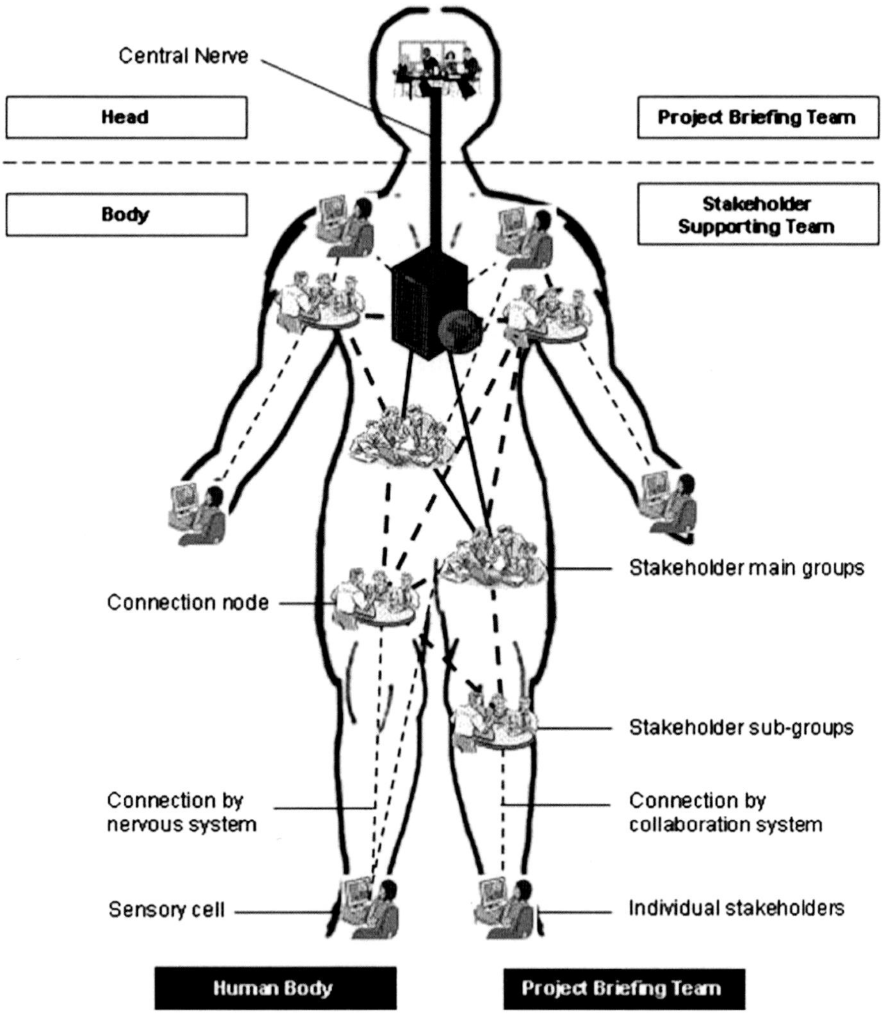

Fig. 26.2 A virtual organisation in briefing Chung et al. [16]

greater stakeholder participation in briefing. Since stakeholders contribute in bringing professional knowledge, experience and creativity to briefing, greater stakeholder participation will increase their inputs and result in more fruitful briefing outputs.

As shown in Fig. 26.2, there are significant similarities in the organisation structure and the communication hierarchy between them. For example, team members (cells) are working closely with in-house colleagues to form a sub-group (tissue) in representing their organisations (e.g. contractors). Various sub-groups are working together to form a main group (organ) in representing their professionals (e.g. engineering consultant group). Several main groups combine to form a functional unit (system) in representing their roles (e.g. construction team). Lastly, various functional units are working together as a virtual organisation (human body) in representing a whole team in briefing (e.g. briefing team). Drawing parallels with the concept of a biological neural network that connects cells, tissues, organs and systems tougher with brains in humans; a computer-based collaborating system is proposed to manage the discussion among team members within this virtual organisation in briefing. The above analogy is only for indication of the holistic interactions rather than for over-simplification of briefing needs [16].

Apart from the above, an 'Integrated Collaborative Briefing Methodology' (INTERCOM) was also developed to translate the CBA approach into a set of actionable methods and job plans for practical use. As shown in Fig. 26.3, the INTERCOM comprises of five components: (i) a value based briefing methodology, (ii) a collaborative briefing job plan, (iii) an integrated briefing team, (iv) a collaborative briefing platform, and (v) facilitation service. The first four components were developed and validated by a group of well experienced multi-disciplinary industry practitioners. The fifth component merits a separate research and development exercise. Research findings reveal that the concept of collaborative briefing approach and the design of the INTERCOM are well supported by the practitioners. In addition, it is concluded that INTERCOM would contribute to improve the briefing process by facilitating team management, enhancing requirement definition and promoting consensus building. It also improves requirement comprehensiveness, decision transparency, decision reliability, and decision satisfaction, as well as the value and quantity of the requirements specified in the brief [15].

In order to extend the research potential of the described study, a new research to study the CBA approach from a new perspective through experimental studies and

Fig. 26.3 Key elements of the collaborative briefing framework

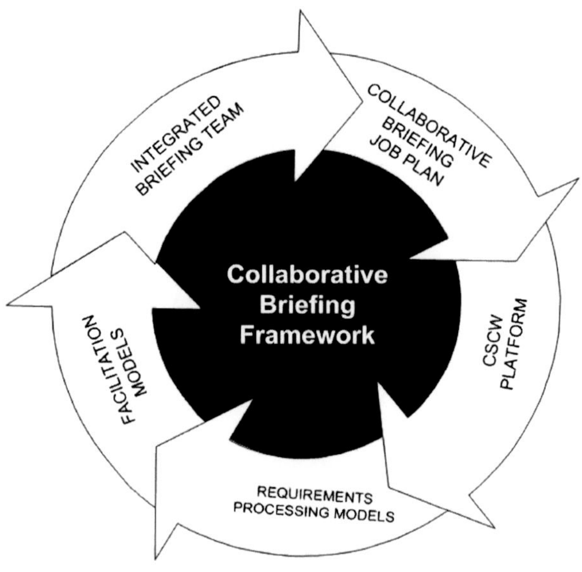

action research and thereby investigate the extent to which we can apply collaboration technology for the purpose of improving briefing output.

26.4 A Proposed Study

26.4.1 Research Objectives

Collaborative Briefing Approach (CBA) facilities group decision making by a large group of multi-stakeholder, which is an important process in construction briefing. This proposed research aims to investigate the extent to which we can apply collaboration technology to improve the briefing output by capturing stakeholder values through the enhanced collaboration between clients and stakeholders in the briefing process. Thus, the objectives of this proposed research are to:

- investigate the extent to which the use of CBA approach can improve the effectiveness and efficiency in the processes of briefing in construction projects;
- investigate the extent to which the use of CBA approach can improve the quality of outputs of briefing; and
- identify critical success factors that lead to effective implementation of CBA approach in briefing.

26.4.2 Research Methodology

Figure 26.4 illustrates the framework for the proposed research. In order to achieve the research objectives, two research methods, namely (i) controlled experimental

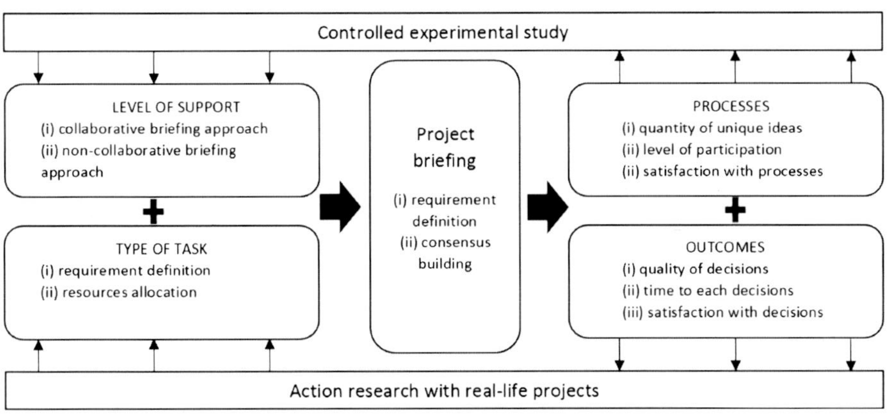

Fig. 26.4 Proposed research framework

Table 26.1 Summary of research methods

Method	Description
Controlled experimental study	A controlled experimental study with 2 × 2 factorial measures will be adopted, under which independent variables are expected to be controlled and thereby the consequent changes in the dependent variables can be observed. In addition, experimental variables can be isolated more easily and their impact can be evaluated more effectively than other methods. In the proposed research, there are two independent variables, namely (i) level of support, and (ii) type of task • The support variable has two levels: CBA approach and non-CBA approach • The task variable also has two levels: requirement definition and resources allocation Statistical test and analysis of variance will be used to examine the likelihood that the means of the experimental and control groups are derived from a single-parent population and can be compared for possible differences
Action research	Real-life construction projects will be conducted to examine the real effect and impact on a number of key issues including the levels of task difficulty and user satisfaction in the research process. It is characterised by involving researchers as part of the situation under exploration. The applicant will play the role of a facilitator to design and deliver the project briefing exercise for these selected projects. Moreover, focus group will be used a supplementary instrument to facilitate the research team to obtain valuable views and insights from participants of the exercises. During focus group meetings, the research team members can ask and adapt questions as necessary, ensure that questions and responses are properly understood by repeating or rephrasing them, and pick up non-verbal cues from the respondents

study and (ii) action research will be adopted and the details of which are summarised in Table 26.1. Illustrates the framework for the proposed research. In order to achieve the research objectives, two research methods, namely (i) controlled experimental study and (ii) action research will be adopted and the details of which are summarised in Table 26.1.

The following hypotheses will be tested and verified in the proposed research:

- H1: Satisfaction with the processes of project briefing for construction projects will vary as a function of the level of support (CBA approach and vs. non-CBA approach) given to the group.
- H2: Satisfaction with the outcomes of project briefing for construction projects will vary as a function of the level of support (CBA approach and vs. non-CBA approach) given to the group.

26.5 Research Contributions

The proposed research is one of the pioneer research studies in exploring the concept of team collaboration in construction briefing. It aims to introduce a shared digital workspace that serves as a collaborative platform and enables all members to work together remotely and asynchronously so as to achieve greater stakeholder participation.

Following the success of the previous study on CBA approach described before, the proposed research will study the approach from a new perspective by using the methods of experimental studies and action research. It is anticipated that this will lead to new knowledge on the real effectiveness of adopting the CBA approach in improving the process and the output quality of construction briefing in a practical manner. Such knowledge will help to justify the real benefits of the approach and thus, determine a "benefit to cost ratio" so as to identify the most suitable project type for using CBA approach. Moreover, a set of critical success factors contributing to the implementation of CBA approach in briefing will be identified in the research. These factors will help to increase our understanding of some hidden factors such as group dynamics, and facilitation, contractual barriers and politics etc. that emerge in real life application. In summary, the proposed research will strengthen the connection between the knowledge of construction briefing and the domain of team collaboration through the "CBA approach".

References

1. Arayici Y, Ahmed V, Aouad G (2006) A requirements engineering framework for integrated systems development for the construction industry. ITcon 11:35–55
2. Arayici Y, Aouad G (2005) Computer integrated construction: an approach to requirements engineering. Eng Constr Archit Manage 12:2
3. Atkin B, Flanagan R, Marsh L, Apapiou A (1995) Improving value for money in construction: guidance for chartered surveyors and their clients. The Royal Institution of Chartered Surveyors, London
4. Austin SA, Baldwin AN, Steele JL (2002) Improving building design through integrated planning and control. Eng Constr Archit Manage 9(3):249–258
5. Barrett PS (1999) Better construction briefing. Blackwell Science, Malden
6. Barrett PS, Hudson CS (1996) Is briefing innovation? In: Proceedings of CIB W65 symposium. Glasgow
7. Barrett PS, Hudson J, Stanley C (1999) Good practice in briefing: the limits of rationality. Autom Constr 8(6):633–642
8. Barrett PS, Stanley C (1999) Better construction briefing. Blackwell Science Ltd., Oxford
9. Blyth A, Worthington J (2001) Managing the brief for better design. Spon Press, London
10. Bouchlaghem D, Rezgui Y (2003) Guest editorial/briefing and design management in construction. Int J IT Archit Eng Constr 1(1):5–7
11. Bouchlaghem D, Rezgui Y, Hassanen M, Cooper G, Rose D (2000) IT tools and support for improved briefing. In: Proceedings of construction information technology 2000, CIB W78 and IABSE. Reykjavik

12. Bowen PA, Pearl RG, Edwards PJ (1999) Client briefing process and procurement method selection: a South African study. Engineering. Eng Constr Archit Manage 6(2):91–104
13. BSI (1995) Performance standards in building—checklist for briefing—contents of brief for building design, the British Standards (BS 7832)
14. Choo HJ, Hammond J, Tommelein ID, Austin SA, Ballard G (2004) DePlan: a tool for integrated design management. Autom Constr 13(3):313–326
15. Chung JKH (2010) Development of a collaborative briefing approach for infrastructure projects in Hong Kong. Department of Civil Engineering. Hong Kong The Universtity of Hong Kong, PhD: 267
16. Chung JKH, Kumaraswamy MM, Palaneeswaran E (2009) Improving megaproject briefing through enhanced collaboration with ICT. Autom Const 18(7):966–974
17. CIB (1997) Briefing the team: a guide to better briefing for clients. Thomas Telford Publishing, London
18. Donaldson T, Preston LE (1995) The stakeholder theory of the corporation: concepts. Evidence, and implications. Acad Manag Rev 20(1):65–91
19. Duerk D (1993) Architectural programming, information management for design. Wiley, New York
20. Duffy F (1987) Change is importance. Architects' J 34–37, 19 and 26 Aug
21. Duffy F, Greenbery S, Myerson J, Powell K, Thomson T, Worthington J (1998) Design for changes: the architecture of DEGW. Watermark Publication, Haslemere
22. Fisher N (1998) A clients project definition tool (CPDT) implementation guide. Reading, The University of Reading
23. Freeman RE (1984) Strategic management: a stakeholder approach. Pitman, Boston
24. Goodacre PE, Noble BM, Murray J, Pain J (1982) Client aid program. Occasional paper, Department of Construction Management, University of Reading. No. 5
25. Green SD (1996) A metaphorical analysis of client organizations and the briefing process. Constr Manage Econ 14:155–164
26. Hansen FE (1994) Change order management for construction projects. Cost Eng 36(3):25–28
27. Hershberger R (1999) Architectural programming and pre-design manager. McGraw-Hill, New York
28. HMSO (1964) The placing and management of contracts for building and civil engineering work (the Banwell report). Ministry of Public Building and Works, London
29. HMSO (1994) Constructing the team: joint review of procurement and contractual arrangements in the United Kingdom construction industry (Latham report). HMSO, London
30. ISO (1994) Performance standards in building—checklist for briefing—contents of brief for building design (ISO 9699)
31. Kamara JM, Anumba CJ (2001) ClientPro: a prototype software for client requirements processing in construction. Adv Eng Softw 32(2):141–158
32. Kamara JM, Anumba CJ (2001) A critical appraisal of the briefing process in construction. J Constr Res 2:13–24
33. Kamara JM, Anumba CJ, Evbuomwan NFO (2001) Assessing the suitability of current briefing practices in construction within a concurrent engineering framework. Int J Project Manage 19(6):337–351
34. Kamara JM, Anumba CJ, N.F.O., E. (2002), Capturing client requirements in construction projects, Thomas Telford, London
35. Kelly J, Duerk D (2002) Construction project briefing/architectural programming. Best value in construction. In: Kelly J, Morledge R, Wilkinson S (eds) Blackwell Science, Oxford, pp 38–58
36. Latham M (1994) Constructing the team: joint review of procurement and contractual arrangements in the United Kingdom construction industry. HMSO, London
37. Macmillan S, Steele J, Kirby P, Spence R, Austin S, Heintz JL (2002) Mapping the design process during the conceptual phase of building projects—collaborative design planning networks. Eng Constr Archit Manage 9(3):174–180

38. MacPherson S, Kelly JR, Male SP (1992) The briefing process: a review and critique. RICS, London
39. Mishra AMD (2013) Applications of stakeholder theory in information systems and technology. Inzinerine Ekonomika Eng Econ 24(3):254–266
40. Newman O (1996) Creating defensible space, U.S. department of housing and urban development, Office of Policy Development and Research
41. Newman R, Jenks M, Dawson S, Bacon V (1981) Brief formulation and the design of buildings. Oxford Polytechnic, London
42. O'Reilly JJN (1987) Better briefing means better buildings. Building research establishment report. Watford, Building Research Establishment
43. O'Reilly JN (1987) Better briefing means better buildings. Building Research Establishment, Garston
44. Othman AAE (2004) Identification, quantification and classification of construction brief development drivers. In: Proceedings of the CII-HK conference 2004 on construction partnering: our partnering journey—where are we now, and where are we heading. Sheraton Hong Kong Hotel, Hong Kong, pp 163–175
45. Othman AAE (2004) Value and risk management for dynamic brief development in construction. In: Proceedings of the CII-HK conference 2004 on construction partnering: our partnering journey—where are we now, and where are we heading. Sheraton Hong Kong Hotel, Hong Kong
46. Othman AAE, Hassan TM, Pasquire CL (2004) Drivers for dynamic brief development in construction. Eng Constr Archit Manage 11(4):248–258
47. Peakman L (2008) Scope for improvement 2008—a report on scoping practices in Australian construction and infrastructure projects. Blake Dawson, Sydney
48. Preiser WFE, Vischer J (2005) Assessing building performance. Elsevier, Oxford
49. Rezgui T, Bouchlaghem D, Austin S (2003) An IT-based approach to managing the construction brief. Int J IT Archit Eng Constr 1(1):25–37
50. Rezgui Y, Body S, Bouchlaghem D, Hassanen M, Cooper G, Barrett P, Austin S (2001) A proposed IT-based approach for managing the construction brief effectively. IN: Proceedings of CIB W78 2001. Mpumalanga, South Africa, pp 5–11
51. RIBA (2000) The Architect's plan of work. Royal Institute of British Architects
52. Salisbury F (1990) Architect's handbook for client briefing. Butterworth Architecture, London
53. Shen GQP, Chung JKH (2006) A critical investigation of the briefing process in Hong Kong's construction industry. Facilities 24(13/14):510–522
54. van Ree HJ, van Meel J (2007) Sustainable briefing for sustainable buildings. In: CIB world building congress 2007. Cape Town, South Africa, International Council for Research and Innovation in Building and Construction (CIB), pp 1011–1023
55. van Ree HJ, van Meel J, Lohman F (2006) Better briefing for better buildings: an innovative modelling tool for specifications management. In: Proceedings of the annual research conference of the royal institution of chartered surveyors. University College London, pp 750–759
56. Yu ATW, Shen Q, Kelly J, Hunter K (2007) An empirical study of the variables affecting construction project briefing/architectural programming. Int J Project Manage 25(2):198–212

Chapter 27
Role and Impact of Business to Government (B2G) Guanxi in Bidding of Infrastructure Projects: A Case in China

Bing Zhang, Zhenggang Huo and Minli Zhang

Abstract Infrastructure projects in China are dominated by government, its entire bidding and tendering process is subject to administrative control, and B2G (Business to government) guanxi has important implications for bidding and tendering activities. This paper investigated professional staff working in bidding and tendering activities of infrastructure projects through guanxi attitude, guanxi value and its roles. It is revealed that the bidding and tendering activities would be very convenient if having B2G guanxi. What's more, the bidders could avoid bidding and tendering risks through B2G guanxi. However, the bidders need spend a lot of time, money and energy establishing and maintaining B2G relationship. Establishing B2G guanxi could through family, friend and intermediary three ways, and the bidders would prefer establishing their B2G guanxi among business activities. B2G guanxi has an impact on the bidding and tendering results and the will of bidding and tendering, in each type of infrastructure project bidding and tendering, B2G guanxi is needed. If having B2G guanxi with senior officials, it would further contribute to

B. Zhang
School of Economics and Management of Tongji University, Shanghai, China

B. Zhang
Civil Science and Engineering, Excel in Construction Management and Construction Corruption Prevention, Yangzhou University, Yangzhou, China

Z. Huo (✉)
Civil Science and Engineering, Expert in Construction Management and New Towns, Yangzhou University, Yangzhou, China
e-mail: 1509389148@qq.com

M. Zhang
Engineering Management Department, Yangzhou University, Yangzhou, China

M. Zhang
The Real Estate Institute of Yangzhou University, Expert in Real Estate Economy, Construction Economy; College of Civil Science and Engineering, Yangzhou University, Yangzhou, China

Y. Wu et al. (eds.), *Proceedings of the 20th International Symposium on Advancement of Construction Management and Real Estate*,
DOI 10.1007/978-981-10-0855-9_27

the successful bidder. Whereas, B2G guanxi violates the principles of fairness and transparency, at last, this paper propose how to foster the competitiveness of construction enterprises and culturing professional ethics to ensure B2G guanxi in a healthy way.

Keywords Infrastructure projects · Bidding and tendering · B2G guanxi · Fraud triangle model

27.1 Introduction

In China, with the increasing emphasis on infrastructure, the current infrastructure investment accounts for 9 % of gross domestic product (GDP) [1]. In the context of rapid development, the number of construction companies has been increasing rapidly. By 2014, there are 81141 construction companies [2]. Facing such a large number of infrastructure investment and such a high competitive market, how to win infrastructure project is the most important thing for each construction company. Considering guanxi could bring several competitive advantages, help in obtaining government permits and assistance in winning projects and resources [3], Ren [4] pointed that in the delivery of infrastructure projects, gaining projects through B2G guanxi (Business to Government guanxi, the guanxi between business person with government official) has become an unwritten rule.

B2G guanxi plays an import role in the bidding and tendering process of infrastructure projects where the bidders establish guanxi with government officials [5]. In China, B2G guanxi has long been recognized as one of the major factors for success when conducting tendering and bidding activities in infrastructure projects, the development and maintenance of B2G guanxi has become a priority for lots of construction companies, because in China, business peasons first strive to build up personal relationships, then business follows [6]. So having good B2G guanxi with government officals would greatly influence the success of winning infrastructure projects. Whereas, there are a large number of cases suggest that B2G guanxi could promote corruption behavior, which always happens between those persons in key positions, furthermore, B2G guanxi is attached at power, it often happens before or as soon as the tendering beginning and the guanxi duration depended on the cost and the profit in the future. To sum up, B2G guanxi is a double-edged sword in the process of tendering and bidding among the infrastructure projects, and B2G guanxi is a highly complex social construct which has its own value (role, effect etc.) in Chinese society.

Despite B2G guanxi has become an important influence factor in infrastructure project bidding and tendering process, and there are lots of studies pay more attention to B2G guanxi. There is still a lack of in-depth knowledge and understanding of B2G guanxi, especially the value, attitude and behavior of B2G guanxi in the infrastructure project bidding and tendering process. To bridge the research gap, the study would discuss why this phenomenon is a pragmatic consideration in terms of "value-attitude-behaviour" model proposed by Homer and Kahle [7], which is a cognitive hierarchy model. According to [7] cognitive hierarchy model, values influence behaviour indirectly through attitudes. The value-attitude-behaviour model therefore implies a hierarchy of cognitions in which the influence theoretically flows from more abstract cognitions (i.e., values) to mid-range cognitions (i.e., attitudes) to specific behaviours [8].

27.2 Research Methods

The research process consisted of three steps. Firstly, a thorough literature review to identify values, attitudes and behavior towards to B2G guanxi in tendering and bidding process of infrastructure projects. Secondly, a questionnaire were designed according to literature review and interview (Table 27.2). Thirdly, a questionnaire survey is described and analyed.

To obtain bidders' opinion of B2G guanxi, a questionnaire survey was conducted, as there are many forms of B2G guanxi among bidding and tendering process of infrastructure projects, Taormina and Gao [9] suggested that in order to achieve a better understanding of guanxi, an empirical investigation is necessary. This research studied B2G guanxi through a questionnaire survey, using a seven point Likert scale, ranging from 1 = total disagree to 7 = completely agree. The questionnaire was first developed based on findings from literature review and consultation with nine top academics who publish widely in this field. After that, a pilot study was conducted with three senior executives in the construction industry, each of which have more than 8 years experience in the industry, to refine and finalize the questionnaire.

In this study, potential respondents were construction professionals who have involved in bidding and tendering process of infrastructure projects with the help from Tongji University and Shanghai Association of Engineering Consulting. The selected cities included Shanghai, Jinan, Hangzhou, Wuxi and Yangzhou, all of which are large cities with massive infrastructure construction these years. The questionnaire survey was conducted in three months. 211 questionnaires were distributed. Finally, 183 responses were received, including 34 invalid ones, Thus 149 valid questionnaires were used for data analysis, representing a response rate of 70,6 %. Among those respondents, 70 % have more than five years experience in the construction industry, nearly 40 % hold senior positions, and 90 % have college degree or above (Table 27.1).

Table 27.1 Demographic profile of respondents

Profile	Categories	Frequency	Percent
Experience	1–5 years	40	27.3
	5–10 years	48	32.0
	10–15 years	31	20.7
	Over 15 years	30	20.0
Position	Staff	91	61.3
	Project manager level	36	24.0
	Department manager level	11	6.7
	Top Manager level	11	8.0
Education	High school or below	15	10.0
	Junior college	43	30.0
	Bachelor's degree	60	40.0
	Master's degree or over	31	20.0
Organization	Contractor	33	23.3
	Quantity survey	14	9.3
	Supervision	43	28.7
	Consultant	59	39.3

27.3 Results and Analysis

The reliability analysis refers to the stability and reliability of the data, which is to test whether the same item could get the consistency of multiple measurements, and coefficient alpha (Cronbach's α) is used for the reliability analysis. In this study, the Cronbach's alpha is 0.932, far bigger than 0.6, reaching the accepted level [10]. This indicates the reliability of seven-point Likert-type scores could be accepted.

27.3.1 Summary Statistics of B2G Guanxi

The mean value of each item is shown in Table 27.2. As the mean scores of most items are over 5.0, which indicates the bidders's value toward B2G guanxi are high, and whether the personal and enterprise's attitude towards B2G guanxi is serious.

For its value, the traditional and instrumental value are all important ones, which indicates that China is a guanxi-oriented society and B2G guanxi is deeply rooted in traditional Chinese culture [11, 12], even in a market economy, B2G guanxi is still an integral part of the Chinese [13, 14]. Most of the respondents also believed that guanxi has become an indispensable part of life and work among the Chinese people. Due to this reason, people focus on ceremony and etiquette, such as weddings, funerals, and festivals, which could be used as means and channels to establish and develop guanxi, and as the principle to maintain and develop guanxi.

Table 27.2 The summary statistics of each item

Questions		Mean	St.d
Traditional value			
Q1	B2G guanxi is common phenomenon current	5.67	1.31
Q2	B2G guanxi is a part of traditional culture	5.68	1.10
Q3	B2G guanxi emphasizes on ceremony and etiquette	5.36	1.19
Q4	B2G guanxi promotes emotion	5.63	1.13
Instrumental value			
Q5	Spending a lot of time and energy to establish and develop B2G guanxi	5.44	1.34
Q6	B2G guanxi makes the bidding easier	5.78	1.25
Q7	B2G guanxi could avoid risks	5.56	1.20
Q8	B2G guanxi has a positive role in construction	4.68	1.58
Personal attitude			
Q9	Whether having strong B2G guanxi determines the bidding will	4.92	1.38
Q10	Good B2G guanxi is important	5.87	1.26
Q11	Help and care could be obtained from B2G guanxi	5.67	1.19
Enterprise' Attitude			
Q12	Getting the bidding because of the B2G guanxi for construction companies	5.05	1.28
Q13	Companies pay more attention to the cultivation and operation of B2G guanxi	5.46	1.18
Q14	Winning infrastructure project by B2G guanxi is in a hign proportion	5.32	1.26
Establish channel			
Q15	Establishing B2G guanxi through government officials' family and good friends	4.99	1.24
Q16	Establishing B2G guanxi through business process	5.56	1.04
Q17	Establishing B2G guanxi through intermediary ways	4.93	1.42
Offical selecting			
Q18	Establishing B2G guanxi with senior officials could be easier getting the bidding	5.62	1.25
Q19	Establishing B2G guanxi with government workers could be easier getting the bidding	5.13	1.27
Q20	Establishing B2G guanxi with offical's agent could be easier getting the bidding	4.60	1.62
Project difference			
Q21	Large projects more need B2G guanxi	5.26	1.41
Q22	Medium projects more need B2G guanxi	5.32	1.17
Q23	Small projects more need B2G guanxi	5.31	1.36

By those ceremony and etiquette, their feeling and trust gradually increase. Besides traditional value, people consider B2G guanxi is an important factor in tendering and bidding process of infrastructure projects.

Furthermore, there are still a variety of irregularities and risks in bidding and tendering process of infrastructure projects. China National Audit Office (CNAO) found 1194 contracts were irregularities in Xinjiang Uygur Autonomous Region Development Projects supported by central government. Until the end of March 2012 (from 2009 to 2012), and since the government possesses too much power in the bidding and tendering process with an arbitrary discretion, so having B2G guanxi with relevant officials could help obtain the bidding and tendering information in advantage and hold dominant position. Most of respondents believed that B2G could reduce these irregularities and illegal act impact on bidding and tendering activity to some extent, and obtained the project under the influence of guanxi.

For B2G guanxi's attitude, there are two parts, that is personal attitudes and enterprise's attitude. Due to its popularity, by the personal view, the factor of good B2G guanxi is important get the highest value while from the enterprise's view, the value of Companies pay more attention to the cultivation and operation of B2G guanxi is highest. At the same time, because the bidders could get help and care through guanxi, what's more, guanxi could bring conveniences, and sc for actors. However, not all people think B2G is positive toward tendering and bidding activities, thus there are different attitudes toward B2G guanxi.

Behavior is the result of value and attitude, due to the bidders pay more attention to B2G guanxi, the behavior of B2G guanxi is popular in the tendering and bidding process of infrastructure projects. For B2G establishing channel, here are similar three guanxi establishing ways. One is establishing B2G guanxi with government officials through officials' family and good friends, second is the way of working together with government officials from the past work, the last but not the least is through an intermediary to establish B2G guanxi, this approach has been revealed from a large number of corruption cases. It could be found from the analysis that, among the three types of B2G guanxi, the bidders mostly want to establish B2G guanxi through business, the following is through family and friends, and the least way is through the intermediary interest way to establish B2G guanxi.

27.3.2 Significant Difference of the Behavior

The development and maintenance of B2G guanxi has become a priority for many bidders, one of the rules in infrastructure for winning projects is establishment of the right guanxi, including the right establishing channel, the right person and selecting the right project, thus the importance of B2G guanxi is its behavior, according the above analysis, this part will conduct the further analysis of the behavior of B2G guanxi by Chi-square test (shown in Table 27.3).

Table 27.3 The chi-square test of different factor

Survey questions	Sum of square	F	P value
Establishing B2G guanxi through government officials' family and good friends	719.72	11.57	0.00
Establishing B2G guanxi through business process			
Establishing B2G guanxi through intermediary ways			
Establishing B2G guanxi with senior officials could be easier getting the bidding	939.18	19.99	0.00
Establishing B2G guanxi with government workers could be easier getting the bidding			
Establishing B2G guanxi with official's agent could be easier getting the bidding			
Large projects more need B2G guanxi	769.43	0.09	0.92
Medium projects more need B2G guanxi			
Small projects more need B2G guanxi			

Different Types of Establishing Channel

The testing whether the three types of B2G guanxi establishing channel exists the significant difference by Chi-square test analysis, the testing result of Pearson Chi-square double-side Sig. value was 0.00, smaller than 0.05, showing the respondents have significant different establishing channel of B2G guanxi in bidding and tendering process of infrastructure projects.

Hwang [15] divided guanxi into three types, which is instrumental guanxi, expressive guanxi and mixed guanxi. The three type of guanxi complied with the demand rule, fair rule, and human rule respectively. B2G guanxi showed more obvious tool characteristics, following the demand rules. In order to establish B2G guanxi, the bidders use various means and methods to establish guanxi with government officials, so there are diversified channels to establish B2G guanxi with government officials. Among all the channels, the respondents mostly want to establish B2G with government officials through business, because in this process, the two sides developed their guanxi from stranger to acquaintances and finally to friend, even entered their own communities, and they prolonged emotional and material inputs, so their guanxi has emotional characteristics with the premise of instrumental factor. Since the two B2G guanxi sides have been trusted and affection, they could easily apply their guanxi among the bidding and tendering process, and the majority of respondents believed that this kind of guanxi did not involve money exchange and other material benefits.

Compared with the B2G guanxi among the business, the B2G guanxi through officials' family, friends and intermediary, showed tool characteristics more obviously. In the bidding and tendering process, the bidders collected information of government officials, established B2G guanxi through those channels, including inviting officials having dinners by intermediary, or even through illegal bribery activities, influenced the government officials' decision through family and friends. While there are no channels, the bidders could get channel from intermediary organizations, after they reached the benefit-sharing agreement, the bidder could get close to the government officials and establish B2G guanxi, which is naked pecuniary interest relationship. In addition, the B2G guanxi through family, friends and intermediary often involved senior officials, and most of those B2G guanxi included pecuniary interest exchange activities, especially for B2G guanxi through intermediary means, showing obvious money-power bartering behavior.

Different Types of Officials

The testing whether the three types of selecting B2G guanxi officials exists the significant difference by Chi-square test analysis, the testing result of Pearson Chi-square double-side Sig. value was 0.00, smaller than 0.05, showing the respondents have significant different selecting officials of B2G guanxi in bidding and tendering process of infrastructure projects.

Due to B2G guanxi is a scare resource and bidders desire establish B2G guanxi with related officials directly, so the willingness of selecting official's agent is lowest. What's more, although establishing B2G guanxi with senior officials could more help secure of obtaining infrastructure projects, not all the bidding and tendering things are processed directly by the senior officials. Compared with the senior officials, establishing B2G guanxi with government workers would be more convenient, and the cost of establishing B2G guanxi could be lower correspondingly. Therefore, in the bidding and tendering process, along with the emphasis the B2G guanxi with senior officials, the bidders would prefer to establish B2G with appropriate government workers.

Different Types of Projects

Then Chi-square test was adopted to test whether there is significant differences among different types of infrastructure project for the need of B2G guanxi for bidders. The Chi-square testing result showed the Pearson Sig. value of double-side test was 0.92, while the acceptable level of the significance is 0.05, so this indicated that the respondents' perception toward the need of B2G guanxi in infrastructure projects is not significant different. From the above analysis, we could see that for all infrastructure projects, it is needed to establish B2G guanxi with government officials in their bidding and tendering process respectively.

The respondents also believed that in bidding and tendering process of Chinese infrastructure projects, no matter which type of the projects, those all need B2G guanxi. Essentially if there is no guanxi among the bidding and tendering process, it could still work only according the bidding and tendering legal system, while if there exists guanxi in the bidding and tendering process, the bidders could get more benefits and play edge ball, easily get infrastructure projects through guanxi.

What's more, emphasizing guanxi has become a trend and informal institution in the bidding and tendering process of infrastructure projects, while there may be no connection between emphasizing guanxi and bidding and tendering legal system. For every type of infrastructure projects bidding and tendering process, guanxi is not only able to facilitate the bidding and tendering, but also to reduce and avoid bidding and tendering risks.

27.4 Conclusion

This research focus on B2G guanxi in tendering and bidding process of infrastructure projects from the value-attitude-behavior model. We may conclude that the bidders value B2G guanxi very much, what's more, B2G guanxi is well recognized by the bidders and construction companies. Considering China is a typical guanxi-oriented society, B2G guanxi has an impact on the entire business, having B2G guanxi could gain information advantage. What's more, government officials in charge of infrastructure projects have arbitrary decision-making power while lack of supervision, having B2G guanxi could enjoy the benefits and interests of the decision power. Thus relying on guanxi has become an informal system.

Accordingly, the bidders and their construction companies have to develop and maintain B2G guanxi with government officials of different levels by different methods in different projects. B2G guanxi could facilitate the bidding and tendering activities, and the bidders could be able to avoid the risks among the bidding and tendering process, so the bidders would establish B2G guanxi with government officials through a variety of ways. Meanwhile, because of the implementation for the construction enterprise qualification management, different qualification of construction companies could contract the required projects, some low-qualified construction enterprises contract some project that law prohibits through B2G guanxi.

As a consequence of the above, although the Tendering and Bidding Law demands that bidding and tendering activities should follow the principle of fairness, openness and justice, but in reality, the bidder is facing fierce competition, and the whole construction enterprises are more and more homogenization and assimilation with the bidding and tendering process being susceptible to be interference by government power. Thus B2G guanxi has become one of the key advantages to win the competition. Additional, because of the confusion among the bidding and tendering activities, administrative power intervention in bidding is often seen, part of bidding and tendering activities become a means for getting money to handful of government officials and unscrupulous bidders.

The existing B2G guanxi in the bidding and tendering process of infrastructure projects have something to do with many things, and generally speaking, there are three main courses for the generating and development of B2G guanxi, which are the imperfect bidding and tendering legal system, the fierce competition in construction and the Chinese traditional culture. So it is needed to take various measures to govern and manage B2G guanxi in the future research.

Acknowledgements This research was funded by the National Natural Science Foundation of China (Grant Nos. 71172107 and 71390523).

References

1. Association RP (2008) America 2050: an infrastructure vision for 21st century America. New York
2. Statistics NBO (2015) China statistical yearbook 2015. China statistics Press, Beijing
3. Warren DE, Dunfee TW, Li N (2004) Social exchange in China: The double-edged sword of guanxi. J Bus Ethics 55(4):355–372
4. Ren Y (2012) Curbing construction project corruption must pay more attention to the tendering and bidding process. People's Daily, 21 Sept 2012 (14)
5. Fan Y (2002) Guanxi's consequences: personal gains at social cost. J Bus Ethics 38(4):371–380
6. Hwang K-K (1987) Face and favor: the Chinese power game. Am J Sociol 92(4):944–974
7. Homer PM, Kahle LR (1988) A structural equation test of the value-attitude-behavior hierarchy. J Pers Soc Psychol 54(4):638–646
8. Milfont TL, Duckitt J, Wagner C (2010) A cross-cultural test of the value–attitude–behavior hierarchy. J Appl Soc Psychol 40(11):2791–2813
9. Taormina RJ, Gao JH (2010) A research model for Guanxi behavior: antecedents, measures, and outcomes of Chinese social networking. Soc Sci Res 39(6):1195–1212
10. Carmines EG, Zeller RA (1979) Reliability and validity assessment. Sage
11. Tian X (2007) Managing international business in China. Cambridge University Press, Cambridge
12. Standifird SS, Marshall RS (2000) The transaction cost advantage of guanxi-based business practices. J World Bus 35(1):21–42
13. Bian Y (1994) Work and inequality in urban China. SUNY Press
14. Yang MMH (1994) Gifts, favors, and banquets: the art of social relationships in China. Cornell University Press
15. Hwang D, Staley B, Te Chen Y, Lan J-S (2008) Confucian culture and whistle-blowing by professional accountants: an exploratory study. Manag Auditing J 23(5):504–526

Chapter 28
Comparative Environmental Performances of Wood Structures from Various Sources in Taiwan

Shenghan Li

Abstract Building materials are factors that cause environmental impacts in construction sector. Among those materials, wood has many benefits in building design in terms of sustainability such as lower energy consumption and lower greenhouse gas emissions in production. Due to insufficient wood resources in Taiwan, wood is imported from many regions of the world. This paper aims to investigate environmental impacts of three different regions that provide wood resources for wood structure construction in Taiwan. Regions have been investigated including West Canada, Pacific Northwest and Southeast in the USA. Methodology of life cycle assessment (LCA) is applied to the research in the analysis with environmental indicators such as energy consumption, global warming potential, acidification potential and air pollution. Results show, firstly, that wood manufacturing process is the major contribution in energy consumption, global warming potential and air pollution in three regions. However, acidification effect is caused greatly from the fact of long marine transportation. Secondly, wood resource from Canada for building structure in Taiwan has the least environmental impacts compared with wood resource from the USA. Therefore, it is suggested that the use of wood from Canada can mitigate environmental impacts toward the goal of sustainability.

Keywords Sustainable construction · Sustainable material · Wood · Life cycle assessment (LCA)

28.1 Introduction

Building construction sector has become an important part in our society, which leads to environmental problems, such as energy consumption, carbon dioxide emissions and other harmful pollutant emissions. Reducing environmental impacts

S. Li (✉)
College of Civil Engineering, Shenzhen University, Shenzhen, China
e-mail: shenghan@szu.edu.cn

Y. Wu et al. (eds.), *Proceedings of the 20th International Symposium on Advancement of Construction Management and Real Estate*,
DOI 10.1007/978-981-10-0855-9_28

in construction filed has been an essential issue related to sustainability. Buildings are constructed by various materials and selecting suitable sustainable materials can lower environmental impacts [1–5]. Among common building materials, many studies have shown that wood is sustainable due to its lower embodied energy consumption and lower carbon emissions in manufacturing process [6]. Besides, wood has another environmentally friendly benefit, i.e., carbon sink, which indicates carbon is stored during wood growth, thus reducing global warming effect [7].

In the construction sector, many researches in the world have shown that structures with wood cause less environmental burdens than other materials. For example, Koch [8] in the USA from 1970s and Buchanan and Honey [9] in New Zealand data from 1980s estimated energy use and CO_2 emissions, finding that environmental impacts reduce if wood material is constructed for building structure. Not far from now, CORRIM found two wood houses that have lower embodied energy and global warming potential than similar design made of steel and concrete [10]. Besides, other studies have quantified ranges of possible use and CO_2 emissions from the manufacture or whole lifecycle of building materials, taking various aspects of lifecycle dynamics into account. Borjesson and Gustavsson [11] went further to estimate the effects of land use and end of life changes of materials, concluding that wood structure buildings have lower energy use and GHGs emissions than concrete structure buildings. Peterson and Solberg [12] found wood construction to cause lower GHGs emissions than non-wood material, depending on material waste management and how forest carbon flows are considered.

In Taiwan, traditionally, building structures are constructed by reinforced concrete material; however, as the issue of material sustainability has been paid attention to, sustainable materials such as wood and bamboo have aroused public concern. It has been found that Tu [13] discovered that RC and steel structure release GHGs 4.2 times and 3.6 times more than wood structure respectively. However, the results were based on using local wood only. Taiwan is a small island with less wood resources, thus importing wood from around the world. According to previous investigation, wood from North America (USA and Canada) account for 40 % of total annual importation [14]. Li [15] investigated environmental burdens including energy consumption and GHGs emissions among three common building structures, wood structure, reinforced concrete structure and steel structure with the wood resource from the USA in the region of Pacific Northwest area (Oregon and Washington States). The results have shown that wood structure with resource from overseas still contribute to less GHGs emissions compared with other two building structures.

Since the resource of wood imports to Taiwan from overseas, it is interesting to investigate which regions that produce and manufacture wood lead to less environmental impacts. This paper will firstly focus on analysing environmental burdens of wood resources from three regions in North America: west Canada (Provinces of British Columbia), Pacific Northwest in the USA (Oregon and Washington States) and Southeast areas in the USA (Texas, Mississippi and Tennessee States). Secondly, with the above estimated data, wood structure in Taiwan from three different wood sources can be analysed to investigate its environmental performances. Besides, only environmental impacts of material production and transportation for construction are considered, and estimation of building operational energy is not included in this research.

28.2 Methodology

Life cycle assessment (LCA) is an environmental methodology for evaluating the impacts of a product or process from its origin to its final disposal [16]. LCA is internationally regarded as a science-based, comprehensive and standardized environmental assessment methodology, used in several sectors, including building industry, manufacturing industry and other applications. In this paper, the methodology of LCA is applied in evaluating environmental performances of wood from different regions to export wood to Taiwan for construction.

28.2.1 System Boundaries

Defining the system boundaries in LCA is not easy simply because the process of manufacturing products is complicated. However, Taiwan imports wood from overseas with several uncertain factors, and thus simplifying system boundaries becomes necessary. The following process of wood product encompasses wood harvesting, road transportation from forest to sawmill, manufacturing, road transportation from sawmill to port and marine transportation from port (local) to Taiwan. Here, we ignore the road transportation on Taiwan because it is a small island with relatively short good transportation. The simplified system boundary can be illustrated in the following Fig. 28.1.

28.2.2 Functional Unit

According to ISO 14040, "the functional unit is a measure of the function of the studied system". The functional unit of wood material in the study is m^3. In addition to buildings, a frequently adopted functional unit is the unitary-usable floor area, sometimes with the whole life span and sometimes with reference to per year. Nevertheless, in the following analysis, floor area (m^2) is adopted for estimation.

28.2.3 Impact Assessment Indicators

LCA results are often seen as the most objective part of an LCA. Since they emerge as a long list of natural resource use and emissions into environment, they must be

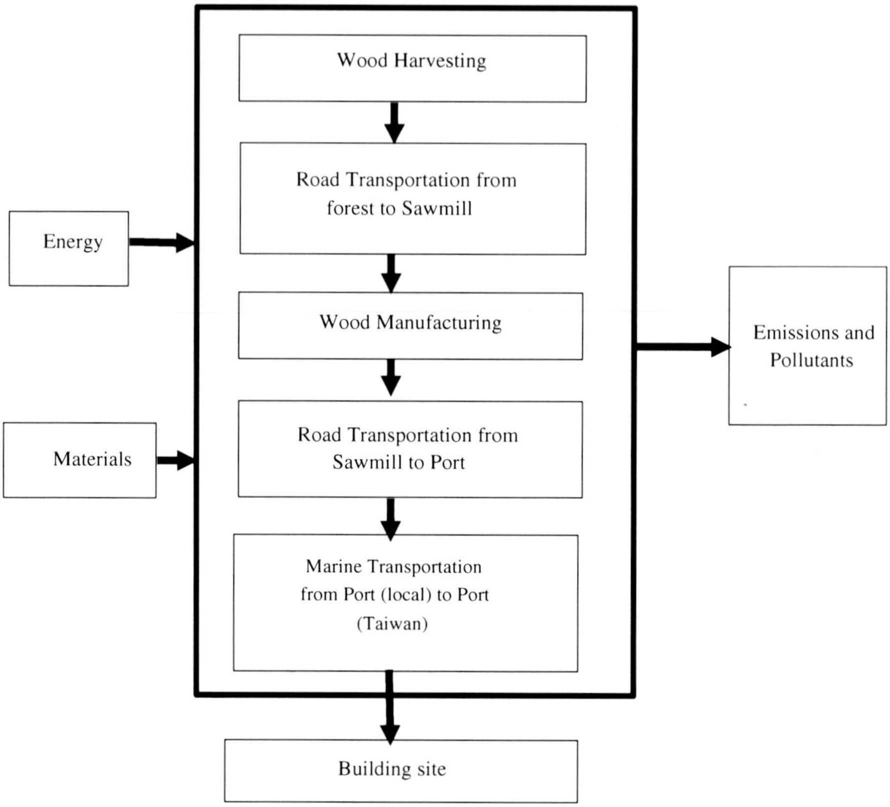

Fig. 28.1 System boundary of imported wood material

converted into understandable and meaningful indicators. And then, it is possible to make practical use of them. Therefore, indicators are chosen to be represented of environmental concern. This method is consistent with the international Standards Organization's (ISO) suggestions for LCA approach, indicating that the impact categories and characterization models should be international accepted. The following shows the indicators in the analysis in this research:

- Energy consumption (EC): indicator of the life cycle primary energy use, which is often expressed as embodied energy consumption;
- Global warming potential (GWP): as an indicators of greenhouse emissions, often referred to as carbon dioxide emissions;
- Acidification potential (AP): as an indication of acidification, for example, SO2 emissions
- Particulate matter (PM): as an indicator of air pollution, which causes human lung disease

28.2.4 Data Collection and Modelling

Data of Wood from Previous Studies

Wood processing goes through many phases. First, the harvesting process is normally divided into the following steps: felling trees, skidding trees to landing area, processing trees to logs, loading and transportation to the process point (sawmill). When wood is manufactured in the sawmill or factory, processes are needed such as sawmilling, chipping, planting and peeling, which need heat for processes such as drying, gluing and pressing, in which fuel oil is used as the major source of thermal energy. Investigations have shown that electricity is the main source of energy in the mechanical wood processing industry, accounting for 40–50 % of the industry's energy needs.

Since wood materials are manufactured in local condition and imported from overseas to Taiwan for construction, environmental impacts should be collected from local regions. Those data can be available from previous studies. In Canada, Athena Institute investigated environmental impacts softwood lumber in the east and west of Canada using LCA in perspective from cradle to gate [17]. The process in analysis includes resource harvesting, resource transportation to sawmill and wood manufacturing. Besides, in the USA, environmental impacts data of wood in Pacific Northwest and Southeast areas can also be obtained from the study [18]. Various wood products, such as lumber, plywood, OSB, and glulam have been investigated and analysed, also including process of harvesting, wood manufacturing and transportation from harvesting sites to manufacturing factories.

In addition, in order to estimate environmental performance of wood structure, the amount of materials used for construction is required. Based on previous study in Taiwan, it has been investigated average use amount of materials in wood structure construction, including wood, concrete and steel [15]. The amount of materials for wood, concrete and steel for wood structure is 0.15 m^3/m^2, 0.17 m^3/m^2, 22.4 kg/m^2, respectively. The functional unit of m^2 is used here to indicate how many materials are needed per meter square in buildings.

Data of Wood from Author's Estimation and Investigation

Road transportation of wood from major sawmills to port may vary greatly; therefore, a proper assumption should be made in order to estimate the environmental burdens due to road transportation effect. Major sawmills have been investigated in three different regions. In the west of Canada, the export port is Vancouver; however, in the USA, the ports in the Pacific Northwest and Southeast regions are located in Tacoma and Houston. Table 28.1 shows the average distance from sawmills to ports. It can be seen that the longest road transportation distance from sawmill to port is in the region of Southeast in the USA (919 km).

Table 28.1 The estimative average distance from sawmill to port in three regions

Canada	Sawmill locations	Average distance from sawmill to Vancouver (km)
	Vanderhoof, Quesnel, Brackendale, Hagensborg	409
USA (Pacific Northwest)	Sawmill locations	Average distance from sawmill to Tacoma (km)
	Portland, Beaverton, Kalama, Seattle	110
USA (Southeast)	Sawmill locations	Average distance from sawmill to Houston (km)
	Beaumont, New Augusta, Nashville, Birmingham, Cordele	919

Table 28.2 Information of heavy duty vehicle and emissions factors

Information of HDV and emissions factors	
Fuel efficiency (L/100 km)	30.89
Fuel heating value (MJ/L)	38.56
Carrying capacity (%)	75
Carrying volume (m^3)	27.3
CO_2 (Kg/L)	2.67
SO_2 (g/km)	0.0672
PM (g/km)	0.113

Inland road transportation is an important factor that leads to environmental problems. There is no global average estimative data concerning the environmental impact of road transportation and each region differs from another. In this paper, information from the Transport Canada Database [19] is taken as providing estimative values. It is assumed that a heavy duty vehicle (HDV, 33,000 lbs) is the main inland wood transportation vehicle type because, based on the author's interviews with truck companies, vehicle of less than 33,000 lbs are not suitable to carry wood due to the large size and the heavy weight of wood. The volume of cargo container is estimated around 40 cubic meters in the HDV and it is assumed 75 % of carrying capacity is used [20]. The basic information of HDV is shown in Table 28.2.

On the other hand, in Taiwan, marine transportation is an important part of its economy due to its running business with other countries and regions. Environmental impacts of marine transportation should not be ignored. Concerning greenhouse gas emissions from marine transportation, the United Nations Framework Convention on Climate Change (UNFCCC) has requested that the International Maritime Organization (IMO) estimates the emissions. Based on the Ocean Policy Research Foundation's figures, the annual total volume of marine transport in bulk carriers is 9.992 × 1012 tonne km, and the total annual fuel consumption is 3.7 × 107 tonnes [21].

Thus, long trip of carrying wood from North America to Taiwan can lead to some certain of environmental impacts. The boundary route of marine transport

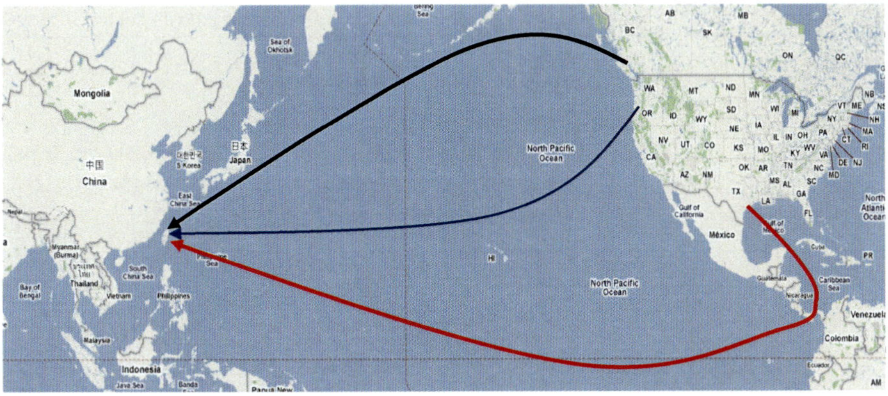

Fig. 28.2 Marine routes of wood transportation from North America to Taiwan

limits only from port to port, i.e., the ports from the countries of wood source to the port of Kaohsiung, which is the largest international port in Taiwan, the destination of marine transportation. Besides, there are thousands of vessel companies sailing from foreign countries to Taiwan and the exact data of vessels is difficult to obtain. However, in the following calculation, one of the largest vessel companies Evergreen Marine Corp. (EMC), which deals with cargo container ships, provides marine information for analysis. The wood marine transportation route can be illustrated in Fig. 28.2.

It has been investigated that the distance from the ports of Vancouver, Tacoma and Houston is 10,376, 10,606 and 19,503 km, respectively. Furthermore, in order to estimate environmental impacts of marine transport, the methodology used by the Network for Transport and Environment (NTM) are adopted. NTM is a non-profit organization initiated in 1993 that aims to establish a common base of values on how to calculate the environmental performance for various modes of transport. As known, all ships are individual with different characteristics. The data provided by NTM is not exact for any given ship, but comprises values measured and calculated over a great number of ships and engines [22]. Container ship with information of cargo capacity (7000 TEU), carrying capacity (80 %) and 75,000 DWT is taken into consideration in the analysis. Engine type has a strong effect on environmental performances during engine operation. Fuel consumption per km and its heating value are 0.163 tonne/km and 41 MJ/kg, respectively. Other information of container ship and emissions factors is shown in the following Table 28.3.

Data of Wood Structure Construction

In order to compare environmental performances of wood structures in Taiwan, the information of the usage of the amount of materials should be known. It has been investigated among several samples of wood houses that to construct wood

Table 28.3 Information of container ship and emissions factors

Information of container ship and emissions factors	
Cargo capacity (TEU)	7000
Carrying capacity (%)	80
Cargo volume (m^3)	25
DWT (tonne)	75,000
Fuel consumption (tonne/km)	0.163
Fuel heating value (MJ/kg)	41
CO_2 (kg/tonne fuel	3110
SO_2 (kg/tonne fuel)	50.3
PM (kg/tonne fuel)	3.65

structure on average requires materials of steel 22.4 (kg/m^2), concrete 0.17 (m^3/m^2), and wood (0.15 m^3/m^2) [13]. Materials of steel and concrete are used to construct the foundations, while wood materials are used for wall, roof, ceiling and pillars. In this paper, environmental impact factors of steel and concrete can be available from previous studies in Taiwan [23]. With this estimation, environmental burdens of materials can be calculated.

For example, concrete production processes include raw material extraction, cement production, and aggregate extraction. Then, a proper mixture of cement and aggregate makes concrete to meet physical standards. It has been estimated that energy consumption and CO_2 emissions of precast concrete per cubic meter are 1581 MJ and 253 kg respectively [23]. Another common building material is steel, which goes through many manufacturing processes, such as ore mining, blasting and casting into steel. Based on the previous studies in Taiwan, we adopt the data of environmental impacts of steel into analysis [24]. It is estimated that 7857 MJ of energy is consumed and 923 kg of CO_2 are released during manufacturing process per tonne of steel.

28.3 Results and Discussion

The results of the LCA based on the impact categories evaluated are presented in this part. Figure 28.3 shows relative percentage of environmental impacts of imported wood from three regions. Energy consumption requirements in manufacturing process account for the major parts, no matter where wood source is. Energy needs in manufacturing wood in the USA are by far higher than in Canada. From the previous study, wood from USA (SE) and USA (PNW) requires energy of 3475 and 3175 MJ/m^3 [18]; however, wood during manufacturing process in Canada only consumes 978.6 MJ/m^3 [17]. Therefore, relative percentage of energy demands of wood manufacturing in the USA account for more than 70 %. The reason may come from the fact that in Canada, manufacturing technique and equipment vary, leading to some difference of energy requirements.

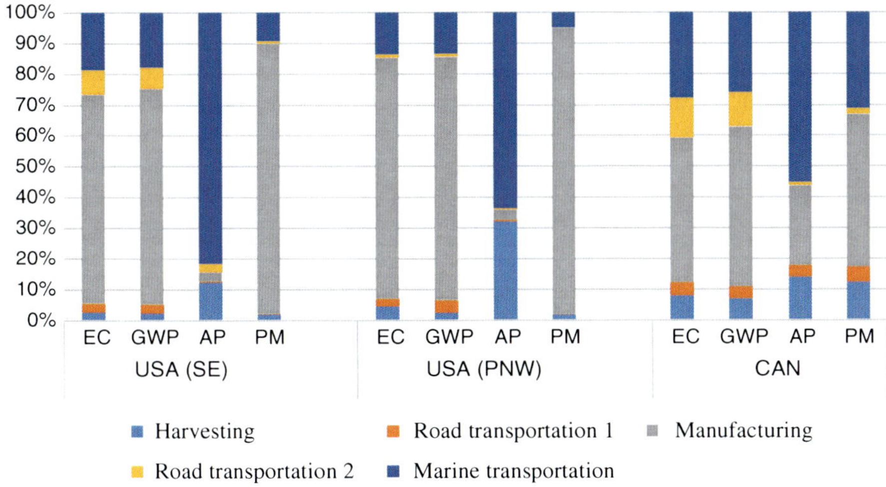

Fig. 28.3 Relative contribution analysis of life cycle of imported wood (road transportation 1: distance from forest to sawmill, road transportation 2: distance from sawmill to port)

In addition, global warming potential (GWP) is an important indicator related to climate change. In the USA, manufacturing process accounts for greater portion in relative percentage of environmental impacts compared with Canadian wood. In the USA, fossil fuel for wood production including coal, crude oil, natural gas contributes to 50 % of total energy needs [18], thus causing more carbon dioxide emissions. In Canada, much renewable energy such as biomass (wood residuals) is used in wood production. As known, biomass is regarded as non-CO_2 emissions or so called carbon neutral because it takes in CO_2 when it grows and releases it when it is burned as fuel. Therefore, wood production in Canada reduces global warming effect.

Acidification is another factor that causes acid rain, which has harmful effects on plants, aquatic animals and infrastructure. Acid rain is caused by emissions of sulfur dioxide and nitrogen oxide, which react with the water molecules in the atmosphere to produce acids. In this paper, sulfur dioxide is used as impact indicator. The results show that among three regions, marine transportation appears to be the major factor that has greatest acidification potential during life cycle process. Due to long transportation route, wood from USA (SE) region releases relatively the most in marine transportation, accounting for 82 % in LCA process, while wood from USA (PNW) and CAN account for 64 and 55 %, respectively. This result is quite similar to the research from Magelli et al. [25], who investigated environmental impacts of wood pellets, which were transported with long distance from Canada to Europe, and discovered that the factor of vessel transportation accounted for the greatest potential for acidification among wood harvesting, production and truck and train transportation. The reason could be the use of the type of fuel. Through the investigation, the engine of vessel uses special heavy residual oil, which has higher contents of sulphur, leading to a great amount of SO_2 emissions during the journey.

With the awareness of this environmental problems, using low-sulphur and distillate fuels for marine transportation is the goal for lessening acidification effect. According to the latest ISO 8217 standard [26], maximum sulfur content in the open ocean is 3.5 % since January 2012. Maximum sulfur content in designated areas is 0.1 % since 1 January 2015.

Human activities, such as the burning of fossil fuels in vehicles, power plants and various industrial processes generate significant amounts of particulates. Particulate matter (PM) has been one essential factor that causes air pollution and human lung disease. In LCA process, wood manufacturing has relatively higher PM values compared with transportation and wood harvesting. It can be noticed that wood from the USA contributes to more than 85 % of total PM values in production due to much use of fossil fuels in electricity energy mix. Coal combusting, as known, pollutes the air and releases particulates the most in comparison with other fossil fuels. Coal accounts for 48 % of total fossil fuels in wood production in the USA [18], while only about 22 % coal of total fossil fuels is used in wood production in Canada [17]. As a consequence of this, manufacturing process in the USA causes higher PM values in relative contribution than in Canada.

As for wood structure construction, factors of energy consumption and global warming potential (the most environmental impact indicators) are taken into analysis of different wood structures from various sources. Figure 28.4 shows energy requirements of wood structures from three regions. Energy consumption of wood materials is higher than concrete and steel due to greater amount of wood use in constructing wood house. Besides, wood from the USA consumes higher energy (1164, 1100 MJ/m^2) than that from Canada (784 MJ/m^2). If Taiwan imports wood from Canada, much energy is reduced compared with wood from the USA. But it can still be noticed that there is no great difference in energy consumption when wood is imported from Southeast or Pacific North West of the USA. The results are quite different from previous study completed by Tu [13] in Taiwan, i.e., wood structure only requires 508.7 MJ/m^2. The reason comes from the fact that previous research lacks considering long trip transportation factor and actual situation of importing wood from overseas. Even so, wood structure still consumes less energy than traditional structures of reinforced concrete and steel, which requires energy of about 2600 and 2100 MJ/m^2, respectively [24].

Fig. 28.4 Energy consumption of wood structures from three regions

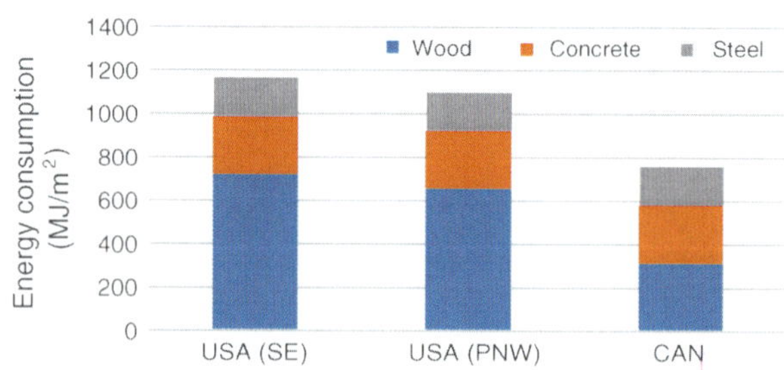

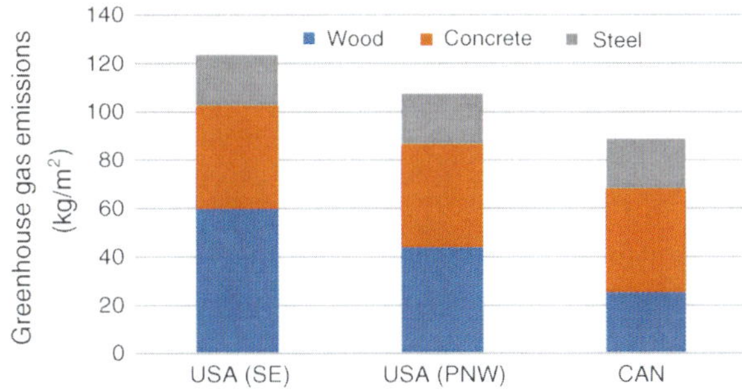

Fig. 28.5 Greenhouse gas emissions of wood structures from three regions

Figure 28.5 indicates greenhouse emissions of wood structures. Wood from Southeast of the USA releases highest CO_2 emissions (122 kg/m^2), compared with wood from PNW (108 kg/m^2) and Canada (89 kg/m^2). Therefore, importing wood from Canada for construction can mitigate global warming potential to some degree. When wood structure is compared with reinforced concrete and steel structures in Taiwan, CO_2 emissions are obviously less because reinforced concrete building releases 340 kg/m^2, much higher than wood structure, while steel structure building releases around 290 kg/m^2. On the other hand, it can be found that the results in this paper are also quite different from the previous study in Taiwan, which indicates wood structure releases only 57.4 kg/m^2 [13] due to ignoring long road and marine transportation from many regions around the world. As known, energy consumption and CO_2 emissions of wood manufacturing process generally are less than other building materials such as cement, steel, glass, brick and so on. Therefore, transportation factor of wood should not be ignored and be taken into account from cradle to gate perspective. As for metal materials, much energy is required for manufacturing, using coal, residual oil and thermal heating. In consequence, transportation factor of metal materials reveals less portion of energy consumption in total life cycle process, and thus transportation might not be the most essential factor that affects the environmental performances in a whole life cycle.

28.4 Conclusions

This study provides a life cycle assessment (LCA) from cradle to gate perspective to investigate energy consumption, greenhouse gas emissions, acidification potential and particulate matter of imported wood to Taiwan for construction from three major regions, Pacific Northwest (PNW) of the USA, Southeast (SE) of the USA and Canada (CAN). LCA is also applied to analyzing environmental performances of energy consumption and global warming potential of wood structure construction in Taiwan. The results indicate wood from the region of SE requires more energy (3475 MJ/m^3) than PNW (3175 MJ/m^3) and Canada (978.6 MJ/m^3). Likewise, CO_2 emissions of wood from SE of the USA release the most

(400 kg/m^3) among three regions. Besides, no matter where wood is imported, marine transportation reveals the major factor of acidification potential during wood life cycle process. For instance, the long marine trip causes wood from USA (SE) region to release SO_2, accounting for 82 % of total life cycle. Particulate matter effect occurs most in the process of manufacturing, especially for wood manufactured in SE and PNW of the USA. On the other hand, environmental performances of wood structure with source from Canada is obviously less than wood source from the USA. Thus, importing wood materials from Canada can reduce global warming effect and energy needs. Finally, compared with reinforced concrete structure and steel structure, wood structure has great benefits in our built-environment.

References

1. Thormark C (2006) The effect of material choice on the total energy need and recycling potential of a building. Build Environ 41(8):1019–1026
2. Venkatarama Reddy BV, Jagadish KS (2003) Embodied energy of common and alternative building materials and technologies. Energy Build 35(2):129–137
3. Abeysundara YUG, Babel S, Gheewala S (2009) A matrix in life cycle perspective for selecting sustainable materials for buildings in Sri Lanka. Build Environ 44:997–1004
4. Gonzalez MJ, Navarro JG (2006) Assessment of the decrease of CO_2 emissions in the construction field through the selection of materials: practical case study of three houses of low environmental impact. Build Environ 41(7):902–909
5. Gao W, Ariyama T, Ojima T, Meier A (2001) Energy impacts of recycling disassembly material in residential buildings. Energy Build 33(6):553–562
6. FAO (Food and Agricultural Organisation of U.N.) (1990) Energy conservation in the mechanical forest industries
7. Micales JA, Skog KE (1997) The decomposition of forest products in landfills. Int Biodeterior Biodegradation 39:145–158
8. Koch P (1992) Wood versus non-wood materials in US residential construction: some energy-related global implications. Forest Prod J 42(5):31–42
9. Buchanan AH, Honey BG (1999) Wood-based building material and atmospheric carbon emissions. Environ Sci Policy 2(6):427–437
10. Thormark C (2002) A low energy building in a life cycle-its embodied energy, energy need for operation and recycling potential. Build Environ, 429–435
11. Borjesson P, Gustavsson L (2000) Greenhouse gas balances in building construction: wood versus concrete from lifecycle and forest land use perspectives. Energy Policy 2(6):575–588
12. Lippke B, Wilson J, Perez-Garcia J, Bowyer J, Meil J (2004) CORRIM: life-cycle environmental performance of renewable building materials. Forest Prod J 54(6):8–19
13. Tu SH (2007) Contribution of carbon sequestration and carbon dioxide reduction by wood construction buildings in Taiwan. Ph.D. thesis, National Taiwan University
14. Li SH (2010) Using wood sustainably in buildings with low environmental impacts in Taiwan. In: Detail design in architecture 9 conference on innovative detailing, pp 81–90
15. Li SH (2011) Environmental impacts of building structures in Taiwan. J Procedia Eng 21:291–29
16. Ortiz O, Castells F, Sonnemann G (2009) Sustainability in the construction industry: a review of recent developments based on LCA. Constr Build Mater 23(1):28–39
17. Athena Sustainable Material Institute (2009) A cradle-to-gate life cycle assessment of Canadian softwood lumber

18. Puettmann ME (2006) Life cycle analysis of wood products: cradle to gate LCI of residential wood building materials. Wood Fiber Sci 37:18–29
19. UTMC (User Guide for transportation emissions calculator) (2012), Canada
20. NTM (Network for Transport and Environment) (2008) Environmental data for international cargo road transport, calculation methods, emission factors, mode-specific issues
21. OPRF (Ocean Policy Research Foundation) (2000) A report on research concerning the reduction of CO_2 emission from vessels
22. NTM (Network for Transport and Environment) (2008) Environmental data for international cargo sea transport, calculation methods, emission factors, mode-specific issues
23. Lin TS, Chang YS, Ou WS (2002) An analysis on environmental load of building material production in Taiwan. J Archit 40:1–15
24. Li SH (2011) Environmental impacts of building structures in Taiwan. J Procedia Eng 21:291–297
25. Magelli F, Boucher K, Bi HT, Melin S, Bonoli A (2009) An environmental impact assessment of exported wood pellets from Canada to Europe. Biomass Bioenergy 33:434–441
26. International Organization for Standardization (2010) ISO 8217. Petroleum products fuels (classF) specifications of marine fuels

Chapter 29
Stakeholder Impact Analysis for Highway Maintenance Management Based on System Dynamics

Hong Zhang, Rongbei Zheng and Fangmin Ren

Abstract It is necessary to study the long-term and dynamic relationships or interactions between the highway and stakeholders such as operator, government department and users in planning the highway maintenance system. The system dynamics method is adopted to model the long-term and dynamic relationships among the highway and stakeholders. The causal loop diagram and stock-flow diagram for the system dynamics model is developed based on the quantitative relationships among relevant variables such as pavement performance, total costs, maintenances policies, threshold to initiate maintenance operations, yearly government investment scheme and subsidy allocation rule, and users' attitude. Application of the proposed system dynamics model is demonstrated through a case study. This study provides an alternative platform to help stakeholders plan and evaluate maintenance polices and relevant management mechanisms from the perspective of a long time period.

Keywords Highway maintenance management · Stakeholders · Long-term and dynamic relationship · System dynamics

29.1 Introduction

Since the first highway was built in October 1988, the highway networks have been developed rapidly so as to possess the longest mileage in the world. Recently, however, the deterioration process of the highway pavement has been greatly addressed. The focus on large scale construction of highway networks in China is being transformed to maintenance management of the highway networks.

The highway operator may consider various maintenance measures with regards to different deterioration degrees of the highway pavement and cause different costs.

H. Zhang (✉) · R. Zheng · F. Ren
Institute of Construction Management, College of Architecture Engineering, Zhejiang University, Hangzhou, China
e-mail: jgzhangh@zju.edu.cn

Y. Wu et al. (eds.), *Proceedings of the 20th International Symposium on Advancement of Construction Management and Real Estate*,
DOI 10.1007/978-981-10-0855-9_29

The highway maintenance fund of China mainly comes from government investment and toll, so the maintenance fund is affected not only by the government but also by the users. Currently, extending of the highway tolling period in spite of loan repay time is being discussed and may be approved to ensure long-term maintenance demand. To make suitable decisions on the highway maintenance management system, it is crucial to evaluate the long-term and dynamic impacts or interactions among the maintenance results or costs of the highway network and stakeholders including operator's maintenance policies, government investment and allocation or supervision mechanism, and users' attitude towards to pavement condition or toll [1].

The current studies on the highway maintenance include optimization of pavement performance from the dynamic perspective [2], two-stage multi-objective optimal allocation of maintenance funds [3], and optimal budget allocation based on optimizing pavement maintenance of network level [4]. However, few studies addressed the long-term and dynamic interactions among multiple stakeholders in the highway maintenance management system. Some studies about stakeholders in infrastructure construction and management, such as fuzzy analysis of stakeholders' public participation satisfaction [5], double analysis of stakeholders' interactions [6], and social performance evaluation of stakeholders [7], could not reflect the long-term and dynamic relationships among the highway and stakeholders. Though some scholars introduced the agent-based simulation of infrastructure asset management activities [8] and the basic framework based on system dynamics [9], further studies on the highway maintenance management are needed based on the regional characteristics.

Therefore, this paper proposes and implements a system dynamics model of the highway maintenance management system, including the quantitative relationships between relevant variables, based on the characteristics of the highway maintenance management in China.

29.2 Analysis of the Highway Maintenance Management System

Pavement performance is the key indicator of the highway condition and is generally indicated by Pavement Condition Index (PCI). Long-term interactions of the traffic load and environment factors lead to deterioration of the highway pavement and decrease of PCI. PCI can also reflect the maintenance results.

The Highway operator is responsible for operation and maintenance of the highway networks. The operator ensures the highway to be used appropriately by implementing certain maintenance measures at suitable time. The operator needs to make decisions o the maintenance policies, such as maintenance measures and required cost, which maintenance measure should be adopted and minimum pavement condition or threshold to initiate corresponding maintenance measure.

The highway maintenance funds in China are mainly composed of government investment and users' toll that has been approved to be collected for life cycle of the highway regardless of the loan repaying deadline. To management the investment and toll rate with regards to the pavement performance and users' satisfaction, the government department needs to setup a relevant supervision mechanism. Therefore, the government department can divide the annual investment into two parts, i.e., direct budget to the highway operator and subsidy that is allocated to the operator for maintenance or used to complement the toll from the users. The allocation rule of the subsidy reflects the supervision mechanism of the government department based on the government satisfaction degree regarding both the pavement performance and users' satisfaction. Increase of the government satisfaction means increase of the government's willingness to support direct subsidy allocation to the operator and simultaneously means decrease of the subsidy for complementing the toll from the users, resulting in increase of the toll rate and users' paid fee. On the contrary, decrease of the government satisfaction means decrease of subsidy allocation to the operator and increase of the subsidy for the toll, resulting in reduction of the toll rate and users' paid fee.

The highway users are the key stakeholder in the highway maintenance management system. One objective of the highway maintenance management is to serve the users well. The highway pavement performance will influence the satisfaction degree of the users and their willingness to pay. Meanwhile, the amount of the toll also affects the satisfaction degree of the users, which is influenced by the government subsidy allocated to complement the toll. Therefore, the total satisfaction degree of the users about pavement condition and paid fee will affect the government's satisfaction degree, investment budget and subsidy allocation [10], correspondingly impact the total maintenance funds and the maintenance policies.

29.3 System Dynamics Model of the Highway Maintenance System

The system dynamics (SD) method able to handle the system issues of high order, nonlinear and multiple feedbacks [11] has been applied to solve various of construction management problems [12, 13, 14]. Causal loop diagram and stock-flow diagram are the core concepts of SD. The causal loop diagram describes the feedback structure of a system through variables and causal link. The stock-flow diagram quantitatively describes the causal relationships among the system variables based on the causal loop diagram.

29.3.1 *Causal Loop Diagram for the Highway Maintenance Management System*

Based on analysis of the highway maintenance management system, the causal loop diagram of correlated variables among the highway, operator, users and government department is established by using the SD software VENSIM. The "+" and "−" in the causal loop diagram mean the positive and negative influences respectively, as shown in Fig. 29.1. There are various causal loops in the highway maintenance system.

Causal loop 1 represents the loop between the variables around the pavement performance and deterioration process, i.e., PCI and deterioration rate. The increase of deterioration rate leads to worse pavement condition. Causal loop 2 represents the loop among the variables including users' total satisfaction, government satisfaction and willingness to support subsidy allocation to the operator, total highway maintenance fund, maintenance measures, pavement performance and users' satisfaction regarding pavement performance. Increase of users' total satisfaction will encourage the government department to allocate more investment, increase the pavement performance and users' total satisfaction. Causal loop 3 represents the loop among the variables including users' total satisfaction, uses' willingness to pay, users' paid fee, total maintenance fund, maintenance measures, pavement performance and users'

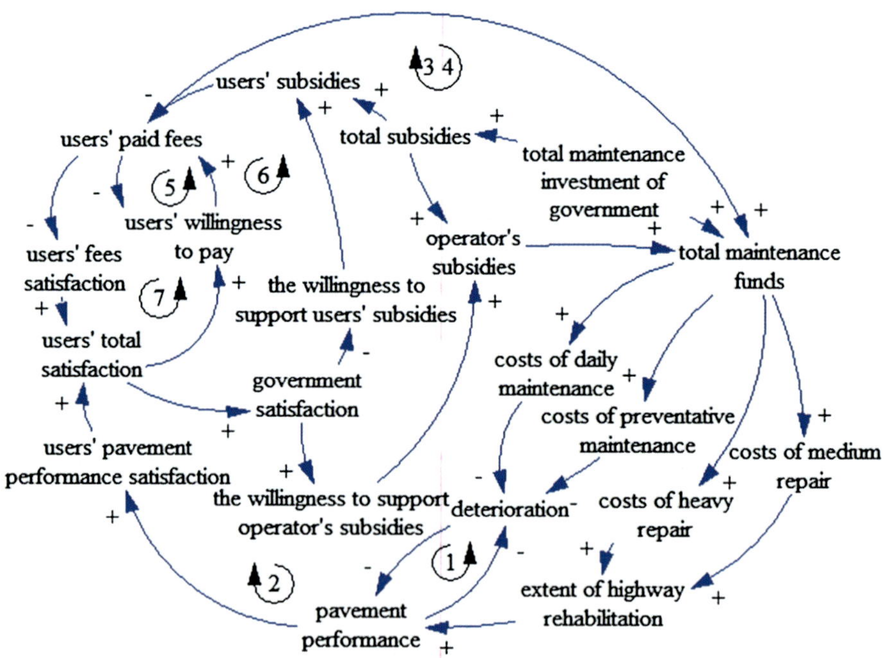

Fig. 29.1 Causal loop diagram for the highway maintenance management system

satisfaction regarding pavement condition. Increase of users' total satisfaction will increase users' willingness to pay, total maintenance fund, pavement performance and users' total satisfaction. Causal loop 4 represents the loop among the variables including users' total satisfaction, government satisfaction and willingness to support subsidy allocation to operator, subsidy allocation for toll, users' paid fee, total maintenance fund, maintenance measures, pavement performance and users' satisfaction regarding pavement condition. Since the total subsidy is fixed, increase of users' total satisfaction leads to increase of government satisfaction and willingness to allocate much subsidy to the operator, and thus simultaneously leads to decrease of the subsidy for toll and increase of the users paid fee. Causal loop 5 represents the loop between the variables such as users' paid fee and users' willingness to pay. Increase of users' paid fee due to decrease of subsidy for toll will decrease users' willingness to pay. Causal loop 6 represents the loop among the variables including users' total satisfaction, government satisfaction, government willingness to support subsidy allocation to the operator or subsidy allocation for toll, users' paid fee, and users' satisfaction regarding toll. Increase of users' total satisfaction will increase government satisfaction and willingness to support subsidy allocation to the operator, thus resulting in decrease of the subsidy for toll and correspondingly increase of users' paid fee. Causal loop 7 represents the loop among the variables including users' paid fee, users' satisfaction regarding paid fee or toll rate, users' total satisfaction and willingness to pay. Increase of the toll rate or users' paid fee will decrease users' satisfaction for toll and decrease users' willingness to pay.

29.3.2 *Stock-Flow Diagram for the Highway Maintenance Management System*

The SD model (Fig. 29.2) for the highway maintenance management system is established by using VENSIM based on the causal loop diagram shown in Fig. 1,

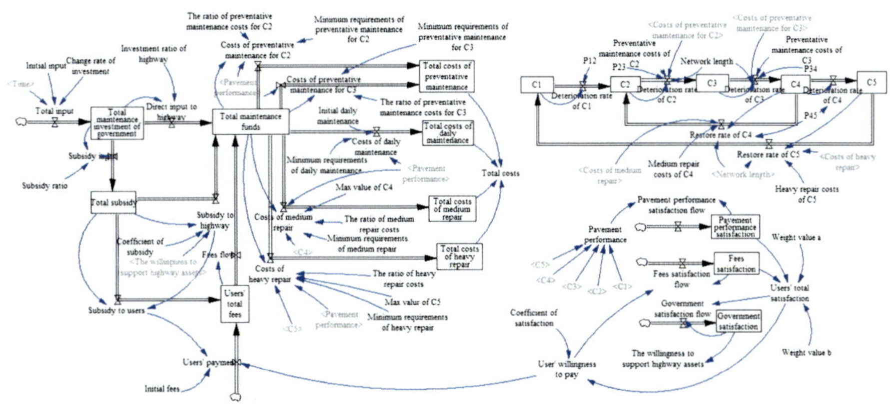

Fig. 29.2 Stock-flow diagram of the highway maintenance management system

which has been verified based on engineering data. This SD model is composed four subsystems that quantitatively describe the causal relationships among the highway and stakeholders such as highway operator, users and government department in the highway maintenance management system.

The subsystem around the highway deterioration process and maintenance activities is based on the Markov model [15]. The highway pavement performance that reflects the integrity or completeness of pavement structure can be classified as five discrete states including excellent, good, fair, poor and very poor, corresponding to different value-ranges of the Pavement Condition Index (PCI), i.e., [100 ~ 85, 85 ~ 70, 70 ~ 55, 55 ~ 40, 40 ~ 0]. Based on the Markov deterioration process model [15], the pavement performance condition can be represented as a set of probabilities for the five states, i.e., [C1, C2, C3, C4, C5], and the sum of the probabilities is equal to 1.0. Using the medium values of the PCI ranges for the five states, i.e., [92.5, 77.5, 62.5, 47.5, 20.0], the PCI for the current pavement performance is the expected value equal to 92.5 × C1 + 77.5 × C2 + 62.5 × C3 + 47.5 × C4 + 20.0 × C5.

The pavement deterioration process can be predicted through the Markov model [15], in which the state transition probability matrix represents the probability of changing performance from one state to another.

$$P = \begin{bmatrix} P_{11} & P_{12} & 0 & 0 & 0 \\ 0 & P_{22} & P_{23} & 0 & 0 \\ 0 & 0 & P_{33} & P_{34} & 0 \\ 0 & 0 & 0 & P_{44} & P_{45} \\ 0 & 0 & 0 & 0 & P_{55} \end{bmatrix} \tag{29.1}$$

When the pavement performance deteriorates, its state will become worse and leads to higher costs for the operator and users. Based on engineering experience, the pavement performance states can only be one of two cases, i.e., remaining unchanged or deteriorating to the next state when no maintenance measures are conducted. In addition, the pavement performance cannot transit from low level states to high level states. Therefore, the state transition matrix can be simplified as Eq. (29.1).

If current pavement performance is [C1, C2, C3, C4, C5] and the state transition probability is P, the pavement performance of the future time (e.g., next year) should be [C1, C2, C3, C4, C5] × P. In the stock-flow diagram of the SD model as shown in Fig. 29.2, the Markov deterioration process model is adopted, where each pavement state is modeled as one stock element and the flow between two adjacent stock elements (e.g., states i and j) represents the transition probability P_{ij}, such as P_{12}, P_{23}, P_{34} and P_{45}.

With regards to the five pavement states, five maintenance measures are respectively considered in this study. Daily maintenance including routine checking and simple maintenance to maintain highway's normal operation is carried out across all states. The preventative maintenance measure is to improve the transition probabilities for good state (C2) and fair state (C3), i.e., reducing P_{23} and P_{34} while increasing P_{22} and P_{33}. The heavy repair can restore the pavement performance from very poor state (C5) to excellent state (C1), while medium repair can restore the pavement performance from poor state (C4) to good state (C2). These maintenance measures will be performed cyclically during the life-cycle of the highway [16].

Based on the pavement conditions and available funds, the highway operator needs to make decisions on maintenance policies, including design of optional maintenance measures such as daily maintenance, preventative maintenance, medium repair and heavy repair, how to adopt these maintenance measures. In addition, the operator needs to setup minimum pavement conditions in terms of PCI for different performance states (e.g., C1, C2, C3, C4, C5) to initiate corresponding maintenance measures. The operator can also determine degrees of various maintenance measures, which causes different maintenance costs.

The subsystem around the users is to model the dynamic relationships among the users' satisfaction regarding pavement condition and toll, users' willingness to pay, government satisfaction, government's willingness to support subsidy allocation for user's toll, user's paid fee and total maintenance funds. The users' total satisfaction is sum of the satisfaction towards the pavement performance and the satisfaction towards paid fee. The significance of the users towards to pavement condition or paid fee is achieved through two weight values. The satisfaction degree regarding pavement condition is proportional to the PCI, while the satisfaction degree for toll is inversely proportional to the increase of toll rate. The users' total satisfaction degree affects not only the users' willingness to pay and the paid fee [17], but also the government's willingness to support the subsidy allocation, and accordingly the total maintenance funds.

The subsystem around the government is to model the relationships of the variables around government department, including investment, subsidy, government satisfaction or willingness to support subsidy allocation to operator or to for toll, users' paid fee, and total maintenance fund. The government department can determine the yearly investment and subsidy. The government satisfaction degree is proportional to the user's total satisfaction degree, while the subsidy allocation to the operator is proportional to the government satisfaction degree. Increase of the allocation to the operator will lead to decrease of the subsidy for users' toll.

29.4 Case Study

Maintenance management of ShenDa (Shenyang-Dalian) highway [18] is adopted to demonstrate the proposed SD model built based on VENSIM. Analysis on the maintenance management over 100 km of the highway is herein considered. Based on the measurements and data analysis, the initial pavement condition is [0.8145, 0.1809, 0.0046, 0.0, 0.0] and the state transition probability matrix before taking maintenance measures are obtained as follows [18]:

$$P = \begin{bmatrix} 0.8145 & 0.1855 & 0 & 0 & 0 \\ 0 & 0.6451 & 0.3549 & 0 & 0 \\ 0 & 0 & 0.5598 & 0.4402 & 0 \\ 0 & 0 & 0 & 0.5602 & 0.4398 \\ 0 & 0 & 0 & 0 & 1 \end{bmatrix}$$

It is assumed that the yearly government maintenance investment for the 100 km highway is 10 million, 50 % of which is directly allocated as maintenance fund and the other 50 % is used as subsidy. Meanwhile, 5 million of toll or users' paid fee is initially planned. The remaining fund from previous years will be used in the following years. Optional maintenance measures are adopted according to the model shown in Fig. 29.2. The cost for each maintenance measure is considered as shown in Table 29.1 based on experience [19].

In order to evaluate the impacts of the stakeholders on the pavement performance and maintenance costs, a series of simulation experiments based on the SD model for the maintenance management system over 60 years have been carried out by changing the variables that are respectively determine by the operator, government department and users. The simulation results are presented and analyzed as follows.

Figure 29.3 describes the pavement performances and total costs over time with different subsidy allocation rules. In this study the subsidy allocation rule is achieved through a coefficient reflecting the proportion size between the government's willingness to support subsidy allocation to operator and the government's satisfaction degree. High value of the coefficient reflects the lean of the subsidy allocation towards the operator, meaning reduction of the subsidy allocation for complementing users' toll, and vice versa. Figure 29.3 shows that the lean of the subsidy allocation towards the operator can lead to increase of the pavement per-

Table 29.1 Cost (RMB/km) for each maintenance measure

Maintenance measures	Daily maintenance	Preventive maintenance for C2	Preventive maintenance for C3	Medium repair	Heavy repair
Cost	60,000	80,000	100,000	500,000	2,000,000

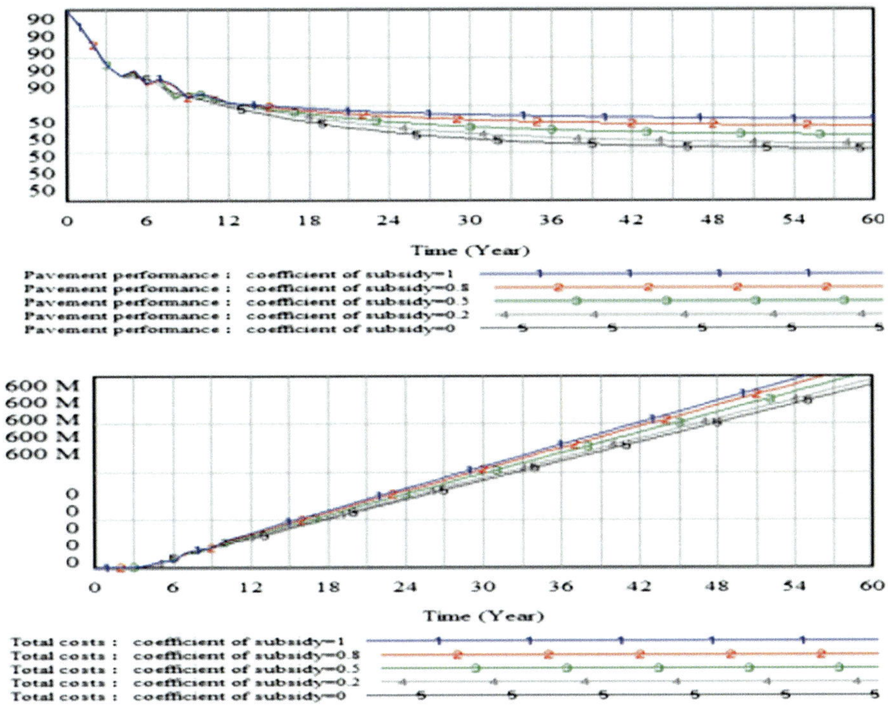

Fig. 29.3 Pavement performances and costs over time with different allocations of government subsidies

formance around 10 years later, though more costs are simultaneously caused. On the other hand, decrease of the subsidy allocation for toll will reduce users' satisfaction for toll and also user's total satisfaction degree. Therefore, the government department should determine suitable subsidy allocation rule so as to achieve dynamic supervision mechanism on the government investment and the toll rate.

Figure 29.4 describes the pavement performances and total costs over time with different minimum pavement conditions or thresholds determined by the operator to initiate daily maintenance, preventative maintenance, medium repair and heavy repair. It is shown that higher minimum pavement conditions or thresholds can lead to better pavement performances, especially during the period from 5 to 30 years, though no obvious differences of the total maintenance costs are caused. Hence, the operator needs to plan suitable maintenance policies including the threshold to

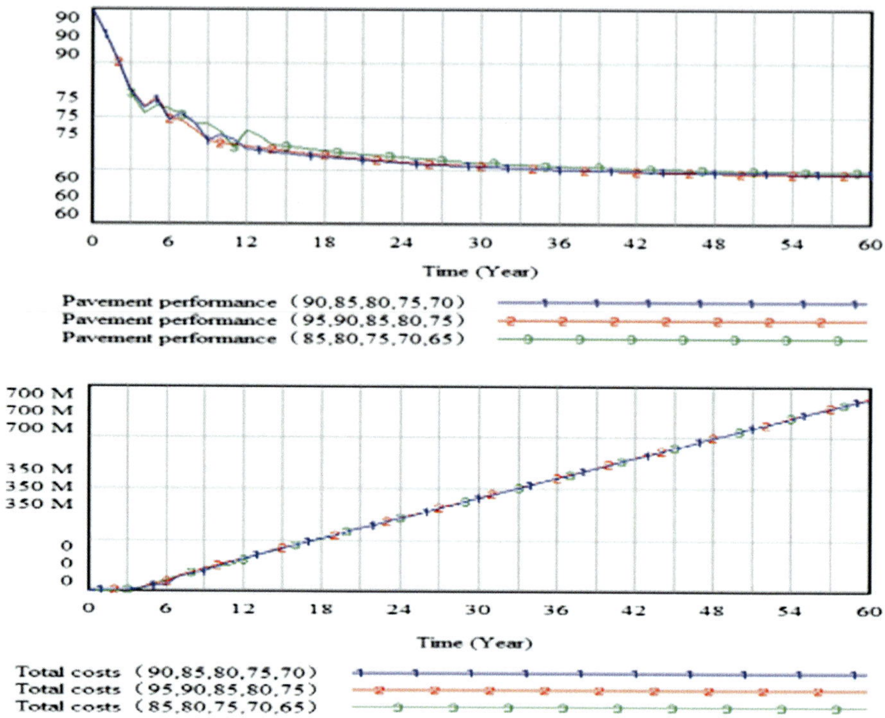

Fig. 29.4 Pavement performances and costs over time for different thresholds to initiate maintenance operations

initiate corresponding maintenance operation for each pavement state in the highway maintenance management system.

Figure 29.5 describes the pavement performances and total maintenance costs over time with different attitudes of the users towards pavement conditions and paid fee. Two weight values a and b are used to represent users' attitudes respectively towards pavement conditions and paid fee. It is shown that the bias of the users towards to pavement condition, i.e., e.g., line 3 with a = 1 and b = 0, leads to better pavement condition and also the total maintenance cost, while the bias of the users towards to paid fee, i.e., e.g., line 2 with a = 0 and b = 1, will decrease pavement condition and also total maintenance cost. Therefore, relative balance between the attitudes towards to pavement conditions and paid fee, e.g., line 1 with a = 0.6 and b = 0.4, will lead to certain tradeoff between the pavement performance and total maintenance cost. Meanwhile, it is demonstrated that good pavement conditions should be provided and toll rate should be supervised reasonably so as to avoid bias of the users.

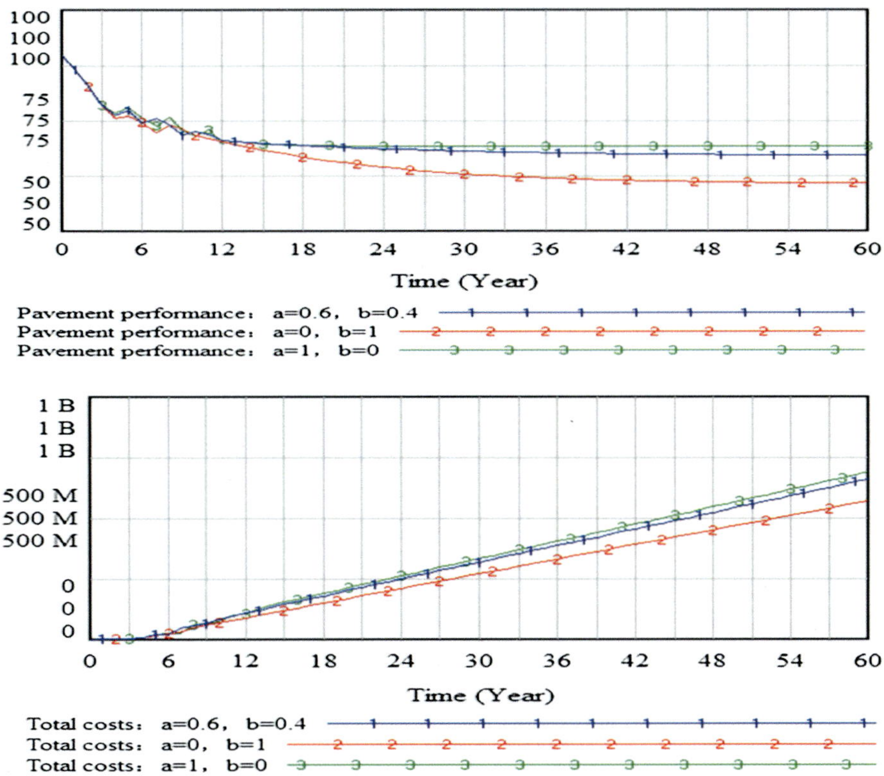

Fig. 29.5 Pavement performances and costs over time with different attitudes of the users

29.5 Conclusion

A SD model has been developed to model the long-term and dynamic relationships or interactions among the highway network and multiple stakeholders including highway operator, government department and users. The causal loop diagram and stock-flow diagram for the SD model have been developed based on the quantitative relationships among the variables around the highway maintenance management system and the relevant characteristics in China. The proposed SD mode has been demonstrated to be able to evaluate operator's maintenance policies such as threshold to initiate maintenance operations, government's yearly investment budget and subsidy allocation rule, and users' attitude towards pavement conditions or paid fee. The SD model is capable of helping stakeholders to make decisions on the long-term highway maintenance plan including maintenance policies and various management or supervision mechanisms on the government investment or toll rate.

References

1. Wu WJ (2009) Research on highway management system, Chang-an University
2. Fallah FS, Rahmandad H, Triantis K et al (2010) Optimizing highway maintenance operations: dynamic considerations. Syst Dyn Rev 26(3):216–238
3. Fwa TF, Farhan J (2012) Optimal multi asset maintenance budget allocation in highway asset management. J Transp Eng 138(10):1179–1187
4. Jesus M, Akyildiz S, Bish DR et al (2011) Network-level optimization of pavement maintenance renewal strategies. Adv Eng Inf 25(4):699–712
5. Li THY, Ng ST, Skitmore M (2013) Evaluating stakeholder satisfaction during public participation in major infrastructure and construction projects: a fuzzy approach. Autom Constr 29:123–135
6. Van der Lei TE, Herder PM (2011) A double analysis of stakeholder interaction in public infrastructure management. Facilities 29(13/14):563–576
7. Doloi H (2012) Assessing stakeholders' influence on social performance of infrastructure projects. Facilities 30(11/12):531–550
8. Osman H (2012) Agent-based simulation of urban infrastructure asset management activities. Autom Constr 28:45–57
9. Osman H, Ali M (2012) Complex systems modeling of infrastructure assets, operators, users, and politicians using system dynamics. Construction Research Congress, pp 2280–2289
10. Zhou Q, Zhou GP, Yu LJ (2009) The evaluation to users' satisfaction of highway service. J Traffic Transp Eng 06:80–86
11. Wang QF (1998) System dynamics. Tsinghua University Press, Beijing
12. Chow J, Kokotovic P (1985) Time scale modeling of sparse dynamic networks. IEEE Trans Autom Control 30(8):714–722
13. Feng L, Zhou J (2010) System dynamics model of engineering management in the mode of partnerring. Syst Eng 08:96–100
14. Cui X, Zhou KC, Cao DB et al (2011) Modeling the system dynamics model of commodity residence in Beijing and the application in early warning. Syst Eng Theory Pract 04:672–678
15. Liu K (2004) The process of practical Markov decisions. Tsinghua University Press, Beijing
16. Deng SY, Pan SP (2012) Prediction and analysis of the best time for the highway preventive maintenance. J Chongqing Technol Bus Univ (Nat Sci Ed) 08:69–73
17. Hu FS (2009) The study of impacts to recommendation intentions and willingness to pay of the visitors from tourism destination image. Zhejiang University, Zhejiang
18. Wang Y, Hu CB, Zhou LY et al (2000) The statistical features study of parameters of highway pavement performance. J Shenyang Architectural Civil Eng Inst 03:182–185
19. Hu CB, Zhou LC, Zhou LY (2000) The study of highway maintenance cost model. J Shenyang Archit Civil Eng Inst 01:7–9

Chapter 30
Integrated Information Management System of Building Materials Based on BIM Technology in Life Cycle Carbon Emissions

Zhenshuang Wang, Yijian Zhao and Xin Ning

Abstract Integrated management of building materials in life cycle carbon emissions is an issue problem in carbon emissions management of low carbon buildings. Considering function demand of carbon emissions integrated management of building materials, the information management system is established with building information modeling (BIM) as technical core and B/S as the network architecture, based on the life cycle theory. This study shows that integrated information management system of building materials carbon emissions can achieve exchange and sharing information, and collaboration work among different participants in different stage of building materials whole life cycle, and realize the integrated management of building materials in life cycle carbon emissions.

Keywords China · Construction industry · Contracts · Competition

30.1 Introduction

Global warming issue has caused widespread concern around the world, IPCC pointed out in the assessment report, Climate warming is primarily the result of human activity and the result of a lot of greenhouse gas emissions [1]. In recent years, carbon emissions have become the focus of global attention. Under the trend of global emissions, the Chinese government has actively take measures to the guarantee of economic development and reduced CO_2 emissions at the same time [2]. On the world conference of climate change in Copenhagen in 2009, the Chinese

---

Z. Wang (✉) · Y. Zhao · X. Ning
Investment Project Management Institute, Dongbei University of Finance and Economics, Dalian, China
e-mail: wangzhenshuang0609@126.com

Y. Wu et al. (eds.), *Proceedings of the 20th International Symposium on Advancement of Construction Management and Real Estate*,
DOI 10.1007/978-981-10-0855-9_30

government has made per unit of GDP by 2020 CO_2 emissions reduced by 40–45 % in 2005. Construction is the most potential industry energy saving and emission reduction. Statistics show that 60 % of the carbon emissions from buildings in a city material production [3, 4]. Therefore, to reduce carbon emissions building materials is the key to our country could live up to its promise. Building materials carbon management involves material production, transportation, construction, maintenance and removal of each stage of the whole life cycle. Li [5] combined the theory of life cycle assessment and had made a evaluation of ready-mixed concrete life cycle environmental impact, established the strength grade of ready-mixed concrete input and output listing. Wang [6] based on the full life cycle assessment theory, constructs the whole life cycle of cement material carbon listing, and the environment coordination of cement material is evaluated.

BIM is the digital expression of the physical and functional characteristics of the engineering project, and the engineering data model of the various relevant information of the construction project is integrated. It provides reliable information and basis for all decision making from the design to the whole life cycle of the whole life cycle, which has the characteristics of visualization, data integration, parameter, data association and so on. It can solve the defects of the traditional project management, and promote the cooperation of project management [7, 8]. BIM can effectively enhance the level of information management system of construction, combining the BIM and RFID technology, GIS technology, control the construction schedule, material and safety management, for the whole life cycle of engineering project management provides a new opportunity. Zhang [9] established BIMDISP system, through the creation and application of the full life cycle BIM platform, using BIM platform integration in the engineering construction cost, quality, safety and other information, and it realized the construction stage of manpower, materials and machinery resources dynamic management and real-time monitoring of the project cost. Hu [10] had established the system which is based on BIM technology mechanical and electrical equipment of intelligent management system. It implements electrical and mechanical equipment installation process of information sharing and operational stage, also for operations staff to provide efficient operational means, in order to ensure all of the subsystems of the mechanical and electrical equipment and its safe operation. This article will the lifecycle carbon emissions from building materials management according to the functional requirements of core based on BIM technology, based on B/S network architecture in whole life cycle of building materials carbon integrated management system, through the information means to promote the whole life cycle of building materials in our country the smooth implementation of carbon management.

30.2 Management System of Carbon Emission of Architectural Material Full Life Cycle

30.2.1 Integration Management of Carbon Emission of Architectural Material Full Life Cycle

Based on the full life cycle assessment theory, divided the whole life cycle of building materials for the production, transportation, construction, maintenance, and demolition of five stages, to determine the whole life cycle of building materials carbon system boundary elements and carbon emissions, as shown in Fig. 30.1. From each stage of whole life cycle of building materials according to the contact of systemic grasp the building materials planning, measurement and control of carbon emissions, to ensure that construction materials at different stages in whole life cycle of different party information sharing and collaborative work. This not only requires different participants in the whole life cycle of building material sharing information integration between different stages, but also between different parties to use the system software shall ensure seamless docking, realize the whole life cycle of building materials carbon integrated management.

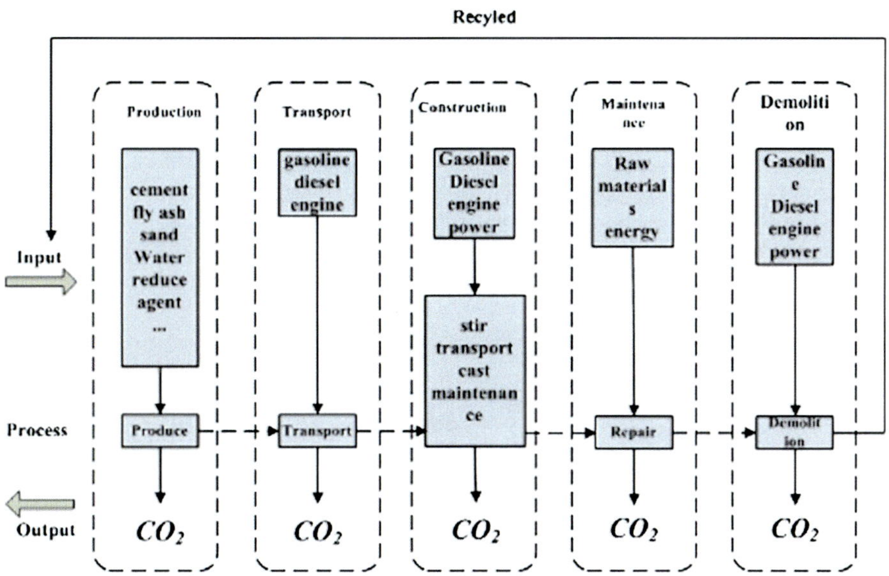

Fig. 30.1 Building materials all life cycle stage of carbon emissions

30.2.2 Functional Requirements of the Information Integration Management System of Carbon Emission of Architectural Material Full Life Cycle

The Information integration management system of Carbon emission of Architectural Material full life cycle is based on the lifecycle of the building materials management, computer aided carbon for building materials information collection, collation, storage, retrieval, transfer, maintenance and use of integrated system. In view of the whole life cycle of building materials of carbon management scope and functional requirements of each participant building materials carbon information integrated management system should have the following functions:

1. Open data information sharing platform. It provides an open data interface, making the whole life cycle of building materials of carbon emissions is different in different stages of the participation in order to get the required data information rapidly and accurately, breaking the fault and information island, realizing the flattening of information flow in the building materials of carbon emissions, improving the whole life cycle of building materials of carbon emissions information transparency, facilitating participants to use building materials carbon integrated information to make decisions.
2. The whole life cycle of building materials application system integration. Low carbon building construction process, involving many parties, different participants all have their own application software, enhance the stages in whole life cycle of building materials the interoperability of different software systems, can effectively control the loss of information between different system interface.
3. Building materials carbon function to the comprehensive information management system application. It required building materials carbon information management system covering the whole life cycle management of scope, system function to become more professional. In order to meet the whole life cycle of building materials of different parties in different stages of the individualized functional requirements.

30.2.3 Building Materials Carbon Emissions in Whole Life Cycle of Information Integration Management Information System Implementation

From the above carbon building materials according to the functional requirements of information integration management information system, implement the information integration management system need to go through the following link:

1. Determine the scope of the building materials of carbon emissions system and the system boundary, the IFC standard, on the basis of the system in the process of whole life cycle of building materials all process related information, classifying information, establishing unified coding system. Using database technology to store data, realize the whole life cycle of building material carbon information storage and fast reading.
2. Through an open data interface and general data exchange standard, implementation model of data sharing and conversion, data storage, data integration management and work together, to realize the integration of the data management, the whole life cycle of building materials in different stages of the different parties and different programming languages application software to realize the integration of the application system.
3. The whole life cycle of building materials carbon information was stored in an integrated and open model, and through the network communication technology, the realization of building materials carbon information transfer and sharing between different parties.
4. Through the BIM platform to implement the intelligent of the building materials of carbon emissions information integrated management, give full play to the initiative of information integration system of auxiliary decision-making management.

30.3 Whole Life Cycle of Building Materials Based on BIM Carbon Information Management System Architecture

30.3.1 The Working Principle of the Information Integration Management System of Carbon Emission of Architectural Material Full Life Cycle

Building materials carbon emissions in whole life cycle of information integration management system is in the network environment, through cost, schedule, carbon emissions and BIM database, in the building materials design, production, transportation, construction, maintenance and removal of different stage, materials manufacturers, suppliers, design, construction, owners and property management room by BIM data layer to obtain the required information, such as building material by integration of professional software measuring the operation cost, schedule, carbon emissions. Then through the API data interface in the IFC format feedback to management BIM components with other software in the effective docking, generate BIM building information model, realize the whole life cycle of building materials of carbon emissions information integrated management, for the various project participants through the information platform construction materials carbon work together. Including building materials 3D visualization work management subsystem, multi participation in collaborative information management subsystem and information intelligent decision subsystem, and finally achieve the different stages of building materials in the whole life cycle of different stages of carbon emissions management and effective integration. It can be calculated after the change in architectural design of carbon emissions data, improving work efficiency, and can carry out a number of programs to improve the efficiency of carbon emissions process management.

30.3.2 Whole Life Cycle of Building Materials Based on BIM Carbon Information Integration Management System Architecture

Based on whole life cycle of building materials management characteristics of carbon emissions data, according to the architecture characteristics of the information model, the project system architecture adopts B/S (browser/Server) mode, the core part of the system function on the Server, through a Web browser installed on the client Server to connect with the Server, using the Web hypertext transfer protocol (HTTP), realize information browsing through a wide area network. Through the network connection between terminal and server, terminal, easy to operation process information users in any place, including the operational layer, exchange layer, model layer and data service layer (as shown in Fig. 30.2).

1. The operational layer. In the operational level, different end users are endowed with different information browsing and editing rights, using Internet Explorer, via Http network hypertext transfer protocol, after get related authentication, enter the system.
2. The exchange layer. Exchange layer main realization model of BIM technology based on terminal equipment and network data transmission and management function, information transmission platform based on LAN or the Internet to build network, establish a "cloud" service platform. At the same time, the terminal equipment such as notebook computers and mobile phones can provide hardware support for BIM technology of data exchange.
3. The model layer. Through various applications with BIM technology as the core to load on the server, through the middleware to accept user's access instructions, then the server process the results back to the user to achieve the function building materials of carbon management.
4. The data service layer. Through the connection middleware, is responsible for the design for translation of instructions and data processing, such as reading, query, deletion, adding operation.

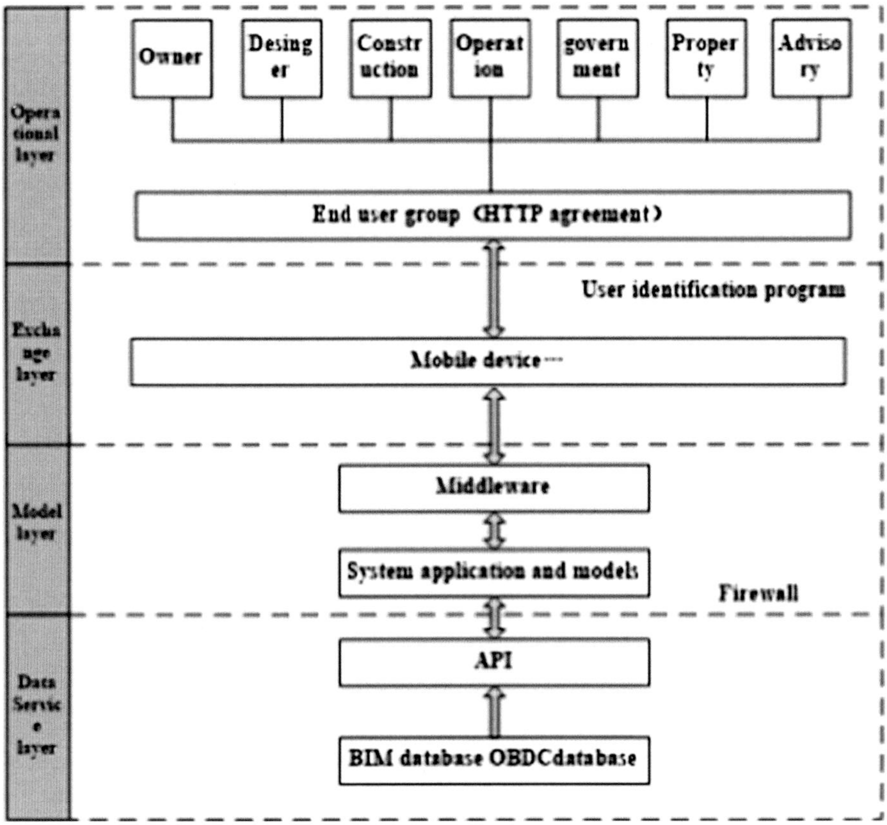

Fig. 30.2 The architecture of information management system of building materials carbon emission

30.4 Information Integration Management System Operating at the Integration of Whole Life Cycle of Building Materials

30.4.1 The Design Phase

Project participants to the minimum carbon emissions in whole life cycle of building materials, construction progress of the simulation based on 3D base model, cost data simulation and carbon emissions measurement, cost data building carbon emissions data, is used to select building materials, construction technology and technical solutions, etc., ensure the whole life cycle of building materials of the carbon emissions quota design, progress and cost control, limitation of BIM technology design value into full play. At the same time, the 3D model and previous reports data are stored in the database.

30.4.2 Stages of Production and Transportation

Building materials manufacturers can pass materials production process simulation, optimize the material formula and preparation technology, lower production costs and carbon emissions in the process of building materials. Through the system, procurement of construction materials can view the required material manufacturer and supplier distribution, product performance parameters, the cost of materials in building and double constraints of carbon emissions, through the system optimization, select materials manufacturers, suppliers, transport mode and transportation route, material inventory optimization, and connecting with the construction field will produce the contract, documents and other automatic classification of load in BIM database query, is advantageous for the building materials management.

30.4.3 The Construction Phase

Reasonable construction units from the perspective of whole life cycle of organization construction, use of BIM information integrated management system and reasonable control of building materials dosage, reasonable arrangement of artificial and mechanical equipment, such as convenience of carbon emissions in the process of construction control, doing fine management. In the process of construction, for engineering design changes, can quickly adjust construction materials of carbon emissions, and associated information, updated in real time, data sharing, the various project participants can be integrated in the information management system, rapid access to data, to make decisions.

30.4.4 Maintenance and Demolition Stage

Building materials through the BIM information integrated management system, from design, production, transportation, construction and maintenance, and demolition of the phase information seamless handover. Property management through the system parameters, and real-time monitoring operational stage maintenance information to judge the service life of the building materials and service state scientific management decision, take timely measures to control the materials of carbon emissions and service costs, and completes the disposal scheme, and will be generated and collected historical data automatically saved in the system database, to provide reference for other projects in the future.

30.5 Conclusions

The Information integration management system of carbon emission of Architectural Material full life cycle is the inevitable trend of the development of low carbon building information in accord with our country's basic national conditions and for future development of construction industry. Make full use of BIM technology power, meet different stages in whole life cycle of building materials of different parties need personalized features, to realize the digitization of whole life cycle of building materials, co-ordinate and fine management, effectively improve the management level of whole life cycle of building materials of carbon emissions. At the same time, the whole life cycle of building materials of carbon emissions information integrated management system still faces many difficulties, however, with the deepening understanding of BIM and BIM development of practical work, the information integrated management system will promote the mature development of low carbon buildings carbon management.

Acknowledgements The research reported in this paper was supported by the Education Department of Liaoning Province, China (No. LZ2014048, W2015130), China Postdoctoral Science Foundation funded project 2015T80259, and DUFE 2015Q15 and BSH201523. This support was gratefully acknowledged.

References

1. Wang ZS, Zhao N, Su HL (2015) Research on coupling coordinative degree of energy-economy-environment. Soft Sci 29(2):33–36
2. Yang G, Liu JX, Wu XY (2014) Carbon reduction targets and effects of emissions policy based on DSGE modeling. Res Sci 36(7):1452–1421
3. Lin XG, Zhou J, Cheng J et al (2012) Research on carbon emissions of building materials products. J Inf Technol Civil Eng Archit 4(3):13–18
4. Wang ZS, Wang LG (2015) Green degree evaluation and certification system of green building material. Constr Econ 1:98–112
5. Li XD, Wang S, Kong XQ et al (2011) Life cycle assessment of environmental impacts of ready-mixed concrete. China Civil Eng J 44(1):132–137
6. Wang ZS, Zhao N (2015) Evaluation of cement materials environmental harmonization based on group AHP and matter element analysis model. J Eng Manage 29(1):48–52
7. Zhang CX (2011) Research on the application of BIM technology in building industry in China. Constr Econ 9:96–98
8. Tang XL, Tian CX (2013) Research on the technology diffusion and application of building information model (BIM). Constr Econ 6:98–100
9. Zhang JP, Yu FQ, Li D (2012) A modeling technology of integrated BIM for building lifecycle. J Inf Technol Civil Eng Archit 4(1):6–14
10. Hu ZZ, Chen XX, Wang L et al (2013) A BIM-based facility intelligent management system. J Inf Technol Civil Eng Archit 5(1):17–21

Chapter 31
Overview of the Application of Social Network Analysis in Construction Engineering and Management Research

Xian Zheng, Yun Le, Albert P.C. Chan and Yi Hu

Abstract Over the last two decades, social network analysis (SNA) approach has gained increasing popularity as an advanced tool in Construction Engineering and Management (CEM) field due to the emerging viewpoint of viewing projects as organizational networks. However, limited studies are available on a systematic review of the use of this tool in CEM field. Therefore, this paper aims to fill this gap by reviewing the SNA papers published in nine selected peer-reviewed journals from 1998 to 2015 to ascertain the state of the art. Based on a three-stage review, 45 papers were identified. It was found that scholars from the United States, the United Kingdom and Australia published the most number of SNA-based papers, which indicated universities (institutions) and scholars from these countries enjoy significant advantage in SNA-based research. The review results provide practical references for future SNA enable applications in CEM research.

Keywords Social network analysis · Construction engineering and management · Review

31.1 Introduction

The last two decades have witnessed the rapid growth of research activities devoted to advance the understanding of social networks [1–3]. The social network (SN) is used to represent a pattern of ties that exist among different entities (nodes) such as individuals and organizations [4]. The concept of SNA was first introduced in the 1930s to tackle the social and political relationships between individuals [5]. Moreno [5] defined SNA as 'a quantitative analytic tool which can study the resources

X. Zheng (✉) · Y. Le · Y. Hu
School of Economics and Management, Tongji University, Shanghai, China
e-mail: sine2588@163.com

A.P.C. Chan
Department of Building and Real Estate, The Hong Kong Polytechnic University, Hung Hom, Hong Kong

Y. Wu et al. (eds.), *Proceedings of the 20th International Symposium on Advancement of Construction Management and Real Estate*,
DOI 10.1007/978-981-10-0855-9_31

exchange among different groups'. Alternatively, it is defined by Haythornthwaite [6] as 'an approach and set of techniques used to study the exchange of resources among actors'. Nowadays, various relationships among individuals and organizations have been investigated in distinct research fields [7], such as economic, political, interactional and affective. Unlike many traditional statistics and data analysis which focus on the attribution of entities, SNA is distinct because it is based on an assumption which pays more attention on the importance of relationships among interacting unites through a set of analytic concepts and methods.

As an advanced technique, SNA has developed over the last two decades and gained increasing popularity in the CEM field due to the increasing recognition that a construction project can be viewed as a network of firms working together for the purpose of the project and thus be analyzed in terms of networks of relationships [8]. As noted by Lin [9], applying SNA to research on construction project management can understand the actual management structures, identify central figures in an organization, and explore the potential threats that cause engineering mistakes.

Literature review is regarded as a useful methodology to gain in-depth understanding of the status quo of a research topic, especially through retrieving from academic journals. As to date, no such critical review has been undertaken regarding application of SNA method in CEM field. Therefore, the aim of this paper is to conduct an overview of the application of SNA techniques in CEM research that has been published in nine selected peer-reviewed journals between 1998 and 2015. A systematic examination of existing publications can help researchers understand the current body of knowledge and stimulate their inspirations for future SNA-based applications in CEM research. Specific objectives of this study are to:

1. What was the coverage of SNA-based research published in CEM journals from 1998 to 2015?
2. What did contributions of countries/regions and institutions to SNA-based research during the same period?

Because of the page limit imposed by the conference, this paper only provides the preliminary findings of the study. Full details of the study will be reported separately in a journal paper to be submitted after receiving feedback from the conference.

31.2 Overview of the SNA

In essence, SNA can describe, visualize and statistically model the ecology of social networks through a set of techniques [10]. In social network, nodes and their behavior are regarded as interdependent rather than independent. Network ties among nodes are defined as flows to transfer resources which can develop into paths. The network models (i.e. ego network and whole network) are regarded as an environment that constrains behavior of individuals (or organizations). Basically,

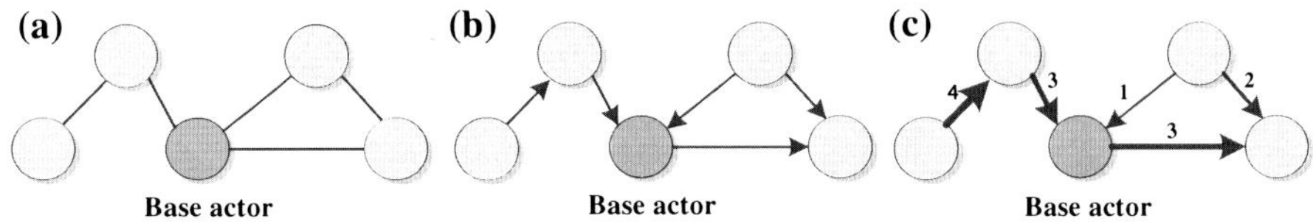

Fig. 31.1 Examples of social network [7]. **a** Undirected network. **b** Directed network. **c** Valued directed network

network can be divided into two types focusing on a tie: nondirectional (symmetric) and directional (nonsymmetric). Figure 31.1a illustrates that a base actor (dark node) links with three neighbors directly and it also links with other one node indirectly through one neighbor. However, the direction of the relationship can be either inward or outward, referring to input and output of actors. Figure 31.1b demonstrates an actor receiving two ties (input) and sending one tie (output), so when it is necessary to investigate the relationships between active and passive actors, directed network is useful. Furthermore, the strength or intensity of each tie can be taken into account. Seen from Fig. 31.1c, valued directed network is a graph in which each tie carries a value [3] and the width of the tie can be represented based on the value. It is commonly admitted that such kind of network can better reflect the reality, such as the different frequency of communication.

SNA enables researchers to focus on connections and interactions in networks by applying mathematical analysis [11]. There are various SNA metrics which are mainly focus on finding out the more central entities than others in a network [12–14], including density, centrality, average distance and diameter. Freeman [15] proposed to assess the central entities by three main metrics of centrality, namely, degree, closeness and betweenness. Degree merely concentrates on direct linkage with a focal entity [16, 17] while the measure of closeness and betweenness assess an entity's contribution to the whole network based on the shortest path and its length as well as location among entities in the network [18]. By determining the scopes of these measures and their relationships, more social network principles and properties have been put forward and applied to understand various social phenomena over the past decades, such as cohesiveness (assessed by density and inclusiveness etc.), component (assessed by cluster and clique etc.), structural position and role (assessed by structural equivalence and structural hole etc.). The development of SNA method has witness a more widespread application in distinct scientific areas, including CEM.

31.3 Research Methodology

This work primarily adopted a structured method advocated by Ke et al. [19] to identify and assess the major outputs of SNA-based research published in peer-reviewed journals. The entire research process included three stages. In Stage

1, two academic databases, the ISI web of knowledge and Scopus, were searched for relevant publications under the "title/abstract/keyword" field of key items "social network analysis" (SNA) and "construction industry". While the use of "construction industry" is too narrow to include all relevant articles, "construction" is used as keyword to search directly by making reference with the journal ranking list of Chau [20] in the CEM area [21].

In Stage 1, the results reflected that the Journal of Construction Engineering and Management (JCEM) and Construction Management and Economics (CME) have published the most SNA-based papers and thus were selected as target journals in Stage 2. Stage 2 conducted a more comprehensive search of all target journals instead of via the search engine. The scope of publication search was scaled down to a time span of 1 January 1998 (when the first peer-reviewed paper employing SNA was published in construction journals) to 1 May 2015. After a brief review of the paper contents, a total of 45 publications were obtained for further analysis. Table 31.1 shows the research result of this study.

In Stage 3, the 45 articles were quantitatively analyzed to determine their contribution by year, country, institution and citation. The scoring method developed by Howard et al. [22] was used to assess the contribution value of each author in multi authored articles. Citations of journal articles were used as a key index to assess research quality [23]. Google Scholar was used to determine the citation status of the journal articles identified. Although Google Scholar provides only an indirect citation report, its powerful search function is a simple yet thorough channel used to acquire such citation reports [24].

Table 31.1 Distribution of selected journal papers

Journal title	Number of selected paper
Journal of Construction Engineering and Management (JCEM)	17
Construction Management and Economics (CME)	12
Journal of Management in Engineering (JME)	5
International Journal of Project Management (IJPM)	4
Building Research and Information (BRI)	2
Project Management Journal (PMJ)	2
Building and environment (B&E)	1
Engineering, Construction and Architectural Management (ECAM)	1
Decision Support Systems (DSS)	1
Total	45

31.4 Discussions of Search Results

31.4.1 Number of SNA-Based Paper Published Annually

Table 31.2 shows the distribution of the identified SNA-based papers over the period of 1998–2015. The literature search work was conducted between May and June in 2015. As shown by Table 31.2, an increasing number of journal papers have adopted SNA techniques since 1998, especially in the last five years, which indicated that there existed a growing trend in CEM research studies with SNA techniques adopted.

31.4.2 Contributions of Countries/Regions and Institutions to SNA Research

By calculation, the score of a specific writer in a multi-authored paper can be obtained. Table 31.3 shows the country/region origins of SNA-based papers, the number of institute/university involved, the number of researchers involved and scores for each country/region. Among the identified countries, the United States published 29 papers and scored top with 15.01. This is understandable many scholars in United States universities in CEM research make a great contribution to SNA application, such as Stephen Mead from Northern Arizona University and Paul Chinowsky from University of Colorado at Boulder. It is also worth noting that merely 16 countries/regions have begun to employ SNA method, which verified that SNA application to civil engineering is still in the early stage [9].

The publications in the nine selected journals witnessed 46 research institutions devoted to SNA-based studies. Table 31.4 shows that seven research institutions were involved in at least three papers. All of them come from the top three countries in Table 31.3, which assert the result in Table 31.3. Overall, the total contribution of researchers from University of Colorado at Boulder, have published 7 papers involving 11 researchers and scoring 6.26, was ranked the highest among all identified universities/institutions, as shown in Table 31.4. University College London and Virginia Polytechnic Institute and State University have contributed 5 and 4 papers respectively, followed by University of New South Wales, Loughborough University, North Carolina State University and Texas Tech University which contribute 3 papers respectively.

Table 31.2 Identified papers using SNA published between 1998 and 2015

Year	98	99	00	01	02	03	04	05	06	07	08	09	10	11	12	13	14	15
Quantity	1	1	0	2	0	0	1	2	2	0	1	1	4	6	4	6	8	6

Table 31.3 Research origin of SNA-based articles published

Country/region	Institution/university	Researchers involved	Total number of papers	Score
United States	12	29	18	15.01
U.K.	9	15	10	9
Australia	5	10	6	5.6
Taiwan (China)	3	4	4	3.32
France	2	4	2	2
China (mainland)	2	6	2	2
Singapore	2	4	2	1.68
Korea	2	3	2	1.04
Chile	1	2	1	1
Israel	1	2	1	1
Italy	2	2	1	1
Bahrain	1	1	1	0.6
Turkey	1	2	1	0.54
Netherlands	1	2	1	0.53
Germany	1	1	1	0.47
Finland	1	1	1	0.21

Table 31.4 Research institutions publishing SNA-based articles in at least 3 papers

Ranking	University/Institution	Country/Region	Researchers	Papers	Scores
1	University of Colorado at Boulder	United States	11	7	6.26
2	University College London	U.K.	3	5	4
3	Virginia Polytechnic Institute and State University	United States	2	4	0.93
4	University of New South Wales	Australia	3	3	2.68
5	Loughborough University	U.K.	5	3	1.36
6	North Carolina State University	United States	3	3	1.36
7	Texas Tech University	United States	1	3	0.7

31.5 Conclusions

SNA technique is increasingly applied in the research area of CEM during the last two decades, but no paper has attempted to draw up a holistic commentary of the existing SNA literature. To fill up this research gap, this paper has conducted a comprehensive review on the SNA-based papers published in nine selected peer-reviewed journals from 1998 to 2015. Based on a three-stage review, 45 papers were identified. In addition, the number of articles published annually, institutional and regional contributions, citations were analyzed and summarized in this study. The analysis results reveal a growing interest in SNA-based research, particularly in the past five years, and three developed countries encompassing the United States, the United Kingdom and Australia contribute most to the development of SNA research in CEM field. Several universities (institutional) and scholars from these countries are most dominant to lead the SNA-based research.

After conducting a comprehensive literature review on the applications of SNA in construction management research, it is recommended that SNA can be adopted in the study of various interdependent relationships and activities. It provides a mathematical quantification of actor roles and positions in the study of iterative processes and transient systems, particularly for comparative purposes [25]. The review of these studies has provided insights for designing future research agendas in CEM sector.

Acknowledgments The writers gratefully acknowledge the National Natural Science Foundation of China (Grant No. 71390523 and Grant No. 71471136), as well as the Joint Supervision Scheme with the Chinese Mainland, Taiwan and Macao Universities from Hong Kong Polytechnic University for providing funding to support this research effort.

Appendix: SNA-Based Papers in Target Construction Journals

No.	Authors	Title of paper	Year	Journal
1	Priven and Sacks	Spanning cultural and geographic barriers with knowledge pipelines in multinational communities of practice	2015	JCEM
2	Lin	An analysis for construction engineering networks	2015	JCEM
3	Aljassmi et al.	Project pathogens network new approach to analyzing construction defects generation mechanisms	2014	JCEM
4	Arriagada and Alarcón	Knowledge management and maturation model in construction companies	2014	JCEM

(continued)

(continued)

No.	Authors	Title of paper	Year	Journal
5	Abbasian-Hosseini et al.	From social network to data envelopment analysis identifying benchmarks at the site management level	2014	JCEM
6	Zhang et al.	Sharing tacit knowledge for integrated project team flexibility case study of integrated project delivery	2013	JCEM
7	Alsamadani et al.	Relationships among language proficiency, communication patterns, and safety performance in small work crews in the United States	2013	JCEM
8	Solis et al.	Hybrid approach to the study of inter-organization high performance teams	2013	JCEM
9	Wambeke et al.	Using Pajek and centrality analysis to identify a social network of construction trades	2012	JCEM
10	Park et al.	Social network analysis of collaborative ventures for overseas construction projects	2011	JCEM
11	Chinowsky et al.	Project organizations as social networks	2010	JCEM
12	Thorpe and Mead	Project-specific web sites friend or foe	2001	JCEM
13	Priven and Sacks	Effects of the last planner system on social networks among construction trade crews	2015	JCEM
14	Chinowsky et al.	Social network model of construction	2008	JCEM
15	Lin	An analysis for construction engineering networks	2014	JCEM
16	Son and Rojas	Evolution of collaboration in temporary project teams an agent-based modeling and simulation approach	2011	JCEM
17	Ning and Ling	Boosting public construction project outcomes through relational transactions	2014	JCEM
18	Lingard et al.	Exploring the link between early constructor involvement in project decision-making and the efficacy of health and safety risk control	2014	CME
19	Sanaei et al.	The influence of generation on knowledge sharing connections and methods in construction and engineering organizations headquartered in the US	2013	CME
20	Alsamadani et al.	Measuring and modelling safety communication in small work crews in the US using social network analysis	2013	CME
21	Ruan et al.	Knowledge integration process in construction projects a social network analysis approach to compare competitive and collaborative working	2012	CME

(continued)

(continued)

No.	Authors	Title of paper	Year	Journal
22	Larsen	Understanding the early stages of the innovation diffusion process awareness, influence and communication networks	2011	CME
23	El-Sheikh and Pryke	Network gaps and project success	2010	CME
24	Hossain	Communications and coordination in construction projects	2009	CME
25	Larsen and Ballal	The diffusion of innovations within a UKCI context an explanatory framework	2005	CME
26	Pryke	Analysing construction project coalitions exploring the application of social network analysis	2004	CME
27	Pryke	Towards a social network theory of project governance	2005	CME
28	Chowdhury et al.	Analysing the structure of public private partnership projects using network theory	2011	CME
29	Loosemore	Responsibility, power and construction conflict	1999	CME
30	Lin and Tan	Performance measurement in the public sector example of the building administration authorities in Taiwan	2014	JME
31	Dogan et al.	Assessing coordination performance based on centrality in an e-mail communication network	2015	JME
32	Wambeke et al.	Task variation and the social network of construction trades	2014	JME
33	Chinowsky et al.	Project network interdependency alignment new approach to assessing project effectiveness	2011	JME
34	Di Marco et al.	Emergence and role of cultural boundary spanners in global engineering project networks	2010	JME
35	Pauget and Wald	Relational competence in complex temporary organizations The case of a French hospital construction project network	2013	IJPM
36	Di Vincenzo and Mascia	Social capital in project-based organizations its role, structure, and impact on project performance	2012	IJPM
37	Brookes et al.	Social processes, patterns and practices and project knowledge management: a theoretical framework and an empirical investigation	2006	IJPM

(continued)

(continued)

No.	Authors	Title of paper	Year	Journal
38	Liu et al.	Evolutionary analysis of the collaboration networks within National Quality Award Projects of China	2015	IJPM
39	Mead	Using social network analysis to visualize project teams	2001	PMJ
40	Di Marco et al.	Exploring negotiation through boundary objects in global design project networks	2012	PMJ
41	Pryke and Pearson	Project governance case studies on financial incentives	2006	BRI
42	Pryke et al.	Resource provision ego-networks in small Greek construction firms	2011	BRI
43	Yang and Zou	Stakeholder-associated risks and their interactions in complex green building projects a social network model	2014	B&E
44	Loosemore	Social network analysis: using a quantitative tool within an interpretative context to explore the management of construction crises	1998	ECAM
45	Hossain	Simulation-based risk network model for decision support in project risk management	2012	DSS

References

1. Lewis K et al (2008) Tastes, ties, and time: a new social network dataset using Facebook.com. Soc Netw 30(4):330–342
2. Liljeros F et al (2001) The web of human sexual contacts. Nature 411(6840):907–908
3. Wasserman S, Faust K (1994) Social network analysis: methods and applications. Cambridge University Press, Cambridge
4. Wambeke BW, Hsiang SM, Liu M (2011) Causes of variation in construction project task starting times and duration. J Constr Eng Manage 137(9):663–677
5. Moreno JL (1960) The sociometry reader. The Free Press, Glencoe, IL
6. Haythornthwaite C (1996) Social network analysis: an approach and technique for the study of information exchange. Libr Inf Sci Res 18(4):323–342
7. Park H et al (2011) Social network analysis of collaborative ventures for overseas construction projects. J Const Eng Manage 137(5):344–355
8. Pryke S, Pearson S (2006) Project governance: case studies on financial incentives. Build Res Inf 34(6):534–545
9. Lin SC (2014) An analysis for construction engineering networks. J Const Eng Manage 141(5):4014096
10. van Duijn MA, Vermunt JK (2006) What is special about social network analysis? Methodology 2(1):2–6
11. Chinowsky P, Diekmann J, Galotti V (2008) Social network model of construction. J Const Eng Manage 134(10):804–812

12. Freeman LC (1977) A set of measures of centrality based on betweenness. Sociometry 40(1):35–41
13. Bonacich P (1987) Power and centrality: a family of measures. Am J Soc 92(5):1170–1182
14. Borgatti SP (2005) Centrality and network flow. Soc Netw 27(1):55–71
15. Freeman LC (1979) Centrality in social networks conceptual clarification. Soc Netw 1(3): 215–239
16. Newman ME (2001) Scientific collaboration networks. II. Shortest paths, weighted networks, and centrality. Phys Rev E 64(1):16132
17. Barrat A et al (2004) The architecture of complex weighted networks. Proc Natl Acad Sci USA 101(11):3747–3752
18. Opsahl T, Agneessens F, Skvoretz J (2010) Node centrality in weighted networks: generalizing degree and shortest paths. Soc Netw 32(3):245–251
19. Ke Y et al (2009) Research trend of public-private partnership in construction journals. J Const Eng Manage 135(10):1076–1086
20. Chau KW (1997) The ranking of construction management journals. Const Manage Econ 15(4):387–398
21. Xiong B, Skitmore M, Xia B (2015) A critical review of structural equation modeling applications in construction research. Autom Const 49:59–70
22. Howard GS, Cole DA, Maxwell SE (1987) Research productivity in psychology based on publication in the journals of the American Psychological Association. Am Psychol 42(11):975–986
23. Hong Y et al (2011) Critical analysis of partnering research trend in construction journals. J Manage Eng 28(2):82–95
24. Hu Y et al (2013) From construction megaproject management to complex project management: bibliographic analysis. http://ascelibrary.org/doi/pdf/10.1061/(ASCE)ME.1943-5479.0000254
25. Pryke SD (2008) Social network analysis. In: Knight A, Ruddock L (eds) Advanced research methods in the built environment. Blackwell, Chichester, pp 171–182

Chapter 32
A Risk Management Approach for Prediction of Contingency Sum for Public-Sector Construction Projects

Terence Lam and Njavwa Siwingw

Abstract A review of the literature supports the inclusion of sufficient contingency to cover risks in construction projects. From the client's point of view, too much contingency may result in a project being aborted or uneconomical and too little may lead to cost escalation if the risk occurs. Risk factors at the construction phase causing cost overruns will be identified and an accurate method for estimation of contingency sum will be determined. Qualitative interviews were conducted with five expert practitioners working in a public works department in Zambia to determine how the contingency sum is estimated and what risk factors are considered. Multiple regression analysis was conducted using cost and risks data collected from 30 building and refurbishment projects recently completed in the department. The qualitative study found that project budget overruns constitute a major issue. The regression analysis results proved that the contingency sum was positively correlated to the estimated contract sum. The qualitative interview results and Pearson correlation coefficient showed there was also a positive correlation between contingency sum and project complexity, although such factor needs to be further reaffirmed by regression analysis using a larger sample size. To enhance project success in terms of cost, it is necessary that project managers should actively implement risk management in projects when calculating the contingency sum. Related project-specific risks should be identified and multiple regression method can be used to predict the contingency fund accordingly.

Keywords Construction projects · Contingency sum · Construction phase · Risk management · Public-sector

T. Lam (✉)
Faculty of Science and Technology, Technological and Higher Education Institute of Hong Kong, Hong Kong, China
e-mail: terencelam@vtc.edu.hk

N. Siwingw
School of Natural and Built Environment, University of South Australia, Adelaide, SA 5000, Australia

Y. Wu et al. (eds.), *Proceedings of the 20th International Symposium on Advancement of Construction Management and Real Estate*,
DOI 10.1007/978-981-10-0855-9_32

32.1 Introduction

Contingency is the percentage of a construction budget that is put aside to accommodate unknown factors and uncertainties connected to the project. In construction projects, according to Demkin and AIA [1] the contingency fund is used as leverage against errors and omissions in construction documents, change in the scope of projects and to pay for unknown conditions. According to Musa et al. [2, p. 42], 'the calculation of the contingency fund is done in many different ways which may depend on the organisation and the complexity of the project'. The contingency fund can be calculated using 'an Architects' advice, a standard percentage of estimated cost based on previous experience with similar projects, or a sum based on an intuitional perceptive of a risk or formal risk analysis. Baccarini [3] contends that most organisations do not have established guidelines to help in the estimation and management of contingency funds and there is a knowledge gap between the use of contingency funds as a leverage against risk in project management with regard to the project management expert's comprehension of contingency funds. This fund is often miscalculated and the result is either an underestimated or overestimated amount [4]. This research focuses on assessing the contingency sum for the pre-tender estimate which has significant impact on the project feasibility and cost control.

Risk factors at the construction phase causing cost overruns will be identified and an accurate method for estimation of contingency sum will be determined. This will result in accurate contingency and hence project budget estimates. This aim of this research is achieved by the following objectives:

- Assess the current methods used for risk assessment in the construction industry.
- Investigate the use of contingency funds during the design and construction phases of a project lifecycle and identify uncertainty factors that affect the amount of contingency provided in the cost estimate.
- Determine an accurate quantitative method for estimating contingency funds by using the risk management approach.

In Zambia, it is common practice to simply apply a percentage of the estimated sum as a contingency value to be used at the project manager's discretion throughout the project life cycle. Data was collected from a pubic works department in Zambia to determine whether this traditional predictive method is accurate, and if not, to identify the risk factors influencing the contingency sum in order to develop a quantitative predictive model.

32.2 Risk Management Concept of Contingency

Makombo [5] defines risks in construction as the existence of possible or actual events which may have a negative or positive outcome on the aim and objectives of the project during any phase of the projects lifecycle. The construction industry constantly faces risks [6, 7]. The significance of risk management in construction projects is to avoid cost overruns and delays in completion of projects due to risks. According to Barrick (cited in [8]) one survey revealed that almost one third of the construction industry clients confirmed their projects normally overran budgets. Furthermore, most of the literature reviewed shows that projects continue to fail due to cost overruns [9, 10]. Cost and time are the two basic elements used in the measure of success in project delivery [11]. Ahuja et al. (cited in [12]) place emphasis on the fact that failure of construction projects is due to inaccurate cost estimates.

There are four response plans that can be allocated to treat the significant risk identified, depending on its rating and risk tolerance, risk appetite and risk threshold of an organisation. PMBOK®Guide [13] identifies that risks can be avoided, transferred, mitigated or accepted. Common practice is that risks that might occur during construction projects are accepted and managed using contingency funds which are incorporated into project cost estimates. According to Günhan and Arditi [14] and Touran [15], contingency funds are allowed to cover three types of risks; 'design contingency' which is established on the various stage of design fulfillment; 'construction contingency' which covers increment in costs that may occur during the construction phase of the project's lifecycle; and 'client's contingency' which covers increases in cost which may arise due to the addition of previously undefined project scope.

It is essential that the contingency sum should be accurately assessed and estimated by the client because it can have a significant effect on the project. First, a high contingency encourages poor cost management, which may cause the project to become too costly and the project may be aborted. Second, a low contingency means an impractical financial environment, which results into having unsatisfactory performance outcome, i.e. cost overrun [16]. This research therefore focuses on assessing the contingency sum for the pre-tender estimate which has significant impact on these two aspects: project feasibility and cost control. This refers to construction contingency and client's contingency allowed for the construction phase of the project. Baccarini (cited in [17]) recommends that the contingency should focus to cover risks including incomplete scope definition, inaccuracy of estimating methods, identified risks and unidentified risks, and that scope changes during the design phase should be excluded.

32.3 Traditional and Regression Methods

The 'traditional percentage' is the most commonly used method in Zambia. A standard percentage is added to the cost estimate, which is normally between 5 and 15 % and is established on preceding experience with projects of the similar nature. Sometimes a figure is added, based on a formal risk management process which in turn identifies and analyses risks in detail and the risk treatment is applied.

Lorance and Wendling (cited in [18]) state that in this method of calculating contingency, an all-inclusive percentage is added to the established estimate. This percentage is commonly developed from instinct, precedent observation and documented information as observed by Lorance and Wendling (cited in [18]). Thompson and Perry (cited in [19]) observed that the method of adding a percentage as a contingency allowance has various shortcomings. The percentage amount is, in all likelihood, arbitrarily arrived at and may be incorrect for a particular project. There is an inclination to enlarge the risk because some estimators have a tendency to provide contingencies for the extreme scenarios. An added percentage will still provide a single amount indicator of the cost estimate, suggesting an extent of certainty which is not justifiable. Since the percentage provides for all risks in the context of cost contingency, it has a tendency to draw attention away from time, performance, and quality risks. Creativity is not encouraged by this method because it allows the process to become routine and mundane, which can increase the chances of certain aspects being overlooked.

Multiple regression is a probabilistic technique in which various independent variables are applied in order to predict some dependent variable of interest [20]. Kim et al. [21] define a regression model as a powerful statistical tool which can be used to estimate costs by analysing and predicting variables that contribute to the final estimate reliability. The research conducted by Thal et al. [22] suggests the following advantages from using the multiple line regression to estimate project contingency funds.

- It significantly reduces contingency shortage for all projects when compared to the traditional percentage.
- It enables the contingency fund to be customised so that it corresponds to project-specific risks.
- It allows justification for having higher contingency funds for high-risk projects
- It prevents trade-offs that may reduce project scope or lengthen construction duration due to lack of funding.

32.4 Factors Influencing Project Contingency

Jimoh and Adama [23] assert that the factors determining contingency funds are: firstly, type of client; secondly, duration of the project; thirdly, type of work (new build or refurbishment); and fourthly, project's location (urban or rural). Owing to the accountability, public-sector officers tends to over estimate the contingency fund in order to make sure that final project cost is within the approved budget. Also construction cost of projects may increase due to prices fluctuations in plant, labour and materials over a longer duration. New build projects may involve more extensive unforeseen foundation works during the construction phase. There may be more claims for direct loss and expenses from the contractor if the project is located at a rural area where transportation may become difficult due to poor road access caused by unforeseen bad weather. McLain et al. [24] identifies geotechnical conditions as another factor that needs to be considered for contingency estimation. According to their study, the management of risks associated with geotechnical site conditions is complicated because in most countries, contracts are awarded before full subsurface investigations are completed. There is a risk that the designs may be inadequate and require adjustments when the actual construction commences.

In the category that Clinton [25] identifies as the economic factors, three factors are identified as determinants for estimating the contingency in construction projects: tax rate, exchange rate and inflation. The exchange rate is notably applicable provided contracting services or other elements of the project are to be procured from other countries. If the exchange rate increases above the level predicted, the cost of the project also increases. On this theme, according to Memon et al. [26], fluctuations in the prices of materials due to inflation is the most important factor that affects construction cost performance. If the estimated duration of the project is long, there is a need to factor for inflation during contingency sum allocation is very important because inflation rate may increase above the predicted level. Furthermore, Clinton [25] observes that tax rates can have a significant effect on gross construction costs because contractors normally have to pay tax for construction services to the central government in most countries and such tax is calculated as a percentage of the project cost. Consequently, this may be considered a risk factor for projects that are implemented over a long period because tax rates are subject to change every year. However, for the purpose of this study, this is a factor which is not considered because all government construction projects in Zambia are not liable to pay tax.

32.5 Research Methods

This research aims to identify the risk factors influencing the contingency sum in order to develop a quantitative predictive model. Based on the literature findings, it was hypothesised that:

- H1: Contingency sum is positively correlated to the type of work: higher contingency for new build and lower contingency for refurbishment.
- H2: Contingency sum is correlated to the location of a project.
- H3: Contingency sum is correlated to the geotechnical conditions.

Three other possible risk factors identified by the literature review were not included in the hypotheses for further examination. 'Type of client' cannot be examined in this research as all projects of interest are funded by the government so the client is common for all. 'Project duration' and the economic factor of 'exchange, inflation and tax rates' were both excluded because the 30 projects being studied were relatively short in their duration and consequently the contracts had not allowed for cost increase arising from these two factors.

A triangulation research method was used to validate the information that was collected from the literature review, qualitative interviews and quantitative analysis. According to Veal [27], the triangulation method comprises the application of more than one research approach in a single study to achieve a broader or more comprehensive interpretation of the topic being investigated. The hypotheses were generated by theory and literature, refined with the findings from qualitative interviews, and objectively tested by quantitative regression analysis.

Qualitative interviews were conducted with expert practitioners. Three project quantity surveyors and three architects were identified for the semi-structured interview on how the contingency sum is estimated. These are professionals working as project managers in a public works department in Zambia, having 6–32 years of extensive experience in estimating contingency funds for building and refurbishment projects. A content analysis was used to identify the risk factors used in practice, which helped refine the hypotheses. Cost and risk factors data were collected from real life projects for the quantitative study. A questionnaire survey was conducted to collect data from 30 building and refurbishment projects recently completed in the public works department. According to Levin and Rubin [28], the sample size that is usually applied in practice is 30, even though in the statistical sense the boundary between large and small sample sizes is taken to be 32. For the quantitative study in this research, 30 projects were used although this was considered to be relatively small.

32.6 Discussion

32.6.1 Qualitative Results

The qualitative interview results are shown in Fig. 32.1 and Table 32.1.

In general the construction of new infrastructure is more costly than refurbishment of existing buildings. According to the participants, this can be attributed to additional costs from activities like external services. The interviews established that construction of new infrastructure required higher contingency sums to address substantial unforeseen events. One senior quantity surveyor emphasised that if these factors are not taken into account, it is likely that the contract sums will be exceeded during the project execution phase due to variations. 80 % of the participants agreed that the project's location (whether rural or urban) is a major risk factor that has to be considered. In Zambia, construction costs usually vary due to distances from the market conditions. According to those who agreed with this factor, projects that are executed in the remote areas of Zambia are more expensive and require a higher contingency sum to be allocated because transporting construction materials, labour and equipment to these areas entails much risk. All interviewees agreed that geotechnical conditions constitute one of the most important factors in contingency sum estimation. According to the participants, determining the actual site conditions for the whole duration of the project is usually not conducted until construction

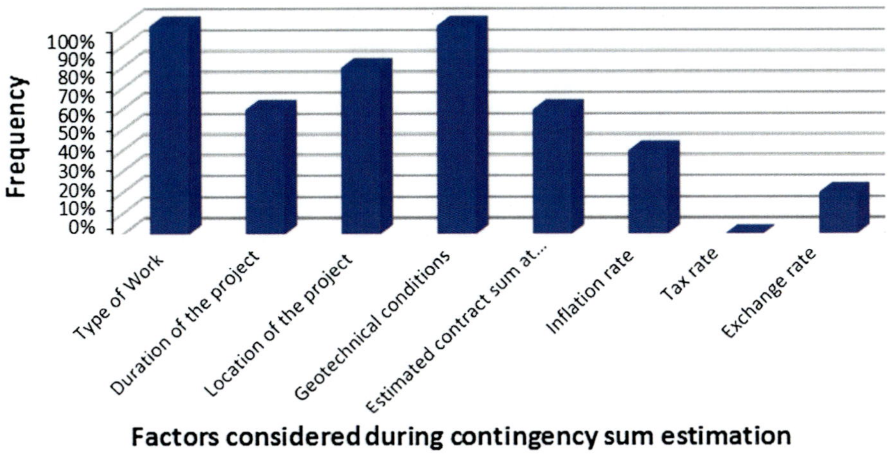

Fig. 32.1 Distribution of factors considered during contingency sum estimation

Table 32.1 Percentage of participants who consider the risk factor to be important

Risk factors	Participants who consider the risk factor in the contingency sum estimation
Type of works	10
Duration of the project	60
Location of the project	80
Geotechnical conditions	100
Estimated contract sum (pre-tender estimate)	60
Inflation rate	40
Tax rate	0
Exchange rate	20

commences. They explained that in practice, difficult conditions are missed during the initial analysis, especially those conditions that may change because of adverse weather conditions. This requires high contingency sums in case unexpected geotechnical conditions like sub-surface bedrock are encountered which may mean redesigning a project. This in itself can be a very costly process.

The interview participants were further asked whether the current 'traditional percentage' method used in the organisation to estimate contingency sums is adequate. This resulted in 60 % of the participants agreeing and 40 % saying it was inadequate. However, they unanimously agreed to there being a need to adopt a method that can accurately and objectively predict the contingency sum for construction projects, as shown by the fact that the contingency sums are usually not adequate.

In addition, interview participants were encouraged to identify other factors that may be considered for contingency estimation. All the interview participants identified 'complexity' of the project as an important factor. According to Baccarini [29] project complexity can be defined by the number of varied components such as criticality of projects, specialist components and interdependence or connectivity, depending on the interpretation of the observer [29]. A further 60 % of interviewees identified the 'estimated project cost' at detailed stage as one of the factors important when considering project contingency sum. They suggested that employing this factor presents a risk because in practice because the estimated projects cost is usually incomplete and therefore this has to be factored in the contingency sum. They suggested the larger and more complex the project, the higher the risk that the estimated projects cost may not be complete for some of the components. Subsequently, they would normally allow for a larger percentage when determining the project's contingency sum. Jimoh and Adama [23] observe that the use of project cost estimates as a deterministic figure for contingency sums is often applied in the construction industry. Some of a project's components may be well defined at the time of estimate while others may be limited and such risk factor must be considered in the cost estimate.

The hypotheses were refined according to the qualitative study findings.

- H1: Contingency sum is correlated to the type of work: higher contingency for new build and lower contingency for refurbishment.
- H2: Contingency sum is correlated to the location of a project.
- H3: Contingency sum is correlated to the geotechnical conditions.
- H4: Contingency sum is positively correlated to the estimated contract sum.
- H5: Contingency sum is positively correlated to the complexity of a project.

32.6.2 Quantitative Results

A multiple linear regression analysis was conducted to statistically estimate the relationships among variables that have been identified: total value of the variations (independent variable) and the five predictor variables. Table 32.2 shows operationalisation of the dependent variable and predictor variables.

Contingency sum had a significant positive correlation with the estimated contract sum as shown by the *p*-value of 0.005 and *t*-value of 5.737. Norusis [30] considers that a R^2 of above 50 % infers a significant correlation between dependent and predictor variables. Hence, the R^2 of 0.542 showed there was a significant correlation between the contingency sum and the estimated contact sum. 54.2 % of the observed variability in contingency sum (dependent variable) can be explained by the estimated contract um (predictor variable). *P*-value for complexity was 0.151 which is close to the significant threshold of 0.05. Type of works, geotechnical conditions and location of projects had insignificant *p*-values > 0.05, thus H1, H2

Table 32.2 Operationalisation of the Variables

Variable		Operationalisation of the variables
Dependant variable	Total variations	Sum of all variations at the final account stage
Predictor variables	Work value (contSum)	Estimated contract sum
	Location (location)	1 = Urban 2 = Rural
	Type of works (typeW)	1 = New 2 = Refurbishment 3 = New and refurbishment
	Geotechnical conditions (geotechnic)	1 = Gravel 2 = Sand 3 = Clay 4 = Silt
	Complexity (complexity)	1 = Low 2 = Medium 3 = High Based on the following: Clarity of scope definition, specialist components, interdependence of elements, criticality of project, project visibility, accountability and type of tasks/activities

Table 32.3 Stepwise regression model analysis

Model	R	R^2
1	0.736	0.542
Model 1 was chosen by SPSS and it included the following variables		
(a) Predictors: estimated contract sum		
(b) Dependent variable: total variations		
Analysis of the chosen regression model		
Significant variable accepted		

Variables	B	SE B	Beta	*t*-value	*p*-value	Collinearity/tolerance
ContSum	0.204	0.036	0.748	**5.737**	**0.000**	0.948

Insignificant variable excluded

	Beta	*t*-value	*p*-value	Partial correlation	Collinearity/tolerance
Constant	−3,985,004,006	−1.336	0.194		
Location	1,882,992,915	1.055	0.302	0.210	0.273
TypeW	124,020,834	0.078	0.939	0.016	0.292
Complexity	1,062,814,739	**1.483**	**0.151**	0.290	0.971

and H3 were rejected by the generalised results. All the results are shown in Table 32.3.

The 'estimated contract sum' with a *p*-value of 0.005 is statistically a significant contributor to the contingency sum's calculation. This result validates the

Hypothesis H1 that 'contingency sum is positively correlated to the estimated contract sum of a project'. This finding corresponds with the discussion on the qualitative study results in Sect. 36.6.1 that the larger and more complex the project, the higher the risk that the estimated projects cost may not be complete for some of the components, and consequently a larger percentage would be allowed when estimating or determining the project contingency sum. Jimoh and Adama [23] suggest that the use of project cost estimates as a deterministic figure for contingency sums is often applied in the construction industry.

The Hypothesis H5 that complexity of a project is positively correlated to the contingency sum may be retained. Pearson correlation coefficients show individual relationships between total cost of variations and the risk factors. The Pearson correlation (r) of 'complexity' risk factor was 0.261, i.e. the second highest Pearson correlation coefficients amongst the five possible influencing factors. Although this value lies in the range of values for a small correlation (0.1–0.29), it is quite close to the range of values for a medium correlation (0.30–0.49), as suggested by Cohen [31]. It is possible that with a larger sample size, a large correlation may occur. According to Pallant [20] the significance of r is strongly influenced by the sample size and a sample size of 30 in terms of the Pearson correlation coefficients is relatively small. In such a small sample size, there may be a correlation between the variables that may not be statistically significant at the traditional p-value of less than 0.05. This is further supported by the results of the qualitative interviews as discussed in Sect. 36.6.1. All the interview participants indicated that project complexity is a factor they apply when determining the contingency sum. This suggests that 'complexity' could be a significant contributor to the estimation. With a larger sample size, complexity can be tested along with other factors using regression analysis.

32.7 Conclusions

Sufficient contingency funds should be allowed to cover risks in construction projects. Too much contingency may result in a project being aborted or uneconomical and too little may lead to cost escalation if the risk occurs. This research therefore focuses on assessing the contingency sum for the pre-tender estimate which has significant impact on the project feasibility and cost control. The assessment should be made with reference to the risk factors involved in construction contingency and client's contingency. However, in the traditional percentage assessment method, there is an inclination to enlarge the risk in order to provide contingencies for the extreme scenarios, especially in the public-sector where officers are concerned about their accountability in cost control.

Results of the qualitative interviews confirmed that project budget overruns is a major issue in construction projects. The regression analysis results proved that the contingency sum was positively correlated to the estimated contract sum. Both the qualitative interviews and the Pearson correlation analysis showed that there was also a positive correlation between contingency sum and project complexity, although such factor needs to be further reaffirmed by regression using a larger sample size. The results confirm that multiple regression method can be used to predict contingency sum accurately and objectively. Such method enables the contingency fund to be customized so that it corresponds to project-specific risks.

To enhance project success in terms of cost, it is necessary that project managers should actively implement risk management in projects when calculating the contingency sum. Related project-specific risks should be identified and multiple regression method should be used to predict the contingency fund accordingly.

While the sample size is relatively small (30), it is considered to be adequate and can produce statistically meaningful results which have provided useful insights into this topic. Nonetheless, it is recommended that further study should be conducted with an expanded sample size so that the results can become much more robust and generalisable to other settings and the private sector. As the projects used for study were relatively short, duration of the project and economic factor of tax rate, exchange rate and inflation rate were not examined in the multiple regression analysis. Further studies should be conducted on longer projects to test out whether these risk factors are significant in influencing the project contingency.

References

1. Demkin JA, AIA (2004) Security planning and design: a guide for architects and building design professionals. Wiley, New York
2. Musa M, Zubairu I, Bala K (2012) Appraisal of the performance of contingency cost provision for building projects in Nigeria. ATBU J Environ Technol 4(1):41–48
3. Baccarini D (2005) Understanding project cost contingency—A survey. In: Conference proceedings of the Queensland University of Technology (QUT) research week international conference, 4–8
4. Baccarini D (2006). The maturing concept of estimating project cost contingency—A review. In: 31st Australasian University Building Educators Association Conference (AUBEA)
5. Makombo HM (2012) The risk management framework for organisations dealing with construction project management in South Africa. Msc thesis, University of Pretoria
6. Idrus A, Nuruddin MF, Rohman MA (2011) Development of project cost contingency estimation model using risk analysis and fuzzy expert system. Expert Syst Appl 38(3):1501–1508
7. Olsson R (2006) Managing project uncertainty by using an enhanced risk management process, Department of Innovation, Design and Product Development, Mälardalen University. Mälardalen University Dissertations no. 34
8. Jackson S (2002) Project cost overruns and risk management. In: Proceedings of Association of Researchers in Construction Management 18th annual ARCOM conference, Newcastle, Northumberland University, UK, pp 2–4
9. Kutsch E, Hall M (2010) Deliberate ignorance in project risk management. Int J Project Manage 28(3):245–255
10. Sharma A, Sengupta S, Gupta A (2011) Exploring risk dimensions in the Indian software industry. Project Manage J 42(5):78–91
11. Doloi H (2012) Understanding impacts of time and cost related construction risks on operational performance of ppp projects. Int J Strateg Property Manage 16(3):316–337

12. Mahamid I (2011) Early cost estimating for road construction projects using multiple regression techniques. Australas J Constr Econ Build 11(4):87–101
13. PMBOK®Guide (2013) A Guide to the Project Management Body of Knowledge (PMBOK Guide), PMI, Newtown Square, PA
14. Günhan S, Arditi D (2007) Budgeting owner's construction contingency. J Constr Eng Manage 133(7):492–497
15. Touran A (2003) Calculation of contingency in construction projects. IEEE Trans Eng Manage 50(2):135–140
16. Ogunlana S, Tabucanon M, Dey P (1993) A methodology for project control through risk analysis: the case of a pipeline project. Managing projects in a borderless world. Pre conference proceedings, IEEE international engineering management conference, 1993. IEEE, pp 18–22
17. Jackson G (2012) Contingency for cost control in project management: a case study. Australas J Constr Econ Build 3(1):1–12
18. Baccarini D (2004) Estimating project cost contingency—A model and exploration of research questions. Departament of Construction Management, Australia
19. Mukhtar MM (2008) An evaluation of contribution of cost contingency towards successful execution of building projects in Nigeria, Building Faculty of Engineering, Ahmadu Bello University, Zaira
20. Pallant J (2013) SPSS survival manual: a step by step guide to data analysis using IBM SPSS. Allen & Unwin, Crows Nest, NSW
21. Kim GH, An SH, Kang KI (2004) Comparison of construction cost estimating models based on regression analysis, neural networks, and case-based reasoning. Build Environ 39 (10):1235–1242
22. Thal AE Jr, Cook JJ, White ED III (2010) Estimation of cost contingency for air force construction projects. J Constr Eng Manage 136(11):1181–1188
23. Jimoh RA, Adama SM (2014) Assessment of contingency sum in relation to the total cost of renovation work in public schools in Abuja, Nigeria. Int J Manag Stud Res 2(10):55–63
24. McLain K, Gransberg DD, Loulakis M (2014) Managing geotechnical risk on US design-build transport projects. Australas J Constr Econ Build 14(12):1–19
25. Clinton AJJ (2015) Factors influencing the final cost of construction projects in Ghana. MSc Thesis, Kwame Nkrumah University of Technology and Science, Kumasi
26. Memon AH, Rahman IA, Abdullah MR, Azis AAA (2014) Factors affecting construction cost performance in project management projects: case of MARA large projects'. Int J Civil Eng Built Environ 1(1):30–35
27. Veal AJ (2005) Business research methods: a managerial approach. Pearson Education Australia/Addison Wesley, South Melbourne, VIC
28. Levin RI, Rubin DS (1991) Statistics for management. Prentice Hall, Englewood Cliffs
29. Baccarini D (1996) The concept of project complexity—A review. Int J Project Manage 14 (4):201–204
30. Norusis MJ (1996) SPSS guide to data analysis. Prentice-Hall, Englewood Cliffs
31. Cohen J (1988) Statistical power analysis for the behavioral sciences. Laurence Erlbaum and Associates, Hillsdale, NJ

Chapter 33
Identification of the Residual Value Risk Factors for Road PPP Projects in China: Questionnaire Survey and Analysis

Yingjie Shao, Jingfeng Yuan and Qiming Li

Abstract With the increasing use of Public–Private Partnership (PPP) mode in road construction, the discrepancy between residual value and agreement of PPP projects in the transition phase has become a prominent problem. Though the risk management of road PPP projects has been lucubrated, systematic research of residual value of them are still in a state of blank. Research described in this paper aims to identify the residual value risk (RVR) factors of road PPP projects and find out the key factors of them. RVR factors of road PPP projects were first identified through the method of literature collection. In this research, 29 RVR factors were identified which were divided into 4 packages. The importance degree of each factor was then presented via descriptive analysis and confirmatory factor analysis based on the data of a questionnaire survey which explored PPP experts' perceptions of 29 factors. The order of the importance degree of each factor indicates that the incidence of residual risks within road PPP projects is strongly influenced by organizational risk, design risk, construction costs, quality of projects, operating costs, quality of service, maintenance costs, market demand and toll pricing. These 29 factors can be used to recognize the risk source and reduce the incidence of residual risks of road PPP projects.

Keywords Residual value · PPP · Road · Risk identification

33.1 Introduction

Over the last twenty years, a major trend in both developed and developing countries has been an increasing level of delegation to the private sector for the provision of public services. However, the discrepancy between residual value

Y. Shao (✉) · J. Yuan · Q. Li
Department of Construction Management and Real Estate,
Southeast University, Nanjing, China
e-mail: 670130216@qq.com

Y. Wu et al. (eds.), *Proceedings of the 20th International Symposium on Advancement of Construction Management and Real Estate*,
DOI 10.1007/978-981-10-0855-9_33

and agreement of the project in the franchise transfer has become a prominent problem.

Though PPP mode and the risk management of it have been lucubrated in academia, systematic research of Residual Value of PPP projects are still in a state of blank. In PPP projects, the Residual Value Risk (RVR) can be defined as the risk that on expiry or earlier termination of the services contract the asset does not have the value originally estimated by government at which the private party agreed to transfer it to government [1]. Bing et al. [2] defined RVR the meso level risk in the study about the allocation of risk in PPP/PFI construction projects in the UK. Chan and Yeung [3] explained RVR that investors overuse the resources like equipment or other technical conditions, etc., which cause insufficient materials and equipment with depreciation at the end of the concession period. As a consequence, it affects the continuous operation of the projects. Leruth [4] found out that Automatic transfer of ownership at the end of the contract transfers the risk of the residual value from the private to the public sector. So that, being able to identify and analyze RVR is very important in at the end of the concession period of a PPP project.

33.2 Road PPP Projects in China

In line with governance trends around the world, a growing number of expressways in China are managed as Public–Private Partnerships (PPPs). Since the late 1990s, the growth in the number of PPPs for roads, especially for expressways programmed in the Chinese National Expressway Network Plan, has speeded up, from 0 to 122 PPP contracts at present.

In China, PPP projects were introduced in the late 1970s as an attempt to encourage the country's reform. With increasing demand for more and better infrastructure, the Chinese government started to apply PPP schemes at a large scale in the 1990s by introducing more foreign investment, especially for water, power, and road projects. The Shen-Da Expressway connecting Shenyang and Dalian is the first expressway in China that was planned in early 1984 and adopted the Build-Transfer (BT) model in its construction. Road projects in PPP mode generally have the characteristics like long implementation cycle, numerous uncertainty risk factors, great economic/technological and environmental risk factors and so on. In the pre-construction, construction as well as pre- and post-franchising period of a project, it will encounter various problems, and these issues will ultimately affect the residual value of the project at the end of the concession period.

33.3 Research Methodology

This paper took road PPP projects as the research objects. RVR factors of road PPP projects were first identified through the method of literature collection. The importance degree of each factor was then presented via descriptive analysis and confirmatory factor analysis based on the data of a questionnaire survey which explored PPP experts' perceptions of 29 factors. Finally the key risk factors were found out according to the order of the importance degree of each factor (Fig. 33.1).

33.4 Identification of the Residual Value Risk Factors in Road PPP Projects

In this paper, 11 literatures about risks of road PPP/BOT/TOT projects were selected, after conducting the comprehensive literature review and content analysis, 29 RVR factors in road PPP projects were identified. The 29 risk factors can be classified into systematic risk factors and non-systematic risk factors. The systematic risk is the out-of-control macro risks from the external surroundings of PPP projects which would occur in all the phases of PPP projects [1]. The non-systematic risk factor is the potential risks form the internal process with the implementation of PPP projects. The non-systematic risk factors were divided into three factor groupings: planning phase risk, construction phase risk, and operational phase risk. The results were list in Tables 33.1, 33.2, 33.3 and 33.4.

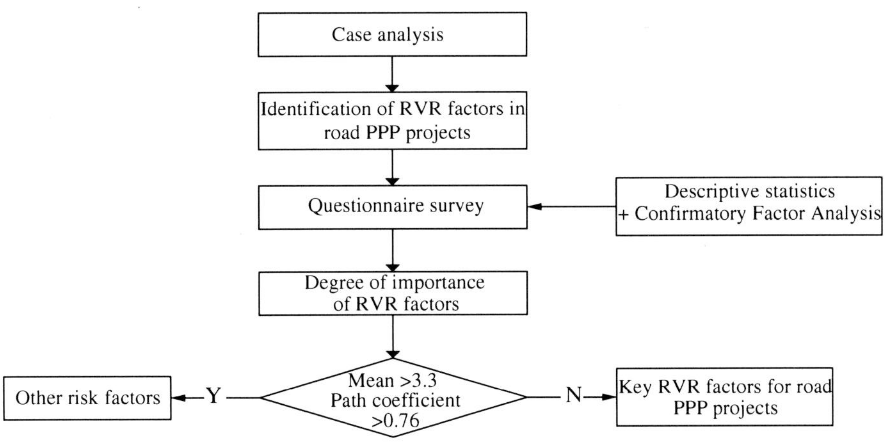

Fig. 33.1 Research methodology

Table 33.1 Systematic risk factors

Code	Risk factors	Description and correlation with RVR	References
F1	Change in Law/policy	In concession period, governments are likely to enact, revise or reinterpret laws or policies, leading to the change of the legitimacy, product service standard, market demand and other factors of the projects. It may cause the project fail or be interrupted	[5–11]
F2	Increase in tax	The financing process of a road PPP project mainly relies on the government's franchise rights, the specific tax policy, price policy and other factors, the change of these conditions will have great influence on the feasibility of the project	[5, 8, 10–13]
F3	Government interference	As the government represents both its own interests and the public's, in the multiple game between stakeholders, the interests of the private is easy to be affected, if the government is very strong	[11, 13–15]
F4	Interest rate fluctuation	A large proportion of the fund of a road PPP project is from bank loans. The fluctuation of the financial market interest rate will have great influences to the project and lead to the increase of the investment	[5, 8, 9, 11, 15]
F5	Foreign exchange fluctuation	Road PPP projects generally cannot earn foreign exchange, construction investment is a bank loan of a currency or several currencies, while operating income is a single local currency. So any kind of foreign exchange will affect the projects' solvency and the actual income	[5, 7, 9, 11]
F6	Inflation	Highway construction need a large amount of cement, asphalt, sandstone, steel and other materials, if inflation causes the rising prices of raw materials and labor wage increasing, it will certainly cause an increase in the total cost of construction	[5, 7, 8, 13]
F7	Public opposition	Public opposition will cause the instability of PPP projects' implementation and management, damage mass base of PPP	[7, 9, 14]
F8	Government/private credit	There are numerous participants in a road PPP project, the contractual relationship is complex. Any party cannot fulfill its responsibilities and obligations of the contract on time, may cause the failure of the project	[7, 10, 14]
F9	Force majeure	Some problems such as collapsed pavement, traffic congestion due to the unreasonably foreseeable natural disasters, may cause the failure of projects or greatly reduce of benefit	[6, 8, 12, 13, 15]

(continued)

Table 33.1 (continued)

Code	Risk factors	Description and correlation with RVR	References
F10	Imperfect contract	Imperfect contracts or unsuccessful contract performance management will lead the project falls into the situation that the two sides wrangle, no matter which party benefits, it will have negative impacts on the project	[13, 14]
F11	Unstable organization	Low efficiency or non-cooperation of organization because of the lack of experience and communication between collaborators, will lead to the project delay or inability to operate	[7, 12, 14]
F12	Stakeholders	If the stakeholders exit unilaterally during the process of cooperation, the project will be significantly affected, and may die on the vine	[14, 15]

Table 33.2 Planning phase risk factors

Code	Risk factors	Description and correlation with RVR	References
F13	Land acquisition	When a long road has to across different provinces, the land acquisition process will be more complex if there is no standardized procedure of it. So that the project will delay	[11, 12, 15]
F14	Approvals and permit	Land acquisition, operating and other procedures require government's approval. If its decision-making efficiency is low, it may lead to high costs and delay of the duration	[9, 11, 13, 15]
F15	Financing risk	The demand of capital investment of road construction is high in China, and laws of road project investment are defective, responsibility and rights cannot be constrained, so the complexity of the project financing and increased risk are likely to cause financial failure	[5, 6, 8, 10, 11]
F16	Design risk	During the construction period, the actual geological structure along the construction sites is bad, construction situation doesn't match the design, causing design change, and it will cause the increase of costs	[6, 7, 9, 10]
F17	Unproven technology	It means that whether the built-up road meet the technical requirements, such as flatness, surface hardness and friction coefficient The loss that unproven technology bring to the project, will lead to costs increasing, project delay and engineering defects	[5, 9, 11]

Table 33.3 Construction phase risk factors

Code	Risk factors	Description and correlation with RVR	References
F18	Construction cost overrun	Any mistakes that design, supervision, or other units of the contractor make is easy to cause cost overrun, result in a shortage of private capital flow side, and affect the quality and post-operation	[5, 7, 8, 10, 15]
F19	Completion risk	Completion risk refers to construction cost overrun, schedule delay, or poor construction quality [5]	[5, 7, 8, 10, 12]
F20	Failure/delay in material/labor delivery	Raw materials or labor supply not in time or in short supply, material suppliers unable to performance, will make the construction costs increase and schedule delay	[9, 14, 15]
F21	Failure to meet performance criteria	If the project fails to meet the requirements of design quality, it will cause project rework or delay and it will increase road maintenance costs, reduce the life of the project and unable to achieve the expected benefit, the project may not operate normally in serious cases	[8–11, 14]

Table 33.4 Operational phase risk

Code	Risk factors	Description and correlation with RVR	References
F22	Operation cost overrun	Operation of transport infrastructure projects involves the maintenance of road, facilities and stations, high operating cost directly affects the cost and income of the project	[5–7, 11, 13]
F23	Shortfall in service quality	Road operating service quality system refers to the organizational structure, procedures and resources which enable customers to achieve safe, efficient, economical, convenient and comfortable road traffic services. Service quality not only affects the government's intention to build the project, but also affects the reputation of the project itself	[7, 9, 11, 15]
F24	Maintenance cost overrun	Road project facilities include anti-drainage systems, lighting systems, fire systems, and road maintenance, etc. All these facilities need maintenance, if maintenance cost overrun, NPV's operation will burden, and may lead to poor maintenance	[6, 7, 9, 11, 15]
F25	Accident in maintenance	Traffic volume of road is very large, it's easy to incur the maintenance accidents. If accidents happen, in addition to the accident itself may cause casualties and property losses, but also the operation of the project will be affected	[11, 14, 15]

(continued)

Table 33.4 (continued)

Code	Risk factors	Description and correlation with RVR	References
F26	Market risk	Traffic prediction is difficult, traffic predictions are always too optimistic, and the actual traffic is often lower than expected, which would have a significant impact on the earnings of NPV	[6, 11, 14, 15]
F27	Competition risk	If there is a competition line in the user area, the traffic forecast accuracy will be seriously damaged, and lead to the lack of the expected return	[7, 11, 12, 14]
F28	Toll risk	Overpriced toll may cause traffic decrease or public opposition, and the project may have to be transferred early	[5, 7, 11, 14, 15]
F29	Environment risk	A road PPP project may damage some of the monuments or cause ground contamination if it has to across the industrial site. To reduce the bad conflict, it has to divagate the line and re-install the existing facilities	[7, 9–11, 14, 15]

33.5 Investigation and Analysis

33.5.1 Questionnaire Survey

A structured questionnaire survey about the significance of RVR factors was conducted among the scholars and experts from different infrastructure types of PPP projects. The questionnaire survey was carried out with the recovery rate of 32.4 % (294 valid questionnaires from all the 907 questionnaires distributed). The content of questionnaire contains two parts: the background info of the respondents and the Likert scale structured questions about the significance of the identified risk factors to RVR: 1-Can be ignored or not important; 2-Maybe important; 3-Important; 4-Very important; 5-Most important. There were 98 questionnaires respondents with transportation experiences from all kinds of infrastructure types. After recycling, we processed the data of the 98 questionnaires from respondents with transportation experiences by SPSS 17.0 and LISREL 8.8.

33.5.2 Descriptive Statistics of Survey Data

Descriptive statistics is the statistics used to describe the observation indexes. In this paper we observed the distribution and the degree of effectiveness of survey data through Mean, Variance, SK, K, etc. of all the risk factors. The results were showed in Table 33.5.

Table 33.5 Descriptive statistics of survey data

Risk factors		Mean	SD	Variance	SK	K
Systematic risk	F1	3.56	1.37	1.878	−0.585	−0.93
	F2	3.35	1.056	1.116	−0.149	−0.736
	F3	3.79	1.262	1.593	−1.03	0.102
	F4	3.21	1.008	1.015	−0.075	−0.692
	F5	2.8	1.121	1.257	0.145	−0.657
	F6	2.99	1.117	1.247	0.111	−0.885
	F7	3.37	1.179	1.389	−0.33	−0.751
	F8	3.48	1.133	1.283	−0.513	−0.31
	F9	3.14	1.193	1.423	0.128	−0.866
	F10	3.8	1.121	1.257	−0.661	−0.269
	F11	3.39	0.938	0.879	−0.471	−0.113
	F12	3.62	0.979	0.959	−0.183	−0.94
Planning phase risk	F13	3.38	1.296	1.681	−0.246	−1.214
	F14	3.55	1.15	1.322	−0.626	−0.419
	F15	3.77	1.242	1.542	−0.958	0.01
	F16	3.36	1.105	1.222	−0.424	−0.387
	F17	3.21	1.018	1.036	0.094	−0.551
Construction phase risk	F18	3.32	1.136	1.291	−0.222	−0.712
	F19	3.46	1.15	1.323	−0.355	−0.736
	F20	3.06	1.003	1.007	−0.125	−0.467
	F21	3.52	1.133	1.283	−0.399	−0.562
Operational phase risk	F22	3.4	1.101	1.211	−0.232	−0.719
	F23	3.33	0.939	0.882	−0.015	−0.645
	F24	3.31	0.901	0.813	−0.563	−0.191
	F25	3.1	1.079	1.165	0.195	−0.561
	F26	3.74	1.143	1.305	−0.582	−0.476
	F27	3.37	1.044	1.09	−0.294	−0.479
	F28	3.45	1.141	1.301	−0.531	−0.273
	F29	3.18	0.978	0.956	−0.177	−0.157

In Systematic risks, the most important factors are 'F3 Government interference', 'F10 Imperfect Contract', and 'F14 Approvals & permit', 'F15 Financing risk' have the most impact in planning phase. In construction phase, 'F21 Failure to meet performance criteria' is the most important risk factor, and next is 'F19 Completion risk', in Operational phase, 'F26 Market risk' and 'F28 toll risk' are the top two factors. SD of all observations are around 1, indicates that respondents' views of the degree of importance of the risk factors are similar. The values of SD of all observations are less than 1 expect F3, and we can say that the data conforms to a Gaussian distribution.

33.5.3 Confirmatory Factor Analysis of Survey Data

Confirmatory Factor Analysis is a powerful tool for explicit hypothesis testing of factor analytic problems. The planning of the analysis is driven by the theoretical relationships among the observed and unobserved variables [16]. In Fig. 33.2, RVR factors are observed variables, and the 4 periods are unobserved variables.

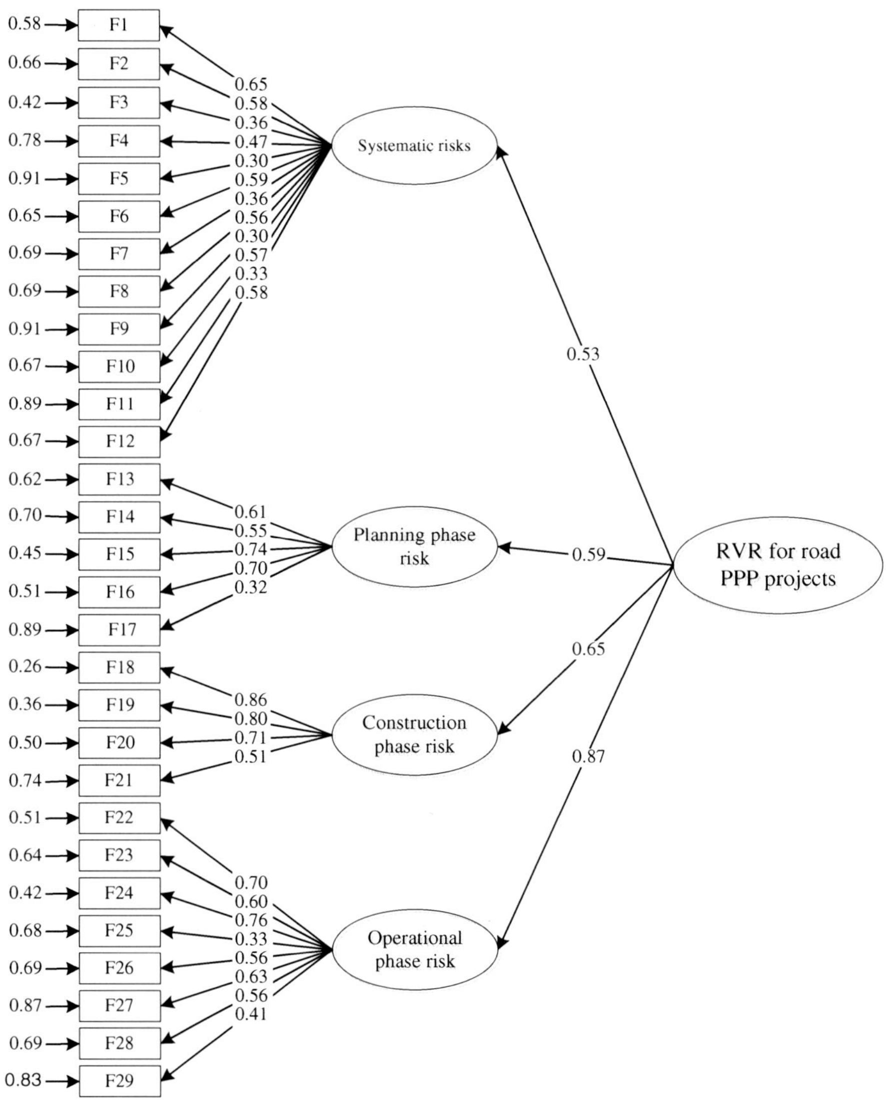

Fig. 33.2 Loading estimates in CFA

CFA parameter estimation uses "maximum likelihood estimation", through CFA, goodness of fit for a specific factor model can be evaluated and also the theoretical framework can be verified. In this paper, CFA were used to test whether measures of a construct are consistent with the authors' understanding of RVR factors of road PPP project and find out the key factors.

The results show that the coefficients $x^2/df = 3.27 < 5$, RMSEA = 0.00 < 0.05, CFI = 0.91 > 0.9, which indicated that the goodness of fit of the model is acceptable. The path coefficient of F1 (0.65) is biggest in systematic risk factors, and so are F15 (0.74) in planning phase risk factors, F18 (0.86) in construction phase risk factors, and F24 (0.76) in operational phase risk factors, which mean they have more contribution degree to their corresponding risk classification than others. The influence coefficient from operational phase risk factors to RVR of road PPP projects (0.87) is the biggest which reflects the effects of this period are greater than others.

33.6 Conclusion

If the risk factor meets the Mean greater than 3.3 and the path coefficient greater than 0.6 at the same time, we identify this factor as a key RVR factor. In summary, the survey received a total of nine key RVR factors: F1, F13, F15, F16, F18, F22, F23, F24, F27.

In this paper 29 RVR factors of road PPP projects were found out through a case study, and a worldwide expert questionnaire survey was conducted to collect the data to analyze the causal relations among the identified risk factors. Descriptive statistics and Confirmatory Factor Analysis were used to test the importance of each factor, and finally 9 key RVR factors were found out. Both the public sector and private sector should attach importance to it and make PPP mode get better development in road projects.

Acknowledgments The authors' special thanks go to all and reviewers of the paper, and to the National Natural Science Foundation of China (NSFC-71001027, 71472037), Social Foundation of Jiangsu Province, China (13GLB005), Program for Outstanding Young Teachers of Southeast University (2242015R30009), and the Fundamental Research Funds for the Central Universities for financially supporting this research.

References

1. Teng TL, Yuan JF, Li QM (2014) Analysis of the formation path of the residual value risk in public private partnership projects based on SEM method. Adv Mater Res 1079–1080: 1126–1130
2. Bing L, Akintoye A, Edwards PJ et al (2005) The allocation of risk in PPP/PFI construction projects in the UK. Int J Project Manage 23(1):25–35

3. Chan APC, Yeung JFY (2011) Empirical study of risk assessment and allocation of public-private partnership projects in China. J Manage Eng 27(3):136–148
4. Leruth LE (2012) Public-Private cooperation in infrastructure development: a principal-agent story of contingent liabilities, fiscal risks, and other (Un)pleasant surprises. Netw Spatial Econ 12(2):223–237
5. Xu Y, Sun C, Skibniewski MJ (2012) System dynamics (SD)-based concession pricing model for PPP highway projects. Int J Project Manage 30(2):240–251
6. Heravi G, Hajihosseini Z (2012) Risk allocation in public-private partnership infrastructure projects in developing countries: case study of the Tehran-Chalus Toll Road. J Infrastruct Syst 18(3):210–217
7. Bagui SK, Ghosh A (2012) Traffic and revenue forecast at risk for a BOT road project. KSCE J Civil Eng 16(6):905–912
8. Wu D, Yin Y, Lawphongpanich S (2011) Optimal selection of build–operate-transfer projects on transportation networks. Transp Res Part B: Methodol 45(10):1699–1709
9. Meduri SS, Annamalai TR (2013) Unit costs of public and PPP road projects: evidence from India. J Constr Eng Manage 139(1):35–43
10. Bagui SK, Ghosh A (2012) Traffic and revenue forecast at risk for a BOT road project. KSCE J Civil Eng 16(6):905–912
11. Ke Y, Wang S, Chan APC (2010) Preferred risk allocation in China's public–private partnership (PPP) projects. Int J Project Manage 28(5):482–492
12. Queiroz C (2007) Public-Private partnerships in highways in transition economies: recent experience and future prosupects. Transp Res Rec J Transp Res Board 1996:34–40
13. Tan Z, Yang H (2012) Flexible build-operate-transfer contracts for road franchising under demand uncertainty. Transp Res Part B Methodol 46(10):1419–1439
14. Akbıyıklı R (2013) Performance assessment of a private finance initiative road project. Transport 28(1):11–24
15. Khan AM (2013) Risk factors in toll road life cycle analysis. Transportmetrica A Transp Sci 9(5):408–428
16. Yuan J, Wang C, Skibniewski MJ (2012) Developing key performance indicators for public-private partnership projects: questionnaire survey and analysis. J Manage Eng 28(3):252–264

Chapter 34
Research on the Application of BIM Technology in Tunnel Project Construction

Lu Zhao, Shihong Zhai, Fuqiang Chen and Fuquan Ji

Abstract In order to improve the efficiency of construction and safety management for tunnel works, BIM (Building Information Modeling) is proposed for application. According to the analysis of the characteristics of railway tunnel works and the actual management demand, BIM-based solutions for construction integrated management are developed, including construction scheduling, visual construction technical clarification, quality control and safety management. The information exchange standard between design and construction has been defined in advance. Information from the design phase about construction components and construction scheduling has been gathered to formulate a 4 dimensional (4D) model. Railway tunnel construction integrated management platform based on BIM is developed independently. During the construction process, actual site data such as process, quality control, safety monitoring and measuring data etc. are continuously collected to add to 4D model. Therefore, the real-time work status of related components can be continuously visualized and compared to the 4D model in the system. Then collaborative management, virtual analysis and risk alarm have been realized. A practical case is also presented to demonstrate the feasibility and validity of the solutions. It is shown that BIM has the potential to improve the construction of tunnel works. The research will benefit tunnel projects which adopt BIM to aid in construction in the future.

Keywords Building information modeling (BIM) · Construction integrated management · Tunnel construction · Railway project

L. Zhao (✉) · S. Zhai · F. Chen · F. Ji
CCCC Second Harbour Engineering Company Ltd., Xicheng, China
e-mail: zhaolu_bee@163.com

Y. Wu et al. (eds.), *Proceedings of the 20th International Symposium on Advancement of Construction Management and Real Estate*,
DOI 10.1007/978-981-10-0855-9_34

34.1 Introduction

As the development of infrastructure facilities, tunnel projects are increasing quickly, especially railway tunnels. Railway tunnel is an essential component of rail transit line, usually located in remote mountain valley where the conditions of construction are poor and the construction process is greatly influenced by hydrogeological. The process of tunnel construction includes complex activities with the following characteristics: limited working space, more risk factors and difficult to manage, etc. Therefore, how to efficient complete tunnel construction become an urgent problem to be solved. The rapid development of information technology provides a new solution method for it.

In order to achieve the scientific and efficient management of railway tunnel construction, Building Information Modeling (BIM) is introduced in the tunnel construction management for the first time. According to the analysis of the characteristics of railway tunnel works and the actual management demand, the tunnel BIM model and 4D construction integrated management technology of railway tunnel construction are researched, and railway tunnel construction integrated management system based on BIM is developed. The information exchange standard between design and construction has been defined in advance, so BIM model of construction phase can easily be established by converting BIM model of design phase. By using BIM model of construction, the construction integrated management can be achieved, including scheduling simulation, management clarification, safety monitoring and quality management, It will provide method, technology, system and application demonstration to railway tunnel project which adopt BIM to aid in construction.

34.2 Basic Concept of BIM

With the in-depth application of IT technology in construction fields, the utilization of BIM has developed and grown significantly. BIM is an engineering data model which integrated variety of construction information, based on three-dimensional digital technology. During the different phases of the life cycle of a facility, different practitioners like the owner, designer and contractor can insert, extract, update or modify information of BIM model at any time to support and achieve the collaborative work. By three-dimensional information transit mode, BIM technology can create a 3D model allowing all practitioners to easy understand, which helps them to intuitive, comprehensive and effective understand the project and reduces the loss of information transit [1]. It can also check design mistakes in advance to provide technical platform and solution method for achieving the integration of design and construction, and effectively solve the existing problem such as poor collaborative, weak integrity in construction. Meanwhile, BIM technology can simulate the construction process to make the actual construction process under control and to achieve integrated project delivery, ultimately maximize the benefit of project.

At present, the application of BIM technology in China is not only in the theoretical research stage, but has entered the practical application stage. And its area has extended from virtual construction and collision detection in the design phase to 4D virtual construction, material tracking, and visual site construction guidance in the construction phase. However, integrated management and integrated delivery based on BIM technology are still in exploring stage. The application of BIM technology in the field of architecture takes priority over in other industry fields, especially in railway. Zhu [2] proposed application of BIM in railway design can be realized phase by phase and step by step. Lu [3] put forward to utilize BIM technology in railway construction projects from the following aspects: project establishment and decision stage, survey and design stage, implementation stage, completion acceptance stage and operation maintenance stage. Li et al. [4] developed 3D design model of Shigu Mountain tunnel to dynamic simulate and statistics quantities. This is the early case of BIM application in railway project. The survey suggests that there is no paper to research on the application of BIM and 4D technology in construction management of the railway tunnel project.

Qingliang Mountain tunnel project in Xicheng Railway is the first BIM application experiment project selected by China Metro Corporation. This paper aim to elaborate the research and application of BIM in this project through descript the application mode of BIM, modeling, delivery standards, management technology, system development and implementation. It provides technical reference and becomes demonstration project to promote the application of BIM technology in railway tunnel.

34.3 Application Architecture of BIM in Railway Tunnel

Combined with the management system of railway projects in China and the actual situation of the pilot project, the most efficient approach, i.e. the developer-driven, the designer and contractor-joint BIM application approach is determined [5]. Application architecture of BIM is shown in Fig. 34.1.

At the beginning of the construction drawing design stage, BIM has been involved in. The developers need to prepare the planning of BIM application, and supervise the designers and contractors to be in charge of their role in BIM application. That is, they should responsible for coordinated management of BIM application. The designers need to develop BIM model of design, and display and analysis tunnel design by BIM technology to detect mistakes, optimize design, and ensure constructability. Then they submit preliminary design documents and BIM model of design to the contractors. Information in the model is consistent with the information in CAD drawing. Combined with the proposed construction program, the contractors need to develop BIM model of construction for simulating and optimizing construction program by BIM technology and virtual technology.

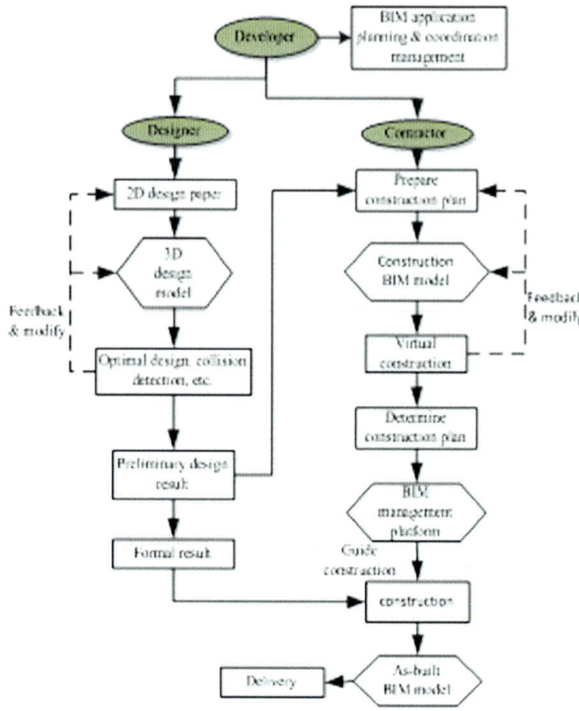

Fig. 34.1 Application architecture of BIM

Meanwhile, they should timely communication with the designers for correcting the mistakes. After all the problems were solved, it is time to develop formal drawings. During the construction phase, BIM management platform is developed by the contractors to guide and assist construction management and to reduce tunnel construction risks. After the project is completed, the construction and as-built BIM model will be delivered to the developers simultaneously.

34.4 BIM Modeling and Design Delivery Standards of Railway Tunnel

34.4.1 Design BIM Modeling

For Railway project, different location has different condition of terrain and geological, with linear characteristics. So, terrain and geological model should be separated from tunnel construction model. The framework is shown in Fig. 34.2. Terrain and geological model is used to support virtual site arrangement and design optimization. And tunnel construction model is used to support construction management and process simulation. After modeling, integrating tunnel

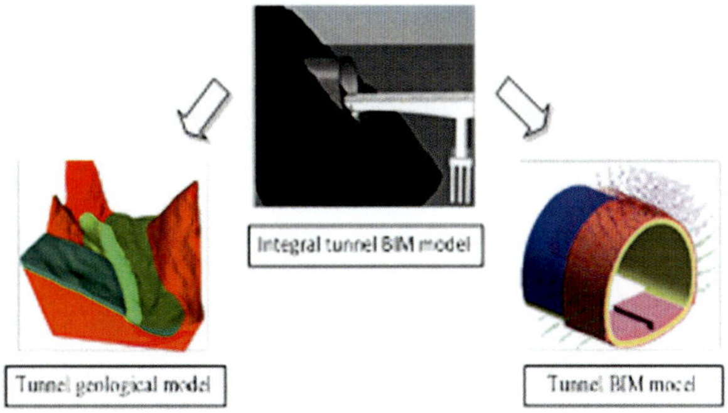

Fig. 34.2 Tunnel BIM model in design phase

construction model with geospatial by mileage of line and relation between geographical coordinates to build near-reality virtual construction environment. Besides, the other reason why separately modeling is the rich of Geological information of tunnel. If only build a BIM model, information is too large to compute.

For Terrain and geological model, 3D terrain model is built based on terrain data, and geological model is built based on tunnel geological mapping date and existing vertical and horizontal section. In order to make the model consistent with the actual situation, using different color, decorative and images to represent geological levels of different ages and using the actual image to form geological levels. Tunnel construction model is built based on design materials and requirements of construction management, and using user-defined clan to represent different components, including the type of component, material, the type and quantity of concrete, the type and quantity of steel, etc.

34.4.2 Design BIM Delivery Standard

In order to make sure information in BIM model same as information in construction drawing, and can satisfy the requirement of subsequent construction management, the designers and contractors make a deal on modeling standard and delivery requirement, like tunnel BIM modeling accuracy, component information granularity, component named and coding rules.

The depth of tunnel BIM model should reach the requirement of construction drawing blueprints, and the accuracy of model should support construction process management. The rule of modeling is shown in Table 34.1.

In addition to covering the geometric information, information granularity of tunnel BIM components also contain construction information, including the grade of concrete, the type of steel, reinforcement radio, and the mileage. In order to make sure the various components name in BIM model same as the name in design drawings, the same type of components is unified named. Information like components name, surrounding rock grade, and lining types should be contained in

Table 34.1 The rule of tunnel BIM modeling

No.	Name of components	Modeling unit
1	Tube; hollow bolt; mortar bolt	Using an pin form of components, and two adjacent pin layout like plum blossom
2	Beam steel frame; grille steel frame	Using a pin form of components, including locking anchor, node connection plate, node bolt
3	Jetting concrete; steel mesh	Using footage length of cycle-excavation, usually 0.5–3.5 m
4	Inverted arch filling; baseplate; inverted arch; waterproof board	Using formworking length of invert trolley or template, usually 6–8 m
5	Ballastless railway foundation; central cover ditch; side ditch groove	Using formworking length of template, usually 6–8 m
6	Arch wall; waterproof board in arch wall	Using formworking length of two-lining trolley, usually 11–12 m
7	Profile design of composite lining tunnel	Consists of multiple circle, commonly 3–4 m

nomenclature of component. As the internal structure of tunnel is not affected by surrounding rock grade and lining types, naming does not need to consider this factor. Nomenclature is shown as follows: components name-surrounding rock grade-lining types.

At the same time, the unique coding system is built, including unit project name, components name, surrounding rock grade, lining types and mileage. The coding rule is shown as follows: the first letter of unit project name-the first letter of component name- surrounding rock grade-lining types-mileage.

34.5 BIM-Based Construction Management of Railway Tunnel

34.5.1 4D Virtual Construction and Optimization

After the Project schedule is imported, 4D construction model is developed through the integration tunnel 3D model and the Project schedule. Visual simulation of the whole construction process is virtual simulated based on 4D model contribute to display tunnel construction procedure and to support the dynamic construction schedule management. Besides, simulation and optimization of important construction procedure and construction method can help to determine the ideal construction planning, and to produce construction animation and image.

34.5.2 Virtual Technical Clarification

Variety of technical clarification files and construction information are integrated with construction BIM model to establish a comprehensive guidance information database. The realization of virtual guidance is shown in Fig. 34.3. To click on any component, information of this component like 2D drawing, materials information, delivery standards, clarifying planning can be checked, which will help construction workers master construction requirements.

34.5.3 4D Construction Schedule Dynamic Management

1. As schedule information is integrated with the model, a status for the construction product is defined as: before construction, under construction, completed on schedule, lag completed. The results are displayed in the model with color metaphors, and the project which completion rate is lower than 95 % is highlighted warning. For manager, they can clearly understand the overall progress and easily get the lag completed part to achieve key monitoring.
2. By real-time recording, uploading, and associating the actual construction information and site photos with the model, construction process is recorded to ensure the integrity of schedule information. When users click on any component, the difference between the actual and the planned will be viewed and then be controlled.

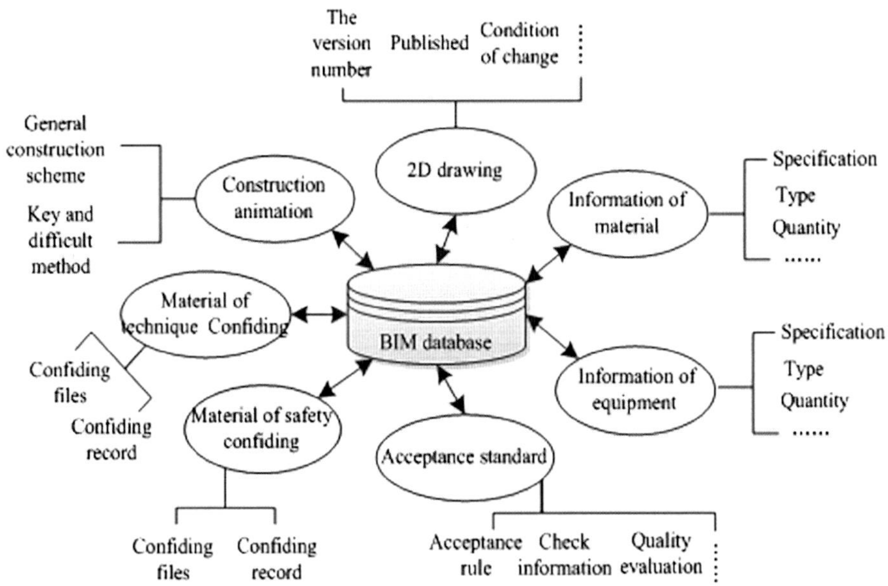

Fig. 34.3 Comprehensive information database of virtual clarification

34.5.4 4D Construction Safety Evaluation and Forewarning Management

1. Safety risk evaluation
   Based on geological conditions like surrounding rock grade, the risk level (high, medium and low) is automatically identify and displayed in the model with color metaphors. It provides guidance for the subsequent construction work.
2. Safety risk forewarning
   Information integration of geological advanced prediction, monitoring, measurement and BIM model is realized. After monitoring information like vault crown settlement or construction information like safety drawing pace exceeds present threshold, it will automatically alarm, and the danger zone in BIM model will be marked red.
3. Safety daily management
   Virtual management of hidden danger is realized, including quickly positioning, looking for hidden danger, checking, and prompting forewarning, etc.

34.5.5 4D Quality Monitoring and Management

1. Information integration of 3D components in BIM model and quality is realized, and the detection results (undetected, qualified, unqualified) are displayed in the model with color metaphors. It is convenient to understand and control the area of quality problem.
2. When users click on any component, the related quality information (like product certification, quality detect files, quality standard files) will be quickly viewed to support site management. It guarantees the quality information traceability.
3. By real-time uploading, connecting quality photos, and quality checking which take components as the main body, participants can realize quality closure management.

34.5.6 Material Management

Statistics amount of various materials according to the conditions and areas to develop material requirement planning, and comparative analysis material consumption.

34.5.7 Project Controlling

By integrating important information like schedule, safe, quality, selecting the outstanding problems in project, and reflecting the trend, project manager can hold this project from macro-perspective.

34.6 BIM-Based Railway Tunnel Construction Management Platform

Based on railway tunnel BIM construction management technology, BIM-based railway tunnel construction management platform is developed. The platform builds virtual construction guidance system by three-dimensional digital technology, based on BIM construction model. All information generated during construction is reflected on the model, forming integrated management environment of railway tunnel construction. The platform takes schedule management as the main line, focus on risk control, and gives consideration to other management. IFC-based BIM data integrated management realizes integration, sharing, and exchange between different software to support accumulation, expansion, integration and application of engineering data generated during construction. The database uses SQL Server to support mass data storage and coordination, consistency and sharing of distributed heterogeneous data. The framework of the platform is shown in Fig. 34.4.

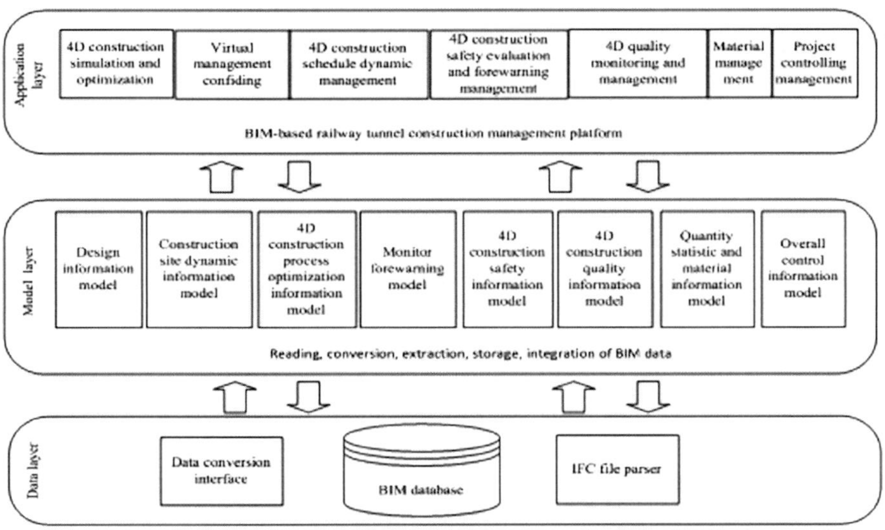

Fig. 34.4 The framework of tunnel construction management platform

Considering on large tunnel construction point, poor network, and inconvenient to carry the computer, this research develop functions of lightweight query and automatic notification of safety and quality based on mobile terminal to support site construction management. Users real-time inspect model and relate technical quality information by smartphone installing Android, and they upload the real-time photos reflected the actual schedule, quality, safety to associate with the related components.

34.7 The Application of BIM in XiCheng Railway Qingliang Mountain Tunnel Project

Qingliang mountain tunnel located in Xi'an—Chengdu passenger dedicated line, and its total length is 12553 m. It through Qinling Mountain, along with water-rich area, unfavorable geology and special geology, so it has many construction risks and is hard to construction. Considering the actual requirement of Xicheng railway Qingliang Mountain tunnel project, refined BIM construction model is built. Through construction process simulation of key process like excavating, lining, inverted arch and key point like junction of tunnel and bridge, efficiency of construction is greatly improved. The simulation is shown in Fig. 34.5.

BIM-based railway tunnel construction management platform has been successfully applied in this project. Next, schedule and safety management will be detail illustrated.

In schedule management, schedule information, schedule photos, and schedule performance can be viewed through overview map (Fig. 34.6). Red represents behind schedule, green represents on schedule, and orange represents under construction. The graphical report of over-excavation can be viewed when double click the selected component in model. The graphical report of two-lining is shown in Fig. 34.7.

In schedule management, tunnel construction risks of different section can be overviewed. During construction, calculating safety drawing pace to forewarn by real-time collecting construction information, including tunnel face, the length of two-lining construction, the length of inverted arch excavation. After imported monitoring data from monitoring equipment, surrounding rock deformation—time cure is automatically generated, and it will automatically alarm when exceeded the warning value. It effectively enhances the level of tunnel construction safety risk monitoring and the ability of process management and control.

Besides, statistics of quantities reduces the workload of checking quantities for construction workers, and provides the reliable basis for material procurement planning. According to simulation of construction process, construction animation or photos are produced to support management confiding. It makes the disclosure

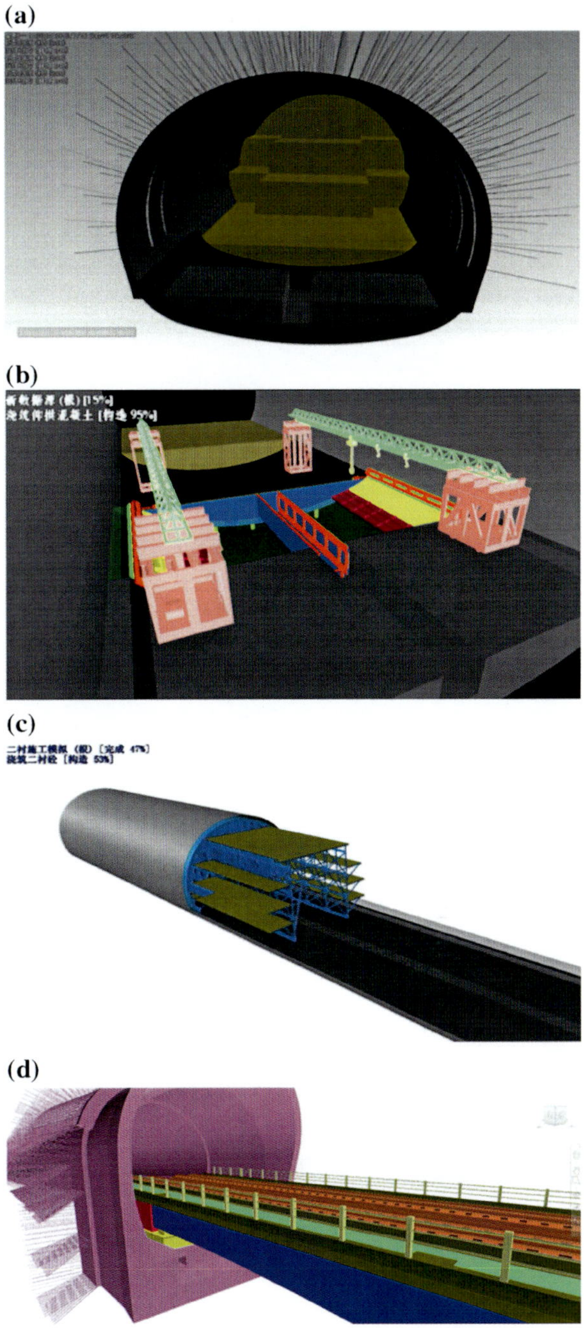

Fig. 34.5 **a** Simulation of seven steps excavation of three stairs. **b** Simulation of inverted arch and filled concrete construction. **c** Simulation of two-lining trolley construction. **d** Simulation of junction of tunnel and bridge construction

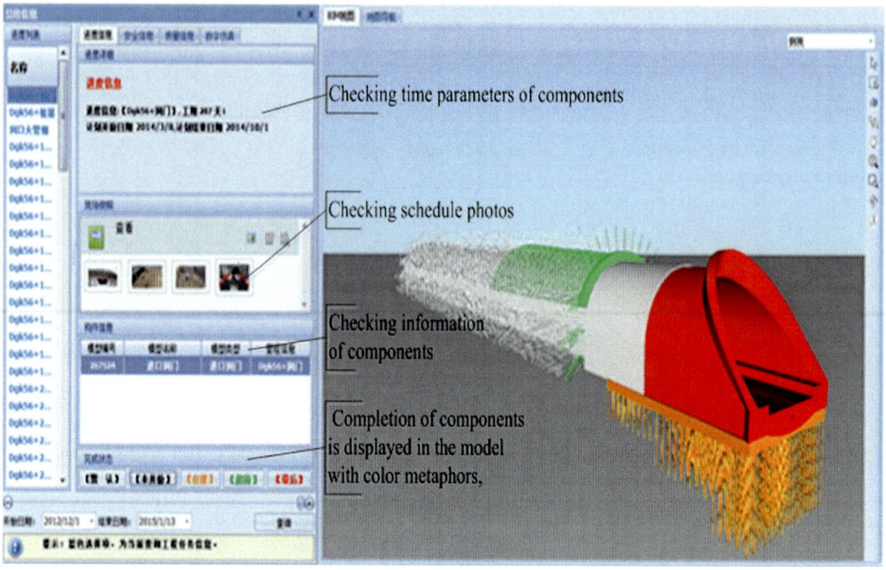

Fig. 34.6 Overview map of tunnel 4D-BIM implementation situation

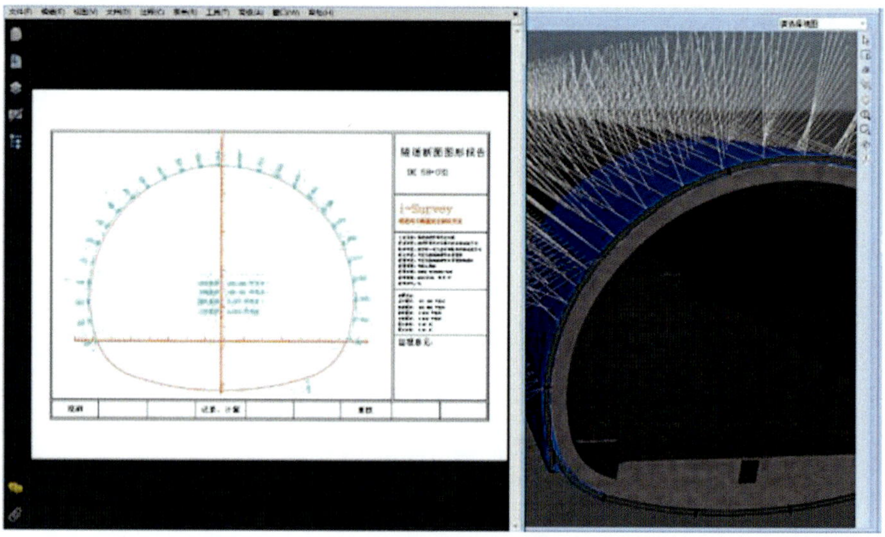

Fig. 34.7 Comparative analysis chart of construction and design of tunnel two-lining

content more intuitive and clearly, and improve the quality of the disclosure, as shown in Fig. 34.8a. Lightweight query also brought great convenience to on-site construction managers. It can quickly access to variety of information to guide construction production, as shown in Fig. 34.8b.

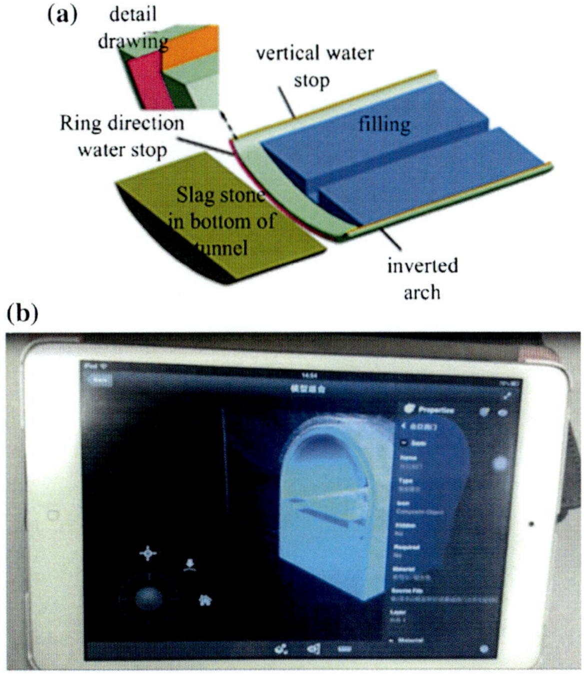

Fig. 34.8 **a** Schematic diagram of tunnel construction joint waterproof. **b** Lightweight query of Smartphone for 4D-BIM

The main problems in application are as following:

1. Slow browsing speed
   The browsing speed of management platform which is developed based on NavisWorks is slow, affecting user experience.
2. Lack of integration
   It is failed to feed rock deformation data back to geological terrain model to simulation analysis tunnel deformation for risk prediction. The application of 3D geological model in safety and quality of construction is affected.
3. Insufficient management support
   It is not fully support the actual management in site, including cost management, subcontract management.

34.8 Conclusion

The applications of BIM in railway tunnel construction project can correct the mistakes of designers much as possible to eliminate the drawbacks caused by the separation of design and construction to a certain extent. And it also can realize digital, visualization, and integrated management of tunnel construction by using rich information in BIM model to improve the capability of risk management.

Meanwhile, developer-driven BIM application mode provides powerful support and assurance to BIM technology. The information exchange standard between design and construction has been defined in advance, so BIM model of construction can easily be established by converting BIM model of design to achieve the construction integrated management. From the above case, we can found that BIM technology provide technical platform for integration of design and construction and refined management of construction site. This shows that BIM technology has great value in the future and needs actively exploration and practice.

References

1. He GP (2011) General of BIM. China Architecture and Building Press
2. Zhu J (2010) Discussion on application of BIM. J Railw Eng Soc 145(10):104–107
3. Lu ZQ (2011) Application analysis of BIM in railway construction projects. Railw Standard Des 24(10):4–7
4. Li JS, Yu Y, Pan T, Tan YJ (2013) Application study of BIM in railway tunnel life-cycle. In: Proceedings of the eighth China workshop on smart city construction technology, vol 24(10), pp 383–390
5. Li H, Guo HL, Huang T, Chen JY, Chen JJ (2010) Research on the application architecture of BIM in building projects. J Eng Manag 24(5):525–529

Chapter 35
Financing Risk Analysis and Case Study of Public-Private Partnerships Infrastructure Project

Zhendong Chen, Jingfeng Yuan and Qiming Li

Abstract This paper focused on establishing a financing risk assessment model and choosing two projects as case studies. Firstly, on the basis of reviewing the relevant studies, this paper identified six financing risk levels by employing the method of content analysis, these were political risks, social risks, financial risks, construction risks and operational risks. And the six risk levels contained seventeen risk factors including government stability, government guarantees, government credit and nationalization etc. Consequently, the article designed an analytic hierarchy structure of PPPs financing risks. By combining project marking sheets with the computed weights of seventeen risk factors, the authors built a financing risk accessing model. After that, an empirical contrast study between Beijing Subway Line 4 and Indian Dabhol Power project was taken. Finally, the paper put forward some suggestions useful to the successful implementation of PPPs projects.

Keywords Public-private partnerships · Financing risk identification · Financing risk evaluation · Analytic hierarchy process · Case study

35.1 Introduction

With the acceleration of urbanization in China, local governments have shifted the focus of investment in infrastructure, resulting an increasing infrastructure investment share of its financial expenditures. In this case, government investment in infrastructure alone can no longer meet its investment needs. PPP, as the overall concept of a series of specific financing models, representing a partnership between private and public sectors to varying degrees, has become an effective infrastructure management worldwide. As an financing model, it can play an effective role to attract private capital into infrastructure projects, making it to be a powerful tool of

Z. Chen (✉) · J. Yuan · Q. Li
Department of Construction and Real Estate, College of Civil Engineering, Southeast University, 2# Si Pai Lou, Nanjing, Jiangsu, China
e-mail: dongzhenchen@126.com

Y. Wu et al. (eds.), *Proceedings of the 20th International Symposium on Advancement of Construction Management and Real Estate*,
DOI 10.1007/978-981-10-0855-9_35

government to coordinate the contradictions between government financial difficulties and infrastructure funding shortages; As a management mode, PPP can introduce advanced concept into public goods and services, accelerate and promote the current political system reform in China [1]. It is precisely because of these characteristics the public sector has been employing PPP model in infrastructure with full expectations [2]. The PPP projects, however, generally have a large resource input, with the characteristics of prolong concession period (generally for 30–40 years) and more stakeholders [3], leading to the PPP project susceptible to political, economic, social and environmental factors [4]. Especially in China, where the PPP mode is still in its infancy, PPP-related laws and regulations are almost blank, which makes China's PPP environment very immature, and prone to risks.

Domestic and international practice experience has also shown that effective management of financing risks is critical for public-private partnerships in achieving their goals. However, it is regrettable that, many infrastructure projects home and abroad with PPP mode failed due to their improper financial risk management. One reason is that PPP financing risks are not merely existing in the traditional project financing stage, but in the entire project process. In general, the financing risk of PPP project can be studied from three aspects including the project participants, the project itself and the external environment. During the process of PPP project financing, the three risk sources are independent or interact with each other, which will have an impact on the financing risk.

To this end, scholars in the field carried out a series of fruitful research. However, it should be pointed out that: although there has been vast amount of literature on financing risks home and abroad, in the PPP project financing, especially in the field of infrastructure financing regarding to PPP features is still poor, therefore, to carry out the research will be a meaningful work. Combined with empirical case studies, the paper will be undertaken PPP project financing risk rating based on Analytical hierarchy process method.

35.2 Literature Review

In general, domestic and foreign research on risks of PPP project focuses on risk identification, risk assessment, and risk sharing. Correspondingly, financing risk research results could also be summed up into three areas. The followings would be focused on financing risk identification and evaluation.

On financing risk levels, scholars tended to focus their research on the hierarchical classification of financing risk with an attempt to achieve the comprehensive recognition of financing risk and to provide convenient conditions for the next phase of risk assessment. Some representative results were as follows. For instance, the financing risk was divided into three levels on the basis of internal and external environments of PPP projects, these were: macro, meso and micro levels [5]. Similarly, PPP project financing risk was divided into project risk and general risk based on the correlation to financing risk of PPPs. Project risk referred to risk caused by micro environment or management of the project; general risk referred to those ones that was not directly related to project itself but related to project macro environment, such as political legal and financial risk, these risks usually had a significant impact on the project [6]. Also, there were researchers divided financing risks of transportation infrastructure PPP projects into four levels including technical risk level, commercial risk level, political risk level, and financing risk level [7]. Besides, some researchers classified financing risk of PPP project into State, market and project levels [8]. Thus, on the basis of pilot works, this article divided PPP financing risks into six dimensions containing factors like political, social, legal, financial ones and construction and operation of the project as well.

In addition to the above work, experts of the field home and abroad continue to refine PPP financing risk into a number of critical risk factors. There was research work on the reasons for project failure which finally inducted 13 major risk factors that may impose a significant negative impact on PPP project by conducting 16 case studies of PPP project, and the paper summarized some common risks of PPP [9]; there was also work on financing risk of water supply projects, according to sources of risks and the contract of the project, the researchers identified a total of eight categories including forty financing risk factors after examining six cases of water supply project in Ghana [10], which had a certain significance for PPP projects, especially water supply projects in China.

On financing risk evaluation levels, since the fact that PPP project had its own characteristics compared with traditional projects, scholars proposed different kinds of financing risk assessment models regarding to the different features of projects. For instance, by employing AHP and fuzzy comprehensive evaluation, the authors conceived financing risk evaluation model aimed for all phase of PPPs including building, operating and transferring stage [11, 12]; also, to obtain the probability of PPPs' financing risk, some researchers employed the method known as Monte Carlo simulation for probability analysis [13]; and some introduced influence diagram method based on Bayesian theory of conditional probability for the assessment of PPP financing risk [14]; besides, there were authors adopted artificial intelligence algorithms for PPP project financing risk assessment [15, 16].

To be noted that, although research work of this field was so much, it seemed there had been a lack of systematic risk identification and risk rating of PPP financing. To bridge the gap, the paper aimed to build a system of PPP project financing risk identification and rating model.

35.3 Methodology

Firstly, in the paper, content analysis method was used for the identification of PPP financing risk factors, then the analytic hierarchy process method was adopted to build the financing risk hierarchy structure chart. On the basis, the authors calculated the weight of seventeen indexes, subsequently, by combining the scoring sheet which was conceived after an extensive literature reading, the paper built up a financing risk assessment model. Meanwhile, two infrastructure of PPP mode were deliberately selected as empirical research. Finally, the paper undertook a comparative analysis of the two cases, and some suggestions on how to cope with financing risk of PPPs were put up.

35.3.1 Content Analysis

As the above mentioned, PPP financing risk can be classified as six levels. Practically, in this study, by searching for "PPP + "risk", "BOT" + "risk", Public private partnership + risk", "financing", "financing risk" and other keywords, the authors got 84 related literature. Finally, 22 SCI articles mostly relevant to the topic was selected. On this basis, the paper selected 17 project financing risk factors of highest frequency of occurrences as the main research. See Table 35.1.

35.3.2 Analytic Hierarchy Process

In this study, the paper constructed the following PPP project hierarchy chart, based on the above financing risk levels and risk factors (Fig. 35.1).

35.3.3 Scoring Sheet

The premise of undertaking quantitative analysis of financing risk is to quantify risk factors. For this, the paper adopted scoring sheet for quantitative evaluation of financing risk.

After an extensive review of the relevant literature and taking in account of the unique situation of China, the authors designed the financing risk scoring sheet (see Table 35.2). Of which two main points were considered: one was variable and the quantization of it. In this study, there were only two possible scenarios of discrete variables: Yes and no, respectively corresponding to highest and lowest point in the scoring sheet. For continuous variables, the paper employed equal division method to divide them into groups corresponding to the each score from the lowest to highest. The second main point was the concretion of the nonfigurative risk. For example, risk factor of government guarantees was quantified by judging the number of guarantee in the concession agreement.

To which should be pay attention was that, the lower score of the project meant the lower level of financing risk. Contrarily, the higher one represented the higher

Table 35.1 Main financing risk factors of PPP project

No.	Risk factor	Risk factor definition
R1	Government stability	Financial risk that New Government does not recognize the funding or some other relating agreement signed by the former one
R2	Government guarantees	Reliability of guarantee agreement of the project signed by the host government
R3	Government credit	Reliability of government to fulfill its direct or indirect financing obligations
R4	Project nationalization	Financial risk caused by the government compulsorily recycling projects
R5	Public opposition	Financial risk caused by public opposition due to damage to its interest
R6	Environmental risk	Financial risk resulting from damage to environment during implementation
R7	Legal perfection	Financial risk caused by low level legislations, poor effectiveness and their conflicts
R8	Change of law	Impact on project financing caused by law or regulation changes
R9	Financing availability	Impact on the project caused by successful or not of funding
R10	Interest rate fluctuations	Negative impact of changes of interest rates on the project financing environment
R11	Exchange rate fluctuations	Impact of changes of exchange rate on international PPP project
R12	Inflation	Project financing risk resulting from financial risk
R13	Project delay	Risks caused by project delay. Such as the extension of loan repayment
R14	Construction overrun	Financing risk resulting from construction cost overruns
R15	Operating cost overrun	Financing risk resulting from overrun of operating cost
R16	Market demand changes	Financing risk resulting from declining market demand during project operations
R17	Price changes	Financing risk resulting from the product/service price changes

level of risk. And in this paper, the lowest score was 0, which meant the project suffered no financing risks. In addition, this scoring sheet was merely designed for the comparison of financing risk level of projects, it would make no sense to weigh the level of financing risk only judging from the score the project.

35.3.4 Weight Calculation

According to the steps of AHP method discussed above, the authors undertook computing the weight of seventeen risk factors. As mentioned earlier, the PPP

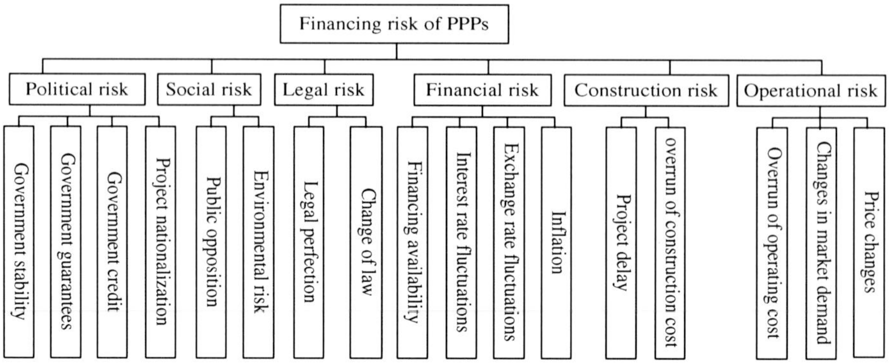

Fig. 35.1 PPP financing risk hierarchy structure

Table 35.2 Scoring sheet of financing risk factors

	Risk factor	0	10	20	30	40	50
C1	Government stability	Government alternation	/	Ever, not much negative influence	/	/	Ever, much negative influence
C2	Government guarantee	/	Over 3 guarantees	1–3 guarantees	None	/	/
C3	Government credit	/	Level 9–10	Level 7–8	Level 5–6	Level 3–4	Level 1–2
C4	Nationalization	No	/	/	/	/	Yes
C5	Public opposition	No	/	/	/	/	Yes
C6	Environmental risk	No	/	Moderate damage	/	/	Severe damage
C7	Legal perfection	Issued PPP law	/	/	/	/	Didn't issue PPP law
C8	Change of law	Not issued harmful law	/	/	/	/	Issued harmful law
C9	Financing availability	Went smoothly	/	Encounter difficulties	/	/	Complete failure
C10	Interest rate fluctuation	No	0–0.5 %	0.5–1 %	1–1.5 %	1.5–2 %	Increasing over 2 %
C11	Exchange rate fluctuation	No negative fluctuation	0–10 %	10–20 %	20–30 %	30–40 %	40 %
C12	Inflation	No	0–3 %	3–6 %	6–9 %	9–12 %	Over 12 %
C13	Project delay	No	1 delay	2 delays	3 delays	/	/
C14	Construction cost overrun	No	/	Moderate	/	/	Severe
C15	Overrun of operating cost	No	/	Moderate	/	/	Severe
C16	Market demand change	No decline of demand	Decline 0–20 %	Decline 20–40 %	Decline 40–60 %	Decline 60–80 %	Decline over 80 %
C17	Change of price	No	Price deceasing	/	/	/	/

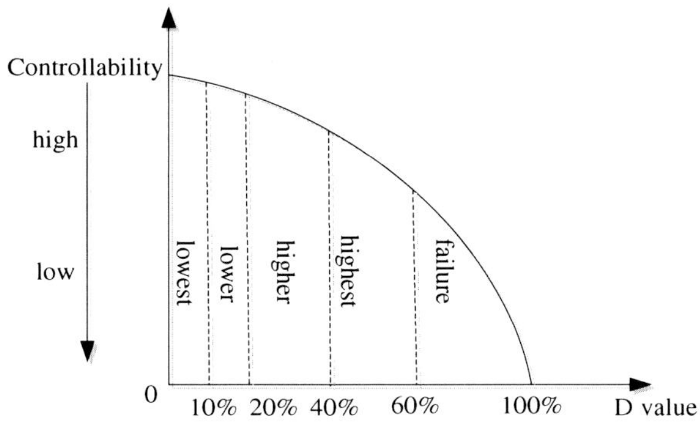

Fig. 35.2 Relationship between D value and controllability of the project

Table 35.3 Weights of PPP financing risk factors and scoring of the two projects

No.	Risk factors	W	S_1	S_2	No.	Risk factor	W	S_1	S_2
1	Government credit	0.139	10	40	10	Government guarantee	0.031	20	20
2	Nationalization	0.129	0	0	11	Construction cost overrun	0.031	0	20
3	Market demand change	0.114	0	50	12	Exchange rate fluctuation	0.029	0	50
4	Change of law	0.089	0	0	13	Inflation	0.027	10	40
5	Legal perfection	0.089	50	50	14	Environmental risk	0.024	0	0
6	Financing availability	0.086	0	20	15	Rice changes	0.02	0	0
7	Government stability	0.083	0	20	16	Interest rate fluctuation	0.016	30	0
8	Public opposition	0.049	0	0	17	Project delay	0.01	0	20
9	Operating cost overrun	0.033	0	50	18				

project financing risk analytic hierarchy chart was shown in Fig. 35.2. Finally, the weight of financing risk factors were computed in Table 35.3.

As can be observed from Table 35.3, experts believed that government credit, nationalization of the project, market demand as well as legal perfection and changes of laws were the top five factors, which occupied about 50 % of total. It was clearly stated that in China, government, laws and the external macro-environment were the most important factors with which should be concerned. This meant implementers of the project should pay much more attention in order to avoid encountering difficulties or the failure of the project.

35.4 Case Study

35.4.1 Beijing Subway Line 4 Project

1. Project profile
Subway Line 4 with total investment of about 15.3 billion Yuan, was consisted of A and B two sections. Section A was mainly about civil engineering and B section about electrical and mechanical equipment. In this project, Beijing Infrastructure Investment Co., Ltd. (BIIC) on behalf of Beijing Municipal Government, was charge of financing and construction of Section A project. And also enjoys the ownership of Section A. While Beijing MTR Corporation Ltd. was responsible for the investment and construction of Section B.

In this project, MTR was responsible for the operation and the maintenance of all facilities except the asset renewal of the hole in vitro. With the recovery investment through the fare revenue and income of business operating in stations, the company would deliver Section B to Beijing Municipal Government gratuitously after 30 years concession period, and Section A should also be returned back to the government [17].

2. Financing risk analysis
The financing risk of Subway Line 4 was quantified base on the relevant literature (see Table 35.3).

3. The scoring of financing risk
The project financing risk score was computed according to Eq. (35.1):

$$Total\,score = \sum_{i=1}^{n} W_i \times S_i \tag{35.1}$$

where: i ∈ {1, 2, 3, 4, …, 17} indicates the subscript of each financing risk factors. Wi represented the weight of financing risk factor of number i; Si represented the score of financing risk factor of number i; Therefore, the financing risk score of Subway Line 4 was computed as follows:

$$Mean = \sum_{i=1}^{17} W_i \cdot S_i = 7.21$$

The score of financing risk of Beijing Subway Line 4 Project was computed as 7.21 points.

35.4.2 Indian Dabhol Power Project

1. Project profile

Dabhol Power Station with its total investment is \$ 2.9 billion was started in 1992. In this project, Enron was Social capital side; Bechtel was contractor and American General Electric Company was the equipment suppliers. Enron set up a special project company responsible for the project—Dabhol Power Company. The Maharashtra State Electricity Board authorized by India's Maharashtra state government signed a 20-year power purchase agreement with Dabhol Power Company in 1993, combined with a "take or pay contract" with DPC. The operation of Phase I started in 1999, however, due to the environment, human rights and other issues. By 2001, the Maharashtra government stopped buying DPC's power, eventually leading to Dabhol Power Station stopped generating electricity [18].

2. Financing risk analysis

The financing risk of Dabhol Power Project was quantified base on relevant literature (see Table 35.3).

3. The scoring of financing risk

According to Eq. (35.1), financing risk score of Dabhol Power Project was computed as followings,

$$Mean = \sum_{i=1}^{17} W_i \cdot S_i = 24.71$$

Therefore, the financing risk score of Dabhol Power Project was 24.71 points.

35.4.3 Results and Findings

For comparison's convenience, we defined financing risk rate, whose value indicated the level of financing risk. In this study, it was represented by symbol D, and which was given by Eq. (35.2):

$$D = \frac{Mean}{Critical\ Value} \times 100\ \% \tag{35.2}$$

where Mean represented the financing risk score of a certain project; Critical Value represented 50 points, the theoretical maximum value of financing risk. According to Eq. (35.2), the D value of any project would be ranging from 0 to 100 %. Clearly, the large D value implied the higher level of financing risks, and it would be more difficult to cope with it in order to achieve the success of the project. Meanwhile, the authors divided the range of D value into several groups which was corresponding to reality (see Fig. 35.2). The magnitude of D value indicated the controllability of financing risks, obviously, Table 35.3 showed that the controllability of the project dropped heavily as the increase of D value.

According to Eq. (35.2), the D values of Beijing Subway Line 4 Project and Dabhol power project were computed as follows:

$$D_1 = \frac{Mean}{Critical\ Value} \times 100\ \% = 14.42\ \%$$
$$D_2 = \frac{Mean}{Critical\ Value} \times 100\ \% = 49.42\ \%$$

According to Fig. 35.2, the financing risk level of Subway Line 4 and Dabhol power project were lower and highest respectively. Actually, the fact had verified the projection: Subway Line 4, as indeed the first PPP project in China, was confirmed to be quite successful, and which was learnt from by numerous implementers of PPP projects for its successful patterns. On the contrary, Dabhol power project ended up with failure because of its huge cost, directly leading to its private partner Enron's bankruptcy. But why the same financing patterns of two projects generated totally different results?

The authors believed that the most reasonable cause was the different mode of financing risk allocation. Specifically, on the Indian government's aspect, the Indian government made several mistakes: First, Dabhol Power project seemed to be much sloppier from its beginning, when the government was eager to attract abroad investment to bridge the financing gap in infrastructure. Under this circumstances, the Dabhol power project was initiated in 1992, following an amendment to the energy law [Electricity (Supply) Act (1948)], which allowed private and foreign investment in the power sector. Second, without any thoughtful consideration, the federal government had provided quite generous warranty for the project in order to attract foreign investment. The problem was that the government would have no ability to keep its permits if the project had encountered difficulties. Besides, the project was awarded through direct negotiation rather than competitive bidding. All of this had made the project confront with a series of systemic risks potentially. On the other, the project's franchisees-now bankrupt Enron Corporation made mistakes too, although the Indian Federal government had made almost the most generous guarantees, the company forgot the independence of Indian state government on its autonomy rights and the ability Federal government keeping its permits. Certainly, there were no pre-research or carefully analysis on the government stability and government credit as well. Therefore, any agreement or guarantee between two parts became just a dead letter.

Different to Dabhol Power project, Beijing Subway Line 4 Project was well prepared from its start. For choosing partners, Beijing Municipal government selected the Hong Kong MTR Corporation as its social partner, having extensive practical experience in construction and operational management of Rail Transit. Thus, this was of great benefit for the smooth implementation of the project on its construction and operation to a significant extent. Secondly, on the aspect of construction risk allocation, project delay risk was assumed by MTR Company which had a strong capacity to handle with construction risks. In addition, the allocation of fare risk was relatively reasonable. Hence, all of this had ensure the success of Beijing Subway Line 4 Project.

35.5 Conclusions

The paper mainly studied the assessment of PPP financing risks. After identifying seventeen key risk factors, the paper found that government credit, project nationalization and changes in market demand were the three most important factors. In addition, the paper developed a financing risk assessment model and chose two typical cases of PPP infrastructure as empirical studies.

Acknowledgments The authors' special thanks go to all and reviewers of the paper, and to the National Natural Science Foundation of China (NSFC-71001027, 71472037), Social Foundation of Jiangsu Province, China (13GLB005), Program for Outstanding Young Teachers of Southeast University (2242015R30009), and the Fundamental Research Funds for the Central Universities for financially supporting this research.

References

1. Li B, Akintoye A, Edwards PJ, Hardcastle C (2005) Perceptions of positive and negative factors influencing the attractiveness of PPP/PFI Procurement for Construction projects in the UK, findings from a questionnaire survey. Eng Constr Arch Manage 12(2):125–148
2. Yuan J, Li Q, Deng X, Li X (2009) A research on drivers in the development of PPP model in China from the perspective of public sector. Constr Econ 1:85–89
3. Tang L, Shen Q, Cheng EWL (2010) A review of studies on Public-Private Partnership projects in the construction industry. Int J Project Manage 28(7):683–694
4. Kim S (2010) Risk performance indexes and measurement systems for mega construction projects. J Civil Eng Manage 16(4):586–594
5. Li B, Akintoye A, Hardcastle C (2005) The allocation of risk in PPP/PFl Construction Projects in the UK. Int J Project Manage 23(l):25–35
6. Ng A, Martin L (2007) Risk allocation in the Private Provision of public infrastructure. Int J Project Manage 25(l):66–76
7. Medda F (2007) A game theory approach for the allocation of risks in transport public private partnerships. Int J Project Manage 25(3):213–218
8. Hastak M, Shaked A (2000) ICRAM-1: model for international construction risk assessment. J Manage Eng 16(1):59–69
9. Qi X, Ke Y, Wang S (2009) Analysis on critical risk factor causing the failures of China's PPP projects. China Soft Sci 5:107–113
10. Ameyaw E, Chan APC (2013) Identifying public-private partnership (PPP) risks in managing water supply projects in Ghana. J Facil Manage 11(2):152–182
11. Jian He (2008) Research on evaluation model of hydroelectric project financing risk based on dynamic multi-level fuzzy evaluation. Guangxi Water Resour Hydropower Eng 2:29–32
12. Li Jie ZP (2011) Fuzzy AHP-based risk assessment methodology for PPP projects. J Constr Eng Manage 137(12):1205–1209

13. Almarri K, Blackwell P (2014) Improving risk sharing and investment appraisal for PPP procurement success in large green projects. Procedia Soc Behav Sci 119:847–856
14. Gong L, Ai P, Liu W (2006) Research on the assessment method of infrastructure project financing risks. Contemp Manager 21:1422–1423
15. Jin X, Zhang G (2011) Modelling optimal risk allocation in PPP projects using artificial neural networks. Int J Project Manage 29(5):591–603
16. Shen J, Wang S (2011) The evaluation of TOT project financing risk based on BP neural networks. Project Manage Technol 9(11)
17. Xie J, Nu Y (2013) The application of PPP model in Beijing Subway Line 2. Financ Acc 4:16–19
18. Zhang H (2002) Beware of project financing risk: analysis the failure of Indian Dabhol Power Project. Case Anal 1:20–28

Chapter 36
The Experiences for Post-disaster Reconstruction in Rural China: The Implications from Data Mining

Yi Peng, Fan Xue and Haijun Bao

Abstract With rapid economic development and increasing devastation of natural hazards, rural China has, and would still have, large amount of post-disaster reconstruction. Many reconstruction cases were reported, especially after the 5.12 Sichuan Earthquake. However, there is a lack of systematic studies on the experiences for post-disaster reconstruction in Rural China. Therefore, the precious experiences gained from one reported case are difficult to deliver to other cases facing natural disasters in future. In order to mitigate this deficiency, this study adopts data mining to systematically investigate post-disaster reconstruction in rural China. Considering news reports have a broader audience than that of academic papers, the source of reconstruction cases is reliable online news reports. The time period for consideration is from 12 May 2008 to 31 May 2015. The keywords for identifying possible news reports are "post-disaster reconstruction" and "rural areas". Preliminary refinement is conducted to exclude the news reports introducing policies rather than reconstruction cases. The similarities and differences of the selected reconstruction cases is mined in terms of manpower, organization, financial source, reconstruction approach (concentration versus in situ), identified experiences, and identified problems. Based on the results, the experience and problem checklist is developed for better experience sharing of post-disaster reconstruction in rural China. The results can also facilitates the government and NGOs to find suitable measures to promote sustainable development in post-disaster reconstruction.

Keywords Post-disaster reconstruction · Rural China · Data mining · Experience sharing

Y. Peng (✉) · H. Bao
School of Urban-Rural Planning and Management, Zhejiang University of Finance and Economics, Hangzhou, China
e-mail: pengyihz@gmail.com

F. Xue
Department of Building and Real Estate, The Hong Kong Polytechnic University, Hung Hom, Hong Kong

Y. Wu et al. (eds.), *Proceedings of the 20th International Symposium on Advancement of Construction Management and Real Estate*,
DOI 10.1007/978-981-10-0855-9_36

36.1 Introduction

China is a country with a highly frequency of natural disasters. According to the database of EM-DAT [1], from 1900 to 2011, there were 681 natural disasters, which resulted in about 12 million deaths, 63 million homelessness, and damages costing nearly 363.43 billion US dollars. Earthquake is one of the most serious natural disasters in China, which has had 1/3 of the devastating earthquakes worldwide even though the land area of China occupies only 1/4 of the world [2]. China recorded an earthquake of magnitude 6.0 or higher 26 times, including earthquake of magnitude 7.0 or higher 7 times [3]. Moreover, it is found that more than 80 % of the earthquakes above magnitude 5.0 occurred in rural areas and brought huge losses [4]. Natural disasters bring about large number of deaths, make countless others homeless, and cause a huge amount of damage in both rural and urban areas. It is difficult to compare the losses between rural areas and urban areas caused by natural disasters as there is no international database of disaster losses that disaggregates data by urban and rural area [5]. However, it is considered that the rural areas face more disadvantages than the urban areas due to insufficient infrastructure, lack of information about natural disasters, and social inequality. For example, housing damage brought about by the 5.12 Sichuan Earthquake in rural areas was much higher than in urban areas of Sichuan Province. As well, it is found that the housing damage is the main loss for the farmers especially in developing countries, such as China [6].

Recent research has focused on post-disaster reconstruction in rural areas of developing countries. Post-disaster reconstruction is a critical, integral part of recovery. Post-disaster reconstruction stresses rebuilding of the physical structures, such as infrastructure and house destroyed or damaged in a natural disaster [7]. Various models have been developed to investigate the reconstruction process [8–10]. The researchers, especially the social researchers, stress social process is important in the recovery after a disaster [11]. Therefore, the recent research moved beyond physical reconstruction, and laid the concern for the social process and sustainable post-disaster reconstruction [12–14]. However, physical reconstruction provides the basis for settling down the victims, restarting industry production, hence facilitating economic recovery, and restoring the environmental function. As physical reconstruction is performed by people, community restoration or reformation such as community participation, and a decentralized approach can be better implemented through means of physical reconstruction [15]. Physical reconstruction is even important for psychological recovery, as the physical building provides the sense of place, which is critical and helpful in achieving psychological recovery [16]. As a result, an increasing number of studies have been conducted on post-disaster reconstruction, particularly in housing reconstruction and relevant aspects.

However, research on an effective mechanism and system to identify, store, and distribute the experiences of housing reconstruction is few. As Hegel criticized that "what experience and history teach is this,-that peoples and governments never have learned anything from history, or acted on principles deduced from it", there is a lack of systematic studies on the experiences for post-disaster reconstruction in Rural China.

In order to mitigate this deficiency, this study aims to systematically investigate experiences of post-disaster reconstruction in rural China through data mining. Section 36.2 introduces the research methods used in this study. Section 36.3 presents the findings from data mining. Thorough discussions have been conducted in this section. Section 36.4 concludes this research with presenting future studies.

36.2 Research Methods

In order to achieve the research aim, the research flow as shown in Fig. 36.1 has been developed.

The established mining aim would guide the whole mining process from defining the source, boundary, and keywords. In addition, the established aim also confines the results interpretation. The aim of the data mining in this study is to systematically investigate the experiences of post-disaster reconstruction in rural China.

There are various types of media recording the experience, such as electronic, paper, audio and video recording. Considering electronic recording is easily

Fig. 36.1 The research flow of this study

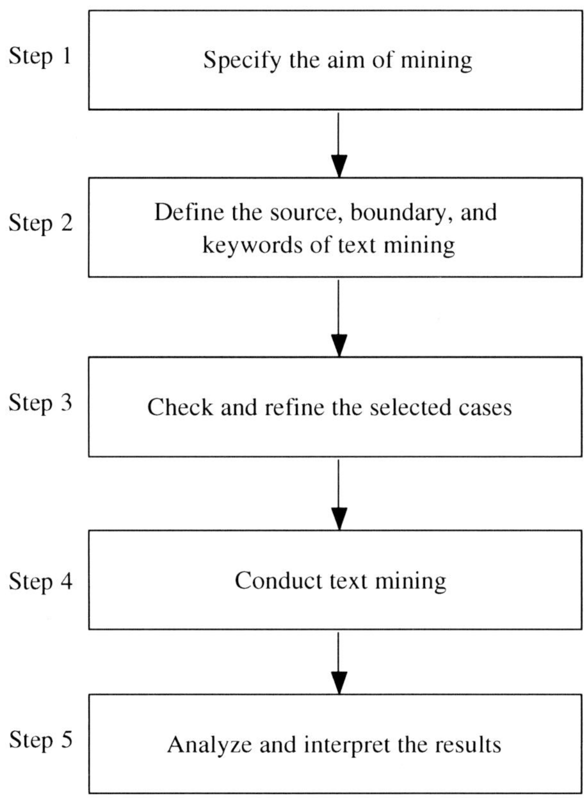

accessed for analysis, this paper confines the analysis to electronic recording. In addition, as news reports have a broader audience than that of academic papers, the source of reconstruction cases is therefore chosen as reliable online news reports. The online website of news reports was chosen from the rank of credible websites, which includes popular websites such as Xinhua Net, Tencent News [17]. Moreover, the time boundary for searching in such sources should be defined otherwise, there would be unexpected ineffective work. This study defines the time period for consideration is from 12 May 2008 to 31 May 2015. Keywords is useful to find the possible cases in the defined sources. The keywords for identifying possible news reports in this study are "post-disaster reconstruction (or post-disaster recovery)" and "rural areas".

By following the first two steps, it is easy to find the possible news reports through searching. Preliminary refinement will be conducted to exclude the news reports introducing policies rather than reconstruction cases. This step is necessary as there are usually some news reports just delivering the formulated policies of post-disaster reconstruction. In addition, the exactly same news reports occurred in different websites was only consider once. It should be noticed that different news reports of the same village is considered as different news reports as different information would be provided in such condition.

After confirming the news reports, data mining would be conducted to select the concerned information in the large pool of news reports. As usually text was used to introduce the reconstruction cases in the electronic news reports, text mining would be used in this study. Text mining, or text data mining, refers to the process of deriving high-quality information from text. Typical text mining tasks include text categorization, text clustering, concept/entity extraction, production of granular taxonomies, sentiment analysis, document summarization, and entity relation modeling [18]. Python was used to develop special programs to mine the reconstruction cases.

Special concerns were paid to the similarities and differences of these cases in terms of manpower, organization, financial source, reconstruction approach (concentration versus in situ), identified experiences, and identified problems. Interpretation of the findings can help to understand what are emphasized in recording such experiences of post-disaster reconstruction in rural China. The understanding of similarities and differences in different reconstruction cases is useful for experience sharing.

36.3 Findings and Discussions

36.3.1 Findings

By following the steps introduced in the Section of research methods, this study found 835 news reports with the keywords "post-disaster reconstruction

Table 36.1 Statistical results of the data mining

Concerns	Number of cases	Key points
Manpower	124	Insufficiency of reconstruction professionals, education after disasters, number of rural victims
Organization	80	Governmental organization improves the reconstruction results, discussion of organization modes suitable for reconstruction, contributions of NGOs, organization process in post-disaster reconstruction
Finance	88	Investment on reconstruction, financial audit, financial support from brother provinces
Reconstruction approach	56 (concentration) 27 (relocation) 11 (in situ)	Benefits of concentrated reconstruction, how to determine the reconstruction approaches, urban-rural coordinated development
Introducing experiences	34	Experiences in terms of organization, site selection, financial sources, housing allocation
Introducing problems	75	Problems in terms of organization, site selection, financial sources, housing allocation

(or post-disaster recovery)" and "rural areas". After excluding the news reports introducing policies rather than experiences and occurring in different websites, there were only 532 useful news reports for analysis. As shown in Table 36.1, not

Table 36.2 Top 10 experiences and problems

Rank	Experience	Problems
1	Transparent and consistent policies	Undetermined property rights of cultivated land generated from consolidating former rural residential land
2	Many matched policies to support post-disaster reconstruction	Incapability of completing project application before reconstruction
3	Sufficient promotions of relevant policies	Insufficient financial support on daily management for CRS
4	Establishing special village affairs board and special supervision board to monitor reconstruction	Little attention to psychological recovery
5	Coordination of governments at different levels	No pre-disaster planning
6	Assessment on degree of damage rather than the economic values of the houses	Little supervision on the housing quality
7	Using local or nearby materials and construction teams	Few risk-coping mechanisms
8	Following rural victims' willingness	Inadequate assessment due to rush of progress
9	Suiting local production and lifestyles	Inadequate infrastructure
10	Adjustable housing and infrastructure planning	Weak basis for future development

all news reports cover our research concerns. Various concerns have been paid in the reconstruction news reports. This is naturally and logically as different journalists usually uses different views to investigate the post-disaster reconstruction. In order to refine the useful information, the information of the first three high frequency was considered as key points in relevant concerns, as shown in the last column of Table 36.1.

This study further analyzes the experiences and problems found in the mined cases. Table 36.2 lists the top 10 experiences and problems respectively.

36.3.2 Discussions

The identified different problems and experiences were useful for experience sharing. The most important experiences mainly lie in available matched policies, successful organization, involving rural victims in decision making, and satisfying rural victims' needs. These issues are valuable considerations and should be considered in future reconstruction. The typical problems include no pre-disaster planning, little supervision on the housing quality, insufficient financial support on daily management for CRS, little attention to psychological recovery [18].

Diversified reasons result in the identified problems above. Typically, some critical activities of developing CRS in post-disaster reconstruction are missing. For example, although there were some emergency measures for responding to natural disasters, no pre-disaster planning was available. As a result, no specific guidelines can be provided to instruct post-disaster reconstruction. In addition, some activities are not so well-designed due to the rush of the project. For example, there is little supervision on the quality of housing and very little assessment on the reconstruction process. This may lead to critical issues if another disaster occurs [19–22].

36.4 Conclusion

This study adopts data mining to systematically investigate post-disaster reconstruction in rural China. Credible news websites were accessed to select the reconstruction cases for analysis. Python was used to develop special programs to mine the reconstruction cases. It was found that various concerns have been paid in the reconstruction news reports. The most important experiences and problems have also been identified for sharing. The results can also facilitate the government and NGOs to find suitable measures to promote sustainable development in post-disaster reconstruction. However, it should be noticed that detailed text analysis has not been conducted due to limit of time. Typical text mining, such as text categorization and text clustering should be conducted in future studies. In addition, the mining source merely includes online news reports, which may also present some bias of the results. Future studies may include the academic papers for mining.

Acknowledgement The work described in this paper was supported by the National Natural Science Foundation of China (71503228).

References

1. EM-DAT (2013) The international disaster database. http://www.emdat.be/database (5 Jan 2013)
2. Zhang YZ (2010) The key issue of new socialist countryside construction is to develop rural economy. J ShanXi Finance Econ Univ 32(2):160 (in Chinese)
3. Na BG, He ZY (2011) Enterprise-village' rural reconstruction financing mode based on CSR perspective. Rural Econ 11:61–64 (in Chinese)
4. Jin L (2008) Some urgent problems on reconstruction after the earthquake of Wenchuan. Planners 7 (in Chinese)
5. Pelling M (2007) Learning from others: the scope and challenges for participatory disaster risk assessment. Disasters 31(4):373–385
6. Cheng FF, Dong XM, Wang SY (2010) Emergency management of Yushu earthquake tests the Wenchuan experience. J Evid Based Med 10(3):157–162 (in Chinese)
7. Quarantelli EL (1999) The disaster recovery process: what we know and do not know from research. http://dspace.udel.edu:8080/dspace/bitstream/19716/309/1/PP%20286.pdf. Accessed on 10 Aug 2010
8. Haas JE, Kates RW, Bowden MJ (1977) Reconstruction following disaster. MIT Press, Cambridge
9. Rubin CB, Popkin R (1990) Disaster recovery after hurricane Hugo in South Carolina. Institute of Behavioral Science, University of Colorado, Boulder
10. Brunsdon D, Smith S (2004) Summary notes from the infrastructure workshop. In: New Zealand recovery symposium, The Ministry of Civil Defence and Emergency Management
11. Hutanuwatr K, Bolin B, Pijawka D (2012) Vulnerability and disaster in Thailand: scale, power, and collaboration in post-tsunami recovery. In: Forces of nature and cultural responses. Springer, Netherlands, pp 69–92
12. Nigg JM (1995) Disaster recovery as a social process. http://dspace.udel.edu:8080/dspace/bitstream/19716/625/1/PP219.pdf. Accessed on 8 Aug 2010
13. Mileti D (1999) Disasters by design: a reassessment of natural hazards in the United States. Joseph Henry Press, Washington DC
14. Miles SB, Chang SE (2006) Modeling community recovery from earthquakes. Earthquake Spectra 22(2):439–458
15. Lyons M (2009) Building back better: the large-scale impact of small-scale approached to reconstruction. World Dev 37(2):385–398
16. Diaz JOP, Dayal A (2008) Sense of place: a model for community based psychosocial support programs. http://www.massey.ac.nz/~trauma/issues/2008-1/prewitt_diaz.htm. Accessed 11 Aug 2010
17. Rank of credible news websites. http://top.chinaz.com/list.aspx?t=247. Accessed on 15 June 2015
18. Berry MW (2004) Survey of text mining. Springer, Berlin
19. Peng Y (2015) A comparison of two approaches to develop concentrated rural settlements after the 5.12 Sichuan Earthquake in China. Habitat Int 49:230–242
20. Peng Y, Shen QP, Shen LY, Lu C, Yuan Z (2014) A generic decision model for developing concentrated rural settlement in post-disaster reconstruction: a China study. Nat Hazards 71 (1):611–637
21. Peng Y, Shen LY, Zhang XL, Ochoa JJ (2014) The feasibility of concentrated rural settlement in a context of post-disaster reconstruction: a study of China. Disasters 38(1):108–124
22. Peng Y, Shen LY, Tan C, Tan DL, Wang H (2013) Critical determinant factors (CDFs) for developing concentrated rural settlement in post disaster reconstruction: a China study. Nat Hazards 66(2):355–373

Chapter 37
The Analysis on the Accounting Methods System of Environmental Costs Focused on the Construction Phase of Construction Projects

Huishu Du, Maomao Ge, Wei Zhang and Yingyue Chen

Abstract Nowadays the construction projects are bringing great economic benefits and value in use for the community while based on the overdraft of the natural resources. So far the scholars seldom take external dis-economy of the environmental costs of the construction projects into consideration, consequently the environmental costs accounting of construction projects lack of a comprehensive and scientific definition as well as a systematic measurement system. Through studying the causes of problems of the environmental costs at construction stage, this paper divides the environmental costs of construction projects into four classes, resources consumption costs, environmental protection costs, pollution treatment costs and environmental loss costs. Then, this paper puts forward an accounting methods system of environmental costs with ABC (Activity-Based Costing method) and FCE (Fuzzy Comprehensive Evaluation method). And this paper also proposes suggestions on the environmental costs management of construction projects. In actual practices, the study results of this paper can help the supervision department of local government strengthen the environmental regulation, meanwhile helping construction firms improve control measures of the environmental cost.

Keywords Construction projects · Environmental costs · Measurement methods

37.1 Introduction

Global economy is now developing at an amazing speed along with the improvement of human society. At the same time, however, the grab of natural resources and damage to the ecological environment have reached an unprecedented point. In these circumstances, construction, as the pillar industry in China, faces server

---

H. Du · M. Ge (✉) · W. Zhang · Y. Chen
School of Construction Management and Real Estate, Chongqing University, Chongqing, China
e-mail: gmm@cqu.edu.cn

Y. Wu et al. (eds.), *Proceedings of the 20th International Symposium on Advancement of Construction Management and Real Estate*,
DOI 10.1007/978-981-10-0855-9_37

environmental problems owing to its resource-intensive feature. In particular, the pollution of air, water, noise, and waste caused at the stage of construction plays the most important role in environmental destruction. Therefore, it is significant to analyze the measurement methods system of environmental costs focused on the construction phase of construction projects which can reflect the cost to protect the environment and the value of the loss caused by environmental damage, and benefit the development of ecological civilization at the same time.

37.2 Research Status

In as early as 1970s, the institutions and scholars had set about to analyze the environmental costs. The theoretical researches on environmental costs in western countries whose focus was on the internal control of enterprise, environmental cost and environmental benefits started earlier with a solid research foundation. At present, the most authoritative definition is the delimitation concluded in the report named Position Statement of Accounting and Financial Reporting for Environmental Cost and Liabilities in 198 by Governmental Experts Group of United Nations International Standards of Accounting and Reporting. This definition emphasized that the environmental costs were supposed to take the environmental responsibility of the enterprise as the center, and take the environmental load caused by the production movement of the corporation and the precautions expenses into account [1].

Domestic research on the environmental costs started late relatively. Research Group of Education Department of Zhejiang Province (2001) holds that the environmental costs are the cost for those methods to cope with the effects on the environment from the management activities. In the article The Research on the Environmental Costs Control based on Circular Economy, Wang et al. [2] insists that from the perspective of circular economy, the environmental costs can be divided into four parts, the prevention costs and treatment costs and for environmental pollution, cost for recycling of waste and the loss of natural resources. Liu et al. [3] brings forward her point in the article The Research on the life-cycle accounting methods of environmental costs that the environmental costs can be classified into internal environmental costs and external environmental costs. Internal environmental costs are those already internalized and related to the environmental pollution treatment and natural resources, while external environmental costs include the loss of human health and ecology health as well as the drop of the productivity of resources.

When it refers to the cost accounting, the opinions of Chinese scholars vary a lot. Wang et al. [4] believe that the environmental costs can be accounted via the method of PLC (product life cycle). Zhang [5] have analyzed the disadvantages of existing environmental cost accounting methods of enterprises before they put forward that the method of ABC (Activity-Based Costing) is an ideal way to account the environmental cost. Zhang and Zhen [6] proposes an accounting model of environmental costs based on the value chain analysis. Guo [7] mentions that the environmental costs of enterprises can be accounted with the three methods, method of ABC (Activity-Based Costing), full-cost accounting method and the method of life cycle, and she mainly analyzes the application of the former two methods.

Now most scholars define the environmental costs and study on the accounting system from the perspective of enterprise. What is more, the focus of their researches tend to be the industrial enterprises such as manufacturing enterprises and coal enterprises, which have a direct impact on the environment. And the researches in view of the construction industry and construction market are relatively few, especially those which take the construction project as the object. Furthermore, Existing definitions about the environmental costs of construction projects are mostly focused on the internal costs, such as the environmental protection costs and pollution treatment costs. They usually do not take external diseconomy of construction projects into consideration, so external environmental costs are often excluded from the cost accounting system in practice.

According to the spirit of ISO14040 series environmental management standard of the International Organization for Standardization, project life-cycle environmental cost equals the potential and prominent influence that project products exerts on the environment during the life cycle. This article classifies the environmental costs of construction projects into four kinds, resources consumption costs, environmental protection costs, pollution treatment costs and environment loss costs under the ecological value theory and externalized theory, based on which this article works out an accounting methods system of the environmental costs of the construction stage in construction projects.

37.3 The Accounting Methods System of the Environmental Costs

When the environmental costs, the environmental protection costs and the pollution treatment costs can easily be measured in the form of a currency, while the measuring of the resources consumption costs and the environmental loss costs is too difficult and complicated to be quantified in terms of currency, as a result of their involving the assessment and analysis on the environment effect, which turns out to be the key problem blocking the improvement of accounting system of environmental costs. Therefore, other methods have to be found to monetize the environmental impact correctly. This article searches for the causes of environmental costs, so as to find out the reasonable components of the environmental costs in construction stage. Then, this article selects suitable and effective accounting methods respectively for different kinds of environmental costs.

37.3.1 Resources Consumption Costs

Huge numbers of various types of building materials are needed during the construction process, and a lot of natural resources and energy are bound to be consumed when manufacturing those building materials, like limestone, clay, sand, iron ore, timber, coal for refining steel and water. The resources consumption costs this article refer to are the costs for the ownership and use right of the natural resources that are consumed to get those materials.

Some of those natural resources are renewable resources, like the forest that provides the timber almost continuously. While the other resources are unrenewable, such as coal and iron ore. On one hand, when the renewable resources are consumed, it has to take a certain amount of time and investment in order to restore them to their original size. During that period, the costs it takes consist of the resources regeneration costs. On the other hand, the unrenewable resources are, in fact, renewable ones whose regeneration cycle is very long. To establish a common accounting model, this article takes all the unrenewable resources as renewable resources to deal with and the regeneration costs equals the resources consumption costs. Based on the above analysis, this article puts forward that the resources consumption costs, or the resources regeneration costs, can be measured as the product of the resource consumption and unit cost of regeneration. The measurement formula of the resource consumption cost is as follows:

$$C_1 = \sum_{i=1}^{n} t_i * (c_i k_i)$$

t_i refers to the quantity of the resource i consumed during the construction stage
c_i refers to the market value of the resource i
k_i refers to the regeneration cost coefficient of the resource i

k_i can be identified with FCE (Fuzzy Comprehensive Evaluation method). The natural resources that the construction projects mainly consume are the following seven, limestone, clay, sand, iron ore, timber, coal and water. The evaluation index set is {regeneration time input, regeneration resources input, regeneration materials input, and regeneration labor input}. The weights are determined by the Delphi method. The comment set is {high, middle and low}. The mark sheet is as Table 37.1 shows.

Because of the regional differences of construction projects, the regeneration cost of the same resources will still differ. According to the result after Fuzzy Comprehensive Evaluation (high, middle or low) aiming at the main natural resources, the corresponding regeneration cost coefficient is set as {a, b and c}, which means k_i = {a, b and c}.

Table 37.1 The FCE mark sheet of regeneration costs coefficient

Main natural resources	Regeneration time input			Regeneration resources input			Regeneration materials input			Regeneration labor input		
	High	Middle	Low	High	Middle	Low	High	Middle	Low	High	Middle	Low
Limestone	v_{11}	v_{12}	v_{13}	…	…	…						
Clay	v_{21}	v_{22}	v_{23}									
Sand	…	…	…									
Iron ore												
Timber												
Coal												
Water												

37.3.2 Environmental Protection Costs

During the construction stage, the environmental protection costs mainly refers to the charge of protection and maintenance of the ecological environment at the construction phase, including greening cost for the construction site, environmental education costs and investment costs of cleaning construction infrastructures. Besides, the environmental protection costs also contain the environmental expenditure, environmental assessment fee and environmental tax payments required (Table 37.2).

37.3.3 Pollution Treatment Costs

Pollution treatment costs are those for reducing the emissions of pollutants and improving the level of environmental pollution at construction stage, such as treatment costs for curing three waste (waste water, waste gas and solid waste) and noise pollution and compensation and fine for the environmental accidents (Table 37.3).

Table 37.2 Activity-based costing details of environmental protection costs

Items of the environmental costs	The drivers of cost	Activity cost pools (ACP)	The environmental cost
Greening cost for the construction site	Green area (m^2)	ACP of greening C_{21}	The environmental protection costs
Environmental education costs	Education cost (yuan)	ACP of environmental management C_{22}	
Investment costs of cleaning construction infrastructures	The expenses of equipment and facilities (yuan)	ACP of cleaning construction C_{23}	
Environmental tax payments	Environmental tax payments (yuan)	ACP of environmental taxes C_{24}	
Environmental expenditure, environmental assessment fee and other fees required by law	The fees of environmental protection (yuan)	ACP of environmental fees C_{25}	

The environmental protection costs, $C_2 = C_{21} + C_{22} + C_{23} + C_{24} + C_{25}$

Table 37.3 Activity-Based Costing details of pollution treatment costs

Items of the environmental costs	The drivers of cost	Activity cost pools (ACP)	The environmental cost
Costs of containment structures	Length of containment structures (m)	ACP of dust reducing and controlling C_{31}	Pollution treatment costs
Costs of watering to prevent the dust	The machine-team of sprinkling cars (times)		
Costs of temporary noise barriers	Noise levels (dB)	ACP of noise controlling C_{32}	
Costs of sedimentation basin construction	Size (m^3)	ACP of waste water treatment C_{33}	
Costs of solid waste transportation	The machine-team of transport vehicles (times)	ACP of solid waste treatment C_{34}	
Compensation and fine for the environmental accidents	Compensation and fine (yuan)	ACP of accident handling C_{35}	

The pollution treatment costs, $C_3 = C_{31} + C_{32} + C_{33} + C_{34} + C_{35}$

37.3.4 Environment Loss Costs

Environment loss cost refers to the substantial damage generating from construction activities to natural environment. Although some protection measurement will be taken and some treatment will be adopted afterwards, environment is still adversely affected by construction activities, including wastewater, waste gas, and solid waste in phase of construction and so on.

A reasonable and effective means of monetization must be found in order to evaluate the cost of environmental damage and destruction. For reason of the complexity and uncertainty of ecological system and the limited techniques, how to measure the environmental loss becomes a problem. Given that "three waste" emission is the source of damage to ecological environment, the calculation of resource loss based on "three waste" can serve as a foundation of the environment loss costs. According to the mass balance theory, substances neither appear nor disappear without reason. So there must be a balance between the total input during construction process and the final output. It can be expressed as following:

$$\begin{aligned}\sum \text{resource input} = \sum \text{building products} + \sum \text{waste gas emission} \\ + \sum \text{wastewater releasing} + \sum \text{solid waste releasing}\end{aligned} \tag{37.1}$$

The resource loss can be determined on the basis of "three waste" emissions which can be known from the transformation of formula (37.1), namely:

$$\begin{aligned}\text{Resource loss } (\mathrm{w_j}) &= \sum \text{waste gas emission} + \sum \text{wastewater releasing} \\ &\quad + \sum \text{solid waste releasing} \\ &= \sum \text{resource input} - \sum \text{building products}\end{aligned} \tag{37.2}$$

The measurement formula of environment loss costs is following:

$$C_4 = \sum_{j=1}^{n} w_j * c_j$$

$\mathrm{w_j}$ refers to the quantity of the resource j lost during the construction stage
$\mathrm{c_j}$ refers to the unit environment loss cost of the resource j, and can be identified as Table 37.4 shows.

The coefficient ρ can be worked out by the analytic hierarchy process, and meet the equation, $\rho_{j1} + \rho_{j2} + \rho_{j3} = 1$. And the unite environment loss cost c_{j1}, c_{j2} and c_{j3} are determined by the Delphi method. Firstly, select 50 representative experts who are familiar with environment costs and authoritative in the field of environment costs for advisor. Then, record the given reference values about c_j from each expert anonymously, and feedback the result to experts after comprehensively analyzing their opinions. Next, experts modified their views through a series of anonymous consultation and eventually get a final conclusion.

37.3.5 Environmental Costs of Construction Projects

According to the above analysis, there is measurement formula of environmental costs of construction projects in construction phase. Environmental costs = resources consumption costs + environmental protection costs + pollution treatment costs + environment loss costs, that is,

$$\mathrm{C} = \mathrm{C_1} + \mathrm{C_2} + \mathrm{C_3} + \mathrm{C_4}$$

Then put all calculation formulas of each cost into the above formula, it can be worked out as follows:

$$\begin{aligned}\mathrm{C} &= \sum_{i=1}^{n} t_i*(c_i k_i) + (C_{21} + C_{22} + C_{23} + C_{24} + C_{25}) \\ &\quad + (C_{31} + C_{32} + C_{33} + C_{34} + C_{35} + \sum_{j=1}^{n} w_j * h_j.\end{aligned}$$

Table 37.4 The evaluation sheet of unit cost and weight of pollution

Main construction materials	Air pollution		Water pollution		Land pollution		Unit environment loss cost (c_j)
	Unite cost (c_{j1})	Weight (ρ_{j1})	Unit cost (c_{j2})	Weight (ρ_{j2})	Unit cost (c_{j3})	Weight (ρ_{j3})	
Concrete							$C_j = c_{j1} * \rho_{j1} + c_{j2} * \rho_{j2} + c_{j3} * \rho_{j3}$
Cement							
Steel							
Brick							
Gravel							
Timber							

37.4 Conclusion

Based on the stage of construction, the methods of classification and measurement system of environment cost was discussed in this paper. With activity-based costing and fuzzy comprehensive evaluation method, eventually there are formulas and accounting methods covering resources consumption costs, environmental protection costs, pollution treatment costs, and environment loss costs over the construction stage of projects. The proposed measurement system of construction projects environmental costs can provide valuable information for relevant Government's supervision departments to check what actions were taken by the construction enterprises over the environment issues and how the effect was. It is also of great meaning for government to strengthen environmental supervision through the construction process and be more severe in punishing those companies who only focus on making profit rarely considering the damaged caused by their business activities. Meanwhile, the use of environmental costs measurement system can facilitate decision-makers in construction corporations to identify all kinds of costs about environment which provides a basis for optimizing the project material input structure and flow, advancing the efficiency of material utilization and analyzing and improving the existing environmental cost control methods.

37.5 Suggestions

Concerning the current problem of environmental pollution caused by construction projects, this paper also puts forward some Suggestions in the following five aspects.

37.5.1 Purchasing Environmentally Friendly Materials at Early Stage of Construction

When gaining raw materials, pay attention that it is better to take advantage of clean energy and environment-friendly materials which do no harm or less harm to environment and those reusable and biodegradable materials. And try to make full use of construction waste to achieve recycling. For example, using hollow brick on a large scale can not only reduce the weight of buildings and construction costs, but also decrease the exploitation of clay minerals and the energy consumption.

37.5.2 Implementing Cleaner Production During Construction Phase

In construction phase, the cleaner production is mainly reflected in clean and efficient production measures. Construction machine should be selected to be in small noise and less volume of exhaust emissions. And those production equipment that is laggard in technology, harmful to environment, and in high resources consumption should be eliminated. The cement processing sites ought to be concentrated and it is needed to build protection wall to block dust and noise effectively because there could be large dust emissions and noise at a high decibel level when cement is produced.

37.5.3 Recycling the Waste in Late of Construction Stage

Turn those redundant raw materials and resources as well as unwanted intermediate products into available raw material. For instance, debris such as broken bricks and concrete can be used in the laying of pavement, and cement bags and other similar objects can be recycled for re-use after reprocessing by manufacturers.

37.5.4 Enhancing the Consciousness of Environment Protection

The government should further spread the news related to environment protection, strengthen the environmental information disclosure of construction projects and construction companies, from which excellent performing ones can be named in recognition of "green building enterprises". Besides, the internal environmental publicity and personal training in enterprises is another efficient way to improve stuff's awareness about protecting environment.

37.5.5 Strengthening Supervision and Assessment

Focusing on environmental protection target and public demand, carry out environmental monitoring mission on construction projects and promote the supervision and assessment work of environmental quality on a regular basis. Furthermore, it is essential to innovate environmental monitoring technology to enhance overall monitoring capability.

References

1. Chen YG (1998) The first international guidelines of environmental accounting and report—the conference proceedings of 15th meeting of the Intergovernmental Experts Working Group for the United Nations international accounting and reporting standards. Accounting Research 05:2–9
2. Wang PC, Dong Y, Su X (2013) The research on the environmental costs control of enterprises based on the circular economy. Ecol Econ 09:116–120
3. Liu SY, Liu YH, Liu T, Huang ZJ (2011) The research on accounting methods of the life-cycle environmental costs. Environ Sci Manag 06:36–38
4. Wang SB, Wang XR, Wang RS (2002) The research on the assessment method of life-cycle environmental costs of industrial products. Shanghai Environ Sci 12:742–744 + 755
5. Zhang YL (2008) The measurement model of environmental cost based on value-chain analysis. Stat Decis 03:58–60
6. Zhang TW, Zhen GH (2008) The accounting of enterprise environmental costs with activity-based costing method. Finance Account Mon 08:38–40
7. Guo XM (2003) The research on the environmental management accounting. Xiamen University Press, 04

Chapter 38
The Level of Crowdedness in Operating Metro Systems in the Selected Cities in China

Liudan Jia, Chenyang Shuai, Liyin Shen and Xianchun Luo

Abstract China has made a great achievement in developing metro systems over last 20 years. There are nearly 22 cities which have built metro systems. These metro systems have been playing an important role in promoting the development of the urban economy and society growth in China. But in the operation of metro system, the problems of crowdedness between passengers have become increasingly serious, due to various reasons, such as scale of metro system and the management skill. Therefore, it is important to understand the level of crowdedness in the operating of metro systems. There is a need for a method to guide us to solve this crowdedness problem. It appears that there is no existing study examining the level of crowdedness in metro system. This paper presents a model for evaluating the level of crowdedness in operating metro system with referring to a number of selected cities in China. This model is called crowdedness evaluation model. The daily ridership (r) and the length of metro system (L) are the main variables for developing the model. 10 metro systems in China are selected to provide data for analysis in this study.

Keywords Metro system · Crowdedness · China · Competition

L. Jia (✉) · C. Shuai · L. Shen
School of Construction Management and Real Estate, International Research Centre for Sustainable Built Environment, Chongqing University, Chongqing, China
e-mail: jld0617@126.com

C. Shuai
e-mail: 1391843874@qq.com

L. Shen
e-mail: shenliyin@cqu.edu.cn

X. Luo
School of Construction Management and Real Estate, Chongqing University, Chongqing, China
e-mail: 516601676@qq.com

Y. Wu et al. (eds.), *Proceedings of the 20th International Symposium on Advancement of Construction Management and Real Estate*,
DOI 10.1007/978-981-10-0855-9_38

38.1 Introduction

Since the first metro line was built in 1863 in UK, nowadays, many countries in the world have developed metro systems to solve the problem of traffic congestion. China has also invested a large amount of resource to construct metro systems. After experiencing the rapid urban development in the past years, the scale of metro systems in China has become the biggest one in the world. It was reported by the China Association of Metros [2] that, by the end of 2014, there were nearly 22 cities which had metro systems in China, and the total length of metro systems has reached to 3173 km with 85 lines. One example is that Beijing benefits from the huge investments on the construction of metro systems. In 2000, there were only just two metro lines with the total length of 54 km in Beijing, with the daily ridership about 1.2 million. But by the end of 2014, the total length of Beijing metro system reached to 604 km including 18 metro lines, and the daily ridership is also increased at 9.53 million. Similarly, the daily ridership of the Shanghai metro systems also increased to 7.84 million in 2014. The metro systems in these two Chinese cities are ranked among the busiest metro systems in the world (http://www.guancha.cn/Project/2013_03_14_131915.shtml).

However, every coin has two sides. With the growing scale of metro systems, the population using the metro systems are also increasing dramatically. If the number of the daily ridership excesses too much than that the metro systems are designed to provide service, the metro systems will be more and more overloaded. Shen et al. [4] pointed out that considering the tradeoff between utility efficiency of metro ridership and crowdedness, the factors of "crowdedness" should be given more attention when a city is in a more developed stage.

In China, there are ten cities that the total length of metro systems have exceed 100 km. The 10 cities can be considered having reached to a more developed stage in view of metro systems development than other cities in China. But as discussed above, the level of crowdedness in operating metro systems is becoming more and more serious. Figure 38.1 shows a common situation in the operation of Beijing

Fig. 38.1 The common situation in using Beijing metro system

metro system. People often have to wait for the arrival of metro vehicles by joining a long queue, with very crowded passengers inside. What is more, this situation is more serious during the peak hours.

However, the crowdedness in using metro systems will reduce the comfort degree of passengers and increase the safety risk and fire risk [3]. Therefore, there is an urgent need to solve this crowdedness problem. Meanwhile the different level of crowdedness of metro systems can be solved by adopting different management methods. The aim of this paper is to build a model to evaluate the crowdedness in using metro systems quantitatively, and find out the level of crowdedness in these major cities in China.

38.2 Research Methods

The research starts with understanding the literatures about existing methods for evaluating the performance of metro systems during the period of operation. Based on the literature study, a model for evaluating the crowdedness in using metro systems will be established. By applying the model, the crowdedness in using metro systems in 10 cities in China will be examined. These cities are representative in terms of their size and importance. They are Shanghai, Beijing, Tianjin, Chongqing, Guangzhou, Shenzhen, Nanjing, Chengdu, Dalian, Shenyang. The research data for these 10 cities are about two aspects: the scale of the metro systems and the daily ridership of metro systems. The resource for retrieving these data are mainly from the two official websites: http://www.ccmetro.com/ and http://mic-ro.com/metro/.

38.3 The Model for Measuring the Level of Crowdedness in Operating Metro System

There is little existing studies for assessing the level crowdedness in using metro systems. Some scholars had studied the methods to depict the performance of Metro systems [1, 5, 6]. In a typical study by Shen et al. [5], ten indicators are identified to depict the performance of Metro systems, as listed in Table 38.1.

As stated early, the aim of this paper is to evaluate the crowdedness in using metro systems. The study by Shen et al. [4] provides important reference in examining the utility efficiency of metro systems. Based on these theoretical studies, this study selects two indicators to indicate the operation performance of metro systems, including daily ridership (r) and the length of metro system (L). The parameter "R" is the ridership per kilometer of the metro system, which can be calculated as:

Table 38.1 Variables for evaluating the performance of metro systems

Variable	Implication
STA	Number of stations
TransSTA	Number of transfer stations
LIN	Number of lines
LEN	Length of system, in km
RID	Daily ridership
FAR	Fare, in EURO
POP	Population of the city, in million
popDEN	Population density of the city, population per km^2
GDP	Per capita GDP of the city
Year	Operation years

$$\mathrm{R} = \frac{r}{L} \tag{38.1}$$

In the model (38.1), the value design for the variable r can be very different from that in the practice. The designed value is denoted as r_d, and that in practice is denoted as r_p.

Considering this, the value of "R" can be presented in two formulas:

$$R_d = \frac{r_d}{L} \tag{38.2}$$

$$R_p = \frac{r_p}{L} \tag{38.3}$$

where r_d is the design value of daily ridership, r_p is the actual value in practice when using metro systems.

This difference between R_d and R_p can be examined by the ratio between the two parameters, as shown:

$$\theta = \frac{R_p}{R_d} \tag{38.4}$$

θ is defined as the level of crowdedness in operating metro systems.

When $\theta > 1$, it can be concluded that the scale of metro system is under supply or the system is overloaded which induces the crowdedness between passengers.

38.4 Research Results

38.4.1 Research Data

By searching for the website of http://www.ccmetro.com/, the value of indicators "r_p" and "L" for the 10 selected metro systems can be collected, and the value of "R_p" can be calculated according, as shown in Table 38.2.

On the other hand, the designed value for r_d in these 10 cities have been obtained through research visit to the Chongqing Metro Management Limited. According to the information provided by the company, the designed value of the operation capacity for every hour per metro line usually ranges from 20,000 to 30,000 persons/h. In this study, we choose 25,000 persons/h as the operation capacity per metro lines. Because different metro systems have different number of metro lines, the designed value of r_d in different metro systems is different. In other words, r_d is different for different metro systems. As a result, the design operation capacity for a specific metro system can be calculated as:

$$R_d = \frac{2.5 * T * Ln}{L} \tag{38.5}$$

In model (38.5), the parameter "L" is the total length of the metro system. The parameter "T" is the operation time of the metro system in a whole day. This study defines that the metro systems are operated from 6:00 a.m. to 11 p.m., so the value of "T" is 17. The parameter "Ln" is the number of metro lines for a specific metro system.

The value of "R_d" for the selected 10 metro systems is calculated shown in Table 38.3.

Table 38.2 Research data of "Rp" for the 10 cities

City	L (km)	r_p (ten thousand)	R_p (ten thousand/km)
Shanghai	643	784	1.2
Beijing	604	953	1.6
Guangzhou	247	610	2.5
Chongqing	202	142	0.7
Nanjing	187	147	0.8
Shenzhen	179	284	1.6
Chengdu	155	78	0.5
Tianjin	147	82	0.6
Dalian	127	25	0.2
Shenyang	114	72	0.6

Table 38.3 Research data of "R_d" for the 20 cities

City	Ln	L (km)	R_d (ten thousand/km)
Shanghai	14	643	0.93
Beijing	18	604	1.27
Guangzhou	9	247	1.55
Chongqing	4	202	0.84
Nanjing	5	187	1.14
Shenzhen	5	179	1.19
Chengdu	2	155	0.55
Tianjin	4	147	1.16
Dalian	1	127	0.33
Shenyang	2	114	0.75

Table 4 Results of for the 10 metro systems

City	θ
Shanghai	1.29
Beijing	1.26
Guangzhou	1.61
Chongqing	0.83
Nanjing	0.70
Shenzhen	1.34
Chengdu	0.91
Tianjin	0.52
Dalian	0.61
Shenyang	0.80

38.4.2 Analysis Results

By applying the collected data to model (38.4), the results of for the 10 selected metro systems is as shown in Table 38.4.

From Table 38.4, the value of these citie—Shanghai, Beijing, Shenzhen and Guangzhou is exceeds 1. The crowdedness are 1.29, 1.26, 1.34 and 1.79 respectively. In the other cities, the value of are all low than 1.

38.5 Conclusion

This study applies the parameter "daily ridership of metro systems" and "the length of the metro systems" to evaluate the crowdedness in using the metro system in China. The crowdedness between the 10 surveyed metro systems varies significantly. According to the overall performance evaluation, the most crowdedness occurs in Guangzhou. Others include the systems in Beijing, Shenzhen and

Shanghai. It signals out that the administrations of the metro systems in Beijing, Shenzhen, Guangzhou and Shenzhen shall adopt certain measures to reduce the pressure of the crowdedness between passengers. In other cities, it appears that their metro systems are underutilized to certain extent. They may consider citizen to choose metro systems as the traffic tools.

The findings from this study provide reference methodology for studying the crowdedness level in using metro systems. The limitation of this study is appreciated that more variables may be necessary to be incorporated into the evaluation model.

Acknowledgments This research is funded by the project of the National Natural Science Fund of China (Project No: 15BJY038).

References

1. Bresson G, Dargay J, Madre JL, Pirotte A (2003) The main determinants of the demand for public transport: a comparative analysis of England and France using shrinkage estimators. Transp Res Part A Policy Pract 37(7):605–627
2. China Association of Metros (2014) Urban rail transit 2014 annual statistical analysis report. Retrieved from http://www.ccmetro.com
3. Jiang S, Sun Y, Du Y (2012) Influence of in-vehicle congestion degree on choice of public transit mode. J Tongji Univ Nat Sci 40(12):1831–1835
4. Shen LY, Jiao LD, He B, Li LC (2015) Evaluation on the utility efficiency of metro infrastructure projects in China from sustainable development perspective. Int J Project Manag 33(3):528–536
5. Shen LY, Jiao LD, Zhou JY, Ren WJ (2014) Evaluation indicators for the efficiency of metro systems from a socio-economic-technical perspective. In: Proceedings of the 18th international symposium on advancement of construction management and real estate, pp 487–496
6. Yu MM (2008) The effects of privatization on return to the dollar: a case study on technical efficiency and price distortions of Taiwan's intercity bus services. Transp Res Part A 42: 935–950

Chapter 39
Study on Regional Spatial Polarization, Regional Imbalance and Functional Complementary in Coastal Megalopolis of China

Xueguang Ma and Peng Dou

Abstract After entering twentieth Century, it becomes more and more serious on regional spatial polarization, regional imbalance development and regional dysfunctional conflicts in Coastal Megalopolis of China (CMC, for short). This paper attempts to discover the characters and rules on spatial-temporal differences, regional imbalance pattern and functional complementary of CMC economic development from the year of 2000 to 2012. First, CMC regional spatial polarization pattern research was carried on by using the system cluster analysis, then followed by spatial imbalance pattern research with concentration index, Gini index, coefficient of variation and the coefficient of range, then investigated regional complementary research by using industrial map to reflect the complementary functions of city function and city group of CMC.

Keywords Coastal megalopolis of China · Regional spatial polarization · Regional imbalance · Functional complementary

39.1 Introduction

In the last century, Jane Gottmann put forward the concept of the megalopolis [1] which became the highest form of the regional urban space organization nowadays. Under the background of globalization, the spatial difference of China's cities has been intensified, and the problems of regional spatial polarization and the unbalanced regional economic development are also caused a wide range of scholars. The research on the division of labor and complementary in academic circles mostly focused on urban function classification. Meijers [2] carried out systematic analysis about the regional complementarity of Randstad of Netherlands, Fleming

X. Ma (✉) · P. Dou
School of Law and Political Science, Ocean University of China,
Qingdao, China
e-mail: hugeomaxg@163.com

Y. Wu et al. (eds.), *Proceedings of the 20th International Symposium on Advancement of Construction Management and Real Estate*,
DOI 10.1007/978-981-10-0855-9_39

diamond district of Belgium and Rhine-Ruhr of Germany. Cowell [3] studied on the complementary and organizational governance of polycentric regions, such as the San Francisco Bay Area of America, Randstad of Netherlands, Emilia-Rome of Italy. Some domestic scholars studied the core cities' functional complementarity of the three metropolitan areas: Beijing-Tianjin-Hebei, Yangtze River Delta and Pearl River Delta. In summary, domestic scholars has made some preliminary results about regional spatial polarization, regional imbalance development and regional function complementation. But there is little empirical analysis of CMC, and the statistical caliber is not uniform. The paper attempts to make up for these deficiencies, carries out the research on spatial polarization, regional imbalance and function complementation of CMC.

39.2 Methodology

39.2.1 Data Resources and Data Measurement

CMC's regional spatial polarization process and quantitative measurement of pattern. Select the annual GDP from China Urban Statistical Yearbook of the cities from 2000 to 2012 as the statistical raw data. Subject to data availability limits, this study's coastal cities do not include coastal areas such as Taiwan, Hong Kong and Macao. Measure of regional imbalance development of CMC. Select the annual GDP and land area from China Urban Statistical Yearbook of the cities from 2000 to 2012. Measure of urban functional division of labor of CMC. Select the Second and third industrial workers from China Urban Statistical Yearbook of the cities from 2003 to 2012 as the statistical raw data.

Based on SPSS analysis software, using the method of clustering analysis, treating the GDP as the basis of classification, selecting the 2000, 2005 and 2010 three years data to explore CMC's pattern and process of regional spatial polarization. Using GDP concentration index, Gini index, difference coefficient and range coefficient for quantitative measurement of regional imbalance development in CMC. Using the correspondence analysis to analyze the functional complementarity of CMC.

39.3 Researching Regions

Related coastal cities of the paper listed in the China Maritime Statistical Yearbook, which involved in 11 provinces (municipalities directly under the central government), totally 53 cities. In order to meet the need of this research, nine coastal urban clusters are included in the study area (includes some inland cities), therefore, there are totally 89 prefecture-level cities. These cities are divided into four scales: single city, provincial, coastal city clusters and CMCs. The CMCs are divided into north CMC (coastal cities in Tianjin, Hebei, Liaoning and Shandong, 17 cities), east CMC (coastal cities of Shanghai, Jiangsu and Zhejiang, 11 cities) and South CMC (coastal cities of Fujian, Guangdong, Guangxi and Hainan, 25 cities). Coastal city clusters are divided into Beijing-Tianjin-Hebei, Shandong peninsula, Liaodong peninsula (namely Bohai Rim Region as one), Yangtze River Delta, Pearl River Delta, Western Taiwan Straits Economic Zone, Southern Fujian Delta(Xiamen, Zhangzhou and Quanzhou area) and the Northern Bay Economic Zone.

39.4 Regional Spatial Polarization in Coastal Megalopolis of China

39.4.1 Characters of Spatial Polarization Pattern

The core cities grow rapidly and become the central and the major driver of the growth in the region. In 2000, there is only Shanghai which ranked first tier cities in CMC, Guangzhou and Shenzhen had been among the first tier cities in 2005, and Tianjin become one of the four first tier cities in 2010 in CMC. The four first tier cities became the leading core of economic development in CMC, and also the mainland China.

Form three growing regions, the Yangtze River Delta, the Pearl River Delta and the Bohai Rim Region. Each of the growing regions has a number of cores to support. Let the core cities to be attraction and radiation center, there are great three growing regions which are Yangtze River Delta, the Pearl River Delta and the Bohai Rim Region in CMC from 2000, within each growing regions also formed a plurality of sub-central cities (such as Hangzhou and Ningbo of the Yangtze River Delta, Dongguan of the Pearl River Delta, Tangshan, Dalian and Qingdao of Bohai Rim Region), all these cities as a whole form a network of regional economic growth spatial pattern which is composed of multiple functional nodes.

Provinces border has formed economic collapse area. We treat the core cities as the center, with circle spread radiation drive around the city to grow in the formation of three major growth areas. At the same time, the distance fading effect of regional economic development still exists which result in economic collapse area among provincial border regions. For example, the Northern Jiangsu area which is the border of Bohai Rim Region and the Yangtze River Delta region, the Northern Bay Economic Zone which lies in the periphery of the Pearl River Delta.

39.4.2 Characters of Spatial Polarization Process

In 2000, Shanghai outshines others in the 53 coastal cities of China, and it is the only and the largest city belonging to the first tier cities. As to the three coastal areas, North China and South China CMC are lack of first tier cities, the former had Tianjin which is only the second tier cities, while the latter showed the pattern of the Guangzhou-Shenzhen dual cities which are the second tier cities. In 2005, the South CMC developed rapidly, Guangzhou-Shenzhen raise together, and became the first tier cities shoulder by shoulder with Shanghai. While the first tier cities of the northern coastal areas are brewing. At the same time, we can also observe that although the northern coastal areas didn't have the first tier cities, it had formed multi-center urban area, with Tianjin-Tangshan, Qingdao-Yantai and Dalian as centers. Eastern China CMC formed distinct levels of coastal urban system comprised of Shanghai, Hangzhou-Ningbo, Nanjing-Suzhou. The Southern China CMC had formed a loose coastal urban system which comprised of Guangzhou-Foshan, Fuzhou-Quanzhou without the third tier cities. In 2010, the most significant change in CMC was that it had formed Tianjin (in north China), Shanghai (in east China), and Guangzhou-Shenzhen (in south China) as centers of the first tier cities.

39.5 Spatial Imbalance in Coastal Megalopolis of China

Firstly, the CMC involves 11 provinces (municipalities) and 53 cities in single city and provincial scale. On single city scale, there is a serious imbalance among cities with the phenomenon of GDP concentrated on some cities. On provincial scale, there are some differences between provinces, but the regional imbalance is not very prominent. With municipal units and provincial units from 2000 to 2012, We can see that the GDP concentration index of municipal units are above 0.3, difference coefficient are more than 110 %, indicating that he phenomenon of obvious unbalanced development among cities. For example, in 2012, GDP ranks the top 5 are Shanghai, Guangzhou, Shenzhen, Tianjin, Hangzhou, the sum of their GDP accounted for 36.24 % of the total of 53 cities. Contrast the provincial unit's concentration index and the difference coefficient with cities', the GDP concentration index of provincial unit below 0.3, the difference coefficient of variation is about 60 %, indicating that the total GDP gap of the 11 coastal provinces is smaller, and that the coastal provinces are relatively more balanced than the single cities.

Secondly, on coastal city clusters scale, there is serious imbalanced development differences between the 81 cities, it has large gap between cities. From the view of the whole urban clusters, the GDP difference between the nine urban clusters is relatively small, but there is the phenomenon of that the GDP is concentrated on a few urban clusters, and the imbalanced development is notable. Within the urban clusters, Beijing-Tianjin-Hebei's internal difference is the most remarkable, Xiamen-Zhangzhou-Quanzhou urban cluster's internal difference is the smallest.

Among the nine urban clusters, Beijing-Tianjin-Hebei's internal difference are the biggest, it is 107.23 %, the second is Liaodong Peninsula city circle, the difference coefficient is 98.33 %. The minimal internal difference coefficient is Xiamen-Zhangzhou-Quanzhou city circle, it is 46.47 %. Shandong Peninsula urban cluster's internal difference coefficient are small as well, it is only 49.92 %.

Finally on CMCs scale, the internal regional development gap of the three CMCs are all great from the view of the city, but the gap between them are small from the regional perspective, GDP's distribution is more balanced. According to the original data, the concentration index of GDP from 2000 to 2012 in the three CMCs is calculated, and we can see the internal regional differences among them. According to the index of the concentration of each year, make the tendency chart of the concentration index with the time. We can see the change of GDP concentration index of the three CMCs in city scale. We can see the CMCs one by one, North China and East China CMC tends to be stable, but the GDP concentration index of the East CMC tends to be decreasing, while the GDP concentration index of the North China CMC is slightly tends to be upward; The wave of the South CMC is slightly larger compared. According to the GDP concentration index of the three sub regions, the three CMCs all have a degree of difference. The imbalanced development phenomenon of South China CMC is the most serious, and the second is the East China CMC, while North China CMC is relatively balanced.

39.6 Functional Complementary in Coastal Megalopolis of China

Make the horizontal comparison among coastal urban clusters, the complementary ratio of Southern Fujian Delta is the maximum area. In 2003, the complementary ratio of Southern Fujian Delta's is the largest, which reached 4.05, then is the Shandong peninsula, after the standardization of the total inertia is 4.01, followed by the Northern Bay Economic Zone, Beijing-Tianjin-Hebei region, Liaodong Peninsula, Pearl River Delta, the Bohai Rim Region, the West Side of the Straits Western Taiwan Straits Economic Zone and Yangtze River Delta region. In 2012, the highest complementary of the region still is Southern Fujian Delta, followed by Shandong peninsula, Beijing-Tianjin-Hebei region, the Northern Bay Economic Zone, Liaodong Peninsula, Pearl River Delta, Yangtze River Delta, Western Taiwan Straits Economic Zone and Bohai Rim Region.

Longitudinal comparison of 10 years, the difference of the CMC is great, and some areas' complementary statistics are increasing, such as Yangtze River Delta, Pearl River Delta, Western Taiwan Straits Economic Zone and Southern Fujian Delta. And some areas' functional complementarity statistics are reducing, such as: the Shandong Peninsula, Liaodong Peninsula, Bohai Rim Region and the Northern Bay Economic Zone. The fastest growing normalized standard value of the regions is the Yangtze River Delta, reached 40 %, followed by Beijing-Tianjin-Hebei region, reached 27.71 %, and the Pearl River Delta region, reached 20.57 %, and the top three are the highest degree CMCs. And then is the Western Taiwan Straits Economic Zone and the Southern Fujian Delta. The standardized total inertia's reducing quantity of Liaodong Peninsula is the most, reached 25.55 %, and the variation range of the Northern Bay Economic Zone is not so notable. The two and three industry economic status of the Shandong Peninsula and the Liaodong Peninsula is less obvious increasingly. The trend of economic role of the city is more homogeneous, and it shows that the reduction of the economic role's complementary.

39.7 Conclusion

The study found that the spatial polarization pattern goes around CMCs, presented circle spread and distance attenuation trend. The process of CMC spatial polarization presents that the absolute degree of polarization are widening while the relative degree of polarization are declining as inverted U-shaped. The study found about spatial Imbalance among 4 scale of spatial unities, such as single city scale, provincial scale, coastal city clusters scale and CMC scale. The results also showed that there are big differences on functional complementary among CMCs during the year of 2003–2012.

Acknowledgements This research supported by National Natural Science Foundation of China Emergency Management Fund Project [Grant No.41440005]. The Project Sponsored by the Scientific Research Foundation for the Returned Overseas Chinese Scholars, State Education Ministry of China [Grant No. 教外司留[2013]1792].

References

1. Bailey N, Turok I (2001) Central Scotland as a polycentric urban region: useful planning concept or chimera? Urban Stud 38(4):697–751
2. Meijers E (2005) Polycentric urban regions and the quest for synergy: is a network of cities more than the sum of the parts? Urban Stud 42(4):765–781
3. Cowell M (2010) Polycentric regions: comparing complementarity and institutional governance in the San Francisco Bay Area, the Randstad and Emilia-Romagna. Urban Stud 47(5):945–965

Chapter 40
Study on the Interactive Relationship Between Cross-Border Trade Development and Construction Land Change—A Case Study of Dongxing City, China

Baokun Liang, Rucheng Lu, Guan Li, Qiuping Li, Qianjing Liang and Ying Lin

Abstract With the development of economic globalization and cross-border economic cooperation, social economy in border areas not only has developed rapidly development in social economy, but also land use has gone through radical changes, especially the use of construction land. To explore and analyze the interactive relationship and interaction between cross-border trade development and change in construction land can be of great help in making scientific and reasonable land supply policy and optimizing the layout of construction land in border areas. By adopting econometric methods like co-integration analysis and impulse response function(IRF), this study probes into total export-import volume of cross-border trade and change of its construction land based on the data of Dongxing city from 1997 to 2012. Results show that there is long-term equilibrium in Dongxing city between cross-border trade development and construction land change that every 1 % increased area of construction land leads to 3.47 % increase in cross-border trade developmental level. And then, the Granger causality is not remarkable, illustrating that the direct pulling effect in between is not significant. It is also found that the impact of cross-border trade development could result in over 34 % of construction land change while construction land change could only lead to about 1 %. It indicates that cross-border trade development can bring about changes in construction land, but not vice versa. Finally, this study suggests that construction

B. Liang · R. Lu (✉) · Q. Li · Q. Liang · Y. Lin
School of Land Resources and Surveying and Mapping,
Guangxi Teachers Education University, Nanning, China
e-mail: lurucheng72@163.com

B. Liang
e-mail: liangbaokun163@163.com

G. Li
Institute of Land Science and Property Management,
Zhejiang University, Hangzhou, China

Y. Wu et al. (eds.), *Proceedings of the 20th International Symposium on Advancement of Construction Management and Real Estate*,
DOI 10.1007/978-981-10-0855-9_40

land growth pattern and cross-border trade development mode in border areas are supposed to improve so as to promote harmonious growth of economy and society.

Keywords Cross-border trade development · Construction land · Co-integration test · Impulse response function · Dongxing city

40.1 Introduction

Border areas are of great strategic significance to defend national sovereignty and territorial integrity, consolidate frontier defense and maintain social stability. As conflicts over "Diaoyu Islands" and "the South China Sea" continue to escalate, border area and their special geopolitical value have become a focus increasingly attracting broad attention around the world. At the same time, within the overall context of economic globalization and regional economic integration, the communication between different countries becomes more frequent, and border areas on both sides of national boundaries grow to be an important space and loading region for international economic exchanges. Cross-border trade, as a unique kind of economic and trade form in border areas, plays an irreplaceable role in promoting international regional economic cooperation, increasing local fiscal revenue, raising the living standard of border people and effectively maintaining stability in border areas. Under such joint effect of special location and economic development pattern, not only social economy of border areas gains rapid, but also the land use, especially that of construction land, goes through profound changes [1]. Therefore, related issues of cross-border trade development and land use change have become a focus of academic research.

So far many experts and scholars have conducted lots of researches on the relationship between construction land change and economic growth from different perspectives such as national and regional scale [2], driving force [3], decoupling analysis [4], contribution between construction land expansion and economic growth [5–7] and by different analytical means. Most researches show that the change of land use, especially construction land change, to a certain extent, reflects requirements of social economic development, and is an inevitable result of high-speed economic development [8]. Economic growth is meanwhile a basic economic factor which drives construction land expansion [6]. For researches on cross-border trade and economic growth, at present academia mainly carries out qualitative analysis from cross-border trade's connotation, characteristics, current situation and so on [7–9], and by using econometric methods like co-integration analysis to discuss on the relationship between cross-border trade and economic growth [10]. There are also some scholars trying to research on trade development and land use change response [11], but most focus on how farm land [12] or ecological land [13] responses to the growth of cross-border trade, while few are about the interrelation between cross-border trade development and construction land.

Therefore, on the basis of above researches, this paper selected a Chinese border city, Dongxing city of Guangxi, as an example, using econometric methods like co-integration analysis, impulse response function and so on, to study on the interactive relationship between cross-border trade development and construction land change of Dongxing city from 1997 to 2012, so as to reveal the inherent coupling mechanism of cross-border trade development and construction land change, and thus functioning as reference when setting land supply policy, optimizing construction land layout and promoting coordinated development of economy and society for border areas.

40.2 Research Methods and Data Processing

40.2.1 Research Methods

With the development of modern econometric methods, the mutual influence and equilibrium relationship between non-stationary variables can be analyzed by mathematical methods like co-integration analysis and Granger causality test and impulse response function and so on [14]. This research adopting relevant theories and methods of co-integration analysis, inspects with dual-perspective observation whether there is a long-term equilibrium relationship and short dynamic causality between cross-border trade development and construction land, and also prove direction and impact strength of the causal effect.

Data Stationarity Test

Before the co-integration analysis, in order to avoid spurious regression caused by instability of time series, stationarity test should be carried out on each variable. This paper uses the ADF (Augment Dickey-Fuller) method to test the stationarity and determine the order of integration. The ADF test model is as follows [15]:

$$\Delta Y_{\mathrm{t}} = \alpha_0 + \beta T + \gamma Y_{t-1} + \sum_{\mathrm{i}=1}^{P} \theta_{\mathrm{i}} \Delta Y_{\mathrm{t-i}} + \varepsilon_{\mathrm{t}} \quad (40.1)$$

In Eq. (40.1), Y_t is the testing variable, Δ is a difference operator, α_0 is a constant term, T is the time trend, P is a lagged value and ε_t is a random error term. The null hypothesis is B_0: $\gamma = 0$, and the alternative hypothesis is B_1: $\gamma < 0$. To accept the null hypothesis B_0 means Y_t has unit root, which indicates that the ordinal variable is a non-stationary variable.

Data Co-integration Test

In order to eliminate the common stochastic trend existing among variables, cointegrafion test is needed to examine whether there is stable equilibrium between variables. There are many kinds of methods to conduct co-integration test. As this paper is to study the co-integration between cross-border trade development and construction land scale, and Engle-Granger two-step procedure (hereinafter referred to as E-G two-step procedure) is often used to test co-integration between two variables, this paper employs E-G two-step procedure in actual research. The basic principle of two-step procedure is to examine the rationality of the regression equation through testing stationarity of the residual error of estimated equation. The residual error model is as follows [16]:

$$Y_t = \alpha + \beta X_t + \varepsilon_t \tag{40.2}$$

In Eq. (40.2), X_t and Y_t are sequences of uniformity integrated. If the residual error is stable, there exists co-integration between the two. Otherwise, there is not.

Granger Causality Test

Co-integration test can only reveal whether there is a long-term equilibrium between variables but cannot analyze whether there exists causal relationship. Therefore, Granger causality test is here adopted to test the causal relationship between variable sequences [17]. Test method is [18]:

$$Y_t = \lambda + \sum_{i=1}^{m} \alpha_i Y_{t-i} + \sum_{i=1}^{m} \beta_i X_{t-i} + \varepsilon_t \tag{40.3}$$

$$X_t = \lambda + \sum_{j=1}^{m} \alpha_j X_{t-j} + \sum_{j=1}^{m} \beta_j Y_{t-j} + \varepsilon_t \tag{40.4}$$

In Eqs. (40.3) and (40.4), λ is a constant term, α_i and α_j are corresponding regression coefficients respectively. If we assume "variable X is not the reason that leads to Y's change", then we can conduct a test on β_i (i = 1, 2, …, m) = 0. If test results reject the null hypothesis, then X is deemed to be the Granger cause of Y. Similarly, to test β_j (j = 1, 2, …, m) = 0 can help determine whether Y is the Granger cause of X.

Impulse Response Function

Impulse Response Function (IRF) can be used to measure variables' dynamic characteristics and Granger causality strength between variables out of the study period, thus dynamically grasping the process of interaction among system endogenous variables [17]. The functional form is as follows [19]:

$$Y_t = \sum_{i=1}^{k} \alpha_{1i} Y_{t-i} + \sum_{i=1}^{k} \beta_{1i} X_{t-i} + \varepsilon_{1t} \tag{40.5}$$

$$X_t = \sum_{i=1}^{k} \alpha_{2i} X_{t-i} + \sum_{i=1}^{k} \beta_{2i} Y_{t-i} + \varepsilon_{2t} \tag{40.6}$$

In Eqs. (40.5) and (40.6), k is the lagged order, ε_{1t} and ε_{2t} are random disturbance terms. If ε_{1t} changes, the value of Y_t will change accordingly and the current Y_t value will affect values of X_t, Y_t in the future, which can thus help to make sure how each variable in the system response to the change of itself and other endogenous variables.

Variance Decomposition

Variance decomposition is one of the important methods to be applied to study table variables' mutual impact and reaction. Variance decomposition model is built to further analyze the contribution that the impact gave rise to endogenous variables' change and to evaluate the importance of different structural impacts, so as to reveal the influence that a variable in the system has on itself and other variables.

40.2.2 Data Sources and Processing

(1) 1997–2012 construction land area is drawn from related statistical data including land use change of Dongxing city over the years and Dongxing City Domesday Book. (2) This study uses the total export-import volume of cross-border trade to measure cross-border trade development level. The data are mainly from Dongxing City Statistical Yearbook (1996–2012); (3) Socio-economic data are derived from Guangxi Statistical Yearbook (1998–2013) and Dongxing City Statistical Yearbook (1996–2012).

In this study, JS is the value of construction land area and BT is the total export-import volume of cross-border trade.

40.2.3 The Study Area

Dongxing city is not only a city where the start point of China's border and the end point of coastline meet, but also the only port city that links China and ASEAN on

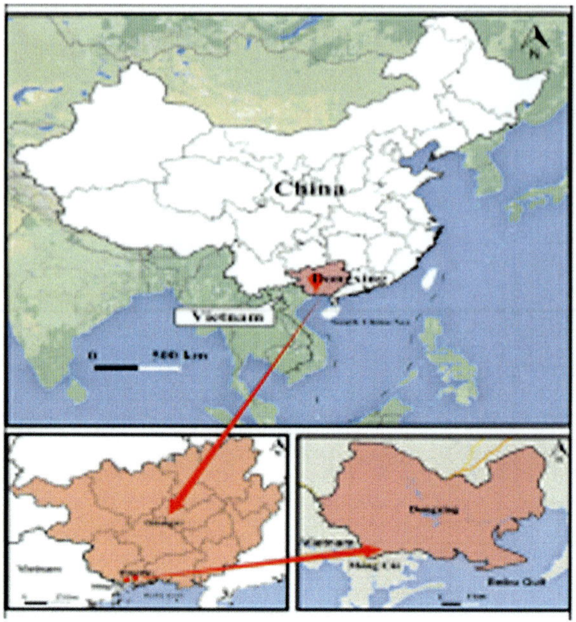

Fig. 40.1 The area chart of Dongxing

land and sea. Its unique geographical location laid a foundation for the rapid development of social economy. From the establishment of Dongxing city in 1996–2013, the total cross-border trade volume to Vietnam reached 103.549 billion yuan with an average annual growth rate of 25 %, showing a trend of steadily rising. It was accompanied by the expansion of construction land. Construction land scale of Dongxing city increased from 2454.21 hm^2 in 1996 to 4858.50 hm^2 in 2012 with an average annual growth rate of 4.36 %, and kept the expanding trend year by year. We can thus conclude that there is an obvious synchronous trend between cross-border trade development and construction land change in Dongxing city. It not only is of good representativeness in border areas, but also provides a solid research foundation for this paper (Fig. 40.1).

40.3 Study on the Interactive Relationship Between Cross-Border Trade Development and Construction Land

40.3.1 ADF Test

In order to improve the degree of fitting of the model and eliminate to some extent the heteroscedasticity phenomena in the data, we calculate the natural logarithm (LN) without altering original variable interactions. The final variables that we

Table 40.1 ADF test of temporal variable

Variable	ADF value	1 % critical value	5 % critical value	10 % critical value	Test form	Conclusion
lnBT	−1.8375	−4.7284	−3.7597	−3.3250	(C,T,0)	Non-stationary
lnJS	−1.2657	−4.7284	−3.7597	−3.3250	(C,T,0)	Non-stationary
D (lnBT)	−4.1537	−4.8864	−3.8290	−3.3630	(C,T,1)	Stationary
D (lnJS)	−3.3667	−4.0044	−3.0989	−2.6904	(C,0,0)	Stationary

Note (1) In terms of test forms, C, T and L in (C,T, L) respectively refer to the intercept, temporal trends and lag order of ADF test. When c or t is 0, there is no intercept or temporal trends involved. (2) The selection of lag period is automatically carried out by EViews6.0 in accordance with SIC and AIC rules

adopt include construction land (lnJS) and the import and export volume of cross-border trade (lnBT). Before the cointegration analysis, we conduct ADF test for lnBT and lnJS. D (lnBT) and D (lnJS) are used to represent the first-order difference of the total import and export volume of cross-border trade and construction land respectively. ADF test is conducted for the calculation of its level of sequence and difference sequence as shown in Table 40.1.

Results show that the test values of lnBT and lnJS are greater than the critical value under 10 % of significance level, which means it failed the test and the original sequence is non-stationary. And the ADF test values of their first-order difference D (lnBT) and D (lnJS) are less than the threshold value under 5 % of significance level, failing the unit root hypothesis that D (lnJS) and D (lnBT) are both stationary sequence. It indicates that during the study, cross-border trade development and the expansion of construction land in Dongxing city are relatively fast with strong temporal trends indicated by data.

40.3.2 Test for Cointegration Relationship

According to the ADF test analysis, *lnJS* and *lnBT* are first-order integrated sequences which are in line with the prerequisites for cointegration analysis. Thus we can further test their cointegration relationship. Regarding *lnBT* as explained variable, *lnJS* as explanatory variable, we use OLS regression to get the cointegration equation between these two variables:

$$lnBT = 3.4715 lnJS - 24.4048$$
$$R^2 = 0.6940 \quad F = 31.7514 \quad DW = 0.4693$$

Table 40.2 Results of ADF test of residuals of regression

Variable	ADF value	1 % critical value	5 % critical value	10 % critical value	Conclusion
e_t	−1.8777	−2.7283	−1.9663	−1.6050	Stationary

In order to examine the rationality of cointegration equation, we use ADF test to test the stationarity of the residual of regression equation. Assume e_t *as* the residual of the above equation, e_t = lnBT − 3.4715lnJS + 24.4048, and then conduct unit root test on its stability. We conclude the results in Table 40.2, indicating that under the 10 % of the significance level, the residual series is stationary.

According to E-G test, there exists co-integration relationship between lnBT and lnJS, which is a long-term equilibrium relation. It also suggests that in such relationships, every increase of the land area for 1 %, the level of cross-border trade development increases for 3.47 %.

40.3.3 Test for Granger Casual Relationship

In order to identify whether there exists a causal relationship between lnBT and lnJS, we test them on the basis of Granger causality test methods. The results are shown in Table 40.3.

According to the test: At 5 % of the significance level, all P-values are greater than critical values, indicating the acceptance of the original hypothesis, namely cross-border trade development and the change in the scale of construction land are not each other's Granger causality. However, the fact that there is no obvious Granger causality does not necessarily mean they do not have a pull effect. Cross-border trade development in Dongxing city leads to local industrial development and human capital growth, indirectly promoting local economic growth in the meantime. On the other hand, the improvement of market environment of cross-border trade and the construction of traffic infrastructure resulting from economic growth will also play a role in promoting cross-border trade development.

Table 40.3 Results of Granger causality test of temporal variable

Null hypothesis	*lnJS* is not Granger causality of *lnBT*		*lnBT* is not Granger causality of *lnJS*	
Lag orders	F statistics	*P*-value	F statistics	*P*-value
1	0.2015	0.6615	4.3773	0.0583
2	0.4142	0.6729	2.3962	0.1465
3	0.1473	0.9277	1.2537	0.3708
4	0.2870	0.8704	0.4792	0.7553

40.3.4 Impulse Response Function

It is necessary to build the VAR models of both lnBT and lnJS first so as to investigate the mutual impact between cross-border trade development and construction land, as well as its corresponding response effect. Meanwhile, we should take into account the degree of freedom of the sample data to establish lag 3-order models of lnBT and lnJS on the basis of the AIC and the SC guidelines as shown in Fig. 40.2. The horizontal axis shows the number of lag periods of impact effect while the vertical axis represents the degree of response effect from dependent variable to explanatory variables. The solid line describes the trend of the variable with impact while the dotted line is twice plus or minus the standard error of the trend.

The IRF graphical results show that when lnJS and lnBT receive an impact sizing one standard deviation of themselves respectively, their responses differentiate largely. When facing an impact sizing one standard deviation of itself, the responses of lnJS vary largely in different stages of the whole impact period. It responds swiftly and strongly in the initial stage, reaching the maximum value in the first period, decreasing continuously afterwards with slight fluctuation during the 4th and 5th period. As for lnBT, it's impact on itself is relatively large with swift

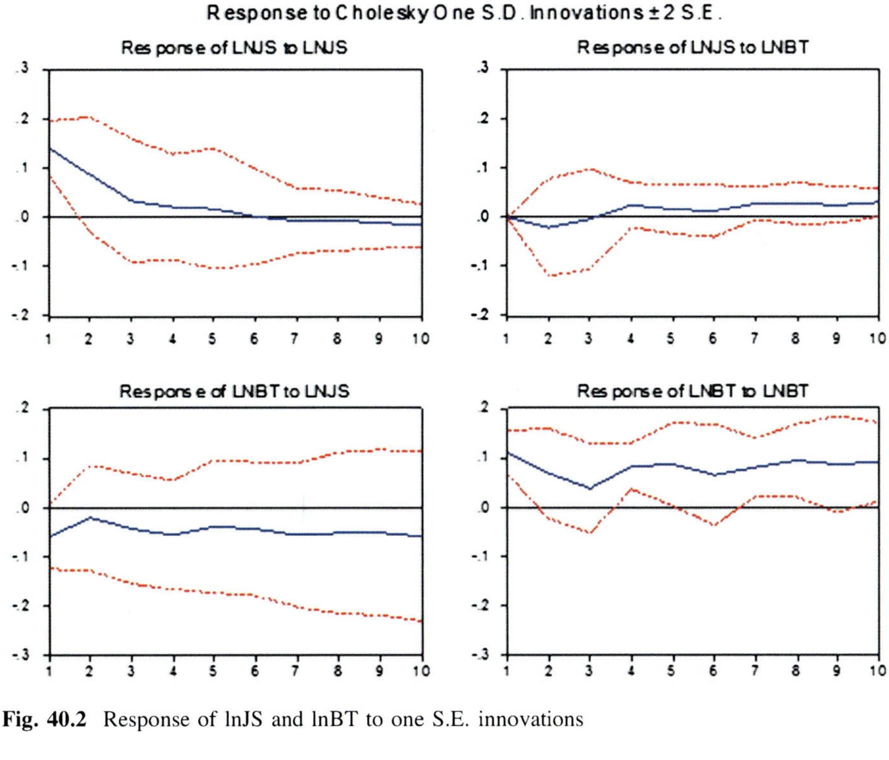

Fig. 40.2 Response of lnJS and lnBT to one S.E. innovations

response, reaching trough in the 3rd period and the peak value between the 4th and 5th period with a slight decrease after the 5th period. Following that, it rises again and maintains a stable trend afterwards. The impact from one standard deviation of lnBT to lnJS is negative and weak at first. It only shows a decreasing trend in the first period and then transforms to positive at the 2nd period, reaching peak value in the 4th period with stable and sound positive response afterwards. The impact of a standard deviation from lnJS brings negative response to lnBT with medium strength, which is relatively stable with fluctuation between the 1st and 10th period.

In conclusion, lnJS responds relatively weakly to the impact of one standard deviation of the variable lnBT; as for the impact from lnJS, lnBT responds relatively strong and quickly, but the degree of fluctuation is weak and negative. In this case, the change of construction land in Dongxing city contributes a little to cross-border trade development while the latter impact more significantly on the former, which is consistent with the actual situation. Take Dongxing city for example, along with cross-border trade development, the demand of infrastructure land will increase because of the improvement of market environment and port construction, resulting in the continuous expansion of construction land. However, economic growth, especially cross-border trade development, is subject to multiple factors of various aspects, such as the international market, macro-economic environment, cross-border trade policies, etc. In this particular context, the influence of construction land change on the development and improvement of cross-border trade is indirectly supportive.

40.3.5 Analysis of Variance Decomposition

When study on VAR model, it is advisable to further analyze dynamic characteristics of the model through building variance decomposition model, which can decompose the changes of each endogenous variable in the system into the impact on other components, providing relatively important information of each random perturbation which influences the endogenous variables [20]. The results of variance decomposition are shown in Table 40.4. The column of Period in the table refers to the forecast period. This study selects 10. The column of S.E. represents the predictive standard error of lnBT and lnJS while figures in lnBT and lnJS columns refer to the contribution level of the equation innovation which takes lnBT and lnJS as dependent variables to the prediction of standard error of each period. The sum of these two in every line is 100.

The results of lnBT variance decomposition based on VAR model reveal that lnBT receives large impact from itself, remaining at around 98 % with slight decline in later periods. In addition, lnJS's contribution to lnBT is 0.97 % in 2nd period, which means 0.97 % of the predictive variance of Dongxing's cross-border trade development level can be explained by the changes of construction land and that in 2nd period the impact degree rises slightly and declines for a certain level over time; as for the variance decomposition results of lnJS, though it is impacted largely

Table 40.4 Results of Variance decomposition of lnBT and lnJS

Period	Variance decomposition of *lnBT*			Variance decomposition of *lnJS*		
	S.E.	*lnBT*	*lnJS*	S.E.	*lnBT*	*lnJS*
1	0.1259	100.0000	0.0000	0.1405	21.4142	78.5858
2	0.1452	99.0343	0.9657	0.1665	27.6684	72.3316
3	0.1561	97.5985	2.4015	0.1699	27.8871	72.1129
4	0.1857	97.9895	2.0105	0.1730	27.3770	72.6230
5	0.2087	98.3145	1.6855	0.1745	26.9924	73.0076
6	0.2230	98.4089	1.5911	0.1750	27.2129	72.7871
7	0.2439	98.4521	1.5479	0.1772	28.9043	71.0957
8	0.2667	98.7016	1.2984	0.1795	30.5785	69.4215
9	0.2853	98.8398	1.1602	0.1815	32.0616	67.9384
10	0.3055	98.8909	1.1091	0.1846	34.3714	65.6286

by its own fluctuation, we cannot ignore the fact that the level gradually goes down afterwards while lnBT's influence on lnJS grows swiftly in the 2nd period with stable and low-speed growth in later periods.

The above analysis indicates that changes in construction land have little contribution to cross-border trade development which is only about 1 % while the contribution of cross-border trade development to construction land changes is relatively high, reaching over 34 % with stable and continuous growth over the forecast period. It points out that currently, cross-border trade development in Dongxing city is majorly influenced by cross-border trade policies put forward by national and regional governments, modes of trade, etc., rather than the expansion of construction land.

40.4 Conclusion

Through the study on the Interactive Relationship between Cross-border trade Development and Construction Land Change of Dongxing city from 1997 to 2012, we can draw a conclusion as follows:

The developing speed of cross-border trade and the expansion of construction land in Dongxing city are relatively quick, the data shows a strong temporal trend; although the study found that there is a long-term equilibrium relationship between them, neither is the Granger causality of each other, and the direct pulling effect in between is not obvious. What's more, though IRF and variance decomposition analysis, we found that the impact response of cross-border trade development which is caused by construction land change is relatively weak, while the reverse impact response is relatively strong, which indicates that construction land change will appear following cross-border trade development, yet will not necessarily promote the latter. The cross-border trade development of Dongxing city is more achieved by driving local industry adjustment, technology upgrade, trade structure change and so on that indirectly promote economic growth. Cross-border trade development has a strong impact on construction land, thus change of the development patterns of cross-border trade will lead to improvement of cross-border trade market environment and port transportation infrastructure land, which then promote construction land change. Therefore, we should pay attention to properly adjusting cross-border trade development way, further optimize import and export commodity structure in frontier trade, and impel optimized structural adjustment of construction land in border areas; on the other hand, ensure the supply of construction land, so as to offer enough support to transformation and upgrading of cross-border trade development way, to ensure the sustainable development of Dongxing city cross-border economy.

References

1. Li L, Mao J, Lu R (2014) Fractal dynamic research on spatiomporal pattern of land use in border area—a case study of Chongzuo City. Res Soil Water Conserv 5:204–209, 215, 341
2. Zhao K, Zhang A, Li P (2011) Driving forces of urban construction land expansion: an empirical analysis based on panel data of provinces. J Nat Resour 8:1323–1332
3. Chen C, Feng C (2010) Driving forces for construction land expansion in China. China Popul Resour Environ 10:72–78
4. Zhong T, Huang X, Wang B (2010) On the degrees of decoupling and recouping of economic in growth and expansion of construction land in China from 2002 to 2007. J Nat Resour 1:18–31
5. Li P, Pu L (2012) Delving into particularity on the relationship between construction land expansion and economic growth in developed region—comparing with the average of the whole China. J Nat Resour 11:1823–1832
6. Jiang H, Qu F (2009) Contribution and response of constructed land expansion to economic growth at different development stages: a case study for Jiangsu. China Popul Resour Environ 1:70–75
7. Ye Y, Zhang H, Xu X et al (2011) Study on the relationship between construction land expansion and economic growth pattern in the Pearl River Delta. Geogra Res 12:2259–2271
8. Huang J, Zhu L, Deng X (2007) The regional difference of construction land expansion and its influencing factors in China. Sci in China (Series D: Earth Sci) 9:1235–1241
9. Shan L (2010) Research on the border trade in Guangxi. Qingdao University, Shandong
10. He C (2012) Growth in western minority areas binary marginal border export trade and economic. J Int Econ Coop 9:52–56
11. René V, Elke S, Geert W et al (2009) The effect of agricultural trade liberalisation on land-use related greenhouse gas emissions. Glob Environ Change 19(4):58–61
12. Sun C, Tang W, Zhou W (2012) Estimation of virtual factors of resource and ecology embedded in grain trade of China. Resour Sci 3:589–597
13. Defries RS, Rudel T, Uriarte M et al (2010) Deforestation driven by urban population growth and agricultural trade in the twenty-first century. Nat Geosci 3(3):178–181
14. Zhou A, Xu H, Tian C (2006) Financial econometrics. Economic press, Beijing, pp 55–84
15. Lu C, Yang Q, Wen F (2010) Cointegration test and granger causality test for the relationship between urbanization and urban land use structure. Scientia Geographica Sinica 4:551–557
16. Ye H, Pu L (2007) Research on the cointegration and causality between Jiangsu province's arable area change and economic growth. J Nat Resour 5:766–774

17. Gao T (2006) Econometric analysis method and modeling: eviews application and instance. Tsinghua University Press, Beijing, pp 106–112
18. Wang S (2003) Macro measuring several frontier geography theory and application. Nankai University Press, Tianjin, pp 86–87
19. Chen L, Long K (2007) Econometric analysis of relationship between cultivated land quantity and economic development. China Land Sci 4:4–10
20. Zheng H, Liu Y, Wang X (2011) Dynamic econometric analysis of the relationship between urbanization and land intensive utilization in China. Resour Environ Yangtze Basin 9:1029–1034

Chapter 41
ABM Based Simulation Research on Construction Waste Management

Zhikun Ding, Yifei Wang and Jinchuang Wu

Abstract Rapid economic growth and urbanization in China lead to extensive construction activities which generate a large amount of construction waste (CW) inducing a great adverse impact on the environment. Therefore, how to implement effective construction waste management (CWM) has received increasing attention from construction stakeholders and researchers in the past few decades. The aim of this study is to provide a simulation model for CWM which can help to reveal the amount of CW generation and quantify the effects of various policy implementations. With the agent-based modeling (ABM) approach, the inter-relations between stakeholder agents and the CWM process are analyzed and a simulation model integrating four subsystems is built. The model is then developed and programmed with Repast Simphony software. The results indicate that the practitioners' awareness and ability of CWM as well as their communications should be further improved. Furthermore, the regulations for the use of low-waste technologies, disposal charging scheme and incentive mechanism should be further refined to reduce landfill, increase the ratio of reuse or recycling. The findings can provide guides and references for predicting CW generation and making effective CWM policies.

Keywords Construction waste · Construction waste management · Agent-based modeling · Repast simphony

Z. Ding · Y. Wang (✉) · J. Wu
Department of Construction Management and Real Estate,
Shenzhen University, Shenzhen, China
e-mail: wangyyyfei@163.com

Z. Ding
e-mail: ddzk@szu.edu.cn

J. Wu
e-mail: jinchuangwu@163.com

Y. Wu et al. (eds.), *Proceedings of the 20th International Symposium on Advancement of Construction Management and Real Estate*,
DOI 10.1007/978-981-10-0855-9_41

41.1 Introduction

With the acceleration of urbanization and urban renewal, huge construction, renovation and demolition activities can be found across China causing the quantities of construction and demolition waste (CDW) to increase at a high speed in the architecture, engineering and construction (AEC) industry. According to statistics, the amount of CDW generated each year has accounted for 30–40 % of the total municipal solid wastes in China, among which more than 100 million tons of wastes are generated due to the construction of new buildings [1]. However, most of the CW is delivered to suburban or rural areas for landfills, which has become an urgent issue due to not only its cost efficiency but also its adverse effect on the environment. Therefore, it is imperative to research how to implement effective CWM to promote the urban sustainable development.

In the past few decades, many researchers devote themselves to the study of CWM by adopting diverse approaches to investigate different topics ranging from CW generation, reduction, reuse, recycling to assessing the effectiveness of CWM, among which quantification of CW generation and CWM policy-making studies have always been the focus. Quantifying CW generation is regarded as a prerequisite for a successful implementation of CWM because knowing in advance the amount of CW generation at the project level or regional level can not only help project managers to develop a reasonable disposal plan and to arrange the on-site temporary piling but also assist government with effective policy-making. Thus various methods have been employed to quantify CW generation at the regional and the project levels. The methods can be classified into six major categories, namely, site visiting method, generation rate calculation method, lifecycle analysis method, classification system accumulation method, variables modeling method and other methods [2]. These studies on CW estimation have paved the way for computer simulation research of CWM. However, these studies don't cover subjective factors affecting CW such as stakeholders' attitude or behavior.

Meanwhile, a series of CWM policies including regulations, codes, and initiatives have been introduced by the government, which mainly focused on the economic and managerial policies. First of all, CW is often the mixture of inert and non-inert materials. Therefore, it is not suitable for reuse or recycling without segregation but generally disposed at landfills. Thus, on-site sorting policy plays an important role in CWM and has great benefits e.g. increasing the rates of reuse and recycling, reducing the cost for waste transportation and disposal, and minimizing the pollution from the huge amount of CW [3]. Furthermore, among the CWM measures proposed in previous studies, the waste disposal charging fee has long been adopted to motivate major project stakeholders to minimize CW generation and increase the recovery of CW materials thereby slowing down the depletion of limited landfill and public filling capacities [4]. Moreover, successful CWM not only requires the government's involvement but also needs other stakeholders' participation. So publicity and education policy was implemented to increase the stakeholders' awareness of CW recycling and environment protection, and improve the waste reduction skills of construction workers [5]. However, the present policies applied in many regions of China are mostly made according to a rule of thumb because it's difficult to quantify how many CW reductions in projects could attribute to the stakeholders' compliance with such policies. And it's almost impossible to conduct an experiment in the real word because of time consuming and high cost. Hence, computer simulation which can quantify the policy implementation effect offers an alternative approach to do experiments.

Based on the above analysis, this paper focuses on developing an agent-based model for better managing CW. The originality of this study includes two aspects: the first is to estimate the CW generation at the project level from the perspective of the stakeholders' interaction; the second is to provide a simulation model which can help to quantify the effect of policy implementation. In following sections, the ABM approach was firstly introduced. Next, the inter-relations between the stakeholder agents and the CWM process are analyzed and a simulation model integrating four subsystems is proposed. Then the modeling and simulation of the proposed model was delineated; the simulation results were analyzed and discussed. Finally, the paper concludes with a summary of the key findings.

41.2 Agent-Based Modeling Method

Agent-based modeling (ABM), also referred to as individual-based modeling (IBM), is a bottom-up modeling method which gains more and more attention for analyzing complex systems and complex adaptive systems over the last few decades. ABM is a computational method that enables researchers to create, analyze, and experiment with models composed of autonomous and heterogeneous agents that interact with each other and their local environment in order to identify the mechanisms that bring about some macroscopic emergent phenomenon of interest. Compared with other modeling methods, ABM has the following advantages [6]: (1) ABM has the ability to describe a complex adaptive system; (2) Macroscopic system and microscopic agent can be effectively integrated; (3) ABM can model agents' behavior and achieve effective description of adaptive agents; (4) Reuse of the model is optional. Based on the above advantages, this method is widely applied in various disciplines including social science, ecology, economics and political science. Recently, ABM has also become a popular technique for modeling construction project management [7].

41.3 Model Development

41.3.1 System Analysis and Description

According to the studies from Hao et al. [8], Yuan et al. [4] and Ding et al. [9], construction waste management system (CWMS) can be viewed as a complex adaptive system because it involved different participants, activities and interactions. The framework of CWMS is constructed to cover four subsystems, namely subsystem of CW generation, subsystem of CW disposal, subsystem of CWM policy, subsystem of CWM assessment (see Fig. 41.1).

1. Subsystem of CW generation is fundamental to quantify CW generated from construction projects so as to establish and analyze CWM. Designers and constructors are critical for CWM due to the application of low waste technologies in design and construction [5] e.g. the adoption of prefabrication. Moreover, the construction agent has the responsibility to collect & sort waste on-site, and the right to determine whether the waste will be dumped in non-designated areas. Therefore, this subsystem mainly involves designer agent, constructor agent as well as the project agent.
2. Subsystem of CW disposal is developed to deal with the CW activities ranging from CW transportation, reusing, recycling to landfill. Different agents carry out the above activities. However, this study uses the disposal agent on behalf of all agents related to the disposal process to simplify the model. It should be noted that CW reusing refers to using the same materials in construction more than

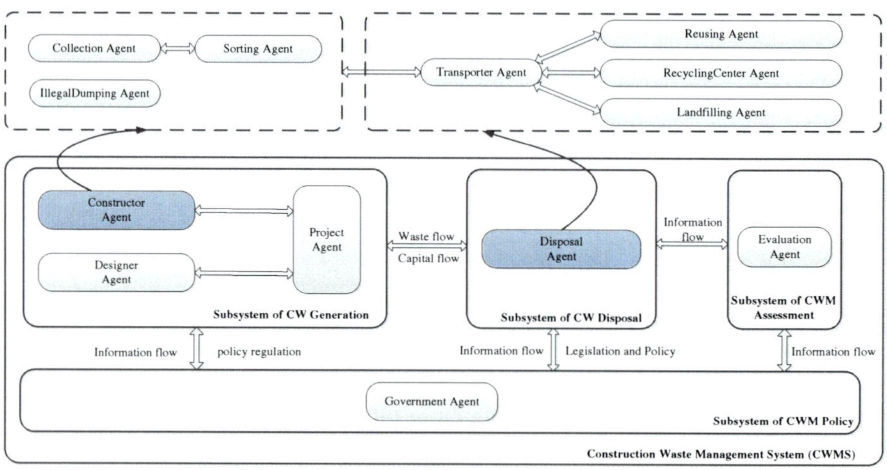

Fig. 41.1 The framework of construction waste management system

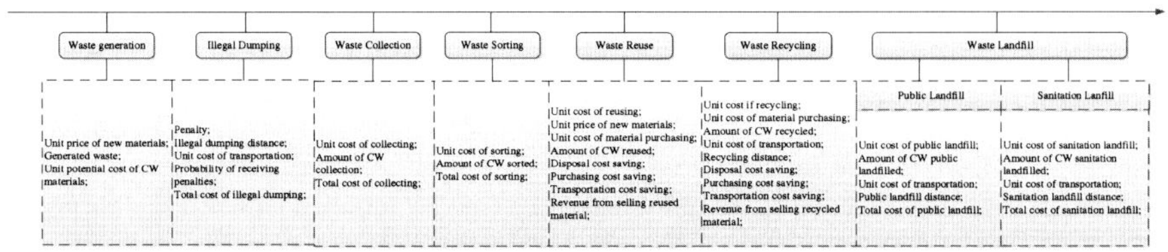

Fig. 41.2 A conceptual model of the CW chain

once i.e. using the material again for the same function and new-life reuse for a new function while CW recycling is the separation and recycling of recoverable waste materials generated from construction activities e.g. the non-inert materials. Nevertheless, when CW can not be reused and recycled, landfill then become the final option for CW disposal e.g. inert materials.

3. Subsystem of CWM policy mainly involves the government agent who plays a supervisory and leading role in CWM such as policy making to regulate the various stakeholders involved in the construction activities. For example, control the illegal dumping behavior through penalty, encourage the reuse and recycling of CW, and motivate designers & constructors to promote the CW reduction management through tax subsides etc.
4. Subsystem of CWM assessment covers the evaluation authority or assessment agencies which can quantify the effects of policy implementation. For instance, the assessment agent in this subsystem can calculate the economic benefits of CWM implementation, and this process involves many variables which are classified by the different construction stages from the perspective of the waste chain (see Fig. 41.2).

41.3.2 Model Formulation and Construction

Computable Model

Based on the above system analysis and agent identification, the next step is to clarify the agents' state & properties and design the system evolution rules. To solve this problem, the following four aspects are considered: (1) the objectives which guide the agents' behavior and influence agents' decision-making. Different agents have different objectives because each agent is a heterogeneous and independent individual. (2) The attributes which distinguish an agent from other agents. As the simulation model runs, agents can update their attribute values according to their rules to better adapt to their environment. (3) The behavior which is the overall observable sum of the agent's actions and state changes. It is an emergent property

Table 41.1 The conceptual model of constructor agent

	Constructor Agent	Constructor Class
Objective	Pursue economic and environmental benefits of CWM to maximize as well as ensure the project quality and complete on time;	Constructor (from agents) onsiteManagementExpectationLevel : Double onsiteManagementLevel : Double AttitudeTowardSortingBehavior : Double SubjectiveNorm : Double PerceievedBehaviorControl : Double illegalDumpingRate : Double sourceReductionRatebyConstructor : Double sortingRate : Double Constructor() step() calculateSourceReductionRatebyConstructor() calculateIllegalDumpingRate() calculateSortingRate() instigate() getAttributes()
Attribute	Source reduction rate by constructor, The level of on-site management, The expectation level of on-site management, Attitude towards on-site sorting behavior, Subjective norm, Perceived behavior control, Illegal dumping rate; Sorting rate;	
Behavior	Source construction reduction management, Determine whether illegal dumping and improve the level of construction reduction management, Collecting waste, On-site sorting, Moving and interaction in the environment;	
Rule	1. If the penalty less than the cost of landfill, then illegal dumping, else not; 2. If the level of construction reduction management below the expectation level, then improve the level, else not; 3. The agents communicate and influence with other agents and their local environment with random movement in the two-dimensional grid space. For example, agent attitude towards construction management change from reduction to non-reduction or conversely; 4. On-site sorting behavior is guided by the Theory of Planned Behavior.	

caused by the interactions between the internal, local, environmental states and the decision rules. Overall system behavior is an emergent property of the interactions between all of the agents' behaviors and the environment. (4) The rules or the "internal models" of agents which describe how states are transformed to actions or new states and could be understood as mechanical decision rules or transformation functions. The rules of agents, either static or dynamic, are usually based on the assumption of rationality or bounded rationality. In the CWMS, it involves a variety of agents who have different objectives, attributes, behaviors and rules. Take the constructor agent as an example for illustration. Its design is presented in Table 41.1.

Model Implementation

Once the computable model is designed, it can be programmed in an appropriate modeling or programming environment. In this research, the model with 3700 lines of source code is implemented in a Java-based modeling environment called Repast symphony 2.2. The model contains three packages and thirteen classes as shown in Fig. 41.3.

For the cwmmodel package, it includes the CWMmodelContextBuilder class extending from DefaultContext<Object> in the class library of Repast Simphony and implementing the interface of ContextBuilder<Object>. The main function of this class is to create the simulation context that involves all of the agents, construct the two-dimensional grid space, add all kinds of agents and control the simulation procedure.

For the cwmmodel.common package, there are three classes namely Constants class, Project Data class and CWMUtils class. Their responsibilities are to define model constants, to store attribute data of the project and to reserve common utility functions in the model. It should be noted that how to read project data into the model is the key to construct the model as a whole. The following codes solve this problem (see Table 41.2).

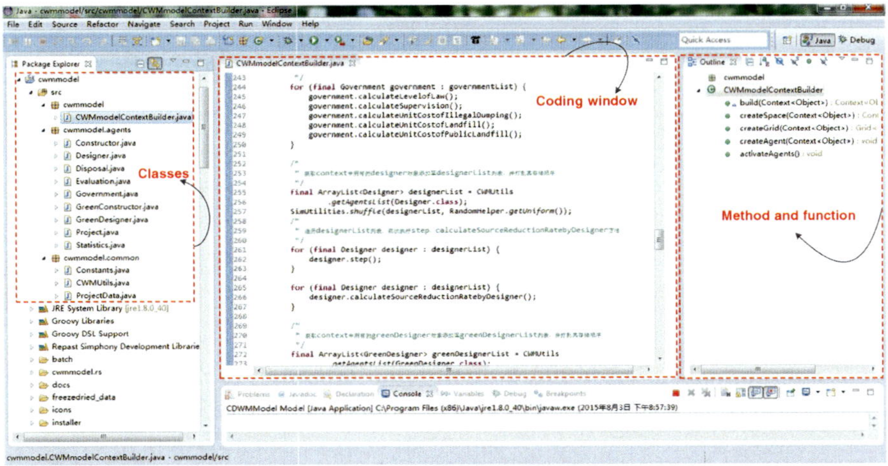

Fig. 41.3 Model implementation

Table 41. 2 Read project data into the model. (ReadDataFile() in CWMUtils class)

<table>
<tr><th>ReadDataFile()</th><th>Classes</th></tr>
<tr><td>//Take ProjectDataFileName as the parameters and return the list containing ProjectData instances
public static List<ProjectData> readDataFile(final String ProjectDataFileName) {
//Create a variable length ret list including ProjectData
final ArrayList<ProjectData> ret = new ArrayList<ProjectData>();
BufferedReader br = null;
try { br = new BufferedReader(new FileReader(ProjectDataFileName));
//Ignore the header of the data file
for (int i = 0; i < Constants.PROJECT_DATA_FILE_HEADER_LINES; i++) {
br.readLine(); }
String line = null;
//Read data in rows
while ((line = br.readLine()) != null) {
final String[] data = line.split("\\s+");
int idx = 0;
//The data in each row is sequentially assigned to projectID、area/startTime、endTime
final String projectID = data[idx++];
final double area = Double.parseDouble(data[idx++]);
final double startTime = Double.parseDouble(data[idx++]);
final double endTime = Double.parseDouble(data[idx++]);
//Create ProjectData by reading the data and add to the ret list
ret.add(new ProjectData(projectID, area, startTime, endTime)); } }
catch (final FileNotFoundException e) { e.printStackTrace(); }
catch (final IOException e) { e.printStackTrace(); }
finally { if (br != null) { try { br.close(); }
catch (final IOException e) { e.printStackTrace(); } } }
return Collections.unmodifiableList(ret); }</td><td>CWMUtils
(from common)
readDataFile()
getGrid()
getSpace()
randomElementOf()
moveTowards()
prob()
getAgentsList()

ProjectData
(from common)
projectID : String
area : Double
startTime : Double
endTime : Double
ProjectData()
getProjectID()
getArea()
getStartTime()
getEndTime()</td></tr>
</table>

For the cwmmodel.agents package, it contains all kinds of agent classes identified in the CWMS including constructor/green constructor, designer/green designer, project, disposal, government and evaluation agent class. All of the agents having their own objectives, attributes, behaviors and rules aforementioned in the computable model. The key decision functions of each agent are presented in Table 41.3. They not only are the framework for all the agent classes but also determine whether the model could run successfully.

Table 41. 3 Sample codes of the agents' decision-making functions

Decision-making Function of the Agents	
(1) Project Agent	(3) Designer Agent
$\mathit{WasteGeneration} = \mathit{WGI} \times \mathit{area} / (\mathit{endTime} - \mathit{startTime})$	$\mathit{SourceReductionRatebyDesigner} = f(\mathit{wasteReductionDesignApplyingLevel})$
$\mathit{WasteGeneration}' = \mathit{WasteGeneration} \times (1 - \mathit{SourceReductionRate})$	$\mathit{WasteReductionDesignApplyingLevel} = f(\mathit{WasteReductionDesignApplyingLevel}, \mathit{WasteReductionDesignExpectationLevel})$
$\mathit{SourceReductionRate} = \mathit{SourceReductionRatebyDesigner} + \mathit{SourceReductionRatebyConstructor}$	$\mathit{WasteReductionDesignExpectationLevel} = f(\mathit{WasteReductionDesignExpectationLevel}, \mathit{InteractionEffect})$
New residential construction: waste generation index (WGI)=37 kg/㎡	(4) Disposal Agent
waste generation classification index (unit: kg/㎡)	$\mathit{wasteCollected} = \mathit{wasteGeneration}' - \mathit{wasteIllegalDumped}$
Concrete=18.7 kg/㎡	$\mathit{wasteSorted} = \mathit{wasteCollected} \times \mathit{sortingRate}$
Brick, block and stone=1.8 kg/㎡	$\mathit{wasteReused} = (\mathit{wasteofMetal} + \mathit{wasteofWood}) \times \mathit{ReusedRate}$
Mortar=1.3 kg/㎡	$\mathit{wasteRecycled} = (\mathit{wasteofMetal} + \mathit{wasteofWood}) \times \mathit{RecycledRate}$
Metal=4.0 kg/㎡	$\mathit{wastePublicLandfilled} = (\mathit{wasteCollected} - \mathit{wasteReused} - \mathit{wasteRecycled}) \times \mathit{wastePublicLandfilledRate}$
Wood=37 kg/㎡	$\mathit{wasteSanitationLandfilled} = \mathit{wasteCollected} - \mathit{wasteReused} - \mathit{wasteRecycled} - \mathit{wastePublicLandfilled}$
(2) Constructor Agent	(5) Evaluation Agent
$\mathit{illegalDumpedRate} = \int (\mathit{illegalDumpedRateIncreace} - \mathit{illegalDumpedRateDecrease})$	$\mathit{economicalCost} = \mathit{costofMaterialbyWasteGeneration} + \mathit{costofWasteIllegalDumped} + \mathit{costofWasteCollected} + \mathit{costofWasteSorted} + \mathit{costofWastePublicLandfilled} + \mathit{costofWasteSanitationLandfilled} - \mathit{costSavingofWasteReused} - \mathit{costSavingofWasteRecycled}$
$\mathit{illegalDumpedRateIncreace} = f(\mathit{costofIllegalDumped}, \mathit{costofDisposal})$	$\mathit{economicalPerformance} = \mathit{economicalCost} - \mathit{economicalCost}'$
$\mathit{costofIllegalDumped} = (\mathit{penalty} + \mathit{illegalDumpingDistance} \times \mathit{costofTransportation}) \times \mathit{probabilityofRecivingPenalty}$	
$\mathit{costofDisposal} = \mathit{costofLandfill} + \mathit{costofTransportation} \times \mathit{landfillDistance}$	
$\mathit{illegalDumpedRateDecreace} = f(\mathit{supervision})$ $\quad \mathit{supervison} = f(\mathit{levelofllaw})$	
$\mathit{SortingIntention} = 0585 \times \mathit{AttitudeTowardSortingBehavior} + 0.188 \times \mathit{SubjectiveNorm} + 0.205 \times \mathit{PerceievedBehaviorControl}$	
$\mathit{SortingBehavior} = 0.935 * \mathit{PerceievedBehaviorControl} - 0.174 * \mathit{SortingIntention}$	
$\mathit{sortingRate} = f(\mathit{sortingBehavior})$	

41.4 Model Simulation and Analysis

41.4.1 Data Collection and Quantification

For quantitative simulation and analysis, all variables and feedback loops in the model have to be firstly quantified which is achieved in this study by collecting data from empirical cases in the construction industry of China. These cases are new residential buildings located in Shenzhen, South China. These project data including project ID, gross floor area, the planned start date and the planned finish date were collected from the open data service platform built by the Housing and Construction Bureau of Shenzhen Municipality.

More related data were collected for the proposed model from extensive literature, specifications, interviews and websites. For example, according to the special planning of construction waste treatment facilities in Shenzhen and the technical code for construction waste reduction in 2012, the transportation distances from construction sites to landfills or recycling facilities, and the CW composition, the rate of waste generation & the waste generation index could be determined. In addition, some parameter values were acquired by the interviews and consultation with industry experts e.g. attitude towards on-site sorting behavior, subjective norm, the level of construction reduction management. After data collection, all variables involved in the model and interrelations between these variables were quantified.

41.4.2 Model Validation

Once all the variables and functions are determined, various tests will be conducted to build up confidence in the model and ensure the model to reflect the real word in a reasonable way. A wide variety of tests, at four levels, have been performed to verify and validate the CWM model, namely recording and tracking agent behavior, single-agent testing, interaction verification limited to minimal model and multi-agent verification [6]. Since the integrated development environment has a built in debugger that allows to go through the code line by line or set "break points", a typical example of the recording and tracking agent behavior by walking through the source code in a debugger mode is illustrated in Fig. 41.4.

41.4.3 Result Analysis and Discussion

Considering the status quo of CWM in Shenzhen and the purpose of this study, the benchmark parameters for the basic policy scenario can be determined by parameter control panel as shown in Fig. 41.5a. There are 806 new residential buildings in this model and the simulation time can reach 336 days when the final project is completed. When all the variables are determined and essential tests for the model verification are finished, simulation experiments can be designed and performed. The following two aspects will be covered in this section: (1) explore the amount of CW generation; (2) explore the cost-benefits of CWM under policy regulation. After the simulation is completed, the output of the model can be analyzed as shown in Fig. 41.5b.

According to the output time series chart and the data set from the system platform, 1.71 million tons of CW will be generated from 806 projects in the base run. However, there will be only 1.21 million tons of CW generated by these projects, saving 30 % million tons, because the interactions between the

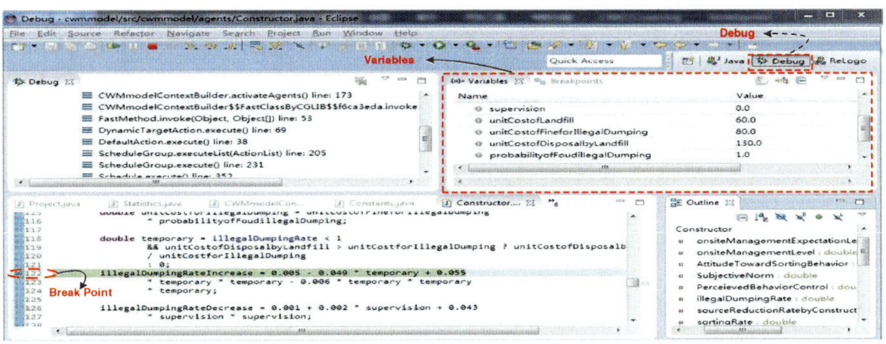

Fig. 41.4 Repast Simphony debug view window

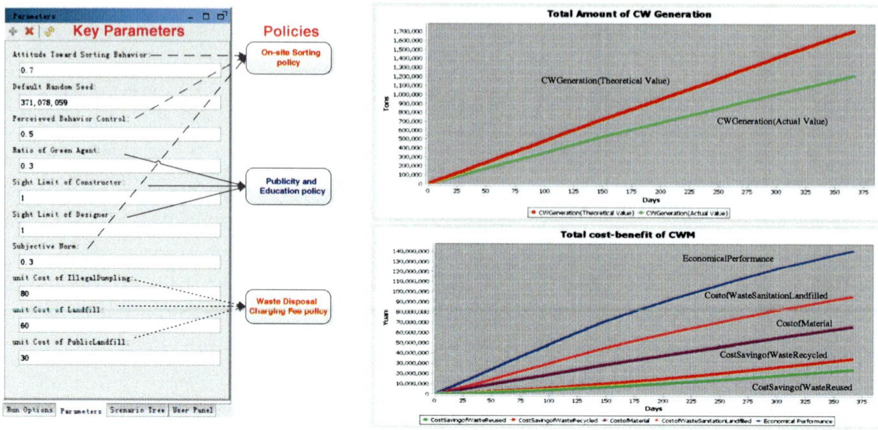

Fig. 41.5 Simulation parameters and output result (the *left* is **a** and the *right* is **b**)

constructors and the designers enables them to improve the level of on-site management and to consider the CW reduction in design. Hence, efforts should be made to promote the communication among all the stakeholders and through the publicity and education to enhance the awareness of CWM [5]. On the other hand, the economic performance, namely the difference between the total cost of CWM (199.96 million yuan) and the total benefit of CWM (58.97 million yuan), shows a continuous growth during the period reaching 140.99 million yuan in the end. Also, the proportion of the cost of waste sanitation landfill and the cost of CW material over the total cost of CWM are 47.81 and 33.05 % respectively. The cost saving of waste reused and waste recycled reach 24.28 million yuan and 34.69 million yuan respectively. Hence, the regulations and policies e.g. the reasonable waste disposal charging fee and the application of low-waste construction technologies should be further promoted to reduce landfill, increase the ratio of reuse or recycling [10].

41.5 Conclusion

The application of an ABM based simulation model for the forecast of CW generation and the outcome quantification of policy implementation presents a new method for improving the performance of CWM. Particularly, the agent-based model integrates all critical variables affecting the economical performance of CWM and is capable of studying interactions underlying these variables from a dynamic perspective. In this way, the cost-benefits of CWM are better revealed. Moreover, the parameter control panel provides a good platform for simulating effects of any given CW policy which allows decision-makers to identify how a particular level of management policy can contribute to the decrease of economic pressure caused by CWM.

The simulation results showed that government should pay attention to the practitioners' awareness of resource saving and environment protection for CWM as well as their ability to conduct good CWM. Moreover, the communications among stakeholders have a positive effect on their attitude toward CWM. Therefore, efforts should be made to publicize the importance of CWM and some essential vocational training should be organized to improve the practitioners' skill and ability [5]. Furthermore, the regulations should focus on the technical recommendations for the use of low-waste technologies in construction and design phases e.g. use of new recycled materials and prefabricated materials. Last but not least, the reasonable waste disposal charging scheme and incentive system for stakeholders should be established to increase the ratio of reuse or recycling and reduce landfill [11].

Further research should concentrate on how to raise agents' intelligence in the model. For example, agents could be provided with a more intelligent learning capability by introducing genetic algorithms, neural networks and other methods. In addition, to take full advantage of building data and to enable the simulation space more close to the real geographical space, GIS with the powerful spatial analysis capabilities could be integrated with ABM.

References

1. Wang JY, Li ZD, Wang XF (2012) Factors analysis on construction waste minimization in the design stage. J Eng Manage 26(4):27–31
2. Wu Z, Yu AT, Shen L, Liu G (2014) Quantifying construction and demolition waste: an analytical review. Waste Manage 34(9):1683–1692
3. Wang JY, Yuan HP, Kang XP, Lu WS (2010) Critical success factors for on-site sorting of construction waste: a china study. Resour Conserv Recycl 54(11):931–936
4. Yuan HP, Wang JY (2014) A system dynamics model for determining the waste disposal charging fee in construction. Eur J Oper Res 237(3):988–996
5. Lu WS, Yuan HP (2010) Exploring critical success factors for waste management in construction projects of China. Resour Conserv Recycl 55(2):201–208
6. van Dam KH, Nikolic I, Lukszo Z (2013) Agent-based modelling of socio-technical systems, vol 9. Springer, Netherlands, p xxvii
7. Knoeri C, Nikolic I, Althaus H-J, Binder CR (2014) Enhancing recycling of construction materials: an agent based model with empirically based decision parameters. J Artif Soc Soc Simul 10(3):1–13
8. Hao JL, Hills MJ, Huang T (2007) A simulation model using system dynamic method for construction and demolition waste management in Hong Kong. Construction Innovation 7 (1):7–21
9. Ding ZK, Wu JC, Wang JY, Wang HT (2014) New horizon for construction waste management research from a complex adaptive system perspective. In: 18th international symposium on advancement of construction management and real estate. Xi'an, China. pp 581–592

10. Zhang XL, Wu YZ, Shen LY (2012) Application of low waste technologies for design and construction: a case study in Hong Kong. Renew Sustain Energy Rev 16(5):2973–2979
11. Saez PV, Merino MD, Gonzalez AS, Porras-Amores C (2013) Best practice measures assessment for construction and demolition waste management in building constructions. Resour Conserv Recycl 75:52–62

Chapter 42
Strategic Direction and Path Selection of Urban Villages Emergence Prevention: A Case Study on Chongqing City

Jiafei Bai, Jirui Yang and Qiulan Sun

Abstract The phenomenon of Urban Villages emerges in the urbanization process of China, which has played a positive role in the process of urbanization. However, Urban Villages cause increasingly prominent problems which constrain urban development and people's living environment. To cope with such problems, this paper firstly reaches the reasons for the emergence of Urban Villages by holistically reviewing and integrating both international and domestic practices of managing and preventing Urban Villages, which includes Urban-rural Dual Management Structure, vague land property, inattentive law enforcement on illegal land using, lacking of policy support and so on, and then brings forward a strategic orientation to prevent their emergences in future, that is Categories Prevention. Secondly, this paper finds out the development trend of Chongqing Urban and Rural Areas between the year of 2016 and 2025, and points out the areas that have the potential to become Urban Villages in future which should be payed close attention by using Landscape Expansion Theory and ArcGIS Buffering Space Technology, taking Chongqing for example. Finally, this paper puts forward five paths or policy recommendations in terms of preventing the emergence of Urban Villages. They are (a) confirming the development timing sequence, renovating villages within Rural-urban Continuum in classification, (b) carrying out central village planning to assure the overall synchronization between countryside and urban development, (c) keeping special village and building intact in case of not affecting city development, (d) dredging and blocking to strengthen urban and rural integration of land management, along with (e) bringing the Urban and Rural Areas into urban management system. This paper primarily contributes to applying ArcGIS Buffering

J. Bai (✉) · J. Yang
School of Economics, Sichuan University, Chengdu, China
e-mail: 249944891@qq.com

J. Bai
Chongqing Engineering Technology Research Center of Land Use and Remote Sensing, Chongqing, China

Q. Sun
Chongqing Xinrong Land and Housing Survey Technology Institute, Chongqing, China

Y. Wu et al. (eds.), *Proceedings of the 20th International Symposium on Advancement of Construction Management and Real Estate*,
DOI 10.1007/978-981-10-0855-9_42

Space Technology to proactively prevent the emergence of Urban Villages, and thereby, it makes an original contribution to providing some meaningful recommendations for making solid urbanization policies in China.

Keywords Urban villages · ArcGIS · Urbanization · Chongqing · Land management

42.1 Background

Winner of the 2001 Nobel Pize in Econonics, the American economist Stiglitz once declared that the high-tech industry in the United States and China's urbanization would be the world's biggest influential things in the 21st century, and deemed that urbanization is one of the biggest challenges confronting the new century of China. Livable is recognized the goal of modern urban construction. Yet Urban Villages is the anti-livable phenomenon occurring in the process of urbanization. Therefore, building livable city must solve Urban Villages problems. Because Urban Villages are the products during the process of urbanization, researching Urban Villages can't leave urbanization.

China's urbanization has entered a rapid development period in the last 10 years, and the rate of urbanization reached 54.77 % in 2014.[1] The number and area of the city are also increasing rapidly. There are 653 cities, covering 49,800 square kilometers by the end of 2014.[2] Our country is still in the period of rapid urbanization, and the urbanization growth rate of China will reach the highest point at around 2030. And after 2030, the urbanization rate will reach about 70 %, then enter a relatively stable stage gradually according to scholars' more consistent judgment.

However, Urban and Rural Areas[3] are turned into Urban Villages from traditional villages on space form, yet in social form, the farmers of Urban and Rural Areas lost their land during the process of urbanization. And farmers who lose their land cannot live their lives depending on farming no longer, thus, they turn to the second or three industry. But restricted by the cultural level and work skills, they can hardly find a new ideal career. Even worse, soaring price of land and house, as well as the obstacle of binary system prompt the residents of Urban and Rural Areas to rush to build illegal houses on a large-scale using the flaws of urban planning.

---

[1]Source: *The National Economic and Social Development Statistical Bulletin (2014)* released by *National Bureau of Statistics.*

[2]Source: *The Urban Construction Statistical Bulletin (2014)* released by *The Department of Housing and Urban-rural Development.*

[3]*Urban and Rural Areas* means the transition zone between urban and rural areas characterized both urban and rural land use properties, also known as peri-urban areas.

They earned the rent of the floating population continuously, while causing chaos of Urban Villages social security, excessive building density, lacking of public facilities, environmental degradation and so on.

As an old says, Preparedness ensures success, unpreparedness spells failure. By the same token, preventing the emergence of new Urban Villages is essential during the urbanization process.

42.2 Purpose and Significance

Although Urban Villages are the products of urbanization, they did not show the tendency of withering away. In contrast, if reorganization of property rights, social security, village development, community culture, planning regulations, construction financing couldn't be well coordinated during the process of urbanization, there will be the emergence of new Urban Villages. This paper intended to find out the reason of the emergence of new Urban Villages through holistically reviewing and integrating both international and domestic practices of managing and preventing Urban Villages, and to put forward coping strategic direction, as well as to prevent policy recommendations in terms of preventing the emergence of new Urban Villages to provide experience for latter transformation of Urban Villages.

There are not only two theoretical significances in this paper, which are revealing the reasons of arising Urban Villages, enriching urbanization theory, but also three practical significances, which are providing recommendations for reconstruction and prevention of Urban Villages during the process of Urbanization, exploring systems, structures, mechanisms of Urban and Rural Harmonious, and improving people's livelihood of Urban Villages.

42.3 Reason Analyzing of Emerging Urban Villages

Urban Village problem is not an accidental, isolated social problem, yet it is a social issue universally emerging during the process of China's urbanization, and there are many similarities with foreign slums.

42.3.1 Binary Systems in Urban and Rural Area Are the Root Cause

China gradually established a highly centralized planned economy system after mid-1950s, and strictly distinguish between the agricultural population and non-agricultural population in terms of account migration, food supply, employment arrangement, welfare and so on, implementing the opposite dual management system between urban and rural areas. In the current regime, although Urban Villages are in the city within the space, yet but the management of these villages have not been fully integrated into the urban management system. And there is a certain vacuum in the management of Urban Villages land use and construction. However, there are no feasible management approaches within the existing division approaches of Rural Homestead, and town planning, construction, management regulations, etc., which put the villages surrounded by the city almost anarchy in terms of construction and management, and it is also root cause of Urban Villages.

42.3.2 Land Tenure Cross Is the Direct Cause

According to Article 23 of Regulation on the Implementation of the Land Administration Law of the Peoples Republic of China (1998), if there is an actual need for using the land outside the scope of urban construction determined by the overall planning for land utilization for such construction projects as energy, communications, water conservancy, mines and military installations involving agricultural land, it shall be handled pursuant to the following provisions: the construction unit shall, on the strength of the relevant approval decumbent of the construction project, file an application for land for construction with the competent department of municipal or county people's government, the competent department of municipal or county people's government shall examine the application, draw up an agricultural land conversion, land requisition plan and land provision plan. This project-oriented way of expropriating land (that is, the constructor expropriated rural collective land according to the project requirements and cost.) split the rural collective land into pieces, causing state-owned land and collective land intersecting. Meanwhile, the vast majority of the rest of the rural collective land are land for house and public facilities and other uses are likely to become Urban Villages. For example, there were 3000 acres of land before in Shuangbei village Shapingba District, but there were 1800 rural collective land to be expropriated because of Metro Line 1, Shuangbei Bridge and State Road Widening and other projects, remaining approximately 1200 acres, of which there were 120 acres left rural constructing. Unfortunately, the phenomenon of Sidaluanjian was serious on the rest of the rural collective land.

42.3.3 Illegal Using of Land Widely Is the Incentive

According to Chongqing Urban-Rural Master Planning (2007–2020), progressively urban expanding and the corresponding gradually expropriating land made a part of the Urban and Rural Areas become city in space, but not in the form of society.

In addition, any development and construction of Urban and Rural Area are prohibited according to Chongqing Urban-Rural Master Planning (2007–2020). That means, the land of Urban and Rural Area are frozen. However, the above lands had obvious advantages in location. So residents said living in Urban and Rural Area developed and constructed themselves without scientific planning driven by economic interests, which bring about widespread phenomenon of illegal land using. And the development with quick benefits, will inevitably resulting in new Urban Villages emerging potentially.

42.3.4 Lacking of Policy Support Is the Main Reason

According to Article 40 of The Law of Town and Country Planning, the constructor or individual shall apply for construction project planning permit if he/she wants to construct projects in the city or the town within the planning area. And when constructor or individual apply for a construction project planning permit, they shall submit the relevant documents of using the land at the same time. However, when the department of land carry on land clearance for constructor or individual, construction land planning license issued by Planning Bureau should be needed. Whereas, when the Planning Bureau conducted construction land planning permit for the constructor or individual, the constructor or individual need to provide the land transfer contract or previews of allocated land. However, transfer or allocation of the relevant provisions do not apply to collective land within urban planning. Thus, lacking of policy support for industrial land within the scope of urban planning makes development of industry in Urban and Rural Area difficulty. For example, although Tongguanyi Town Jiulongpo District solved the problem of development of rural industrial (that is, people, places and money problem) well by the way of Increasing and Decreasing Project, there are about 100 million yuan of investment without land clearance due to corporate ownership of land cannot be handled timely, affectting the normal development of industry.

42.4 The Prediction of Focus Region Prevention in Chongqing Urban Villages

42.4.1 Range Determining

The range of key prevention village of future Urban Villages within Chongqing Urban and Rural Area is located between 2014 Built-up Area Boundary and the scope of Chongqing Urban and Rural Overall Planning (2007–2020).

42.4.2 Determine the Proper Extension Scale

This paper gets an average annual expansion rate of each unit relative area of the space unit by computing unit area unit (1 km^2). That is the annual average of urban land use expansion percentage a unit area of the space unit during the period of study, and the relative annual extension rate is calculated by the follow formula:

$$AGR = \frac{UA_{i+n} - UA_i}{n} \times 100\,\%$$

In this formula, *AGR* indicates *Annual extension Rate* (or *Annual Growth Rate*), UA_{i+n} and UA_i indicate respectively the area of urban land use in the i + n year and the i year, and n is the time in units.

At same time, the paper carries out a processing of partitioning *Built-up Area* within *Center City of Chongqing* in order to get a strong spatial directivity between the expand scale of urban space and speed characteristics. And, then it takes the of Chongqing (*Guanyinqiao*) as the origin, and divides the research area into eight regions according to 8 azimuths, which are north, northeast, east, southeast, south, southwest, west, northwest on the basis of generating new *Built-up Area* spatial distribution of different time.

After obtaining the above data, the paper calculated out the total area of *Built-up Areas* landscape patches of the eight areas in *Center City District* within each period of beginning and end studying time by using *ArcGIS Statistical Functions*, and completed the calculation of *Built-up Areas* landscape changes amount in different of time (as is shown in Table 42.1). It can be seen the development rapidly during the year of 2010–2014 from table. And the plaque area of Urban *Built-up Areas* have increased from 219.91 to 380.02 km^2, with the average annual growth of 32.02 km^2, mainly growing in north; northeastern and northwestern, which means that the development trend of the city is north.

Table 42.1 The table of change situation of patch scale changes of built-up downtown (2010–2014) unit: km^2

Change situation	2010–2014			
	Beginning	Terminal	Total variation	Annual change
Northeast	32.77	59.2	26.43	5.29
Southeast	18.73	27.58	8.85	1.77
Northwest	26.58	47.28	20.7	4.14
Southwest	36.6	54.86	18.26	3.65
North	29.65	66.22	36.57	7.31
East	18.08	29.3	11.22	2.24
South	30.27	54.68	24.41	4.88
West	27.23	40.88	13.65	2.73
Total	219.91	380.02	160.11	32.02

42.4.3 The Expansion Trend Forecast of Built-up Area

The paper forecasts that the Main City of Chongqing will extend at a rate of about 50 km^2/year before 2025 according the scale of Built-up Area during the year of 2010–2014, combining with socio-economic development of Chongqing, the Chongqing General Land Use Planning (2007–2020) and Chongqing Urban and Rural Overall Planning (2007–2020). And there will be 250 km^2 of urban growth scale during the year of 2016–2020. Then, it matches the expansion rate of Built-up Area averagely, and concludes that each direction of Built-up Area extend outward at the same speed, by using ArcGIS Overlay and Buffering Function. Basing on this, the paper finds out that if the existing Built-up Area extend outward 2 km, there will be 310 km^2 of urban growth scale during the year of 2016–2020 totally by conducting buffer analysis on the border of 2014s Main City, deducting relevant areas located on Chongqing General Land Use Planning (2007–2020). Therefore, the scale of Built-up Area during the year of 2016–2020 is the town planning range which is formed by the existing Built-up Area extending outward 2 km.

Whereas, the scale of Built-up Area during the year of 2021–2025 is the rest area after deducting the existing Built-up Area and the scale of Built-up Area during the year of 2016–2020 relevant areas from Chongqing General Land Use Planning (2007–2020). The paper reaches the Map of Development Trend for Chongqing Urban Villages, as is shown in Fig. 42.1.

42.4.4 Determine the Land Use Situation of Rural-Urban Continuum in the Future

After the Chongqing core area Built-up Area extended range be forecasted during 2016–2020 and 2021–2025, using the overlay analysis function of ArcGIS, the situation of land use in the Rural-urban Continuum in the future decade can be analyzed through the combination with the present land use map of Chongqing core area.

42.4.5 The Key Prevention Villages of Rural-Urban Continuum in the Future

This paper analyzed the space expansion mode in different periods of the Chongqing Rural-urban Continuum and brought from that there are mainly three modes of landscape space expansion within Rural-urban Continuum, which are filling-up style, edge style and enclave type by combining the theory of landscape expansion, the future trend of urban development, the social and economic condition, and the investigation, reconnaissance and consulting for the land. Then, chose the key prevention area in the future decade through the application of Landscape Expansion Theory and ArcGIS Buffering Space Technology (as is shown in Fig. 42.2).

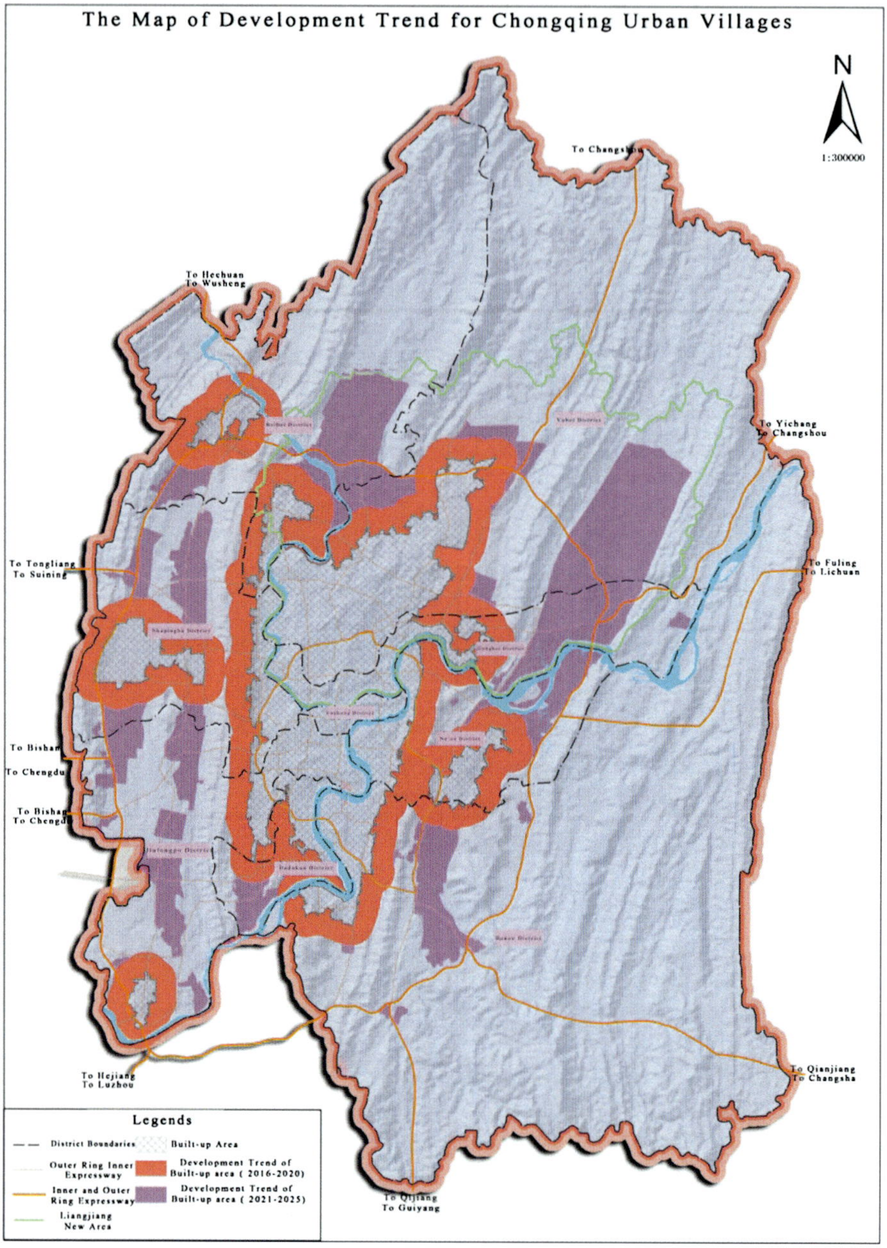

Fig. 42.1 The map of development trend for Chongqing urban villages

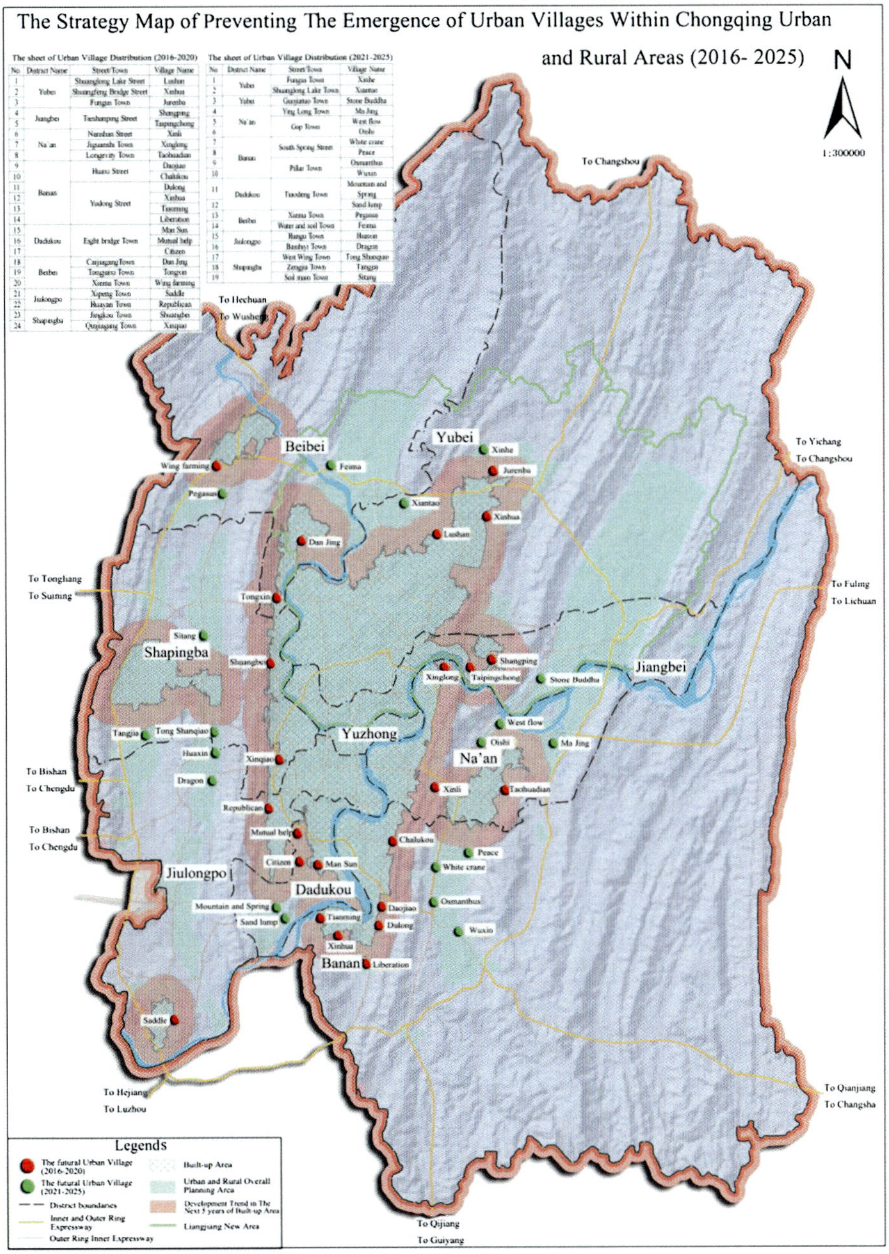

Fig. 42.2 The strategy map of preventing the emergence of urban villages within Chongqing urban and rural areas (2016–2025)

42.5 Proposals

42.5.1 *Confirming the Development Timing Sequence, and Renovating the Villages Within Rural-Urban Continuum Classifiedly*

The local government should deal with the villages within the planned range into two cases according to urban development needs in the framework of Chongqing Urban-Rural Master Planning (2007–2020): the recent changes to the of the city.

1. Villages will be changed into city recently. The local government should forecast a reasonable scale of farmers' housing resettlement, and plan as a whole, make arrangements for resettlement land and resettlement housing construction in advance to make the demolition work organization orderly.
2. Villages will be transformed into urban construction land long-term but exist short-term. May explore to combine with The Second Ring Era Large Residential Area Planning of Chongqing Main City, bring the project area farmers directly into the large settlements within the Second Ring Era through regulating rural settlements with the policy of Increasing and Decreasing. Distract land development rights in short-term at point blank range to ensure the urban farmers' income of the Rural-urban Continuum in the near future, avoiding the malignant pattern of changing from Welt Construction[4] to Urban Villages, and then to transformation, achieving Rural-urban One Step Transformation.[5]

At the same time, it may be appropriate to solve the housing problems of the urban-rural migrant workers on the base of the current public rental policy.

42.5.2 *Carrying Out Central Villages Planning to Assure the Overall Synchronization Between Countryside and Urban Development*

The central village planning should do well focused on the following points, adhering to the principle of long-term-view, reasonable guide, and advance planning: Firstly, integrating resources and creating special industries, as well as fostering economic growth to make Rural-urban Continuum to be Dual Typical

[4]Welt Construction mainly refers to the phenomenon of the construction of affordable housing for rent on the border near the town planning homestead collective construction land.

[5]There are two ways for *Step Transformation*. Firstly, bringing the farmers within the project area directly into the city through the policy of increase and decrease. Secondly, compensate for losses to farmers' short-term development interests through the policy of increase and decrease.

consisting of promoting urban modernization and rural modernization, in accordance with the requirements of the overall urban planning and coordinating the construction requirements of surrounding area planning to the center of the village, as well as adapting to the trend of regional economic and social development and urbanization. Secondly, strictly controlling the total amount of land development and development intensity to per capita land use indicators locating in a reasonable range. Thirdly, guiding the villagers to improve the construction way and urban consciousness. And eventually, reaching the goal of rational layout in Central Villages, orderly construction and development, optimized environmental quality and villagers enjoying a happy life through scientific planning and construction guidance as well as efficient and standardized management. Thus, bring Central Villages planning to assure the overall synchronization between countryside and urban development, making Central Villages to be new specific area of urbanization.

42.5.3 Keeping Special Village and Building Intact in Case of not Affecting the Development of Urban

The paper recommends retain and develop the featured countryside within Urban and Rural Areas under the circumstances of affecting the development of urban, on base of scientific planning and layout. For instance, the government may combine rural tourism and agricultural tourism and leisure, rural festivals, etc. together to form the rural tourism format mainly themed as farmhouse, country hotels and so on by integrating tourism resources, to enrich the connotation of rural tourism and promote the sustainable and coordinated development of rural tourism and leisure agriculture. Putting urban and rural together closely to let the majority of both urban and rural masses enjoy high quality of life of city. Simultaneously, give prominence to Bayu culture style during the process of construction, forming a beautiful landscape of urban suburbs.

42.5.4 Strengthening the Land and Housing Management of Urban and Rural Areas

The paper suggested that the government should adopt the approach of combining guidance when dealing with the problem of illegal land use of Urban and Rural Areas actively and steadily, and it is not only necessary to strengthen the management, to protect the reasonable needs of housing, but also necessary to carry out comprehensive remediation activities to strengthen the management of illegal land using. Land and Housing Department should lead relevant departments to organize jointly launched special campaigns as soon as possible. According to Several

Opinions About Registration and Certification of The Rural Collective Land Right Verification (Land and Resources issued [2011] No. 178), land and housing departments indeed the right registration certification against the ultra-area homestead in accordance with different historical stages. The paper suggested that the government should identify the history and status of land using for illegal homestead and the land of collective construction, and should give the approval and registration certification for the land complying with Overall Land Use Planning, Town Planning and related land policy. And resolutely punished if not.

42.5.5 Bringing Urban and Rural Areas into Urban Management System

There is a prevalent problem that the municipal infrastructure of Urban and Rural Areas is weak, and the government should implement appropriately tilted fiscal policy not only being in line with the principle of laying the foundation for the future of the Urban and Rural Areas but also holding the goal of optimizing the environment and enhancing the development potential of Urban and Rural Areas, and then accelerate to make up for the gap of urban fringe infrastructure, to ensure the development of fringe infrastructure in moderately overdeveloping. Meanwhile, public service between multiple groups should be implemented Seamless Connection for Chongqing's multi-center, multi-group model of development to avoid Villages between groups. Based on the status of social and administrative institutional of Urban and Rural Areas, the paper suggested that bring the urban fringe into the urban management system, and implement urban management villages through the introduction of urban lifestyles and management, if the conditions are ripe, to avoid producing Villages because of different management systems.

References

1. Chen S (2010) Spatial morphology of city village harmonious transformation in great cities of Central and Western China. Chongqing University, Chongqing, p 1
2. Smart A, Deng Y (2007) Informal path-illegal construction in mainland China and Hong Kong
3. Wang XW (2002) Subculture discussion of city villages. Anhui Build 2

Chapter 43
Experience, Lessons of India's Urbanization and Its Warnings to China

Yan Wang and Jiafei Bai

Abstract The urbanization in India is an excessive one. Its urbanization speed is not in line with the industrialization level and economic development; and poor population quality of those who migrate from rural areas into urban areas results in poor quality and low efficiency of urbanization development; excessively rapid urbanization also leads to severe social problems. The solution to excessive urbanization in India requires scientific layout and planning, adherence to the path of ecological civilization and intensiveness and high efficiency, emphasis of urban infrastructure construction, and adherence to coordinated development between urbanization, industrialization and urban-rural integration. All these are important lessons to the urbanization process in China.

Keywords India · Urbanization · Experience and lessons

43.1 Introduction

Urbanization is a process of economic and social progress characterized by constant concentration of population and social productive forces from rural areas to urban areas in the process of economic development. It is often regarded as an important mark measuring the social-economic development of a country or a region. In such a process, rural population is turned into urban population, and non-agricultural economic activities are constantly gathered in urban spaces. The constant improvements in urbanization are accompanied by the rising number of cities, the expanding size of cities and urban industries, and improving urban quality. China and India share many things in common, such as their history, national conditions, and economic basis after they achieved independence respectively, although they

Y. Wang (✉) · J. Bai
School of Economics, Sichuan University, Sichuan, China
e-mail: 109280741@qq.com

Y. Wu et al. (eds.), *Proceedings of the 20th International Symposium on Advancement of Construction Management and Real Estate*,
DOI 10.1007/978-981-10-0855-9_43

show obvious differences in urbanization process and economic development, with China far exceeding India. The Indian government is still puzzled by various problems caused by urbanization in spite of its efforts to solve them over the past decades. It is from various efforts made by the Indian government that we found China and India share many similar problems in the urbanization process. Therefore, we can draw on some important implications and experience from the urbanization process in India.

43.2 Overview of the Urbanization Development in India

As a country with the second largest population in the world and the most potential emerging economy, India has attracted worldwide attention with its development in recent years. However, India has always been slow in the development of urbanization. Its overall urbanization rate, which amounted to 17.9 % in 1960, was less than 20 % in the 1960s. Since the 1970s, the growth rate of urban population increased from the annual average of 2.1 % in the 1960s to over 3 % with the urbanization process. Since the 1990s, India has accelerated urbanization process, with the growth of urbanization rate of 25.7 % in 1991 to 28.7 % in 2005, which was less than 30 %. During the 14 years from 1991 to 2005 when Rao government conducted reform, the urbanization rate grew accumulatively to nearly 3 % points, with an annual average growth rate of 0.21 % points (Table 43.1). It was not until 2011 that urbanization rate reached 31.2 %, exceeding 30 % [1].

The urban geographical boundary has always been constantly expanded along with the rapid urbanization process. The corresponding urban construction has been in full swing as well. From 2001 to 2011, the urban construction was accelerated against a backdrop of rapid urbanization process. The number of urban agglomerations increased from 384 in 2001 to 475 in 2011, up 23.7 %. The number of large cities and towns increased from 1362 in 2001 to 3894 in 2011, up 186 %. The number of designated towns showed a slight increase of 6.36 % from 2001 to 2011 (Table 43.2).

Table 43.1 Urbanization rate in India

	Population urbanization rate (%)	Urban population (unit: 10,000)
1960	17.9	7784
1991	25.7	22,287
2005	28.7	31,415
Average annual growth rate from 1960 to 2005	0.24	525
Average annual growth rate from 1991 to 2005	0.21	652

Table 43.2 Changes in the numbers of urban agglomerations, cities and towns in India [2]

Types of cities and towns	Number of 2001	Number of 2011	Growth rate (%)
Urban agglomerations	384	475	23.7
Large cities and towns	1362	3894	185.9
Designated towns	3799	4041	6.37
Villages	962	981	1.98

The geographical boundary of urban area in India is increased constantly with the urbanization process. The average geographical boundary of urban area was expanded from 38,509.28 km^2 in 1961 to 78,199.6 km^2 in 2001. Of all the cities, the geographical boundary of urban area of level-I cities was expanded from 8174.29 km^2 in 1961 to 30,984.69 km^2 in 2001 on the average (Table 43.3). In the urbanization process in India, the growth rate of Level-I cities is faster than that of the cities as a whole. The main reason for such a phenomenon is that India's urban planning structure takes the form of an inverted pyramid, which is mainly reflected by the imbalance in the layout of urban space development [3]. A small number of extra-large cities and large cities are distributed in a relatively independent way in some regions, where they obtain high-speed development. However, these cities neither interact with nor drive the development of small and medium-sized cities. As a result, the population size in large cities cannot be put under control, while small and medium-sized cities show slightly insufficient development. Therefore, India governments later lowered the growth speed of Level-I cities intentionally. For example, the growth rate of the geographic boundary of cities as a whole increased from 1.19 % in the period of 1961–1971 to 2.05 % in the period of 1991–2001. In contrast, the area of Level-I cities increased in an opposite direction, with its growth rate decreasing from 3.64 % in the period of 1961–1971 to 2.58 % in the period of 1991–2001 (Table 43.3).

Table 43.3 Expansion of geographic boundary of urban area [2]

Year	Area (1000 m^2)		Annual compound growth rate (%)		Number of urban agglomerations and that of cities and towns
	Cities	Level-I cities	Cities	Level-I cities	
1961	38,509.28	8147.29	–	–	2657
1971	43,336.77	11,689.53	1.19	3.64	3081
1981	52,380.58	17,176.5	1.91	3.92	3891
1991	63,836.01	24,021.97	2	3.41	4615
2001	78,199.66	30,984.69	2.05	2.58	5161
2011	–	–	–	–	7935

43.3 Main Problems in the Urbanization Process in India and the Reasons Behind Them

43.3.1 Main Problems Caused by "Great-Leap-Forward" Type of Urbanization in India

There exists to some extent the phenomenon of "great-leap-forward" type of development in India. Such a type of development leads to an excessively rapid speed of urbanization, which is exceeding the necessary speed at which the city can sustain industrialization and economic development. That's why we say the urbanization in India, in essence, is arguably a type of "excessive urbanization." As farmers flow into cities in excessively large numbers and at an excessively rapid speed, the number of cities, the scope of urban space and the population size all increase at an unreasonably high speed, which may give rise to severe urban and rural diseases. The disordered expansion of population in large cities leads to the collapse of urban basic services, and results in a series of problems with regard to housing, slum, use of water, infrastructure and life quality. The disordered urban life and the poor rural life coexist, which make it difficult to maintain vibrant momentum of development in many cities in India.

43.3.2 Main Reasons for the Formation of "Great-Leap-Forward" Type of Urbanization in India

First, expansion of urbanization does not play the role of supporting industrialization and economic development. The manufacturing industry not only produces a large number of employment opportunities for labor forces, but also provides employment opportunities of higher income than those in rural areas. The most important reason behind the shift of population from rural areas into urban areas is the necessity of pursuing more highly-paid jobs than those in rural areas. After gaining independence, India also implemented the five-year plan of giving priority to developing heavy industries and basic industries. Such a plan promoted to some extent the establishment and development of India's modern industry. The gathering of both state-run industrial sectors and privately-run industrial sectors in some cities, to some extent, helped improve the urbanization level in India. However, the agricultural development lagged behind for a long period of time, which severely affected the development of India's industry. That's why the service industry always occupies a fairly high proportion of India's GDP. For example, in 2013, the tertiary industry accounted for 67 % of India's GDP, apparently higher than those of industry and agriculture, which were 13 % and 20 % respectively. Such a large proportion of the tertiary industry, coupled with the green revolutionary that effectively promote the development of modern agriculture, causes unemployment of rural populations while increasing grain outputs. The surplus laborers in larger numbers are in the state of hidden unemployment. These laborers are seemingly engaged in agricultural or non-agricultural activities. In fact, the labor supply far exceeds labor demand [3] ,which promotes to some extent the concentration of poverty population from rural areas into urban areas.

Second, in addition to the factors playing the role of "pushing" the development of urbanization, there is the power playing the role of "pulling" the development of urbanization. The biggest difference between India and China in terms of urbanization lies in the fact that India does not have the household registration system. As India did not undergo social revolution, the private ownership of land is preserved. When combined with other traditional social systems in India, the private ownership of land has some negative impact on urbanization [4]. The co-ownership of larger family together with the system of equal division of family properties among many sons give rise to the progressive decrease in the land owned by each household. Since the land outputs cannot support the whole family, the rural laborers move to the urban area. This is mainly caused by the "pushing" role played by the rural poverty, which is also evidenced by the large numbers of slums in Indian cities. Such a kind of urbanization is arguably driven to some extent by poverty, not by population migration.

Third, explosive growth of population becomes an important factor of promoting the accelerated development of urbanization in India. The factors contributing to urban population growth include: natural growth of urban population, migration of population from rural areas to urban areas, and redefinition of geographic boundary of urban area. Of all the sources of urban population growth in India, natural growth of urban population is the main pulling force. During the period of 1971–1981, the natural growth of urban population accounted for about 41 % of all the increase in urban population growth in India, the migration of population from rural areas to urban areas and the expansion of the geographical boundary of urban construction contributed to about 36 % of such increase. During the period of 1981–1991, the proportions of the above two factors were 60 and 22 % respectively. Therefore, natural growth of urban population is the main factor of promoting urbanization in India [5]. According to the result of a national consensus, India had a total population of 360 million in 1947 when it attained independence. In spite of the measures it took, the Indian government failed to control the population growth. By 2011, India had a total population of 1.24 billion. The population growth rate decreased from 1.82 % in 1997 to 1.37 % in 2011, while the population density grew from 324.71 people per square kilometer in 1997 to 417.56 people per square kilometer in 2011, showing an increase of 92.85 people per square kilometer [6].

To deal with the "great-leap-forward" style of urbanization, the India government has taken many measures, e.g., formulating the overall planning of urban development, constructing large industrial parks, transforming slums in cities, improving living conditions in rural areas, developing township enterprises, reducing appropriately the birth rate in urban areas, and controlling the number of people flowing into cities. The results of implementing these policies, however, did not seem to be satisfactory.

43.4 Experience and Lessons of Indian Urbanization to China

India took roundabout courses and accumulated some experience in its urbanization process. Based on our analysis of the models, process and measures taken to develop urbanization in India, we can draw on the experience and lessons of India in the following respects.

First, healthy urbanization should be developed in a coordinated way with industrialization and urban-rural integration. In India, the reason why a large number of rural people flow into cities is that it is impossible to provide employment opportunities for these people in rural areas, instead of rural people coming into cities for pursuing better development opportunities. As cities receive people from rural areas in a passive way, there appear a large number of slums in cities. To cope with this problem, the India government has promulgated a series of laws and measures to guarantee the development and employment in rural areas, including National Rural Employment Guarantee Act (NREGA) and National Rural Employment Guarantee Scheme (NREGS). In many practices of urbanization in China, since lopsided emphasis is placed on the development of the regions surrounding cities while the interactions between cities and their surrounding areas, rural areas in particular, are ignored, there appears imbalance between urban and rural development. As China's new type urbanization is developed following the principle of taking people first, it is imperative for such urbanization and agricultural modernization to complement each other. Therefore, it is necessary to shift the urban planning model from the one that simply focuses on city and town planning to the one that highlights urban-rural integration. The focus should be put on developing small towns and cities for the purpose of driving rural social-economic development, increasing farmers' income, gradually narrowing the gap between the rural areas and the urban areas, and achieving the urban-rural development on an equal footing and in a coordinated way. In addition, the comprehensive supporting reforms with the reform of household registration system as the focus should be initiated as soon as possible, so as to grant the same treatment to rural people flowing into urban areas as urban people in terms of employment, social guarantee, education, medical care and housing guarantee.

Second, scientific layout can promote the coordinated development of large, small and medium-sized cities. Due to historical reasons, India has long been emphasizing the development of a small number of large coastal industrial cities while ignoring the development of small and medium-sized cities and small towns. As a result, a large number of people are highly concentrated in large cities, the urban layout becomes imbalanced, and there arise such problems as the appearance of a large number of slums and the lack of infrastructure. China should draw on the experience and lessons of India when developing its new-type urbanization. In the new round of making urban planning and layout across the country, China should, by following the planning of national major function zones, design the layout of city agglomerations in a scientific way, and form a development pattern in which large cities occupy leading position, small and medium-sized cities play supporting role, county towns serve as links and small towns act as basic units. Efforts should be made to promote the scientific positioning of each city or town, reasonably divide the functions of cities, develop cities in a coordinated way, enable cities and towns at different levels to connect with and complement each other more than before, so that all units at different levels of city agglomerations form a cascaded development pattern, in which large, small and medium-sized cities and small towns develop in a synergetic way. China shall work hard to embark on a path of new type urbanization suitable for its national conditions.

Third, China should attach importance to strengthening urban infrastructure construction and providing government-subsidized housing including public rental housing. India paid a heavy price for the backward construction of its urban infrastructure. So it has taken a series of measures to enhance the construction and management of urban infrastructure in the development of urbanization. In contrast, China is better and advanced with regard to the construction of its urban infrastructure. But China is still facing many challenges which severely restrict the promotion of urbanization, including traffic congestion and urban water-logging. China should further strengthen the construction of urban infrastructure, focusing on improving the weak links in such construction, so as to enhance the comprehensive urban carrying capacity. China should also improve the management of urban infrastructure, and by relying on modern methods and technologies, construct smart infrastructure so as to bring the service functions of cities into full play. In addition, providing government-subsidized apartments is an important project of people's livelihood. Over the past decades, the Indian government has taken many measures to provide houses for poor urban populations. China should draw on the good practices of India in solving its housing problem for urban residents, and constantly promote the construction of government-subsidized housing projects. Sufficient importance should be attached to the spatial layout of government-subsidized housing projects in its urban planning. It should observe the principle of "balanced layout, convenient traffic, complete supporting facilities, environment suitable for living", and follow the thinking of "localized concentration and extensive dispersion" to coordinate the construction of government-subsidized housing projects with industrial development and spatial expansion. By making reasonable layout of government-subsidized housing projects, China should guide the development of such projects to be consistent with the overall planning of urban land use and the overall urban planning, so as to avoid the over-concentration of low-income groups of people, a phenomenon which would lead to social differentiation and the forming of quasi-slum. China should draw on the advanced experience of India in transforming slums and constantly promote the rebuilding of urban shanty towns and urban villages, so as to avoid the phenomenon of "juxtaposition of high risers and low shanties."

Fourth, China should adhere to the requirements of ecological civilization and intensiveness and high efficiency in developing urbanization. It should not follow the path of high energy consumption, high pollution and low output when developing new-type urbanization. Instead, it must follow the path of ecological civilization and intensiveness and high efficiency. In addition, China should draw on the successful experience of Bangalore in planning and constructing the city by following the requirements of ecological civilization, formulate clear objectives and guiding principles of building ecologically-friendly city, enhance the awareness of ecological-friendliness among citizens, and build ecologically-friendly cities step by step by adopting measures in light of local conditions. In India, a large amount of farm lands and forests disappeared in the process of urban sprawling, which caused severe damages to India's ecological system. For example, large quantities of high-yield land disappeared when the city of Allahabad was expanded. The increase in the number of slums severely damages the urban landscape. In China, urban development land area per capita is 130 m^2, which far exceeds the average level of 82.4 m^2 in developed countries and that of 83.3 m^2 in developing countries. This shows an extensive model of developing urbanization in China [9]. In this respect, China should draw on the lessons of India. In light of the present situation of land resources in China, particularly the shortage of cultivated land, China should accelerate the comprehensive transformation of old cities, optimize urban spatial layout, adjust industrial mix, follow the path of using the urban land in an intensive and economic way, and realize the shift of focus from "incremental adjustment" to "stock adjustment" in the provision of land, so as to improve the efficiency in using urban land.

References

1. Zhaoyi L (2013). Annual report on India's national report (2011–2012). China Social Sciences Academic Press
2. Tripathi S (2013) An overview of India's urbanization, urban economic growth and urban equity. MPRA Paper No. 45537, p 6
3. Shihai S (1992) Urbanization in India and its characteristics, South Asian Studies, Issue No. 4, pp 9–17
4. Hongsheng W (2006) Inquiry into and analysis of the slow progress in the urbanization in India, collection of the theses of the 4th forum on modernization study in China, pp 65–71
5. Peilin L (2010) Characteristics, experience and lessons of urbanization in India, urban and rural construction, Issue No. 10, pp 76–78
6. Aibai A, Xinmei K, Bibo W (2015) Shift of surplus rural laborers in the process of urbanization in India and its implication to china. World Agriculture, Issue No. 2, pp 27–32

Chapter 44
Research and Application of Indexes System of Comprehensive Benefit of Large Urban Underground Space Development

Jingkuang Liu and Yunyi Liao

Abstract With the rapid growth of China's national economy and the speed development of urbanization, the development and utilization of underground space in the city has achieved some results. Therefore, the research on the rational utilization and standard management of underground space attracts more and more attention. However, in developing the underground space, the aggregation effect of it can not be realized because of the lack of scientific and reasonable planning, unsound coordination mechanism and relatively independent development project. This paper, based on the experience of other countries in development and utilization of underground space, constructs overall efficiency evaluation index system. Focusing on underground space demonstration project in Zhujiang New Town of Guangzhou, it applies this index system to cases. The establishment of overall efficiency evaluation index system can promote the better development and utilization of underground space in our country.

Keywords Urban underground space development · Index system · Development and construction · Fuzzy comprehensive evaluation

44.1 Introduction

In the 21st century, urban environmental problems are relatively serious, including air and noise pollution, low rate of green space, traffic congestion and public facilities shortage, which hinders the development of underground space. However, in some cities, the utilization of underground space is restricted to shopping malls, entertainment venues, and the development density is large. The initiative of developers is very high because of high economic benefits. People's understanding

J. Liu (✉) · Y. Liao
School of Business Administration, Guangzhou University, Guangzhou, China
e-mail: ljkgowell@163.com

Y. Liao
e-mail: 515603002l@qq.com

Y. Wu et al. (eds.), *Proceedings of the 20th International Symposium on Advancement of Construction Management and Real Estate*,
DOI 10.1007/978-981-10-0855-9_44

of underground space in the aspect of economy is limited to construction cost and they hold that the development of underground space is uneconomic and unreasonable. Therefore, it is necessary for us to establish the overall efficiency evaluation index system of underground space in the city to construct a more perfect system for big cities, which can make the utilization of underground space more scientific and reasonable. This paper, based on the experience of other countries in development and utilization of underground space, constructs overall efficiency (including economic benefits, social benefits, environmental benefits, etc.) evaluation index system of underground space in big cities through analytic hierarchy process and assessment of fuzzy math, and it applies this index system to the underground space demonstration project in Zhujiang New Town of Guangzhou. The construction of overall efficiency evaluation index system of underground space contributes to the expansion of financing channel for the development of underground space, the realization of unified planning, unified management, and orderly development of underground space, and the improvement of social benefits and economic benefits, thus promoting better development and utilization of underground space in China.

44.2 Construct the Overall Efficiency Indicator of Urban Underground Space

44.2.1 Construct the Evaluation Index System of Urban Underground Space

The Selection Principle of Evaluation Index System

Comprehensive evaluation refers to make a comprehensive evaluation of a thing or phenomena influenced by many factors. In constructing the whole evaluation system, the establishment of the index system is the basis. The selection of index system should meet the following principles: (1) comprehensiveness, the combination of qualitative and quantitative; (2) systematicness: fully and comprehensively reflecting the comprehensive level of the project; (3) representativeness, reflecting the current situation and future development direction of underground construction; (4) operability, convenient for real calculation and on-site management; (5) dynamicity, effectively reflecting the current situation and the construction level in different stages; (6) standardability: quantitative index originates from standardability. In order to reflect the current management situation of urban underground space comprehensively and objectively, the evaluation index system from micro-level to macro-level should be established to evaluate the whole project, so as to provide realistic basis and reference for the promotion of new technology in actual project. Through system analysis and induction, the comprehensive evaluation system of underground space demonstration project is shown in Table 44.1.

Table 44.1 Overall efficiency evaluation index system of underground space

Goal layer	Criterion layer	Sub-criterion layer 1	Sub-criterion layer 2
Comprehensive benefits of urban underground space	Economic benefit A1	Direct benefit B1	All payback period C1
			Rate of total investment change C2
		Indirect benefit B2	Market requirement C3
			Technological competitiveness C4
			Data review C5
			Project quality C6
			Engineering safety C7
			Duration of a project C8
	Social benefit A2	Project demonstrativeness B3	Advancement of new Technology C9
			Innovativeness of new technology C10
			Maturity of new technology C11
			Application of new technology C12
		Project practicability B4	Personnel placement C13
			Influences on surrounding residents C14
			Influences on surrounding transportation C15
	War readiness benefit A3	Project sociality B5	Disaster-resistance and civil air defense C16
			Utility rate of space C17
			Use function benefit C18
		Ecological environment B6	Underground water protection C19
			Ground subside control C20
			Original soil protection C21
			Influence of underground relics C22
			Soil surface environmental impact C23
	Environmental benefit A4	Cultural environment B7	Utility rate of land C24
			Historic preservation C25
			Supersound disturbance C26
		Civilized construction B8	Technical cleaning C27
			Materials and equipment placed C28
			Technical energy conservation C29
			Public hygiene C30
			Non-renewable resources input C31
			Traffic safety warning C32
			Garbage cleaning C33

Determine Evaluation System

The determination criterion of evaluation system should be specific, credible, flexible and operable. It can be made under single condition or multi-conditions. As shown in Tables 44.2 and 44.3.

The expert grading method is used to determine the grade of qualitative evaluation index that can not be directly quantified. Firstly, each index is divided into 6 grades: excellent, good, middle, bad, qualified and unqualified. The quantitative score of each grade is 45, 85, 75, 55, 65 and 95 respectively. Then, each expert marks the grade of evaluation index element according to the content of examination of evaluation index. Finally, the score value of this evaluation index is calculated

Table 44.2 Determination criterion of rate of change in total investment in the period for recovery of investment

No.	Evaluation factor	95	85	75	65	55	45
C1	All payback period	≤10	(10, 15]	(15, 20]	(20, 25]	(25, 30]	>30
C2	Rate of total investment change	≤−0.15	(−0.15, −0.05]	(−0.05, 0.05]	(0.05, 0.15]	(0.15, 0.20]	>0.20

Note The unit of all payback period is year
Rate of total investment change = (full cost of real project − planned investment)/planned total investment * 100 %

Table 44.3 Determination of project quality

Level	Condition ①	Condition ②	Condition ③	Logical relationship that the condition meets
	Qualification rate of project checkpoint related to new technique P19 (%)	Qualification rate of all inspection projects P2 (%)	Changes in process rating P3	
95	≥10	>6	The level improves and new technique has improving effects	① or ② or ③
85	[5, 10]	[3, 6]	No change	(① or ②) and ③
75	[2, 5]	[1, 3]	No change	((① or②) and ③) or no correlation
65	(0, 2]	[0, 1]	No change	(① or ②) and ③
55	(−5, 0]	(−3, 0]	No change	(① or ②) and ③
45	≤−5	<−5	The level reduces and new technique has negative effects	① or ② or ③

according to the following formula: quantitative score value of the qualitative index = $\sum$ Selected grade coefficient of each expert/The number of experts.

Determine the Weight of Evaluation Index System

Weights are non-dimensional values, which can reflect the relative size of the value in number set, and they are used to weigh the importance degree in practical application.

In the evaluation problem, weights can be divided into two categories according to the source of source information: one is subjective weight determination, and its source information comes from expert consultation, namely the use of expert knowledge and experience, such as expert consultation method and analytic hierarchy process; another is objective evaluation method, and its source information comes from statistical data themselves, such as principal component analysis and artificial neural network analysis. Criterion weight can be set through artificial neural network analysis, and the weight of sub-criterion layer can be set through analytic hierarchy process. These methods study weight problem from different angles, but they also have some limitations. In this paper, the analytic hierarchy process is used to determine the weights. According to the nature of problems and established goals, it divides the problem to be solved into different component elements which are for hierarchy clustering combination according to the influencing and subjection relationship between the factors to form an ordered hierarchical structure. The relative importance of factor in each layer is determined through the comparison between any two means. Then, the mathematical method is used to determine the weight of relative importance order of all factors in the same layer. The combination weight of relative importance of each factor in the lowest layer (index layer) against the highest layer (general goal) is obtained according to calculating the weigh value of relative importance of factors in each layer to get evaluation basis.

44.3 Integrative Efficiency Evaluation of Underground Space in Cities

44.3.1 Fuzzy Comprehensive Evaluation Steps of Underground Space in Cities

Since the comprehensive benefit of urban underground space project is a complicated concept, the benefit is fuzzy. And its related factors are more complex. On the one hand, some indicators can not be described by the exact number and they are just fuzzy concepts; on the other hand, there is not a corresponding function relation between the changes in various factors and the overall effectiveness, so it is impossible to establish exact mathematical model. Therefore, the fuzzy theory is introduced to evaluate the comprehensive benefit of urban underground space project [1]. Basic steps of fuzzy comprehensive evaluation are:

1. Establish evaluation set. The evaluation criterion is divided into several grades according to the concrete conditions, and the evaluation set is: $V = [V_1, V_2, \ldots V_n]$
2. Establish the membership relationship to obtain fuzzy matrix.The questionnaire statistic method is used in this paper. The public and experts mark the evaluation factor. The scores are collected and normalized to get fuzzy evaluation matrix. $R = \begin{bmatrix} R_1 \\ R_2 \\ \vdots \\ R_n \end{bmatrix}$, Among which, $R_i = [r_{i1} \quad r_{i2} \quad \ldots \quad r_{in}]$ is the fuzzy evaluation vector of the ith element.
3. Fuzzy comprehensive evaluation. Make blurring operation for the weight vector of evaluation factor and fuzzy evaluation matrix to obtain fuzzy evaluation results. $B = U \cdot R[b_1 \ b_2 \ \ldots \ b_n]$.

44.4 Case Application

Zhujiang New Town CBD underground space is located in Zhujiang New Town CBD of the new axis in Guangzhou, and the main part of the project is supporting facilities for public service, including underground multi-functional commercial facilities, automatic transport system for passenger and surficial central landscape square. The gross floor area of three stories underground is 754,000 m^2, and the overall floorage is 440,000 m^2. It was launched in 2006 and completed before the Asian games in 2010, with a total investment of about 3.5 billion yuan, and it is currently the largest and the most important development and utilization project of underground space in Guangzhou.

44.4.1 Integrative Efficiency Evaluation of Underground Space Project Based on AHP

1. Determine the weigh of index factor. AHP is used in this paper to calculate the relative importance general ranking weigh of each index factor in the lowest layer against that in the highest layer. The main calculation steps are shown below.
2. Construct the judgment matrix. Based on expert consultation, field practice and the author's opinion, the relative importance of each factor in the same layer in this evaluation model was judged. And these judgements are expressed by values after introducing appropriate scale, namely, the judgment matrix. One-nine scaling method is used to construct the judgment matrix of each factor in the same layer.

3. Single hierarchical arrangement and consistency check. The square root method is used to determine the relative importance ranking weight of each factor in the same layer against a certain factor in last layer, W_B, W_C, and the feature vector is calculated, which is called single hierarchical arrangement. To ensure the satisfactory consistency of judgment matrix, check consistency is necessary. When the random conformance rate, CR meets the formula: CR = CI/RI < 0.1, single hierarchical arrangement has satisfactory consistency. RI is average random consistency index; CI is coincident indicator, determined by $CI = \frac{\lambda_{max} - n}{n-1}$ (n is the order of judgment matrix)

This index system involves many indexes. In order to make calculation easy, the model is input into the AHP program block through the MCE software made by Du et al. [2] to conduct single hierarchical ranking, display evaluation structure and educe some tables (Tables 44.4, 44.5, 44.6 and 44.7). At the same time, each single hierarchical ranking weigh, W is provided and the consistency of each judgment matrix is proved to be satisfactory.

The total weight vector of index factors of integrative efficiency evaluation index system of Zhujiang New City underground space project is recorded as $W_a = (W_1, W_2, W_3, \ldots W_{32}, W_{33})$. Through calculation we know that $W_a = (0.0417, 0.2083, 0.0060, \ldots, 0.00050, 0.0035)$. The weight of index factor shows the importance of the factors to the objective evaluation. Therefore, we can know that B1 (Direct benefit) has the most significant influence on target (Economic benefit) evaluation; the influence of C28 (Material equipment placed) on target evaluation is minimum.

Table 44.4 Destination layer—criterion layer judgment and consistency check

Comprehensive benefits of urban underground space demonstration project	Economic benefit A1	Social benefit A2	War readiness benefit A3	Environmental benefit A4
Economic benefit A1	1	1	1	1
Social benefit A2	1	1	1	1
War readiness benefit A3	1	1	1	1
Environmental benefit A4	1	1	1	1
Single-layer weight	0.2500	0.2500	0.2500	0.2500

Note 1max = 4; CI = 0; RI = 0.58; CR = 0 < 0.1

Table 44.5 A1-B judgment and consistency check

Economic benefit A1	Direct benefit B1	Indirect benefit B2
Direct benefit B1	1	3
Indirect benefit B2	1/3	1
Single-layer weight	0.7500	0.2500

Note 1max = 2; CI = 0; RI = 1E-6; CR = 0 < 0.1

Table 44.6 Reckoning of B-layer single ranking and hierarchy general ranking and consistency check

Index layer	A1	A2	A3	A4	Comprehensive weight Wi
	WA1 = 0.2500	WA1 = 0.2500	WA1 = 0.2500	WA1 = 0.2500	(i = 1, …, 8)
B1	0.7500				0.2500
B2	0.2500				0.0833
B3		0.1634			0.0545
B4		0.5396			0.1798
B5			0.2970		0.0990
B6				0.5936	0.1978
B7				0.2493	0.0831
B8				0.1571	0.0524

$CI = \sum W_{AK} \cdot CI_K = 0.0105$; $RI = \sum W_{AK} \cdot RI_K = 0.3867$; $CR = CI/RI = 0.0272 < 0.1$

44.4.2 *Fuzzy Evaluation of Comprehensive Benefits*

Integrative efficiency evaluation index system of Zhujiang New City underground space project is a three-level four-layer index system. These evaluation indexes are fuzzy. According to the characteristics of the index model, we treat the evaluation of sub-criterion layer 2 against sub-criterion layer 1 as the first-level evaluation, the evaluation of sub-criterion layer 1 against goal layer as the second-level evaluation, thus constructing a two-level three-layer fuzzy synthetic evaluation model

Construct Evaluation Factor Set and Determine the Weigh

Fuzzy comprehensive evaluation method is an effective way to make a comprehensive evaluation of things affected by many factors. To evaluate the comprehensive benefits of Zhujiang New City underground space project, a corresponding evaluation factor set should be established, and the evaluation factor set of each layer should be written according to the weights of sub-criterion layer 1 and 2 determined by analytic hierarchy process.

$W = (0.25\ 0.08\ 0.05\ 0.18\ 0.10\ 0.20\ 0.08\ 0.05)$; $W_1 = (0.17\ 0.83)$; $W_2 = (0.07\ 0.13\ 0.03\ 0.22\ 0.48\ 0.07)$; $W_3 = (0.20\ 0.44\ 0.27\ 0.09)$; $W_4 = (0.64\ 0.10\ 0.26)$; $W_5 = (0.64\ 0.10\ 0.26)$; $W_6 = (0.16\ 0.10\ 0.29\ 0.06\ 0.39)$; $W_7 = (0.67\ 0.23\ 0.10)$; $W_8 = (0.16\ 0.04\ 0.25\ 0.06\ 0.33\ 0.10\ 0.07)$.

Construct the Evaluation Set

The evaluation set is $V = [V_1\ V_2\ \ldots\ V_n]$. The evaluation set defined by the evaluation model of this project is V = [excellent, good, middle, bad, qualified and

Table 44.7 Reckoning of criterion layer 1—criterion layer 2 (C layer)single ranking and hierarchy general ranking and consistency check

Index layer	B1	B2	B3	B4	B5	B6	B7	B8	Comprehensive weight Wi
	WB1 = 0.2500	WB2 = 0.0833	WB3 = 0.0545	WB4 = 0.1798	WB5 = 0.0990	WB6 = 0.1978	WB7 = 0.0831	WB8 = 0.0524	(i = 1, 2, …, 8)
C1	0.1667								0.0417
C2	0.8333								0.2083
C3		0.0717							0.0060
C4		0.1290							0.0107
C5		0.0299							0.0025
C6		0.2229							0.0186
C7		0.4770							0.0397
C8		0.0696							0.0058
C9			0.1990						0.0108
C24							0.6738		0.0560
…							…		…
C33								0.0664	0.0035

$CI = \sum W_{AK} \cdot CI_K = 0.0434$; $RI = \sum W_{AK} \cdot RI_K = 0.6529$; $CR = CI/RI = 0.0665 < 0.1$

unqualified] and they are assigned. In this paper, the comprehensive benefits of Zhujiang New City underground space project are divided into 6 grades and they are assigned.

Determine Subordination Relationship to Obtain Fuzzy Matrix

The sub-criterion layer evaluating sub-factor set U_i to evaluation set V is regarded as a fuzzy mapping to determine fuzzy evaluation matrix. In order to obtain essential data, aiming at index system of comprehensive benefits of Zhujiang New Town underground space project, the author adopts choice questionnaire and surveys the engineering technicians who are engaged in planning, design and management in the project. Eighty questionnaires were given out and seventy-two were collected. Each index was evaluated according to the evaluation set to obtain the following fuzzy evaluation matrix and the first comprehensive evaluation was conducted.

$$R_1 = \begin{bmatrix} 0.10 & 0.42 & 0.38 & 0.05 & 0.05 & 0.00 \\ 0.12 & 0.24 & 0.22 & 0.30 & 0.07 & 0.05 \end{bmatrix};$$

$$R_3 = \begin{bmatrix} 0.52 & 0.29 & 0.10 & 0.09 & 0.00 & 0.00 \\ 0.48 & 0.23 & 0.14 & 0.14 & 0.01 & 0.00 \\ 0.42 & 0.20 & 0.16 & 0.18 & 0.04 & 0.00 \\ 0.42 & 0.28 & 0.14 & 0.08 & 0.08 & 0.00 \end{bmatrix}$$

$$R_2 = \begin{bmatrix} 0.16 & 0.42 & 0.36 & 0.06 & 0.00 & 0.00 \\ 0.32 & 0.26 & 0.13 & 0.20 & 0.09 & 0.00 \\ 0.34 & 0.38 & 0.13 & 0.14 & 0.01 & 0.00 \\ 0.46 & 0.24 & 0.20 & 0.10 & 0.00 & 0.00 \\ 0.42 & 0.32 & 0.12 & 0.08 & 0.06 & 0.00 \\ 0.10 & 0.11 & 0.36 & 0.25 & 0.14 & 0.04 \end{bmatrix};$$

$$R_5 = \begin{bmatrix} 0.42 & 0.28 & 0.12 & 0.10 & 0.07 & 0.01 \\ 0.46 & 0.24 & 0.11 & 0.09 & 0.08 & 0.02 \\ 0.48 & 0.23 & 0.10 & 0.12 & 0.06 & 0.01 \end{bmatrix}$$

$$R_4 = \begin{bmatrix} 0.62 & 0.24 & 0.12 & 0.02 & 0.00 & 0.00 \\ 0.41 & 0.21 & 0.16 & 0.18 & 0.04 & 0.00 \\ 0.68 & 0.22 & 0.10 & 0.00 & 0.00 & 0.00 \end{bmatrix};$$

$$R_7 = \begin{bmatrix} 0.43 & 0.34 & 0.06 & 0.14 & 0.03 & 0.00 \\ 0.48 & 0.32 & 0.10 & 0.10 & 0.00 & 0.00 \\ 0.38 & 0.24 & 0.16 & 0.08 & 0.08 & 0.06 \end{bmatrix}$$

$$R_6 = \begin{bmatrix} 0.45 & 0.38 & 0.08 & 0.08 & 0.01 & 0.00 \\ 0.12 & 0.16 & 0.20 & 0.24 & 0.18 & 0.10 \\ 0.14 & 0.18 & 0.38 & 0.15 & 0.09 & 0.06 \\ 0.12 & 0.16 & 0.14 & 0.32 & 0.14 & 0.12 \\ 0.16 & 0.10 & 0.22 & 0.30 & 0.12 & 0.10 \end{bmatrix};$$

$$R_8 = \begin{bmatrix} 0.40 & 0.33 & 0.15 & 0.10 & 0.02 & 0.00 \\ 0.13 & 0.25 & 0.38 & 0.14 & 0.08 & 0.02 \\ 0.48 & 0.26 & 0.15 & 0.08 & 0.02 & 0.01 \\ 0.40 & 0.32 & 0.14 & 0.08 & 0.04 & 0.00 \\ 0.10 & 0.24 & 0.35 & 0.17 & 0.08 & 0.06 \\ 0.40 & 0.30 & 0.14 & 0.05 & 0.03 & 0.08 \\ 0.32 & 0.40 & 0.13 & 0.06 & 0.08 & 0.01 \end{bmatrix}$$

$B_1 = W_1 * R_1 = (0.12\ 0.27\ 0.25\ 0.26\ 0.07\ 0.04)$; $B_2 = W_2 * R_2 = (0.37\ 0.29\ 0.17\ 0.11\ 0.05\ 0.00)$
$B_3 = W_3 * R_3 = (0.47\ 0.24\ 0.14\ 0.14\ 0.02\ 0.00)$; $B_4 = W_4 * R_4 = (0.61\ 0.23\ 0.12\ 0.03\ 0.01\ 0.00)$
$B_5 = W_5 * R_5 = (0.44\ 0.26\ 0.11\ 0.10\ 0.07\ 0.03)$; $B_6 = W_6 * R_6 = (0.27\ 0.18\ 0.28\ 0.22\ 0.12\ 0.07)$
$B_7 = W_7 * R_7 = (0.44\ 0.33\ 0.08\ 0.12\ 0.03\ 0.01)$; $B_8 = W_8 * R_8 = (0.31\ 0.28\ 0.22\ 0.11\ 0.05\ 0.05)$

After the second comprehensive evaluation, we draw the conclusion:

$$R = \begin{bmatrix} B_1 \\ B_2 \\ B_3 \\ B_4 \\ B_5 \\ B_6 \\ B_7 \\ B_8 \end{bmatrix} = \begin{bmatrix} 0.12 & 0.27 & 0.25 & 0.26 & 0.07 & 0.04 \\ 0.37 & 0.29 & 0.17 & 0.11 & 0.05 & 0.00 \\ 0.47 & 0.24 & 0.14 & 0.14 & 0.02 & 0.00 \\ 0.61 & 0.23 & 0.12 & 0.03 & 0.01 & 0.00 \\ 0.44 & 0.26 & 0.11 & 0.10 & 0.07 & 0.03 \\ 0.27 & 0.18 & 0.28 & 0.22 & 0.12 & 0.07 \\ 0.44 & 0.33 & 0.08 & 0.12 & 0.03 & 0.01 \\ 0.31 & 0.28 & 0.22 & 0.11 & 0.05 & 0.05 \end{bmatrix}$$

$B = W * R = (0.34\ 0.25\ 0.19\ 0.16\ 0.06\ 0.03)$, After normalization, $B = (0.33\ 0.24\ 0.18\ 0.16\ 0.06\ 0.03)$
"95, 85, 75, 65, 55 and 45" are assigned to "Excellent, Good, Middle, Bad, Qualified and Unqualified". Through $B = Wa * R$, we learn that the uniformizable comprehensive benefit evaluation result of Zhujiang New Town underground space is 80.30 and the comment rate is "Middle".

44.5 Conclusion

There is a growing emphasis on the development and utilization of underground space, and people gradually realize the advantage and potential of the underground space to expand urban space capacity. Before the construction of urban underground space project, scientific economic argumentation must be made. In terms of economic benefits, people should take not only the cost, including construction cost, maintenance and management cost, but also the social and environmental benefits into consideration, that is, the overall efficiency evaluation index system should be established to evaluate the efficiency, optimize the project and program decision and promote the development and utilization of underground space in cities.

References

1. Jian Y, Wu LX, Du LQ (2005) On the index system for city underground space developing-utilization capacity evaluation. Urban Stud 12(5):47–52
2. Du D, Pang QH, Wu Y (2008) Modern comprehensive evaluation method and case selection, 2nd edn. Tsinghua University press
3. Wang Y (2012) Research on quantitative assessment of disaster-prevention benefit for urban underground space development. Chin J Undergr Space Eng 8(sup1):1560–1566
4. Luo ZQ, Liu WP, Liu XM et al (2007) Benefit analysis of urban underground space. Chin J Undergr Space Eng 3(1):5–8
5. Kong LX, Shen RF (2007) Evaluation of sustainability for development of underground space in urbs. J Nat Disasters 16(1):119–123
6. Godard JP (2004) Urban underground space and benefits of going under ground. World tunnel congress 2004 and 30th ITA general assembly. Singapore, pp 22–27

Chapter 45
An Integrated Research Design for Developing a Holistic Sustainable Decision Making Framework in Regenerating Chinese Cities

Wenli Dong, Jamie Mackee and Michael Mak

Abstract It has often been recognised that urban planning and regeneration play a crucial role in the achievement of urban sustainability. However, problems still exist with regards to the approaches for decision making in urban regeneration of Chinese cities. Significant limitations are identified and discussed in the context of sustainability, such as the reductionism within many of the approaches and the lack of holism in the evaluation. The identified deficiencies provide the motivation for the development of a new framework which can integrate the hierarchical structure and holistic utilisation of sustainability in the built environment. The framework helps decision makers identify the sustainability aspects involved in an urban planning project, guiding them in the evaluation on the basis of a number of problem solving methods. This paper describes the development of the research methodology. It is structured into four main sections. In order to address the research question and select the appropriate research methods for achieving the research objectives, the opening section provides information on the contemporary context of the study area. Existing inductive modelling approaches and their suitability to this research are discussed. Then the following section provides the research objectives and a justification for the chosen strategy. Both the selection of the research theory and the immediate context of the study are discussed. Subsequently, the logic linking the research design to the research objectives is considered. Finally, the detailed methodology for building an integrated sustainability evaluating framework will be introduced. Case studies are adopted to show the benefits of the framework in the urban projects and across different design processes. Practical applications of the framework are used to test the validity of

W. Dong (✉)
College of Engineering and Architecture, Zhejiang University, Zhejiang, China
e-mail: wenlidong@zju.edu.cn

J. Mackee · M. Mak
School of Architecture and Built Environment, The University of Newcastle, Newcastle, Australia

Y. Wu et al. (eds.), *Proceedings of the 20th International Symposium on Advancement of Construction Management and Real Estate*,
DOI 10.1007/978-981-10-0855-9_45

this approach in different cultural regions and using different local practices. The resulting framework will provide a significant step forward in understanding and evaluating the built environment in the context of a sustainable urban development in China.

Keywords Research design · Decision making · Urban regeneration

45.1 Introduction

Entering the 21st century and distancing itself from its early planned economy, there are now new economic and social imperatives in China. However, in the last three decades, Chinese cities have confronted serious problems of balancing fast, intensive economic and urban development with achieving sustainability. The planning of eco-cities and urban regeneration is becoming prevalent in China to meet this challenge for sustainable urban configuration and envision for sustainable living. While the global regeneration approaches that are practiced today are founded on Euro-centric traditions, many of the urban theories and sustainability evaluation methods applied to Chinese cities have been appropriated from Western models of urban development [1]. Nonetheless, whether these urban developments, either new development or redevelopment, will provide actually sustainable living environment in practice, requires the evaluation on their sustainability of decision making. The evaluation measures these developments towards the yardstick that featured with low carbon emission, efficient energy and resource, low consumption and harmonious society, being culturally sensitive to their users and catering the needs of habitants. A complex system to manage ourselves, individually and collectively is required [2].

However, the cultural sensitiveness of the decision making in sustainable development is attributed to the value-based nature of sustainability, while values can vary over time and between cultures. There has recently been a movement toward indigenous approaches to sustainability supported by local knowledge. While the approaches developed in the west are appropriate in a scientific sense, there are limitations to how Eurocentric theories deal with the relationship between cultural values and the urban environment [1]. Eurocentric approaches are based on reductionist philosophical foundations, while the Asian philosophical traditions are more organic, holistic and systemic. Traditional reductionist methods of analysis which breakdown and isolate the component parts bring the risk of fragmented decision making with potential unforeseen consequences. Systemic, holistic, goal-directed and hierarchical assessments of sustainability are urgently needed. The philosophical experiences of Europe which has provided the basis for the current urban sustainability evaluation theories and practices in the West are inappropriate in China. An alternate framework that establishes a more holistic classification system for sustainability criteria and indicators in the evaluation of the built environment at a local planning level is needed to evaluate urban regeneration in the context of Chinese cities. Sustainability based on indigenous philosophy recognises holistic relationships and incorporates cultural values, which are often more holistic in nature. Indigenous philosophy provides a way of adding new issues to current evaluation practices.

45.2 Review on Modelling Approaches for Sustainable Decision Making

45.2.1 The Quantitative Multi-criteria Methods

In this section, the establishment of integrated qualitative models for evaluating sustainable development are explored. They are significantly different in regards to the structure of the framework, as well as the methodology that are employed. Blanc simplifies the construction of a composite index into the stepwise selection, scaling, weighting and aggregation of the variables [3]. Palme adapts the Lifecycle Assessment (LCA) method in the development of environmental sustainability indicators for municipal water systems [4]. The underpinning methodology for the development of composite indices is also stepwise. These steps use simple techniques and qualitative assessments [5, 6]. However, the weakness of this method resides in the lacking of strength in the cross-case analysis. Moreover, Kinderytė suggests not only quantitative but also qualitative indicators are recommended because of the difficulties to quantify certain aspects of sustainability [7].

In the current development of multi-criteria assessment of sustainability, indicators are becoming not just a 'best practice', but 'standard practice' in many different contexts. The challenges of growing complexity and the demand for simplicity are causing dilemmas for researchers [8]. As the complexity of sustainability is becoming widely acknowledged, the monitoring and management of such large amounts of information is increasingly challenging the development of indicator sets. Meanwhile, there is a demand for sustainability frameworks that are simple, can be used in public education and administration and are comprised of schemes which are comprehensible, elegant and effective, without compromising the underlying complexity.

Among the attempts to address the above challenges, Zavrl, Žarnić and Šelih propose a two-step method for assessing sustainability where the initial procedure is to identify the impact of the sustainability level of the building [9]. Then, the indicators are aggregated according to their influence on individual sustainability aspects. Atkisson and Lee Hatcher also develop an aggregation, scaling and presentation methodology which clusters indicators and assessment scores into four quadrants of the compass [8]. The proposed framework displays the indicator and informs comprehensive sustainability in the form of systemic compass. The inherent risks, on the other side, are censured on the indicator selection, scale strategy, ethics and data availability.

45.2.2 The Statistical Reference Between Sustainability Indicators

Either theoretical or statistical reference between indicators is crucial for obtaining an aggregate with a clear meaning and an objective estimation while avoiding a disparate combination. The determination of weights assigned to the indicators makes the assessment method applicable in the local context, helps to confirm the competences of the involved actors, as well as identifies the weak points in the procedure [9]. Otherwise, several indicator sets suffer from the inconsistency caused by the mixing of indicators without a common unit or known equivalency between them.

Rather than validation of sustainability aspects, Zavrl checks whether the evaluation of criteria and determination of indicators ranking is feasible [9]. Blanc devises a novel weighting scheme in the construction and expression of indicators according to a coherent framework issuing from the Driver- Pressures- State-Impacts-Responses (DPSIR) system [3]. Hemphill utilises a methodology of hierarchical modelling in the aggregated weighting system for evaluating sustainable urban regeneration [10]. The Delphi technique and multi-criteria research develop an effective means of weighting among the key attributes of sustainable urban regeneration in accordance with their relative importance. However, Blanc discovers that three types of weakness and two primary sources of errors can be identified within the existing aggregating indicators [3]. The lack of coherence in the choice of indicators, the lack of a common unit for aggregation and the missing impact level in the use of the DPSIR framework, the weakness in the choice of the systematic assessment level, and the mixed indicators from different levels within the same category, all lead to ambiguous interpretations and to a risk of double-counting. Kinderytė argues that assessment according to the standardised methodology is only the first step since it can display only fundamental problems [7]. Budget allocation processes were also used as a weighting method. The systemic analysis of existing sustainability indicators systems, the determination of sustainability indicators and their significance were determined by the experts, which is repeated to improve indicators and weighting coefficients. However, the experts' survey has questioned the need to assign weights to indicators. Instead, determining appropriate indicators for evaluation is perceived as the most critical step [11].

45.2.3 The Theoretical Reference Between Sustainability Indicators

The questioning of the statistical weighting between sustainability indicators and their weakness and errors, draws attention to the theoretical references between sustainability indicators. Such philosophical underpinning, as opposed to mechanic weighting, means that integrated models for evaluating sustainable development are worth investigating [1, 12, 13].

The approach used by Plessis [1] is a theoretical exploration to define a worldview by the study of ecological and social articles. The themes include conceptual framing, method used, aspects studied, research design, objectives and case studies, as well as how the system and its properties are defined and studied [1]. Despite its focus on the theoretical exploration, which has limited implications for the qualitative model establishment, this method could be applied to the literature review on the existing sustainability assessment methods and traditional philosophies.

Based on Dooyeweerd's philosophical idea, Lombardi establishes a framework for evaluating sustainable development at a policy level, its values, how it is applied in a management context, how it is implemented and its time horizon [14]. It reveals the three hierarchies of assessment including the evaluation framework, the aggregated indicators at all levels and assessment methods which are linked to the framework with a directory. Several issues are supposed to be addressed in constructing the proposed framework, including the dimension, space, functions, accessibility, environmental compatibility, historical and cultural significance, feasibility, visual appeal, flexibility, adaptability, institutional sustainability, interest, concern and the time horizon [13].

Unlike the indicators and assessment methods of the framework in Lombardi's theory, the properties of the system in Plessis's theory share similarities to eastern philosophies [1, 14]. The system stresses not only the entities that make up the system, but also the interactions and the linkages between them. The behaviour of the system adopts the dynamics, adaptability, transformability and self-organisation of the internal activities to clarity the attributes of the system. Plessis described the city as a complex and adaptive system characterised by holarchical and panarchical structured noosystem, as well as symbolic construction. All these statements reflect the spirit of eastern philosophy.

Case studies draw attention to the context of the cases and the capacity to interpret casual links. They play a crucial role in developing theory using multiple sources of evidence. They are relevant to theory and strategies which integrate multiple research designs [15]. This research strategy is adopted by researchers of integrated theories. The methodology in Lombardi's theory includes case studies for validation and applications. They check the completeness, duplication, consistency and internal logic of the model. The cases in Brandon and Lombardi focused on identifying relevant evaluation criteria, stakeholders' views and synthesising quantitative indicators, whereas the additional case in Brandon and Lombardi highlights the interrelations and links in decision-making, which is one of the key features in holistic theories [12]. This change represents the increased awareness of the holistic features of the framework in addition to the previous focus on complex structure and stakeholders' views.

RESCUE used the case studies are employed for the mutual purposes of developing an evaluation procedure and finding out the best local practices. In this instance doing so led to the development of an analytical sustainability framework [16]. This analysis method is valuable for the qualitative analysis in this research. The applicability check of indicators (ACI), strengths, weaknesses and gaps analysis (SWG), good or best practice analysis (GBP) and transferability analysis (TA), which are used to find the best practice for brownfield redevelopment, are also useful to identify the best practice for urban regeneration.

45.2.4 The Theory of the Qualitative Model Development Cycle

Qualitative research is reflective of the routine life of individuals or organisations. It emphasises natural settings and focuses on interpretation and explanation of how the respondents make reflections of their own circumstances. It involves the use of various tactics and other aspects of qualitative research strategy, such as holistic, open ended questions, the researcher as a measurement device and interpretation through words [17]. The researcher obtains a holistic overview of the circumstances under the logic, explicit and implicit rules for qualitative research [18]. This research primarily draws on interviews and archival strategies which lead to the division of content analysis and reasoning.

With a vertical time series, Miles and Huberman shows the steps involved in qualitative data analysis, including the initial study design, conceptualisation, data collection, coding, visualisation and forming conclusions [19]. However, the development of a model is an iterative rather than a linear process which is composed of the setup phase, the conceptualisation phase, the model construction phase, the evaluation phase, the use phase and finally the reporting phase [20]. Every action can result in a revision of an earlier step. For example, in the case when the assessment test fails, the calibration needs to be improved for adjustment.

During the setup phase for the development of a model, the research problem and its context needs to be described. It is from here that the objectives and requirements of the model can be derived. The purpose and the intended functionality are defined from the objectives. It is at this beginning stage, the requirement to evaluate the model is already identified for defining the functionality of the model. The building of the conceptual model defines the model structure which includes the processes and the relationships [20]. The mutual influence of a conceptual framework and research questions start this process and both lead to plans for sampling and instrumentation [19]. The data availability and a description of the system need to be examined for the assumptions about the theoretical development and simplifications of raw data. The decisions about representation and structure of data are therefore determined in this process [20]. This data analysis typically begins with back effects which provide feedback to the sampling plan and conceptual framework itself. The interacting between initial coding scheme and the research questions develops the interim contribution for this research and lead this research into the next step of within-case analysis. Both descriptive and explanatory conclusions can be fed back for verification and revision [19].

The real construction of the model includes both the building of the model and its calibration, which assesses and improves the model parameters and constructs to essentially make the model fit with the available data [19]. However, in the process described in Miles and Huberman, a significant gap is the lack of explanation about how the coding could evolve into descriptive displays and an explanatory model [19]. Carney develops a 'ladder of logical abstraction' which fills in the gap by integrating the data into an explanatory framework in a series of ascending levels [21]. It begins by trying out coding categories on a text. Then it moves to identify themes and trends, to test propositions and findings, which aims to formulate the complex structure for integrating the information into an explanatory framework. The data transforms as the information is condensed, clustered, sorted and linked through the process [19].

The validation phase deals with the value of the results, which proves the functionality and targets as proposed in the initiating of the model. The committed assessment essentially means that a model has adequate achievements for a targeted utility. The process of validation is similar to the establishment of the model [20]. The entire cycle of within-case analysis will be repeated to derive cross-case conclusions in the study of multiple cases.

45.3 Research Justification

45.3.1 Nature of Research

The ultimate research problem in this research is: how can indigenous Chinese philosophical traditions support a more appropriate evaluation methodology in decision-making for sustainable urban regeneration in China? The research strategy is described as 'the way in which the research objectives can be questioned' [22]. Given this question, the listed research objectives provide a brief summary of the research framework. The general aim of this study includes the following sub-objectives for this research: Objective A reviews the latest research of sustainable urban developments and evaluation methodologies for understanding the fundamental rationale and highlighting the gaps (literature review). Objective B develops theoretical evaluation framework and indicator matrix for analysing the city in a holistic approach and developing analytical approaches for case studies (conceptual framework). Objective C analyses urban strategies at both urban and district level and improves the holistic sustainability evaluation framework from previous case study analyses (case studies and building models). Objective D implements the framework to external examples of sustainable urban practice and validate the evaluation framework (application and validation).

Understanding the nature of this proposed research will help to determine what research strategies and which research methods are appropriate for this study. The following statements propose the main nature of the proposed research and each required approach. First, the research focuses on the contributions that can be made from Taiji Knowledge to sustainability evaluation, where exploring the principles and relationships between Taiji and western evaluation concepts of the relationships between human and nature will be a focus. Therefore, a deep understanding of indigenous knowledge and western practice is required for establishing a conceptual framework. Second, it is intended to develop a conceptual framework suitable for Taiji knowledge to be applied to evaluate the sustainability of urban design practices. This research will apply and refine the model based on an analysis of the case studies. Therefore investigation into the selection of an appropriate qualitative analysis method is required. Thirdly, exploratory analysis of urban regeneration strategies and establish a conceptual sustainability evaluation model based on the case studies. Thus, the method for the cross-case study and methods for qualitative data analysis need to be explored. Finally, an appropriate validation technique is required to implement and validate the research. Based on the conceptual framework and refined model, a prototype model will be developed to evaluate the urban design. The validity of the conceptual framework needs to be approved by the validation of the prototype model. The proposed prototype model needs to be operationalized to demonstrate the assessment outcomes visually. Therefore the selection of appropriate measurement methods is also required.

45.3.2 Research Design

In order to achieve the objectives, the following methodology will be applied using the selected case studies. Cross-case analysis is the chief method of deepening understanding of the generic processes that occur across individual cases. The cases are selected to ensure variety and opportunities to learn. The two-fold complexity resulted from both sustainable issues and urban systemic issues in one research require strong research methodology that is informed from more than one research methods. Thus, both statistical and qualitative methods will be used in this analysis. In the quantitative analysis, factor analysis will be used for exploring spatial patterns related to the total covariance among the indicators. The multivariate analysis and clustering will be used for factor analysis, using PASW Statistics17 to evaluate the significance of indicators and make quantifiable comparisons between cities. The results of the factor analysis can explain the total covariance of quantitative indicators in terms of a smaller number of underlying factors. In the qualitative analysis, the clustering analysis will be applied to identify clusters of the practices. NVIVO 9.0 is used for qualitative data and coding.

In order to achieve the research objectives, several steps are needed.

1. Objective A firstly requires an overview of the methods and problems in decision making of urban regeneration and design. This includes the identification of the factors which are essential in relation to sustainable urban development in the West and China, as well as a critical analysis of current assessment methods and planning evaluation approaches in the West and China.

2. A philosophical framework within the built environment to understand sustainability is developed. This framework provides a richer, integrated and holistic view of the problem, compared to Western systems. This stage includes a detailed analysis of the characteristics of sustainability in the Chinese background and recognition of their interrelationships on the basis of the Taiji theory of the hierarchical structure and internal forces. The development of a theoretical framework to understand sustainability which is based on the Chinese indigenous theory is presented. It also sets up the evaluation framework and criteria matrix and develops the qualitative criteria and quantitative indicators for evaluation.
3. The development of a conceptual framework in urban planning and design, taking into account the issues recognised in the previous step by incorporating the developed theoretical framework. The steps start with the initial quantitative studies, indicate the reasoning behind the case selection, takes into account the issues recognised previously and the results of the quantitative analysis. Then, this research develops a number of questions from the theoretical framework to assess sustainability, which is recognised from previous approaches that linked the principles and assessment of sustainable development. The semi- structured interviews were conducted about the cases with the designer, manager, experts and NGOs on sustainability. The establishment of an evaluation framework and sustainability factors in urban planning is based on the integrated structure, and the stakeholders' viewpoint identified on the basis of real world case studies; Illustration of the framework for the evaluation of sustainability in urban regeneration decision making takes the form of a Taiji compass which is convenient for Chinese decision makers to understand and evaluate sustainability in local contexts. It will be used to check whether a development proposal meets the criteria of sustainability in the built environment, providing a guide for decision making.
4. The validity of the framework demands the assessment of the framework based on its applications to two case studies which are related to different decision making process and planning situations. In addition, the study suggests future developments and further research in regard to urban sustainability paradigm and the practical applications of the framework.

In this research, both statistical and qualitative research methods will be used. The literature review and theoretical framework are primarily using the secondary qualitative resources from journals and books. The preliminary data analysis before data collection is the quantitative methods using the secondary statistics, whereas after the data collection, qualitative case studies become the chief method in pilot study and building the conceptual model.

The selected methodology must fit the overall framework of the study and fulfil its objectives. The aim of this study is to find appropriate evaluation methodology in decision-making for sustainable urban regeneration in China using indigenous Chinese philosophical traditions. Six main stages adapted from Miles' qualitative data analysis process and Carney's ladder of logical abstraction form the skeleton of the research process. Methodologically, the research will be following the steps of building and using data management system, building and using documentation scheme, repackaging and aggregating the data, developing and testing propositions to construct an explanatory framework, within case analysis and generating implications for theory, policy, actions.

45.3.3 The Research Process

In consideration of the nature of the research problem and the research design, there are various research stages that require the appropriate research methods.

From the starting point, the conceptual framework is used for underpinning the city information for evaluation. Thus building and using the data management system is essentially the process of building and using the conceptual framework itself. With the research question and conceptual framework, it initially creates an instrumentation plan for the research overall, and then the sampling plan for data collection. To gain access to cases is a separate module and requires ethics approval. The data collection for primary data and the data transcription method will be introduced. Meanwhile, the secondary data, mainly quantitative data, will be analysed using multivariate analysis and clustering so as to write preliminary analysis and summaries.

The transcribed data are free–flowing text as a proxy for experience. It can be coded either by Key Words in Context (KWIC) or schema analysis. The former could be conducted through word count, but not necessarily; the latter needs to develop a coding scheme first. Both of them aim to establish codebooks which are the basis of the indicators of framework.

Repackaging and aggregating the data is a transitional stage for the preparation of model building. One aspect is searching for relationships in the data by writing analytical memos. The other is the calibration which involves finding out where the emphases and gaps in the data are. This could lead to the applicability examination of indicators, and the strengths, weaknesses and gaps analysis in the pilot study.

Developing and testing propositions to construct an explanatory framework firstly require testing hypotheses and reducing the volume of data for analysis of trends in it, then delineating the deep structure. The former step starts with the crosschecking of preliminary findings, before the matrix analysis of major themes in data. Cluster analysis is the further exploration of the major themes in order to display the matrix more visually in NVIVO. The later step is a process of synthesis, by integrating the data into one explanatory framework.

Within-case analysis aims to apply the refined descriptive model and validate the explanatory displays. For the cross-case analysis, this is an iterative process, to extract and verify conclusions. The final stage is to produce implications for theory, policy and actions.

In the above a brief outline of the development of a model is given. The six main steps, as well as their constituent stages form the ladder of logical abstraction. The process is iterative in various stages for the purpose of checking, complementing and improvement. The actual methods to implement this process are described in the methodology following section.

45.4 Conclusion

This paper focuses on the methodological approach for an in-depth research of the holistic framework for evaluating sustainability of urban regeneration in China. This paper describes the research strategy, methods and techniques which are employed to collect and analyse the data in order to answer these research questions. The justification of the adopted research strategy was highlighted. The research design was discussed in detail including the processes of case study analysis and qualitative analysis. The latter implies a series of actions taken to answer the question of how to build a qualitative model. This approach closely follows the case study research strategy described by Miles and Huberman and the theory induction process described by Carney. The paper concluded with discussions of the detailed stages of data collection, data analysis, modelling and validation.

References

1. Plessis CD (2009) An approach to studying urban sustainability from within an ecological world view. University of Salford, Salford
2. Peet J (2000) Being fully human and creating a better future: sustainable development from an integrated systems perspective. Paper presented at the Baltic 21 workshop on sustainable development. Sigtuna Foundation, Sigtuna, Sweden
3. Blanc I et al (2008) Towards a new index for environmental sustainability based on a daly weighting approach. Sustain Dev 16(4):251–260
4. Palme U et al (2005) Sustainable development indicators for wastewater systems—researchers and indicator users in a cooperative case study. Resour Conserv Recycl 43(3):293–311
5. Carvalho SD et al (2009) Application of a sustainability index for integrated urban water management in southern african cities: case study comparison—Maputo and Hermanus. Water SA 35(2)–144–151
6. Castells M (2000) The rise of the network society 2nd edn, vol 1. Blackwell Publishers, Oxford
7. Kinderytė L (2010) Methodology of sustainability indicators determination for enterprise assessment. Darnaus vystymosi Rodiklių Nustatymo Metodika Jmonėms Vertinti 52(2):25–31
8. Atkisson A, Hatcher RL (2001) The compass index of sustainability: prototype for a comprehensive sustainability information system. J Environ Assess Policy Manage 3 (4):509–532
9. Zavrl MŠ et al (2009) Multicriterial sustainability assessment of residential buildings. Daugiakriterinis Gyvenamuju Namu darnos Vertinimas 15(4):612–630

10. Hemphill L et al (2002) An aggregated weighting system for evaluating sustainable urban regeneration. J Property Res 19(4):353–373
11. Phillips LD, Costa CAB (2005) Transparent prioritisation, budgeting and resource allocation with multi-criteria decision analysis and decision conferencing. The London School of Economics and Political Science, London
12. Brandon PS, Lombardi PL (2005) Evaluating sustainable development in the built environment. Blackwell Science, Oxford
13. Iuliis MD, Brandon P (2010). The time horizon in the evaluation of sustainable development. Paper presented at the 1st international conference on sustainable urbanization 2010. Hong Kong, China
14. Lombardi PL (1999) Understanding sustainability in the built environment: a framework for evaluation in urban planning and design. (Unpublished doctoral dissertation). T.I.M.E. Institute, Department of Surveying, University of Salford, Salford
15. Yin RK (2009) Case study research: design and methods, 4th edn. Sage Publications, Thousand Oaks
16. Creswell JW (2007) Qualitative inquiry & research design: choosing among five approaches, 2nd edn. Sage, Thousand Oaks
17. Wolcott HF (1990) Writing up Qualitative Research. Sage Publications, Newbury Park
18. Miles MB, Huberman AM (1994) Qualitative data analysis: an expanded sourcebook, 2nd edn. Sage Publications, Thousand Oaks
19. Vliet JV (2006) Validation of land use change models- a case study on the environment explorer. (Unpublished masters thesis). Wageningen University and Research Centre, Wageningen
20. Carney YF (1990) Collaborative inquiry methodology. The University of Windsor, Windsor, Ontario
21. Naoum SG (2007) Dissertation research and writing for construction students, 2nd edn. Elsevier Butterworth-Heinemann, Amsterdam
22. Mak MY (2009) Scientific Feng Shui: application of Feng Shui knowledge to preliminary building design evaluation using knowledge-based expert systems approach. VDM Verlag. Dr. Müller, Saarbrucken

Chapter 46
The Impact of Urbanization on Investment of Real Estate Industry in China

Yan Liu and K.W. Chau

Abstract During the period of 1995 to 2014, the urbanization rate in Mainland China increased from 29.04 to 54.77 %, with an annual increase of 1.28 %. Meanwhile, the real estate market also boosted dramatically and worked as a driving force for the national economy. The interaction between urbanization and development of housing markets has long been investigated by researchers. Theoretically, the urbanization process should boost the demand for various properties and nurture more investment in this sector. For another, real estate industry works as a foundation and provides a supporting environment to the urbanization process. Previous research mainly focuses on qualitative policy analysis instead of direct quantitative analysis and the majority discuss about the housing price. This paper would like to fill in the knowledge gap by exploring the relationship between urbanization and annual real estate investment. Based on the macro data in China over the past 20 years, empirical study is carried out, with application of Unit Root Test, Co-integration Test and Granger Causality Test. The findings suggest that there is a long-term equilibrium and a one-way granger relationship between urbanization rate and annual real estate investment. The findings are useful for understanding the sustainable development of real estate industry and forecast future property demanding caused by the urbanization process.

Keywords Urbanization · Real estate investment · Granger causality test · Co-integration test

---

Y. Liu (✉)
Department of Urban Management, University of Electronic Science and Technology of China, Chengdu, China
e-mail: liuyanhit2009@gmail.com

K.W. Chau
Department of Real Estate and Construction, The University of Hong Kong, Pok Fu Lam, Hong Kong

Y. Wu et al. (eds.), *Proceedings of the 20th International Symposium on Advancement of Construction Management and Real Estate*,
DOI 10.1007/978-981-10-0855-9_46

46.1 Introduction

The real estate industry in Mainland China (hereafter China) has experienced great development and expansion during the past decades. Three variables are selected in Fig. 46.1 to present the trend: REA stands for the "Annual Property Saleable Area (million Sq.m)", while REI is the "Annual Real Estate Investment (billion)"and REP refers to "Average Commercial Property Price (RMB/Sq.m)". From the figure, we can see that REA maintains a steady rising during 1995–2014, and REP increased gradually especially after the year 2003. The amount of annual investment (REI) demonstrates a more dramatic increase, from the amount of 314.9 billion in 1995 to 9503.56 billion in 2014.

Meanwhile, cities have been dramatically developed and the rapid urbanization process brings about significant economy changes in China. The term urbanization originated from the Latin word 'urbs', which means city. This is normally described as "the population shift from rural areas to urban areas" or "the gradual increase in the proportion of people living in urban areas". Urbanization has been highly relevant to modernization, industrialization and rationalization, and may cause social, economic and environment changes in the society. The level of urbanization is widely measured by the percentage of population living in the urban area, which is.

During the period of 1995 to 2014, the urban residents in China raised from 351.74 million to 749.16 million while the urbanization rate increased from 29.04 to 54.77 %, with an annual increase of around 1.28 %. The actual percentage change of urban and rural population can be seen from the Fig. 46.2. Ever since 2010, the urban percentage exceeds that of rural population.

The urbanization process brings great change to city development and there is apparent interaction between urbanization and real estate industry. The urbanization process has boosted the development of real estate industry via great increase of housing demand. Meanwhile, real estate industry also works as a foundation and

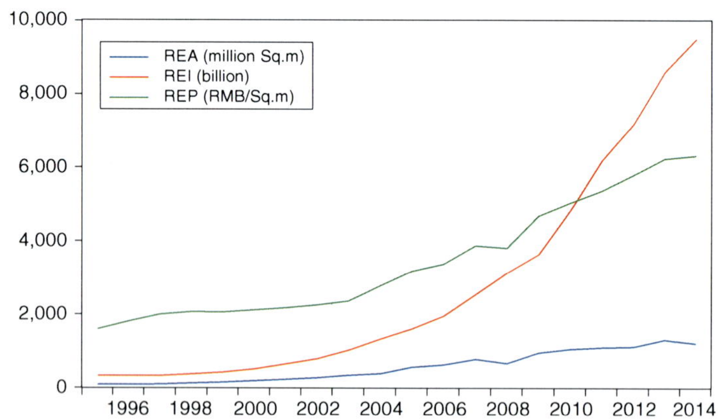

Fig. 46.1 Development of real estate industry during 1995–2014

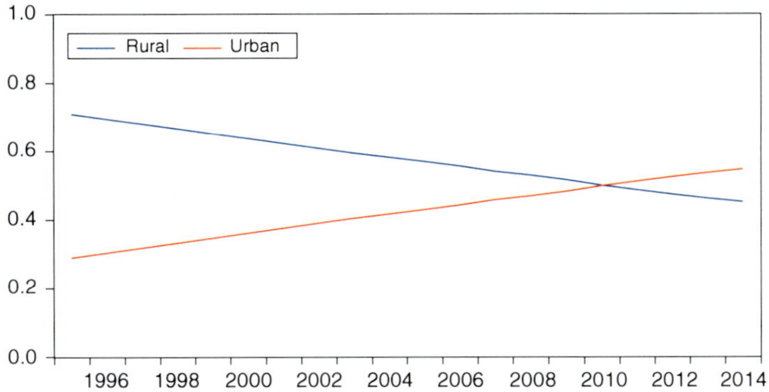

Fig. 46.2 Population percentage change during 1995–2004

provides a supporting environment to urbanization process. Interaction from the supply and demand theory can reveal the theoretical relationship between them. Therefore, this study aims to investigate the impact of urbanization on the real estate industry in China from a supply and demand perspective.

To carry out the aforementioned issue, the rest of the paper is organized as follows. After the brief introduction, literature review is addressed in Sect. 46.2, and then follows with research methods in Sect. 46.3, afterwards, Sect. 46.4 comes up with results and findings, and final Sect. 46.5 draws the conclusion.

46.2 Literature Review

Current research mainly focus on housing price, housing related policies, or the casual relationship between real estate market and the stock market, see [1, 2], etc. Little has been done with the urbanization issue. This research will fill this research gap and link the interaction between real estate market and urbanization. Because the real estate market is quite regional and various in different countries, and also considering the specialty of the economic policy and environment in China, mainly Chinese market related research will be the focus.

Zhang and Zheng [3] analyze the real estate demand based on the 2003–2007 panel data, with use of STATA. They quantify the living demand with variables of price, income and urbanization degree and find that the urbanization process motivates the commercial housing demand greatly. The elasticity for urbanization is 5.961, which means that when urbanization rate increases by 1 %, the housing demand will increase by 5.861 %. Wang [4] studies the correlation between urbanization and real estate market demands, and find that saleable real estate area will rise by 5.21 % when the urbanization rate increases by 1 %. The finding reveals the quantitative relationship, but the choosing of saleable are as the independent variable may cause biases. Chen et al. [5] examine how rapid urbanization and massive rural-urban migration affect housing dynamics in China from 1995 to 2005. They find that due to the different migration status in coastal areas and inland areas, the housing market varies. When making related policies, the labor migration and urbanization should also be considered.

Many past research study the real estate related supply and demand factors either from a macro or a micro perspective, few concentrate on the direct relationship between urbanization and real estate industry. This research will focus on the direct impact of urbanization on development of real estate industry. Some previous studies selected the saleable area (REA), or the real estate price (REP) to stand for real estate industry development. But a more detailed factor can better increase the accuracy and liability, so the direct annual real estate investment (REI) will be selected in this study as it will avoid certain biases and can better represent the industry development trend.

This study will focus on the granger causality relationship between real estate investment and urbanization rate particularly, with a hypothesis that "Urbanization Rate is the Granger-cause of Real Estate Investment". Because the urbanization process will attract more and more citizens working and living in the urban areas, this will foster greater demanding for various types of properties. This increase of demand then will motivates more supply and thus need more money to invest in real estate sector. Therefore, urbanization rate is assumed to be the Granger-cause of real estate investment.

46.3 Research Methods

The theoretical methodology will be introduced in this section and follow such sequence: "Unit Root Test"-"Co-integration Test"-"Granger Causality Test".

46.3.1 *Unit Root Test*

For time series variables, mostly of which are financial and economic attributes, if they are non-stationary, the Ordinary Least Squares (OLS) estimation and some diagnostic indicators, e.g. R-square value and t-statistic may be biased [6]. Therefore, before any further analysis, it is important to test whether the independent and dependent variables demonstrates the same order on stationarity. In this study, ADF method will be selected for time series stationary test.

In order to test unit root, Dickey and Fuller [7] developed the following function:

$$\Delta Y_t = \alpha + \delta_t + \gamma Y_{t-1} + \sum_{i=1}^{k} \beta_i \Delta Y_{t-i} + \varepsilon_t \quad t = 1, 2, \cdots, T$$

where:

Y = The variable under test;
α = A drift;
δ_t = A time trend;
k = The integer lag length;
ε_t = The disturbance term

If the estimated coefficient of the lagged variable, γ is significantly less than 0, the null hypothesis that Y contains a unit root ($\gamma = 0$) should be rejected, vice versa. Specifically:

$$\begin{cases} H_0 : \gamma = 0 \\ H_1 : \gamma < 0 \end{cases}$$

46.3.2 Co-Integration Test

Non-stationary variables will induce spurious estimation results. While there also exists exception if the linear combination of the non-stationary variables demonstrates the stochastic trend and produces stationary residuals. This can be referred that the set of variables are co-integrated. Under such circumstance, the regression result can be regarded as reasonable and reliable. The co-integration equation can be interpreted as a long-run equilibrium relationship between two time series variables [8]:

$$Y_t = \alpha + \beta X_t + \varepsilon_t$$

The residual ε_t is tested for stationarity with use of ADF test. If it is stationary, the two series are defined as co-integrated, which means there is long-term equilibrium between the two series. Johansen test will be adopted is this study.

46.3.3 Granger Causality Test

The Granger Causality test is to examine how much of the current variable Y can be explained by past values of Y and whether adding lagged X can improve the explanation ability. If the coefficients of the lagged X's are jointly significant, the null hypothesis of no granger causality between X and Y is rejected. That is to say, Y is granger caused by X or X is the granger cause of Y, vice versa.

$$Y_t = \sum_{i=1}^{n} \alpha_i Y_{t-i} + \sum_{i=1}^{n} \beta_i X_{t-i} + \mu_{it}$$

$$X_t = \sum_{i=1}^{n} \alpha_i X_{t-i} + \sum_{i=1}^{n} \beta_i Y_{t-i} + \mu_{it}$$

The null hypotheses is that $H_0{:}\beta_1 = \beta_2 = \cdots = \beta_k = 0$, which means that X does not granger cause Y. The granger causality test can help identify whether there are bidirectional, one-way or no causal relationship between variable X and Y.

Particularly, this study aims to investigate the relationship between urbanization process and annual real estate investment in China. The annual amount of real estate investment $(Y{:}REI)$ will be selected as dependent variable, and annual urban rate $(X{:}UR)$ will be independent variable.

46.4 Empirical Analysis

46.4.1 Data Description

To conduct the empirical study on the hypotheses that urbanization rate (UR) is the granger-cause of real estate investment (REI), both variables will be included and the sample range will start from 1995 to 2014. All the data can be got from the China Statistical Yearbook. The following Table 46.1 shows the general data description of both variables.

The natural logarithm form of both variables instead of the raw data will be selected for a better representation of the percentage change of certain economic attribute. This form can directly represent the elasticity of the variable. Thus, the following estimations will all base on LnREI and LnUR.

Table 46.1 Raw Data Description

Attributes	UR	REI
Mean	0.422832	27533.37
Median	0.423750	14533.75
Maximum	0.547704	95035.61
Minimum	0.290404	3149.020
Std. Dev.	0.080753	29550.33
Skewness	−0.049589	1.120187
Kurtosis	1.796419	2.929509
Jarque-Bera	1.215371	4.186871
Probability	0.544610	0.123263
Sum	8.456631	550667.5
Sum Sq. Dev.	0.123900	1.66 E + 10
Observations	20	20

Table 46.2 Augmented Dickey-Fuller test for variables LnREI, LnUR

Variables	(c, t, k)	ADF test statistic	Test critical values			Test result
			1 %	5 %	10 %	
LnREI	(c, t, 0)	−4.468233[b]	−4.532598	−3.673616	−3.277364	Stationary
LnUR	(c, 0, 0)	−12.69613[c]	−3.831511	−3.029970	−2.655194	Stationary

Notes c denotes including a constant
t denotes including linear trend
k denotes lag lengths
[a]10 % level of significance, [b]5 % level of significance, [c]1 % level of significance

46.4.2 Unit Root Test

Unit root test helps to determine whether the time series variable is stationary and the stationarity is the prerequisites before any further analysis. As shown in Table 46.2, the ADF test indicates that both series of LnREI and LnUR are stationary. Therefore, those two time series variables can be further analyzed.

46.4.3 Co-Integration Test

The empirical result of JJ co-integration test is presented in Table 46.3, including the trace and maximum eigenvalue statistics. The result indicates that both tests reject the null hypothesis of no co-integration existing between LnREI and LnUR at the 1 % significance level. This suggests a long-run equilibrium between the two variables.

46.4.4 Granger Causality Test

The empirical test of whether there is causality relationship between them is shown in Table 46.4. The analysis indicates that for both lag 2 and lag 3 circumstances, the

Table 46.3 Result of Co-integration Test

Hypothesized No. of CE(s)	Eigenvalue	Trace statistic	0.05 critical value	Prob.[b]
Unrestricted cointegration rank test (trace)				
None[a]	0.872429	48.24698	15.49471	0.0000
At most 1[a]	0.462757	11.18349	3.841466	0.0008
Unrestricted cointegration Rank Test (maximum eigenvalue)				
None[a]	0.872429	37.06348	14.26460	0.0000
At most 1[a]	0.462757	11.18349	3.841466	0.0008

Note Trace test indicates 2 cointegrating eqn(s) at the 0.05 level
Max-eigenvalue test indicates 2 cointegrating eqn(s) at the 0.05 level
[a]denotes rejection of the hypothesis at the 0.05 level
[b]MacKinnon-Haug-Michelis (1999) p-values

Table 46.4 Result of Granger Causality Test

Pairwise granger causality tests			
Null Hypothesis	Obs	F-statistic	Prob.
Lags: 2			
LNUR does not granger cause LNREI	18	31.3328	1.E-05
LNREI does not granger cause LNUR		1.67354	0.2256
Lags: 3			
LNUR does not granger cause LNREI	17	12.9828	0.0009
LNREI does not granger cause LNUR		0.63762	0.6078

results reject the null hypothesis of "LNUR does not Granger Cause LNREI" at the 1 % significance level. However, both estimations cannot reject the other hypothesis that "LNREI does not Granger Cause LNUR". This suggests the variable LnUR is the granger cause of LnREI, which means that the urbanization rate is one of the driven forces for real estate investment and help explain the trend of the real estate industry development.

46.4.5 Discussion

From the above analysis, we can see that past urbanization rate can help better explain (or predict) the investment in real estate sector. Urbanization is a granger cause of real estate investment. This can mainly be understood that urbanization increases demanding greatly which in turn will motivate huge real estate supply and therefore attract more investment.

The reason why urbanization process can serve as an incentive for more investment in real estate industry can mainly be explained by that urbanization is a driven force for increasing the demand. Certain rising of demand can be detailed in the following four aspects.

Firstly, urbanization will increase the "living demand". With more and more people move to the urban area, more houses are needed to meet human's basic living requirement. It's the fundamental and base for daily life. And thus will motivate more demand for residential real estate.

Secondly, this is also caused by increase of "productive demand" which refers to larger demand for offices or industrial properties to guarantee the producing activities. As more and more people move from rural areas to urban areas, they will try to work in the cities as well. The job market therefore will need more working spaces.

The third aspect can be called as "consuming demand", as citizens moving from rural to urban area, their lifestyle will also change gradually. Increasing dining, shopping and leisure demand will thus motivate the expanding for those leisure or entertainment properties.

The last motivation is "investment demand". As known that property is not only consumer goods, it's also investment goods that people can directly make profit through selling or renting out a property. Driven by speculating pursuit, the demand of properties will therefore be increased.

46.5 Conclusions

This paper investigated the granger causality relationship between the real estate investment (REI) and urbanization process (UR) in China during the period of 1995 to 2014, from the supply and demand perspective. The raw data showed great development of both sectors in the past decades. Both time series variables are stationary and demonstrate long-term equilibrium. This indicates the similarities between them and their common responses to general macro-economic and political changes. This study further finds causal relationship between LnREI and LnUR, which means that the including of urbanization rate can help better explain and predict current real estate investment.

Significances of this paper are in two folds: first, this study can help better reveal the relationship between urbanization and real estate industry from a quantitative analysis base which is a great improvement. Especially under the new challenges of real estate industry development, to learn more about the urbanization process and better serve the modern urban sectors is of great importance. Secondly, the choosing of real estate investment as the dependent variable differs from existing research and this will provide a direct observation and reference to the capital market. Whether the interaction varies in different geographic areas or including of other macro-economic attributes help better explain the real estate industry will be the possible future study areas.

References

1. Kapopoulos P, Siokis F (2005) Stock and real estate prices in Greece: wealth versus 'credit-price' effect. Appl Econ Lett 12(2):125–128
2. Okunev J, Wilson P, Zurbruegg R (2000) The causal relationship between real estate and stock markets. J Real Estate Finance Econ 21(3):251–261
3. Zhang YH, Zheng DC (2009) Study on the relationship between urbanization and real estate market demands. Spec Zone Econ 12:277–279
4. Wang WT (2010) Analysis of the impact of urbanization on Chinese real estate market demands. Mod Property Manage 4:21–23
5. Chen JH, Guo F, Wu Y (2011) One decade of urban housing reform in China: urban housing price dynamics and the role of migration and urbanization, 1995–2005. Habitat Int 35(1):1–8
6. Granger CW, Newbold P (1974) Spurious regressions in econometrics. J Econometrics 2 (2):111–120

7. Dickey DA, Fuller WA (1979) Distribution of the estimation for autoregressive time series with a unit root. J Am Stat Assoc 74:427–431
8. Ling T, James R (2000) Causality in real estate markets: the case of Hong Kong. J Real Estate Portfolio Manage 6(3):259–271

Chapter 47
Analysis on the Urbanization of Key Ecological Function Area—A Case Study on Tibetan Areas of Sichuan

Xia Lei

Abstract Urbanization is the necessary way to achieve modernization, and is also important way to prompt economic growth. On condition of the construction of main function area, the experience about urbanization from the areas with high level of urbanization such as optimizing development area and key development area cannot apply to key ecological function area. This paper analyzes how the construction of key ecological function area influences the urbanization theoretically. Then it takes Tibetan areas of Sichuan as example to analyze how the construction of key ecological function area influences the urbanization practically. It concludes that the whole level of urbanization of Tibetan areas of Sichuan is low but unbalanced in counties; the development level of urbanization mainly matches the development of economy, but is influenced by its industrial structure; the urban population is small and the designated towns are few. And the construction of key ecological function area influences the Tibetan Areas of Sichuan by limiting the urban form, affecting the orientation of urban function, and affecting the layout of urban system. In order to promote the urbanization of Tibetan Areas of Sichuan, it suggests to cultivate leading industry, optimize the layout planning of urban and rural system, increase fiscal support, balance the rural and urban development, and promote the ecological immigration steadily.

Keywords Urbanization · Construction of key ecological function area · Tibetan areas of Sichuan

47.1 Introduction

Urbanization is the necessary way to achieve modernization, and is also important way to prompt economic growth. Since 1980s, China has started its urbanization and made great progress. However, the traditional way to develop urbanization has

X. Lei (✉)
School of Economics, Sichuan University, Chengdu, China
e-mail: 83114605@qq.com

Y. Wu et al. (eds.), *Proceedings of the 20th International Symposium on Advancement of Construction Management and Real Estate*,
DOI 10.1007/978-981-10-0855-9_47

caused some problems. [1] On condition of that, the government issued the Main Function Area Planning which aims to restrict the land usage to harmonize the development of population, economy and resource. [2] According to the National Main Function Area Planning and Sichuan Main Area Planning, Tibetan Areas of Sichuan whose urbanization still stays at low level is in the limited development area and mainly plays the part of key ecological function area. [2, 3] For large-scale and high-intensity development of industrialization and urbanization is forbidden in key ecological function area, the urbanization of Tibetan Areas of Sichuan can't follow the traditional way the other developed areas have experienced. How to properly prompt the development of urbanization to achieve modernization in Tibetan Areas of Sichuan is not only important for the development of it, but also helpful for the development of other key ecological function area and minority areas.

The urbanization level of Tibetan Areas of Sichuan is still low and unbalanced, and this process should be prompted. [4, 5, 6] According to the analysis on the relationship between urbanization and supporting industry, in order to speed up the urbanization of ethnic areas in Sichuan, it should depend on the supporting industry to raise productivity in rural land and narrow the gap of income between urban and rural areas. [7] For the development of population urbanization synchronize with the development of economy, and the population mainly accumulates in small towns, completing the infrastructure in small town, cultivating leading industries of designated town, as well as supporting the poor are the way to prompt the urbanization of Tibetan Areas in Sichuan. [8] For the development of urbanization of Tibetan areas are different from east and middle China, the drive force of urbanization in Tibetan areas is decided by both economy and social cultural environment, the urbanization of Tibetan areas should pay attention to their own characteristics to develop plateau characteristic economy such as tourism, completing public service and infrastructure, and developing green towns, by integrating tradition and modernization [4].

The existing literatures analyze the current situation of Tibetan Areas of Sichuan and the drive force of their urbanization, and also realize it should go on a special way which is much different from the other areas to prompt urbanization. However, these literatures don't concentrate on the effluence of the construction of key ecological function area, which is the key factor for their urbanization.

47.2 Theoretical Analysis

According to Northam Curve, a country's or region's trail of urbanization will follow a rule which presents like a flat "S" curve. [9] Although this rule is concluded with western countries' experience, the urbanization in most regions of China also obey it. [10] Nevertheless, the precondition is that the development of urbanization will not affect the ecological environment negatively, and the actions such as expanding urban land and developing the secondary industry are the optimal way to develop urbanization for most areas. However, it does not match the key ecological function area. In these areas, the ecological system has degraded to a certain extent but is very important and relates the ecological security of large-scale areas or even the whole country. It means that large-scale and high-intensive urbanization may not only worsen the ecological environment locally, but also influence the natural environment all over the country; and the prohibition of large-scale and high-intensive urbanization may slow the local economic and social development but benefit the national ecological environment. That is to say, the cost and revenue of urbanization includes national and local cost containing economic, social, and environmental cost and revenue. And the cost of traditional way to achieve urbanization in these areas is higher than the national revenue and the local economic and social revenue is lower than the national ecological cost. On contrary, the national revenue of prohibiting large-scale and high-intensive urbanization is higher than the local cost for the national ecological revenue is higher than the economic and social cost.

The construction of key ecological function area affects the urbanization directly and indirectly. As we know that, the construction of key ecological function area requires to prohibit large-scale and high-intensive urbanization and industry. Therefore, it directly affects the urbanization by limiting the development of urban land which is the part of urbanization. In another way, it also affects the urbanization indirectly by limiting the development of industry. For urban industry is the absorber of population, the prohibition of high-intensive industry means industry cannot absorb enough rural population to help realizing urbanization, and it also means that the fund of infrastructure is hard to be financed by industry. So the limit of industry narrows the methods to realize population urbanization and the ways to finance the infrastructure fund. And it is described in Figure 47.1.

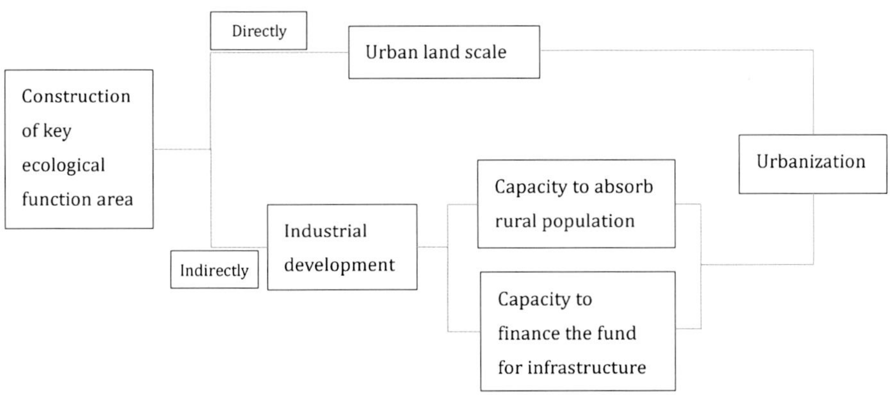

Fig. 47.1 How the construction of key ecological function area influences the urbanization

47.3 Analysis on the Urbanization of Tibetan Areas of Sichuan

47.3.1 Research Object and Data Resource

Tibetan Areas of Sichuan consists of "two autonomous prefectures and one county", it includes Ganzi Tibetan Autonomous Prefecture which consists of 18 counties and Ngawa Tibetan and Qiang Autonomous Prefecture which consists of 13 counties and Muli County which is administered by Liangshan Yi Autonomous Prefecture. All these data are from Sichuan Statistical Yearbook or calculated based on it except the data about designated towns which are cited from the government websites, respectively.

47.3.2 Brief Introduction to the Urbanization of Tibetan Areas of Sichuan

Tibetan Areas of Sichuan locates at west Sichuan Province and east Tibet Plateau. It covers 250.1 thousands square kilometers, which occupies 51.57 % of the land of whole Sichuan. [11] At the end of 2013, the total residential population here is 2181.1 thousands, and it is the second largest Tibetan settlement in China. The urban population here is 626.3 thousands and the urbanization rate is 28.71 %. At the same year, the GDP totaled ¥45.97 billion yuan consists of primary industry totaled ¥8.73 billion yuan, secondary industry totaled 21.45 billion yuan and tertiary industry totaled ¥15.80 billion yuan; And the proportions of the three industry are 18.99, 46.65 and 34.36 %. In addition, the per GDP of it is ¥21077.44 yuan. The competitive industries of these towns are mainly tourism, mining industry, and trade industry. In general, the development of Tibetan Areas of Sichuan is low.

47.3.3 The Characteristic of the Urbanization of Tibetan Areas of Sichuan

The Whole Level of Urbanization Is Low but Unbalanced in Counties

The urbanization rate of the whole Tibetan Areas of Sichuan is 28.71 %, which means that it stays at the early stage of urbanization. Whereby, 10 counties' urbanization rate is lower than 20 %, which occupies 31.25 % of the 32 counties. And there are 7 counties whose urban population is less than 10,000, and the urban population of Derong is only 4900.

Although the urbanization of it is low, it is unbalanced in these areas. Kangding whose urbanization rate is the highest of 49.55 is 37.05 % higher than Xinlong whose urbanization rate is only 12.50 %. And the urbanization rate of Kangding ranks 35 in 184 counties in the whole province, which means that it ranks the top 20 % of the whole province. However, there are 21 counties whose urbanization rate are lower than 30 %, and it occupies 7 of the bottom 10 counties of the whole province.

The low urbanization rate indicates that increasing population urbanization rate is the key work when prompting urbanization. But the unbalanced urbanization means that the method to prompt the urbanization of it should be various according to different counties.

The Development Level of Urbanization Mainly Matches the Development of Economy, but Is Influenced by Its Industrial Structure

The development level of urbanization mainly matches the development of economy, the higher the per capita GDP, the higher the level of urbanization. However, the industrial structure influences the urbanization significantly. For example, Jiulong's urbanization rate is 18.54 % which ranks only 26 in 32 counties, but its per capita GDP achieves ¥42,761 yuan which ranks 2 in 32 countries. The great different between these two indexes is caused by its industrial structure. Secondary industry including mining industry and hydropower industry develops well in Jiulong. For the supporting industries are capital-drived industry, the well-developed secondary industry does not bring in good development of urbanization. The GDP ratio of the three industries 1:8.79:1.73 also reflects it. On contrary, Litang's urbanization rate is 34.38 % which ranks only 7 in 32 counties, but its per capita GDP is only ¥11,799 yuan which ranks 25 in 32 countries. For the tertiary industry including tourism develops well in Litang and it is labor-drived industry, it absorbs many labors and its urbanization level is comparatively high. And the GDP ratio of the three industries 1:0.44:1.18 also reflects it.

In general, the development of secondary industry decides the development of economy, and the tertiary industry deices the development of urbanization. How to balance the development of economy and urbanization will be the key issue to be considered.

The Urban Population Is Small and the Designated Towns Are Few

The urban population of Tibetan Areas of Sichuan is small. There is only one county-level city Kangding whose urban population exceeds 50,000, and there are 7 counties whose urban population are less than 10,000. The bottom 10 of the whole province are all from Tibetan areas. In addition, the designated towns are few, too. Wenchuan covers 8 designated towns which is the top of these 32 counties, and there are 15 counties which only cover 1 designated town.

The research indicates that, a town will generate agglomeration effect only if its urban population exceeds 50,000. [11] With the view of scale effect, the urbanization of Tibetan Areas of Sichuan should be promoted. However, for the total population of it are small than east and middle China, the urbanization of it will be different from them, too.

47.4 The Influence of the Construction of Key Ecological Function Area on the Tibetan Areas of Sichuan

The construction of main function areas confirms different areas' function about economy and ecology. [2] The Tibetan Areas of Sichuan are all oriented as key ecological function area. It means that its ecological system is weak and its ecological function is important, the carrying capacity of its natural resource and environment is weak; so that there is no condition to conduct large-scale and high-intensive urbanization and it should take increasing productivity capacity about ecological product as the prime task and limit large-scale and high-intensive industrialization and urbanization. [1] It changes the traditional no-differentiated strategy about urbanization, and has significant influence on the development of urbanization for most areas of China. [11] On condition of that the urbanization of Tibetan Areas of Sichuan which is comparatively backward stays at the early stage, the orientation of key ecological function area causes that this area cannot follow the others' experience. The construction of key ecological function influences the urbanization of it in these ways described below.

47.4.1 The Construction of Key Ecological Function Area Limits the Urban Form

For the construction of key ecological function area prohibits large-scale and high-intensive urbanization and industry, the expansion of urban land should be slow, the development of industry is strictly limited, and with the sequence, the urban population which depends on the land scale in a certain extent and its construction which depends on industrial construction in a certain extent has been limited.

From 2009 to 2013, the rate of primary industry, secondary industry and tertiary industry which changes from 1:1.56:1.53 to 1:1.97:1.77. Although the percentage of secondary industry raises a little, the percentage of secondary industry is still low for it is less twice of the primary industry and is little more than tertiary industry. Accompanying with the little change of the rate of three industries is the little change of the urban scale. Although the urban resident population raises from 482.4 thousands to 626.3 thousands, and the urbanization rate raises from 23.42 to

28.71 %, the urbanization rate is still low than 30 % and the average annual growth rate is just 5.22 %. In addition, these urban population allocate at 32 counties, there is only 65.9 thousands urban population in the biggest administrative region Kangding which upgrades to be a county-level city from a township, and there is only 4.6 thousands urban population in the smallest county Derong. The limit of urbanization and industry causes Tibetan Areas of Sichuan be hard to develop its urban scale and will keep stay at the form of small-town.

47.4.2 The Construction of Key Ecological Function Area Affects the Orientation of Urban Function

For the construction of key ecological function area prohibits large-scale and high-intensive urbanization and industry, the development of industry is strictly limited. Besides the mining industry and power industry which can be developed properly, the secondary industry is not encouraged. At the same time, it promotes to develop ecological industry which belongs to primary industry and tertiary industry. Therefore, the construction of key ecological function area causes the Tibetan Areas of Sichuan develop towards ecological region and exert ecological function.

As it discussed above, since the issue of the Main Function Area Planning in 2010, the rate of primary industry, secondary industry and tertiary industry which changes from 1:1.56:1.53 in 2009 to 1:1.97:1.77 in 2013. Although the secondary industry occupies the biggest proportion of GDP, the primary industry and tertiary industry occupies 62.25 % of it. Hereinto, the primary industry and tertiary industry mainly contains planting industry, trading industry and tourism industry which include the ecological industry or could be transited to ecological industry easily. Therefore, the construction of key ecological function area causes Tibetan Areas of Sichuan be hard to be industrial town but possible to develop ecological industry and exert the ecological function.

47.4.3 The Construction of Key Ecological Function Area Affects the Layout of Urban System

For the construction of key ecological function area prohibits large-scale and high-intensive urbanization and industry, the way of connection among these towns is narrowed, and ecological resource will be the main reason to organize these towns.

With the limitation of landform and traffic condition, the towns of Tibetan Areas of Sichuan stay spotty distribution along the artery traffic, but not form the pole-axis pattern. As the production of ecological products are the most important task for these towns, they should realize ecological town may be the final form of them and the Tibetan Areas of Sichuan should develop as a whole. For the existing counties are small and the urban scale may keep steady, small town will be the main form of the whole urban system. In addition, some counties and county-level city such as Kangding, Maerkang, Jiuzhaigou, Wenchuan, Maoxian, Luding and so on which occupy advantages such as nature resource or administrative resource will be the leaders in the urban system. Thus, it will form the ecological urban layout with network pattern.

47.5 Conclusion

The construction of key ecological function area will influence the urban form, urban function and urban layout. And in order to develop the urbanization well, it should concentrate on these facts.

47.5.1 Cultivating Leading Industry to Promote Urbanization

The development of leading industry is the drive for sustainable urban development, and the base of urbanization is the transfer of industry from rural area to urban area which should beyond the transfer of population. [6] Basing on the existing resource and orientation of key ecological function area, the Tibetan Areas of Sichuan should cultivate the ecological tourism. Firstly, the development of ecological tourism will help the construction of key ecological function area. Secondly, the ecological tourism is a labor-intensive industry, which helps to absorb the surplus rural labor to promote urbanization. Thirdly, the employment threshold of ecological tourism is comparative low, which matches the reality that the education background and labor skill of the people in these areas are comparative low.

47.5.2 Optimizing the Layout Planning of Urban and Rural System

According to the nature resource and environment, existing population scale, industrial layout, and the construction of key ecological function area, it should optimize the layout planning of urban and rural system. It should confirm the different functions of different towns and lead the rural people and production factors transfer to the urban areas. It also should pay attention to some big counties to cultivate central towns which will drive the development of other towns around.

As the quantity of designated town is comparative few, it should strive to cultivate more designated towns to promote the urbanization.

47.5.3 Increasing Fiscal Support

For the construction of key ecological function area benefits the whole country, and the Tibetan Areas of Sichuan do sacrifice some economic benefit to satisfy the whole country, the government should provide them some subsidies to balance its cost and benefit. Particularly, the Tibetan Areas of Sichuan is undeveloped, it needs external support to finance the necessary fund to construct infrastructure which is the base of urbanization. The fiscal support could be eco-compensation, transfer-payment, or central specialize fund, and so on.

47.5.4 Balancing the Rural and Urban Development

The development of urbanization is not only the development of towns, but also the development of country. For the Tibetan Areas of Sichuan should concentrate on developing small towns which connect the country closely, the development of rural land will influence the development of small towns deeply. Therefore, it should promote the development of rural land and primary industry to balance the rural and urban development. It should increase the investment on rural land and primary industry, and improve their infrastructure to form country-level traffic. It should concentrate on the transfer of agriculture to form modern ecological agricultural chain. It also should pay more attention to rural vocational training and education to help the farmers improve their skill. Finally, it will realize the balance of rural and urban development.

47.5.5 Promoting the Ecological Immigration Steadily

For the prohibition of large-scale and high-intensive urbanization and industrialization, the ability to absorb the rural population is limited. In addition, for the serious environment, the land carrying capacity is limited, and the increasing population brings it much pressure. Therefore, the capacity of local immigration is limited, long distance immigration should be considered. However, for the obvious ethnic characteristics, it is hard for them to integrate into the other places. So the ecological immigration should be considered carefully. It should not only consider the economic reasons, but also should consider the social, religionary, ethnic and psychological reasons.

References

1. Net X (2014) National new-type urbanization plan (2014–2020). http://news.xinhuanet.com/politics/2014-03/16/c_119791251_2.htm
2. State Council of PRC (2010) Main function area planning. http://www.gov.cn/zwgk/2011-06/08/content_1879180.htm
3. Sichuan provincial People's Government (2013) Sichuan main area planning [OL]. http://www.sc.gov.cn/10462/10883/11066/2013/4/23/10258501.shtml
4. Xueping L, Bo Ding (2015) Studying on the development path of new-type urbanization of tibetan areas—A case study on Ganzi County, Tibetan areas of Sichuan. J Southwest Univ Nationalities (Humanity Soc Sci) 36(2):110–114
5. Hong W, Pengzhen C, Lilong H (2013) Discussing on New Towns in Ethnic Regions—Take Aba Tibetan as an example 150(2):97–100
6. Maoying S (2010) Study on the urbanization of minority area—A case study on Tibetan areas of Sichuan. J Ethno-National Econ 31(10):136–140
7. Wen Z, Taibi J (2008) Empirical studying on urbanization progress and development of leading industries at Sichuan Ethno Area. J Southwest Univ Nationalities (Humanity Soc Sci) 203(7):147–151
8. Maoying Shen (2010) On the issues about population urbanization in minority nationality areas—Taking Sichuan Tibetan area as example. J Tibetan Stud 123(5):112–120
9. Xuehui D, Weibo H (2012) Northam curve can not be copied to study the urbanization in China, J Hebei Univ Econ Bus 33(4):22–25, 51
10. Shengzu G, Hua L, Shance Y (2010) Urbanization as an engine for expanding domestic demand and sustaining economic growth. Chin J Popul Sci 30(3):2–10
11. Guiwen L (2010) Developmental way of county urbanization from the perspective of major functional regions 30(2):194–199

Chapter 48
Study on Social Management System Reform in Urban Fringe Area Based on Urban Service Boundary

Teng Su and Yuzhe Wu

Abstract This paper selected the water supply and police fire protection in Sandun Town, Hangzhou as the research object, and the Urban Service Boundary (USB) as the core theory, the Institutional Analysis and Development framework (IAD framework) as the train of thought. Through studying on the public goods and services supply, the policy recommendations were put forward to solve the social problems in urban fringe area; moreover, social management system reform in urban fringe area was explored based on the Urban Service Boundary. Through research, this paper found that the local government, enterprises and residents had differently cognitive to the supply of public goods and services such as water and police fire protection, which could lead to the necessity of information communication among the three, and the communication cost bigger. The distance from urban center, population density and revenue together made Sandun Town have no advantage to incorporate into the USB. Moreover, the design of system rules isn't sound in Sandun Town. And then, this paper put forward the following recommendations: perfect the communication mechanism, broaden the supply channels and change the role of actors; strengthen renovating urban fringe area and improve the conditions and attributes of community; optimize the design of system rules, and make them coordinate and promote each other. Finally, based on USB-IAD framework applied to urban fringe area, this paper reached the following conclusions: what made the social problems in urban fringe area such as public goods and services supply arise and be solved? The direct cause was the actors' interactive action in the action situation enclosed within an action arena; the conditions and

T. Su · Y. Wu (✉)
Department of Land Management, Zhejiang University, Hangzhou, China
e-mail: wuyuzhe@zju.edu.cn

T. Su
e-mail: suteng0813@zju.edu.cn

Y. Wu et al. (eds.), *Proceedings of the 20th International Symposium on Advancement of Construction Management and Real Estate*,
DOI 10.1007/978-981-10-0855-9_48

attributes of community, system and rules were restraining factors; and the root cause was the Urban Service Boundary which could interact with the conditions and attributes of community, system and rules.

Keywords Urban fringe area · The urban service boundary · The institutional analysis and development framework · Social problems

48.1 Introduction

Since reform and opening up, the control of population flow was gradually let go of, and a large number of rural residents, including peasant-workers, came into urban. Combined with the urban boundary expansion, a lot of people populated urban fringe area. As a result, many social problems appeared such as household registration management, public services, floating population, and land development and management.

This paper selected the water supply and police fire protection in Sandun Town as the research object, and the Urban Service Boundary (USB) as the core theory, the Institutional Analysis and Development framework (IAD framework) as the train of thought. Through studying on the public goods and services supply, the policy recommendations were put forward to solve the social problems in urban fringe area; moreover, social management system Reform in urban fringe area based was explored based on Urban Service Boundary.

48.2 Theoretical Basis and Analysis Framework

48.2.1 The Urban Service Boundary (USB)

Urban Service Boundary was the boundary of the public infrastructure service, which could decide supply public infrastructure service or not. Within the USB, public infrastructure service could be enjoyed, and beyond not. USB limited (or guided) urban growth through refusing to supply public infrastructure service with the areas beyond the USB [1, 2].

One word, the USB policy was practiced to limit urban unchecked growth and spread, and then improve the efficiency of urban management, realize economies of scale [3], because urban managers deemed that it was diseconomy to develop the area beyond the USB.

48.2.2 The Institutional Analysis and Development Framework (IAD Framework)

As Ostrom's research focus, the Institutional Analysis and Development framework (IAD framework) [4] was committed to explain how external variables including the conditions and attributes of community and rules, influenced the outcomes of autonomous governance in public pool resources policy, through the actors' interactive action in the action situation enclosed within the action arena. At the same time, through the results of policy evaluation and reflection on the existing operation situation, it optimized the existing system of rules.

In the IAD framework, every problem was a part of complicated public management and social problems meanwhile could be analyzed as a relatively independent part. Through the theoretical research with the help of IAD framework, we could think macroscopically, and also pay attention to the relationship between the part and whole.

48.2.3 The USB- IAD Framework Applied to the Urban Fringe Area

This paper used the USB as the core theory, the IAD framework as the train of thought, and built the USB-IAD framework applied to the urban fringe area.

As shown in Fig. 48.1, we could explain the mechanism of action how the social problems in urban fringe area arose and were solved with the help of the following propositions: First, the actors' interactive action in certain action situation enclosed within the action arena formed the internal motive power of the social problem solving in urban fringe area such as public goods and services supply, and it played a leading role. Second, the external variables including conditions and attributes of community, system and rules formed the exterior motive power, which could affect the internal motive power to work, and it played the role of restraining factors. Tried, the USB was very important to the social problem solving, which could affect the inner motive power through affecting the exterior motive power.

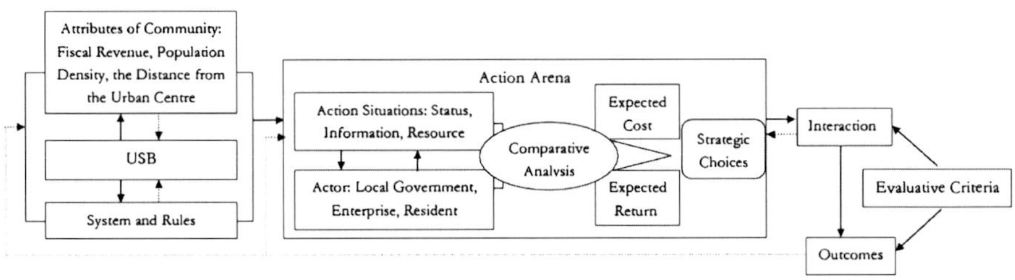

Fig. 48.1 The USB-IAD framework applied to urban fringe area

48.3 Research Cases and Data Investigation

48.3.1 Selection and Overview of the Research Object

This paper selected Sandun Town as the research object. As the capital of Zhejiang province, Hangzhou was a developed coastal city, and the size didn't reach the level of the large cities such as Beijing, Shanghai, which provided the possibility of analogical research because of the large number of cities with the similar size to Hangzhou.

As one of the towns in Xihu District, Sandun Town was located on the northwest fringe of Hangzhou and reached out to Xiangfu Subdistrict on the east, Jiangcun Sub-district on the south, Liangzhu and Cangqian Towns on the northwest. There were not only urban areas such as Sanba and Xialongyu Communities, but also rural areas such as Hualian and Shanlian Villages in Sandun Town. Xuan-Hang Railway Line played the rough boundary between urban and rural, to the southeast of which was urban, and the northwest was rural. Our research object was mainly concentrated in the urban-rural fringe areas including Tangwangtou, Dagangqiao, Chufangdou, Houchengqiao Communities, etc.

48.3.2 Selection and Overview of Social Issues in the Research Area

This paper selected the water supply and police fire protection to research. One hand, it was very important for people's daily life and protecting their life and property safety; on the other hand, compared with other problems such as pension and floating population management, the water supply and police fire protection was easier to be researched and measured, which made the data and information used in the theoretical analysis more real, scientific and reliable.

The water supply in Sandun Town was mainly provided by Xiangfu Waterwork, and administered by Sandun Branch, which both belonged to Hangzhou Water Industry Group Limited, and located in Gongshu District. Xiangfu Waterwork undertook the water supply for the whole north block with the supply capacity of 250,000 tons per day. Sandun Branch administered the area with 77.7 km^2, which reached out to the Beijing-Hangzhou Grand Canal on the east, Yuhang Tong River on the south, Cangnan town in Yuhang District on the west and Xuan-Hang Railway on the north. The fire protection in Sandun Town was in the charge of the nearest Special Force Team 2 Squadron called Xiangfu Squadron, which was responsible for the fire protection of Xiangfu Subdistrict in Gongshu District, Sandun Town in Xihu District and Liangzhu Town in Yuhang District.

48.3.3 Data Collection and Preliminary Analysis

On the water supply in Sandun Town, we had the following discoveries: In recent years, the floating population increased quickly, which led the water supply not enough and the water pressure insufficient in the water peak period at night, such as in Dagangqiao Communities. One day in April 2013, Xiangfu Waterwork's major water pipe to the north of the city burst, and the water pressure insufficient even no water appeared. In those days, this major water pipe was only two meters deep across the Shixiang Road. But along with the development of urbanization, the Shixiang Road broadened, became elevated road and was mat for 4 meters high, which made it become a 6-meter-deep major water pipe that was the deepest in Hangzhou. At the same time, motor vehicles especially heavy duty trucks shot up, and the major water pipe broke down under the crush, although it was only 15–20 years while as a general rule, the service life period of an underground water pipe was about 50 years. About the police fire protection in Sandun Town, we had the following discoveries: In many fire cases recently, it was Xiangfu Squadron that received reports and executed rescue execution in general case. When the fire was larger, there would be the help from Daguan, Jiangcun and Xihu Squadron.

In preliminary analysis, we could draw the following conclusions: First, the water supply and police fire protection in Sandun Town could maintain residents' daily life demand, though located on the urban fringe area. Second, with the development of urbanization, the water supply and police fire protection in Sandun Town had gradually showed the trend that couldn't adapt to the economic and social development demand. Xiangfu Waterwork, Sandun Branch and Xiangfu Squadron were all located in Gongshu District, which caused the geographical differentiation of supply and demand. Third, there were many reasons that the water supply and police fire protection couldn't maintain residents' daily life demand, but the root cause was that Sandun Town especially the north-west vast region lay beyond the Hangzhou Urban Service Boundary, where it was not economic to provide the public infrastructure service in theory.

48.4 Social Management System Reform in Urban Fringe Area Based on the IAD Framework

48.4.1 Analysis on Actors' Action

Actors' Cognitive Contrast

In the action arena, Sandun Town, actors included the local government, enterprises and residents. Through analyzing the cognitive contrast among the three sides, as

Table 48.1 The cognitive contrast among the local government, enterprises and residents

	The local government	Enterprises	Residents
Supply status	Relatively satisfied	Relatively satisfied	Still had potentials to be improved
Importance degree	Important	Very important	Very important
Potential risk	Not big	Some risks but not big enough to affect people's normal life	Some risks and accidental sudden situation may affected their normal life
The first subject of liability	Residents and enterprises	Residents and enterprises	The local government
the evaluation of supply and demand	Basically meet residents' daily demand	Basically meet residents' daily demand	Monopoly profits in enterprises and the local government

shown in Table 48.1, the differences were mainly manifested in some respects such as supply status, potential risk, the first subject of liability, the evaluation of supply and demand, etc.

Analysis on Actors' Action

Actors' action or strategy selection was influenced by the certain action situation enclosed within the action arena. The action situation and the above-mentioned differences led to different actions of the local government, enterprises and residents.

As a manager or coordinator, the local government's action was influenced by public opinion, cost benefit and discount rate. For example, if the public opinion was more inclined to requiring government to solve the water supply and police fire protection problems, the local government would have to make the balance of negotiations tilt to the residents under pressure. And in consideration of the current government performance evaluation mechanism which gave priority to economic development in China, the local government paid less attention to the public goods and services supply whose investment cycle was long and effect was slow.

As providers, the enterprises' action was influenced by cost benefit, internal specification and discount rate. In particular, the ultimate goal of social enterprises was the profit maximization, and the expected cost and return would be important decision-making factors.

As demanders, the residents' action was influenced not only by the severity and cognition of the water supply and police fire protection problems, but also the expected cost and return. For example, residents would pay more attention to the water supply and police fire protection problems where the problems were relatively serious, and would pay more attention to the public infrastructure service in fringe area if they perceived the water supply and police fire protection more deeply.

48.4.2 Analysis on External Variables

Analysis on Conditions and Attributes of Community

The conditions and attributes of community included many factors such as the geographical position, economic and social development situation, population density, residents' cultural level, cognition to the water supply and police fire protection, superior governments' attention, etc. For example, governments and enterprises would pay more attention and investment to the central area because of the low cost and to the relatively developed area with a large population density because of the high return. Furthermore, governments and enterprises also had to pay more attention and investment to the area under pressure where the residents had higher cultural level, more deeply understanding of public services, and stronger consumer consciousness.

This paper selected the population density, the distance from the urban centre and the fiscal revenue to analyze the conditions and attributes of Sandun Town. As shown in Table 48.2, we selected Xiangfu Sub-district as comparative study to make the research more considerable, which was also in the north edge of Hangzhou.

According to the role of indicators to the USB, it would bring the area with a large population density and high economic development level nearer the urban centre into the boundary, while exclude the area with a small population density and low economic development level far from the urban centre outside the boundary. Therefore, Sandun Town didn't have enough comparative advantage to get the superior government's more attention compared with Xiangfu Sub-district. In other words, Sandun Town had no advantage to incorporate into USB.

Table 48.2 The contrast between Sandun town and Xiangfu sub-district

Indicators	Sandun town	Xiangfu sub-district
Area	37.93 square kilometers	18 square kilometers
Population size	Permanent 63,000, and external 50,000	Permanent 227,000, and external 80,000
Population density	2979 per square kilometer	5706 per square kilometer
Distance from the urban centre	5.5 km	6 km
Fiscal revenue	283 million yuan	477 million yuan
Per capita fiscal revenue	2504.42 yuan/person	4644.60 yuan/person

Analysis on System and Rules

The public goods and services supply of urban fringe area was affected by the institutional environment. Good system and rules could not only ensure and promote supplies, but also reduce social running cost.

The government departments associated with the water supply and police fire protection in Sandun Town included the General Management Office, the Urban Management Office and the Social Affairs Management Office. But there were no clear provisions about the water supply and police fire protection in the function descriptions. Visibly, Sandun Town didn't have special department to manage the local water supply and police fire protection, which might cause the situation of multiple or no management.

Look from the rules and regulations, Hangzhou government had issues relevant rules and regulations aimed at the water supply and police fire protection, such as Hangzhou Urban Water Supply Management Regulations, and the Implementation Opinion on Further Implement the Responsibility System for Fire Control Safety. However, these rules and regulations were trying to detailed, but they still needed further specific and practicable to implement. Moreover, the investigation found that most residents wasn't clear the rules and regulations about the water supply and police fire protection, although they all believed the government should have some about them.

48.4.3 Findings

Problem Description

According to the field investigation and theoretical analysis, about the water supply and police fire protection in Sandun Town, this paper had the following findings:

First, according to actors' action analysis, there were some cognitive differences which brought the demand to communicate. And for the cognitive differences about the first subject of liability, the communication would cost larger and be ineffective. In addition, the relatively single mode without various sources and competition was likely to lead to the supply lack or ineffective.

Second, according to the analysis on the conditions and attributes of community, the demand and conditions of the water supply and police fire protection would be stronger with the better-playing social service function, if the distance from the urban center was shorter, the population density was larger, and the revenue or money was more. Sandun Town had no advantage to incorporate into USB. However, with the groundbreaking of the North Hangzhou New Residential Zone, Sandun Town's regional advantages would be more obvious, and it would be brought into the USB with economic social development and population increase.

Third, according to the analysis on system and rules, Sandun Town didn't have special department to manage the local water supply and police fire protection, which might cause the situation of multiple or no management. On the other hand, most residents wasn't clear the system and rules, not only because there was no supporting measures for the system and rules put into practice, but also for Hangzhou government never brought Sandun Town into its USB in thought and action, and then the system and rules became non-binding in Sandun Town.

To sum up: Because of excluded outside Hangzhou USB, the public infrastructure service in Sandun Town was incomplete compared with the main urban zone, which further affected the floating population to Sandun Town and the development of public transportation. Then the disadvantage of the distance from urban centre and the limit of economic social development were unable to change, the system and rules became non-binding and couldn't play their roles in Sandun Town. The non-mature conditions and non-perfect rules caused the social problems, through affecting actors' action in the situation enclosed within the action arena. One word, why did the water supply and police fire protection problems in Sandun Town exist? In the final analysis, it was because Sandun Town was excluded outside the Hangzhou USB and couldn't get the same treatment as urban.

Policy Suggestions

In order to solve the water supply and public fire protection problems in Sandun Town, this paper put forward the following policy recommendations:

First, perfect the communication mechanism, broaden the supply channels and change the role of actors. Cognitive differences affected supplies, so establishing a perfect communication mechanism would make for clarifying the object of liability, and realize the standardization of the public goods and services supply in urban fringe area. The three sides should all open their mind and increase the supply channels of public goods and services. The local government shifted from a manager to a server, enterprises and residents built and held on the customer awareness, only then could the supply efficiency of the public goods and services increase.

Second, strengthen renovating urban fringe area and improve the conditions and attributes of community. With the groundbreaking of the North Hangzhou New Residential Zone as an opportunity, vigorously develop social economy and ensure the water supply and police fire protection by plenty of money and fiscal income. Gain superior government's attention through attracting the floating people and the population increase. Develop the public infrastructure service especially public transport to shorten the time distance from urban center. Only in those ways, could the action relationship among the local government, enterprises and residents be affected, and then the water supply and public fire protection would be optimized.

Third, optimize the design of system rules, and make them coordinate and promote each other. According to the system in the collective level such as Hangzhou Urban Water Supply Management Regulations, and the Implementation Opinion on Further Implement the Responsibility System for Fire Control Safety, design the supporting measures in the operation level for the system and rules put into practice. At the same time, Hangzhou government should bring Sandun Town into its USB in thought and action, and then the system and rules would become influential and binding.

All in all, the most important measure to solve the water supply and police fire protection problems in urban fringe area was the system and rules and reversing the idea. In other words, bring Sandun Town into Hangzhou city development strategy plan as an organic part, and place it in Hangzhou USB.

48.5 Conclusions

Through the research of the water supply and police fire protection in Sandun Town, this paper drew the following conclusions:

First, the actors' interactive action in the action arena was the direct cause for the social problems in urban fringe area such as public goods and services supply arising and solved. Therefore, it had great significance for solving the social problems to establish a perfect communication mechanism, strengthen the communication between the local government and residents, broaden the supply channels, change the role of local government, and set up resident consumer awareness.

Second, the conditions and attributes of community, system and rules which could affect the action decision made by the actors in the action arena were the restraining factors for the social problems in urban fringe area such as public goods and services supply arising and solved. Therefore, it would make much difference in solving the social problems to strengthen renovating urban fringe area, improve the conditions and attributes of community, optimize the design of system rules, and make them coordinate and promote each other.

Third, the USB influenced the different actors' action in the action arena, through influencing the conditions and attributes of community, and it could be seen as the root cause for the social problems in urban fringe area such as public goods and services supply arising and solved. Therefore, placing it in Hangzhou USB was the most important measure to solve the water supply and police fire protection problems in urban fringe area.

In short, what made the social problems in urban fringe area such as public goods and services supply arise and be solved? The direct cause was the actors' interactive action in the action situation enclosed within an action arena; the conditions and attributes of community, system and rules were restraining factors; and the root cause was the USB which could interact with conditions and attributes of community, system and rules.

Acknowledgements The authors appreciate the financial support by the National Natural Science Foundation of China (Grant No. 71373231), and the helpful work from the organizing committee of CRIOCM2015 International Symposium on Advancement of Construction Management and Real Estate.

References

1. McGinnis MD (2011) An introduction to IAD and the language of the Ostrom workshop: a simple guide to a complex framework. Policy Stud J 39(1):169–183
2. Pendall R, Martin J, Fulton WB (2002) Holding the line: urban containment in the United States. Center on Urban and Metropolitan Policy, the Brookings Institution
3. Miao ZZLY (2010) The patterns of urban containment and its impacts to urban development. Mod Urban Res 1:015
4. Zhu D, Liu D (2005) From urban management to urban service: thinking on human-oriented mode of urban administration. In: Urban planning forum (vol 6, p 009)

Chapter 49
Analysis of Health Factors Influencing Construction Workers' Operating Efficiency Based on Structural Equation Model

Miaomiao Wang, Jingfeng Yuan and Qiming Li

Abstract With the rapid development of construction industry, the requirements for architectures and construction workers have been improved, thus enterprises pay more attention to workers' production efficiency. Evidence has suggested that poor health status may lead to decreased productivity. Structural equation model is introduced to find out the main factors influencing efficiency by exploring how the factors of health status and social network influence work performance. The model is built by determining latent variables and the interaction relationship. Observable variables are found through literature and a questionnaire is made to get data. Finally, run the model with Amos and conclude that the major factors affecting their efficiency are self-protection consciousness, the frequency of physical examination, leisure time and anxious feelings. Moreover, the built model can be in common use for general construction projects while the construction companies run the model with their data and then they can set out to improve productivity with the latest results.

Keywords Structural equation model · Physical and psychological health · Social network · Operating efficiency

49.1 Introduction

Recently, real estate industry develops rapidly, leading construction industry as its foundation to get fast development. Meanwhile, construction workers' health and safety continue to be a huge concern for the construction industry. Previous

M. Wang (✉)
Department of Construction and Real Estate, Southeast University, Nanjing, China
e-mail: 505869564@qq.com

J. Yuan · Q. Li
School of Civil Engineering, Department of Construction and Real Estate, Southeast University, Nanjing, China

Y. Wu et al. (eds.), *Proceedings of the 20th International Symposium on Advancement of Construction Management and Real Estate*,
DOI 10.1007/978-981-10-0855-9_49

researchers have focused on the reasons of health and safety problems and corresponding solutions are undertaken to handle possible conditions.

Time, quality and cost are three main control points in project management. It not only depends on management by engineers, but also relates to worker performance. If they concentrate on their work, time and quality of projects would be guaranteed and in consequence cost of projects would be under control. So enterprises should pay more attention to their efficiency.

In order to control workers' operating efficiency conveniently, this paper introduces structural equation model to explore how the factors of health status influence work performance.

49.2 Literature Review

Over decades, researches related to workers' health and safety are various and cover different aspects of the subject. However, most of the reports are focused on a certain aspect of health and safety problems and usually organized by professional medical organizations or related government department to concentrate on a specific disease or mental health. Few of them research problems related to construction workers from a general perspective. Additionally, comparatively fewer researches concerned workers' performance.

Laura investigated the effects of a worksite health promotion intervention on work ability and work performance, and the results showed that the intervention was not successful in improving work-related measures [1]. Wang studied psychological health of workers and proved it connected closely with a positive social network [2]. Thus, this paper adopts generalized concept of health and tries to find out the main factors influencing work performance while keeping workers as usual.

49.3 Research Methodology

Commonly, questionnaire survey is usually used to find out the main factors that affects health status of workers. The result deviation is difficult to be measured because of all kinds of uncertainties during the research process. Questionnaire survey can only give a brief overview of factors and it is unable to reflect internal relationships. Thus this paper introduces structural equation model to explore concrete degree of influence that factors of health status have on work efficiency.

Structural equation model analyzes the relationship between variables based on covariance matrix, including the measurement model and the equation model. The measurement model describes the relationship between latent variables and observed indexes while the structural model describes the relationship between the latent variables.

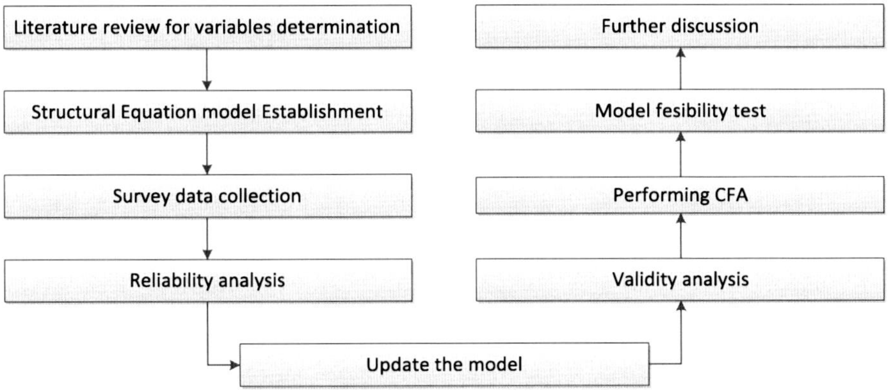

Fig. 49.1 The methodology adopted in this study

Structural equation model can process multiple dependent variables at the same time and allow measuring error exist in both dependent variables and independent variables [3]. Statistical analyses were performed, using the SPSS 17.0 software by multiple methods including Cronbach's alphas and so on. A confirmatory factor analysis (CFA) was performed using AMOS 7.0 software to test the correlation of the theoretical model with the survey data and to clarify the relationships among the variables. The organization of methodology adopted is shown in Fig. 49.1.

49.4 Modeling

49.4.1 Determination of Latent Variables

The most authoritative definition of occupational health is defined by the international labor organization and the world health organization joint professional committee in 1950. It says that occupational health should aim to promote and keep the physiological status, psychological status and social status of workers in different industries at its best [2]. It also claims that occupational health should prevent workers from being damaged by hazard factors in work environment and make work environment suitable for their physiology and mentality.

In this case, physical health, psychological health and social relation network are defined as latent variables. Limitation of recognition due to lack of observable variables could be avoided only through related estimate (namely add a new observable variable), so new latent variable called work efficiency is introduced to quantitatively describe influence on workers' performance. The concrete relationships between the latent variables are shown as below (Table 49.1).

Table 49.1 Designed path structure and the basic path hypothesis

Designed path structure	Basic path hypothesis
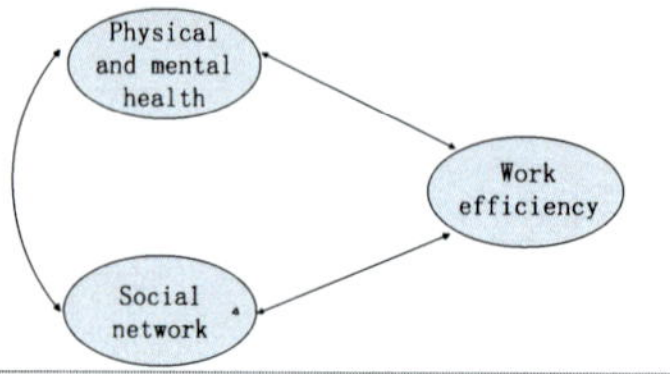	1. The physical and mental health has influence on work efficiency 2. Social network has influence on work efficiency 3. The physical and mental health and social network have influence on each other

49.4.2 *Determination of Observed Indexes*

Variables of physical health are selected from six previous researches about physical and mental health of construction workers [2, 4–8]. Statistics indicate that the first six factors affecting physical health are self-protection consciousness, labor intensity (namely daily working hours), rest time, frequency of physical examination, sleep time, and working years.

The definition of mental health divides individual psychological problems into three categories: developing psychological problems, adaptive psychological problems and obstacle psychological problems. Variables of psychological health could arise from symptoms in SCL-90. However, the respondents are all construction workers and do not have enough time and knowledge to distinguish the factors. Then the occurrence frequency of anxious and upset mood is set as the observed variable. Every latent variable in structural equation model needs at least three observable variables. Only one observed variable could not meet the requirement, so the two latent variables physical health and mental health are merged into physical and mental health.

Social network includes all the social interaction. In terms of construction workers, their social network mainly includes interaction with government, companies and individuals, such as regular check by government department, protection appliance provided by companies, specific occupational health lectures and themselves social status.

Table 49.2 Corresponding model variables table

Latent variables	Physical and mental health	Social network	Work efficiency
Observed variables	Self-protection consciousness (a1) Daily working hours (a2) Rest time (a3) Frequency of physical examination (a4) Sleeping time (a5) Working years (a6) Occurrence frequency of anxious and upset mood (a7)	Frequency of check by government (a8) Social status (a9) Protection appliance provided by companies (a10) Specific occupational health training (a11)	Quality of finished work (a12) Work progress (a13) Work attitude (a14)

Time, quality and cost are three key control points in project management. Observed variables corresponding to work efficiency are work progress, the quality of finished work and work attitude. While latent variables and observed variables have been confirmed, all model variables are shown in Table 49.2.

49.5 Survey on Factors

49.5.1 *Questionnaire Survey*

As the principle indicates that a problem cannot correspond to multiple observed variables while multiple problems can correspond to the same observed variable, problems were set up one by one for each observed variable. Investigation was conducted on site with random. Gender, native place, type of work, age and other variables are set to avoid homogeneity of samples. All the problems use Likert 5 measurement for the convenience of statistics analysis and factor analysis.

Taking Gorsuch's opinion for reference [9], the best proportion of items and interviewees is 1:5 and the total number of samples shall not be less than 100.

In this survey, 150 questionnaires were sent out and 133 valid samples were taken back. The missing value was coped with column deletion method, that is, once there is a missing value in a record then deletes the record. Finally, 116 records were left and the effective rate was 77.3 %.

49.5.2 *Consistency of Survey Data*

A questionnaire involves multiple factors and each factor corresponds to multiple data of investigation, whether each data describes the corresponding factor objectively needs further test. So consistency analysis of data is necessary and it usually includes reliability test and validity test.

Reliability represents the consistency and stability of scale. Consistency mainly represents internal relationship between problems and inspects whether each problem test the same content. Stability refers to consistency of repeat measurement results of the same subject in different time. Test the reliability of existing 14 observed variables and the reliability coefficient is 0.910. The corresponding project total statistical scale is shown in Table 49.3.

Table 49.3 Project total statistical scale

Factors	a1	a2	a3	a4	a5	a6	a7
Cronbach's alpha if item deleted	0.894	0.908	0.902	0.895	0.909	0.919	0.905
Factors	a8	a9	a10	a11	a12	a13	a14
Cronbach's alpha if item deleted	0.898	0.908	0.899	0.897	0.901	0.902	0.896

Reliability reflects consistency and stability of results through Cronbach alpha which ranges from 0 to 1. More closely the value approaches 1, higher the internal consistency of data is.

As Table 49.3 shows, the Cronbach's alpha if item deleted of working years is more than 0.910. It indicates that the reliability coefficient will increase if the item is deleted, that is, the item has a low correlation with other factors. While the factor working years is deleted, the reliability coefficient of left 13 variables is 0.919. The obvious increase in reliability coefficient shows the removal of working years is rational.

Validity test mainly rely on Kaiser-Meyer-Olkin inspection and Barlett test of sphericity of valid data. It is generally believed factor analysis can be carried out when the value of KMO is more than 0.5 and the Barlett coefficient is below 0.05. In this survey, the value of KMO is 0.854 and the Barlett coefficient is 0.000. It shows the coefficient matrix is not an identity matrix and the least factors extracted can explain most of the variance. Overall validity of the questionnaire is high and confirmatory analysis of factors can be continued.

49.6 Operation and Analysis of Model

49.6.1 Confirmatory Analysis of Factors

Draw the model shown in Fig. 49.2 in AMOS and import questionnaire data from SPSS, and the path coefficients have shown in Table 49.4.

Before model evaluation, it is necessary to figure out whether all the parameters estimated from the model have statistical significance. It is evaluated by critical

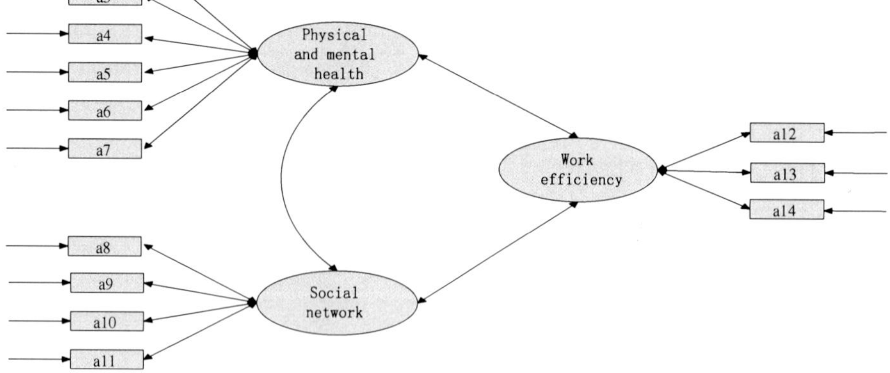

Fig. 49.2 Built structural equation model

Table 49.4 Coefficient estimation of parameters

			Estimate	S.E	C.R.	P	Standardized estimate
Work efficiency	←	Physical and mental health	0.310	0.079	3.922	***	0.574
Work efficiency	←	Social network	0.300	0.122	2.469	0.014	0.364.
a1	←	Physical and mental health	0.863	0.065	13.349	***	0.883
a2	←	Physical and mental health	0.339	0.058	5.885	***	0.500
a3	←	Physical and mental health	0.772	0.080	9.603	***	0.727
a4	←	Physical and mental health	1.000				0.800
a5	←	Physical and mental health	0.268	0.081	3.305	***	0.301
a7	←	Physical and mental health	0.410	0.065	6.285	***	0.532
a8	←	Social network	1.087	0.115	9.453	***	0.759
a9	←	Social network	0.189	0.090	2.091	0.037	0.191
a10	←	Social network	0.837	0.084	9.931	***	0.782
a11	←	Social network	1.000				0.785
a12	←	Work efficiency	1.000				0.803
a13	←	Work efficiency	0.988	0.116	8.549	***	0.722
a14	←	Work efficiency	1.213	0.124	9.747	***	0.836

*** $P < 0.001$

ratio (CR) value in AMOS, equivalent to the value of t test. While the absolute value of CR is more than 2.58, the parameters are estimated at 0.01 significant level.

As shown in Table 49.4, physical and mental health has more effects on work efficiency rather than social networks for it is not significant at 0.01 significant level. In this research, all observed variables corresponding to physical and mental health have obvious influence on it. Particularly, a1, a4 and a3 have a great influence on physical and mental health because their coefficient all reach 0.7, followed by a7 and a2 for their coefficient all reach 0.5. Therefore, self-protection consciousness, frequency of physical examination and rest time are main factors affecting worker health, meanwhile daily working hours and their psychological status cannot be neglected.

Among all the factors corresponding to social network, influence coefficient of social status is only 0.191. It indicates that workers have few obstacles during social interaction. Maybe it is due to selected large scale projects which contains different organizations and is similar to a small society. Thus it is not likely to become social isolation or less communication. The coefficients of other three factors, a8, a10 and a11 are all close to 0.8, a relatively high value.

Table 49.5 Indirect coefficient estimate to work efficiency

			Indirect estimate
Work efficiency	<—	Physical and mental health	0.574
Work efficiency	<—	a1	0.507
Work efficiency	<—	a2	0.287
Work efficiency	<—	a3	0.417
Work efficiency	<—	a4	0.459
Work efficiency	<—	a5	0.173
Work efficiency	<—	a7	0.305
Work efficiency	<—	a8	0.276
Work efficiency	<—	a10	0.285
Work efficiency	<—	a11	0.286

The coefficients of a12, a13 and a14 are up to 0.7. It indicates that it is feasible to test work efficiency with these three factors.

From the perspective of enterprises, probably the key point that they pay attention to is how the factors finally impact work efficiency. The degree of influence on work efficiency needs to be calculated and the formula is as the following: indirect influence of factors a1 to a11 on work efficiency = (the path coefficient of factors a1–a11 on physical and mental health or social status) *(the path coefficient of physical and mental health or social status to work efficiency)

All factors are calculated one by one except those in Table 49.4 whose critical ratio value did not reach 2.58. The calculation results are shown in Table 49.5.

As is shown in Table 49.5, physical and mental health has considerable influence on work efficiency. The four factors a1, a3, a4 and a7 may be major reasons for it.

49.6.2 *Performing CFA to Analyze Data*

Model fitting indexes are statistical indicators which evaluate the fitting degree between structural model and data. Different model fitting indexes can evaluate models in multiple aspects, including the complexity of model, sample size, relativity and absoluteness and so on.

Using CFA, the adequacy of a model is evaluated using a number of goodness-of-fit indices(GIFs), which reflect the correlation between the hypothesized statistical model and the actual data set [10]. Common GIFs include the chi-square (χ^2) statistic, comparative fit index (CFI), and the root mean square error of approximation (RMSEA) [11]. A summary of the recommended benchmarks for GFI is presented in Table 49.6. All measured value are mainly acceptable and meet the requirements. So the model is feasible and all above analysis proves to be effective.

Table 49.6 Index and Recommended Values

GFI	Measured value	Recommended value
χ^2/df	4.8	2.0–5.0
CFI	0.896	>0.9
RMSEA	0.077	<0.1

49.7 Conclusions

Import data from questionnaire survey to the established structural equation model and get actual degree of impact that the factors affecting worker health have on their work efficiency through quantitative analysis. The results show that the main factors affecting worker health are self-protection consciousness, frequency of physical examination and rest time, followed by anxiety and daily working hours. Moreover, self-protection consciousness, frequency of physical examination, rest time and anxious mood play a major role in affecting work efficiency.

Thus construction enterprises should specially pay attention to these four aspects and care more about workers. Enterprises should organize regular physical examination and arrange their daily working hours and rest time reasonably. Additionally, training to enhance their self-protection consciousness and appropriate leisure activities should be arranged. With small invention to prevention, ensure quality of work to guarantee the success of the whole project.

Moreover, the built model can be in common use for general construction projects while the construction companies run the model with their data and then they can set out to improve productivity with the latest results.

Acknowledgments The authors' special thanks go to all and reviewers of the paper, and to the National Natural Science Foundation of China (NSFC-71001027, 71472037), Social Foundation of Jiangsu Province, China (13GLB005), Program for Outstanding Young Teachers of Southeast University (2242015R30009), and the Fundamental Research Funds for the Central Universities for financially supporting this research.

References

1. Viester L, Verhagen EALM, Bongers PM, van der Beek AJ (2015) The effect of a health promotion intervention for construction workers on work-related outcomes: results from a randomized controlled trial. Int Archives Occup Environ Health 88(6):789–798
2. Wang JH (2011) Psychological health of construction workers and related social supporting network. Soc Psychol Sci 26:152–155
3. Hou JT, Wen ZL, Cheng ZJ (2004) Structural equation model and its application. Science Education Press, Hong Kong
4. Ma T, Qian ZQ, Liang JZ (2008) Investigation of occupational health knowledge, attitude and behavior of construction workers. Career Health 24(21):2260–2261
5. Ma MY Survey of construction workers integrated situation in Shenzhen. J Shenyang Architectural Univ 12(3):281–285

6. Wang D, Wang YY (2011) Survey and thinking of labor and living conditions of construction workers. Ind Discuss 11(5):18–20
7. Yang XM, Liu SR, Guo X (2012) Survey and analysis of basic rights and interests safeguard situation of construction workers. Sci Technol Inf 3:222–223
8. Guo YS (1996) Present situation and methods of health education of urban construction workers. Primary Health Care in China, pp 47–48
9. Wu ML (2010) Questionnaire analysis—operation and application of SPSS. Press of Chong Qing University, Chong Qing
10. Sanders RD, Allen DN, Forman D, Tarpey T, Keshavan MS, Goldstein G (2005) Confirmatory factor analysis of the neurological evaluation scale in unmedicated schizophrenia. Psychiatry Res 133(1):65–71
11. Yuan JF, Wang C, Miroslaw JS, Li QM (2012) Developing key performance indicators for public-private partnership projects: questionnaire survey and analysis. J Manage Eng 28:252–264

Chapter 50
An Evaluation of Project Management System of Public Construction Sector in Shenzhen, China

Chengke Wu, Xiao Li, Shirong Li and Bo Xu

Abstract As one of the most developed cities in China, Shenzhen has experienced a rapid urban construction in the last 30 years and features a highly modernized city now. Since 2002, to facilitate constructions of larger and more complex public projects, Shenzhen government founded the Public Works Department (PWD) to centralize management of all public construction projects except transportation and water sector. After operation, the PWD of Shenzhen has significantly reduced projects durations and corruptions, increased quality, gain of investment and safety in the last decade. However, there are problems existing in the system which mainly include the lack of communication among authorities and between authorities and PWD, poor project decision-making methods, inadequate and obsolete construction management and lack of implementation of innovations in AEC sector. These gaps need to be bridged to improve project management of public projects further. This paper however, focus on regulatory reforms to facilitate permitting processes, and discuss possible methods of reform taking references from international literatures.

Keywords Public project construction management · Regulatory reforms · Public works department

50.1 Introduction

Public construction project includes all fixed assets which are financed by government entities and constructed within the administrative region of Shenzhen government. Main public projects in Shenzhen are constructions of non-profit projects which focus on protecting environments, enhancing coordination and simultaneously development, improving industrialization of new techniques and

C. Wu (✉) · X. Li · S. Li · B. Xu
Faculty of Construction Management and Real Estate,
Chongqing University, Chongqing, China
e-mail: 404835780@qq.com

Y. Wu et al. (eds.), *Proceedings of the 20th International Symposium on Advancement of Construction Management and Real Estate*,
DOI 10.1007/978-981-10-0855-9_50

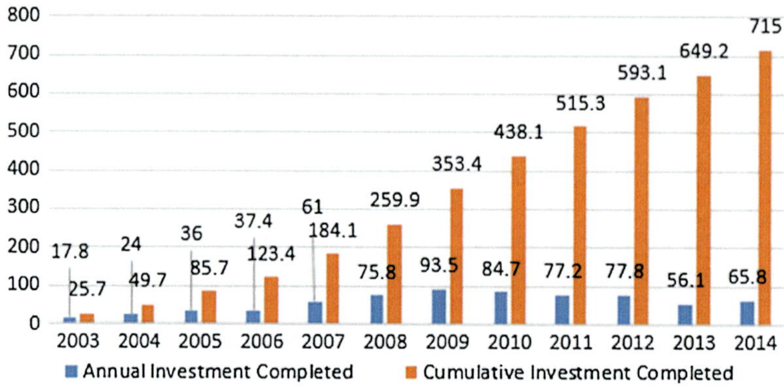

Fig. 50.1 Increase of completed investments in last decade

innovations. To address problems in traditional construction delivery system and adapt new situations, Shenzhen founded the Public Works Department (PWD) to implement centralized management for public constructions except for transportation and water projects. PWD has professional stuff and are responsible for management from feasibility analysis until handover. After 11 years' development, PWD has completed 75 billions' investment. See Fig. 50.1.

More importantly, the professional skills and centralized management system reduce schedule and cost while ensure quality. Now, many other cities began to learn from Shenzhen and to set up the same mode. However, problems exists. For some of them, solutions are under development and progresses can be observed, and for the rest, innovations and references are necessary. In the following sections, these problems of 2 categories will be discussed, then possible solutions will be revealed, taking references of good practices of construction regulation reforms from countries like Canada, New Zealand.

50.2 Problems of Institution Structure

The basic idea of institution structure of construction management system in Shenzhen is that PWD is at municipal level and is responsible for projects financed by municipal government while sub-district authorities also have their own construction offices managing projects within particular sub-districts invested by local political subdivisions. However, their duties are the same—act as the "owner" of projects, manage feasibility analysis, conceptual design, bidding process, construction, and projects completion and then transfer them to end users. Problems In these complicate relations and hierarchies, are listed below.

50.2.1 Vague Definition of Liabilities of PWD

As mentioned above, currently PWD is the technically "owner". However, it's the end users who prepare initial projects proposals and decide all functions of projects. In other words, the department has no corresponding authorities as an owner. Especially in the early decision-making stage, PWD has much less voice due to that all functions or requirements are set by end users.

50.2.2 Problems of the Role of Projects End Users

End users play critical roles in project life cycle. First, they provide demands of functions of projects. Second, while PWD is accountable for obtaining all approvals from multiple authorities, end user will offer help during the process. Third, end users operate projects after handover and need to manage facilities. There are lots of interactions between PWD and end users which significantly affect performances of projects. Problems in this field include: First, each project is independent according to its individual end users, which reduce the effect of economy scale; Second, given some end users are not completely formed during construction of projects, it's impossible to provide clear demands or requirements, leading to frequent changes increasing costs and durations; Third, liabilities of maintenance and defects are hard to define for lack of clear procedures and regulations.

50.2.3 Poor Communication Between PWD and Sub-District Construction Departments

Because central PWD has no direct authorities over smaller construction departments, local departments function independently without communication channels between the 2 levels, which results in several negative effects. First, Experiences gained can't be shared; Second, projects may face different political requirements and standards which leads to confusions; Third, isolated project management can't take advantage of economy of scale.

50.2.4 Poor Co-Ordination Among Authorities

Multiple authorities are involved in the process of getting permits for public projects. Main authorities include development and Reform Commission (DRC), Financial Department, Urban Planning Department, Ministry of Construction and so on. The main problems here is multiple management. Due to each authority has its own purposes and liabilities, the aims of their supervisions, the ways they inspect projects and the degree they investigate all vary. As a results, inspection outcomes may contradict to each other, causing serious delay of construction and cumbersome, repetitive procedures. Furthermore, the lack of an overall co-ordination organization which cover all authorities means PWD has no superior department to resort to.

50.2.5 Co-Ordination Problems Between PWD and Authorities

It can be concluded that the structure of public project construction management system is a matrix, with PWD lying on the horizontal axis and authorities lying on vertical axis. Given that these authorities are directly authorized by Chinese central government, while PWD is local governmental agency in Shenzhen, the matrix here is a weak one, which would hamper construction process:

(a) No flexibility in permitting process: Current permitting process implement require the same standards and procedures for almost all public projects irrespective of their status, risks, complexity or functions. If different permitting procedures can be implemented according to types of projects, it will no doubt significantly increase effectiveness by saving unnecessary time.
(b) Permitting deviates from market: One feature of authorities is that they are not specialized in AEC sector and not as familiar with construction process, especially innovations as PWD do. The standards which authorities adopt usually seriously lags and deviates from market.
(c) No updating of ideas: PWD has better understanding of market and more professional stuff, so can provide more reasonable targets, budget estimates or investment plans. However, the thoughts and ideas of authorities are not updated according to the new system. They are still prone to cut budget, set additional requirements, making targets hard to be achieved.

50.2.6 Problems in Supervision Mechanism

All public projects are under supervisions from multi-disciplines like Bureau of Housing and Urban and Municipal Audit Bureau. Some deficiencies of them can hinder construction of public projects:

(a) Supervision power are highly separated: Highly separated supervision and lack of an overall supervision system can reduce effectiveness and contradict the original ideas of centralized management. PWD has to spare limited resources to deal with frequently and randomly inspections.

(b) Inadequate transparency and communication: In various permitting and supervision processes, authorities often only provide outcomes, no reasons or explanations are informed. Given this, PWD can't fix problems without knowing reasons behind.
(c) Unequal rights and responsibilities: Supervision authorities are not accountable for any goals of projects, lack of construction skills and knowledge, but they have superior powers. In many cases, supervision authorities issue unauthorized or inappropriate commands but do not take risks.

50.3 Problems of Management Mechanism

The following section will go through public project process from initial decision-making stage to finale completion acceptance to reveal these problems.

50.3.1 Problems in Decision-Making Stage

End users, PWD and Development and Reform Commission (DRC) are 3 main players. Effectiveness and reasonableness of divisions of liabilities and co-ordination among the 3 parties are essential. End users would provide project proposal, then get the approvals from DRC, after which PWD is involved. Liabilities of PWD in this stage include compiling feasibility analysis and obtaining approvals. Problems include:

(a) Unclear demands of functions: As has been discussed, end users hardly provide clear descriptions of project functions. On the other hand, PWD can reduce potential changes but it has few voice in decision-making stage.
(b) Poor Communication with DRC: DRC is prone to cut project scale and investment volume because standards that DRC uses are macroscopic or average thus doesn't reflect perspective demands of a modernized city. Without communication, it's not possible for PWD to help DRC fix these deficiencies and PWD has to take risks to meet strict or even unreasonable requirements.
(c) Inadequate scientific decisions: Project investments is significantly affected by construction technical plans. Without specialized knowledge of PWD, the validity of decisions is not ensured. Competence of PWD is not recognized and utilized by authorities. Instead, authorities intervene too much in detailed technical plans which violate their macro-management roles.
(d) Lack of social participation: To gain the maximum benefit and to increase fairness and transparency, social participation, including both experts and citizens are indispensable. On the one hand, given that current public projects involve more techniques, it's necessary to bring in third parties which is

specialized in these fields to conduct comprehensive investigation. On the other hand, public participation in government invested projects are statutory but now there is no appropriate arrangements.

(e) Lack of life cycle versions: The aims of public projects should be the maximum benefits of whole project life cycle instead of concentrating on one-off investment. Currently, authorities emphasize on saving investment of construction stage and lack of life cycle version taking long-term functions, operation and savings into consideration. In consequence, techniques like saving energies, increasing qualities are not adopted due to cut investments, then long-term costs increase.

(f) Permitting decisions contradicts centralized management: Traditional permitting based on individual project not only hinder realization of economy of scale, but also the optimization of decision-making. With centralized management, PWD can prioritize public projects to gain maximum benefits. However, if decision-making process is still dependent on individual projects, relations among these projects would be competitive rather than cooperative to achieve their own profits or aims, making collaborative construction contributing to whole public not possible.

50.3.2 Problems in Design, Bidding and Construction Stage

Generally speaking, PWD can complete construction in a relatively transparent, effective way and meet targets of quality and schedule. However, deficiencies still exist:

(a) Construction methods contradict centralized management: Centralized management, is the key to increase entire benefits of public projects. However, the idea is not combined into construction management system, leading to several consequences: (i) lack of strategic consulting that provides suggestions to project groups; (ii) temporary facilities that are used in one project can't be utilized again in another project; (iii) batch bidding or strategic partnership can't be applied.

(b) Consulting services not fully utilized: Despite that more consulting companies emerge recent years, they are not able to play an efficient role. Main problems here are: (i) Consulting services are highly fragmented. Feasibility consulting, engineering consulting, cost consulting, extra, are all separated and PWD has to co-ordinate with all of them. There is no individual or team that can overall plan, and be links between PWD and consulting teams. (ii) Design is not in line with construction. Design system is not suitable for current market situation. First, no design parties are involved in feasibility analysis, causing large deviations of feasibility from reality; second, primary design, whose main purpose is to obtain estimated budget, is based on fixed ratio price which can't reflect Shenzhen market; Third, Design parties who are most familiar with detailed drawings are excluded from bidding and have very limited effects afterwards.

(c) Poor adoption of value engineering and easy construction: Value engineering should have been utilized in decision-making actions through entire project life cycle, but its potential has not been recognized yet. Easy construction, on the other hand, can encourage design parties take more construction factors into account and combine these separated sector together. Unfortunately, there is no such mechanism in AEC sector of Shenzhen

(d) Problems during bidding stage: Recent years, 2 main approaches of evaluating biddings are adopted in Shenzhen, "evaluated-lowest-quotation-as-bid-winner" and "comprehensive evaluation method". Drawbacks are found in both of them. The former is a relatively impartial approach which select bidders with the lowest price based on all other requirements are met. However, due to inconsistences between design and construction and ambiguous, unscientific designs, there can be unreasonable low bids since bidders have more opportunities to claim afterwards. The latter is to select the best bidder based on a comprehensive evaluation of technical abilities, business competence and reputations. It turned out to be impractical because it provide rooms for corruption and illegal actions like surround-bidding. Recently, Shenzhen government adopted another method of bidding by separating evaluation process and determination where bid evaluation committee only complete evaluations and let owners to decide the winner.

50.3.3 Problems During Facility Management

After completion and handover, projects are out of control of PWD and end users are responsible for facility management. Due to that management of PWD doesn't cover operation stage, it's easy to cause disputes if any construction deficiencies or quality problems are left. In addition, inadequate necessary knowledge of facility management of end users may result in insufficient utilization of project functions and inappropriate maintenance, increasing life-cycle costs.

Current alternative that end users adopt is to outsource facility management to relating companies. However, the market of facility management companies is not mature and under a disorder competition. Complex public projects entail highly skilled stuff and techniques which many companies don't have. The solution is also simple—allocate more resources in terms of fiancé and trained persons to PWD and make PWD responsible for life-cycle facility management.

50.3.4 Obstruction to Promote Innovation

For the 2 major revolutions in AEC sector, namely, green building and BIM, government should take responsibility to promote them through typical public projects. However, there are some challenges that Shenzhen government has to face: First of all, lack of incentives, including both financial encouragement and relating guidelines; Second, lack of co-ordination, few private third parties conduct specialized evaluations or supervisions; Finally, lack of platform for information exchange, no co-ordination platforms supporting innovations like BIM, cloud techniques.

50.4 Address Existing Problems and Improve Performance of PWD

To improve construction management of PWD, possible approaches will be discussed in this section with international literatures, including regulatory reforms and developing BIM-based E-tendering.

50.4.1 Simplify & Improve Permitting Procedures

By some estimates, 60–80 % of building projects in developing economics are undertaken without proper permits and approvals [1]. Studies found that formalities before construction are the most time-consuming and costly. Countries, both developed and developing, are implementing various methods of regulation reform to improve permitting systems. More or less, PWD can learn. Major 3 stages can be summarized: (a) A broad discussion stage which helps central institution formulate sound policy objectives. (b) A legal foundation stage focusing on the rationalization, grouping, and modernization of building codes. (c) An implementation phase, with a carefully planned sequencing of the particular reforms and best practices. Following contents will elaborate approaches further:

(a) Central government initiates reforms: Actions of Shenzhen central government in the beginning stage include publishing objectives of reform to entire society; and organizing forums or workshops with all stakeholders to develop diagnostics of their respective regulatory systems. More importantly, a Central Reform Committee should be established to which PWD could resort when it faces impediments from authorities. For one thing it develops an overall co-ordination party, enhancing communications, for another, it protects interests of both developers and public, by preventing schedule and quality from inappropriate intervention of authorities.

(b) Introduce building codes into construction law: Building codes, which cover extensive range like green building performance, energy saving, structural efficiency, public health, safety and even inspection or approval requirements establish common points of reference between regulator and industry practitioners [2]. Furthermore, building codes generally outweigh traditional fixed standards for they can be modified and updated in constant frequency reflecting market demands and political requirements correctly. Especially given that current public projects become more complex and multi-disciplinary, introducing building codes into law is essential to enhance equality and transparency and reduce disputes. In addition, building codes should be objective- or performance- based to avoid inhibiting innovations since they only set targets but don't determine the way in which to comply codes.
(c) Enhance co-ordination between departments: To improve communication between PWD and sub-district departments, central government should establish specialized department to create and maintain the channels which would better be part of or subordinate to the Central Reform Committee. For instance, in New Zealand, Department of Housing and Building established a particular organization acting as joint regulatory role between central agency and local smaller departments. After implementation, PWD, in a higher level, will constantly has feedbacks and information from all sub-district departments
(d) Measure service qualities of authorities: CRC is responsible for monitoring service qualities of authorities by monitoring performance of permitting authorities using multiple indicators. By doing so, on the one hand, unnecessary, repetitive and cumbersome permitting procedures would be identified and evaluated; on the other hand, if any reforms are implemented CRC could monitor the progress and take control actions.
To measure performance of authorities, effectiveness and efficiency are 2 major outcomes. Effectiveness is defined as the extent to which the building regulatory system achieves compliance with regulatory requirements. Efficiency, on the other hand, indicates how efficiently the system achieves the effectiveness measures and includes indicators like time and cost required to get approvals, number of procedures and level of transparency. In New Zealand, authorities that fail in measurements will simply lose the right to issue building permits [3], and an adjacent or even private agency will take the places. Despite implementation of such high competitive system is unrealistic in Shenzhen, a link between leadership of authorities and service performance could still be developed to enforce authorities to improve.
(e) Involve private inspection organizations: Resorting to private entities which have both techniques and fully trained professionals is a high cost-effective solution. Take New Zealand as an example, any private organizations can also be permitting authorities as long as they meet requirements set by central government then get registrations. These private authorities would evaluate

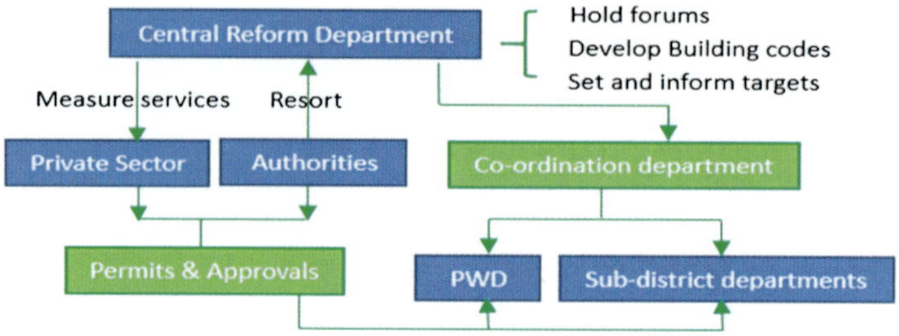

Fig. 50.2 Basic structure after reforms

particular fields like Green Building Codes or anti-seismic analysis. In Ontario, the government also provided a new status called "Registered Code Agencies" to encourage private practitioners to do design reviews and site inspections. In fact, these actions introduce a degree of competition between building officials and private sector practitioners.

(f) Enhance professional skills of authorities: Central government should provide necessary professional trainings to authorities Reforms is not to privatize permitting sector. Official authorities still have superior powers, so expertise is essential for these authorities being guiders rather than barriers. Tests should be set to evaluate outcomes of training programs and training outcomes should be linked to the measurement of leadership of relating authorities.

(g) Create reform insurance system: Sound insurance system is also necessary to support reforms mentioned above. Central government should encourage insurance firms provide liability insurance for designers, contractors, governmental authorities and private inspection practitioners by offering tax reduction benefits or other preferential policies. In this reform process, the role that insurance companies play can give another guarantee and supervision of both private and official authorities. In addition, it helps local authorities to reduce their aversion to risks and their tendency to stall the reform process. In summary, the basic structure is shown in Fig. 50.2.

50.5 Discussion

Problems and deficiencies in the first part of this paper are based on official analysis of structure and mechanism of construction management of public projects in Shenzhen, which can objectively reflect real situations. Solutions refer to international literatures in Canada, New Zealand. Major Considerations here are whether these innovations can be implemented given political and economic environments of Shenzhen. According to policies and attitudes of government in recent years, the perspective of reforms and improvements of public construction industry is bright. First of all, as a modern city which developed from reforms, both public and government are able to better recognize the potential and benefits of reform, making implementation easier. Second, the government began to simplify permitting processes by regulating time, number of approvals of permits since 2011, which means problems have been aware and initial actions have been taken. With innovations like BIM and ICT which can facilitate communication and interoperability, the reforms will surely have a bright perspective.

50.6 Conclusion

Since establishment of Public Works Department, much costs and time of public projects are saved due to centralized management and specialized knowledge of PWD. Despite these achievements, PWD needs to address problems like poor communication, cumbersome permitting, inefficient bidding and other problems. This paper discussed several possible solutions, referring to international literatures to simplify permitting processes, mainly include private sector introduction, building code establishment and service performance measurement and PWD should constantly learn from countries which successfully implemented reforms.

Acknowledgements We gratefully acknowledge the financial support from Chongqing Graduate Student Research Innovation Project (CYS14036)

References

1. World Bank Group (2014). Dealing with construction permits. Doing Business. Retrieved from http://www.doingbusiness.org/reports/global-reports/~/media/GIAWB/Doing%20Business/Documents/Annual-Reports/English/DB14-Chapters/DB14-Dealing-with-construction-permits.pdf
2. International Finance Corporation, World Bank, Multilateral Investment Guarantee Agency (2013). Good Practices for Construction Regulation and Enforcement Reform: Guidelines for Reformers. Investment climate. World Bank Group, Washington, DC. Retrieved from https://openknowledge.worldbank.org/handle/10986/16612
3. Moullier T (2009) Reforming Building Permits Why Is It Important and What Can IFC Really Do? IFC Advisory Services

Chapter 51
Comparative Analysis of the Indicator System for Guiding Smart City Development

Shiju Liao, Xi Chen, Yan Qian and Liyin Shen

Abstract Smart city has been gaining increasing attention among academia and practitioners as a new technology-based solution to meet the city disease challenges. A number of smart city indicator systems have been introduced to guide the practice of smart city initiatives. This paper examines the effectiveness of these typical existing indicator systems based on a benchmarking framework of smart city indicators, which is built on a comprehensive understanding of the definition and the connotations of smart city. Eight representative indicator systems both in China and abroad are collected and compared. The result of this study reveals that the existing smart city indicator systems have limitations for guiding the practice of smart city initiatives.

Keywords Smart city · Indicator system · Comparison · Indicator effectiveness

51.1 Introduction

According to the WDI (World Development Indicators) database of World Bank, the global average proportion of urban population had reached 53 % in 2013 [30], and this figure was predicted to be 70 % by 2050 [28]. The rapid growth of

S. Liao (✉) · Y. Qian · L. Shen
School of Construction Management and Real Estate,
International Research Centre for Sustainable Built Environment,
Chongqing University, Chongqing, China
e-mail: liaoshiju@cqu.edu.cn

Y. Qian
e-mail: cqqianyan@cqu.edu.cn

L. Shen
e-mail: shenliyin@cqu.edu.cn

X. Chen
Department of Real Estate and Construction, The University of Hong Kong,
Hong Kong SAR, China
e-mail: chenx90@hku.hk

Y. Wu et al. (eds.), *Proceedings of the 20th International Symposium on Advancement of Construction Management and Real Estate*,
DOI 10.1007/978-981-10-0855-9_51

population and increasing urbanization has generated a large number of city disease problems, such as energy shortage, environment pollution, traffic congestion and increasing social inequality [23]. In this circumstance, smart city has been gaining increasing attention as a new technology-based solution to achieve the sustainable development of cities [7, 11, 18].

There have been numerous examples of cities designated as smart cities in recent years. The IT Strategic Headquarters in Japan launched "I-Japan Strategy 2015" in 2009 [17], and focused in three priority areas for striving to create a citizen-driven and vibrant digital society, including electronic government and electronic local government, healthcare and health, and education and human resource. The government of Singapore issued a document entitled "A Vision of an Intelligent Island: IT2000 Report" in 1992 with the vision of becoming an "intelligent island", and "Intelligent Nation 2015" (iN2015) and "e-government 2015" in 2006 and 2011 respectively, with the aim of developing the info-communications sector and structuring a co-operation-oriented government [15, 24]. Other typical smart city initiatives include: the Dubuque project in American, the Stockholm Intelligent Transportation System in Sweden, the U-Korea and Seoul IPTV government in Korea, the Multimedia Super Corridor, and others (ICF 2015). In China, the number of smart city pilots introduced by Ministry of Housing and Urban-Rural Development has amounted to 290 by 2015 [21].

Whilst the initiatives of smart city are booming, there is no general consensus on the meaning of the term of "Smart City" among different parties including practitioners and academia except the pervasive use of Information and Communication Technologies (ICT). Most of the smart city cases reported on the internet are about specific aspects and focus on the construction of ICT infrastructures. In the process of development of smart cities in China, many problems have been revealed, such as treating the smart city initiatives as image projects, paying attention to the construction process of ICT infrastructure but neglecting the application, and isolated island of information among different city departments [26]. These problems to a large extent are due to poor understanding on the meaning of smart city and what contributes to the success of smart city initiatives. Therefore it is hard to guide policy makers and planners to build an effective smart city. Smart city indicator system is considered as an effective tool to provide guiding information for decision making and city planning for implementing the principle of smart city. Nevertheless, an indicator system is valuable only if the system is effective for application. In other words, the system can represent accurately the target and the connotations of smart city. Therefore, it is important to examine the effectiveness of the existing indicator systems in order to improve the quality of indicator systems for guiding the practice of smart city.

Various scholars and research institutions have been making efforts to develop various smart city indicator systems. For example, in promoting broadband economy as part of smart city, Intelligent Community Forum [19] defined six factors that determine a community's competitiveness in the broadband economy, including broadband, knowledge workforce, innovation, digital equality, sustainability and advocacy. Giffinger et al. [14] developed a smart city indicator system containing six dimensions to rank the city competitiveness of European medium-sized cities, including smart economy, smart people, smart governance, smart mobility, smart environment and smart living. The China's Ministry of Housing and Urban-Rural Development (MOHURD) and Ministry of Industry and Information Technology (MIIT) introduced smart city indicator systems respectively in 2012 from the perspective of their departmental functions. There are still other institutions and researchers who have produced various smart city indicators [8, 9, 10].

The above discussions show that various smart city indicator systems have been introduced throughout the world. However, it appears that little research has been done in examining the effectiveness of these established indicators. In fact, the effectiveness of these systems can be examined through a comparative analysis between these systems. This paper attempts to address this research gap by establishing a benchmarking framework of smart city indicator systems to compare the typical established systems, in order to improve the quality of indicator systems and their application effectiveness.

51.2 Research Methodology

The research starts with comprehensive literature review for establishing proper understanding on the definition and meaning of smart city. As existing smart city indicator systems include different dimensions, the literature review helps identity which dimensions should be counted in a smart city indicator system. Based on this, a benchmarking framework of smart city indicators will be established for further comparing the existing indicator systems.

Secondly, by means of internet search and library search, extensive literature surveys are conducted to identity and collect the representative existing smart city indicator systems. Standardization work is processed to ensure the comparability between these surveyed systems.

In the third part of this research, comparisons between the benchmarking framework and individual surveyed systems are examined in order to study the effectiveness of these selected systems.

51.3 Establishment of the Benchmarking Framework of Smart City Indicator Systems

51.3.1 Principle of Smart City

Smart city is still in conceptual stage, and there is still no general consensus on the definition of smart city [18]. A number of research literatures presented different definitions of smart city from various perspectives, and a number of typical definitions are shown in Table 51.1.

Different definitions highlight different aspects. Some definitions stress the construction and application of ICTs or computing technologies [8, 9, 27]. The definition of smart city given by Natural Resources Defense Council (NRDC) regarded the adjective word "smart" as efficient, sustainable, equitable, and livable. Giffinger et al. [14] considered that smart city was built on the smart combination of endowments and activities of self-decisive, independent and aware citizens. Caragliu's (2011) approach defined a city as smart city when the investments can fuel sustainable economic growth and a high quality of life. Rios [25] saw smart city as a city that can give inspiration and share culture, knowledge, and life. Cohen [8, 9] opined that smart city is a city which use ICTs to support innovation and low-carbon economy.

51.3.2 Dimensions of Smart City Benchmarking Framework

There is no agreement on the definition of a smart city, but a number of main dimensions of a smart city have been proposed. Giffinger et al. [14] suggested that a smart city contains six main components, including smart economy, smart people, smart governance, smart mobility, smart environment and smart living. Chourabi et al. [6] identified eight success factors of smart city initiatives, including (1) management and organization, (2) technology, (3) governance, (4) policy, (5) people and communities, (6) the economy, (7) built infrastructure, and (8) the natural environment. Lombardi et al. [20] presented the performance indicators of smart city in five cluster: smart governance (related to participation); smart human capital (related to people); smart environment (related to natural resources); smart living (related to the quality of life); and smart economy (related to competitiveness).

Table 51.1 Definitions of smart city

Definition	Reference
• A city striving to make itself "smarter" (more efficient, sustainable, equitable, and livable.) (*Available from http://smartercities.nrda.org*)	NRDC
• A city well performing in a forward-looking way in economy, people, governance, mobility, environment, and living, built on the smart combination of endowments and activities of self-decisive, independent and aware citizens	[14]
• The use of Smart Computing technologies to make the critical infrastructure components and services of a city—which include city administration, education, healthcare, public safety, real estate, transportation, and utilities—more intelligent, interconnected, and efficient	[29]
• A city "connecting the physical infrastructure, the social infrastructure, and the business infrastructure to leverage the collective intelligence of the city"	[16]
• A city combining ICT and web 2.0 technology with other organizational, design and planning efforts to dematerialize and speed up bureaucratic processes and help to identify new, innovative solutions to city management complexity, in order to improve sustainability and livability.	
• A useful definition is to call a city "smart" when investments in human and social capital and traditional (transportation) and modern infrastructure (ICT-based) fuel sustainable economic growth and a high quality of life, with a wise management of natural resource, through participatory government"	[5]
• A city that gives inspiration, share culture, knowledge, and life, a city that motivates its inhabitants to create and flourish in their own life	
• Smart cities use information and communication technologies (ICT) to be more intelligent and efficient in the use of resources, resulting in cost and energy savings, improved service delivery and quality of life, and reduced environmental footprint—all supporting innovation and the low-carbon economy	[9]

There are still other definitions and explanations on smart city [22, 3]. In fact, there are other terminologies similar to smart city, such as digital city, intelligent city, sustainable city, low-carbon city, and others. The indicators included in these systems also provide important reference for establishing the benchmarking framework.

Based on the above, this paper proposed a five-dimension benchmarking framework for smart city indicators. These dimensions are discussed as follows:

ICT infrastructure

The accessibility, availability and quality of the ICT infrastructure are fundamental to a smart city's development [14]. ICT infrastructure includes wireless infrastructure (fiber optic channels, Wi-Fi networks, wireless hotspots, kiosks) [1], service-oriented information systems [2].

Economy

Economy is the impetus to promote the development of smart city, and also the main driver of the smart city initiatives. If a city has a high competition index of economic development, it will be in advantageous position to the development of smart city [6]. Smart economy includes factors about economic competitiveness, such as innovation, entrepreneurship, trademarks, productivity, flexibility of the labor market and the integration in the national and global market [6].

Governance and policy

Accounting the research of Forrester, the core of smart city initiatives is smart governance [4]. Chourabi et al. [6] summarized main governance factors, including collaboration, leadership and champion, participation and partnership, communication, data-exchange, service and application integration, accountability, and transparency. On the other hand, policy context is critical to use ICTs systems in appropriate ways. A government cannot transfer to a smarter one without a normative drive of policy [12]. As there are no specific literatures discussing the "policy" factor in the context of smart city, findings in e-government research can help understand this factor. García and Pardo [13] examined the success factors of e-government, including legal, regulatory, institutional and environmental challenges.

People

People is the main factor determining the change to a smart city because the role of citizens is the principal part in the process of construction and development of smart city. Chourabi et al. [6] summarized the main factors related to people in smart city, including digital divide, information and community gatekeepers, participation and partnership, communication, education, quality of life and accessibility.

Natural environment

The mission of smart city is to solve the city disease problems, and to achieve the sustainable development of cities. So it is important to better manage natural resources and protect environment in a smart city.

51.3.3 Establishment of the Benchmarking Framework of Smart City Indicator Systems

Based on the above analysis, a five-dimension benchmarking framework is established, containing 40 specific indicators located in various categories, as shown in Table 51.2.

Table 51.2 The benchmarking framework of smart city indicator system

Dimension	Category	Indicators
ICTs infrastructure (SCI1)	Infrastructure	WIFI coverage
		Broadband network coverage
		Next generation broadcasting (NGB) coverage
		Rate of fiber to the home (FTTH)
		Average internet speed
	Service-oriented information systems	Urban public information platform
		Data base
		Information security
Economy (SCI2)	ICTs application	Level of E-commerce
	Productivity	GDP
	Industrial structure	Industrial planning/industrial upgrading
		Input and development level of information industry
	Innovation	Innovation/entrepreneurship/patents
	International cooperation and communication	Integration in the global market
Governance and policy (SCI3)	Laws and regulations	Smart city development planning
		Security laws and regulations
		Operational guidance and supervision
		Financial planning and continuous support
	Institution change	Institutional framework
		Operation security system
	ICTs application	E-government
	Participation and partnership	Public participation in decision making
	Service and application integration	Government service system
		Government administrative efficiency index
	Transparency	Transparency
People (SCI4)	Participation and partnership	Degree of public participation in political/social life
	Communication	Public awareness/internationalization/diversification
	Quality of urban living	Quality of urban living (convenience, security, etc.)
	Education	Education level
		Continuous learning
		Training and recruitment of information industry talents
	ICTs application	Level of IT application
		Digital divide

(continued)

Table 51.2 (continued)

Dimension	Category	Indicators
Natural environment (SCI5)	Environment protection	Quality of natural environment (Air, Soil, Water)
		Contamination discharge
		Public participation in environmental protection
	Energy saving and greenhouse reduction	Energy efficiency
		Renewable energy use
	ICT application	Environmental monitoring
		Informatization level of environmental protection

51.4 Data Collection

Through extensive literature surveys by means of internet search and library search, eight typical existing smart city indicator systems established by different institutions are identified for further analysis in this paper, as shown in Appendix 1.

There are differences between the eight smart city indicator systems in the definition of individual indicators. Similar indicators may be classified into different categories. In order to ensure the comparability between the eight systems and benchmarking framework, standardization on the indicators are processed as follows:

- Standard definition is used to solve the problem that indicators in different systems assume same meaning but quoted in different names.
- For those indicators which are in different names but measuring the same context of smart city in a same system, a combined term is applied.
- Some indicators are outdated and considered unnecessary, and they are removed from the list.
- The standardized indicators are allocated into the five dimensions of benchmarking framework built in the former chapter.

After the above processes, the comparison chart is derived to exhibit the comparisons between the benchmarking framework and the surveyed individual systems, as shown in Table 51.3.

Table 51.3 The comparisons between the benchmarking framework and the surveyed individual systems

Dimension	Cl.	Indicators	S1	S2	S3	S4	S5	S6	S7	S8
SCI1	S0	WIFI coverage	✓	–	–	✓	✓	✓	✓	–
		Broadband network coverage	✓	✓	✓	✓	✓	✓	✓	✓
		Next generation broadcasting (NGB) coverage	–	–	–	✓	–	–	–	✓
		Rate of fiber to the home (FTTH)	–	–	–	–	✓	✓	✓	–
		Average internet speed	–	–	–	–	✓	–	–	–
		Urban public information platform	–	–	–	✓	–	–	–	✓
		Data base	–	–	–	✓	–	–	–	✓
		Information security	–	–	–	✓	–	–	–	✓
	Others	internet penetration rate	–	–	–	–	✓	–	–	–
		3G network coverage	–	–	–	–	✓	–	–	–
		Rate of smartphone usage	–	✓	–	–	–	–	–	–
		Average computer no. of one family	–	–	–	–	✓	–	–	
		Application of internet of things (IOT)	–	–	–	✓	✓	–	–	✓
		Application of cloud computing technology	–	–	–	–	–	–	–	✓
		Information transformation of key fields	–	–	–	–	–	–	✓	–
		State key laboratory for smart city	–	–	–	–	–	–	✓	–
		Smart grid technology and equipment application	–	✓	–	–	–	–	–	–
SCI2	S0	Level of E-commerce	✓	–	–	–	–	✓	✓	–
		GDP	–	✓	–	–	–	–	✓	✓
		Industrial planning/industrial upgrading	–	–	✓	✓	–	–	–	–
		Input and development level of information industry	✓	–	–	–	–	✓	✓	✓
		Innovation/entrepreneurship/patents	✓	✓	✓	✓	–	–	✓	–
		Integration in the global market	–	✓	–	–	–	–	–	–

(continued)

Table 51.3 (continued)

Dimension	Cl.	Indicators	S1	S2	S3	S4	S5	S6	S7	S8
		Flexibility of the labor market	–	✓	–	–	–	–	–	–
		Trade barriers	✓	–	–	–	–	–	–	–
		Trademark	–	✓	–	–	–	–	–	–
		Synthetic energy consuming per 10,000 rmb of GDP	–	–	–	–	–	–	✓	✓
SCI3	S0	Smart city development planning	–	–	–	✓	✓	✓	–	✓
		Security laws and regulations	–	–	–	✓	✓	–	–	✓
		Operational guidance and supervision	–	–	–	✓	–	–	–	–
		Financial planning and continuous support	✓	–	–	✓	✓	–	–	–
		Institutional framework	✓	–	–	✓	✓	✓	–	–
		Operation security system	✓	–	–	✓	✓	✓	–	–
		E-government	✓	–	✓	✓	✓	✓	–	–
		Public participation in decision making	–	✓	–	✓	✓	–	–	✓
		Government service system	–	–	–	✓	–	–	–	✓
		Government administrative efficiency index	✓	–	–	–	–	–	✓	–
		Transparency	–	✓	–	✓	✓	✓	–	✓
	Others	Smart city forum/conference	–	–	–	–	–	✓	–	–
		Funds investment of E-government construction	–	–	–	–	–	–	✓	–
		Cooperative application system	–	–	–	–	–	–	✓	–
		Public service application penetration rate	–	–	–	–	–	–	✓	–
SCI4	S0	Degree of public participation in political/social life	–	✓	–	–	–	–	–	✓
		Public awareness/internationalization/diversification	–	✓	–	–	–	–	✓	–
		Quality of urban living (convenience, security, etc.)	–	–	–	–	–	✓	✓	–
		Education level	✓	✓	–	–	–	✓	✓	✓

(continued)

Table 51.3 (continued)

Dimension	Cl.	Indicators	S1	S2	S3	S4	S5	S6	S7	S8
		Continuous learning	–	✓	–	–	–	–	–	–
		Training and Introduction of information industry talents	–	✓	✓	–	–	–	–	✓
		Level of IT application	–	–	–	–	–	✓	✓	✓
		Digital divide	–	–	✓	–	–	–	–	–
SCI5	S0	Quality of natural environment (air, soil, water)	–	✓	–	–	–	–	–	–
		Contamination discharge	–	✓	–	–	–	✓	–	–
		Public participation in environmental protection	–	✓	–	–	–	–	–	–
		Energy efficiency	–	✓	–	–	–	✓	–	✓
		Renewable energy use	–	✓	–	–	–	✓	–	–
		Environmental monitoring ratio	–	–	–	–	–	✓	–	–
		Informatization level of environmental protection	–	–	–	–	–	–	–	✓
	Others	Intelligent use of resources	–	–	–	–	–	–	–	✓
Others		Transportation	✓	✓	✓	✓	✓	✓	–	✓
		Medical system	✓	✓	–	✓	✓	✓	–	✓
		Housing	✓	✓	–	✓	–	–	–	–
		Education	✓	✓	✓	✓	✓	✓	–	✓
		Environment protection	–	–	–	✓	–	✓	–	–
		Social security	–	–	–	✓	✓	–	–	✓
		Community management	–	–	–	✓	–	✓	–	✓
		Energy	–	–	–	✓	–	✓	–	✓
		Territory	–	–	–	✓	–	–	–	–
		Emergency management	–	–	–	✓	–	–	–	–
		Public security	✓	✓	–	✓	–	✓	–	✓

(continued)

Table 51.3 (continued)

Dimension	Cl.	Indicators	S1	S2	S3	S4	S5	S6	S7	S8
		Logistics	–	–	–	✓	–	–	–	–
		Household	–	–	–	✓	–	–	–	–
		Finance	–	–	–	✓	–	–	–	–
		Water supply/drainage/lighting/gas	✓	–	–	✓	–	–	–	–
		Culture and sports	–	✓	–	✓	–	–	–	–
		Tourism	–	–	–	–	✓	–	–	–
		Employment	–	–	–	✓	–	–	–	–
		Urban planning	–	–	–	✓	–	–	–	–

* "✓" Indicates that this indicator is adopted in the concerned indicator system
"—" Indicates that this indicator is not adopted in the concerned indicator system

51.5 Discussion

51.5.1 Comparison Analysis on the Consistency Between Benchmarking Framework and the Selected Systems of Smart City Indicators

From the information in Table 51.4, it can be seen that the degree of inconsistences between the benchmarking system and the eight selected systems are different. For sake of further analysis, the number of dimensional indicators of benchmarking framework adopted in the eight individual systems is counted (which is denoted as "N"). And ratio of variable "N" to the total number of indicators belong to each individual dimensions of benchmarking framework (which is denoted as "T") is also calculated, which is denoted as "α". The results are shown in Table 51.4.

It is observed from Table 51.5 that only three indicator systems (S2, S6, S8) have adopted to certain extents the benchmarking indicators across all the five dimensions, with the overall ratio value of consistence of 40, 43, 45 %. Other five systems (S1, S3, S4, S5, S7) do not adopt indicators under the dimension of natural environment (SCI5) into their systems, which is in line with the aim of smart city. The systems S4 and S5 do not include the indicators of smart people dimension (SCI4), which is important in the smart city initiatives. It is interesting to note that S5 give no consideration to the indicators under economy dimension. (SCI2).

Table 51.5 demonstrates that these existing indicator systems do not reflect some focal issues relating to smart city. The insufficiency of these systems weakens their effectiveness in application for guiding the practice of smart city initiatives.

51.5.2 Focused Indicator Areas of Each Individual Systems

By referring to the Table 51.4, the focuses of each individual systems can be examined by counting the ratio (β) of the number of adopted indicators under different dimensions (N1) to the total number of adopted indicators in individual systems, the results is shown in Table 51.5.

Table 6 indicates that the indicators in the systems S2, S6 and S8 are relatively evenly distributed among the five different dimensions. Indicators in S3 and S7 is distributed homogeneously in four dimensions except the natural environment dimension (SCI5). S1, which is related to the IT background of IBM, puts more attention on the dimensions of technology (SCI1), governance (SCI2), economy (SCI3) and others (ICT applications in different city field). S4 emphasizes those indicators related to governance and policy (SCI3) and others (the smartness of different fields of cities, such as smart transportation, smart medical system, smart

Table 51.4 The level of consistency between benchmarking framework and the selected systems

System dimension	S0	S1		S2		S3		S4		S5		S6		S7		S8	
	T	N	α (%)	N	α (%)	N	α (%)	N	α (%)	N (%)	α (%)	N	α (%)	N	α (%)	N	α (%)
SCI1	8	2	25	1	13	1	13	6	75	4	50	3	38	3	38	5	63
SCI2	6	3	50	3	50	2	33	2	33	0	0	2	33	4	67	2	33
SCI3	11	5	45	2	18	1	9	10	91	8	73	5	45	1	9	5	45
SCI4	8	1	13	5	63	2	25	0	0	0	0	3	38	4	50	4	50
SCI5	7	0	0	5	71	0	0	0	0	0	0	4	57	0	0	2	29
Total	40	11	28	16	40	6	15	18	45	12	30	17	43	12	30	18	45

Table 51.5 The number of indicators distributed in different dimensions

System dimension	S1		S2		S3		S4		S5		S6		S7		S8	
	N_I	β (%)	N_I	β (%)	N_I	β (%)	N_I	β (%)	N_I	β (%)	N_I	β (%)	N_I	β (%)	N_I	β (%)
SCI1	2	12	2	8	1	13	6	17	9	43	3	13	5	28	6	21
SCI2	4	24	5	20	2	25	2	6	0	0	2	8	5	28	3	11
SCI3	4	24	2	8	1	13	9	26	7	33	5	21	4	22	5	18
SCI4	1	6	5	20	2	25	0	0	0	0	3	13	4	22	4	14
SCI5	0	0	5	20	0	0	0	0	0	0	4	17	0	0	3	11
Others	6	35	6	24	2	25	18	51	5	24	7	29	0	0	7	25
Total	17	100	25	100	8	100	35	100	21	100	24	100	18	100	28	100

finance, et al.). It can be considered that the system S4 focuses on the smart city performance in urban planning and management. The indicators in S5 are concentrated in information technology, ICTs infrastructure, the level of ICTs application, and the construction and application of information supporting system, emphasizing on the investigation to the ability of technical support of cities. The indicator systems of S4 and S5 are introduced by MOHURD and MIIT respectively. The compositions of indicators in S1, S4 and S5 reflect the phenomenon that the administrations or commercial organizations count more smart city indicators which are related to their functions or business coverage.

51.6 Conclusion

An effective smart city indicator system is an essential tool to diagnose and promote the development of smart city. There are a number of existing smart city indicator systems either in China or abroad, but their effectiveness has not yet be reported. Comparison analysis on the existing systems can help to produce an effective indicator system for guiding the initiatives of smart city. In order to conducting this comparison, this paper structured a benchmarking framework containing five dimensions: ICTs infrastructure, economy, governance and policy, people and natural environment. The comparison was conducted accordingly between the eight selected smart city indicator systems and the framework.

The finding from the study shows that the surveyed systems S2, S6 and S8 are more consistent with the benchmarking framework, and the other systems do not take into account of some dimensions to certain extents. Furthermore the focuses of each individual systems are different, which reflect the characteristics of departments or institutions who introduced the corresponding systems.

The inconsistence and different focuses among these smart city indicator systems indicate that the understandings on smart city between different scholars and research institutions are different, and are limited to their own perspectives. In fact, an indicator system based on incomprehensive understanding on smart city is undoubtedly lack of effectiveness to assist the effective practice of smart city.

The ineffectiveness of these existing indicator systems reflects the requirements for an effective indicator system when a specific city is concerned. The benchmarking framework presented in this paper provides a reference for further study to develop valid indicators for supervising the development of smart city in a particular city.

Acknowledgements This paper is funded by the Key Project of the National Natural Science Fund of China (Project No: 15AZD025).

Appendix 1: Lists of Smart City Indicator Systems

No.	Name	Publication date	Publication instruction	Assessment range	Dimensional indicators	No. of second-tier indicators	No. of bottom level indicators
S1	IBM Smart City Evaluation Criteria and Elements	2008	IBM Institute for Business Value (IBV)	Global	City service, citizen, economy, transportation, communication, water supply and energy	28	50
S2	Indicators for Ranking the Smartness of European Medium-Sized Cities	2007	Vienna University of Technology	European Union	Smart economy, smart people, smart governance, smart mobility, smart environment and smart living	31	74
S3	Smart Taiwan Performance Index	2008	Executive Yuan of Taiwan	Province	Broadband networks, creative industry, e-government, real world application, digital divide, talent development	–	9
S4	Smart City (District, Town) Pilots Assessment Indicator System (Trial Version)	2012	Ministry of Housing and Urban-Rural Development of the People's Republic of China	National	Security system and ICT infrastructure, construction and living, smart management and service, smart industry and economy	11	57
S5	Smart City Evaluation Index System (Draft)	2012	Ministry of Industry and Information Technology of the People's Republic of China	National	Smart preparation, smart management and smart service	9	51
S6	Smart Pudong Assessment Indicator System (2.0)	2012	Government of Pudong New Area in Shanghai	District	Infrastructure, social management and public services, economy, smart living, citizens' subjective perception, urban soft environment	18	37

(continued)

(continued)

No.	Name	Publication date	Publication instruction	Assessment range	Dimensional indicators	No. of second-tier indicators	No. of bottom level indicators
S7	Smart Nanjing Assessment Indicator System	2011	Nanjing Information Center	Municipality	ICT infrastructure, smart industry, smart service and smart citizens	–	22
S8	Smart Ningbo Assessment Indicator System	2012	Ningbo Academy of Smart City Development	Municipality	ICT infrastructure, governance, living, industry, citizen, environment	19	40

References

1. Al-Hader M, Rodzi A, Sharif AR, Ahmad N (2009) Smart city components architicture. In: International Conference on IEEE Computational Intelligence, Modelling and Simulation, 2009. CSSim'09. pp 93–97
2. Anthopoulos L, Fitsilis P (2010) From online to ubiquitous cities: the technical transformation of virtual communities. In: Technological and legal issues. Next generation society. Springer, Heidelberg, pp 360–372
3. Anthopoulos LG, Vakali A (2012) Urban planning and smart cities: interrelations and reciprocities. In: The future internet, Springer, Berlin, Heidelberg, pp 178–189
4. Belissent J (2011) The core of a smart city must be smart governance. Forrester Research Inc., Cambridge
5. Caragliu A, Del Bo C, Nijkamp P (2011) Smart cities in Europe. J Urban Technol 18(2):65–82
6. Chourabi H, Nam T, Walker S, Gil-Garcia JR, Mellouli S, Nahon K, choll J (2012) Understanding smart cities: an integrative framework. In: 45th Hawaii international conference on IEEE system science (HICSS), pp 2289–2297
7. Coe A, Paquet G, Roy J (2001) E-governance and smart communities a social learning challenge. Soc Sci Comput Rev 19(1):80–93
8. Cohen B (2012a) What exactly is a smart city? Co. Exist
9. Cohen B (2012b) The top 10 smart cities on the planet. Co. Exist 11
10. Dedao GU, Wen Q (2012) Study on the construction of evaluation index system of China's smart city. Future Dev (10):79–83
11. Eger J (2003) Smart communities: becoming smart is not so much about developing technology as about engaging the body politic to reinvent governance in the digital age. Urban Land 60(1):50–55
12. Eger JM, Maggipinto A (2010) Technology as a tool of transformation: e-cities and the rule of law. In information systems: people, organizations, institutions, and technologies. Physica-Verlag HD, pp 23–30
13. García JR, Pardo TA (2005) E-government success factors: mapping practical tools to theoretical foundations. Gov Inf Q 22(2):187–216
14. Giffinger R, Fertner C, Kramar H, Kalasek R, Pichler-Milanovic N, Meijers E (2007) Smart cities-ranking of european medium-sized cities. Vienna University of Technology
15. Ha H (2013) E-Government in Singapore: critical success factors. E-Government success around the world: cases, empirical studies, and practical recommendations, pp 176–197
16. Harrison C, Eckman B, Hamilton R, Hartswick P, Kalagnanam J, Paraszczak J, Williams P (2010) Foundations for smarter cities. IBM J Res Dev 54(4):1–16
17. Headquarters IS (2009) I-Japan strategy 2015. Striving to create a citizen-driven, reassuring & vibrant digital society
18. Hollands RG (2008) Will the real smart city please stand up? Intelligent, progressive or entrepreneurial? City 12(3):303–320
19. Intelligent Community Forum (2006) Intelligent community indicators. Retrieved from https://www.intelligentcommunity.org/index.php?submenu=Research&src=gendocs&ref=Research_Intelligent_Community_Indicators&category=Research&link=Research_Intelligent_Community_Indicators
20. Lombardi P, Giordano S, Farouh H, Yousef W (2012) Modelling the smart city performance. Innov Eur J Soc Sci Res 25(2):137–149
21. MOHURD (2015) Announcement of the list of 2014 national smart city pilots. Retrieved from http://www.mohurd.gov.cn/zcfg/jsbwj_0/jsbwjjskj/201504/t20150410_220653.html
22. Nam T, Pardo TA (2011) Conceptualizing smart city with dimensions of technology, people, and institutions. In: Proceedings of the 12th annual international digital government research conference: digital government innovation in challenging times, pp 282–291 (ACM)
23. Neirotti P, De Marco A et al (2014) Current trends in smart city initiatives: some stylised facts. Cities 38:25–36

24. Ng P (2011) Embracing emerging technologies: the case of the singapore intelligent nation 2015 vision. IGI Global, Hershey
25. Rios, P. (2012). Creating "the smart city". Doctoral dissertation
26. Shengzu Gu, Yangjian Wu, Ri Liujiang (2013) Problems in the development of smart city in China and their solution. China Soft Sci 2013(1):6–12
27. Toppeta D (2010) The smart city vision: how innovation and ICT can build smart, "liveable", sustainable cities. The Innovation Knowledge Foundation, Think
28. Un-habitat (2010) State of the world's cities 2010/2011: bridging the urban divide, Earthscan
29. Washburn D, Sindhu U, Balaouras S, Dines RA, Hayes NM, Nelson LE (2010) Helping CIOs understand "smart city" initiatives: defining the smart city. Its drivers, and the role of the CIO. Forrester Research, Cambridge
30. World Bank (2014). Urban population (% of total). Retrieved from http://data.worldbank.org/indicator/SP.URB.TOTL.IN.ZS?display=default.

Chapter 52
SNA Based Identification of Key Factors Affecting the Implementation of Emission Trading System (ETS) in Building Sector: A Study in the Context of China

Xiangnan Song, Liyin Shen, Michael C.H. Yam and Zongnan Zhao

Abstract Emission Trading System (ETS) has been promoted as a tool for providing financial and cost-effective incentives to the GHG emitters to implement emission reduction measures. The ETS has been applied in many energy-intensive industries internationally. However, its implementation in building sector appears encountering many challenges due to the unique characteristics of buildings. This study attempts to identify and analyze the key factors within the context of China. Research data are from semi-structured interviews with a selected group of ETS experts from official and academia. Fifteen representative factors have been identified from extensive content analysis and validated with experts' opinions. Among these representative factors, some are key factors playing critical and essential roles. This paper applies the Social Network Analysis (SNA) to find the key factors. Finally, it is found that there are four factors considered as the most important influence factors, including "government", "emitters", "the scattered energy consumption and carbon emission in building sector", and "decentralization of property rights in building sector". The findings from the study provide valuable reference for decision-making in promoting ETS in those developing countries such as China where carbon emission in building sector contributes considerably to environmental pollution.

X. Song (✉) · L. Shen · Z. Zhao
School of Construction Management and Real Estate, International Research Centre for Sustainable Built Environment, Chongqing University, Chongqing, China
e-mail: 1018500021@qq.com

L. Shen
e-mail: shenliyin@cqu.edu.cn

Z. Zhao
e-mail: 270705489@qq.com

M.C.H. Yam
Department of Building and Real Estate, Hong Kong Polytechnic University, Hung Hom, Kowloon, Hong Kong
e-mail: bsmyam@polyu.edu.hk

Y. Wu et al. (eds.), *Proceedings of the 20th International Symposium on Advancement of Construction Management and Real Estate*,
DOI 10.1007/978-981-10-0855-9_52

Keywords Emission trading system (ETS) · Building sector · Social network analysis (SNA) · Key factors · China

52.1 Introduction

The majority of the world reached an agreement that the climate change is mainly caused by the atmosphere warming as a result of the greenhouse gases (GHG) emissions released into the atmosphere. And energy consumption in building sector is commonly appreciated as a major contributor to the global warming and environmental unsustainability as it produces large amount of GHG emissions [1, 24]. Being at the stage of rapid urbanization, China is experiencing unprecedented growth in building demand. With the high population densities and the unprecedented rate in economic development, China has become the largest GHG emitter worldwide, and its building sector is the second largest energy consumer and carbon emitter in China [8, 26]. Therefore, it is widely recognized that any attempt to mitigate the carbon emission in the world must ultimately have to include China.

The Kyoto Protocol, which was issued by the United Nations Framework Convention on Climate Change (UNFCCC) members in Tokyo of Japan in 1997, has been stimulating the development of ETS at the national and international levels. The implementation of the Kyoto Protocol has effectively limited the emissions of carbon dioxide and other greenhouse gases because the rights of carbon emissions have been limited significantly according to the Kyoto Protocol. However, carbon emissions are inevitable in reality as we pursue economic development. Furthermore, the costs for reducing carbon emissions are different between different industries and enterprises. Therefore, some industries have spare rights for carbon emissions, and some do not have, which provides the possibility of the right exchange between different emitters. In line with this, the mechanism of carbon trading has been spawning internationally recently.

The effectiveness of ETS has been well accepted in many economics sectors such as electricity, power, and other sectors [7, 39]. Previous policy instruments for addressing emission reduction in building sector range from mandatory measures, subsidy schemes, and various voluntary schemes [30]. Greiner and Lieberg [12] pointed out that whilst the existing instruments are not able to provide incentives for emission reduction, ETS mechanism can provide a viable alternative to the existing policy framework. Li and Colombier [22] opined that the establishment of emission trading system in building sector provides a more efficient way of low-carbon transition through effective resource allocation by marketing mechanism. Nishida and Hua [25] detailed the key factors affecting trading program settings in applying the Tokyo Cap-and-Trade Program (TCTP), which is the first ETS for emission reductions from urban buildings introduced in 2010. Lam et al. [21] introduced the ETS to property and facility managers to improve energy efficiency and reduce carbon emission in building sector. It is considered that the advantages of marketing mechanism embodied in carbon trading system will be brought into full play in achieving the objective of energy saving and emissions reduction if the carbon trading instrument is used properly for building sector. However, existing ETS systems have little application in building sector. For example, as the largest multinational implemented carbon trading system in the world, the EU-ETS has included many economic sectors but expect for building sector. This happens to other typical ETSs, such as the Regional Greenhouse Gas Initiative (RGGI) in USA, the Chicago Climate Exchange (CCX) in USA, and the New South Wales Greenhouse Gas Abatement Scheme (NSW GGAS) in Australia [28].

Some attempts have been devoted in applying ETS to building sector. For example, the TCTP in Japan, which is considered one of the influential ETSs globally, is the first emission trading system mainly targeting at emission reduction from buildings [25]. The Chinese government has also contributed efforts in introducing ETS in its building sector. The Ministry of Housing and Urban-Rural Development (MOHURD) has proposed three carbon trading schemes, targeting for nonresidential buildings, the existing residential buildings in heating regions, and heat supply facilities respectively [12]. These initiatives have been playing driving force in promoting the ETS for controlling the emissions in the Chinese building sector. In line with this development, some local building industries have tried applying ETS to building sector. There are seven ETS pilots in China, Beijing, Shanghai, Tianjin, Shenzhen, Chongqing, Guangdong, and Hubei. Five of these ETSs have covered building sector to different extents, including Shenzhen, Tianjin, Beijing, Shanghai, and Guangdong [38]. However, the application of ETS for buildings in China is still in its initial stage and has not seen many fruits.

There are many factors contributing to the successful implementation of ETS. Particularly, in building sector, the implementation of ETS appears encountering more challenges due to the unique characteristics of buildings, and these challenges are more prominent in those developing countries such as China [3, 21, 33]. For example, Jiang and Tovey [18] pointed out that the key barriers preventing the adoption of carbon reduction strategies in large commercial buildings in China include financing barriers, technological barriers, infrastructure barriers, and management barriers. Ren et al. [29] identified four main factors affecting the implementation of ETS in the Chinese building sector, namely, scattered ownership of buildings, relatively small quantities of emission reduction from individual buildings, lack of effective statistical data about carbon emissions, and high transaction costs associated with quantifying, certifying, marketing, selling and transferring the emission reductions.

The above discussions show that there is little existing research conducted comprehensive factors analysis on the ETS application in building sector within the context of China. Therefore, a detailed and systematically analysis of the factors and their relative importance is a significant research content. This study adopts Social Network Analysis (SNA) to build a key factors identification model through factor analysis, relationship analysis, and matrix analysis on the ETS factors. The results are expected to provide valuable reference for adopting appropriate measures to promote the ETS in the Chinese building sector successfully.

52.2 Research Methodology

This paper employs SNA to identify key factors affecting the implementation of Emission Trading System (ETS) in building sector by exploring their structural characteristics.

Network analysis, derived from graph theory, attempts to describe the structure of relations (displayed by links) between given entities (displayed by nodes), and applies quantitative techniques to produce relevant indicators and results for studying the characteristics of a whole network and the position of individuals in the network structure [32]. Social Network Analysis is an interdisciplinary methodology developed mainly by sociologists and researchers in social psychology, further developed in collaboration with mathematics, statistics, and computing that led to a rapid development of formal analyzing techniques which made it an attractive tool for other disciplines like economics, management or marketing [6, 36]. In recent years, the idea and principle of this method are extended to identify key elements and factors within a complex network [11, 13, 16] and has been get highly recognized. In order to achieve the aim of this paper, the research works are planned as follows:

Factor analysis-identification of representative factors system.

Selecting and confirming a comprehensive and representative factors system is usually considered the first and vitally important step for further identification of key factors. Content analysis is a proven effective method for collecting and analyzing data to identify research problems from literatures, official documents, websites, databases, and other various existing studies, especially in sociological research field [4, 20, 35]. In view of this, content analysis is conducted to a wide range of literatures, news reports, websites, and Chinese government documents to formulate a comprehensive list of factors. However, process of identifying the factors through content analysis has limitations, such as high correlation and synonym between the identified factors [37]. For example, the factors "legal system" "legal infrastructure" and "rule of law" are synonymous. Therefore, semi-structured interviews with the attendance of a group of selected experts who have knowledge of ETS and the practice of the Chinese building sector are organized to select the representative factors.

Relationship analysis-Establishment of Adjacency Matrix for representative factors.

Adjacency matrix is used to present the contextual relationships between the identified representative factors. The contextual relationship is usually established through collecting experts' opinions [2].

According to the principle of SNA method, the qualitative relationships between factors can be transformed into a square matrix, called "Adjacency Matrix", in which the description of the relationship will be expressed by the degree of mutual influence between various factors. On the other hand, the direction of "influence" needs to be considered in filling the Adjacency Matrix. For example, Factor "α" affects Factor "β", but Factor "β" may not affect Factor "α", and the degree of

Factor "α" affects Factor "β" maybe not the same as Factor "β" affects Factor "α". The rules for defining this influence are elaborated as follows:

All the rows and columns in the Adjacency Matrix represent each factor in turn.

The value of (i, j) entry denotes the influence degree of Factor "i" to Factor "j", and similarly the value of (j, i) entry denotes the influence degree of Factor "j" to Factor "i".

The influence degree will be assigned with 4-level value, including no relations (0), weak relations (1), moderate relations (2), and strong relations (4).

Matrix analysis-identification of key factors.

Matrix analysis based on the established Adjacency Matrix is the most important step for accurately identifying the key factors affecting the implementation of ETS in the Chinese building sector. Key ETS factors refer to those factors which play the leading role in the ETS factors network. And they are not vulnerable to be interfered by other factors and possess more capacity to influence the other factors at the same time. In other words, the leading position indicates the centrality of networks, which is the core concept in the application of Social Network Analysis (SNA) [5]. This study employ the SNA method to conduct the matrix analysis.

Centrality in network analysis is a quantification of the relative importance of a node within a graph network. The centrality of a node in a network is a measure of the structural importance of the node. The three most widely used centrality measures are Freeman's degree, closeness, and betweenness [5, 9].

"Degree" centrality is measured as the number of direct ties that involve a given node [10]. Due to the Adjacency Matrix established is asymmetric, the degree centrality will be denoted by two variables: "in degree (ID)" and "out degree (OD)". In degree and out degree for a given node fi is calculated as:

$$\mathrm{ID}(\mathrm{fj}) = \sum_{i=1}^{M} f(i,j) \tag{52.1}$$

$$\mathrm{OD}(\mathrm{fi}) = \sum_{j=1}^{N} f(i,j) \tag{52.2}$$

where ID(fj) represents the "in degree" of column j, and the M is the total number of row in the established Adjacency Matrix. And OD(fi) represents the "out degree" of row i, and the N is the total number of column in the Adjacency Matrix.

However, the ID and OD only measure the immediate ties of an individual node in the network, and they fail to consider the indirect ties. On the other hand, the degree centrality only represent partial centrality. So, this study intends to employ closeness centrality (Cc) to measure the extensity of influence over the entire network for a given node. Closeness centrality often emphasizes the distance of a node to others in the network by focusing on the geodesic distance from each node to all others. Nodes with high closeness centrality can presumably connect and interact with other nodes more easily and directly in a network.

The closeness centrality (Cc) of a node is defined as:

$$\mathrm{Cc(fi)} = {}^{1}\Big/{\textstyle\sum_{n=1}^{N}} d(fi, fj) \tag{52.3}$$

Here, the count d(fi, fj) denotes the geodesic distance between nodes i and j, which is defined as the length of the shortest path between the two nodes Closeness centrality of a node i is the inverse of the sum of the geodesic distances from node i to all the other nodes in the network (1 to N). Similarly, due to the Adjacency Matrix established is asymmetric, the closeness centrality will be denoted by two variables: "in closeness centrality (ICc)" and "out closeness centrality (OCc)", respectively, based on inward and outward connections, although their calculation are based on the same formulas (52.3).

Key factors often have direct impact on other factors, but not be interfered by other factors, so their "in degree" are relatively lower and "out degree" are higher. Accordingly, the key factors in this research have: higher "out closeness centrality" (OCc) and higher "out degree" (OD).

52.3 Results

In the first stage of this research, based on a wide range of literatures, news reports, websites, and Chinese government documents, the research team formulated a comprehensive list of factors affecting the implementation of Emission Trading System (ETS) in building sector, which is shown in Table 52.1.

Semi-structured interviews were then organized to select and confirm the representative factors by referring to the factors in Table 52.1. In order to select sufficient and effective interviewees, the research team carried out extensive investigations from December 2014 to April 2015. As a result, the above research efforts have led to approaching 58 candidate interviewees. Finally, the research team managed to get the supports from 7 of these experts for interview discussions. The detail information of these 7 experts is shown in Appendix A. These 7 experts were invited to judge which factors in Table 52.1 are important and should be considered in the context of the Chinese building sector. In addition to the information of Table 52.1, a number of open-ended questions were also presented to these 7 interviewees, including "Do you think what are the representative factors in Table 52.1" "Are there any synonymous factors in Table 52.1?" "Are there any factors in Table 52.1 neglectable?" "Are there other important factors missed in Table 52.1?"

Following the semi-structured interview discussions, a list of 15 factors are selected as representative ETS factors as shown in Table 52.2. The selection of these 15 factors is based on the principle of "The minority is subordinate to the majority". In other words, a factor is selected if four or more experts agree that it is representative.

Table 52.1 The comprehensive ETS factors based on content analysis

Factors identified	Factors identified	Factors identified
Zhang [40] • Government laws • Administrative measures • Trading platform • The threshold of coverage • Permit allocation • Monitoring, reporting and verification (MRV) • Penalty for noncompliance • Independent verification bodies	Han et al. [14] • Emission measurement • Legal infrastructure • Permit allocation system • Rules (trading rules, emission accounting and verification rules) • MRV • Accountability system for non-compliance • Administrative capacity	Jiang et al. [17] • Coverage and scope • Cap on emissions • Allowance allocation • Monitoring, reporting, and verification (MRV) • Penalty for non-compliance • Supervision and management • Marginal abatement cost • Government supervision and intervention
Shen [31] • Carbon related expertise and experience • Penalty system • Capped and uncapped enterprises • Carbon intermediaries • Financial reward • Government attitude to the ETS	Jotzo [19] • Policy instruments • Size threshold for inclusion in ETS • Coverage of the carbon market • Transaction costs • Incentive mechanism • Permit allocation • Emission data collection and analysis	Peng [27] • Weak emissions statistics in the Chinese building sector • Lack of reliable technical and economic data and information related to Chinese building energy consumption and carbon emissions • Lack of trade methodologies and standards

Table 52.2 The 15 representative factors

Factors
Government (**F1**)
Emitters (**F2**)
Independent consulting organizations (**F3**)
Technology for building energy saving and emission reduction (**F4**)
Statistics of carbon emissions in the Chinese building sector (**F5**)
Monitoring performance on carbon emissions in the Chinese building sector (**F6**)
The scattered energy consumption and carbon emission in building sector (**F7**)
Decentralization of property rights in building sector (**F8**)
Cost for carbon emission reduction in the Chinese building sector (**F9**)
Legal system for ETS (**F10**)
Coverage scope of ETS (**F11**)
Carbon emission baseline for various types of buildings (**F12**)
MRV systems (**F13**)
Incentive mechanism (**F14**)
Penalty mechanism (**F15**)

Table 52.3 The Adjacency Matrix between 15 representative factors (A)

A	F1	F2	F3	F4	F5	F6	F7	F8	F9	F10	F11	F12	F13	F14	F15	OD
F1		2	2	1	2	3	2	2	1	3	1	1	2	3	3	28
F2	1		1	3	3	2	2	1	3	1	1	1	1	1	1	22
F3	1	1		0	1	2	0	0	1	0	0	0	1	0	0	7
F4	1	0	0		0	1	1	1	2	0	1	1	0	0	0	8
F5	1	0	1	0		0	0	1	0	0	2	2	0	0	0	7
F6	1	0	0	0	0		1	1	0	0	2	2	0	0	0	7
F7	3	2	2	1	2	2		1	1	1	3	3	2	2	1	26
F8	3	2	2	1	2	1	1		1	2	3	3	1	1	1	24
F9	0	0	0	0	0	0	1	1		0	2	1	0	0	0	5
F10	1	3	2	1	2	2	2	1	1		2	2	1	1	1	22
F11	0	2	1	2	1	1	1	1	0	1		1	1	1	1	14
F12	0	2	1	2	1	1	1	1	0	0	2		1	1	1	14
F13	1	2	3	1	2	3	1	1	1	0	1	1		0	1	18
F14	0	2	1	1	1	1	1	0	1	1	1	1	1		1	13
F15	0	2	2	1	1	1	1	0	1	1	1	1	1	1		14
ID	13	20	18	14	18	20	15	12	13	10	22	20	12	11	11	

In order to get the Adjacency Matrix between the 15 representative factors, the seven experts selected above were approached further through emails at later stage to assess the contextual relationships between the 15 representative ETS factors by conducting the pairwise comparison of the factors. As a result, the degree of contextual relations between the 15 representative factors are expressed in the Adjacency Matrix, as shown in Table 52.3.

Table 52.3 shows the mutual influence degree, which are based on the value of influence degree with the help of the experts' evaluation using the 0–3 points system. Figure 52.1 reveals the connectivity status of each factor. It can be seen that there exist difference of the OD, ID, and connective lines, but these indicators just only explain the mutual relations between each other and have not taken into account the influence of one factor to the system. Therefore, it is not very reasonable to identify key elements according to the "degree" centrality.

In order to identify key factors more accurately and comprehensively, this study further calculated "closeness centrality" of each factor based on the Adjacency Matrix. The OCc and ICc of each factor were calculated with the use of SNA software "Ucient 6". The calculation result is shown in Fig. 52.2.

According to the study of Huang and Xiong [16], the key factors are those factors that having high "out closeness centrality" ($90 < OCc \leq 100$) and relatively lower "in closeness centrality" ($70 < ICc < 89$). Based on this, it can be seen from the Fig. 52.2 that "government" (F1) (OCc = 100, ICc = 73.684), "emitters" (F2) (OCc = 100, ICc = 77.778), "the scattered energy consumption and carbon emission in building sector" (F7) (OCc = 100, ICc = 82.353) and "decentralization of property rights in building sector" (F8) (OCc = 100, ICc = 87.500) are key factors

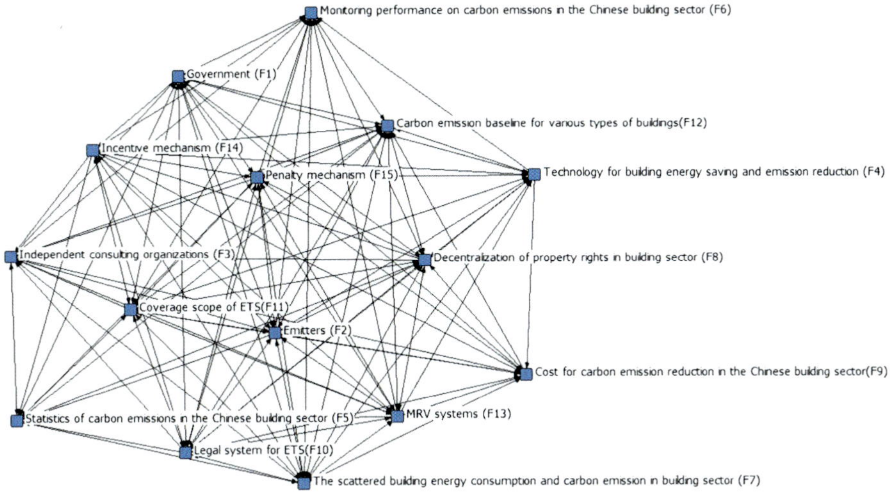

Fig. 52.1 Map of relations between each factor

		1 inFarness	2 outFarness	3 inCloseness	4 outCloseness
11	Coverage scope of ETS(F11)	15.000	16.000	93.333	87.500
12	Carbon emission baseline for various types of buildings(F12)	15.000	17.000	93.333	82.353
6	Monitoring performance on carbon emissions in the Chinese building sector (F6)	16.000	23.000	87.500	60.870
7	The scattered building energy consumption and carbon emission in building sector (F7)	16.000	14.000	87.500	100.000
3	Independent consulting organizations (F3)	17.000	22.000	82.353	63.636
8	Decentralization of property rights in building sector (F8)	17.000	14.000	82.353	100.000
5	Statistics of carbon emissions in the Chinese building sector (F5)	17.000	23.000	82.353	60.870
2	Emitters (F2)	18.000	14.000	77.778	100.000
4	Technology for building energy saving and emission reduction (F4)	18.000	21.000	77.778	66.667
13	MRV systems (F13)	18.000	16.000	77.778	87.500
9	Cost for carbon emission reduction in the Chinese building sector(F9)	18.000	24.000	77.778	58.333
1	Government (F1)	19.000	14.000	73.684	100.000
15	Penalty mechanism (F15)	19.000	16.000	73.684	87.500
14	Incentive mechanism (F14)	20.000	16.000	70.000	87.500
10	Legal system for ETS(F10)	21.000	14.000	66.667	100.000

Fig. 52.2 Results of OCc and ICc generated by using the Ucient6.0 software

affecting the implementation of emission trading system (ETS) in the Chinese building sector. And this result is consistent with the conclusion presented in Table 52.3, which is based on the degree centrality.

52.4 Discussion

In the previous section, four key factors have been identified that affect the implementation of ETS in the Chinese building sector, namely "government" "emitters", "the scattered energy consumption and carbon emission in building sector", and "decentralization of property rights in building sector".

"Government", as the ETS designer, supervisor and policies provider, government provides policies and makes decisions through verifying carbon emission, subsidizing the application of carbon emission reduction technology, allocating carbon emissions quotas to carbon emitters, imposing penalties on non-compliance of emission reduction regulation, and regulating the practice of ETS market [34]. Previous studies have appreciated that the leadership role by government will determine to a large extent the success of the implementation of ETS market in China [19, 34]. So, it should be given the highest consideration to the Chinese government to enhance her governing capacity to build an effective ETS market in the Chinese building sector.

"Emitters", as the ETS targeted objects, make their decisions for energy technology selection, carbon credit transactions and commodity marketing planning by minimizing the losses caused by the ETS policy. This factor plays a linkage role, with receiving the influence and actions from the other factors, and in turn, exerting effects on those dependent factors. For example, emitters will be affected by the factor of scattered energy consumption which determines the emitters' emission reduction behavior and performance [15, 21, 23]. In turn, emitters will affect other factors such as the technology for building energy saving and emission reduction, statistics of carbon emission. It can be seen that these linkage factors affect all participants, so a joint effort from all participants should be made to address these factors.

"The scattered energy consumption and carbon emission in building sector" and "Decentralization of property rights in building sector", as the unique features of building sector, are too scattered in comparing with other industries such as power industry, materials industry. The implementation of ETS in building sector encounters its own challenges due to the unique characteristics of the industry [3, 21, 33]. So these two factors should be treated as strategic issues in promoting the ETS in the Chinese building sector.

52.5 Conclusion

This study provides greater insight into the key factors affecting the implementation of ETS in the Chinese building sector and distinguishes the key factors from typical factors. This key factors include "government", "emitters", "the scattered energy consumption and carbon emission in building sector", and "decentralization of property rights in building sector".

The findings from this study provide valuable insights on the influence factors affecting the implementation of ETS in the Chinese building sector. For example, understanding of the key factors and the importance degree in connecting with the actual conditions can help the top-level authorities and policy designers make effective policies and decisions to guide the ETS in building sector develop smoothly and effectively. From the perspective of building owners and business companies, understanding of these factors will help them develop and promote new technologies, professional talents and other more targeted measures to meet the requirements of ETS. For relevant management departments of building sector, understanding the key influencing factors will move them to produce specific standards and regulations in the field of building energy saving and emissions reduction to mitigate those barriers.

Acknowledgments This research is funded by the Key Project of the National Natural Science Fund of China (Project No: 15AZD025).

References

1. Al-Sallal KA (2014) A review of buildings' energy challenges. Int J Environ Sustain (IJES) 3
2. Attri R, Dev N, Sharma V (2013) Interpretive structural modelling (ISM) approach: an overview. Res J Manage Sci 2319:1171
3. Bartels C, Berlin K, Cavey EA, Desiderio M, Lesmes S, Raines FD, Sahadi RJ (2005) System and method for residential emissions trading. Google Patents
4. Berelson B (1969) Content analysis for the social sciences and humanities. Addisson-Wesley, Reading
5. Borgatti SP (2005) Centrality and network flow. Soc Netw 27:55–71
6. Cantner U, Graf H (2006) The network of innovators in Jena: an application of social network analysis. Res Policy 35:463–480
7. Demailly D, Quirion P (2008) European Emission Trading Scheme and competitiveness: a case study on the iron and steel industry. Energy Econ 30:2009–2027
8. Eom J, Clark L, Kim SH, Kyle P, Patel PL (2012) China's building energy use: a long-term perspective based on a detailed assessment. PNNL report, PNNL-21073
9. Freeman LC (1977) A set of measures of centrality based on betweenness. Sociometry 35–41
10. Freeman LC (1979) Centrality in social networks conceptual clarification. Soc Netw 1:215–239
11. Ge X (2014) Key element identification in cooperative technological innovation risk on social network analysis 316–319
12. Greiner S, Lieberg K (2011) Carbon trading in the Chinese building sector. Greenhouse Gas Market Rep 2011:26–29
13. Guo Y (2012) Influence factors analysis of co operative innovation for aeronautical enterprises based on social network analysis. J Ind Technol Econ 7:011
14. Han G, Olsson M, Hallding K, Lunsford D (2012) China's carbon emission trading. An overview of current development. Stockholm: FORES/Stockholm Environment Institute. Online at: http://www.sei-international.org/publications
15. Hou J, Wu Y, Liu Y (2014) Research on the marketization ways of existing public buildings'. Energy Effi Retrofit 21:10001–10005
16. Huang Y-L, Xiong Y (2010) Key element recognition of regional innovation environment based on social network analysis. J Beijing Univ Posts Telecommun (Soc Sci Ed) 2:019
17. Jiang JJ, Ye B, Ma XM (2014) The construction of Shenzhen's carbon emission trading scheme. Energy Policy 75:17–21
18. Jiang MP, Tovey K (2010) Overcoming barriers to implementation of carbon reduction strategies in large commercial buildings in China. Build Environ 45:856–864
19. Jotzo F (2013) Emissions trading in China: principles, design options and lessons from international practice. Centre for Climate Economics & Policy, Crawford School of Public Policy, The Australian National University
20. Krippendorff K (2012) Content analysis: an introduction to its methodology. Sage
21. Lam P, Chan EHW, Yu ATW, Cam WCN, Yu JS (2014) Mitigating climate change in the building sector. Facilities 32:342–364
22. Li J, Colombier M (2010) Economic instruments for mitigating carbon emissions: scaling up carbon finance in China's buildings sector. Clim Change 107:567–591

23. Liu X, Wu Y, Chen X (2013) An overview of building emission trading system development in China. Urban Dev Stud 20:64–69
24. Nejat P, Jomehzadeh F, Taheri MM, Gohari M, Majid MZA (2015) A global review of energy consumption, CO_2 emissions and policy in the residential sector (with an overview of the top ten CO_2 emitting countries). Renew Sustain Energy Rev 43:843–862
25. Nishida Y, Hua Y (2011) Motivating stakeholders to deliver change: Tokyo's Cap-and-Trade Program. Build Res Inf 39:518–533
26. Oberheitmann A (2012) CO_2-emission reduction in China's residential building sector and contribution to the national climate change mitigation targets in 2020. Mitig Adapt Strat Glob Change 17:769–791
27. Peng M (2013) Analysis on the status and feasibility of the emission trading system in the Chinese building sector. Constr Sci Technol 16:57–61
28. Perdan S, Azapagic A (2011) Carbon trading: current schemes and future developments. Energy Policy 39:6040–6054
29. Ren H, Lu Y, Cai W, Xie Q (2013) Study on framework of buildings' carbon emissions trading in China. Urban Dev Stud 20:70–76
30. Shen L, He B, Jiao L, Song X, Zhang X (2015) Research on the development of main policy instruments for improving building energy-efficiency. J Cleaner Prod
31. Shen W (2014) Chinese business at the dawn of its domestic emissions trading scheme: incentives and barriers to participation in carbon trading. Clim Policy 1–16
32. Shih H-Y (2006) Network characteristics of drive tourism destinations: an application of network analysis in tourism. Tour Manag 27:1029–1039
33. Swan LG, Ugursal VI (2009) Modeling of end-use energy consumption in the residential sector: a review of modeling techniques. Renew Sustain Energy Rev 13:1819–1835
34. Tang L, Wu J, Yu L, Bao Q (2015) Carbon emissions trading scheme exploration in China: a multi-agent-based model. Energy Policy 81:152–169
35. Weber RP (1990) Basic content analysis, quantitative applications in the social sciences, vol 19. Sage Publications, Inc, Beverly Hills, CA, pp 24–26
36. Yao X, Yang Y, Fu YJ, Li YL, Wu GS (2014) Hot issues detection on Weibo based on social network analysis. Adv Mater Res Trans Tech Publ 1818–1825
37. Yu ATW, Wu Y, Zheng B, Zhang X, Shen L (2014) Identifying risk factors of urban-rural conflict in urbanization: a case of China. Habitat Int 44:177–185
38. Zhang D, Karplus VJ, Cassisa C, Zhang X (2014) Emissions trading in China: progress and prospects. Energy Policy 75:9–16
39. Zhang F, Guo Y, Chen X (2011) Research on China's Power sector carbon emissions trading mechanism. Energy Procedia 12:127–132
40. Zhang H (2013) Emissions trading and China: learning by doing and implications for policy making. Aust Environ Rev 28:690–694

Chapter 53
Economic Thinking on the Development of Tourism Real Estate in Poyang Lake Ecological Economic Zone

Qunhong Liu, Qian Yang and Shengyun Niu

Abstract With the rapid development of tourism and real estate, tourism real estate becomes a new economic growth gradually. It has the vital significance that tourism real estate develops in Poyang Lake ecological economic zone by using ecological resources. And based on the demand and supply analysis of tourism real estate's development, this paper explain some economic effects about its developing from these aspects: driving the tourism economic growth in the area, solving the increasing of local resident's income, improving the regional real estate's development, and achieving the interaction between ecological economy and industrial economy. This paper further probes into how to develop, such as integrating the tourism real estate resources, building the ecological brand of tourism real estate, and formulating the laws and special planning of tourism real estate.

Keywords Poyang lake ecological economic zone · Tourism real estate · Economic thinking

53.1 Introduction

In recent years, with the development of China's rapid economy, the improvement of people's living standards and the traffic modernization, a large demands for tourism products have emerged. Tourism real estate provides special facilities for tourists, which is part of the tourism products. In the contest of that real estate, industry has become an important pillar of the national economy. Tourism real estate develops vigorously, whose form of development and operation gradually derive to property right hotel, timeshare hotel, scenic residential, leisure and eco-resorts and other models, in addition to hotels and resorts in traditional development methods [1].

Q. Liu (✉) · Q. Yang · S. Niu
Institute of Urban Construction, Jiangxi Normal University, Nanchang, China
e-mail: liuqh6678@163.com

Y. Wu et al. (eds.), *Proceedings of the 20th International Symposium on Advancement of Construction Management and Real Estate*,
DOI 10.1007/978-981-10-0855-9_53

The State Council officially approved the Poyang Lake ecological economic Zone planning that taking the construction of Poyang Lake ecological economic zone as a national strategy in December 12, 2009. Jiangxi Province will focus on building “Ten industrial base”, including eco-tourism base in the future around Poyang Lake area “three districts, one platform” for the overall positioning. With tourism real estate and tourism closely linked, building eco-tourism base is inseparable from the support of tourism real estate, both interactively complement and promote each other. The development of tourism real estate in Poyang Lake ecological economic zone is adaptation to local conditions, which occupies one third of Jiangxi and relies on its unique ecological advantages in resources. This paper expounds the economic functions for the development of tourism real estate and its countermeasures through the analysis of the supply and demand of the development of tourism real estate in Poyang Lake ecological economic zone.

53.2 Demand Analysis of Tourism Real Estate's Development in Poyang Lake Ecological Economic Zone

53.2.1 *The Arrival of Leisure Time, Large-Scale Demands for Holidays Generated*

The arrival of leisure time, the tourism real estate industry has brought about the development opportunity. Firstly, it is reflected in the transformation of the concept and diversification of tourism purposes. People's traditional tourism concept has gradually evolved into a leisure vacation tourism, and the purpose of tourism is also showing a trend of diversification with income increasing. There are rich ecological resources in Poyang Lake ecological economic zone such as a variety of rare birds, aquatic organisms and wetland resources. Building tourism real estate based on ecological resources effectively meet the people's demands for leisure travel.

Furthermore, according to the World Leisure Organization prediction that when the per capita GDP of a region reached 2000 US dollars, leisure demands are increasing rapidly and the leisure options are diversified; when per capita GDP reached 3000 US dollars, large-scale demands for holidays generated [2]. A series of data show that Jiangxi's per capita GDP in 2009 has surpassed \$2000, per capita GDP of Jiangxi Province in 2014 was about \$5677, the national GDP per capita was about \$7485. It indicates that leisure demands will grow rapidly and show the character of diversity, which has formed a large-scale holiday needs in nationwide scope. Coupled with the cities and counties in the region strengthened the current infrastructure, dedicated to improving the appearance and transport, tourism real estate in Poyang Lake ecological economic zone will be greatly developed.

53.2.2 *People Income's Increase, the Improvement of Tourism Demand Level*

In recent years, with the constant increase of people's income of our country, the standard of living is gradually improving in all regions.

As shown in Table 53.1, from 2007 to 2013, the household consumption expenditure by region is increasing in Jiangxi [3], which shows that tourism is also rising.

Tourism spending and incomes indicate that with increasing of people's travel desire and ability, tourism demand level gradually improved. As a result of the tourism, real estate is a kind of tourism demand products, people's demand level for tourism real estate is also constantly improved. According to Maslow's theory of demand level and features of demand level for tourism products, demand for tourism real estate can be divided into three levels: (1) Basic needs, including life needs, security needs and the needs of tourism real estate products; (2) Innovation needs, it is that people change their original living environment, content and methods, it is also a pursuit for the new living environment, content and manner, which mainly refers to enjoying new products and services for tourism real estate endowed with local culture, customs, spirit thought. (3) Communication needs, it is about to communicate with others in tourism real estate, sharing ideas and culture from different places. The development of tourism real estate in Poyang Lake ecological economic zone is adhering to the trend of the times, meeting people's needs of the improvement of living standards and spirit and culture. It complies with a social trend of improving tourism real estate demand level.

Table 53.1 Household consumption expenditure by region

Region	2013	2012	2011	2010	2009	2008	2007
Nanchang	31,072	28,051	24,975	20,261	18,788	18,057	14,848
Jingdezhen	16,137	14,637	13,489	10,835	8519	7997	6994
Pingxiang	10,732	10,323	9660	8764	8199	7977	6758
Jiujiang	11,121	10,115	8925	7680	6897	6035	5018
Xinyu	14,496	13,465	12,326	11,169	9520	10,227	7239
Yingtan	11,445	10,480	9512	9069	10,494	6816	7536
Ganzhou	7308	6604	5912	5555	4780	4295	3914
Jian	8056	7196	6329	5468	4782	4394	3760
Yichun	9750	8810	7866	6815	5830	5192	4397
Fuzhou	6830	6007	5177	4320	4005	3987	4517
Shangrao	7319	6759	6195	5102	4686	3946	3648

53.2.3 The Development of Ecological Tourism Brings Secondary Demand

The development of traditional tourism follows the management thought and methods of industrial revolution to develop and utilize tourism resources, as a result of that the ecological environment has been severely damaged. Over the past 20 years, with the enhancement of human environmental consciousness, ecological tourism with the theme of "approach to nature, close to nature, return to nature" is growing rapidly. According to the Poyang Lake ecological economic zone planning, Jiangxi will focus on the development and management of green tourism and build ecological tourism demonstration base for the development and construction of tourism real estate. According to the guidance function of supply on demand, the demand for ecological tourism will increase. Wherefore, tourism real estate, as a kind of tourism products and its reliance on eco-tourism, will become a secondary demand generated by the demand for eco-tourism.

53.3 Supply Analysis of Tourism Real Estate's Development in Poyang Lake Ecological Economic Zone

53.3.1 The Supply Status of Tourism Real Estate in Poyang Lake Ecological Economic Zone

The tourism industry has taken shape in Poyang Lake Ecological Economic Zone. It relies on the mature scenic spots such as Lushan, Sanqingshan, LungFuShan, Wuyuan, Jingdezhen and others. The tourism real estate which provides accommodation and leisure for tourists is also developed, such as Lushan Sunshine Hotspring Resort, Wuyuan Cloud River villas. However, the current majority of tourism real estate development is based on the development of mature resorts, while ignoring the economic role of other natural ecological resources.

As shown in Table 53.2, the income of Jiangxi's tourism industry is not high from 2007 to 2013, this restricts the development of tourism real estate because that tourism is an important support for industry development. Overall, the tourism real estate in the Poyang Lake region is insufficient in effective supply.

Table 53.2 Jiangxi tourism income over the years

Year	Total tourist income (billion Yuan)	Accounting for total national tourism income (%)	Accounting for the province's GDP (%)
2007	463.67	4.23	8.43
2008	559.38	4.83	8.63
2009	675.61	5.20	8.83
2010	818.32	5.21	8.66
2011	1105.93	4.92	9.45
2012	1402.59	5.42	10.83
2013	1896.06	6.43	13.22

53.3.2 The Conditions and Advantages of the Increase of Tourism Real Estate Supply in Poyang Lake Ecological Economic Zone

Resource advantages: Poyang Lake Ecological Economic Zone not only has rich history and culture and red tourism resources, but also has exceptionally eco-tourism resources, including 2 World Heritage Sites, 2 World Heritage nominations, 2 World Geological Parks, an important International Wetland, 5 National Heritages, 2 national historical and cultural cities, 5 China excellent Tourism Cities, 8 State-level Scenic Spots, 5 National Nature Reserves, 22 National Forest Parks, a Chinese strong tourism county and 2 Jiangxi strong tourism counties. These are the basis of tourism and tourism real estate development.

Policy advantages: Poyang Lake ecological economic zone is essentially a policy area, Jiangxi, as an undeveloped province, enjoys preferential policies granted by the central government, and Poyang Lake ecological economic zone planning is an important opportunity. After planning approval, governments at all levels are actively coordinating and accelerating the construction of infrastructure, and in accordance with the development plan. Jiangxi Province will focus on creating eco-tourism base in the future so that it become a strong driving force of tourism real estate development.

Location advantages: Poyang Lake ecological economic zone is located in the central area, but it is the only ecological economic zone to be surrounded by the Yangtze River Delta economic zone, Jingjiang city circle, Wuhan city circle, Tan economic zone, the Pearl River Delta economic zone and the Western Taiwan Straits economic zone. It is able to expand its eco-tourism and tourism potential demand for real estate through taking all advantages of its own conditions and surrounding economic zones to achieve complementary resources to become a direct economic hinterland accepted radiation.

Traffic advantage: Poyang Lake Ecological Economic Zone has the Gan River connecting to the Yangtze River, which has formed a preliminary traffic architecture. It still strives to build a integrated and efficient transport system of joining the north and south, linking up the west and east. Such as railways construction, including Jiujingqu railway, Anhui-Jiangxi railway, XiangPu railway, Hengchaji railway, Hangzhou south railway; high-speed construction, including Ganzhou to Chongyi, Fengxin to Tongu and other 12 highway, and Dexing to Nanchang, Jiujiang Yangtze River Bridge; aviation construction, including Changbei airport, Jinggangshan airport, Mingyueshan airport and Sanqingshan airport. The improvement of traffic conditions provides a favorable guarantee for the development of tourism real estate.

53.4 The Economic Role of Tourism Real Estate Development in Poyang Lake Ecological Economic Zone

53.4.1 The Development of Tourism Real Estate Can Guide the Tourism Demand and Promote the Growth of Tourism Economy

The development of tourism real estate in Poyang Lake ecological economic zone has two aspects: on the one hand, it should be developed in accordance to local conditions. To be specific, tourism real estate development and construction should based on natural geographical features and ecological resources; on the other hand, the sources of the real estate include transformation for the original place in addition to the development of tourism, it is possible to reform the original place and make appropriate promotional strategies, which plays a guiding role for tourism demand and its secondary demand. With the multiplier effect, tourism real estate can ultimately promote tourism growth.

53.4.2 Development of Tourism Real Estate Can Effectively Solve the Problem of Increasing the Income of Local Residents

First of all, the tourism real estate development and construction can be taken as the project to attract investment, and actively called for local residents to participate in the project construction and management to solve the problem of employment; secondly, the tourist destinations in the Poyang Lake ecological economic zone relying on eco-tourism resources is basically in suburban or rural areas, where need land transfer for the development of tourism real estate, so it can use of transfer to solve the employment problem of the residents.

53.4.3 The Development of Tourism Real Estate Can Improve the Real Estate Industry in the Region

Tourism real estate development in the eco-tourism base means that the eco-tourism base has become the geographical expansion of the base of the real estate enterprises. The policy regulations of the state for the housing market provide an opportunity for the Poyang Lake eco economic zone tourism real estate development, because in the context of leisure and tourism, real estate enterprises and social capital will flow into the tourism real estate industry. In addition, tourism real estate development, transformation and product diversity can promote product structure adjustment of regional real estate market. In short, the development of tourism real estate can improve the real estate industry in the region.

53.4.4 Actively Protect the Ecology, Achieve Ecological Development of the Industrial Economy

The development of tourism real estate in Poyang Lake ecological economic zone will inevitably damage the local natural environment. In order to meet its development orientation, eco-tourism as the slogan of tourism real estate projects must pay attention to ecological protection and restoration. In project development of ecological protection, actively protect the ecology with the consciousness of green low-carbon and the integration of eco-tourism resources to achieve ecological development of the industrial economy.

53.5 Thinking on the Development of Tourism Real Estate in Poyang Lake Eco Economic Zone

53.5.1 Integration of Eco-tourism Resources and Tourism Real Estate Resources

The existing eco-tourism resources should be integrated with tourism real estate resources and establish eco-tourism and tourism real estate management department in Poyang Lake ecological economic zone to conduct a comprehensive investigation for ecological resources, break time and space constraints, with which to explore more and more excellent resources; develop appropriate tourism real estate projects exploiting or creating traffic conditions; gradually realize the inter cooperation of regional projects and achieve the resource sharing and advantage compensation

53.5.2 Formulate Relevant Laws and Regulations, and Promote the Healthy Development of Regional Tourism Real Estate Market

The phenomenon of people returning to nature, and appreciating nature have promoted the tourism real estate development, while it can also result in destruction to the natural environment. At present, there is a lack of ecological tourism and tourism real estate in the legal system that has weakened the bottom line of the natural environment protection, which has caused contradictions between economic development and ecological conditions. Thus there is necessary to formulate relevant laws and regulations, and achieve the legalization of industrial development to actively guide eco-tourism and tourism real estate industry into the healthy development track, effectively avoid the environmental problems and social problems arising from the development process. Ultimately, it could achieve the unity of economic benefits, ecological benefits and social benefits.

53.5.3 Strengthen Ecological Marketing Positioning, Building the Ecological Brand of Tourism Real Estate State

Ecology is a brand of tourism real estate. Now people's environmental consciousness has been awakened and the requirements of real estate products have been increased. The development of green, low-carbon products of tourism real estate not only meets the strong desire to protect the environment and ecology of the community, but also can reduce the damage to the ecological situation in the project area. In the times of rapid development of tourism and real estate, in order to cope with the fierce market competition, it must be properly positioned to build their own brand. Ecological resources are the most unique advantage in Poyang Lake ecological economic zone which should be fully utilized and take eco-marketing positioning to achieve brand effect.

53.5.4 Formulate Special Planning of Tourism Real Estate Development

The location selection and layout of tourism real estate should focus on tourism resources because it is the basic condition of tourism real estate. Due to the constraints of traffic, location and many other factors, tourism real estate must be combined with town planning, tourism planning and land use planning to formulate scientific and perfect tourism real estate development plan. Only the planning first can achieve the rapid and healthy development of the entire industry.

References

1. Yang J, Su C (2015) Empirical research on ecological and cultural tourism real estate development model. J Ecol Econ 01:147–149
2. Zhan J, Yu X (2008). The rise and demand characteristics of leisure vacation tourism real estate. J Econ Forum 2:62–65
3. Jiangxi Bureau of Statistics (2014) Jiangxi statistical yearbook. China Statistics Press

Chapter 54
The Calculation and Prediction of the Non-equilibrium Degree of the Beijing Commercial Housing Market

Hong Zhang, Shuai Gao, ZhiYi Shu and Yang Zhang

Abstract The non-equilibrium degree is the ratio of the difference effective supply to effective demand with the actual transaction volume in that condition to the imperfect market, no sensitive price system. It can be more sensitive to reflect the state of the housing market fluctuations between the effective supply and demand. This paper constructed a non-equilibrium model of Beijing city commercial housing market on the basis of Lessoning the theory on non-equilibrium model. Further to combine with the consideration of various Commodities' supply and demand factors, This paper constructed the system dynamics model of commercial housing market in Beijing city from the theory of comprehensive system view, verified the validity of the model through the simulation empirical study, measured the non-equilibrium degree of Beijing commercial housing market consistenting with the reality of the situation, Meanwhile, predicted the non-equilibrium degree of commercial housing market in Beijing from 2013 to 2018, thus provided a new sensitive index for early warning of the real estate market.

Keywords Commercial housing market · Effective demand · Effective supply · Disequilibrium degree · System dynamic model

54.1 Introduction

The non-equilibrium degree is the ratio of the difference effective supply to effective demand with the actual transaction volume in that condition to the imperfect market, no sensitive price system. It can be more sensitive to reflect the state of the

H. Zhang · S. Gao (✉) · Z. Shu
Center for Urbanization and Industrial Development,
Tsinghua University, Beijing, China
e-mail: sparkle1979@163.com

Y. Zhang
Department of Property Management, Beijing Forestry University, Beijing, China

Y. Wu et al. (eds.), *Proceedings of the 20th International Symposium on Advancement of Construction Management and Real Estate*,
DOI 10.1007/978-981-10-0855-9_54

housing market fluctuations between the effective supply and demand. If the non-equilibrium degree became smaller, the market situation was closer to the ideal equilibrium state and the commercial price would also close to its value. Thus, the non-equilibrium degree could be used to forecast whether the market effective supply and effective demand are equilibrium or not. However, as effective demand and effective supply that is effected by others factors are unobservable variables. The influence factors of price with the demand and supply formatted a complex circulatory system, as a result, the non-equilibrium haven't been used as a warning index before. Most of previous studies are focus on using models to test history data for some static analysis report. They didn't consider about the dynamic change, there is also no researchers try to use this index in commodity market prediction.

Based on the lacking of existing research, this paper take non-equilibrium degree model as the basis, use system dynamics theory and take Beijing commodity market as a research object, Because Beijing commodity market is the representative of national market, its fluctuation can affect the national economy. This paper try to use system to analysis the factors that affect non-equilibrium degree, create the system dynamics model of non-equilibrium degree, then provide effective policy basic advice with this index.

54.2 Literature Review

The ideal supply-demand means there is no quantitative restriction in an equilibrium market, people in this market can pursue maximal benefit by unlimited trade and these trades compose the demand and supply. In reality market, effective demand and effective supply are affected by many other factors, for example, the market structure's flaw and frictions. Thus, overproduction and undersupply might happen. Generally, the market is non-equilibrium, it is hard to reach Pareto optimality condition, which is non-equilibrium against Walrasian Equilibrium, Charemza and Quandt established a Hyperbola Regress model of non-equilibrium formula [1] (54.1) to calculate the non-equilibrium degree.

$$\begin{cases} S_t = f(P_t, Y_t) \\ D_t = f(P_t, X_t) \\ Q_t = 1/2 * (D_t + S_t) - 1/2 * \sqrt{(D_t - S_t)^2 + 4 * r^2 * D_t * S_t} \\ Z_t = (D_t - S_t)/Q_t \end{cases} \tag{54.1}$$

In formula (54.1), S_t = effective supply in a given year, D_t = effective demand in a given year, Q_t = trade volume of product in a given year, Z_t = non-equilibrium degree in a given year, P_t = price of product, X_t = variable set of demand, Y_t = variable set of supply, r = convergence level of micro markets, the lower r is, the lower trade friction emerge, the Macro Market also more closer to market conform to shortage Rule. At the same time, $r > 0$ is normal phenomenon of market, thus, for any r($r > 0$), there is $Q_t < \min(D_t, S_t)$ and $\lim_{r \to 0} Q_t = \min(D_t, S_t)$.

Many scholars made outstanding contribution on empirical research using the non-equilibrium theoretical model. For example, Li [2], Wang [3] classified the non-equilibrium situations. Wang [4] set up a dynamic analysis model of real estate demand-and-supply market. However, most of these studies are confined to econometric model static analysis of supply and demand in commodity market. Nevertheless, the supply and demand is each other's essential prerequisites, also a dynamic adjustment mode in reality market. Furthermore, the purpose of empirical research is forecast future accurately. Apparently, there are no many studies about dynamic Analysis and forecast. It is because there are some lacks in econometric model static, thus, we need some new research methods combine with theory to forecast the non-equilibrium degree.

System Dynamics is a subject that analyzes research Information and fed back to science. Because the system function determined by system Structure, we can use the causal models to design the feedback loop and use system Dynamics software to simulate the variation of non-linear model and complex time-varying systems, it provides a method to measure non-equilibrium degree. System Dynamics possess many incomparable advantages than econometrics in real estate market non-equilibrium degree analysis. This article build the non-equilibrium of Beijing real estate market model, use both econometrics method to analyze the statistical data, then use the System Dynamics to confirm those data. In addition, forecast the non-equilibrium degree of Beijing. Then provide some effective policy basic advice by this index.

54.3 Theoretical Analysis and Model Specification

54.3.1 Influence Factors of Real Estate Market Analysis

According to the contraction model [5] built by and others and cost theory of Industrial Economics. This paper select commercial housing price, resident's disposable income, and quantity of urban population and market expectation of purchaser as factors that affect effective demand housing market. On the other side, this paper chose commercial housing price, cost of land, cost of construction, and installation and market expectation of developers as factors that affect effective supply.

Influencing Factors of Commercial Housing Demand Analysis

Average disposable income: If average disposable income raise, purchasing power will also increase. Beside, because residence is both investment and consumer good, as a matter of course it becomes an important options that people distribute their wealth. This paper chose (inc_t) index and analyze its effect against effective demand.

Quantity of urban population: According to contraction model, quantity of urban population is positive correlation with commercial housing demand in both investment and consumer parts. Thus, we select population of household registration $(Population_t)$ as a proxy variable.

Influencing Factors of Commercial Housing Supply Analysis

Influencing factors of commercial housing market supply are based on cost theory. Commercial housing costs are composed of different kinds of real estate development cost. Based on normal production and business operation activities, calculate the actual cost during real estate development process including land cost, compensation for demolition, cost of construction and installation and tax.

The factors of land supply: Because land is unrenewable and scarce, its cost continue to rise, furthermore, our government monopolize the land market in our country, they adjust and control the land use rights by auction and give local government a fixed amount land-transferring fees. In order to prevent overdevelopment and use land designedly, government must set higher land price, then it will restrain the commercial housing supply. We use as a proxy variable in this article.

The factors of construction and installation cost: The price of building Materials also increase apparently in recent years, such as steel, cement and labor cost are keep rising. By measuring, the construction and installation cost accounts for 40 % of total cost of real estate development, it will affect commercial housing supply. This paper use as a proxy variable of construction and installation cost.

Influencing Factors of Both Commercial Housing Demand and Supply Analysis

Commercial housing price: The price influence is act on both demand and supply, this paper select P_t as a proxy variable. First of all, according to price determinism, commercial housing supply is determined by its price, the goal that developers improve supply is pursuing profits, it also mean higher price contain higher profits.

The effect of price against demand is that price racing make purchaser's burden serious and makes them feel overwhelmed with purchasing, thus, the racing of house price probably restrain the commercial housing demand.

Market expectation of developers and purchaser: If developers predict that housing price will continue rising and make more profit, it will stimulate the housing supply. Besides, the market expectation of purchaser will affect commercial housing demand. Generally speaking, people always buying up not buying down, for example, housing price in Beijing continue rising, the risk of house purchase also become higher, However, because location of Beijing has many advantage just like better service, good infrastructure and optimistic economy situation, people might look forward to more appreciation. But, excessive housing prices go beyond payment capacity of most purchaser, those people might wish for housing price falling.

Through the above analysis, housing market expectations affects both housing demand and supply situation distinctly. This paper select P_{t+1} as a proxy variable of housing market expectation.

54.3.2 Model Construction

Based on non-equilibrium theory model, housing market equilibrium is temporal and relative, however, non-equilibrium is regular and absolute. As a result, we build the model with many possible factors as follows:

$$\begin{cases} S_t = b_0 + b_1 Earth_t + b_2 Cost_t + b_3 P_t + b_4 P_{t+1} + \varepsilon_{st} \\ D_t = population_t * (a_0 + a_1 inc_t + a_3 P_t + a_4 p_{t+1}) + \varepsilon_{dt} \\ Q_t = 1/2 * (D_t + S_t) - 1/2 * \sqrt{(D_t - S_t)^2 + 4 * r^2 * D_t * S_t} \\ Z_t = (D_t - S_t)/Q_t \end{cases} \tag{54.2}$$

In this formula, D_t is effective commercial housing demand, S_t is effective commercial housing supply in given time (non-observed quantity), P_t is commercial housing price in given time (observed quantity), P_{t+1} is commercial housing price in postpone given time, Q_t is commercial housing trading volume in given time (observed quantity), inc_t is average disposable income in given time, $Population_t$ is quantity of urban population in given time, $Earth_t$ is land price in given time, $Cost_t$ is cost of construction and installation in given time of Beijing commercial housing market, ε_{dt} and ε_{st} are random perturbed variables.

Base on non-equilibrium theory model of Beijing commercial housing market and existing papers, combine with statistical data of Beijing commercial housing market, and simplicity of the model, then build the System Dynamics model. This paper chose 17 index variable simulate cause-and-effect relationships in Beijing commercial housing market and external environment effects against Beijing commercial housing market. As in Table 54.1, begin with level variable, affirm input variable and output variable and analyze their influence factors and form,

Table 54.1 System dynamics model primary variable of Beijing commercial housing market

Level variable	Auxiliary variable
Commercial housing demand area	Demand accomplishment, new commercial housing supply, population, average disposable income, commercial housing price, previous commercial housing price, supply accomplishment, new increase supply area, construction and installation's cost, land price, comprehensive cost, commercial housing sales area, supply-demand ratio, price impact factors of supply-demand ratio, non-equilibrium degree measurement
Commercial housing supply area	

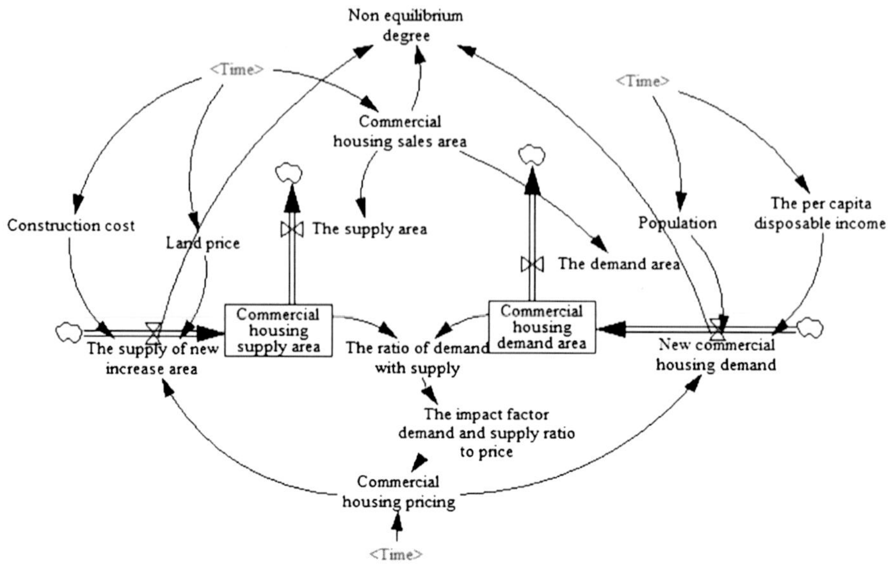

Fig. 54.1 System flow chart of Beijing commercial housing market

some factors like population, average disposable income impacted by policy, we use external call to control. On the other hand we use non-equilibrium theory model as basis and affirm the relationship between the variables by fitting, regression and table function.

From System evolution point of view,controlling principles of non-equilibrium housing market can use Fig. 54.1 to explain.

54.4 Empirical Analysis

54.4.1 Data Sources

In consideration of availability and others factors, this paper use Beijing commercial housing market date during 1994–2011 as study sample, those date collected from Beijing Municipal Bureau of Statistics and CSMAR. Furthermore, we use consumer price index for the urban residents in 1994 as benchmark to adjust inflation effect. In order to eliminate heteroskedasticity, we use logarithmic to each variable. Finally, we use average disposable income ($\ln inc_t$), land price ($\ln Earth_t$), construction and installation's cost ($\ln Cost_t$), commercial housing price ($\ln P_t$), average commercial housing ownership ($\ln \lambda_t = \ln(D_t/Population_t)$), commercial housing sales area ($\ln Q_t$).

System dynamics simulation use 2011 as base year, calculate interval is one year. The relation between these variables seeing below:

Commercial housing supply area = INTEG (new housing supply − supply accomplishment, INTIAL)
Commercial housing demand area = INTEG (new housing demand − demand accomplishment, INTAL)
Supply accomplishment = demand accomplishment = commercial housing sales area
Comprehensive cost = (construction and installation's cost + land price per floor area) * (1 + 10 %) * (1 + 30 %) (According to housing empirical data estimation)

Others relation between these variable estimated by effective demand and supply non-equilibrium estimation.

54.4.2 *Empirical Analysis*

Equilibrium Estimation of Effective Demand and Supply

Suppose that market clear, estimate effective demand equation and effective supply equation by the method of OLS. First, use stationary against each macro variable. According to ADF unit root test results as Table 54.2, variables are I(1) sequence, we also use Lagrange test to affirm whether there is the correlation item random disturbance. So we can ensure model is set correctly.

Table 54.2 The sequence and differential sequence ADF unit root test results

Sequence	Test type	ADF statistics	1 % critical level	5 % critical level	*P* value	Stationary
$\ln Q$	(C, T, 0)	−0.152	−4.380	−3.600	0.992	N
$\Delta \ln Q$	(C, T, 0)	−4.601	−4.380	−3.600	0.001	Y
$\ln Earth$	(C, T, 0)	−1.946	−4.380	−3.600	0.630	N
$\Delta \ln Earth$	(C, T, 0)	−3.953	−4.380	−3.600	0.010	Y
$\ln \cos t$	(C, T, 0)	−3.172	−4.380	−3.600	0.090	N
$\Delta \ln \cos t$	(C, T, 0)	−3.705	−4.380	−3.600	0.022	Y
$\ln P$	(C, T, 0)	−1.316	−4.380	−3.600	0.8838	N
$\Delta \ln P$	(C, T, 0)	−4.110	−4.380	−3.600	0.0061	Y
$\ln inc$	(C, T, 0)	−1.350	−4.380	−3.600	0.8749	N
$\Delta \ln inc$	(C, T, 0)	−3.929	−4.380	−3.600	0.0111	Y
$\ln \lambda_t$	(C, T, 0)	−4.372	−4.380	−3.600	0.9934	N
$\Delta \ln \lambda_t$	(C, T, 0)	−4.180	−4.380	−3.600	0.0024	Y

Note In test (C, T, L), C = constant tern, T = time trend, L = lagged variable

According to unit root test results, then use co-integration theory put forward by test co-integration relationship between variables. In the first place, use regression test the effective demand model, then use ADF unit root test, we found that residual sequence is stationary, thus, it is long-term stability between variables. Results of regression are as follows:

Demand equation:

$$\begin{array}{llll} \ln \lambda_t = 2.65 & * \ln inc_t - 1.09 & \ln P_t - 1.84 & \ln p_{t+1} \\ \quad (5.49) & \quad (-1.18) & \quad (-2.13) & \end{array}$$

$Adj - R^2 = 0.689$, further can be simplified as

$$\ln Q_t = 2.65 * \ln inc_t - 1.09 \ln P_t - 1.84 \ln p_{t+1} + \ln population_t \tag{54.3}$$

On the test results inc_t, P_t and P_{t+1} are significant with t statistics, adjust R^2 are also significant, imitative effect is good. Then test co-integration relationship between variables, use regression test the effective demand model, then use ADF unit root test, we found that residual sequence is stationary, thus, it is long-term stability between variables.

Supply equation:

$$\begin{array}{llll} \ln S_t = -0.504 & \ln Earth_t - 1.697 & \ln Cost_t + 0.509 & \ln P_t + 0.510 \ln P_{t+1} \\ \quad (-2.69) & \quad (-3.19) & \quad (0.40) & \quad (0.41) \end{array} \tag{54.4}$$

$$Adj - R^2 = 0.99$$

On the test results, all variables are significant with t statistics, adjust R^2 are also significant, imitative effect is good.

Non-equilibrium Estimation of Effective Demand and Supply

Because of time series is only 17 years, we use simple trading volume equation that r = 0 and use NLS [6, 7] size up Hyperbolic aggregation equations [8]:

$$Q_t = 1/2 * (D_t + S_t) - 1/2 * \sqrt{(D_t - S_t)^2 + 4 * r^2 * D_t * S_t} \tag{54.5}$$

NLS model is estimated by iterative method, specific results as Table 54.3.

Base on significance of T statistics, P value of effective demand and supply variables are under or close to 5 % critical level, regression equation has good explain ability. Thus, we can build commercial housing market non-equilibrium econometric model as

Table 54.3 Trading volume equation parameters estimate result

	Explanatory variable	Coefficient estimates	T statistics	*P* value
Supply equation	$\ln Inc_t$	3.85^{**}	2.87	0.014
	$\ln P_t$	-1.868^{**}	1.92	0.043
	$\ln P_{t+1}$	-2.347^{*}	−2.10	0.058
	$\ln Population$	1	–	–
Demand equation	$\ln Earth_t$	-0.525^{*}	−5.663	0.076
	$\ln Cost_t$	-2.055^{***}	−0.922	0.000
	$\ln P_t$	0.600^{***}	4.224	0.001
	$\ln P_{t+1}$	0.686^{***}	4.997	0.001
	$Adj-R^2 = 0.995$		D.W = 1.74	

*=10 % critical level
**=5 % critical level
***=1 % critical level

Effective demand equation:

$$\begin{aligned}\ln D_t &= 3.85 * \ln inc_t - 1.868 \ln P_t - 2.347 \ln P_{t+1} + \ln Population_t \\ &= \ln \lambda_t = 3.85 * \ln inc_t - 1.868 \ln \ln P_t - 2.347 \ln P_{t+1}\end{aligned} \quad (54.6)$$

Effective supply equation:

$$\ln S_t = -0.525 \ln Earth_t - 2.055 \ln Cost_t + 0.600 \ln P_t + 0.686 \ln P_{t+1} \quad (54.7)$$

Transactions equation

$$Q_t = \min(D_t, S_t)$$

According above results, from the effective demand Eq. (54.6), average demand and average disposable income are positive correlation, when average disposable income increase 1 % average demand will also increase 3.8 %, that is to say, economic development in our country makes effective demand raise. Higher disposable income elasticity coefficient reflect that people willing to buy a house. However, housing price rising and anticipation of house prices rising are both depressed the demand, housing price increase every 1 % will lead to 2.347 % reduce of housing demand. Thus, in order to stimulate housing demand, we need to control housing prices reasonably. Also, from the effective supply Eq. (54.6), if we can improve people's living standards and upgrade peasant consumption condition, then housing purchase pressure of people will be relieved, finally demand will increase and Home Ownership Scheme could be achieved.

On the other hand, from the supply Eq. (54.7), Beijing commercial housing supply is negative correlation with land price and construction and installation cost, but, it is positive correlation with housing price and anticipation of house prices rising. For example, when land price increase 1 %, housing supply will decrease

0.525 %, also, when construction and installation cost increase 1 %, housing supply will decrease 2.055 %.This show that cost factor is the main factor that determine housing supply.

The enthusiasm of developers to develop commercial housing will be restrained by cost factor, and then it will reduce the housing supply. Besides, hosing price rising will stimulate the developers to increase housing supply, that is "capital chasing profit". When hosing price increase 1 %, housing supply will also increase 0.6 %.

When the house anticipation prices increase 1 % housing supply will increase 0.686 %. Obviously, housing price and anticipation of house prices rising can bring more profit ratio to developers and also increase effective supply.

Through those empirical analysis above, it provide us theoretical basis and variable relationship of System dynamics. Based on model, we can use System dynamics simulation on Beijing commercial housing market.

Simulation Models Results and Effectiveness Test

This paper use Vensim PLE to simulate Beijing commercial housing market for year long, based on the system dynamic model and the relationship between variable of non-equilibrium model. The calculation results during 2001–2011 are as shown in Table 54.4. System dynamic simulation results usually compare with historical data and calculate relative error to test relative result of model $e_i = \hat{y}_i - y_i/y_i, y_i$ = real value index in given year, $\hat{y}_i$ = predictive value in given year. By observing the representative indicators, we found that error in most years are less than 10 %, show that the mode simulate the change trend of Beijing commodity house market well.

Table 54.4 Beijing commercial housing market system dynamics model simulation result

Index	Commercial housing sales area (million square meters)			Commercial housing price (Yuan)		
Year	Real value	Predictive value	Error (%)	Real value	Predictive value	Error (%)
2001	1205	1175.24	2.47	2674.37	2625.70	1.82
2002	1708.3	1697.20	0.65	2563.26	2525.32	1.48
2003	1895.8	1828.31	3.56	2543.63	2500.64	1.69
2004	2472	2287.09	7.48	2686.57	2640.36	1.72
2005	2803.2	2668.09	4.82	3286.59	3315.84	−0.89
2006	2607.6	2626.64	−0.73	4298.33	4338.73	−0.94
2007	2176.6	2185.74	−0.42	5857.22	5810.95	0.79
2008	1335.4	1289.46	3.44	5990.2	5551.72	7.32
2009	2362.3	2058.51	12.86	6757.65	6496.80	3.86
2010	1639.5	1445.71	11.82	8504.33	8140.34	4.28
2011	1440	1487.23	−3.28	7629.25	7874.91	−3.22

Data source Beijing bureau of statistical

Beijing Commercial Housing Market Non-equilibrium Degree Calculation and Prediction

Base on system dynamic model and in same system structure, we can build Beijing non-equilibrium econometric index as Fig. 54.2.

From the relation between demand and supply in Non-equilibrium degree, Beijing commercial housing market from 1994 to 2012 can be divided three stages:

Stage1: During 1994–2000, the non-equilibrium degree are all minus, it means that Beijing commercial housing market was supply exceeds the demand. Because the housing market hadn't marketization in our country, there's no demand and housing supply is excess.

Stage2: During 2000–2006, the non-equilibrium degree are upright, it means that Beijing commercial housing market was demand exceeds the supply. Since state department released "No. 23" file and declared housing commercialization in 1998, Beijing commercial housing market development increased rapidly. The non-equilibrium returned to 0 on 2000, then the Ministry of Land and Resources announced the "issued 11" tender auction listing transferring state-owned land use rights provision, it pushed up the land price and restrained the supply, so the demand couldn't be satisfied.

Stage3: During 2006–2011, the non-equilibrium degree is in a trend fluctuations, in this stage, the government introduced a series of policies to tight the housing consumption and regulate the real estate market, for example "Article 11" increased the down payment of second house to 40 %. Intensive real estate regulation policy and price continue rising during 2006–2011 both restrained the effective demand.

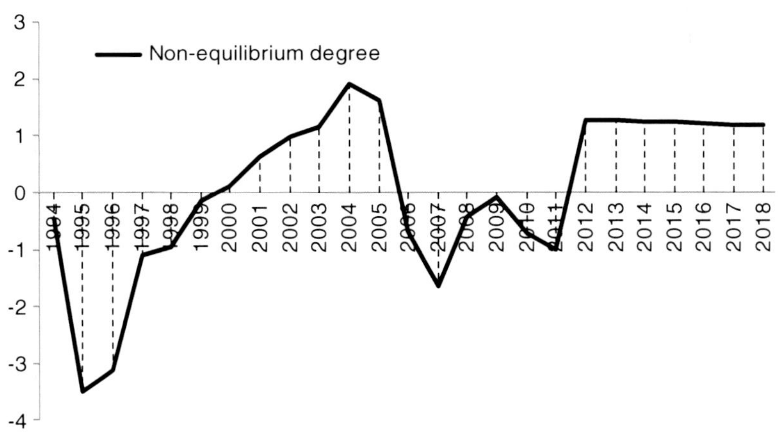

Fig. 54.2 Beijing commercial housing market non-equilibrium degree during 1994–2018

Through the analysis of the above three stages, it shows that the non-equilibrium degree index is no different with Beijing commercial housing market demand and supply. We can further predicts that the non-equilibrium degree in 2013–2018 are upright, On behalf of a period of time in the future with the same model will remain in situation that demand exceeds supply. But it is amplitude remain at about 1, it also means that commercial housing market demand and supply are in a relatively balance situation.

54.5 Conclusion

Base on non-equilibrium econometric theory mode and System dynamics theory, we use system dynamics model to summarize and analysis Beijing commercial housing market. This mode simulates the operation of commercial housing market in the last 17 years. The results of non-equilibrium degree index and imitation are consistent, they reflect the Beijing commercial housing market supply and demand situation. Further, we predict the Beijing commodity house market non-equilibrium degree in next five years, it seems that the effective demand and effective supply will enter the stable stage, demand will exceed supply and non-equilibrium degree will maintain 1. Thereby, offer advice against Beijing real estate policy formulation and implementation.

The deficiency of this paper is that in this article does not discuss the most advantage of system dynamics "Simulation experiments of policy", for example, the impacts from change of the population and the change of per capita disposable income to the commercial housing market.

In addition, while we made a beneficial attempt in System dynamics model analysis, it still not be enough, some causal relationships between variables unable to reveal.

Acknowledgments We gratefully acknowledge the financial support of the National Natural Science Foundation of China (Grant No. 71373143).

References

1. Charemza W, Quandt RE (1982) Models and estimation of disequilibrium for centrally planned economics. Rev Econ Stud 49:109–116
2. Li YN (2000) Chinese economy on disequilibrium. Guangdong Economic Press, pp 63–68
3. Wang X, Deng S (1997) The disequilibrium analysis of the cash supply and demand in China. Syst Eng Theory Pract 9:9–14
4. Wang SH (2009) The empirical analysis of the disequilibrium supply about housing security system in China. Econ Prob 3(4):41–44
5. Fortura P, Kushner J (1986) Canadian intercity house price differentials. J Am Real Estate Urban Econ Assoc 14(4):525–536

6. Li ZN (2010) Econometrics. Higher Education Press, Beijing, pp 63–74
7. Ji LC (2005) The disequilibrium real estate market. Economic Management Press, Beijing, pp 72–102
8. Wang J, Gao T (2004) The dynamic analysis of the demand and supply function of the real estate. Chin Soft Sci 4:69–74

Chapter 55
Study on Premium Level of School District Housing from the Perspective of Mass Discrepancy—Take Shenzhen as Example

Botong Song, Yaning Li and Ying Yan

Abstract Nowadays, concerning the problem that how to gain the biggest profit for the buyers of the school district housings, the scholars at home and abroad have not yet taken the housing prices difference into consideration. It is believed that the analysis of housing prices difference is more meaningful than that of absolute figures. This paper is going to take the housing prices difference among well-known and common schools as the objective and it contains three parts: ① exploring the quantitative relationships between the premium rate of the school district housing and common housing prices fluctuation, and educational levels of different schools; ② setting up the HPM (Hedonic Price Model) of the premium rate of the school district housing; ③ researching the influence degrees of commodity house price volatility and mass discrepancies of different school districts on the premium rate of the school district housing quantitatively. It is concluded that mass discrepancy makes a significant influence on the premium rate of different school district housings through empirical studies on the popular school districts in Shenzhen. Moreover, strategies of investment and purchasing of school district housings as well as suggestions on school hierarchical planning are put forward in this paper to meet requirements of those buyers.

Keywords The premium rate of the school district housing · Boundary fixed method · Hedonic price model · School district · Shenzhen · Investment

55.1 Introduction

School district housing refers to the real estates or the house properties located in targeted area set by schools to enroll students living there in accordance with the principle of entering nearby schools without exam by following rules made by education sectors or schools [1]. The big difference in price obviously exists

B. Song · Y. Li (✉) · Y. Yan
School of Shenzhen University, Shenzhen, China
e-mail: liyaning@email.szu.edu.cn

Y. Wu et al. (eds.), *Proceedings of the 20th International Symposium on Advancement of Construction Management and Real Estate*,
DOI 10.1007/978-981-10-0855-9_55

between school and non-school district housing with larger premium space in the former. Though the price of the school district housing is on the rise, its investors do not decrease in number. Some are for the famous school degree, while some are for pure investment. The right purchasing opportunity of school district housing changes as investor's different requirements. Therefore, a series of problems concerning the premium level of the different school district housing come to arise.

Researches on the influences of domestic public goods on housing price mainly adopt the HPM [2], generally focusing on basic education, landscape facility, park, rail transit, etc. separately but with less attention to a number of public goods at the same time [3]. At present, many domestic and foreign scholars haven't made any systematic researches on the premium of the school district housing. And most of the previous studies were concentrated on the influencing factors of surrounding schools on the residence price. For instance, the foreign scholars such as Black [4], Dougherty et al. [5], Gibbons et al. [6] took students' test scores, average education expenditure, student-staff ratio, housing transaction price and features into the consideration to point out that schools have great influence on the price of their nearby residential houses; what's more, the domestic scholars such as Shi [7], Yang [8], Shi [9] drew a similar conclusion. Based on that, it is meaningful to make researches on school district housing premium (price difference between school district housing and commercial housing) in this paper.

On the basis of the previous studies, this paper will take foreign dominant methods as reference, namely, the boundary fixed method [10], and to build the HPM of school district housing premium and then empirically study the relationships between premium rate of school district housing and price fluctuation of common residential housing, school mass discrepancy as well as the school district characteristics. Lastly, the appropriate school district housing investment and purchasing strategies as well as suggestions of school scale planning will be put forward based on the modal analytical results in accordance with the common residential housing's price fluctuation to meet different requirements of various pursuers.

55.2 The Establishment of the Hedonic Price Model

In the empirical analysis of HPM, the semi-logarithmic model is more reasonable [11]. This paper will analyze the school district housing's premium rate and the change rate of the commercial housing price as well as the quantitative relations of mass discrepancy among different schools. And the adopted HPM (55.1) is as follows:

$$\begin{aligned}\mathrm{PremiumRate_{it}} &= \beta_0 + \mathrm{ChangeRate_{it}} + \sum_{j=1}^{m} \gamma_j k_j + \\ &+ \sum_{i=1}^{n} \alpha_i z_{it\_jt} + \mathrm{Landprice_{it\_jt}} + \alpha_i + \varepsilon_{it}\end{aligned} \tag{55.1}$$

In the above equation:

$\mathrm{PremiumRate_{it}}$	stands for the premium rate of the ith popular school district housing chosen as samples;
$\mathrm{ChangeRate_{it}}$	stands for the change rate of the commercial housing price of the ith popular school district chosen as samples;
Ki	stands for a series of architectural features of housing estates, including the age of building, plot ratio, green coverage rate and scale (the total number of households);
Z_{it_jt}	stands for a series of mass discrepancies among famous and common schools at the time of t, which contain primary schools' levels, student-faculty ratio, higher vocational teachers' ratio, the enrollment rate of junior high school and students' average area;
$\mathrm{Landprice_{it\_jt}}$	stands for the benchmark land price difference between school district housing i and commercial housing j at the time of t;
ε_{it}	stands for stochastic error term;
β_0	stands for the constant quantities influencing price other than the hedonic variable;
α_i	stands for some unobserved features, including regional condition, architectural structure, architecture planning layout, etc.;
$\gamma_1, \gamma_2, \gamma_3, \ldots \gamma_m$; $\beta_1, \beta_2, \ldots, \beta_n$	stands for parameters to be estimated;

55.3 Empirical Researches

According to the HPM based on the time interactive items, The paper will make empirical researches on the influences of the change rate of commercial housing price and the mass discrepancy among different school districts on the school district housing's premium rate.

55.3.1 The Delimitation of Research Areas

Situated in the ancient island areas in Shenzhen, the three districts, Fu Tian, South Mountain and Luo Lake concentrating high qualified educational resources, will be chosen as the research areas in this paper. With earlier development, mature market, active second-hand housing trading market, perfect municipal facilities and fully furnished school, hospital and other governmental units, those three districts are the best to be chosen as research samples. What's more, academic degree must be taken into consideration when choosing school samples. Therefore, the stage of compulsory education (including primary school and junior middle school) will be the focus, without involving kindergarten, senior high school, higher or vocational education. In view of the investment-oriented demands of newly-built houses, unprivileged favor of "admission into the nearest schools" with excellent educational resources when renting houses, famous schools with longer history and more older buildings around schools, the writer consider it is the best to take second-hand houses data as samples. With respect to the assessment of elite school and common school, it will be ensured on the basis of the comprehensive education level presented by the metropolitan-level Education Bureau and various schools' official websites, as well as the school ranking arranged by the public.

55.3.2 Samples Illustration

The research objects involve twenty-one sample points in six school districts. Every sample point is composed of a paired school district housing and common commercial housing situated along and within the distance of 800 m [12] of the same boundary. There totally have 42 housing estates, equally split between school district housing and common commercial housing. And the samples are mainly multiple-storey and high-storey housing estates with the exception of villa and security houses. The samples of districts are listed as follows: seven sample points in two popular school district areas in Fu Tian District, three sample points in one key school district area in Luo Lake District and eleven sample points in three key school district areas in South Mountain District.

55.3.3 Data Sources and Variable Selection

In this paper, the data of the empirical research are collected from three sources: Second-hand Housing Information System of Real Estate Assessment Center in Shenzhen, Shenzhen electronic map from www.Chachaba.com, www.sz.fang.com, www.sz.gov.cn and other authoritative webs. The time span of the data is from the first quarter of the year 2010 to the fourth quarter of the year 2013.

Table 55.1 Independent variables

Category	Name of variable
Architectural feature	The age of building (from the house completion date to the date of setting the price), the scale of community (the total number of households), property fee, per household's parking space
School feature	The poll grading of primary school, the distance from primary school to the center of housing estate, the floorage of primary school building, primary school's student-faculty ratio, primary school's ratio of higher vocational teachers, the enrollment rate of junior middle school (the ratio of admission into the four elite senior high schools in Shenzhen)
Neighborhood feature	The price replacement of school district housing by the commercial housing located oppositely along the boundary
Quarterly time trend variable	The initial quarter is the first quarter of 2010 recorded as 1. The next quarter is increased one unit on the basis of the previous unit progressively as quarter goes on

Main micro influencing factors of real estate will be divided into three aspects, location, neighborhood and architecture [13]. According to the specific situations of Shenzhen and the previous related research [14], this paper's independent variables are shown in Table 55.1.

55.3.4 Analysis of School District Housing's Premium Rates

Descriptive Analysis

The Table 55.2 is the descriptive statistics of selected variables. And this table shows that there have 336 valid samples in total, and the mean value and standard deviation of model variables meet the requirements. Besides, the samples are in good quality.

Parameter Estimation

The paper uses the software Stata to conduct the regression of panel data, as well as to carry on the estimation of the mixed OLS, Fixed-effects ("FE" for short), and Random-effects ("RE" for short). Each model's regression conclusion, variance analysis and regression coefficient analysis are listed in the Table 55.3.

Table 55.3 presents the results based on the mixed OLS estimation. And the data adopted are 21 paired panel data, which are collected from six hot school district areas in Fu Tian, Luo Lake and South Mountain Districts in Shenzhen ancient island areas. We can get the Eq. (55.2):

Table 55.2 Descriptive statistics

Variables	Mean value	Standard deviation	Minimum	Maximum	Observed number
Explained variable					
Premium rate of school district housing	0.1316	0.0251	0.0705	0.2594	N = 336
Explanatory variable					
1. Change rate of commercial housing price	0.9672	0.0571	0.0981	0.4424	N = 315
2. Architectural factors of housing estate					
Housing estate's scale (the total number of households)	831.01	1221.0	110.00	6045.0	N = 336
Per household's parking space	0.5437	0.3186	0.0783	1.4285	N = 336
Plot ratio	3.4563	1.8951	1.0000	8.8000	N = 336
Green coverage rate	0.4250	0.1245	0.2500	0.6900	N = 336
Age of building	11.309	4.7151	4.0000	26.000	N = 336
Property fee	3.0309	0.0151	1.0000	9.8000	N = 336
3. Locational factors					
Benchmark land price spread	38.050	230.40	−498.00	696.00	N = 336
4. School factors					
Difference in enrollment rate of junior high school between ES&CS	10.865	11.056	−8.4300	38.490	N = 336
Difference in the distance to school between ES&CS	−231.61	431.75	−1763.0	430.04	N = 38
Difference in student-faculty ratio between ES&CS	−0.0006	0.0068	−0.0111	0.0153	N = 38
Difference in higher vocational teacher ratio between ES&CS	0.0892	0.1561	−0.1951	0.4204	N = 38
Difference in per student occupied area between ES&CS	2.8706	3.4864	−2.9741	10.060	N = 38
Difference in ranking between ES&CS	0.3809	0.5762	0	2	N = 38

Note ES&CS stands for elite primary school and common primary school

Table 55.3 Regression results of school district housing's premium rate

Explained variables (premium rate)	(1) FE	(2) RE	(3) OLS egression
Change rate of commercial housing price	0.348*** (0.0822)	0.347*** (0.0816)	0.409** (0.127)
Housing estate's scale		0.001* (0.000)	0.000*** (0.000)
Per household's parking space		0.109** (0.215)	0.153** (0.0512)
Plot ratio	−0.0307 (0.0221)	−0.0350* (0.0176)	−0.050*** (0.00581)
Green coverage rate		0.0862** (0.498)	−0.839*** (0.118)
Benchmark land price difference	−0.000 (0.000)	−0.000 (0.000)	−0.000 (0.000)
Difference in enrollment rate of junior high school between ES&CS	0.168** (0.002)	0.165** (0.002)	0.154*** (0.001)
Age of building	−0.004 (0.012)	−0.006* (0.009)	−0.007** (0.002)
Difference in the distance to school between ES&CS		0.001** (0.000)	0.0001*** (0.000)
Difference in poll grading between ES&CS	0.000 (0.000)	0.001 (0.000)	0.000*** (0.000)
Difference in student-faculty ratio between ES&CS		−3.736 (6.758)	−1.467 (1.936)
Difference in higher vocational teachers' ratio between ES&CS		−0.092* (0.024)	−0.105* (0.091)
Difference in per student occupied area between ES&CS		−0.014* (0.014)	−0.008* (0.004)
Difference in ranking between ES&CS		0.273** (0.088)	0.290*** (0.027)
Constant	0.036** (0.160)	0.270** (0.330)	0.211** (0.151)
N	336	336	336
Intragroup R2	0.52	0.672	0.467

Note Numbers in the bracket stand for the standard error; *** for 1 % notable level; ** for 5 % notable level; *stands for 10 % notable level

$$\begin{aligned}
&\text{School District Housing's Premium Rate} \\
&= 0.270 + 0.347 * \text{Change Rate of Commercial Housing Price } + 0.109 \\
&\quad * \text{Per Household's Parking Space} + 0.035 \\
&\quad * \text{Plot Ratio} + 0.0868 * \text{Green Coverage Rate} + 0.165 \\
&\quad * \text{Difference in Ranking between ES\&CS} - 0.006 * \text{Age of Building} - 0.092 \\
&\quad * \text{Difference in Higher Vocational Teacher Ratio between ES\&CS} + 0.014 \\
&\quad * \text{Difference in Per Student Occupied Area between ES\&CS} + 0.273 \\
&\quad * \text{Difference in Ranking between ES\&CS} + 0.001 \\
&\quad * \text{Different in the Distance to School between ES\&CS} + \mu_{it}
\end{aligned} \tag{55.2}$$

Model Test

1. Hausman Test. The calculation result of Stata12.0 is as follows:

$$\text{chi2}(10) = (\text{b} - \text{B})'\left[(\text{V}_\text{b} - \text{V}_\text{B})^{(-1)}\right](\text{b} - \text{B}) = 2.40\text{Prob} > \text{chi2} = 0.9539$$

   The test value of Hausman is 2.40 and p value is 0.9539 > 0.000, which conform to the original hypothesis. The acceptance of it means that the estimated results of RE are superior to that of FE. In fact, the RE model is suited to the condition of estimating the overall situation based on samples.
2. F Test. The calculation result of Stata12.0 is as follows:

$$\text{F test that all u}_\text{i} = 0 : \text{F}(20, 281) = 40.10 \quad \text{Prob} > \text{F} = 0.0000$$

   It can know that F value is notably tested, and there has no existence of the original hypothesis of individual effects in the rejected model, which illustrate that FE is superior to the mixed OLS.
3. LM Test. The calculation result of Stata12.0 is as follows:

$$\text{Test}: \ \text{Var(u)} = 0 \quad \text{chibar2}(01) = 671.69 \quad \text{Prob} > \text{chibar2} = 0.0000$$

   As shown in the calculation result, P value is notably greater than 0. The rejection of the original hypothesis, presenting the unobserved random variable μ_i as 0, illustrates that RE is superior to the mixed OLS.

To sum up, RE is superior to FE and mixed OLS and the estimation result should be regarded as the final result of this empirical model.

55.3.5 Results Analysis

The Table 55.3 shows that among fourteen independent variables, there are ten variables having notable impacts on the dependent variables. And the specific analysis is shown as follows:

1. The commercial housing price roughly keeps in pace with the change rate of school district housing's premium rate.
   It makes possible to provide the participants of school district housing market with some decision-making basis according to the fluctuation range and trend of commercial housing price.
2. The difference in school ranking, enrollment rate of junior high school between ES&CS and per student occupied area have notable influences on the premium rate of school district housing.

The difference in ranking between ES&CS having largest influence on school district housing's premium, which reflects that parents' education-core thought plays an important role in the choice of school district housing. And it also boosts the price rise of school district housing situating around the elite primary schools.

The difference in enrollment rate of junior high school between ES&CS shows that parents pay more attention to the dispersion of enrollment rate between ES&CS, and it also gives a reasonable explanation of developers' behaviors in introducing elite school, which has high enrollment rate, into the common school district with the stunt of "key school district housing". The bigger the difference in enrollment rate between ES&CS becomes, the wider the premium space of school district housing gets.

The difference in per student occupied area between ES&CS shows that parents pay more attention to the area of school building and the source number of students. And it is probably due to the fact that the area of school building represents school's financial strength to some degree in Shenzhen Special Economic Zone. A good school cannot be separated from the support of abundant educational appropriations, and per student occupied area can also influence students' comfort level during school attendance.

55.4 Conclusions

1. The fluctuation of commercial housing price and the mass discrepancy among schools have notable influence on the premium rate of school district housing.
2. When the change rate of commercial housing price increases 1 %, its corresponding school district housing's premium rate increases 0.347 %; When the difference in ranking between ES&CS increases one level, the corresponding school district housing's premium rate goes up 0.273 %; When the difference in enrollment rate of junior high school between ES&CS increases 1 %, the school

district housing's premium rate goes up 0.165 %; When the difference in per student occupied area between ES&CS decreases 1 %, the school district housing's premium rate increases 0.014 %.

3. As for the pure investment-oriented investors, they should buy the school district housing when the school district housing premium is at its lowest price and sell it when at its highest price from the perspective of getting the maximum benefit; For the developers, they should develop the school district housing and put into market quickly when the school district housing premium is at its highest price. And they should also introduce the educational theme into the development of new houses, and pay more attention to the nearby peer schools quality, such as the index of enrollment rate. If the difference in enrollment rate becomes higher, the enrollment rate of school district housing also gets higher. Then the developers can get the maximum benefit.
4. As for the supervision departments of real estate market, they should join hands with the education departments to stabilize the premium rate of school district housing. They should pay more attention to the hierarchical planning of schools and strengthen the regulation of real estate market. Due to the fact that the mass discrepancy among schools, such as the enrollment rate and the difference in ranking, has positive impact on school district housing's premium rate, which means that mass discrepancy expansion can lead to the greater room for school district housing premium. And it can result in the consequence that school district housing price goes against the main keynote of market regulation and even the society-level differentiation may appear.

References

1. Wang Xi GY, Han Zhang (2010) A research on price mechanism of school district housing in nanjing old urban district. CO-Operative Econ Sci 12:10–13
2. Shao F, Zhang X (2007) An analysis of the influencing factors of Shanghai housing price based on the hedonic price model. Econ Tribune (12):9–13
3. Shi Y, Zhang R (2010) Temporal-spatial impact effects of large-scale parks on residential prices: exemplified by the Huangxing park in Shanghai. Geogr Res 29(3):510–520
4. Black SE (1999) Do better schools matter? Parental valuation of elementary education. Q J Econ 577–599
5. Dougherty J, Harrelson J, Maloney L, Murphy D, Smith R, Snow M, Zannoni D (2009) School choice in suburbia: test scores, race, and housing markets. Am J Educ 115(4):523–548
6. Gibbons S, Machin S, Silva O (2013) Valuing school quality using boundary discontinuities. J Urban Econ 75:15–28
7. Shi W (2013) The radiation effects of urban compulsory education resource on housing price. Harbin Institute of Technology
8. Yang S (2013) The study of impact of educational facilities on housing price. Zhejiang University
9. Shi Y, Wang Y (2014) The impacting mechanism of housing prices in the school districts in Shanghai City. China Land Sci 12:47–55

10. Ries J, Somerville T (2010) School quality and residential property values: evidence from Vancouver rezoning. Rev Econ Stat 92(4):928–944
11. Wang X (2006) The research of housing characteristic price in Shanghai based on hedonic model. Tongji University
12. Hu W, Zheng S, Wang R (2014) The capitalization of school quality in home value: a matching regression approach with housing price-rent comparison. China Econ Q 13(3):1195–1214
13. Wen H, Jia S (2004) Housing characteristics and hedonic price: analysis based on hedonic price model. J Zhejiang Univ (Eng Sci) 38(10):101–105
14. Wen H Jia S (2006) Market segment and hedonic price analysis of urban housing. J Zhejiang Univ (Humanit Soc Sci) 36(2):155–161

Chapter 56
Research on the Purchase Decision and Determinants of Public Rental Housing Tenants in Chongqing

Xu Ma, Yuhong Pan and Yi Zhan

Abstract The purchase decision of the Chongqing first group tenants staying in public rental housing for five years will certainly affect it's sustainable development. In the paper, we found that the 63 % tenants would like to buy public rental housing through the questionnaire results from three public rental housing communities in Chongqing. Based on the survey data, the Logistic model is used to get the significant influence factors. Finally, the suggestions are put forward according to the situation in order to give government some sustainable development ideas.

Keywords Public rental housing · Logistic model · Purchase decision · Influence factor

56.1 Introduction

The public rental housing of Chongqing is the earliest to start and the largest in scale nationwide. There have been 39 public rental housing communities which are under construction or completed in Chongqing on the end of September 2014, Property management and operation management of Chongqing public rental housing are completely funded and controlled by government and market share is not allowed to get involved in management. Public rental housing lease is about 5 years in Chongqing. After the first five years, the tenants can make a decision whether or not to buy the house which they are live in. As for the the tenants, they can buy the house at market price by one-off payment or installment payment. Once the buyers do not want to stay in the public rental housing after buying it, the houses can only be bought back by Chongqing government.

As for the construction of public rental housing nationwide, the biggest challenge is financing. According to the 2010 government plan, Chongqing government strive to build 40 million square meter public rental housing in the three years,

X. Ma (✉) · Y. Pan · Y. Zhan
School of Chongqing Jiaotong University, Chongqing, China
e-mail: 871753136@qq.com

Y. Wu et al. (eds.), *Proceedings of the 20th International Symposium on Advancement of Construction Management and Real Estate*,
DOI 10.1007/978-981-10-0855-9_56

which require 1200 hundred million fund. The principal and interest will be paid back by rental income of public rental housing, sale and rental income of business stores and sale income of public rental housing. It is predicted that government will make more than 400 hundred million income if they sale 30 % public rental housing. In conclusion, the sale revenue of public rental housing will influence the fund and process of public rental housing, which will influence the government goal of 'people enjoy their rights to live'.

According to the current situation, the first five years of Chongqing public rental housing communities is coming soon one after another and it is necessary to discuss the purchase decisions of the public rental housing tenants in Chongqing. As for the government, the public housing sale revenue can repay construction fund of the public rental housing, reduce the pressure of government and ensure sustainable development of public rental housing. When it comes to tenants, they can get the their own houses in the city, which will increase their sense of belonging and well-being. In this paper, the three typical public housing communities in Chongqing are as the research objects and binary logistic model is used to analyse the significant factors influencing the tenants to buy public housing in order to provide advice to government for the sustainable development of public rental housing.

56.2 Research Hypothesis

The influencing factors which influence the purchase decisions of the public housing tenants were found by reviewing the literature at home and aboard and were divided into two categories, are the characteristics of public rental housing tenants and the characteristics of public rental housing current situation respectively.

56.2.1 The Characteristics of Public Rental Housing Tenants

The characteristics of public rental housing are the fundamental element to the purchase decisions of public rental housing tenants. Western economics said that rational people can make full use their limited resources to maximize utility, profit and social benefits. So low and middle income groups also choose where to live according to their own characteristics. Their main characteristics mainly include monthly income, family population changes, employment group categories.

Monthly Income

A family's ability to pay depend on the family income, which will affect the family the choice of buying housing. Now many scholars such as Hassan [4], Zhou [5], Jing [6] and so on all confirmed this view. During the five years, the income level of some public rental housing tenants have increased significantly, some low-income groups may become middle or high income groups, so they can buy commercial housing to solve the housing problem, which will improve the quality of life and cast off status label of public rental housing tenants. Therefore, hypothesis H1: high monthly income is negatively related to the probability of purchasing public rental housing; H2: high monthly income is positively related to the probability of purchasing public rental housing.

Family Population

Yang [7] and some other domestic scholars put forward: the change of family population structure is one of the important factors which affect the housing requirements, especially on the different class and specifications, etc. General area of public housing is about from 35–80 m^3, when a family population increase, living cost of every individual will be saved, but the living space will be crowded. When a family population decrease, living cost of every individual will increase. Based on the situation, a rational family is likely to choose a more suitable living space. Therefore, hypothesis H3: a family whose population increase or decrease will give up to buy public housing; H4: a family whose population increase or decrease will choose to buy public housing.

Employment Group Category

Peng and Fu [8] and some other scholars concluded that Chongqing public rental housing group mainly can be divided into three categories: (1) employee who work in the urban area and can not afford the commercial housing or some who suffer living problem. (2) fresh employee who graduate from colleges and vocational schools. (3) employee who move from rural area to the city. According to age, the employment groups can be divided into the group of young, middle-aged and the old-aged. Future development direction of young group can not be predicted, while middle-aged group working environment is relatively stable and they tend to more stable living environment. Therefore, hypothesis H5: purchase decision of public rental housing tenants will decrease or increase along with the labor group categories. Youth group will choose to buy public housing; H6: on the contrary.

56.2.2 *The Characteristics of Public Rental Housing Current Situation*

The characteristics of public rental housing current situation are also the important factors influencing the public rental housing tenants to buy public rental housing. The characteristics of public rental housing current situation mainly include evaluation of management mechanism, evaluation of employment distance, payment methods and understanding about public rental housing policy.

Evaluation of Management Mechanism

In recent years, domestic scholars are actively studying community management mechanism of public rental housing. Based on the view of Yan [9], Chen Yongfan, Zheng Xiaoman scholars, there are still many problems such as defective management institutions, lacking support group for disadvantage groups, low community culture quality, poor property management innovation and so on which result in the difficulty in handling social affairs and low degree of satisfaction. The quality of public rental housing community management mechanism will affect the satisfaction degree of public rental housing tenants and affect purchase decision directly. Therefore, hypothesis H7: purchase decision of public rental housing tenants will increase or decrease along with the quality of the management mechanism, and improve the community management mechanism positively related to the probability of purchasing public rental housing; H8, on the contrary.

Evaluation of Employment Distance

Based on 2014 Zeng [10] survey data, the average age of tenants who live in 'minxin garden' public rental housing communities in Chongqing is 38 years old, the biggest proportion is young and middle-aged group who are under 35, followed by retirees who is more than 60 years old. For the majority of employment population, convenient transportation can reduce the tenants time cost and transportation cost. Therefore, hypothesis H9: purchase decision of public rental housing tenants increase or decrease along with the distance between the the public rental housing and working places. The probability of purchasing public rental housing will increase when the distance is relatively close; H10 on the contrary.

Payment Methods

Aihua LI et al. [11] pointed out that based on the research of the urban housing purchasing power, we can analyze whether the purchasing power of residents match the housing price of real estate market. This method is also suitable for the study of public rental housing. Government regulate that the public rental housing tenant can purchase the public rental housing which they live in at cost price after the first five years. They can buy the house at market price by one-off payment or installment payment. If they choose one-off payment, they do not need to pay rent anymore, but if they choose installment payment, they have to pay the rent for the non-payment house area. Some public rental housing tenants who is not insufficiency of purchasing power will not purchase the public rental housing. Therefore, hypothesis H11 tenants who can pay for public rental housing will purchase the public public housing; H12 on the contrary.

Understanding About Public Rental Housing Policy

Zhen and Daochi [12] and some other scholars put forward that the urban low and middle income earners will consider more about future changes of the housing market when they face with the instability and unpredictability of public rental housing policy. Now the government policy and future evolution trend is not yet clear, which increase the concern of public rental housing tenants and affect tenants the decisions of purchasing public rental housing. Therefore, hypothesis H13: purchase decision of public rental housing tenants increase or decrease along with understanding about public rental housing policy. The probability of purchasing public rental housing will increase when tenants get a better understanding about the public rental housing policy; H14 on the contrary.

56.3 Data Resource and Empirical Research

56.3.1 Data Resource

This study combine the questionnaire survey method and field interview method. In field investigation stage, the data is obtained through random questionnaire of three Chongqing public rental housing communities, are 'minxinjiayuan', 'kangzhuangmeidi', 'liangjiangmingju' respectively, in order to understand the purchase intention of public rental housing tenants. The data is dealt by SPSS software to analyze the significance of the influencing factors. In this paper, the data was collected through questionnaire survey from October to December in 2014. There are 300 questionnaires we hand out in total, and we receive 251 questionnaires back. Finally, there are 227 questionnaires as request after dealing with the 251 questionnaires. These samples are representative through the summary and test.

56.3.2 Model Specification

Logistic regression model which is a mathematical statistical method is widely used in research. In this study, if the dependent variable Y = 1, it means that public rental housing tenants are willing to buy public rental housing, if Y = 0, it means that they give up purchasing public rental housing. The dependent variable is dichotomous variable. The binary logistic model is used in this paper to analyze the factors affecting the purchase decision of the public housing tenants.

$$\text{given:}\ \mathrm{P_i} = \mathrm{F}(1/\mathrm{Y_i}) = \frac{1}{1+\mathrm{e}^{-y}},\quad \text{then:}\ \frac{p_i}{1-p_i} = \mathrm{e}^{y},\quad \mathrm{Y_i} = \ln\frac{p_i}{1-p_i}$$

$$\text{assume:}\ \mathrm{Y_i} = \beta_0 + \beta_1\chi_1 + \beta_2\chi_2 + \cdots + \beta_n\chi_n + v_i$$

$$\text{then:}\ \ln\frac{p_i}{1-p_i} = \beta_0 + \beta_1\chi_1 + \beta_2\chi_2 + \cdots + \beta_n\chi_n + v_i$$

Then there is a linear model,

$$\text{Logit P} = \ln\frac{p_i}{1-p_i} = \beta_0 + \beta_1\chi_1 + \beta_2\chi_2 + \ldots\ldots + \beta_n\chi_n + v_i$$

Among them, the Pi is the probability of purchasing public rental housing, 1-Pi mean the probability of not purchasing public rental housing. β0 is constant, βi is the variable coefficient, vi is Random perturbation terms, Xi is the influence factors which are obtained in the research hypothesis. Are monthly income (X1), family population (X2), employment group category (X3), evaluation of management mechanism (X4), evaluation of employment distance (X5), payment methods (X6) and understanding about public rental housing policy (X7) respectively. n is the amount of the independent variables and n is seven in this paper.

56.4 The Regression Analysis of Model

According to the Table 56.1, monthly income, employment group category and payment methods are virtual variables, Thus we can infer that: with the increase of family income, the smaller possibility of purchasing public rental housing; Middle-aged and old-aged group have different purchase decision from youth group, middle-age and old-age groups are more likely to buy public rental housing, and youth groups tend not to buy public rental housing. As for the payment methods of purchasing public rental house, the public rental housing tenants who are unable to pay are less likely to purchase public rental housing. Evaluation of management mechanism, evaluation of employment distance, understanding about public rental housing policy are 4 independent variables, and the Wald index of the four independent variables were: 0.305, 17.401, 8.873, 17.401. The significance of

Table 56.1 The significant of the independent variable

		B	S.E.	Wald	df	顯著性	Exp (B)
步驟 1	X_1 (income decrease 0–500)	0.446	0.227	5.824	1	0.016	1.562
	X_1 (remain stable)	0.171	0.402	1.060	1	0.303	1.187
	X_1 (income increase 0–500)	−0.427	0.959	0.464	1	0.496	0.652
	X_1 (income increase more than 500)	−1.625	0.451	8.000	1	0.005	0.197
	X_3 (middle-aged group)	0.129	0.106	11.389	1	0.001	1.138
	X_3 (old-aged group)	0.762	0.266	10.771	1	0.001	2.143
	X_7 (installment payment)	1.161	0.590	3.876	1	0.049	3.192
	X_7 (can not pay for the house)	−1.478	0.239	25.953	1	0.000	0.228
	X_2 (family population)	0.240	0.887	0.305	1	0.581	1.271
	X_4 (management mechanism)	−1.659	0.309	17.401	1	0.000	0.190
	X_5 (employment distance)	−0.887	0.316	8.873	1	0.003	0.412
	X_6 (understanding about public rental housing policy)	0.113	0.840	0.160	1	0.689	1.119
	常數	−1.488	1.272	0.919	1	0.350	0.226

the evaluation of employment distance and evaluation of employment distance reach 0.05. Namely, the two variables has a significant influence on whether to buy public rental housing. The coefficient of X4, X5 is negative, namely hypothesis H7 and H9 are true. Namely if improve the community management, shorten the employment distance, more public rental housing tenants will purchase the public rental housing. Because the X2, X6 was not significant, so it can not predict whether the family population change and understanding about public rental housing policy have an influence on purchasing public rental housing.

56.4.1 Logistic Regression Model Analysis

According to the result of the logistic regression model, we reach the conclusions are as follows.

1. The increase of family income has a significant negative effect on the purchasing public rental housing. Namely, the higher income the public rental housing tenants earn, the smaller probability of purchasing public rental housing. The main reason is that public rental housing tenants will transfer to commercial residential building from public rental housing to improve the living condition with higher income. Some people who live in public rental housing for a long time will be stigmatized, which effect their social activities. So there is a

negative correlation between the increase of family income and purchasing public rental housing.

2. Employment category has a significant positive influence on purchasing public rental housing, namely, the middle-aged and old-aged employment group are more likely to purchase public rental housing. The main reason is that the future development direction of youth group can not be predicted. And many fresh graduates are bound by the general social concept that they should buy the house before getting a marriage and the house can not be too small which lead many graduates lost the freedom of selecting public rental housing. Compared to the youth group, the marriage of the middle-group and old-aged group are more stable and have stronger ability to pay. So there is a positive correlation on employment category and purchasing public rental housing.
3. Evaluation of management mechanism has a significant negative influence on purchasing public rental housing, namely, the worse community management mechanism, the lower probability of purchasing public rental housing. There is a close relationship between the Community management mechanism and the life quality of public rental housing tenants. Poor management mechanism have a bad effect on daily life of public rental housing tenants and make a bad impression on the public rental housing tenants, which lead them to give up purchasing public rental housing. Therefore, there is a negative correlation on evaluation of management mechanism and purchasing public rental housing.
4. Evaluation of employment distance has a significant negative influence on purchasing public rental housing. Namely, the longer distance between public rental housing and working places, the lower probability of purchasing public rental housing. Employment distance is the Geographical factors which will a significant effect on employment condition and living cost of public rental housing tenants. If the public rental housing is located in remote area, it can not meet the employment needs. Even if, they have to pay for more time cost and transportation cost. And the public infrastructure is poor in remote area. So, there is a negative correlation on evaluation of employment distance and purchasing public rental housing.
5. Payment methods has a significant negative influence on purchasing public rental housing. Namely, The weaker ability to pay, the lower probability of purchasing public rental housing. The public rental housing tenants who have a strong ability to pay will pay by a lump-sum payment or installment, while the public rental housing tenants who have week ability to pay are less likely pay for the public rental housing. Instead, they are tend to rent the public rental housing. Therefore, there is a negative correlation on payment methods and purchasing public rental housing.
6. Family population do not have a significant influence on purchasing public rental housing. Namely, the main reason is that the family member number do not change significantly. As for a family, the size of public rental housing is quite appropriate. And the data is limited. So it can not predict the correlation between family population and purchasing public rental housing.

7. Understanding about public rental housing policy do not have a significant influence on purchasing public rental housing. Studies have shown that Chongqing public rental housing tenants generally don't understand the policy of public rental housing, the main reason is that the policy and future evolution trend of Chongqing public rental housing is not clear. And the data is limited. So it can not predict the correlation between understanding about public rental housing policy and purchasing public rental housing.

56.5 Conclusion

In this study, It can be seen that the more family income earned, the younger the employment group is, the less probability of purchasing public rental housing, the better the community management system is, the shorter the employment distance is, the more probability of purchasing public rental housing. It can not predicted whether the family population and understanding about public rental housing policy have a influence on purchasing public rental housing.

At present, the purchase policy of public rental housing policy in chongqing mainly mada according to the suppliers, which lack the investigation and study of demanders. With the increasing scale of public rental housing construction, the government should pay attention to the needs of public rental housing tenants and make an effort to improve basic public services in the community in order to coordinate the 'rent' and 'purchase' and promote sustainable development. According to current situation of public rental housing, government can take some specific measures to improve the public rental housing construction. For example: (1) establish specialized agencies with clear responsibilities to ensure that the service and management work effectively to improve satisfaction of public rental housing tenants (2) increase public transport nodes and the subway entrances around the public rental housing communities, shorten the employment distance between public rental housing and working places; (3) promote community employment through the active guidance to improve the ability to pay of public rental housing tenants, the government can take appropriate housing subsidies to the people who have difficult to pay; (4) the data also shows that Chongqing public rental housing tenants generally lack of the understanding of specific policies of purchasing public rental housing. Government should clear housing policy orientation and future evolution trend to eliminate the concerns of public rental housing tenants.

References

1. Jansen SJT (2012) What is the worth of values in guiding residential preferences and choices. J Housing Built Environ 27(3):273–300
2. Lee SW, Myers D (2003) Local housing-market effects on tenure choice. J Housing Built Environ 18(2):129–157
3. Chen D, Yu Q, Yu X (2011) Preferences and the influencing factors of housing tenure choice: an empirical study of Hangzhou. J Zhejiang Univ Technol (Soc Sci) 10(1):1–6
4. Hassan GF (2012) The enabling approach for housing supply: drawbacks & prerequisites-Egyptian experience. Alexandria Eng J 1:421–429
5. Zhou L (2014) Consumption attribution of housing subsidies and the factors—logistic test based on a survey data of the low—income rent allowance policy in Shanghai. Shanghai Econ Rev (4):90–98
6. Jing J, Liu Z, Man JY (2010) International experiences of low-income housing policy and its implications to China: demand-side subsidy policies. Urban Stud 17(6):56–63
7. Yang X, Xu D (2011) Research on population structure change and housing demand of developing countries. Res Dev 2:84–87
8. Peng X, Fu G (2012) Research on management of new public rental housing communities. China Market 20:47–50
9. Yan J (2013) Service and management of public rental housing communities—based on survey of 'minxingjiayuan' public rental housing in Chongqing. Theoretical Exploration 02:82–85
10. Zeng D, Quan L (2014) The matching issue about living and employment of public rental housing tenants—based on Chongqing. Urban Problems (02):88–93
11. Aihua LI et al. (2006) Study on the housing power of urban residents. Journal of Management Sciences in China 9(5):8–17
12. Zeng Z, Qiu D, Li F, Li X (2012) Public rental housing consumption intention of university graduates and its influencing factors—an empirical study based on Chongqing 34(10):124–130
13. Maynard K (2012) Application logistic regression analysis, 2 ed. Due press
14. Deng H, Ke F (2014) Logistic regression analysis of factors influencing the public housing demand. Stat Decis 14:100–104
15. Gibler KM, Kananen J, Tyvimaa T (2014) The relationship between the determinants of rental housing satisfaction and considering moving in Finland. Property Manage 32(2):104–124
16. Lai AWY, Lai WM (2013) User's satisfaction survey on building maintenance in public housing. Eng Constr Architectural Manage 20(4):420–440

Chapter 57
The Evaluation on the Suppliers of Prefabricated Housing Components Based on DEMATEL Method

Yi Zhan, Jiyuan Liu and Xu Ma

Abstract An evaluation method on the suppliers of prefabricated housing components was proposed based on DEMATEL. Firstly, combing with the characteristics of construction industry, the establishment of evaluation system about suppliers was used to solve the problem of effective choice. Secondly, considering the logical relationship of factors in the evaluation system, a method of evaluation based on DEMATEL was used to calculate the degree of correlation among factors, which is used to modify the linguistic assessment and evaluate the suppliers of prefabricated housing components in the perspective of multi-stakeholder. Finally, the feasibility and effectiveness of this method was demonstrated through the case analysis. The result demonstrates that this method can contribute to the selection of suppliers.

Keywords DEMATEL method · Prefabricated housing components · Suppliers

57.1 Introduction

With the development of housing industrialization, more construction enterprises recognize the importance of the prefabricated housing system. Meanwhile, the number of suppliers of prefabricated housing components as an intermediate link in the industry chain is increasing. Nevertheless, compared with the traditional construction industry chain, the selection method about suppliers of prefabricated housing components is different. Although the traditional chain mode has a good operation under the sound law environment and industry standard, it cannot satisfy requirement of housing component for housing industrialization. At the present, industrialized modes of production and installation will promote new supply modes, but an efficient evaluation standard is lacking in the selection to supplier of prefabricated housing component.

Y. Zhan (✉) · J. Liu · X. Ma
School of Chongqing Jiaotong University, Chongqing, China
e-mail: 917636654@qq.com

Y. Wu et al. (eds.), *Proceedings of the 20th International Symposium on Advancement of Construction Management and Real Estate*,
DOI 10.1007/978-981-10-0855-9_57

On the other hands, as far as a large project of prefabricated housing, the problem that is interest coordination among multi-stakeholders is difficult to be solved. Professor Qiping Shen considered that the management of stakeholders is the key to resolving the conflict of interest and achieving the benefit maximization in the construction project [1]. However, in the process of supplier selection, the results of selection from various stakeholders are different, and there is a direct or indirect relationship among the evaluation factors usually. If the relationship is ignored, error will be caused in the evaluation result. In order to solve this problem that is hard to select appropriate suppliers of prefabricated housing components effectively from various stakeholders, this paper puts forward an evaluation method based on Decision-Making test and Trial Evaluation Laboratory (DEMATEL) principle. Firstly, combined with the characteristic of supplier selection from manufacturing industry, this paper establishes the comprehensive evaluation system which is suitable for the suppliers of prefabricated housing component in the perspective of stakeholders. Secondly, the matrix model of DEMATEL is used to quantitatively analyze the influence factors based on the multi-stakeholder and indicate the relationship between each factor and one another, which can evaluate the suppliers objectively.

Finally, this paper demonstrates the feasibility and effectiveness of the DEMATEL method in the selection of suppliers by case analysis and provides the scientific evaluation methods and suggestions for the stakeholders, so as to make suppliers and buyers as well as industry chains of pre-fabricated housing gradually move towards the direction of sustainable development.

57.2 The Present Research Situation on Supplier Evaluation System

Since 1996, according to the questionnaire survey, the American scholar Dickson [2] summarized 23 criterions of evaluation for supplier performance including quality, delivery, etc. Weber [3] analyzed 74 literatures of suppler selection and found that prize, on-time delivery rate and quality have significant influences on supplier selection. In China, Guo and Lu [4] considered that the basic evaluation indexes on suppler selection include technical level, product quality, supply capacity, price, geographical position, etc. Professor Cao [5] considered that the evaluation factors of real estate partner can be divided into four systems: management capabilities, staff capacity, operational capacity and finance cost. Lianfa and Jialing [6] summarized the indexes of product advantage and operation ability in the green building supplier's perspective. It can be seen that the most of supplier selection literatures were concentrated in industry, trade economy and other areas, in the construction field, most supplier selection studies were based on traditional construction mode. However, the researches on supplier of prefabricated housing components were less.

Although most of areas had a certain degree of similarity in the establishment and classification of supplier evaluation, for example, the supplier evaluation criterion focused on product quality, price and delivery term, various fields still had their own significant characteristics. If the current mode of supplier selection directly is applied to Suppliers of Prefabricated Housing Components, it will lack the adaptability and can not well reflect the characteristics of prefabricated housing components as well as can not assess the prefabricated housing components' usability, safety and others. Therefore, the establishment of an appropriate evaluation system is particularly necessary.

57.3 Establishment of the Evaluation System

The nature of prefabricated housing system is making use of the industrialized production and installation methods to construct residential projects. This new form makes the past on-site construction work to be completed in the industrial workshop of standardization, then assembled and installed in the construction site by means of transportation, which can achieve the project objectives as well as meet the constraint conditions including time, quality, cost and the customer demand.

As can be seen, this new construction method is similar to manufacturing industry. Therefore, this paper, based on reading literatures about manufacturing and construction [7, 8], and analyzes their characteristics, and combines with housing performance to evaluate the technical standard, etc. [9, 10], and establishes a suitable evaluation system about the suppliers of prefabricated housing components in order to cover the possible factors that should be considered when suppliers are selected and evaluated. This evaluation system consists of 28 index factors in seven aspects including quality factor, price factor, production and delivery factor, product performance factor, comprehensive factor of suppliers, and service factor, financial factor. The quality factor includes product qualification ratio (F1), quality certification (F2), quality management level (F3) and technological level (F4). The price factor includes product price (F5), product cost (F6), logistics cost (F7) and product price fluctuation (F8). The production and delivery factor includes delivery term (F9), accuracy rate of delivery (F10), on time rate of delivery (F11), flexible production capacity (F12), flexible delivery capacity (F13) and market share (F14). Product performance factor includes product applicability (F15), product safety (F16), product durability (F17) and product diversity (F18). Comprehensive factor of suppliers includes research and development capability (F19), design capability (F20), development situation of suppliers (F21) and employees' comprehensive capability (F22). Service factor includes capability of after-sales service (F23), capability of technology service (F24) and problem treatment capability (F25). Financial factor includes situation of assets and liabilities (F26), financing capability (F27) and status of capital operation (F28).

57.4 The Evaluation Method Based on DEMATEL

57.4.1 Establishment of Evaluation Sets

Firstly, set F = {F1, F2, …, Fn}(n = 28) is described as 28 index factors. Then, according to evaluation system, the linguistic assessment set S is established, S = [0 (no influence/very bad), 1 (slight influence/bad), 2 (general influence/medium), 3 (serious influence/good), 4 (highly influence/very good)]. Set S is described respectively as the stakeholders' evaluations about index system and suppliers.

57.4.2 The Normalized Treatment Based on DEMATEL

DEMATEL method was called Decision-Making and Trial Evaluation Laboratory, which is a factor analysis tool based on matrix operations. It is proposed by Gabus and Fontela [11, 12]. By using the DEMATEL method, the relationship and the influence of each factor were evaluated quantitatively and then achieved the perfect result of evaluation.

At present, the DEMATEL method has been more widely used in various fields [13–15]. As far as the selection problem about suppliers of Prefabricated Housing Components, the complicated relations of factors, for example the direct and indirect influence among factors, also can be treated by DEMATEL method. Therefore, this method can clarify the internal or external relationship of factors and the relatively better suppliers can be selected precisely for stakeholders. The concrete calculation steps as follows:

1. Suppose that f is the number of stakeholders who participate in evaluation. The kth direct evaluation matrix from stakeholders is denoted as. It can be calculated by Eq. (57.1)

$$X_k = \left[x_{kij}\right]_{n\times n} = \begin{array}{c} \\ C_1 \\ C_2 \\ \vdots \\ C_n \end{array} \begin{array}{c} \begin{array}{cccc} C_1 & C_2 & \cdots & C_n \end{array} \\ \begin{bmatrix} - & x_{k12} & \cdots & x_{k1n} \\ x_{k21} & - & \cdots & x_{k21} \\ \vdots & \vdots & \ddots & x_{k21} \\ x_{kn1} & x_{kn2} & \cdots & - \end{bmatrix} \end{array} \tag{57.1}$$

$$x_{ij} \in S;\ i,j = 1, 2, \ldots, n;\ k = 1, 2, \ldots, f$$

2. The direct influence matrix, which is given by stakeholders, is determined by arithmetic average and the calculation as follows.

$$x_{ij} = \frac{1}{f}\left(\sum_{k=1}^{f} X_{kij}\right) \quad i,j = 1,2,\cdots,n; \;\; k = 1,2,\cdots f \tag{57.2}$$

3. The direct influence matrix is converted into which denotes the normalized direct influence matrix by using the normalized treatment. The concrete calculation steps as follows.

$$y_{ij} = x_{ij} / \max_{1 \le i \le n} \sum_{j=1}^{n} x_{ij} \tag{57.3}$$

4. The comprehensive influence matrix is obtained by Eq. (57.4), and t_{ij} indicates the direct influence and indirect influence degree of the factor i on the factor j.

$$T = \lim_{\lambda \to \infty} (Y^1 + Y^2 + \cdots + Y^{\lambda}) = Y(I - Y)^{-1} \tag{57.4}$$

5. The degree of influence R_i, degree of being influenced C_i, centrality degree e_i and the reason degree f_i can be calculated by the comprehensive matrix T. The concrete calculation steps for variables as shown in Eq. (57.5).

$$\begin{aligned} R_i &= \sum_{j=1}^{n} t_{ij}, \quad i = 1,2,\ldots,n \\ C_j &= \sum_{i=1}^{n} t_{ij}, \quad j = 1,2,\ldots,n \\ e_i &= R_i + C_i \\ f_i &= R_i - C_i \end{aligned} \tag{57.5}$$

In the Eq. (57.5), the degree of influence R_i indicates the influence extent of factor i on other factors. The degree of being influenced C_i indicates the influenced extend of the factor i on other factors. Centrality degree e_i is the sum of the degree of influence and the degree of being influenced, which indicates the factor's importance in the evaluation system. The reason degree f_i is the difference between degree of influence and degree of being influenced, which reflects the function of the factor, such as direct influence or indirect influence, and the logical relation among various factors. If the reason degree is greater than 0, which means that the factor has a major influence on other factors. If the reason degree is less than 0, it implies that this factor is under the influence of other factors.

57.4.3 The Distribution of Index Weight

The distribution of index weight reflects the relative importance of each index factor in order to evaluate suppliers objectively and precisely. Through the sequence of reason degree, the index weights were reallocated and the calculation method of weight z_i as follows.

$$z_i = \frac{e_i}{\sum_{i=1}^{n} e_i}, \quad i = 1, 2, \ldots, n; \quad \sum_{i=1}^{n} z_i = 1 \tag{57.6}$$

57.4.4 The Selection and Evaluation of Suppliers

The matrix is established according to the evaluation set S, and m denotes the number of suppliers who will be evaluated. Then the final evaluation result h_q is equal to index weight z_i multiplied by each evaluation of suppliers h_{iq}, the calculation formula as follows.

$$h_q = \sum_{i=1}^{n} (z_i \times h_{iq}) \quad i = 1, 2, \ldots, n \quad q = 1, 2, \ldots, m \tag{57.7}$$

57.5 Case Analysis

We can suppose that a large prefabricated housing construction project is going to purchase multiple prefabricated components. After extensive investigation and screening, three prefabricated suppliers are selected in total and will be evaluated by DEMATEL method. Meanwhile, three suppliers are described by A, B, C respectively.

57.5.1 The Weight Distribution of Evaluation System

In order to demonstrate the feasibility of the DEMATEL method and ensure the reasonable allocations of weights, in this paper, 5 experts including project manager, supervising engineer, professor who major in construction field, et al., were interviewed. After two weeks, 5 evaluation results about suppliers and evaluation system were received. Combined with the calculation steps above, the reason degree e_i and weight z_i can be obtained. The evaluation on index factors as listed in Table 57.1.

Table 57.1 Five values of each factor

Index	R	C	e	f	Weight (%)	Index	R	C	e	f	Weight (%)
F_1	0.68	0.93	1.61	−0.24	3.62	F_{15}	0.45	0.82	1.27	−0.36	2.85
F_2	0.93	0.62	1.55	0.31	3.49	F_{16}	0.51	0.77	1.28	−0.27	2.88
F_3	1.09	0.57	1.66	0.52	3.72	F_{17}	0.51	0.77	1.28	−0.27	2.88
F_4	1.05	0.73	1.77	0.32	3.98	F_{18}	0.48	0.47	0.95	0.02	2.13
F_5	1.22	1.47	2.69	−0.25	6.06	F_{19}	1.11	0.53	1.65	0.58	3.70
F_6	0.61	0.85	1.46	−0.23	3.28	F_{20}	1.00	0.53	1.54	0.47	3.45
F_7	0.97	0.43	1.39	0.54	3.13	F_{21}	1.12	1.64	2.77	−0.52	6.21
F_8	0.27	0.28	0.55	−0.02	1.24	F_{22}	1.30	0.23	1.53	1.07	3.43
F_9	0.63	0.75	1.38	−0.12	3.09	F_{23}	0.27	0.40	0.67	−0.13	1.51
F_{10}	0.62	0.80	1.43	−0.18	3.20	F_{24}	0.86	0.53	1.38	0.33	3.11
F_{11}	0.62	0.82	1.44	−0.20	3.24	F_{25}	1.24	1.07	2.32	0.17	5.20
F_{12}	0.56	0.88	1.43	−0.32	3.22	F_{26}	0.71	0.80	1.51	−0.09	3.40
F_{13}	0.82	0.89	1.71	−0.07	3.85	F_{27}	0.60	0.82	1.41	−0.22	3.17
F_{14}	0.90	1.80	2.71	−0.90	6.09	F_{28}	1.11	1.05	2.16	0.06	4.85

57.5.2 The Evaluation and Selection on Suppliers of Prefabricated Housing Components

According to the information of investigation, the modified evaluations on suppliers are obtained by using Eq. 57.7. The final evaluation results are hA = 2.84, hB = 2.49, hC = 1.81. Therefore the perfect suppler of prefabricated housing components is A. The sequence of priority selection is A > B>C.

57.5.3 Analysis of Result

As far as the suppliers of prefabricated housing components, the self-analysis and summary for suppliers also can be obtained through the comprehensive analysis on Table 57.1. The sequence of centrality degree implies that the attention of stakeholders or customers focus on development situation of suppliers (F21), market share (F14) and product price (F5), which indicates that these factors should be the main object of attention for suppliers. On the other hand, the factors of high influence degree are employees' comprehensive capability (F22) and research and development capability (F19), which mainly affect on other factors. It reflects that the comprehensive ability of research and the quality of staff can significantly improve the other key factors, and explains that the technology of prefabricated housing components still does not get the approval of people. Especially in real estate market, this new construction mode is rarely applied except government construction project. By using this analysis, the suppliers of prefabricated components can grab the attention of customers effectively and continue to reform themselves, which can facilitate the sustainable development of prefabricated housing system for demanders as well as suppliers.

57.6 Conclusion

This paper combines with the characteristics of prefabricated housing system, and establishes an evaluation system on suppliers of prefabricated components that satisfies the requirements for housing industrialization. Under the evaluation system, the matrix analysis model of DEMATEL is used to calculate the weight of each factor in order to provide an objective basis for stakeholders or customers, which can avoid possible risk due to the decision-making misplay and reduce the implicit cost in construction.

On the other way, the advantages of using the DEMATEL method are: (1) There are considerations about not only the unidirectional relationship but also the interactive relationship among the factors. (2) The clear relationship about direct influence as well as indirect influence among factors are reflected. It can avoid the information loss and misinterpretation when the information of evaluative language transforms into mathematical model, which may make the results of evaluation more comprehensive and objective. (3) The reduction in the number of system factors can cancel a part of complicated calculation steps such as consistency check, which simplifies the process of calculation and enhances the application for decision maker in practice.

In a word, the selection on suppliers of prefabricated components serves as an important link in the industry chain of prefabricated housing system, is one of problems what we have to face in the development of housing industrialization. With the continuous development of housing industrialization, more and more unsolved problems might be appeared because of the birth of new manufacturing technology and method in construction field. It is also an inevitable process the human society and civilization must experience in development. Therefore, the establishment of evaluation system on suppliers of prefabricated housing components and scientific evaluation method are the necessary means which can reduce the mistakes in decision-making process and take preventive measures, and insure the construction industry turning to intensive mode from extensive mode of production.

References

1. Sheng Q, Yang J (2010) Research on stakeholder management framework in construction projects. J Eng Manage 4(24):412–419
2. Dickson GW (1996) An analysis of vendor selection systems and decisions. J Purchasing 2 (1):5–17
3. Weber CA, Current JR et al (1991) Vendor selection criteria and methods. Eur J Oper Res 50 (1):2–18
4. Guo S, Lu Z (2008) Logistics and supply chain management. China Machine Press, Beijing, pp 81–82
5. Cao J, Gao X (2005) Fuzzy syntheses evaluation of supply chain partners of real estate. J Tongji Univ 33(6):843–847
6. Lianfa R, Jialing C (2011) The selection of green suppliers based on VIKOR. Stat Decis 21:62–64 (in Chinese)
7. Xu X, Li R, Wu X (2011) Research on the model of supplier selection in supply chain collaboration. J Zhejiang Univ Technol 3(5):550–554
8. Zheng X, Jia L (2013) A study on the selection of construction component suppliers in the process of housing industrialization. East China Econ Manage 27(10):93–97
9. GB/T 50362-2005 (2006) Housing performance to evaluate the technical standard. China Building Industry Press, Beijing
10. JG J1-2014 (2014) Technical specification for precast concrete structures. China Building Industry Press, Beijing
11. Herrera F, Martinez L (2000) A 2-tuple fuzzy linguistic representation model for computing with words. IEE Trans Fuzzy Syst 8(6):746–752
12. Fontela E, Gabus A (1976) The DEMATEL observer, DEMATEL 1976 report. Battelle Geneva Research Centre, Switzerland Geneva
13. Suo W, Fan Z, Feng B (2012) A method for vendor selection in IT services outsourcing based on multifactor correlation analysis. Ind Eng Manage 6(17):1–6

14. Li K, Qi B, Wang H (2013) The analysis of constraint factors in prefabricated housing based on DEMATEL, pp 49–51 (in Chinese)
15. Gao P, Lu X (2014) Theory and application study on extended DEMATEL method based on interval number. Oper Res Manage Sci 23(1):44–50

Chapter 58
Measuring Price Differentials Between Large and Small Housing Units: The Case of Hong Kong

Ling Li and K.W. Chau

Abstract In Hong Kong, the private residential property prices have undergone significant changes over the last two decades which can be mainly divided into three phases: the price level grew nearly 105 % between year 1993 and 1997, then followed by a radical decline of 66 % during 1997–2003, and then a fourfold surge since 2003. Over this time span, there was increasing divergence of price changes in the luxury (large units) and mass (small units) residential markets. One possible explanation is that large units are preferred by investors and therefore more sensitive to changes in investment demand, whilst buyers of small units are mostly owner-occupiers and thus mainly affected by changes in consumption demand. As house price appreciation plays an important role in household's decisions of when, where and how "much" house to buy, this study aims to investigate the main factors contributing to the price differentials between large and small housing units. We hypothesize that the changes in price differentials of large and small housing units are primarily caused by changes in investment and consumption demand. We test this hypothesis using empirical data from Hong Kong and find that capital flow and market sentiment have a stronger impact on the prices of large units than that of small units since they can induce more investment demand for housing. Instead, marriage rate, divorce rate, unemployment rate and mortgage ceiling are closely related to the housing consumption demand and are found to affect the prices of small units more.

Keywords Price differential · Large housing units · Small housing units · Investment demand · Consumption demand

---

L. Li (✉) · K.W. Chau
Department of Real Estate and Construction, University of Hong Kong, Pok Fu Lam, Hong Kong
e-mail: u3002298@connect.hku.hk

Y. Wu et al. (eds.), *Proceedings of the 20th International Symposium on Advancement of Construction Management and Real Estate*,
DOI 10.1007/978-981-10-0855-9_58

58.1 Introduction

House price appreciation plays an important role in households' decisions of when, where and how "much" house to buy, and whether to default on their mortgages, or whether to sell out it. Although the average change in housing prices is related to changes in "fundamentals" such as national and local macroeconomic variables, not all houses in a market appreciate at the same rate. The residential market of Hong Kong is such an example. As showed in Fig. 58.1, the changing price differentials between large (Class D&E) and small housing units (Class A) are noteworthy which demonstrate that they appreciate at very different rates across time.

The primary focus of this study is to investigate the reasons for such price differentials by size stratification. We draw on one thing for guidance, that is, the dual role of housing as an investment and consumption good. Theoretical work by [1] suggests that investment demand for housing is more sensitive to wealth and income, but consumption demand is more sensitive to demographic variables. Since they are driven by very different factors, and if they also aims differently, for instance, invest in larger units but consume smaller units, then relative changes between investment demand and consumption demand can be captured by the price differentials stratified by size. This conjecture seems plausible in Hong Kong, where there are predominant investors in the large housing market but vast majority consumers in the small housing market. On one hand, investors can benefit from the higher price appreciation of larger units, which are always located in popular areas with convenient transport and attractive neighborhood environment. Such land is extremely limited and property price is very high. Undoubtedly, land price becomes the main contributor to the high price whereas construction cost is rather stable and accounts for much less of the property value. And this higher land leverage can further benefit investors with less information asymmetry as information about land quality is more transparent than building quality [2]. On the contrary, soaring property price in Hong Kong has caused the homeownership to be beyond the reach

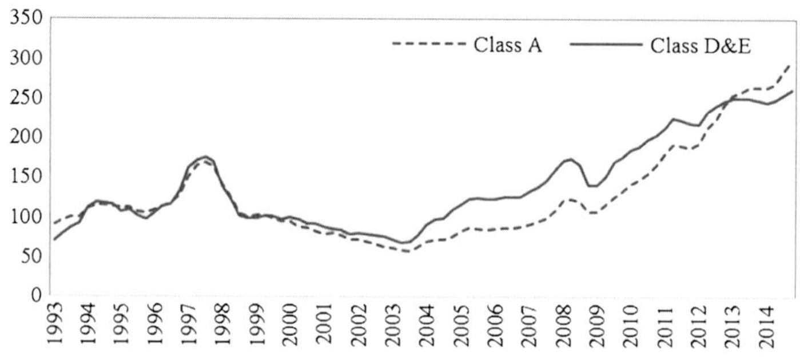

Fig. 58.1 Price movements of residential housing by size (the residential units in Hong Kong are divided by reference to floor area into five groups: Class A (saleable area less than 40 m^2), Class B (saleable area of 40–69.9 m^2), Class C (saleable area of 70–99.9 m^2), Class D (saleable area of 100–159.9 m^2), and Class E (saleable area of 160 m^2 or above))

of many aspiring households. Even the smallest units are not easily affordable—it will take a household 10 years to afford a 30 m^2 unit without any other expenditure.[1] The general residents are forced to look at small units only. These provide sufficient theoretical reasons for the property market of Hong Kong to be segmented by investment demand and consumption demand as large and small housing market.

Further support comes from the real world. The most developed area with high land leverage, Hong Kong Island (HKI), do possess a higher proportion of large units. Compared to the other two districts, Kowloon and New Territories, HKI is generally accepted as the historical, political and economic center and enjoys the most prosperity, including the land price. It owns much more large units and few small units (accounting for 53 % of the stock of Class D&E units but only 30 % of the stock of Class A units). This implies that large units tend to be located in areas with higher land leverage and more small units in less developed zones. Further, the large housing market also bears a lower occupancy rate but a higher vacancy rate. Houses for the purpose of investment should be either rented out for rental income or left vacant waiting for price appreciation, while those for basic consumption are always occupied by owners. And we find among the newly completed units from 1985 to 2013, large units have a much lower occupancy rate (Class D: 66.08 %; Class E: 40.97 %) than small units (Class A: 82.4 %). And they are also characterized by a higher vacancy rate (Class D: 6.9 %; Class E: 8.2 %), which fluctuates a lot, whilst that for small units remains rather stable at only 3 %. This shows that consumption demand for small units is quite low but large units which are mainly purchased for investment purpose are more likely to be left vacant for price appreciation. And the high volatility is due to frequent short-term speculative activities. Overall, large units are more located in developed areas with higher land leverage, and characterized by lower occupancy rate and higher vacancy rate, whereas small units tend to be situated in regions with lower land leverage, and more occupied by owners instead of being rented out or left vacant. And these are observable evidence for our argument that the purchase of large units is driving by investment demand, whereas that of small units by consumption demand.

So it is reasonable to account for the different appreciation rates of large and small housing units by relative changes in investment and consumption demand. We test it with the Hong Kong data from 1986 to 2014, and the results show changes in capital flow and market sentiment which represents investment demand have a stronger impact on prices of large units. Instead, marriage rate, divorce rate, unemployment rate, and mortgage ceiling, which are closely related to the housing consumption demand, affect the prices of small units more. In other words, we show that factors with an unequal effect on housing investment demand and consumption demand can contribute to the changing price differentials between large

[1] Raw data from Rating and Valuation Department and Census and Statistics Department, Hong Kong.

and small housing units. And this is because the large and small housing market are driven by investment and consumption demand, respectively.

The rest of the paper is structured as follows. After hypotheses development, we will introduce the empirical test and analyses. Concluding remarks are provided at the end.

58.2 Hypotheses

Based on the assumption that large units are mainly purchased for investment and small units are for consumption, the price differentials between large and small housing units should capture relative changes between investment and consumption demand. Since consumption and investment demand for housing are driven by various forces, collectively or respectively, factors that have unequal influences on the two, for instance, more impact on consumption demand but minimal impact on investment demand, should contribute to the price differentials stratified by size. We divide these factors into two groups, one is to lead changes in investment demand but have relatively less influence on consumption demand, and another is to induce consumption demand mainly. Other factors that can affect the price differentials are also taken into account as control variables.

58.2.1 Investment Demand Drivers

Investment demand depends on a number of factors, such as interest rates, costs, risks and opportunities. This study considers three representatives: capital flow, market sentiment, and stamp duty.

Interest rate is an effective measure for investment demand. A downward interest rate suggests a sufficient capital supply and investment is encouraged, whilst an upward interest rate indicates less capital supply and investment activities are discouraged. So large movements of capital can exert a significant impact on the property market by affecting investment activities. In Hong Kong, the exchange rate of the Hong Kong dollar to the United States dollar has to be maintained at a rate of 7.8 under the Linked Exchange Rate System (LERS) since 1983. The Hong Kong Monetary Authority is authorized to buy and sell Hong Kong dollars (HKD) to absorb any impact that both capital inflows and outflows may have on the linked exchange rate. When large amounts of capital flow into Hong Kong, it has to sell HKDs to stabilize the HKD, resulting in an increasing aggregate balance in the banking system, which makes it easier to borrow. The price of capital, that is, the Hong Kong Inter-bank Offer Rate (HIBOR) will fall. Conversely, when large amounts of capital flow out of Hong Kong, it has to buy HKDs. The monetary base thus contracts and HIBOR will rise. Taking advantage of the LERS system, the capital flow in Hong Kong can be better captured by introducing the U.S.-corresponding London Inter-bank Offer Rate

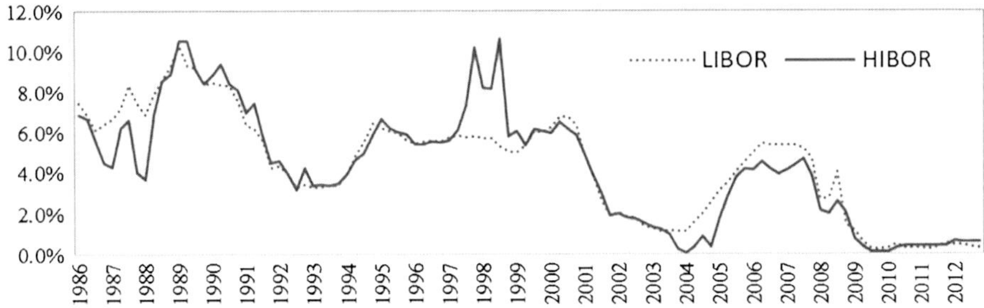

Fig. 58.2 Movements of HIBOR and LIBOR over the period of 1986Q1–2012Q2 (*Source Hong Kong Monthly Digest of Statistics*, Census and Statistics Department, Hong Kong)

(LIBOR). Under normal conditions, due to arbitrage activity, interest rate in Hong Kong should follow closely that in U.S. Capital flow from the market of a lower interest rate to the higher one, and equivalent HIBOR and LIBOR should be expected. However, referring to the movements of HIBOR and LIBOR in Fig. 58.2, we find some deviations. It implies that the process of arbitrage in interest rates doesn't always work well. During the financial crises period from 1997 to 1999, HIBOR remains higher than LIBOR, which suggests a lost confidence in HKD. Substantial capital flowed out the Hong Kong market. On the contrary, HIBOR has been persistently higher than LIBOR since 2003. Possible explanation is the economic development of China [3] which attracts a large amount of capital driven by speculation in the potential appreciation of renminbi. So the direction of capital flow can be captured by such deviations. That is, a higher HIBOR-to-LIBOR ratio reveals capital outflow, whilst a lower ratio represents capital inflow. Since residential market is a popular investment in Hong Kong, the changing ratio can indicate corresponding changes in the investment demand for large housing units. A negative link is thus formed between this ratio and price of large units. So our capital flow hypothesis is developed as:

H1.1 The changes of the HIBOR-to-LIBOR ratio will have a stronger negative impact on the prices of large housing units than those of small housing units, ceteris paribus

Stock price is very sensitive to economy changes. The huge trading volume and value in the stock market make it an ideal indicator of the investment sentiment. Conversely, investing in the housing market will take much longer time to make a deal due to heavier transaction cost. Higher stock price suggests a positive investment sentiment and more investment activities will happen in the residential market. The Hang Seng Index (HSI) which is a stock market index including the largest and most liquid stocks listed on the Main Board of the Stock Exchange of Hong Kong plays such a role. The constituent stocks are classified into four sub-indices by reference to distinct sectors. Property sector is one of them, and the index is Hang Seng Properties Sub-index (HSIP). It is based on the performance of dominant investors and developers in Hong Kong and this makes it a better indicator of investment sentiment. An increase in HSIP reflects positive attitude toward the market while a decline means lacking of confidence. So our market sentiment hypothesis is developed as:

H1.2 The changes of Hang Seng Properties Sub-index will have a stronger positive impact on the prices of large housing units than on those of small housing units, ceteris paribus

Increases in transaction costs will dampen the investment demand due to the decline in expected capital gains. Stamp duty is one of such transaction costs. Hong Kong has a long history of depending on property taxes. In fact, at least one-third of the government revenue comes from land property-related tax. The one imposed on property transactions is ad valorem stamp duty with a nonlinear rate schedule. The idea is that consumers of lower-value properties tend to be relatively poor and it is neither cost-effective nor socially desirable to collect tax from such group. But those traders of higher-value properties are considered to be less tax-sensitive and more capable of paying the tax. This means large units will be associated with higher stamp duty. Since 1980s, the Government has modified the stamp duty schedule eight times and the last two revisions (April 2010 and March 2013) are of vital importance with an aim to curb the heated market. The continued surge in property prices despite the current global and local economic slowdown since 2008 is a cause for such revisions. The Government precipitates a buildup of exuberance in the housing market and one solution is to increase the stamp duty. So they doubled the stamp duty rate, called Double Stamp Duty (DSD). Besides this, another two stamp duties are imposed to further cool down the market. Too many investment and speculative activities are always thought to be responsible for the surging property price. So a Special Stamp Duty (SSD) which applies to residential properties resold within 3 years after purchase is proposed (Nov. 2010) to increase the transaction cost of short-term investment and expect for diminished prospect for quick profits from speculating in properties. Moreover, in light of the increasing share of residential housing taken up by non-local buyers, Buyer stamp duty (BSD) is introduced subject to any person except a Hong Kong Permanent Resident, and it is charged at a rate of 15 % on all residential properties, on top of DSD and SSD. This is to ensure that the basic consumption demand of local buyers will be accorded priority amidst the tight supply in the residential market. If the large housing market is dominated by investors whose willingness to invest are heavily dampened by these stamp duty tools, conceivably it will be more affected than the small housing market and results in a lower price appreciation. So a lower price differentials between large units and small units is expected. This leads to two stamp duty hypotheses:

H1.3a An increasing gap between the stamp duty of large and small housing units will lead to a lower price differentials between large and small housing units, ceteris paribus

H1.3b The price differentials between large and small housing units will be lower after the implementation of BSD and SSD, ceteris paribus

58.2.2 Consumption Demand Drivers

Consumption demand for housing mainly comes from population growth, changing social patterns which see a shift from extended to nuclear families, and increasing affluence which generates higher aspirations for home ownership and better living conditions. In this study, marriage and divorce rates, unemployment rate, and mortgage are taken into account.

Population growth is one of the main drivers for housing consumption, including the demand of buying a house and renting one. The purchase demand is more affected by changes in marriage rate and divorce rate in the district. This is because it is a very popular idea in Hong Kong that owning a house is the precondition for the success of a marriage. An increase in marriage rate can induce more such purchase demand. Instead of staying with their families, new couples always choose to live away from their original home. This is supported by a decreasing average household size in Hong Kong, from 3.9 in 1982 to 2.9 in 2014. If majority of these new couples aim for small units to meet their basic consumption demand, a higher marriage rate can contribute to a lower price differential between large and small units. The marriage rate hypothesis is motivated as:

H2.1 Changes in marriage rate will negatively affect the price differentials between large and small housing units, ceteris paribus

Similarly, a growing number of divorces can also exert a strong impact on the housing market. According to [4], six to seven years after the divorce the extra demand for housing is about one-third of the number of divorces in any given year in Netherlands. The housing market, in particular the small market, will see a growing consumption demand from the divorced. This demand increase will contribute to a lower price differentials between large and small units. And our divorce rate hypothesis is formed as:

H2.2 Changes in divorce rate will negatively affect the price differentials between large and small housing units, ceteris paribus

Employment affects the housing demand through well-known channels. Declining unemployment raises consumer confidence and willingness to undertake long-term commitments as they obtain a stable income flow for their mortgage payment [5]. Employment alleviates the wealth constraint on down payments, especially buyers of small units [6]. Rising employment can also encourage tenure shifting from rental to first-time homeownership [9] and population migration to more prosperous regions [7]. Consequently, unemployment rate changes will exert a relatively greater impact on buyers of small units than buyers of large units. This is consistent with a higher income elasticity for lower income homebuyers than for higher income homebuyers since we assume buyers of small units with a relatively lower income compared with those of large units [9]. Declining unemployment will lead the price differentials between large and small units to converging, and this motivates the unemployment hypothesis:

H2.3 Changes in unemployment rate will positively affect the price differentials between large and small housing units, ceteris paribus

A mortgage ceiling is the maximum loan-to-value ratio for a mortgage borrower. A lower mortgage ceiling requires a homebuyer to possess a larger amount of wealth for down payment, and thus, has a negative impact on housing consumption demand. This is because consumption demand predominantly comes from buyers of small units and they rely more heavily on mortgage borrowing. In contrast, a housing investor buying a large unit is less concerned about a mortgage ceiling, but more concerned with the risk and return features of a housing investment compared to those of other investments [8]. The price differentials between large and small units will be enlarged with a high mortgage ceiling. So our mortgage hypothesis is expressed as:

H2.4 Mortgage ceiling will negatively affect the price differentials between large and small housing units, ceteris paribus

58.2.3 Control Variables

Five control variables are identified from the literature (mainly from [8, 9]). If higher-income households always shop for large units, while lower-income households for small units, the income gap and inflation gap between the two groups is expected to enlarge the price differentials between large and small units. A higher vacancy rate will increase a sellers' expected length of sale, then reduce the reservation price and eventually the sale price [10]. So we expect the vacancy gap between large and small units to negatively affect their price differentials. So does supply gap. Lack of supply will drive up the product price, whereas over-supply will take the price down. Last is public housing supply. Increased supply of public housing may divert some demand for the small units, though the limited supply and its non-transferable nature suggest that they are not good substitutes. We expect a positive impact of the public housing supply on the price differentials between large and small units.

58.3 Empirical Tests

58.3.1 Models and Variables

To test our hypotheses, one regression is developed and variables and other details are listed in Table 58.1. The study period is from 1986 to 2014 determined by data availability, and we use quarterly data. Three season dummies are included to capture seasonal difference. According to H1.1, $\gamma 1$ should be negative as capital

Table 58.1 Variables

Variables	Coefficient	Expected relationship	Lag	Data source
Dependent variable				
Price differential (*PL_PS*)				Rating and valuation department
Investment demand variables				
The Hibor-to-Libor ratio (*HI_LI*)	γ_1	−	0	Hong Kong Monetary Authority
Hang Seng Properties Sub-index (*HSIP*)	γ_2	+	1	Census and Statistics Department
Ad valorem stamp duty differential (*STDL_STDS*)	γ_3	−	0	GovHK: Stamp Duty Rates
Special Stamp Duty (*SSD*)	γ_4	−	0	GovHK: Stamp Duty Rates
Consumption demand variables				
Crude marriage rate (*CMR*)	γ_5	−	0	Census and Statistics Department
Crude divorce rate (*CDR*)	γ_6	−	0	Census and Statistics Department
Unemployment rate (*UR*)	γ_7	+	1	Census and Statistics Department
Mortgage ceiling (*MC*)	γ_8	−	0	Hong Kong Mortgage Corporation
Control variables				
Inflation gap (*CPIL_CPIS*)	γ_9	+	0	Census and Statistics Department
Income gap (*INL_INS*)	γ_{10}	+	0	Rating and Valuation Department
Vacancy gap (*VL_VS*)	γ_{11}	−	1	Rating and Valuation Department
Supply gap (*SL_SS*)	γ_{12}	−	5	Census and Statistics Department
Public housing supply (*PUS*)	γ_{13}	+	5	Rating and Valuation Department

outflow will contribute to lower prices of large units. And $\gamma 2$ should be positive by reference to H1.2. High investment sentiment leads to higher prices of large units. Both $\gamma 3$ and $\gamma 4$ are expected to be negative predicted by H1.3a&b. The stamp duty increase will dampen the investment for large units and thus lower price. H2.1&2 predict a negative sign for both $\gamma 5$ and $\gamma 6$ since higher marriage rate and divorce rate can induce more consumption demand for small units and hence higher price. H2.3 expects a positive $\gamma 7$, while H2.4 a negative $\gamma 8$. Lower unemployment and higher mortgage ceiling can encourage housing consumption and result in higher price of smaller units. Among the control coefficients, $\gamma 9$, $\gamma 10$, and $\gamma 13$ are expected to be positive, whereas $\gamma 11$ and $\gamma 12$ to be negative.

Measurement of variables:

1. PL_PS = ln (PL/PS). PL and PS are the price indexes of large (Class D&E) and small housing units (Class A), respectively.
2. HI_LI = ln (HIBOR/LIBOR). The 3-month rate of HIBOR and LIBOR are used. HIBOR is only available from 1982 to 2012, and the rest (2013–2014) is estimated by reference to HIBOR fixing.[2]
3. STDL_STDS = ln (STDL/STDS). STDL and STDS are the ad valorem stamp duty payments for Class D&E units and Class A units, respectively, and they are estimated by reference to the average unit price in each class. And the average sizes of a Class A, D and E unit are assumed to be 30, 130, and 160 m^2, respectively. STDL and STDS are calculated by reference to the ad valorem stamp duty rate in each time period and unit price.
4. SSD is a dummy variable and equals 1 if the SSD is implemented; otherwise, 0. Since the period of BSD is covered by SSD and the implementation period is very short, only a SSD dummy is used.
5. UR is a dummy variable. And it is identified through comparing the real unemployment rate and the natural unemployment rate which is calculated as the average rate over years from 1981 to 2014. UR equals 1 if the real unemployment rate is larger than the natural unemployment; otherwise, 0.
6. CMR (CDR) is the marriage (divorce) rate per 1000 population and calculated as the ratio of the number of marriages (divorces) to the total population in thousand.
7. CPIL_CPIS = ln (CPIC/CPIA). CPIC and CPIA measure the inflation rates affecting the high-income and low-income households, respectively, and they are compiled based on the expenditure patterns of households in the relatively high and relatively low expenditure ranges.
8. INL_INS = ln (INL/INS). INL and INS are average income of households living in large units (Class D&E) and small units (Class A), respectively.
9. MC is a dummy variable and equals to 1 if the mortgage ceiling is higher than 0.7; otherwise, 0.

[2]HIBOR fixing is daily HKD Interest Settlement Rates and it is fixed by reference to market rates for HKD deposits in the Hong Kong interbank market, namely, HIBOR. It is established since 1996.

10. VL_VS = ln (VL/VS). VL and VS are the respective vacancy rates of Class D&E and Class A. A time lag of one quarter is set to avoid endogenous problem.
11. SL_SS = ln (SL/SS). SL and SS are the respective supply of Class D&E and Class A. And SL_SS is assumed to lead the price differentials by 5 quarters. This is because after the acquisition of the consents to commence work, it will always take developers 8 quarters to complete and they do the presales 3 quarters in advance. So 5-quarter lag is plausible.
12. PUS is the ratio of the number of public housing units with consents to commence work to the total units including both private and public sectors with consents to commence work. Similarly, 5-quarter lag is assumed.

58.3.2 Empirical Results

All the variables have to be stationary to obtain a reliable OLS result. We apply the Augmented Dickey-Fuller (ADF) test and the results show only HI_LI and SL_SS are stationary. The first difference of the rest variables are examined through the ADF test again and all turn out to be stationary. So the first difference of the nine variables, ΔPL_PS, ΔHSIP, ΔSTDL_STDS, ΔCMR, ΔCDR, ΔCPIL_CPLS, ΔINL_INS, ΔVL_VS, and ΔPUS take their places in the regression. Table 58.2 shows the regression result. Our hypotheses are all confirmed except the stamp duty one, and R2 turns out to be a high level. This means we do provide some important explanations for the changing price differentials between large and small housing units.

Table 58.2 Results of the regression

Dependent variable: *PL_PS*; Method: least squares; Sample (adjusted): 1987Q2–2014Q4; Observations: 111

Variable	Coefficient	t-Statistic	Variable	Coefficient	t-Statistic
C	0.046027	4.077351	*ΔCPIL_CPIS*	0.339351[b]	2.212810
HI_LI	−0.019562[a]	−3.448507	*ΔINL_INS*	0.046557	1.301459
ΔHSIP	0.051353[a]	3.442775	*ΔVL_VS*	−0.055452[b]	−2.208830
ΔSTDL_STDS	0.001224	0.425313	*SL_SS*	−0.041625[b]	−2.207203
SSD	−3.98E-05	−0.005021	*ΔPUS*	0.001816	0.330660
ΔCMR	−0.024117[c]	−1.906085	$SEASON_1$	−0.036637[b]	−2.208625
ΔCDR	−0.252061[c]	−1.957763	$SEASON_2$	−0.027462[b]	−2.116742
UR	0.012759[b]	2.129297	$SEASON_3$	−0.032963[b]	−2.850443
MC	−0.029080[a]	−4.324688			
R-squared	0.486570		Adjusted R-squared		0.399178

Note [a]Denotes significance at 1 % level, [b]significance at 5 % level, and [c]significance at 10 % level

Table 58.3 Results of the Jarque-Bera normality test and the White Hetero skedasticity test

Normality test: Jarque-Bera				Heteroskedasticity test: White			
Skewness	−0.147	Kurtosis	3.136	F-statistic	0.583	Prob. F (16,94)	0.889
Jarque-Bera	0.785	Probability	0.785	Obs*R-squared	10.023	Prob. Chi-Square (16)	0.865

As expected, HI_LI has a significantly negative effect on PL_PS. An increase in the ratio of HIBOR-to-LIBOR indicates capital outflow and investment demand for large units reacts with decline. This contributes to a lower price of large units. Conversely, ΔHSIP has a significantly positive effect on PL_PS. When the stock market performs well, it signals strong confidence for the market and encourages more investment activities. However, a positive and insignificant effect of ΔSTDL_STDS on PL_PS is showed, and this might be caused by the rough calculation. And also a negative and insignificant effect of SSD is revealed. The implementation period might be too short to provide sufficient statistics. The four variables representing the influences of consumption demand on price differentials all obtain expected signs. Both ΔCMR and ΔCDR have a significantly negative effect on PL_PS, though only at 10 % level. Increases in marriage rate and divorce rate can bring new consumption demand for small units. One thing deserves further discussion is the effect of divorce rate which takes immediately according to the result in this study while [4] find a gap of six to seven years before such increases take effect. UR positively affect PL_PS at 5 % level, while MC exerts a negative effect at 1 % level. The implication is that a decreasing unemployment rate or a higher-than-7 % mortgage ceiling can induce more consumption demand for small units and contribute to a lower price differential between large and small units. Among the control variables, ΔCPIL_CPIS, ΔVL_VS, and SL_SS show significantly expected signs at 5 % level, whereas ΔINL_INS and ΔPUS have expected signs but not significant.

To ensure the plausibility of our regression, we check residuals' normality and homoscedasticity. The Jarque-Bera test is used to check the normality based on skewness and kurtosis and the White test is applied to test the homoscedasticity. As showed in Table 58.3, we cannot reject the assumption of normal distribution and constant variance for the residuals.

58.4 Conclusion

This study examined the contributors to the changing price differentials between large and small housing units. We argue that the property market of Hong Kong is segmented as small and large housing markets by consumption demand and investment demand. Any factors that have unequal effects on investment demand and consumption demand can account for the varying price differentials. We test eight such variables using data of Hong Kong from 1986 to 2014 and find a significant effect from six of them. Capital flow and market sentiment have a stronger impact on the prices of large units than that of small units since they can induce more investment demand for housing. Instead, marriage rate, divorce rate, unemployment rate and mortgage ceiling are closely related to the housing consumption demand and are found to affect the prices of small units more. We expect that these results about house price appreciation stratified by size will be of interest to investors and homebuyers, and also should be an important input to housing policy.

References

1. Ioannides YM, Rosenthal SS (1994) Estimating the consumption and investment demands for housing and their effect on housing tenure status. Rev Econ Stat, 127–141
2. Wong SK, Yiu CY, Chau KW (2012) Liquidity and information asymmetry in the real estate market. J Real Estate Financ Econ 45(1):49–62
3. He D, Leung F, Ng P (2008) How do macroeconomic developments in Mainland China affect Hong Kong's short-term interest rates? Macroeconomic Linkages between Hong Kong and Mainland China
4. Dieleman FM, Schouw RJ (1989) Divorce, mobility and housing demand. Eur J Popul 5 (3):235–252
5. Böheim R, Taylor MP (2002) Tied down or room to move? Investigating the relationships between housing tenure, employment status and residential mobility in Britain. Scott J Polit Econ 49(4):369–392
6. Linneman P, Wachter S (1989) The impacts of borrowing constraints on home ownership. Real Estate Econ 17(4):389–402
7. Gabriel SA, Shack-Marquez J, Wascher WL (1992) Regional house-price dispersion and interregional migration. J Hous Econ 2(3):235–256
8. Chau KW, Lam HM (1998) Investment and consumption demand for housing in Hong Kong. In: The Third Asian Real Estate Society conference (p 14)
9. Smith LB, Ho MH (1996) The relative price differential between higher and lower priced homes. J Hous Econ 5(1):1–17
10. Wheaton WC (1990) Vacancy, search, and prices in a housing market matching model. J Polit Econ, 1270–1292
11. Huang W (2013) Price movements of large and small housing units in Hong Kong: an empirical investigation. Doctoral dissertation, The University of Hong Kong

Chapter 59
The Development and Differentiation Trend of Real Estate Market in the Yangtze River Delta Urban Agglomeration

Ling Zhang, He Wang and Shenghua Jia

Abstract This paper examines the development prospect of urban real estate market with the evolution law of Urban Agglomerations. To grasp the evolution trend of Yangtze River Delta and the dynamic characteristics of its real estate market, this paper was carried out from three dimensions, namely the overall evolution dimension hierarchical evolution dimension and the classification evolution dimension. By analyzing data about city development and real estate market in Yangtze River Delta and comparing it with the similar data at home and abroad, this text concluded that the Yangtze River Delta has developed into mature stage;its real estate market has great development potential and will polarize more obviously among different levels and categories; the progress and divergence trend for real estate market depends on the evolution process of the Urban Agglomerations.

Keywords The yangtze river delta · Urban agglomerations · The real estate market divergence

59.1 Introduction

Being as the pioneer of Reform and Opening-up policy and enjoying high reputation for its economic prosperity, Yangtze River Delta always has the best economic foundation the highest marketization degree the most prosperous culture and

L. Zhang
Real Estate Research Center, College of Civil Engineering and Architecture, Zhejiang University, Hangzhou, China

H. Wang (✉)
College of Civil Engineering and Architecture, Zhejiang University, Hangzhou, China
e-mail: wanghelotus@163.com

S. Jia
Real Estate Research Center, Zhejiang University, Hangzhou, China

Y. Wu et al. (eds.), *Proceedings of the 20th International Symposium on Advancement of Construction Management and Real Estate*,
DOI 10.1007/978-981-10-0855-9_59

balanced social development in China, its development potential and direction influence the national economy a lot just like a vane. Meanwhile, the real estate industry, an important part of national economy, received much concern in recent years. So it happens to be a very intriguing issue of what direction will the real estate market in the Yangtze River Delta be heading in. Therefore, this paper will set out from the evolution trend of urban agglomeration in Yangtze River Delta, then go deep into the different directions of the real estate market in the Yangtze River Delta in all the levels and types.

59.2 Overall Evolution Trend of Real Estate Market in Yangtze River Delta

59.2.1 Assessment of the Development Stage of Yangtze River Delta Urban Agglomeration

There is an obvious regularity in the spatial evolution of urban agglomeration, and some scholars divided that progress into five stages according to the spatial composition, industrial relation, regional structure and core function, which are the embryonic stage, the initial stage, the middle stage, the mature stage and the advanced stage [1]. We can measure and further evaluate the stage of the urban agglomeration evolution from three dimensions: spatial structure feature, economic development and social development [2].

Spatial structure feature. The Yangtze River Delta urban agglomeration is composed of three metropolitan areas including Shanghai, Hangzhou and Nanjing. Shanghai being the lead, Shanghai-Nanjing and Shanghai-Hangzhou as development axes, there generates an urban network which contains Shanghai, Nanjing, Hangzhou, Suzhou, Ningbo and other cities as the center. Being deemed as China's political, economic and cultural center, the Yangtze River Delta urban agglomeration act as the national growth center and takes the function of contacting with the outside world.

Economic development. According to Cheney's [3] industrialization stage theory and Clark's [4] labor transfer law, the Yangtze River Delta urban agglomeration has been at the mature industrialization stage with its per capita GDP of $9917 and three industries' proportion of 10, 48, 42 % respectively, which is close to the value of American Boswash urban agglomeration in 1980.

Social development. The social development level can be measured from four aspects, the urbanization rate, the population quality, the basic public service and the ecological civilization, in which the Yangtze River Delta urban agglomeration shows an obvious dominance in China. Though there is still a big disparity with developed countries, but it is close to the level when they were in the mature stage of urban agglomeration (as shown in Table 59.1).

Table 59.1 Comparison of social development between yangtze river delta and developed areas

Indicator	The yangtze river delta 1	Developed countries in their mature stage 2	Developed areas 3
Urbanization rate	66 %	73.7 %	82 %
College entrance rate	27 %	53 %	94 %
Number of doctor per million people	18	19	24.2
Energy consumption per unit GDP (tons of standard oil/million dollars)	2.70	3.12	1.66
Carbon emissions per unit GDP (ton/million dollars)	13.7	16.9	3.77

Note The indicators 'college entrance rate', 'energy consumption per unit GDP' and 'carbon emissions per unit GDP' in 1 column take the Chinese national data, and they take American data in 1980 and 2010 in 2 column and 3 column, the international data comes from the World Bank Database

According to the three dimensions above, we could draw the conclusion that the Yangtze River Delta has been in the mature stage of Urban Agglomeration development already. At the same time, the Yangtze River Delta's input in innovation and public service is obviously above the national average these years, and the construction of Shanghai free trade zone will further strengthen its leading position in the reform and opening up territory. Besides, the reform dividend will gradually release, which will certainly accelerates the Yangtze River Delta's pace in catching up with the international standard.

59.2.2 The Overall Development Trend of Real Estate Market in the Yangtze River Delta

With huge development potential, the actual space for the development of Yangtze River Delta's real estate market lies on the urban agglomeration's ability to restructure and upgrade.

In the Yangtze River Delta urban agglomeration's evolution toward advanced stage, the service industry, especially the real estate's proportion in its economic structure will see a great improvement. As for now, the real estate's added value in Yangtze River Delta accounts for barely 6 % of its GDP (so is it for its core city Shanghai) while other developed cities like the United States, Japan, Canada and United Kingdom take a proportion of 12.8, 11.2, 16 and 22.4 % respectively.[1] Difference as there may be in the statistics caliber, it still makes sense in believing that the Yangtze River Delta real estate has a great development potential.

[1]Data source: 《The development of international metropolis and its enlightenment to Beijing》, International metropolis real estate development research group, 2011, p. 4.

It is worth noting that, the growth of the proportion mentioned above in the Yangtze River Delta is slowing down after 2005, presenting an almost saturation state. This reflects a limited development in the real estate industry under the current industrial structure and institutional framework, and only through transformation and upgrading, can the Yangtze River Delta's urbanization level be greatly improved and the connotation of the real estate be enriched, which will eventually turn the preceding development potential into growth in real terms.

Speed and magnitude matching with the development of urban agglomeration, the increase of the proportion of commercial property and high-quality residential will be durative.

The gap of nonresidential building's proportion between the Yangtze River Delta and national average is widening year by year, which is closely related to the Yangtze River Delta's higher proportion of service sector than the nation and its active transition to service industry. In fact, the U.S. commercial real estate didn't reach saturation until its proportion was 30 %, compare to that, the Yangtze River Delta still have considerable development potential since its commercial real estate proportion is less than 15 %. With the continuous increase in the proportion of the third industry hereafter, all kinds of commercial real estate are likely to witness further prosperity.

The Yangtze River Delta residential building shows an obvious trend to take high-end route, which is highlighted by the higher proportion of its villa and high-grade residential sales than the national average. Such phenomenon consists with the Yangtze River Delta's appeal to numerous high-level talents and the induced high demands for upgraded residential. Although the high-grade housing's share is declining under the supply regulation recent years, its posting rising price suggests a firm demand for quality housing. Given a transformation in regulation strategy, the high quality residential sales' percentage will stage a comeback out of question.

With continuous rise in tertiary industry's proportion and persistent gathering of high level personnel, there is a considerable development space for commercial real estate and quality housing in Yangtze River Delta, especially when the urban agglomeration achieves a smooth transformation and upgrading. Because under that circumstances, the improvement of industrial structure and economic quality will release huge effective demand for commercial real estate and quality housing, thus their proportion will see a new round of promotion.

59.3 Hierarchical Evolution Trend of Real Estate Market in Yangtze River Delta

There is an obvious hierarchical structure in urban agglomeration. As the result of the interaction of concentration and diffusion effect in the evolution of urban agglomeration, it illustrates the hierarchical distributions of structure, function, level and other development state of different cities in the urban agglomeration [5].

Table 59.2 Hierarchical division of the yangtze river delta urban agglomeration

Hierarchy	City type	City
The first tier	Cosmopolis	Shanghai
The second tier	Metropolis	Nanjing ningbo suzhou wuxi hangzhou
The third tier	Big city	Jiaxing nantong changzhou wenzhou taizhou jinhua xuzhou yancheng shaoxing
The fourth tier	Medium city	Yangzhou taizhou zhenjiang quzhou lishui zhoushan lianyungang huaian suqian huzhou
The fifth tier	Small city	Kunshan zhangjiagang changshu taicang cixi yiwu yuyao ling'an zhuji and other counties

59.3.1 Hierarchical Structure of Urban Agglomeration

Urban flow intensity reflects the economic and social bonding strength between one city and the other in the area [6]. Urban flow intensity can be calculated by the urban outward functional capacity and urban functional efficiency, the former is measured by the number of employees in the 8 major export service sectors in the downtown area of each city, the latter is measured by GDP per capita, and the calculated ranking results of each city[2] is similar to the result of urban economic rank in Yangtze River Delta from Zhang XuLiang (2011) [7]. Thus catalogue the 25 prefecture cities into 4 tiers and list the counties and county-level cities as fifth tier, as shown in Table 59.2.

59.3.2 Hierarchical Differentiation Trend of the Urban Agglomeration

The development state of each tier is codetermined by its agglomeration degree and development level, and the agglomeration degree plays a more important role to influence the future.

Agglomeration degree. This index is measured by the urban expansion rate and its proportion in the total amounts of urban agglomeration in the last 5 years. The former is reflected by net migration rate or growth rate of population, while the latter is tested by GDP share, fixed asset investment share and total export share.

Development level. This index is reflected by the city wealth level and structure effect, the former includes the per capita income level, the per capita consumption level and saving level, while the latter includes industrial structure (the proportion of the third industry), population structure (the proportion of college graduates and employees) and consumption structure (proportion of education and entertainment expenses).

[2]The calculation data of urban flow intensity is not listed here due to space limitations

Development state. On account of the different meaning and unit between indexes of concentration ability and development level, this paper takes the deviation standardization method to grade each index, namely uses the 10 points system, where the minimum value of certain index is 1, the maximum is 10, and the other is calculated through the interpolation method. Plus the 0.6 weight of agglomeration degree and 0.4 weight of development level, we can get the composite score for each hierarchy is 9.8, 5.5, 3.2, 1.7, 1.4 respectively, with clear hierarchical difference.

Future trend. According to the composite development score of each hierarchy in 2000, 2005, 2008 and 2011, we can first calculate the ratio of development state score of each hierarchy to the whole Yangtze River Delta average and observe the spatial features of hierarchical difference; and then calculate the ratio difference between these years to see the relative growth rate of each hierarchy. It's found that the hierarchical development difference has increased than decade ago. To be specific, the discrepancy was slight between the first tier and the second tier while more apparent between the second tier and the third, fourth tier. Besides, the third tier and the fourth tier showed a relatively stable relation while the difference between the fourth and the fifth tier was significantly reduced. On the whole, the evolution trend is featured with promotions in both ends and gaps inside all the tiers.

59.3.3 Hierarchical Differences in Urban Agglomeration's Real Estate Market

There are discrepancies of various extents in market size, commercial housing prices, land market structure and commercial housing use structure of real estate market in the five hierarchies.[3]

Housing price. It's found that the first tier has a huge sales volume and a slowdown growth, the second tier has a continuous enlargement in both the size and share while a large increase is witnessed in the third 、the fourth and the fifth tier. All the hierarchies show an obvious differential except for the more and more identical third and fourth tier. Meanwhile, the commercial housing price in all hierarchies has a differential over 30 % except for the fourth and fifth tier.

Market structure. The proportion of urban industrial land in the fourth and fifth tier has increased obviously, the fourth tier achieved a jump from 28.7 % in 2008 to

[3] Data in this section comes from the housing index database. The first and second hierarchies take the average value of all the cities they include, the rest hierarchies only take the average of some representative cities due to limited data sources, the third hierarchy takes Shaoxing, Wenzhou, Xuzhou, Nantong and Changzhou; the fourth hierarchy takes Zhenjiang, Yangzhou, Taizhou, Lianyungang, Zhoushan and Huaian; the fourth hierarchy takes Kunshan, Jiangyin, Ninghai, Haimen, Cixi, Yuyao, Fenghua, Linhai, Zhangjiagang, Changshu etc.

60.1 % in 2012(especially the continuous increase in the recent three years), the fifth tier went up to 62.9 from 54 %. The land structure change in these two hierarchies reflects the trend for industrial manufacturing industry's diversion into the fourth and fifth tier. Though drop gradually, the proportion of commercial housing sales is still more than 85 % in all hierarchies, this signifies the structural characteristic of the commodity housing's leading role in property market under the impetus of rapid urbanization and improvement of living standard. Apart from that, the proportion of office sales in the first and second tier is significantly higher than other hierarchies, the 6.8 % average proportion in five years shows a strong demand for commercial real estate.

59.4 Classification and Development Trend of the Real Estate Market in the Yangtze River Delta

The differences in development patterns among different cities is decided by their status, function, industrial division and interaction in the urban agglomeration [8]. In general, cities in the Yangtze River Delta Urban Agglomeration can be divided into four types: metropolis,large and middle-sized city, small industrial cluster city and small urban satellite city.

59.4.1 Metropolis

Metropolis is developed from the megacities in the first and second tier, pulling and driving the regional economic development [9]. Its future development pattern should be "structure upgrading and function improvement", propelled by innovation ability and function level. Cities of this kind is basically featured by: rapid expansion in scale, continuous improvement in function, sustained prosperity in central city, expansion from city to suburb, high level of public service, strong purchasing power of residents, further talent aggregation and further convergence of hi-tech industry, finance, logistics, culture and education industry.

The real estate market of metropolis is characterized by the following performances: (1) Housing market remains prosperous, rigid and upgrade demand remain strong, demand for quality housing is huge and the house price of scarce lots in core city keeps up with world city. (2) There is great development space for office and business real estate like office building, financial center, shopping center, hotel, urban complex and headquarters base. (3) Cultural entertainment, creativity and exhibition real estate will see more opportunities in metropolis.

59.4.2 Large and Middle-Sized City

Large and middle-sized city involved in the third and fourth tier has a relatively sound urban function, connects the other cities in the urban agglomeration. City of this kind has the ability to gather regional economic activities in its area on the one hand, on the other hand, it undertakes some shift industries from central city. So its development pattern is "scale economy and function improvement", driven by gathering ability and undertaking capability.

The real estate market features of this city type includes: (1) The upgrading demand from the native people and rigid demand from the immigrant boost a stable growth in housing market, but demand for high-end property is limited. (2) The industrial park and warehousing estate will achieve rapid growth because of their strong industrial cluster ability and location advantage. (3) There is still some demand for commercial property in the third tier while very limited in the fourth tier. (4) Some cities with tourism resources will achieve certain development in real estate of tourism and cultural creative.

59.4.3 Small Industrial Cluster City

Small industrial cluster city in the fifth tier mainly comes from the county-level city driven by industrial agglomeration and takes an advantage in industry and cost. Such cities have the basic urban functions, but it's difficult to promote their city level and city quality. By building up leading industry, some of them can form a specific industrial cluster and develop their service industry around the superior industry. Its future development is partly leaning on the agglomeration and upgrading of superior industry and partly relies on the industrial transfer in central city, and forms a development pattern of "agglomeration economy".

The real estate market of this kind is featured by: (1) A limited scale of housing market especially the demand for high-end residential, the potential demand for house comes from the rural population brought by industrial agglomeration and new urbanization. (2) Industrial estate, commercial office property and professional wholesale market estate are relative prosperous and potential. (3) Influenced by environment and competition, the fluctuation in leading industry's boom degree will cause a risk of volatility in real estate market. Generally speaking, housing market demand in these cities has a lot of policy flexibility, the industrial and commercial estate revolving around the superior industry is guaranteed with development potential.

59.4.4 *Small Urban Satellite City*

Small urban satellite city in the fifth tier is often distributed in the periphery of metropolis and undertakes some metropolitan functions. As an organic extension of the metropolis, the key to its development is to give full play to its regional and cost advantages.

The real estate market features of this city type includes: (1) A great property demand from young first-time home buyer out-of-town. (2) Convenience to develop low-density and high-grade residential as well as leisure and aged-care building for those cities who take an advantage in natural environment. (3) Supporting commercial property for large residential area or other functional area will spring up. (4) Improper development program, construction pace or auxiliary transportation facility can induce a risk of imbalance in real estate market. In summary, there is a rapid growth for rigid demand housing, aged care architecture and low-density building, hence the auxiliary commercial property is also guaranteed with great potential.

59.5 Conclusion

Research shows that the Yangtze River Delta urban agglomeration has stepped into mature stage and has great potential in the overall real estate market. According to the real estate development law of development countries and the Yangtze River Delta's appeal to talent flow, the proportion of commercial property and high-quality residential will increase continuously and its price will further gap away from the national average. There are great difference and obvious differentiation among cities from different hierarchies (sorted by economic level) and different types (sorted by development pattern) in real estate market.

Furthermore, the actual development space for the real estate market in Yangtze River Delta depends on the evolution process of the urban agglomeration, and the market differentiation in real estate will continue along with the evolution of the Yangtze River Delta urban agglomeration. In case that the Yangtze River Delta urban agglomeration stagnates in the mature stage for a long time in future, the industry and labor structure of all level cities will reach a steady, the difference of market structure and price among cities in different hierarchies will stabilize or increase slightly. In another case, if the Yangtze River Delta urban agglomeration achieves transformation and upgrading in the future, marked with rapid improvement of real urbanization level and substantial breakthrough in industrial upgrading. The industry and labor structure of cities in different hierarchies will differentiate a lot, the real estate market scale of all the cities especially the lower hierarchy one will have a large growth space and be identified with a synchronous improvement in both quantity and quality. Besides, the difference of market structure and price among cities in different hierarchies will continue to expand. As for whether the

Yangtze River Delta urban agglomeration can enter into the advanced stage, it is decided by the intensity, breadth and depth of reform and opening up as well as the transform and upgrade process and independent innovation capability, full of uncertainties.

For the city government and local real estate department, it's necessary for them to figure out the urban specialization and orientation under the framework of urban agglomeration development, planning and guiding for a steady and sound development of the real estate market. With respect to the real estate enterprises, they should fully realize the differences between cities in Yangtze River Delta and pay more attention to the differentiation trend of property market in cities from all hierarchies and types, and then seek for suitable investment opportunity and evade risk effectively according to the respective feature of different cities.

References

1. Fang C (2010) Theory and practice on sustainable development of urban agglomeration in China, Science Press
2. Luo S (2012) The interaction among cities and industrial development within a metropolitan area,Shanghai people's Publishing House
3. Cheney JT (1989) Comparative study on industrialization and economic growth,Shanghai Triple Bookstore
4. Changle Xu (2013) Research on the transformation and development of urban agglomeration in the yangtze river delta,Shanghai people's Publishing House
5. Guoxi Wu, Jinchui Liu, Fuzhong He (2008) Study on urban spatial evolution and the regional organization of zhong-yuan urban aggolomeration. Urban Dev. Stud. 2:35–38
6. Na Li (2011) Study on spatial contact and integration of urban agglomeration in yangtze river delta. Areal Res. and Dev. 30(5):72–77
7. Xuliang Z, Yuemin Nin (2011) Urban economic contacting and internationalization spatial development strategy in yangtze delta area urban agglomeration. Econ. Geogr. 31(3):353–359
8. Kaiming C (2010) Distribution structure and function of the urban system–an empirical study based on the yangtze river delta. Res. on Dev. 3:11–15
9. Xiaocheng L (2012) The economic analysis on city industry transformation and the space scale upgrade. Comment on prod. 3:59–64

Chapter 60
Empirical Study on Influence Factors of Capital Structure of Chinese Real Estate Listed Companies

Hong Zhang, Linjun Li and Jiawei Chen

Abstract This paper will make theoretical analysis on influence factors of capital structure of Chinese real estate listed companies and put forward hypotheses to be tested, then make empirical analysis with a panel data model, based on a sample of 48 Chinese real estate list companies from 2005 to 2014. The results show that among company characteristic factors, company size and growth ability have significantly positive effect on capital structure, profitability and asset liquidity have significantly negative effect on capital structure, operation capacity, collateral value of assets and non-debt tax shield show no significant effect. In terms of company governance factors, ownership concentration shows significant negative correlation with capital structure, while managerial ownership, board size and independent directors have no significant effect. Among macro-economic factors, supply of money has significantly positive effect on capital structure, interest rate and inflation have no significant effect.

Keywords Capital structure · Real estate · Panel data

60.1 Introduction

At present, economic development of China has entered into the new normal, development tendency of real estate has presented some new change, the golden age of real estate has ended, and growth rate slowed down, high consolidation showed up, and the return to rational prices appeared. In this macro-background,

H. Zhang (✉) · L. Li · J. Chen
Center for Urbanization and Industrial Development, Hang Lung
Center for Real Estate, Tsinghua University, Beijing 100084, China
e-mail: zhannie@tsinghua.edu.cn

Y. Wu et al. (eds.), *Proceedings of the 20th International Symposium on Advancement of Construction Management and Real Estate*,
DOI 10.1007/978-981-10-0855-9_60

how real estate understands and adapts to the new normal has become a question deserving discussion. As a typical capital-intensive industry, financing modes combination of real estate has an important impact on its development, so it is very necessary to study on financial selection of real estate. As is known to all, capital structure is an important sign to reflect financial selection, therefore, study on influence factors of capital structure of real estate listed companies has important significance. Exploring this issue can provide real estate listed companies with some advice to make rational financing decision, it also can help government make policies on real estate market regulation, and then promote the healthy development of real estate industry.

There are a lot of research results about influence factors of capital structure of listed companies, Ozkan [1] uses panel data analysis and GMM estimation to study on UK companies, and concludes that current liquidity, current profitability, non-debt tax shield, growth ability have negative impact on capital structure, previous profitability has positive impact on capital structure, and company size shows no significant effect. Hong and Shen [2] find that company size and profitability have significant effect on capital structure, and company equity, growth ability and industry factor cannot influence capital structure. Zhong and Zhu [3] conduct empirical research in the way of multivariate regression analysis, results indicate that total assets growth rate, shareholders' equity turnover rate show positive correlation with capital structure, net profit margin, turnover of total capital, liquidity ratio have negative effect on capital structure, riskiness of the company, external liquidity ratio, company size, tax shield characteristics and industry characteristics have no significant correlation with capital structure. Guan and Zhang [4] analyze the influence factors of capital structure of mainland real estate companies listed in Hongkong, conclude that profitability have significant negative effect on capital structure, and growth ability show positive correlation with capital structure. Using GMM model, Liu and Song [5] study influence mechanism of company characteristic factors on capital structure of real estate companies, and find that growth ability, the unique of the products correlates positively with capital structure, profitability, company size, liquidity and asset tangibility have negative effect on capital structure.

Comprehensively, most of existing study on influence factors of capital structure of real estate listed companies are proceeding from the angle of company characteristic factors, just a few of them take company governance factors and macro-economic factors into consideration. Therefore, for these existing relevant research, influence factors are not considered completely, this paper will synthesize company characteristic factors, company governance factors and macro-economic factors, use a sample of 48 Chinese real estate list companies from 2005 to 2014 to empirically analyze the influence factors of capital structure, and then the study come to the conclusions.

60.2 Theoretical Analysis and Hypotheses to Be Tested

60.2.1 Definition of Capital Structure

Capital structure is formed by different financing methods adopted by enterprises, various combination of all kinds of financing method determine enterprise capital structure and its change.

The measure of capital structure depends on the purpose of research, this paper will put emphasis on the proportion of different financing methods of real estate listed companies, so the ratio between debt financing and all financing will represent capital structure, in other words, the asset-liability ratio is adopted.

60.2.2 Summary of Influence Factors

According to the results of literature research, the summary of influence factors of real estate listed company capital structure is presented as Table 60.1.

60.2.3 Influence Factors Analysis and Hypotheses to Be Tested

Company Characteristic Factors

1. Company size

The size of company is bigger, the expected future cash flows of company are more stable, so the anti-risk ability of the company will be stronger, and the company will be more possible to choose debt financing. We expect that company size have positive effect on leverage ratio.

2. Profitability

According to pecking order theory, the financing pecking order is internal financing, debt financing and equity financing. More profitable company will have

Table 60.1 Influence factors of real estate listed company capital structure

Type of influence factors	Company characteristic factors	Company governance factors	Macro-economic factors
Influence factors	Company size, profitability, operation capacity, growth ability, asset liquidity, collateral value of assets, non-debt tax shield	Ownership concentration, managerial ownership, board size, independent directors	Supply of money, interest rate, inflation

more retained earnings for internal financing, and then the asset-liability ratio declines. So we assume that profitability has negative correlation with asset-liability ratio declines.

3. Operation capacity

Companies with strong operation capacity are more efficient on using asset and capital, they need less capital for operation, so they are less possible to use debt financing. Therefore, it is assumed that operation capacity will show negative impact on asset-liability ratio.

4. Growth ability

Companies with strong growth ability have great potential for development, and because initial shareholders are optimistic on companies' future, they may tend to use debt financing which cannot threaten their right of control, so growth ability has positive influence on asset-liability ratio.

5. Asset liquidity

Asset liquidity is the capacity to quickly monetize assets without loss, the higher the level of asset liquidity is, the stronger debt paying ability is. We expect that asset liquidity has positive correlation with capital structure.

6. Collateral value of asset

In the corporate asset structure, if there are more assets suitable for collateralizing, credit capacity of the corporate will be stronger. Therefore, in this paper, we expect that collateral value of asset shows positive correlation with capital structure.

7. Non-debt tax shield

The trade-off theory holds that when the tax rate is high, companies are more likely to utilize the tax offsetting effect of debt interest to lower taxes, so they will choose debt financing. According to Chinese accounting standard and taxation regulations, depreciation can be deducted as cost expense before taxes. Therefore, when non-debt tax shield plays a great role, companies will less rely on tax saving through debt interest. As a consequence, we assume that non-debt tax shield will show negative correlation with capital structure.

Corporate Governance Factor

1. Ownership concentration

The ownership concentration is higher, means that strong stock holders has a high degree of correlation with economic benefits of the company, and strong stock holders voting power is bigger. According to Agency theory, strong stock holders are inclined to use debt as a tool to restrain the management. So we suppose that ownership concentration has positive correlation with capital structure.

2. Managerial ownership

The management ownership is higher, the overlap ratio between strong stock holders and the management is higher, and this will weaken board of directors' supervision on the management, then debt, which can be used to restrain the management, is less. We expect that management ownership will have negative impact on asset-liability ratio.

3. Board size

The more the number of board members is, the small the ownership concentration is, and then reduce the restraint the management feel, so the management use the debt less, we expect that board size will show negative correlation with asset-liability ratio.

4. Ratio of independent directors

The ratio of independent directors is higher, the restraint board put on the management is bigger, which may increase the debt. So this paper suppose that ratio of independent directors has positive effect on capital structure.

Macro-economic Factors

1. Supply of money

When monetary policy is easy, supply of money will increase, which is good for expanding the credit scale, and then companies are more possible to get loans. So we expect supply of money has positive impact on asset-liability ratio.

2. Interest rate

When interest rate gets higher, the debt financing cost will increase, which make the companies who need money are less inclined to get loans. Therefore, we expect interest rate will show negative impact on asset-liability ratio.

3. Inflation

Generally speaking, inflation will make the nominal capital cost higher, but due to interests can be tax shield, inflation will make actual debt cost lower, which may encourage companies to get debt financing. So, we expect that inflation shows positive correlation with capital structure.

60.3 Empirical Study

After analyzing the influence factors of capital structure of real estate listed companies, this paper will test the hypotheses based on empirical study with real data, and will offer objective evidences to analyze the influence factors of capital structure.

60.3.1 Sample Selection

We select real estate listed companies from the A-share market, and after rejecting the companies who operated abnormally and the companies from which we cannot get enough data, we finally select 48 real estate listed companies with date from 2008 to 2014, and all data are from Resset and CSMAR.

60.3.2 Primary Election of Indicators

According to existing study and data availability, combined with above hypotheses, we select indicators for all influence factors as Table 60.2 shows.

60.3.3 Determination of Indicators

In the primary selection of indicators, to guarantee the comprehensiveness of indicator system, we selected more than one indicators for some influence factors, but for indicators expressing the same influence factor, they may have repetitive information to some extent. Therefore, this paper will use statistical method to select one representative indicators for every influence factor, then comprehensiveness and representativeness of the indicator system can both be taken into consideration.

If some influence factors only have one indicator, then we choose the only one to be representative indicator; and for the influence factors which have more than one indicators, we will firstly compute the correlation between these indicators and capital structure, presented by Pearson's correlation coefficient, and then we choose the indicators with the largest correlation coefficients in the highest level of significance as the representative indicators, and the selection of representative indicators is shown as Table 60.3.

Table 60.2 Primary election of indicators of influence factors

Influence factors	Primary elected indicators
Company size	Logarithm of total assets
	Logarithm of operating revenue
Profitability	Return on assets
	Return on equity
	Sales margin
	Operating profit margin
Operation capacity	Inventory turnover
	Fixed asset turnover
	Total assets turnover
Growth ability	Total assets growth rate
	Operating revenue growth rate
Asset liquidity	Liquidity ratio
	Quick ratio
Collateral value of asset	(Fixed assets + inventory)/total assets
Non-debt tax shield	Depreciation rate
Ownership concentration	The shareholding ratio of the top five major shareholders
Managerial ownership	Executives ownership
Board size	Number of directors
Ratio of independent directors	Ratio of independent directors
supply of money	Growth rate of supply of broad money
Interest rate	Loan rate for one to three years
Inflation	Consumer price index

60.3.4 Empirical Test

Panel data analysis is used in the process of empirical study. And the regression model is constructed as formula 60.1.

$$L_{it} = \alpha_i + X_{it}\beta + \mu_{it}, \quad i = 1, 2, \ldots, 48, \quad t = 1, 2, \ldots, 7 \tag{60.1}$$

In formula 60.1, L_{it} is asset-liability ratio, X_{it} represents all the influence factors of capital structure, the subscript i and t means the number i company and number t year.

In the panel data we used, T = 7, N = 48, obviously it is short panel data, so we don't need to think about the problem of panel autocorrelation in the analyzing process. As for the model selection, we do the Hauseman test and get a P value at 0.002, which means rejecting the null hypothesis, namely, random effect model is better than fixed effect model, therefore, we use the fixed effect model for analysis, the regression result of panel data is shown in Table 60.4.

According to the test results from Table 60.4, we can conclude that:

Table 60.3 The selection result of representative indicators

Influence factors	Primary selected indicator	Pearson's correlation coefficient	Significance	Representative indicator
Company size	Logarithm of total assets	0.4484	***	Logarithm of total assets
	Logarithm of operating revenue	0.4212	***	
Profitability	Return on assets	−0.1528	***	Sales margin
	Return on equity	−0.1012	**	
	Sales margin	−0.1808	***	
	Operating profit margin	−0.1634	***	
Operation capacity	Inventory turnover	−0.2255	***	Inventory turnover
	Fixed asset turnover	0.0863	**	
	Total assets turnover	−0.1464	***	
Growth ability	Total assets growth rate	0.0828	*	Total assets growth rate
	Operating revenue growth rate	0.0075		
Asset liquidity	Liquidity ratio	−0.3377	***	Quick ratio
	Quick ratio	−0.3625	***	

Note In terms of the expression of significance, *** means 1 % significance level, ** means 5 % significance level, and * means 10 % significance level

In the company characteristic factors, company size, profitability, operation capacity and asset liquidity have significant impact on capital structure, and growth ability, collateral value of asset and non-debt tax shield have no significant impact on capital structure. And company size, profitability, operation capacity show the correlation with capital structure as the above hypotheses expect, asset liquidity shows the correlation with capital structure which is contrary to the hypothesis. After analysis, we realized that asset liquidity can affect capital structure through different ways, Lipson and Morta [6] founds that if a company's asset liquidity is stronger, then this company is more likely to use own money, instead of debt, which can decrease the leverage level. And from this perspective, the negative correlation between asset liquidity and asset-liability ratio can be explained.

In terms of corporate governance factors, only ownership concentration shows significant effect on capital structure. The effect of Ownership concentration on capital structure is different from that in the hypothesis, one possible explanation for this result is that when external financing is necessary, agents for strong shareholders and the management prefer to choose equity financing which has no economic constraint and has no effect on status of strong shareholders, for example,

Table 60.4 The empirical study result of influence factors of capital structure

Influence factors	Variables	Coefficient	*P* value	Significance (10 % level)
Company size	Logarithm of total assets	0.0991569	0.000	Significant
Profitability	Sales margin	−0.1102692	0.000	Significant
Operation capacity	Inventory turnover	−0.0137922	0.066	Significant
Growth ability	Total assets growth rate	0.0173100	0.236	Non-significant
Asset liquidity	Quick ratio	−0.0901638	0.000	Significant
Collateral value of asset	(Fixed assets + inventory)/total assets	0.0316834	0.583	Non-significant
Non-debt tax shield	Depreciation rate	0.0258384	0.923	Non-significant
Ownership concentration	The shareholding ratio of the top five major shareholders	−0.1534609	0.032	Significant
Managerial ownership	Executives ownership	−0.1216183	0.910	Non-significant
Board size	Number of directors	0.0026478	0.407	Non-significant
Ratio of independent directors	Ratio of independent directors	−0.0918156	0.274	Non-significant
Supply of money	Growth rate of supply of broad money	0.2897769	0.013	Significant
Interest rate	loan rate for one to three years	0.8855415	0.370	Non-significant
Inflation	Consumer price index	0.0008291	0.772	Non-significant

allotment of shares, and Guan and Zhang [4] founds that the company with higher stock concentration is more likely to use this method.

In the macro-economic factors, supply of money has significant effect on capital structure, and interest rate, inflation show no significant impact on asset-liability ratio. The analysis results show that supply of money has positive significant effect on capital structure, which is consistent with our hypothesis.

60.4 Conclusions and Suggestions

Conclusions from our study are as following: in the company characteristic factors, company size, growth ability show positive correlation with capital structure, profitability and asset liquidity have negative effect on capital structure, operation capacity, collateral value of asset, non-debt tax shield have no significant impact on capital structure; in terms of corporate governance factors, ownership concentration has significant negative effect on asset-liability ratio, managerial ownership, board size, ratio of independent directors show no significant correlation with capital structure; as for macro-economic factors, supply of money has significant positive effect on capital structure, interest rate and inflation both have no significant impact on asset-liability ratio.

According to our conclusions, capital structure of real estate listed companies is closely related to company size, profitability, growth ability, asset liquidity, ownership concentration and supply of money. Therefore, combined with those influence factors' effect on capital structure, real estate listed companies can adjust their capital structure to achieve optimization, then promote the companies. And for the government, related departments can adjust the capital structure of the whole real estate industry through macroeconomic policies, and then make real estate industry develop healthily.

References

1. Ozkan A (2001) Determinants of capital structure and adjustment to long run target: evidence from UK company panel data. J Bus Fin Account 28(1-2):175–198
2. Hong X, Shen Y (2000) Empirical analysis on influence factors of capital structure of Chinese listed companies. J Xiamen Univ 03:114–120
3. Zhong H, Zhu J (2006) Empirical study on capital structure of Chinese real estate listed companies. Construct Econ 10:34–36
4. Guan M, Zhang Y (2011) Empirical study on capital structure of real estate companies—taking Hong Kong-listed companies from the Mainland as the example. Spec Zone Econ 01:132–133
5. Liu p, Song Y (2014) Study on influence factors of capital structure of listed companies—taking Chinese real estate companies as the example. Econ Vision, 01:25–27
6. Lipson ML, Mortal S (2009) Liquidity and capital structure. J Fin Mark 12(4):611–644

Proceedings of the 20th International Symposium on Advancement of Construction Management and Real Estate

Yuzhe Wu · Sheng Zheng · Jiaojiao Luo
Wei Wang · Zhibin Mo · Liping Shan
Editors

Proceedings of the 20th International Symposium on Advancement of Construction Management and Real Estate

Volume II

Springer

Editors
Yuzhe Wu
Department of Land Management
Zhejiang University
Hangzhou, Zhejiang
China

Wei Wang
Department of Land Management
Zhejiang University
Hangzhou, Zhejiang
China

Sheng Zheng
Department of Land Management
Zhejiang University
Hangzhou, Zhejiang
China

Zhibin Mo
Department of Land Management
Zhejiang University
Hangzhou, Zhejiang
China

Jiaojiao Luo
Department of Land Management
Zhejiang University
Hangzhou, Zhejiang
China

Liping Shan
Department of Land Management
Zhejiang University
Hangzhou, Zhejiang
China

ISBN 978-981-10-0854-2 ISBN 978-981-10-0855-9 (eBook)
DOI 10.1007/978-981-10-0855-9

Library of Congress Control Number: 2016935985

Printed on acid-free paper

This Springer imprint is published by Springer Nature
The registered company is Springer Science+Business Media Singapore Pte Ltd.

Acknowledgements

The symposium gratefully acknowledges the support of organizations below:

Conference Organizer

The Chinese Research Institute of Construction Management (CRIOCM), Zhejiang University

Conference Executive Organizer

School of Public Affairs, Zhejiang University
College of Civil Engineering and Architecture, Zhejiang University
Land Academy for National Development, Zhejiang University
Institute of Land Science and Real Estate, Zhejiang University
Institute of Construction Management, Zhejiang University
Institute of Urban and Rural Planning Theories and Technologies, Zhejiang University

Conference Co-organizers

Enjoy-town Investment Management (Hangzhou) Co., Ltd.
Department of Real Estate and Construction, The University of Hong Kong
Department of Building and Real Estate, The Hong Kong Polytechnic University
International Research Center for Sustainable Built Environment, Chongqing University
School of Urban-rural Planning and Management, Zhejiang University of Finance and Economics

Academic Committee of Civil Engineering Management in Zhejiang Province
Journal of Urban Management
International Journal of Construction Management

Advisory Committee (In Alphabetical Order by Last Name)

Organizing Committee

Chairman

Co-chairman

Executive Chairman

Secretary-General

Executive Secretary

Members (In Alphabetical Order by Last Name)

Prof. Haoying Han, Zhejiang University
Dr. Lifeng Han, Macau University of Science and Technology
Dr. Jane J.L. Hao, Ryerson University, Canada
Dr. Dezhi Li, Southeast University
Prof. Xinhai Lu, Huazhong University of Science and Technology
Dr. Wilson W.S. Lu, The University of Hong Kong
Dr. Yi Peng, Zhejiang University of Finance and Economics
Dr. Chengshuang Sun, Harbin Institute of Technology
Dr. Vivian W.Y. Tam, University of Western Sydney, Australia
Prof. Hui Wang, Zhejiang University
Prof. Zhigang Wu, Nanjing Normal University
Dr. Bo Xia, Queensland University of Technology, Australia
Dr. Michael C.Y. Yam, The Hong Kong Polytechnic University
Prof. Kunhui Ye, Chongqing University
Dr. Mingxuan Yu, Renmin University of China
Dr. Hongping Yuan, Southwest Jiaotong University
Prof. Wenze Yue, Zhejiang University
Prof. Hong Zhang, Tsinghua University
Dr. Xiaoling Zhang, City University of Hong Kong
Prof. Zhenyu Zhao, North China Electric Power University
Prof. Yimin Zhu, Louisiana State University, USA
Dr. Jian Zuo, University of South Australia, Australia

International Academic Committee

Chairman
Prof. Liyin Shen, Chongqing University
Prof. Martin Skitmore, Queensland University of Technology, Australia

Members (In Alphabetical Order by Last Name)
Prof. Haijun Bao, Zhejiang University of Finance and Economics
Prof. Abdol R. Chini, University of Florida, USA
Prof. Stuart Green, University of Reading, UK
Prof. Miklos Hajdu, Szent Istvan University, Hungary
Prof. Xianjin Huang, Nanjing University
Prof. Xiangmu Jin, Zhejiang University
Prof. Per-Erik Josephson, Chalmers University of Technology, Sweden
Prof. Craig Langston, Bond University, Australia
Prof. Qiming Li, Southeast University
Prof. Jian Lin, Peking University
Prof. Anita M.W. Liu, The University of Hong Kong
Prof. Guiwen Liu, Chongqing University
Prof. Weidong Liu, Zhejiang University
Prof. Hualou Long, Chinese Academy of Sciences

Prof. S.P. Low, National University of Singapore, Singapore
Prof. Ping Lv, Renmin University of China
Prof. Mona Shah, National Institute of Construction M&R, India
Prof. Geoffrey Shen, The Hong Kong Polytechnic University
Prof. Jonathan Shi, Louisiana State University, USA
Prof. Qian Shi, Tongji University
Prof. Dalu Tan, Sichuan University
Prof. Jiayuan Wang, Shenzhen University
Prof. Shouqing Wang, Tsinghua University
Prof. Xueqing Wang, Tianjin University
Prof. Xiaolong Xue, Harbin Institute of Technology
Prof. Donglang Yang, Xi'an Jiaotong University
Prof. Saixing Zeng, Shanghai Jiao Tong University
Prof. Qing Zhu, Southwest Jiaotong University
Prof. George Zillante, The University of Adelaide, Australia
Prof. Patrick X.W. Zou, Swinburne University of Technology, Australia

Conference Secretariat

Dr. Sheng Zheng, Zhejiang University
Ms. Lu Song, Zhejiang University
Ms. Jiaojiao Luo, Zhejiang University
Ms. Liping Shan, Zhejiang University
Ms. Wei Wang, Zhejiang University
Mr. Zhibin Mo, Zhejiang University

Contents

Chapter 61
Research on Selection of Logistics Supplier in the Process of Housing Industrialization

Yongfei Pan and Junwei Wang

Abstract In recent years, a variety of fields of manufacturing industry has been relatively mature, but there are few study on the logistics supplier of prefabricated housing components. The choice of the logistics supplier is unique because of the particularity of existing residential prefabricated. Based on Data Envelopment Analysis, a logistics supplier performance evaluation model is presented in this paper. And the latest date of 10 logistics companies will be served as the samples for empirical research in our country to provide a reference for relevant decision-making departments.

Keywords Housing industrialization · Data envelopment analysis · Logistics supplier · Prefabricated components

61.1 Introduction

In 2006, the Ministry of Construction proposed the "national housing industrialization base in the pilot scheme" policy, which open a new model of the housing industrialization development. In the report on the work of the State Council in 2014, Li Keqiang, China's deputy Prime Minister, said that the need to promote the housing industrialization at many times. Yu Zhengsheng, chairman of the CPPCC also specially held a meeting to discuss housing industrialization problems and noted that not only the housing industrialization is needed, but also the process of housing industrialization must be promoted vigorously. With the further promotion of housing industrialization policy our country, local governments have also

Y. Pan (✉) · J. Wang
School of Management, Chongqing Jiaotong University,
Chongqing 400074, China
e-mail: 1522850625@qq.com

Y. Wu et al. (eds.), *Proceedings of the 20th International Symposium on Advancement of Construction Management and Real Estate*,
DOI 10.1007/978-981-10-0855-9_61

introduced polices to encourage and guide the real estate enterprises to build prefabricated housing. On November 10th 2014, for example, Shenzhen issued the "guidance on accelerating the Shenzhen housing industrialization", which pointed out that the industrialization of residential construction should adopt the prefabricated building system in which the prefabricated rate should be above 15 % and assembly rate should above 30 %, meanwhile the prefabrication should be gradually improved [1]. Prefabricated house is the important symbol of housing industrialization and all kinds of residential components in an industrial mode will be assembled at the construction site to component the residential building. Industrialized production mode greatly improves the flexibility of enterprise production and has been used widely, but the changing of the mode proposes stricter requirements on logistics enterprises suppliers. In addition, there are many serious challenges especially the large prefabricated, such as internal and external wall panel (PC), laminated floor, balcony, stair nose etc., and transportation vehicle selection, high limits of the bridges and culverts along the bridge, objects match of different size [2] and supply timeliness of prefabricated parts and the choice of logistics suppliers. Therefore, scientific selection and optimization of prefabricated logistics suppliers are the important guarantees for promoting the process of housing industrialization.

61.2 Literature Review

At present, the research on prefabricated housing both at home and abroad focuses on the application prospect of components, cost analysis, the economic scale effect and the selection of component suppliers, the research on selection and optimization of logistics supplier in the process of housing industrialization is still relatively few. For other industries, the supplier selection models of logistics are more mature. Wang and Zhang [3], viewed that the rail freight logistics supplier selection is important to effectively recognize the high-quality suppliers for railway departments, so carried out a selection research in logistics supplier of certain railway freight departments using projection pursuit model; Shen [4], used AHP combining with the characteristics of the water conservancy construction project, studied the theoretical analysis and calculation of the logistics suppliers' evaluation factors, and realized the logistics supplier selection about water conservancy project; Chou and Chang [5], took the method fuzzy SMART to characterize expert's opinion and evaluate and select the logistics supplier, showed some advantages in terms of expressing the expert opinion; Jackson and Boyd [6], analyzed the various factors which affect the comprehensive ability of the logistics supplier through Logistic Regression Analysis (LRA), this analysis seems simple and easy to operate.

Based on the research results of the selection of logistics supplier in different fields, the prefabricated housing supplier selection in the process of transportation research has certain reference value. Due to prefabricated component's big size, varied shape, the high demand of integrated appearance (surface roughness, no distortion, no fracture, no lack of edge off angle, etc.) [7], they increased the difficulty in managing the prefabricated components in the process of transportation, this is one of the reasons which is different from the logistics supplier selection in other fields. Therefore, research on selection and optimization of prefabricated logistics supplier is an important link of promoting housing industrialization with more fast developing economy in our country. Now using prefabricated housing logistics supplier as the research object, this paper chooses the reasonable metrics and models to evaluate and make relevant recommendations.

61.3 The Establishment of Supplier Evaluation Index and Data Collection

61.3.1 The Selection of Indicators

Now the study of selection and optimization of logistics supplier in the process of housing industrialization is relatively less at home and abroad, and the experiences which could borrow from are not much. But the management style of prefabricated components in prefabricated houses about purchasing, transportation, warehousing features both the construction and logistics [8]. So this paper collected the related references and the financial statements of listed logistics company, as shown in Table 61.1

According to statistics in Table 61.1 and the index data about listed logistics company, considering the available assessment information and industry characteristics, this paper selects selection index system of logistics supplier from the prefabricated houses. The financial indicators are divided into cost indicators and efficiency indicators combining the requirements of the DEA method, it will take cost index as the input indicators, efficiency indicators as DEA output indicators. It is divided into seven indicators specifically: cost of sales, management fees, research and development expenses, financial expenses, operating income, basic earnings per share, the weighted average return on assets.

61.3.2 Data Collection and Pretreatment

According to our country each big listed annual report data of logistics enterprises in 2013 and the characteristics of the DEA evaluation model, the first is

Table 61.1 Relevant authority literature information about the DEA model

Authors	Research object	Indicators 1	Indicators 2
Barros and Dieke [9]	31 airports in Italy (2001–2003)	(1) Labor costs (2) construction investment (3) operating costs(not including wage costs)	(1) The plane number (2) the number of passengers (3) general cargo (4) the admissibility of revenue (5) BV aviation (6) BV applicable business department
Lam et al. [10]	11 airports of Asia-Pacific international (2001–2005)	(1) Labor input (2) the scale of investment in fixed assets (3) trade (4) soft inputs (including service, external purchase of goods and commodities)	(1) Freight amount (2) the number of flights (3) the number of passengers
Deng et al. [11]	8 listed logistics companies in Shenzhen and Shanghai (2001-2006)	(1) Fixed assets (2) the total amount of wages of workers (3) operating costs (not including the worker salary)	Total pre-tax profits
Li and Ma [12]	26 listed logistics companies in Shenzhen and Shanghai (2008)	(1) Total assets (2) the main operating costs (3) the number of employees	(1) Main operating income (2) the net profit
Ma [13]	12 listed logistics companies (2011)	(1) The strength of sales staff (2) the strength of advertising investment (3) cost of sales strength	(1) Main business revenue growth (2) return on net assets (3) inventory turnover (4) accounts receivable turnover
Ding and Qu [14]	24 listed logistics companies (2010–2012)	(1) Total number of enterprise employees (2) the scale of investment in fixed assets (3) main business cost	(1) Net come of enterprises (2) the main business income (3) equity net profit margin

preprocessing the evaluation data of listed logistics enterprises preprocessed to eliminate the effect of the peak data. As follows [15]: cost of sales divided by 10^3, the management fees divided by 10^4, research and development expenses divided by 10^4, financial expenses divided by 10^3, operating income divided by 10^5, statistical data are shown in Table 61.2.

Table 61.2 Pretreatment parameters of logistics companies about important input and output indicators

Logistics company	WL1	WL2	WL3	WL4	WL5	WL6	WL7	WL8	WL9	WL10
Decision unit	DMU1	DMU2	DMU3	DMU4	DMU5	DMU6	DMU7	DMU8	DMU9	DMU10
Cost of sales	2.45	1.12	3.69	1.49	0.59	0.00	16.45	5.15	0.80	0.50
Management fees	0.73	0.97	1.79	1.07	21.51	13.13	2.87	5.58	1.48	2.27
Research and development expenses	0.40	0.00	0.22	0.00	3.04	7.67	1.07	9.41	0.99	0.00
Financial expenses	0.06	0.73	12.97	1.88	8.57	1.86	0.77	0.44	0.96	0.62
Operating income	4.28	1.29	3.05	1.84	28.16	16.63	27.84	7.86	5.65	2.45
Basic earnings per share	0.32	0.12	0.18	0.10	0.23	0.64	0.48	0.32	0.03	0.62
The weighted average return on assets	10.01	4.01	5.73	3.73	10.76	8.67	6.12	9.80	2.21	11.36

Source Annual reports of major listed logistics companies of 2013

61.4 Establishment of the Model

For the construction industry, selecting logistics suppliers is a question about optimization from variety of factors, it is a multi-objective decision making. The DEA method which is different from many other analysis methods has simple structure and is easy to use, it rules out lots of subjective factors and has strong objectivity. It can evaluate the relative effectiveness of multi-input and multi-output indicators, and has the absolute advantage, without any weights assumptions, the weight of each input and output is not based on the evaluator's subjective judgments, but the optimal weighs of the actual data obtained from the decision [16]. The first is to establish the corresponding evaluation system, collecting relevant data, use MATLAB tools to calculate the DEA model, so suppliers can be divided into two categories in order to realize the optimization, simultaneously do some optimal adjustment of the invalid supplier, so that more listed logistics companies could become the high quality of prefabricated logistics suppliers.

The method of Data Envelopment Analysis (DEA) was founded in 1978 by Charnes et al. [17]. It is a new method of system analysis which base on multiple index input and multiple index output to the same type of units (departments) and take relative effectiveness or benefit evaluation. In the DEA method, there are N listed logistics companies, namely N decision-making units (DMU), each DMU has m types of "input" and s types of "output". x_j is the first j a decision-making unit output, y_j is the first j a decision-making unit output, s^+, s^- are inputs and outputs of slack variables and surplus variables. Through the dual model to further introduce slack variable s^+ and s^- remaining variables, obtain the following model:

$$\begin{cases} \min \quad \theta \\ s.t. \sum_{j=1}^{n} \lambda_j x_j + s^- = \theta x_0 \\ \sum_{j=1}^{n} \lambda_j y_j - s^+ = y_0 \\ \lambda_j \geq o, \ j = 1, 2, \ldots, n \\ \theta \text{ 无约束} \\ s^+ \geq 0, s^- \geq 0 \end{cases} \tag{61.1}$$

Set the optimal solution for $\lambda^*, s^{*+}, s^{*-}, \theta^*$, the following conclusions:

1. If $\theta^* = 1$, and $s^{*-} = 0$, $s^{*+} = 0$. At this time decision-making unit j_0 is the DEA effective. The production of decision making unit j_0 is effective in technology and scale.
2. If $\theta^* = 1$, in this case, the decision unit j_0 is weak DEA effective.
3. If $\theta^* < 1$, in this case, the decision unit j_0 is not a valid DEA.

61.5 Empirical Analysis

This paper selected 10 companies listed logistics as research subjects, involving all the data from listed companies' annual report of 2013. Using MATLAB tools, the operation results are shown in Fig. 61.1.

61.5.1 Efficient Results Analysis of DEA

From the analysis results based on MATLAB, it can be seen that the overall efficiency value of these 10 listed logistics companies reflect the achieved maximum possible level of real output radio under the current logistics technology. Known by $\theta^* = 1$, and $s^{*-} = 0, s^{*+} = 0$, decision making units, such as 1, 2, 4, 5, 6, 7, 9 and 10, are DEA efficient, and effective decision-making units comprise more than 80 %, so it is indicate that the production factors of 8 listed logistics companies in 2013 have reached optimal combination, and obtained the best output effect. In an economic sense, it also suggests that these eight decision-making units

```
lambda =

    1.0000    0.0000    0.1672    0.0000    0.0000    0.0000    0.0000    1.3998    0.0000    0.0000
    0.0000    1.0000    0.0000    0.0000    0.0000    0.0000    0.0000    0.0000    0.0000    0.0000
    0.0000    0.0000    0.0000    0.0000    0.0000    0.0000    0.0000    0.0000    0.0000    0.0000
    0.0000    0.0000    0.0000    1.0000    0.0000    0.0000    0.0000    0.0000    0.0000    0.0000
    0.0000    0.0000    0.0000    0.0000    1.0000    0.0000    0.0000    0.0000    0.0000    0.0000
    0.0000    0.0000    0.0000    0.0000    0.0000    1.0000    0.0000    0.1124    0.0000    0.0000
    0.0000    0.0000    0.0550    0.0000    0.0000    0.0000    1.0000    0.0000    0.0000    0.0000
    0.0000    0.0000    0.0000    0.0000    0.0000    0.0000    0.0000    0.0000    0.0000    0.0000
    0.0000    0.0000    0.0000    0.0000    0.0000    0.0000    0.0000    0.0000    1.0000    0.0000
    0.0000    0.0000    0.3274    0.0000    0.0000    0.0000    0.0000    0.0000    0.0000    1.0000

s_minus =

    0.0000    0.0000    0.6307    0.0000    0.0000    0.0000    0.0000    0.0000    0.0000    0.0000
    0.0000    0.0000    0.0000    0.0000    0.0000    0.0000    0.0000    1.2185    0.0000    0.0000
    0.0000    0.0000    0.0000    0.0000    0.0000    0.0000    0.0000    4.8445    0.0000    0.0000
    0.0000    0.0000    7.1590    0.0000    0.0000    0.0000    0.0000    0.0000    0.0000    0.0000

s_plus =

    0.0000    0.0000    0.0000    0.0000    0.0000    0.0000    0.0000    0.0000    0.0000    0.0000
    0.0000    0.0000    0.1029    0.0000    0.0000    0.0000    0.0000    0.1999    0.0000    0.0000
    0.0000    0.0000    0.0000    0.0000    0.0000    0.0000    0.0000    5.1864    0.0000    0.0000

theta =

    1.0000    1.0000    0.5717    1.0000    1.0000    1.0000    1.0000    0.6659    1.0000    1.0000
```

Fig. 61.1 MATLAB runs results based on DEA model

are effective in scale and technology. This is closely related to the business management level and the production operation mode, they can be used as the excellent logistics suppliers of the prefabricated housing in the process of transportation management, and establish long-term cooperative relationship with the enterprise.

61.5.2 Effective Projection Analysis of Non-DEA

The total efficiency value of decision-making units 3 and 8 are less than 1 and close to 0.5, and belong to the non-DEA effective, indicating that the logistics efficiency of listed companies WL3 and WL8 is low, resources which these two companies invested are not fully utilized, and outputs are not reached optimization. The company did not emphasis on resource utilization enough. There may be several reasons: first, because of conditions such as funding, policy, environment got constrains, lack of professional management and advanced technology and equipment, it result in a waste of resources when the logistics investment; second, there exist internal management problems, poor management did not make full use of resources, ultimately affect the output, and so on.

61.5.3 Effective Adjustment and Optimization of Non-DEA

With the advancement of housing industrialization, there are more and more demand for prefabricated logistics suppliers. In order to improve the operational efficiency of listed logistics companies and to optimize resources, it can adjust the amount of input resources and benefit output so that all decision-making units could reach DEA effective, to provide more excellent suppliers for the transportation of prefabricated components and logistics management of inventory. Take an example from WL3, the non-DEA effective data of the listed logistics company is analyzed to optimize, WL8 similar.

WL3 logistic company input and output indicators result in the DEA effective in MATLAB.

Given:

$$\lambda_1 = 0.0000,\ \lambda_2 = 1.6720,\ \lambda_3 = 0.0550,\ \lambda_4 = \cdots = \lambda_7 = \lambda_9 = \lambda_{10} = 0,\ \lambda_8 = 0.3274;$$
$$s_1^- = 0.6307,\ s_2^- = 0.0000,\ s_3^- = 0.0000,\ s_4^- = 7.1590;$$
$$s_1^+ = 0.0000,\ s_2^+ = 0.1029,\ s_3^+ = 0.0000;$$
$$\theta_7 = 0.5717.$$

Adjustment scheme:

See Table 61.3.

Table 61.3 Non-DEA efficient data projection analysis of WL3 logistics company

Enter before adjustment		After entering adjustment	Enter before adjustment		After entering adjustment
Cost of sales	3.69	3.69 × 0.5717 − 0.6307 = 2.11	Operating income	3.05	3.05
Management fees	1.79	1.79 × 0.5717 = 1.02	Basic earnings per share	0.18	0.18 + 0.1029 = 0.28
Research and development expenses	0.22	0.22 × 0.5717 = 0.13	The weighted average return on assets	5.73	5.73
Financial expenses	12.97	12.97 × 0.5717 − 7.1590 = 0.26	–	–	–

By calculation, the decision-making unit 3, that is WL3 company, need to adjust input and output. If make decision-making unit 3 DEA effective, it need to reduce the cost of sales: from 5.15 down to 3.43. In practice, it should control the input and output of resources, establish the corresponding supervision mechanism, improve the system of transfers and other expenses, reduce the management fees accordingly to 2.49. In the aspect of management, it should do pre-planning, specify the use conditions and limits of cost with associated systems, it also can examine the effect on cost from the process and the input and output, introduce a lot of the professional management personnel, so as to reduce the cost. Research expenses also have greatly reduced from the original 9.41 for 1.42, the reasons may be that in the process of research and development of various resources and charges are not reasonable effective use, less returns, resulting in the gratuitous waste of resources. Financial expenses should be reduced to 0.29, may reduce bank lending, reduce the use of checks and other measures to achieve. For the output variable, operating income and the weighted average return on assets do not need for adjustment, basic earnings per share from 0.32 to 0.52, which makes the WL3 company costs (input) and benefits (output) get the maximum utilization of resources, so as to improve the efficiency of logistics of the company. That is the decision-making unit 8 (WL8), similar to the adjustment method, not to repeat.

61.6 Conclusions and Recommendations

There are many factors which influence the prefabricated logistics supplier selection. In this paper, as for the prefabricated housing building, DEA is used to build the prefabricated logistics supplier selection index system to evaluate the input and output indicators and to optimize a non DEA efficient logistics supplier adjustment, so as to realize logistics management performance evaluation of listed companies. Based on the above conclusion, some recommendations are as follows.

Firstly, as for the manufacturers, suitable high quality logistics suppliers can control from the source material input of prefabricated houses, guarantee the quality of prefabricated, enable manufacturers to gain credibility for their products to expanding the sources of income. In view of this, on the basis of trust between each other, the choice of logistics provider should consider the enterprise management, technical measures, customer satisfaction and other factors to establish long-term quality assurance channel and to ensure that the prefabricated of prefabricated houses in every link to get reasonable use, so as to promote the process of housing industrialization.

Secondly, as for developers, they start from the index data of the listed logistics company and give full consideration to the enterprise economic operation conditions, analysis the company' input and output rates of optimal allocation of resources, and assess the credibility of logistics enterprises, take the location and the number of service points into account at the same time in order to select the right partners, it forms a certain degree of the interests of the consortium and realizes the integration of industrial chain in prefabricated houses.

Thirdly, as for the construction, choosing logistics suppliers correctly can guarantee prefabricated components reach their destinations timely and quickly. So in the research of cost index and efficiency index of listed logistics enterprises, analyzing whether the enterprise is in the choice of transport and stacking of prefabricated, considering product reinforcement and protection, it must ensure the integrity of prefabricated parts. It is related to the project quality, safety and progress, is a decisive factor in the process of the prefabricated housing promotion, affecting the construction of cost benefit essentially.

Fourthly, as for the listed logistics enterprises, it should strengthen the management and control, improve the operational efficiency, improve the optimization and management level of enterprises continuously, sum up the experience in sales, management, development, finance and other investment factors and continue to exploit advantage, at the same time make greater progress in revenue, earnings per share, net assets yield output factors, grasp the market opportunities, improve the enterprise's core competitiveness.

References

1. http://precast.com.cn/index.php/news_detail-id-1177.html
2. Venselaar M, Gruis V, Fenne V (2015) Implementing supply chain partnering in the construction industry: work floor experiences within a Dutch housing association. J Purchasing Supply Manage 3:1–8
3. Wang X, Zhang C (2013) Based on projection pursuit model of third party logistics supplier selection for railway freight transportation research. Tech Methods, 10
4. Shen X (2014) Research on selection of logistics supplier in water conservancy construction projects. Proj Manage Technol, 07(12)
5. Chou SY, Chang YH (2008) A decision support system for supplier selection based on a strategy-aligned fuzzy SMART approach. Expert Syst Appl 34(4):2241–2253
6. Jackson JD, Boyd JW (1988) A statistical approach to modeling the behavior of bond rates. J Behav Econ 17(3):173–193
7. Li Y (2011) Talk about concrete prefabricated construction. Heilongjiang Traffic Sci Technol, 11
8. Zheng X, Jia L (2013) Research on selection of construction parts supplier in the process of housing industrialization. East China Econ Manage, 10
9. Barros CP, Dieke PUC (2007) Performance evaluation of italian airports with data envelopment analysis. J Air Transp Manage 13(4):184–191
10. Lam SW, Low JMW et al (2009) Operational efficiencies across Asia Pacific airports. Transp Res Part E 45(4):654–665
11. Xueping D, Wang X, Ng ASF (2008) Analysis of logistics enterprises in our country total factor productivity. Syst Eng 26(6):1–7
12. Li Y, Ma Z (2012) A modified model of DEA and its application in listed companies of logistic. J Beijing Jiaotong University, 04
13. Ma X (2013) Study on performance evaluation of listed logistics enterprise. Tech Methods, 03:078

14. Bin D, Huimin Q (2014) Analysis of operational efficiency and influencing factors of China's listed logistics company. Manage Modernization 4:74–76
15. Zhou T, Liu M, Zhang X (2014) Empirical analysis on performance evaluation of listed logistic enterprises in China. Railway Transport Econ, (8):10–14
16. Du D, Pang Q, Wu Y (2008) Modern comprehensive evaluation method and case selection. Tsinghua University Press, Beijing
17. Charnes A, Cooper WW, Rhods E (1978) Measuring the efficiency of decision making units. Eur J Oper Res (2):429–444

Chapter 62
Australian Institutional Investors and Residential Investment Vehicles

Chyi Lin Lee, Graeme Newell and Valarie Kupke

Abstract A lack of institutional investor involvement in the private rental residential sector is a structural weakness in the Australian housing rental market. To encourage institutional investment in the private rental market, several residential investment vehicles such as REITs have been introduced in the US and internationally. Despite Australian REITs being the second largest REIT market in the world, no residential REIT vehicle is available in Australia. Therefore, it is not only essential to assess the attitudes of Australian institutional investors regarding housing investment, but also residential investment vehicles. A survey of Australian institutional investors concerning residential property investment was conducted in August-September 2014. The results showed that the lack of well-structured residential investment vehicles and low returns were seen as critical issues in the residential property market. In addition, the most desirable features for an effective residential investment vehicle were being managed by an experienced manager, a diversified portfolio by location and delivering stable income returns with low debt. The implications of the findings are also discussed.

Keywords REITs · Unlisted residential funds · Private rental market · Residential investment vehicles · Institutional investment and Australia

C.L. Lee (✉)
School of Business and Urban Research Program,
Western Sydney University, Penrith, NSW, Australia
e-mail: c.lee@westernsydney.edu.au

G. Newell
School of Business,
Western Sydney University, Penrith, NSW, Australia

V. Kupke
School of Commerce, University of South Australia,
Adelaide, SA, Australia

Y. Wu et al. (eds.), *Proceedings of the 20th International Symposium on Advancement of Construction Management and Real Estate*,
DOI 10.1007/978-981-10-0855-9_62

62.1 Introduction

A clear decrease in housing affordability among Australians has been observed in recent years [18]. The recent report from the Council of Australian Governments confirmed that only a relatively small proportion of homes sold were actually affordable to low-income households [7]. With increasing housing stress in the home purchase market and a limited supply of housing, many potential home purchasers have to seek private rental accommodation. This also pushed up the demand for private market properties [2, 27]. The demand for private rental properties has increased significantly in recent years. In 2014, it was estimated that the Australian private rental market included 2.4 million rental dwellings with 6 million residents across Australia [1]. In 2011–2012, almost 25 % of Australian households rented privately [1]. More importantly, the market grew by 11 % over 2001–2006 [26]. These figures have clearly demonstrated the significant role of the private rental market in the Australian housing sector. To meet the demand of private rental properties, institutional investments in the residential sector would play a critical role in enhancing the supply of rental properties [21].

However, the lack of institutional investor involvement is a structural weakness in the Australian residential rental market, which is dominated by small and individual investors [5, 20, 21, 25, 27]. Indeed, the majority of rental stock in Australia is owned by individual investors over many decades. This is further confirmed by the low level of market share concentration in this market in which the largest four residential operators (e.g. Defence Housing Australia) only account for 5 % of market share [12]. One of the reasons of the lack of institutional investor involvement in this sector can be attributed to the absence of an effective residential investment vehicle. To encourage institutional investment (e.g.: key pension funds) in the private rental market, several residential investment vehicles have been launched in the United States (US) and internationally. In the US, residential real estate investment trusts (REITs) have been introduced. As at June 2014, there were 16 US residential REITs accounting for $95 billion (A$102 billion) in market cap. This sees residential REITs as the second largest REIT sub-sector, being 15 % of the US REIT market [23]. These residential REITs predominantly invest in apartments, with a focus on mobile, affluent markets and growth cities. As well as apartments, some US residential REITs invest in student accommodation and manufactured homes [24]. Importantly, US residential REITs have also significantly enhanced the supply of privately rented properties in which US residential REITs own and manage over 2,300 residential properties with over 644,000 apartment units [23]. Residential investment vehicles have also been successfully introduced in Canada, Japan, Singapore and France. There were 12 residential REITs in the above mentioned markets with a total market capitalisation of A$23.2 billion at June 2014. These 12 residential REITs have nearly 187,000 residential units [4, 11].

Despite Australian REITs being the second largest REIT market in the world (only exceeded by US REITs) [17], no residential REIT vehicles are available in

Australia. Therefore, it is not only essential to assess the attitudes of Australian institutional investors regarding housing investment, but also suitable residential investment vehicles. A large number of papers have examined institutional investment in the residential property sector. That literature, which is reviewed in the literature section, has largely considered institutional investors' attitudes towards residential property investment. No study has been undertaken to examine institutional investors' attitudes regarding residential investment vehicles. More specifically, the important features of an effective residential investment vehicle have not been identified.

Therefore, this study aims to fill in this gap in the literature by examining a number of key elements concerning institutional investment in the private rental residential market. More specifically, it assesses the attitudes of Australian institutional investors concerning residential property investment in general, and residential investment vehicles in particular. This study contributes to the literature in a number of ways. Firstly, this is the first study to investigate institutional investors' perceptions regarding residential investment vehicles. This study not only provides some insights into the perceptions of institutional investment towards residential investment, but also residential investment vehicles. This will assist policy makers and property fund managers to develop an effective residential investment vehicle. More importantly, an effective residential investment vehicle is essential to encourage institutional investment in housing, in light of Australian institutional investors normally not investing directly in acquiring their own property assets [3]. Therefore, an effective residential investment vehicle could enhance the involvement of institutional investors such as Australian superannuation funds in the private rental residential sector. Secondly, the extent of institutional investment in the Australian housing sector is also gauged for the first time. The findings will be of great interest to policy makers and researchers. Specifically, this would enable policy makers to access better informed and robust findings on the institutional investment in the privately rented sector. In turn, an enhanced policy can be formulated to encourage an increased level of institutional involvement in this important property sector.

The remainder of this paper is organized as follows. The following section provides a literature review on institutional investment in housing markets. Section 62.3 discusses the methodology and data. Section 62.4 investigates institutional investors' attitudes regarding residential investment vehicles via a survey. The important features of an effective residential investment vehicle are also identified. The final section provides concluding comments.

62.2 Literature Review

The role of institutional investors in housing markets has been intensely debated in the housing literature. In general, most studies confirm that the level of institutional investment in residential property is marginal. Numerous previous studies have attempted to examine institutional investors' attitudes towards residential property investment. A UK study reported that many institutional investors in the UK considered that, in principle, private renting ought to be an attractive investment [8]. Nevertheless, they were very reluctant to enter into the private rental market for a wide range of reasons, such as small lot size of investment, poor and greater costs of housing management than commercial property, illiquidity, low volatility of rent, low returns, companies' images and other factors. These difficulties see the private rented sector as a much more risky investment sector compared with commercial property (e.g.: office, retail, residential). Consequently, a higher rate of return is required by institutional investors as a compensation for the perceived higher risk level of the private rented sector. Comparable findings have also exhibited by several studies [6, 10, 14, 15, 21] in which low rental yield, high taxes, poor liquidity, higher investment risk, reputational risk, management issues, a lack of sufficient scale, the dearth of market information, poor housing management and political risk have been widely recognised as the key obstacles for institutional investing in residential properties.

Recently, a study has re-examined these issues and identified the barriers of Australian institutional investors to invest in residential properties [21]. It identified five key outstanding deterrents based on the recent experience of institutional investors. These factors are the high risk profile of residential property, track records of property managers, scale of investment in residential property, liquidity of housing investment and political risk. In the UK, a survey of 48 institutional investors also identified the obstacles for investing in residential properties [13]. Low income yield has been cited as the most important barrier for UK institutional investors to invest in residential property. Other critical deterrents are insufficient market size, lack of liquidity, reputational risk, management issues, lack of sufficient scale and political risk. Interestingly, Housing Investment Trusts had been suggested as an important potential vehicle for attracting large scale institutional investment into private renting [10]. To encourage institutional investment in private rented housing, the lack of suitable residential investment vehicles appear as the key important investment barrier [9]. More recently, real estate investment trusts (REITs), a tax transparent investment vehicle, have been seen as an effective vehicle to encourage institutional investors to invest in residential properties [21]. However, a survey of institutional investors' attitudes towards residential property as an investment vehicle in Switzerland, the Netherlands and Sweden also found that institutional investors in these countries invest in residential property directly [22]. The study also suggested that institutional investors' exposure to indirect residential property investment is minor. This refutes the view of the absence of an indirect investment vehicle in housing is a major obstacle to institutional investment in the private rented sector. In addition, a recent survey also showed that, in the UK, direct ownership remains the most popular method of holding residential property by institutional investors, representing almost 60 % of all residential asset value [13]. Listed property companies have not been seen as a good strategy by UK institutional investors to gain exposure in residential property. The principal rationale for investing in residential property is the returns profile. Although the abovementioned studies found that Swiss, Dutch, Swedish and UK institutional investors may prefer to invest in residential properties via direct ownership compared with indirect ownership, this could be attributed to the lack of an appropriate residential investment vehicle in these markets. In these markets, listed property companies are the major indirect property players. Importantly, listed property companies are mainly engaging in property development activities instead of managing rental properties. Therefore, these may not suit the needs of institutional investors who would only like to invest in the residential rental properties market. Conversely, in the US, residential REITs have been recognised as successful investment vehicles. In addition, Australian institutional investors such as superannuation funds do not typically invest directly in acquiring their own property assets. Therefore, an effective residential investment vehicle could be seen as the important strategy to boost institutional investment in this important sector.

Overall, institutional investors do not play a significant role in residential property investment. Residential investment vehicles have been argued as an appropriate vehicle to encourage institutional investment in housing. As can be seen, no papers have as yet considered the desirable features for an effective residential investment vehicle.

62.3 Data and Methodology

To assess Australian institutional investors' attitudes regarding residential property investment, a survey of Australian institutional investors was conducted in August-September 2014. A questionnaire was designed to examine the perceptions of institutional investors towards residential property investment [22]. More specifically, desirable features for effective residential investment vehicles were also surveyed. There were 61 superannuation funds identified from the Australian Prudential Regulation Authority. Superannuation funds were selected on the basis of specific property or CIO in house still being available, rather than those superannuation funds that relied heavily on asset consultants. 23 property investment companies were identified from the 2014 Australian Property Funds Industry Survey that was published by Property Investment Research. These 23 property investment companies are key property investors. Smaller funds with residential property investment were also included in this survey. In total, 84 surveys were distributed.

With 26 respondents, this saw a survey response rate of 31 %, comprising superannuation funds (62 %) and property investment companies (35 %). The total value of the respondents' portfolio was $147 billion, comprising superannuation funds (A$136 billion) and property investment companies (A$11 billion). A wide range of AUM values for these superannuation funds, with an average AUM of A $8.5 billion, up to A$29 billion for the largest superannuation fund has been evident. The total value of the respondents' property portfolio was A$25 billion, comprising superannuation funds (A$13 billion) and property investment companies (A$11 billion). This saw a wide range of property AUM values for these superannuation funds, ranging from A$65 million to A$3.3 billion, with the average property AUM being A$827 million. This sees the respondent superannuation fund profile matching the typical 5–10 % allocation to property seen for Australian superannuation funds and in most mature markets. The average property AUM for the property investment companies was A$931 million, ranging from A$180 million to $3 billion. Survey respondents were at senior executive levels and had considerable experience in the property industry. For the superannuation funds, it was an average of 14 years experience and for the property investment companies, it was an average of 24 years experience. Both the significant AUM of the respondents and their significant experience in property further add credence to the integrity and rigor of the survey responses. The survey questions were largely based on a previous survey for European institutional investors in residential property [22]; with the questions varied to account to local differences. The specific information received in the survey related to:

1. Factual information concerning their portfolios
2. Rating the importance of specific factors; this was done using a 5-point rating scale comprising "unimportant = 1", "less important = 2", "important = 3", "very important = 4" and "critical = 5".
3. Questionnaires were sent to respondents who are at the level of "Managing Director" or "General Manager" or "Fund Manager" or "Chief Investment Officer". This approach was designed to benefit the study by ensuring a high level of reliability. In addition, the respondents have daily exposure to the fund's management, decision-making process, portfolio management and performance measurement.

The survey results are analysed with frequency analysis. Importantly, a survey has been viewed as the effective way to assess the perceptions of human and have been widely used to gauge property investors' perceptions in the property literature [16, 19, 22, 26].

62.4 Results and Discussion

62.4.1 Current Property Types in Portfolio

Retail property (96 % of respondents), office property (92 %) and industrial property (88 %) were seen as the most popular property sectors in the Australian institutional investor property portfolios. Lesser emphasis was seen for the land (42 %) and hotel/motel (19 %) sectors. Residential property was only seen to be in the portfolio of 8 % of institutional investors; this represented only two of the responding funds.

62.4.2 *Factors Influencing the Decision to Invest in Residential Property*

The importance of the various factors influencing an institutional investor's decision to invest in residential property is shown in Table 62.1. This section was completed by all funds surveyed (# = 26), not just those with residential property in their portfolios. The main factors influencing their decision to invest in residential property were seen to be total expected return (4.27 out of 5), potential for capital appreciation (3.88), cash flow (3.73), risk diversification (3.42) and inflation hedging (3.08). These factors reflect the importance attached by funds to performance analysis and how an asset contributes to the overall portfolio. In particular, these factors rated more highly than the more qualitative factors such as tax benefits (2.62), socially responsible investment (2.38), matching liabilities (2.38) and government subsidies (2.15). This is further emphasized in 85 % of respondents seeing total expected return as "critical/very important", with 62 % seeing cash flow as "critical/very important".

62.4.3 *Issues Relating to Residential Property Investment*

Table 62.2 indicates the importance of specific issues relating to residential property investment by institutional investors. All specific issues rated highly, particularly the lack of well-structured residential investment vehicles (4 out of 5), low returns (3.84), lack of management expertise (3.72) and poor market information (3.52). The focus on lack of well-structured residential investment vehicles and low returns are fundamentally important to institutional investors and highlight their reluctance to see residential property as a key ingredient in their property portfolios. This is

Table 62.1 Importance of factors influencing an institutional investor's decision to invest in residential property

Factor	Average score	Critical/very important (%)
Total expected return	4.27	85
Cash flow	3.73	62
Potential for capital appreciation	3.88	58
Risk diversification	3.42	50
Inflation hedging	3.08	31
Portfolio regulations	2.92	8
Lack of other investment opportunities	2.50	15
Socially responsible investment	2.38	15
Tax benefits	2.62	12
Other government subsidies	2.15	15
Match against liabilities	2.38	15

Table 62.2 Importance of residential private rented sector investment issues

Issue	Average score	Critical/very important (%)
Lack of well-structured residential investment vehicles	4.00	77
Low returns	3.84	69
Lack of management expertise	3.72	58
Poor market information	3.52	54
Poor liquidity	3.51	54
Rent control	3.48	46
Small lot size and poor quality	3.44	50
Tenancy regulation	3.36	50

further highlighted where 77 % of institutional investors saw the issue of the lack of well-structured residential investment vehicles as "critical/very important" and 69 % of respondents saw the issue of low returns as "critical/very important"; with lower levels of "critical/very important" evident for the other issues relating to residential property investment by institutional investors. This has also been emphasized in previous studies regarding residential property in an institutional investor's portfolio.

62.4.4 Desirable Features for Effective Residential Investment Vehicles

77 % of institutional investors considered that a well-structured residential investment vehicle would encourage institutional investment in the residential private rented sector. Given the importance attached to the issue of the lack of well-structured residential investment vehicles, Table 62.3 indicates the importance of desirable characteristics for an effective residential investment vehicle. Top priority is given to the investment vehicle needing to be managed by an experienced manager (4.35/5), followed by diversified portfolio by location (3.92) and focus on delivering a stable income stream (3.88). This clearly reflects the "people" dimension to property funds management and the key role of the fund manager, suitable risk management strategies (i.e. diversification by location) and the importance of income returns. This is further highlighted with 92 %, 81 % and 69 % of respondents respectively seeing these three desirable features as being "critical/very important". Other desirable characteristics seen as being of much less importance included diversification by property type (i.e.: various residential sub-sectors) (3.27), investing in social housing/affordable housing (2.35) or top-end residential properties (2.04) and the need for liquidity by listing on the stock market

Table 62.3 Importance of desirable characteristics for an effective residential investment vehicle

Characteristics	Average score	Critical/very important (%)
Managed by an experienced manager	4.35	92
Diversified portfolio by location	3.92	81
Focused on delivering a stable income stream	3.88	69
Low gearing level (debt usage)	3.31	46
Minimize volatility of portfolio returns	3.38	42
Large scale size	3.42	46
Focus on maximising capital gains	3.58	50
Diversified portfolio by property types	3.27	35
Investing in social housing and/or affordable housing	2.35	15
Investing in middle-end residential properties	2.60	12
Investing in top-end residential properties	2.04	0
Listed on stock exchange (liquid asset)	2.15	12

(e.g.: REIT) (2.15). This highlights the focus on unlisted property investment vehicles for institutional investors to achieve their residential property exposure.

62.4.5 *Potential Problems of Investing in Residential Investment Vehicles*

Whilst recognizing the desirable characteristics needed for residential investment vehicles, the institutional investors also recognized the potential problems with investing in residential investment vehicles. Table 62.4 indicates the importance attached to these potential problems. The most significant potential problems were

Table 62.4 Potential problems of investing in residential investment vehicles

Problem	Average score	Critical/very important (%)
Low returns	4.15	77
Poor market information	3.73	69
Low quality portfolios	3.77	65
Poor liquidity	3.38	54
Regulatory restrictions	3.31	54
Resistance from the board of directors	3.04	31
Not considered as a popular investment vehicle	2.77	27

seen to be low returns (4.15/5), low quality portfolios (3.77) and poor market information (3.73). These are clearly important issues, as seen with 77 %, 65 % and 69 % respectively seen as "critical/very important" by the superannuation funds. These problems clearly relate to property portfolio quality and property information issues, rather than the broader external and internal factors of regulatory restrictions (3.31) and resistance from the Board of Directors (3.04), or not being considered as a popular investment vehicle (2.77) for institutional investors.

62.5 Concluding Comments

While Australian superannuation funds have significant commercial property portfolios, the typical lack of exposure to residential investment by Australian superannuation funds in their property portfolios is highlighted. Thus, this study aims to examine the attitudes of institutional investors in relation to residential property investment in general and residential property investment vehicle in particular. A survey of Australian institutional investors regarding residential property investment was conducted. There are several important findings from this study. Firstly, critical factors influencing Australian institutional investors' decision to invest in residential property have been identified. Return and risk of housing and portfolio considerations have been seen as critical factors. In addition, over 75 % of institutional investors agreed that a well-structured residential investment vehicle would increase institutional investor interest in residential property investment. Secondly, the most desirable features for an effective residential investment vehicle were being managed by an experienced manager, a diversified portfolio by location and delivering stable income returns with low debt. Lastly, institutional investors acknowledged that low returns, poor market information and low quality portfolios are the potential problems with residential property investment vehicles. In particular, the results presented a range of significant challenges and opportunities if we are to see increased levels of residential property in Australian institutional investor property portfolios in the future. Clearly, establishing a well-structured residential investment vehicle is an essential step to enhance institutional investment in residential properties. Ongoing research is required to identify enabling strategies that can be used to increase the level of residential property in Australian institutional investor portfolios. The finding is not only critical for Australian policy makers, but also international policy makers, including Chinese policy makers to enhance the institutional investment in residential property and enhance the supply of rental properties.

References

1. ABS (2011) Household income and income distribution, Australia, 2009–2010. Canberra, Australian Bureau of Statistics
2. AIHW (2011) Australian's welfare 2011. Canberra, Australian Institute of Health and Welfare
3. APRA (2014) Annual statistical bulletin: june 2014. Canberra, Australian Prudential Regulation Authority
4. APREA (2014) Asia Pacific REIT statistical compendium: june 2014. Singapore, Asia Pacific Real Estate Association
5. Berry M (2000) Investment in rental housing in Australia: small landlords and institutional investors. Housing Studies 15(5):661–681
6. Berry M, Whitehead C, Williams P et al (2004) Financing affordable housing: a critical comparative review of the United Kingdom and Australia, Final Report 72. Australian Housing and Urban Research Institute, Melbourne
7. COAG (2010) National Affordable Housing Agreement: baseline performance report for 2008–2009, Canberra, Council of Australian Governments pp 1–158
8. Crook ADH, Hughes J, Kemp PA (1998) Housing investment trusts and the returns from residential lettings. J. of Property Res. 15(3):229–248
9. Crook ADH, Kemp PA (1999) Financial institutions and private rented housing, Joseph Rowntree Foundation, pp 1–4
10. Crook ADH, Kemp PA (2002) Housing investment trusts: A new structure of rental housing provision? Hous. Stud. 17(5):741–753
11. EPRA (2014) EPRA research: monthly statistical bulletin September 2014. Belgium, European Public Real Estate Association, Brussels
12. IBISWorld (2015) IBISWorld industry report L6720: real estate services in Australia. Melbourne, IBISWorld
13. IPF (2014) Institutional attitudes to investment in UK residential property. Investment Property Focus, London, pp 22–24
14. Lawson J, Gilmour T, Milligan V (2010) International measures to channel investment towards affordable rental housing Research Paper. Melbourne, Australian Housing and Urban Research Institute
15. Lawson J, Milligan V, Yates J (2012) Housing supply bonds- a suitable instrument to channel investment towards affordable housing in Australia?, Final Report No. 188, Melbourne, Australian Housing and Urban Research Institute
16. Lee CL (2010) Use of derivatives by Australian property funds. Pacific Rim Property Research Journal 16(2):151–170
17. Lee CL, Lee ML (2012) Hedging effectiveness of REIT futures. J. of Property Investment and Finance 30(3):257–281
18. Lee CL, Reed R (2014) The relationship between housing market intervention for first-time buyers and house price volatility. Housing Studies 29(8):1073–1095
19. Lee CL, Reed R, Robinson J (2008) An investigation on the risk perceptions of Australian property fund managers. Pacific Rim Property Research Journal 14(2):199–221
20. Milligan V, Phibbs P, Fagan K et al (2004) A practical framework for expanding affordable housing services in Australia: Learning from experience, Final Report 65. Melbourne, Australian Housing and Urban Research Institute
21. Milligan V, Yates J, Wiesel I, et al (2013) Financing rental housing though institutional investment-volume 1: outcomes from investigated panel, Final Report No. 202, Melbourne, Australian Housing and Urban Research Institute
22. Montezuma J (2006) A survey of institutional investors' attitudes and perceptions of residential property: the Swiss. Dutch and Swedish cases. Housing Studies 21(6):883–908
23. NAREIT (2014) REIT watch, Washington D.C., US, National Association of Real Estate Investment Trusts

24. Newell G, Fischer F (2008) The role of residential REITs in REIT portfolios. J. of Real Estate Portfolio Manag. 15(2):129–139
25. Newell G, Lee CL, Kupke V (2015) The opportunity of residential property investment vehicles in enhancing affordable rental housing supply, Positining Paper No 166. Melbourne, Australian Urban and Housing Research Institute
26. Rogelberg SG, Stanton JM (2007) Introduction: Understanding and dealing with organizational survey nonresponse. Organ. Res. Methods 10(2):195–209
27. Wulff M, Reynolds M, Arunachalam D et al (2011) Australia's private rental market: the supply of, and demand for, affordable dwellings, Final Report No 168. Melbourne, Australian Housing and Urban Research Institute

Chapter 63
Evaluation of the Sustainability Asian Housing Market Based on Multiple Criteria Assessment

Xiang Li

Abstract The aim of this research is to assess the sustainability of the housing market of four countries in Asia. We find socio-economic indicators of Singapore, China, Japan and Malaysia to do comparison. For this purpose we take example by Land Use Policy (42(2015)642–651) and simplify the method it used. In order to visualize the results we plot the figure by using Matlab. The results include five criteria groups, which are general economic, housing stock, housing affordability, population and social conditions and housing quality. The analysis of ranking and assessment results allows recommendations to be made for improving the indicators in order to increase housing market sustainability. A case study presents a practical application of the proposed methodology. It introduces the calculation results, conclusions and recommendations based on an assessment of four countries to forecast the housing sustainability of Asian.

Keywords Asia · Housing sustainability

63.1 Introduction

Housing is accompanied with human since we born to the world. Thus its future development influence human's future living. We agree with Tim Iglesias, who views housing issues through five comprehensive paradigms: housing as a human right means that adequate, safe, and affordable housing is critical to proper human development. Housing as an economic good means that substantial capital gains and losses occur regularly, as housing is mostly financed, produced and distributed by the private market. Housing as a home means rights and privileges affecting safety, freedom, and privacy. This includes access to and tenure in safe, decent housing for all people. Along with these main paradigms, housing is providing social order and is one competing land use [1].

X. Li (✉)
Department of Mathematics, University of Macau, Macau, China
e-mail: arthurlx94@hotmail.com

Y. Wu et al. (eds.), *Proceedings of the 20th International Symposium on Advancement of Construction Management and Real Estate*,
DOI 10.1007/978-981-10-0855-9_63

A lot of attention is focused on tenure forms: "As on the macro level, the policy theory of market correctives means that political decisions on tenure forms are crucial. The dominant policy theory says it is not for the state to decide how citizens should be housed, but it may be for the state to set up guarantees that citizens have a real opportunity to find decent housing in the market at a reasonable cost. This is why housing tenures should be seen as the most important political instruments of housing provision as welfare state policy" [2].

In China, the real estate is developing at a rapid speed in the last decades. The price of the private using house raised tremendous, for example, in Beijing. The price has raised over three times than it was ten years ago. Since the housing market change rapidly, is housing market will keep on developing? When will it end? These question are further analyzed in this research.

63.2 Literature Review

From 1960 Japan economics keeps on applying Export-oriented economic growth model, foreign trade surplus increase a lot since the early 1980s. As a consequence, JPY faces a great appreciation pressure. After the Plaza Accord in 1985, resulting in the JPY exchange rate rose from 250 yen against the US dollar quickly to 120 JPY in the next five years. At the same time, in order to prevent the decline in exports, the Bank of Japan performed expansionary monetary policy. Therefore, from 1986 to 1990, Japan's average annual growth rate of the money supply (M2) was 16.7 %, significantly faster than the historical average level. Money supply growth is completely beyond the growth of real GDP, which plant a probability of collapse of Japan economic in the future. Another noteworthy background, because of the stagnancy of world market and the pressure of international competition, Japanese companies merely maintain sales and production by lowering commodity prices, leading to a rapid decline in profitability in the manufacturing sector. The decreasing microscopic level of efficiency of Japanese companies reducing the real economy demand for money supply. Under the effect of excess liquidity and real economic downturn, excess funds are not being absorbed by the real economy, but attracted to the area of virtual economy. Thus in the whole 1980s, money supply far exceeded money demand. In 1980, the supply demand ratio exceeded two. As a result, the exceeding money had to find a new investment area. Then, when the excess liquidity into real estate and stock market, resulting in a price soaring in all the land, including residential land, commercial land and industrial land. And in 1985–1990, real estate index rose from 65.19 to 147.8, increasing 126 % with the average annual growth rate of 25 %. Facing the growing accumulation of bubble, the Japanese government was very worried. However, because of fearing the damage to the real economy, the negative impact of the US stock market bubble busted in 1987 and other unfavorable international economic environment. They hesitated on doing macroeconomic policy adjustments. By the time in late 1989, the Japanese government finally realized the seriousness of the problem. Ministry of

Finance was forced to use the interest rate instruments, from May 1989 to August 1990, a total of 350 points interest rates was raised and made the rapid rise in the Japanese discount rate to 6 %. At the same time, the authorities also banned real estate loans issued by financial institutions, and forced banks to return the funds gradually. In 1991, Japan's economic bubble burst, the stock and real estate prices began to fall, the Japanese economy has entered a recession for 10 years.

We also look at the historical trends in household credit of Singapore. Household credit is a major proportion of credits in Singapore, accounting for about 60 % the proportion of bank loans to non-bank customers. According to the statistics agency,mortgage almost accounts for two of three of the housing credit. But in recent years, after suffering continuous economic change and series of economic crisis, including government intervention in the real estate bubble crunch in May 1996; The Asian financial crisis between July 1997 and September 1998; Technology bubble burst from 2000 to 2002; SARS in 2003 and fluctuations in oil prices between 2003 and 2006. During this time, Singapore economic also have to face the competition pressure from China, India as well as other emerging competitors. Thus households become more conservative to household credit. Household debt as a percentage of GDP falls from 97 % in 2003 to 77 % in 2006. However, the uniqueness of Singapore, such as the increase of high-income immigrants, decline of the labor market and the decrease of interest rate, will influence the development of household credit. And will further influence the sustainability of the real estate. Since in 2006, Singapore announced a relaxation of immigration policy: attracting more foreigners to settle in and allow more residents to become Singapore citizens. Moreover, because of the current account surpluses as well as a large number of large capital inflows, Singapore decrease the interest rate to 3 %, and the real estate will have a new growth. For example in the center of Singapore city, the rental fee for a small flat is 1200SGD, which in the last year, this number was only 900SGD.

And in Malaysia, which is a developing country just like China, has been through a financial crisis in 1997. Therefore there are some experience can be adopted. Since the 1990s, in order to pursue a higher growth rate, Malaysia invested excessively, credit expanded blindly and absorbed large quantity of foreign capital. As the financial system is incomplete and lack of financial supervision, a lot of money influx in real estate, which caused the domestic real estate bubble. There is some identical phenomenon. First, before the crisis, followed by the investment and credit expansion, the quantity of demand in real estate remained high; the price also went up rapidly. And after the crisis, price went down dramatically and housing vacancy rate rise. Second, with low interest rate, a large number of bank loans went into the real estate industry, which in the end became bad debts. Therefore, a good macro-control is needed and sound regulatory system should be used in financial system.

In China, the price of real estate has been increasing continuous from 2004. From the beginning of 2006 to the end of 2010, the amplitude of the increase is about 28.19 %. Only in 2008, because of the international financial crisis, the price has a temporary small decline. As the real estate is developed, the housing vacancy rate goes up. Although the definition in China is a little bit different from international countries, the number still reflect a potential problem. For example in Beijing, this number attained 27.16 %, much higher than 20 % of which is considered as a dangerous level. Meanwhile,the Housing price-Income ratio in Beijing has attained 31 on average, according to international convention, at present, the ratio between 3 and 6 times as a reasonable interval of real income, and housing consumption accounted for the proportion of household income should be lower than 30 %. What else, house price to rents ratio is the ratio between the housing price and the total monthly rent. Housing price rent ratio is used to judge whether there is a real estate bubble. And it is used to judge whether a region has investment value. In general the number should be between 200 and 300. If the number is greater than 300 refers to a bubble and lower means good potential of investment. In Beijing, the housing price rent ratio is 315 on average, the highest which is 700 (Jinghua Times 2009.11).

This paper aims to use some index such as the ratio mentioned above to compare the housing sustainability of four countries in Asia.

63.3 Development of System of Criteria for Housing Sustainability Evaluation

In order to do the comparison we should include as many criteria as possible. However, with the limit of data approach we mainly find the data from the Statistical yearbook of each country. We divided indicators into five groups: general economic, housing stock, housing affordability, population and social conditions, and housing quality. We found in total of 21 different criteria and form a matrix that is if there are new criteria we can add it into the ordinary one. The system of criteria is presented in Table 63.1. [3–6].

63.4 Assessment Methodology

Followed by Land Use Policy (42(2015)642–651), we used the same method of Multiple Criteria Decision Making. There are four stages according to get the significance and priority of alternatives (in this case, countries).

Table 63.1 List of data

General economic indicators	Gross domestic product (GDP) per capital
	Unemployment rate
	Inflation rate
Housing stock indicators	Number of household
	Housing owner rate
	Agricultural acreage (%)
	Residential construction as % of GDP
Housing affordability indicators	Share of housing cost in disposable income (%)
	Representative interest rates of new residential loans
	Average number of dependents as employer
	Housing price index
	House price to rents ratio (capital)
	Total outstanding residential loans to GDP ratio
	Housing price-income ratio (capital)
	General government expenditure for housing and amendments
	Population density
	Housing vacancy rate
Population and social conditions	Employers in real estate
	The gene coefficient
Housing quality indicators	Average households size
	Average useful floor area per person

Stage 1.
Find $\hat{x}_{ij}$. Where x_{ij} is the data we found previously. N is the number of countries, which are 4 in this case. q_i is the significance of ith criterion.

$$\hat{x}_{ij} = \frac{x_{ij} \cdot q_i}{\sum_{j=1}^{n} x_{ij}}; \quad i = \overline{1,m}; \quad j = \overline{1,n};$$

Stage 2.
Find S_{+j} (S_{-j}). Since we only find the average value of each criteria, therefore $S_{+j} = S_{-j}$. And similarly, $S_+ = S_-$.

$$S_{+j} = \sum_{i=1}^{m} \hat{x}_{+ij}; \quad S_{-j} = \sum_{i=1}^{m} \hat{x}_{-ij}; \quad i = \overline{1,m}; \quad j = \overline{1,n}.$$

$$S_{+} = \sum_{j=1}^{n} S_{+j} = \sum_{i=1}^{m} \sum_{j=1}^{n} \hat{x}_{+ij};$$

$$S_{-} = \sum_{j=1}^{n} S_{-j} = \sum_{i=1}^{m} \sum_{j=1}^{n} \hat{x}_{-ij}; \quad i = \overline{1, m}; \quad j = \overline{1, n}.$$

Stage 3.
Find Q_i. The significance (efficiency) of comparative alternatives.

$$Q_j = S_{+j} + \frac{S_{-\min} \cdot \sum_{j=1}^{n} S_{-j}}{S_{-j} \cdot \sum_{j=1}^{n} \frac{S_{-\min}}{S_{-j}}}; \quad j = \overline{1, n}.$$

Stage 4.
Find N_j. Determining the priority order of alternatives. The greater the Q_j, the higher is the efficiency of an alternative. In order to visually assess alternative efficiency we calculated the utility degree N_j. The degree of utility is determined by comparing the alternative analyzed with the most efficient alternative.

$$N_j = \frac{Q_j}{Q_{\max}} \cdot 100\,\%$$

63.5 Case Study, Results and Discussion

Different Asian countries have different housing policies and history, different economic development and different housing respective. Data for the assessment was acquired from each countries 2010s statistics yearbook. We do realize that deficiencies exist in all countries, but we assume, that rules for international database formation are uniform.

Figure 63.1is the general calculation result. The blue line is the total multiple criteria result and others are all categorized indicators.

During the analysis the normalization of decision making matrix was performed and weighted decision making matrix was constructed by using the formula. Further, based in the matrix data, we calculated significances Q_j and utility degree. I determined weights of criteria in each group; other experts if needed can change them.

Next I would like to talk about some aspects that may influence each countries behaviors during the calculation.

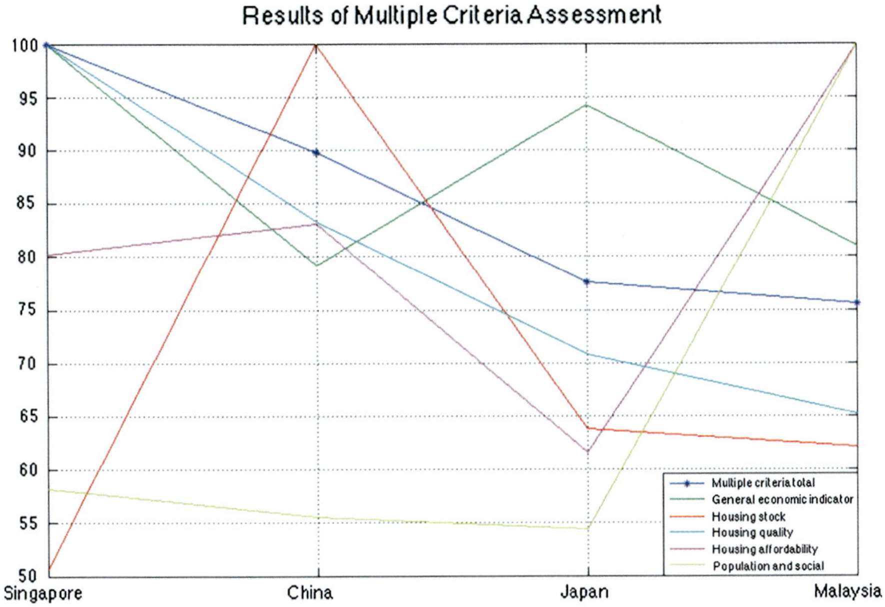

Fig. 63.1 Results of multiple criteria assessment

63.5.1 Housing Stock

Housing stock market reflects part of the whole development of the housing market. Since it is an enormous and complicated task to do the analysis in details. Thus we only look through the data from Bloomberg. I pick a total of four indexes from each countries. They are TPX500 (Japan), SSE180 (China), FBMEMAS (Malaysia) and FSTAS (Singapore) [7]. By comparison of the indexes within one year we can see that China has the greatest fall but still in the leading position. Japan follows and the rest two have almost the same index. However, if we look through the movement within five years, Japan takes the first position and has a stable increase. China, though once catch up with Japan with a rapid rise, it falls hugely in the beginning of June. Malaysia has a much smooth development while Singapore suffers a consistent decline.

One year comparison of four index

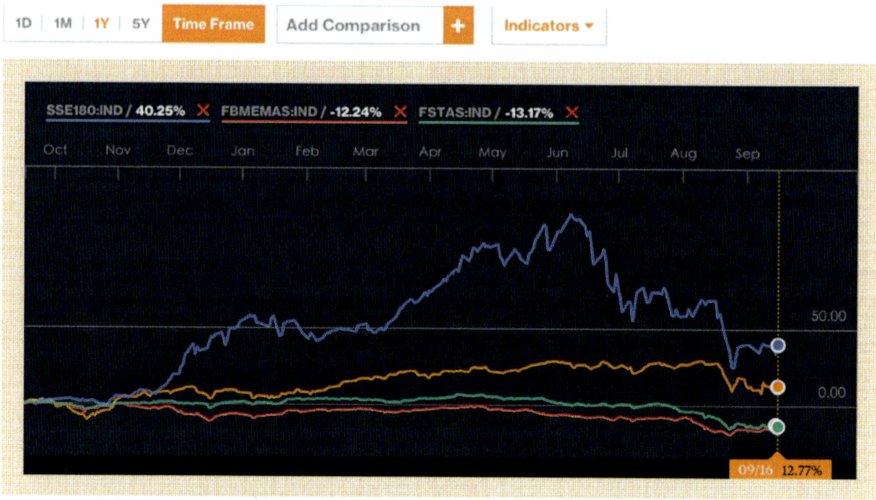

Five year comparison of four index

63.5.2 Housing Affordability

Housing affordability relates to many aspects such as income and housing price. And both aspects have other factors that matters with. Hence through my limited knowledge I cannot give an inclusive analysis. But still I want to talk about my opinions. Maybe affordability cannot only be judged by data. If the necessity of

Multiple criteria total	100%	89.75%	77.55%	75.63%
General economic indicators	100%	79.20%	94.20%	81.00%
Housing stock indicators	50.70%	100%	63.86%	62.10%
Housing affordability indicators	100%	83.33%	70.84%	65.21%
Population and social conditions	80.29%	83.08%	61.59%	100%
Housing quality indicators	58.17%	55.53%	54.38%	100%

Fig. 63.2 The accuracy results of each countries

owning a house exceeds the burden of afford the loans from a house. Then people will scarify other assumptions in order to improve the affordability. Like Chinese parents will purchase a house nearby the school so that to convenient their children.

63.5.3 Housing Quality

Housing quality mainly depend on the time when the house is built. And it will deteriorate as time going on. Moreover, the development speed of a country also contribute to the quality of the house which means the skills of construction, the skills of production and so on. Sometimes the quality depends the geographic location of the house, like in Japan where earthquakes happens easier and in China it is not.

Figure 63.2 are the accuracy results of each category. From left to right followed as Singapore, China, japan and Malaysia.

By all groups of criteria, the most sustainability housing is in Singapore, but it also takes the last place in the Housing stock criteria. The less sustainability housing is Malaysia, but the gap between its and Japan are not large. In all groups, China performed at a medium level. As it has the largest population as well as the land area, then if the indicator is by capita, the result will be normal obviously.

We do realize that industrial countries with mature economies and housing markets cannot ne strictly compared to economies in transition. But all the countries have more or less suffered the housing bubbles and economic downturns [8]

63.6 Conclusions

In conclusion, after looking up many materials I have a feeling that the sustainability of housing mainly depend on the policy or future design of the country or the certain area. The real estate, by all means, contributes to the GDP, or at least account a large part of it. So that it is the policy that decide the sustainability of housing. For example, Chinese government control the construction land as a commodity that will sell to the real estate company whenever they need to stabilize the economy or increase GDP. It control the price of the house by limit the amount of the available land. In Singapore, the government want to decrease the venture activity and thus it cut the loans which time period is over 35 years. In Malaysia, for instance, its capital where have the lowest housing price among all other capitals in Asia countries and according to its future plan, it aims to build a modern city that will give a wondrous increase in its real estate as well as other aspects.

When purchase housing is considered as an investment, then the sustainability of the housing market will become less sustainable. Since everyone hope that the price will increase forever so is their wealth and this disobey the regularity of economy. Look back at 1980s, Japan's real estate market has a fascinating development. All most all the individuals, companies has devoted to the market. The country really had an enjoyable time when the price kept rising. However, when the bubble exploded the economy falls to the bottom. Hence the real sustainable way is how to control the bubble effectively, on one hand it will increase the economy quickly but inconsistently. On the other hand, if it cannot be treated suitably, it will pulls back the economic in decades.

In China, the government want to move the capital from the type of investment to consumption so that reduce the bubble of the real estate market. Meanwhile increase the housing affordability of the population. This is the most sustainable way of develop the housing market. We need the development of housing market, but we do not need a growth too fast to control.

References

1. T. Iglesias. Housing Paradigms. Selected Works of Tom IglesiasUniversity of San Francisco, USA (2012)
2. B. Bengtsson. Housing politics and political science D.F. Clappham (Ed.), et al., The SAGE Handbook of Housing Studies, SAGE Publications Ltd., London (2012), pp. 206–229
3. Japan Statistical Yearbook 2013
4. Yearbook of Statistics Singapore 2013
5. Statistics Yearbook Malaysia 2011
6. China Statistical Yearbook 2013
7. Bloomberg (http://www.bloomberg.com/quote)
8. A. Kaklauskas, L. Kelpsiene, E.K. Zavadskas, D. Bardauskiene, G. Kaklauskas, M. Urbonas, V. Sorakas Crisis management in construction and real estate: conceptual modeling at the micro-, meso- and macro-levels Land Use Policy, 28 (2011), pp. 280–293
9. Housing policies in Singapore: Evaluation of recent proposals and recommendations for reform Sock-Yong Phang,* David Lee,*Alan Cheong,** Kok-Fai Phoon* and Karol Wee*

Chapter 64
Cadre's Local Embeddedness and Land Law Enforcement: Evidence from China

Lin Li and Xiaoling Zhang

Abstract Illegal land use is a threat to land administration as it reveals the weaknesses of the system. In order to ensure the management of land, it's necessary for government to take some measures and law enforcement is a feasible alternative. Law enforcement enables government to detect as well as penalize illegal land use and therefore is indispensable for effective land administration. This paper investigates land law enforcement in the Chinese context and finds that cadre's embeddedness to some extent results in the dynamics of this activity. In this way, this paper argues that a political perspective is needed when studying land law enforcement in contemporary China.

Keywords Land administration · Law enforcement · Cadre · Embeddedness

64.1 Introduction

Over the past few decades, China experiences dramatic changes in a variety of areas and one example is the rapid urbanization. In 1978, the proportion of urban population is less than 18 % while in 2013 this proportion has exceeded 50 %. Urbanization greatly reshapes and restructures the landscape throughout the whole country and land is a crucial component which is deeply involved in this process. In addition to land development in many cities, illegal land use has been another notable issue that emerges and persists during Chinese urbanization. According to the statistics from China Land and Resources Almanac and China Land and Resources Statistical Yearbook, 166,042 cases concerning illegal land use are uncovered in 1999, involving 28,674.82 ha of land. In 2012, the number of cases drops to 61,821 whereas the area of land involved increases to 32,026.18 ha.

L. Li · X. Zhang (✉)
Urban Research Group, Department of Public Policy, City University of Hong Kong, Kowloon Tong, Hong Kong
e-mail: xiaoling.zhang@cityu.edu.hk

Y. Wu et al. (eds.), *Proceedings of the 20th International Symposium on Advancement of Construction Management and Real Estate*,
DOI 10.1007/978-981-10-0855-9_64

Given the severity of illegal land use, central government in China has developed some strategies in order to promote the enforcement of land laws, but whether they would pay off still remains a question. As Chinese government is a multi-hierarchy organization, therefore the integration between governments at different levels greatly affects the success of enforcement. Studies of Chinese politics have demonstrated that there are at least two reasons why this integration may collapse and fail to bring about favorable outcomes. One reason has something to do with flawed monitor mechanism, which means that superior government is unable to inspect the behavior of its subordinates in a comprehensive and timely manner [1]. Another reason is related to the huge discretion lower level governments possess when governing the territories, namely they are authorized to determine the timing and the way of implementing policies within their own jurisdictions [2, 3]. These two reasons jointly contribute to the dynamics of hierarchy system, as well as the significance of local government. In this regard, we argue that law enforcement in China is a political process where local government's autonomy shall be taken into account.

Following this argument, this paper pays special attention to cadres in local government and explores the effects they have on law enforcement in land management arena. By taking provincial governor as an example, this paper attempts to uncover how local embeddedness shapes his or her motivation and influences government's performance on law enforcement.

64.2 Literature Review

64.2.1 Local Government and Land Development in Contemporary China

Many scholars point out that urbanization in China is quite different from that in western countries as it is deeply influenced by local government and its economic and fiscal incentives. Lin [4], for example, argues that landed urbanization in the Chinese context is greatly driven by local government's motivation to gain revenue from land development and land related income has accounted for a considerable proportion of local government's total income. Ye and Wu's [5] study lends support to this argument. Based on prefecture level data, the two scholars find that percentage of built up area is positively associated with proportion of land leasing revenue, implying that land finance has been an important momentum driving the expansion of urban areas.

Given the significance of land finance, some studies further explore its institutional foundation. Generally, they argue that land finance is local government's strategy to cope with fiscal deficit resulting from the reconfiguration of fiscal arrangement in the middle 1990s [6]. Specifically, lower level governments may be more fiscally disadvantaged as their superiors have the authority to formulate details of the arrangement and therefore are able to enjoy a larger sharing of the revenue [7]. The importance of fiscal arrangement has been confirmed by various studies. Ye and Wang's [8] research, for instance, indicates that sub provincial government's reliance on land revenue is negatively related to the extent of revenue decentralization within a province, meaning that institutions do affect the evolution of land finance in contemporary China.

As an important foundation of land finance, land apportionment system in China has experienced fundamental reform, namely from an administrative dominant system to a market oriented one [9]. Nevertheless, the development of land market varies in different regions throughout China, requiring a close examination of the driving forces of this marketization reform. Among them, local government may be one of the most crucial forces. Deng [10], for example, studies land market reform in China based on provincial level data and reveals that city density, a proxy of intergovernmental competition, has a negative effect on administrative allocation. In this sense, local government could also be an advocator and facilitator of pro-market reform, as long as its power and ensuing benefits are preserved or even promoted with the progress of the reform.

However, land market may be more complicated than it seems to be, since there are several different ways through which local government could lease land use right, including tender, auction, negotiation and so forth and there is no guarantee that local government will necessarily choose the way which is more open, transparent and competitive. A consequent question is therefore: what is the pattern of local government's choice? Tao et al.'s [11] study provides a possible answer in suggesting that municipal government strategically applies land leasing to get fiscal revenue, namely leasing land to manufacturing industries through negotiation so as to lower their land costs and assigning land to commercial and residential sectors by auction in order to get a reasonable price. Though this strategy of land distribution may help local government gain economic benefits and reduce fiscal pressure in the short term, the lack of institutionalization in land market will eventually result in some serious and far reaching consequences. Cartier [12], for instance, notes that the absence of formal rules governing transfers of land use right may finally undermine local state's capacity to obtain land leasing income while enable individual officials to gain personal interests during land transactions.

Extant studies of local government's behavior in land development reveal its profit-making motive and deepen our understanding of land politics in contemporary China. However, it should be noted that these studies are only a part of the story as the role of local government is multifaceted. Practically, local government is an important component of the hierarchy system and is responsible for enforcing laws concerning land use within its own jurisdiction. Nevertheless, how local government carries out such a duty and which forces shape its performance have not been fully explored to date, implying that there is still much work to do to obtain a complete image of land management in the Chinese context.

64.2.2 Law Enforcement in Land Administration and Determinants of Its Outcomes

Law enforcement is an important part of land administration through which government could punish those who violate laws or regulations and maintain the order of land use. Therefore, scholars have been paying great attention to this issue. Some of them conclude that the effects of restrictions have increased due to the development of economy, technology and management tools [13], whereas others find that in some developing and underdeveloped countries such as Brazil and Ghana, defiance of regulations is an even more prevalent phenomenon [14, 15]. Thus, a following question would be: what are the determinants that affect law enforcement in land management?

Some studies argue that legislation itself is a crucial factor influencing its enforcement. When exploring farmland decline in China, Lichtenberg and Ding [16] find that farmland loss persists despite the existence of administrative restrictions, implying the necessity of adjusting these restrictions so as to promote their effectiveness. Buitelaar et al. [17] study of Netherlands' planning law also argues that adaption to local institutions, both formal and informal, would greatly determine the extent to which legislation can achieve its intended aims. Moreover, a successful legislation should not only reactively respond to current situation, but also in some way foresee the trends in the coming future, otherwise de facto land use will still deviate from preset arrangements as a consequence [18].

Additionally, institutional configuration could also be an obstacle preventing laws from achieving their initial targets [19]. A study of Laos, a Southeast Asian country that is experiencing rapid urbanization, indicates that master planning fails to control urban sprawl in Vientiane partly due to the lack of supporting institutions that are indispensable for the implementation of such plans [20]. A similar situation could be found in Ghana. Boamah et al. [21] study, for example, reveals that political interference is an important source of enforcement failure in this country, as politicians and superior authority sometimes intervene in law enforcement and therefore inevitably undermine the impartiality of local authorities. In some more industrialized and developed countries such as the United States, however, institutional configuration also matters. By exploring planning practice in Washtenaw County, Loh and Sami [22] argue that lacking state intervention in some way leads to the ending of regional planning in that county.

Apart from institutional reasons, organizational defects also deserve our attention [23]. In some underdeveloped countries such as Nigeria, departments concerned often suffer from a shortage of manpower, funding and necessary tools and thus are unable to fully enforce laws at the local level [24]. In addition to the shortage of resources, inappropriate division may impede effective enforcement as well. Rizzo's [25] study, for instance, demonstrates that the complex administration structure in Qatar, where diverse departments are involved in the same affairs but in charge of different aspects, is an important factor that hinders the implementation of planning. Meanwhile, resource consuming procedure is also a notable force that undermines the efficacy of laws concerning land use. In a study of land management system in Tanzania, Kironde [26] concludes that lengthy administrative procedure discourages land users' willingness to comply with restrictions and therefore worsens the outcomes of land use regulations.

Although the above studies reveal structural variables that steer law enforcement in land administration, they to some extent overlook the complexity of this process. It is often the case that laws could be intensively and effectively enforced in some regions and periods, while fail to exercise equivalent influence in other regions and periods. Rodrigues-Filho et al. [27], for example, note that election and resulting administrative instability bring about periodically lax implementation of forestland preservation policies in the Amazon area. In this sense, it is necessary to uncover the dynamics of law enforcement by exploring the factors that drive spatial and temporal variation and the mechanisms through which they make a difference.

64.3 Cadre's Embeddedness and Its Implications for Law Enforcement

Cadre's embeddedness indicates the association between an individual cadre and a given locality and its strength is determined by a set of factors. Among all the factors, this paper mainly concentrates on characteristics of cadres and regards tenure, age and domestic origin as the most important ones.

Tenure refers to the time a cadre has been acting as the governor of a given province. Generally, cadres with longer tenure are more embedded in the locality as they have more chances to interact with various actors in the territory and therefore are more likely to establish and strengthen relationships, formal or informal, with these actors. Scholars have been making efforts to explore the effects of long tenure and associated relationships on cadres' acts and the quality of governance and they argue that such effects could be negative on some occasions. Zhou [2, 3], for instance, claims that long tenure would make cadres deeply involved in interpersonal networks and thus undermine their neutrality when dealing with public affairs. Landry [28] also sheds lights on the adverse influences of long tenure. As he suggests, long tenure may lead to some severe problems such as cadres' misconduct, including abuse of power or even corruption. In this regard, we propose that the increase of tenure will weaken cadres' motivation and result in a worse performance on law enforcement.

Age implies the time a cadre can hold a position with substantive power at most before he or she retires. With the practice of cadre retirement in the 1980s, age has been an important determinant of cadres' political mobility with elder ones being less likely to get promoted [29, 30], which then weakens their association with and accountability towards superiors. In this way, elder cadres are more embedded in the jurisdiction than their younger counterparts. Studies have uncovered how this time horizon and related prospect shape cadres' behavior. Zhong [31], for example, argues that cadres who are elderly and are less likely to achieve upward mobility tend to put their own interests prior to responsibilities and engage in rent-seeking activities. Nie and Jiang's [32] study of work safety management in coalmine industry in China echoes with this argument by concluding that younger cadres are more committed to their statutory duties and thus are able to exhibit better performance on the duties concerned than their elder fellows. Therefore, we propose that the increase of age would reduce cadres' incentive and lead to a laxer law enforcement.

Domestic origin indicates whether a cadre mainly worked in the locality he or she is currently governing. In general, cadres with domestic origin are more embedded in the territory than those who spent most of their career in other places due to the local identity stem from past working experience. As a result, these cadres incline to put more emphasis on local interests while leave superiors' concerns aside if they cannot satisfy both sides at the same time. Literatures have provided empirical evidence for this argument. Bai et al. [33], for instance, explore determinants of economic specialization in China and find that cadres who have closer association with the local society are more prone to local protectionism. Huang and Sheng's [34] study of macroeconomic management also implies the effect of such association by demonstrating that cadres who are more integrated with a given jurisdiction perform worse on inflation control. In this way, we propose that domestic origin is likely to undermine cadres' compliance and bring about lower level of task accomplishment concerning law enforcement.

64.4 Data and Methodology

This study involves 30 provincial level jurisdictions in mainland China except Tibet and the time span is from 1999 to 2012.

Dependent variable of this study is percentage of land settled cases involve ("settled percentage" hereinafter) and is calculated according to the following formula:

$$pit = \frac{sit}{fit + uit - 1}$$

where means settled percentage of province in year, denotes the area of land involved in settled cases, is the area of land involved in filed cases and implies the area of land involved in unsettled cases in last year. Data concerning the area of land involved in settled cases, filed cases and unsettled cases come from China Land and Resources Almanac and China Land and Resources Statistical Yearbook.

Explanatory variables of this study are cadre's tenure, age and domestic origin. Tenure is measured in the span between the year a cadre takes governor post and the year of study. Age refers to the distance between a given year and the year of birth. Domestic origin is a dichotomous variable where "1" implies that a cadre

Table 64.1 Descriptive statistics of the variables

Variable	N	Mean	Std. dev	Min	Max
Settled percentage (%)	420	81.59	17.09	16.26	100
Tenure	420	3.114	1.895	1	10
Age	420	57.86	4.198	43	65
Domestic origin	420	0.469	0.500	0	1
GDP (a hundred billion yuan)	420	6.669	6.476	0.240	36.22
Proportion of revenue to expenditure (%)	420	52.15	18.69	14.83	95.09
Population (million people)	420	43.14	26.10	5.100	105.9

mainly served at the locality he or she is currently appointed and "0" means that a cadre mainly worked at other places before he or she is assigned governor position. Information of explanatory variables could be found in cadres' biographies through the Internet. Control variables of this study comprise GDP at 1999 constant price as an indicator of economic development, proportion of public revenue to expenditure as an indicator of government's fiscal pressure and number of population by the end of the year as an indicator of population size. Data of control variables come from China Statistical Yearbook and China Population and Employment Statistics Yearbook (see Table 64.1).

Two-way fixed effects model is used in this study for statistical analysis:

$$yit = \beta xit + \omega i + \kappa t + \varepsilon it$$

where represents dependent variable, indicates independent variables consisting of both explanatory and control variables, implies entity fixed effects and implies time fixed effects, while is the error term.

64.5 Results

In this part, we present regression results of two-way fixed effects model in different specifications. Details are shown in Tables 64.2 and 64.3.

In model 1, model 2 and model 3, the effects of explanatory variables are examined respectively without any control variables. The coefficient of tenure is negative and significant at 5 % level, implying that increasing tenure would lower settled percentage. Therefore, this result lends support to the argument that longer tenure undermines local government's performance on law enforcement. However, neither the coefficient of age nor that of domestic origin is significant, which disconfirms the hypotheses concerning these two factors. In other words, age and domestic origin fail to exercise considerable influences on cadres' preferences on this occasion. In model 4, model 5 and model 6, we include one of the control variables, namely regional GDP and its squared term. The coefficient of tenure

Table 64.2 Results of two-way fixed effects regression (Model 1 to Model 6)

	Model 1	Model 2	Model 3	Model 4	Model 5	Model 6
Tenure	−0.815**			−0.842**		
	(0.324)			(0.322)		
Age		−0.265			−0.259	
		(0.198)			(0.210)	
Domestic origin			−0.582			−1.126
			(2.551)			(2.675)
GDP				2.169**	2.071**	2.035**
				(0.872)	(0.878)	(0.866)
GDP^2				−0.0570***	−0.0554***	−0.0554***
				(0.0170)	(0.0173)	(0.0173)
Revenue to expenditure						
Population						
Constant	86.03***	98.98***	84.09***	80.52***	93.34***	79.16***
	(2.297)	(11.00)	(2.891)	(2.710)	(11.99)	(3.074)
Observations	420	420	420	420	420	420
R-squared	0.091	0.086	0.082	0.116	0.111	0.107

1. Robust standard errors in parentheses
2. ***$p < 0.01$, **$p < 0.05$, *$p < 0.1$

remains negative and significant at 5 % level, with the magnitude changing only slightly. Meanwhile, the coefficients of age and domestic origin are still insignificant and do not vary greatly in magnitude even after controlling economic development. Thus, it seems that tenure does make a difference whereas age and domestic origin are unable to exert significant effects.

In order to test the robustness of above argument, we also add other control variables one by one. In model 7, model 8 and model 9, proportion of public revenue to expenditure is included and we can find that the results are similar. The coefficient of tenure is negative and significant at 1 % level, with its magnitude increasing only a little. Nevertheless, the coefficients of age and domestic origin remain statistically insignificant yet even though fiscal pressure of local government is taken into account. In model 10, model 11 and model 12, we add all control variables and the results do not change very much. The coefficient of tenure remains significant and negative and its magnitude varies only marginally. However, the coefficients of age and domestic origin are still not significant with the magnitude of the former being slightly larger when population size is controlled. In this way, we argue that the increase of tenure does matter to the decrease of settled percentage while age and domestic origin do not significantly lead to such variation and this argument is quite robust among different model specifications.

Table 64.3 Results of Two-way fixed effects regression (Model 7 to Model 12)

	Model 7	Model 8	Model 9	Model 10	Model 11	Model 12
Tenure	−0.911***			−0.909***		
	(0.310)			(0.310)		
Age		−0.262			−0.272	
		(0.207)			(0.203)	
Domestic Origin			−1.546			−1.462
			(2.737)			(2.748)
GDP	2.019**	1.938**	1.889**	2.010**	1.930**	1.882**
	(0.831)	(0.862)	(0.835)	(0.827)	(0.857)	(0.834)
GDP^2	−0.0547***	−0.0534***	−0.0534***	−0.0513***	−0.0496***	−0.0501***
	(0.0160)	(0.0167)	(0.0164)	(0.0161)	(0.0167)	(0.0162)
Revenue to Expenditure	0.238	0.197	0.213	0.285	0.250	0.258
	(0.167)	(0.174)	(0.180)	(0.185)	(0.195)	(0.197)
Population				−0.351	−0.396	−0.340
				(0.366)	(0.358)	(0.373)
Constant	67.48***	82.56***	67.55***	79.26***	96.41***	78.97***
	(9.259)	(15.12)	(9.332)	(12.68)	(12.22)	(12.96)
Observations	420	420	420	420	420	420
R-squared	0.121	0.114	0.111	0.122	0.116	0.112

1. Robust standard errors in parentheses
2. ***$p < 0.01$, **$p < 0.05$, *$p < 0.1$

64.6 Conclusion and Discussion

Studies concerning land laws have been continuously emerging in the past few decades, covering a wide variety of topics. Among all the topics, law enforcement is one of the few that receive great attention. The reason may be that enforcement, as many scholars argue, does have significant implications. For one thing, enforcement, as a form of implementation, plays the role of turning blueprint into reality [35]. For another thing, enforcement, in a broader context, makes up a crucial part of state capacity, indicating the extent to which state could ensure the obedience to rules [36]. In China, land law enforcement has drawn central government's attention and relevant legislations have been enacted. As early as 1989, State Bureau of Land Administration has issued Interim Regulation of Handling Illegal Land Use, stating various details of land law enforcement such as the responsibilities of departments concerned, requirements for case filing and etc. At the end of 1995, Regulation of Handling Illegal Land Use is issued, with some revision of the interim version. In 2014, Ministry of Land and Resources issues Regulation of Administrative Penalties in Land and Resources Management, which covers land as well as mineral resources. Given the importance of land law enforcement, both theoretical and practical, this paper focuses on this topic and explores it within the Chinese context.

As there are diverse forces shaping land law enforcement, so a comprehensive examination of them is beyond the reach of this paper. Therefore, we put more emphasis on political factors by underlining the role of local cadres play during this process. Cadres, as extant studies suggest, are influential actors in Chinese politics who greatly determine the outcomes of policies or programs at the local level [37]. This is even true for the reform era, since cadres successfully sustain and strengthen their influences. In this sense, we propose that cadres matter to land law enforcement. By studying the effects of cadres' embeddedness on subnational government's performance, we to some extent confirm this argument. Specifically, the significant and negative relationship between settled percentage and cadres' tenure implies that longer tenure does impair the performance of local government. Meanwhile, the coefficients of age and domestic origin are both statistically insignificant, meaning that these two variables only have marginal effects. Given these findings, we are able to conclude that land management in general and land law enforcement in particular are deeply rooted in institutional context where cadres could make a difference.

Moreover, this paper also lends support to some practices of cadre management such as term limit. In 2006, General Office of the Chinese Communist Party Central Committee issues Interim Rules for Leading Cadres' Term Limits. According to this document, each term lasts for five years and cadres cannot be nominated for a position if he or she has spent two consecutive terms on that position, implying that cadres could hold the same post for at most 10 years. Meanwhile, one cannot be appointed as leading cadres at a given level if he or she has been holding a leadership position at that level for 15 years. Although this document also states that term limits of minority cadres working in national autonomous areas could be relaxed with approval, it does impose stringent restrictions on cadres' term and therefore in some way shape their behavior pattern. Some studies have revealed that term limits and frequent leadership turnover, together with performance-based evaluation and highly competitive upward mobility, have resulted in cadres' preferences for short-term achievements and superficial or even unsustainable solutions [38]. This behavior pattern could be an unintended consequence of term limits, but it does not fundamentally challenge the necessity of such limits. Rather, it implies that term limits should be carefully designed so that the limits would not be a cause of expedients while still exerts the effect of ensuring cadres' integrity.

References

1. Zhan JV (2009) Decentralizing China: analysis of central strategies in China's fiscal reforms. J Contemp China 18(60):445–462
2. Zhou LA (2010) Incentives and governance: China's local governments. Cengage Learning Asia Pte. Limited, Singapore
3. Zhou X (2010) The institutional logic of collusion among local governments in China. Modern China 36(1):47–78
4. Lin GC (2014) China's landed urbanization: neoliberalizing politics, land commodification, and municipal finance in the growth of metropolises. Environ Plan A 46(8):1814–1835
5. Ye L, Wu AM (2014) Urbanization, land development, and land financing: evidence from Chinese Cities. J Urban Aff 36(S1):354–368
6. Lin GC, Zhang AY (2014) Emerging spaces of neoliberal urbanism in China: land commodification, municipal finance and local economic growth in prefecture-level cities. Urban studies, 0042098014528549
7. Tian L (2015) Land use dynamics driven by rural industrialization and land finance in the peri-urban areas of China: "the examples of Jiangyin and Shunde". Land Use Policy 45:117–127
8. Ye FZ, Wang W (2013) Determinants of land finance in China: a study based on provincial-level panel data. Aust J Publ Adm 72(3):293–303
9. Ho SP, Lin G (2003) Emerging land markets in rural and urban China: policies and practices. China Q 175:681–707
10. Deng FF (2003) China's urban land reform, urban productivity, and local government behavior. Eurasian Geogr Econ 44(3):210–227
11. Tao R, Su F, Liu M, Cao G (2010) Land leasing and local public finance in china's regional development: evidence from prefecture-level cities. Urban Stud 47(10):2217–2236
12. Cartier C (2001) 'Zone Fever', the Arable land debate, and real estate speculation: China's evolving land use regime and its geographical contradictions. J Contemp China 10(28):445–469
13. Long Y, Gu Y, Han H (2012) Spatiotemporal heterogeneity of urban planning implementation effectiveness: evidence from five urban master plans of Beijing. Landscape Urban Plan 108 (2):103–111
14. Ditt EH, Knight JD, Mourato S, Padua CV, Martins RR, Ghazoul J (2008) Defying legal protection of atlantic forest in the transforming landscape around the Atibainha reservoir, South-Eastern Brazil. Landscape Urban Plan 86(3):276–283
15. Boamah NA (2013) Land use controls and residential land values in the offinso south municipality, Ghana. Land Use Policy 33:111–117

16. Lichtenberg E, Ding C (2008) Assessing farmland protection policy in China. Land Use Policy 25(1):59–68
17. Buitelaar E, Galle M, Sorel N (2011) Plan-led planning systems in development-led practices: an empirical analysis into the (lack of) institutionalisation of planning law. Environ Plan A 43 (4):928–941
18. Pethe A, Nallathiga R, Gandhi S, Tandel V (2014) Re-thinking urban planning in India: learning from the wedge between the de jure and de facto development in Mumbai. Cities 39:120–132
19. Verburg R, Rodrigues Filho S, Lindoso D, Debortoli N, Litre G, Bursztyn M (2014) The impact of commodity price and conservation policy scenarios on deforestation and agricultural land use in a frontier area within the Amazon. Land Use Policy 37:14–26
20. Sharifi A, Chiba Y, Okamoto K, Yokoyama S, Murayama A (2014) Can master planning control and regulate urban growth in vientiane, Laos? Landscape Urban Plan 131:1–13
21. Boamah NA, Gyimah C, Nelson JKB (2012) Challenges to the enforcement of development controls in the Wa municipality. Habitat Int 36(1):136–142
22. Loh CG, Sami N (2013) Death of a planning department: challenges for regionalism in a weak mandate state. Land Use Policy 32:39–49
23. Smith H, Jenkins P (2015) Trans-disciplinary research and strategic urban expansion planning in a context of weak institutional capacity: case study of Huambo, Angola. Habitat Int 46:244–251
24. Arimah BC, Adeagbo D (2000) Compliance with urban development and planning regulations in Ibadan, Nigeria. Habitat Int 24(3):279–294
25. Rizzo A (2014) Rapid urban development and national master planning in Arab Gulf Countries. Qatar case study. Cities 39:50–57
26. Kironde JL (2006) The regulatory framework, unplanned development and urban poverty: findings from Dar es Salaam, Tanzania. Land Use Policy 23(4):460–472
27. Rodrigues-Filho S, Verburg R, Bursztyn M, Lindoso D, Debortoli N, Vilhena AM (2015) Election-driven weakening of deforestation control in the Brazilian Amazon. Land Use Policy 43:111–118
28. Landry PF (2008) Decentralized authoritarianism in China: the communist party's control of local elites in the post-mao era. Cambridge University Press, Cambridge, New York
29. Manion M (1992) Politics and policy in post-mao cadre retirement. Chin Q 129:1–25
30. Li Y (2014) Downward accountability in response to collective actions: the political economy of public goods provision in China. Econ Transit 22(1):69–103
31. Zhong Y (2003) Local Government and politics in China: challenges from below. M.E. Sharpe, Armonk, New York
32. Nie H, Jiang M (2011) Coal mine accidents and collusion between local governments and firms: evidence from provincial level panel data in China. Econ Res J 6:146–156
33. Bai CE, Tao Z, Tong SY (2008) Bureaucratic Integration and regional specialization in China. Chin Econ Rev 19(2):308–319
34. Huang Y, Sheng Y (2009) Political decentralization and inflation: sub-national evidence from China. Br J Polit Sci 39(02):389–412
35. Jones CO (1984) An introduction to the study of public policy, 3rd edn. Brooks/Cole Publishing Company, Monterey, California
36. Mann M (1984) The autonomous power of the state: its origins, mechanisms and results. Eur J Sociol 25(02):185–213
37. Kung JKS (2014) The emperor strikes back: political status, career incentives and grain procurement during China's great leap famine. Political Sci Res Methods 2(02):179–211
38. Eaton S, Kostka G (2014) Authoritarian environmentalism undermined? local leaders' time horizons and environmental policy implementation in China. China Q 218:359–380

Chapter 65
An Empirical Analysis on the Reasons of Indemnification Housing Encounter Cold—Take a Case of Guangzhou Longgui Garden Project

Lin Chen, Tao Tao, Jianhui Tan and Ziping Zhang

Abstract Based on the demand and tenant intention survey of Guangzhou Longgui garden indemnification housing project, an empirical analysis is made in this paper. Firstly, using the qualitative and quantitative research method, the real living situation and the housing security demands of the low income groups are explored in this paper; Secondly, combined with the experience and lessons of other cities in the indemnification housing project construction and allocation process, the reasons of the indemnification housing project may receive cold reception are analyzed, including that the publicity is inefficient, location is remote, lack of community infrastructure and the higher entrance standard of applications and etc. Finally, based on above all, relative suggestions are made in this paper, including strengthening the publicity, optimizing the project sites, improving community infrastructure, reducing application threshold, increasing the number of rental subsidy, increasing income standard, and let more migrant workers be included into the housing security protection scope.

Keywords Housing security · Public rental housing · Indemnification housing encounter cold · Longgui garden project

L. Chen · J. Tan (✉)
Business School, Guangzhou University, Guangzhou, China
e-mail: tjhgzu@126.com

L. Chen
e-mail: 2450289765@qq.com

T. Tao · Z. Zhang
Civil Engineering School, Guangzhou University, Guangzhou, China
e-mail: 718023860@qq.com

Z. Zhang
e-mail: 729637094@qq.com

Y. Wu et al. (eds.), *Proceedings of the 20th International Symposium on Advancement of Construction Management and Real Estate*,
DOI 10.1007/978-981-10-0855-9_65

65.1 Introduction

In recent years, together with the increasing supports from the government to the housing security work, more and more straitened people are included into the housing security range, and the housing difficulty problems are relieved in certain degree. But at the same time, while the public-rental housing construction projects are finished in many places and the houses are put into use, in some cities the public-rental housing is receiving cold reception when the housing just comes to the market. The vacancy of indemnification housing will not only leave the public resources unused, it will also limit the further improvement of the housing security policies. For this reason, in this paper, an empirical study is done based on the survey of the tenant intents to the public-housing project of Longgui Garden made by the government as the indemnification housing project and will soon be put into the market, with questionnaires from those families which are qualified before 31st July, 2014, to wait for indemnification housing, as well as those families which have handed in their applications and are waiting for the approval of their qualifications. The purpose of this paper is to find out the real reasons why the public-rental housing would meet with cold reception, so that strategic suggestions may be made for the similar public-rental housing projects, for the better leasing effects and social benefits.

65.2 Review of Relevant Domestic Researches and Practice in Other Cities

65.2.1 Domestic Researches

Domestic scholars and experts have made relevant studies and researches of the reason why the indemnification housing may meet with cold reception. Some scholars are of the opinion that the vacancy of large quantity of indemnification housing is due to the main reason that current system frame has no encouragement to the assignment and supply of indemnification housing. Some local governments let the indemnification housing be vacant temporarily is to make the construction speed faster than the supply speed in the indemnification housing, to avoid arguments in the assignments [1]. Also some scholars point out that there is a gap between the indemnification housing application conditions and the conditions which the demanding people really have. The access system is not reasonable which forms unmatched supply and demand in the structure, so that large quantity of demanding groups are not covered by the security [2]. Some scholars mention that vacant indemnification housing is a waste of land resources, but it shows that possibly the indemnification housing development mode may have problems [3]. Xie and Liu [4] are of the opinions that the construction of indemnification housing should not be blindly increasing the housing quantity, but should practically understand the demands of the indemnified people, speeding up the planning and construction of supporting facilities in transportation, education, and commerce, etc., and combine them with the market management mechanism of the constructions of public-rental housing, etc. Yang [5] thinks that the main reason of the "high vacancy rate" is due to the inconvenience of transportation conditions of public-rental housing, and the shortage in public facilities. Usually, as the lands in city centers are in saturation, most governments have to build the indemnification housing in suburban areas, city edges. Therefore, the indemnification housing has the common points of insufficient transportation basic facilities, including incomplete markets, schools and shops, etc. [6]. Applicants removing into indemnification housing not only for the purpose of residing, but also for a better live, with community environment as an important referent factor. Many factors such as remote locations, incomplete public facilities, poor housing quality, etc., are impressing the selection willing of the indemnified target groups [7]. Since the new round of large scale indemnification housing construction, the central government would appraise local governments only by the operation rates and the finishing rates of indemnification housing, and the occupancy rates are not included. Then it has been a hidden rule that quantity is more important than quality, in certain respects [8].

The study and research results of domestic scholars and experts in the problems of why indemnification housing would meet with cold reception have made firm foundation for the continuous studies, and are worth learning. This paper would, on the foundation of the studies of so many scholars and experts, continue the empirical study, as the supplement and verification of the relevant viewpoints.

65.2.2 The Phenomenon of Public-Rental Housing Encounter Cold in Relevant Cities

While the public-rental housing in some cities in China are completed and put into the market for use, in some cities, earlier or later, the public-rental housing is meeting cold reception in the applications, typically in cities including Dongguan, Foshan and Shenzhen, etc. The reason behind such phenomenon are worth contemplation.

From the data of Dongguan Municipal Bureau of Housing and Urban we may see that till September, 2011, in Dongguan totally 2629 flats of indemnification housing had completed, among which there were 846 flats of public-rental housing, economically affordable housing 1783 flats. So far 297 flats of indemnification housing have been assigned, among which there are 140 flats of public-rental housing, and 157 flats of economically affordable housing, and 2332 flats of remaining housing are waiting to be assigned, among which, 706 flats of public-rental housing and 1626 flats of economically affordable housing. The indemnification housing assigned is less than 20 %. For specific assignments please refer to Table 65.1.

Table 65.1 Assignments of Dongguan indemnification housing (till Sept. 2011)

Flats completed		Flats assigned		Flats to be assigned	
Public-rental housing	Economically affordable housing	Public-rental housing	Economically affordable housing	Public-rental housing	Economically affordable housing
846	1783	140	157	706	1626
Total: 2629		Total: 297		Total: 2332	

Data source Dongguan Municipal Bureau of Housing and Urban

Foshan Municipal Bureau of Housing and Urban has the data that till 18th October, 2013, in Foshan, 8450 flats of public-rental housing had been completed and totally 5100 flats are assigned, remaining almost 40 % vacant. Large quantity of public-rental housing became "None-rental" housing. Of which, in Haibei Plaza of Nanhai District, in Phase one project totally 640 flats public-rental housing of indemnification nature had been completed and totally removed in residents and successful applicants were of 210 families, only 100 families had removed in (Data source: Foshan Municipal Bureau of Housing and Urban).

In August 2013, the assignments of Longyueju, which was the largest public-rental housing project in Shenzhen, entered into the last stage. For the launched 8250 flats as the first batch of public-rental housing sources, from the date of the announcement of assignment, totally 25815 copies of online application forms were received. Totally 10352 families were qualified as primary candidates and standby candidates. But when the first batch applicants had visited the sample flats, about 45 % applicant families selected to give up the submitted materials. Totally written application materials delivered by 5683 families were actually handled, with the actual handling rate of 55 %, in which, 4527 families were from social groups, and 1156 families were from talent groups. By the primary verification and re-verification, totally 5435 families were qualified, in which, 4344 families were from social groups and 1091 families were from talent groups, qualification rate was 96 %. 2815 flats of housing source are remained embarrassingly as more than 30 % applicant families have given up their applications. For specific assignments please refer to Table 65.2.

From above indemnification housing meeting with cold reception in relevant cities, we may see that comparing with the huge quantity of housing security demands of the society, the indemnification housing meeting with cold reception is obviously not due to the construction volume is too much and the supply is exceeding demands. But still it is an actually existing problem, worth deepened

Table 65.2 Assignments of Shenzhen largest public-rental housing project (till Aug. 2013)

First batch launched flats	Families applying primarily online	Families screened and qualified	Remaining families after visiting sample flats	Families after verifications	Indemnification housing rent out flats
8250	25815	10352	5683	5435	2815

Data source Shenzhen municipal bureau of housing and urban studies. Though in different cities such phenomena may have different reasons, but something in common should exist. Based on this point, in the following paragraphs, I am going to study the real reasons why the indemnification housing would meet with cold reception, with an empirical study method, based on the public-rental housing tenants intention survey of Guangzhou Longgui Garden Project, understanding the real ideas and specific demands of the research objects.

65.3 Research Process and Important Findings

65.3.1 Basic Information About Longgui Garden Project

Longgui Garden located in Longgui Town of Baiyun District of Guangzhou, is the largest indemnification housing community under construction in Guangzhou. The project occupies the total area of 347.5 thousand square meters of land, total construction floor area is more than 1.07 million square meters, with plot ratio of 3.4, planned to have residents of about 30 thousand headcount. When the project is completed, it will provide totally 12,000 flats of indemnification housing to Guangzhou (mainly as public-rental housing products, of 61.56 %, 7576 flats). Of which, the phase 1 of project was completed at the end of 2014, bringing more than 5000 flats of indemnification housing as the supply volume. Longgui Garden Project is about 20 km far to the downtown area of Guangzhou, currently it is about 3 km far to the nearest Metro Line 3 Longgui Station, and two public buses have stops near that project. Now the Metro station going through the project, and large sized schools, hospitals, etc., as public facilities are under construction and they are estimated to be completed in 2–3 years.

65.3.2 Investigation Method and Basic Situations of Respondents

In this study, the survey used the method of questionnaires. With the helps from government authorities, totally 10718 families were investigated by up-to-door visits. Such families were qualified before 31st July, 2014, to wait for the effective housing indemnification, or were those families who had submitted their public-rental housing applications and were under verifications. Of which, 8454 families were effectively qualified, and 2264 families were under verifications. Totally 8848 copies of the questionnaires were collected for the Longgui Garden housing demands, with the ratio of 82.6 %. Of which ineffective questionnaires were 138 copies, and effective ones were 8710 copies, as 98.4 %. Effective questionnaires were from Yuexiu, Liwan, Haizhu, Baiyun, Huangpu, Panyu, Huadu and Luogang, etc., totally 9 administrative districts, among them Yuexiu, Liwan and Haizhu districts were separately 27.4, 28.8 and 32.3 %, totally 88.5 %.

65.3.3 Findings from the Quantitative Survey

Nearly Half of the Families Had no Idea of the Project

The survey showed that the visited families generally didn't understand Longgui Garden Project. 49.2 % families stated that they had no idea of it. 35.1 % stated that they knew something about it, and only 15.2 % said that they understood that project. It was quite clear that for similar large size indemnification housing projects, the advertisement work should be further reinforced, as the understanding degrees of the people would play important influence to them in the leasing willingness, and would play active functions in lowering the vacancy risks of the projects.

The Understanding Degrees of the Visited People Played Important Influence to Their Leasing Willingness

Relevant analyses of the leasing willingness and the understanding degrees of the visited people to Longgui Garden Project showed that the more they understood that project, the stronger their leasing willingness would be. In the survey, the visited people who understood the project were more inclined to accept the leasing, those people who knew something about that project were more inclined to possible acceptance, while those who knew nothing about Longgui Garden might possibly refuse the leasing. Therefore, strengthening the advertisement of the project to make people more understand the project details would help raising the leasing willingness of the straitened people.

From the Demands in Districts, the Districts of Yuexiu, Liwan, Haizhu, Tianhe and Baiyun Had Demands of Higher Quantities and Higher Ratios

The statistic result showed that in the effective 8710 copies of questionnaires, only 2214 families clearly stated that they would accept leasing Longgui Garden, the percentage was 25.4 %, those clearly stated refusal were 4201 families, the rate was 48.2 %, and the hesitating ones were 2295 families, as 26.3 %.

Those families who would accept Longgui Garden were mainly concentrated in the old downtown districts where there were more straitened people. Of the 2214 families who would accept leasing, 716 families were in Yuexiu District, as 32.3 %, Liwan District 580 families, as 26.2 %, Haizhu District 652 families as 29.4 %, Tianhe District 122 families as 5.5 %, Baiyun District 104 families as 4.7 %, Huangpu District 15 families as 0.7 %, Panyu District 3 families as 0.1 %, Huadu District 15 families as 0.7 %, Luogang District 7 families as 0.3 %. The statistic result is showed in Table 65.3.

Table 65.3 Demands distribution in districts

Adm. district	Yes		No		Hesitating		Total
	Families	Percent (%)	Families	Percent (%)	Families	Percent (%)	
Yuexiu	716	30.0	1042	43.7	625	26.2	2383
Liwan	580	23.1	1159	46.2	769	30.7	2508
Haizhu	652	23.2	1469	52.3	688	24.5	2809
Tianhe	122	9.0	196	46.6	103	24.5	421
Baiyun	104	35.7	115	39.5	72	24.7	291
Huangpu	15	8.3	147	81.2	19	10.5	181
Panyu	3	42.9	3	42.9	1	14.3	7
Huadu	15	17.6	56	65.9	14	16.5	85
Luogang	7	28.0	14	56.0	4	16.0	25
Total	2214	25.4	4201	48.2	2295	26.3	8710

From the demands in districts, the headcount of people willing to lease Longgui Garden was 716 in Yuexiu District, the headcount as the top 1, as 30.0 % of the effective investigated families of 2383 in that district. Samely, in other districts where there were more families demanding housing, Liwan District had the rate of 23.1 %, Haizhu District 23.2 %, while Baiyun District 35.7 %.

Refusal Reasons Were Mainly the Location of Project Was not Meeting the Demands

The survey data (for this issue the question was with multiple choice, multiple choice analysis method was adopted) showed that, the reason to refuse leasing Longgui Garden was mainly that the location was not in line with the demands. The visited families had the general idea that it was too far away from downtown area, 86.3 % families had such idea. There were other reasons such as having no idea of that project, as 17.7 %, complaining shortage in public facilities, as 11.4 %, untold reasons were 6.8 %. The result of analyses is showed in Fig. 65.1.

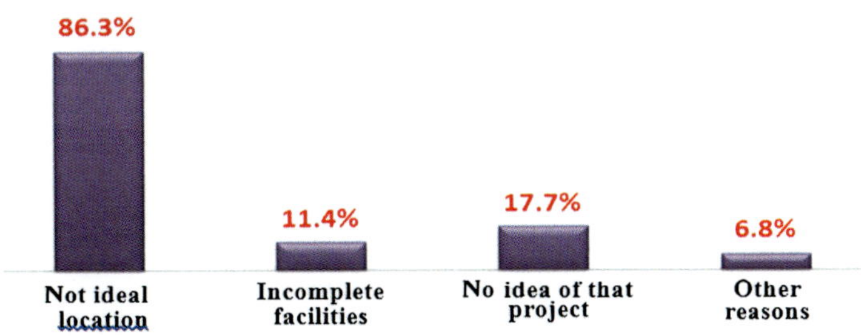

Fig. 65.1 Reasons why refuse to accept leasing

By crossover frequency analysis it was more obvious that those applicant families who clearly stated refusal of Longgui Garden were mainly due to the reason of giving up that the location was not an ideal one to them, the percentage of such idea was as high as 92.1 %. Other reasons such as having no idea of that project, incomplete facilities and other reasons were of much lower percentages.

For those families who stated that they might choose Longgui Garden, at the same time they might not choose it. 69.4 % families provided their reasons of not choosing Longgui Garden. If these families who stated at the time of survey that they "might choose", finally giving up Longgui Garden, the reasons would be: Not the good location, as 75.6 %, not understanding the project, as 30.2 %, incomplete facilities, 13.0 %, other reasons, 5.6 %.

Even in the families who stated acceptance of Longgui Garden there would be the possibility of final given up. 15.1 % of such families provided their refusal reasons that if they finally gave up that project, the reasons would be: not ideal location, as 67.1 %, no idea of that project as 21.9 %, incomplete facilities, as 14.4 %, other reasons, 8.1 %. The result of analyses is shown below in Table 65.4.

The other reasons reflected by the visited people for refusal of leasing Longgui Garden were mainly: (1) too far away from working location or schools of kids, inconvenient transportation and high transportation costs, which would make their living burdens even heavier; (2) incomplete public facilities for living and transportation, causing inconvenience in living and outdoor activities; (3) too far away to the registered hospitals appointed by the medical care insurance, if there were patients in family, not easy for them to go to hospitals; (4) the families had aged members living in downtown areas, if the applicants lived in Longgui Garden they would be unable to take care of aged family members; (5) the room areas of the project were too small, not practical, while the rentals were too high; (6) they would rather give up the security manner of providing housing to them with rental payable, and they preferred to have rental allowances or to live in directly managed public housing.

Table 65.4 Analyses of reasons to refuse Longgui Garden

			Reasons to refuse Longgui Garden				
			Not ideal location	Incomplete facilities	No idea of that project	Others	Total
Whether Longgui Garden would be selected?	Yes	Families	224	48	73	27	334
		Percent (%)	67.1	14.4	21.9	8.1	100
	No	Families	3714	423	503	290	4034
		Percent (%)	92.1	10.5	12.5	7.2	100
	Hesitating	Families	1205	207	481	90	1592
		Percent (%)	75.6	13.0	30.2	5.6	100
	Total	Families	5143	678	1057	407	5960
		Percent (%)	86.3	11.4	17.7	6.8	100

Analyses of Supply and Demand Matching Showed that the Supply Quantities of All Types of Flats in that Project Exceeded Current Demanded Quantities

Analysis results showed that in the families stating acceptance of Longgui Garden leasing, most were one-person families, as 47.4 %, totally 1049 families; less than that were families of 3 members, 608 families, as 27.5 %; single parent families of mother and son, or father and daughter, were 282 families, as 12.7 %. Other two member families were 179 as 8.1 %, families with 4 or even more members were 69, as 3.1 %. The analysis result is in Fig. 65.2.

The total quantity of indemnification housing in Longgui Garden Project which would launch soon was 5161 flats, which was far more than the 2214 families who clearly stated their demands in this survey. In the flat types of the indemnification housing launched by Longgui Garden, quantities of different flat types were sufficient to meet the demand of the people. For specific available quantities please refer to Form 3–3. In the flats, one bedroom and one bedroom with one sitting room type had 2082 flats, as 40.3 % of the total quantity, which might be sufficient for one person families or two member families. Other types of flats were also in sufficient supply, the types of two bedrooms and one sitting room, and three bedrooms and on sitting room were far more than the quantity of the types prepared for families with 3 or even more members. The analyses of supply and demand matching showed that the supply quantities of all types in this project were more than current demands. Flat types and available quantities are shown in Table 65.5.

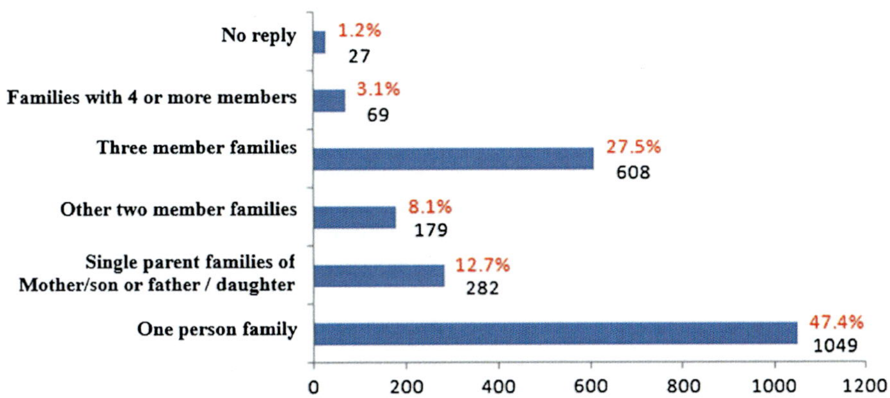

Fig. 65.2 Structure of family security headcount

Table 65.5 Available quantity statistics of flat types in Longgui Garden

Flat type	Flats	Percent (%)	Building areas (m^2)
Single room	124	2.4	35.9–36.27
1 bedroom 1 sitting room	1958	37.9	31.1–41.6
2 bedrooms 1 sitting room	1443	28.0	44.67–51.4
3 bedrooms 1 sitting room	1636	31.7	60.72–63.32
Total	5161	100	

65.4 Conclusion

From above empirical study we may see that the main reasons why the applications for indemnification housing of projects similar to Longgui Garden Project were meeting with cold reception including the advertisement is insufficient, the information communication channels are not smooth, many people with housing difficulties do not understand the policies and the information about the housing security, and the psychological distances are increasing; the location of the indemnification housing is far away from downtown area of the city, the transportation is inconvenient, and of higher costs; ambient living facilities, medical facilities and educational facilities of the indemnification housing are not sufficient, causing inconveniences in living; the threshold to apply for indemnification is too high, admittance conditions are strict; the indemnification housing is of narrow coverage, mainly for registered permanent residents, large quantity of migrant labors have not been included into the security range, etc.

Against such questions and their reasons, to explore the effective measures, following suggestions are raised, including Strengthening advertisement, increasing understanding to the projects by the people, shortening the psychological distances; Optimizing the project location selections, and completing transportation facilities; Notice the actual demands of the security targets, completing the public facilities in the indemnification housing communities; Broadening the housing indemnification considerations, increasing rental allowance and standards; Reasonably expand the range, heightening the qualification standards of income limits, including more difficult families with household registers and migrant labors into the range of indemnification

While commodity housing flats are highly priced, the issues of insufficiency of the small flat type housing sources available to lease and the limited paying abilities, etc., are bringing housing problems to large quantity of middle and low income people. On the other hand, in some cities the launched indemnification housing projects are meeting with cold reception embarrassingly. How to make good use of the available resources, and to match the supply and demand, to make the indemnification housing policies favoring more demanding people, are worth consideration. Last but not the least, this paper is a case study empirically based on Longgui Garden Project, but the conclusion and suggestions are of general meanings.

Acknowledgements Guangdong Provincial Natural Science Fund project (S2013010013533); Guangdong Province; Philosophy and Social Science Planned Project (GD13CGL04).

References

1. Li X (2013) Whether public-rental housing can really work. J Econ Syst Reform 10:142–147
2. Yun H, Zheng J, Sun Y, Gong Z, Tian Y (2012) Living with housing, people go first—an empirical study of public-rental housing plan and construction in Tianjin. J City Plann 05:13–19
3. Wang H (2011) Risks existing in the indemnification housing constructions in China and the measures suggested. J China Real Estates 22:71–74
4. Xie D, Liu J (2012) Discussions on the operation modules of indemnification housing. J China Real Estates 05:53–55
5. Yang J, Huang X (2012) A typical case study on public-rental housing construction. J Econ Issues 01:50–54
6. Chen S, Fan Q (2014) Later period management of public-rental housing and the solutions–based on the sample of Wuhan City. J City Issues 09:81–86
7. Qing F (2010) Zero transaction—public-rental housing. China Real Estates 7:8–9
8. Lin W (2009) Practical significance of development of public-rental housing and the analysis of prospects. J City Rural Constr 10:59–63
9. Lu P, Zhen H (2010) Unifying the establishments of housing security systems in the integrated developments of cities and villages. J City Rural Constr 01:123–133
10. Zeng D, Wen X (2012) Study on comparisons of domestic and foreign operation and management systems in public-rental housing. J City Devel Study 10:80–85
11. Xie D, Liu J (2012) Exploring the public-rental housing management in Chongqing. J China Real Estates 09:67–69
12. Chen W, Zhang R (2013) Re-consideration of "Hongshan" module of public-rental housing in Wuhan. J China Real Estate 15:39–42
13. Lin S (2013) Study on the public-rental housing meeting with cold reception phenomenon. J China Real Estates 15:39–42
14. Meng W, Liu X (2011) Study on the pricing mechanism of the rentals for city public-rental housing leasing. J Pricing Theor Pract 12:66–70
15. Zhang J (2013) Research and study on the vacant rates of indemnification housing in Guangzhou. Degree paper of Guangzhou, South China University of Technology

Chapter 66
Indemnification Housing Construction Task Assignment Model Building—An Empirical Study Based on Guangzhou

Jianhui Tan, Ziping Zhang, Lin Chen and Tao Tao

Abstract After reviewing the practice of indemnification housing construction tasks sharing between city and districts level in Guangzhou, analyzing the mechanism and the problems of the current mode, this paper develops the regression models of random effects and fixing effects to ensure the reasonability and enforceability of social housing construction task assignment between different administrative levels. On the basis of model analysis, relevant policies and suggestions are raised.

Keywords Indemnification housing · Task assignment · Regression model

66.1 Introduction

In recent years, in China, the housing security is getting more and more emphasized, practically solving the housing problems of quite a lot of straitened people. But in a short period and in some cities, the tasks of indemnification housing are relatively heavy, and the construction volume organized only by the city level authorities is hard to satisfy. Therefore some construction job must be assigned to districts or counties level to ensure that the construction tasks of indemnification housing can be completed smoothly. From 2011, Guangzhou has begun the trial of indemnification housing construction tasks shared by the city and districts. In three years of practice,

J. Tan · L. Chen (✉)
Business School, Guangzhou University, Guangzhou, China
e-mail: 2450289765@qq.com

J. Tan
e-mail: tjhgzu@126.com

Z. Zhang · T. Tao
Civil Engineering School, Guangzhou University, Guangzhou, China
e-mail: 729637094@qq.com

T. Tao
e-mail: 718023860@qq.com

Y. Wu et al. (eds.), *Proceedings of the 20th International Symposium on Advancement of Construction Management and Real Estate*,
DOI 10.1007/978-981-10-0855-9_66

certain achievements are obtained, whereas obstacles are met as well. The main problem is the issue of how to effectively establish the assignment model shared by the city and districts for the construction tasks. Against this background, building a suitable tasks assignment model via empirical study is very necessary.

66.2 Literature Review

Construction of indemnification housing is an important measure to realize the central government's policy of "home to live in". During the period of the 12th 5 Year Plan, the target of constructing 36 million flats may basically solve the problem of insufficient supply of the indemnification housing [1]. The research shows that construction of indemnification housing has practically helped many straitened people [2]. But behind the completion of the indemnification housing construction, many problems still exist.

In the construction of indemnification housing, insufficient land and shortage of funds are two main problems which are seriously hindering the progress and effects [3]. Studies and experiences in other countries show that when the funds for social housing construction are insufficient, the better and effective solution is to use funding leverage, to absorb social capitals or revitalize stock assets [4–6]. In China, many scholars are of the opinion that leverage of financing with administrative funding should be sufficiently utilized, the indemnification housing construction and the assets of indemnification housing construction should be securitized and capitalized, and the model of public-private cooperation should be used in financing [7–9]. Meanwhile, combining with the actual situation in China, scholars also raised many suggestions, such as establishing special plan for indemnification housing construction and take it as an emphasized aspect in evaluating the governmental achievements, etc. [10–12].

Many studies show that either funding issue or land issue, the sources usually may be summarized as problem of upper-level system. The appraisal mechanism and land finance have seriously affected the progress and quality of the completion of the construction tasks [13].

Wang [14] believed that under current financial decentralization and land finance system arrangement, local governments may have resistant and gambling behaviors to the indemnification housing construction. Wang [15] has studied the behavior changes of local governments in the indemnification housing construction with the theories of reasonable selections and new institutionalism, etc., and he has found that the behavior changes of local governments are mainly limited by the financial system, the authority structures of different levels of governments, and the relationship of local benefits, etc. Yan Rong has studied the behavior logic of the local government in the social housing construction and found that local governments have to accept the current political incentive system arrangement. For completing the quota planned by the central government, with the background of insufficient of usable funding, local government would select to build the social housing aggregating and by large scale at the city edges [16].

On the other hand, the completion of progress and the efficiency of the social housing construction tasks are affected by the issues of un-clarified authorities or unequal rights and responsibilities among different administration levels. The rights and responsibilities of the government in the housing security should be further clarified, and the complete legal systems of social housing should be built up [17, 18].

66.3 Indemnification Housing Construction Task Assigning Model Building

66.3.1 Practice and Problem Study

By massive empirical researches, discussions with personnel engaged in housing security departments of different levels of governments in Guangzhou, site visiting of typical indemnification housing projects in Guangzhou, detailed information and data are obtained. It is found that the current practice has really lessened the pressure of housing security work in the city level of Guangzhou. However, there still exists some problems and short comings, such as: (1) In some districts the enthusiasm of self-decided construction of indemnification housing is not high generally. (2) Difficulties are existing in some projects' construction, funding and management. (3) Quantity over quality, i.e. most districts merely focus on the completion of the indemnification housing construction task, and the quality is not assured.

The questions of how to further activate the enthusiasm of the districts, how to determine the right proportions of the distribution between the city level and the district level, have become the main issues troubling all the parties. In order to solve the problems, the indemnification housing construction tasks assigning model will be tried to build up.

66.3.2 The Changes of the Assigning Situation Between the City and the Districts for the Indemnification Housing Construction from 2011 – 2014

Gradual Changes from Mainly by the City to Sharing Mutually by the City and the Districts

In recent years, the mode of indemnification housing construction in Guangzhou changed from mainly responsible by the city to the mode of mutually sharing by the city and the districts, it is shown in the transfer of the housing security tasks to the districts step by step. According to the data provided by Guangzhou Housing Management Bureau (see Table 66.1), affected by the policies made and the

Table 66.1 2011–2014 Indemnification housing construction tasks quantity and percentage in districts and (unit: flats, %)

	2011		2012		2013		2014	
	Quantity	Percentage	Quantity	Percentage	Quantity	Percentage	Quantity	Percentage
Whole city	81,457	100	36,200	100	16,736	100	11,682	100
City level	54,500	66.91	26,370	72.85	10,967	65.53	5800	49.65
Yuexiu	500	0.61	200	0.55	100	0.60	100	0.86
Haizhu	1500	1.84	500	1.38	200	1.20	300	2.57
Liwan	3000	3.68	100	0.28	300	1.79	200	1.71
Tianhe	1500	1.84	500	1.38	300	1.79	100	0.86
Baiyun	1300	1.60	200	0.55	100	0.60	200	1.71
Panyu	4000	4.91	1000	2.76	500	2.99	400	3.42
Huadu	1800	2.21	500	1.38	336	2.01	200	1.71
Huangpu	1500	1.84	500	1.38	300	1.79	200	1.71
Luogang	6100	7.49	2000	5.52	1000	5.98	1000	8.56
Nansha	2300	2.82	2100	5.80	1733	10.35	882	7.55
Conghua	1000	1.23	730	2.02	400	2.39	300	2.57
Zengcheng	2457	3.02	1500	4.14	500	2.99	2000	17.12

Source of data Guangzhou housing management bureau

Table 66.2 Shared between city and districts of the tasks 2011–2014

Time (years)	2011	2012	2013	2014
Proportion between city and districts	2.02	2.68	1.90	0.99

demand of the housing security development, the magnitude of inclination of indemnification housing construction tasks in districts is getting up year by year. By the building coefficient Γ the proportion of tasks between the city and the districts is measured:

$$\Gamma = yt/(1 - yt) \quad (66.1)$$

y_t proportion of indemnification housing construction task assignment of city level itself in year t (of which t = 2011…2014)

The smaller Γ value shows the heavier tasks at the district level layer. Table 66.2 shows that the Γ value drops from 2.02 to 0.99.

The Changes of Construction Situations in Districts Are of Unbalanced Feature

The 12 districts in Guangzhou have different situations in population, land area, GDP volumes, economic development levels, housing security object quantities, housing policies, even in personnel systems, and there exists unbalance. It is hard to compare the task difficulties among the districts, and the differences are not favorable to the correct reflection of development in indemnification housing construction tasks. To solve this problem, the comparison of task assignments in 2011–2014 in districts may be indicative. As all the districts during certain period would have relatively stable indexes, a coefficient δ is introduced which reflects the comparison between the current year construction task and that one in the previous year.

$$\delta = X_t/X_{t-1} \quad (66.2)$$

Xt the indemnification housing construction task volumes in the districts (of which t = 2012…2014)

The larger δ value shows the larger task pressure in the district level. For details refer to Table 66.3.

Table 66.3 shows that the pressure of completing the construction task in different districts are quite different. Yuexiu, Haizhu, Baiyun, Luogang and Zengchen have the value of $\delta \geq 1$, which shows the larger pressure, and it will directly cause the tension relationship between the city and the districts. Therefore,

Table 66.3 Changes of district task volumes in 2012–2014

District	Yuexiu	Haizhu	Liwan	Tianhe	Baiyun	Panyu	Huadu	Huangpu	Luogang	Nansha	Conghua	Zengcheng
2012	0.4	0.33	0.03	0.33	0.16	0.25	0.27	0.33	0.32	0.91	0.73	0.59
2013	0.5	0.4	3	0.3	0.5	0.5	0.66	0.60	0.5	0.83	0.55	0.33
2014	1	1.5	0.67	0.33	2	0.8	0.60	0.66	1	0.51	0.75	4

Table 66.4 Basic indexes of districts in 2011

District	Task volume	Indemnification housing applied volume	Personnel setup	Financial income (×10 k)
Yuexiu	500	11,354	9	389,459
Haizhu	1500	3431	11	355,403
Liwan	3000	6999	2	368,868
Tianhe	1500	1700	6	422,968
Baiyun	1300	1153	2	392,671
Panyu	4000	245	8	708,448
Huadu	1800	47	1	593,931
Huangpu	1500	148	1	131,334
Luogang	6100	50	1	996,043
Nansha	2300	366	4	312,543
Conghua	1000	64	9	236,285
Zengcheng	2457	251	8	466,573

Table 66.5 Basic indexes of districts in 2012

District	Task volume	Task pcts %)	Indemnification housing applied volume	Personnel setup	Financial income (×100 million)	Completion ratio in 2011 (%)
Yuexiu	200	0.55	16,992	9	43.26	190
Haizhu	500	1.38	3106	11	37.8	100
Liwan	100	0.28	1490	2	39.99	262
Tianhe	500	1.38	502	6	49.16	113
Baiyun	200	0.55	771	2	44.5	133
Panyu	1000	2.76	75	8	79.96	100
Huadu	500	1.38	101	1	57.51	100
Huang pu	500	1.38	172	1	12.19	100
Luo gang	2000	5.52	30	1	109.89	100
Nansha	2100	5.8	259	4	109.89	100
Cong hua	730	2.02	34	9	46.73	136
Zeng cheng	1500	4.14	109	8	26.18	149

considering the actual difficulties of the districts, only reasonably assigning the indemnification housing construction tasks may guarantee the long running and sustainability of the indemnification housing construction.

Table 66.6 Basic indexes of districts in 2013

District	Task volume	Task pcts (%)	Indemnification housing applied volume	Personnel setup	Financial income (×100 million)	Completion ratio in 2012 (%)
Yuexiu	100	0.6	13,924	9	46.88	100.0
Haizhu	200	1.2	2745	11	46.63	100.0
Liwan	300	1.79	6805	2	40.76	138.0
Tianhe	300	1.79	1425	28	58.18	100.4
Baiyun	100	0.6	849	10	51.95	218.8
Panyu	500	2.99	322	12	72.98	112.4
Huadu	336	2.01	38	1	66.43	103.8
Huangpu	300	1.79	281	2	17.15	103.5
Luogang	1000	5.98	20	1	109.81	74.6
Nansha	1733	10.35	75	4	52.58	101.5
Conghua	400	2.39	26	10	29.67	111.8
Zengcheng	500	2.99	40	13	62.99	108.5

Note The data in Tables 66.4, 66.5 and 66.6 are from 2012–2013 "Guangzhou City Statistic Yearbook" and actually reported data by the districts

66.3.3 Building up the Task Assignment Estimated Model of the Districts

Specific consideration is as follows: the influence mechanism of independent variable indexes to the task assignments with dependent variables would be estimated with some important indexes of each district, the estimates would be made with LSDV, to estimate the task assignment volumes of each district. Following indexes are selected, according to the principles of equal rights and responsibilities, including "financial income", "application volume", "completion of previous year", "personnel setup", etc., such independent variables are all continuous variables, the direct relationship between some key independent variables and tasks assigned with dependent variables, and based on it, the indemnification housing construction task assignment model will be built. The tasks assigned between the city and the districts are decomposed as follows:

$$Y_T = y_C + y_D \tag{66.3}$$

$$\lambda = y_C / y_D \tag{66.4}$$

T total
C city
D district

Based on the data of 2011–2013, it is tried to discuss the causality between the variables, so as to estimate the trends of district task assignments in 2015, based on the data of the districts in the years, without consideration of inter-functioning items and interfering items among the districts, estimating with LSDV. The dependent variables of model A may use the proportion between the task volume and the task, the calculation result shows that the model with task volume as dependent variables is obviously better than the model with dependent variables as task proportion. The model is built as follows:

$$y_{Di} = \alpha x_i + \beta x_2 + \delta x_3 + \gamma x_4 + \varepsilon \quad (66.5)$$

y_D Assigned task volumes of districts;
α Coefficient of housing security application quantity of each district;
β Coefficient of housing security clerk setup quantity of each district;
δ Coefficient of previous year task completion ratio;
γ Coefficient of current annual financial income;
ε Intercept term.

The data are taken into relevant treatment, including treatment of missing values, if the financial income will be taken by logarithm to meet the demand of normality, the result will be as shown in Table 66.7:

F statistics and relative P value from combined inspection of model (a) are separately 3.49 and 0.07, which show that the parameters are obvious wholly on the statistic 10 % level. F statistics and relative *P* value from combined inspection of model (b) are separately 13.45 and 0.0009, showing that the model is relatively obvious. The F statistic relative *P* value (0.0009) is inspected to see whether the fixed effect is obvious, in order to compare model (a) and (b), Hausman inspection method is used. According to the principles of "large value by random and small

Table 66.7 Task sharing model

	Random effect model (a) Assigned task volume	Fixed effect model (b) Assigned task volume
Applied volume	−0.267	0.554
	(−1.42)	−0.71
Clerk setup	12.91	−19.92
	−0.71	(−0.51)
Logarithm of financial income	47.25**	36.53
	−3.18	−2.12
Previous year completion ratio	−141.5	−30.58
	(−0.64)	(−0.21)
Constant	492.9	−164.8
	−1.42	(−0.37)
N	23	23
R2		0.666

Note Relationship between significance level and standard deviation: * $p < 0.05$, ** $p < 0.01$, *** $p < 0.001$

value by fixed", the result shows the P value is 0.12, for this reason fixed effect model (b) is refused and random effect model (a) selected. Model (a) result shows that the influence of financial income to the tasks shared by the city and the district is highly obvious in statistics, each increase of one unit in the financial income, the task assigned volume will be added with about exp (47.25) ≈ 29, other variables are not obvious in statistics, and are available according to model (a).

The housing tasks assigned total volume of the district layer is:

$$y_D = y_C/\lambda$$

$$y_{Di} = -0.267x_i + 12.9x_2 - 141.5x_3 + 29x_4 + \varepsilon \quad (66.6)$$

Once the total volume of task of Guangzhou is determined by the "fixed proportion" between the city and the districts, the total task volumes of city and district layers may be calculated. By formula (66.6) the specific task volume of the specific district may be calculated, and the value of ε (intercept) depends on the total task volume of the district layer.

66.4 Model Analysis and Suggestions

66.4.1 Model Analysis

We can see that the model above only reflects the obvious positive relationship between the annual task volume and the financial incomes of the districts, but there is no obvious relationship with the annual application volume. From this point of view, the model's explanatory power is relatively limited, and further discussion is needed.

On the other hand, a sound assignment model bases on certain preconditions such as equal rights and responsibilities, having right of resources distributions, multiple housing security products (including rental allowance). It is feasible only when above conditions are available. At the same time, it should be realized that there is no unified standard of social housing construction task assignment proportion between the city and the districts. The assignments of construction task between the city and districts are, in fact, the transfer of public affairs and responsibilities among different governmental management layers.

66.4.2 Relevant Suggestions

Based on literature and empirical studies, combining with the actual situation in Guangzhou, some suggests are raised as follow. Firstly, municipal government may provide different incentive measures for different kinds of districts, for example, raising the rental allowance standard for poor in old downtown districts, such as

Yuexiu, Liwan, whereas providing supportive policies to new districts to help them fulfilling the construction tasks. Secondly, municipal government should refine housing security tasks by distinguishing different demands, and reasonably determine the city and district sharing proportions. Thirdly, learning the experience of Qingdao, the authority may introduce the quota trades mechanism of indemnification housing to renew the concept of "sharing". At last, not only city level but also districts level should strengthen inter-department coordination, especially reinforce the construction of grass-root organizations.

66.5 Conclusion

The current practice of the indemnification housing construction task shared by the city and the districts in Guangzhou has obtained certain achievements, but there are still many difficulties and obstacles. In comparing with different models, the random effect regression model of indemnification housing construction task assignment was chosen. It provides important foundation for reference in the refining of the social housing construction task assignment between the city and districts layers for Guangzhou and the cities of same type. Some suggestions such as adjusting the limited resources and refining the original rules for rights and responsibilities in different administration layers are provided as well.

Acknowledgements This research is supported by Guangdong Province Natural Science Fund Project(S2013010013533); Province Philosophy and Society Science Planned Project (GD13CGL04)

References

1. Li YS (2013) Indemnification housing construction and social management renovation. J Hebei 32(4):111–117
2. Chen HS, Li ZG (2014) Construction and evaluation of Guangzhou housing and indemnification housing. J Trop Geogr 34(6):823–830
3. Liao HY (2012) Public-rental housing land supply and financing mode case study. J Econ Rev 3:77–80
4. Priemus H (2003) Dutch housing associations: current development and debates. J Hous Stud 18:327–351
5. Engel E, Fischer R (2012) The basic public finance of public-private-partnerships. J Eur Econ Assoc 11(1):83–111
6. Fabozzi J (2001) The handbook of fixed income securities. McGraw Hill, Monograph of New York
7. Ba SS, Wang ZF (2010) Funding sources, system reform and international experience: from public-rental housing. J Reform 3:80–84
8. Guo MJ, Wang Y (2011) Study of modes of financing in the city indemnification housing construction. J Financ Study 11:36–39

9. Yang H (2013) Discussion of financing modes for Chinese indemnification housing—with reference of Japanese experience. J Financ Study 10:77–80
10. Wang JQ (2014) International experience and China's road—non-profit seeking organization participating construction and management of indemnification housing. J Adm Forum 6:78–83
11. He YB (2010) Experience reference for indemnification housing and the choices of development modes in China. J Econ Issue Discov 6:164–170
12. Xu DY, Liu XG (2012) Current existing problems and influence factor analysis for indemnification housing land planning. City Issues 2:97–101
13. Cheng DT (2013) Research of motive and mechanism of indemnification housing construction by local government under dual system for housing lands. J Zhejiang 4:149–157
14. Wang GX (2013) Financial decentralization incentive and land finance—internal logic and its adjustment of the indemnification housing. J Cent Finan Univ 95:1–5
15. Wang YP (2012) On the system restraint to the changes of local government behaviors—with indemnification housing as the policy behavior observation area. J Explor 4:63–65
16. Yan R (2014) Indemnification housing construction—local government behavior logic. J Modern Econ Explor 10:13–17
17. Sun SJ, Sun J (2013) Completing social indemnification housing system in China—from and angle of the government. J China Adm Manage 10:57–60
18. Wang LZ, Li F, Chang HH (2014) Government coordinating behavior study in the indemnification housing policy in China. J China Adm Manage 2:102–106

Chapter 67
Study on the Foreign Investment of Chinese Real Estate Enterprises According to the Eclectic Paradigm

Liangliang Yuan and Yousong Wang

Abstract With rapid economic growth in the process of reform and opening up, China's real estate has become one of the pillar industries of China's national economy, but on the other hand, at present, because of the domestic regulation on real estate industry, Chinese real estate enterprises need to bear more pressures and risks than ever: the pressures of 1 and resources and bank credit, and business risks. According to the Eclectic Paradigm, this paper analyzed the present situation of foreign investment in China and characteristics, in addition it put forward related suggestions in two aspects on how to control the risk and improve the core competitiveness of the enterprise. The discussion of the paper will offer a reference for the foreign investment of Chinese real estate enterprises.

Keywords Real estate enterprises · Internationalization · Foreign investment · Eclectic paradigm

67.1 Introduction

In recent years, the growth of the global economy has slowed down. The developing countries have become the main power to drive a new round of world economic recovery, and the economic growth in many countries is unbalanced [1]. In the process of economic globalization, Chinese real estate enterprises start to seek the investment district more favorable to them through foreign investment [2]. A report from Jones Lang LaSalle showed that the foreign investment of Chinese real estate enterprises surged in 2014, more than $165 billion and that it was the first time for it to surpass the domestic investment [3]. Under the background that

L.L. Yuan (✉) · Y. Wang
School of Civil and Transportation, South China University of Technology, Guangzhou, China
e-mail: yuanliangliang@gzmtr.com

L.L. Yuan
Construction Department, Guangzhou Metro Corporation, Guangzhou, China

Y. Wu et al. (eds.), *Proceedings of the 20th International Symposium on Advancement of Construction Management and Real Estate*,
DOI 10.1007/978-981-10-0855-9_67

our country tightens the regulation of real estate market, more and more domestic real estate enterprises have paid more attention to foreign investment. In many countries, the requirements of foreign-funded companies in the aspect of policy are very demanding out of trade protection. However, there are many problems in the foreign investment of Chinese real estate enterprises, such as, low purchasing power of overseas market, the lack of cost advantages, insufficient experiences in international real estate development, low capital operation ability and low financing level. Real estate is featured by fixed location, long-term service life, long development cycle and great cost, which makes the international operation of real estate different from other industries. Few of enterprises producing general consumer goods succeeded in foreign investment, let alone real estate enterprises. The foreign investment of Chinese real estate enterprises is still at the exploring stage. We should pay enough attention to the risk in foreign investment of Chinese real estate enterprises for it is the industry with the biggest risk. Foreign investment should be treated with caution. Therefore, this paper further analyzes the problems of the foreign investment of Chinese real estate enterprises according to the Eclectic Paradigm and the research conclusions has positive influences on promoting the internationalization development of Chinese real estate.

67.2 Literature Review

Domestic and foreign scholars have analyzed enterprises' foreign investment from different angles. There are many foreign representative studies. For example, American scholar, Hymber [4] illustrated the motivation of multinational enterprises' outward foreign direct investment through monopolistic competition theory in western micro economics and proposed Monopolistic Advantage Theory [4]; British scholars, Buckley and Casson [5] illustrated the motivation of international direct investment and proposed The Internalization Theory; professor from America Harvard University, Vernon [6] proposed Product Life Cycle Theory, which explains the motivation, opportunity and locational choice of American enterprises' postwar foreign investment; Dunning [7] proposed Eclectic Paradigm and it synthesizes Hymber's Monopolistic Advantage Theory, Buckley and Casson's The Internalization Theory and Dunning's Area Superiority Theory [6]. Dunning's [7] Eclectic Paradigm holds that for multinational enterprises wanting to be engaged in foreign investment, ownership advantage, regional advantage and internalization advantage are three basic factors that they should have. The ownership specific advantage includes technological advantage, scale advantage, organization and management ability advantage and financial advantage of enterprises, which are superior to that of enterprises in countries where the foreign investment is made. The internalization advantage refers to the advantages caused by the internalization use of assets owned by multinational enterprises. The regional advantage means multinational enterprises' advantage in location choice after they have ownership specific advantage and internalization advantage, that is, some area has object advantage or preferred location in the aspect of economic development. The Eclectic Paradigm is appropriate not only for the developed countries but also for the developing countries. For example, Malaysian scholars, ABDUL-AZIZ and AWIL [8] studied the international development of Malaysian real estate enterprises.

Compared with foreign research, China's domestic research has just started. Chinese scholars believe that the research emphasis on foreign investment should be put on how to develop enterprises into multinational enterprises. The specific research content is divided into 3 aspects of the choice of international road, comparison of entrance mode into foreign market and international business strategy [9]. After analyzing main investment theories, Chen [10] chose the Eclectic Paradigm of International Production to analyze the strategic choice of Chinese enterprises' foreign investment. Based on the framework of Eclectic Paradigm of International Production, Shao et al. [11] established the index system of foreign investment projects from the perspective of ownership advantage, internalization advantage and location advantage, and analyzed and evaluated the investment risk of foreign direct investment through multi-level fuzzy comprehensive evaluation method. To some extent, these researches can explain and guide the traditional foreign investment of multinational enterprises. However, there is bigger risk in foreign direct investment of Chinese real estate, and domestic research is limited. Therefore, combining our special environment and development characteristics of real estate enterprises, this paper makes a deep analysis of foreign investment of Chinese real estate according to the Eclectic Paradigm.

67.3 Research Method

In order to get more valuable information, based on the existing domestic and foreign theoretical literature, this paper adopted structured interview. Fifty managers from Country Garden Holdings Company Limited and Wanda were selected for interview, including 7 senior managers, 15 middle managers and 28 junior managers. The structured interview was adopted to collect information. The questions included the qualified investment conditions of foreign real estate market, foreign real estate market investment decision-making, specific investment projects, investment risk, development and construction problems and investing and financing model.

67.4 Analysis of Foreign Investment of Chinese Real Estate

The determinant of foreign investment of Chinese real estate is explained according to the interview conclusion and the Eclectic Paradigm.

67.4.1 Ownership Advantage

The ownership advantage is one of the three key elements in the Eclectic Paradigm, which explains the reason for enterprises to make foreign investment. The ownership advantage is the precondition of foreign investment. Real estate enterprises that make foreign investment should analyze their advantages at first. Then, they should further consider investment strategy and investment destination. Whether the real estate enterprises should make foreign investment or not depends on whether they have ownership advantage. To know their advantages, real estate enterprises must have a precise market positioning and suitable development strategy. Whether and when enterprises should make foreign investment must conform to the development strategy and their characteristics. In addition, they should clearly know their customers and service object. Moreover, they should know policies, exchange rates, taxes related to real estate in different countries. Take tax as an example. The tax policy in Korea is very preferred. Reinvestment can be deducted after paying income tax. Policies in different countries are different. Of course, high tax rate should also be taken into consideration. Finally, they should make good use of professional service organizations in overseas market where division is thinning and systematic. In developed countries, there are many professional core organizations, which can help you to raise funds, lock customer base and market sales. You can get accurate report by paying them a little money and they are basically credible. Despite all this, enterprises should analyze the effectiveness of the report. If the real estate enterprise has his advantages, whether internal or external advantages, it can consider making overseas direct investment. Then, it should determine operating mode and investment strategy according to the locational factor of host country. In addition their own advantages, enterprises can also borrow other advantages.

In recent years, domestic real estate market develops rapidly, and the overall management level of real estate enterprises improves. Many famous domestic real estate enterprises, such as Vanke, Wanda and Country Garden Holdings Company Limited are competitive in the world. These mature real estate enterprises have strong real estate development capacity. The development of domestic real estate enterprises and the increased overall strength is a powerful guarantee of successful foreign investment.

67.4.2 Location Advantage

Through the analysis of the ownership advantage, enterprises have preliminarily known foreign investment strategy and operating mode, and they still need to investigate the location advantage that has significant influence on foreign investment. Location factors mainly refer to the factors influencing investment way and cost in host country, such as human geography, customs, resources, labor costs, market demand, investment policies and other factors. Real estate enterprises should choose areas where they can give full play to enterprise advantage and allocate resource efficiently. Enterprises must make deep investigation of each area and market in host country, and determine investment area after making detailed feasibility study and analysis. Only by selecting suitable investment area according to their advantages and investment strategy, can the foreign investment of real estate enterprises succeed. The location advantage in host country plays a decisive role in the foreign investment of Chinese real estate enterprises. For example, the Golden Bay Project developed by Country Garden Holdings Company Limited takes a full consideration of location advantage. It takes only 3 h to go to Kuala Lumpur from Guangzhou by airplane, which makes more and more Chinese regard Malaysia as new investment hot spot. According to the interview, although the politics, economy and cultural environment in Malaysia are total different from that in China, Country Garden Holdings Company Limited have copied the domestic rapid development strategy. It took only 4 months for it from applying for the project to opening the demonstration plot. The sales rate reached 60 % within 7 months since it was assessable to the land. Chinese speed brought by Country Garden Holdings Company Limited shocks the local people. The freehold, unlimited purchasing, low down payment and ten-year residence visa of Country Garden Holdings Company Limited give a deep impression on countrymen.

67.4.3 Internalization Advantage

After making investment decisions according to ownership advantage and location advantage, enterprises should also consider whether they have the ability to allocate resource in host country efficiently, that is, whether they can obtain higher profit through internalization advantage and whether they can deal with possible problems and risks. Since each country has different customs, legal risks and language environments, enterprises should make full preparation to deal with inevitable conflicts, otherwise, they can not achieve expected results and will suffer losses, and the market competitiveness will be influenced. Before making foreign investment, they must make a detailed survey, predict the possible problems after investment, weigh risk and earnings, look for solutions and implement investment plan after full preparation to minimize the investment risk.

67.5 Discussion and Recommendation

With the gradual deepening of Chinese real estate enterprises layout on a global scale, the analysis of and response to foreign investment risk of Chinese real estate enterprises start to attract more attention. Many ministries in our country introduce files to encourage domestic enterprise to make foreign investment to express their support. This paper discusses the measures and recommendation to promote the international development of Chinese real estate enterprise in two aspects of improving their competitive power and risk control in investment process.

67.5.1 Improve the Competitive Power of Real Estate Enterprise

Through questionnaire survey and data analysis, Abdul-Rashid ABDUL-AZIZ and Roslinda ALI discuss and study the entrepreneurship of Malaysian and adventive (Chinese) real estate developer from the perspective of demographic statistics, and they draw the conclusion that the most striking divergence between local and adventive entrepreneurship lies in innovativeness, perspectiveness and risk propensity and that the common characteristics include possessiveness, internal control, independence and tolerance [8]. In making foreign investment, real estate entrepreneurs should be innovative when they consider capital, human and technology, and accurately position the investment demand related to land and real estate development. Based on above viewpoints, the overseas market development of Chinese real estate enterprises should generally follow 8 steps, including seeking and selecting investment opportunity, refining investment plan, feasibility research, contract negotiation, official signature of cooperation agreement, construction, putting into use after completion and real estate assets management. If real estate enterprises want to survive and develop in overseas investment, they must pay attention to the changes in enterprises themselves and external environment, and find a strategic link in the value chain to foster the core competitiveness.

67.5.2 Strengthen the Management of Foreign Investment

In order to effectively reduce the investment risk, the investment environment should be analyzed at first to avoid political and legal risk. Investment environments include politics, economic development level, the level of friendliness to our country, the foreign exchange control and the industry investment restrictions. The politic risk in foreign investment mainly depends on the diplomatic relations between two countries and the supporting level of officials in investment country. It is important for Chinese real estate enterprises to select right investment destination. The real estate is localized business after all. The real estate enterprises that make foreign investment must make full preparation in the aspects of policy, law, market demand, financing environment, political system and customs. For them, local policies and regulations are the primary source of risk. The changes in policies may influence the schedule of project and further arrangement of investment plan. The real estate enterprises should treat these changes as normal thing and solve them by adjustment instead of avoiding them.

67.5.3 *Increase Investing and Financing Method of Overseas Market*

Firstly, enterprises should train and recruit team members who know Chinese enterprise needs and understand the rules of international game. These people can play the role of intermediary organization, especially legal adviser, and determine financing channel according to the investigation of business and legal environment of the host country, making full preparation for foreign investment. Secondly, Chinese real estate enterprises should build a financing platform suitable for both domestic and foreign financing, increase investing and financing manner and ensure the source of funds. As a capital-intensive industry, the real estate is strongly dependent on finance. The key to solve financial risk lies in the full development of real estate capital market. For Chinese real estate enterprises, it is necessary to realize double insurance in financing channel. The double-platform suitable for both domestic and foreign financing channel can ensure the source of funds and avoid the risk caused by single domestic or foreign financing channel.

67.5.4 *Look for Appropriate Partners*

At present, the investment modes of foreign projects developed by Chinese developers are different, because the domestic and foreign developments are totally different and each country has its own characteristics. In the initial period of foreign investment, it is better for Chinese developers to look for local partners to reduce operating risk. They can well know local laws, construction, tax, market situation and the social relations through their partners, which can help them to avoid detours and mistakes. They can form a development group having the advantages of information, technology, management, talent and capital by establishing the joint venture company to improve the success rate of project. According to this interview, we know that Vanke cooperated with Tishman Speyer, the local large-scale real estate developer when it developed project in San Francisco. In this cooperation, the marketing rights are still possessed by the local developer, so Vanke could not determine the sales method and housing price. Sales and price should be determined through consultation and agreement of the parts involved, but the local developer has greater decision-making power.

67.6 Conclusion

This paper analyzes the ownership advantage, location advantage and internalization advantage of the foreign investment of Chinese real estate enterprises according to the Eclectic Paradigm and puts forward the related suggestion on risk control, enterprise competitiveness and investment and financing method. The research results show that when the domestic enterprises prepare for foreign investment, they must have a clear understanding of overseas market and themselves to make full preparation, realize and avoid possible risk in advance, and escort for foreign investment by evaluating local political and economic situation, cooperating with local enterprises and purchasing insurance. The conclusion in this paper is of great reference value for the foreign investment of Chinese and other developing countries' real estate enterprises.

References

1. Shu JX (2010) The golden times for expanding international market. Chinasoe 9:38–39
2. Wang JY (2010) The operation environmental analysis and prospect forecast of current real estate market of our country. Mod Econ Res 4:9–13
3. He XY (2014) Parsing the investment direction of Chinese enterprise overseas real estate. China Nat Conditions Strength 6:55–57
4. Hymber SH (1970) The multinational corporation and the law of uneven development. In Bhagwati JN (ed) Economics and world order. Macmillan, London, pp 113–140
5. Buckley PJ, Casson M (1976) The future of the multinational enterprise. Holmes and Meier Publishers, NY
6. Vernon R (1966) International Investment and International Trade in the Product Cycle. Int Executive 8(4):16–16
7. Dunning JH (1988) Explaining international production. Unwin Hyman, London
8. Abdul-Aziz AR, Awil AU (2010) Examining the internationalisation of Malaysian housing developers using the eclectic paradigm. Int J Constr Manage 10(4):75–99
9. Xiong W, Xiong Y (2008) The comprehensive influences of the system on FDI—on the basis of amended international production compromise theory. Reformation Strategy 23(6):17–20
10. Chen RA (2007) The eclectic paradigm and strategic choice of overseas investment by Chinese compan. Prod Res 8:96–98
11. Shao YG, Guo X, Yang ND (2008) The investment risk study for foreign dired investment based on eclectic theory. Soft Sci 22(9):41–45

Chapter 68
Research on the Evaluation of Indemnificatory Housing Residential Environment—in the Case of Guangzhou

Fan Wu, Xiaoya Su, Suning Zhai, Mingjie Dai, Wenyao Huang, Kaihang Lin and Yanhong Zhu

Abstract The development of indemnificatory housing is the important policy to improve people's living conditions, and promote the harmonious society. However, there are numerous residential environment problems in the current indemnificatory housing in China. Therefore, analyzing and evaluating these problems are very important to promote indemnificatory housing development. This paper studies the evaluation methods of indemnificatory housing environment in recent years and structures the evaluation models. Combining the features of indemnificatory housing residential environment evaluation, this study selects Analytic Hierarchy Process (AHP) to construct residential environment evaluation index hierarchy and determine the index weight through questionnaire survey in Guangzhou, based on household demands. This research also elaborates the score mechanism of the target neighborhoods by using geographic information system (GIS) and expert ratings method, and then establishes indemnificatory housing residential environment evaluation model.

Keywords Indemnificatory housing · Residential environment · AHP · GIS

68.1 The Indemnificatory Housing Residential Environment

As a special housing for low-income groups and family with financial difficulties, offered by country or region, indemnificatory housing has the characteristics of welfare, guarantee object specificity, quasi-public goods, government-led [1].

F. Wu (✉)
Department of Construction Management, School of Civil Engineering and Transportation, South China University of Technology, Guangzhou, China
e-mail: ctfanwu@scut.edu.cn

X. Su · S. Zhai · M. Dai · W. Huang · K. Lin · Y. Zhu
School of Civil Engineering and Transportation, South China University of Technology, Guangzhou, China

Y. Wu et al. (eds.), *Proceedings of the 20th International Symposium on Advancement of Construction Management and Real Estate*,
DOI 10.1007/978-981-10-0855-9_68

Since the reasons that the economic development in China is unbalanced, the supply and management system of indemnificatory housing is immature, and indemnificatory housing has some disadvantages itself, the indemnificatory housing is showing a tendency of 'remote, centralized, large-scale', leading to a series of problems [2]. Taking Guangzhou as a case, the environmental issues of indemnificatory housing are mainly as follows:

(1) Indemnificatory housing in Guangzhou mostly located in Panyu, Huadu, Huangpu, Baiyun and other suburbs where traffic facilities and public resources are poor. This situation will not only limit the improvement of residents' life quality, increase the problem of unfair distribution of medical and educational resources, but also widen the economical gap between the rich and the poor, easily lead to residential segregation and intergeneration transmission of poverty [3].
(2) Due to the features of indemnificatory housing, the government often directly run the construction or contract to general contractors as construction enterprises, which, however, not only lead to an very possible crisis of construction funds and a slow pace of the promotion of indemnificatory housing construction, but can also cause the shortage of green landscape and entertainment facilities [4].

68.2 Research Methods

Based on relevant literature on residential environment of indemnificatory housing, the common research methods used by scholars are summarized in Table 68.1.

It should be noted that there is no single method to make perfect evaluation of indemnificatory housings residential environment up to now, as the evaluation involves numerous impact factors. This paper tries to make best use of the methods stated above and bypass the disadvantages, to study the impact factors of indemnificatory housing residential environment.

68.3 The Development of Evaluation Index System

68.3.1 The Determination of Research Methods

This research is about establishing indemnificatory housing assessment model, and analyzing the objective environment. The biggest difficulty is to compare social factors (soft environment) which are hard to quantify including sense of security and friendliness of neighborhood relations with physical factors (hard environment) such as traffic network and infrastructure. Therefore, we use the Analytic Hierarchy

Table 68.1 Research methods summarization

Literature analysis (document regression method)	On the basis of collecting and reading literatures in China and abroad, this method summarizes the theory and method of residential evaluation and the theory, which provides theoretical basis and technical support for the evaluation research in the article	
Questionnaire survey	A field research method. By making questionnaires pointing at the key points of tasks, some data, like residents' satisfaction and detail opinions of residential area built environment, which meets the reality and provides analysis basis for following data transacting and analyzing is gained	
Mathematical statistics method	AHP	AHP makes the evaluation model hierarchical. Each hierarchy is made of different factors, and combines in accordance with their correlations and affiliations, and there comes the evaluation model
	Structural equation model	SEM is a family of statistical methods designed to test a conceptual or theoretical model. Some common SEM methods include confirmatory factor analysis, path analysis, and latent growth modeling. SEM has been widely used in fields like tourism geography and traffic research, etc.
	GIS	Combined with other mathematical analysis, get final evaluation results by grid computing on GIS platform
	Principal component analysis	PCA is a branch of multivariate statistical analysis. By means of orthogonal transformation, former random vectors are transferred into new random vectors whose components are irrelevant. Variance measures information and reduces new random vectors' dimension. By making appropriate value function, the system will get further transformation
Comprehensive evaluation method	Data envelopment analysis	The DEA, a model named by the creator's name—CR model, not only evaluates and sorts relative efficiency of all decision making units of the same type, but it can further more analyze the reasons why each decision making unit is non-DE valid and bring ways to improve it, and therefore, important information for decision are taken to the decision maker
	Fuzzy evaluation	Fuzzy evaluation is on the basis of fuzzy mathematics. On one hand, it can evaluate and sort objects according to comprehensive score. On the other hand, it can rank objects under the maximum subordination principle according to the value on fuzzy evaluation set

Table 68.2 The hierarchy model of the residential environment evaluation impact factors

Level 1	A. Residential environment evaluation model of the indemnificatory housing			
Level 2	B1 community facilities	B2 housing and community	B3 traffic and location	B4 community atmosphere
Level 3	C11 medical and hygienic facilities	C21 housing quality	C31 public transport	C41 security
	C12 education facilities	C22 environmental quality	C32 private transport	C42 community attachment
	C13 retail		C33 downtown distance	C43 neighborhood
	C14 recreation	C23 property management	C34 commuting	
	C15 landscape		C35 traffic cost	C44 density

Process (AHP) to quantify the social factors and then calculate the impact of each factor on Guangzhou security room residential environment, and finally establishment a relatively complete evaluation system.

68.3.2 Determination Evaluation Index and Hierarchicalization

The Residential Environment Evaluation Model of the Indemnificatory housing (REEMI) is developed according to the relationship between the impact factors. The evaluation model is also based on the general thinking of the indemnificatory housing living conditions of hard and soft environment and its characteristics, and the work and living needs of the residents. The hierarchy model is as follows (Table 68.2).

The evaluation model consists of three levels: A, purpose level; B, category level; C, factor level. The category level includes Community Facilities, Housing and Community Quality, Traffic and Location, and Community Atmosphere. The contents of the impact factors are stated as follow:

B1 Community Facilities

C11 the index ‘Medical and Hygienic Facilities’ is used to measure the sufficiency of surrounding hospitals, clinics and other hygienic facilities. C12 the index ‘Education Facilities’ is used to measure the sufficiency of the surrounding educational facilities and institutions, such as kindergartens, primary schools, secondary schools, libraries, etc. C13 the index ‘Retail’ is used to measure the sufficiency of surrounding shopping and dining stores for living convenience, such as shopping mall, supermarket, fresh market, fast food shops, restaurants, etc. C14 the index 'Recreation' is used to measure the sufficiency of recreational facilities nearby the community, such as playground, KTV, gym, theater, stadium, etc. C15 the index 'Landscape' is used to measure the sufficiency of surrounding open green spaces and other landscaping amenities, such as garden, squares, parks, greenway, etc.

B2 Housing and Community Quality

C21 the index 'Housing Quality' is used to measure the quality of the housing. A high housing quality means the residents in this apartment have good living experience and comfortable staying. This index could be judged by the housing durability, appearance, ventilation and lighting, heat insulation, etc. C22 Environmental Quality index is used to measure the environmental conditions supplied by housing project community. This index could be judged by the air quality, noise levels, green rate, surrounding pollution, etc. C23 Property management index is used to measure the management quality of a housing project. This index could be judged by the building and facility maintenance, community clean, community security, etc.

B3 Traffic and Location

C31 Public Transport Network index is used to measure the convenience for residents to make use of public transport travel. C32 Private Transport Network index is used to measure convenience for the residents to make use of private car travel to accessing roads and parking. C33 Downtown Distance index is used to measure the distance to the city center and the accessibility of housing project. C34 Commuting index is used to measure the convenience for residents commute to work. C35 Traffic Cost index is used to measure the housing residents' cost of daily traffics.

B4 Community Atmosphere

C41 Security index is used to measure the sense of safety felt by the residents in the residential perimeter. C42 Community attachment index is used to measure the intense degree of adoring the community by the residents. C43 Neighborhood index is used to measure the relationship of the neighbors. A harmonious neighborhood gets higher index score. C44 Density index is used to measure the residential density in communities whether to allow residents to feel comfortable.

68.4 Determination of the Index Weight

To reduce excessive interference of subjective factors and to enable researchers to obtain a more scientific and effective evaluation system, this paper adopts AHP evaluation to analyze the weights of indemnificatory housing residential environment, using questionnaire to obtain evaluation data pair wise comparison judgments. By calculating the feature vectors of the judgment matrix, this study calculates the index weights of level B and level C in the evaluation model, and ultimately determines the index weights of the entire evaluation system model. Since it is difficult in actual operation to use the 9 scale method of comparison matrix developed by the AHP, the paper make some improvement to the original AHP which is reduced to 5 scale applications to simplify the difficult judgment.

In order to obtain evaluation index weights which are based on the needs of residents, this study conducts random questionnaires in two representative indemnificatory housing communities, Longgui and Ansha, while the former is a large newly built indemnificatory housing community located in the suburbs and the latter is a mid-sized community located in downtown. In the 55 collected questionnaires, 33 questionnaires are valid. The investigators explained the goal of our survey and the questions with respondents to ensure that respondents are familiar with their residential environment, and can express their views correctly in the questionnaire. The age composition of all respondents is evenly distributed, which reflects the opinion of residents across age groups. According to the requirements of AHP scoring method, we use five-scale method for each index factor pair wise comparison, the judgment matrix and mean weights of each factors of Level 2 and 3 are then generated (Table 68.3).

68.4.1 Analysis of the Environmental Impact Factors

- **Level B categories analysis:**

From the results of Table 68.3 we can find that the Community Facilities, accounting for 44.08 %, are most valued in the eyes of residents, followed by Community Atmosphere (26.37 %), Housing and Community Quality (21.33 %), while the weight of Traffic and location (8.25 %) is lower than its expectation.

It is generally believed that the traffic and location of housing is very important, while there is a saying ‘location, location, location’. However, the result of this survey is opposite to that. Since quite a part of respondents work nearby or being unemployed, they consider the facilities nearby are more important, while they can work, shop, or see a doctor near their community, the demand of traffic network or center location is relatively low. This result indicates that the ‘job-housing’ balance is very necessary in the indemnificatory housing planning. For low-income residents without higher education, job opportunities and living facilities nearby the neighborhood are most needed.

Table 68.3 Mean weights of residential environment factors

Level 1	Level 2	Weight	Level 3	Weight
A residential environment evaluation model of the indemnificatory housing	B1 community facilities	0.4408	C11 medical and hygienic facilities	0.4870
			C12 education facilities	0.1620
			C13 retail	0.1594
			C14 recreation	0.0668
			C15 landscape	0.1248
	B2 housing and community quality	0.2133	C21 housing quality	0.6331
			C22 environmental quality	0.2372
			C23 property management	0.1297
	B3 traffic and location	0.0825	C31 public transport	0.4071
			C32 private transport	0.0600
			C33 downtown distance	0.2096
			C34 commuting	0.2382
			C35 traffic cost	0.0852
	B4 community atmosphere	0.2637	C41 security	0.4749
			C42 community attachment	0.1820
			C43 neighborhood	0.2145
			C44 density	0.1285

It should be noticed that in the comparison of 4 categories of level B in the REEMI, Community Atmosphere is the second important category for the respondents. This result indicates that the security and the sense of belonging are highly weighted by indemnificatory housing residents, while often being underestimated by the developers or governments. The planning and operation of an indemnificatory housing project should consider residents' mental needs more, by offering more public places for communication, better property management, warmer humanistic cares and more harmonious community atmosphere. According to Maslow's hierarchy of needs, Physiological needs are the physical requirements for human survival. With their physical needs relatively satisfied, the individual's safety needs take precedence and dominate behavior. The weights of 4 categories are well correspond to Maslow's theory.

Table 68.4 Weights and order of impact factors in Level C

Order	Factors	Weights
1	Medical and hygienic facilities C11	0.2147
2	Housing quality C21	0.1350
3	Security C41	0.1252
4	Education C12	0.0714
5	Retail C13	0.0703
6	Neighborhood C43	0.0566
7	Landscape C15	0.0550
8	Environmental quality C22	0.0506
9	Community attachment C42	0.0480
10	Density C44	0.0339
11	Public transport C31	0.0336
12	Recreation C14	0.0294
13	Property management C23	0.0277
14	Commuting C34	0.0197
15	Downtown distance C33	0.0173
16	Traffic cost C35	0.0070
17	Private transport C32	0.0050

- **Level C factors analysis:**

The order and weights of all impact factors in level C are shown as follows (Table 68.4).

It's showed that the top three factors most concerned by indemnificatory housing residents are Medical and Hygienic Facilities (21.47 %), Housing Quality (13.50 %) and Security (12.52 %), while the downtown distance, traffic costs and private transport network are relatively insignificant. It could be easily understood that the Medical and Hygienic Facilities is essential for the residents, since the one of the goals of indemnificatory housing policy is to offer shelters for low-income people, such as the elderly and disabled. The convenience to acquire medical care is very important for the residents in demand, so the density of hospital and health service network should be reasonable, the accessibility of medical care should be improved and the quality of health care should be enhanced.

Housing quality is the second important factor concerned by indemnificatory housing residents, which indicates that the well-designed, high quality-built, and orderly operated housing is highly demanded by the residents to guarantee their living quality. However, in recent years, the housing quality problems such as leakage, cracking, lighting and ventilation problems, and noise in indemnificatory housing projects happens occasionally, consequently public confidence in the indemnificatory housing mechanism sags. This result indicates that a life-cycle quality safeguard mechanism should enter the development to enhance and guarantee the living quality of indemnificatory housing residents.

As the third important factor, security is also highly concerned by the residents. To strengthen the sense of safety, the reasonable planning and professional property management including building maintenance and security system are very important. Also, the sublet problem which causing the disordered management should receive attention. In addition, education is also concerned by residents. To avoid the poor extension between generations and enhance the education level of next generation, adequate and quality education institutions and facilities should be allocated near the indemnificatory housing communities. It is often criticized that the lack of nearby educational facilities leads long commute distance and time and high risk while commuting for the children and youths who live in the indemnificatory housing.

Factors associated with soft environment, such as the neighborhood relationship, community attachment and density, are concerned after basic needs about living facilities, housing quality and security, which is identical with Maslow's theory that after the living demands are satisfied, people need to meet their demands of love and belonging. The construction of harmonious community should be attended by the government.

68.4.2 Construction of the Evaluation Model

After the confirmation of the evaluation index and its weights, the evaluation model is obvious. For a target community, the score of physical environment in the categories of Community Facilities, Housing and Community Quality, Traffic and Location is estimated by GIS technology, while the score of social environment is valued by Experts Grading Method (EPM). The corresponding base data layer will be obtained by using ArcGIS software and the GIS data source of the target community could be acquired by thematic maps (bus stops, subway map, maps of public service buildings, etc.) and planning drawings. The score of target community is obtained by accumulating all scores of level B categories, while the score of a category is the summation of every score of level C index under it. The score level C index could be calculated by multiplying the score of an index by its weight. The higher score a community gets the better residential environment it offers. A high score of a category also means a quality environment in this category the community offers.

68.5 Conclusion

This paper reviews the evaluation methods of the indemnificatory housing residential environment in recent years and structures the evaluation models. Combining the features of indemnificatory housing residential environment evaluation, this study selects AHP to construct residential environment evaluation index hierarchy and determine the weight of index through questionnaire survey in Guangzhou, based on household demands. By using AHP, the social factors such as security and community attachment are quantified and evaluated. The weights of 17 impact factors in 4 categories are evaluated by the residents in indemnificatory housing communities in Guangzhou. The survey results indicate that in the 4 categories, the Community Facilities, accounting for 44.08 %, are most valued in the eyes of residents, followed by Community Atmosphere (26.37 %), Housing and Community Quality (21.33 %), while the weight of Traffic and location (8.25 %) is lower than its expectation; in the 17 factors, Medical and Hygienic Facilities (21.47 %), Housing Quality (13.50 %) and Security (12.52 %) are most concerned by indemnificatory housing residents. The results are then analyzed.

As a pilot study, this paper determines the evaluation index and hierarchicalizes them into 3 levels, finds out the weights of index and category, and also elaborates the grading mechanism of the target neighborhoods by using GIS and EPM, and then establishes indemnificatory housing residential environment evaluation model. Exploring the residential environment of indemnificatory housing is significative in real world. The model could be used to evaluate the residential environment of indemnificatory housing and indicate the aspects needed to be improved. This model needs to be further consummated by expansion of survey data and experience of evaluation implementation.

References

1. Dong S (2015) China's construction of indemnificatory housing: status, problems and route selection. Teach Thought Polit Study 5:53–57
2. Zhu D, Zhang J, Wang X (2014) Evaluation indexes and method of influences of affordable housing's residential space differentiation based on GIS and MAS. Mod Urban Res 5:22–26
3. Li Z, Ren Y, Li L (2014) A case study of everyday life within the affordable housing community of Jinshazhou, Guangzhou. Architect J 2:12–16
4. Chen Z, Liu Y, Zhao T, Wang Y (2015) The status quo of China's affordable housing construction funds. Hous Ind 5:20–23

Chapter 69
Research on Demand of Replacing to Public Rental Housing Tenants in Chongqing

Jiyuan Liu, Yi Zhan and Xu Ma

Abstract There are more and more public rental housing residents in Chongqing with the development of public rental housing construction. However, because of marriage, fertility, change in jobs, educational issues, living with old parents, a few public rental housing tenants have to reconsider the objective public rental housing factors according to the variable personal requirements. Those tenants have to replace their public rental housing with new units of different sizes or of different communities to solve their problems. But it is hard to fulfill their exchange renting needs especially changing public rental housing among communities due to current policy. As the result of the situation, public rental housing resources should make a rearrange and reassignment to better meet the needs of public rental housing tenants and to make a contribution to the sustainable development of public rental housing. The data was collected by random questionnaire survey of three public housing communities of Chongqing and bidirectional market matching theory will be used to build models to analyse the problem. Based on the results of the research, suggestions and countermeasures will be put forward to improve the allocation mechanism of public rental housing.

Keywords Public rental housing · Replace the rent · Bidirectional market matching theory

69.1 Introduction

Comfortable and convenient living environment is the basic premise of human's living and development. In recent years, there are more and more large-scale public rental housing construction in Chinese mainland. The main aim is to solve the

J. Liu (✉) · Y. Zhan · X. Ma
School of Chongqing Jiaotong University, Chongqing, China
e-mail: 297312214@qq.com

Y. Wu et al. (eds.), *Proceedings of the 20th International Symposium on Advancement of Construction Management and Real Estate*,
DOI 10.1007/978-981-10-0855-9_69

housing problems of low-income families. According to the 2010 government plan, Chongqing government strive to build 40 million square meter public rental housing in the next three years to solve the housing problems of 2 million people. There have been 39 public rental housing communities which are under construction or completed in Chongqing on the end of September 2014, the number of the public rental housing tenants is around 156 hundred families, about 400 thousand population. As a result, not only do people care about planning problem of public rental housing, but also they pay more attention on the later operation management of public rental housing. Fresh employee who graduate from colleges and vocational schools and employee who move from rural area to the city are the major population of the public rental housing tenants who face many difficulties such as job changes, marriage, giving birth to a baby and living with their parents. Inconvenient living environment will lead to a lot of problems. In order to adapt to the change of living and working conditions, they need to adjust the house type and location of public rental housing. For this reason, they demand to replace the previous public rental housing. however, the policy is imperfect at this stage, public rental housing tenants can only look for the right public rental housing from a large number of public rental houses and system can only screen out only one target which match the requirements of public rental housing tenants, meanwhile, public rental housing tenants will miss many public rental housing information which is close to the target. This is not a scientific and reasonable mechanism which leads a lot of tenants lost their precious chances to replace public rental housing.

This study combine the questionnaire survey method and field interview method. In field investigation stage, the data is obtained through random questionnaire of three Chongqing public rental housing communities named Minxinjiayuan, kangzhuangmeidi, Liangjiangmingju respectively. According to the survey, it can be found that a large number public rental housing tenants are suffer housing problems which cannot be solved timely and effectively. The paper is based on the analysis of Chongqing public rental housing current situation and problems to solve the problem of replacing public rental housing by Shapley and Gale [1–3] market two-way matching theory in order to achieve the public rental housing redistribution efficiently.

69.2 Current Situation and Problems of Public Rental Housing Development in Chongqing

The public rental housing of Chongqing guarantee a wide range of people with low barriers to join, low rent money and good public rental housing location. Beginning in 2010, a large number of urban residents called "sandwich class" [4] moved into public rental housing gradually. Over time, a series of questions came up. According to the survey, the surrounding facilities of public rental housing communities are often relatively backward or even some of the surrounding is undeveloped at all. In this case, a large number of public rental housing residents choose to find employment, go to school, go shopping at someplace near the main urban centers. So far, the first batch of tenants who live in public rental housing have rented for 5 years. Most of them are college graduates and migrant workers. They have to get married or live with their parents or change their job frequently. First, the increase of household population will lead to the decline of per capita living space; if somebody move out from a set of public rental housing due to some work reasons, the rest tenants would share more rent money and it is also a waste of public resources. Secondly, changing jobs will inevitably cause workplace change [5]. It will takes a long time from extremely remote communities to the work site. Especially in the case of changes in the work, this kind of jobs-housing spatial [6] mismatch will cause substantial increase in commuting costs. So in this case a group of tenants require to replace location [7] and size of their public rental housing.

According to Chongqing's public rental housing management system, public rental housing tenants can change flats by their own under the supervision of public rental housing owners and operators. But this approach is not very effective. The reason is that most tenants are not able to master accurate, timely and comprehensive information. After missing the optimal selection, they can not find the similar one to substitute the target houses.

69.3 Explanation of Data from the Questionnaire Survey

69.3.1 Introduction of the Questionnaire Survey

The questionnaire, which last about one and half month including design and data collection, is targeted at three security housing communities that have occupancy rate in main urban area in Chongqing. There are three ways adopted to gather information, which are paper questionnaire, telephone interview and Internet interview question. Ultimately, 491 replies have been collected, which 540 ones are distributed in total. The aim of this questionnaire is to understand how many residents who live in the security housing community want to move out and the reason why they want to move out. In the meantime, this paper also tries to ascertain the cause why it is difficult for them to move out.

69.3.2 Basic Information About the Questionnaire

There are 154, 166, 171 replies collected from communities of Minxinjiayuan [8], Kangzhuangmeidi and Liangjiangminju separately. Through sorting out and analyzing data in depth, some valuable information have became clear. In terms of the intention of moving out, 20.8 % of residents from Minxinjiayuan, 17.5 % from Kangzhuangmeidi and 22.2 % from Liangjiangminju have this kind of will. The rest of the residents, however, are contented with the status quo. In this case, if giving the consideration to the whole main urban area that has 39 security housing communities, the amount could be huge.

Generally, there are three reasons causing householders want to move out the community by analyzing the data based on the questionnaire. First of all, the increase in population of family is the primarily important reason. According to the survey from city council, most of residents living in security house are migrant worker and graduates from technical institute and secondary technical college. From one side, initially, almost migrant workers move into city, leaving their old parents, spouse and young children in their hometown. After a few years, they could earn more money and enjoy a more stable life, which becomes possible to bring their whole families into city and live together. By another side, a part of graduates living in the community has to face the issue about their marriage and the living condition for their baby. It cannot be avoided for both the worker and the graduates that living area is too small to meet every family members' need if they chooses to live in the community continually. As a consequence, the highest priority for them is "Great for Small", which means they need to change the unit with only one bedroom into the one with two bedroom at least. According from the survey, there are 32 leaseholders in Minxinjiayuan that hold the will of change, which 75 % of them with the reason of the increase of population. The figures in Kangzhuangmeihu and Liangjiangminju are 65.5 and 68.4 %. From these data, it could be obvious that the increase of family population is the most important reason causing the need of change house, which should get priority for local government.

Secondly, changes in works also could be the reason why people want to move out. In China, both migrant workers and new graduates are low-income people, compared with those residents that born and live in the city all the time and change their work more frequently. It is true that only 20 % migrant workers have a stable job, which means that 80 % of them experience the situation of job-hopping more often. Similarly, around 70 % students that graduate from university no more than three years are job hopper when they just enter work arena, which because they are risk-takers, like try new things and, in some extent, are lack of patient and long-term plan for their career. Obviously, changes in work can cause changes in work places. In this circumstance, leaseholders have to face the problem of the unbalance between working and living, which can tighten their financial budget because they need to pay more on commutation. Living under this situation for long time, it is harmful them both physically and psychologically. Moreover, it is also quite common for them to share apartment with others. However, some roommates have to move out because of job-hopping, which means person who left has to afford extra rental fee and suffer more pressure that could be avoid. Besides, it also a huge waste for public resource. The main need for this group of people is to change to other community or change to a bigger unit. Based on this survey, 18.8 % of leaseholder that wants to move out their community from Minxinjiayuan holds the

Table 69.1 Survey statistic

Public housing community	Amount of questionnaire	The will of changing house		The percentage of changing housing in total resident (%)
		Yes	No	
Minxinjiayuan	154	32	122	20.8
Kangzhuangmeidi	166	29	137	17.5
Liangjiangmingju	171	38	133	22.2

Table 69.2 The reasons of replacing rent

Reason / Community	Increase of population (%)	Changing job (%)	Other reasons (%)
Minxinjiayuan	75	18.8	6.2
Kangzhuangmeidi	65.5	27.6	6.9
Liangjiangminju	68.4	23.7	7.9

reason of job-hopping. The figures in Kangzhuangmeidi and Liangjiangminju are 27.6 and 23.7 % respectively. It is obviously that this fact cannot be ignored.

At last, according to survey, a small part of leaseholder move out because of other issues, like education, infrastructure and the quality of dwelling house, which occupies 6.2, 6.9 and 7.9 % in Minxinjiayuan, Kangzhuangmeidi and Liangjiangminju separately. These facts also should be taken into account refer to government management (see Tables 69.1, 69.2).

69.4 Using Bidirectional Market Matching Theory to Solve Replacing Rent

69.4.1 The Bidirectional Market Matching Theory

Bidirectional market matching theory was proposed by the Gale and Shapley 1950s, the United States began to arrange work for medical students in what they called "national resident matching project" in this way and achieved success. The greatest significance of the Gale-Shapley algorithm is that as a matchmaker of these men and women, we don't need to calculate stable marriage match and we don't even need to know everyone's preference. It is enough to organize a matchmaking activities for them according to the algorithm. All we have to do is to tell the candidates the algorithm process in the form of rules of the game. Then the game will automatically make everyone satisfied after marriage match. For example, we use the letters a, b to name men, women are named with numbers 1, 2. Assuming that every man prefer woman 1, but woman 1 prefer man b and woman 2 prefer male a. If we match a-1, b-2 respectively, then man b and woman 1 are more like each other. So such marriage collocation is unstable. However, if we change them into a-2, b-1, such collocation is stable. The best solution, of course, is everyone obtaining the best results which are impossible in the real world. So Gale-Shapley algorithm is not to provide every candidate the best matching results but to give participants a stable matching scheme.

69.4.2 The Bidirectional Market Matching Theory Used in the Process of Replacing Rent

There are two market players set W and Y, where $W = \{w_1, w_2,\ldots, w_n\}$ represents n applicant of replacing rent in the collection W; $Y = \{y_1, y_2,\ldots, y_m\}$ indicates m sets of public rental housing in the collection Y and government make decisions on behave of those houses.

$Z(w) = y_1, y_2, \ldots, w, \ldots, y_m$ represents the replacing rent applicants' order of preference. Ranking the top express preference while ranking the bottom means replacing rent applicants would rather give up the application than replacing rent.

$Z(y) = w_1, w_2,\ldots, y, \ldots, w_n$ represents order of preference of house. Ranking the top express preference while ranking the bottom means house would rather be empty than be applied.

$Z = \{Z(w_1), Z(w_2),\ldots, Z(w_n), Z(y_1), Z(y_2),\ldots, Z(y_m)\}$ indicate the preference collection of all the participants, so (W, Y, Z) is the bidirectional matching market.

Each applicant apply for the public rental housing which rank first in their list. The government on behalf of the application object rejects unacceptable applicants and retain the best applicant sorted in a number of applicants. In the next round of application, if there is an applicant who is closer to the top then we should keep the later one and eliminate the previous candidate. And each applicant who is rejected could continue to pick the next application object. Application object repeats the above step. Thus after several cycles of the above steps when there are no replacing rent applicants being refused, the applicants either have finished match or are turned down by the acceptable application object in the end. Of course, it is reasonable for the replacing rent applicants to start with applying for the best apartment. And it is also in the interest of the government to pick the better applicants constantly. Therefore, we will consciously abide by the rules of the game without worrying about someone fakes his preference. The way that the tenants apply for the flats before the government accept or refuse those applications is obviously more beneficial to the applicants according to studies. In fact, there are often more than one stable matching, but the above algorithm can guarantee that every applicant obtains the public rental housing which is the best of all the possible stable matching schemes, while an application object is given to the tenant who is worst of all the possible stable matching schemes.

Of course, there must be a sufficient number of public rental housing available for the applicants to match if the we want to implement above-mentioned scheme effectively. The replacing rent applicant could change his public rental housing directly for another set which belongs to the original community according to the relevant regulations of Chongqing. However, it is extremely difficult if someone has the idea to trade his public rental housing for another set of public rental housing which belongs to other communities. Because the applicants have to participate in the next round of screening. On the one hand, they are easier to master the housing information concerning their own communities. On the other hand, the people who are eager to remove to a new community must experience screening process again which is complicated and time-consuming. As the result, the majority of tenants choose to apply for alternative public rental housing in their original community. Now applicants have artificially divided public rental housing communities into several isolated groups. But if we put them into one big collection of application objects, then the number of stable matching will increase as a geometric series finally. This approach greatly increase the probability and efficiency of successful applicants in rent. It also reduce the waste of social resources caused by idle public rental housing. Existing mechanism of replacing rent emphasize on equity of allocation process. However, this screening process does not fully take the diverse needs of the applicants into account, which is very prone to creating unstable matching, resulting in replacing rent again. The bidirectional market matching theory not only fully take different needs of the applicants into consideration, but also give the government an opportunity that regulator could provide public rental housing for those who are more urgent to change their flats.

69.5 Conclusion

Chinese Public rental housing starts relatively late than some other countries and there are many drawbacks in the later management operations. Tenants cannot meet the replacing rent demand timely. In the view of the current situation, the paper will put forward the suggestions and measures as follows.

69.5.1 Improvement of Replacing System

The government is the main undertaker of Chongqing public rental housing construction and operation management and the replacing system cannot be improved without government support. The government should make rules and laws to support replacing system and to provide better service for the public rental housing tenants. At present, the replacing system is time-consuming and the success rate is low, meanwhile, the public rental housing resources cannot be allocated reasonably, which would require the government to integrate related resources and simplify replacing process to increase efficiency.

69.5.2 Strengthen the Management of Returning Public Housing

There are many irregularities such as cheat, rascal and transfer happened in the operation process of public rental housing, which lead to unfair allocation of public rental housing resources and some public rental housing tenants who suffer difficulty indeed can not meet their requirements timely. On the one hand, the government should ask the public rental housing tenants who have a improvement in economic conditions and is no longer serve for the public rental housing tenants to return the public rental housing and provide they preferential policies when they buy economically affordable housing, price-limited housing or commercial residential building. On the other hand, government should publish the irregularities such as cheat, rascal and transfer severely. For example, at the beginning, the government can increase the rent to stimulate they return the public rental housing as soon as possible. As for some public rental housing tenants who refuse to correct their errors, the government can give them tickets and ask the law-enforcing department force they to return public rental housing.

69.5.3 Being Targeted to Solve Different Replacing Demand

Replacing demand is caused by family population increase, job change, children school-entering. The main house types (about 80 % of total) of Minxinjiayuan, Kangzhuangmeidi, Liangjiangminju are single rooms or one living room apartments. Family population increase is the main reason why public rental housing want to replace their previous public rental housing because of living with parents or marry and give birth to babies. Application objects of most applicants who want to replace their previous public rental housing are two rooms one hall or three rooms one hall which is relatively scarce. This is the main reason why the success rate is low. Therefore, government should gradually increase the percentage of large house types to meet the growing needs of public rental housing tenants. When it comes to the problems of public rental housing such as job changes and children school-entering, the government should speed the infrastructure construction process of public rental housing and improve road transport network for public rental housing tenants. Government should also improve health-care education and economy around the public rental housing area to make public rental housing tenants to have a access to those resources.

69.5.4 Increase the Intensity of Public Housing Construction

No matter how much public rental housing replacing efficiency is improved, the main way to address the root cause is owning enough public rental houses to support the process in order to process smoothly [9, 10]. Although Chongqing have developed rapidly in the five years, there are still a large number of low-income population who is waiting for queue lottery do not live in the public rental housing. Therefore, the government should devote greater effort to the construction and planning of public rental housing under the premise of stability development of other aspects in order to solve burning issues of public rental housing tenants [11, 12]. Certainly, the construction fund from government is quite limited, but the public rental housing which is a public welfare program have a strong demand for the fund. It is not practical to solve all problems by government direct investment. It is a scientific choice to make full use of the funds from society and folk to build public rental housing. The government can consult the stock system reform experience of state-owned enterprises to set up joint venture company which is controlled by government. The government can provide preferential policy to attract funds of all aspects to develop public rental housing.

Certainly, the difficulties of replacing public rental housing can not be solved at once and it require the patience from both the government and public rental housing tenants, with the development of economy, society, replacing public rental housing will eventually be solved one day.

References

1. Gale DL, Shapley S (1962) College admissions and the stability of marriage. Am Math Mon 69:8–15
2. Roth AE, Marilda A, Oliveira S (1990) Two-sided matching: a study in game modeling and analysis. Cambridge University Press, Cambridge
3. Roth AE, Peranson E (1999) The redesign of the matching market for American physicians: some engineering aspects of economic design. Am Econ Rev 89:750–778
4. Tan J, Li M (2012) Research on the demand characteristics of the "Sandwich Class" segments—a case study in Guangzhou. Constr Econ 4:83–87
5. Cai Y, Yan Z, Liu Z (2011) Research on spatial difference and influencing factors of separation of workplace and residence. J geogr 66(2):157–166
6. Li M, Zhou Y, Hu P (2013) Welfare loss study of the jobs-housing spatial mismatch of residents in ensuring housing communites. Urban Dev Stud 20(10):63–68

7. Preston V, Mclafferty S (1999) Spatial mismatch research in the 1990s: progress and potential. Reg Sci 78:397–402
8. Pan Y, Ju X (2012) Applicable study of AHP method on site strategy of public housing—a case of "Minxinjiayuan" in Chongqing. Constr Econ 3:57–60
9. Xu L (2012) Mixed supply model for public rental housing: a strategic for enforcement. Urban Stud 19(8):54–59
10. Zeng D, Wen X (2012) Comparative study of domestic and international operation management system in public rental housing. Urban Stud 19(10):80–85
11. Yang Y (2014) Pushing the housing security system based on public rental house: coming from Germany experience. Urban Dev Stud 21(2):77–82
12. Lv P, Xiu D, Li S (2013) The theoretical and practical significance of mutual property rights mode used in affordable housing system. Urban Dev Stud 20(2):20–24

Chapter 70
Risk Assessment on the Development and Utilization of Government Housing Information Resources

Xiaoqing Cui, Qinlei Li and Yanfang Gao

Abstract Government Housing Information Resources (GHIR) have been accumulating wealth with the development of China's housing market. Further studies on the risk of GHIR development and utilization is needed in order to develop and utilize these resources safely and efficiently. Risk factors are identified according to the method of structures breakdown, in accordance with the lack of practice cases, Bayesian network model for risk factors is established basis on the expert investigation, the influence weight of risk factors is determined by using hierarchical analysis method, the comprehensive value of risk factors is calculated. The three main risk factors are: the imperfect regulation of government information resources development and utilization, the inadequate ability of information development companies, and the absence of a sharing policy for government information.

Keywords Government housing information resource · Risk of development and utilization · Risk assessment · Bayesian network

70.1 Introduction

The definition of Government Housing Information Resources development and utilization (GHID&U) is the collection, processing, storage, exchange, sharing, reprocessing, recreation, and reuse of the housing information resources owned by government [1]. As a systematic engineering process, GHIR development and

X. Cui (✉)
Department of Engineering Management, Shandong Jianzhu University, Jinan, China
e-mail: cxq6706@sdjzu.edu.cn

Q. Li
Qingdao Hisense Real Estate Co., LTD, Qingdao, China

Y. Gao
School of Engineering Management, Shandong Jianzhu University, Jinan, China

Y. Wu et al. (eds.), *Proceedings of the 20th International Symposium on Advancement of Construction Management and Real Estate*,
DOI 10.1007/978-981-10-0855-9_70

utilization maybe affected by uncertainty factors such as information disclosure and sharing, capital, information technology, and government behavior. Consequently, the incremental development and diversified utilization of government housing information resources is difficult to achieve. Adopting targeted risk coping strategies, strengthening the understanding and evaluation of risk factors in science, can improve the enthusiasm of all parties involved in the development and utilization of information resources and the subsequent value of the data information set.

At present, research on the risk to the development and utilization of Government Housing Information Resources is inadequate. Related studies include; research on the mode of government housing information resources industrialization development and utilization by Xiaoqing Cui [2], and research on risk assessment on development and utilization of government housing information resources by Qinlei Li [1], who listed risks to the Government Housing Information Resources development and utilization and built the system of risk control. Other relevant research has mainly focused on the risk to government information resources development and utilization, and the risk to library and information resources development and utilization.

In respect to the security of government information, Hongxia Wang [3] analyzes the risk categories of building an electronic government from three aspects including information security, marketing, and management. Shuai Sun [4] distinguishes E-government risks solely from a network of 5 security layers (physical, network, system, application, and management), and separately lists risk prevention countermeasures. Xianghong Meng [5] proposes 7 key components to E-government information security risk. In the aspect of risk management regarding the government information technology project, Surong Zheng [6], Jimei Xie [7] and Yi Zhou [8] fully researched risk management in relation to the government information technology project to a macro level, summarizing 10 aspects of risk sources; analysis of risk guarantee system comprehensively including security system, unified planning and management of information technology projects, increased resource sharing and integration in longitudinal and transverse dimensions, giving consideration to professional training, speeding up the construction of related laws and regulations systems (rent-seeking punishment, information protection, information property protection and public), pushing bidding rules for system hardware and software, strict control of cost effectiveness according to the risk identification-Risk Management Theory. In the field of library and information resources development and utilization, Hui Pan [9] introduces the concept of risk management in digital library information resources construction projects of school, identifies 6 risk factors including: budget, quality, technology, basic environment, security, and intellectual property rights. Lupeng Du [10] discusses adequately the connotation of construction risk in the development of library information resources, analyzes the details of the risk according to the risk classification standards (macro occurrence area, the specific reasons, characteristics of the micro performance), then designs a risk list and provides reliable risk identification for digital library construction. In addition, Yixin Huang [11] summarizes the risk factors related to information systems software development projects, which mainly comprise scale risk, demand risk, technology risk, management risk, intellectual property risk factors and so on.

In conclusion, research on the risks of government housing information resources development and utilization is inadequate, the risk factors list is simpler, experts surveyed form is unitary, available references focused mainly on qualitative analysis of risk management at the macro level. Therefore, in considering the utilization characteristics of government housing information resources, a risk assessment model is constructed based on Bayesian Networks and Analytic Hierarchy Process. The main risk factors of information resources development and utilization are identified in order to promote the government housing information resources development and utilization.

70.2 Building a Risk Evaluation Model of Government Housing Information Resources Development and Utilization Based on the Bayesian Networks and Analytic Hierarchy Process

70.2.1 The Overview of Bayesian Network Technology

Bayesian Networks (BN) is a network of probability theory, with the characteristics of flexible input and output data, dealing with small samples and incomplete data, using expert advice, which do not depend on historical data. At present, it has become one of the most effective theoretical models in the field of artificial intelligence uncertain knowledge representation and a chain of reasoning. In the fields of financial risk [12, 13], engineering project risk [14], information security risk and risk of software requirements [15], etc., a large number of practical cases and applications are produced. BN consists of a Bayesian network structure and Conditional Probabilities Table (CPT). In the Bayesian network structure, nodes represent random variables, directed edge represents the causal relationship of two variables between nodes, the starting node of a directed edge is the parent node and the other node is the child node. In addition, each child node has a conditional probability table which indicates the causal relationship between the node and its parent node quantitatively, the conditional probability table will be expressed by the prior probability if the node doesn't have a parent node.

BN is based on Bias's theorem, and uses the Bias formula to realize the reasoning process.

$$P(A_i|B) = \frac{P(B|A_i)P(A_i)}{P(B)} = \frac{P(B|A_i)P(A_i)}{\sum_{i=1}^{n} P(B|A_i)P(A_i)} \tag{70.1}$$

where, $P(A_i)$ is the prior probability and $P(A_i|B)$ is the posterior probability.

The prior probabilities are divided into two categories. First, the objective prior probability is calculated accurately from the past historical materials and data. Second, the subjective prior probability is illation by the experts based on knowledge and experience in the absence or lack of historical information [16].

70.2.2 The Building Process of a Bayesian Network Model

In general, the build process of a Bayesian Network Model mainly comprises four steps: confirming the network node from the identification of risk factors→confirming the domain of the network node→building a Bayesian network structure→determining the conditional probability tables→model reasoning.

The method for confirming the network node is that, in accordance with the method of work breakdown structures (WBS), risk can be subdivided into risk types and risk factors. The target risk is expressed by "A", risk types are expressed by "A_i" which are intermediate nodes; risk factors are expressed by "R_i" which are evidence nodes.

The method for confirming the Network node domain is that, establishing two status values for each node by using expert knowledge, represent the risk realization either "yes" or "no", namely the status values set for {1, 0}. The reason for taking two status values is that respondents easily understand and fill in the questionnaire. In addition, if more status and the survey data are not enough, the result of many zero values could be predicted [17].

The method for building a Bayesian Network structure is based on the corresponding relationship of risk factors between the evidence node, the intermediate node, and the target node. Then build the Bayesian network structure, as shown in Fig. 70.1.

The method for determining the conditional probability tables is that, the Bayesian network node conditional probability tables can be confirmed through expert research in the cases of less practical cases or lack of historical data. In order to unite the probability value and possibility of risk, the paper uses the seven description language classification of risk probability [18]. Probability of statements represent the probability of the scope of value takes the median of the probability scope represented by the description language respondents choose [19]. Calculating the weighted average of each risk probability in all questionnaires to generate the risk probability value.

70.2.3 Comprehensive Evaluation Model of Risk Based on the Analytic Hierarchy Process

Comprehensive risk evaluation is based on statistics of risk probability "P", the effective weight "E" of the development and utilization risk determined by analytic

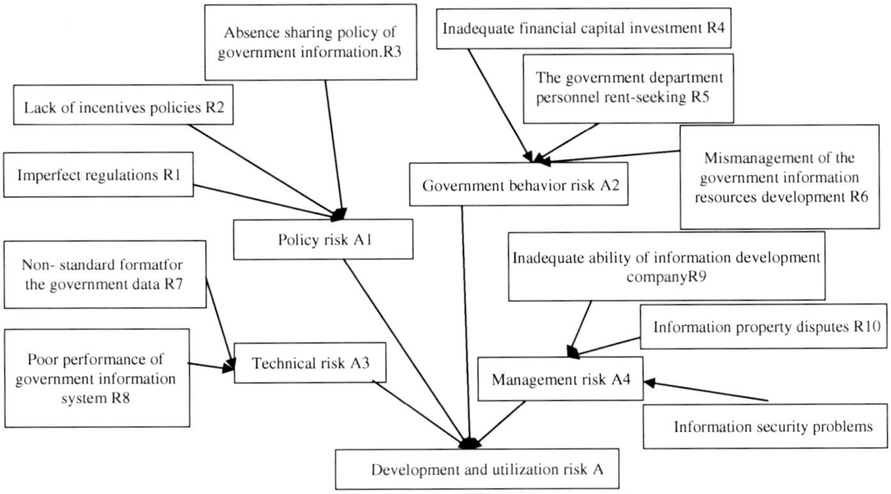

Fig. 70.1 Bayesian network structure of risk assessment on development and utilization of government housing information resources

hierarchy process (AHP), and then based on the definition of risk, $R = P * E$, generates a comprehensive risk value for each risk factor. First, change the Bayesian network structure of government housing information resources development and utilization risk into a hierarchical structure, the target risk is the target layer "*T*", the intermediate layer is composed of the intermediate nodes, the bottom layer "*F*" is made up of various risk factors. Second, construct the comparison matrix, calculate weight vectors and through the combination test, obtain the weight. In the present situation, the higher the risk evaluation value, the greater influence degree of the risk factors [20].

70.3 Risk Assessment of Government Housing Information Resources Development and Utilization

70.3.1 Risk Identification of Government Housing Information Resources Development and Utilization

In the early stage of the study, work stage division and risk factor decomposition are simpler. According to the method WBS and the characteristics of information project development, the activities of the process of government housing information resources development and utilization can be divided into 4 stages comprising 11 main specific work contents. The 4 stages include: positioning and design of information products and services, information resources

Table 70.1 Work table of government housing information resources development and utilization

Work stage	Work content
Positioning and design of information products and services	Survey the user (government, enterprise and individual) requirements, determine the function, content of information products and services, etc.
	Design implementation plan according to the information products and services positioning
Information resources acquisition	Housing related government departments collect basic information data
	The competent government department audit, collect and upload data
Information resources development	Government departments integrate, process and handle government basic information data
	Government departments transfer part of the housing information data to non-governmental organizations
	The company of providing information products and services process and develop information, etc.
Utilization of information resources or the products/service	Government disclose the part of housing information or provide information products and services
	Sharing information resources or information products and services among government departments
	Enterprise supply information products and services market-oriented or monetization
	Users collect or purchase information, products or services

acquisition/transport/transfer, information resources value-added development, and utilization of information resources and the products. Refer to Table 70.1.

The process of government housing information resources development and utilization is affected by many factors. According to the activities of government housing information resources development and utilization and subject matter expert interview results, 4 categories of risk, 11 key risk factors, and consequences description when risks happened are summarized and shown in Table 70.2.

70.3.2 Building a Bayesian Network Structure of the Government Information Resources Development and Utilization

According to the method for determining network nodes, 4 risk types of the government housing information resources development and utilization such as policy, government behavior, technical and management risk, are used as intermediate nodes, expressed by the "$A_1 \rightarrow A_4$". 11 key risk factors are used as evidence nodes, respectively represented by "$R_1 \rightarrow R_{11}$". The risk of government housing

Table 70.2 Risk description of government housing information resources development and utilization

Risk categories	Name of the risk	Description of the risk consequences
Policy risk	Imperfect regulations	The market-oriented practice of the development and utilization has obstacles, the government is very conservative. The development and utilization is lack of motivation
	Lack of incentives policies	Departments related government housing have low will to collect information and value-added development
	Absence sharing policy of government information	Phenomenon of information island, housing information resources value is low, less government departments share information to the society, narrow scope of information products and services
Government behavior risk	Inadequate financial capital investment	Data acquisition work of government housing information is restricted, government has not the ability to develop housing information resources, then cause waste of resources
	The government department personnel rent-seeking	The government departments use permissions to monopoly housing information resources, and cause rent-seeking, caused vicious competition between the housing information service enterprises
	Mismanagement of the government information resources development	The company of providing information products and services use housing information which is transmitted from government illegally, cause misleading for social investment or purchase behavior, affect social and economic order
Technical risk	Non-standard format for the government data	The collected data from different regions or departments cannot be unified, it is difficult to integrated utilize
	Poor performance of government information system	It is hard to meet collection efficiency and the quality requirements of the housing data
Management risk	Inadequate ability of information development company	Caused information products and services position inaccurately, development technology level is low, cannot provide high quality and personalized needs, enterprises operate difficultly
	Information property disputes	Property rights of housing information data from the government and housing information products and services provided by enterprises is violated, copied and transferred illegally
	Information security problems	The government internal database system is violated or stolen

information resources development and utilization is a target node, represented by "A". Based on the corresponding relationship between risk factors and the evidence node, the node and destination node among them, the Bayesian network structure is constructed as shown in Fig. 70.1.

70.3.3 Risk Probability Assessment Value of the Government Housing Information Resources Development and Utilization

Bayesian network node conditional probability tables can be confirmed by expert investigation in situations where there are insufficient practical cases or a lack of historical data. The research objects are groups who have knowledge and or are undertaking research on the subject matter, such as teachers in colleges and universities, the staff of government departments and offices, the staff of information products and services development companies. 11 portion valid questionnaires including 4 from college teachers, 3 from the staff of government departments, 4 from the staff of information products and services development companies have been won.

Risk assessment on the occurrence probability has been done with the help of Bayesian network analysis software GeNIe. The first step is inputting the risk probability and conditional probability of Bayesian network node from survey data statistics into the software, then obtaining Bayesian reasoning results, that is the probability of risk occurrence of government housing information resources development and utilization, as shown in Table 70.3.

From Table 70.3, at present, the risk of government housing information resources development and utilization that is the probability of the risk of the target node A, is 0.72. It appears that the risk to the government housing information resources development and utilization is relatively large. Among evidence nodes, the occurrence probability of R1 (Imperfect regulations), R6 (Mismanagement of the government information resources development) and R3 (Absence sharing policy of government information) is relatively high. It indicates that policy-level system design and mandatory regulation are incomplete, so strengthening the construction of legal systems should be the focus in the future.

70.3.4 Impact Weight of Government Housing Information Resources Development and Utilization of Risk

Impact weight of government housing information resources development and utilization refers to the relative importance of the 11 risk factors ("$A_1 \rightarrow A_{11}$") to the target node ("A"), and the sum of weighted values is 1. The impact of risk factors is

Table 70.3 Risk probability table of government housing information resources development and utilization

Risk ID	Name of the risk	Probability of occurrence
R1	Imperfect regulations	0.823
R2	Lack of incentives policies	0.598
R3	Absence sharing policy of government information	0.689
R4	Inadequate financial capital investment	0.521
R5	The government department personnel rent-seeking	0.517
R6	Mismanagement of the government information resources development	0.690
R7	Non-standard format for the government data	0.624
R8	Poor performance of government information system	0.481
R9	Inadequate ability of information development company	0.635
R10	Information property disputes	0.573
R11	Information security problems	0.669
A1	Policy risk	0.67
A2	Government behavior risk	0.60
A3	Technical risk	0.64
A4	Management risk	0.74
A	Development and utilization risk	0.72

Data sources survey data from experts and GeNIe output result

calculated by using the AHP method, in accordance with the following steps: (1) The different risk factors are listed in the comparison matrix. (2) According to the 1–9 scale method for two comparison. (3) Calculate weighted value by the method of square root, then normalized processing and calculates the consistency index of CR. This study has investigated six experts, construct the comparison matrix, calculate the hierarchical structure matrix with the aid of Excel formula editor and the automatic calculation function. The effective weighting of risk factors to government housing information resources development and utilization is shown in Table 70.4

70.3.5 Comprehensive Evaluation Value of the Risk of GHIRD&U

On the basis of calculating the risk probability ("*P*") and risk effective weight ("*E*") of government housing information resources development and utilization, according to the definition of risk $R = P * E$, comprehensive evaluation value of each risk factor can be obtained [21]. Refer to Table 70.4:

From Table 70.4, comprehensive evaluation value of R_1, R_9 and R_3 are higher, these three factors are the biggest factors influencing the degree of risk, not only

Table 70.4 Comprehensive risk evaluation of government housing information resources development and utilization

Risk ID	Name of the risk	Probability of occurrence(P)	Effective weight(E)	Risk value	Sorting
R1	Imperfect regulations	0.823	0.354	0.291	1
R2	Lack of incentives policies	0.598	0.067	0.040	5
R3	Absence sharing policy of government information	0.689	0.155	0.106	3
R4	Inadequate financial capital investment	0.521	0.015	0.008	10
R5	The government department personnel rent-seeking	0.517	0.008	0.004	11
R6	Mismanagement of the government information resources development	0.690	0.054	0.037	6
R7	Non-standard format for the government data	0.624	0.065	0.041	4
R8	Poor performance of government information system	0.481	0.028	0.013	9
R9	Inadequate ability of information development company	0.635	0.174	0.110	2
R10	Information property disputes	0.573	0.030	0.017	8
R11	Information security problems	0.669	0.051	0.034	7

Note values are taken after the decimal point three data. *Sources* expert survey data

having a high probability of occurrence but also influence weight. So these are the key factors in limiting government housing information resources development and utilization, which mainly involves the government system construction and company capacity of information development enhancement. Therefore, three areas of work are needed to be undertaken. First, the government should increase the relevant institutional arrangements in the whole industry chain, such as the housing information resource collection, transfer, the development and utilization, give full play to the government main body role. Then, greater disclosure and sharing of government housing information resources, unifying the housing information format standard across the relevant departments avoids departmental information island. Moreover, information development companies, who are the key participants, should pay more attention to market demand, design types, and content of value-added products and services effectively, serving investors and consumers in the entire real estate market preferably.

70.4 Conclusion and Expectation

The risks to government housing information resources development and utilization can be divided into four categories: policy, government behavior, management, technology risk, comprising 11 risk factors. Through risk evaluation Bayesian network model of government housing information resources development and utilization, a comprehensive risk evaluation value can be obtained. The three main risk factors of imperfect regulations, inadequate ability of information development companies, and the absence of a sharing policy of government information demonstrate that the government plays an important role in the control of risk.

Due to industrialization, the level of the government housing information resources development and utilization is too low and sample data is limited. Bayesian network conditional probability table is based on expert knowledge from investigation. This study lacks practical data. However in the future, with government housing information resources development and utilization increasing, and with the help of Bayesian network convenient infinitely expand functions, risk assessment based on BN and AHP methods will become an effective tool to control risk.

References

1. Li Q (2013) The study on risks of government housing information resources development and utilization. Shandong Jianzhu University, Jinan
2. Cui X, Hou Y (2012) The industrialization of development mode research on the city government housing information resources. Shandong jianzhu University, Jinan
3. Wang H (2003) The risk and prevention mechanism in the electronic government affairs. J Hefei Univ Technol 8:944–946
4. Sun S (2004) Government information security risk analysis and prevention countermeasures in electronic government. Econ Manag BBS 24:64–65
5. Meng X (2009) Review on E-government information security risk management research. E-government 8:67–70
6. Zheng S (2004) How to improve the efficiency of the E-government construction in China from E-government high risk. Office Inf 1:12–14
7. Xie J (2006) Risk analysis and control of government information project. Sci Technol Manag Res 7:182–184
8. Zhou Y (2009) Research on government information assets and operational strategy. Inf Stud Theor Appl 6:18–21
9. Hui P (2011) Risk management of school digital library information resources construction project. Libr Theor Pract 3:66–67
10. Du L (2012) Risk management and research of library information resource construction. Sci-Tech Inf Dev Econ 8:18–21
11. Huang Y (2012) Theory of risk management on information system software development projects. Inf Res 12:88–90
12. Alexander C (2001) Bayesian methods for measuring operational risk. Reading: ISMA Center of the University of Reading, pp 1–22

13. Bo C, Wang Z (2008) Risk management of commercial bank operation based on Bayesian network. Financ Theor Pract 1:43–46
14. Yang Y (2008) Fault tree and Bayesian network used in the risk analysis study on levee breaches. Dalian University of Technology, Dalian
15. Si Y (2008) Research of assessment technology on information security risk. Guizhou University, Guizhou
16. Selma, Yaoting Z (2005) An introduction to econometrics Bayesian inference. The MIT Press, Shanghai
17. Tao S (2009) Electric power engineering project risk management research based on Bayesian network. Shenyang University of Technology, Shenyang
18. Patt AC, Schrag DP (2003) Using specific language to describe risk and probability. Clim Change 61(1):17–30
19. Wang T, Song D (2010) Assessment method of construction safety risk probability based on Bayesian network. China Civ Eng 43:385–391
20. Lu L (2009) Risk assessment technology research on the software requirement based on Bayesian network. Nanjing University of Science and Technology, Nanjing
21. Tang A, Wang R, Hu C (2010) The application of Bayesian networks in software project risk assessment. Comput Eng Appl 7:62–65

Chapter 71
Study on the Mechanism and Countermeasures of Community Deterioration of Public Rental Housing—Case from Chongqing

Lin Wang, Lifang Huang, Long Yin, Xiaoman Zheng and Hui Xu

Abstract Past two years, in order to solve the housing difficulties of low-income groups, Chongqing has increased efforts in building public rental housing. According to the related experience of United Kingdom and United States, a lot of the construction of public rental housing could lead to a series of social problem and community deterioration. From the four dimensions of the city residential price screening mechanism, the government policy guidance mechanism, the dynamic circulation mechanism between communities, and the acceleration effect of cumulative circulation, this article researches the drive factors on public rental housing community coming up Community degradation. It finds that community dynamic degradation results in "high-end people fled and low-end crowd gather", but through time disappeared, the degree of degradation of low-end community will aggravate constantly, and bring a series of social problems. In order to effectively prevent the risk of Community degradation, some suggestions should be available, such as choosing the rational arrangement and location of public rental housing, building a good cultural atmosphere of the community and strengthening good property management etc.

Keywords Public rental housing · Community escape · Community deterioration · Mixed-income housing

L. Wang (✉) · L. Huang · L. Yin · H. Xu
School of Construction Management and Real Estate,
Chongqing University, Chongqing, China
e-mail: wangidill@163.com

X. Zheng
Chongqing Real Estate College, Chongqing, China

Y. Wu et al. (eds.), *Proceedings of the 20th International Symposium on Advancement of Construction Management and Real Estate*,
DOI 10.1007/978-981-10-0855-9_71

71.1 Introduction

Public rental housing is refers to the government investment and provide the policies support, limited by model area and preferential rent standard, and supply of affordable housing to the family which match condition [1]. In order to solve the housing problem of Middle-Low income masses of housing difficulties in the city, and achieve the goal of "home to live in" which put forward in party's 17th report, Chongqing will construct 40 million square meters of public rental housing for the low-income groups and the "sandwich layer" housing problem in 2010–2015.

Only in 2010, a total of 13 million square meters of public rental housing has been constructed in main district of Chongqing such as Yuanyang, Caijia, Xiyong, Microelectronics Park, and 12 suburb areas such as Wanzhou district, Fulin district, Qianjiang district. When the Chongqing public rental housing is constructing in full swing, the state of affordable housing projects coordination group assign the 10 million units of affordable housing construction tasks to the local government through signed the goal responsibility on 24th February, 2011. Compared to the construction quantity of 5.8 million units in 2010, it has increased by 72 %. The construction of such a large scale of public rental housing and other affordable housing in Chongqing and even the national regions, which reminds people of the large-scale and rapid construction of public rental housing bring a series of social problems in western developed countries such as US.

Serious Slumdog problem caused the extensive concern of the public and social care, tried to America government to establish the Federal Housing Administration in 1934, and made the US began to build public housing. Because the land is expensive and difficult to get, it forced the public housing to high-rice and site in the suburbs. It is worth mention the Pruitt-Igoe Housing, which has a good plane, space, including the beautiful park, the newest amusement and rest square and a large number of greenery landscapes around the building. After one and a half year, the property services and warranty become bad to worse, and the community consisted of the whites, blacks, the poor, the disabled, the mental patient, became a hotbed of theft, destruction, crime, and the industry suffered serious damage. After 17 years, the Pruitt-Igoe Housing has no one to live, and been razed to the ground by the government [2].

What kind of impact on the city development that the construction of China's affordable housing will be? The problem has attracted the attention of many scholars. For example, Jiao Yixue appointed the construction of affordable housing intensified the urban residential space differentiation, and it once informed will produce a series of catastrophic social problem [3]. Liu Zhilin and Li Jie declined China's affordable housing construction may lead to problems of poor crowd gather and residential segregation [4].

71.2 The Analysis of Three Dynamic Mechanisms of Public Rental Housing Degradation

Community degradation is refers to the community of high-income people continue to leave, low-income people continue to come into, community residents' social and economic status are declining or maintain in low level, community peripheral become bad to worse, the level of property management is more and more low, community policing problem is more and more serious, unemployment maintain in high level, and it would spread to the surrounding communities, and even spread to the city, resulting in the degradation phenomenon of the city. Thus, in this part, mainly through analysis the three mechanisms led to public rental housing community degradation. The three mechanisms are housing price filtering mechanism, guiding mechanism in the government policy, and the flow mechanism of the crowed between similar and different communities. Meanwhile, analysis the cumulative acceleration mechanism how accelerate the public rental housing degradation.

71.2.1 The Housing Price Filtering Mechanism in the Market Mechanism

The key factor of housing price led to low-income crowed gather is that their income is not enough to support they select a house with good condition, and the determinants of housing price not only the influence of the macroeconomic background, but also other factors, such as: location, transportation, huxing design, peripheral matching, the level of community property management. Under the background of dominated by real estate developer, resource scarcity leads to the price of the house which has good lactation, transportation, complete functional space of housing, mature of support of surrounding commercial, medical, education, and high level of property management, will be relatively high. On the contrary, the community which has location disadvantage, single unit, surrounding facilities defective, because of its low price to attract low-income people to purchase and rent. Rejected by high housing price, low-income groups had to move to the place such as suburbs, junction between urban and rural areas, which the housing price relatively low. Housing price has intention or no intentional separate high-income group from low-income group, and informed relatively closed two groups. According to residential differentiation theory, it brings a series of negative social effects, such as increase the contradiction between high-income and low-income class, the unfair distribution of social resources, deprived of living space [5].

71.2.2 The Guiding Mechanism in the Government Policy

The perfect function of market cannot be make everyone has own housing, the government's affordable housing policies can help low-income people to solve the housing problem. But, to some extent, this mechanism made the spatial agglomeration of the low-income groups sharpened. Because China's affordable housing has strictly access system, only the crowed satisfied with condition to be eligible to live in. Thus, under the filtering effect of this mechanism, the low-income crowed gather in the city.

The smallest of seven public rental housing area that start construction in main district of Chongqing in 2010 accommodates 43,000 people. Under the action of government filtering standards, the big and high construction will directly lead to crowds with relatively low of economic income level, social status, education level into the same community. It will intensify the information of community differentiation. It has significant difference compared public rental housing with ordinary commercial housing community.

71.2.3 The Flow Mechanism of the Crowed Between Similar and Different Communities

According to the housing price, the paper defines the low density residential such as villas, and high-rise luxury as high-end housing, defines the ordinary high-rise residential as the ordinary commercial housing. Because in the dominated of the government, public rental housing is considered to be the representative for meet the housing of housing of the low-income groups, so it is defined as the third of housing. Meanwhile, consider factors of house type, peripheral supporting facilities, environment condition, high-end housing and ordinary commercial housing can divided into high-end, midrange, general. Chongqing public rental housing system had effectively integrated the low-rent housing and affordable housing system, realized integrative security. Therefore, according to the security object, from top to bottom the potential tenants of public rental housing is divided into transitional household, ordinary tenants, and low-rent home in the housing filtering mode. Ordinary tenants is mainly refers to security objects of first type and a small amount of the third type. The transitional household is mainly refers to security objects of the second type, it mainly consists of colleges and vocational schools. Because they had accepted higher education, the possibility of their income level rise is larger, and the possibility of leave public rental housing is larger also. The formation of community through above two mechanisms is not still. When the economic level of a resident rise obviously, it will appear the phenomenon of "up filtering" if improve the environment becomes a urgent demand, namely choose more high-end housing again; on the contrary, when the economy level declined obviously, and cannot afford the mortgage, it will appear the phenomenon of "off filtering", namely select to sell existing housing and choose relatively low-end housing. The happen of "up-off filtering" made different housing types appear the phenomenon of "high-end people fled and low-end crowd gather".

71.2.4 The Cumulative Acceleration Mechanisms

Under the function of the above three mechanisms, it will appear Matthew effect as constantly circulate. For the high-end housing, it will gather more and more people who have high income and social status, and its commercial value will improve also. But for public rental housing, the nature determines gathering a large number of low-income groups at the beginning of the cycle. As further accumulation of low-income group, the property management becomes bad to worse; the consciousness of community residents to maintain the community environment is relatively low, the community environment becomes worse to worst; and also because of the limitation of community residents' economic capacity, the surrounding commercial is difficult to development. As the time passing, it will appear the effect of cumulative acceleration: for the transitional household, they will choose to leave the community to rent the housing in ordinary commercial housing. And the third groups that Chongqing public rental housing introduces will be avoided. The vacancy for public rental housing will attract a large number of lower income people to live in. Thus, the diversity of income and population of the public rental housing will constantly to reduce. And in the case of market economic fluctuations, the low-income groups are impacted largest. It will lead to the phenomena that they refused to pay the rent and property management fee. These phenomena had the conduction effect: once a tenant refused to pay the rent, other tenants will follow. In a result, the whole community reached the point of nobody management, and gathered eroticism, gambling, drug, high crime rate and unemployment rate is the synonymous for the community.

Therefore, through the above analysis, under the background of market economic, the selection of housing price made the urban community had the trend of spontaneous differentiation, combined with the conduct of affordable housing security system, it will increasing homogeneity communities gathering. In the process of gathering, the dynamic filtering mechanisms—the flow mechanism of the crowed between similar and different communities and Cumulative acceleration mechanisms—increased the community differentiation. In addition, as the effect of Cumulative acceleration mechanism, it will further accelerate the community differentiation, and may spread to the surrounding communities. In a result, it will leading to the urban differentiation.

71.3 The Analysis of the Reasons of the Dynamic Process of Community Differentiation

The possibility of the differentiation of public rental housing community which is the low-end housing in the filtering mechanism is largest, and the phenomenon of community fled is more obvious. Expect the influence of economic level, what other reasons will aggravate the occurrence of this phenomenon?

71.3.1 The Decline of Building Quality and Community Property Management Level

The quality and community property management level of the public rental housing directly affect person that has middle-income or in secondary status whether he/she would escape from the community. Due to the proprietary rights of public rental housing belong to the government not the tenant, the tenant takes care of the indoor environment and the community public implementation is much less than the commercial housing community, it increases the difficulty to management the public rental housing community. With the aging of public rental housing community, the cost of daily maintenance and repair increase progressively also, and the charge of the property management is low, it will lead to increased difficulties of property management of public rental housing community. Capered to the commercial housing, public rental housing is a government project; so the possibility of rent-seeking behaviors is digger, and it is hard to guarantee the quality of design and construction. For transition households, they will choose to rent smaller house in other commercial housing community when the economic ability did not changed, and will initiative fled the public rental housing community when their economic ability promoted.

71.3.2 Traffic Inconvenience and Peripheral Supporting Is Imperfect

If the lactation of public rental housing choose in the relatively remote and traffic inconvenience, the cost of traffic will increase. The traffic cost includes the cost of time and economic. For the low-income group, traffic cost is the primary factor of living cost need to consider. If the traffic cost exceeds the scope of the low-income groups' pay, they will choose to rent the house in urban village or other commercial housing community. It can cause the security failure, and public rental housing is idle.

When the peripheral supporting such as commercial, medical and education is deficiency (not enough), the possibility of the transition groups considered to rent public rental housing will be smaller. Such as education, Chinese parents want their children can receive a good education, which requires has a relatively good school near the public rental housing community. Most of the transition tenants are young, and many of them have no child or the child not reach the school age, when the child arrived and the education supporting of public rental housing community not satisfy people's demand, for their child they will choose to leave the community. Most of public rental housing tenants are low-income, the level of their consumption is limited, the peripheral commercial is difficult to development, it can appear high vacancy rate, even there has public commercial supporting, then those who have the ability to pursue the demand of high material and culture will choose to leave.

71.3.3 Social Psychology and Social Discrimination

Compared to the ordinary commercial housing community, public rental housing community is easier to form "social" for the differences in the nature, and become the synonymous with "poor". The in-group and out-group theory proposed by American sociologist William Graham Sumner, it verified the gap and discrimination exist in different classes. The member of the in-groups showed help each other, unity and friendly; correspond to the in-groups, the out-groups are that the people not joined, the member of it always showed discrimination and an unfriendly [6]. The stratification of housing causes the isolation between different levels of housing community, added to high and low, strong and weak of social stratum, so it made the high-income and strength groups have more abilities and means to discriminate the low-income and disadvantaged groups. And other commercial housing community as relatively strength group, it treats public rental housing community as different foresight, then the public rental housing tenants will be affected, he/she has the sense of inferiority, and it also have a negative effect on their children. Meanwhile, the traditional Chinese culture emphasizes "equalitarianism" would prejudice low-income groups against high-income groups in social psychological, and high crime rate and order confusion are connected by high-income groups with low-income groups. In addition, the residents that live in the other ordinary commercial housing near public rental housing face public rental housing every day, there will be a pressure of unsafe in psychological, gap of huge the poor makes the contradiction between low-income groups and high-income groups highlighted, and estrangement will become big to bigger.

71.4 The Prevention Measures of Risk

The mass and concentrate of the construction of public rental housing, through a longtime, probably 10 years, 20 years or even longer, due to the low-income crowed gather and the community environment worsening, public rental housing community may appear the degradation phenomenon, typical similar phenomenon of "urban village" and "slum", high unemployment and crime rate. The serious social problem—such as low education level, social status, income level—of public rental housing will affect the development of the surrounding community, and even lead to the price of around housing appear concave phenomena. At last, the government had to sell or demotion of public rental housing. So it is particularly important to take measures to prevent the risk. According to the specific condition of China and Chongqing, the paper puts forward several suggestions.

71.4.1 Combined with the Location of the Public Transportation

The location of public rental housing close to public transportation, such as bus station, subway and light rail, and the distance of the location is better within 10–15 min from the subway station in walk, it can minimize the traffic cost of the tenants. Low-income groups affected by market volatility is bigger, in order to prevent the economic level of low-income groups fall further, the government can make matching corresponding industry or industrial park nearby, it is benefit for the residents in employment.

71.4.2 Adopting Mixed Residential Mode When Planned

Mixed living mode is referred to organic integrate the groups of different incomes, classes, social status into a residential community. The social distance between low-income residents and other income residents in the mixed living mode is less than homogenous residential mode [7]. But it should avoid too much difference when adopt mixed living mode, public rental housing should moderate mixed with ordinary commercial housing, and it is better not to mixed with high-end housing, such as villa.

Chongqing public rental housing is mass concentrate constructed, although according to mixed living mode to design it in the stage of planned location, it is possible to make the location of planned area near ordinary commercial housing, and the proportion is 1:3. But it is within the scope of the overall planned and did not spread to every community, such as the construction of public rental housing in 2010, the smallest area contains 43,000 people, it is equivalent to build a small city that consisted by low-income people in Chongqing. Compared to the international common practices "small centralize, large dispersion", there is still a certain gap.

71.4.3 *Shared the Infrastructure with Surrounding Residential Communities*

It should configure the infrastructure of education, medical, commercial, leisure and sports cross communities, and makes the education, medical and business services of public rental housing integrated with other communities. The function of infrastructure is a link medium, and it can effectively connect public rental housing community with ordinary commercial housing community, which can increase the exchanges and communication between different communities.

71.4.4 *Guarantee the Quality and Property Management Level of Public Rental Housing*

Try to avoid the rent-seeking behavior and guarantee the quality of the construction of public rental housing, and effectively avoid the rapid aging and shabby of public rental housing. Compared to ordinary commercial housing community, the property management of public rental housing community is more difficult, it should try to select the property management company that has the corresponding qualifications and the ability of capital, at the same time, the government can put the appropriate subsidies in the operation and maintenance stages of public rental housing.

71.4.5 *Increasing the Construction Area of the Small Family, Reducing the Cost of Residents to Preventing None Dwell in the Public Rental Housing Community*

The construction area of Chongqing public rental housing families that construct in 2010, generally between 30 and 80 m^2, the proportion of the area of the family below 50 m^2 is more than 60 %, the proportion of 50–60 m^2 is controlled with 25 %, and the 60–80 m^2 is 15 %. According to the condition of the first time to apply for public rental housing, the masses' demand of small families such as single rooms, one room one hall, and two rooms one hall is bigger, so the government should increase the construction area of this family to meet the demand of people and "marketable", only which can truly benefit people.

Although the water, electricity and gas meter of the interior decoration engineering of Chongqing public rental housing has installed to the door, the house has adopted security door, kitchen and toilet has carried on the simple packaging, but lack of the necessary simple furniture, in order to moving in, the lessee must buy furniture and electrical appliances such as bed, wardrobe by himself. Thus increasing occupancy costs of the lessee, in the same conditions, the potential lessee can completely rent the housing with simple furniture in the urban villages or other ordinary commercial housing community. So the government can be equipped with simple furniture, reducing the cost of the lessee check-in, and attract residents rent, to prevent the situation that no swell or low rent rate happens.

71.4.6 Cultivating the Spiritual Idea and Values of Community and Improving the Homeownership Rate of Public Rental Housing

Chongqing government set that the lessee can buy public rental housing after the full five years, and the sale price of public rental housing is based on the comprehensive cost. The policy is to encourage the tenants to buy public rental housing and improve the homeownership rate of public rental housing, let low-income groups really have their own homes. At the same time, the regulation of this policy can enhance the residents' protection consciousness of public rental housing, and it will bate the trend of community environment degradation. There is a community in Brooklyn, New York, which built in 1974 and more than 30 different nationalities live together, they put forward the concept of life is "We live together in harmony living the good live", and the development planning of the community is "safe, unity, good education, fully rest". Public rental housing community can built own community concept and values to make the residents have the sense of the community, and take good care of the home they live.

Acknowledgements Project Supported by National Social Science Fund of China (Grant No. 11XZZ001); Project Supported by Chongqing University Postgraduates' Innovation Project (Grant No. CYB15005).

References

1. Chongqing Public Rental Housing Information Network. http://www.cqgzfglj.goy.cn/
2. Wang G (1998) Study on the origin and development of the American public housing. Time Archit 3:62–65
3. Jiao Y (2007) Exploration on the mixed building mode of public housing aiming at habitation integration. Urban Stud 14(5):57–61
4. Liu Z, Li J (2010) Public rental housing policy: comparative international experience and its implications for China. Mod Urban Res 12:21–26

5. Yang S, Wang C (2006) A sociological study on the differentiation of the urban residential space in Shanghai. Society 6:117–137
6. Song L (2010) The sociological analysis of social discrimination. Theor Horiz 7:193–194
7. Tian Y, Li D, Bi X (2006) Study on the mixing-incoming housing and its feasibility. Archit J 4:36–39

Chapter 72
The Impact of Changing Business Taxes on the Rationality of Housing Price

Jingming Liu and Zhanqi Wang

Abstract This paper examines the effects of business taxes policy on real estate prices, using mathematical model to compare and analyze the effects of two business taxes reforms on real estate ROI, and then do research in the relationship between the change of business taxes policy and market price mechanism, compared the influence of two taxes policies on the rationality of the housing price. Through the study found that the five year business taxes policy is lower than two years in the rate of return on investment and the rate of housing price growth, which is better to curb the rapid growth of housing prices and keep it in reasonable.

Keywords Business taxes · Housing price · Return on investment (ROI)

72.1 Introduction

March 30, 2015, The ministry of Finance, the taxes The State Administration of Taxation released a file named "Notification concerning the policy Adjusting in the Business Tax of individual housing assignment" (Taxes [number[2015]39]), which conclude the latest adjustment in business taxes policy. If individuals purchase housing less than two years, full business taxes should be levied when it is on sell, but ordinary housing which keep more than 2 years (including 2 years) will be exempted from business taxes when it sells. It is different from non ordinary housing which business taxes should be levied according to the gap between sells income and purchase price of housing.

In early 2011, according to the notice of "Circular of the Ministry of Finance and the State Administration of Taxesation on the adjustment of the business taxes

J. Liu (✉)
Department of Public Administration, China University of Geosciences, Wuhan, Hubei, China
e-mail: 347826460@qq.com

Z. Wang
School of China University of Geosciences, Wuhan, China

Y. Wu et al. (eds.), *Proceedings of the 20th International Symposium on Advancement of Construction Management and Real Estate*,
DOI 10.1007/978-981-10-0855-9_72

policy for the transfer of the individual housing" [No. 12 (fiscal 2011)], ordinary business taxes changed to 5 years which is different to the policy published in 2015. This policy's order is to curb the fast growth in housing market, but now the business taxes payment period is two years, what is the influence of the business taxes on the housing price, this paper will discuss from the following several aspects.

72.2 The Hypothesis of the Involved Factors in Real Estate Transactions

In the process of real estate transactions, a short time of investment is the best choice for the investors to achieve capital appreciation and risk shifting. But it plays an opposite role in the housing market price stability. Housing is not only investment goods and more is a necessities, the tendency of the housing price closely related to social stability and harmonious.

Due to the implementation of the bank's housing loan policy, the majority of buyers in China are used to use mortgage loans in purchase. The difference of the years in the loan will affect the interest rates. In order to estimate the loan interest costs cautiously, according to the proportion of the existing housing bank loans in China, if loan duration is one year, when it sells the home loan interest rate is 5.35 %. If loan period is two years, the home loan interest rate is 5.91 %. if loan period between two years and five years, the loan interest rate is 6.45 %. The loan ratio is 6.61 % when loan term is more than five years. All lending rates will be calculated according to the simple interest, but not the compound interest [1].

Assuming that the holder of the housing cost M yuan to purchase a housing, and the own fund ratio share b %, which mean the rest of $(1 - b)$ % will loan from bank. The housing rent occupy q % in the total housing price. In order to simplify the calculation process of the model, assume that the housing rent will not be changed in the whole time, the income of the rent is M times q. But most of the housing holder own short period, if the rental contract within 2 years, we can assume that the rent ratio remain unchanged. Judging from past experience, it is generally believed that the income from rent about 25 years can approximately equal to real estate prices, which could calculate the rent ratio about 3–4 % within two years of real estate holding period. In order to estimate the level of return on real estate investment cautiously, assuming that the level of the rent is q1 about 3 %.

If the contract within five years, according to prudence principle, the commodity price rise in recently 5 years is about 3 % every year. If 25 years rent incomes is equal to housing price, calculate the ratio in housing rent with compounding formula. The proportion of rent income in housing price is about 2.74 %, that is, the price of the rent levels is q2 about 2.74 %. T stands for housing holding period, and the average annual increase of the holding period is p, the business taxes rate is s, the current business taxes rate is 5.5 %.

72.3 The Structure of the Model

Based on the assumption of above variables, calculate real estate ROI in two taxes period respectively, assume period is N years (N = 2 or 5), and then structural average annual return on investment model from less than N years or more than N years [2].

(1) When T < N years the annual rate of return on the investment model as follow [1] (72.1):

The annual rate of ROI Y

$$= \sqrt[T]{\frac{M \times (1+p)^T \times (1-s) + M \times q \times T - M \times (1-b) \times (1+R \times T)}{M \times b}} - 1$$
$$= \sqrt[T]{\frac{(1+p)^T \times (1-s) + q \times T - (1-b) \times (1+R \times T)}{b}} - 1 \tag{72.1}$$

The investors initial investment of real estate is the amount of $M \times b$; When sells real estate, the cash income after deduction business taxes is $M \times (1+p)^T \times (1-s)$; The rent income in holding period of real estate is $M \times q \times T$, which neglect the capital reinvestment, and do not consider the relevant taxes on rental income; The loan amount of principal and interest after sell the property is $M \times (1-b) \times (1+R \times T)$.

(2) When $T \geqq N$ Ordinary housing and non ordinary housing have the following two kinds of models to calculate the annual rate of return on investment.
The first case, the real estate is ordinary housing, after N years when sell the housing will not pay for the business taxes, the formula as following (72.2):

The annual rate of ROI Y

$$= \sqrt[T]{\frac{M \times (1+p)^T + M \times q \times T - M \times (1-b) \times (1+R \times T)}{M \times b}} - 1$$
$$= \sqrt[T]{\frac{(1+p)^T + q \times T - (1-b) \times (1+R \times T)}{b}} - 1 \tag{72.2}$$

The cash income is $M \times (1+p)^T$, when sell the real estate, no business taxes.
Second cases, The real estate is a non ordinary housing, after N years (N = 2 years or N = 5 years) when it sells, we should pay for the taxes based on the gap between profit and normal. The formula is as follows (72.3):

The annual rate of ROI Y

$$= \sqrt[T]{\frac{M \times (1+p)^T - [M \times (1+p)^T - M] \times s + M \times q \times T - M \times (1-b) \times (1+R \times T)}{M \times b}} - 1$$

$$= \sqrt[T]{\frac{(1+p)^T \times (1-s) + s \times q \times T - (1-b) \times (1+R \times T)}{b}} - 1 \tag{72.3}$$

72.4 Theoretical Verification

On the basis of the above model and the existing conditions of the current loan interest rate, rent rate in housing price and business taxes, assuming that the price rise to a certain extent, to examine the relationship between the change in business taxes policy [3], the investment holding period, self funding ratio and the return on investment.

72.4.1 *The Impact of the Annual Rate of ROI on Self Funding Ratio*

In the first case, we set the rise of housing price to a certain extent, according to the ROI model, examine the relationship between the business taxes, investment holding period, self funding ratio and the annual rate of return on investment respectively.

The article chooses Shanghai as the research area, in the past 6 years, the second-hand housing price index from 2297 to 2941 with persistently increasing.

From Table 72.1, we can calculate the mean rate in the housing price growth, it is about 5.2 % every year. In order to calculate the average rate of ROI cautiously, we take known number from YUE Juan- LI. And based on our result, we assumed the increased rate of housing price is five percents.

Table 72.2 shows the impact of the change in self funding ratio to the return on investment. The smaller the proportion of own funds is, the higher return on investment in real estate, when self funding ratio reach to 70–80 %, the return on investment is flattening out [4]. If use two years business taxes policy, in the first year it can effectively inhibit the growth of return on investments in real estate, but

Table 72.1 Second-hand housing sells price index in recent five years in Shanghai

	2009	2010	2011	2012	2013	2014
Sells price index	2297	2538	2575	2583	2607	2941
Growth rate (%)		10.5	1.45	0.31	0.92	12.81

Table 72.2 The impact of the change in self funding ratio to the return on investment

	0.30 (%)	0.40 (%)	0.50 (%)	0.60 (%)	0.70 (%)	0.80 (%)	0.90 (%)	1.00 (%)
One year without taxes	14.18	11.98	10.65	9.77	9.14	8.66	8.29	8.00
Two years without taxes	12.51	10.86	9.85	9.18	8.70	8.33	8.05	7.82
Three years without taxes	10.46	9.40	8.75	8.31	8.00	7.76	7.58	7.43
Four years without taxes	10.23	9.20	8.56	8.13	7.82	7.59	7.41	7.26
Five years without taxes	10.20	9.15	8.51	8.07	7.75	7.51	7.32	7.16
Six years without taxes	9.84	8.88	8.29	7.88	7.59	7.36	7.19	7.04
1 year (taxes)	−5.07	−2.46	−0.90	0.14	0.89	1.44	1.88	2.22
2 years (taxes)	3.14	3.80	4.19	4.45	4.64	4.77	4.88	4.97
3 years (taxes)	4.21	4.68	4.96	5.15	5.28	5.38	5.46	5.52
4 years (taxes)	5.81	5.83	5.85	5.86	5.87	5.87	5.87	5.88
5 years (taxes)	6.83	6.56	6.40	6.29	6.22	6.16	6.11	6.08
After 2 years (no taxes)	11.05	9.91	9.22	8.75	8.42	8.17	7.98	7.82
After 2 years (spread tax)	10.20	9.26	8.70	8.32	8.05	7.84	7.68	7.56
After 5 years (no taxes)	9.94	8.98	8.39	7.99	7.70	7.48	7.30	7.16
After 5 years (spread tax)	9.24	8.44	7.95	7.61	7.37	7.19	7.05	6.93

after the second year, the rate of ROI is still positive though it pay for business taxes. But compared to 2 years without taxes policy, it has greatly reduced the return on investment, and effectively control the growth of housing price, preventing the existence phenomenon of real estate investment speculation.

If use the five years business taxes policy, in the first five years, compared to non collection of taxes, collection taxes could be better to control the increase ratio in investment, and the change in self funding ratio pay small impact in the fourth years and fifth years. What's more, It could better control the growth of return on investment in real estate.

Compared to the analysis in two differences taxes collection methods, levy taxes for five years will be better to control the return on investment than two years. Compared to two years taxes policy, five years business taxes is stronger to reduce the increase of the housing price. But if the market is tighten market, two years taxes policy will be better mobilize the housing market activity. When levy taxes in different way, two years taxes policy will be slightly higher in the rate of return on investment than of five years taxes policy.

In summary, five years taxes policy can be better to curb the growth of housing prices, and restrain the existence in housing investment speculation. But two years taxes policy can be good to mobilize the activity of the market in the process of the growth of housing prices [5] (Fig. 72.1).

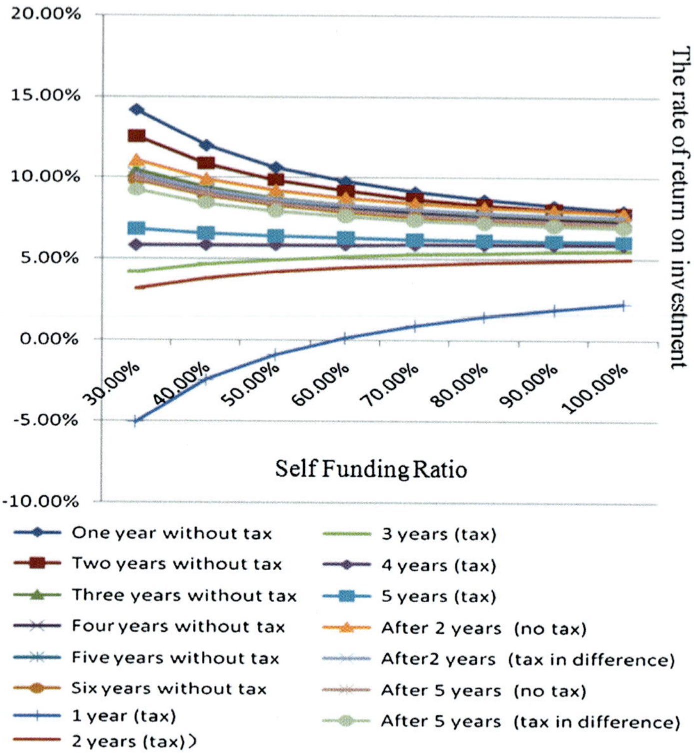

Fig. 72.1 Distribution map of the change in self funding ratio to the rate of return on investment

72.4.2 The Impact of the ROI on the Housing Price

The relationship between the growth rate of return on investment and housing price growth is closely. The paper construction formula based on the above model, then conversion of the return on investment rate. Take the rate of return on investment as independent variable, and the growth ratio of housing prices as dependent variable. To analysis the impact of the growth rate of return on investment to the rate of housing prices.

(1) When T < N years, the growth of housing price model as follow (72.4):

$$\begin{aligned}&\text{The growth rate of housing price p}\\&=\sqrt[T]{\frac{(Y+1)^{T}\times M\times b+M\times(1-b)\times(1+R\times T)-M\times q\times T}{M\times(1-s)}}-1\\&=\sqrt[T]{\frac{(Y+1)^{T}\times b+(1-b)\times(1+R\times T)-q\times T}{(1-s)}}-1\end{aligned}\tag{72.4}$$

The investors initial investment of real estate is the amount of M, assuming that the proportion of buyers of self funding ratio is 0.5, b = 0.5. Y is the growth rate of real estate investment.

(2) When T ≧ N (N = 2 years or N = 5 years), Ordinary housing and non ordinary housing have the following two kinds of models to calculate the growth of housing price.

The first case, the real estate is ordinary housing, after N years when sell the housing will not pay for the business taxes, the formula as following (72.5):

$$\text{The growth rate of housing price p} = \sqrt[T]{\frac{(Y+1)^T \times M \times b + (1-b) \times (1+R \times T) \times M - M \times q \times T}{M}} - 1 = \sqrt[T]{(Y+1)^T \times b + (1-b) \times (1+R \times T) - q \times T - 1} \tag{72.5}$$

Y is the growth rate of real estate investment. b is the ratio of self funding ratio equal to 0.5.

Second cases, The real estate is non ordinary housing, after N years (N = 2 or 5 years) when it sells, it should pay for the taxes based on the gap between profit and normal. The formula is as follows (72.6):

$$\text{The growth rate of housing price p} = \sqrt[T]{\frac{(Y+1)^T \times M \times b + M \times (1-b) \times (1+R \times T) - M \times q \times T + \left[M \times (1+p)^T - M\right] \times s}{M}} - 1 = \sqrt[T]{\frac{(Y+1)^T \times b + (1-b) \times (1+R \times T) - q \times T}{(1-s)}} - 1 \tag{72.6}$$

According to the conversion model above, it impressed the impact of the real estate ROI to the growth of housing price. The shown reveals in Fig. 72.2 and Table 72.2.

From Fig. 72.2 we can come to a conclusion that the growth rate of return on investment in real estate is much higher than the growth rate of housing price. In the forms of 6 years without taxes, even if the rate of return on investment in real estate is negative [6], the ratio of housing prices is still positive. This shows the importance of the business taxes for the development of housing prices.

In 2 years business taxes policy, when the rate of ROI is −5 %, the ratio of the housing price is still positive. This shows that the short-term investment will bring the growth of housing prices. When the rate of ROI is 5 %, the growth rate of housing price is about 2.5 %, when housing prices rise to 3 %, it is possible to bring 6 % return on investment, huge returns will stimulate the market to increase

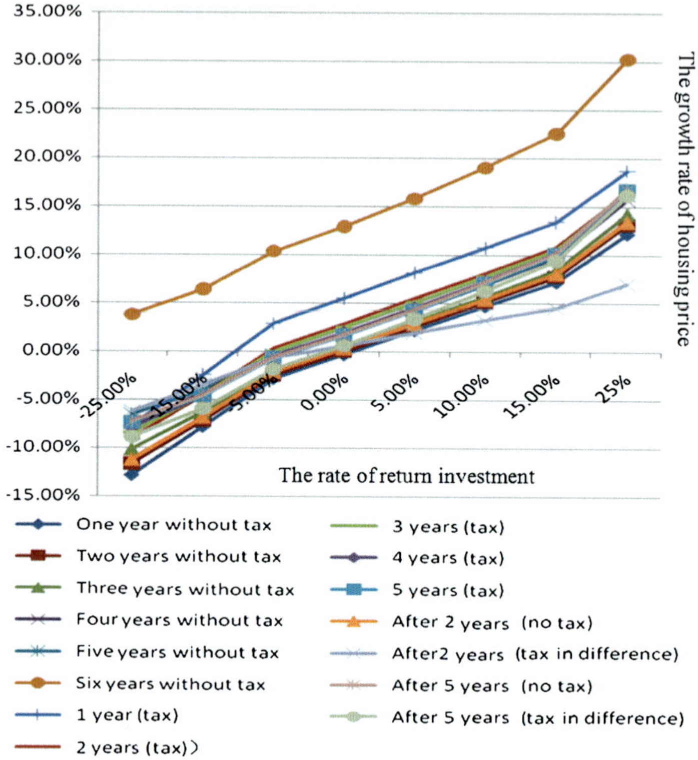

Fig. 72.2 Distribution map of return on investment in real estate to the growth of housing price

investment, so the two years business taxes is failed to play a good role in the control of the housing price growth.

In the 5 years business taxes policy, when the return rate of investment is 0, the growth rate of housing price is positive. It shown that the growth rate of housing price is slightly less than the rate of return on investment. When the return rate of investment is 5 %, the growth rate of housing price is almost the same as that of the rate in return of investment [7]. Compared to 2 years business taxes policy, five years business taxes policy will be better control of the growth rate of housing prices.

72.5 Judgment in the Rationality of Housing Price

The judgment of the reasonable housing price use the housing price income ratio method.

In general, the growth rate of disposable income will not be doubled in one year. But due to the existence of price changes and inflation, the growth rate of

Table 72.3 Effects of the return on investment in real estate to the growth of housing price

	−25.00 %	−15.00 %	−5.00 %	0.00 %	5.00 %	10.00 %	15.00 %	25 %
One year without taxes	−12.83	−7.82	−2.83	−0.32	2.18	4.68	7.18	12.18
Two years without taxes	−11.66	−7.24	−2.51	−0.05	2.49	5.08	7.72	13.15
Three years without taxes	−10.14	−6.34	−1.93	0.48	3.02	5.67	8.43	14.24
Four years without taxes	−7.49	−4.40	−0.47	1.81	4.27	6.91	9.71	15.75
Five years without taxes	−6.58	−4.02	−0.41	1.79	4.24	6.93	9.83	16.20
Six years without taxes	3.73	6.33	10.32	12.87	15.77	19.00	22.51	30.25
1 year (taxes)	−7.75	−2.46	2.83	5.48	8.12	10.77	13.41	18.70
2 years (taxes)	−9.13	−4.58	0.28	2.82	5.43	8.09	10.81	16.40
3 years (taxes)	−8.43	−4.56	−0.06	2.40	4.98	7.69	10.50	16.42
4 years (taxes)	−7.98	−4.67	−0.49	1.91	4.51	7.27	10.20	16.49
5 years (taxes)	−7.41	−4.62	−0.73	1.62	4.23	7.07	10.12	16.75
After 2 years (no taxes)	−11.27	−6.87	−2.16	0.30	2.83	5.41	8.04	13.46
After 2 years (spread tax)	−6.21	−3.43	−0.73	0.59	1.90	3.19	4.46	6.96
After 5 years (no taxes)	−7.30	−4.51	−0.64	1.70	4.30	7.13	10.17	16.80
After 5 years (spread tax)	−8.93	−5.96	−1.86	0.59	3.30	6.23	9.37	16.17

disposable income is a positive value. Assume that the growth rate of the residents disposable revenue is X ($X < 0$) [8]. When the above data divided by disposable income growth rate in the same time, due to the denominator consistent, all of the rate less than 0, then $1/X > 1$, that is to say Housing price income ratio and the ratio of housing prices to a positive ratio relationship.

Assuming the return rate of investment is 5 %, according to the model of the growth of housing price. Two years business taxes policy and five years of business taxes policy will bring the impact of changes in housing prices as Table 72.3. When have no business taxes, even if the growth of housing price is negative. The return on investment still could get 5 %. Compared two schemes with taxes and non taxes, taxes will slow down the growth of real estate income, and five years taxes policy is better than two years business taxes policy for reduce the growth of housing price income ratio.

72.6 Conclusion

Business taxes can effectively control and slow down the growth rate of housing prices, reduce speculation or investment in real estate market. But compared with two different policies of the business taxes system, the risk of investment in two years is lower, so most investors will choose sell real estate after two years, so that the implementation of the policy is not obvious. Five years business taxes system compared to two years taxes system can be better inhibit the growth of housing prices.

(1) Two different taxes system plays an important role in the regulation of real estate market. If need to activate the market. We should take two years taxes system or cancel business taxes system, because in the short term can greatly reduce the return on investment properties, but due to the short time, the low risk and the effect is not significant for the price of housing. If you want to better reduce the real estate increase rate, two years of taxes system will be a better choice flexibly and effectively for the control of real estate prices.
(2) If not levy real estate business taxes, real estate return on investment will be far greater than the rate of growth rate, which is more strongly to stimulate the purchase of housing, leading to the imbalance of real estate market, which is easily to cause the phenomenon of bubble in residential market. Although the business taxes can better control the return on investment, to prevent speculators use smaller proportion of self funding to get more benefits. But when the proportion of self funding less than 0.5, the return on investment is greater than the rate of return on housing prices. So the state should consider the proportion of self funding in the collection of business taxes, development of the real estate industry rationally.
(3) Due to the growth rate of price and housing price income ratio is closely related, the control of housing prices for the control of the housing bubble is essential. Because the real estate industry's investment rate of return is almost 2 times the proportion of housing growth, so work hard to improve the rate of return on investment in real estate prices is very important.

References

1. YUE Juan-LI, GUAN Hong-XI (2005) The effects of sales tax on housing estate price. Finance and Stock 6:159–162
2. Wang YH, Ma YH (2006) The effect of contrary to business taxes control prices. J Taxes Res 09:46–47
3. Sun J, Ma YH (2010) The impact of business taxes on the volume and price of second-hand housing, the real estate market in Shanghai. J Era Financ 09:58–59
4. Jiang BL, Xiao BH (2009) Monetary policy, bounded rationality and low price of housing credit. J Jiangxi Soc Sci 4
5. Yin B (2004) On housing market regulation and development. J Econ Latitude Longitude 4
6. Wang Y, Cai XJ (2004) The further analysis of net present value and the research of its modification algorithm. J Quant Econ Technol Res 12:72–77
7. Vafeas N (2003) Board structure and the informativeness of earnings. J Financ Econ 19:139–160
8. Smith BF, Amoako-Adu B (1995) Relative price of dual classshares. J Financ Quant Anal 30:223–239

Chapter 73
Incentives Selection and Reputation of Government in the Supply of Public Rental Housing in China: A Household Survey Data

Ping Lv and Bo Zang

Abstract The enforcement and pricing problems of the effective supply of public products have always been a hot academic topic, but the basic research paradigm is always principal-agent framework of central and local governments. This paradigm has neglected the ordinary households' preferences which are the actual public demand, which is the argument foundation of our study. So using the public rental housing as the analysis object, the households' preferences as the foundation for analyzing the government behavior, through principal-agent dynamic game with imperfect information and signal game with mixed strategy, we research which levels of incentives should the central government choose and how does the reputation effect of local government influent the long-term supply of public rental housing. Results show that under the premises that the local government pursue the maximization of achievements utility and economic utility, the enforcement of the supply of public rental housing significantly related to the levels of central government's incentives, strong incentives have marginal amplification effect and weak incentives have marginal narrow effect. When information is so incomplete that households cannot distinguish the quality of public rental housing in advance, the equilibrium rent will be influented by the government reputation, that is the construction quality of public rental housing. The government reputation is closely related to the gap between the rent determined by the local government and the rent that households willing to pay.

Keywords Public goods · Government behaviors · Rent gaps · Households' preferences · Reputation effects

---

P. Lv · B. Zang (✉)
Department of Land and Real Estate Management, Renmin University of China, 59 Zhongguancun Street, Beijing 100872, China
e-mail: zb2416650@163.com

Y. Wu et al. (eds.), *Proceedings of the 20th International Symposium on Advancement of Construction Management and Real Estate*,
DOI 10.1007/978-981-10-0855-9_73

73.1 Introduction

Public rental housing ("PRH" for short) is a good that is dominated by the state and will be rented for low-income groups. This good includes two main properties, one is that it can not be provided by markets and must be supported by the government-led, the other is that it has limited competitiveness because for the low-income groups. From these two properties, public housing is "quasi-public good", which is a kind of collective consumer products can not be found dispersed spontaneous solution products through market access [1]. Some scholars [2–5] concern the influence factors of public goods supply, and think that the central and local government's preference is the most important factor. How to study this factor? Most scholars [6–9] use the principal-agent model and some scholars [10–12] use mechanism design theory to study the implementation of public goods supply. From China's actual situation, building 36 million sets of PRH has been defined by Chinese government in 2011, so PRH has become the important examination of government performance. So the government behavior is needed to study deeply.

However, the basic research paradigm to study the government behavior in the supply of PRH is always principal-agent framework of central and local governments and has neglected the households' preferences which are the actual public demand [13–15]. So our contribution is propose a framework that contains two game relationship in the supply of PRH, one is a principal-agent dynamic game (central and local government) with imperfect information, which discusses which incentives selection should central government take. The other is a signal game (local government and low-income household) with mixed strategy, which discusses the local government's reputation.

The paper is set out as follows. The next section discusses the relationships between the supply objectives of PRH and households' preferences, to give a evaluation foundation and key variables that reflect governments performances and behaviors in the supply of PRH. Section 73.3 discusses evaluation frameworks of governments behaviors, one is unsupervised principal-agent game of central and local government, the other is signal game under mixed strategy of local government and households, and then proposes two propositions. Section 73.4 is to test these two propositions and to provide a unified framework to explain the government behaviors in the supply of public goods. Section 73.5 discusses some suggestion on which incentives should the government select and what governance structure should the government make in the supply of quasi-public goods or public goods.

73.2 Key Variables of Governments Behaviors and Data

73.2.1 Evaluation Foundation of Governments Behaviors

Tiebout model [16, 17] points out that residents' display preferences[1] will bring transaction costs, so the price of public goods can not be determined by the general equilibrium under the neoclassical market frameworks. Under the premise of asymmetric information and local governments with financial constraints, the rent is defined according to the period of cost recovery of PRH and low-income urban households budget constraint, and the household only choose 'accept' and 'refuse' passively. So this is a typical problem of dynamic game with incomplete information. The rent of PRH (r_j) is related to the players' strategy selection and payoff, government and household, so it is the first variable to study the players' behaviors. Meanwhile, the influence factors related to the input costs of local government and the households' display preferences are the location of PRH (l_j), convenient degree of transportation (t_j), housing structure (s_j), living environment (c_j) and quality of construction (b_j). Here these factors are named 'the quality of PRH' collectively, which is expressed as Q_j, where $Q_j = f(l_j, t_j, s_j, c_j, b_j)$. Accordingly, we get two basic variables to evaluate the performance of governments in the supply of PRH, r_j and Q_j.

73.2.2 Data

Samples are selected by stratified sampling, and the survey are carried out through one-to-one open-ended interview. Firstly we choose those 10 communities which occupancy rate are more than 90 %. Secondly, we choose communities by the properties of living groups to avoid property homogenization. Finally, we choose communities by the distance from the nearest large-scale business centers. Through these three steps, five communities, 1–5 are selected. The main groups are graduates in Community 1; Community 2, low-income households in the cities; Community 3, foreign migrant workers without special technologies; Community 4, industrial and technical workers; Community 5, graduates, foreign migrant workers, industrial and technical workers and the low-income households.

The main contents of questionnaires are whether the living groups are satisfied with the PRH and which the main reasons for being dissatisfied with. 1300 survey questionnaires were distributed and 1180 are acquired, of which 1146 are valid. The

[1]The costs of display preferences are transaction costs actually, and they are produced by residents' flowing between different communities and making a selection. The costs of display preferences includes information gathering costs, transportation costs and migration costs, and higher this cost is, the lower the efficiency of resource allocation is when other conditions remain unchanged.

effective ratio is 97.12 %. Assuming sampling precision is similar to random sampling, sampling error of this research is ±1.467 % at the 95 % confidence level.

73.3 Theoretical Analysis

73.3.1 *Incentive Choices: The Principal-Agent Relationship Between Central and Local Government*

The supply quantities of PRH are decided by central government, and then allocated to local governments in China. Both governments are typical principal-agent relationships which belong to dynamic game. Under the principal-agent model without supervision, assuming that the central government has two incentives, one is strong incentive, w_i^H; another is weak incentives, w_i^L. Assuming that the total amount of PRH construction defined by the central government is 1, the amount of PRH construction carried out by the local government i is q_i, and $q_i \in (0, 1]$. As the central government can not to supervise the local government's effort, so w_i is a function of q_i, and $0 < w_i^L < w_i^H$. Correspondly, high and low effort is e_i^H and e_i^L respectively, and $0 < e_i^L < e_i^H \leq 1$. Assume that there is no fixed costs, unit variable cost of building is c_i, the rent level of PRH determined by local government is r_i, and $r_i \in (0, 1]$. During the construction of high-quality PRH, the probability of high effort (q_i^H) in selected by local government is P_1, and the probability of low effort (q_i^L) is $1 - P_1$. Similarly, in the construction of low-quality PRH, the probability of q_i^H is P_2, the probability of q_i^L is $1 - P_2$, and $0 \leq P_2 < P_1$. The objective function of local government is:

$$\begin{aligned}
&\underset{w_i, e_i}{Max}\ U_i^w + U_i^r \\
&s.t.\ U_i^w = P_1(w_i^H - e_i^H) + (1 - P_1)(w_i^L - e_i^H) + (1 - P_2)(w_i^H - e_i^L) + P_2(w_i^L - e_i^L) > 0 \\
&P_1(w_i^H - e_i^H) + (1 - P_1)(w_i^L - e_i^H) > (1 - P_2)(w_i^H - e_i^L) + P_2(w_i^L - e_i^L) \quad \text{(IC)} \\
&P_1(w_i^H - e_i^H) + (1 - P_1)(w_i^L - e_i^H) > 0 \quad \text{(PC)} \\
&U_i^w = r(q_i^H + q_i^L) - c_i \cdot q_i^H - c_i \cdot q_i^L
\end{aligned} \tag{73.1}$$

The first-order conditions of Formula (73.1) are:

$$e_i^H (e_i^H)' = (w_i^H)'(1 + P_1 - P_2) + r - c_i \quad \text{and}$$
$$(e_i^L)' = (w_i^L)'(1 - P_1 + P_2) + r - c_i$$

where, $(w_i^H)'$ and $(w_i^L)'$ are strong and weak marginal incentive respectively, and $(e_i^H)'$ and $(e_i^L)'$ are marginal construction efforts of high-quality and low-quality

PRH. Since the PRH is a quasi-public good, the unit rents are equal to unit costs roughly. So we get $(e_i^H)' \cong (w_i^H)'(1+P_1-P_2)$, and $(e_i^L)' \cong (w_i^L)'(1-P_1+P_2)$. Since $1+P_1-P_2>1$ and $0<1-P_1+P_2<1$, we get:

Proposition 1 *Supposing local governments to pursue maximization of political achievements and economic benefits, the supply of PRH is significantly associated with the incentives*[2] *selection by the central government. Strong incentive can not only increase the scale of high-quality construction, but also increase the speed of high-quality construction, which is called marginal amplification effects. Weak incentive increases the scale of high-quality construction although, but shows marginal narrowing effects.*

73.3.2 Government Reputation: Incomplete Information Games Between Local Government and Household

The objectives and implementation of PRH policy are the first test of the policy sustainability, the second test is the operation and management of PRH project, in which rent level is an important variable that relates to the all subjects' benefits. Supposing during the construction of PRH, the capacity of local government is η, and can be classified into high (*H*) and low (*L*) respectively. The probability of *H* is $P(\eta=H)$, the probability of *L* is $P(\eta=L)$, and the capital and probability are common knowledge of low-income households and local governments. Local government knows clearly its own ability, and selects a scale of construction q. There is competition in the PRH allocation, and if two households observe the scale of high-quality PRH, they will apply simultaneously and provide the rent r. Local government accept the apply of household with more difficult and stable source of income. If the degree of difficulty and income stability is the same, the priority will be selected randomly by lottery.

Supposing there is no fixed costs of PHR construction, local government's benefit is $r-c(\eta,q)$, where $c(\eta,q)$ represents the marginal cost of a unit of PRH construction by the local government with the ability η. The PRH household benefit is $y(\eta,q)-r$, where $y(\eta,q)$ represents PHR productivity of local government. If a household do not live in the PHR, its benefit is 0. Because the scale of PRH construction is a signal to reflect the local government's capacity of policy implementation, this can be seen as a model of two-stage signal game with incomplete information.

In this game, local government with high capacity select q_H^*, which is the scale of high-quality construction, local government with low capacity make a random

[2] Incentives are a broad concepts in here that contain incentive standard, contents and forms, meanwhile they consist of positive (reword) incentives and negative (punitive) incentives.

selection between q_H^* and q_L^* with the probability π and $1-\pi$ respectively, and $\pi \in [0,1]$. The Bayes' rule as follows:

$$P(H|q_H^*) = \frac{P(H)}{P(H) + [1 - P(H)]\pi} \quad \text{and} \quad P(L|q_H^*) = 1 - P(H|q_H^*),$$
$$P(H|q_L^*) = 0, \quad P(L|q_L^*) = 1$$

Assuming that high-capacity local government select q_H^*, because low-capacity local government make random selection between q_H^* and q_L^*, a unit of rent of high-quality and low-quality PRH are $r^*(q_H^*)$ and $r^*(q_L^*) = r^*(L)$ respectively, and they must be met:

$$r^*(L) - c(L, q_L^*) = r^*(q_H^*) - c(L, q_H^*) \tag{73.2}$$

where, $q_L^* \in \arg\max\left[r^*(L) - c(L, q_L^*)\right]$, $q_H^* \in \arg\max\left[r^*(q_H^*) - c(L, q_H^*)\right]$. That is the low-capacity local government will get the same benefit, when it choose the two scale of construction.

Given local government's mixed strategy and household's judge from above, household's balanced strategy is to give the rent $r^*(L)$, when the scale of construction is q_L^*. When the scale is q_H^*, the rent will be:

$$r^*(q_H^*) = P(H|q_H^*) \cdot y(H, q_H^*) + P(L|q_H^*) \cdot y(L, q_H^*), \quad r^*(L) = y(L, q_L^*),$$
$$s.t.\ 0 < r^*(L) \le r^*(q_H^*) \le \bar{r}$$

where, $\bar{r}$ is the market rent. Regardless of q_H^* or q_L^*, the rents in the equations are equal to productivity or expected productivity. Under the Formula (73.2), the mixed strategy is the local government's best response, when given the household policy. Therefore, both sides are sequential rational, which is a perfect Bayesian equilibrium with mixed strategy.

Thus, we get:

Proposition 2 $\pi \in [0,1]$, $q_H^* > 0$, $q_L^* > 0$, then $\frac{\partial r^*(q_H^*)}{\partial \pi} < 0$.

The economic meaning of Proposition 2 is: When information is imperfect and household can not distinguish the quality of PRH in advance, equilibrium rent will be affected by the probability of 'camouflage', which is low-capacity local government pretend to high-capacity. The bigger the probability, the smaller the equilibrium rent. That is, when the rent determined by local government is significantly higher than the rent that household willing to pay, there is possibility that low-quality PRH pretend to be high-quality.

73.4 Evidence and General Discussion

73.4.1 Evidence of Proposition 1

Currently, it is still difficult to obtain a large sample of data about PRH from the statistical departments in China, so the second best option is to look for the typical cases, such as the A city, in real world. The quality of PRH is obtained by a comprehensive Index which is calculated by three steps. Firstly, calculating proportions of 'unsatisfied' variables, such as locations of PRH, convenient degrees of transportation, housing structures, living environments and qualities of construction, in total by each communities, and assigning the proportions to X_i. Secondly, calculating weights of each variables through pairwise comparison matrix. Thirdly, calculating the comprehensive Index to reflect the quality of PRH, and the calculation formula is (Table 73.1). Because of the random sampling, so a large number of survey data can reflect the qualities of PRH.

The central government adopted strong incentives to the supply of PRH in 2010, and issued some big policies, such as the "Notices on Stable and Healthy Development of the Real Estate Markets" in January 2010, and "Guidance on Accelerating the Development of Public Rental Housing" in June 2010 and so on. Meantime, the central government set up a mission that 36 million sets of PRH construction, and let the local governments to complete a certain part in every year and complete all at the end of 2015. Since then, large-scale constructions of PRH are launched by local governments. In A City, construction to completion time of community 1 and 2 are between 2010 and 2011 years, and there are 15,000 sets of PRH with high-qualities in the community 1 and 18,000 in the Community 2. The proportions of high-qualities PRH in all have increased from 87.09 to 88.54%, which showing marginal amplification effects.

Completion time of community 3, 4 and 5 are between 2011 and 2013 years, the amount and speed of constructions of high-qualities PRH decreased respectively, which shows the marginal narrowing effects, although the central government issued "Administrative Measures on Public Rental Housing" and "Notices on Unified Managements of Public Rental Housing and Low-Rent Housing". during this period. Because these policies are 'neutral', and they will diminish their effects before changing them to 'punitive' [8, 18].

73.4.2 Evidence of Proposition 2

The rent gaps are characterized by the households' evaluations to the government rents (Table 73.2), and the qualities of PRH are continue to use the results in Table 73.1. We found that the bigger proportions of "rents are little higher" and "higher", the lower the qualities of PRH. The correlation coefficient of 'little higher' proportions and low-qualities is 0.445 (p value is 0.000) and the correlation

Table 73.1 Qualities of PRH of each communities in A city

Index	Weight	Community 1	Community 2	Community 3	Community 4	Community 5
Without large shopping mall	0.0582	3.66 %	3.89 %	5.96 %	11.85 %	12.36 %
Without top three hospital	0.0285	12.20 %	7.22 %	13.91 %	8.06 %	9.55 %
Without key middle school	0.1356	5.69 %	6.39 %	7.95 %	6.64 %	10.67 %
Inconvenient traffic	0.3085	6.50 %	6.11 %	16.56 %	19.43 %	17.42 %
Irrational structure	0.0582	6.91 %	2.78 %	2.65 %	1.42 %	1.12 %
Poor environment	0.1856	22.76 %	14.72 %	27.15 %	27.49 %	22.47 %
Poor quality of construction	0.2254	21.95 %	23.89 %	15.23 %	32.70 %	39.33 %
Comprehensive index	1	87.09 %	88.54 %	84.44 %	79.63 %	79.09 %
Construction to completion time	–	Mar 2010–Nov 2011	Nov 2010–Dec 2011	Sep 2011–Nov 2012	Sep 2012–Oct 2013	Dec 2012–Dec 2013

Note The percentages are the proportions of each variables to characterize the qualities in total samples of each communities

Table 73.2 Regression results of rent gaps and households' behaviors

y	Variable	Valuation	MNL		MNP	
			Coefficient	S.E.	Coefficient	S.E.
r	Rent gap	Lower = 1; reasonable = 2; higher = 3; much higher = 4	0.283** (0.000)	0.068	0.314** (0.000)	0.077
huk	Hukou	Urban = 1; rural = 2	1.574** (0.000)	0.282	1.120** (0.003)	0.372
age	Age	≤21 = 1; 22–35 = 2; 36–49 = 3; 50–60 = 4; ≥60 = 5	−0.507** (0.000)	0.11	−0.583** (0.000)	0.126
mar	Marriage	Single = 1; married = 2; divorced = 3; widowed = 4	0.028 (0.727)	0.079	0.078 (0.389)	0.090
wor	Employment	Enterprise = 1; government = 2; self-employed = 3; unemployed = 4; retired = 5; migrant worker = 6; graduate = 7	−0.066 (0.237)	0.056	0.595* (0.051)	0.304
edu	Education	Primary school or below = 1; middle school = 2; high school (vocational school) = 3; undergraduate (junior college) or above = 4	0.620** (0.000)	0.163	0.296 (0.201)	0.232
hea	Health condition	Well = 1; chronic = 2; serious illness = 3; disabled = 4	−1.426** (0.000)	0.286	−1.561** (0.000)	0.321
inc	Annual household income	≤15,000 = 1; 15,000–30,000 = 2; 30,000–50,000 = 3; 50,000–80,000 = 4; 80,000–100,000 = 5; ≥100,000 = 6	0.089 (0.189)	0.068	1.155** (0.018)	0.486
*huk * edu*	Hukou × Education	Interaction	−0.409** (0.000)	0.095	−0.226* (0.089)	0.133
*wor * inc*	Employment × Income	Interaction	0.008 (0.654)	0.018	−0.230** (0.035)	0.109
*age * hea*	Age × Health	Interaction	0.345** (0.000)	0.075	0.429** (0.000)	0.091
α	Constant	–	1.699** (0.014)	0.69	−0.526 (0.676)	1.259
N			1146		1146	
Other statistical index			R^2 = 0.092,D – W = 1.801, F = 10.435 (0.000)		R^2 = 0.081, D – W = 1.648, F = 8.964 (0.000)	

Note: (1) MNL means multinomial logistic regression model, and MNP means multinomial probit regression model. (2) ** and *represent 5% and 10% significance levels, respectively coefficient of 'higher' proportions and low-qualities is 0.527 (p value is 0.000). These results indicate that the rent gaps can indeed reflect 'shoddy' quantities and qualities on overall, which indirectly proves the correctness of Proposition 2. In order to reflect the relationship between rent gaps and policies sustainability of PRH, namely the existence of a government reputations, we construct regression model as follows:

$$y = \alpha + \beta_1 \Delta r + \beta_2 huk + \beta_3 age + \beta_4 mar + \beta_5 wor + \beta_6 edu + \beta_7 hea + \beta_8 inc + \beta_9 huk \cdot edu + \beta_{10} wor \cdot inc + \beta_{11} age \cdot hea$$

where, y is households' behaviors after tenancies are expired, and it is discrete variable, to relet = 0, to rent market housing = 1, and to purchase market housing = 2. Δr is rent gaps, and other variables are relevant control variables. Meanings and values of all variables are shown in Table 73.2.

Table 73.2 shows that the larger rent gaps, the more households are reluctant to rent PRH, that is they are more likely to rent market housing. This results show that when rent gaps are going larger, the households are likely to leave out the PRH and return to the market, which reflects 'organization failure' partly. When rent gaps going larger continually and reach a threshold, all households are will to leave out the PRH, which is 'organization failure' completely. Once organization failures occur, they will sent out signals that there are 'shoddy' in the supply of PRH and the negative reputations of the local governments. And the negative reputations may led PRH policies to be non-sustainable.

73.5 Conclusions

We study governments behaviors in the supply of PRH through households' preferences under a game theoretical framework. This framework includes the unsupervised principal-agent game of central and local government, and the signal game under mixed strategy of local government and households. And the using households' preferences of PRH are measured by the comprehensive Index that consists of locations, convenient degree of transportation, housing structure, living environment and quality of construction and the rent gaps to study governments behaviors in the supply of PRH.

Results from the evidences test respectively the two propositions. First, supposing local governments to pursue maximization of political achievements and economic benefits, the supply of PRH is significantly associated with the incentives selection by the central government. Strong or punitive incentive can not only increase the scale of high-quality construction, but also increase the speed of high-quality construction, which is called marginal amplification effects. Weak or reword incentive increases the scale of high-quality construction although, but shows marginal narrowing effects. Second, when information is imperfect, equilibrium rent will be affected by the probability of 'shoddy', which is low-capacity local government pretend to high-capacity, and the bigger the probability, the smaller the equilibrium rent. Once organization failures occur, they will sent out signals that there are 'shoddy' in the supply of PRH and the negative reputations of the local governments. And the negative reputations may lead to 'organization failure', that households are likely to leave out the PRH and return to the market and PRH policies are going to be non-sustainable.

As long as performance evaluation are the main of central government's incentives in China, and there are differences in marginal returns when local governments take different behaviours, local governments will have incentives to 'shoddy' to deliver a strong executive capacities. In this sense, the analytical framework of this paper is suitable for analysis of local governments behaviours in the supply of other public goods and in the process of executing major policies in China.

Acknowledgements This research was supported by the National Natural Science Foundation of China (Grant # 71173223), the Fundamental Research Funds for the Central Universities, and the Research Funds of Renmin University of China (No. 10XNI011) and the Outstanding Innovative Talents Cultivation Funded Programs 2014 of Renmin University of China.

References

1. Samuelson PA (1954) The pure theory of public expenditures. Rev Econ Stat 36(4):387–389
2. Blackley PR, Deboer L (1987) Measuring basic wants for state and local public goods: a preference independence transformation approach. Rev Econ Stat 69(3):418–425
3. Besley T, Coate S (1997) An economic model of representative democracy. Q J Econ 112 (1):85–114
4. Besley T, Coate S (2003) Centralized versus decentralized provision of local public goods: a political economy approach. J Public Econ 87:2611–2637
5. Economides G, Philippopoulos A, Vassilatos V (2014) Public, or private, providers of public goods? A dynamic general equilibrium study. Eur J Polit Econ 36:303–327
6. Besley T, Ghatak M (2007) Retailing public goods: the economics of corporate social responsibility. J Public Econ 91(9):1645–1663
7. López R, Galinato GI (2007) Should governments stop subsidies to private goods? Evidence from rural latin America. J Public Econ 91(5–6):1071–1094
8. Stephens M (2005) An assessment of the British housing benefit system. Eur J Hous Policy 5 (2):111–129
9. Nie H, Li J (2013) Collusion and economic growth: a new perspective on the China model. Soc Sci Electron Publishing 1(2):18–39
10. Laffont JJ, Martimort D (2005) The design of transnational public good mechanisms for developing countries. J Public Econ 89(2–3):159–196
11. Maniquet F, Sprumont Y (2010) Sharing the cost of a public good: an incentive-constrained axiomatic approach. Games Econ Behav 68(1):275–302
12. Saijo T (1991) Incentive compatibility and individual rationality in public good economies. J Econ Theor 55(1):203–212
13. Guccio C, Pignataro G, Rizzo I (2014) Do local governments do it better? Analysis of time performance in the execution of public works. Eur J Polit Econ 34:237–252
14. Bowles S, Hwang SH (2008) Social preferences and public economics: mechanism design when social preferences depend on incentives. J Public Econ 92(8–9):1811–1820

15. Speer J (2012) Participatory governance reform: a good strategy for increasing government responsiveness and improving public services? World Dev 40(12):2379–2398
16. Tiebout CM (1956) Exports and regional economic growth. J Polit Econ 64(2):160–164
17. Tiebout CM (1956) A pure theory of local expenditures. J Polit Econ 64(5):416–424
18. Umezawa M (2012) The replacement principle for the provision of multiple public goods on tree networks. Soc Choice Welf 38(2):211–235

Chapter 74
Option Game Model of Low-Carbon Residential Investment Decision-Making Under Imperfect Competition

Kai Guo, Dezhi Li and Yanchao Chen

Abstract The low-carbon residential project is of great uncertainty and always in long-term, the investment on these projects requires flexible decision basis and long-term decision strategy. According to the variation of the influencing factors in a certain project, the investment decision can be analyzed and adjusted by using the real option method. This paper studied the process of investment decision in low-carbon residential projects based on the real option theory and the research in the market. The method of investment decisions under the market of imperfectly competition is also explained, identified the value function of developers and the relationship between development value and development threshold. It also pointed out the best time and way for developers to make the investment. The conclusion of the paper emphasized the competitive advantage of the call option and offered a strong decision basis for the decision makers.

Keywords Low carbon residence · Real option · Imperfect competitive market · Investment decision

74.1 Introduction

Residential housing projects have always been the main part of the construction industry in our country, it is essential of the economic development and social stability to put great effort on developing the residential projects. Meanwhile, under the background of that the economic and social development are increasingly constrained by the shortage of energy and the environment, vigorously enforcing energy conservation and emissions reduction measures is very imperative. Therefore, considering from the long-term interests of enterprises', it is necessary for real estate developers to get into low-carbon residential market, and corre-

K. Guo (✉) · D. Li · Y. Chen
Department of Construction and Real Estate, Southeast University, Nanjing, People's Republic of China
e-mail: 220141127@seu.edu.cn

Y. Wu et al. (eds.), *Proceedings of the 20th International Symposium on Advancement of Construction Management and Real Estate*,
DOI 10.1007/978-981-10-0855-9_74

spondingly the investment strategies become the new central issue. Especially, the low-carbon residential housing projects are still in the starting phase, the market has defects in many ways [1]. It has more uncertainty compared to traditional residential housing. Thus, how to properly handle this uncertainty is now the key problem to developers' successful investment. The traditional Discounted Cash Flow method has large limitation [2] to the highly uncertain project, does not apply to investment decisions of low carbon residential project. Therefore, the investment strategy this paper proposed is based on real option theory [3], allowing the developers adjust the strategy in view of the changing market situation. And according to the research on residential housing projects, which are under imperfect competition market, this paper pointed out the value function for developers and the right time to invest, which would be significant about making decisions of investment in low-carbon residence.

74.2 Decision Foundation

74.2.1 Research of Low-Carbon Residential Market

In order to accurately comprehend the low-carbon residential market, this paper firstly investigates the uncertain factors concerning the low-carbon residential prices and the corresponding influential degree, so as to figure out the more reasonable investment decision methods, providing reliable consulting for developers. Taking the Eastern China area as an example, this paper made a study on customer characteristic, general preference, preference of low-carbon, and price situation, then concluded that:

(1) Compared to ordinary residence on the same level, low-carbon residence has more useful function and investment value, which greatly improve the owner's expectation for the low-carbon projects. This makes low-carbon residence has more value and higher risk resistance capacity, thereby increasing the specific requirements.
(2) Customers, who once had been in low-carbon housing, are more loyal to this kind of residence. Because that low-carbon housing project itself has relatively completed configuration and positioned in middle, high grade customers, specific requirements of low-carbon residential projects will become less. This means, compared to ordinary residence, the specific requirements of low-carbon residence have more uncertainty.
(3) With the incremental cost of low-carbon technology, the low-carbon residence can easily improve the specific requirements and reduce the entire cost in the long-term operation. It has been found from the survey, whether residents are willing to pay a premium for low-carbon, that nearly three quarters of people are willing to accept extra cost for living in low-carbon housing (showed in Fig. 74.1), this means the price is accepted by most customers for being higher

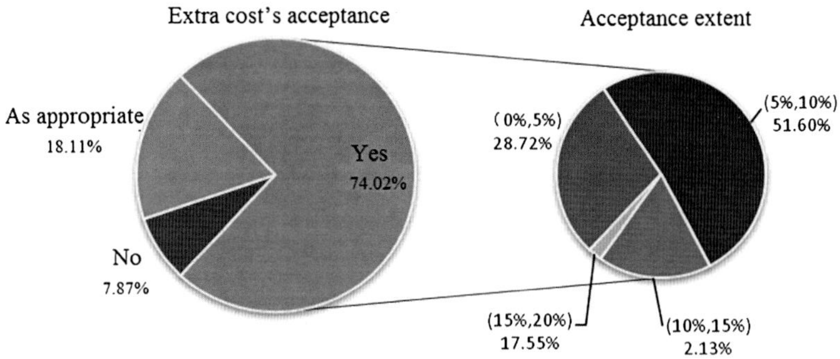

Fig. 74.1 Customers' response to extra cost

than ordinary residence. This makes developers actively reduce costs in order to get the biggest profit.

(4) Compared to ordinary residential projects, low-carbon residences not only have customers' preference, but also the incremental cost in low-carbon technology, this asymmetry embodies the discrepancy between cost and demand of low-carbon residences, meanwhile, it affects the investment timing and profitability of enterprises.

It is can be easily seen from researches above, low-carbon residential market is influenced by many uncertainty factors, thus it is very essential for investors to know how to handle these uncertainties. The following will be a hypothesis based on above conclusions, in order to study the investment decision-making process.

74.2.2 Market Hypothesis

In reality, because of the scarcity of land and the emerging of the low-carbon housing and other reasons, low-carbon residential market, or even the entire real estate market, embodies the feature of imperfect competition, this competition has an influence on the value of the projects and the timing for developers entering the market. Especially for low-carbon residence market, this kind of new emerging segment, there are usually only two or three components competing with each other, thus it forms the new competition environment of duopoly. Calculation model under this market environment better simulate the low-carbon residential developers long-term investment decisions and market behavior, this will be very helpful for entrepreneurs to make decisions of long-term strategies. Then, in view of the market survey, specific conditions are assumed as:

(1) In the duopoly market, there only exists two developers, the one this paper studies is called low-carbon residential (LCR) developer, the other is called

competitor. Known from above analysis, there are double asymmetry of cost and demand between LCR developer and the competitor.

(2) The value of low-carbon residential projects is measured by market rent. Learned from the survey, the sale prices of low-carbon residence are strongly affected by human factor, difficult to be predicted and described clearly by random process. However, the market rent not only reflects the real demand of low-carbon residence, but also conforms to the principle of random process. This paper studies the decision-making behavior under the same market environment, thus, using market rent rate or using sale price of low-carbon residence is the same difference.

(3) According to the market rule, rental rate obeys a downward sloping inverse demand function with increasing demand elasticity, showed as:

$$R_i(t) = G_{\mathrm{i}}(t)K(t)D[Q(t)] \quad i = 1, 2 \tag{74.1}$$

where, $R_i(t)$ is the rental rate of developer I at time t; Q(t) is the supply quantity of low-carbon residence at time t; D(Q) is the inverse demand function of market; Q(t) = q1(t) + q2(t), and $D'(Q) < 0$, $D''(Q) > 0$. To simplify the analysis, this paper assumes that the two developers have the same share of the market, i.e. q1(t) = 1; q2(t) = 1; Q(t) = 1, 2, and $D(1) > D(2)$.

$G_i(t)$ represents the influence on investment decision, directly or indirectly affected by national macroeconomic regulation and policy.

In addition, rental rate is also under the influence of uncertainty of exogenous housing rental market, this kind of uncertainty is embodied as demand shock wave. $K(t)$ is the market demand shock, obeying geometric Brownian motion:

$$dK/K = \alpha dt + \sigma dw \tag{74.2}$$

where α is expected grow rate of K, σ is the corresponding volatility, dw is the increment of standard wiener process.

(4) Assuming that the incremental cost of low-carbon technology is the only difference between the two developers, no other aspects can cause any cost variance. Assuming that the cost of low-carbon residential enterprise is I, the competitor's is kI, because of the cost of low-carbon residence is higher than of ordinary one, so $0 < k \leq 1$.

(5) Assuming that the time the two developers construct the projects are all T.

(6) Determining the difference of competitiveness between developers. Before study their strategies, it is necessary to conform which one of the two developers is in the edge. However under the asymmetry of demand and cost, especially when customers' need for low-carbon residence has strong correlation properties (i.e. both of the competitors are ordinary developers, the decision-making investor will have more share of the market, but it also has to pay higher cost), this makes it not clear which one of the competitors is in the edge of the market. However, it is certain that the developer possessing the advantage position has lower development threshold than the competitor, thus

the development threshold value of them (K_F^1) can be compared. The formula is: $\frac{K_F^1}{K_F^2} = \frac{G_1(t)I}{G_2(t)kI} = \frac{G_1(t)}{G_2(t)k}$. Assuming that the reciprocal of this formula is γ, then γ is the difference of competitiveness between the developer and its competitor, it is the integrated reflection of demand advantage and cost advantage of them. Clearly, when $\gamma > 1$, it means the decision-making developer has more obvious advantage on competition than the competitor.

When $\gamma = 1$, it means there is no difference of competitiveness, no useful strategies, i.e. the two developers will permanently go parallel. For their respective development value $J_i^S(K)$, there is always $J_S^i(K) = nJ_S^j(K)$, $(n > 1)$, the development value of high specific requirement is always n − 1 higher than that with low specific requirement, where n is the demand rate of their specific requirement or cost ratio. When $\gamma < 1$, the decision-making investor has no advantage, thus it is not necessary to develop at this timing. According to research above, low-carbon residence has more advantage in usefulness and investment value, that means low-carbon developers are in more competitive position than the competitor, thus it is assumed in the following discussion that $\gamma > 1$.

74.3 Decision-Making Process

74.3.1 Determination of Value Function of Developers

Developer i ($i = 1, 2$), relative to developer j ($j = 3 - i$), has three possible strategic action: (1) before developer j, developer i invests, becoming the market pioneer; (2) after developer j, developer i invests, becoming the market follower; (3) at the same time, developer i and j invest. When one of the developers enters the market, this investment behavior will have an influence on the market demand, thus affecting the market rental rate, thus it will have a leverage in the other developer's strategy on entering the market, thereby, studying the market behavior and strategic action of the two developers will need option game theory. Reverse tracing method is often used to dynamically study this situation, that means assuming one of the developers has already put investment into the market, then analyzing the follower's value and threshold value, and then analyzing the prior one's development value. So it is not difficult to draw the following conclusions [4]:

(1) Follower's development value is:

$$J_F^i(K) = \begin{cases} \left(\frac{I}{\beta-1}\right)\left(\frac{K}{K_F^i}\right)^{\beta}, & K < K_F^i \\ G_i(t)A_2 - I, & K \geq K_F^i \end{cases} \tag{74.3}$$

where

$$A_2 = \frac{KD(2)e^{-(\rho-\alpha)T}}{\rho - \alpha}$$

$$\beta = \frac{1}{2} - \frac{\alpha}{\sigma^2} + \sqrt{\left(\frac{\alpha}{\sigma^2} - \frac{1}{2}\right)^2 + \frac{2\rho}{\sigma^2}} > 1$$

$$K_F^i = \frac{\beta(\rho - \alpha)I}{G_i(t)(\beta - 1)D(2)} e^{(\rho-\alpha)T}$$

This is to say when the pioneering developer has already invested, the best strategy of the follower is to invest at timing T_F^i, where is the first time that $K(t)$ is equal or more than K_F^i.

(2) Pioneering developer's value is:

$$J_L^i(K) = \begin{cases} G_i(t)A_1 + \frac{G_i(t)}{G_j(t)} FkI - I, & K < K_F^j \\ G_i(t)A_2 - I, & K \geq K_F^j \end{cases} \quad (74.4)$$

where

$$A_1 = \frac{KD(1)e^{-(\rho-\alpha)T}}{\rho - \alpha}$$

$$F = \frac{\beta[D(2) - D(1)]}{(\beta - 1)D(2)} \left(\frac{K}{K_F^j}\right)^\beta$$

(3) Parallel developers' value is:

$$J_s^i(K) = G_i(t)A_2 - I \quad (74.5)$$

74.3.2 Judgment of Decision-Making Behavior

According to the value function curve of the two developers in the preemption area [5], it can be easily described that the relationship between development value and threshold value in the low-carbon residential market, shown in Fig. 74.2.

Where (K_L^2, K_P^2) is the preemptive area, it means: if the initial demand shock located in this area, it is possible for both developers to be the pioneering one, this will more intuitively reflect the decision-making behavior of both parties.

Because of k $(0 < k \leq 1)$ is uncertain constant, the intersection of the curve and axis J is moving in (0, I), i.e. the curve will change as it moves, in actual

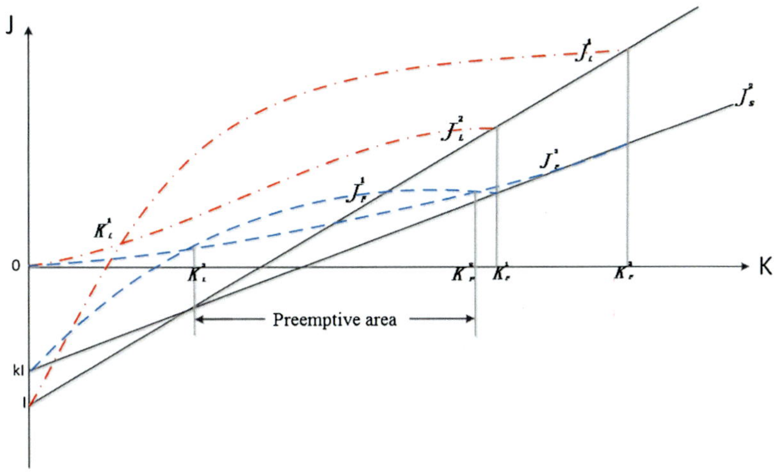

Fig. 74.2 Value function curve of the two developers in the preemption area

decision-making process, the development threshold value can be calculated from k.

Assuming that both the two developers can be aware of the above relationship of value function, development value and development threshold value, i.e., this is a complete information dynamic game process. They will make their decisions of investment behavior in view of the difference of competitiveness and the initial situation of the market. The former decides their order to invest, the latter decides the optimal investment strategy, i.e. the balance investment strategy under the asymmetry of demand and cost. Based on this analysis, investment behaviors are as follows:

(1) When the two developers are all low-carbon developers, the difference of their competitiveness is relatively small, so they both are likely to be the pioneering developer. There exists a preemption area (as shown in Fig. 74.2), however, their specific development strategies depend on the market situation. Specific investment behaviors are as follows: K_M^1 is monopoly development threshold, K is market demand.

① When $0 < K < Min(K_L^2, K_M^1)$, decision-making developer will, at the timing $Min(K_L^2, K_M^1)$, develop instantly, the competitor will follow when the market demand is up to K_F^2;

② When $K_M^1 \leq K < K_L^2$, decision-making developer develops instantly, the competitor will not follow until K_F^2;

③ When $K_L^2 \leq K < K_P^2$, mixed development strategy, i.e. how to develop depends on the best development probability of the two developers within a certain time;

④ When $K_P^2 \leq K < K_F^2$, decision-making developer develops instantly, the competitor will not follow until K_F^2;

⑤ When $K \geq K_F^2$, both of them develop instantly [6]; When the developers are ordinary residential developers, benefiting from the advantage of low-carbon residence, it will make the one who develops low-carbon residence in a dominant position, monopolizing the entire low-carbon market. Then, the development strategies will depend on the market demand.

① When $0 < K < K_M^1$, both of them will not develop, but when the demand is over K_M^1, the low-carbon market is profitable to low-carbon developers;

② When $K_M^1 \leq K < K_F^2$, the low-carbon market is profitable to low-carbon developers;

③ When $K \geq K_F^2$, both low carbon residential developers and ordinary residential developers can invest at this time.

74.3.3 Decision-Making Analysis

In the above part, this paper defined the value function of developers, studied the relationship of development value and development threshold, stated the equilibrium strategies, then analyzed the actions that developers how and when to invest. Then, some other factors which may affect decision-making will be further analyzed next:

(1) It can be easily seen from Fig. 74.2, no matter who has the competitive edge, the slope of value function of the developer with higher specific requirements is always bigger. This means once the demand is strong, exceeding the intersection of the two value function of the two developers, then the developer with higher specific requirements always possesses more investment value. That is to say, when the real estate market is in rising period, it will be more profitable to develop projects with higher specific requirements; when the demand of the real estate market is in a slowdown period, to develop stable requirement projects is less risk and more profitable.

(2) The critical difference is very important to the game behaviors of the two developers. This difference determines their competitive position, i.e. determining that which of the developers is in the advantageous position of the market. In addition, the difference also determines whether the disadvantageous developer can take preemptive action. But apparently, in any case, the developer in the advantageous position has more profit, which mostly comes from monopoly because of first enter.

(3) When both of the two developers are low-carbon developers, they could all benefit from the mutual development cost savings. Reducing the cost is one of

the primary methods for developers to obtain more profit. As to low-carbon residential developers with high cost, reducing the cost under the critical difference helps the developer firstly enter the market and possess more share. As to low-carbon developers with low cost, they can gain high monopoly profits from kicking the competitors out of the market by reducing the cost more. Apparently, reducing the cost is the primary strategy for low-carbon developers.

(4) When both of the two developers are ordinary developers, improving the competitive difference degree is optimal strategy for their long-term operation. This kind of competitive difference is the synthesis of demand difference and cost difference.

(5) When both of the two developers invest into low-carbon residence, the direct and indirect effects from national macroeconomic regulations and policies are the same to them, no matter the effects are positive or negative to the low-carbon market. All they have to do is to accurately measure the regulations and policies. However, when one of the two developers invests into ordinary residence, no matter the national regulations and policies are positive or negative, developing the low-carbon market is always more profitable due to the high quality of low-carbon residence and the pursuit of low-carbon life.

74.4 Conclusion

This paper provided investment making-decision method of low-carbon residential development. It clearly analyzes the timing for investors entering the low-carbon market under imperfect competition. Through defining the value function of developers and studying the development value and the development threshold curve, it not only can be seen that the earlier they enter the market, the bigger the critical difference; The higher the uncertainty, the smaller the critical difference; The higher the rental rate, the smaller the critical difference; And the higher the discount rate, the bigger the critical difference. Furthermore, it concluded that how big the difference of competition is plays a very important role in investment making-decision of low-carbon residence. Therefore, when developers consider entering the low-carbon residential area, it is necessary to realize the long-term influence of national policy on low-carbon residence and the call option making the owner in the more competitive position brought by developing low-carbon market, this business strategy of low-carbon residence values a lot more than the strategy of ordinary residence.

Acknowledgments This research is jointly supported by National Natural Science Foundation of China (No. 71301024) and Ministry of Education of the People's Republic of China (No. 20120092120019 & No. 13YJC790067).

References

1. Zhang SL, Zhao F (2010) Operation mode of low-carbon construction market in China. Constr Econ 328(2):50–52
2. Wang A (2010) Comparison of real asset valuation models: a literature review. Int J Bus Manag 5(5):14
3. Kiriyama E, Suzuki A (2004) Use of real options in nuclear power plant valuation in the presence of uncertainty with CO_2 emission credit. J Nucl Sci Technol 41(7):756–764
4. Hou CX, Song YF, Ma N (2010) Real estate investment decision-making research using real option. J Eng Manag 24(4):447–451
5. Liu T (2007) Investment decision-making theory of real estate using real option. Shanghai Jiaotong University, pp 109–128
6. H YB, Zhang M (2010) Life-cycle cost analysis of residence from low-carbon perspective. Ecol Econ 232(11):109–113

Chapter 75
Lessons Learnt from Delivering Oil and Gas Projects in Australia: An Empirical Research

Gonzalo Rafael Chávez Abaunza and Konstantinos Kirytopoulos

Abstract The engineering and construction companies working in the oil and gas industry treat lessons learnt as secrets of the trade. Therefore, they rarely publish relevant registers for professional or educational purposes. Aim of this research is the investigation of current issues in oil and gas industry through capitalising on experience coming from practitioners (lessons learnt). The research tool comprises a semi-structured questionnaire that was built around the project management subject groups as these are described in the ISO21500 standard: Guidance on Project Management as well as the PMBOK. Nine Australian oil and gas project managers with experience ranging from 10 to 20 years, were interviewed, and their personal experiences were investigated through cluster analysis. Relevant ideas among interviewees were grouped together forming captured lessons learnt that are presented under four overarching themes, namely (a) Integration and Quality, (b) Human Resources, Communications and Stakeholders, (c) Risk and (d) Procurement, Scope, Time and Cost. In conclusion, this empirical research presents a particular insight to the Australian oil and gas construction projects and extends the limited repository of scholar publications for lessons learnt from this particular industry.

Keywords Project management · Oil and gas · Lessons learnt · Engineering and construction projects

75.1 Introduction

In accordance with the Project Management Body of Knowledge (PMBOK), every project manager should record lessons learned at the closing stage of a project, once the scope has been completed [1]. The importance of capturing the knowledge

G.R.C. Abaunza · K. Kirytopoulos (✉)
School of Natural and Built Environments, University of South Australia,
Adelaide, Australia
e-mail: Konstantinos.Kirytopoulos@unisa.edu.au

Y. Wu et al. (eds.), *Proceedings of the 20th International Symposium on Advancement of Construction Management and Real Estate*,
DOI 10.1007/978-981-10-0855-9_75

created through the execution of projects is vividly given by Buttler et al. [2] who state that "lessons learnt are one way to retain experience and knowledge in project-based organisations, helping them to prevent reinventing the wheel or repeating past mistakes". Bischoff [3] suggests that lessons learnt should be promoted as a means for continuous improvement in an organisation, not limiting the relevant workshops when things go wrong, but also recording steps to ensure things go as planned.

However, according to both knowledge and project management literature, it seems that lesson learnt processes do not take place for most projects and organisations [4]. The reason why, may relate to the complexity of capturing and communicating the generated information from the individual projects and relevant participants to the company level. In order to do so, specific frameworks need to be developed, including cognitive and situated learning theories [5].

The engineering and construction companies working in the oil and gas industry are probably an exception to the rule as they often capture lessons learnt. However, they rarely publish these registers for professional or educational purposes. Most times, such knowledge repositories are considered secrets of the trade. The companies expect to get competitive advantage out of them and also avoid loss of reputation when managerial or technical errors are recorded in the process. As a result, the research problem raised here is that publicly available information is scarce and scholars do not have a sound basis of literature to build upon.

This paper reports the outcomes of an empirical research that provides several examples of issues that the Australian engineering and construction companies operating in the oil and gas industry currently face. The main issues are classified against the project management subject groups as per ISO [6], and directions for potential solutions are offered. In a nutshell, this research gives insights into lessons learnt from the industry projects and highlights the importance of documenting project's lessons learnt.

The rest of this paper is organised as follows. Section 75.2 presents the literature review related to capturing knowledge from the delivery of oil and gas construction projects. Section 75.3 describes the research method as well as the detailed aims and objectives of the study. Section 75.4 elaborates on the research findings and Sect. 75.5 discusses these results in relation to existing literature. Section 75.6 concludes the paper and offers paths for further research.

75.2 Literature Review

The research was complemented with a comprehensive literature review and the most interesting and relevant papers are presented hereafter. The literature review supported the importance of capturing and documenting professional experiences when managing engineering and construction projects in the oil and gas industry.

The oil and gas industry was practically established in Australia with the first offshore well in Bass Strait at 1965 and since then it "has delivered enormous benefits in the form of exports earnings, domestic economic activity, employment and investment… and… created and sustained a globally respected, innovative technology, services and manufacturing sector" [7]. Projects related to this industry are usually classified as medium to large construction projects. A posteriori analysis of projects and lessons learnt have been discussed in the literature. Especially for large projects in Australia, Hatcher et al. [8] claim that "Concerns have been raised about our capacity constraints, productivity, and the cost and efficiency of the delivery of major projects". These issues of large, complex projects are similar across industries and usually result in cost overruns and delivery delays. Some of the issues raised in Hatcher et al. [8] relate to salary expectations, low productivity and the lack of qualified professionals. The finding is further supported by the skills shortage report by the Department of Education, Employment and Workplace Relations [9]. The report states that Australia needs petroleum engineers, geologists and geophysicists, as well as engineering trade workers like fitters, mechanics, etc.

Other issues related to construction projects in Australia relate to the tax regime, environmental assessments, labour and material costs and duplication of infrastructure [10]. In their research, Marlow and Jang [10] use lessons learnt from recent Australian oil and gas projects to foster the relevant industry in Canada. The authors followed a qualitative approach, presenting examples of lessons learnt from the construction of LNG projects in Australia (Gladstone in particular). The research aim for the project undertaken by Marlow and Jang [10] was to highlight these lessons in order to assist with the approval of similar LNG projects in British Columbia.

In a more managerial context, Badiru and Osisanya [11] present a global approach to the execution of oil and gas projects using the PMBOK knowledge areas. They discussed the particularities of this type of projects for some knowledge areas with a particular focus on schedule control and forecasting. Badiru and Osisanya [11] foresaw that "the challenges the oil and gas projects will continue to face, fall in the following categories: technical, managerial and human resources". The research expanded on similar responses from the participants, depending on their local experiences.

Following the managerial focus but this time at a higher level, Crawford and Koh [12] used "a pragmatic perspective" and "sequential multimethod approach" to evaluate portfolio management across six Australian organisations delivering projects for multiple industries. Especially for the case relating to the energy sector, findings show how the understanding of portfolio management can improve project delivery.

Common ground of the sources examined in the literature was the importance of capturing and using lessons learnt along with the perception that lessons learnt literature is limited.

75.3 Research Method

Aim of this research is the investigation of current issues in engineering and construction projects in the oil and gas industry through capitalising on experience coming from practitioners. The approach followed is of qualitative and inductive nature while the underpinning philosophy leans against interpretivism.

Creswell (cited in Carrillo et al. [13]) suggests using qualitative research approach to knowledge management and learning investigations, due to the abstract nature of the topics. The research tool comprises a semi-structured questionnaire that was built around the project management subject groups (knowledge areas) as these are described in the ISO21500 standard: Guidance on Project Management [6] as well as Project Management Body of Knowledge—PMBOK [1]. Fink (cited in Jansen [14]) recommends "qualitative survey analysis for the exploration of meanings and experiences… as… qualitative survey is the study of diversity (not distribution) in a population". The questionnaire used had one section for each ISO21500/PMBOK subject group/knowledge area: Integration, Scope, Time, Cost, Quality, Human Resources, Communications, Risk, Procurement and Stakeholders.

As part of data analysis, lessons were grouped in a lessons learnt register and scored when different PMs provided similar responses. In order to present unbiased results, the top scoring issues (those that achieve the highest consensus amongst participants are presented in this paper.

In total, nine Australian oil and gas project managers with experience ranging from 10 to 20 years, were interviewed, and their personal experiences were investigated through cluster analysis. Relevant ideas among interviewees were grouped together forming captured lessons learnt that are presented under four overarching themes.

75.4 Research Findings

75.4.1 Construction Business License

The interviews contacted offered significant insights on the issues that the engineering and construction companies face when delivering oil and gas projects. Approximately 50 distinct lessons learnt/issues were captured during the empirical research. Table 75.1 presents those that have a consensus which is higher than 50 %. That is, 5 or more out of 9 interviewees raised them as important for the industry.

Human resources management and integration management knowledge areas had the highest number of "high consensus" lessons learnt/issues while lessons related to the use of lessons learnt processes, the use of appropriate software tools for scheduling, the meetings as a means of communication and the importance of

Table 75.1 Issues with high consensus amongst project participants

Lessons learnt/issues	Consensus (% among participants)	Integration	Scope	Risk	Time	Quality	Procurement	Human resources	Communications	Stakeholders	Cost
Lessons learnt sessions take place	89	×									
Primavera or MS project for project planning depending on project budget	89				×						
Progress reported against established baseline	56				×						
Site remoteness significantly impacts cost	56										×
Scope changes are formally managed (defined processes)	56		×								
Often, scope statements are not well presented	67		×								
Project delays due to poor quality	56					×	×				
Existing risk registers (formal process)	56			×							
Difficulty to retain qualified professionals due to salary expectations	67							×			

(continued)

Table 75.1 (continued)

Lessons learnt/issues	Consensus (% among participants)	Integration	Scope	Risk	Time	Quality	Procurement	Human resources	Communications	Stakeholders	Cost
Staff frequent turnovers leads to inefficiencies	56				×			×			
Professional engineers supervising and signing off documentation should be registered at a national level	67							×			
Regular (weekly) meetings is the main form of communication	89								×	×	
Interdisciplinary coordination mainly happens through weekly progress meetings	56	×							×	×	
Meetings are formally minuted—communicated to attendees	78								×	×	
Communications files are handled through document management systems	56								×		

(continued)

Table 75.1 (continued)

Lessons learnt/issues	Consensus (% among participants)	Integration	Scope	Risk	Time	Quality	Procurement	Human resources	Communications	Stakeholders	Cost
Professionals are conscious on environmental safety issues and believe that the processes are sufficient	78			×						×	
Project Managers need to be aware of all the relevant knowledge areas (ISO21500/PMBOK)	89	×						×			
PM concepts from other Industries might be the same but specific knowledge of the O&G industry is also required	67	×						×			
Engineering background will be better suited for a PM position in the O&G industry	56	×						×			
	Totals	5	2	2	3	1	1	6	4	4	1

project management standards knowledge for those managing projects in the industry received the highest consensus.

The most important of lessons learnt/issues that received the highest consensus among interviewees are organised below in four areas, namely (a) Integration and Quality, (b) Human Resources, Communications and Stakeholders, (c) Risk and (d) Procurement, Scope, Time and Cost.

75.4.2 Integration and Quality

The main research finding in this area was that the interviewed project managers valued quality as more important than cost and schedule. Oil and gas companies will have different ways to deal with projects delivered late or over budget; however, the final users will always complain if a project does not meet the specified objectives.

The interviewed project managers also indicated that the value of lessons learned is recognised in most organizations. They have either participated in lessons learned workshops or searched through lessons learned registers. The participants expressed the difficulty of recording knowledge during project execution because of everyday project commitments and then at project completion, most of the team has disbanded or the project has run out of money to hold a close-out lessons learned workshop.

75.4.3 Human Resources, Communications and Stakeholders

Participants confirmed that the Australian oil and gas companies often struggle to retain qualified professionals. The interviewed project managers explained that this may be caused by staff mobility, higher salary expectations than other industries, limited resource pool, and competition with larger oil and gas hubs like Middle East, North America and Eastern Europe, amongst others.

Almost all research participants considered that professional registration for engineers needs to be adopted at a national level. This would improve the quality of work for oil and gas projects.

75.4.4 Risk

According to the research participants, the Australian oil and gas industry is risk-aware. The project managers should be aware of scope changes deriving from risk review meetings like safety in design, job hazard analysis, safety management studies, hazard and operability reviews, safety integrity level/layers of protection analysis and others.

Most research participants have experienced the inefficiencies associated with handovers and learning hours for new personnel coming up to speed on projects that are half-way through execution. A significant risk that the project managers should consider in their project estimates are the down-time and rework hours associated with staff turnover.

75.4.5 *Procurement, Scope, Time and Cost*

The Australian oil and gas industry has stringent quality assurance requirements, which should be clearly spelled out in the contracting arrangement when using an overseas company. This could be a cost-effective measure for the project manager to transfer risk and use a modularised approach to the design. The project manager will need to consider language barriers, time zone differences, need for third party inspections and pre-qualification amongst other issues in the procurement management plan. One participant mentioned that overseas vendors didn't understand the Australian oil and gas market, because unless it is fully aligned with European or American standards, the Australian market is relatively small for them to do a location-specific version of their product.

Some of the participants have used an overseas engineering consultant, fabricator or supplier. They agree that it is very important to clearly define the scope and provide as much documentation as possible. Communication has been a barrier between Australia and countries in Asia because of language and time difference. The project managers may need to allow supervision, third party inspections or regular video conferences in order to measure progress. They will then need to consider the logistics and lead times involved with the shipment of products from Asia to the remote oil and gas construction sites.

Two research participants expressed their concerns for the lack of integration between the schedule and the cost tracking packages. Furthermore, almost all research participants declared that the cost tracking information recorded by an accounting package (i.e. SAP, Oracle and others) is not project management-oriented. They often end doing data downloads to an MS Excel spreadsheet, in order to manipulate the data and present it via pivot tables and graphs, as part of their cost accrual and monthly progress reports.

75.5 Discussion

From the participant's responses, it was suggested that for small to medium oil and gas construction projects in Australia, a project engineer is better suited to combine both project management and technical roles cost effectively. This finding aligns with Plummer [15] stating that "The project engineer uses engineering skills to bring judgment to decisions and to understand the arguments and rationale that other engineers present". The oil and gas project manager needs to understand the legal and technical implications of the decisions taken throughout the execution of a project.

Another aspect that the industry's project managers should pay attention to is safety. The research participant's all agree that the Australian oil and gas industry has stringent risk assessments, environment, health and safety (HSE) procedures and strict compliance to Australian standards as well as industry practices. In particular, one participant mentioned that in oil and gas projects, safety is more important than quality, then we have cost and schedule. As in Lester [16] "Apart from the conventional responsibilities for time, cost and performance/quality, the PM must ensure that all safety requirements and safety procedures are complied with. Serious accidents do not only have personal tragic consequences, but can destroy a project or indeed a business overnight". This statement captures the high HSE standards applied across Australian oil and gas sites.

For small and medium oil and gas projects, the participants confirmed that the scope statements are often incomplete and need to be improved. One option suggested to overcome this issue is that the project manager gets directly involved in defining the project's objectives, business justification and project need, in order to have a clear idea of the project's scope and then transfer this intention into the design team. The selection of an engineering consultant should then be based on an objective evaluation of proposals that add value to the project and will design a cost effective solution. The project manager will need to then identify risks as well as mitigation strategies, then constantly revisit the risk register as the design progresses. Eweje et al. [17] also agree to this conclusion: "The stronger the project manager's sense of control is over risk exposures, the more likely his or her project information feed system will create better long-term strategic value".

The research also confirmed that "Onshore, the challenges are related to remoteness and a lack of infrastructure. Here the domestic industry has sufficient experience and it is more a matter of time and money" [18]. According to the interviewees, the remoteness of the Australian sites has a significant impact on the costs associated with materials and labour, as well as the logistics required to deliver them. On top of that, the lack of skilled resources [9] is another important issue in the industry. This is coupled by the fact that frequent turnover amongst the oil and gas professionals is considered one of the highest in the nation. According to the interviewed project managers, staff needs to be offered secure career opportunities and complex challenging work. These professionals need to maintain technical skills via continuing professional development, courses, seminars and classroom training as required. Therefore the senior management needs to consider all these training needs and human resource issues in the execution plan and contingency allowance.

75.6 Conclusions and Further Research

This research project has concluded on practical issues that the engineering and construction companies operating in the oil and gas industry face, specifically in Australia, along with relevant recommendations from project practitioners. The research aim was addressed by the compilation of a list of lessons learnt/issues currently relevant to the industry that has been gathered through empirical research including nine Australian oil and gas construction project managers with experience ranging from 10 to 20 years.

The main lessons learnt that were revealed through this research are as follows. In regards to integration and quality management, it seems that the oil and gas industry, unlike other industries, shows a clear preference to safety and quality compared to time and cost while implementing projects. Lessons learnt, as part of the integration area are highly valued, although capturing them is challenging. Considering human resources, the main challenge that led to significant issues in the past is the lack of and retention of qualified staff at all levels. It is not uncommon for staff to get experience in Australia and then set for more rewarding markets like the Middle East and North America. The risk knowledge area in the industry is mainly focusing on safety. Another crucial lesson learnt is the inefficiency associated with handovers and learning hours for new personnel coming up to speed on projects that are half-way through execution. Procurement is also a major issue in the industry. The Australian oil and gas companies follow stringent quality assurance requirements, which should be clearly spelled out in the contracting arrangements when using an overseas company. Communication has also been reported as a barrier between Australia and countries in Asia because of language and time difference. Careful consideration of logistics and lead times involved with the shipment of products from Asia to the remote Australian oil and gas sites is the main concern for time management.

In conclusion, this empirical research presents a particular insight to the Australian oil and gas industry and envisages to extend the limited repository of scholar publications of lessons learnt from this particular industry.

This research project has not presented how lessons learnt are managed at an oil and gas program or portfolio level, neither has it evaluated the project management practices required for oil and gas multibillion dollar mega-projects, using engineering, procurement and construction (EPC) arrangements. However, it has created the ground for further relevant research to take place. It is suggested that the further research should focus on the topics identified here as the main issues in the industry.

References

1. Project Management Institute (2013) A guide to the project management body of knowledge —PMBOK guide, 5th edn. Project Management Institute, Pennsylvania (USA)
2. Buttler T, Lukosch S, Lindstaedt S, Granitzer M (2012) Rethinking lessons learned capturing —using storytelling, root cause analysis and collaboration engineering to capture lessons

learned about project management. In: Knowledge management and knowledge technologies: proceedings of the 12th international conference, ACM, New York

3. Bischoff B (2010) Building a lessons-learned culture, PMI, viewed 09 September 2014, <http://www.pmi.org/Knowledge-Center/ ~ /Knowledge%20Shelf/Bischoff_2010.ashx>
4. Duffield S, Whitty SJ (2015) Developing a systemic lessons learned knowledge model for organisational learning through projects. Int J Project Manage 33(2):311–324
5. Kitimbo, Dalkir (2013) Learning from project experience: creating, capturing and sharing knowledge. Knowl Manage 12(4):59–74
6. ISO (2012) ISO 21500 guidance on project management, international organisation for standardisation (ISO). Geneva, Switzerland
7. Austrade (2013) Oil and gas—industry capability report. Australian Trade commission, commonwealth of Australia, viewed 27 September 2014, <www.austrade.gov.au/ArticleDocuments/2814/Oil-and-Gas-ICR.pdf.aspx>
8. Hatcher C, Pisarski A, Zolin R (2013) 'Leading, large, complex projects', AIPM's project manager, August–September 2013, p. 39
9. Department of Education, Employment and Workplace Relations (2013) 'Skill shortages list—Australia', Skill shortages, Australian Government, viewed 27 September 2014, <http://pandora.nla.gov.au/pan/129492/20131002-0817/foi.deewr.gov.au/system/files/doc/other/skillshortagelistaus.pdf>
10. Marlow I, Jang B (2014) The LNG race: the lessons Canada can learn from Australia. The globe and mail, gladstone—Vancouver, viewed 15 April 2014, <http://www.theglobeandmail.com/report-on-business/international-business/the-lng-race-the-lessons-canada-can-learn-from-australia/article17946509/?page=1>
11. Badiru A, Osisanya S (2013) Project management for the oil and gas industry: a world system approach. Taylor and Francis corporation, Boca Raton, viewed 17 August 2013, <http://reader.eblib.com.au.ezlibproxy1.unisa.edu.au/(S(k50foe40t2ikfvwltdtoceun))/Reader.aspx?p=1114004&o=110&u=%2f0V8DkPD6bpd6HzyIOiCZQ%3d%3d&t=1376739738&h=F654A87D0AF0982C4CC1327F9746509995D8FC6A&s=9551000&ut=318&pg=1&r=img&c=-1&pat=n>
12. Crawford L, Koh A (2012) Portfolio management: the Australian experience. PMI's Project Manage J 43(6):33–42
13. Carrillo P, Ruikar K, Fuller P (2012) When will we learn? Improving lessons learned practice in construction. Int J Project Manage 31(1):567–578
14. Jansen H (2010) The logic of qualitative survey research and its position in the field of social research methods. Forum Qual Soc Res 11(2) viewed 09 September 2014 <http://www.qualitative-research.net/index.php/fqs/article/view/1450/2946>
15. Plummer F (2007) Project engineering: the essential toolbox for young engineers. Butterworth-Heinemann, Burlington MA (USA)
16. Lester A (2007) Project management, planning and control: managing engineering, construction and manufacturing projects to PMI, APM and BSI standards, 5th edn. Butterworth-Heinemann, Jordan Hill (GBR)
17. Eweje J, Turner R, Müller R (2012) Maximizing strategic value from megaprojects: the influence of information-feed on decision-making by the project manager. Int J Project Manage 30(6):639–651
18. Bradshaw M (2010) A new energy age in Pacific Russia: lessons from the Sakhalin oil and gas projects. Eurasian Geogr Econ 51(3):330–359

Chapter 76
Development of a Smart Work Site for Early Warning of Heat Stress

Xin Liu and Xiangyu Wang

Abstract Heat stress is a serious and important issue to be addressed during the work process in even moderate environments let alone under extreme heat and exertion situations. It can lead to deaths and injuries caused by heat stroke. The Western Australia's resource industry sites in remote areas expose to hot and sometimes humid conditions, where the probabilities of experiencing heat stress injuries are more likely. Traditional safety management of heat stress mainly depends on the workers' awareness and behaviours. This project aims to develop a smart work site to manage workplace heat stress proactively, by integrating the functionality of Geographic Information System and Building Information Modelling. Cisco smart sensors will be applied to monitor the environmental conditions of the smart work site. The location of these smart sensors will be optimised by suitable spatial sampling methods. 3D spatial modelling will be used to simulate the 3D environmental conditions for the entire site. Furthermore, the locations will be monitored by Radio Frequency Identification Devices. This real-time monitoring and early warning system will safeguard the well-being of workers exposed to heat stress. The method developed in this study can also be applied to areas of smart city and smart campus for healthcare solutions.

Keywords Heat stress · Early-warning · Geographic information system · Building information modelling

X. Liu (✉) · X. Wang
Australasian Joint Research Centre for Building Information Modelling, Curtin University, Bentley, WA 6102, Australia
e-mail: xin.liu@curtin.edu.au

X. Wang
Department of Housing and Interior Design, Kyung Hee University, Seoul, Republic of Korea

Y. Wu et al. (eds.), *Proceedings of the 20th International Symposium on Advancement of Construction Management and Real Estate*,
DOI 10.1007/978-981-10-0855-9_76

76.1 Introduction

Workers are often exposed to severe environmental heat stress on the work site, especially in summer in extreme environments as evidenced by deaths and injuries caused by heat stroke [1]. Heat stress can result in workers' inattentiveness, which may lead to further tragedies like drowning or impact by falling or moving objects. To prevent heat-related incidents on the site, a series of fundamental practice notes and guidelines have been promulgated [2]. The performance of traditional safety management methods regarding heat stress mostly depends on the specific regulations of workers' behaviours, which is normally not under the control of site safety managers.

Most previous studies on workplace accidents merely focus on where hazards may occur and how to intervene, while the fact is that for heat stress related injuries the real situation of the work site could be much more complicated, and precautions can be difficult to take. The main objective of this research is to develop a smart work site that is able to provide real-time monitoring and early warning of potential heat stress.

76.2 Heat Stress Management

76.2.1 Heat Stress Among Mine, Oil and Gas Workers

Workers employed in outdoor occupations such as surface mining, construction and farming, are recognised as high-risk groups, especially those workers with high frequency, duration or intensity of physical activity [3].

Although miners normally have a high quality of acclimatisation, heat illnesses are still experienced [3]. Therefore, further cooling or early warning methods are required to be developed in heat stressful occupations, especially for the resource companies in remote and hot north of Australia.

76.2.2 Determinates of Heat-Related Illness or Mortality

Risk factors of heat stress can be categorised as intrinsic, such as age, low level of physical fitness, medical conditions, medications, drug and alcohol abuse, lack of

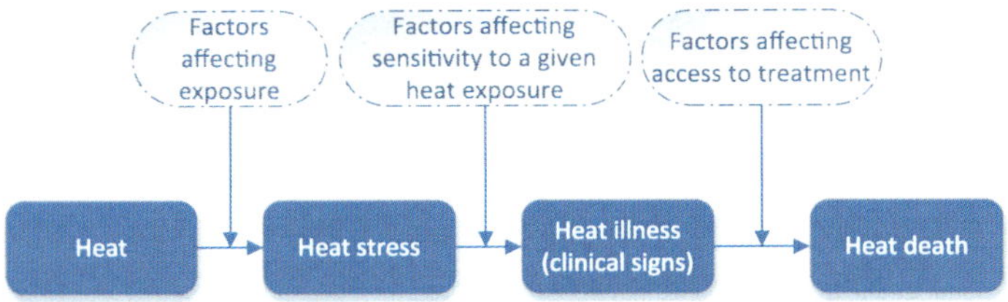

Fig. 76.1 Points along the causal chain from heat exposure to heat death

acclimatisation, dehydration; and extrinsic, such as high temperature and/or humidity, reduced air movement, working near radiant heat sources, contact with conductive heat sources [2].

Heatwave has also been identified as a significant factor for heat-related mortality [2]. However, the effect of a heatwave event will be dependent on a number of factors. They include heat wave magnitude, timing in season, population experience of heat wave events, and public health responses [4].

The building condition that workers or people working or living can be another determinate of heat stress. Furthermore, some socio-economic factors can matter. For example, individuals with low incomes are more chronic disease or other medical risk factors, which will increase the risk of mortality during a heat wave [5]. Figure 76.1 illustrates the causal chain from heat to heat death.

76.2.3 Risk Assessment and Control

In an occupational context, the risk of heat strain is usually identified by many indices. Heat stress indices use environmental and behavioural parameters to model the thermal strain experienced by the body. According to Brake and Bates [6], in excess of 50 heat stress indices have been developed since the 1920s. The most commonly measured environmental variables include air temperature, relative humidity, air velocity, and radiant heat. A further two variables included in heat stress indices relate to individual behaviour: clothing levels and metabolic rate (referring to the heat generated by the body's activity).

Brake and Bates [6] further argues that heat stress indices are overly conservative to account for such uncertainties. However, indices of heat stress such as the Wet-bulb globe temperature (WBGT) and the heart rate provide useful screening tools as part of preliminary risk assessments. In addition, although much is known about the relationship between built environment and health, the evaluation of health condition on the site becomes crucial, considering the complex environment and intensive work condition.

76.3 Development of Smart Work Site for Early Warning of Heat Stress

Research has already been undertaken to assess the significance of heat stress on the safety and productivity of outside workers employed in environments subject to extreme weather conditions [2].

To achieve the goals of this research, several tasks need to be accomplished, namely, spatial sampling of smart sensors; 3D spatial interpolation of environmental factors; integrating the spatial infrastructure with the smart sensor infrastructure; linking real-time environmental data to the modelling system and developing a management platform of early-warning of potential heat stress. In this study, Woodside Pluto LNG worksite has been selected as study area to test the developed method (Fig. 76.2).

76.3.1 Optimization of the Smart Sensors' Location

The significance of and spatial variation of environmental factors on the work site will be determined first to optimise the spatial sampling of smart sensors. Traditional spatial sampling methods, considering the significant variation of environmental conditions in the work space, are not suitable for this complex situation.

Transitional spatial sampling methods include: random sampling, stratified sampling and systematic sampling [7]. Neither random sampling nor systematic sampling can easily account for the spatial variations of the environmental condition on the working site. While, stratified sampling normally requires good knowledge about the study area, which is not easy to get. Furthermore, not all of these methods can guarantee that the sampling is spatially balanced. Therefore, a more effective method is required.

Fig. 76.2 Study area

The Kriging method was applied to exam the spatial relationship of the historical 2D heat stress index (HSI) on the site and interpolate the HSI:

$$\hat{Z}(s_0) = \sum_{i=1}^{N} \lambda_i Z(s_i) \tag{76.1}$$

in which, $Z(s_i)$ is the historical statistical information at location i, λ_i is the weight at the location i, s_0 is the location to be estimated and N is the number of value used to predict $\hat{Z}(s_0)$. And, the 12 locations with maximum standard error of environmental condition were chosen as sampling point:

$$\text{Sampling} \quad (s_i) = \max_{i=1:N} ste(\hat{Z}(s_0 - s_i)) \tag{76.2}$$

The optimised sampling locations can better capture the spatial variation of the environmental condition and represent the work condition for heat stress (Fig. 76.3).

76.3.2 3D Environmental Condition Monitoring and Early Warning for Heat Stress

The digital 3D model of the work site has been developed by Geography Information System (GIS) and Building Information Modelling (BIM). The Kriging method has been applied to interpolate the 2D environmental condition by the information from sampling points. While, sensors have also been placed on each level of the buildings. Therefore, 3D environmental condition can be monitored by these sensors (Fig. 76.4).

The positioning system has also been integrated into this digital 3D model. It can monitor the indoor and outdoor location and status of workers by applying Radio Frequency Identification (RFID) devices. The ArcSever works as the platform and engine for the tracking system (Fig. 76.5).

The early warning management system has been used to analyse, display data and give suggestions on work site heat stress problems. The early warning of heat stress was built based on the Bayesian Networks (Fig. 76.6). The quantified relationships in the network have been calculated by training historical data. The threshold in this warning system is expressed as the high likelihood of heat stress.

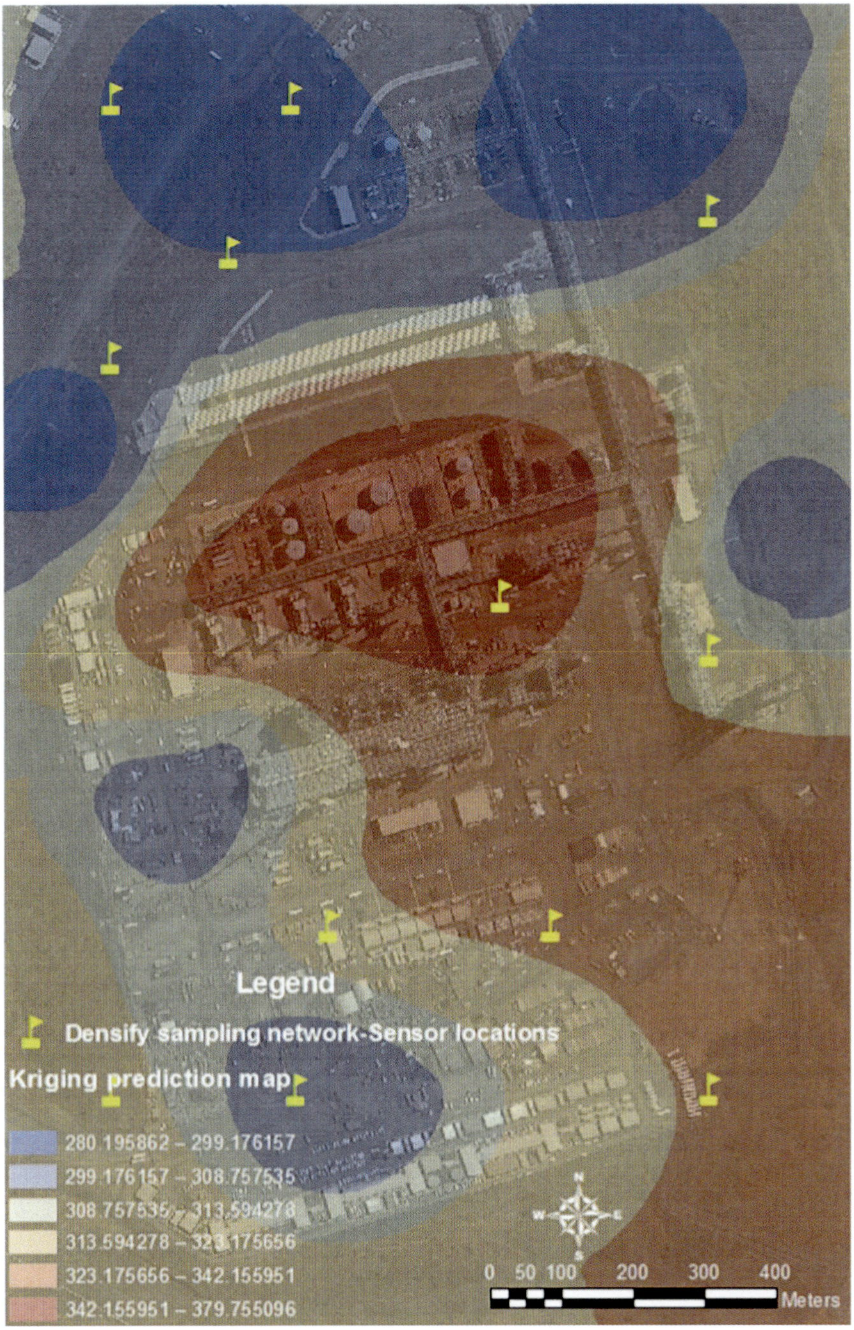

Fig. 76.3 Optimised positions of the sensors for heat stress environment monitoring

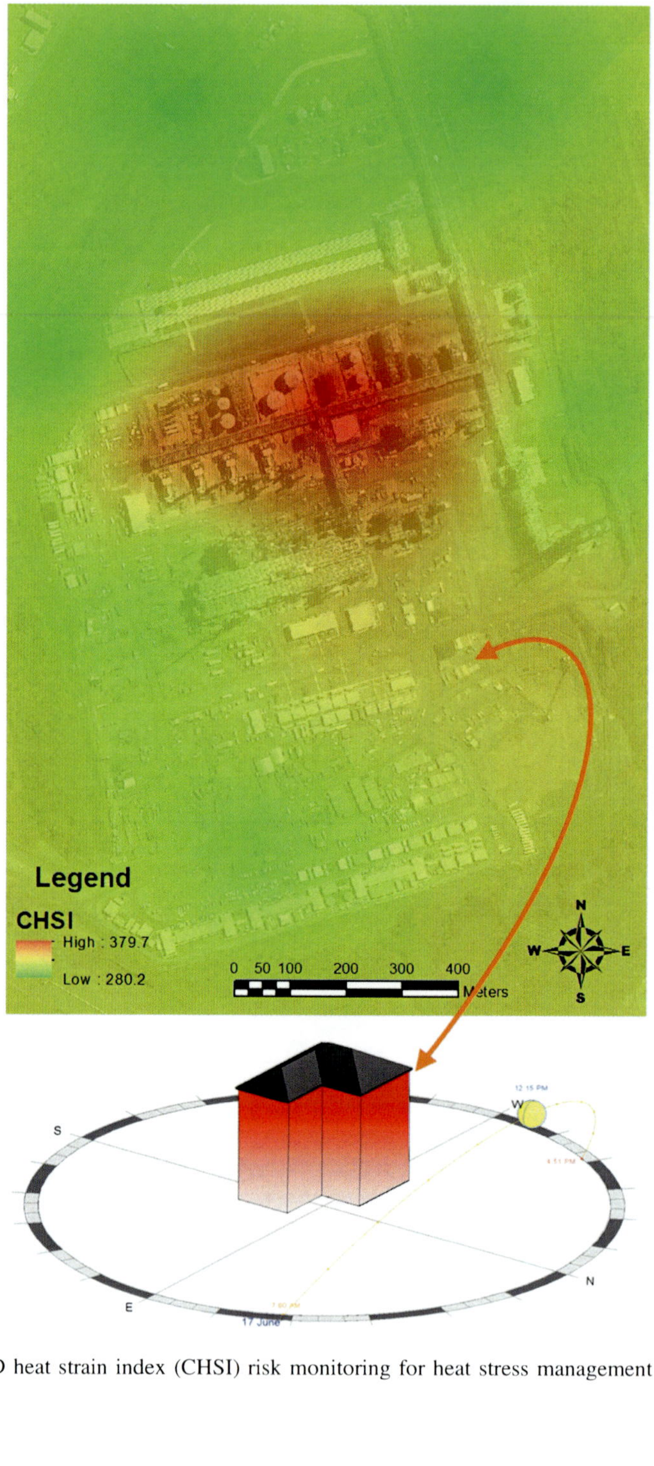

Fig. 76.4 3D heat strain index (CHSI) risk monitoring for heat stress management

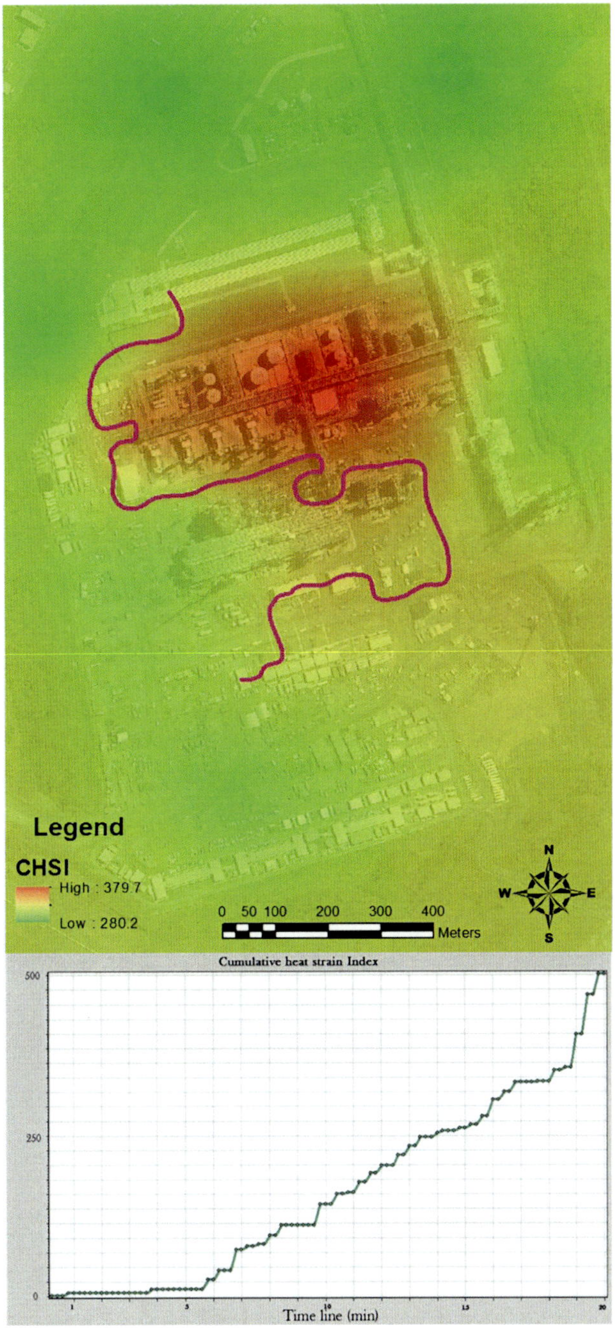

Fig. 76.5 Cumulative heat strain index increases with real-time tracking on worker's position

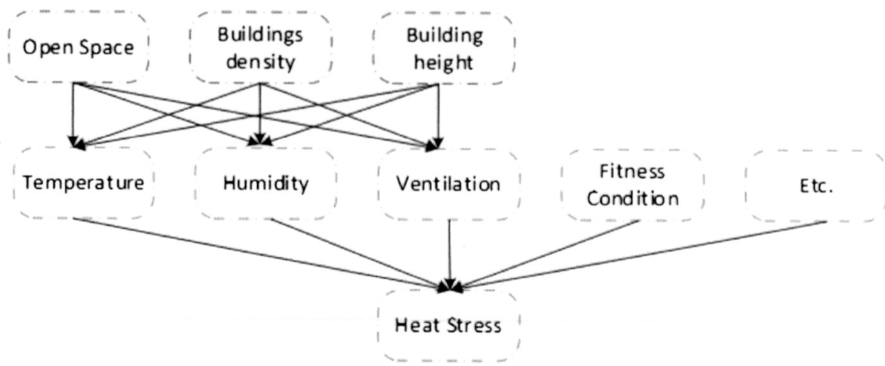

Fig. 76.6 Simplified Bayesian networks relationship of heat stress

76.4 Conclusion

Heat stress is a well-known occupational hazard, and climate change together with the increased frequency and intensity of extreme heat events has made risks more severe and widespread. A smart work site for early warning of heat stress is capable of detecting the extent of heat stress that workers are confronted with. It would be a useful way to protect workers from the hazardous effects of heat stress. The workers and site managers will be alerted once certain thresholds are exceeded. Appropriate interventions could be implemented to prevent injuries and deaths.

This system can be further improved by evaluating the results with the meteorological parameters and physiological parameters. The "heart rate monitoring" via a smart bracelet on workers will be incorporated into this system. Heart rate (HR) is the safest index because it is the earliest response of physiological strain [8]. International Standard Organization (ISO) has identified HR as an indicator of heat strain. Furthermore, this research can be improved by further considering (1) a more reasonable relationship between environmental factors and heat stress for Bayesian Network; (2) a real 3D spatial infrastructure information and environment monitoring; (3) individual differences in the response to heat stress.

References

1. Chan AP, Yi W, Chan DW, Wong DP (2012) Using the thermal work limit as an environmental determinant of heat stress for construction workers. J Manage Eng 29(4):414–423
2. Kovats RS, Hajat S (2008) Heat stress and public health: a critical review. Annu Rev Public Health 29(1):41–55
3. Maté J, Oosthuizen J (2012) Global warming and heat stress among Western Australian mine, oil and gas workers. In: J Oosthuizen (ed) Environmental health emerging issues and practice. InTech, Croatia, pp 289–305

4. World Health Organization (2003) Heat-waves: impacts and responses. Briefing note for the fifty-third session of the WHO Regional Committee for Europe, Vienna, Austria
5. Michelozzi P, De Donato F, Bisanti L, Russo A, Cadum E, DeMaria M, D'Ovidio M, Costa G, Perucci C (2005) The impact of the summer 2003 heat waves on mortality in four Italian cities. Euro Surveill (Eur Commun Dis Bull) 10(7):161–165
6. Brake R, Bates G (2002) A valid method for comparing rational and empirical heat stress indices. Ann Occup Hyg 46(2):165–174
7. Wang J-F, Stein A, Gao B-B, Ge Y (2012) A review of spatial sampling. Spat Stat 2:1–14
8. Bernard TE, Kenney WL (1994) Rationale for a personal monitor for heat strain. Am Ind Hyg Assoc 55(6):505–514

Chapter 77
Construction Project Bidding Research Based on Game Theory and Target Cost Pre-control Management

Ning Xu, Qiuyue Xu and Qiming Li

Abstract Construction project bidding behavior is affected by the value of the interests of the game participants. Information openness and resource selection directly affects the results of the bidding construction projects. From the perspective of the target pre-control costs, bidding management system to be established under the target cost. This will form a construction project bidding management model under the target cost pre-control based on Game Theory. Game Theory and the target cost control are applied to the pre-bidding decisions, which will make bidding more proactive. Bidding conclusion will be more reasonable.

Keywords Construction project · Game theory · Bidding · Target cost · Pre-control management

77.1 Introduction

When determining the cost of a construction project enters stage tender of choice is very important. These decisions directly related to the cost of construction projects, as well as new bilateral cooperation agreement. Participants of construction project bidding negotiations and consultations for price. This reflects the interests of the game kind of value. It is a combination of management, science and art [1, 2].

The earliest of the ancient "Art of War" is a game theory monographs. According to the definition of the National Science and Technology Terminology Committee announced, game theory is a mathematical method of competition and

N. Xu (✉) · Q. Xu · Q. Li
Department of Construction Management and Real Estate, School of Civil Engineering, Southeast University, Nanjing, China
e-mail: xun1973@163.com

Q. Xu
e-mail: xnqyyj@163.com

Q. Li
e-mail: njlqming@163.com

Y. Wu et al. (eds.), *Proceedings of the 20th International Symposium on Advancement of Construction Management and Real Estate*,
DOI 10.1007/978-981-10-0855-9_77

cooperation in the decision-making process, study the competition participants how to fight for the best interests of the behavior of decision should be made based on information analysis and the ability to judge, behavioral interactions between multiple decision-making body and balance each other, so that a profit or utility maximization game theory research involving multiple disciplines of geography, ecology and resource science. In theory, game theory is a theory of rational form of interaction between actors [3, 4].

77.2 Construction Project Bidding Behavior Based on Game Theory

People often unilateral research projects bidding behavior of both sides from the system and management. But the practical point of view, construction project bidding issues related to the bidding of two decision-making bodies of both sides, and their mutual influence of decisions. Pursue their own interests for their own benefit maximization, forming a bidding game on both sides to determine the price, namely bidding game.

77.2.1 Interests of the Game in the Construction Project Bidding

In bidding activities, there is value of the game on construction projects price between bidders bidding. Bidding people from their own interests want to project to the lowest price, and choose the best bidder in their favor. But the bidders are from their own interests want the project to be carried to the highest price, and want to select the most beneficial to their own tender. Constraints of these acts are selective bilateral asymmetry and freedom of information. Asymmetry refers to both sides bidding degree of information disclosure, and freedom selectivity refers bidding parties on the other side of the freedom of choice. Another important assumption is that the tender bidders who meet the constraints of all potential bidders who do not communicate with each other information, that do not know each other each other's ideas and opinions in detail (as shown in Fig. 77.1).

In the open tender, the tender announced construction projects complete information, but the bidder disclosed information only once in a public tender pre-qualification stage. On the other hand, in their own interests, the tender will have unrealistic fantasies of its low-price report, hoping to get the maximum benefits in multiplayer auction. Of course, people do not understand the tender bidders also contributed to its construction team hopes to get that low price and good management experience, and the team previously worked best, good reputation of the bidder, reduce management costs.

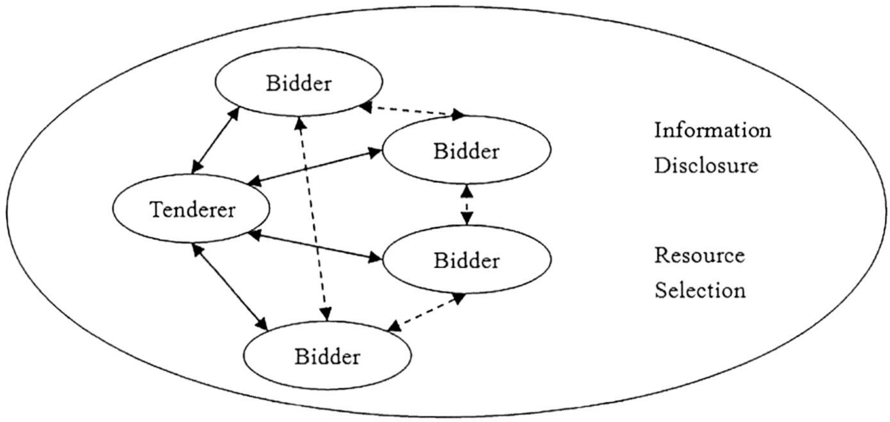

Fig. 77.1 Construction project bidding Game

Bidders bidding stage due to the asymmetry of information between each other, unable to understand each other's thoughts and tender human intentions. But their self-interest, the tender will want offer the higher the better. Bidding people want to find their partners in the open competition the team, and would like to tender a partner who is honest, and preferably previously worked with reputable tenderer.

77.2.2 *Construction Project Bidding of the Basic Elements of the Game*

Participants: all parties involved in the construction project bidding, including decision-makers (refers to one of the game in the first decision, the parties are often based on their own feelings, experiences, and the apparent condition of taking a directional action), antagonist (in game two game action lags behind that person, and policy makers to make decisions basic con, and his action is delayed, the default, passive, but eventually dominated).

Competitive information: All project information, competing parties involved in the project historical information data can be collected.

The degree of resource selection: parties involved in the project to compete with each other freedom of choice.

The results of the game: move the final results of the tender.

Equilibrium: the tenderer optimal bidding strategy.

In the bidding process of the construction project, the tender as the lead decision-makers in the game makes bidding decisions first. This directly affects the following game results. As a competitor, the bidder's strategy relies on the merits of the policy decision makers choose, tender man active in decision-making party. And in the subsequent bidding process, under different tender, bidding participants have different initiative, there will be different results.

Table 77.1 Open bidding Game results

Open tender	Information disclosure		Resource selection		Game results
Tenderee	Full disclosure	Passive	Uniqueness	Passive	The lowest price, the satisfied with the winning bidder
Bidder	Non-full disclosure	Initiative	Perfect competition	Initiative	The lowest price, the winning bidder is not satisfied
					Lower prices, more satisfied with the winning bidder

77.2.3 *Bidding Game Results*

In the open tender, the tender is fully released as decision maker construction project information. The bidders are required for competitive bidding and other bidders were non-full disclosure of information, confidential corporate data retention. On the information openness tender man party passive, active party bidders. Different levels of the parties mutually selected to participate in the bidding process of the project competition, the tender bidders who choose to open, fair and non-discriminatory behavior, have initiative, but without the right to choose the bidder on the project, either winning or no bid has passive. Open bidding Game results are shown in Table 77.1.

77.2.4 *Invite Bidding Game Results*

In the invitation to tender, the tender as a decision maker announced the construction project information entirely. Bidder by the tenderer from the familiar or recommended or highly professional people choose potential bidders. Therefore, the bidder's information is completely open and understanding. The tender and bidders have peer initiative in degree in information disclosure,. Different levels of the parties mutually selected to participate in the bidding process of the project competition. There are people in the choice of the tender bidders certain selectivity, has the initiative, but not the right to choose the bidder on the project, winning or not winning, the more passive. Invite bidding Game results are shown in Table 77.2.

As can be seen from the above analysis, information and resources to select the degree of openness is the impact of construction projects bidding game equilibrium outcome of the main factors. For tender people, the higher the degree of external information disclosure, the easier to get the real item price. For bidders, the external information openness is a very cautious problems. Asymmetric information on the tenderer is a big obstacle. Similarly, the higher the degree of resource selection, the

Table 77.2 Invite bidding Game results

Open tender	Information disclosure		Resource selection		Game results
Tenderee	Full disclosure	Peer	Uniqueness	Passive	Lower prices, more satisfied with the winning bidder
Bidder	Full disclosure	Peer	Imperfect competition	Initiative	The higher the price, the satisfied with the winning bidder

tender project more easily get a lower price after full competition, and the winning bidder to obtain a satisfactory risk is also higher. For bidders, the higher the degree of resource selection means lower offer. If the parties information symmetry and rational bidding, his best result of the cost of the lowest bid. When the increase in the number of bidders, its offer has a tendency to approach the cost. So let more people participate in the bidding tender interests' demands, which is consistent with reality.

77.3 Target Cost Control for Pre-bidding

To achieve cost management front and truly full cost control, we need to improve the regulation of target costs and dynamic costs. We should form a pre-control cost in the bidding stage, and build a complete cost management system (as shown in Fig. 77.2).

77.3.1 Cost Control Objectives and Achieve Cost Known and Controllable

Target cost is based on the real estate business market conditions, combined with the project management plan, pre-determined based on the expected selling price and profit targets, and achieve cost targets after efforts. Target cost stepwise refinement, such as land version, boot version, the program version, construction plates, adjust edition. Target cost version of the program is the implementation of standards. Target cost is broken down to the relevant departments to establish liability and costs linked to the performance appraisal system. In addition, the goal of determining the cost of the approval, does not allow arbitrary adjustment. To change the time, we need to have a strict examination and approval procedures. This reflects the seriousness of the target cost.

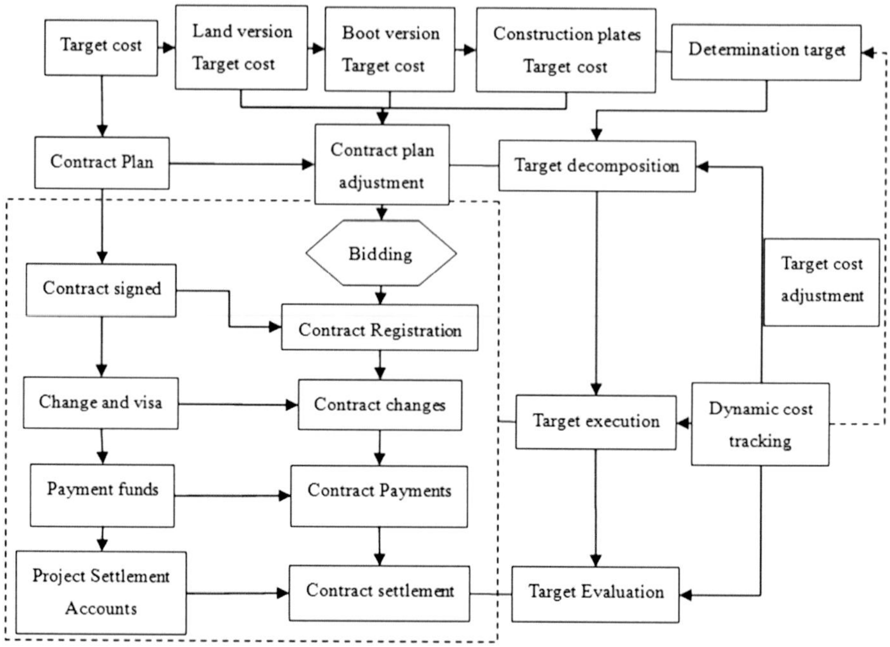

Fig. 77.2 Bidding management under the target cost pre-control

77.3.2 Cost Targets Regularly Review and Analysis, Dynamic Cost Control

We can establish a regular monthly cost reports with dynamic cost of the project review meeting mechanism to ensure timely and cost data refresh. At the same time, we need to establish cost overruns early warning and trip mechanism, which can be clearly the program of measures over time the cost of early warning indicators and mandatory targets.

The purpose of project cost control is to strengthen the real estate cost management of the entire process, and avoid risks. Through the above cost management system set up, we can achieve cost management control requirements in advance, and realize the full cost of the project known and controllable, effectively prevent project cost risk [5, 6].

77.4 Bidding Control Points

Construction project bidding is a typical characteristic of the game with a competitive activity. Research trick bid construction projects based on game theory shows that different tender, scaling rules will result in the bidder's tender offer different strategies, different bidding efficiency. In the construction project bidding process, the tender should be ready to work in the following areas.

77.4.1 Information Balance Control

The role of information in the bidding game is very important. Information asymmetry will cause "adverse selection" and "moral hazard." Information asymmetry in the short term will be beneficial for the party, but it will eventually destroy the entire market. Game two sides need to make efforts to seize the high ground information.

77.4.2 Selection Control of Resources

The selection of bidder resource is another very important constraint. The tender will face interest groups, it's the nature of the bidding decision making is significant. Selection of bidders in different ways to produce different bidders will be binding. Full competition is very important, which makes sense of crisis bidders doubled, to reduce the cost of purposes. Game leads to cooperation, but can not guarantee to continue cooperation. The cost of cooperation could be built on the basis of personal interests above the damage. From the perspective of future earnings analysis, the future is bound by the contract specifications of the winning bidder.

77.4.3 Management Control System

Standardized system of bidding avoids damage to the interests of the event, to lay a solid foundation for the future development of relations. First, the strict implementation of the process is very important. Second, set the scaling formula directly affect the winning bidder's choice. Third, not restrict competition among bidders. Construction project bidding information should be made public, resource selection should be fair, procedures should be fair and transparent. Construction project bidding activities shall follow the openness, fairness, impartiality and honesty.

77.5 Conclusion

For the construction project bidding to obtain higher returns, we must pay attention to strategies and techniques bidding. We will use game theory to bidding decisions, research the optimization problem about the tender and the bidder of projects bidding, from the perspective of the target pre-control costs, establish bidding management system under the target cost. These will make bidding more dynamic, initiative and maneuverability, the bidding is more reasonable conclusion. This will have profound philosophical implications.

References

1. Jiang W (2004) Game theory in project bidding application. Ind Technol Econ 23(1):58–60
2. Xu W, Yang H (2005) Based on Game theory construction project bidding. Infrastruct Optim (5):36–38, 41
3. Zhang W (1996) Game theory and information economics. Shanghai People's Publishing House, Shanghai, pp 258–261
4. Zhou L, Ann J (2007) Study bidding based on game theory. Hebei Eng Univ (Nat Sci) 01:95–98
5. Zhang C (2010) Real estate projects target cost control theory and practice to explore. Tianjin University
6. Zhang Z (2007) The main phase of real estate development project cost control research and application of. Southwest Jiaotong University

Chapter 78
Reframing Public Private Partnerships Through 'Performance' Contracting

John Douglas Thomson

Abstract Many governments and aid agencies are of the view that the use of 'performance' regulations could better meet their demands for 'value'. The use of 'performance' regulations is now mandated by some national building codes. Despite this, there has been a struggle with policy implementation. At the implementation level, owners and contractors seem as yet unready and less than willing to use 'performance' specifications, so use of 'prescriptive' specifications prevails. In addition, there is no universally accepted, legally binding 'performance' contract. This research finds that the use of 'performance' specifications by an owner to express 'value' requirements can be responded to by tenderers using 'prescriptive' specifications. The use of an interpretative methodology, a case study and interviews/correspondence with various stakeholders demonstrates the validity of this research, which could now be extended to public private partnership projects. Pre-contract award, critical success factors include the use of a 'performance' contract and a competitive tendering process which favours innovation, with tender selection based on 'value' rather than lowest price. Post contract award, these arrangements provide opportunity for owner or contractor to be innovative within the contract structure.

Keywords Performance · Prescription · Value · Public private partnerships · Contract · Specification

78.1 Introduction

At a policy level, many governments and aid agencies have been of the view that the use of 'performance' regulations could equitably meet both its and private sector demands. But in response to these demands, there has been little progress with 'performance' regulation implementation. At the implementation level, industry

J.D. Thomson (✉)
Graduate School of Business and Law, RMIT University, Melbourne, Australia
e-mail: doug.thomson@rmit.edu.au

Y. Wu et al. (eds.), *Proceedings of the 20th International Symposium on Advancement of Construction Management and Real Estate*,
DOI 10.1007/978-981-10-0855-9_78

seems unready or unable to interpret the more sophisticated 'performance' regulations or the derivative 'performance' specifications. Continued industry use of 'prescriptive' specifications almost universally prevails, and industry training remains largely around the use of 'prescriptive' regulations and specifications. Another difficulty is that there is as yet no universally accepted 'performance' contract.

This research is conceptual and was undertaken to examine if there is a process which meets an owner's use of 'performance' specifications to achieve better 'value' outcomes, while continuing the use of 'prescriptive' specifications by industry. The concept is then considered for application to public private partnership projects.

78.2 Early Industry Involvement

Much recent research is supportive of early industry involvement in new product design and development. Lopez and Love [16, p. 585] found that 'design errors can adversely influence project performance' and 'were found to not significantly vary with the procurement method or project type used'. Design adjustments taken at a later time were 'directly responsible for considerable costs driven by changes that could have been avoided emerging at later stages of a project. Late changes are far more costly and time consuming than when they are made early on in a project' [12]. The 'ability to influence the final characteristics of a project is highest at the start of a project and decreases as the project progresses towards completion … the cost of changes and correcting errors typically increases substantially as the project approaches completion' [17, p. 17, 8–11]. Levitt and Mahalingam [14] suggest that an owner's project planning ability to influence project outcomes falls rapidly once the project begins (the inverse of the cost S-curve)—when outcomes become known, all project funds have already been expended and no influence remains. Despite this research, there is little evidence of early industry involvement [5, 25, 26]. A recent survey by Chen and Manley [5, Table 1] of 320 Australian public and private sector alliance or collaboration contracts, only 8 of these used early industry involvement and only 27 early contractor involvement. Their research supports the early engagement of industry experts early in the pre-contract award concept, design and development stages.

78.3 National Construction Codes and 'Performance' Regulation

Some national construction codes such as that of Australia [1] now mandate the use of 'performance' regulations. These do not dictate how to achieve required results as 'prescriptive' regulations do, but allow for any acceptable proposal to be approved. Because industry's heritage is with 'prescriptive' specifications, the culture is difficult to change. To accommodate existing culture, national construction codes such as the Australian Building Codes Board [1] provides for 'deemed to satisfy' prescriptive solutions that may be used. This inclusion seems contrary to the intention behind the mandatory use of 'performance' regulations. For example, a 'performance' specification for natural lighting for housing is stated as 'a habitable room must be provided with windows so that natural light, when available, provides a level of illuminance appropriate to the function or use of that part of the building' (Performance Requirement P2.4.4). Such 'performance' specification does not set a minimum distance from windows to an allotment boundary as does the 'deemed to satisfy' prescriptive specification which states 'in a Class 1 building, a required window that faces a boundary of any adjoining allotment must not be less than a horizontal distance of 900 mm from that boundary' [1]. Despite the mandatory requirement for the use of 'performance' based specifications, the 'deemed to satisfy' option gives an alternative to those who prefer to continue to use 'prescriptive' specifications, which is most of the industry.

78.4 'Performance' Contract Terms and Conditions

The use of 'performance' specifications is not new. In 1920, the Supreme Court of the United States, (1920), United States V. Atlantic Dredging Co, W.B. Brooks, Agent, 253 U.S. 1 (1920), decided on the 26 April, 1920 that 'where one agrees to do, for a fixed sum, a thing possible to be performed, he will not be excused or become entitled to additional compensation, because unforeseen difficulties are encountered.' This decision is conditional upon 'a fixed sum' and a 'thing possible to be performed'. More recently and in keeping with the 1920 ruling, the Australian Attorney General's advice on the use of 'performance' specifications and terms and conditions for inclusion in any 'performance' contract was that 'any inaccuracy or mistake however arising shall neither affect the contractor's obligation to complete the work under the contract nor entitle the contractor to payment of any extra moneys whatsoever; after acceptance of the contractor's tender, the contractor shall expeditiously prepare and complete all other documentation; no approval, direction or assistance given to the contractor in respect of specification or designs or other data produced shall relieve the contractor of responsibility under the contract for the correctness of all such designs, drawings, specifications and other data created or supplied for the purposes of the contract; the contractor warrants the sufficiency and fitness for its purpose of all designs, drawings, and specifications prepared pursuant to the Contract for use in the execution of the work; and the contractor shall not be entitled to claim from the Principal any damages, loss, or expense or extra costs incurred by the contractor in respect of cause or delay'. This provides the basis for a legally binding 'performance' contract.

78.5 Methodology

Theory building from case studies does not rely on previous literature nor prior empirical evidence—case study theory building is a bottom up approach such that the specifics of the data produce generalizations of theory [7, 28, 29]. For this research, there are no published case studies, and little research. The method therefore is focused on theory building from one pilot case initially and the concepts derived from new ideas. This conceptual research was directed toward the development of a testable hypothesis and theory which may be generalizable across settings such as public private partnership projects. The process is designed to be iterative, novel, and testable with constructs that can be measured and a model that can be proven false [7, 28, 29].

78.6 The 'Performance' Contract Concept

The 'performance' contract concept can be applied by an owner using 'performance' specifications to describe 'an end result, an objective or standard to be achieved'. This leaves the determination of how to reach the 'performance' proposed to tenderers (Stuyvesant Dredging Co. v. United States, 834 F.2d 1576 [Fed. Cir. 1987] in Calvert et al. [3]). This challenge facilitates innovative solutions from the marketplace which can be expressed in 'prescriptive' terms. Responding 'prescriptive' designs from tenderers will 'detail the materials to be employed and the method in which the work is to be performed. The contractor is then required to follow them as one would a road map and without deviation' (L.L. Simmons Co. v. United States, 412 F.2d 1360 [Ct. Cl. 1969] in Calvert et al. [3]). Calvert et al. [3] indicate that the principal consequences of the use of 'performance' specifications are risk related. That is, there will be no owner's warranty of the sufficiency of plans and specifications, liability for design failures is shifted from the owner to the contractor, and risks arising in 'performance', such as the risk of unforeseen conditions or necessary changes, are shifted to the contractor. The contractor is not entitled to recover the cost of changing to the correct means or methods later, if initial selection of the means or methods was wrong. The contractor assumes responsibility for the total design build (Fig. 78.1).

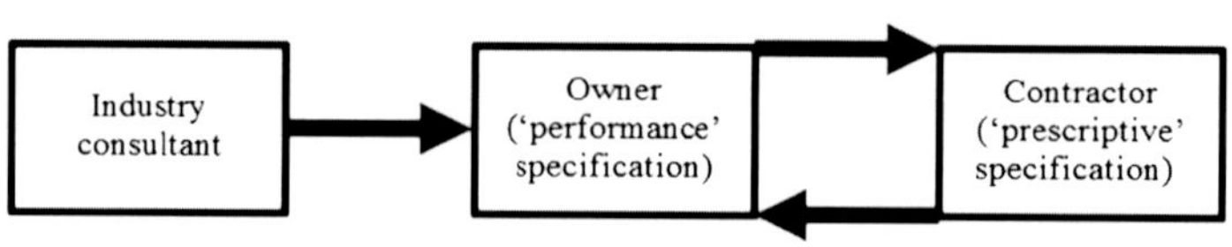

Fig. 78.1 The 'performance' contract model

78.7 Design Rationalization

In the mid 1970s, Australian defence (amongst others) became interested in the use of 'design to cost' targets which gave 'greater attention to cost control during the design and documentation stages of projects, the implementation of an accommodation building cost per person and a more flexible approach in the application of Services Scales and Standards of Accommodation. A 'building cost per person targeting system was developed, and compared with the cost of building from the private sector which provided a similar standard of accommodation' [2, pp. 1–2]. By 1990, little if any progress had been made and defence again recognised the value of examining opportunities for rationalising the design process for common facility types such as living in accommodation [6, p. 1]. Defence anticipated that design rationalization would provide a significant increase in the value for money obtained [21]. This led to the development of a number of defence accommodation types and 'design to cost' targets for each accommodation type. Typical of these 'design to cost' targets was that developed for the SNCO self contained units (Fig. 78.2).

The 'inclusions' were a wardrobe, writing table and cupboard, bookshelf, bed with storage under, lighting and power points, TV aerial fitting, hand basin, shower, toilet, a box storage space of 0.6 m^2, shared laundry facilities between four with washing machine, wash trough, clothes dryer and ironing board. 'Exclusions' were stairwells, lift wells, external landings, external access ways, connecting walkways between buildings, car parks or fire escapes which were subject to additional 'design to cost' pricing once the site was known.

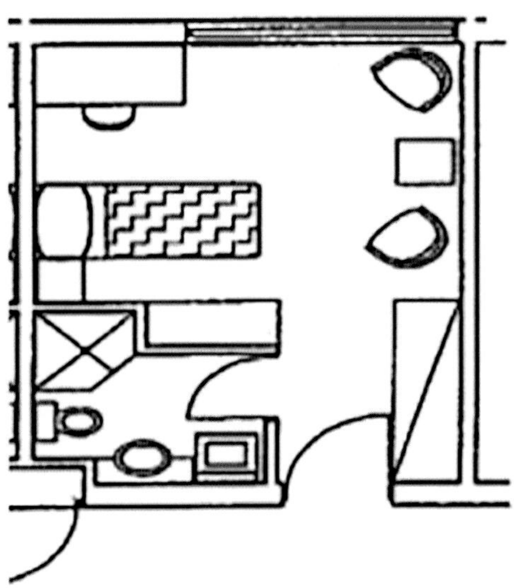

Fig. 78.2 Senior NCO self contained unit. *Source* Defence Scales and Standards of Accommodation

78.8 The Preliminaries

The 'design to cost target' 'inclusions' were calculated by quantity surveyors to be $30,250 per person for a room of 21 m^2 floor area for a single storey, self contained accommodation of full masonry construction in a temperate climate (Fig. 78.1). The balance of all other works 'exclusions' were to be estimated according to the selected site and other environmental factors. For a site at HMAS Cerberus, Victoria 'exclusions' were estimated by quantity surveyors to be AU$6875 per person, that is, for the 'balance of all other works'. The total 'design to cost' target for the 24 person accommodation facility therefore became AU$891,000 (i.e. [$30,250 + $6875] × 24). As a 'fixed sum', this could be used to enable direct comparison between the use of 'performance' specifications and the 'prescriptive design and construct' specification model usually used by the defence owner for this type of project. Using its preferred 'design and construct' procurement model, it was further estimated by quantity surveyors that the pre-contract award period for this project would take the owner four months to process i.e. development of the business case and its approval, functional and detailed design brief drawings and specifications, approvals, confirmation of the selected procurement process, invitation and assessment of companies registering interest, calling and assessment of tenders, post tender discussions, approval of the selected tenderer, contract discussions and award. The contract award to contract completed period was estimated by the same quantity surveyors as eight months, based on the many previous similar defence 'design and construct' contracts.

78.9 Pilot Case Study—Fixed Price, Fixed Period, 'Performance' Specification

To measure the 'performance' contract model against the 'prescriptive design and construct' model, the defence owner agreed to make available a test case [7, 18, 28, 29]. This was the HMAS Cerberus accommodation facility in Victoria for 24 senior sailors, using for comparison purposes the same capital cost of AU$0.891m and the same post contract award construction period of eight months as estimated for the 'design and construct' process for this project. Early on, the owner acknowledged it lacked current industry knowledge and experience, so competitively engaged an industry consultant to assist it through the entire pre and post contract award processes. With assistance from the consultant and quantity surveyors, the owner determined a seven page 'performance' specification, and used as a fixed contract price the 'design and construct' AU$0.891m, and as a fixed contract period the 'design and construct' 8 months before going to market. Within these boundaries, tenderers were required to respond to the owner's seven page 'performance' specification with a 'prescriptive' specification providing details of financial soundness, resources, schedules, designs, specifications, construction programs and cash flow charts. Post contract award, the consultant was to be engaged in mediating between the owner and the contractor. For conflict of interest reasons, the consultant was not permitted to bid on the contract and was subject to a confidentiality agreement with the owner.

78.10 The Event

During the pre-tender award period, the owner and consultant prepared the 'performance' specification and 'performance' contract, and the owner called for expressions of interest. 49 were received and reduced to five based on those with the best financial capacity, resource capacity, professional indemnity, quality of subcontractors and designer's experience. The five each submitted innovative tenders and were given the opportunity to brief the owner on their proposals the day after tenders closed. Because the construction period and price were fixed for all bids, tender assessment could be made on the basis of 'value' only. This meant all tenderers had to innovate their 'value' offered, not only to meet the owner's 'performance' specification, fixed period (8 months) and fixed price (AU$0.891m) but also to be more competitive and innovative than the other four tenderers. After presentations by each tenderer, offers were assessed by the owner and consultant, and the contract was awarded the following day. The owner's pre-contract award 'performance' contracting process took two months less than the four months of the owner's usual 'design and construct' process. Both owner and tenderers benefitted from this fifty percent reduction in pre-contract award process time and the reduced transaction costs [27] (Table 78.1).

Tenderer 1 provided much better quality than the 'design to cost' prescriptive specification required while meeting the fixed price and period. The design provided 24.8 m^2, 3.8 m^2 more floor area than the minimum prescribed, with conveniently located individual store rooms and car parks. Tenderer 2 provided an area of 21 m^2 as prescribed, but did not submit design drawings or renderings with qualifications as to the general conditions of contract, did not include site or common

Table 78.1 Comparison of 'prescriptive' and 'performance' costs and times

Pre and post contract award	'Prescriptive' specification transaction costs and times	'Performance' specification transaction cost and times
Pre-contract award (concept to contract award)	$133,650 4 months	$49,874 2 months
Post-contract award (contract award to contract completion)	$89,100 8 months	$50,961 7 months
Comparison of time and cost	$222,750 12 months	$99,835 9 months

services costs or design costs, and specified that the engagement would be a project management engagement. Tenderer 3's submission provided a floor area of 21.2 m^2, but there was no corridor carpet provided, a poor quality landscape plan, a conflicting life cycle cost plan, an unsympathetic car park and architecture, lack of exhaust fans in bathrooms, no mention of security and below average internal décor with quality assurance costs to be added. Tenderer 4's cost at $1,127,174 exceeded the fixed cost of $891,000, so demonstrating a lack of adherence to the owner's fixed cost requirements. Tenderer 5 met the new product design to cost target of $891,000 and provided locality concept, configuration, master plan, and room layout drawings and other design details but lacked detailed design drawings and renderings, cash flow details, landscape plan and funding, life cycle costing schedule, but no area schedule showing how the requirements of the 'performance' brief were to be met. Post tender submission presentations and discussions with each of the tenderers was undertaken to confirm their tender and the owner's assessment. No submissions or changes to the tenders were permitted subsequent to the close of tenders.

Table 78.2 Tender evaluation summary

Item	Tenderer 1	Tenderer 2	Tenderer 3	Tenderer 4	Tenderer 5
Performance measures	Trade offs	Trade offs	Trade offs	Trade offs	Trade offs
Design component cost	34,000 (3.8 %)	55,200 (6.2 %)	80,000 (9.0 %)	57,888 (5.1 %)	50,000 (4.6 %)
Building works	562,770 (63 %)	485,600 (55 %)	551,440 (62 %)	645,380 (57 %)	543,401 (61 %)
Mechanical	20,010	In 'electrical'	In 'bldg wks'	In 'bldg wks'	12,000
Electrical	45,140	98,200	In 'bldg wks'	78,648	62,000
Site services	40,580	87,000	20,000	92,340	67,024
Landscaping and planting	18,500	7000	5000	22,148	
Furn/fittings/effects	66,000 (7.4 %)	61,000 (6.8 %)	58,239 (6.5 %)	74,580 (6.6 %)	65,616 (7.4 %)
Ancillary	2000	In preliminaries'	101,041	No advice	No advice
Preliminaries	102,000	97,000	75,280	156,190	90,959
Total ($0.891)	891,000	891,000	891,000	1,127,174	891,000
Floor area (21 m^2)	24.8	21	21.2	TBA	TBA
Architectural rating	1	2 (−)	2 (+)	5	2
Overall Rating	Excellent	Good (−)	Good (+)	Poor	Good

78.11 'Performance' Contract Trade-Offs

The 'performance' contract required tenderers to list the allocation of funds to the various requirements and provide the basis for the owner to assess the trade offs made by each tenderer. The owner's 'performance' contract trade-off tender evaluation summary is shown in Table 78.2.

Table 78.2 provides insight into how design trade-offs and resource allocations were made by each of the tenderers when the key 'performance' indicators of price and period are fixed, as well as the competitive, innovative and transparent nature of the process. Tenderer 1 was awarded the contract and innovatively completed construction in seven months, one month less than contracted, and for \$0.892m, AU \$1000 more than the contracted fixed price. The winning design is shown in Fig. 78.3.

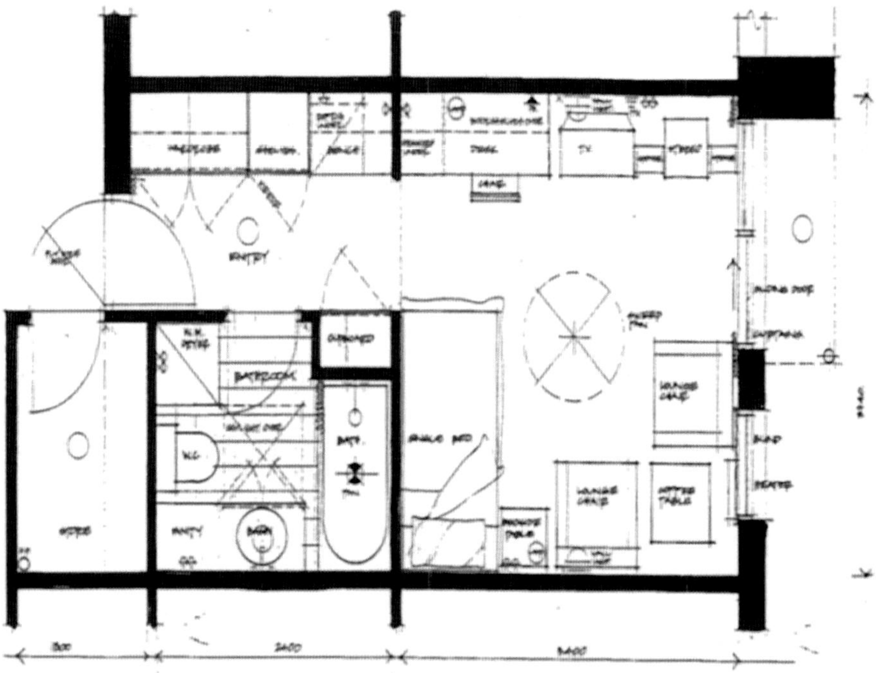

Fig. 78.3 Tenderer 1 winning design

78.12 Application to Public Private Partnerships

This 'performance' contract concept may now be applied to public private partnerships and other projects in a similar way. The owner will prepare a 'performance' specification in conjunction with an industry consultant if necessary, and fix the contract cost and time. The owner will then call for registrations of interest, followed by selection of the three to five best qualified responses. These will be invited to submit 'prescriptive' specification tenders and a contract awarded on that basis. Further research is to be carried out on a pilot public private partnership test case project.

78.13 Conclusions

This research examines the concept of an owner's use of a 'performance' specification, a fixed price, a fixed period and a legally binding contract. The 'performance' specification is determined by the owner with the assistance of an industry expert in advance of going to market. This pre-contract award diagnosis of the owner's 'performance' requirements reinforces the need for an owner to consider engaging an industry expert early in the design and development process so as to better understand the 'performance' required and the associated cost and time involved. This places greater demands on an owner to truly understand the 'thing possible to be performed'. Making the fixed contract time and fixed cost criteria known to the market as part of the owner's request for tender levels the playing field for the tenderers. Tenderers can then respond to the owner's 'performance' specification on the basis of 'value' with a 'prescriptive' specification, and this is held through the 'performance' contract. With time and cost fixed, this leaves only the 'value' component for tender assessment by the owner, which significantly simplifies the tender assessment process so reducing transaction costs and times for both owner and tenderers. The appointed contractor in developing its own 'prescriptive' specification should have a thorough knowledge its contents. The process effectively removes underbidding on price and the follow on problems underbidding often causes. An owner's use of 'performance' specifications provides a useful framework for early industry involvement in project definition and development before going to the marketplace. In response, a contractor will be required to 'do, for a fixed sum, a thing possible to be performed' and 'will not be excused or become entitled to additional compensation because unforeseen difficulties are encountered'. Future research will apply this concept to a public private partnership project.

References

1. Australian Building Codes Board (2014) Understanding the BCA's performance requirements. National Construction Code Training Program Module 2, Canberra
2. Australian Government (1976) Services sleeping accommodation design to cost targets. Department of Construction, Central Office, Hawthorn, Defence Section, April 1976
3. Calvert CA, Gavin PC, Hamilton KH (1996) Performance specifications: the issues, the problems and some solutions. ABA forum on the construction industry twentieth anniversary. Crider, Calvert & Bingham, P.C. Albuquerque; Wickwire Gavin, P.A. Minneapolis; Stafford Frey Cooper, Seattle; USA
4. Chen L, Manley K (2014) Validation of an instrument to measure governance and performance on collaborative infrastructure projects. J Constr Eng Manage (American Society of Civil Engineers)
5. Department of Administrative Services (1990) Other ranks single living in accommodation rationalization schematic design proposals. Australian Construction Services, Canberra, Sept 1990
6. Eisenhardt KM (1989) Building theories from case study research. Academy Manage Rev 14 (4):532–550 (Academy of Management)
7. Gambatese JA, Hallowell M (2011) Enabling and measuring innovation in the construction industry. Constr Manage Econ 29(6):553–567
8. Gambatese JA, Hallowell M (2011) Factors that influence the development and diffusion of technical innovations in the construction industry. Constr Manage Econ 29(5):507–517
9. Gannon MJ, Smith NJ (2011) An effective outline business case to facilitate successful decision making. Constr Manage Econ 29:185–197
10. Gibson GE, Hamilton MR (1994) Analysis of pre-project planning effort and success variables for capital facility projects. Rep. Prepared for Construction Industry Institute, University of Texas at Austin, Austin, Texas USA
11. Hwang B-G, Ho JW (2012) Front end planning implementation in Singapore: status, importance, and impact. J Constr Eng Manage 138:567–573
12. Levitt RE, Mahalingam A (2002) Predicting and mitigating institutional costs in global projects. Department of Civil and Environmental Engineering, Stanford University, USA
13. Lopez R, Love P (2012) Design error costs in construction projects. J Constr Eng Manage 138 (5):585–593. doi:10.1061/(ASCE)CO.1943-7862.0000454
14. PMI (2008) A guide to the project management body of knowledge (PMBOK), 4th edn. Project Management Institute, PA, USA
15. Soy SK (2006) The case study as a research method. University of Texas at Austin
16. University of Cambridge (2010) A brief guide to value for money (summarised from HEFCE website accessed 29 June 2014)
17. Victorian Government (2011) Tendering, evaluation and acceptance. Capital Projects and Service Planning, Department of Health
18. Victorian Government (2014) Alliance and traditional contracting. Department of Treasury and Finance, Melbourne
19. Williamson OE (1998) Transaction cost economics: how it works; where it is headed. De Economist 146(1):23–58
20. Yin RK (2003) Case study research design and methods, 3rd edn. Sage Publications, CA
21. Yin RK (2003) Applications of case study research, 2nd edn. Sage Publications, California

Chapter 79
Research on the Reformation and Innovation of Public Works Management System on the Experience of Shenzhen

Jianpu Wang, Xiao Li, Yue Teng, Shirong Li and Bo Xu

Abstract For a long time, there have existed a variety of problems in the management of public projects including overinvestment, slow construction progress, poor quality and corruption. Although many efforts have been done for a long time, the essential problems have not been solved at present. The main reason can be explained by no comprehensive measures from the aspects of system, the legal system and the mechanism at the same time. According to the experience of developing countries, this paper put forward the management of specialized organizations could be brought into guarantee the centralization, specialization, unification and investment efficiency of public projects. Then an innovative framework of public project management system has been built based on the centralized management system of Municipal Public Works Bureau in Shenzhen, which is a typical model of reform in China.

Keywords Public works · Management system · Centralization management · Reformation · Innovation

79.1 Introduction

With the rapid development of China's economy and urbanization, the Chinese government increase in pushing forward the construction of the engineering project to promote the coordinated development of public works, economic development and social management. And all kinds of public works got great development.

Bo Xu—Chongqing Graduate Student Research Innovation Project (CYS14036)

J. Wang (✉) · X. Li · Y. Teng · S. Li · B. Xu
Construction Management and Real Estate School, Chongqing University, Chongqing, China
e-mail: wjp910212@163.com

Y. Wu et al. (eds.), *Proceedings of the 20th International Symposium on Advancement of Construction Management and Real Estate*,
DOI 10.1007/978-981-10-0855-9_79

However, the main management pattern of domestic public works is decentralized management, namely, a variety of functions such as construction, management and operation are part of dispersed project implementation main body. Under this management pattern, the problems for the management's professional restrictions and not familiar with the law of professional construction are ubiquitous. The project goals are often difficult to control, which led to a variety of problems including overinvestment and slow construction progress [1, 2].

Since 2002, to solve the disadvantages of traditional management pattern of public works, the Shenzhen government explored the specialization of centralized management pattern and set up Shenzhen Municipal Public Works Bureau to in charge of the implementation organization and management of government investment project, and carried out the agent construction system for the first time in domestic attempt and practice. Currently, there are two kinds of patterns of the agent construction system, which is the market agent construction system and public institutions agent construction system. However, the market agent construction system can't solve the problems which public works suffers under the traditional management pattern such as the lack of efficiency and regulation. Besides, there are no substantial differences between the public institutions agent construction system and the unified public project implementation and management carried by government-affiliated institutions directly [3–5]. Therefore, the Shenzhen government definitely set up professional public institution to in charge of the implementation and management of public works, and the Bureau of Public Works of Shenzhen Municipal is the main and representative institutions.

By investigating and interviewing 24 works departments of Municipal Public Works Bureau, we find following problems existing in the implementation process,

- The responsibility and role positioning of Municipal Public Works Bureau is not clear, and it didn't get complete government legal authorization. Therefore, the professional ability of Municipal Public Works Bureau can't exert completely for the low voice in the decision making and using stage.
- The coordination interaction mechanism has not established between different works departments, which impede the experience sharing. The system advantage of centralization management can't exert completely.
- The management interface between works departments and facility management departments is arranged imperfect.
- There actually exist some problems of overlapping functions and work coordination between works departments and the administrative departments.
- The regulative function of administrative department is too scattered, opaque, and excessive formalization of tender supervision.

Therefore, it's necessary to carry on the reformation and innovation in the existing management system to perfect the construction management of public works.

79.2 Analysis of the Management System of Public Works in Developed Countries

In the course of public works management, government owner uses the taxpayers' money instead of its own, which may be driven for the increase of capital expenditure and given the higher risks of corruption; meantime, it can hardly rationally keep the one investment and whole life-cycle cost and performance in balance, if government owner doesn't serve as the major role of whole life-cycle management. To solve the problems of this kind, US government sets the owner as a real estate developer engaged in running the government real estate property, and refers it to the comparable real estate market for the purpose of specialized business. In which: (1) It prices the government property with reference to the market, and establishes its internal property leasing market; operation funds of government owner is mainly sourced from its income from property leasing and management, and appropriate such funds to recover construction investment and keep the financially sustainable business running; (2) Government sector writes the costs of utilizing various properties into the fiscal budgets for each item user, rather than directly appropriates these costs to the government owner, then such item user pays to the government owner in the form of rent and service charge, which in turn allows them to exert the necessary influence on the government owner by their money [6]. While in Britain, public works construction depends on government action, and public works management is enforced by government sector as a owner. British government holds public works in decentralized management, which means that government sector as the project launcher at its will organizes the implementation and management of construction project [7]. Germany has formed the "Staatliche Hochbauverwaltung". As a specialized organ, it is held responsible to assure the construction task with national investment completed as per the normative standard, provided that public interests are suited [8].

In the centralized unified pattern, centralized unified organ is not the end-user, which may ignore the end-user needs because of bureaucracy, causing the under-performance of construction quality in respect of applicability; meanwhile, government owner in the centralized unified pattern acts as the monopolistic government property business entity, which may give rise to the unreasonable monopolistic pricing. To that regard, US government resorts to (1) Introduces the competitive mechanism. When government owner fails to offer the products and service to the satisfaction of project user at the market level or higher, project user may be entitled to obtain the relevant products and service. For this reason, GSA (General Service Administration) will always refer its leasing and property service to the market price and deliver some preferential terms to keep its competitiveness. (2) Separate its revenue from expenditure. Although GSA theoretically runs at own revenue and expenditure, it has no right to dispose its income, but all expenditures will be under the congressional oversight. Therefore, all investment and construction activities are subject to two constraints, i.e.: public construction funds and congressional approval. Obviously, this will also contain GSA from seeking sector benefits by increasing the price in its own discretion. (3) Keep transparency. When the pricing of government leasing and property service is disclosed in entirety, the comparability between its service price and market price will be uncovered, which in turn contains the unreasonable monopoly profits. If GSA makes public its data ranging from budgets, income, various service and pricing, it will constitute the strong social supervision on GSA, and impose stress on project user in denial of GSA service provided that GSA service price is examined to be reasonable. German government may commission other tasks to any third party, private company or legally independent user in need [9], other than: providing land for construction use, signing an agreement, making all necessary decisions, defining user requirement, funding and paying. British government has formed the instruction and supervision organ for the purpose of public works management: OGC (Office of Government Commerce) and MPA (Major Projects Authority), which is primarily held responsible to lay down the norm, stipulation and standard with a view to public works management; instruct the government project owner; oversee the construction activities and results of government project in alignment with the highest level of government [10]. In British government, Crown Commercial Service is the service provider of the centralized management of government public works (to accept the commissioning by government project owner) [11].

79.3 Framework Design for the Reformation and Innovation of Public Works Management System—Taking Shenzhen as an Example

Since 2002, Shenzhen has been establishing the centralization management, with Bureau of Public Works of Shenzhen Municipal at the core, for the government invested constructions. However, there are still all kinds of problem about the system. If these problems can't be solved promptly and efficiently, it will definitely affect the government's benefit on the investment of construction programs. Meanwhile, annual investment amount the government put on fixed assets is still at a high level. Like in 2012, the total amount of Shenzhen government invested on the fixed assets reached at 39.65 billion yuan, taking up 21 % of the total financial expenditure that year. For taking an efficient management on such a big amount of fixed assets, it is necessary for us to mirror the developed countries' successful experience by setting up a department which is independent, strong in professions and also able to coordinate horizontal departments in government. Bureau of Public Works of Shenzhen Municipal has professional management, but is not equipped with all the three abilities mentioned above. So reformation and promotion should be made on the bases of the existed one, a new political authority, which is called Shenzhen Municipal Public Works Administration Bureau, should be found for the aim at the management of public works. This professional department will supervise in the whole life cycle of government public works.

79.3.1 The Organizational Function of Municipal Public Works Administration Bureau

For accomplishing functions both in management and execution, a double organizational structure should be set in the Municipal Public Works Administration Bureau. As shown in the Fig. 79.1, the management should be cored on public works administration through external coordination and internal supervision to formulate standards and reformations, as well as the organization for the whole process of the government public works management. During which the Municipal Public Works Administration Bureau interact with other departments by means of horizontal coordination, popularizing technologies and offering decision support at the beginning of a project, etc. Meanwhile it will monitor the government-affiliated institutions by implementing supervision, standards formulation, personnel management and cost controlling. The government-affiliated institutions unit can be treated as an institution under the control of Municipal Public Works Administration Bureau, and it's job is to carry out the specific execution works for the government public works by purchasing management, construction management or some other functions connected with project management, then finished the tasks about project insurance and facility operation and maintenance management.

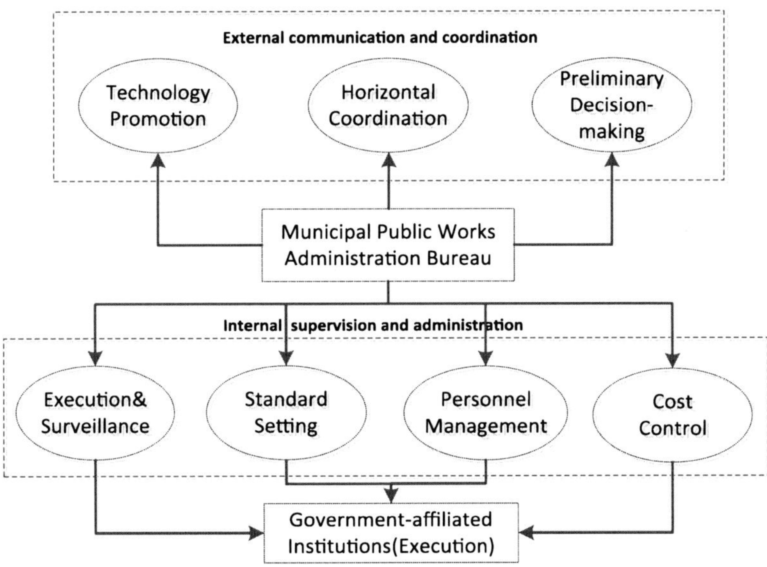

Fig. 79.1 The basic organization functions of Municipal Public Works Administration Bureau

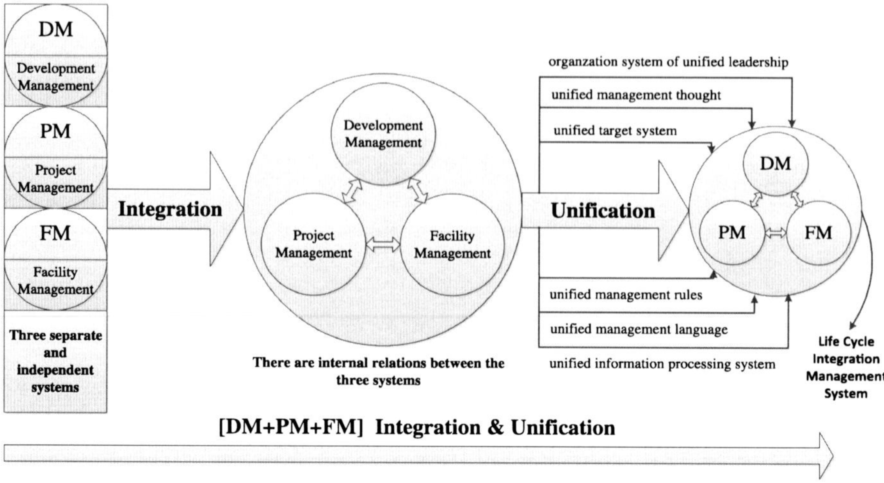

Fig. 79.2 Life cycle management system of construction projects

79.3.2 *The Administrative Functions of Shenzhen Public Works Administration Bureau*

According to the theory of vertical integration division of public service and the life cycle management theory of construction project, for the purpose of improving the investment profit and the appreciation of construction project, the Municipal Public Works Administration Bureau must be responsible for all the management activities took place during the three main periods of a construction project's life cycle, which are DM-Development management, PM-Project management and FM-Facility management. Works as the Fig. 79.2.

Decision-Making Support for Public Works

The traditional decision-making process for the public works mainly consists of project proposals, project requirement collection and project feasibility analysis and some other works. But there are still lots of problems during the execution, such as the professional restrictions occurred when the users doing the formation job and the limited personnel in Municipal Public Works Bureau and Approval Department. To solve this kind of problem, firstly, we need to found an interdepartmental agency who has strong professions, high authorities and is able to mix differing views, and this agency is named Decision-making Committee of Public Works. As is shown in Fig. 79.3, the committee should be charged by the municipal leaders, and can be organized by National Development and Reform Commission, including Municipal Public Works Administration Bureau, Planning and Land Administrative Departments, Construction and Management Department, Finance Department,

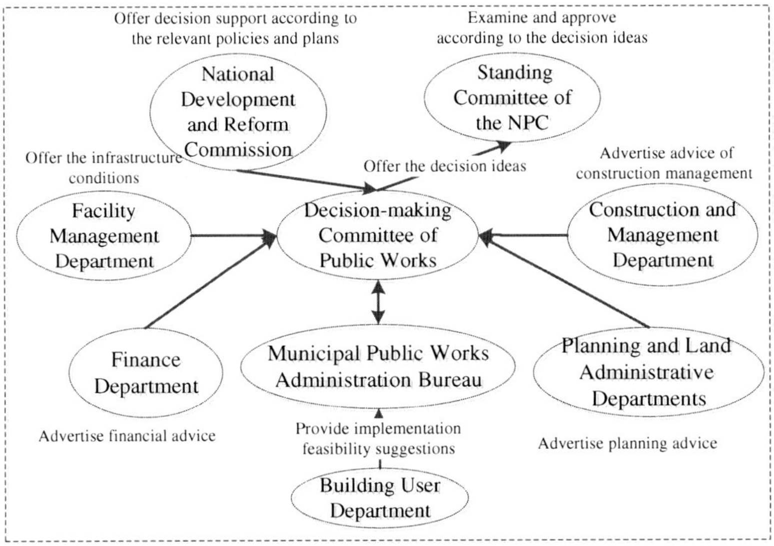

Fig. 79.3 The working relationship of Decision-making Committee of Public Works

Facility Management Department, the building user and it's department in charge, etc. Considering the existing system the reformation could be made step by step. For example, we can classify investment programmes by taking the invested amount, criterion, scale, influence etc. into consideration, and make decision upon the classification. The decision making consists of preliminary decision and final decision. Preliminary decisions should be made by different decision-making organs to different projects, like the preliminary decision for major project and important projects should be made by Decision-making Committee of Public Works, while for middle and small-sized project should by the National Development and Reform Commission, which means there is no need to go through the decision-making committee. The final decision and approval for all projects should be made by the Municipal People's congress.

Execution of Public Works

Municipal Public Works Administration Bureau can authorize its government-affiliated institutions the functions connected to construction, and the main task for it would be communicating with other departments, formulating standards, supervising investment and inspecting its subordinate institution. During the period of execution, in order to make the legal basis for project management activities of the subordinate institutions, also in order to facilitate the management, project assessment and performance assessment a series of working criteria should be made, so that can make it more convenient to limit their performance as well as guide their work. Such as the development of project management manual, technical standards or work processes, etc. The main means for supervising the investment of subordinate institutions can include into before, performing and after three periods. Pre control mainly through the brand and strategic partners to make the bidding price to reach the best, and the budget review; the control during the performing of the project is mainly the analysis and control of the settlement, the project supervision, and the major changes in control. Post control, which is to evaluate and control the final accounts.

Facilities Management of Public Works

An essential function that should be given to Municipal Public Works Administration Bureau is to organize the activity for Property management. The Public Works Administration Bureau will play two roles, one is construction unit, the other is facilities management unit, to make the facilities management related team get involved in the project life cycle as soon as possible, it can make a smooth transition from the implementation period to maintenance period. Centralization management can take full advantage of modularization, obtain facilities management services which is more competitive on price and reduce maintenance costs. Centralization management can benefit for playing advantages in professions, to make a management team who is more professional and also more suitable for public works for the improving of the efficiency of management and assure the quality of its maintenance. What's more, centralization management can also lower the risk of corruption.

79.3.3 The Organization Structure and Management Object of Municipal Public Works Administration Bureau

Matrix organization structure is a new module of organization, it is especially suitable for the system whose main task of management is construction objects. The intersection, namely a specific construction project of a matrix organization structure, has both vertical and horizontal orders, which will easily leads to contradictory orders. Every year the Municipal Public Works Administration Bureau has to face with mounts of execution of various projects, which means that there will be many project leaders in the organization of different projects horizontally. It is necessary to carry out the unified and standardized management of different projects, that is to realize the standardization, normalization, and therefore, it is necessary to take the organization structure of the vertical (Fig. 79.4). The Municipal Public Works Administration Bureau can only take management on public works, and the core of which is government investment. There are mainly two purposes of government investment, one is that the majority of the projects invested by the government is

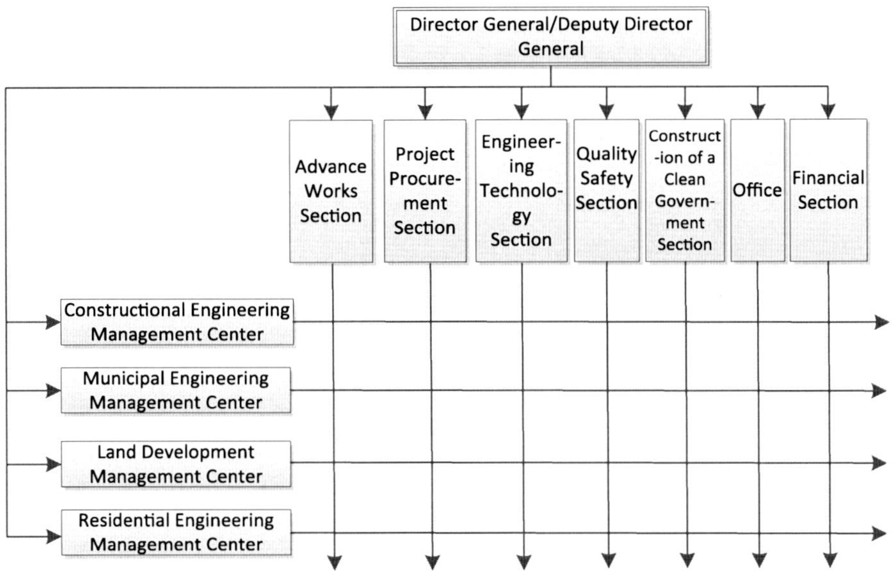

Fig. 79.4 The organizational chart of Municipal Public Works Administration Bureau

aimed at providing public products for our society. In other words, providing projects which is needed by the majority of the society and non-profit. These projects have high social benefit (Positive external economic benefits), if they were invested by the society it will take a long time to make profit, so the investors tend to be not enthusiastic. These projects also include some infrastructure projects which are difficult to recover the investment or have a longer recovery period. The other is for some confidential construction projects which cannot openly use social capital

79.3.4 The Personnel Structure of Municipal Public Works Administration Bureau

The Public Works Administration Bureau should consist mainly of professionals and part of the administrative personnel. The proportion of professional quantity occupy no less than two-thirds, and should cover the whole life cycle of construction project management's full relevant professional. Related disciplines including but not limited to: planning, architectural, structural, geotechnical, electrical, construction, cost, equipment and data processing and information technology, etc. According to the experience of the developed countries, the institution of public works management has huge personnel. Thus on the basis of controlling the

Table 79.1 The melioration of the business relationship between Municipal Public Works Administration Bureau and the related administrative department

Departments		Traditional functions			Reformation and innovation	Municipal Public Works Administration Bureau
		Decision-making phase	Implementation phase	Operation and maintenance phase	Function orientation	Relationship
National development and reform commission		The examination and approval of projects and fund scale	Project inspection	Post project evaluation	Ambiguous and repeated orientation in the examination and approval of projects and fund scale, project inspection and post project evaluation	Share in decision making, different division
Professional Construction and Management Departments	Bureau of housing and Urban	The examination and approval of construction permits	Whole process supervision		The functions are repeat with Public Works Administration Bureau and should be separated	Municipal Public Works Administration Bureau take charge of the examination and approval of construction permits and whole process supervision of public works

(continued)

Table 79.1 (continued)

Departments		Traditional functions			Reformation and innovation	Municipal Public Works Administration Bureau
		Decision-making phase	Implementation phase	Operation and maintenance phase	Function orientation	Relationship
	Bureau of Planning and Land Administration	Planning control index and land determining boundary	Land use and planning	Restrict commercial activities	Submit to Bureau of Planning and Land Administration on the site selection and planing issue	
	Other engineering departments such as water affairs and road, etc.		Whole process supervision		Maintenance of water affairs and road	Stepwise bring into the Municipal Public Works Administration Bureau to do a centralized management
	District Public Works Administration Bureau				Owner guidance, information sharing, project classification implementation	
	Other examination and approval departments (fire control, environmental protection, etc.)	The examination and approval of each specialized			Simplify the approval process	

(continued)

Table 79.1 (continued)

Departments		Traditional functions			Reformation and innovation	Municipal Public Works Administration Bureau
		Decision-making phase	Implementation phase	Operation and maintenance phase	Function orientation	Relationship
Financial Bureau		Appropriation according to investment plan	Change of fiscal fixed assets		Supervise the use of funds and rationality	Supervise the Municipal Public Works Administration Bureau
Audit Bureau		Supervise the implementation of budget	Final account audit	Audit of financial revenue and expenditure	Comprehensive audit in dealing with finance, quality, safety, environment of invested projects	Audit Bureau organizes specialized organizations to form the inspection team, Municipal Public Works Administration Bureau prepares material at fixed period

number of professional personnel in every specialty,in view of the requirement for each department is professional dynamic allocation, and establish the staff appraisal and incentive mechanism.

79.3.5 The Business Relationship Between Municipal Public Works Administration Bureau and the Related Administrative Departments

Our government public works decision-making and implementation process are supervised by National development and reform commission (formulation, approval, bidding and other comprehensive supervision), financial (appropriated funds and financial management), the audit (application of funds), construction committee (such as quality and safety and market trading regulation), supervision, discipline inspection commission and other departments, and major construction project audit ombudsman's office. In the concrete implementation, it shows many phenomenons. Such as regulatory fragmentation, rights and liabilities divided not clear, cooperating not harmonious between departments and responsibility implementation not reaching the designated position. Many departments are in the tube, but the effect not beautiful. Lack of the whole process of the project and a full range of unified and strict regulation. So in this management system designed for the following improvement and relationship orientation (Table 79.1).

79.4 Conclusions and Recommendations

To sum up, On the basis of the centralized management of Bureau of Public Works of Shenzhen Municipal, setting up the management system of the innovation of public works management system also need the constantly practice and summary. But it can be a two-step in general: the first step to optimize, improve the Municipal Public Works Administration Bureau related functions, increase under the administration of design management, cost control, quality inspection, facilities maintenance and other technical support department. Water, traffic engineering still belong to the competent administrative department for water, transportation management. The second step is to stay after the completion of the Municipal Public Works Administration Bureau, and with reference to the Hong Kong development mode, deepening the reform of the government public project management system in the city, the government investment in public works, water, traffic construction functions all integrated into one department or commission.

Suggest that all levels of government investment in public works in the city of Shenzhen divide in secondary management, the municipal public engineering construction department is responsible for, the municipal financial investment in public works. District public engineering construction department is responsible for district-level financial investment in public works. The street agency of public works management organizations are cancelled. City, district public works management institutions belong to the municipal, district secondary government management. There are no subordinate relations, but in the aspects of policies and regulations, technical standards, information sharing maintain close relations.

References

1. Huang M (2013) Problems and the methods to deal with in government investment project management. Contemp Econ 12:76–77
2. Chen H (2012) Problems analysis and countermeasures of government investment project management. Manager' J 6:49
3. Hao J, Yin Y (2003) Research on the management of USA government investment project. Technoecon Manage Res 3:91–92
4. Zeng X (2015) The orbit of the construction of foreign public project and its revelation. Nanjing J Soc Sci 3:95–101
5. Dequan Li (2002) Management system of government investment project in developed countries and zones. Constr Econ 6:8–12
6. Zhu J, Liu G, Yin Y (2002) International practice of government investment project management mode. Chin Consult Eng 11:24–25
7. Hao J, Liu J (2004) Fiscal investment project management in the developed countries and zones. Water Resour Hydropower Eng 35(8):83–85
8. Wu X (2008) Analysis and reference of government investment project management system in developed countries. Spec Zone Econ 7:100–101
9. Yan M, Yan L, Yin Y (2011) A study of institutional innovation of centralized management mode for government investment projects: a perspective of transaction cost analysis. China Civil Eng J 44(9):130–138
10. Nan Z (2012) The construction agent system of government investment project Management Model Analysis and Countermeasure Research, Xi'an University of Science and Technology
11. Su Y, Yin Y (2011) An investigation of agent-construction management system under the situation of centralized construction agent—case of the practice of Bureau of Public Works of Shenzhen Municipality. J Shenyang Jianzhu Univ, Soc Sci 13(1):53–57

Chapter 80
Analysis and Agent Modeling of Metro Passengers' Unsafe Behavior Based on Theory of Planned Behavior

Wenjie Sun and Jingfeng Yuan

Abstract Metro is a significant part of modern urban traffic system. But in recent years, metro accidents happen occasionally. Passengers' unsafe behavior is a significant cause. This research focuses on factors and influence path of unsafe behavior and tries to take passengers' attributes as the breakthrough point to propose appropriate suggestion to facilities improvement and evacuation plans. The theory of planned behavior is the fundamental theory of this research. Questionnaire design and its analysis are all based on five variables from theory of planned behavior and another four variables introduced from literature. Difference of personal attributes on unsafe behavioral intention and influence factors were analyzed. Structural equation model was used to analyze the influence path of factors. Passengers' personal attributes were transferred into speed parameters used in simulation. Hall and platform were used as two main simulation scene in Anylogic. Time and number of passengers in queue are recorded in whole progress of taking metro and evacuation. After all those analysis, respondents' opinions were shown on implementation intention, risk awareness and factors of unsafe behavior. Focusing group with specific attributes was pointed out. Influence path of factors was found by structural equation model. Suggestion to facilities improvement was proposed and verified. Appropriate evacuation measurements were proposed to increase the evacuation efficiency.

Keywords Metro passengers · Unsafe behavior · TPB · Agent modeling

80.1 Introduction

Metro is a significant part of modern urban traffic system. By its unique advantages, metro performs an increasingly important role in urban economic life. But in recent years, metro accidents happen occasionally. Passengers' unsafe behavior is a

W. Sun (✉) · J. Yuan
Department of Civil Engineering, Southeast University, Nanjing, China
e-mail: jeffreyswj@163.com

Y. Wu et al. (eds.), *Proceedings of the 20th International Symposium on Advancement of Construction Management and Real Estate*,
DOI 10.1007/978-981-10-0855-9_80

significant cause. This research focuses on factors and influence path of unsafe behavior and tries to take passengers' attributes as the breakthrough point to propose appropriate suggestion to facilities improvement and evacuation plans. This research include questionnaire data analysis by SPSS17.0, structural equation model by AMOS and agent modeling by Anylogic. Results were analyzed to improve metro station service and to decrease occur of passengers' unsafe behavior.

80.2 Literature Review

The theory of planned behavior is an extension of the theory of reasoned action made necessary by the original model's limitations in dealing with behaviors over which people have incomplete volitional control.

As in the original theory of reasoned action, a central factor in the theory of planned behavior is the individual's intention to perform a given behavior [1]. The theory of planned behavior postulates three conceptually independent determinants of intention [2]. The first is the attitude and refers to the degree to which a person has a favorable or unfavorable evaluation or appraisal of the behavior in question [3]. The second predictor is a social factor termed subjective norm; it refers to the perceived social pressure to perform or not to perform the behavior [4]. The third antecedent of intention is the degree of perceived behavioral control which, as we saw earlier, refers to the perceived ease or difficulty of performing the behavior and it is assumed to reflect past experience as well as anticipated impediments and obstacles.

Overseas research focuses more on unsafe behavior in the field of transportation, especially in the field of pilots' unsafe behavior. Boudrifa et al. identified the influence factors of road users' unsafe behavior and motivation, through questionnaire survey on drivers. And he found out the most often and the most dangerous unsafe behavior of road users. Friedman's rank test method is applied to rank each road users' factors and motivation [5, 6].

80.3 Methodology

80.3.1 The Extended Theory of Planned Behavior

Along with further researches, scholars found out that quite a large portion of variance in the study of intent hasn't been explained. It is necessary to add new variables in order to improve the predictive ability of the whole model. For this reason, there are four assumptions introduced here as variables to make some improvements for the model of TPB: (1) Risk propensity; (2) Outer perception; (3) Past behavior; (4) Passengers' personal attributes.

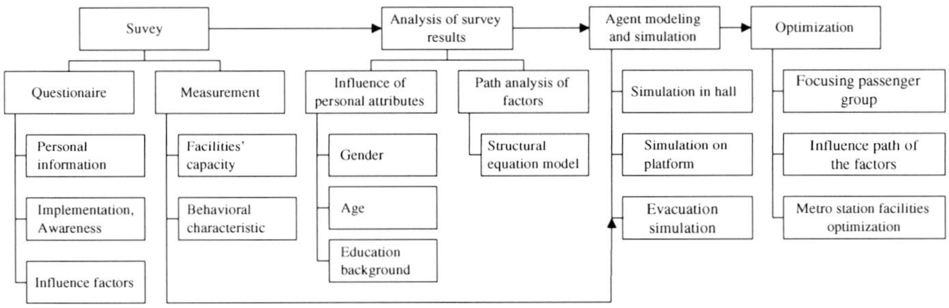

Fig. 80.1 Roadmap

80.3.2 Research Design

The whole roadmap for this research is shown in Fig. 80.1.

The research was conducted by questionnaires. The questionnaire was divided into three parts. The first part was about personal information, including gender, age, education background and the frequency of taking a subway. The second part was about respondents' implementation and awareness of unsafe behavior. In this part, 16 kinds of typical unsafe behaviors were listed according to the literature review and relevant case studies. The third part was about the influence factors of unsafe behavior. According to the research above, attitude, subjective norm, perceived behavior control, risk propensity, outer perception, past behavior and behavioral intention were designed as variables. All questions were used in the form of 5 points of the Likert scale. The respondents were asked to explain their degree of agreement from 1 (strongly disagree) to 5 (strongly agree).

80.3.3 Influence of Personal Attributes

An independent-samples T test by SPSS17.0 was conducted to analyze difference of gender on unsafe behavioral intention and influence factors. A one-way ANOVA and multiple comparisons by SPSS17.0 were conducted to analyze difference of education on unsafe behavioral intention and influence factors. A one-way ANOVA and multiple comparisons by SPSS17.0 were conducted to analyze difference of age on unsafe behavioral intention and influence factors.

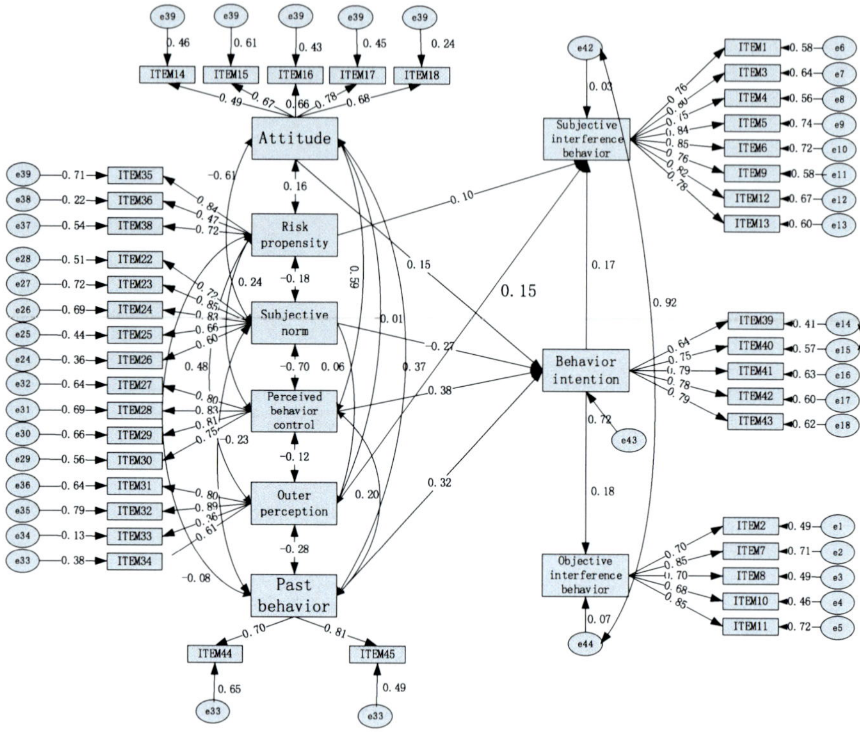

Fig. 80.2 Influence path of factors

80.3.4 Influence Path Analysis of Factors

To make sure of the unidimensionality of variables, this paper make every question as an measurement index of one variable. Figure 80.2 shows the final fitting structural model of our research.

80.3.5 Passenger Simulation in Hall of Zhujianglu Subway Station

The completed environment model of hall of Zhujianglu Subway Station established by Anylogic is illustrated in Fig. 80.3. Data of passengers' characteristic was used to set the parameters of passengers. Simulation of taking a subway was shown in Fig. 80.3.

Each passenger's time and number of passengers in the queue were recorded through the simulation.

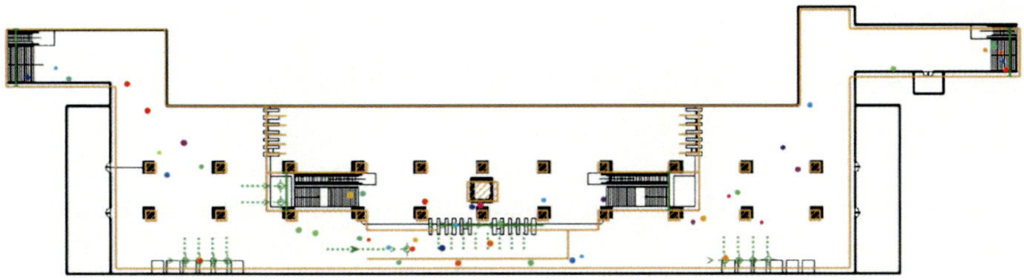

Fig. 80.3 Simulation in hall

80.3.6 Passenger Simulation on Platform and Trains

The completed environment model of platform and trains was shown in Fig. 80.4. Data of passengers' characteristic was used to set the parameters of passengers. The simulation was shown in the following figure. Time of each process of taking on and off trains were recorded through the simulation.

80.3.7 Passenger Evacuation Simulation

Different conditions of passengers were set to simulate the evacuation under emergency. The 3D working sketch was shown in Fig. 80.5. Evacuation time of each conditions was recorded through the simulation.

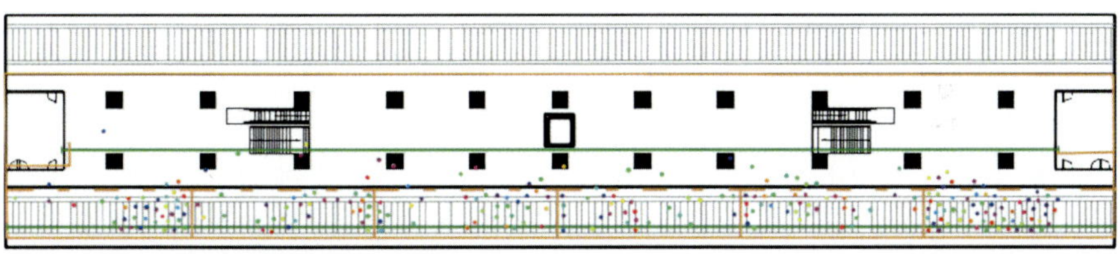

Fig. 80.4 Simulation on platform

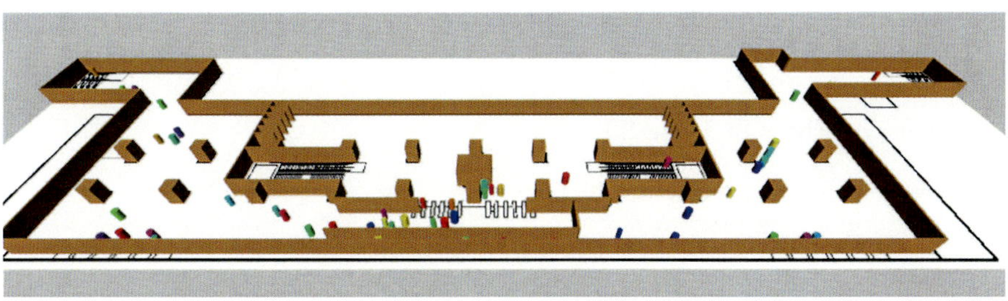

Fig. 80.5 Evacuation simulation

80.4 Results

80.4.1 Results of Influence of Personal Attributes

From the results of those analysis, several results were concluded as follows. Gender has significant difference on attitude towards unsafe behavior. Passengers with various education levels are significantly different in unsafe behavior, attitude, perceived behavior control, outer perception and risk propensity. Passengers in different ages are significantly different in unsafe behavior, attitude, perceived behavior control, outer perception and risk propensity. Generally speaking, females have more tendencies to unsafe behavior; Passengers with lower education level have stronger intension to conduct unsafe behavior; Young and middle-aged have stronger intension on unsafe behavior.

80.4.2 Results of Passenger Simulation in Hall of Zhujianglu Subway Station

1. Walk time of passengers and queuing number of each facility are revealed in the figure below. Severe congestion's happening after 20 m simulation, which made the simulation out of reality, so part of the data in the figure is blank.
2. The density of passengers increases in the metro entrance and gradually leads to be steady.
3. Queuing line will generate at ticket office after partial passenger flow comes. As time goes on, passengers will gradually become crowed because the lack of ticket capacity.
4. With the coming of passenger flow, service is insufficient and speed of evacuation is not enough, which results in long queuing line and affects the order and comfort of station.
5. Congestion is easily formed in front of ticket gate because of the lack of gate number, insufficience of gate service, and the concentration of gate (Table 80.1).

80.4.3 Results of Passenger Simulation on Platform and Trains

Time data was show in Table 80.1, 80.2.

Table 80.1 Simulation Results

Projects	Value
Maximum of walk time	–
Minimum of walk time	107
Mean value of walk time	–
Maximum number of queuing people in front of ticket machine	9
Minimum number of queuing people in front of ticket machine	3
Mean number of queuing people in front of ticket machine	5
Mean number of queuing people in front of artificial ticket	3
Maximum number of queuing people in security checkpoint	15
Minimum number of queuing people in security checkpoint	6
Mean number of queuing people in security checkpoint	9
Maximum number of queuing people in ticket gate	–
Minimum number of queuing people in ticket gate	6
Mean number of queuing people in ticket gate	–

Table 80.2 Simulation Results

Serial number	Number of passengers getting on	Number of passengers getting off	Time (ordered)	Time (disordered)
1	50	50	4.2	6.3
2	100	100	9.8	14.6
3	150	150	14.3	20.8
4	200	200	21.5	28.0
5	250	250	27.7	37.4

80.4.4 Results of Passenger Evacuation Simulation

Evacuation results were displayed on Figs. 80.6, 80.7 and 80.8.

80.5 Discussion

80.5.1 Optimized Analysis of Hall of Zhujianglu Subway Station

1. Canceling security fence and increasing a security facility.
2. Increasing two gates and separating the gates' location will relieve the congestion.
3. Increasing two ticket machines in the south and north of the hall respectively.
4. Increasing an artificial ticket window in the south location.

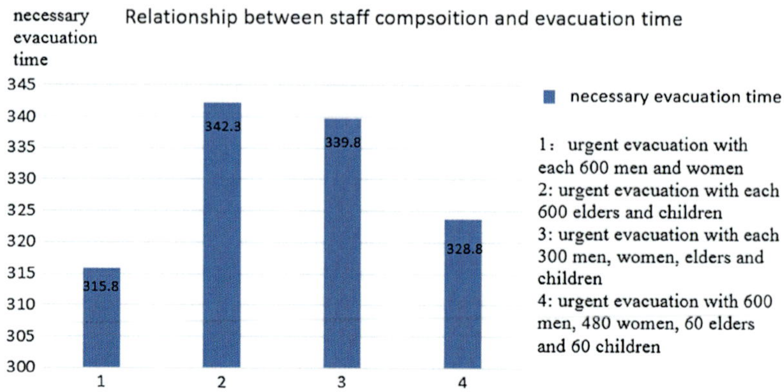

Fig. 80.6 Simulation results

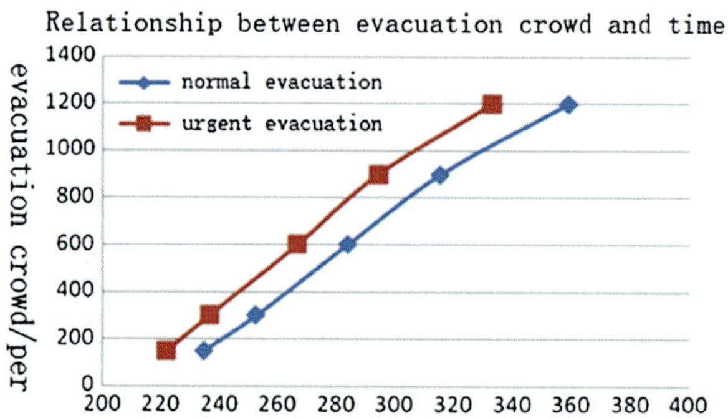

Fig. 80.7 Simulation results

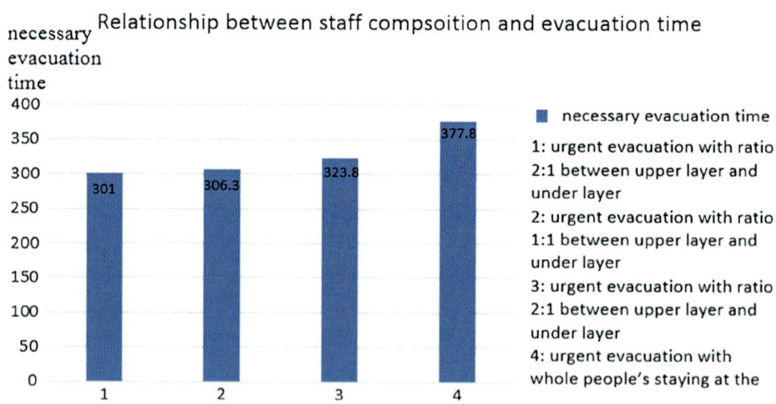

Fig. 80.8 Simulation results

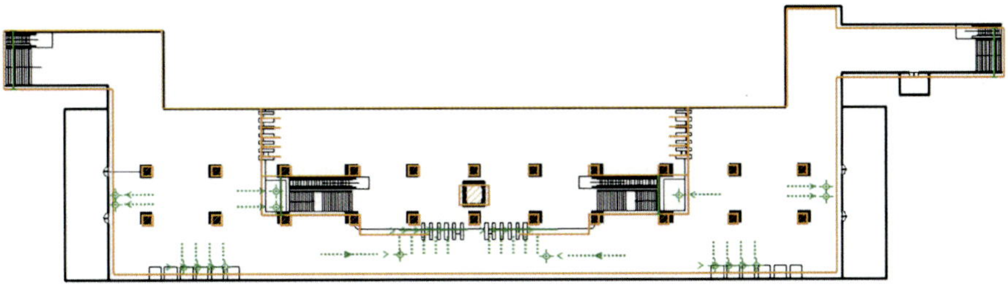

Fig. 80.9 Optimization

Table 80.3 Optimization Result

Projects	Value(before optimization)	Value (after optimization)
Maximum of walk time	–	568
Minimum of walk time	107	93
Mean value of walk time	–	281
Maximum number of queuing people in front of ticket machine	9	6
Minimum number of queuing people in front of ticket machine	3	2
Mean number of queuing people in front of ticket machine	5	4
Mean number of queuing people in front of artificial ticket	3	2
Maximum number of queuing people in security checkpoint	15	9
Minimum number of queuing people in security checkpoint	6	3
Mean number of queuing people in security checkpoint	9	6
Maximum number of queuing people in ticket gate	–	8
Minimum number of queuing people in ticket gate	6	3
Mean number of queuing people in ticket gate	–	4

After optimization, the environment model is shown in Fig. 80.9.

After optimization, from Table 80.3, traffic capacity was greatly improved.

80.5.2 Optimized Analysis of Platform and Trains

1. As the number of passengers increased, time lag between ordered and disordered condition was rising faster. Time change faster and longer. It indicates that on this occasion, force between passengers and environment raised faster and stronger and jam at doors was more severe.
2. Number of passengers at peak time is between 200 and 250. The needed time is between 21.5 and 27.7 s both of which are longer than the time doors open (19 s). The longer needed time is, the more passengers' unsafe behavior would occur.

80.5.3 Optimized Analysis of Passenger Evacuation Simulation

Firstly, the required safety egress time of old people and children in station is 27 s longer than that of male and female. As a consequence, the elderly and children need more attention in safe evacuation. Secondly, the more passengers stay on the lower floor, the more safety egress time is required. Hence, the staff should firstly lead passengers to the upper floor and ensure the evacuation passageway unblocked. Thirdly, evacuation time limit is 360 s. The design of this station satisfied the requirement. As the number of passengers grows, the time difference between normal and emergency evacuation becomes more significant. Problems of bottlenecks of exit and evacuation passageway are increasingly serious.

80.6 Conclusion

1. Respondents' opinions were shown on implementation intention, risk awareness and factors of unsafe behavior.
2. Focusing group with specific attributes was pointed out.
3. Influence path of factors was found by structural equation model.
4. Suggestion to facilities improvement was proposed and verified.
5. Opening time of doors of metro trains was suggested to extend in a certain scope and verified through simulation.
6. Appropriate evacuation measurements were proposed to increase the evacuation efficiency.

Acknowledgments The authors' special thanks go to all and reviewers of the paper, and to the National Natural Science Foundation of China (NSFC-71001027, 71472037), Social Foundation of Jiangsu Province, China (13GLB005), Program for Outstanding Young Teachers of Southeast

University (2242015R30009), and the Fundamental Research Funds for the Central Universities for financially supporting this research.

References

1. Fogarty GJ, Shaw A (2010) Safety climate and the theory of planned behavior: towards the prediction of unsafe behavior. Accid Anal Prev 42(5SI):1455–1459
2. Clarke S (2006) Safety climate in an automobile manufacturing plant—the effects of work environment, job communication and safety attitudes on accidents and unsafe behaviour. Pers Rev 35(4):413–430
3. Goncalves S, Da Silva SA, Lima ML et al (2008) The impact of work accidents experience on causal attributions and worker behavior. Saf Sci 46(6):992–1001
4. Samantha Jamson MWRB (2008) Developing a driving Safety Index using a Delphi stated preference experiment. Accid Anal Prev 40:435–442
5. Deborah J, Tunnicliff BCWK (2012) Understanding the factors influencing safe and unsafe motorcycle rider intentions. Accid Anal Prev 2012(49):133–141
6. Dahlen ER, Martin RC, Ragan K et al (2005) Driving anger, sensation seeking, impulsiveness, and boredom proneness in the prediction of unsafe driving. Accid Anal Prev 37(2):341–348

Chapter 81 The Life Cycle Cost Forecast of 220 kV Transmission Line Project Based on LS-SVM

Jing Fu, Guiwei Shao, Zhaoxia Gao, Zhongfu Qin, Xiaolong Lei and Liqing Meng

Abstract The influence factors of transmission line engineering construction are complex. The historical data is limited. It is a typical small sample and nonlinear problem. Aiming at the shortage of the traditional regression method in solving small sample problem, this paper constructs a new model of cost prediction based on least square support vector machine. Through the analysis of historical projects and consulting the relevant indicators to choose index of correlation which affect the project cost as the original input set, a new input set is obtained by using principal component analysis. The results show that the new model is more prominent in the small sample learning, which can help the cost analysis and forecast of the line project. The research findings can support Inspection Efficiency Assessment in Transmission Line Projects.

Keywords Power engineering cost · Principal component analysis · Least square support vector machine

81.1 Introduction

The influence factors of transmission line engineering construction are complicated, and the number of historical projects is limited, which makes the prediction of transmission line engineering cost is a typical small sample and nonlinear problem, which brings difficulty to the investment management. Therefore, it is urgent to build

J. Fu · G. Shao
China Electric Power Research Institute (CEPRI), 15 Xiaoying E Rd, Haidian, Beijing, China

Z. Gao
Zhejiang Electric Power Corporation, State Grid Corporation of China, No.8, Huanglong Road, Hangzhou 310007, China

Z. Qin (✉) · X. Lei · L. Meng
College of Civil Engineering and Architecture, Zhejiang University, Zhejiang, China
e-mail: lxl5358949@126.com

Y. Wu et al. (eds.), *Proceedings of the 20th International Symposium on Advancement of Construction Management and Real Estate*,
DOI 10.1007/978-981-10-0855-9_81

a new model of project cost forecast for small sample and nonlinear problem, in order to improve the investment management capability of the transmission line project.

At present, many domestic and foreign experts and scholars have carried out in-depth research on the prediction model of project cost, but most of the research are focused on the traditional statistical learning methods, such as linear regression [1], parameter estimation [2], artificial neural network [3], etc. However, the traditional statistical learning method is the pursuit of empirical risk minimization, so it needs a large number of samples, which will lead to learning problems and lead to poor generalization ability.

Statistical learning theory (SLT) [4] is a machine learning theory developed by Vapnik. It is a training samples that is studied under the condition of limited machine learning rules of discipline. Support Vector Machine [5] (SVM) is based on the development of the theory, and introduces the structural risk minimization principle and the kernel function. Compared with the traditional machine learning, support vector machine is good at the small sample learning and nonlinear problems [6], and in many scientific research fields have shown a good generalization ability. Therefore, the study of support vector machine has become a hot spot in the world. Many of them are for the improvement of SVM algorithm itself, such as block processing algorithm, fixed working sample set algorithm, SMO algorithm [7]. Least square support vector machine (LS-SVM) [8, 9] is an improved algorithm of SVM, which is used to solve the dimension disaster of support vector machine. And it can further improve the training speed and accuracy of learning machine.

In this paper, based on general design and general cost manual for transmission line of Zhejiang Electric Power Corporation and the historical engineering of Zhejiang province and the relevant expert opinion to Select the indexes of the main body of transmission line project. The main components are selected as the input set of the new model. Through principal component analysis to reduce the dimension and eliminate the correlation between the indicators to get a small amount of comprehensive index. The small amount of comprehensive index as a new input set input based on least square support vector machine to achieve rapid and accurate prediction of transmission line project cost purpose.

81.2 Principle of Principal Component Analysis and Least Square Support Vector Machine

81.2.1 Principal Component Analysis

There is a high degree of correlation between the various indicators which affect the cost of transmission line project, and it has the overlap of the sample information. Component analysis Principal (PCA) [10] can be used to reduce the dimension of the original variables and get less comprehensive variables, which can overcome the curse of dimensionality of SVM. The above mentioned variables can represent the information of the original variables as well as to eliminate the correlation between each other.

81.2.2 Support Vector Machine and Least Square Support Vector Machine

Support vector machine (SVM) theory is developed from the optimal classification line [11]. The optimal classification line can not only separate the two types of samples, but also get the maximum classification interval. The optimal classification line is increased to a multi dimension space. Finding the optimal classification surface is a two time planning problem. The support vector machine is used to search the optimal classifier, and the linear map of the original space can be classified by using the kernel function, and the fault tolerance is introduced by using the loss function.

LS-SVM is an improved algorithm of SVM. And the support vector machine is improved in the process of optimizing problem. The different loss function is selected, the inequality constraints are changed into equality constraints, so the optimization problem of LS-SVM is:

$$\begin{aligned} &\min\phi(\omega,\varepsilon) = \tfrac{1}{2}\|\omega\|^2 + \tfrac{1}{2}\gamma\sum_{i=1}^{l}\varepsilon_i^2 \\ &\textit{subject to}\cdots y_i((\boldsymbol{\omega} * x_i) + b) = 1 - \varepsilon_i, \quad i = 1,\ldots,l; \quad \varepsilon_i \geq 0 \end{aligned} \tag{81.2.1}$$

Define the Lagrange function:

$$L(\omega, b, \varepsilon; \alpha) = \frac{1}{2}\|\omega\|^2 + \frac{1}{2}\gamma\sum_{i=1}^{l}\varepsilon_i^2 - \sum_{i=1}^{l}\alpha_i(y_i(\boldsymbol{\omega} * \varphi(x_i) + b) - 1 + \varepsilon_i) \tag{81.2.2}$$

Make $\omega, b, \varepsilon_i, \alpha_i$'s partial derivative is equal to 0, and eliminate ω, ε:

$$\begin{pmatrix} \mathbf{0} & \mathbf{Y}^{\mathbf{T}} \\ \mathbf{Y} & \mathbf{Z}\mathbf{Z}^{\mathbf{T}} + \frac{1}{\mathbf{r}}\mathbf{I} \end{pmatrix}\begin{pmatrix} b \\ \boldsymbol{\alpha} \end{pmatrix} = \begin{pmatrix} \mathbf{0} \\ \mathbf{I} \end{pmatrix} \tag{81.2.3}$$

$Z^T = [y_1\varphi(x_1), y_2\varphi(x_2), \ldots, y_l\varphi(x_l)]$, $Y^T = [y_1, y_2, \cdots y_l]$, $I^T = [1, 1, \ldots, 1]$, $\alpha^T = [\alpha_1, \alpha_2, \ldots, \alpha_l]$. And according to the Mercer condition [12], the existence of nuclear function:

$$\psi(\mathbf{x_i}, \mathbf{x_j}) = \varphi(x_i)^T\varphi(x_j) \tag{81.2.4}$$

Therefore, the decision function of least squares support vector machine is:

Table 81.1 Characteristic value and contribution rate of principal components

Principal component	Characteristic value	Contribution rate (%)	Cumulative contribution rate (%)	Principal component	Characteristic value	Contribution rate (%)	Cumulative contribution rate (%)
Z_1	4.139	24.3472	24.3472	Z_{10}	0.7023	4.1313	90.1935
Z_2	1.9126	11.2505	35.5977	Z_{11}	0.5437	3.1983	93.3918
Z_3	1.5984	9.4021	44.9998	Z_{12}	0.3533	2.0783	95.4701
Z_4	1.4957	8.798	53.7978	Z_{13}	0.3269	1.9231	97.3932
Z_5	1.4297	8.41	62.2078	Z_{14}	0.2135	1.2557	98.6489
Z_6	1.2386	7.286	69.4938	Z_{15}	0.1618	0.952	99.6009
Z_7	1.105	6.5	75.9938	Z_{16}	0.0675	0.3972	99.9981
Z_8	0.924	5.4353	81.4291	Z_{17}	0.0003	0.002	100
Z_9	0.7876	4.6331	86.0622				

$$y(x) = sign[\sum_{i=1}^{l} \alpha_i y_i \psi(\mathbf{x_i}, \mathbf{x_j}) + b] \tag{81.2.5}$$

Kernel function $\psi(\mathbf{x_i}, \mathbf{x_j})$ has a lot of settings, we chose RBF radial basis function in this study [11]:

$$\psi(\mathbf{x_i}, \mathbf{x_j}) = \exp\left(-\frac{|x - x_i|^2}{\sigma^2}\right) \tag{81.2.6}$$

σ^2 is the parameters of the kernel function, and as the same as γ that can find in the MATLAB through the parameters optimization.

81.3 Index Selection

The influence factors of transmission line engineering construction are complicated, which are influenced by different technical conditions, natural conditions and economic conditions, which make the cost analysis face difficulties. In this study, the great impact of the project cost on the construction cost of the transmission line based on Zhejiang province is selected as the input index. In addition, this study selects the unit cost of transmission line engineering (million yuan/km) as the output set. Through the above analysis, this study selects 17 indicators as the input set and the output set:

Input set = {the ratio of five terrain condition, unit tower (base/km), unit resistance angle tower (base/km), ire splitting number, single wire cross-sectional area (mm^2), unit wire quantity (t/km), wire price (yuan), ice (mm), wind speed (m/s), unit volume (m^3/km), unit concrete volume (m^3/km), unit base steel (t/km), basic steel price (yuan)} [13, 14]. Output set = {unit cost (million yuan/km)}.

Explanation: the base of the unit index is the line length (converted to Dan Xianlu).

81.4 Model Simulation Analysis

81.4.1 Case Description

The sample of this study is 2010–2012, the voltage level is 220 kV in Zhejiang province, which uses the new standard of the iron tower, the new standard of ice and the three way of the overhead line project. Through the data collection and sorting, and combined with experts' opinion, after removing a part of line length which is less than 1 km, we get a total of 87 samples, including 80 samples as training samples and 7 samples as the test sample in the new model simulation analysis.

81.4.2 Principal Component Analysis

Using MATLAB to deal with the standard processing of the original input set. The principal component analysis get the characteristic values, the contribution rate and the cumulative contribution rate which are shown in Table 81.1.

From the above table we can see that the eigenvalues of the 17 principal components are more than 0, which meet the requirements of the principal component analysis. The characteristic values and the contribution rate of each principal component are decreased from Z_1 to Z_{17} in turn. The contribution rate of Z_1, $q_i = 24.3472$ %, which contributes 24.3472 % to the original index information. The general cumulative contribution rate is $Q_i \in (80, 95$ %) [14], which is considered that the main components of the former i can be used to instead of the original index information. In this study select the first 8 principal components ($Q > 80$ %), the first 9 principal components ($Q > 85$ %) and the first 10 principal components ($Q > 90$ %) are tested to determine the best principal components, and to avoid the curse of dimensionality. The study will not select more than 10 main components for the test.

81.4.3 Model Training and Prediction of LS-SVM

In view of the adjustable parameters of LS-SVM model, we select initial parameters: the penalty coefficient = 10, the kernel function of the width coefficient = 0.1, in the MATLAB for training and prediction. And the forecast values of the 7 test samples and actual values are based on the first 8, 9 and 10 principal components in

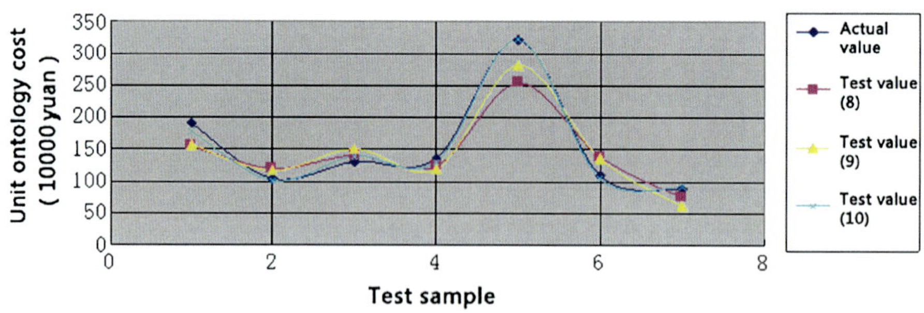

Fig. 81.1 The prediction effect of LS-SVM model based on the first 8, 9 and 10 principal components respectively

Table 81.2 The error of the predicted value and actual value from the test sample (select the first 10 principal components)

Sample serial number	Actual value (unit cost/million yuan)	Predictive value (million yuan)	Absolute error (million yuan)	Relative error (%)
1	190.5	178.1	−12.4	−6.5092
2	104.2	100.8	−3.4	−3.2630
3	131.1	139.6	8.5	6.4836
4	136.4	130	−6.4	−4.6921
5	321.4	322.4	1	0.3111
6	109.8	104.7	−5.1	−4.6448
7	88.87	92.36	3.49	3.9271

MATLAB using the function image to be combined into a demonstration as shown in Fig. 81.1.

Note: the yellow line represents the actual output value of the test sample, the red line represents the predicted output value of the test sample, and the vertical coordinate represents the unit body cost of the sample.

By the image above : select the first 10 principal components ( ) as the new input set can obtain the best fitting results. The error of the test sample and the actual value is shown in Table 81.2.

From the above chart, the relative error of the prediction is −6.5092 %, the minimum is 0.3111 %. The new model has good performance and stability at controlling relative error and the simulation analysis is very good.

81.5 Summary and Outlook

In this study, the cost forecasting model of transmission line project based on LS-SVM that is from the training and testing of the historical project in Zhejiang province. Through simulation analysis of the model, we find that the new model has good performance and stability at controlling relative error and the simulation analysis is very good. It is convenient for the evaluation and prediction of the technical and economic personnel, and it has a high popularization value.

Acknowledgments The research findings can support Inspection Efficiency Assessment in Transmission Line Projects.

References

1. Huiwen W, Jie M (2007) Predictive modeling method for multivariate linear regression. J Beihang Univ 33(4):500–504
2. Li Y, Xie B, Peng Q (2009) Study of past, present and future. Data (7):58–60

3. Yang X, Chen T (1994) Inherent advantages and disadvantages of artificial neural networks. Comp Sci 21(2):23–26
4. Vapnik VN (1998) Statistical learning theory. John Wiley, New York
5. Vapnik VN (1995) The nature of statistical learning theory. Springer, New York
6. Jiayuan Z, Yun Y, Hengxi Z (2004) Study on small sample prediction based on optimized least square support vector machine. J Aviat 25(6):565–568
7. Liu H, Ma S (2002) Research status of support vector machine. Chin J Grap 7(6): 618–623
8. Yan W, Shao H (2003) Comparison and application of support vector machine and least square support vector machine. Control Decis 18(3):358–360
9. Suykens JAK, Vandewalle J (1999) Least squares support vector machine classifiers. Neural Process Lett 3:293–300
10. Jincheng Fan, Changlin Mei (2002) Data analysis. Science Press, Beijing
11. Bai P, Zhang X, Zhang B et al (2008) Support vector machine theory and engineering application examples. Xi'an Electronic Science and Technology Press, Xi'an
12. Cristianini N, Shawe-Taylor J (2000) An introduction to support vector machines and other kernel-based learning methods. Cambridge University Press, Cambridge, pp 30–34
13. Ling Y, Yan P, Han C, Yang C (2012) Cost forecasting model of transmission line project based on BP neural network. China Power, 45(10): 95–99
14. Peng G, Yu J, Wei J, Yang G (2009) Feature extraction and small sample learning of power engineering cost forecasting model. J Chongqing Univ 32(9):110–1104

Chapter 82
Based on Hypergame Analysis of Construction Disputes

Zhongfu Qin, Xianrong Wei and Liqing Meng

Abstract Construction disputes in the construction industry is common and inevitable phenomenon. Under the condition of information asymmetry, hypergame situation often appear in the construction disputes. This paper discussed the reason of misunderstanding, and how to deal with the misunderstanding. In this paper, the author gives the preference vector determination method and the dynamic development mechanism of hypergame. It has practical significance for the development of the construction project disputes and equilibrium solution.

Keywords Construction disputes · Hypergame · Misunderstanding · Preference vectors

82.1 Introduction

Due to its various type of work, organization, confused between each main purpose can be said to be inevitable in construction engineering [1]. If it is like the village development, old city rebuilding and widening roads, housing demolition, urban infrastructure construction, large-scale engineering projects such as construction of water conservancy facilities, it would be easier to produce dispute. When disputes appear in the construction projects, information exchange is not completely between the players, no matter it is due to subjective or objective reasons。.

Game theory in the study of the complex phenomenon of the dispute process, through constant theoretical development and applied exploration has achieved outstanding results. Kassab et al. [2] applied graph model to construction project of the dispute, using construction disputes chart model study the interaction between owners and contractors, and considering the decision making participants with their

Z. Qin (✉)
Department of Zhejiang University, Zhejiang, China
e-mail: qinzhongfu@zju.edu.cn

X. Wei · L. Meng
School of Zhejiang University, Zhejiang, China

Y. Wu et al. (eds.), *Proceedings of the 20th International Symposium on Advancement of Construction Management and Real Estate*,
DOI 10.1007/978-981-10-0855-9_82

decision ideology, in the process of formation model, allows larger parameter selection of freedom [3].

Qin [4, 5] considered the preferences, strategy, uncertain factors in the players, the changing structure of the conflict and power asymmetry, and so on, which make disputes models and algorithms gradually closer to complex disputes in reality. But there still exist the following problems to be further expand and deepen:

1. The hypergame theory is the effective tool for dispute under the condition of incomplete information, but there is still no system complete theoretical framework in the research as so far, such as how to extract the basic elements of hypergame from the reality of the dispute in accordance with certain rules, how to avoid and deal with the misunderstanding, how to obtain more effective equilibrium solution with the misunderstanding exist [6].
2. For super countermeasures preference vector is obtained, There is still no a unified and effective analysis method to obtain preference vectors of hypergame. The preferences of players are uncertain in hypergame, it is difficult for one player to clearly recognize the opponent's preference order. How to obtain the more accurate preference vector of opponent's is needed to study [7].

82.2 The Reason of Misunderstandings Occur in Hypergame

Misunderstanding each other in disputes mainly because of information asymmetry [8]. Information asymmetry refers to in the market economy activity, the mutual information between corresponding economic individuals are not uniform, asymmetric distribution state. Because there are differences between all kinds of personnel who understanding of relevant information. The people who grasp the information is sufficient staff, often have a more advantageous position, and on the contrary, poor information of personnel, are at a comparative disadvantage. Namely this information of different between players called information asymmetry.

The reason of information asymmetry can be divided into objective and subjective factors, as shown in Fig. 82.1. Objective factors are outside the players consciousness, do not rely on the spirit of the players and exist independently, and subjective opposite; Subjective factors are from players own reason who lead to lack of information. The common representation of the information asymmetry is the lack of information and error information, information collection and provide process is likely to lead to the occurrence of such conditions.

Objective factors mainly reflected in the process of information collection, incomplete information and error caused by communication barriers, one player mistakenly cognitive things, the difference in the status of players and the asymmetric resources of players.

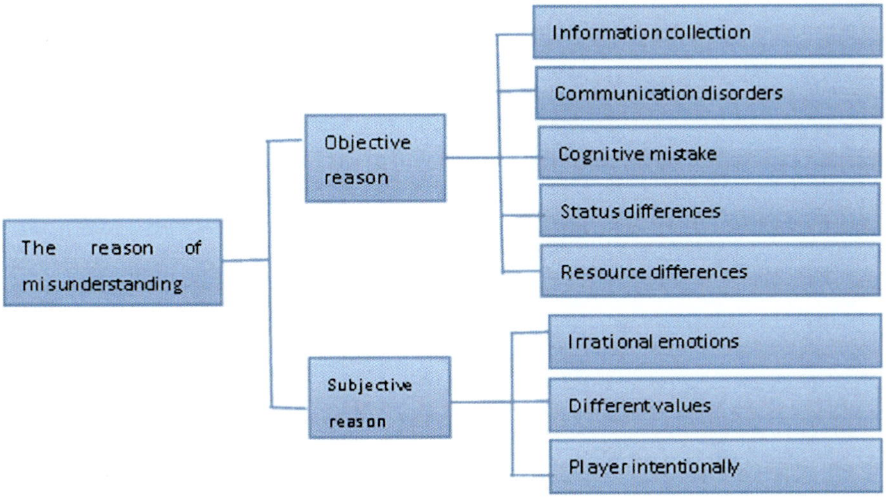

Fig. 82.1 The source of misunderstanding in hypergame

Subjective factors [9] mainly reflects in the players who have different values and irrational emotions, as well as the players themselves deliberately concealing or deliberately leaked some false information, etc.

82.3 Deal with Misunderstandings in Hypergame

When players are in a hypergame game, if players do not understand the opponent, or aware having misunderstanding to their opponents, or due to the unexpected ending or himself don't satisfied with the result, or feel the presence misunderstanding in the game, then he will have intrinsic motivation to change. Players can improve their situation from the following three aspects, as showed in Fig. 82.2.

82.3.1 Learning by Himself

1. Cognitive himself
   Self cognitive in psychological cognition is a kind of more advanced cognitive ability. The psychological cognition is a process of infinite, it will change and development as the individual experience, memory, and change of the surrounding environment. So in the dynamic development of dispute, in some cases, players may position errorly to oneself. Then players could improve the position to himself.

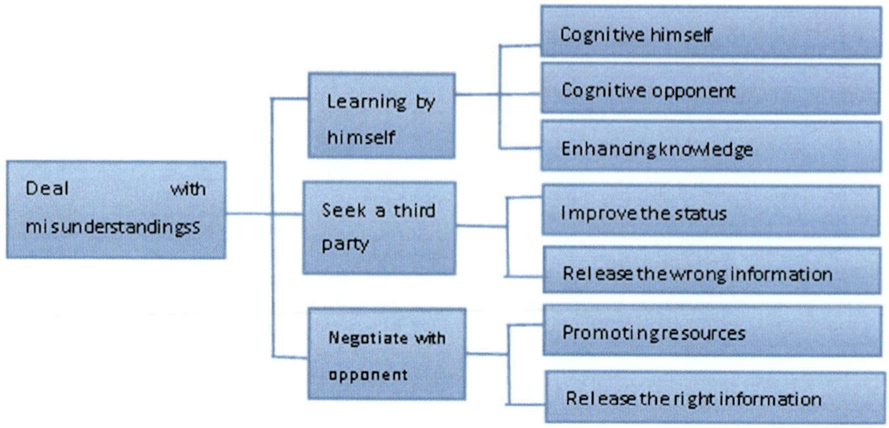

Fig. 82.2 Deal with misunderstandings in hypergame

2. cognitive opponent
   Cognition is different from the perception, it have a choice, actively, systematically to consciousness, but not receive all the information from the environment. Cognition is the rational cognition stage, it is the stage to try to know the enemy and himself, it asked to choose information constantly, feedback, processing and handling. Through the analysis of competitors, such as swot analysis, comprehensive analysis the opponent's situation, then find more real preferences of opponents.
3. improve knowledge
   In the process of the whole disputes, players should constantly improve their knowledge level, including cognitive ability, the knowledge learning involving disputes, analyse the entire dispute situation, etc., in order to stand in the active position in the process of dispute.

82.3.2 Seek a Third Party

1. Improve status
   In the process of dispute, the difference of status can affect both sides of the strategy choice.
   The position in a strong conditions may put forward a little harsh requirements in the dispute, but the weak side will be fewer choices. But, if a player is weak, then he can seek opportunities to improve their social status, change the status of inequality, strive for more choices for himself.
   But in the actual situation, change their position is often more difficult, such as under the condition of migrant workers unpaid wages, migrant workers is

always vulnerable groups. At this time they can improve their positions by seeking the help of a third party, such as government or the media. Through the support of the government, media, public opinion pressure to elevate the status of migrant workers, so they can get more fair position, and ultimately achieve their desired salary.

2. Promoting resources
   Players can get better social relations resources, more complete and accurate information resources, increase investment in their time resources and economic resources, improve the resource they share in dispute. The extent of resources occupied can change the strategy choice of players.

82.3.3 Negotiate with Each Other

1. Release the wrong information
   If there is a lose-lose strategy, and the opponent's loss may be greater than oneself, if taken, opponents will be afraid of, but considering their own lose he also will not necessarily take their losses. But he still can take some measures to let the other side think the actions will be taken on our side, so as to make some concessions, won the other strong conditions for their own.
2. Release the right information
   If your opponent errorly understand yourself, then yourself never satisfactory equilibrium solution, so you must think of some way to get your opponents know your real thoughts. As the typical story of the battle of the sexes, if the girl go against the wish of her heart choose to watch the boxing match with boys, so the boy would think that the girl's hobby is the same with his own hobby, so every time when they go out to entertainment, the chance of watching the boxing match will become high, but the girl get pleasure effect from the activity is very low, so the girl can be in the appropriate time to show the boy of her hobby is actually to see the musical, so that the boys can't misunderstanding the girl. Thus in the later dates it's avoided that the girl will be misunderstanding by the boy.

82.4 The Method of Fuzzy Analysis

This article uses the fuzzy analysis method to obtain the opponents of preference vectors.

82.4.1 The Basic Idea of Fuzzy Analysis

Set $Q = \{q^1, q^2 ..., q^m)$ be m practical outcomes of the n people conflict model. $A^{j} \in F(Q)$ express the outcomes fuzzy set of player j (j = 1, 2, …, n), namely μ_{Aj}: $Q \times Q \rightarrow [0, 1]$. R^j is a fuzzy relation in Q, represented as the matrix $R^j = (r^{il})^{m \times m}$. If $r^{il} + r^{1i} = 1(\forall i \neq 1)$, say R^j is complementary.

For $q^{\in}Q, q^i, q^{1\in}Q$, Make r^{il} said the preferences extent that q^i priority than q^1. For $q^{\in}Q$, directly give $\mu_{Aj}(q)$ is difficult, but for q^1, $q^{2\in}Q$, it's easy to compare $\mu_{Aj}(q^1)$ and $\mu_{Aj}(q^2)$. For q^i, $q^{1\in}Q$, let r^{il} reflect the degree of priority q^i than q^1. Assuming that (1) $0 < r^{il} < 1$, i, $1 = 1, \ldots, n$; (2) $r^{il} + r^{1i} = 1$, $\forall i \neq 1$. (2) reflect that for love, the q^i and q^1 together as a whole (That is to say, only compare between A and B, the degree of preference for A is 0.7, then B must be 0.3. In addition, if A and A itself, this value would be 1) [10].

82.4.2 The Establishment of the Fuzzy Matrix

Next is to establish matrix R by comparing one and the other. The expression of mathematical formula is very troublesome. Here we use the intuitive example to express, it is exactly the same meaning.

$$
\begin{array}{c}
\quad\; \begin{array}{cccccccccc} 1 & 2 & 3 & 4 & 5 & 6 & 7 & 8 & 9 & 10 \end{array} \\
\mathbf{R}^{ij} = \left\{ \begin{array}{cccccccccc}
1.00 & 0.70 & 0.25 & 0.55 & 0.70 & 0.25 & 0.60 & 0.70 & 0.30 & 0.60 \\
0.30 & 1.00 & 0.10 & 0.25 & 0.60 & 0.20 & 0.30 & 0.40 & 0.25 & 0.40 \\
0.80 & 0.90 & 1.00 & 0.60 & 0.70 & 0.55 & 0.60 & 0.80 & 0.55 & 0.65 \\
0.45 & 0.75 & 0.40 & 1.00 & 0.65 & 0.20 & 0.55 & 0.70 & 0.30 & 0.60 \\
0.30 & 0.40 & 0.30 & 0.35 & 1.00 & 0.20 & 0.40 & 0.60 & 0.30 & 0.45 \\
0.75 & 0.80 & 0.45 & 0.80 & 0.80 & 1.00 & 0.60 & 0.70 & 0.55 & 0.65 \\
0.40 & 0.70 & 0.40 & 0.45 & 0.60 & 0.40 & 1.00 & 0.30 & 0.35 & 0.60 \\
0.30 & 0.30 & 0.20 & 0.30 & 0.40 & 0.30 & 0.70 & 1.00 & 0.20 & 0.40 \\
0.70 & 0.75 & 0.45 & 0.70 & 0.70 & 0.45 & 0.65 & 0.80 & 1.00 & 0.80 \\
0.40 & 0.60 & 0.35 & 0.40 & 0.55 & 0.35 & 0.40 & 0.60 & 0.20 & 1.00
\end{array} \right\}
\end{array}
\qquad (82.1)
$$

Take the first row for example, R_{11} is said comparing outcome 1 and outcome 1's preferences, and the result is $1.R_{12}$ is said comparing outcome 1 and outcome 2's preferences, get the result of 0.7 (assume value). By the preceding definition we can know the value of R_{21} will be 0.3 ($R_{ij} + R_{ij} = 1$). The matrix of each line and so on. These Numbers are assuming, of course, if in the engineering practice, we must pass a written survey or similar approaches such as expert evaluation to determine the participants for these preference values of outcomes.

Table 82.1 Player i's preference vector matrix

Outcomes number	1	2	3	4	5	6	7	8	9	10
Player i	(5)	(10)	(1)	(6)	(8)	(3)	(4)	(9)	(2)	(7)

82.4.3 The Calculation of Subordinate Function

Next we can use two simple methods to get A's membership function (namely preference vector), as follows:

1. the minimum method: c $\mu_{Aj}(q_i) = \min_{l \neq i} r_{il}$, $i = 1, 2 \ldots n$, Inc is minimum value of the top five lines of upper triangle on the matrix R, and after the five lines of the lower triangular on the matrix R.
2. the average method: The values $\mu_{Aj}(q_i) = \min_{l \neq i} \sum_{l=1}^{m} r_{il}/m$, $i = 1, 2 \ldots n$, which is the average of the top five lines of upper triangle on the matrix R, and after the five lines of the lower triangular on the matrix R.

By any of the above two methods we can get $a_j = (\mu_{Aj}(q_1), \mu_{Aj}(q_2), \ldots.$ $\mu_{Aj}(q_m))$. Order the minimum value (or average) in row. Take the matrix above as an example. With minimum method, we can get $a_j = (0.2, 0.1, 0.55, 0.2, 0.1, 0.45, 0.40, 0.2, 0.45, 0.2)$, which can be found with the same size Numbers (0.2 and 0.2), at this time, we can reuse mean method to compare the rows where these numbers in, sorting. Average method: $A_1 = 0.51$; $A_4 = 0.50$; $A_5 = 0.39$; $A_8 = 0.36$; $A_{10} = 0.43$; $A_6 = 0.49$, $A_9 = 0.65$. Use 1.2.3... 10 instead of specific value. 1 representative the largest value, 10 representative the minimum value. Finally we can get the players preference of sorting table, as shown in Table 82.1.

From what has been discussed above, we get preference vector matrix under the fuzzy preference circumstances. The following, only use the partial countermeasures method, we can get the final stability analysis results of disputes.

82.5 Dynamic Development Mechanism of Hypergame

Conflict is the game between both sides who have interest disputes. Construction conflicts disputes just about both sides who have interest disputes. Each game in the process of dispute is a process of conflict analysis, and the hypergame is a specialization of conflict analysis.

Each result of the game in construction disputes can be divided into satisfied and unsatisfied to players, can also be divided into unexpected and expected. Then sum up four conditions are expected satisfied, unexpected satisfied, expected unsatisfied, unexpected unsatisfied. Player's attitude toward outcomes decide whether continue the game. If player accept this result, the game is over, but if the players do not accept this result, it will continue.

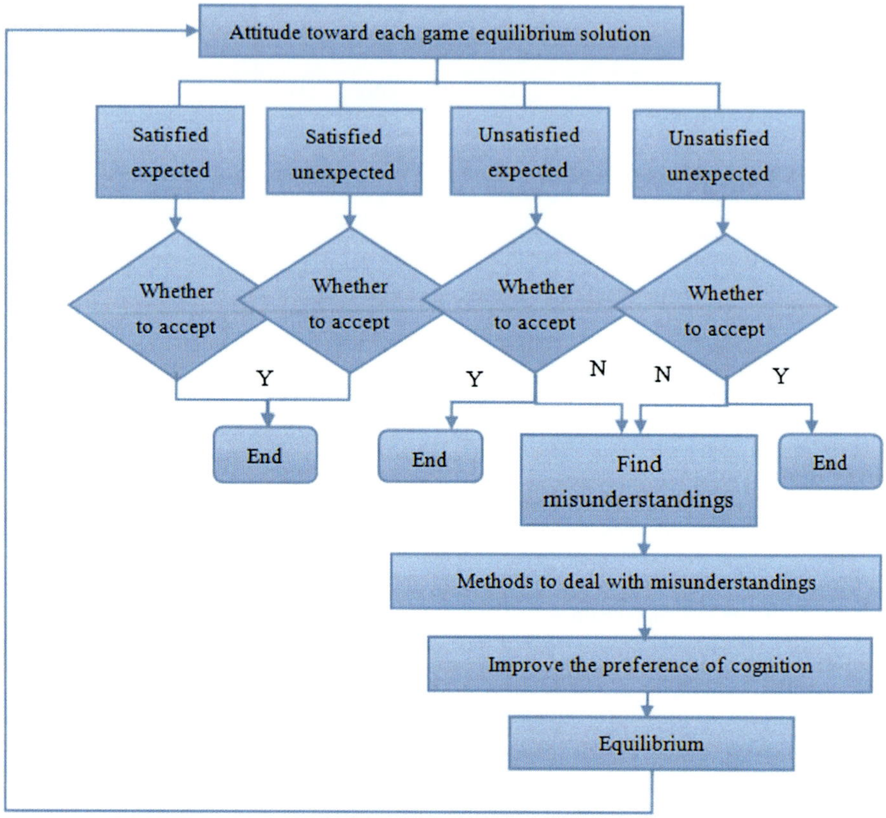

Fig. 82.3 Dynamic development mechanism of hypergame

According to the actual situation, if the game result is satisfactory, players don't have the motivation to improve the result whether it's expected or unexpected, players can choose to accept the current result; For the unsatisfied result, the player may accept the ending, but there is big motivation to improve the outcome that they are not satisfied with the current result, what's more, the motivation of unexpected outcome than expected is bigger.

After player decided not to accept the current, first, analyze the cognition to the opponent, according to the description of the second chapter, find the cause of the misunderstanding to the opponent, so as to find a way to deal with the misunderstanding, then player can form a new cognitive to opponents, and improve the cognition to the opponent preferences, get a new equilibrium solution. The attitude player toward the new equilibrium solution determine whether the next improvement occur. If player choose to improve the current result, then the game enter the circulation, until player accept the equilibrium, the game over. As shown in the Fig. 82.3.

82.6 Conclusion

This paper discussed the reason of misunderstanding, and how to deal with the misunderstanding. In this paper, the author gives the preference vector determination method and the dynamic development mechanism of hypergame. It has practical significance for the development of the construction project disputes and equilibrium solution. This paper does not give the example to verify the theory, but the author will improve the example.

References

1. Kumaraswamy MM (1998) Consequences of construction conflict: a Hong Kong perspective. J Manage Eng 14(3):66–74
2. Kassab M, Hipel K, Hegazy T (2006) Conflict resolution in construction disputes using the graph model. J Constr Eng Manage-Asce 132(10):1043–1052
3. Xu HY, Hipel KW, Kilgour DM (2008) Preference strength and uncertainty in the graph model for conflict resolution for two decision-makers. IEEE
4. Zhongfu Q, Chunhua K, Xue Z (2010) Conflict analysis under different power. J Modern Bus 08:280–283
5. Hua F, Qin Z (2010) Incentive algorithms and the development of decision support system. Zhejiang University
6. Sasaki Y (2008) preservation of misperceptions–stability analysis of hypergames. In: proceedings of the 52nd annual meeting of the ISSS-2008. Madison, Wisconsin 3(1)
7. Haojie Y, Yexin S, Yong Q (2013) Hypergame analysis with players' behavior preference. Comput Digital Eng 3:410–413
8. Bennett PG (1977) Toward a theory of hypergames. Omega 5:749–751
9. Wang Y, Wang H, Liu Y (2006) The properties and function of positive emotions. J Capital Normal Univ (social science edition) 1:168
10. Wang P, Han L (1988) Application of fuzzy mathematics. Beijing economic institute publishing house, Beijing, vol 8

Chapter 83
Lessons Learned from Managing a Remote Construction Project in Australia

Andrew Hay, Jian Zuo, Sangwon Han and George Zillante

Abstract There has been a growing demand on developments in remote areas. Apart from enjoying the associated benefits, architectural, construction and engineering (AEC) firms also face unique challenges to manage remote construction projects. This study adopted a case study approach to investigate the challenges and strategies of managing remote projects. The results showed that human resources, material supply and communication present most significant issues in remote construction projects. To overcome these issues, contractors are encouraged to engage local trades and suppliers. These results provide a useful reference to industry practitioners to manage remote projects.

Keywords Remote construction · Case study · Australia

A. Hay
Mossop Construction + Interiors, Hindmarsh, Australia
e-mail: Andrew.Hay@mossop.com.au

J. Zuo (✉)
School of Architecture and Built Environment; Commercialisation and Innovation Centre (ECIC), The University of Adelaide, Adelaide, SA 5005, Australia
e-mail: jian.zuo@adelaide.edu.au

S. Han
Architectural Engineering Department, The University of Seoul, Seoul, Korea
e-mail: swhan@uos.ac.kr

G. Zillante
School of Architecture and Built Environment, The University of Adelaide, Adelaide, SA 5005, Australia
e-mail: george.zillante@adelaide.edu.au

Y. Wu et al. (eds.), *Proceedings of the 20th International Symposium on Advancement of Construction Management and Real Estate*,
DOI 10.1007/978-981-10-0855-9_83

83.1 Introduction

Due to the recent resources boom, Australia has responded with a massive increase in mining investment through the private sector and therefore an accompanying surge in mining and construction employment [1]. Many of the new mines that have developed during the boom were planned as large-scale operations that required large workforces requiring houses, community services and transport facilities. The Australian government made it a requirement that the mining companies had to provide, or make a large financial contribution to the infrastructure of the mining operation as part of the mining lease [2]. This contribution was not only major roads, railways and ports but social infrastructure such as streets, houses, schools, hospitals and recreation facilities. Therefore in the mid-1980s, many mining companies started shifting to fly-in/fly-out type situations to commute their labor force instead of constructing townships. Therefore Australia has not seen a new mining town developed since Roxby Downs was constructed in the late-1980s [2]. With the recently approved Olympic Dam expansion, the township of Roxby Downs is expected to grow to 10,000 people, thus requiring changes to Roxby Downs and Northern Australia's regional communities in the form of increased demand for accommodation and services [3]. Other townships in the area are also expected to see a population growth and this again stresses the need for services and accommodation. It can therefore be seen that wherever there are mining projects underway, construction is required to facilitate the process. This has created significant demands for the architectural, construction and engineering (AEC) firms to manage remote projects.

83.2 Literature Review

A remote project is one in which a significant part is carried out in a location remote from the owners office [4]. Remoteness can be described as the 'physical distance' from the construction site to the required resources needed for the construction process [5]. These resources may include the proximity of the project team for the ease of consultancy or availability and delivery times of materials. Therefore construction projects that are at some distance away from the major cities can pose difficulties for the construction process especially when time, cost and quality are key factors to a successful project. While the management of these contracting firms may be confident in completing conventional projects that they are experienced and successful in, it is important that any contractor recognizes the risks, special characteristics and any problems that may be generated by remote projects [4].

For the majority of construction companies, their projects are located in major cities where they have access to skilled labor, materials, close proximity to job sites and therefore regular observation of progress, and reduced costs. However, working in regional areas creates difficulties to the company and project team in terms of skill shortages, restricted access to materials and advice, fly-in/fly-out issues, psycho-social problems with staff such as increased strain on families, and on-site productivity issues mainly due to the absence of the project manager [6, 7].

In order to better manage these issues, attempts have been made by the companies to implement strategies such as fly-in and fly-out systems where the labor commutes to their workplace for a period of time and then returns to their residence place after the period of work is complete. There are great advantages for companies who choose to travel to and from remote project sites as an alternative to establishing permanent infrastructure [7]. This is particularly relevant to mining areas where it is advantageous to fly-in and fly-out labor instead of developing housing and supermarkets to accommodate the new labor population. Australian companies preferred fly-in/fly-out arrangements to permanently relocating staff [6]. However, their findings also suggested that for construction workers, personal issues were reported to be the greatest challenge for them when working in remote locations. In contrast, workers usually express satisfaction with fly in operations and do not appear to have any long term difficulty adjusting to work schedules which involve spending lengthy periods away from home [8]. In considering this it is acknowledged that the primary burden of adjustment to fly-in operations falls on the wives and families of the workers, rather than the workers [8].

Recent studies have suggested that the construction industry must embrace new ways of working in order to sustain competitive advantage and to meet the high demands of clients [9]. Therefore strategies have been developed in an effort to better manage construction projects and in particular remote construction projects. These include tools such as Advanced Electronic Project Management Systems (AEPMS) and Web-based Project Management Systems (WPMS) [6]. The use of information technology (IT) has been suggested to increase the quality of documents, increase the speed of work, lead to better financial control and communications of projects, allow faster access to data, and reduce the amount of errors in documentation [10].

Although some of the problems with remote project management could potentially be avoided if web based systems were used, many of the problems in remote construction still do not benefit from the use of web-based systems [11]. For example, difficulties with site personnel recruitment and the transportation of materials, in addition to unforeseen problems, such as possible shortages in labor and the breakdown of equipment are all issues that web based systems would be able to quickly alert involved parties of, but cannot assist in resolving the problems. This idea is also supported by Shelbourn et al. [9] who believe that approaches that are solely based on the use of information technology will not be successful unless the organizational and people issues are considered. It is also considered that many individuals, in particular the older parties involved in the industry are apprehensive when confronted with technology and a change to the old processes [9].

Similarly, contractors may experience the difficulty of finding adequate labor. Due to the construction industry being a labor intensive industry, the availability of local labor is critical. However, due to the high and low periods of construction, this often means that there is always a labor shortage or a labor surplus. Due to the fluctuating nature of the industry, the hiring of unqualified labor at higher costs can cause issues when labor is of high demand. This can have even larger effects when considering labor sourcing for remote construction projects as it can be hard enough to source adequate labor for remote projects. The management of this labor (and in particular subcontractors) is often critical to the success of any construction project whether it is remote or not.

A common problem in the construction industry is the availability of material and ensuring materials are on site at the right time during construction. One of the biggest contractor-related delays in construction was material management problems [12]. They also outlined that the delay in the materials is often a contributory cause to cost overruns in construction projects and that the main cause of these delays were found to be organizational weaknesses, suppliers defaults, government regulations and transportation delays. This can be evident during the construction of remote projects due to the vast distances that the building materials may have to go to reach the construction site. Therefore it is critical to ensure that the management of material is closely monitored. Site material management is defined as "allocation of delivery, storage, and handling, spaces and resources for the purposes of supporting the labor force and minimizing inefficiencies due to congestion and excess material movement" [13]. However, they went further to also outline that proper management involves storage, identification, retrieval, transport and construction methods which can be linked to safety, productivity and schedule performance.

Due to the remoteness of some projects, quality can be hard to manage at times. This is due to the unlikelihood of meeting normal expectations of a city project; receiving immediate supplies, readily available spare parts, experienced advice, and the abundance of well qualified labor. This therefore places extra pressure on the construction process [14].

83.3 Case Study: The Ceduna Hospital Project

An in-depth case study was undertaken to investigate the challenges and strategies to the management of a remote construction project. The case study approach is adopted in this paper because it is an effective approach for 'why' and 'how' research questions. The case study allowed for a more detailed understanding of the entire remote construction process and outlined the specific challenges that occurred on the selected remote construction project.

The Ceduna Hospital was selected as the case example studied in this research. The project was constructed by a construction company based in Adelaide, South Australia and is relatively new to working on remote construction projects. There is a large distance from the small township of Ceduna to the capital city of Adelaide where the construction company is located. Due to this large distance, a significant part of the project is carried out in a remote location far from the construction company's office and difficulties are likely to be encountered.

Three key project participants were interviewed to solicit their professional opinions on challenges and strategies for managing remote projects. Their profiles

Table 83.1 Profiles of Interviewees

Interviewee	Organization	Title	Experience (years)
A	Contractor	Project manager	14
B	Fire service subcontractor	Project manager	12
C	Plumbing subcontractor	Project manager	7

are shown in Table 83.1. This is followed by a review of project documentations. Each interview took around 1.5 h. Content analysis was employed to identify merging themes.

83.4 Findings

The Ceduna Hospital redevelopment project was announced by South Australian government in 2008. The redevelopment project consisted of two stages, which includes construction of a new Health Care Centre, new Aboriginal health service facilities, new high level aged care facility and the redevelopment of acute services. The redevelopment is seen as a major positive to the community of Ceduna and surrounding townships as people can now access the hospital facilities without traveling large distances.

In October 2008, construction started by an Adelaide construction company, who worked closely with the architects, engineers and cost consultant, all of which were also located in Adelaide. The construction took 19.5 months with the construction being split into two stages. Stage one which consisted of the 3500 m^2 new hospital build was handed over to the client on March 20th, 2011 and the stage 2 the 1800 m^2 refurbishment of the existing building was successfully handed over to the client on the August 24th, 2011.

The construction team consisted of one project manager, estimator and contract administrator based in the Adelaide office and a site manager, safety supervisor and another contract administrator based on site. All site staff was from Adelaide and was permanently relocated to Ceduna for the duration of the project. The project manager would regularly fly in and fly out of Ceduna based on meeting requirements with the project team or the construction team. Over 40 different trades where contracted to complete the project and over 600 workers were inducted onto the job site by the head contractor.

83.4.1 Challenges for Managing Remote Projects

Interviewee A outlined that the two most important challenges relating to the Ceduna project were trade management and site coordination as it was critical to have the site coordination under control which leads to the good trade management well before they arrived on site. As trades could invoice up to $2000 for a period of two days which included labor, travel and living away expenses from Adelaide, it was critical that the trades were given notice of when they were required and what was required of them on site. Therefore, a building program was produced regularly throughout the project to give updates to the trades of where the project was up to and what had to be completed in the next 2–4 week period. This was critical for the trades as some of them would stay in Ceduna for larger periods such as 1–4 months while others may have only stayed in 2–4 week periods. This meant that the trades were given good direction of when they were required onsite. This was also confirmed by interviewee B as he had subcontracted all of his labor to Adelaide based trades therefore he relied heavily on regular information from the head contractor and also his tradesman on site on how the project was progressing and what was required.

Material management was also seen as a challenge as deliveries had to be scheduled in stages to ensure that there was always enough material on site but at the same time not too much due to space restrictions. Large deliveries would often come from Melbourne or Perth to Adelaide and would then have to be couriered to Ceduna which takes time and required pre-planning. Deliveries from Perth were a frustration as the delivery would drive straight past Ceduna to Adelaide.

Labor sourcing for the project was not seen as a major challenge. All of the main subcontractors were Adelaide-based with the exception of the earth worker/site worker as no Adelaide subcontractors were willing to bid the project. Due to the large amount of Adelaide-based subcontractors, problems with minor work being completed was commonly seen as an issue. Due to the subcontractor not being onsite all of the time, local contractors were called upon to complete minor work and small defects in order for the site to keep running and the subcontractor's that were onsite going. As interviewee A stated:

> We exhausted the local trades and sometimes this was seen as beneficial to the subcontractors as the cost back charged from their contract was minimal compared to the financial cost of the Adelaide subcontractor having to organize their Adelaide labor due to other commitments, travel and accommodation.

Subcontractors such as the plumber, electrician and the plasterboard fixer had larger contracts and extensive work loads and were therefore required onsite for longer periods of time. In this instance, they would often rent houses for a period of time until the work was completed on site. A tactic used by some contractors was to have their material and work vehicles couriered to Ceduna over the weekend by truck and then fly to Ceduna on Monday to be ready for their period of work. This meant that the subcontractor was not having to pay his tradesmen wages for the time spent travelling to and from site. The plane ticket may have been expensive but a day was still not lost when the tradesmen time could be best spent completing the contractual works. A lesson learned quickly from the head contractor's project manager was that if the subcontractor stated they would be on site on Monday it was not entirely true as the eight and a half hour drive from Adelaide would eliminate what time they would actually spend on site. Therefore Monday was always seen as a lost day.

Onsite morale was seen as a challenge for the remote project as the site staff often found that the tradesmen would get short tempered and frustrated after a period of three weeks. For the productivity of the project, they were often told to leave the site and go home to have a short break with their families and come back refreshed. It was seen as very important that a positive relationship was built between the site staff and the subcontractors. Therefore, offsite activities such as fishing and lawn bowls were used to build good relationships and often competiveness onsite.

The construction program was critical to the success of the project. Soon after awarding the contract, the client wanted the head contractor onsite early which caused a few problems with programming and pre-construction planning. During the early stages of the project, constant questions and requests for information (RFI's) were issued to the consultants to clarify any issues or discrepancies in the documentation. The subcontractors soon built a positive relationship with the site team and the consultants which helped the project progress in the early stages. This was especially important with all the technical questions from the services contractors.

Another challenge that occurred and was mentioned by interviewee A was that they often had to pay a premium to get trades who specialized in remote work and they have higher costs. During the tender negotiations with the subcontractors, it was noted and often promised by the subcontractors that they would have one or the most experienced project manager overseeing the project. However, the senior project managers having extensive experience often do not wish to relocate to a remote setting due to family commitments and/or lack of interest. Therefore, younger inexperienced project managers would be placed in charge of the project. This sometimes caused a few issues onsite as their lack of experience caused errors and mistakes. Therefore, as stated by interviewee A, "you don't always get the best team but you do get a team that is interested and wants to be there". It was also noted during the tender negotiations that subcontractors would only allow for a certain amount of visits to site. This really was a critical point as extra visits were sometimes required to complete different sections of work and handover different zones to the client. The completing of defects also became an issue at times as trades were reluctant to visit the site just for some minor defect work.

83.4.2 Construction Process Affected Due to Remoteness

Interviewee B being a subcontractor project manager outlined the importance of making the most of the allocated trips to the remote construction site. This is to ensure that no time is wasted on behalf of the subcontractor as well as to not waste the other trades time on site. It is important to maximize efficiency on site as there is no point in doing a job if you are going to make a loss. Good communication is required with the site staff to understand what stage the construction is currently at in order to ensure the site and other subcontractors are ready for their arrival. For example, if only three visits were allowed during tender time, then it is critical that all of the work can be completed during the three visits because the money spent on travel and accommodation can be expensive.

Interviewee A highlighted the importance of preloading materials onsite before the subcontractor arrives and also while they are completing their work. In one instance, the plasterboard fixers arrived onsite before the plasterboard and were forced to install ceiling tiles while they were awaiting the plasterboard to arrive. This was not favorable to the process of the construction as the services' subcontractors had not finished their above ceiling works. Therefore, all damaged tiles by the services contractors were not seen as their fault and were the responsibility of the plasterboard fixer.

As briefly mentioned above, due to the remoteness of the project, the completing of defects were seen as a challenge as subcontractors were reluctant to travel to Ceduna to complete minor defects. Therefore, some of the subcontractors were made aware of local trades that could complete their work and therefore save on travel and accommodation.

83.4.3 Management Tools and Techniques

All interviewees highlighted the importance of the construction program. For interviewee A, a workable program was critical which was signed off by the subcontractors to ensure that they could meet their deadlines. This alongside the implementation of liquidated damages in their contracts was a solid way to start of the construction process and ensures that the subcontractors agreed to the set dates. Then, during the construction process, the construction program was constantly updated to reflect the progress onsite. This alongside the issuing of the program in different formats and sizes helped to keep the entire project on track. On top of this, an eight week program was always on the whiteboard in the site office to ensure that the site staff could focus on the coming weeks ahead.

Regular site meetings are crucial to any construction project. However, for remote construction projects, the implementation of a successful meeting with all parties involved can be difficult. Throughout the Ceduna project, fortnightly project managers meeting were held with all of the subcontractor project managers. This was based in Adelaide due to the majority of Adelaide-based trades and their project manager's location. A weekly meeting was also held on site with the site team and the subcontractor's responsible person as well as a daily occupational health safety and welfare pre-start meeting. On top of these, a joint project managers and responsible persons meeting should be held onsite in order for everyone to understand where the project is up to and inform every one of all the current issues. However, this needs to be discussed at tender time to ensure that the subcontractor has allowed for the number of visits for these meetings by their project managers. Only one meeting was ever held where the project managers, responsible people and the project consultants were involved. However, this did prove beneficial since a large number of issues were discussed and resolved.

Communication was highlighted by the interviewees as one of the most important management tools for the project. As stated by interviewee C, good communication is vital in order to understand where the project is currently at, what issues there are and how they can be resolved. Interviewee A noted this also and to ensure that everyone was made aware of the projects progression, emails would be sent a month in advance to the subcontractors to give them clear direction of what was expected and when they were required to be on site. Interviewee C noted that a positive of the project was that the subcontractors were never left waiting for information as requests for information were promptly followed up by the site team. A communication plan was developed that outlined the entire projects contacts, therefore the relevant information could be sought from the correct person quickly and efficiently.

In addition to emails and telephone calls, two forms of web-based management called Aconex and Grazer were used on the project. As a result, interviewee A noted that only 5–7 % of communication was sent by email. For the Ceduna project, it was used as a form of document management control which was a positive as it reduced the number of emails and acted as an information hub. The Aconex system also allows information to be sent and accessed by certain people. Therefore, the information is relevant and targeted to the correct people. The second web based tool was Grazer which was used for the submission of manuals. Subcontractors were required to allow about $3000 at tender for training and using the Grazer process. Interviewees suggested that the Grazer system provides a streamlined way of managing the handover of documentation.

83.5 Conclusions

The Ceduna Hospital Redevelopment, a remote construction project had numerous challenges/difficulties. Firstly, the head contractor's project manager and various subcontractors had to manage fly-in and fly-out system while the majority of construction site staff was permanently relocated. Secondly, material management was identified as a challenge since constant supply and delivery had to be maintained to fuel the construction process. Third challenge was labor management. Getting the contracted labor onsite at the right time required careful programming. Also, labor was hard to get back to site and complete minor defects. Lastly, maintaining onsite workers' moral is identified as another challenge.

In addressing the challenges mentioned above, various management tools and techniques were implemented during the project. Firstly, it is important to constantly update the construction program in order to reflect the progress and changes onsite. It is effective to deliver the program in different formats and timeframes to get a better understanding of the construction process and to always be looking in advance. Secondly, it is essential to have regular site meetings with the consultants, the subcontractor's project managers and their site responsible person. Lastly, communication systems including emails, telephone, communication plan and web-based project management system would enhance the efficiency of remote construction projects.

Future research opportunities exist to conduct an industry-wide survey to investigate the effectiveness of the strategies to manage remote projects. Similarly, an international benchmarking study will be beneficial to develop a best practice guide for remote project management.

References

1. Ewing R, Fenner S, Kennedy S, Rahman J (2007) Recent productivity outcomes and Australia's potential growth' economic roundup, viewed 28/08/11. http://www.treasury.gov.au/documents/1304/PDF/Economic%20Roundup%20Winter%202007.pdf#page=53
2. Australian Bureau of Statistics (2000) The Australian mining industry: from settlement to 2000, Cat 8414.0
3. BHP Billiton (2011) Olympic dam expansion EIS—Roxby downs and regional communities, viewed 13/11/11. http://www.bhpbilliton.com/home/aboutus/regulatory/Documents/Olympic%20Dam%20Supplementary%20EIS/Information%20Sheets/Olympic%20Dam%20EIS%20Roxby%20Downs%20And%20Regional%20Communities.pdf
4. Moselhi O, Poulton C (1991) Planning and control of remote projects. J Manage Eng 7(1):83–97
5. Kestle L, London K, Bodger P, Storey B (2002) The development of a typology for remote construction projects. In: The 18th annual ARCOM conference. Association of Researchers in Construction Management, London, 2–4 September 2002
6. McAnulty S, Baroudi B (2010). Construction challenges in remote Australian locations. In: 26th annual ARCOM conference, Leeds. Association of Researchers in Construction Management, UK, 6–8 September 2010
7. Storey K (2001) Fly-in/fly-out and fly-over: mining and regional development in Western Australia. Aust Geogr 32(2):133–148
8. Houghton DS (1993) Long-distance commuting: a new approach to mining in Australia. Geogr J, 281–290
9. Shelbourn M, Bouchlaghem NM, Anumba C, Carrillo P (2007) Planning and implementation of effective collaboration in construction projects. Constr Innov Inform Proc Manage 7(4):357–377
10. Nitithamyong P, Skibniewski MJ (2004) Web-based construction project management systems: how to make them successful? Autom Constr 13(4):491–506
11. Sidawi B (2010) The use of advanced electronic management systems to manage remote projects in the kingdom of Saudi Arabia. In: 26th annual ARCOM conference, Leeds. Association of Researchers in Construction Management, UK, 6–8 September 2010, pp 633–642
12. Sambasivan M, Soon YW (2007) Causes and effects of delays in Malaysian construction industry. Int J Project Manage 25(5):517–526
13. Thomas HR, Riley DR, Messner JI (2005) Fundamental principles of site material management. J Const Eng Manage 131(7):808–815
14. Harris F, McCaffer R, Edum-Fotwe F (2006) Modern construction management. Wiley, New York

Chapter 84
Factors Influencing Transaction Costs in Construction Projects: A Critical Review

Xiao-Hua Jin, Guomin Zhang, Yongjian Ke and Bo Xia

Abstract The management of large-scale construction projects involves the efficient and effective management of financial resources. The physical cost of a construction project governs the estimation of construction budgets and forecasts. However the Transaction cost economics theory in the construction industry aims to promote the relationship and impact of transaction related costs in construction projects. These costs are related to the economic exchange of a product or service and differ from the actual physical production costs of a construction project. Previous academic studies have identified transaction costs and the factors of influence. However a significant gap in the research area is the analyses of the influence procurement methods have on transaction costs in construction projects. Procurement methods dictate the contractual obligations of the parties and the method in which the project is delivered and essentially represent different types of transactions, thus it is assumed that procurement methods influence transaction costs in construction projects. In this paper, transaction costs are categorized into pre and post contract. The identified factors that are expected to influence transaction costs are associated with predictability of owner's behavior, predictability of contractor's behavior, project management efficiency, and uncertainty in the

X.-H. Jin (✉)
Program of Construction Management, School of Computing, Engineering and Mathematics, Western Sydney University, Penrith, NSW, Australia
e-mail: xiaohua.jin@westernsydney.edu.au

G. Zhang
School of Civil, Environmental and Chemical Engineering, RMIT University, Melbourne, VIC, Australia

Y. Ke
School of Architecture and Built Environment, Faculty of Engineering and Built Environment, University of Newcastle, Callaghan, NSW, Australia

B. Xia
School of Civil Engineering and Built Environment, Faculty of Science and Engineering, Queensland University of Technology, Brisbane, QLD, Australia

Y. Wu et al. (eds.), *Proceedings of the 20th International Symposium on Advancement of Construction Management and Real Estate*,
DOI 10.1007/978-981-10-0855-9_84

transaction Environment. It is also expected that different procurement systems have various levels of impact on transaction costs. Empirical research is proposed to further test the prepositions in this paper.

Keywords Transaction cost · Procurement · Influence factors · Construction projects

84.1 Introduction

The theory of transaction costs in the construction industry aims to promote the relationship and impact of transaction related costs in construction projects. Transaction costs are related to the economic exchange of a product or service however these costs differ from the actual production costs. In the construction industry transaction costs are related to the process of procuring a project but are not related to the actual production costs of the project and are costs associated with economic exchange (a transaction). The theory demonstrates this relationship and has been applied to the construction industry in order to analyze the effect transaction costs may have.

The theory of transaction costs economics and its application in the construction industry portrays the understanding that in addition to the physical production and labour costs required to deliver the final product there are also transaction costs. The primary concept is that in addition to the costs of production there are also transaction costs between the parties [1]. These costs although may be directly related to the construction process are different in the sense that production costs are the costs of transforming inputs into outputs, whereas transaction costs arise from economic exchange [2].

In the construction industry there are various types of transaction costs, which occur during different stages of the project lifecycle. Essentially due to the theory of transaction costs economics being closely related to the contractual exchange of products and services, the types of transaction costs are distinguished prior to the contract being awarded and after a contract has been awarded and works have commenced. The transaction cost economics theory portrays the concept that individual firms and market factors can influence the market environment and thus transaction costs in the construction industry. The influence that individual firms and market factors can have on a market environment are referred to in the industry as factors that influence transaction costs [1]. These factors of influence are directly related to the influence individual firms and the market environment can have on transaction costs in the construction industry.

Procurement methods in the construction industry define the nature of the transaction environment, which in turn influences transaction costs in the construction industry. However in order to understand how procurement methods influence transaction costs, a link must first be established between procurement systems and transaction costs in the construction industry. The relationships between procurement methods, contractual arrangements and tendering procedures have not been extensively analyzed in the construction industry and thus this creates a need to further analyze this area of study. Contractual arrangements in the construction industry are influenced by the procurement strategy in which the project will be delivered [3]. Thus this aspect portrays the fundamental relationship between procurement systems and transaction costs as the procurement strategy dictates contractual arrangements, which influences transaction costs in the construction industry [4]. The need to further identify the relationship between procurement systems and transaction costs is required in order to give construction professionals a better understanding of how procurement systems can influence transaction costs and implement the appropriate strategies in order to reduce this affect.

The research aims to comparatively analyze the affect that procurement systems have on the transaction environment, which influences the various factors of influence and thus transaction costs in construction projects. In this paper, the literature associated with transaction costs and their influential factors is reported. The possible impact by procurement systems is also discussed.

84.2 Transaction Costs in the Construction Industry

In the construction industry there are various types of transaction costs, which occur during different stages of the project lifecycle. The types of transaction costs in the construction industry have been split up into two categories. These are referred to as pre and post contract transaction costs [5]. The reason for the categorization of transaction costs is to clearly differentiate between the different types of transaction costs and when they occur in the project life cycle. Essentially due to the theory of transaction costs economics being closely related to the contractual exchange of products and services, the types of transaction costs are distinguished prior to the contract being awarded and after a contract has been awarded and works have commenced. The types of transaction costs identified as pre and post contract transaction costs are broken down and further evaluated.

84.2.1 Pre Contract Transaction Costs

Pre contract transaction costs are incurred before a transaction takes place between two or more parties and essentially include the costs associated with initiating the project; these costs are borne by the owner before the construction contract is signed. Pre contract transaction costs are identified as and broken down into the following categories, initiation costs, preliminary design costs, negotiation and contracting costs and feasibility study costs [6]. Initiation costs are associated with the process of initiating the project and include team selection, proposal evaluations and start up meetings. Preliminary design costs involve developing early design documentation in order to assist the development approval and feasibility process they include preliminary design drawings and design workshops. Feasibility study costs involve evaluating the environmental, financial feasibility and profitability of the project; these costs are associated with information search and assessment. Negotiating and contracting costs are a major contributor to pre contract transaction costs and are associated with the process of awarding or winning a project. They include preparing bidding documentation, approvals and commercial negotiations [5]. Negotiating and contracting costs are not just worn by the owner and contractor, unsuccessful tenderers also have to wear these costs and are essentially considered transaction costs as a result of a potential to engage in economic exchange. During the pre contract stage [6] differentiate between external costs such as technical, legal and financial advice and in house costs such as project preparation costs. Ultimately pre contract transaction costs in the construction industry are identified as initiation, feasibility study, preliminary design and contraction negotiation costs.

84.2.2 Post Contract Transaction Costs

Post contract transaction costs are associated with the costs incurred after the contract has been signed and before the entire transaction has been completed. All contractually obliged parties may potentially wear these costs. Post contract transaction costs are associated with the set up and running costs of the contract governance structure in which monitoring is assigned and disputes are resolved [7]. Post contract transaction costs are once again broken down into sub categories. These include monitoring and control costs, dispute resolution costs and implementation costs [5]. Monitoring and control costs include the process of ensuring the terms and conditions of the contract are met and both parties are upholding their responsibilities under the contract documents. Dispute resolution costs occur when conflict between the parties arises and is associated with the costs of resolving these conflicts. This may include the costs of mediation, arbitration and litigation. Post contract transaction costs arising from contract disputes and litigation are costly both in direct costs such as lawyers, claim consultants and delays to productivity and to indirect costs such as degeneration of working relationships and the consequences of mistrust between the parties [2]. Implementation costs are the most common type of post contract transaction costs and essentially involve the day-to-day implementation of the contractual obligations and include the daily contract administration, administering claims and change orders [5]. In essence post contract transaction costs are identified as monitoring and control costs, dispute resolution costs and implementation costs. These costs arise as a result of the process of ensuring the final product is delivered in accordance with the contractual documentation and are directly related to the costs incurred during the pre contract stage in order to develop, administer and negotiate the very contractual documentation used to govern the relationship between the contractually obliged parties.

84.3 Factors that Influence Transaction Costs

The transaction cost economics theory portrays the concept that individual firms and market factors can influence the market environment and thus transaction costs in the construction industry. The influence that individual firms and market factors can have on a market environment are referred to in the industry as factors that influence transaction costs [1]. These factors can be identified as predictability of the owner and contractors behavior, Project Management efficiency and uncertainty in the transaction environment [8]. These factors of influence are directly related to the influence individual firms and the market environment can have on transaction costs in the construction industry.

84.3.1 Predictability of the Owner's Behavior

Predictability of the owner's behavior indirectly influences transaction costs as it reduces uncertainty in the transaction environment and increases the efficiency of project management [5]. It is essential to build a good working relationship with the owner as its behavior and actions can indirectly influence transaction costs through uncertainty and the ability to project manage. This relationship is of particular importance in situations where gauging the owners response and behavior is critical during communication. The owner's behavior is essential in reducing uncertainty during the transaction in order to enable efficient and effective management of the entire construction process. The owner's experience in similar type project is essential in order to reduce change orders and ensure efficient and effective decision-making. This is also linked to organizational efficiency and the ability to produce maximum output. Relationships with other parties are also of particular importance as cooperation and stability between all parties is essential in reducing transaction costs. Payment on time is also an important aspect as this ensures uncertainty is reduced and cash flow is maintained which ultimately satisfies the financial requirement of all parties involved. This reduces disagreements and thus transaction costs [5].

84.3.2 Predictability of the Contractor's Behavior

The predictability of the contractor's behavior can influence transaction costs, this concept is portrayed as the more predictable the contractor's behavior the less frequent transaction costs are incurred and vice versa. In the instance that a contractor is well established and performs professionally and ethically it is likely that it will not explore opportunistic behavior, such as issuing unnecessary claims, thus reducing transaction costs [5]. The predictability of the contractor's behavior can have a positive affect on project management efficiency as a stable relationship with subcontractors and clients has a positive influence. The predictability of the contractor's behavior affects transaction costs directly and indirectly through it's influence on project management efficiency [5]. The predictability of the contractor's behavior influences transaction costs through the issuing of opportunistic and unnecessary claims, bidding behavior, relationships with subcontractors, clients and experience. These aspects all have the ability to affect the behavior of the contractor and thus influencing transaction costs.

84.3.3 Project Management Efficiency

The quality of project management has a large impact on project productivity, quality and rework. Rework in particular is estimated to be greater than 10 % of the total project cost [5]. Project management efficiency has a large impact on transaction costs in construction projects as the efficient management of the project reduces the frequency and affect of transaction costs in construction projects and vice versa for inefficient management of the construction process [8]. The quality of decision making, communication, conflict management and technical experience and leadership all affect the efficient management of the construction process and influence transaction costs in construction projects. This factor of influence is related to the predictability of the contractor's behavior as the behavior and efficient project management are directly related. This is evident in the influence the contractors behavior has on efficient project management and the effect efficient project management has on the contractor's behavior [1]. Uncertainty in the transaction environment also reduces project management efficiency, whereas predictable owner and contractor behavior enhance project management efficiency, thus influencing transaction costs. Efficient project management and a reduction in transaction related costs may be achieved by effective leadership, efficient communication and conflict management, through efficient project management the frequency and impact of transaction related costs can be effectively reduced through the above mentioned qualities [5]. The process of project management is a major factor that influences transaction costs in the construction process as it is directly related to a representation of the contractors behavior and provides a platform for opportunistic practices, thus influencing transaction costs in construction projects.

84.3.4 Uncertainty in the Transaction Environment

The concept of uncertainty in the transaction environment is related to the contractual documentation outlining the responsibilities and obligations of each party. This relates to documentation such as design drawings, specifications, scope of works and any other contract documentation [5]. The ability of these documents to accurately portray the responsibilities and obligations of each party influences transaction costs in the construction industry as any deviation to these contract documents may result in potential variations and disputes under the provision of the contract. Certainty in the transaction environment indicates that the scope of the project is well defined and the design documentation is clear and complete. Uncertain contract documentation is a dominant influence on the transaction environment however this is not only attributed to its direct influence but also the contractor's behavior in response [5], this is linked to opportunism and portrays the understanding of how the contractors behavior can play a major role in influencing transaction costs in the construction industry.

Uncertainty in the transaction environment will always occur especially during the execution of large and complex construction projects [9]. The ability to establish a well-defined project scope and accurately portray the project requirements on all design documentation will effectively contribute to the reduction of transaction costs in the construction industry. In this aspect reducing transaction costs can be achieved through the reduction of project uncertainty by ensuring the design documentation is complete and a well-defined scope is developed. The link between transaction costs and the uncertainty in the transaction environment conveys the influence this aspect has on transaction costs in the construction industry and is an ongoing influence factor [10].

84.4 Procurement Systems' Impacts on the Factors

Procurement methods in the construction industry define the nature of the transaction environment, which in turn influences transaction costs in the construction industry. However in order to understand how procurement methods influence transaction costs we must first establish a link between procurement systems and transaction costs in the construction industry. The relationships between procurement methods, contractual arrangements and tendering procedures have not been extensively analyzed in the construction industry and thus this creates a need to further analyze this area of study. Contractual arrangements in the construction industry are influenced by the procurement strategy in which the project will be delivered [3]. Thus this aspect portrays the fundamental relationship between procurement systems and transaction costs as the procurement strategy dictates contractual arrangements, which influences transaction costs in the construction industry [4]. The need to further identify the relationship between procurement systems and transaction costs is required in order to give construction professionals a better understanding of how procurement systems can influence transaction costs and implement the appropriate strategies in order to reduce this affect.

The link between procurement systems and the transaction environment is evident through the contractual arrangements dictated by each project delivery method. Each procurement system has a degree of ambiguity, uncertainty and risk [1]. For example in the case of a design and construct contract, design documentation is not fully complete and thus the project is defined by a scope of works which aims to portray all the project requirements, however in relation to the level of complete design documentation there is a higher degree of risk associated for both the developer and contractor [11]. The procurement system often dictates how the project will be constructed from a documentation point of view. It is important to ensure all documentation is coordinated in order to adequately portray the projects requirements in line with the project delivery method [11]. The link between procurement systems and the transaction environment conveys the level of influence procurement methods have on transaction costs in the construction industry. The nature of contractual arrangements such as variations, extension of time and contingent claims all differ according to the procurement strategy and level of design documentation, thus influencing transaction costs in the construction industry [12].

There is a degree of uncertainty and ambiguity associated with each procurement system influencing the requirement to identify and evaluate how these systems influence transaction costs in the construction industry. In most cases the level of design documentation, specification, size and complexity of the project dictates the procurement selection process. Through the incorporation of transaction costs analysis in the procurement selection process, emphasis could be placed on developing governance mechanisms, such as formal contracts, rules and regulations, termination agreements, warranties and appropriate strategies such as longer-term agreements and alliances [12]. By developing appropriate governance mechanisms it may be possible to reduce the risk associated with procurement uncertainties, reducing the instances of opportunism, bargaining and asset specificity issues, thus reducing the impact of transaction costs. The thinking behind such strategies is the elimination of contract coordination, management and motivation costs, ultimately reducing the affect of transaction costs [12].

The procurement system relevant to the particular project influences the transaction costs incurred at these stages for example in the traditional procurement system the design is relatively complete before construction begins and a contract is executed, thus the tender prices submitted accurately reflect the project deliverables because the project scope is well defined, thus reducing transaction costs in the pre contract phases. In relation to a design and construct procurement system the majority of design documentation is incomplete, this creates a degree of uncertainty and ambiguity, which causes difficulty during cost prediction and estimation [3].

84.5 Conclusion

The management of large-scale construction projects involves the efficient and effective management of financial resources. Apart from the physical cost of a construction project that governs the estimation of construction budgets and forecasts, transaction related costs in construction projects should be studied and promoted because these costs are related to the economic exchange of a product or service and differ from the actual physical production costs of a construction project. Previous studies have identified transaction costs and the factors of influence. However a significant gap in the research area is the analyses of the influence procurement methods have on transaction costs in construction projects. Procurement methods become the focus of this study is due to the fact that they dictate the contractual obligations of the parties and the method in which the project is delivered and essentially represent different types of transactions. It is therefore claimed in this paper that procurement methods influence transaction costs in construction projects. This paper identifies transaction costs and the factors of influence through a literature review. In this paper, transaction costs are categorized into pre and post contract. The identified factors that are expected to influence transaction costs are associated with predictability of owner's behavior, predictability of contractor's behavior, project management efficiency, and uncertainty in the transaction Environment. It is also expected that different procurement systems have various levels of impact on transaction costs.

Future research has been planned to examine the impact of procurement methods on transaction costs in construction projects by comparatively analyzing the levels of influence of the identified factors between projects adopting different procurement methods based on a data set collected through a questionnaire survey. The findings of this research are expected to assist construction professionals to identify and gain a clearer understanding of transaction costs in construction projects. This will allow professionals to better forecast the costs associated with the transaction prior to its occurrence. This will also enable an identification of transaction costs during the tender and legal documentation process to ensure the "transaction" does not negatively influence the interests of one party more than the other. Identifying the influence of transaction costs and the impact this may cause will allow professionals to consider their actions and opportunities. Through a comparative analysis of the procurement methods professionals are able to identify which procurement method is more efficient, effective and suits their interests in specific relation to their business and the associated construction project. Through conducting the research the ambiguity and controversy surrounding the theory of transaction cost economics in relation to construction projects will be reduced by clearly portraying its nature, definition, influence and overall impact on construction projects.

References

1. Winch G (1989) The construction firm and the construction project: a transaction cost approach. Constr. Manag. and Econ. 7(4)
2. Winch G (2001) Governing the project succcess, a conceptual framework. Constr. Manag. and Econ. 19(8)
3. Hackett M, Robinson I, Statham G (2007) The aqua group guide to procurement, tendering and contract administration. Oxford; Malden, Mass, Wiley-Blackwell

4. Dudkin G, Välilä T (2006) Transaction costs in public private partnerships: a first look at the evidence. Competition and regulation in network industries: CRNI. 1(2)
5. Li H, Arditi D, Wang Z (2013) Factors that affect transaction costs in construction projects. J. Constr. Eng. and Manag. 139(1)
6. Soliño AS, Gago de Santos P (2009) Transaction costs in PPP transport infrastructure projects: comparing procurement procedures. J. Eur. Manag. 19(3)
7. Williamson OE (1985) The Economic Institutions of Capitalism: Firms. Relational Contracting, Free Press, New York, Markets
8. Li H, Arditi D, Wang Z (2012) Transaction-related issues and construction project performance. Constr. Manag. and Econ. 30(2)
9. Jin X-H, (2010) Determinants of efficient risk allocation in privately financed public infrastructure projects in australia. J. Constr. Eng. Manag. 136(2)
10. Jin X-H, (2011) A model for efficient risk allocation in privately financed public infrastructure projects using neuro-fuzzy techniques. ASCE J. Constr. En. and Manag. 137(11)
11. Melese F, Franck R, Angelis D et al (2007) Applying insights from transaction cost economics to improve cost estimates for public sector pruchases: the case of u.s. military acquisition. J Int Public Manage 10(4)
12. Angelis D, Dillard J, Franck BGC, et al (2008) Measuring transaction costs in dod acquisition programs, In Acquisition Sponsored Research Reports, Naval Postgraduate School, Graduate School of Business and Public Policy, Monterary CA

Chapter 85
Construction Schedule Risk Analysis Based on Complex System Theory: Methodology and Empirical Study

Xiaoxiao Xu, Jiayuan Wang, Wenke Huang, Huanyu Wu and Yaning Li

Abstract With the development of China's society and economy, a large number of construction projects are being actively carried out. Project schedule, as one of the major concerns in construction projects, is always affected by high uncertainties and risks brought by the substantial amount of human and nonhuman factors. Therefore, more prominence should be given to proposing effective measures to minimize the losses caused by schedule risks. Although a certain progress has been made from the previous studies, the lack of dynamics and uncertainty analysis lead to results different from reality. This study aims to propose a novel model for analyzing construction project schedule risk. On the basis of complex systems theory, this paper builds a system dynamics model, and encapsulates the system dynamics model into discrete event simulation to form "work module" which can be combined according to the requirements of different projects, Monte Carlo simulation is also employed to analyze the uncertainty of risks. In addition, a WeChat platform is established for collecting schedule risk data. Finally, the present model is validated by a case study after the structure test and behaviors test. The results showcase that: (1) System dynamics can effectively deal with the complex problems of project schedule from the perspective of system, which includes rework, quality problems, the allocation of resource, productivity, decision-making delays, schedule pressure, etc. (2) It greatly improves the applicability of the model to combine system dynamics and discrete event simulation together. (3) Monte Carlo simulation enables the model to obtain the distribution of real-time planned duration in the process of construction, and to be suitable for managers with different risk attitudes and risk-bearing capacity to make decisions.

Keywords System dynamics · Discrete event simulation · Monte carlo simulation · WeChat

---

X. Xu · J. Wang (✉) · W. Huang · H. Wu · Y. Li
School of Civil Engineering, Shenzhen University, Shenzhen, China
e-mail: wangjy08@qq.com

Y. Wu et al. (eds.), *Proceedings of the 20th International Symposium on Advancement of Construction Management and Real Estate*,
DOI 10.1007/978-981-10-0855-9_85

85.1 Introduction

Nowadays, with the prosperous of China's economy, all kinds of projects are being actively carried out. As one of the pillar industries of the national economy, the construction industry is also booming. Construction is an integrated, systematic and complex social activity. The complexity of a construction project is contributed by the substantial amount of human and nonhuman factors involved in the project. The interaction between factors leads to high uncertainties and risks [1]. Project schedule, as one of the major concerns in construction project management, could be significantly affected by the uncertainties and risks. Current practice of construction risk management relies heavily on experts' experience. However, an individual's beliefs, attitudes, judgments, and feelings may affect his risk perception. Also, professionals can hardly conduct an extensive risk analysis in the early stage due to the uniqueness of every project.

Being aware of the aforementioned situation, this paper aims to propose a novel conceptual model for construction schedule risks analysis by means of complexity science and to provide insights for its utilization for risks prevention and control.

85.2 Methodology and Data

85.2.1 Model Summary

System dynamics, originated by Forrester in the 1960s, is a science which focuses on the structure of complex systems and the relationship between function and dynamic behavior based on feedback control theory and computer simulation technology. System dynamics has been used in a wide variety of applications, such as economic development, military system, energy and resources, urban planning and ecological environment.

Discrete event simulation has great advantages and application prospects in the design and analysis of the dynamic, complex and interactive construction system [2], which is regarded as an effective tool in the quantitative analysis of the events in the whole building facilities life [3]. Discrete event simulation mainly focuses on that the state variables are discrete changed and the system simulation process is driven by discrete event [4].

Monte Carlo simulation is an effective tool for risk analysis. Considering the threats and opportunities and selecting the probability of occurrence of different variables are two important characteristics of Monte Carlo simulation [5]. Monte Carlo simulation may become increasingly important statistical tools to help risk assessors accessing a risk of uncertainty [6].

85.2.2 Study Area

China Union Technology Building, located on the C17-2 ground in Dayang stone village of the High-tech zone, is being built by CMCU. The expected period of construction is nearly 3 years from 1 December 2013 to 31 December 2016. This building is an office and ancillary facility. It has 26 floors, of which 5 floors underground and 21 floors on the ground. The project uses the frame-core wall structure form and its design working life is 50 years. In addition, the construction design grade of this project is grade 1 and the total number of indoor parking spaces in this building is 480.

85.3 Complex System Modeling

85.3.1 Construction of System Dynamics Model

Based on the research of Nguyen and Ogunlana [7], Lee et al. [8] and Wang [9], this paper divides system dynamics model into construction process subsystem, resource subsystem, schedule target subsystem and project performance subsystem. Subsystems are connected to and associated with each other. The system dynamics model is shown in Fig. 85.1.

For example, if quality problems have been found, rework and schedule delay may happen. Therefore, project performance subsystem is interrelated with

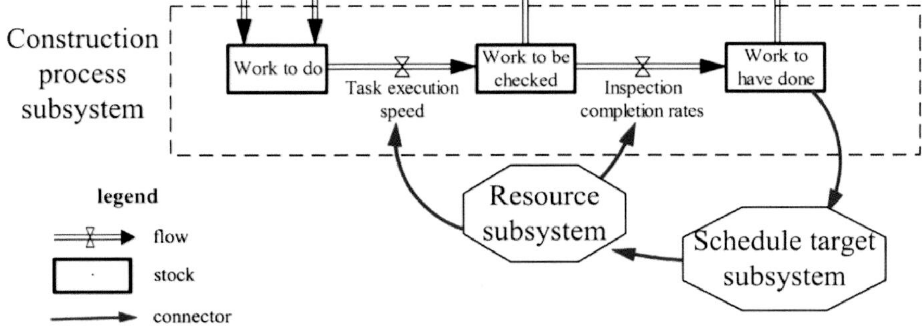

Fig. 85.1 System dynamics model

schedule target subsystem. On the other hand, project performance subsystem is also interrelated with construction process subsystem as getting work done faster may increase the likelihood of the occurrence of project quality problems.

85.3.2 Construction of Discrete Event Simulation Model

A major feature of the discrete event simulation is modeling from micro-level. A project work unit is built by discrete event simulation. In order to achieve the target, the property of a basic unit should be defined. For example, earthwork construction, drilling, geothermal pipe channel construction and infrastructure construction are set as work unit, logical relationships between the work units (including preceding activity, following activity and bonding relation), duration, start time, working hour requirement and resource requirement should be defined. In discrete simulation process, the start time and duration of work unit will change with the impact of schedule risk, which provides data support for system dynamics model.

85.3.3 Building the Interactive Board Comprised of System Dynamics, Discrete Event Simulation and Monte Carlo Simulation

A model that could only be applicable to a particular project can hardly be considered as successful. One of the objectives of this research is to build a model with a high level of university. To achieve this, a completely new concept is adopted in this study: packaging a system dynamics model into a module of the discrete event simulation (namely a job of the model) to constitute a concrete "work module". Different work modules could be combined in order to get a schedule risk systematic analysis model which suits the requirement and characteristics of different construction projects. The process is shown in Fig. 85.2.

85.3.4 Building a WeChat Platform to Get Schedule Risk Information

People may not always take a computer at site, but they will always have a cell phone on hand. It is more convenient for a person to collect real-time data by cell phone than by computer [10]. In order to obtain schedule risk information effectively, a WeChat platform, as shown in Fig. 85.3, which can be accessed by construction project stakeholders is established. Any project stakeholder with a

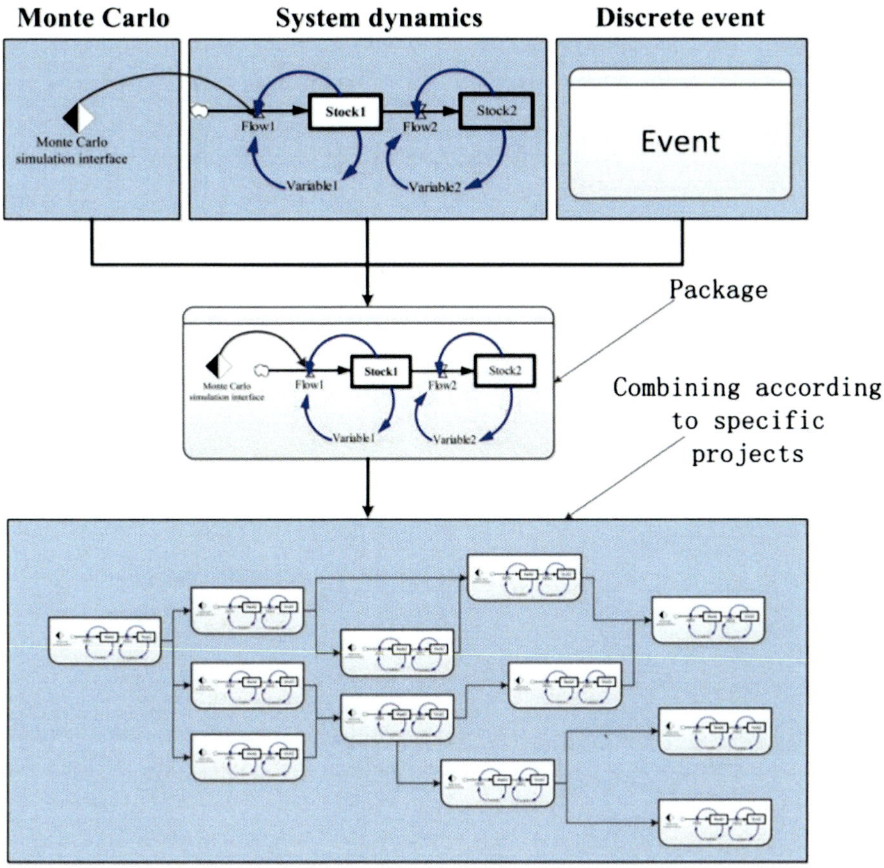

Fig. 85.2 The schematic diagram of model combination

WeChat account can enter the platform and input the construction information (including the schedule risk data in the process of construction). These data will be sent into database and form massive data. By data mining, the probability distribution of each schedule risk can be obtained. The probability distribution will then be inputted into the system model for simulation. Through such processes, users are allowed to forecast the impact from risks to their project more accurately.

85.3.5 *Model Test*

To test the validity of the model, model structure test and model behavior test are adopted.

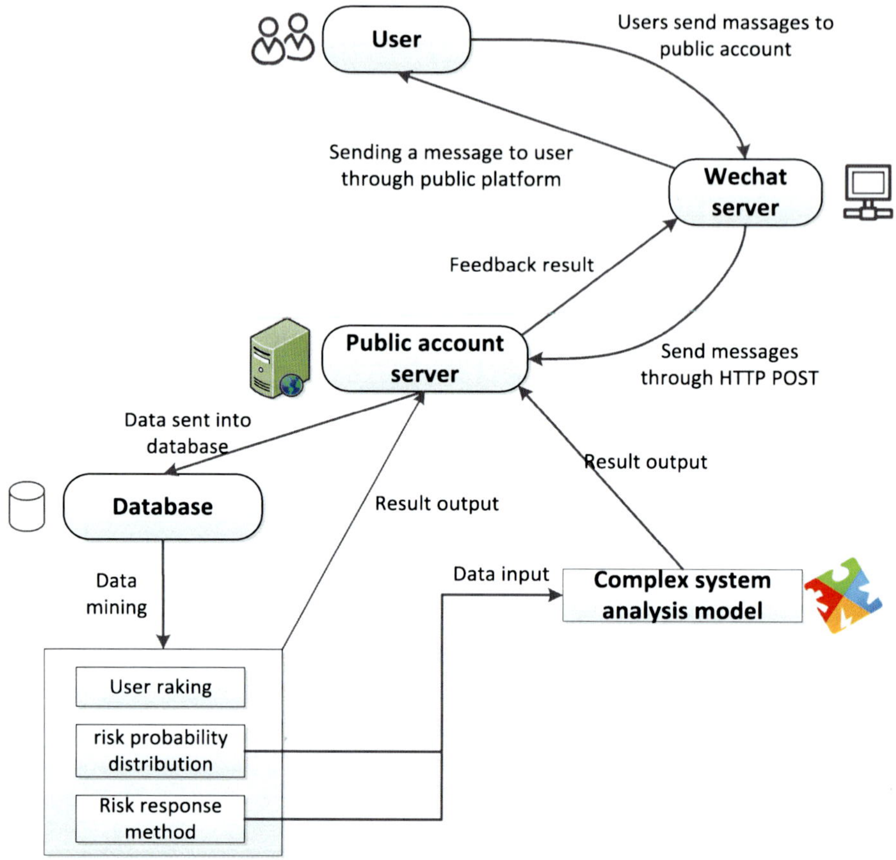

Fig. 85.3 The overall framework of WeChat platform

Model Structure Test

Model structure test includes direct structure tests and structure-oriented behavior tests [11, 12]. Direct structure test contains structure confirmation test, parameter confirmation test, boundary adequacy test and dimensional consistency test [11, 13]. Through a symposium of experts, system dynamics model is properly modified and in line with direct structure test. Structure-oriented behavior rest involves extreme-condition test, behavior sensitivity test and integral error test. The purpose of extreme-condition test is to test whether system dynamics model is consistent with common sense under extreme condition. We set all the variables associated with schedule risks to be 0 and find the output conform to actual situation, which prove the model meets the requirement of extreme-condition test. The role of behavior sensitivity test is to identify sensitive parameters in system dynamics model. By sensitive analysis, we find all sensitivity parameters are in permitted range, which indicate the model in line with the requirement of behavior sensitive test. Integral error test is aiming at determining whether model is sensitive to difference step and integral mode. By the way of changing difference step and integral mode, we find the output change a little, which demonstrate the model meets the requirement of integral error test.

Model Behavior Test

Historical data comparison analysis is adopted for model behavior test. The common practice is to check whether the simulation results are in line with the corresponding historical data [14]. Considering that it is not practical to demonstrate all the output variables due to the lack of historical data, we choose real-time planned schedule as indicators. The model simulated variable value is compared to its historical data. The average error of each variable is 0.0045 %, and the variable whose relative error is less than 5 % accounts for 100 %, which indicates the validity of our model in agreement with the requirement of model behavior test [15].

85.4 Case Study

One of the most important indicators to measure project delay is the real-time planned schedule. The longer the project is postponed, the more influential the schedule risk is. By simulation, we can attain the dynamic changes of schedule risk, as shown in Fig. 85.4. The horizontal axis in Fig. 85.4 indicates "Time" and the vertical axis represents "real-time project schedule". Curve 1 means the planned schedule, which means the real-time project schedule equals the planned schedule without risk. Once risks occur (as risk 1 and risk 2 in Fig. 85.4), the real-time project schedule increases. Afterwards, the real-time planned schedule will decline

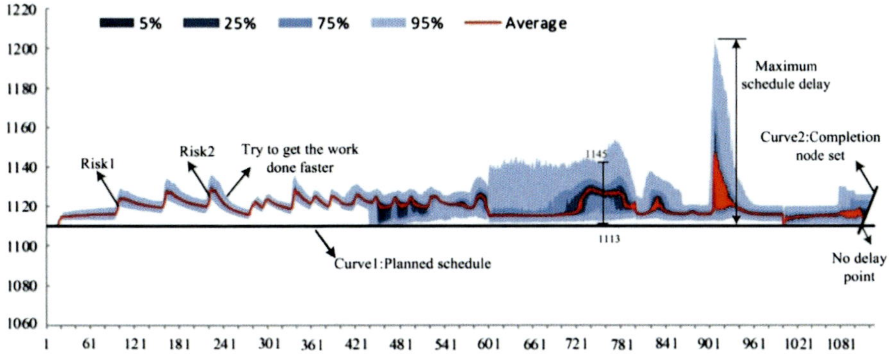

Fig. 85.4 Risks dynamic change

and constantly approach curve 1. Due to the uncertainty of risk, system model will simulate hundreds or even thousands of times and will generate thousands curves. By probability statistics of the value on the curves, we can get the distribution of real-time planned schedule in everyday under the action of risks. Take the 761th day for example, the real-time planned schedule has 95 % chance between 1113 and 1145 and has 75 % chance between 1120 and 1130. The difference value between the ordinate of each curve and curve1 indicate schedule delay. We can see from Fig. 85.4 that the risk which impacts the most occurs at the 902th day and it causes the real-time planned schedule reach to 1200 days. The end point of each curve constitute curve 2 (completion node set) which is a straight line with a slope of 1 (because only when the time equals to real-time planned schedule will the project finish). The curve, the end point of which is (1110, 1110), indicates the schedule is not postponed. Apart from this, all the curves are postponed. Project manager can import the identified risks into the model in different times according to their needs, and they can get: (1)the impact of risks on the project at the same time; (2)the impact of different risk on the project at the same time; (3)the impact of different combinations of risk on the project. Due to the fact that the output can express the uncertainty of risks, the decision maker with different risk attitude can make different response program according to the output.

Further data mining can be conducted for the impact of schedule risk according to output data from risk dynamic change chart. As shown in Fig. 85.5, the horizontal axis represents the risk occurrence time and the vertical axis represents the average of real-time planned schedule which is the ordinate of red curve in Fig. 85.4. Each "bubble" in Fig. 85.5 represents a risk, the radius of which indicates the schedule fluctuation. The bigger the radius is, the more significant the fluctuation is. The following information can be obtained from Fig. 85.5: (1)average schedule delay, the ordinate of each bubble's center indicates the average schedule delay under a certain risk; (2)the impact of the same risk happens at different times, for example, rework occurs at the 110th day, the 171th day and the 589th day

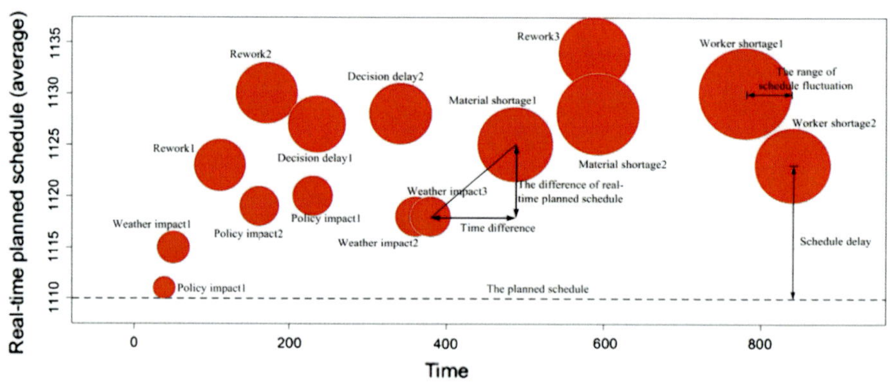

Fig. 85.5 Risk measure

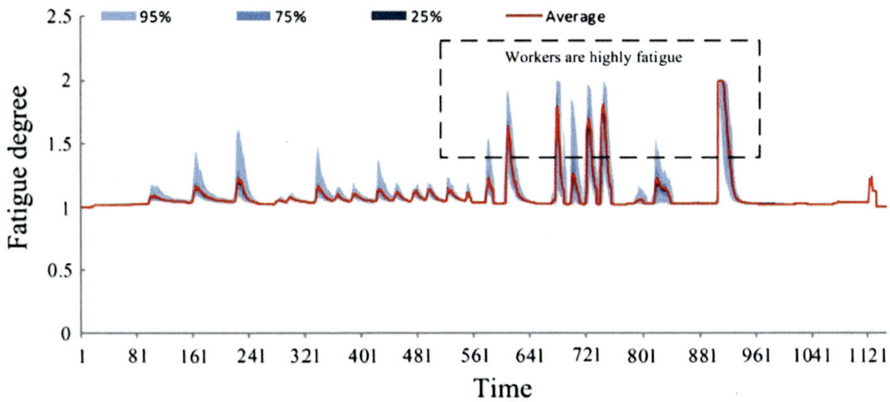

Fig. 85.6 Fatigue degrees in the process of construction

respectively (Fig. 85.5 only intercepts a part of the risks). By comparing the ordinates and radius of the three "bubbles", the impact of risk occurring at different time can be obtained (rework1 < rework2 < rework3); (3)The comparison of the effects of any two schedule risk. Any two schedule risk (e.g. weather impact3 and material shortage2) in Fig. 85.5 can constitute a right triangle. The length of horizontal angle side of right triangle represents time difference, namely the time difference between two risks. In Fig. 85.5, weather impact3 happens 108 days earlier than material shortage2. It indicates that two risks happen at the same time when the time difference is zero. The length of vertical angle side of right triangle represents the difference of real-time planned schedule, for example, the schedule delay caused by material shortage2 is 8 days more than the schedule delay caused by weather impact3.

In addition to the result of project schedule, we can also attain the change of other variables over time, such as fatigue degree, as shown in Fig. 85.6. Worker's fatigue degree is between 1 and 2. 1 shows fatigue does not exist, while 2 represents worker has extreme fatigue. Overtime is the main reason for fatigue. We can see from Fig. 85.6 that fatigue exist slight fluctuations in the first 500 days. From the 500th day to 1000th day, workers repeatedly experience extreme fatigue. In the case of extreme fatigue, construction workers most likely to make a mistake which will lead to schedule delay to a certain extent and create a vicious cycle.

85.5 Conclusion

With the help of complex system theory, this study succeeds in proposing a novel conceptual model and demonstrating its application for construction project schedule risk analysis. However, there are still shortcomings in the process of the analysis concerning the influence of construction project schedule risks. For example, interactions between environment, safety, cost and schedule have not been considered in the analysis. Similarly, interactions between risks have not been taken into account. It is the writers' opinion that future studies can be developed in the following directions: (1) Combining the model in this study with BIM. Each task can be connected with architectural models in BIM model library making it possible to conduct system simulation and 3D virtual building simultaneously. (2) Considering cost, schedule, quality, environment and safety, and building a comprehensive system analysis model. (3) Collecting real-time data in the construction process more effectively with the help of the Internet of things.

Acknowledgements We thank National Natural Science Foundation of China (No. 71272088) for their financial support of this research.

References

1. Choudhry RM, Aslam MA, Hinze JW, Arain FM (2014) Cost and schedule risk analysis of bridge construction in Pakistan: establishing risk guidelines. J Constr Eng Manage 140(7)
2. Lu M (2003) Simplified discrete-event simulation approach for construction simulation. J Constr Eng Manage 129(5):537–546
3. Martinez JC (2009) Methodology for conducting discrete-event simulation studies in construction engineering and management. J Constr Eng Manage 136(1):3–16
4. Law AM, Kelton WD, Kelton WD (1991) Simulation modeling and analysis, vol 2. McGraw-Hill, New York
5. Rezaie K, Amalnik MS, Gereie A, Ostadi B, Shakhseniaee M (2007) Using extended monte carlo simulation method for the improvement of risk management: consideration of relationships between uncertainties. Appl Math Comput 190(2):1492–1501
6. Hayse JW (2000). Using monte carlo analysis in ecological risk assessments. Argonne National Laboratory, USA, Oct 27
7. Nguyen LD, Ogunlana SO (2005) Modeling the dynamics of an infrastructure project. Comput Aided Civil Infrastruct Eng 20(4):265–279
8. Lee S, Han S, Peña-Mora F (2009) Integrating construction operation and context in large-scale construction using hybrid computer simulation. J Comput Civil Eng 23(2):75–83
9. Wang Y (2011) The research on system dynamics simulation-based tome control methodology for complex construction project. Tongji university, China
10. Lee H, Park H, Kim J (2013) Why do people share their context information on social network services? A qualitative study and an experimental study on users' behavior of balancing perceived benefit and risk. Int J Human–Comput Stud 71(9):862–877
11. Barlas Y (1996) Formal aspects of model validity and validation in system dynamics. Syst Dyn Rev 12(3):183–210
12. Barlas Y, Kanar K (2000) Structure-oriented behavior tests in model validation. In 18th international conference of the system dynamics society, Bergen, Norway, pp 33–34
13. Lee S, Peña-Mora F (2007) Understanding and managing iterative error and change cycles in construction. Syst Dyn Rev 23(1):35–60
14. Li Z, Shen GQ, Alshawi M (2014) Measuring the impact of prefabrication on construction waste reduction: an empirical study in China. Resour Conserv Recycl 91:27–39
15. Maddala GS (1986) Limited-dependent and qualitative variables in econometrics. Cambridge university press, USA

Chapter 86
The Application Scenarios of Smart Construction Objects (SCOs) in Construction

Yuhan Niu, Weisheng Lu and Diandian Liu

Abstract The primary aim of this study is to investigate the application scenarios of smart construction objects (SCOs). SCOs are construction resources (e.g. machinery, device, and materials) that are made "smart" by augmenting them with technologies conferring autonomy, awareness, and the ability to interact with their vicinity. The research starts from a brief review of recent developments of smart technology in different industrial sectors including construction. Based on the definition and properties of SCOs, interviews and site visits are conducted to investigate how SCOs could be applied under different scenarios of the construction industry. Perspectives for future studies are proposed in order to fully realise their potentials. The research encourages a wider adoption of SCOs and smart technologies in improving current construction practices. It also provides academia with a platform for further exploring the innovative uses of SCOs in construction.

Keywords Smart construction objects · Smart technology · Information management · Application scenarios · Construction

86.1 Background

The construction industry is calling for changes with the help of technology. A series of influential construction industry reports such as the Latham Report [1], the Egan Report [2], and the Tang Report [3], have been published to elucidate

---

Y. Niu (✉) · W. Lu · D. Liu
Department of Real Estate and Construction, The University of Hong Kong, Pok Fu Lam, Hong Kong
e-mail: yuhanniu@connect.hku.hk

W. Lu
e-mail: wilsonlu@hku.hk

D. Liu
e-mail: liudd@connect.hku.hk

Y. Wu et al. (eds.), *Proceedings of the 20th International Symposium on Advancement of Construction Management and Real Estate*,
DOI 10.1007/978-981-10-0855-9_86

some of the deep-rooted problems in construction, e.g. adversarial attitudes, fragmentation, discontinuity, and labor shortage. Technical and managerial innovations have been quested for solving these problems at an industry level [4]. Therefore, new construction management technologies and concepts, particularly smart construction, relating to Sensing technologies, Augmented Reality (AR), Building Information Modelling (BIM), Internet of Things (IoT) have emerged against this background.

The construction industry shall be more intelligent through the adoption of technologies as mentioned above. The tenet underpinning this thread of research is that human beings, with their intelligence and cognitive abilities, are central decision-makers; in the construction process, determining the use of construction resources such as materials and machinery. Whilst this people-centric decision-making model makes sense in construction, it must also be acknowledged that human beings are not infallible when it comes to processing information and making informed decisions [5]. In this context, human intelligence shows deficiencies (such as being slower and more error-prone) when compared with artificial intelligence (AI) [6].

To meet the even-heightened demand of information management for future construction, smarter technologies or decision-making models that deviating from traditional people centric ones could be reasonably envisaged. Ideally, the construction resources (e.g. materials, precast components and machinery) should be made "smart" by augmented them with ubiquitous smartness including sensing, processing and communication abilities so as to enhance information management for construction. Therefore, to meet the challenge in construction industry, Smart Construction Object (SCO) is introduced as a way to embed "intelligence" into construction recourses in order to assist construction management.

The primary aim of this study is to investigate the potential application scenarios of SCOs in construction. The rest of the paper comprises of five sections. In Sect. 86.2, the developments of smart objects in different industrial sectors including construction are reviewed. Definition and properties of SCOs are presented in Sect. 86.3. Based on data collected from site visits and interviews conducted for construction personnel on site, four typical scenario cases that illustrate the potential applications of SCOs are demonstrated with their respective property diagrams in Sect. 86.4. SCOs under some of the proposed scenarios have been under development and testing. Therefore, next in Sect. 86.5, current development of SCOs and future studies are briefly introduced for fully realizing the potentials. Finally, conclusions are drawn.

86.2 Literature Review

86.2.1 Smart Objects

The trend to implant smartness to everyday objects is originated from the concept of "ubiquitous computing", which was first put forward in the Computer Science Laboratory (CSL) of the Xerox Palo Alto Research Center (PARC) in early 1988. The idea of ubiquitous computing is to shift the paradigm of "one person-one desktop computer" to a new form where computers are spread ubiquitously and invisibly into the everyday objects of our lives [7]. Ubiquitous computing operates as a rationale for the trend of transplanting "smartness" into everyday artifacts, which can be defined as "a non-computational physical entity with established purpose, appearance and use in everyday experience" [8]. Smart objects emerge with the purpose to provide added-values of information processing and exchanging to everyday object while its original appearance and functions will not be restricted or compromised [8]. Instead of the invention of new artifacts with computational intelligence, everyday artifacts could be augmented with the ability of computer to gather, process and exchange digital information. The concept of "smart objects" is developing along with their unique properties as shown in Table 86.1.

A consensual typology of SOs is yet to be agreed but efforts have been paid to developing it in various industries. To improve living quality of human beings, smart wall is designed to convey information and atmosphere when perceiving different people passing by [13]. To help occupants escape from fire, a smart fire response guide (FireGuide) is developed to alert occupants as well as to advise the fastest safe escape route [14]. In transportation industry, a driver assistance systems (DAS) was designed to assist inexperienced drivers and to indicate dangerous situation ahead with the help of intelligent vehicles and smart road infrastructure [15]. In medical industry, the coat of doctor is made smart with a wearable user interface that could be controlled by gestures [16]. The smart coat could provide updated information of patient by interacting with RFID tags on patients and smart bedside display [16].

Table 86.1 The properties of smart objects (SOs)

Properties of smart objects (SOs)	
Contact-awareness	The ability to perceive surrounding environment [8, 9], and to provide condition information [9–11]
Interactivity	The ability of ad hoc information sharing [8, 10] and communication with people/other smart objects [9, 11, 12]
Autonomy	The ability to conduct self-directed actions [9]
Traceability	The ability to discover where they are [11] and provide unique identification [10]
Record-keeping	The ability to describe what happens to them in the past [11]
Representativeness	Possessing programming model and functions inside smart objects [12]

86.2.2 Smart Objects in Construction

Although the development of smart objects have been continuously progressing in other industries, the concept and technology of smart object can hardly be indiscriminately imitated from other industries to the construction industry. Most construction works are project-based with varying workflows. Unlike production-based industries, there is neither standard procedure nor prototype for mass production for the construction industry. At the meantime, the construction work is communication-intensive. Under the contracting and layers of sub-contracting system, manifold parties need to cooperate and exchange updated information. Thus it is assumed that there is no routine assembly line in construction. In addition, construction work is large in scale and non-reversible, for which flexible and interoperable solutions are commonly in need. Therefore, smart objects that are tailor-made for construction are needed to cater for the specialty of the construction industry.

In the construction industry, while not directly using the term "smart object", there are a few studies proposing some devices augmented with smartness to facilitate construction work. To facilitate construction work in different stages and for different purposes, these studies have addressed one or more properties that smart objects possess, including the autonomy to act, the awareness to capture real-time information and the ability to communicate.

For supply chain management, the focus is to utilize the sensor network and auto-ID tags such as RFID tag to provide real-time information. The significance of real-time information in construction supply chain management have been proposed and tested through simulations [17]. The just-in-time delivery enabled by RFID and sensors network in the case study of Shin et al. [18] proves to enhance the time efficiency by 32 % compared to the traditional method. Cho et al. [19] develops a vertical material delivery framework for a liftcar, also relying on the RFID and sensor network.

For on-site construction, the SO-related research mainly covers studies on automation and context-awareness. Contour Crafting (CC) is an example for autonomously fabricating various kinds of superstructures of building [20]. Another way to autonomously construct the superstructure of buildings is to use robots, which are controlled by computers [21] or other kinds of stimulus on site [22]. On the other hand, context-aware information and real-time data that are captured on-site are used to facilitate on-site management, such as providing context-specific information to workers [23], locating construction components on-site [24] and updating simulation models [25].

In construction facility management and safety management, there have been prototypes of smart objects where equipment or plants are augmented with additional functions. For example, it has been suggested that rented plants be equipped with a pay-per-use unit to calculate rental cost based on actual use [26]. For worker health and safety, methods of detecting the vibration of construction tools have been developed [9]. With context-awareness, a collision-prevention method is also proposed to assist crane operators to avoid collision in operation [27].

Despite these research efforts, SCOs and their definition, properties, applications, representations, and prospects have never been systematically explored. Single or scattered smart objects that have been proposed are not enough to exert the full potential of smart objects. It is necessary to steer toward a panoramic and interconnected smartness in construction industry. To do so requires examination of the inheritable properties of SOs from other industries and incorporation of properties addressing the heterogeneity of construction under the construction application scenarios.

86.3 Smart Construction Objects (SCOs)

A Smart Construction Object (SCO) is a step towards the ubiquitous computing and ubiquitous smartness in construction contexts. SCOs are construction resources (e.g. materials/precast components and machinery) that are made "smart" by augmented them with sensing, processing and communication abilities so as to enable better decision-making in construction [28]. In earlier study of Niu et al. [28], there are three core properties for a SCO to function, which are awareness, communicativeness and autonomy. As demonstrated in Fig. 86.1, each of the properties can be developed to different levels. Each type of the awareness, communicativeness and autonomy are summarized in Table 86.2.

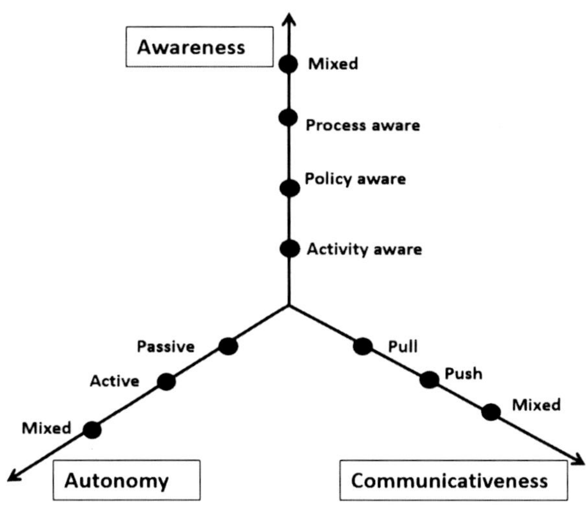

Fig. 86.1 SCO property diagram outlining the three core properties of SCOs

Table 86.2 Properties of SCOs

Awareness	*The ability of SCOs to sense and log the real-time condition of SCOs and the surrounding environment*
Activity-aware	To understand and make record when certain type of activity or event is triggered
Policy-aware	To understand to what extent the real-time condition or activity comply with rules and regulations
Process-aware	To understand and recognize the workflow and transition between construction activities
Mixed	To have more than one type of above awareness
Communicativeness	*The ability of a SCO to share information with managerial personnel or other SCOs*
Pull	To provide information on requests
Push	To proactively send updated information or make alert in a regular interval
Mixed	To have both pull and push communicativeness
Autonomy	*The ability of SCOs to alert people for actions or to take autonomous actions*
Passive	To make alerts to people and to assist people in making decision and taking actions
Active	To take self-directed actions proactively based on change of conditions
Mixed	To have both passive and active autonomy

86.4 Application Scenarios of SCOs

Utilizing the properties of a SCO, the smartness that the SCO contains could optimize working procedures and performance in the construction industry under various circumstances. In this study, pilot interviews are conducted to collect information on current needs and problems faced by the construction industry. Based on data collected from interviews, the potential applications of SCOs are proposed in four scenarios, using the SCO property diagram. The proposed application scenarios are then calibrated by site visits and further discussions with construction personnel, which are introduced with the SCO diagrams as follows.

86.4.1 Component Tracking

Tracking and checking construction materials/components is extremely important in construction. The procedure could be enhanced by embedding smartness into transported materials, transporting vehicles, and checking points to different extents, respectively. The material supply chain of the construction industry is complex but vitally important to on-site work management and cost control. Multiple parties including the suppliers, material distributors, project managers and workers on-site are involved in the transporting process, where the materials might be stored in several transfer depots temporarily. High cost and serious delay may incur when either there is a delay by shortage of materials or on-site warehouse runs out of space by overly stockpiled inventory. Therefore, in time logistic information is of key importance to inventory management and site management.

One option is to augment each component or each piece of material to be SCOs with awareness so that they can sense and push the information of their real-time location all the way. However, considering the massive quantity of construction materials, another more cost-efficient way is also proposed. Materials or prefabricated components could be tagged individually or in batch with smart tags (e.g. RFID) with unique ID and production information. Then, smartness is embedded to the vehicles and possibly the transfer depots. Smart construction vehicles could read and log the information at each time of loading and unloading by activity awareness. Besides, when entering and leaving a transfer depot, the tracking record is updated by active autonomy. Meanwhile, as each transportation record is logged into the memory unit of a SCO. The track-record could be pulled anytime so as to facilitate the regular inventory checking at any time required. Compared to the amount counted by human each time, the SCO log would provide more objective proof to reduce potential disputes by discrepancies. Unlike traditional practices, SCOs have awareness and autonomy, and can interact with the vicinity (Fig. 86.2).

86.4.2 Safety Management

Condition monitoring using SCO may contribute to the safety control on construction site. For example, on construction sites there are environment factors that may harm the health of workers, such as Particulate Matter (PM), noise, vibration

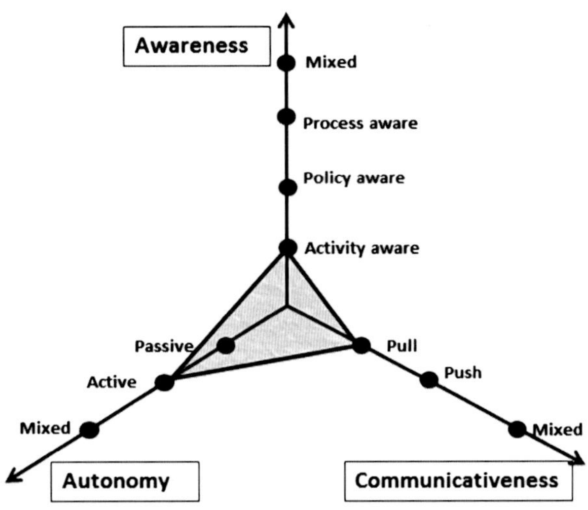

Fig. 86.2 SCO property diagram for component tracking scenario

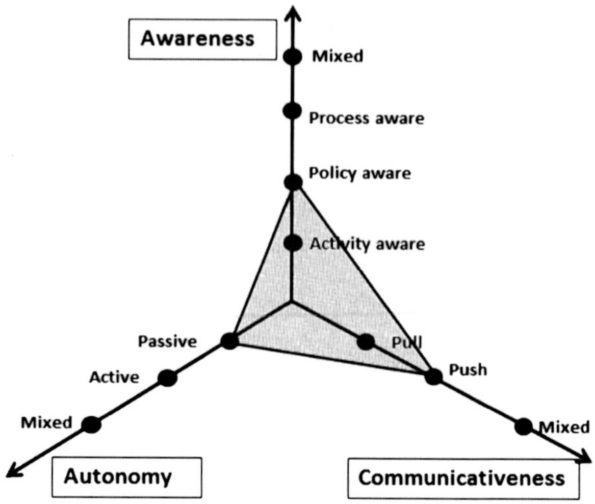

Fig. 86.3 SCO property diagram for safety management scenario

etc. Some of the factors are non-perceptible for human beings. Threshold below the maximum human bearing limits can be input into smart tools, helmet or other wearable devices. With policy awareness, these SCOs could sense the condition and changes in environment. Below the threshold, SCO just push the real-time condition to user interface for normal monitoring. Once the threshold is broke, SCO automatically make alert people for further action (Fig. 86.3).

86.4.3 Procedure Guiding

The process awareness is extremely useful to guide delicate workflow on site, such as on-site assembly of prefabricated components and cooperation between machines. The workflow consists of activities that are linked by transition point [12]. At every transition point, SCO may choose and redirect the workflow based on information from sensors or human input.

Taking the on-site installation of prefabricated components as an example, smart prefabricated component may guide tower crane in the assembly process to locate the precise assembly location and to avoid possible clashes, by sending the information of real-time position to the tower crane on requests. After the smart prefabricated components getting the destination location, the feedback will trigger the smart tower crane entering into the next procedure, i.e., searching for next target component to be installed. The communications between the smart prefabricated components and the smart tower crane relies on them pulling information from each other. By active autonomy, the smart tower crane would automatically find the optimum route and hoist the components from initial storage area to designated installation area. When the component arrives destination successfully, an alert will be sent back to the control center using passive autonomy (Fig. 86.4).

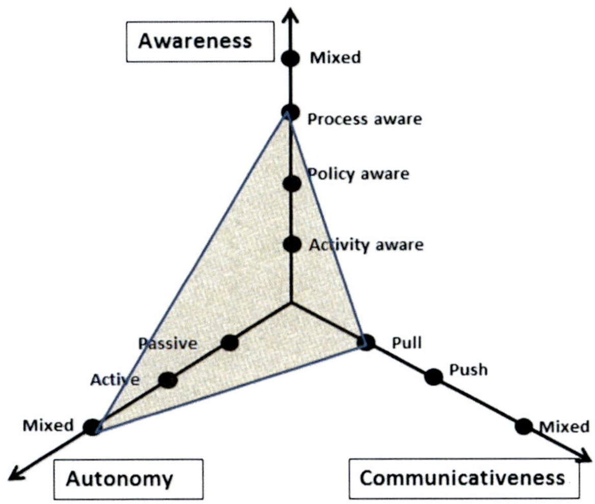

Fig. 86.4 SCO property diagram for procedure guiding scenario

86.4.4 *Facility Management*

Condition monitoring using SCO may also contribute to the facility management. Tools and plants augmented with activity awareness could record the actual operating duration and frequency. By recording the actual using rate of tools and plants, it could assist the estimation of the service life of device. For some plants that may impose danger when getting scrapped, such as tower crane, an alert will be triggered when the usage reaches certain times. So the operator may examine, repair or change a new device before any hazard occurs. Meanwhile, estimation of the service life of device could also assist the cost control. A pay-per-use unit proposed by Fitton et al. [26] is to augment tools or plants with activity awareness to record the usage. Pulling out actual usage then could be adopted as a new way to calculate rental cost [26].

Building maintenance serve as a crucial part for the life-cycle facility management. By integrating smartness to some building components that are unapproachable by human after construction, it may provide information that can hardly be obtained before. For example, once the construction work is completed, the condition of piles and foundations can be hardly monitored by human. Under this condition, policy awareness to sense real-time locations could be embedded into piles and foundations that are driven below the ground where the foundation base is poor or subject to frequent earthquakes. When the shifting or settlement effect reaches threshold, timely alert will be made to offer sufficient time for maintenance or evacuation. For monitoring purpose, the SCOs under the building maintenance scenario may possess a mixed type of communicativeness. On normal working condition, the SCOs push the real-time status of building in a regular interval. When property manager would like to look into the status record at particular time spot, the track-record can also be pulled out (Fig. 86.5).

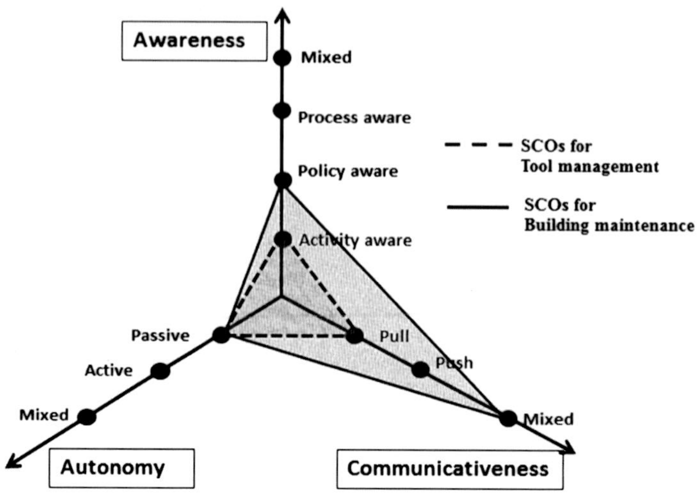

Fig. 86.5 SCO property diagram for facility management scenario

86.5 Discussion

The four typical scenarios discussed above provide a picture of potential applications of SCOs and their smart properties. Among these scenarios, the prototype of SCOs for component tracking and installation procedure guiding are currently under a pilot development and testing. Testing and application feedbacks will be presented and discussed in future studies. Nevertheless, the application scenarios are by no means exhaustive. There are numerous construction scenarios, on-site or off-site, requiring the augmented capabilities of sensing, processing, computing, networking, and reacting to alleviate human beings' incapability in decision-making. With probable increasing needs, the application of SCOs might be further explored for additional level of details. It is also envisaged that SCOs that with dynamic combinations of properties would be tailor-made to cater for different working conditions.

86.6 Conclusions

Smart Construction Object is proposed to be the evolved generation of traditional construction object and the basic element t in future construction. Augmented with the properties of awareness, communicativeness, and autonomy, SCOs are capable of sensing, communicating, and automatically taking actions in addition to performing its original function. This paper envisages the promise for SCOs to be implemented in the construction industry. The application scenarios investigated in this study assist researchers to understand the essence of SCO, and help the application designers to select context to implement the smartness and use of SCO. In the effort to substantiate the claim, we are working on the technical aspects of the three core properties of SCOs with deeper understandings in the context of construction industry. A scenario-based pilot study of SCO is under development. The working cases with SCO application details are envisaged to be presented in future studies.

Acknowledgements The work is supported by HKU Post Doctor Fellowship (PDF)/Research Assistant Professor (RAP) Scheme. Project title: BIM-and IoTs-oriented Smart Construction Objects (SCOs): The Elementary Particles of Future Construction.

References

1. Latham M (1994) Constructing the team. Industry Review of Procurement and Contractual Arrangements in the UK Construction Industry HMSO, London
2. Egan J (1998) Rethinking construction: the report of the construction task force. Department of the Environment, Transport and the Regions, London
3. Tang H (2001) Construct for excellence: report of the construction industry review committee. Construction Industry Review Committee, Hong Kong
4. Ellmann S (2008) Management of complex projects: invisible structures, coordination and recommendations for management. Paper presented at the proceedings of the 22nd IPMA world congress, Roma
5. Reason J (2000) Human error: models and management. BMJ 320(7237):768–770
6. Sterman JD (1989) Modeling managerial behavior: Misperceptions of feedback in a dynamic decision making experiment. Manage Sci 35(3):321–339
7. Weiser M, Gold R, Brown JS (1999) The origins of ubiquitous computing research at PARC in the late 1980s. IBM Syst J 38(4):693–696
8. Beigl M, Gellersen HW, Schmidt A (2001) Mediacups: experience with design and use of computer-augmented everyday artefacts. Comput Netw 35(4):401–409
9. Kortuem G, Alford D, Ball L, Busby J, Davies N, Efstratiou C, Finney J, White MI, Kinder K (2007). Sensor networks or smart artifacts? An exploration of organizational issues of an industrial health and safety monitoring system. Springer, Berlin
10. López TS, Ranasinghe DC, Harrison M, McFarlane D (2012) Using smart objects to build the internet of things. IEEE Internet Comput (to appear, 2013)
11. Mattern F (2003) From smart devices to smart everyday objects. Paper presented at the proceedings of smart objects conference, pp 15–16
12. Kortuem G, Kawsar F, Fitton D, Sundramoorthy V (2010) Smart objects as building blocks for the internet of things. Internet Comput IEEE 14(1):44–51
13. Streitz NA, Rocker C, Prante T, van Alphen D, Stenzel R, Magerkurth C (2005) Designing smart artifacts for smart environments. Computer 38(3):41–49
14. Li Y, Feng L, Qiao L, Li Y, Kong S, Yi Y, Qin W (2010) FireGuide: a context-aware fire response guide for the building occupants. Smart Sens Context, 1–14
15. Fuchs S, Rass S, Lamprecht B, Kyamakya K (2007) Context-awareness and collaborative driving for intelligent vehicles and smart roads. Paper presented at the 1st international workshop on ITS for an ubiquitous ROADS
16. Cheng J, Bannach D, Adamer K, Bernreiter T, Lukowicz P (2008) A wearable, conductive textile based user interface for hospital ward rounds document access. Smart Sens Context, 182–191

17. Min JU, Bjornsson HC (2008) Agent-based construction supply chain simulator (CS2) for measuring the value of real-time information sharing in construction. J Manage Eng 24(4):245–254
18. Shin TH, Chin S, Yoon SW, Kwon SW (2011) A service-oriented integrated information framework for RFID/WSN-based intelligent construction supply chain management. Autom Constr 20(6):706–715
19. Cho CY, Kwon S, Shin TH, Chin S, Kim YS (2011) A development of next generation intelligent construction liftcar toolkit for vertical material movement management. Autom Constr 20(1):14–27
20. Khoshnevis B (2004) Automated construction by contour crafting-related robotics and information technologies. Autom Constr 13:5–19
21. Louis J, Dunston P, Martinez J, West SMD (2014) Automating construction operations using discrete event simulation models. In: Construction research congress 2014, pp 1043–1052
22. Werfel J, Petersen K, Nagpal R (2014) Designing collective behavior in a termite-inspired robot construction team. Science 343(6172):754–758
23. Aziz Z, Anumba CJ, Ruikar D, Carrillo PM, Bouchlaghem DN (2005) Context aware information delivery for on-site construction operations. In: Proceedings of the 22nd CIB-W78 conference on information technology in construction, institute for construction informatics. Technische Universitat Dresden, Germany, CBI Publication, vol 304, pp 321–332
24. Torrent DG, Caldas CH (2009) Methodology for automating the identification and localization of construction components on industrial projects. J Comput Civil Eng 23(1):3–13
25. Akhavian R, Behzadan AH (2012) An integrated data collection and analysis framework for remote monitoring and planning of construction operations. Adv Eng Inform 26(4):749–761
26. Fitton D, Sundramoorthy V, Kortuem G, Brown J, Efstratiou C, Finney J, Davies N (2008) Exploring the design of pay-per-use objects in the construction domain. Smart Sens Context, 192–205
27. Hwang S (2012) Ultra-wide band technology experiments for real-time prevention of tower crane collisions. Autom Constr 22:545–553
28. Niu Y, Lu W, Chen K, Huang G, Anumba C. Submitted to Journal of Computing in Civil Engineering

Chapter 87
A SCO-Based Tower Crane System for Prefabrication Construction

Diandian Liu, Weisheng Lu, Yuhan Niu and Hongdi Wang

Abstract Prefabrication construction has been increasingly applied to the construction industry. Automation and intelligentization of construction site and construction objects has become the fore for the development of the prefabrication construction. In this study, a SCO-based tower crane system, as the core resource site on the critical path of prefabrication construction, is proposed to make the cranes more intelligent and efficient. A set of smart units are applied to the tower cranes and prefabricated components to make them into Smart Construction Objects (SCOs), enabling the functions of awareness, communicativeness and autonomy. The smart tower crane could "talk" with prefabrication components through sensors and real-time wireless connections. A system operation flow is developed for fetching and placing target components. The SCO-based tower crane system can provide the operation instruments, real-time traceability of the components, and necessary warning. The system is being tested in a pilot project, which is delivered through BIM and prefabrication construction. It is envisaged that the SCO-based tower crane system could significantly improve the safety performance and operation efficiency of the tower crane operations in prefabrication construction.

Keywords Prefabrication construction · SCO · Tower crane · Safety

87.1 Background

Prefabrication is hallmark development of the industrialization of the construction industry. Different from traditional cast in situ method, precast components are produced and tested in the factory, before they are assembly on-site. Prefabrication is widely perceived as a superior construction technology to traditional construction method in aspects such as including the improvement in environmental impacts

D. Liu (✉) · W. Lu · Y. Niu · H. Wang
Department of Real Estate and Construction, The University of Hong Kong, Pok Fu Lam, Hong Kong
e-mail: liudd@connect.hku.hk

Y. Wu et al. (eds.), *Proceedings of the 20th International Symposium on Advancement of Construction Management and Real Estate*,
DOI 10.1007/978-981-10-0855-9_87

through reduction in waste and dust [1], the increased worker safety through reduced exposures to hazardous operations [2], and the increment in whole-life performance and profitability [3].

Due to its merits, prefabrication has been adopted in Hong Kong for more than two decades, mostly in in public housing projects [1]. In 2002, the precast components accounted for approximately 17 % of the total concrete volume used in public housing projects [4]. The precast elements cover a wide range of construction elements including facades, staircases, parapets, partition walls, semi-precast slabs and, more recently, volumetric precast bathrooms and kitchens [2]. The advancement of prefabrication techniques has contributed significantly to a dense urban environment like Hong Kong, economically and environmentally [1].

However, problems associated with the prefabrication construction are also reported. For example, high transportation cost has led to that overall prefabrication construction around 2 % more expensive than using traditional technologies. It requires more carnage. When working with large amount of on-site prefabrication work, installation is often complicated especially when working under very congested floor layout at a high altitude [5]. Although innovation in techniques has improved the prefabrication construction [1], the innovations mainly focus on the casting technique. It is desired that prefabrication could be improved by continuously developing innovative technologies such as automation and robotics.

To improve the construction by the means of automation has been proposed since decades ago [6]. Using robots to replace existing labor work and existing construction machinery is the main trend. There are robots like mobile crane that are designed for brick-laying and masonry work [7]. Using Contour Crafting (CC) system provides a way to fabricate a whole piece of hollow concrete structure on site by automatic computer control [8]. There are also breakthrough technologies for prefabrication construction. For example, a gantry is augmented with robotic intelligence to automatically assemble prefabricated component on site [9], although it is designed for low-rise buildings.

Putting robots onsite as "helpers" for construction work might be a way to drive the technological development in the construction industry. Alternatively, to augment the existing on-site construction machinery and equipment with "smartness" for conducting traditional work might be another way without radical change in routines. In this study, a framework to develop a smart tower crane system is proposed to assist the on-site prefabrication construction work. Based on the concept of Smart Construction Objects (SCOs), the tower crane and prefabricated components in this system are augmented with the ability to sense, communicate and act autonomously. The remainder of the paper comprises six sections. Section 87.2 reviews the concept of SCO and defines the properties of SCOs in the smart tower crane system. Section 87.3 elaborates the rationale for setting up a digitized 3D site to support the system. Section 87.4 describes three components in the smart unit. A workflow of the system is introduced in Sect. 87.5. Section 87.6 discusses the prospects and future research areas presented by the smart tower crane system, and conclusions are drawn in Sect. 87.7.

87.2 SCO and Smart Crane System

To meet the challenge in construction industry, Smart Construction Object (SCO) is introduced as a way to embed "intelligence" into construction recourses in order to assist construction management. A SCO is proposed as a step towards the ubiquitous computing in the construction context [10]. In earlier study of Niu et al. [10], SCOs are defined as follows:

> Construction resources (e.g. machinery, tools, device, materials, components, and even temporary or permanent structures) that are made "smart" by augmenting them with sensing, processing and communication abilities so that they have autonomy and awareness, and can interact with the vicinity to enable better decision-making.

The smart tower crane system is one application example of SCOs, where traditional tower cranes and prefabricated components are augmented into SCOs. The sensing ability of SCO is addressed by awareness, which enables the prefabricated components to sense their real-time locations. Besides, by applying the communicativeness, a connection could be established between prefabricated components and the smart crane. Through the connection, prefabricated component could exchange information with the smart crane including the real-time coordinates, destination coordinates, and other geometric parameters. Based on the information captured by and transmitted from the smart prefabricated components, the smart tower crane could fetch and relocate these components with autonomy.

87.3 Smart Unit

To obtain the awareness, communicativeness and autonomy as described for SCOs, each of the prefabricated component and the crane are augmented with a smart unit. The core components of the smart unit are the location tracking module, the microcontroller, and the Bluetooth module.

87.3.1 Location Tracking Module

The awareness to sense real-time locations is achieved by the location tracking module in the smart Unit. A Global Positioning System (GPS) tracker and an Inertial Measurement Unit (IMU) work complementarily as the location tracking module for the smart tower crane system.

The GPS has been widely used in navigation and surveying application due to its capability to provide a 24-h worldwide positioning service at a relatively low cost [11]. Relying signals from on the satellites, the accuracy of GPS has been made possible in a 10-m level of change [12]. Although the accuracy is comparatively high for road transportation positioning, it is still not enough for positioning of single building component. Thus a more precise positioning system for short distance is needed to help calibrate the result given by GPS system. Moreover, GPS may suffer from signal masking and multipath errors in area shielded by dense buildings or trees [11]. Hong Kong Island is typical city canyon area where only around 50 % of the test area is receptive of adequate GPS signals in the study of Lu et al. [11]. Therefore, GPS system alone is not enough for location tracking in the smart tower crane system.

Not like GPS that relies on satellites for positioning, an IMU is self-contained and it requires no external motion signals for positioning [11]. The IMU is reliable and stable while navigating under conditions of external disturbance [13]. The IMU consists of a gyroscope to sense angular rates, an accelerometer to sense acceleration and other sensors such as the magnetic sensor so that data from separate sensors could fuse into a single, optimum estimation [14]. After the acceleration of gravity is subtracted, the remaining accelerations would then be double integrated over time to determine the displacement of the IMU relative to a known starting point [15]. Since location tracking using IMU is based on integration against time of movement, even a small change of distance can be calculated. Thus IMU could offset the accuracy problem of GPS. On the other hand, with the increase of tracking time and distance, the IMU will suffer from drifting problem by integration [11, 14, 16, 17]. The tracked locations may deviate from the real-time locations over time if IMU is used alone. Thus IMU is usually used with other sensors or systems, such as the optical navigation sensor [17], for navigation and positioning purpose. Therefore, for the smart tower crane system, GPS is needed to calibrate the IMU result at regular time interval.

87.3.2 Microcontroller

The microcontroller is vitally important for realizing autonomous control. The application of microcontroller is widely adopted in autonomous robot control [18, 19]. Likewise, acting like the 'brain' in every SCO, the microcontroller centrally manages the sensors, communicating devices, and the actuators in the Auto-In system. Data generated from the location tracking module would be compiled and interpreted by the microcontroller before communication. Besides, in traditional construction process, the motion of jib, trolley and hoisting rope of the tower crane are manually controlled by the operator. In the smart crane system, the motions of these components will be guided by instructions sent from the microcontroller, which are calculated and provided by the algorithms programmed inside.

87.3.3 Communication Module

Communicativeness of SCOs is achieved by the communication module in the smart unit. Bluetooth is chosen for the smart tower crane system. Compared to other wireless communication technologies, Bluetooth is preferred because low-energy Bluetooth could perform lower power consumption [20]. It usually takes no less than one month for prefabricated components to be the manufactured, delivered and installation on-site. In the Hong Kong-Shenzhen cross-board cooperation for prefabrication construction, the transportation of prefabricated components may take longer time. Most wireless connection modules such as Wi-Fi module may largely consume the battery before the on-site operation. Comparatively, the low-power Bluetooth may support the smart prefabricated components for longer duration up to 6 months. Meanwhile, depending on the power class of the device, the typical working distance of Bluetooth ranges from 10 to 100 m [11], covering the hoisting range of the tower crane.

Besides, the Bluetooth wireless module provides a low-cost way for information exchange among mobile devices and provides connectivity of these devices to the Internet [21]. To complete the system operation, the smart tower crane needs to be connected by servers or handheld devices. Therefore, the Bluetooth module is preferred in the smart tower crane system.

87.4 The Digitized 3D Site

To establish a digitized three dimensional (3D) site with three mutually orthogonal axes is the first and foremost step for the SCO-based tower crane system. The purpose of digitizing the site is to assign each location within the site with three a unique triaxial (x, y, z) coordinate. X-axis and y-axis forms the plane that are parallel to the ground. Z-axis is parallel to the direction of gravity. Together, x-axis, y-axis and z-axis form a 3D cubic grid that is large enough to contain the whole site and buildings to be erected. Therefore, anywhere on site including the temporary storage area, vehicle parking area, construction area and so on can be located by corresponding coordinates. Digitizing the whole construction site area is fundamental for SCOs to achieve location awareness. The coordinate system serves as the standard of site location information. After SCOs sense their real-time locations within the site, the information will be compiled in a (x, y, z) format. When SCOs need to communicate the location information between each other or update location information back to database, the location information is also conveyed in the (x, y, z) format.

The establishment of the digitized 3D site is based on site plans. The z-axis is set referring to elevation plans while x-axis and y-axis are set on site layout plans. Since the location tracking and route calculation are based on relative displacements, the base point (0, 0, 0) can be set at the center or any angle of the cubic grid. The cubic grid is large enough to contain the irregular-shaped site. Each façade could be matched with a (x′, y′) value as their destination location value in the designated installation location. For facades on ground level, the z′ values are set as 0.

For facades on above levels, the z′ values are set according to floor levels in elevation plans. If project has been designed with Building Information Modeling (BIM) models, the model can be directly matched with the 3D cubic grid so that each objects in the BIM model carries the matched (x′, y′, z′) value as destination locations.

87.5 System Operation Flow

As shown in Fig. 87.1, the system block diagram specifically demonstrates the dynamic interactions among three elements of the SCO-based tower crane system, including the smart crane, smart prefabricated component and the information center.

The information center serves as a database to support the system, containing information of the digitized 3D site as well as information of designated installation locations for every prefabricated component. The destination location (x′, y′, z′) of each smart prefabricated component, which is extracted from 3D digitized site from the information center, will be firstly input into its smart unit during prefabrication

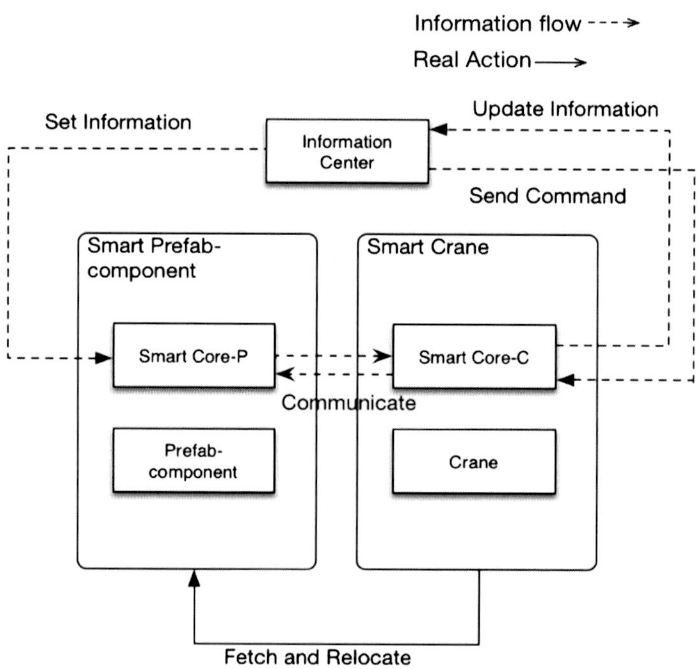

Fig. 87.1 The block diagram of the smart crane tower system

process. Besides, the smart prefabricated component could also carry other information such as weight, height, and installation requirements. The database of the information center is shared by on-site servers and can be accessed by handheld devices.

When smart prefabricated components have been delivered on site, the smart crane will be triggered into operation mode by a command from servers in the site control center, or from handheld devices. A set of smart unit will also be augmented to the tower crane so that the smart tower crane could communicate with the smart prefabricated components and act autonomously. When smart prefabricated components enter into the temporary storage area, they are configured to the discoverable mode. Since each smart component carries a unique ID, the smart crane could identify each smart component by its ID.

The smart crane will then establish the communicative Bluetooth connection with smart components. After the smart tower crane validates the ID of the each component, the component would send back its current location (x0, y0, z0) and destination location (x′, y′, z′) to the crane. By using the current location (x0, y0, z0) and the destination location (x′, y′, z′), the crane is enabled to calculate the optimum route for fetching and relocating the target component by calculating the rotation degree of the jib, parallel moving distance of the trolley and vertical moving distance of the hook block.

Basically, there are three steps for the smart crane to reach the targeted components. First, the jib of the smart crane will rotate in a 360° plane that is parallel to the ground until pointing the targeted component. Second, the trolley will move forward or backward along the jib based on the parallel moving distance calculated. Finally, the crane will move the hook block up and down by adjusting the hoisting rope. Similarly, by reversing and repeating the above three steps, the crane could relocate the target component to the destination location after fetching it. During the fetching and relocating operation of the smart tower crane, the smart prefabricated component could sense the replacement of its location though awareness, which is achieved by the microcontroller and the location tracking module in the smart unit. The IMU could provide the real-time acceleration and rate of angular change. These data would then be converted into the 3D coordinate format as the real-time location (xi, yi, zi) by the microcontroller, which will be calibrated by the GPS outputs.

Once the real-time location (xi, yi, zi) matches destination location (x′, y′, z′), the smart prefabricate component would alert the smart tower crane, suggesting that the component is successfully arriving at the designated installation location. Afterwards, the smart crane will update current location (xi, yi, zi), which should be equal to the designated destination location (x′, y′, z′) theoretically, of the smart prefabricated component to the information center. At this time, the information of current location (xi, yi, zi) could be regarded as the as-built information of the smart fabricated component, which is potential for future use for the purpose of maintenance and facility management. The full process of the operation could also be demonstrated through the schematic diagram as Figs. 87.2, 87.3 and 87.4.

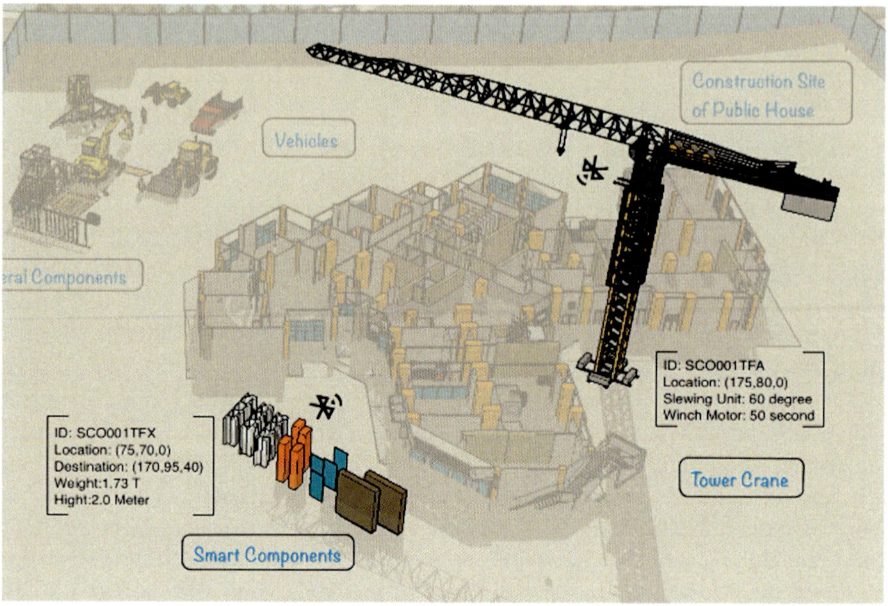

Fig. 87.2 Communication establishment

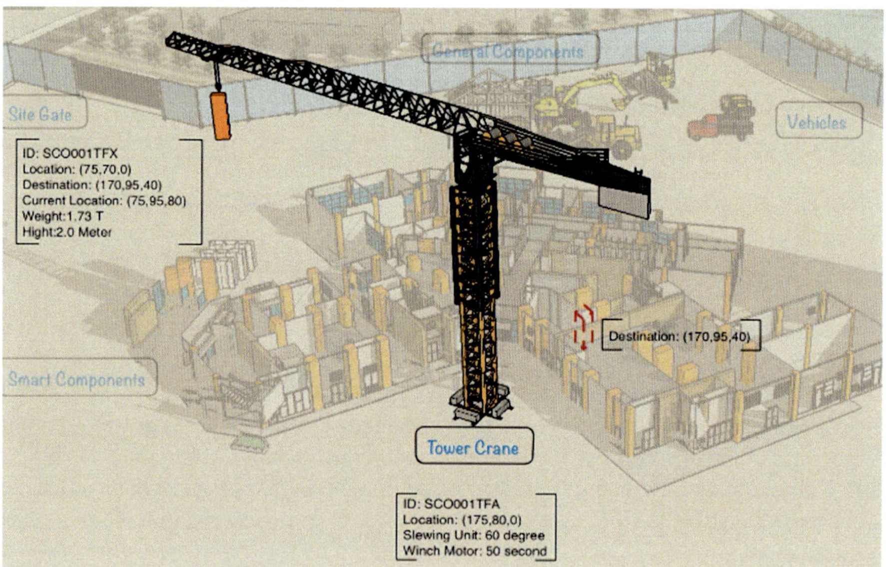

Fig. 87.3 Autonomous installation

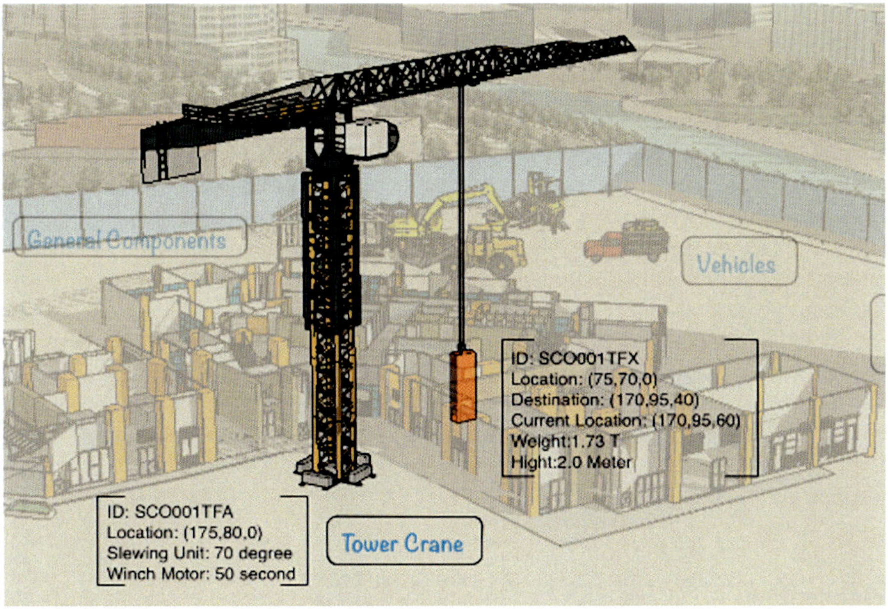

Fig. 87.4 Destination confirmation

87.6 Discussion

The smart tower crane system is envisaged to improve the prefabrication construction by autonomous action-taking to free labor from site operations. The reduction in human involvement is potential to enable better safety control and to decrease human-centric interventions and possible errors. Besides, the object-to-object communicativeness presented by the smart tower crane and the smart prefabricated component lays the foundation towards Internet of Things in the construction industry.

As the smart tower crane system is being developed and tested in a pilot study in a prefabrication construction project in Hong Kong, the effectiveness and efficiency of this system is yet to be demonstrated. The empirical performance of the smart tower crane will be presented in future study. Besides, more construction objects including machines, equipment and on-site components are potential be augmented into SCOs under various circumstances to step towards a panoramic smart construction industry.

87.7 Conclusion

This paper presents the key concepts, technologies, and the framework for developing smart tower crane system for prefabrication construction. The fundamental concept underlying this study is the concept of smart construction objects (SCOs), which are construction objects that are augmented with sensing, communicating and autonomous abilities. A smart unit is embedded into the traditional tower crane and prefabricated components to make them into SCOs. To establish a reference location system, the construction site would be digitized with 3D cubic gridlines so that each location could be assigned with a unique coordinate. Based on the SCOs and location referring to the 3D digitized site, the on-site installations could be preceded autonomously by the smart tower crane system.

Acknowledgements The work is supported by HKU Post Doctor Fellowship (PDF)/Research Assistant Professor (RAP) Scheme. Project title: BIM-and IoTs-oriented Smart Construction Objects (SCOs): The Elementary Particles of Future Construction.

References

1. Jaillon L, Poon CS (2009) The evolution of prefabricated residential building systems in Hong Kong: a review of the public and the private sector. Autom Constr 18(3):239–248
2. Lu W (2013) Enhancing housing production in Hong Kong through BIMatizing offshore prefabrication. Build J, 12–15
3. Venables T, Barlow J, Gann D (2004) Manufacturing excellence: UK capacity in offsite manufacturing. The Housing Forum, London
4. Chiang YH, Chan EHW, Lok LKL (2006) Prefabrication and barriers to entry—a case study of public housing and institutional buildings in Hong Kong. Habitat Int 28:1–18
5. Wong WM (2000). Prefabricated construction in Hong Kong. Construction and Contract News. Issue No.3, Hong Kong
6. Balaguer C, Gambao E, Barrientos A, Puente EA, Aracil R (1996) Site assembly in construction industry by means of a large range advanced robot. In: 13th international symposium on automation and robotics in construction, Tokyo, pp 65–72
7. Heintze J, Teerhuis PC, Weiden A (1996) Controlled hydraulics for a direct drive brick laying robot. Autom Constr 5(1):23–29
8. Khoshnevis B (2004) Automated construction by contour crafting—related robotics and information technologies. Autom Constr 13(1):5–19
9. Balaguer C, Abderrahim M, Navarr JM, Boudjabeur S, Aromaa P, Kahkonen K, Atkin B (2002) FutureHome: an integrated construction automation approach. Robot Autom Mag IEEE 9(1):55–66
10. Niu Y, Lu W, Chen K, Huang G, Anumba C, Submitted to Journal of Computing in Civil Engineering
11. Lu M, Chen W, Shen X, Lam HC, Liu J (2007) Positioning and tracking construction vehicles in highly dense urban areas and building construction sites. Autom Constr 16(5):647–656
12. Ochieng WY, Sauer K (2002) Urban road transport navigation: performance of the global positioning system after selective availability. Transp Res Part C: Emerg Technol 10(3): 171–187
13. Koruba Z, Tuśnio J (2009) A gyroscope-based system for locating a point source of low-frequency electromagnetic radiation. J Theor Appl Mech 47(2):343–362

14. Madgwick SO, Harrison AJ, Vaidyanathan R (2011) Estimation of IMU and MARG orientation using a gradient descent algorithm. In: 2011 IEEE international conference on, rehabilitation robotics (ICORR), pp 1–7. IEEE
15. Titterton D, Weston JL (2004) Strapdown inertial navigation technology. IET, 17
16. Huang C, Liao Z, Zhao L (2010) Synergism of INS and PDR in self-contained pedestrian tracking with a miniature sensor module. Sens J IEEE 10(8):1349–1359
17. Hyun D, Yang HS, Yuk GH, Park HS (2009) A dead reckoning sensor system and a tracking algorithm for mobile robots. In: IEEE international conference on mechatronics. ICM 2009, pp 1–6
18. Caprari G, Balmer P, Piguet R, Siegwart R (1998) The autonomous micro robot "alice": a platform for scientific and commercial applications. In: Proceedings of the 1998 international symposium on paper presented at the micromechatronics and human science. MHS'98
19. Crespi A, Badertscher A, Guignard A, Ijspeert AJ (2005) AmphiBot I: an amphibious snake-like robot. Robot Auton Syst 50(4):163–175
20. Gomez C, Oller J, Paradells J (2012) Overview and evaluation of Bluetooth low energy: an emerging low-power wireless technology. Sensors 12(9):11734–11753
21. Bisdikian C (2001) An overview of the Bluetooth wireless technology. IEEE Commun Mag 39(12):86–94

Chapter 88
Quantization on Social Cost of Large-Scale Construction Project Based on Emergy Analysis Method

Hong Zhou, Bing Li, Qiang Shen and Shasha Yu

Abstract For some of the major projects, there are always numbers of people involved and huge environmental and social impact. Traditional thinking has always been focused on the development of economic and financial evaluation of the project construction, while ignoring the social benefits of the project. Considering all aspects of society has become the new problem researchers are facing. Currently the measurement of social cost is still in an early stage. This article defines the social structure for large-scale projects' social analysis. The social costs system is on the basis of the social components. Combined with improved Rabel *J*. Burdge social evaluation system, we use the feature that emergy could be a standard of various value forms, establish a quantitative system of social costs, and for some variables of social assessment we present a quantitative method. We did the empirical study of Gulei PX program by field investigation and statistical calculations. At last we come to a conclusion that Gulei PX program have social risks. The potential problems as well as the solutions are pointed out.

Keywords Emergy · Social cost · Social cost assessment · Quantitative

88.1 Introduction

In recent years, with the constant social conflicts caused by large-scale construction project in China, the first case of Xiamen "PX" event in the public participation has raised the thinking of social impact assessment in the large-scale construction projects assessment. What is more, the result of some engineering whose purpose was the development of the local economic and social, is the fall of the society. Drug addiction and gambling were prevalent in the region after Yunnan Manwan

---

H. Zhou (✉) · Q. Shen · S. Yu
School of Architecture and Civil Engineering, Xiamen University, Xiamen, China
e-mail: mcwangzh@xmu.edu.cn

B. Li
School of Management, Xiamen University, Xiamen, China

Y. Wu et al. (eds.), *Proceedings of the 20th International Symposium on Advancement of Construction Management and Real Estate*,
DOI 10.1007/978-981-10-0855-9_88

Hydropower Station was built. So the social stability risk assessment of major projects has become a necessary part of the project in the early stage since 2013. In our country, the social conflict problem occurred in the practice of large engineering projects illustrates the importance and inadequacy of social evaluation. Namely, accuracy is not enough and social evaluation is a mere formality. From 2000, the social impact assessment has gradually turned to the comprehensive evaluation of public participation and quantitative research. Emergy, this scale has been widely used in the field of environmental ecology, and extended to the economic field since Odums proposed it in 1996. And in recent years it has received the attention of social and management discipline. Ruan et al. [1] applied the emergy analysis method to regional labor transfer, and put forward that the regional potential energy was the driving force of labor transfer. From the biosphere perspective, the social system is a part of the biosphere, which is the largest ecosystem. From this point of view, the heterogeneity dimension of the social system can be quantified by the scale of emergy, thus to get quantitative evaluation of the impact of project on the social system. In this paper, the emergy analysis method was used to evaluate the social impact in order to quantitatively explore social evaluation, and PX Gulei project was taken as an example.

88.2 Social Evaluation Theory and Method

88.2.1 Social System Definition

Social system is “any kind of human action, and all kinds of time and process associated with such actions constitute a relatively independent general system” [2]. “Social impact assessment variables” refer to the measurable changes in the relationship among people, community and society, which are caused by policy changes or development projects. Social impact assessment needs to be conducted before making decision [3].

In this paper, the narrow definition of the environmental system of the environmental economist Mr. Li Jinchang is adopted: From an environmental point of view, all natural resources are the elements of the environment and this environment is the natural environment. So the scope of the social system can be distinguished, which includes human and the activities and contact between them. Among them, the community infrastructure provides a place for people’s activities as a part of the social system (Table 88.1).

Table 88.1 Social system structure definition table

Classification	Detailed description
People in community	Specific social individual
Community facility	Construction degree of communal facilities
	Improvement situation of science and education facilities: school, library, etc.
	Perfection situation of sanitary facilities:hospital, clinic
	Entertainment: network, television and other community activities, etc.
	Other facilities
Social atmosphere	Personal emotion, mood, psychological states
	Social atmosphere, collective emotion
	Interpersonal relationship
Information intelligence culture	Social and cultural information
	Advanced technical expertise
	Intelligence level and education status
	Cultural heritage

88.2.2 *Social Impact Assessment Index System*

According to Labelle's theory, there are currently twenty-eight social impact assessment variables. Categories and definitions are shown in the following table, and the quantitative and qualitative indicators can be distinguished (Table 88.2).

Table 88.2 Social evaluation variables and definitions

Classification	Variables	Evaluation method
Population influence	1. Demographic change	Quantitative
	2. Temporary workers' flow	Quantitative
	3. The seasonal (tourism) emergence of residents	Quantitative
	4. Relocation of individuals and families	Quantitative
	5. Differences in age, gender, race and nationality	Qualitative
Community system	6. Forming the attitude to project	Qualitative
	7. Activities of interest groups	Qualitative
	8. Changes in the local government structure and size	Qualitative
	9. Any comprehensive planning activities	Qualitative
	10. Industry diversification	Quantitative
	11. Living/family income	Quantitative
	12. Strengthen economic inequality	Qualitative
	13. Changes in employment equality of minority groups	Qualitative
	14. Changes in employment position	Quantitative

(continued)

Table 88.2 (continued)

Classification	Variables	Evaluation method
Community in transition	15. Emergence of external institutions	Qualitative
	16. Cooperation between institution	Qualitative
	17. Emergence of new social strata	Qualitative
	18. Local commercial/industrial center change	Qualitative
	19. The emergence of residents on weekends (tourism)	Qualitative
Impact of individual and family	20. Disturbance of daily life and activity pattern	Qualitative
	21. Differences in religious and cultural activities	Qualitative
	22. Changes in family structure	Qualitative
	23. Interference of social networks	Qualitative
	24. Perception of public health and safety	Qualitative
	25. The change of leisure opportunity	Quantitative
Residential infrastructure needs	26. Changes in community infrastructure	Quantitative
	27. Land acquisition and land allocation	Qualitative
	28. The effects on suppressive culture, history, religion and archaeological sites	Quantitative

Table 88.3 The supplement of social impact assessment variable

Classification	Variable	Evaluation method
Mutual adaption between engineering and community	29. Implementation situation of project expectations	Qualitative
	30. Expected difference between social repercussions and the actual situation	Qualitative
Public participation and policy transparency	31. Plans for public participation	Quantitative
	32. Ways of public reflection	Quantitative
	33. Policy openness	Qualitative

The social impact of the project is evaluated by these twenty-eight variables in detail from five perspectives. That social risk should be taken into consideration in the aspect of social mutual adaption is pointed out in "Outline and description of the preparation of the social stability risk assessment report of major fixed asset investment projects". In China, citizen participation is increasingly valued. After questionnaire and expert interview, five variables of two categories were added to Labelle's social evaluation variables (Table 88.3).

88.3 Emergy Analysis Method of Social Impact Assessment

Expert scoring, social survey, fuzzy mathematics and other theories are widely used in Chinese social assessment. These methods are more biased in qualitative analysis and social evaluation is a mere formality. Basic assumption of emergy analysis method is that the sun is the energy source of the earth. This is the basis of the quantitative social cost. The variables of the social system can be converted by the conversion rate of solar emergy. The magnitude of the emergy transformity reveals the fundamental reasons for the differences in the value of the different resources energy, commodity labor and technical information [4]. In social impact assessment, the input and output of the calculation of the social cost not only consider the specific goods, but also the value of labor culture and other valuable parts. The basic formulas are (Fig. 88.1):

$$\text{solaremergy(sej)} = \text{energy(J)} * \text{emergytransformity(sej/J)}$$
$$\text{solaremergy(sej)} = \text{material(g)} * \text{emergytransformity(sej/g)}$$
$$\text{solaremergy(sej)} = \text{value} * \text{emergytransformity(sej/\$)}$$

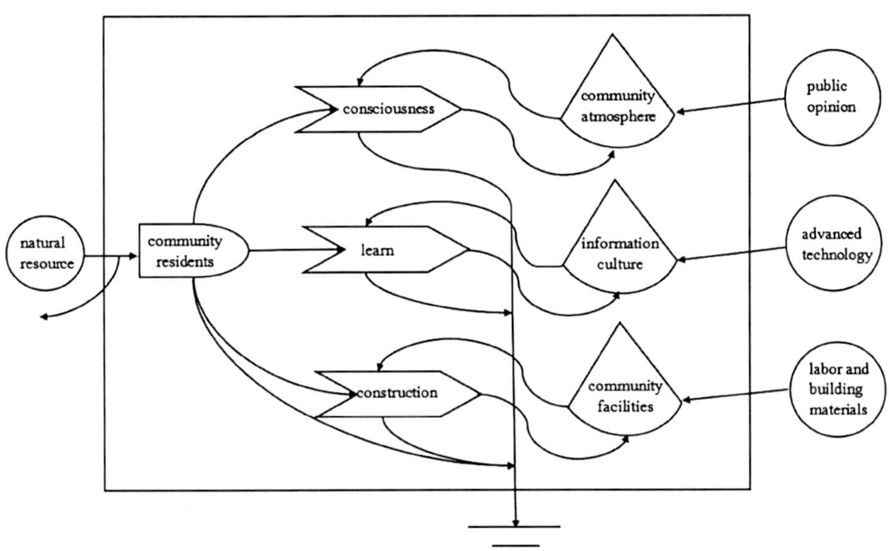

Fig. 88.1 Social emergy system diagram

88.4 Social Impact Emergy Evaluation of PX Project

88.4.1 PX Chemical Project

Xiamen Haicang PX project relocated from Xiamen to Zhangpu Gulei peninsula triggered the first Chinese public participation "walk" event. There used to be rich natural resources and vigorous aquaculture, planting, and other ecological industries. There also used to be three nature reserves. Since PX petrochemical project settled, Gulei peninsula has changed into petrochemical base, planning to extend and develop the petrochemical industry chain and related supporting industries (Figs. 88.2, 88.3 and 88.4).

Fig. 88.2 Seaside condition of Xingzai village

Fig. 88.3 Dismantling condition of Ponei village

Fig. 88.4 Rafts for abalone culture and fishers' working condition

88.4.2 Social Cost Emergy Calculation

Empirical research data came from field survey (March 21–22, 2014, Gulei Xingzai Village, Ponei Village, Mingda wharf, etc.), questionnaire and interviews with villagers and other means to understand the specific situation of the project. The calculated data was also taken from the 2013 Statistical Yearbook of Fujian Province [5], field survey and questionnaire of subject, planning documents of Gulei PX project, "emergy analysis of ecological economy system" [6] and related network literature. Data collection of construction project evaluation is a meticulous work and has a great influence on the quantitative calculation of social cost and the objectivity of social evaluation. It not only includes the above, but also includes the network media reports, and data of similar projects before project. The aforementioned thirty-three social impact assessment variables of seven categories consider the social impact brought about by the implementation of the project from different angles respectively. The twelve quantitative indicators of quantifiable part can be treated as follows:

1. Demographic change: the gap between emergy of resident moving out of and moving in the community, the calculation should consider the different cultural background or the level of labor. Moving in is positive, while moving out is negative.
2. Temporary workers' flow: the emergy difference between temporary workers coming out and coming in was classified and calculated by different levels of labor. Moving in is positive, while moving out is negative.
3. The emergence of seasonal residents: the difference of seasonal population change may be desirable for the change in annual average or monthly average before and after the project construction. The emergy transformation rate is preferable to average individual emergy value. If the most seasonal residents are tourists in community in the past, this can also be represented by the change of tourism output value.
4. Resettlement of individuals and families: when the population moving out is basically residents of the demolition, the emergy value of residents of the

demolition is equal to emigration population emergy value of social cost according to different educational background or different levels of labor to classify and calculate.

5. Industry diversification: the emergy value of new private enterprises.
6. Living/family income: calculating the emergy value corresponding with average monthly income of normal family living above the poverty line, calculating the change before and after the project.
7. Changes in employment position: the product of the number of occupations and the corresponding emergy value before and after project change can be taken as the total community emergy value, comparing its change. A representative value of the payment of the unit labor time can be taken as the position emergy.
8. The change of leisure opportunity: the emergy value corresponding with the change of the entertainment facilities.
9. Changes in community infrastructure: the increase or decrease of infrastructure in the community should be credited to the emergy value of optimization and construction of infrastructure in community brought about by project construction on the basis of the social cost of demolition of infrastructure.
10. The effects on suppressive culture, history, religion and archaeological sites: the loss of emergy value in the community of historical and cultural heritage is due to the lack of proper protection. The construction of the project will not bring the community historical and cultural heritage in general, so this item is in line with the social cost.
11. Plans for public participation: multiply the number of government propaganda, the average number of participants, the estimated value of the average amount of knowledge every time to get emergy value.
12. Ways of public reflection: multiply the number of public reflecting issues, the value of the issue, the number of citizens to participate to get emergy value.

88.4.3 Emergy Quantization on Social Cost

Research results showed that the sustainable employment period that residents were not disturbed was expected to about two decades, that long-term planning period to 2020, about total ninety-nine years including demolition, construction, operation, and ecological restoration, so we take ninety-nine years as calculation period. The calculation range was PX project planning range, and the number of calculation people involving the number of relocation people reached 9857. According to the research results of Gong [7], in 2008, the total amount of emergy value (U) in Fujian Province totalized 3.12E+23sej, including renewable resources with 3.48E +22sej, nonrenewable resources with 2.09 E+21sej, nonrenewable energy with 1.71E+23sej and monetary flow with 1.04E+23sej. Nonrenewable energy sources included thermal power with 4.34E+22sej and auxiliary energy with 1.28E+23sej (fuel energy with 1.0723sej, fertilizer with 1.78E+22sej, pesticides, plastic sheeting

Table 88.4 Labor force background-Emergy transformity [6]

Labor force background	Emergy transformity/(sej/person)
College student	6.00E+07
Undergraduate	7.33E+07
Postgraduate	3.43E+08
Expert	1.03E+09
Ordinary worker	1.24E+07

and agricultural machinery with 3.85E+21sej). GDP of Fujian in 2008 was $1.56E+11 and calculating the emergy-money ratio was 2.00E+12 (sej/$). Emergy-money ratio was: [current exchange rate was 6.234 (¥/$)]

$$2.00\text{E}+12\ (\text{sej}/\$)/6.234\ (¥/\$) = 3.208\text{E}+11\ (\text{sej}/¥)$$

Emergy transformity of people from deferent labor force background is as follows (Table 88.4).

The Quantization on Social Emergy Loss at Planning and Demolition Stage

Public participation and community quality in the planning phase of the project did not significantly change. System emergy change started from demolition phase.

1. Local residents moving out:

   (1) Individual education and cultural level: According to the results of the questionnaire, the demolition involved 9857 residents in 2306 families. Obtained by the survey, the cultural level of the residents moving out consisted of 6604 college students, 3253 ordinary workers. The emergy of residents moving out was:

   $$\text{E1}' = 6604 \times 6.00\text{E}+07 + 3253 \times 1.24\text{E}+07 = 4.366\text{E}+11(\text{sej})$$

   In the ninety-nine years of the whole life cycle, considering population growth, according to Fujian Statistical Yearbook 2012, the natural population growth rate was 0.701 %. The emergy of the residents moving out considering the population growth was:

   $$\text{E1} = 4.366\text{E}+11 \times (1+0.701\,\%)99 = 8.718\text{E}+11(\text{sej})$$

(2) Decrease of personal income: according to research results, the economic losses were estimated:
Current average personal monthly income: Y = 1078.130 (¥/person); monthly income expected (discounting in calculate period): Y = 1000.000 (¥/person)
The total emergy of economic losses:

$$\begin{aligned}E2 &= (1078.130\,¥/(\text{person} \cdot \text{month}) - 1000.000\,¥/(\text{person} \cdot \text{month})) \\ &\quad \times 12\ \text{month/year} \times 99\ \text{years} \times 9857\ \text{people} \times 3.208E + 11\text{sej}/¥ = 2.935E + 20(\text{sej})\end{aligned}$$

(3) Increase of daily life consumption:
According to survey results, the new life cost: 222.500 (¥/person · month)
The emergy of new life cost:

$$\begin{aligned}&222.5(¥/\text{person} \cdot \text{month}) \times 3.208E + 11(\text{sej}/¥) = 7.138E + 13(\text{sej}/\text{person} \cdot \text{month}) \\ &E3 = 7.138E + 13(\text{sej}/\text{person} \cdot \text{month}) \times 9857 \times 12 \times 99 = 8.359E + 20(\text{sej})\end{aligned}$$

(4) Increase of health cost:
New health cost: 23.000 (¥/person · month)
The emergy of new health cost: 23 (¥/ person · month) × 3.208E+11 (sej/¥) = 7.378E+12 (sej/person · month)

$$\begin{aligned}E4 &= 7.378E + 12(\text{sej}/\text{person} \cdot \text{month}) \times 9857 \times 12 \times 99 \\ &= 8.640E + 19(\text{sej})\end{aligned}$$

(5) Resettlement of Individuals and families:
Relocation compensation, land compensation: relocation and resettlement policy referred to appendix. Sea area, facilities and seedling compensation: the total compensation for the damage to biological resources during the construction phase of the project added up to 101,640 yuan. Sea area, facilities and seedling compensation totalized 475 million yuan.
According to the survey, after the completion of the relocation, purchase, decoration and other resettlement, relocation compensation did not have too much surplus, and therefore effect of the resettlement of individuals and families counted as (475 + 10.164) million, equivalent to:

$$E5 = (475 + 10.164)E + 4\,¥ \times 3.208E + 11(\text{sej}/¥) = 1.556E + 18(\text{sej})$$

2. Emergy loss of tourism:

(6) Tourism loss
Gulei community had certain tourism revenues in the past. After the impact of the plant, the tourism industry simply could not continue to develop, thus the project impact on tourism is reflected on the decreased tourism output. According to statistics, in 2008 the total tourism revenue of

Zhangpu reached 368 million yuan. Gulei tourism output value approximately accounted for one sixth of GDP in Zhangpu, emergy involved was calculated as follows:

$$(2.05 \times 2 \times (1 + 9\%)) \times (1 - 1.0999)/(1 - 1.09)/6 = 41971.478\ (\text{million})$$
$$E6 = 3.68 \times 108 \times 99\,¥ \times 3.208E + 11\text{sej}/¥ = 1.169E + 22(\text{sej})$$

3. Community Infrastructure: project construction more occupied wasteland and farmland, and the destruction of the social infrastructure was not significant. The communities before was backward, of no libraries, research institutions and other facilities, here negligible.
4. Social atmosphere: After investigation, although there were some problems during construction and reaction from residents as well, the possibility of eruption of Gulei community collective protests approached zero, thus the emergy of social atmosphere was zero. Community crime rate was low, and not considered as a key factor, and because of the lack of information, here negligible. Former communities did not have cultural heritage, therefore negligible.
5. Information intelligence culture: the original community did not have outstanding and special cultural skills, and people's income was mainly from fishing, aquaculture, thus the emergy loss of social technologies could be approximately valued as the local per capita income, from the above calculation:

$$\begin{aligned} E7 &= 1078.130\,¥/(\text{person} \cdot \text{month}) \times 12/\text{year} \times 99\ \text{years} \\ &\quad \times 9857\,\text{people} \times 3.208E + 11\text{sej}/¥ = 4.050E + 21(\text{sej}) \end{aligned}$$

The original community did not have cultural heritage, so the emery loss of cultural heritage could be recorded as zero.

The Quantization on Social Emergy Loss at Construction Stage

(7) Wages of construction workers: the total investment of Gulei PX project was 13.784 billion yuan. According to the budget targets, for industrial buildings, labor cost accounts for 13–16 % of the total investment, adopting 14.5 %:

$$\begin{aligned} E8 &= 1.38 \times 1010\,¥ \times 14.5\% \times 3.208E + 11\text{sej}/¥ \\ &= 6.419 \times 1020(\text{sej}) \end{aligned}$$

(8) Industry diversification: the emergy value of new private corporations. According to the survey, local residents reflected that due to the construction of the project, the past private restaurants and shops partly closed and that due to the entry of foreign workers and their families, some new private shops

opened. The overall change was not big, thus industry diversification did notperforme significantly.

The Quantization on Social Emergy Loss at Operation Stage

6. Staff moving in:

   (9) Demographic change: the condition of technical staff immigrating was as follows: 600 new jobs, recruiting a number of highly educated technical workers, the level of staff education counted in the following Table 88.5.

$$\text{Emergy of immigrant popuation} = 99 \times 6.00\text{E}+07 + 393 \times 7.33\text{E}+07 + 53 \times 3.43\text{E}+08 + 148 \times 1.24\text{E}+07 = 5.476\text{E}+10(\text{sej})$$

   Checking Fujian Statistical Yearbook 2012, the natural population growth rate was 0.701 %, and the operation period was 20 years, therefore changes occurred in the 20 years was:

$$\text{E9} = (5.476\text{E}+10) \times (1 + 0.701\%)20 = 6.287\text{E}+10(\text{sej})$$

   (10) Wages of production workers: According to the 2014 national average wage slips, 3060 yuan per capita in Zhangzhou

$$\text{E10} = 3060\,¥/\text{person} \times 693\,\text{people} \times 3.208\text{E}+11\text{sej}/¥ \times 20 = 1.361\text{E}+19(\text{sej})$$

   (11) Workers' health cost: the change of working cost in economic zone in two decades was:

$$\text{E11} = 7.378\text{E}+12 \times 12 \times 20 \times 9857\text{sej} = 1.745\text{E}+19(\text{sej})$$

   (12) Temporary workers' flow: Gulei community before was relatively isolated and of few temporary workers. After construction, skilled workers flew into the community. There was no obvious temporary workers' inflow. This item was not included.

Table 88.5 Background of labor force moving in number

Labor force background	Number/person
College student	99
Undergraduate	393
Postgraduate	53
Expert	0
Ordinary worker	148

88.5 The Analysis of Calculation Result

After calculation, considering the whole life cycle, the social cost of Gulei PX program was 1.72E+22(sej), equivalent to 53.616 billion yuan. Social cost was relatively high. The components of each group are shown in Fig. 88.5 (Table 88.6).

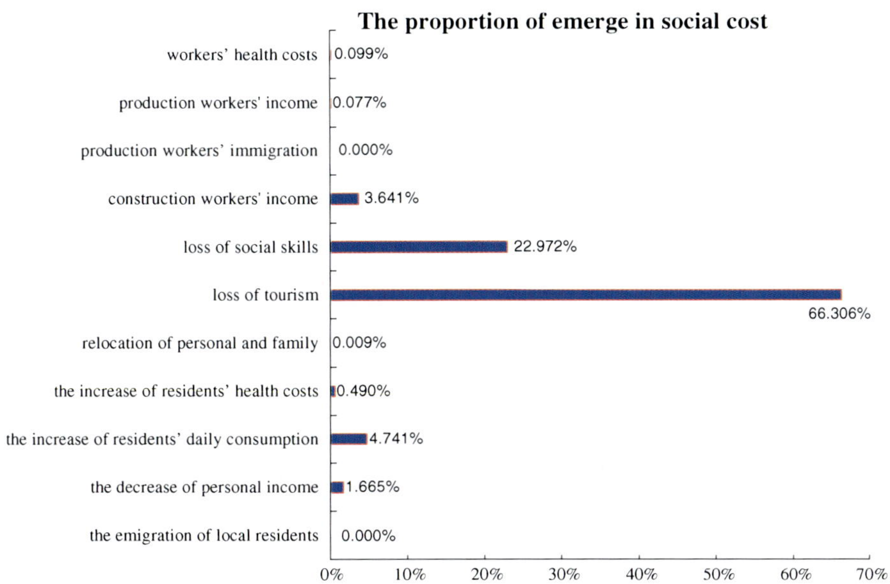

Fig. 88.5 The proportion of social costs

Table 88.6 The proportion of emerge in social cost

Period	Project	Solar emerge(sej)
Demolition stage	The emigration of local residents	8.718E+11
	The decrease of personal income	2.935E+20
	The increase of residents' daily consumption	8.359E+20
	The increase of residents' health costs	8.640E+19
	Relocation of personal and family	1.556E+18
	Loss of tourism	1.169E+22
	Loss of social skills	4.050E+21
Construction stage	Construction workers' income	6.419E+20
Operation stage	Production workers' immigration	6.287E+10
	Production workers' income	1.361E+19
	Workers' health costs	1.745E+19
Total		1.72E+22

88.6 Conclusion

In this paper, a social assessment system was established for the assessment of large-scale projects in social perspective. On the basis of field survey of the Zhangzhou Gulei PX program, the social cost and social evaluation of PX project based on the emergy method were discussed preliminarily. The Gulei Peninsula where the PX project is located used to be dependent on abalone farming aquaculture, fishing industry. Technology passed by from one generation to another was faced with the loss problem after the introduction of the project. The problem of the employment of indigenous people became a prominent social problem. In the process of project construction in Gulei town, the main emergy loss was the loss of the tourism industry, which accounted for 66.31 % of the total emergy loss. The second was the social skill loss, accounting for 22.97 % of the total energy loss. Daily consumption of residents increased, accounting for 4.71 % of the total energy loss. After the introduction of PX project, the living consumption of local residents increased and health costs also increased. And it caused the great social skill loss and the traveling industry loss. All those directly reflected the negative social effects of the project. Because the PX project data was not open, the results obtained in this paper were only a general verification method that could be used in the mutual comparison of similar projects. If the exact value is used for engineering evaluation and decision making, detailed information will be needed.

Acknowledgements The work is supported by by National Natural Science Foundation of China (Grant No.71271180 and G71272078) and Undergraduate student basic funding of Xiamen University) (CXB2014113).

References

1. Ruan P, Wu Y et al (2005) The emergy analysis and thinking of regional labor transfer. J China Popul Sci, supplement:44–47
2. Ding D (2005) Luhmann and his "social system theory". J World Philos 5:34–38
3. Burkee LJ (2011) The concept process and methods of social impact assessment. China Environmental Science Press, Beijing
4. Statistical Yearbook of Fujian Province, 2012
5. Lan S, Qin P, Lu H (2002) Emergy analysis on ecological-economic systems. Chemical Industry Press, Beijing
6. Zhang Y (2008) Emergy value and value evaluation on Fujian province forest resources system based on emergy theory. Fujian Normal University, Natural Geography Major, Fuzhou
7. Gong R (2011) Study on the accounting of green GDP in Fujian province from the perspective of emergy analysis. University Of Agriculture and Forestry In Fujian

Chapter 89
An Investigation of the Latent Barriers to BIM Adoption and Development

Ke Chen, Weisheng Lu, Yi Peng, Linzi Zheng, Yuhan Niu and Steve Rowlinson

Abstract The building information modeling (BIM) has been under aggressive promotion by governments throughout the world. By contrast, the Architecture, Engineering, and Construction (AEC) industry shows reluctance to adopt BIM in its projects. Two questions are immediate: why the market hesitates with BIM adoption; and whether the governmental compulsory BIM implementation is regarded as reasonable. Previous literature is in-depth but is distant from the market, which have failed in analyzing the basic reasons lying behind. This study aims to provide an alternative explanation. Based on the assumption that any market activity is to maximize economic profits, this study concluded that (1) since BIM adoption involves opportunity costs, stakeholders would choose to adopt BIM at the market-determined time; (2) due to the social costs/benefits of using BIM, the market-determined timing of BIM adoption tends to be delayed, the governments' enforcing BIM hence is reasonable; and (3) the compulsory requirements on clarifications of the responsibilities and the corresponding benefits among AEC disciplines reinforces the importance of governments in promoting BIM implementation.

K. Chen · W. Lu · Y. Peng · L. Zheng (✉) · Y. Niu · S. Rowlinson
Department of Real Estate and Construction, The University of Hong Kong, Pok Fu Lam, Hong Kong
e-mail: xiaopeng1900@gmail.com

K. Chen
e-mail: leochen@connect.hku.hk

W. Lu
e-mail: wilsonlu@hku.hk

Y. Peng
e-mail: pengyihz@gmail.com

Y. Niu
e-mail: yuhanniu@connect.hku.hk

S. Rowlinson
e-mail: hrecsmr@hku.hk

Y. Wu et al. (eds.), *Proceedings of the 20th International Symposium on Advancement of Construction Management and Real Estate*,
DOI 10.1007/978-981-10-0855-9_89

Keywords Building information modeling (bim) · Government intention · Profit-maximization · Latent barriers

89.1 Introduction

The productivity of the construction industry has lagged that of many other industries [15], which equally being threatens and chances. The construction industry is facing the paradigm to innovate [20], and modern information technologies make this opportunity available. Particularly, building information modeling (BIM) is with evident benefits facilitating the project management in terms of better coordination and information management for decision making [6]. Nevertheless, BIM has been introduced to Architecture, Engineering, and Construction (AEC) participants for more than two decades, but its adoption rate is still lethargic. The current construction industry remains disciplines-fragmented and interoperability-inadequate. These obstinate defects, together with the increasing attention of both academics and industry professionals on how to successfully improve BIM, suggest the market's reluctance to adopt BIM, and highlight the inquiry into reasons lie behind.

Nowadays, there are evidences stating that the AEC industry is under governmental pressure to adopt BIM. Governments have been the aggressive promotors of BIM in many countries and regions, including Mainland China, Hong Kong SAR, Singapore, the UK, the USA, Netherlands and many other areas [12]. Even more ambitiously, BIM will soon become the mandatory requirement in public construction projects [21]. Under such pressure, undesired but widely observed "half" and "fake" BIMs are emerging in terms of clumsy 3D digital building images with insufficient or without information at all, simply aiming to cope with the mandatory regulation made by the government.

On observing the unwanted consequences of governmental regulation of using BIM, an immediate impression may lead to an argument that the political pressure is deemed to be wasteful in terms of twisting the resource allocation in the society. However, significant social benefits generated from using BIM seem to rationalize the government's propelling BIM. Two research questions hence are developed: (1) why the construction industry resists adopting BIM; (2) whether the governmental regulation of using BIM can be regarded as reasonable.

This study provides an alternative explanation from the economic view. Based on the assumption that any market activity is to maximize economic profits. This study concluded that (1) since BIM adoption involves opportunity costs such as capital cost, labor training cost, and possible project delay, stakeholders would choose the most profitable cash flow with BIM being introduced at the market-determined time, after combining the labor and capital inputs and the trade-off between earning more first and investment more first within a project life cycle; (2) due to the social benefits of using BIM, including knowledge spread and environmental improvements, the market-determined BIM timing tend to be delayed compared with when the benefits of BIM could have been thoroughly digested by individual stakeholders, the governments' enforcing BIM hence is reasonable for the market to adopt BIM at the social-desired time; and (3) the compulsory requirements on clarification of the responsibilities and the corresponding benefits among AEC disciplines reinforce the governmental action in terms of contract arrangements to enable BIM's effective usage. Both economic modeling and qualitative analyses are conducted to fulfill the research purpose, and the Cobb-Douglas production function is employed as an illustrative device.

Limited by page, the tasks accomplished in the present paper have not provided quantitative evidence, which is suggested to be a potential future extension. In the remaining contents, Sect. 89.2 provides the literature review. Section 89.3 presents the economic modeling of the market-determined timing for BIM adoption, and discussions of the affecting factors. Section 89.4 develops the reason why the governments' enforcement of BIM could be reasonable. Section 89.5 is the conclusion of this paper.

89.2 Literature Review

In its most basic definition, BIM is a virtual representation of both physical and functional characteristics of a facility, forming a reliable resource of information about a facility throughout its life cycle [18]. More than two decades have passed ever since the introduction of BIM technology to the AEC industry. However, the adoption rate of BIM has been very low prior to the governments' aggressive promotion [10], the BIM performance and its effects on improving the construction management efficiency turned out to be unsatisfactory or even lethargic [1].

Extensive literatures have proposed possible reasons to explain the market's reluctance to adopt BIM or achieve full capacity of BIM, among those are: there was insufficient "initiative" for the market to use BIM, or even the "initiative" was available, there was no sufficient training [4]; the AEC industry was by nature fragmented, which was contrary to the spirits of BIM [13]; the preparations for BIM adoption across different disciplines of the construction industry were inconsistent [13]; and the BIM-related technologies were still immature [14]. Compared with these studies, Holzer provided a relatively impressing idea to address the market's resistance towards BIM, arguing that the lack of clarity in the allocation of roles, responsibilities and benefits was the very reason [11].

Although these explanations are enlightening from their own perspective, research on the BIM adoption is frequently done at a distance from the market situation, and thus becomes more or less superficial. The distance from the real market hinders the possibility to analyze fundamental reasons such as innovation purpose for maximizing commercial profits or its effects on a firm's competencies. The research gap exists in that, the opportunity costs involved in the processes of the BIM-based innovation are largely ignored.

89.3 Modeling the Timing of BIM Adoption by the Market

The more productive an industry is, the more the goods can be produced in a sustainable fashion, i.e. a higher productivity leads to a higher economic growth rate. With stable constraints of technologies and regulations, the law of diminishing marginal productivity holds within the production function of any industry. The following deduction of the timing of BIM adoption is structured by the Cobb-Douglas [7], based on maximizing the net profits of BIM subjected to capital input.

The economic benefits of using BIM mainly involve reducing re-work and saving time [16, 17]. Although both the re-work reduction and time saving can be regarded as the reduction in unit cost of building output in terms of monetary value, it is generally recognized that the immediate innovators are not able to realize positive profits due to the outlay to implement BIM. Such outlay within a project life cycle is substantial, which includes: (1) labor training cost; (2) opportunity cost of adapting existing workflows to BIM-aided process; (3) equipment cost on buying and administrating new software and hardware; (4) negotiation fee to procure the required collaboration, integration and interoperability; and (5) the concern of hindering the future free contracting among individual AEC stakeholder, which would have been caused by (4). The substantial initiation inputs of BIM adoption, if could be invested elsewhere, is very possible to generate higher profit than implementing BIM within a certain period. Consequently, to potential BIM users, the profit motive is not strong enough for them to adopt BIM at an earlier time.

Prior to further modeling the timing that private stakeholders adopt BIM, several assumptions should be firstly set and the definitions of variables in the model are meanwhile given. Based on the paradigm, effects of changes in some of the assumptions will be examined to provide detailed discussion on the rationale of governments' actions to promote BIM.

(a) The economic benefits of BIM, if adopted by the entire AEC industry without any delay, are equal to reducing unit cost of building production by k dollars.
(b) Assumed there are no social benefits of using BIM, and no benefits will pass to the consumers or others. Consequently, the unit cost saving—k dollars—equals the profits obtained by an individual BIM adopter, i.e., the BIM adopter is the only beneficiary.
(c) Following the previous assumption, the reduction of unit cost will not affect the final price of the building market.
(d) The benefits of BIM take effect on its implementation.
(e) The total outlay to implement BIM equals to I dollars, including the accrued interest during the wear-in period until everything is settled for well use of BIM.
(f) The market demand for the construction products is X_0 at time t_0, which is increasing at a constant rate p as time passing by, and hence $X_t = X_0e^{pt}$ exists.

(g) The market discount rate r is instantaneous and constant for any type of resource.
(h) Controlling the labor input and productivity unchanged.

Provided that it is dependent on the individual's own choice to adopt BIM at the certain time in order to maximize profits, stakeholders should consider that the substantial inputs for BIM may be more profitable for alternative usage. Controlling the labor factor unchanged and suppose an individual earning stream introducing BIM at time t, the present value of net profits of using BIM at time t_0 is the discounted stream of BIM benefits over the project lifecycle minus the discounted value of total input for implementing BIM, as expressed by Eq. (89.1):

$$\pi_0 = \left(\int_t^{\infty} kX_0 e^{-(r-p)^{\tau} d\tau} \right) - Ie^{-rt} \tag{89.1}$$

Notice that π_0 is impossible to grow infinitely, thus, $p < r$ holds, and Eq. (89.1) can be rewritten as

$$\pi_0 = \frac{kX_0 e^{-(r-p)t}}{r-p} - Ie^{-rt} \tag{89.2}$$

To maximize π_0, there is $\frac{d\pi_0}{dt} = 0$. Solving the differentiation equation, the optimal root of timing to generate the maximum present value of net BIM profits can be derived and expressed by Eqs. (89.3) and (89.4) respectively:

$$t_m = \frac{lnr + lnI - \ln(kX_0)}{p} \tag{89.3}$$

$$\pi_0(m) = I\frac{p}{r-p} \tag{89.4}$$

As shown in Eq. (89.3), the optimal timing for maximizing individuals' profits to adopt BIM are dependent on the market-determined discount rate, the growth rate of market demand, and the unit cost saved by employing BIM. The upper limit of the net profits brought by BIM is also determined by the market conditions, as well as the capital cost involving in activating BIM. Postulated by individual rationality, the lower limit of the present value of net BIM profits is zero. Replace π_0 by zero in Eq. (89.2), the solution of time is:

$$t_{zero} = \frac{\ln(r-p) + lnI - \ln(kX_0)}{p} \tag{89.5}$$

As shown in Fig. 89.1, any early or late adoption of BIM other than t_m would bring up undesired resources allocation, and any BIM adoption earlier than t_{zero} would have made the present value of the net profits of using BIM as negative

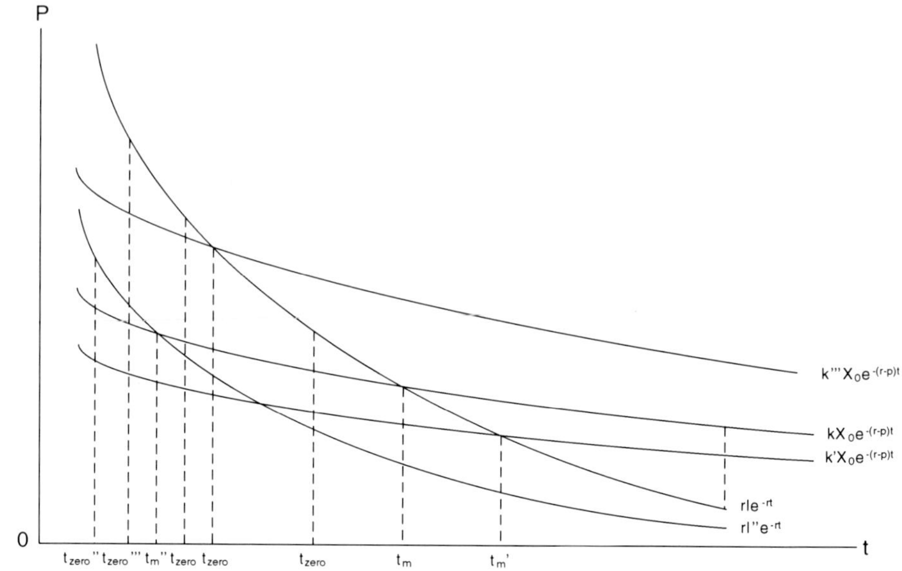

Fig. 89.1 The timing of BIM adoption

investment, because the capital invested in BIM would generated higher present value if could have been invested in alternative usages.

89.4 The Rationale of the Governmental Promotion of BIM

Given affecting factors known, the optimal time t_m to adopt BIM can be calculated, which are defined by the BIM's capability to reduce the unit cost of building production, the labor training cost, and the market conditions, including the prevailing discount rate and the growth rate of market demand. If releasing the assumption (b) and (c), it suggests that the profits derived from cost saving are not necessarily appropriated to the sole private BIM adopter because BIM has significant social benefits [3, 8]. Therefore, the total benefits of BIM are greater than what the individual stakeholder has received. Provided that this difference between the social benefits and private benefits can be internalized by the interactions among different stakeholders such as contractual arrangements, there would have no disincentive for them to use BIM. However, if the internalization of a certain parts of total social benefits remains too costly for market forces to cope with, a very likely consequence is that, the timing of BIM adoption would be delayed to the point which is much later than t_m. Governmental actions in terms of direct or indirect monetary subsidy and contractual regulation respectively, are obviously necessary [5].

H1: The government subsidy in various terms to promote the market to adopt BIM is expected to positively relate with the social benefits derived from the use of BIM.

89.4.1 Information of Previous BIM Successes

The information of successful experiences and previous practices will decrease I (total outlay to implement BIM) of the following BIM adopters. As shown in Fig. 89.1, a decrease in I will drive the market to adopt BIM much earlier than before. Therefore, if that information is available, the market would be likely to adopt BIM comprehensively. However, due to the competitive pressure, it suggests that the spread period for BIM to generate cost-saving profits may not sustain as long as originally expected, which causes one of the major disincentives for individual stakeholder to take the lead and adopt BIM voluntarily.

Wherever successful experiences of BIM implementation are observed, various aspects of these observable successes, such as strategy, innovate timing, BIM software and etc., will be copied by other potential BIM adopters in their pursuit of commercial profits. This "rough-ready-imitation" process of BIM adoption is particularly crucial in the present stage for the construction industry. First, identifiable criteria for BIM-aided decision-making are still in absence. Perhaps the autonomous intelligent management will be realized in future, but to date, numerous decision makings during the project life cycle still rely on the managers' experience. Second, the "trial-and-error" processes are not yet available. Determined by its own characters, any improvement on BIM technology that is in practical requirement is dependent on the experimental knowledge of "trial-and-error", which must be derived from the actual applications of BIM. Third, factors affecting the well use of BIM are multiple and complex. A BIM can only be said as in effect when the interoperation and cooperation among different stakeholders have been arrived. Thus, in addition to internal factors affecting the use of BIM by a single stakeholder, the external complexity has to be attended. Experimental knowledge on how to deal with the external relation for achieving the efficient cooperation, together with how to deal with internal adaptive process, hence is valuable for following BIM adopters.

Above all things, a given piece of information of pervious BIM success is by nature a commodity. It might be expected that, if this information commodity can be traded in, and if the payments received from selling successful experiences by the leading adopters could be equal with or even larger than the losses in cost-saving benefits due to change in market price, the aspect of externality can hence in internalized, and the government subsidy can be waved. However, the information commodity is with peculiar attributes of in-appropriability and uncertainty. The typical difficulties of marginal charging appear in the potential trading process of the information commodity—the value of information, if on sale, would be sharply decreased. Consequently, it is extremely costly for one to search the market in order to find the individual who bidding the information highest. Witnessed by the bulk of research has been carried on outside the industrial system, such as universities or institutions, the innovative knowledge could have been too costly to be traded on the open market by individual transactions in the absence of special legal protection [2]. For the pervious BIM success as a typical innovative knowledge, it is difficult for pioneers to charge the following BIM users, they would be taken advantages of without any payments.

In sum, the in-appropriability of producing information of BIM success hinders the market's reluctant to adopt BIM substantially. In this sense, the government subsidies, for example, in indirect terms by commercializing the BIM success information, or in a direct manner to incent the leading adopters, is rational to drive the market back to the desired timing to introduce BIM.

89.4.2 Social Benefits of BIM

The abilities of BIM of reducing rework and improving efficiency of construction project are obviously able to reduce the waste and pollution to the environment as well as in the global ecosystem, therefore, the BIM adoption brings benefits in the environmental sense compared with the traditional way of construction. From this perspective, an early timing of BIM adoption, even earlier than t_{zero}, is very likely to be socially desired, when extending the outcomes of efficient resource allocation from commercial benefits to the wellbeing of the society as a whole. Constrained by commercial purpose, any timing of BIM adoption earlier than t_{zero} will make the present value of resource allocation negative. However, taking environmental improvements into account, situations perhaps would be very different.

For analytical consistency, suppose the implicit value of environmental improvements is able to be digested into the market price. It is equal with an outwards move of the diagram of $kX_0e^{-(r-P)t}$. An immediate implication shows clearly that, an early adoption timing of BIM, perhaps even earlier than t_{zero}, is very likely the actual optimal time from the perspective of the social wellbeing as a whole. However, it is too costly for individual stakeholder to charge side payments from the public who would have been benefited from improvements in living environments compared with before. The benefits of BIM to the environment are much more costly to be internalized within the private market compared with the discussion in the above section. Reasons lie behind are obvious: (1) the beneficiaries are difficult to count, (2) the value of charging is difficult to calculate, (3) the agreements on charging are almost impossible to arrive, and (4) the enforcement cost of charging hence is infinitely high. In other words, the construction industry, as pursuing commercial purposes, is not obliged to provide and maintain the built environments and therefore needs to be paid its part in achieving the goals. Again, the government action on subsidizing the adoption of BIM by the AEC industry is hypothesized to be rational.

H2: The government regulation in terms of contract innovation is expected to be critical to promote the market to employ BIM in its thorough effects.

The property rights of BIM, which enables the owner to derivate profits from using, developing, and freely transferring it, is one of the most critical problems for the market to adopt BIM. As having been widely agreed, in order to thoroughly use BIM, the transparency of open data access and the transfer of information are fundamentally required. However, with respect to the AEC industry, they are completely new to all participants.

A BIM involves in the lifecycle of a building project, which contains multiple constitutes, over which the property rights need to be clearly delineated, including the design, the technological aspects (hardware and software) and the information collected. Further, the property rights package over each BIM constitution has multiple attributes, including the right to use, the right to make revisions, the right to derive benefits, and the right to transfer. To date, there is a lack of clear understanding of the responsibilities of different stakeholders during the BIM adoption and implementation process [9].

Practical beings that people are, most of the time compromises have to be made. In the traditional way of construction management, each party is responsible for specified parts, and able to keep one's information under control with relatively small efforts. Accompanied by the ease to control, it is difficult to access the information of other stakeholders. Negotiation always is involved, and each time of negotiation indicates an implicit value of side-benefits.

With the presence of BIM, it is apparently a trade-off between sharing the information and controlling the information. A BIM is a digital model with all project-relevant information. Even there are multiple files designed for different stakeholders, the advanced technologies make it easy to connect one file with the other. Actually, this connection is compulsorily needed according to the original intention of BIM. Therefore, it has a major influence on the whole property rights debate that, with the valuable information contained in BIM which is much easier for another stakeholder to access than before, the implicit value of information trading is hence much more difficult to charge. It is therefore obvious that the property rights disputes are more difficult to internalize.

The widely existing public ownership of BIM does nothing good to solve the defined problems. For example, the NYC Department of Design and Construction [19] takes the property rights of BIM saying that:

> …DDC holds ownership of the BIMs including all inventions, ideas, designs, and methods contained within the model. This includes, but is not limited to; the content submitted as part of the BIMs itself…

This definition of BIM property rights by nature equals with that nothing has been addressed on how to pay different stakeholders with correspondence to their responsibilities. Provided that a stakeholder is not sure about how many benefits they would derive from implementing and developing a BIM, it is very unlikely for the stakeholder to improve BIM more than the minimum required by the government. A paradox here is that, while BIM aims at addressing the interoperated and collaborated project management, unspecified benefits and responsibilities among various construction disciplines tend to tear the BIM-aided management for building life cycle apart. In this situation, the government innovative regulation is hence needed to firstly initiate such underwritten contracts to promote the BIM adoption, as well as to ensure that projects could deploy auditable processes on reliable interoperable information operational project life cycle, by specifying the responsibilities and the corresponding benefits of different disciplines.

89.5 Conclusion Remarks

Innovation adoption must be a risky process due to the perpetual uncertainty, in which the outputs cannot be predicted perfectly from the inputs. Individual stakeholder in the construction industry is often profit-maximizing. Without considering the opportunity costs involved and adopting BIM abruptly, the adopters may not able to make a commercial fortune. This is, to a certain degree, counter to market pressure where short-term goals and returns are seen as essential.

With the presence of uncertain market and the social benefits of BIM, the social cost and the private cost separate. The in-appropriability and uncertainty prevent the individual stakeholder from performing adequately, delaying the commercially beneficiary timing for BIM adoption, and failing to fulfill the social desires. This problem arises when the government finds it necessary to engage in the market activities. In this study, the government actions in terms of direct or indirect subsidies are hence suggested. Besides, one cannot go any further without firstly clarifying the individual's responsibility and specifying the profits with corresponding to one's inputs when investigating problems related with corporative works. Therefore, governmental innovative regulations in terms of contract inventions are also suggested as a possible choice to promote BIM in the modern construction industry. Limited by page, how the suggested government actions can come into effect has not articulated with details. A game model of budget constrained incentive mechanism could be one choice to effectively determine the value and objects of government subsidies. Finally, it should be aware that, any government intervention should always be careful to be addressed. Constraints with respect to control the scope of political intervention must be clearly predefined.

References

1. Arayici Y, Coates P, Koskela L, Kagioglou M, Usher C, O'Reilly K (2011) BIM adoption and implementation for architectural practices. Struct Survey 29(1):7–25

2. Arrow K (1962) Economic welfare and the allocation of resources for invention. In: The rate and direction of inventive activity: economic and social factors. Princeton University Press, Princeton, pp 609–626
3. Azhar S (2011) Building information modeling (BIM): trends, benefits, risks, and challenges for the AEC industry. Leadersh Manag Eng 11(3):241–252
4. Bernstein PG, Pittman JH (2004) Barriers to the adoption of building information modeling in the building industry. Autodesk Building Solutions
5. Buchanan JM, Stubblebine WC (1962) Externality. Economica 29(116):371–384
6. Chen K, Lu W, Peng Y, Rowlinson S, Huang GQ (2015) Bridging BIM and building: from a literature review to an integrated conceptual framework. Int J Project Manage 33(6):1405–1416
7. Cobb CW, Douglas PH (1928) A theory of production. Am Econ Rev 18(1):139–165
8. Deutsch R (2011) BIM and integrated design: strategies for architectural practice. Wiley, Hoboken
9. Eastman C, Teicholz P, Sacks R, Liston K (2008) BIM handbook: a guide to building information modelling. Wiley, Mississauga
10. Gu N, Singh V, Taylor C, London K, Brankovic L (2010) BIM adoption: expectations across disciplines. Handbook of research on building information modeling and construction informatics: concepts and technologies, 501–520
11. Holzer D (2007) Are you talking to me? Why BIM alone is not the answer [online]. Available from: https://epress.lib.uts.edu.au/conferences/index.php/AASA/2007/paper/view/48
12. Hong Kong Construction Industry Council (HKCIC) (2013). Final draft report of the roadmap for BIM strategic implementation in Hong Kong's construction industry [online]. Available from: http://www.hkibim.org/?p=2771
13. Johnson RE, Laepple ES (2003) Digital innovation and organizational change in design practice. CRS Center, Texas A and M University
14. Khemlani L (2004) The IFC building model: a look under the hood. AECbytes feature, 1–10
15. Koskela L (1992) Application of the new production philosophy to construction (No. 72). Stanford University, Stanford, CA
16. Lu WS, Fung A, Peng Y, Liang C, Rowlinson S (2014) Cost-benefit analysis of building information modeling (BIM) implementation in construction projects by demystifying the time-effort distribution curves. Build Environ 82:317–327
17. Lu WS, Peng Y, Shen QP, Li H (2013) A generic model for measuring benefits of BIM as a learning tool in construction tasks. ASCE J Constr Eng Manage 139(2):195–203
18. National Institute of Building Sciences (NIBS) (2007) National building information modeling standard Part-1: overview. Principles and methodologies. US National Institute of Building Sciences Facilities Information Council. BIM Committee
19. NYC Department of Design and Construction (DDC) (2013). BIM Guidelines [online]. Available from: http://www.nyc.gov/html/ddc/downloads/pdf/DDC_BIM_Guidelines.pdf
20. Nour M (2007) Manipulating IFC sub-models in collaborative teamwork environments. In: Proceedings of the 24th CIB W-78 conference on information technology in construction
21. Zeiss G (2013) Widespread adoption of BIM by national governments [online]. Available: http://geospatial.blogs.com/geospatial/2013/07/widespread-adoption-of-bim-by-national-governments.html

Chapter 90
Study and Realization of Core Calculation and Interactive Mode of Construction Project's Cost and Schedule Dynamic Control System

Haoshuai Qiao, Hong Zhou, Li Su and Yi Sun

Abstract BIM (Building Information Modeling) technology has brought a global construction industry revolution. But in our country, we still fall behind with foreign countries in BIM software innovation, lacking of independent software platform and the core technology. In order to realize building software independent innovation, we must explore the core technology of BIM software. This study tries to find out the possible way for BIM software to calculate and interact, and tries to use GUI to compile BIM application framework, employing Matlab as a powerful tool for computing. The central idea of this program is to demonstrate the relation between the images and real-time data with controlled primary data and timeline. These images and data are extracted from a certain building, showing the stage and rate of the construction. The program mainly consists of the control part, the cost function and the graphic function. The control part is a slider representing timeline, allowing users to control time as a variable and view the full construction cycle when dragged from one end to the other. The cost function generates different kinds of value, which are needed to be observed and compared in the process of construction management, from time variable controlled by the slider, with expected cost, critical point of material usage and consumption rate as type-ahead parameters. The graphic function generates real-time construction image from time variable, with parameters such as expected critical point of construction schedule and the construction sequence of building structure. These graphics visually express building information changes. The program has constructed a core operation of BIM software which has broad prospects for development. By adding functions such as automatic material price search, process flow diagram drawing, flow construction plan generating, this software can be applied in all fields of construction. Using more advanced programming language, it can be perfected to a high level, practical and durable BIM software with native core technology.

Keywords Cost schedule · Dynamic control · Core calculation · Interactive mode

---

H. Qiao · H. Zhou (✉) · L. Su · Y. Sun
School of Architecture and Civil Engineering, Xiamen University, Xiamen, China
e-mail: mcwangzh@xmu.edu.cn

Y. Wu et al. (eds.), *Proceedings of the 20th International Symposium on Advancement of Construction Management and Real Estate*,
DOI 10.1007/978-981-10-0855-9_90

90.1 Introduction

90.1.1 Background and Meanings of This Research

BIM is the abbreviation of Building Information Model which integrates all the information needed in construction process, including geometry information, geographic information, the information of the properties and quantities of each architectural component and so on. Through digital simulation, we can simulate the whole life cycle of construction projects, from virtual construction, collaborative construction, cost calculation to management operation and maintenance.

As a building information model, BIM can compute and express the information more efficiently. In the traditional design method, engineers use 2d drawing papers to express 3d architectural designs which they imagined in their mind. Construction workers build 3d architectural entity according to 2d drawing papers. The most striking feature of this process is invisible, and rework along with hidden danger can be caused by design errors. In terms of information calculation, the advantages of parametric modeling is that BIM can collaboratively organize invisible information with accuracy.

If the cost dimension is added into 4D BIM model, it will form 5d BIM model. 5d model integrates the factors such as quantities, schedule, cost, construction organization. Through the integration of the information, 5d models can be used for simulating the construction process and calculating the costs of project at any time, so as to realize the dynamic management for the cost of construction project [1]. It will help customers to manage project efficiently and improve the quality of project.

In the actual construction process, a lack of reliable data often causes budget overruns. Cost accounting is an indispensable part of construction. When we add the cost dimension into 4D models to form 5D models, we meet a lot of problems which are caused by a few characteristics of cost accounting, such as the large amount of data and related sectors, complicated consumption and money payment. It is these features that make the development of the 5d model meet certain obstacles. But with the improvement of software, the development from 4d models to 5d models gradually matures, some practical 5d software which are applied for cost accounting come into use, and the accuracy of cost accounting is also improved.

Most of cost accounting software can only process files in specific formats, and cannot deal with 2D drawings of non-native vector accurately. They are not good combinations of BIM and cost budget. In addition, different subfields of building often use different software to establish models. For example, Revit software is used in the construction field, Tekla software is used in the field of steel structure design, and Civil software is often used in the area of bridge design and so on. Due to different design software used in different subfields, there is trouble dealing with statistic data in a whole pattern. So to design a budget software supporting all the file formats of different design software is very valuable. This can greatly promote the development of the 5d models, forming a complete accounting system and improving the accuracy of cost accounting [2].

90.1.2 Content, Method and Technology Roadmap

This article is aimed at simulating and realizing the core mode of operation of 5d BIM software which is referred as 3D model construction with cost and process management. After the data analysis and digital simulation, we compile the program which can realize the effect of 5D BIM software on Matlab GUI interface. To achieve the effect, we joined the time dimension through the function of the time axis and the cost dimension through the cost control function on the basis of BIM 3d model. Through the program, we can compare the construction simulation and real progress to realize the management of cost, controlling the schedule and cost of projects (Fig. 90.1).

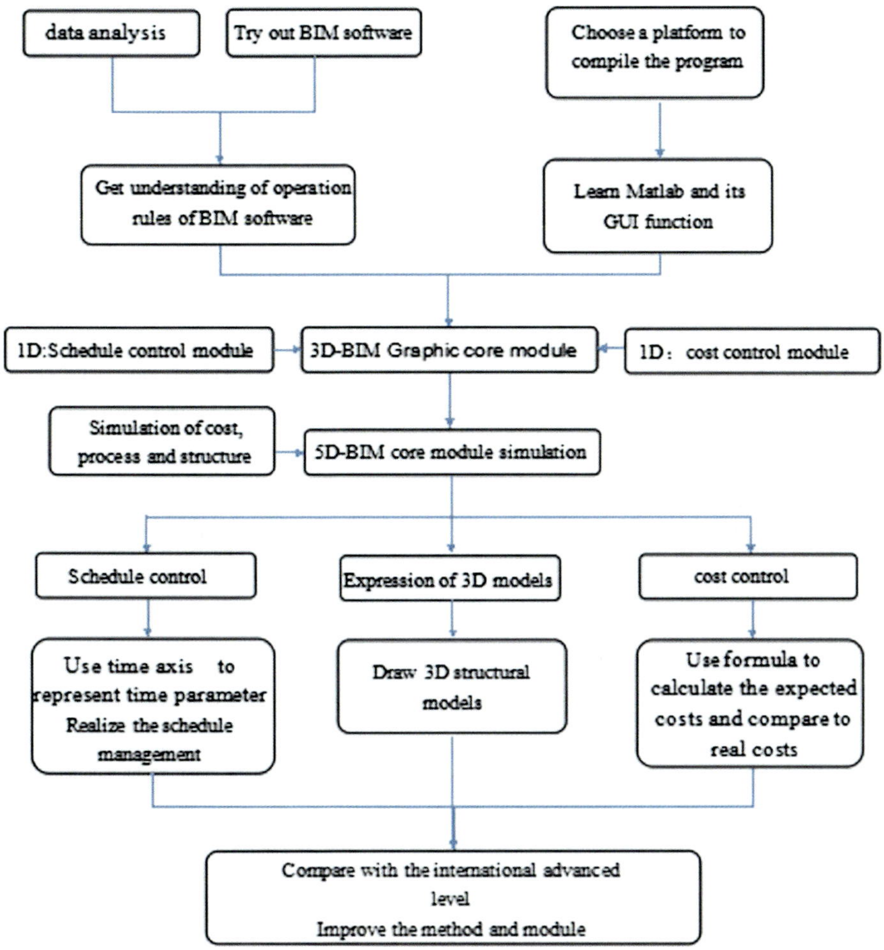

Fig. 90.1 Technology roadmap

90.2 Research and Development of Schedule Management

90.2.1 Expected Functions of Schedule Management

The four-dimensional construction model based on BIM technology is accomplished by introducing time axis as the forth dimension to the existing three-dimensional designing model. Usually it is expressed with the aid of computer technology by the four-dimensional digital model, but not simply equivalent to generating simulated pictures of construction process using computer virtual reality technology. 4D construction technology, by contrast, requires a more profound integration of time and space elements. It displays the space elements on the dimension of time, so as to formulate a reasonable construction plan and determine the construction sequence, duration and cooperative relationship of the construction process, significantly improve the efficiency and accuracy of project construction, and even play a key role for project quality.

90.2.2 Schedule Management Implementation Method

Structural Schedule Management Implementation

Import Images Generated by Third-Party Programs

As BIM software, image is the intuitive way to convey information. Due to technical limitations, a five-storey building construction process is used as an example. Here we use the "cell" function to provide read and storage space in the memory for

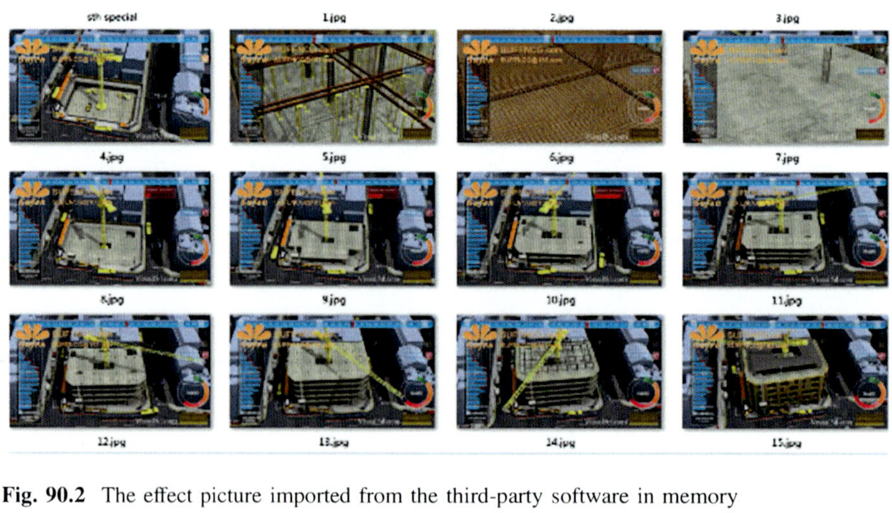

Fig. 90.2 The effect picture imported from the third-party software in memory

15 pictures of key-process points of control taken by a third-party program, and they are stored in the Matlab Current Directory of images folder in the hard disk, waiting for further changes of the timeline parameters to achieve orderly calls respectively. Users can use the "imread" function to add file name, and show each control point image in axes1 in accordance with the time axis, completing the requirement and expectation of the final image expressing effect of the program (Fig. 90.2).

Using the Matlab Graphics Generate Images

One of the goals this program aiming to achieve is to control the construction schedule be comparing the real-time model (axes2) to the forecast model (axes1) directly.

Using the progress of point-line-plane—body in the program, we graphical code the cube, which is the basic element of the structure, and store it as a primary sub function "Rec.m" and three secondary sub functions "x0.M", "y0.M", "z0. M". These three secondary functions adopted mesh grid function, according to the different time parameter and the location and shape of a cube, draw the three different surface of the cube, namely, X, Y, Z axis three surface, the color and shape of the points can be set by the user. After completion of surface rendering, it can be combined with the primary function to form the cube, then the cube can be shown in axes2, together with other cubes to form the whole structure (Fig. 90.3).

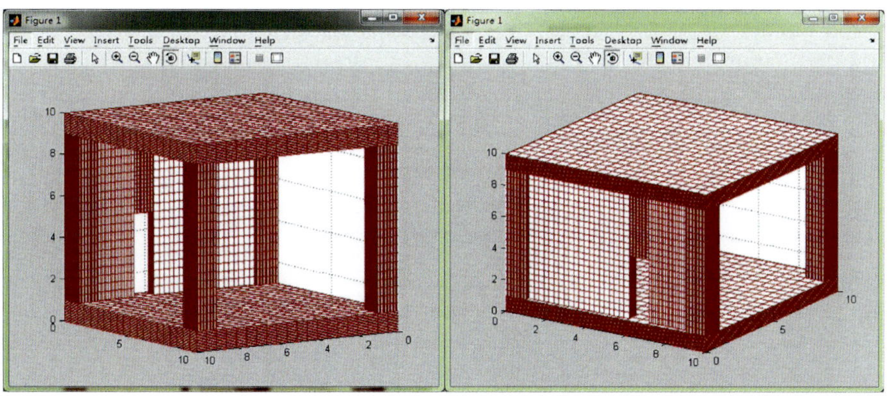

Fig. 90.3 3D graphs generated by matlab

Progress Schedule Management Implementation

In this program, the controlling bar at the bottom of the window is representing time. Its value is set from zero to one, as the progress of the percentage of the said work. All the image expression and parameters calculation are controlled by it.

Total duration(cost) control: After reading the simulation schedule file, drag the scroll bar, the program will read the controlling bar value, then multiply it by the total time limit for a project (total cost) and round the result, get the amount of time (cost) the project has already spent. Using judgment statement such as "if" with basic operations it can be calculated that the project should be in which stage, the remaining time limit for a project, the amount of spent cost and so on a number of general parameters.

Accurate daily management: in this paper, we add seven control points in "Sta. TXT" as examples to express certain amount of values a project is expected to reach when a certain point of time is coming, such as "expected cost", "expected reinforced usage", "expected concrete consumption", "expected numbers of stories", using the basic law of linear interpolation to express the relationships between these values and time schedule, and displays it on the right side of the data display area, for construction supervision personnel to input real-time data and acquire comparative analysis from the program, so as to make informative decisions.

90.2.3 Achievements

By adjusting the time axis parameters, we can accurately get the corresponding period number of the expected data, the model of the progress and the overall expected parameters for the entire duration of construction, joint by the real-time progress comparing module, we can achieve the goal of schedule management (Fig. 90.4).

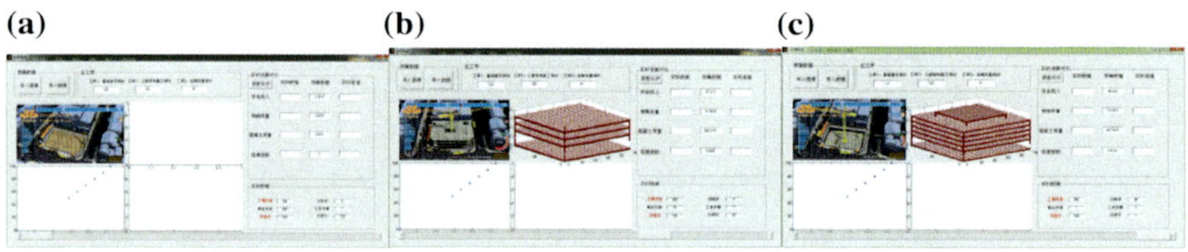

Fig. 90.4 **a–c** Effect pictures of schedule management

90.2.4 Results Analysis

Analysis

Time axis function changed a number of different related parameters, thus achieving the goal to reflect the variation of schedule, cost and 3D graphic model along with the progress of the project.

Structural Schedule Management Development Prospects

As a powerful numerical analysis software, it is difficult for Matlab to be used to make complex 3D images. For further development, we can use.Net, C, Java to develop professional level graphics program, it will greatly improve the simulation level of the images.

If we combine the program with more direct and simple ways of image interactive expression, and replace "Rec. M" with model building of parameter entry dialog box and directly depicting tools such as brush, it will achieved the basic sketching needs for BIM software.

Progress Schedule Management Development Prospects

1. The "time" parameter can be expressed by modes other than scrolling bar. Adding numerical entry mode can help the operator to adjust "time" to a precise point in time. We can also add an automatic increasing time module, so the overall preview of the project can be done by a click on a button.
2. The "time" parameter can be further refined. The current unit of this parameter is day, after perfection we may use more detailed time units such as "12 h", "h", "15 min" even "min" to achieve fine management.
3. The "phase of project" can be input by the user. The current project generally conclude building construction phases as "foundation construction period", "supper-structure construction period", "later perfection period", if there are any inconsistencies or other special construction process buildings, this program will not be able to apply. With fixed parameters "e8", "e9" and "e10" to represent the above three periods in the existing program, it's impossible to change the construction phases directly in the operation interface. We can try to use sequence, set up special "edit" frame in the operation window for operations such as assignment, this improvement will greatly improve the applicability of the program, from the general building construction to the entire building construction, and even to large equipment production can use this program as an "Information Model" to monitor and manage.
4. The categories of fine management variations can be input by the user. This program only lists the most common categories such as "reinforcement",

"concrete" instead of all the variations we probably need to manage in the whole construction process. The requirement can be met by the same method described in "3".

5. The function setting of the law of the numerical changing. In this program, we set the law of changing as control points collocate with linear interpolation, but it's not always like this in the real process of construction. The solution for this problem is also the method described in "3".

90.3 The Research and Development of Cost Management

90.3.1 Expected Functions of Cost Management

Cost management uses the timeline as control core, makes every effort to display all kinds of resources consumption level, consumption trend in the project and the comparison of real-time data and expectation, for the project managers to better understand the status quo, and adjust the resource usage to achieve a better cost control and management.

90.3.2 Cost Management Implementation Method

Expected Data Import

By selecting and reading records from the "Sta. TXT" file, which contains the expected cost and schedule of the project, the program imports all the information of this building to Workspace, waiting for the time parameter to call.

The Display and Comparison of Cost

The program will read the related parameters which are previous imported from the "Sta. txt" file according to the value of the time parameter, and then show them in the numerical display area on the right side and images in axes3.

90.3.3 Achievements

It get some achievements: (1) Import of the related parameters in the project; (2) Show different construction phases; (3) Make comparison.

Shown in the following Fig. 90.5.

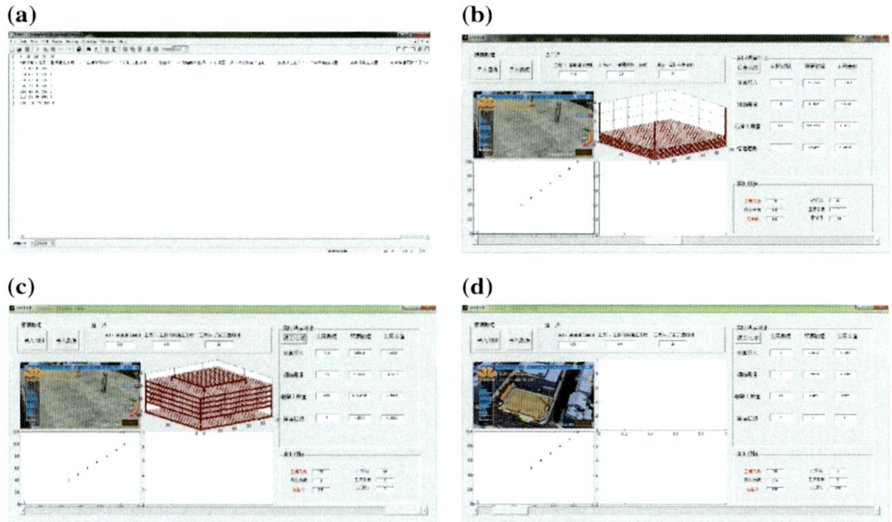

Fig. 90.5 **a–d** Effect pictures of cost management

90.3.4 *Results Analysis*

Analysis

In accordance with the time parameter variation, program calls the cost controlling points from the data base, and shows the budget of each item after calculation, finally shows them in graphs, successfully conveying the expected data. At the same time, the program uses real-time data input by supervisors of the project to generate the differences between the real-time data and expectation and the trend of the real cost, successfully conveying the actual data. In conclusion, this program will realize the two major functions of cost management.

Development Prospects

1. Building information input relies on the analysis summary of project supervisors as required.
   In the process of further perfect of the program, we can use the method of directly importing files from structural designing soft such as Autocad and project cost management software instead of manual entry.
   Further more, we can try to integrate the 3D drawing ability and information storage management of designing and management software, using the Internet to get various materials and human resource quotation, realizing automatic mapping and budgeting.

2. For the function key "error comparison", we would like to display the real-time data entered in this category in a more intuitive way with the images of budget in Axes3 image display area at the same time, express the changing trend of the current cost and reflects the differences with the budget data at the same time. In addition we can also add new parameters in the data contrast areas, such as supplemented the fourth column "catch-up schedule", that is to determine the time and material needed to fix up the error based on the actual error acquired by previous calculation and the analysis of the schedule of the same procedure.

90.4 Analysis of the 5D-BIM Simulation System

This program has successfully expressed the core operation mode of 5D-BIM program, combined time, cost and 3D model together. The core thought is to control the whole system with time parameter, and to display every single steps of a specific project with graphs and data. We have accomplished the purpose of writing a core framework for 5D-BIM software.

Nonetheless, there is still much space to develop in many aspects, such as the numerical input method, numerical simulation and prediction, graphing method, degree of graphic strips. The design of the interface can also be more hierarchical. Matlab is chosen to be the programming platform due to the limited ability of the author, but it is not the most ideal one. Some of the algorithm is not sufficient enough, it may increase the running time of the program. To resolve these problems is the next step of this program.

As a demonstration program, it has elaborated the authors' idea of 5D-BIM programming for programmers, and provided a huge space to enrich and perfect.

90.5 Conclusions

This program joined time parameter and cost parameter which can be regulated to the 3d model, greatly improving the performance of the program. But the effect of time parameter needs more professional analysis and simulation to reflect the real situations more accurately.

In this demo, we set up a total of four "Axes" to express "desired image effect", "simple function method of building models" and "dynamic data expression and comparison". And the fourth "stay white" Axes can express different effects according to different needs.

Along with the development of computer technology, more intelligent, automated production line will come into use to replace human labor force. This kind of software can not only instruct the computer to complete the construction of the mechanized assembly process, it also can reflect to the operator the progress of the project at the same time, which is convenient for the supervision and administration.

Acknowledgements This article is supported by STITP: Research of Construction Project's Cost and Schedule Dynamic Control Based on BIM (DC2013038) and National Natural Science Foundation of China (Grant No. 71271180).

References

1. Z Z (2015) Research of project cost management bases on BIM-5D. Inner Mongolia coal Econ 04:45–47
2. Chen X (2013) Design and realization of 5D model in building management software. Shanghai

Chapter 91
Project Organization Structure and Design Coordination in Architecture and Engineering Design Projects: An Advanced Structural Contingency Perspective

Rong Zhang and Anita M.M. Liu

Abstract Peter Morris calls for attention to the front-end in construction project. Design stage is an important part in the front-end of construction project. Due to lack of coordination amongst different professionals in design stage, information inconsistency (e.g. clash) amongst different disciplines (e.g. architecture, structure engineering) results in delay and cost overrun in construction stage. How to design project organization structure in design team in order to improve design coordination performance is concerned. Structural contingency theory argues that there is no best organization structure. The effect of organization structure on performance depends on task uncertainty. However, organization structure does not react sensitively to contingency change. But project organization structure, as a temporary organization structure is much easier to be adjusted. Coordination is an information processing activity to deal with task uncertainty. It is postulated that project organization structure affect project performance by facilitating or prohibiting information processing to reduce task uncertainty. Incorporating structure-conduct-performance perspective into traditional structural contingency theory, an advanced structural contingency framework is developed to investigate the relationship between project organization structure and project performance. It is postulated that, when task uncertainty changes its level, performance restores by organization structure-conduct adapting to fit new information processing requirement. A case study is conducted to test the framework. The advanced framework helps to understand how project organization structure affects project performance from a more systematical and process view.

Keywords Project organization structure · Task uncertainty · Design coordination · Structural contingency · Information processing · Fit

---

R. Zhang (✉) · A.M.M. Liu
Department of Real Estate and Construction, Faculty of Architecture, The University of Hong Kong, Porkfulam Road, Hong Kong
e-mail: srzhang@hku.hk

Y. Wu et al. (eds.), *Proceedings of the 20th International Symposium on Advancement of Construction Management and Real Estate*,
DOI 10.1007/978-981-10-0855-9_91

91.1 Introduction

As a classical topic in organization study, the research interest on organization design starts from early 1900s, bursts after 1960s and begins to decline in the middle of 1980s [1]. Recently, there is a resurgence of research on organization design [2]. On one hand, the rate of environmental changing is more rapid than ever since. Organizational context is more complex than existed when organization design theory firstly emerged. Triggered by complex and dynamic situations, there is a growing demand for robust theorizing and empirical research on new forms of organization design [2]. One the other hand, it is recognized that design is a central problem of management scholarship and practice [2]. Therefore, how organization should be designed in order to make organization works effectively is still an important topic today, as indicated by Burton et al. [3]. As the booming property market declines, the architecture and engineering design (AED) firms in China is in transition period, from extensive pattern to refinement. In the background, this study investigate how project organization structure design affect project performance in AED firms.

A key word in discussing organization design is organization structure. Organization structure is internal differentiation and patterning of relationships amongst major components within organization [4] (p. 51). Mintzberg's [5] view on organization structures from coordination perspective is applied in this study: the structures of an organization can be defined simply as the sum total of the ways in which its labor is divided into distinct tasks and then its coordination is achieved among these tasks. In this study, the coordination of information exchange between different discipline teams are concerned, because common problems in design coordination are due to insufficient and inadequate information exchange [6].

Mainstream organization theory is based upon the assumption that organizations are or should be permanent [7]. Fairly recently, the concept that project organization is a temporary organization is well acknowledged [8, 9]. Project-based firms are those firms that operate as permanent organizations and carry out their main activities in projects and where projects are the primary so called "unit of production [10]. AED firms are project-based firms. On one hand, there is permanent organization structure in AED firm; on the other hand, project is operated with temporary organization structure. The characteristic of project organization in AED firms will be discussed following.

91.2 Project Organization Structure of AED Firms

As for permanent organization structure, functional organization is most common organization structure in AED firms in China. In functional structured firms, designers are grouped by discipline (e.g. architecture department). The head of discipline department is functional manager. Some large AED firms have cell

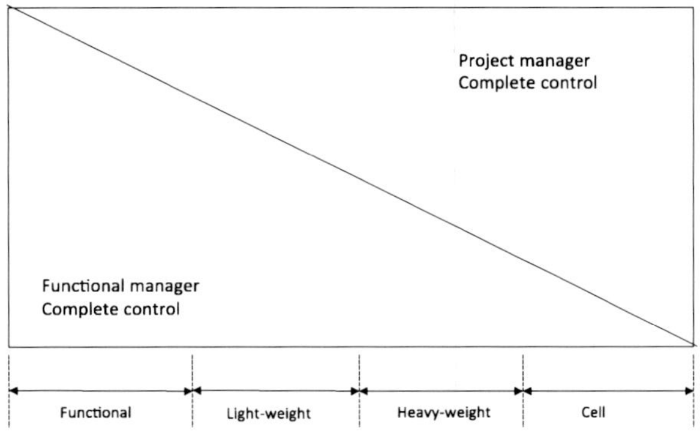

Fig. 91.1 Project team structure in architecture and engineering firms (*Source* developed from Winch (2010), Fig. 15.1)

organizations embedded in. Cell organizations specialize in one type of building (e.g. stadium, hospital). In cell organization, project managers have complete autonomy. It is equipped with all disciplinary professionals.

When project comes in, a temporary matrix organization is set up with a project manager and with a number of designers from functional departments. In terms of control power balance between project manager and functional manager, temporary project team organizations in AED firms are classified into four types: functional structure, light-weight structure, heavy-weight structure and cell. Although some tension and even conflict between project manager and functional manager is inevitable in matrix structure [9] (p. 412), people have to make matrix structure functional since it has becoming more dominant in modern AED firms (Fig. 91.1).

Matrix organization is the predominant project organization structure to operate project in AED firms. P-form organization [10] is a type of heavy weight matrix organization. The role of project as organization forms and of project managers as important "agent" in the capability building process is emphasized in P-form organization. Söderlund and Tell [10] highlight the fact that project managers and project directors played a fundamental role in the growth of the company since 1950. Consistent with the world trend, many AED firms in China are in the process of transform to P-form organization.

Project organization structure in design team is a temporary structure. There are two reasons for temporality. Firstly, project has definite start and end point. The temporality is reflected not only in production but also in organization [11]. From production perspective, temporality is shown by there are a beginning, an input-transform-output process, and an ending. From organization perspective, project organization is set up for a specific task/temporary assignment. After the task is completed, the project organization is diminished. Individuals in project team have other "homes" before, during and after being involved in a temporary organization [7]. Secondly, the human resource requirement is dynamic rather than stable in the design process. In this study, the "structural" characteristics of project organization structure are focused: size, span of control, and flat/tall hierarchy, spatial complexity [12, 13]. At the same time, centralization between supervisor and subordinate is also concerned.

91.3 Design Team as an Information Processing System

Organizations can be seen as information processing system [14, 15]. Whilst information processing perspective from general organization and management theory is relevant and useful to project management, before it is able to provide insights which contribute to making project organizations more effective, it is needs to be carefully distilled and developed [8].

Winch [9] put forward that project could also be viewed as information processing system. As such, it is argued that project organization in design stage could be viewed as information processing system under three assumptions. Firstly, project organizations are open social systems rather than closed systems that deal with work related uncertainty. Secondly, the basic function of project organization structure is to facilitate the effective collection, processing and distribution of information. Thirdly, project organization is comprised of sub units (e.g. architecture team, BIM consultant team). Individual is the basic information-processing unit in sub unit. This perspective shifts unit of analysis to individual in a discipline team. This study is consistent with the trend in organization study in two aspects: one is there is an increasing sense of focus from the industry as a whole to individual behavior [16]; the other is that there is an increasing emphasizes the context.

Malone [17] defines coordination activity as additional information processing activity performed when multiple actors pursue a goal that a single actor pursing the same goal would not perform. Activity in organization could be classified into production activity and coordination activity [17]. Design coordination activity refers to additional information processing activity performed when people from different disciplines work together to complete a building design. Design coordination activity includes information seeking from programming or people, communication (discussion or negotiation), and information transfer.

91.4 Perspective on Organization Design

91.4.1 Structural Contingency Perspective

The inquiry on how organization should be designed in order to enhance performance could date back to 1950s, when structural perspective was applied by early researchers in organization studies [20, 21]. Mintzberg (1979) [18, 19] comments that, during the era of classical organization theory, a good structure was based on rules and a rigid hierarchy of authority with spans of control no greater than six. Classical organization theory is a type of universalistic theory of organization [20] (p. 3). The central difficulty in classical organization theory is its preoccupation with discovering categorical "principles" of organization, applicable unqualifiedly to all organizations at all times. For example, classical management argues that maximum formalization and specialization lead to good organization performance [20, 21]. Many researchers tested the "principles" in classical organization theory in different contexts (e.g. health and welfare organizations, community colleges, brokerage firms, manufacturing firms) [22–24]. However, the results seemed to be contradicting. Dalton et al. [25] reported a comprehensive review on the empirical studies from 1950s and 1970s on relationship between organization structure and performance. Inconsistent empirical results showed that the relationship between organization structure and performance had not been convincingly demonstrated by classical organization theory.

Later on, research finds that there is no best way to organize. Researchers begin to consider the fit between structure and internal contingencies (e.g. technology, size) as well as external contingencies (e.g. environment) [4, 26–28]. This approach is called structural contingency perspective. In an abstract level, structural contingency perspective says the effect of organization structure on organization performance depends upon contingencies [20] (p. 6). In the absence of a sense of contingency, no assessment can be made of the extent to which cited best practice is transferable to any other firm [16] (p. 5). A well-acknowledged proposition of structural contingency approach is to which extent a coordination mechanism predicts performance depend on a third contingency factor. They predict what structure would be more efficient rather than what structure would be observed. In this study, task environmental uncertainty will be investigated as a contingency factor.

Although structural contingency theory has been challenged by other theories, such as institutional theory, population ecology, configuration theory and complexity theory (refer to Donaldson (2001), Chap. 6, pp. 161–179 and Van de ven et al. (2013) for a detailed discussion) [22], it continues to enjoy conceptual richness and validity [2, 20]. In additional, based on Kuhnian perspective, Qiu et al. [29] and Donaldson and Luo [30] state the necessity for persisting with existing paradigms/established theory rather than seeking new theories for their own sake. Van de Ven et al. [2] also support this idea. Underlying this is the notion that progress in organizational theorizing is made through persisting with existing theory rather than looking for new ones. Because paradigm continuity, elaboration, and extension develops theory, adding to its complexity.

91.4.2 Structure-Conduct-Performance Perspective

Structure-conduct-performance perspective finds its roots in organization economics. The theory argues that firms derive competitive advantages by responding to the characteristics of the industry in which they compete [31, 32]. Firms pursue strategies in response to market conditions, which alter firm conduct to positively impact the level of profits earned. Recently, researcher in management field applies this perspective [33].

Traditional structural contingency approach to organization design neglects the process how organization structure interacts with organization performance. Sometimes, it treats process as part of structure. For example, Donaldson (2001, p. 9) [20] used the word "structure" to indicate coordinate mode in reviewing Van de ven et al.'s [34] study on task contingency. While, coordination mode (e.g. scheduled meeting) is an information processing process. The process is consisted of a series of actions, or in another word, information processing activity.

91.4.3 An Advanced Structural Contingence Perspective

Structural contingency perspective points out it is contingency change (e.g. task uncertainty) that initiate the structural change. Donaldson (2001) [20] describes it as structural adaption to regain fit (SARFIT). However, organization structure does not react sensitively to contingency (e.g. task uncertainty). Donaldson (2001) [20] explains that there is time lag for structure to adapt to contingency. However, project organization structure as a temporary organization structure, is much easier to be adjusted. In additional, it is postulated that actions in process react more sensitively to contingency change. Coordination is an information processing process which could be adjusted timely to suit task uncertainty. A project organization structure is initially in fit, having a series of actions that fit its existing level of the contingency variable. Fit positively affects project performance. However, when contingency variable changes its level, there is a misfit between coordination and contingency variable if project organization retains its existing practice to coordinate. In turn, the misfit leads to lower project performance. When project performance is lower than acceptable level, the project organization structure has to make adaptive change. The adaptive change is to adjust coordination action to fit its

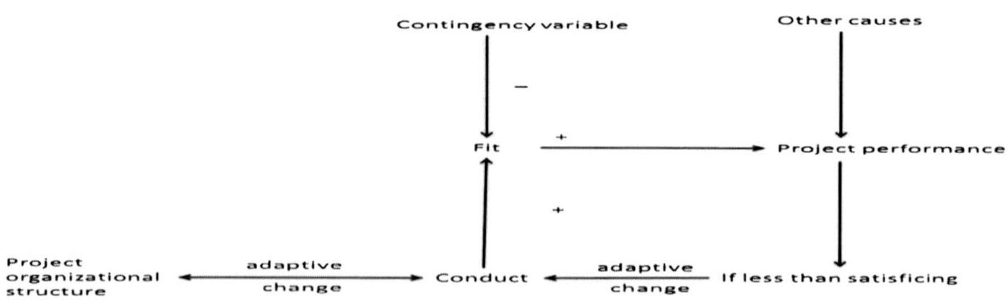

Fig. 91.2 The advanced contingency theory of project structural adaptation to regain fit (*Sources* developed from Donaldson (2001), p. 12, Fig. 1.1)

new contingency level. The new fit restores performance. The modification of coordination action affects structure and vice versa. The framework is showed in Fig. 91.2. Combining structure-process-performance perspective and structural contingency perspective helps to understand how project organization structure affects project performance from a more systematical and process view.

91.5 Research Design

Case study is an appropriate research strategy for exploring these how and why questions [35]. In this research case study is applied to as an explanatory one to explain how project organization structure and interdisciplinary team information processing activity interact to fit changing task uncertainty, in order to maintain performance. The criteria of selecting projects are: (1) ongoing project which is near to the end; (2) there is adaption process in the project. In the end, a school project from an AED firm in China, corporation alpha, is selected.

Corporation Alpha was set up in 1990s at the background of fast development of Chinese residential market. Over 20 years, Corporation Alpha has become one of the largest AED firms in China. It had branch offices all over around China. The basic organization structure is functional department by disciplines. They operate project with light matrix organization.

The first empirical study of this project was conducted in June 2014. It consisted of a group meeting with project manager, lead architect, lead plumping engineer, lead mechanical engineer and lead electrical engineer; and in depth interview with project manager and lead architect. The second of empirical study of this project was conducted in December, 2014, and a follow up interviews were conducted after eight months. A thirteen-mouth gap allowed for a degree of longitudinal insight. Face-to-face interviews were conducted with project manager, lead architect, lead plumping engineer, lead mechanical engineer, lead electrical engineer, lead quantity surveyor and most of designer and engineers.

91.6 Case Study

The analysis of design coordination applies the framework in Fig. 91.2. In this case, contingency is task environmental uncertainty and conduct is information processing. Initial state, the change of task uncertainty, the adaption of project organization structure will be discussed. Architecture team will be further discussed as examples to investigate the relationship between project organization structure, information processing and design coordination performance.

91.6.1 Project A

Project A was a school project. It included teaching buildings, library, canteen and dormitory buildings. The height of the tallest building was 33.8 m. The Gross Area was 80513 m^2. It was a public school project. Government provided the construction funding. Hence, the auditing procedure was quite strict. The project proposal was approved in 2006. The feasibility study was approved in 2010. In 2012, Corporation Alpha was awarded the design contract for the project though tender and bid.

Initial State

Initially, the project was treated as a normal project. The project was located in City A. The location of design team was fragmented in different locations in City B. The task of design started from 2012.

Increased Uncertainty from Task Environment

DRC rejected the first submission of construction budget. Construction budget was calculated based on shop drawings. Both the Gross Area and construction budget were over the number in approved feasibility study. Mainly two factors led to the over budget. Firstly, the aimed capability of the school increased from 2000 to 3000. As a result, the Gross Area was increased. Secondly, there was a two year gap between the approval of the project feasibility study and commence of design. During the two years, material price and labor price increased a lot.

Design team in City B was asked to do optimization to reduce construction budget. It seemed that deign project had two clients: bureau of public works (BPW) and the school management team (SMT). That made task environment more complicated. BPW was responsible for the construction management of the school project. The end user was SMT. SMT had the motivation to make the campus more beautiful and more comfortable. In tender and biding stage, architects tried to satisfy the requirement of SMT. At the same time, BPW gave pressure to design team to reduce construction budget. According to regulation, construction budget that exceeding project investment estimation for more than 10 % than feasibility study would be rejected. The first construction budget exceeded for about 50 %. SMT insisted to keep the school capability was 3000 students. It was not practical to control the budget overrun within 10 %. But what did they do was, on one hand they gave pressure to design team to do design optimization to reduce construction budget, and on the other hand they expected that the second submission could be granted even it exceed more than 10 % through bargaining with DRC. The reason for that solution was partly related with lack of experience to know the practical range of cost deduction by design optimization, and partly because of Chinese culture, that was flexible. The "flexible" culture induced people to disobey regulation, or play in the gray area. To sum up, the task environmental for design team was in high uncertainty.

The Signal of Misfit from Performance

The project design schedule was delayed. Design team in City B conducted design optimization to reduce cost. BPW submitted construction budget for the second time. Again, it was rejected by DRC. It required intensive communication between client and design team, design team and cost estimation team. As design team and client were in different cities, geographic distance prohibited smooth information processing and increased coordination cost. Initially, the period of design production was set up as six months. More than one year passed, the project was still in stuck. There should be some change.

Project Organization Structure and Information Processing Adaptation Process to Regain Fit

Hence, top management team in corporation alpha decided to assign new design team in City A to replace the team in City B in order to facilitate information processing. However, the architects and structure engineers in City A was not enough. In the end, three architects and one structure engineers from City C moved to City A to supplement architecture team and structure team in City A. Altogether, there were 25 staffs in the project team from six discipline teams. Project manager was an architect by training. The project organization chart was as in Fig. 91.3.

New project team brought changes in information processing. Besides frequency meeting with client, with team members were located near each other, information

Fig. 91.3 Project organization chart. *Notes* (1) *AT* architecture team; *SET* structure engineering team; *MET* mechanical engineering team; *EET* electrical engineering team; *PET* plumping engineering team; *QST* quantity surveying team; (2) *staff from City C 3. *Name with underlying* denotes discipline leader

processing between discipline teams were largely increased by all means: inter-discipline group meeting, face to face meeting, et al.. "In detailed design stage, we often have group meeting when meeting across design changes. We discuss the solution together. For example, plumping engineering asks me how much money it will cost by they plan pipe routine in such a manner. I will calculate and tell him shortly after the meeting. "Said the lead quantity surveyor. Geographical location increased the feasibility of communication between architects, engineers and quantity surveyors. They could learn more through discussion and knowledge sharing. Project manager could call for ad-hoc face-to-face meeting with discipline team easily, because they are located near each other. Enabled by sufficient information exchange, they conducted three times or re-design and three times of design optimization within three months.

Reduced Task Uncertainty from Environment

After re-design and design optimization, the construction budget was still over around 50 %. Client got to know that they could not find gray area to play. Finally, they decided to submit a new feasibility study based on increased capability of 3000 students. The second project feasibility study was approved by DRC with a slight cutting down of gross area and budget (around 6 %). This time, client asked design team to do design optimization strictly according to the new approval. It was good news to design team, because the object was clear.

91.6.2 Architecture Team

Project Organization Structure

Besides lead architect, project manage monitored a lot in architecture team. Her span of control was 10 within the firm. It exceeded normal range. After a period of time of working with other teammates in City A, architects from City C asked to return to branch office in City C. They thought they were familiar with the project. After they went back to city C, spatial complexity increased in architecture team. As a result, their information-processing channel changed.

Change of Information Processing Channel

Intranet was set up within one branch office in a city. There was online database for project documents in the intranet. Most of time, they share design information by uploading and downloading design information from the online database. However, the speed was very low if staff in city C to upload or download design drawing to/from branch office's intranet in city A. Architects in city C had to exchange design information both with architecture discipline and with other disciplines. In that situation, the lead architect acted as the coordinator for information exchange. If architects in city C deliver drawings to other disciplines, they gave them to the lead architect, and vice versa. Sometimes, lead architect could help them in uploading or downloading aimed design drawings. No matter what approach was used; the information exchange effectiveness was reduced. Project manager had to fly to city C to coordinate. That increased coordination cost.

Design Coordination Performance

In the end, the performance of architecture team was qualified but not quite good. The project manager was required to evaluate the performance of architecture team. She gave a score of six (the highest score was ten). The reason included: (1) Three teammates from City C used different size of symbol in drawing. It reduced the consistency amongst architecture drawings. In additional, the level of detail in drawing by architects in city C was not deep enough. (2) Designers were lack of the concept of cost control.

91.7 Conclusion and the Way Forward

Recognizing the strength and weakness of existing paradigm of structural contingency theory, this research seeks to improve it in two ways. Firstly, early organizational structural contingency studies emphasized on characteristics of organization structure and strategy, but tended to overlook how work was done [2]. This study tries to fill the gap by adding behavior into the framework to investigate how design coordination task is done. Secondly, another criticize is that fit in structural contingency theory is operationalized as a static equilibrium. Static fit could not reflect the dynamic views of adaptive fitness with evolving organizational landscape [34, 35]. Based on Donaldson's (2001) [20] SARFIT framework, this study incorporates structure-conduct-performance perspective to explain the project performance is a dynamic fit between structural conduct adaption process and task uncertainty.

Although ICT technology claims that it could facilitate remote coordination and communication, physical distance is still a big barrier in design coordination. (1) some staff are reluctance to accept web-based communication, especially the older staff; (2) face-to-face communication is required to deal with problem with high equivocality; (3) the speed of internet is not satisfactory. Further study could investigate to which extent ICT could facilitate remote design coordination.

References

1. Barry D (2011) Re-designing organization design. In: Design business conference, Barcelona
2. Van de Ven AH, Ganco M, Hinings C (2013) Returning to the frontier of contingency theory of organizational and institutional designs. Acad Manag Ann 7(1):393–440
3. Burton RM, Obel B, De Sanctis G (2011) Organizational design: a step-by-step approach. Cambridge University Press, Cambridge
4. Thompson JD (1967) Organizations in action. McGraw-Hill, New York
5. Mintzberg H (1979) The structuring of organizations: a synthesis of the research. University of Illinois at Urbana-Champaign's Academy for Entrepreneurial Leadership Historical Research Reference in Entrepreneurship
6. Hegazy T, Khalifa J, Zaneldin E (1998) Towards effective design coordination: a questionnaire survey 1. Can J Civ Eng 25(3):595–603
7. Lundin RA, Söderholm A (1995) A theory of the temporary organization. Scand J Manag 11 (4):437–455
8. Walker A (2002) Project management in construction, 4th edn. Blackwell Science Ltd., London
9. Winch GM (2010) Managing construction projects, 2nd edn. Wiley, Singapore
10. Söderlund J, Tell F (2010) The P-form organization: contingencies, characteristics, and challenges. Unpublished mimeo
11. Turner JR, Müller R (2003) On the nature of the project as a temporary organization. Int J Project Manag 21(1):1–8
12. Campbell JP et al (1974) The measurement of organizational effectiveness: a review of relevant research and opinion. Navy Personnel Research and Development Center, San Diego
13. Daft RL (2010) Organization theory and design, 10th edn. Cengage learning
14. Cyert RM, March JG (1992) A behavioral theory of the firm, 2nd edn. Blackwell, Englewood Cliffs
15. Tushman ML, Nadler DA (1978) Information processing as an integrating concept in organizational design. Acad Manag Rev 3(3):613–624
16. Winch GM (1992) The Implementation of CAD/CAM systems. In: The metalworking industries, in faculty of social studies. University of Warwick, Coventry
17. Malone TW (1998) What is coordination theory? In: National science foundation coordination theory workshop. Massachusetts Institute of Technology Massachusetts, Cambridge
18. March JG, Simon HA (1958) Organizations. Wiley, New York
19. Simon HA (1976) Administrative behavior, vol 3. Cambridge University Press, Cambridge
20. Brech EFL (1957) Organisation: the framework of management. Longmans, London
21. Donaldson L (2001) The contingency theory of organizations. Sage Publications, Thousands Oaks
22. Blau PM, Heydebrand WV, Stauffer RE (1966) The structure of small bureaucracies. Am Soc Rev 179–191
23. Child J (1974) Managerial and organizational factors associated with company performance part I. J Manage Stud 11(3):175–189
24. Hage J, Dewar R (1973) Elite values versus organizational structure in predicting innovation. Adm Sci Q 279–290
25. Dalton DR et al (1980) Organization structure and performance: a critical review. Acad Manag Rev 5(1):49–64
26. Galbraith JR (1977) Organization design. Addison-Wesley, Reading
27. Lawrence PR, Lorsch JW (1967) Differentiation and integration in complex organizations. Adm Sci Q 1–47
28. Van de Ven AH, Walker G (1984) The dynamics of interorganizational coordination. Adm Sci Q 598–621
29. Qiu J, Donaldson L, Luo BN (2012) The benefits of persisting with paradigms in organizational research. Acad Manag Perspect 26(1):93–104

30. Donaldson L, Luo BN (2014) The Aston programme contribution to organizational research: a literature review. Int J Manag Rev 16(1):84–104
31. Bain JS (1956) Barriers to new competition: their character and consequences in manufacturing industries, vol 329. Harvard University Press, Cambridge
32. Caves RE, Hall P (1964) American industry: structure, conduct, performance. Prentice-Hall, Englewood Cliffs
33. Ralston PM et al (2015) A structure–conduct–performance perspective of how strategic supply chain integration affects firm performance. J Supply Chain Manag 51(2):47–64
34. Van de Ven AH, Delbecq AL, Koenig Jr R (1976) Determinants of coordination modes within organizations. Am Soc Rev 322–338
35. Yin RK (2013) Case study research: design and methods. Sage publications, Thousands Oaks
36. Becker B, Gerhart B (1996) The impact of human resource management on organizational performance: progress and prospects. Acad Manag J 39(4):779–801
37. Drazin R, Van de Ven AH (1985) Alternative forms of fit in contingency theory. Adm Sci Q 514–539

Chapter 92
Research on Profit Distribution of IPD Project's Participants Based on Cooperative Game Theory

Yue Teng, Xiao Li, Huiping Li, Shirong Li, Bo Xu and Jun Wang

Abstract Developing IPD (Integrated Project Delivery) in projects requires the operable implementation scheme that can improve productivity and earn more profits. And it is important to provide incentives tied to achieving project goals. This paper suggested cooperative game theory as the method for analyzing profit distribution in IPD project and determined the designer, construction contractor, owner, BIM consultant as the key participants according the degree of importance in IPD projects. The main purpose of this paper is to: (1) Building the framework of BIM platform and using information sharing to guarantee the achievement of IPD project. (2) Exploring the profit distribution model of IPD according to various cooperative game theoretic solution methods. (3) Using the Shapley value to achieve the distribution of profits and the FCE (Fuzzy Comprehensive Evaluation) methods to quantize the risk of each index, while the AHP (Analytic Hierarchy Process) was used to give the weight. (4) Exploring the application of the identified allocation rules based on the case study of Autodesk Inc.

Keywords Profit distribution · IPD · Cooperative game theory · BIM · Shapley value

92.1 Introduction

With the participants of engineering projects being segmented, inefficient and competitive, simultaneously each participant only perform its own functions to maximize its own interests [1]. Government and enterprises have started to embrace Design-Bid-Build (DBB) since the 1940s in the United Stated. But due to the

Y. Teng (✉) · X. Li · H. Li · S. Li · B. Xu
School of Construction Management and Real Estate, Chongqing University, Chongqing, China
e-mail: 604758622@qq.com

J. Wang
School of Built Environment, Curtin University, Perth, Australia

Y. Wu et al. (eds.), *Proceedings of the 20th International Symposium on Advancement of Construction Management and Real Estate*,
DOI 10.1007/978-981-10-0855-9_92

bilateral contract relationship among owner, contractor and designer; it generated numerous problems, such as high levels of fragmentation, long construction period and high costs of collaboration [2]. In 1980s, a delivery method known as Project Partnering (PP) was used to achieve the participants consensus objectives by building an project team that involved early. It has improved the control of process, cost and quality. But the restraint of contract and the timeliness of information sharing are still existing problems [3]. With the development of information technology and the prominent shortage of traditional delivery methods, a new project delivery method that was seek to rebuild trust, commonly referred to as Integrated Project Delivery (IPD) [4].

92.2 Construction Business Environment in China

The AIA [5] defines IPD as, a project delivery approach that integrates people, systems, business structures and practices into a process that collaboratively harnesses the talents and insights of all participants to optimize project results, increase value to the owner, reduce waste, and maximize efficiency through all phases of design, fabrication, and construction. Profit distribution rewards "what's best for project" behavior, such as by providing incentives tied to achieving project goals [6]. These goals may vary but are usually associated with cost, schedule, and quality metrics commonly used to measure project success. Making fair profit distribution methods is the key of IPD success. The following reward sharing methods are found in literature: Based on value, Incentive pool [7], Innovation and outstanding performance, Performance bonuses, and Profit sharing [8]. As each IPD participant has its own interest, introducing the cooperative game theory into profit distribution is very important.

Cooperative game theory has been applied in engineering projects, mainly concentrated in the profit distribution of Project Partnering [9] and Public-Private-Partnership (PPP) [10], project management and Franchise determination method of BOT projects [11]. Cooperative game theory emphasizes the collective rationality, efficiency, fairness and equality rather than individual's rationality and Individual optimal decision [12]. In cooperative game, if agreement, promise or threat is fully binding and can be enforced, and there should exist distribution rule of Pareto Improvement within the consortium [13]. In terms of the solution of Cooperative game model, Shapley [14] used the axiomatic method to give the Shapley Value of profit distribution solution, and established the core of profit distribution. Objection and counter-objection was put forwarded by Aumann and Maschler [15] to regard the A-M bargaining set as the distribution solution. The basic conditions of cooperative game solution are information sharing and binding contract. In order to realize the information sharing and binding contract, this paper introduces the framework of BIM platform.

92.3 Profit Distribution Model

92.3.1 *The Realization of the Basic Assumptions*

In the IPD projects, the key point to solve the problem of distribution is to allocate extra profit, rather than the average profit. It is necessary to satisfy the following assumptions when the use of cooperative game theory:

(1) All participants take the maximization profit distribution strategy;
(2) All participants are fully trusted and there is the necessary information sharing;
(3) It should establish multiple agreements to restrain the participants.

The essence of enterprise is the pursuit of profit maximization that will fit the first point. Then this paper put forward a framework of BIM platform (Fig. 92.1) to realize the binding agreements of participants and information exchange (Fig. 92.2).

92.3.2 *The Basic Model of Cooperative Game Theory*

Assume that N is the set of participants. $S \in N$ is one of alliances in N. $V(s)$ is the characteristic function which is defined in the set of alliances and represents the profit of alliance S. $V(i)$ represents maximum utility between the participant i and others in the process of the game.

92.3.3 *Basic Model of Shapley Value*

In game theory, it needs to define the quantitative results for a given model. The representative concept in solution is Shapley value. Compared to other solutions, Shapely value has unique value. It explains the contribution margin to alliance from participants in multi-party cooperative game.

Set x_i as the profit of participant i. The profit that Alliance participant i should get is equal to the average marginal contribution of participant i. According to the theorem of Shapley:

$$x_i = \sum_{s \in N} \frac{(|s| - 1)!(n - |s|)!}{n!} [v(s) - v(s - i)] \qquad (92.1)$$

$|S|$ represents the scale of alliance S. ! stands for factorial. $[v(s - i)]$ represents the profit without participant i. Thus $[v(s) - v(s - i)]$ can be regarded as the contribution of participant i to alliance S. Assuming $K(s) = \frac{(|s|-1)!(n-|s|)!}{n!}$. Then

$$x_i = \sum_{s \in N} K(s)[v(s) - v(s - i)] \tag{92.2}$$

92.3.4 Profit Distribution Model of Modified Shapley Value

Assuming x_1, x_2, x_3, x_4 stands for profit of designer, owner, construction contractor and BIM consultant before modified. In fact, basic Shapley model does not consider

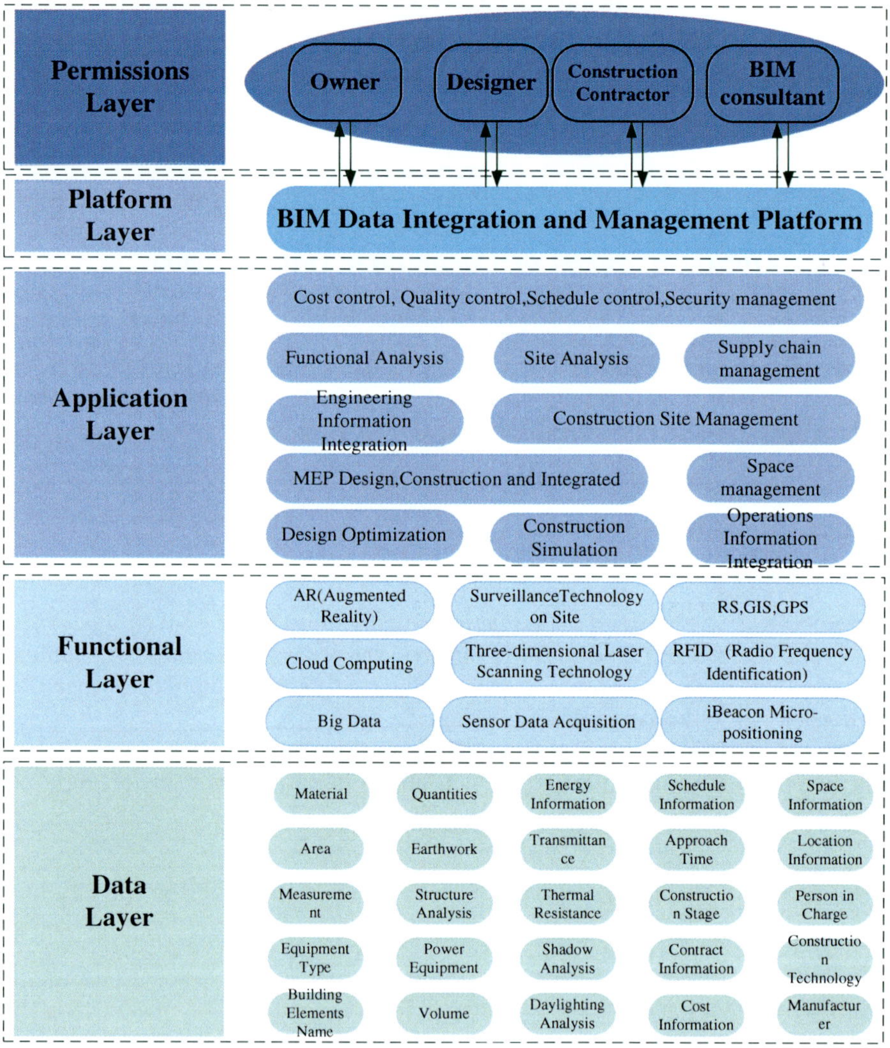

Fig. 92.1 Framework of BIM platform

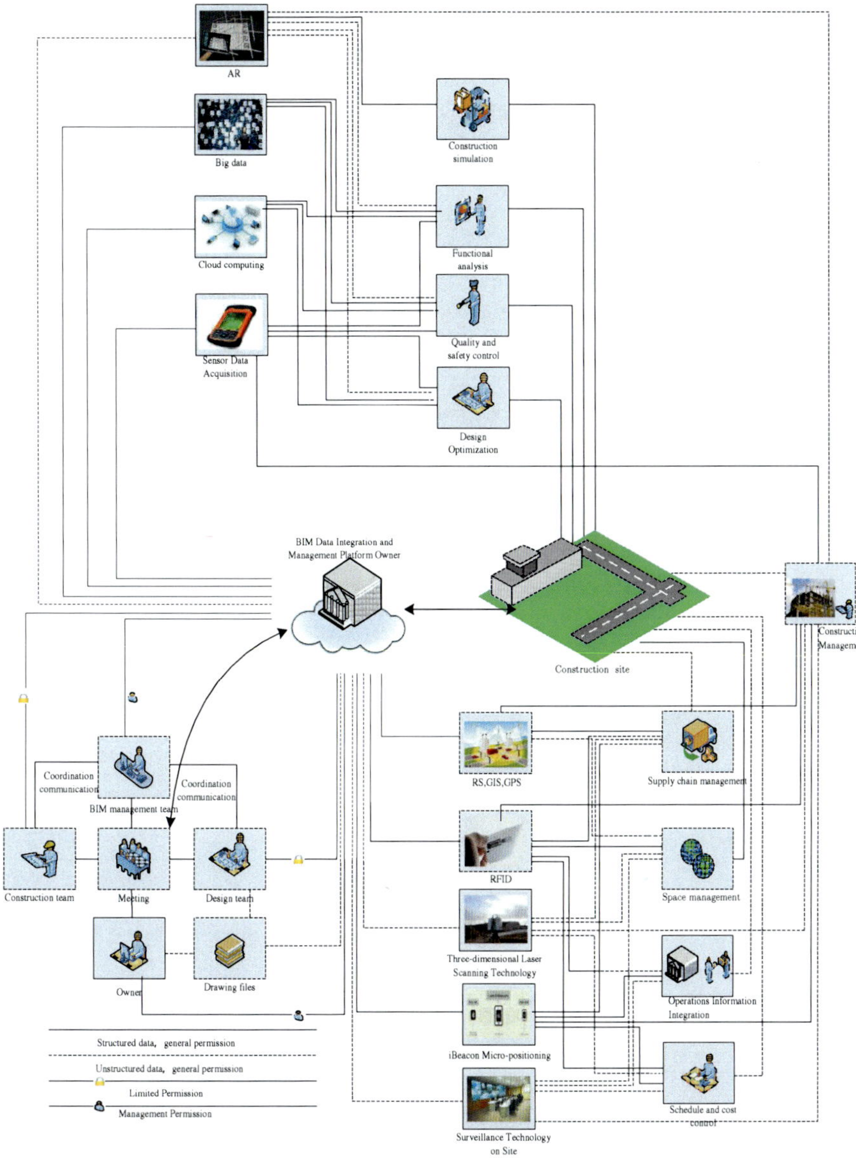

Fig. 92.2 Framework of Information Sharing

the actual risk of participants. That is the default for the risk borne by participants is equal: $1/n$. In actual IPD operation, the risk of participants is complex and varied. This paper considered four types of risks.

- Operation Risk: It includes human resource risk r_1, information resource risk r_2, technology risk r_3, equipment failure risk r_4.
- Financial Risk: It includes liquidity risk r_5.
- Profit Risk: It includes cost variation risk r_6, payment credit risk r_7, reasonable design r_8.
- Market Risk: It includes interest rate fluctuation risk r_9, policies and regulations r_{10}.

Risk evaluation methods usually have genetic algorithm, neural network and the fuzzy comprehensive evaluation (FCE) method, analytic hierarchy process (AHP) method. IPD projects have some risk that cannot quantize. FCE is suitable for the project that is difficult to define and quantize the risk. But FCE is difficult to unify the multiple indexes into a single evaluation index. AHP can give the weight to the indexes and unify the indexes. This paper adopts the combination of AHP and FCE, the main steps are as follows:

① Determine the decision target X.
② Establish evaluation index system U, If X has m influence factors, $U = \{u_1, u_2, \ldots u_m\}$.
③ Determine the weight set,assume the weight of $u_1, u_1, \ldots u_m$ is $w_1, w_1, \ldots w_m$, weight vector can be built by using AHP.
④ Establish evaluation V. The factors in evaluation set represent magnitude of risk. Assuming $V = \{v_1, v_2, v_3, v_4, v_5\}$, $v_1 \sim v_5$ represent magnitude of risk from low to high. It gives evaluation to the u_i in the U and gets evaluation matrix R_i of u_i.

$$R_i = \begin{bmatrix} r_{11} & \cdots & r_{1n} \\ \cdots & \cdots & \cdots \\ r_{m1} & \cdots & r_{mn} \end{bmatrix} \tag{92.3}$$

The comprehensive evaluation of u_i: $B_i = w_i' \times R_i$, comprehensive fuzzy rating matrix R combined by $R = (B_1, B_2, B_3, B_4)$.
⑤ Determine the comprehensive evaluation matrix R and weight vector W,

$$B = W \times R \tag{92.4}$$

Take the normalized process to B. $B' = \left[b_1', b_2', \ldots b_n'\right]$, according to maximum membership principle, the normalization of the maximum is the risk factor s_i.
⑥ Participants undertake the actual risk s_i, the difference between actual risk and average risk is $\Delta s_i = s_i - \frac{1}{n}$, $\sum_1^n \Delta s_i = 0$. So it needs to modify each participant's profit, $v(s)$ is the total profit and it can get $\Delta s_i \times v(s)$. The final value of profit distribution is $x_i' = x_i + \Delta s_i \times v(s)$.

92.4 Case Validations

92.4.1 *Brief Introduction of Case*

Autodesk Inc., a company (owner) that creates design software for the AEC industry. The company decided to put those goals forward with two of its own projects. The Waltham project is a 55,000 ft^2, three-story interior tenant improvement that uses all of the space in a new speculative office building near Route 128 in Boston's technology corridor. KlingStubbins (designer), Autodesk software (BIM consultant) and Tocci (construction contractor) were chosen because of their qualifications, familiarity with the local market, BIM and willingness to abide by a "true" IPD agreement. This paper uses the case data from AIA case studies to validate the profit distribution model.

92.4.2 *Shapley Value*

This paper builds the cooperative game model $\{N, v\}$, of which the collection of IPD participants is $N = \{1, 2, 3, 4\}$ (1 = Designer, 2 = Owner, 3 = Construction, Contract 4 = BIM Consultant).

① Each Participant Expected Profit in Traditional Project Delivery
If all the participants only get the average profit of industry, the distribution value of extra profit is 0, namely V(1) = 0, V(2) = 0, V(3) = 0, V(4) = 0 (Table 92.1).

② Expected Profit of Multiple Participants Alliance
If participants collaborate with each other, they can get extra profit. Thus the expected profit of multiple-participants alliance adopts the rising rate based on addition of single participant expected profit. Because the rising rate will not affect the distribution rate of each participant, this paper assumes Bi-participants alliance could increase the 10 % profits and Tri-participants alliance could increase the 20 % profits. Alliance can allocate the extra profit and get Table 92.2.

Table 92.1 Participants basic information (unit: million USD)

Participants	Average profit rate of industry (%)	Expected cost	Expected profit
Designer	2.50	1231	32
Owner	11.10	14530	1814
Construction contractor	7.09	12223	1048
BIM consultant	5.80	1089	67

Table 92.2 Multiple-participants profit distribution (unit: million USD)

Bi-participants alliance	V (1, 2)	V (2, 3)	V (1, 3)	V (1, 4)	V (2, 4)	V (3, 4)
Expected profit	2030.6	3148.2	1188	108.9	2069.1	1226.5
Distributable profit	184.6	286.2	108	9.9	188.1	111.5
Tri-participants alliance	V (1, 2, 3)	V (1, 2, 4)	V (1, 3, 4)	V (2, 3, 4)		
Expected profit	3472.8	2295.6	1376.4	3514.8		
Distributable profit	578.8	382.6	229.4	585.8		

③ Maximum Profits of IPD project
According to the expectation, this IPD project would increase 40 % profits compared to traditional way. Thus, expected profit would lead the total profit to 4145 million dollars and the distributable profit V(S) = 1184 million dollars. If allocate the profit by average, the profit of each participant is 296 million dollars. It is not good for the simulation of participants working enthusiasm. This paper uses the random variable of expected profit $R(N) \sim U[2961, 4145]$.

All the collections including designer, Owner, Construction contractor, BIM consultant are as in Tables 92.3, 92.4, 92.5 and 92.6.

Table 92.3 Profit distribution value of the designer

T_1	1	1 ∪ 2	1 ∪ 3	1 ∪ 4	1 ∪ 2 ∪ 3	1 ∪ 2 ∪ 4	1 ∪ 3 ∪ 4	1 ∪ 2 ∪ 3 ∪ 4
$v(s)$	0	184.6	108	9.9	578.8	382.6	229.4	1184
$v(s-1)$	0	0	0	0	286.2	188.1	111.5	585.8
$v(s) - v(s-1)$	0	184.6	108	9.9	292.6	194.5	117.9	598.2
s	1	2	2	2	3	3	3	4
$K(s)$	1/4	1/12	1/12	1/12	1/12	1/12	1/12	1/4
$K(s)[v(s) - v(s-1)]$	0	15.38	9	0.83	24.38	16.21	9.83	149.55

Table 92.4 Profit distribution value of the owner

T_2	2	1 ∪ 2	2 ∪ 3	2 ∪ 4	1 ∪ 2 ∪ 3	1 ∪ 2 ∪ 4	2 ∪ 3 ∪ 4	1 ∪ 2 ∪ 3 ∪ 4
$v(s)$	0	184.6	286.2	188.1	578.8	382.6	585.8	1184
$v(s-1)$	0	0	0	0	108	9.9	111.5	229.4
$v(s) - v(s-1)$	0	184.6	286.2	188.1	470.8	372.7	474.3	954.6
s	1	2	2	2	3	3	3	4
$K(s)$	1/4	1/12	1/12	1/12	1/12	1/12	1/12	1/4
$K(s)[v(s) - v(s-1)]$	0	15.38	23.85	15.68	39.23	31.06	39.53	238.65

Table 92.5 Profit distribution Value of the construction contractor

T_3	3	1 ∪ 3	2 ∪ 3	3 ∪ 4	1 ∪ 2 ∪ 3	1 ∪ 3 ∪ 4	2 ∪ 3 ∪ 4	1 ∪ 2 ∪ 3 ∪ 4
$v(s)$	0	108	286.2	111.5	578.8	229.4	585.8	1184
$v(s-1)$	0	0	0	0	184.6	9.9	188.1	382.6
$v(s) - v(s-1)$	0	108	286.2	111.5	394.2	219.5	397.7	801.4
s	1	2	2	2	3	3	3	4
$K(s)$	1/4	1/12	1/12	1/12	1/12	1/12	1/12	1/4
$K(s)\,[v(s) - v(s-1)]$	0	9	23.85	9.29	32.85	18.29	33.14	200.35

Table 92.6 Profit distribution value of the BIM consultant

T_4	4	1 ∪ 4	2 ∪ 4	3 ∪ 4	1 ∪ 2 ∪ 4	1 ∪ 3 ∪ 4	2 ∪ 3 ∪ 4	1 ∪ 2 ∪ 3 ∪ 4
$v(s)$	0	9.9	188.1	111.5	382.6	229.4	585.8	1184
$v(s-1)$	0	0	0	0	184.6	108	286.2	578.8
$v(s) - v(s-1)$	0	9.9	188.1	111.5	198	121.4	299.6	605.2
s	1	2	2	2	3	3	3	4
$K(s)$	1/4	1/12	1/12	1/12	1/12	1/12	1/12	1/4
$K(s)\,[v(s) - v(s-1)]$	0	0.83	15.68	9.29	16.5	10.12	24.97	151.3

Therefore, the distribution profit of designer is $x_1 = 225.18$, the distribution profit of owner is $x_2 = 403.38$, the distribution profit of construction contractor is $x_3 = 326.78$, the distribution profit of BIM is $x_4 = 228.68$. Each participant gets the both average profit and extra profit.

92.4.3 Modified Shapley Value

By using FCE and AHP, it can build the risk evaluation index of the Shapley model. It takes the owner's risk measurement as example. The decision-making objective is X_2, and owner has four risk factors. Namely, $m = 4$, $U_{\text{owner}} = \{u_1, u_2, u_3, u_4\}$.

① Comparison matrix A is constructed by AHP. This paper uses expert decision making and 1–9 scaling method to get the weight set $W_{owner} = \{0.24, 0.30, 0.29, 0.17\}$. After the consistency check, it gets CR = 0, IR = 0.89 and it meets the consistency check.

② Establish the evaluation set to evaluation the factors in U_{owner}. Taking the market risk u_{owner4} as example, Choosing 10 experts to score the interest rate

Table 92.7 Risk evaluation of the owner

A	B	Evaluation				
		1	3	5	7	9
Operation risk	Human resource risk r_1	0.1	0.1	0.4	0.4	0
	Information resource risk r_2	0.1	0.4	0.5	0	0
	Technology risk r_3	0.4	0.3	0.3	0	0
	Equipment failure risk r_4	0.9	0.1	0	0	0
Financial risk	Liquidity risk r_5	0	0	0	0.2	0.8
Profit risk	Cost variation risk r_6	0	0	0.3	0.2	0.5
	Payment credit risk r_7	0	0	0.3	0.3	0.4
	Reasonable design r_8	0	0	0.1	0.5	0.4
Market risk	Interest rate fluctuation risk r_9	0	0	0.1	0.2	0.6
	Policies and regulations r_{10}	0	0	0.2	0.4	0.4

fluctuations in market risk and choose first high five. The membership of them is 5/10 = 0.5, then it can get the risk evaluation Table 92.7.

Then it gets $R_{owner4} = \begin{bmatrix} 0 & 0 & 0.1 & 0.2 & 0.6 \\ 0 & 0 & 0.2 & 0.4 & 0.4 \end{bmatrix}$

Based on the AHP it can build market risk with the weight vector $w'_{owner\,4} = [0.56, 0.44]$. It meets the consistency check and the comprehensive evaluation of risk factor u_{owner4}:

$$B_{owner4} = w'_{owner\,4} \times R_{owner4} = [0.56, 0.44] \cdot \begin{bmatrix} 0 & 0 & 0.1 & 0.2 & 0.6 \\ 0 & 0 & 0.2 & 0.4 & 0.4 \end{bmatrix}$$
$$B_{owner4} = \begin{bmatrix} 0 & 0 & 0.14 & 0.29 & 0.51 \end{bmatrix}$$

Similarly, it can get the other comprehensive evaluation $B_{owner1}, B_{owner2}, B_{owner3}$ of owner and build the fuzzy comprehensive rating matrix

$$R_{owner} = (B_{owner1}, B_{owner2}, B_{owner3}, B_{owner4})$$
$$R_{owner} = \begin{bmatrix} 0.235 & 0.09 & 0.125 & 0.255 & 0.295 \\ 0 & 0 & 0 & 0 & 1 \\ 0 & 0 & 0.15 & 0.28 & 0.51 \\ 0 & 0 & 0.106 & 0.2 & 0.588 \end{bmatrix}$$

③ Finally it can get the owner's risk rating:

$$B_{owner} = W \times R = [0.48, 0.07, 0.26, 0.19] \cdot \begin{bmatrix} 0.21 & 0.22 & 0.38 & 0.19 & 0 \\ 0 & 0 & 1 & 0.2 & 0.8 \\ 0 & 0 & 0.24 & 0.32 & 0.44 \\ 0 & 0 & 0.14 & 0.29 & 0.51 \end{bmatrix}$$

$$B_{owner} = [0.1018 \quad 0.1050 \quad 0.3402 \quad 0.2416 \quad 0.2689]$$

After the normalization processing, it gets $\bar{B}_{owner} = [0.0963 \quad 0.0993 \quad 0.32170.22850.2543]$.

And get the risk parameter 0.3217 of owner according to maximum membership principle. Similarly, it can get the risk parameter of other participants: designer 0.2506, construction contractor 0.3025, BIM consultant 0.2483. After the normalization processing get the risk factor:

$$\Delta s_{de} = 0.0269, \Delta s_{owner} = 0.0364, \Delta s_{cc} = 0.0194, \Delta s_{bim} = -0.0289$$

④ Thus, the ultimate distribution profit value of owner is:

$$x_2' = x_2 + \Delta s_{owner} \times v(s) = 403.38 + 0.0364 \times 1184 = 446.478$$

Similarly, it can get the other participants distribution profit value:

$$x_1' = 257.03, x_3' = 349.75, x_4' = 194.462$$

92.5 Discussion and Conclusions

There exist some obstacles in the IPD implementation including multi-party contract, liability waivers and transparent financials. But stakeholders of IPD project most care about the profit distribution. This paper suggested cooperative game theory as the method for analyzing profit distribution in IPD project and determined the designer, construction contractor, owner, BIM consultant as the key participants according the degree of importance in IPD projects. The cooperative game theory has numerous successful case validations and theory models in solving profit distribution. To facilitate cooperative game theory in solving profit distribution of IPD projects, this study introduced the Shapley value within risk factors to build the profit distribution model. The risk that considered in the model includes operation risk, financial risk, profit risk and market risk. In order to quantize the risk value in IPD projects, this paper put forward using the FCE to quantize the risk and AHP can give the weight to the indexes to unify the multiple indexes into a single evaluation index. But there always exist some assumptions to satisfy, such as information sharing and contract constraint. To meet the assumptions, this study establishes the framework of BIM platform and information sharing.

Acknowledgements We gratefully acknowledge the financial support from Chongqing Graduate Student Research Innovation Project 2015 (CYS15002) and Chongqing Graduate Student Research Innovation Project 2014 (CYS14036).

References

1. Lichtig WA (2006) The integrated agreement for lean project delivery. Constr Law 26(3)
2. Construction Industry Institute (2009) Value of collaboration. UT Austin
3. Lahdenperä P (2012) Making sense of the multi-party contractual arrangements of project partnering, project alliancing and integrated project delivery. Constr Manag Econ 30(1):57–79
4. Matthews O, Howell GA (2005) Integrated project delivery: an example of relational contracting. Lean Constr J 2(1):46–61
5. Guide AIA (2007) Integrated project delivery: a guide. American Institute of Architects, CA
6. Sive T (2009) Integrated project delivery: reality and promise, a strategist's guide to understanding and marketing IPD. Society for Marketing Professional Services Foundation
7. Lichtig WA (2006) Integrated agreement for lean project delivery. Constr Law 26:25
8. Brady T, Davies A, Gann DM (2005) Creating value by delivering integrated solutions. Int J Project Manag 23(5):360–365
9. Lazar FD (2000) Project partnering: improving the likelihood of win/win outcomes. J Manag Eng 16(2):71–83
10. Scharle P (2002) Public-private partnership (PPP) as a social game. Innov Eur J Soc Sci Res 15(3):227–252
11. Shen LY, Bao HJ, Wu YZ et al (2007) Using bargaining-game theory for negotiating concession period for BOT-type contract. J Constr Eng Manag 133(5):385–392
12. Schmidt C (ed) (2003) Game theory and economic analysis: a quiet revolution in economics. Routledge
13. Suijs J, De Waegenaere A, Borm P (1998) Stochastic cooperative games in insurance. Insur Math Econ 22(3):209–228
14. Shapley LS, Shubik M (1954) A method for evaluating the distribution of power in a committee system. Am Polit Sci Rev 48(03):787–792
15. Aumann RJ, Maschler M (1964) The bargaining set for cooperative games. Adv Game Theor 52:443–476

Chapter 93
Business Expanding and Strategic Decision-Making Innovation for Real Estate Valuation Companies Based on Big-Data

Yan Xiao and Lin Xu

Abstract With the penetration of large data on real estate and related industries, real estate appraisal industry is not only faced with technical challenge of processing huge amounts of data, but also with the innovation of strategic decision-making. This paper starts with the current situation of the development of big data, and basing on the analysis of Five Forces model, it analyzes the threats of real estate valuation enterprises under the background of big data competition. It is suggested that massive valuation, automatic valuation and the innovation of valuation technology are the main methods for real estate valuation enterprises to expand their business. Real estate valuation enterprises should make innovations in strategic decision-making model, developing model, marketing model as well as industry cooperation and resource integration, so that they can build up advantages in the future competition.

Keywords Big data · Real estate valuation · Five forces model · Business expanding · Strategic decision-making

93.1 Introduction

In may 2011, The McKinsey Global Institute (MGI), McKinsey & Company's (a global consulting agency) business and economics research arm published a report named *Big data: The next frontier for innovation, competition, and productivity*. In this report, it pointed out that data have become a torrent flowing into every area of

Y. Xiao (✉)
Department of Management, Chongqing Technology and Business University, Chongqing, China
e-mail: xiaoyan@ctbu.edu.cn

L. Xu
School of Economics and Management, Urban Vocational College of Sichuan, Chengdu, China

Y. Wu et al. (eds.), *Proceedings of the 20th International Symposium on Advancement of Construction Management and Real Estate*,
DOI 10.1007/978-981-10-0855-9_93

the global economy. Research shows that data can create significant value for the world economy, enhance the productivity and competitiveness of companies and the public sector and create substantial economic surplus for consumers [1]. Big data create value in the ways of creating transparency, segmenting populations to customize actions, replacing human decision making with automated algorithms as well as innovating new business models, products, and services.

As Chinese real estate industry's golden decade ends, the entire real estate industry has to face with future strategic transformation and the innovation of the business models. As a key link in real estate industry, real estate appraisal industry is faced with more complex forces and interests. Technically, it is faced with business innovation challenge of dealing with torrents of data. Under this circumstance, to explore a solution based on big data is the most important means for China's real estate appraisal enterprises to upgrade their business and improve efficiency and productivity. The era of big data is also the right time for real estate appraisal enterprises.

93.2 The Definition and Application of Big Data

Because of different concerns, scientific and technological enterprises, research scholars, data analysts, and technical practitioners have different definitions of big data. As early as 2001, Doug Laney, an analyst of META (presently Gartner) defined challenges and opportunities brought about by increased data with a 3Vs model, i.e., the increase of Volume, Velocity, and Variety, in a research report [2]. Companies like Microsoft [3] and IBM [4] described three attributes of big data with a "3Vs" model: volume, velocity and variety. In the "3Vs" model, Volume means, with the generation and collection of masses of data, data scale becomes increasingly big; Velocity means the timeliness of big data; Variety indicates the various types of data, which include semi-structured and unstructured data such as audio, video, webpage, and text, as well as traditional structured data.

In 2010, Apache Hadoop defined big data as "datasets that could not be captured, managed, and processed by general computers within an acceptable scope." On the basis of this definition, in May 2011, McKinsey & Company announced Big Data as the next frontier for innovation, competition, and productivity [5].

Also in 2011, an IDC report defined big data as "big data technologies describe a new generation of technologies and architectures, designed to economically extract value from very large volumes of a wide variety of data, by enabling the high-velocity capture, discovery, and/or analysis" [6]. With this definition, characteristics of big data may be summarized as four Vs, i.e., Volume, Variety, Velocity and Value. Such 4Vs definition was widely recognized since it highlights the meaning and necessity of big data. This definition indicates the most critical problem in big data, which is how to discover values from datasets with an enormous scale, various types, and rapid generation. In 2013, Pat Saporito added another V to the attributes of big data, i.e., veracity.

Table 93.1 Big data applications

Application categories	Sectors
Goods	Manufacturing Construction Natural resources Computer and electronic products Real estate, rental, and leasing Information
Services	Transportation and warehousing Retail trade Administrative, support, waste management, and remediation services Accommodation and food services Other services (except public administration) Arts, entertainment, and recreation Finance and Insurance Professional, scientific, and technical services Management of companies and enterprises
The regulated and public	Government Educational services Health care and social assistance Utilities

Source McKinsey Global Institute analysis

Digital data is now everywhere—in every sector, economy, organization and user of digital technology. Big data is now relevant for leaders across every sector, and consumers of products and services stand to benefit from its application. In the past decade, big data has been widely used in the supplies of goods, services and the regulated and public (see Table 93.1).

According to a report from International Data Corporation (IDC), in 2012, it is estimated that by 2015 the service market of big data technology will gain value of 16.9 billion dollars, compared with 3.2 billion dollars in 2010 [7]. With an annual growth rate of nearly 40 %, big data creates value far higher than other industries including the high-technology sector. And from the angle of management, application of big data in enterprises to support business analysis and decision making, is also a hot topic of business schools, as the trend has shown in the European and American business schools [8].

93.3 Big Data's Influence on Real Estate Appraisal Enterprises

Real estate appraisal is mainly to measure the value of real estate and forecast the price level, which means that the quality and quantity of data will directly affect the accuracy of the assessment. Data is vital to real estate appraisal enterprises because

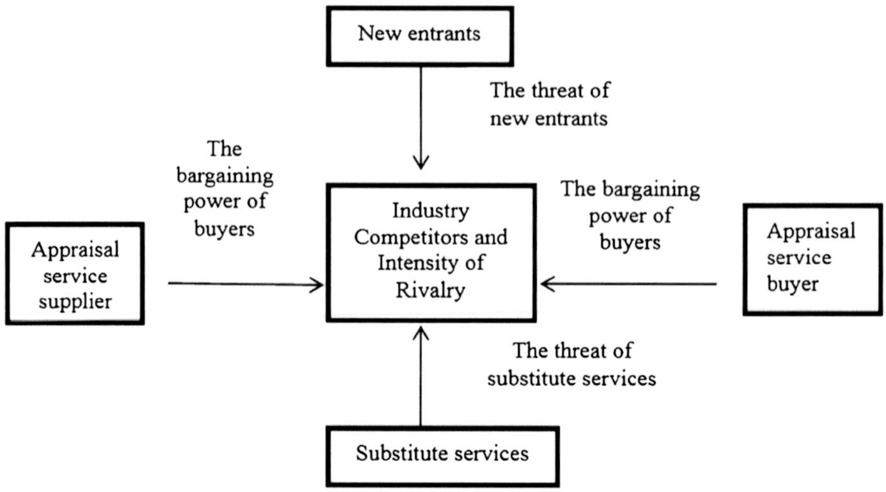

Fig. 93.1 Michael Porter's five forces model

real estate appraisal industry has obvious characteristics of big data. The era of big data will have far-reaching influence on real estate appraisal business.

One of the universally accepted frameworks for the organization of information about industries is the 'Five Forces that Drive Industry Competition' Model developed by Michael Porter. Porter's model assumes competition is the background to industry structure and a firm's performance. The model uses the widely accepted definition of industry as a collection of firms with the capacity to provide services that are close substitutes to each other. According to Porter's definition, real estate valuation industry is a collection of companies offering valuation services that satisfy the same basic customer need for real property appraisal.

The intensity of competition within the valuation industry, in terms of Porter's Model, is established by the generic five forces of competition shown diagrammatically in Fig. 93.1.

93.3.1 Industry Competitors

A real estate valuation company's closest competitors are considered to be those organizations that serve the same basic customer needs. Up to now, there are more than 5000 real estate appraisal companies in China that compete for data. The sources of their data include external accesses such as government statistics, professional research institutes and consultant agencies, as well as lots of first-hand data they collect. But due to the particularity of real estate industry, either the information of macro policy or micro policy such as the adjustment of the interest rate will have a significant impact on the real estate market. With the advent of the era of big data, a large number of products and customer information can be obtained from the websites. More and more real estate evaluation enterprise build their own database, through the real-time data mining, image mining, text mining, semantic mining, more structured method of data mining, machine learning, and so on, much first-hand information can be obtained from the network, radio, television and other media. It would be reasonable to argue that the intensity of rivalry between competing appraisal firms will differ due to big data. It will be the competition of data accuracy and efficiency.

93.3.2 The Threat of New Entrants

The threat of entry into an industry depends on the barriers to entry that are present, coupled with the reactions from existing competitors. If the barriers are high and/or the newcomer can expect sharp retaliation from competition, the threat of entry to an industry is low [9]. The major barriers to entry to the real estate valuation industry can be summarized as government policy through regulation and licensing of professionals, the costs of professional indemnity insurance and customer loyalty.

These barriers can be considered as relatively minor in the era of big data, because most of the information needed for their real estate evaluation can be barrier-free access. This makes the threshold of entering real estate appraisal industry lower than before. Therefore, some consultants, and even some network operators related with real estate are engaged in real estate appraisal market. At present, some consultants are providing appraisal services for banks and other financial institutions. Another example is the property tax base assessment market, which will be bigger than mortgage market. At present, software companies provide a solution for property tax assessment using their desktop appraisal software. Traditional real estate appraisal institutions are losing their superiority if new entrants keep coming into this market, which will make the market more competitive in the future.

93.3.3 The Threat of Substitute Services

Substitute services restrict the potential returns of appraisal industry by placing limits on the prices that firms in the industry can profitably charge. The real estate valuation industry has recently been experiencing the effect of substitute services in the residential sector. The principal substitute service has been the adoption of desktop valuations. Desktop valuation models provide an assessment of the value of a property without the need for a physical inspection by utilizing desktop access to extensive real estate databases. They are sometimes referred to as automated valuation models. Automated valuation model (AVM) is the name given to a service that can provide real estate property valuations using mathematical modeling combined with a database.

Appraisers, investment professionals and lending institutions use AVM technology in their analysis of residential property. AVMs are increasingly used by mortgage lenders to determine what a property might be worth in order for them to lend against the valuation. Besides, AVMs are particularly useful in assessing the value of a property portfolio. Using AVMs can also be useful for valuing an individual property where the provider can deliver a suitable level of accuracy.

Perhaps the most important feature in this trend is the underlying rationale for the substitute service of a desktop valuation. The buyers of desktop valuations, with respect to the residential sector, are seeking a service that will reduce their risk position, increase efficiency and reduce their costs in processing loans. As a result, they are looking for a standardized loan valuation process that reduces costs. Clearly, given the structure of the valuation industry and the significance of residential mortgage activity within the industry, this competitive force will continue to be an important influence on future directions for the valuation industry.

93.3.4 The Bargaining Power of Suppliers

Generally speaking, the power of suppliers is related to the number of suppliers and the specific inputs they provide. If there are few suppliers of a critical input then the power of the supplier as a competitive force is strong. For real estate appraisal suppliers, the most critical input is professional labor, i.e., valuers, which would typically form 50 % or more of the total expenditure of a firm. Suppliers of this input, professional labor, are not necessarily restricted or collectively organized and it is therefore postulated that they are not a threat in terms of a competitive force. As discussed before, the advantages of using AVMs over traditional chartered appraisers/surveyors are that they save time, money and resources, thus lowering the cost of valuing a property. Many AVMs can be made and used with little cost, so more choices in valuation methodology are also possible. It is claimed that unlike traditional surveyor valuations, AVM outputs do not suffer from the same fraud risk although certain providers can have their systems manipulated intentionally or otherwise if property features are incorrectly entered. AVMs remove the human element from the valuation process and rely on computer automation so as to remove human bias and subjectivity [10, 11]. A key evaluation method or no intellectual property is in the hands of one or a few agencies, institutions to provide products of alternative is higher, appraisal institution in the market competition, the price war is still the main method. However, in the era of big data, with huge amounts of data, customers do not have to consult professional appraisers, as long as data processing can clearly be understood by the market. Therefore, in the long run, the suppliers' bargaining power will fall further.

93.3.5 The Bargaining Power of Buyers

The bargaining power of the buyers is presently a much more significant force in determining present and future valuation industry structure. For the buyer's bargaining power, with the development of the entire real estate industry, the number of real estate appraisal services buyers will increase, but on the other hand, the number of appraisal service suppliers is also increasing. At present the real estate appraisal services have a strong homogeneity. Especially in recent years, Chinese government strengthens the regulation of real estate industry. The fastest growing stage for China's real estate industry has gone, followed by decreasing demand of real estate appraisal services. Big data era will bring more and more competitors. Therefore, different from the real estate market, in the real estate appraisal market, the buyers' bargaining power will be stronger and stronger.

93.3.6 Summary

Big data provide strong support for real estate industry to develop. It benefits realty enterprises to implement diversified investment through data re-mining for potential value. Therefore, real estate appraisal companies should make revolutionary changes in the way of collecting and using the big data, so as to gain new revenue growth opportunities in the age of big data.

With the inevitable growth of desktop valuations for residential mortgage lending, if appraisers and the firms providing valuation services are to prosper, they need to find out a solution, utilizing new technologies to enhance their valuation services. If desktop valuation services add value to the valuation industry's clients, then the provision of the service is inevitable and is best provided by qualified valuers capable of identifying possible valuation anomalies and interpreting market trends: the skills in GIS technology, the availability of more comprehensive and rigorous market data, risk management approaches to the mortgage market and statistical valuation methodologies.

The nature of the real estate valuation industry means that it will continue to be impacted upon by the bargaining power of the buyer and the threat of substitute services. Large volume buyers of valuation services will continue to explore ways of achieving the same or better levels of lending risk at lower costs. As appraisers, members of professional bodies and firms, they must embrace changes that technology and economic advances bring to their professional lives. They must lift the professionalism of the valuation practice to more than simple mass production of valuations at lowest cost, to a service which clients see as adding to the business process and for which they are prepared to pay a fee worthy of the expertise of the property professional.

93.4 Real Estate Appraisal Enterprises' Business Expanding in the Era of Big Data

With the advent of the era of big data, the traditional valuation activities and services are unable to meet the customers' diversified needs. Therefore, real estate appraisal industry should combine professionals with the big data in all areas, so as to realize the automation of real estate appraisal work, and innovate valuation methods.

93.4.1 Massive Valuation

At present, many real estate appraisal projects are involved with massive appraisal method, such as real estate market analysis in a certain area or a city, comprehensive damage assessment of houses in a certain area, hedonic price model analysis of a certain type of real estate, assessment of macro policy's impact on real estate industry and so on. However, under the condition of existing technology level, many real estate appraisal companies are limited to manual information management and analysis stage. They do not have their own data management platforms and are difficult to get access to data such as the government tax data and bank housing loans.

In the process of modernization in our country, the government will gradually open the data. Data sharing in the future will become a trend in the era of big data and will be paid more and more attentions by the government. For example, the federal government of America provided 500,000 types of data to private sectors. In fact, the International Association of Assessing Officers (IAAO) proposed a method of computer-assisted mass appraisal (CAMA) by combing public data and accumulated historical data. This method has been introduced in some districts of America [12]. Real estate appraisal companies will also benefit from this trend, and the most direct impact is that it is possible to use massive processing method in real estate appraisal. At present, different industries all over the country are engaged in the developing and research of computer-aided mass appraisal technique and system. In the future, who makes a breakthrough in the development of mass appraisal techniques, who will be able to achieve advantages in the future competition.

93.4.2 Automated Valuation

Automation is another big direction for real estate appraisal enterprises using big data. The ultimate aim of big data analysis is to turn data into information, turn information into knowledge, and turn knowledge into wisdom. Nowadays, more and more real estate companies start to launch automatic real estate assessment system, which can offer a range of vital property information to customers. Prepare useful property reports, value estimates, verify information and conduct valuable research and highly targeted marketing is no longer a difficult problem for clients. Take RP Data Company, a leading property data solution used by property professionals in Australia and New Zealand, as an example. Based on the Auto Val and Fusion Val AVMs, RP Data publishes an average of 3000 valuation reports. RP Data can even deliver a market value estimate of every available residential property in Australia. It uses statistical techniques to model comparable physical characteristics and recent sale prices of nearby properties and suburb characteristics.

Traditional real estate appraisers should learn information technology and know how to do data analysis. Through the analysis of massive amounts of data about real estate appraisal, they can set up mathematical models and design new products and services.

93.4.3 Valuation Methods Renovation

There are three traditional real estate appraisal methods: sales comparison approach, the cost approach and the income approach, but these three methods have their disadvantages and big data can effectively make up the defects of these methods. For instance, in the application of cost method, the index of depreciation rate is mainly based on the experience of the appraiser, which is very subjective. But now with big data, valuers can scientifically extract this index by analyzing the massive amounts of data and establishing the relationship between the market value and replacement cost. As for the income approach, the key of appraisal is to estimate the capitalization rate and rent (cash flows) and both these data are difficult to get. However, based on the analysis of large data, valuers can accurately determine the changing trend of the rent and capitalization rate. Likewise, when valuers use sales comparison approach, the number of the comparable sales is often too limited to meet the needs of appraisal, because even two apartments in the same building will be very different in value because of the natural lighting, orientation of the house or even interior decoration.

In the era of big data, valuation enterprises can make the selection of comparable sales very easy by type in elements such as a street address and zip code in the database management programs, it will provide comparable sales information up on up to 20 comparable properties to help the valuer analyze the value of the subject property.

93.5 Appraisal Enterprises' Innovation of Strategic Decision-Making

The era of big data will not only influence the size of the data which enterprises process, but also the way of thinking of enterprise manager, the enterprise's management system and the transformation of business model. Since it is still the early stage in the era of big data, for enterprises, the early they make a strategic decision through innovation of business expanding, the earlier they will get advantages in the future competition.

93.5.1 Innovation of Strategic Decision-Making Model

Enterprise's strategy innovation is the fundamental factor that will decide the direction of the enterprise. However, strategic decision-making process is highly complex and dynamic involving a large number of factors inside and outside the organization, as well as information collection, information costs, uncertainty and conflicts. Under the background of big data, the complexity and areas or structure involved with strategy decision are obviously different from traditional for enterprises. It is time for real estate appraisal enterprises to do big data in innovative investment and change the ways of acquiring information. It is new to both the supervisors and ordinary employees and neither of them have experience, therefore, decision-makers should update both software and hardware, and the corresponding cost will also increase [13].

Real estate appraisal enterprises should invest in the construction of database and the development of software programs, to provide support for strategic decision-making innovation. The strategy innovation, on the other hand, also faces enormous risks, because of the innovation of strategic decision demands for higher performance of information collection, information integration, information ability and rationality. At the same time, some enterprises may not take the big data for granted, so that they may have limited investment of money, energy and time, which will greatly reduce the accuracy of the strategic decision. Therefore, under the background of big data, the establishment of large data thinking is very important, which need real estate appraisal enterprises to update their thinking and ideas.

93.5.2 Innovation of Developing Model

Under the background of big data, due to the progress of information system and data processing technology, the traditional real estate appraisal business will be greatly expanded. The technology of mass appraisal and automatic valuation will be mature and well known, and the application of these new valuation technologies will be more accurate and reasonable. In the era of big data era, the golden rule is the application. Real estate enterprises should seize this rare strategic opportunity, integrate real estate data and provide all kinds of digital products and additional services as a commercial program. In the future, real estate appraisal industry will expand its business chain from single real estate appraisal to finance and real estate assets management, and even areas such as housing pricing, so that real estate appraisal enterprises will complete their transformation from the downstream of real estate value chain to the upstream–real estate consulting services provider.

93.5.3 Innovation of Marketing Model

In fact, there are over-developed real estates in some medium and small cities in China. Therefore, how to use big data to promote sales is crucial. In addition, traditional marketing models are ineffective as the popularization of electronic commerce is changing the consumption manners of the Chinese.

On the one hand, we should use the database and turn human real estate appraisal and assessment services into automatic assessment. On the other hand, we also should provide personalized custom services based on the individual needs of customers. Data resources are very important for real estate appraisal enterprises to raise competitiveness. The huge and perfect sources of data ensure precise customer location and the effective marketing. Accordingly, enterprises can implement precise marketing relying on information system. They can build the customer data system based on the possession of big data to categorize the customers, then extract useful information by big data mining for the precise marketing.

93.5.4 Strengthen Industry Cooperation and Resource Integration

Real estate appraisal is an integrated system, and there are many variables besides valuation technology and methods. However, real estate enterprises have their own advantages in forecasting the price trend and understanding the macro policy of the country. Therefore, they can make full use of their technology advantages and strengthen the exchange and collaborative construction with enterprises in the industry of database, data exchange and sharing, and technology development. If the estate enterprises are willing to share data, they can cooperate with the third-party, joining up with the market players like developers and home services and customers, fully demonstrating the advantage of big data.

93.6 Conclusion

The big data has become a type of significant strategic resource for estate appraisal enterprises to enhance competitiveness. Nevertheless, since the big data is just beginning, the practice on this field is incomplete, facing lots of challenges and threats, such as data collecting and processing technique. Some Chinese real estate enterprises have started to make huge innovations and achieved great success in the applications of big data. Further research might focus on the application of the big data in some typical real estate appraisal companies, studying the massive valuation methods and technology in detail. The applications of big data in appraisal enterprises will deepen related research. Meanwhile, scientific research will be conducive for estate enterprises at the era of big data.

References

1. A special report on managing information: data, data everywhere, The Economist, February 25, 2010; and special issue on "Dealing with data," Science, February 11, 2011
2. Laney D (2001) 3-d data management: controlling data volume, velocity and variety, META Group Research Note
3. Meijer E (2011) The world according to linq. Commun ACM 54(10):45–51
4. Zikopoulos P, Eaton C et al (2011) Understanding big data: analytics for enterprise class hadoop and streaming data. McGraw-Hill Osborne Media
5. Chui M, Brown B, Bughin J, Dobbs R, Roxburgh C, Byers AH (2011) Big data: the next frontier for innovation, competition, and productivity. McKinsey Global Institute
6. Chen M, Mao SW, Liu YH (2014) Big data: a survey. J Mob Netw Appl 19(2):171–209
7. Gantz J, David R (2011) Extracting value from Chaos. Sponsored by EMC Corporation, pp 1–10
8. Parry M (2011) A 'moneyball' approach to college. The Chronicle of Higher Education
9. Porter, Michael E (1980) Competitive strategy. The Free Press, A Division of MacMillan Publishing Co. Inc., New York
10. Downie ML, Robson G (2007) Automated valuation models: an international perspective. Council of Mortgage Lenders, London, pp 32–33. ISBN 1-905257-12-0
11. Mitropoulos A, Wu W, Kohansky G (2007) Criteria for automated valuation models in the UK. London, pp 1–2. ISBN 1-905257-12-0
12. Jensen DL (2011) The use of cross-validation in CAMA modeling to get the most out of sales. J Prop Tax Assess Adm 8(3):19–40
13. Du DY, Li AH, Zhang LL (2014) Survey on the applications of big data in Chinese real estate enterprise. J Procedia Comput Sci 30:24–33

Chapter 94
A Framework of Developing a Big Data Platform for Construction Waste Management: A Hong Kong Study

Xi Chen, Weisheng Lu and Shiju Liao

Abstract Big data has shown great potentials in improving management discretion in many areas. The applications of big data in areas such as finance, computer science, health care and medical science have made continued success. Despite of big data's potentials, its applications in construction waste management (CWM) are still in infant stage. In order to embrace these prospects, this research proposes a platform to apply big data technologies in CWM, by focusing on the CWM status of Hong Kong. This study first presents a framework of the big data platform by describing the data collection, storage and analysis involved in the roadmap of CWM in Hong Kong. The existing unintentionally generated big dataset in CWM in Hong Kong is used to examine the availability of the proposed platform. It was found that this platform could make government, industry and other CWM stakeholders benefit from proper cooperation. Though this study focused on the CWM of Hong Kong, it may act as a driving force to stimulate the adoption of big data in CWM across countries where construction waste is growing as an urgent issue that concerns government, industry, academia, and all the stakeholders.

Keywords Construction waste management · Big data · Platform · Hong Kong

94.1 Introduction

Construction waste has become a serious problem that hinders the sustainable development worldwide. The U.S. Environmental Protection Agency (EPA) estimated that 136 million tons of building-related construction and demolition (C&D)

X. Chen (✉) · W. Lu
Department of Real Estate and Construction, The University of Hong Kong, Hong Kong SAR, China
e-mail: chenx90@hku.hk

S. Liao
School of Construction Management and Real Estate International Research Centre for Sustainable Built Environment, Chongqing University, Chongqing, China

Y. Wu et al. (eds.), *Proceedings of the 20th International Symposium on Advancement of Construction Management and Real Estate*,
DOI 10.1007/978-981-10-0855-9_94

debris was generated in 1996 [1]. The wastage rates within the construction industry could reach 10–15 % in UK [2]. In Australia, C&D waste accounted for a large proportion of the industrial solid waste ended landfills [3]. In Hong Kong, the statistics show that solid waste ending up in landfills reached 14, 311 ton per day (tpd) in 2013, of which 25 % or 3591 tpd was from construction activities estimated by Hong Kong Environment Protection Department (HKEPD) [4]. Construction waste therefore places tremendous pressure on the valuable landfill space in this highly condensed city. Hong Kong is running out of landfill space far earlier than expected, and the existing landfills will be exhausted one by one by 2020 if waste levels continue to increase at current levels [5]. To pursue construction waste reduction, reuse and recycling (3Rs) researchers in CWM raised measures, such as on-site sorting [6, 7], prefabrication [8–10], and selective demolition [11, 12]. Innovations are therefore urgently needed to effectively implement the agenda in CWM on the basis of the traditional strategies.

The emerging big data has become a reality with a variety of prospects. Big data is defined as things one can do at a large scale that cannot be done at a smaller one, to create a new form of value in living, working, science and industry by changing markets, organizations, relationship between people, and more [13]. This technology is becoming the frontier for innovation, competition and productivity [14, 15]. It has been argued that an organization should foster the data-driven decision-making culture for management revolution to improve management performance [16]. Big data have been proved promising in various disciplines, including medical science [17], ecological science [18] and business [19]. Despite the application success of big data in so many fields, the application of big data in improving the performance of CWM is still in infant stage. Therefore, it is necessary to develop a platform that makes CWM take the advantages of big data technologies.

This research proposes a platform to apply big data technologies in CWM. This paper will present the framework of the big data platform by focusing on the data collection, storage and analysis based on the roadmap of CWM in Hong Kong. The existing unintentionally generated big dataset in CWM and related databases in Hong Kong will be used to examine the availability of the proposed platform.

94.2 A Framework of Big Data Platform in CWM

This study will focus on the collection, storage and analysis of big data in CWM in Hong Kong. First, the CWM roadmap in Hong Kong will be presented to demonstrate the occasions where big data can generate. This study will then describe the framework of big data platform by focusing on the requirements of the basic stages of big data application in CWM, including data collection, storage and analysis.

94.2.1 The CWM Roadmap in Hong Kong

Lu and Tam developed a roadmap based on the present CWM status in Hong Kong (see Fig. 94.1) [20]. This roadmap generally describes the process of transferring construction waste from construction site to government waste reception facilities. In Hong Kong, the waste materials generated on construction sites are dealt with 3Rs for using up their value in principle. Then, on-site sorting activities should be conducted to separate 'materials without value' to non-inert and inert waste materials, the rest of which are named mixed construction waste. The three types of waste are sent to corresponding three types of waste reception facilities, namely landfills, public fill reception facilities, and off-site sorting facilities. Within these facilities, the off-site sorting facilities will separate the mixed waste to non-inert and inert materials, which are finally sent to landfills and public reception facilities.

94.2.2 Data Collection

It is much easier to discover patterns in structured data than unstructured data or semi-structured data [21]. The necessary priority for data collection is to establish a scheme, wherein every construction project is demanded to report the project details to an organization at the design, construction, operation and demolition phases. Developers and contractors are clearest on these details, which are suggested fully recorded by government as structured data. Since 2005, Hong Kong Environmental Protection Department has enacted the Construction Waste Disposal Charging Scheme (CWDCS) as legislation to demand contractors open an account number for a construction project, which may send construction waste to disposal facilities managed by government [22]. This scheme recorded information of every construction project with a billing account, which is far from detailed though certain patterns can be discovered from the existing recorded information, including contract sum, site address, department, type of construction work, detail of construction work, and remarks. The good thing is they are formed as structured data filed and

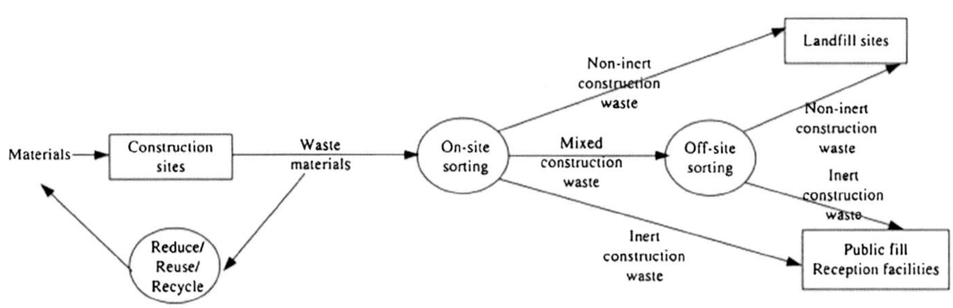

Fig. 94.1 The roadmap of CWM in Hong Kong developed by Lu and Tam [20]

stored by EPD, while some information (e.g., floor area of building projects, green building/not green building, developers, building information modeling (BIM) applied/BIM not applied) useful for analyzing the correlations between CWM and various features has not been recorded in this way. Actually, these details must be available in different organizations, and for unified analysis of data in CWM, they are regarded as unavailable and unstructured data. To address the availability of big data analysis, it is necessary for government to set a well-planed scheme should be set for demanding as much structured data composed of very detailed information contributed by stakeholders as possible to get ready for the future analysis of construction waste.

With the ambition and scheme to collect big data in CWM, the next thing is to define the data sources. The roadmap reveals the route of construction waste, where data may generate if sensors are implanted in the right positions. At present, there are some principles to improve CWM performances, so that the proposed platform can conduct data collection with certain purposes. First, the on-site waste 3Rs are crucial for the saving materials [23], thus should be encouraged with incentives. On-site sorting has benefits including increasing the rates of reuse and recycling, reducing the cost for waste transportation and disposal, prolonging the lifespan of landfills designed for receiving non-inert construction waste, and lessening the pollution resulted from the huge amount of construction waste [6, 7]. Sensors or manual supervision should be placed to construction site to measure the quantity and types of on-site sorting and materials conducted with 3Rs if a construction project want to apply an award or competition organized by government, such as 'Annual 3Rs Best Award', where CWM performance ranking presents reputation of construction companies. Then, the construction waste is sent to the designated facilities. In the transportation process, the routes and timelines of vehicles may be recorded for analysis for afterward route optimization. The facilities should also record the features of every vehicle of construction waste as structured data for afterward analysis. Also under the CWDCS, EPD have recorded such data since 2005, which is somehow useful but far less than enough to discover much of the value. The data recording should be as systematical and detailed as possible [22]. The data can be collected for real-time and afterward analysis for the purposes that people presently attempt to achieve or are not yet aware of now but might be meaningful someday.

94.2.3 Data Storage and Analysis

Objects, process and services of the CWM system can generate an explosive growth of many types of new data to be saved and analyzed. This platform needs to find a way forward with its storage infrastructure to facilitate the big data whether it's in the cloud or within the data center. People nowadays are living in a real-time world, therefore the data storage and analysis advance wave upon wave in a very short time period. A power data analytics infrastructure is required to make big data fast, affordable and available for decision makers in some real-time cases. For example, if the routes of vehicles carrying construction waste show that a large number of vehicles are approaching one of the disposal facilities, at that time, the big data platform should inform the other waste haulers who is ready to go to that facility to change their schedule in case of joining a very long queue. Another example is the allocation of vehicles to construction site according to the distances, availability and service attitude of drivers. In this data collection, storage and analysis process, this platform creates transformational value in form of new services, which can be providing convenience to stakeholders in CWM. The nowadays data storage technologies developed by companies, such as IBM have become very mature to deal with the proposed big data generated in CWM in Hong Kong.

The big data generated in the proposed platform are used to discover patterns and correlations using the big data analytics. Big data of waste generation in this study may possibly be used to build models in CWM. Such models (e.g., the waste generation amount vs. construction time model) are very useful in planning the CWM for new projects with proper features input, so that a consultant can make a reasonable waste management plan before the commencement of a new project. With the data growth, the models can be reshaped and more accurate in prediction. These models can also be used to calculate the lifespan of the existing landfills and other facilities, and estimate the necessity of building new facilities to facilitate the large amount of construction waste. The value of big data in sustained promotion of CWM can only be realized by hybrid persons who not only have skills in big data technologies, but also know how to tackle problems and communicate with people about these problems, solve them and then figure out the way to transfer the solutions into real scenarios.

94.2.4 Stakeholders of the Big Data Platform

The development of the big data platform aims to bring values to the environment and different groups of people. The most important stakeholders in CWM unavoidably should be government, who manage this platform, and construction companies. Although government has to pay for the establishment of such a platform, the benefits from such a system can be very promising evidenced by the following examples. Stimulating the 3Rs and on-site sorting can reduce the amount of construction waste so as to slow down the environmental deterioration contributed by construction waste. With incentives to stimulate 3Rs on site, the construction companies, which well performed 3Rs, may gain many benefits. The most important benefit is saving cost for materials purchases. On the other hand, Poon et al. suggested that at the first stage, waste producers only need to pay 50 % of the landfilling cost (HK$ 55/t, US$ 7/t), but in the subsequent stages, the charge will be increased to cover the full construction and operation costs of the landfill sites [6]. Saving waste disposal cost is another benefit for construction companies. The amount of construction waste received from so many construction sites, the necessity of landfills and other facilities. There are some added benefits for construction companies such as gaining reputation from excellent waste management, followed by more business opportunities. In order to implement the on-site CWM to make sure the availability of the big data platform, government would be better to provide education for contractors, on-site supervision, and easily understanding website with relevant knowledge.

Meanwhile, other stakeholders such as waste haulers, big data professionals, CWM professionals, and residents near landfills may also benefit from this big data platform. For example, this big data platform may provide convenience for construction waste haulers by monitoring their transportation routes. Also, the government can avoid waste illegal dumping activities with haulers under supervision of this real-time tracking system of this platform. The platform may also provide professionals in big data and CWM with career opportunities. People who live in this compact city may worry less about of the building and extension of landfills and other waste disposal facilities near their homes.

94.3 Availability of the Big Data Platform in CWM

Section 2 described a framework of developing the big data platform in CWM. To examine the availability of the platform, this study will raise some examples that can demonstrate big data generated from such a platform have great potential, by analyze the construction waste disposal records recorded by EPD in 2011, 2012, 2013 and 2014 combined with an 'account detail' database. The existing big data are not generated from the proposed big data platform, but would become a part of the big data from the platform once it is established. In the proposed framework, this study has raised some potential usages of this platform in providing services and improving environment. Since some of the values are already discovered and published, these studies will be used as examples to demonstrate the availability of this platform.

Chen et al. used more than a million waste disposal records generated in 2011 to benchmark the performance of CWM in Hong Kong [24]. Waste generation rate is a key performance indicator for interpreting the waste management performance. The benchmarks can be used as a baseline for measuring the good or bad performances of CWM. With the benchmarks, the contractors of future projects may constrain their waste generation amount to a reasonable value. Government can also use the benchmarks to improve the waste disposal charging system. Another example is Lu et al. compared the CWM performances between private and public sectors using the big data in 2011 and 2012, totally more than two million construction waste disposal records of about 4000 construction projects conducted in Hong Kong, and found there is a notable disparity of construction waste management (CWM) performance between the public and private sectors [25]. This also provides references for government and sectors to improve their strategies in CWM. The use of big data generated from the platform can also be predicting the future waste generation. Chen et al. developed a universal S-curve for predicting the waste generation of new projects based on the waste disposal records in 2011, 2012, 2013 and 2014 [26]. The S-curve equation of a new project in Hong Kong can be expressed by inputting the characteristics, including contract sum, location, public and private nature of the project. This makes waste management plan easier and more accurate.

94.4 Conclusion

This study proposed a framework of developing a big data platform in CWM in Hong Kong by considering the CWM roadmap, data collection, and data storage and analysis. The framework of the platform can provide government and industry with a better understanding of their performance in CWM so as to achieve sustained improvement. Since different regions may have different roadmaps for CWM, the framework is only applicable to those regions whose construction waste is systematically managed by government. The study then examined the availability of the platform by referring to the previous examples of using the existing big data. It is found that this platform will contribute to environmental protection and benefit a large group of stakeholders, including government, contractors, professionals and general publics. This study only listed some examples of benefits from the platform. Future studies are suggested to figure out more value that can be discovered from the big data platform, so as to demonstrate the importance of adopting such a platform in CWM.

References

1. U.S. EPA (1998) Characteristics of building-related construction and demolition debris in the United States. U.S. Environmental Protection Agency. Report prepared by Franklin Associates, Rep. No. EPA 530-R-98-010
2. McGrath C, Anderson M (2000) Waste minimizing on a construction site. Build Res Establishment Digest 447
3. Lingard H, Gilbert G, Graham P (2001) Improving solid waste reduction and recycling performance using goal setting and feedback. Construct Manage Econ 19(8):809–817
4. HKEPD (2015) https://www.wastereduction.gov.hk/sites/default/files/msw2013.pdf. Accessed on 11 June 2015
5. HKEPD (2015) http://www.epd.gov.hk/epd/english/environmentinhk/waste/waste_maincontent.html. Accessed on 09 Sept 2015
6. Poon CS, Ann TW, Ng LH (2001) On-site sorting of construction and demolition waste in Hong Kong. Resour Conserv Recycl 32(2):157–172
7. Wang J, Yuan H, Kang X, Lu W (2010) Critical success factors for on-site sorting of construction waste: a China study. Resour Conserv Recycl 54(11):931–936
8. Jaillon L, Poon CS, Chiang YH (2009) Quantifying the waste reduction potential of using prefabrication in building construction in Hong Kong. Waste Manag 29(1):309–320

9. Lu WS, Yuan HP (2013) Investigating waste reduction potential in the upstream processes of offshore prefabrication construction. Renew Sustain Energy Rev 28:804–811
10. Tam VW, Hao JJ (2014) Prefabrication as a mean of minimizing construction waste on site. Int J Constr Manage 14(2):113–121
11. Kourmpanis B, Papadopoulos A, Moustakas K, Kourmoussis F, Stylianou M, Loizidou M (2008) An integrated approach for the management of demolition waste in Cyprus. Waste Manage Res 26(6):573–581
12. Poon CS, Yu ATW, See SC, Cheung E (2004) Minimizing demolition wastes in Hong Kong public housing projects. Constr Manage Econ 22(8):799–805
13. Mayer-Schönberger V, Cukier K (2013) Big data: a revolution that will transform how we live, work, and think. Houghton Mifflin Harcourt
14. Manyika J, Chui M, Brown B, Bughin J, Dobbs R, Roxburgh C, Byers AH (2011) Big data: the next frontier for innovation, competition, and productivity. Report by McKinsey Global Institute
15. Gobble MM (2013) Big data: the next big thing in innovation. Res-Technol Manage, January–February
16. McAfee A, Brynjolfsson E (2012) Big data: the management revolution. Harvard Bus Rev 90:60–66
17. Murdoch TB, Detsky AS (2013) The inevitable application of big data to health care. JAMA 309(13):1351–1352
18. Hampton SE, Strasser CA, Tewksbury JJ, Gram WK, Budden AE, Batcheller AL et al (2013) Big data and the future of ecology. Front Ecol Environ 11(3):156–162
19. Chen H, Chiang RH, Storey VC (2012) Business intelligence and analytics: from big data to big impact. MIS Quart 36(4):1165–1188
20. Lu W, Tam VW (2013) Construction waste management policies and their effectiveness in Hong Kong: a longitudinal review. Renew Sustain Energy Rev 23:214–223
21. McCallum A (2005) Information extraction: distilling structured data from unstructured text. Queue 3(9):48–57
22. HKEPD (2015) http://www.epd.gov.hk/epd/misc/cdm/scheme.htm. Accessed on 09 Sept 2015
23. Wu Z, Shen L, Ann TW, Zhang X (2015) A comparative analysis of waste management requirements between five green building rating systems for new residential buildings. J Cleaner Prod
24. Chen X, Lu W, Ye M, Shen L (2015) Construction waste generation rate (WGR) revisited: a big data approach. In Proceedings of the 19th International symposium on advancement of construction management and real estate. Springer, Heidelberg, pp 843–854
25. Lu W, Chen X, Ho DC, Wang H (2015) Analysis of the construction waste management performance in Hong Kong: the public and private sectors compared using big data. J Cleaner Prod
26. Chen X, Lu W, Peng Y (2015) The S-curve for forecasting construction waste generation. In: Proceedings of International conference on solid waste. 12–23 May 2015, Hong Kong

Chapter 95
Assessment Model of the Effectiveness of Construction and Demolition Waste Reduction Measures: A Case of Shenzhen

Ting Wang, Jiayuan Wang, Dong Wu and Yu Gao

Abstract With the continuous development of urbanization, a large number of construction and demolition wastes (CDW) caused by the construction projects have become increasingly prominent, and CDW reduction is one of the effective means to improve the status quo. In recent years, CDW reduction has been one of research focuses in both industrial and academic areas. However, previous researches mainly pay attention to reduction measures while the effectiveness of these measures is seldom. In addition, most of the existing studies use static methods, lacking of consideration of the mutual influences between the various factors in the reduction system. Therefore, based on system dynamics theory, the paper developed a dynamic model to assess the effect of the key measures for CDW reduction. Firstly, on the basis of literature review, this paper identified 11 key measures mainly from the aspects of management, techniques and stakeholders' motivation. Secondly, a model to evaluate the effectiveness of waste reduction measures was established via Vensim software. Finally, the real data of Shenzhen were used to conduct a case study and a scenario simulation analysis. The simulation results show that the model is valid and robust which can be used for quantitative assessment of the effectiveness of CDW reduction measures and serve as a valuable decision support tool to management departments.

Keywords CDW · Reduction · Effectiveness · System Dynamics

T. Wang · Y. Gao
School of Civil Engineering, Shenzhen University, Shenzhen, China

J. Wang (✉)
Department of Civil Engineering, Shenzhen University, Shenzhen, China
e-mail: wangjy08@qq.com

D. Wu
School of Ecology and Environmental Science, East China Normal University, Shanghai, China

Y. Wu et al. (eds.), *Proceedings of the 20th International Symposium on Advancement of Construction Management and Real Estate*,
DOI 10.1007/978-981-10-0855-9_95

95.1 Introduction

As known, new construction, expansion, renovation as well as other construction processes has generated a lot of CDW. According to statistics, the total amount of CDW generated each year in China is accounted for 40 % of total municipal solid waste [1] while America annually produces about 143 million metric tons of CDW [2]. A large number of CDW not only consumes a lot of raw materials, pollutes natural environment, but also causes huge economic losses. Kulatunga et al.'s [3] research shows that in the United States, 25 % of the national annual consumption of wood and 40 % of the national annual consumption of sand, stone are consumed by construction industry.

As the increasing amount of CDW, over the past 20 years, some experts have made a lot of beneficial research and study in an effort to reduce CDW generation and minimize their social and environmental impacts. It is worth mentioning that the CDW reduction is also the most environmentally friendly, which it is in the top-level of the construction waste management "3Rs" (reduction, reuse, recycling) principles [4] and has become one of the focuses in the field of CDW management [2]. Hao et al. [5] from the aspects of design stage, technical reform and construction technology of Hong Kong CDW reduction measures have been identified, and some useful suggestions are proposed based on the research. Through interviews and field research, Li [6] have made the investigation on the implementation CDW reduction measures on construction site, and points out measures such as improving workers' technology education training, field classification and sorting, can effectively reduce CDW. Through a survey of construction workers in Shenzhen, Zhu et al. [7] points out that it is the lack of awareness of waste reduction education and technical training, as well as incentives from the owners who are vital to support to CDW reduction that impede construction workers from actively carrying out reduction. In terms of design and construction technology, with the case study of HK area, Jaillon et al. [8] found using prefabricated construction technology can reduce about 52 % of waste. At the same time, the modular design can reduce the waste generation during welding process [9], and reducing design changes can effectively avoid CDW brought by repeated construction [10]. In terms of the management of policies, Kautto et al. [11] carried on a comparative study on the situations before and after levy CDW landfill tax in Finland. The results show that the landfill tax can be effective in promoting the construction enterprises' reduction initiatives. The ineffective governmental supervision of the implementation of CDW management policies also hinder the effective implementation of the reduction measures [12].

As mentioned above, previously studies are relatively fruitful on the CDW reduction measures but effectiveness assessment in the field of CDW is seldom discussed, especially the effectiveness of reduction measures. Yuan et al. [2] proposed an assessment model of CDW management policies based on system dynamics, and they noted that this simulation model can effectively help policy-makers clearly understand the impact of different management policies in reduction system, thereby decision-makers can develop more scientific and sustainable measures. Tam [13] assessed the effectiveness of waste management plan in Hong Kong via questionnaires and structured interviews, and studies have shown that lack of economic incentives and high indirect costs are major obstacles for the implement of policies.

The main purpose of this paper is to assess the effectiveness of major reduction measures in CDW reduction system. Firstly, we identified the key measures of CDW reduction. Secondly, a modeling and conducted simulation via Vensim software is developed. This article not only identifies the key factors influencing the effect of CDW reduction, but also proposes a model to help decision-makers better understand under different management policies so that make more scientific and effective management measures.

95.2 System Dynamics (SD) Methodology

SD was first founded by MIT Professor Forrester in 1956, it is a research and simulation method to solve the complex problems with dynamic feedback based on a feedback control theory, computer technology, structural-functional analysis [14]. CDW management system itself is a complex system. Moreover, coupled with a handle highly nonlinear dynamical systems, higher-order, more variable, more feedback solving skills as well as qualitative and quantitative characteristics, make the system dynamics widely used in the CDW management research in recent years. For example, CDW management policy analysis [15], environmental and economic impact assessment of CDW landfill [16], as well as CDW recycling plant site selection [17].

The system dynamics modeling typically have four steps as shown in Fig. 95.1. It should be pointed out that the causal loop diagram and stock-flow diagram are

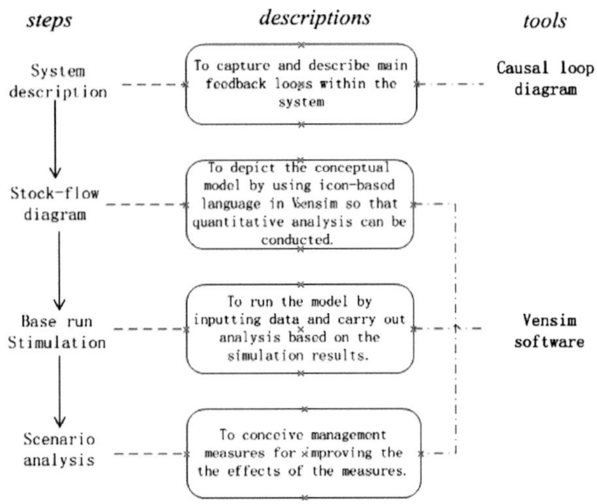

Fig. 95.1 The development steps of system dynamics modeling

essentially the same and the difference is that the relationship outlined with arrows and words between by the former one, while the latter is more complicated which depicts the quantitative relation among all variables with equations and computer programs [14].

95.3 Model Development

95.3.1 Key Measures Identification

Base on related literature summarized in this paper, 11 key measures in terms of CDW reduction measures were identified, namely Management A (on-site classification and sorting, CDW landfill charging fee, supervision of CDW policy), Stakeholders' Reduction Motivation B (reduction education and technical training on construction workers, reduction education and technical training on designers, governmental subsidies policy, enterprises' economic incentives to employees) as well as Design and Construction Techniques C (use of prefabricated component, design changes reduction, use of metal templates, modular design), see in Table 95.1

95.3.2 Stock Flow Diagram

As mentioned before, stock-flow diagram depicts the relationship between variables more comprehensively. According to identification of various key measures

Table 95.1 CDW reduction key measures

Code	Measures	Key references
M_{A1}	On-site classification and sorting	[6]
M_{A2}	CDW landfill charging fee	[11]
M_{A3}	Supervision of CDW policy	[12]
M_{B1}	Reduction education and technical training on construction workers	[7]
M_{B2}	Reduction education and technical training on designers	[5]
M_{B3}	Governmental subsidies policy	[5]
M_{B4}	Enterprises' economic incentives to employees	[7]
M_{C1}	Use of prefabricated component	[8]
M_{C2}	Design changes reduction	[10]
M_{C3}	Use of metal templates	[6]
M_{C4}	Modular design	[9]

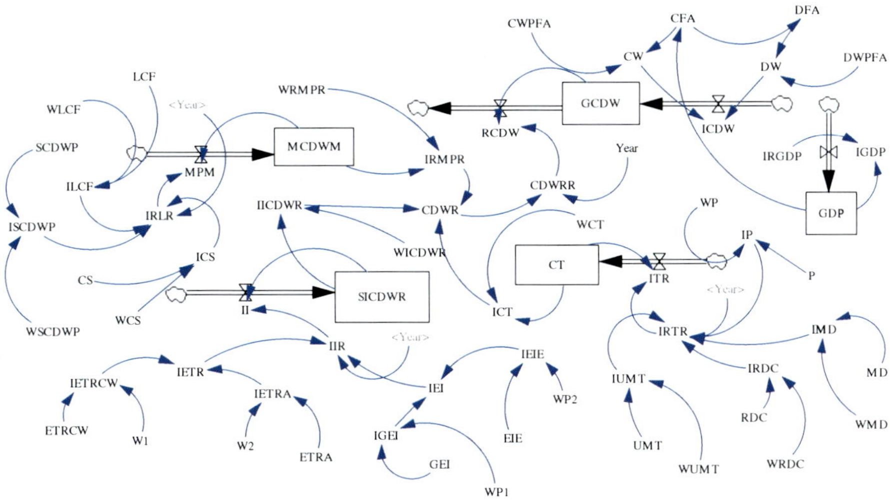

Fig. 95.2 A stock-flow diagram for assessment the effectiveness of CDW reduction

(Table 95.1), this paper developed the stock flow diagram via software Vensim (see Fig. 95.2). There are five stocks, six flows, 26 auxiliary variables and 27 constants in model and the details of the variables are referenced in Appendix 1.

95.4 Model Simulation and Results Analysis

95.4.1 Data Collection and Quantification

As for qualitative analysis and simulation, stock flow diagram of all variables should first be quantified. In this paper, there are mainly two different channels to quantify data. Parts of the data were from the statistics, the governmental reports and literature, such as the construction floor area. Assumed that Shenzhen annual construction floor area has a linear relationship with GDP [17], Shenzhen 2005–2012 year with construction floor area and GDP data of Shenzhen from 2005 to 2012 (as shown in Table 95.3) [18]. The construction floor area was worked out via Excel.

$$\text{Construction floor area} = 1.4939 * \text{GDP} - 1107.1$$

According to the "Shenzhen City Master Plan (2010–2020)" [19], GDP growth rate was set as follow: 2015–2020 year 9 %, 2021–2030 year 7.5 % (Table 95.2).

As some variables in this model are difficult to obtain from existing studies, so they were obtained through the questionnaire. In this research, 50 copies were

Table 95.2 Construction floor area and GDP statistics in Shenzhen [18]

Years	Construction floor area (ten thousand m^2)	GDP (billion)
2005	5063.23	4950.91
2006	5163.59	5813.56
2007	5866.51	6801.57
2008	5394.61	7786.79
2009	5690.62	8201.32
2010	5980.3	9581.51
2011	8328.6	11505.53
2012	9731.9	12950.06

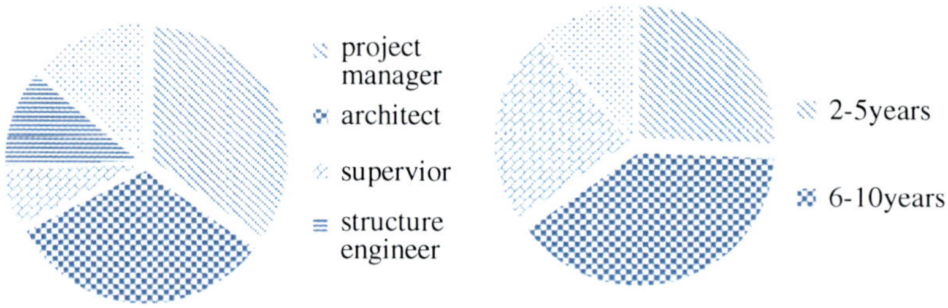

Fig. 95.3 **a** Source of respondents **b** work experience

distributed in the form of an electronic questionnaire (http://www.sojump.com/jq/4342954.aspx), and the details of the questionnaire are shown in Fig. 95.3a, b.

The degree of importance of the reduction measures in the questionnaire was divided into five levels, 1 represents "not important", 2 represents "more important", 3 represents "important" and 4 represents "relatively important" and 5 represents "very important". The average level of importance for each measure is as follow:

$$\mathrm{u}_i = \frac{1 * n_1 + 2 * n_2 + 3 * n_3 + 4 * n_4 + 5 * n_5}{n_1 + n_2 + n_3 + n_4 + n_5}$$

Among them, the weight of each sub-measure is calculated as follows:

$$\mathrm{w}_i = \frac{u_i}{\sum_{i=1}^{j} u_i}$$

which shows the impact weight of measure i; j indicates the existence of measure j in the module.

Meanwhile, in order to convert qualitative variables (such as supervision of CDW policy, use of prefabricated component, etc.) into quantitative factor and input them into model, the following formula is processed:

Table 95.3 The impact weight and value of CDW reduction measures

Rank	Measures	The average level of importance	Weight	Value
1	M_{B4}	4.12	0.264	0.78
2	M_{A3}	4.05	0.358	0.76
3	M_{C1}	3.99	0.266	0.74
4	M_{C4}	3.97	0.265	0.72
5	M_{B3}	3.93	0.252	0.73
6	M_{B2}	3.78	0.243	0.71
7	M_{A1}	3.77	0.332	0.69
8	M_{B1}	3.76	0.241	0.68
9	M_{C2}	3.60	0.241	0.65
10	M_{A2}	3.51	0.310	0.53
11	M_{C3}	3.42	0.228	0.60

$$\mathrm{d}_i = \frac{n_1 * 0 + n_2 * 0.25 + n_3 * 0.5 + n_4 * 0.75 + n_5 * 1}{n_1 + n_2 + n_3 + n_4 + n_5}$$

where represents the value of quantitative measures i.

As mentioned above, the impact weight of each measure and quantitative value calculate are shown in Table 95.3.

95.4.2 Model Validation

After being quantified, the validation of model should be tested to ensure this model can reflect the real world in a meaning way. This study selected three common tests, namely boundary adequacy test, dimension consistency test and extreme conditions test. Finally, this model has past all tests mentioned above.

95.4.3 Results of Base Run

Considering that the implementation of CDW reduction measures would take some time, so the simulation time was set for 10 years (2015–2024). After inputting the data of Shenzhen into model, relevant simulation outputs are shown in Fig. 95.4a, b.

Figure 95.4a shows the simulation result of CDW reduction rate. As shown in Fig. 95.4a, the increase of CDW reduction rate at almost exponential, and from the simulation period of beginning to sixth year, growth rate is relatively stable, but in the late simulation (seventh—final) period, it witness a significant increase. As implementation of CDW reduction measures and the process usually take a long

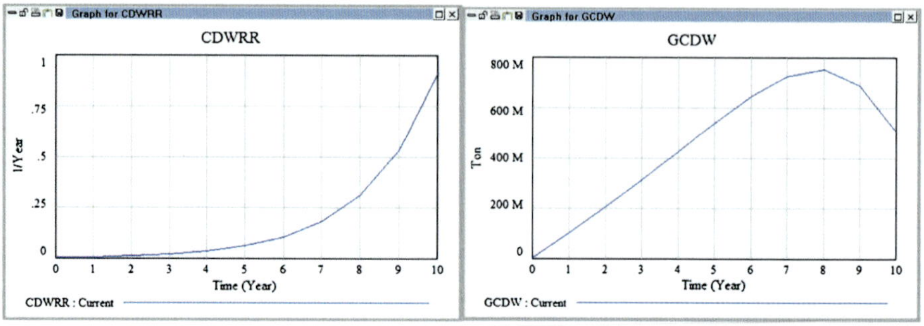

Fig. 95.4 **a** CDW reduction rate simulation results **b** CDW generated simulation results

period time, in the early simulation time, the effectiveness of reduction measures may not fully appeared. However, in the middle and later periods, the effectiveness of the measures gradually appears so that CDW reduction rate increases significantly.

Figure 95.4b shows the simulation result of CDW generation. As can be seen, the generation amount of CDW increases first and then decreases. During the early 8 years, the amount of CDW continues to increase, peaking at 76.74 million tons in 8 year, and followed by a decline, reaching to 50.3 million tons at the end of simulation. This may be that the early stage of CDW reduction measures is not taken good in the early stage, but in later, CDW reduction rate witnesses an upward trend since the effect of reduction measures reflect on the generation amount of CDW. Table 95.4 lists the simulation result of effectiveness of different measures on CDW reduction. In the final, CDW reduction rate reaches to 0.917, including the impact of management measures on CDW reduction (0.054), the impact of stakeholders' motivation on CDW reduction (0.804) and impact of design and

Table 95.4 Simulation results of effectiveness of different measures on CDW reduction

Years	The impact of management measures on the CDW reduction	The impact of stakeholders' motivation on CDW reduction	The impact of design and construction techniques on CDW reduction	CDW reduction rate
1	0.00055	0.00590	0.00055	0.00700
2	0.00092	0.01019	0.00092	0.01203
3	0.00153	0.01759	0.00156	0.02068
4	0.00255	0.03037	0.00261	0.03553
5	0.00424	0.05243	0.00439	0.06106
6	0.00706	0.09052	0.00738	0.10496
7	0.01176	0.15625	0.01241	0.18042
8	0.01959	0.26973	0.02087	0.31019
9	0.03263	0.46564	0.03506	0.53333
10	0.05435	0.80382	0.05895	0.91712

construction techniques on CDW reduction (0.589). Obviously, as values shown in Table 95.4, the impacts of measures on CDW reduction are ranked as follows: Stakeholders' motivation > design and construction techniques > CDW management measures. This result shows that the stakeholders' motivation is the most important factor to promote the CDW reduction, which echoes with findings from previous studies, such as Yuan et al. [4] and Zhu et al. [7].

95.5 Scenario Analysis

In theory, the effectiveness of reduction is more significant if promoting CDW reduction measures. However, what the extent of effectiveness on CDW reduction still need to be simulated. Therefore, three different scenarios have simulated and demonstrated in this paper. As revealed above, the two measures, namely, "enterprises' economic incentives to employees" and "governmental subsidies" could influence the CDW reduction in a more efficient way. Therefore, they are selected to simulate how changes in such measures could influence the CDW reduction. As a result, three scenarios are developed, which are:

Scenario 1 (S1): this scenario is concerned with how changes of the enterprises' economic incentives would affect the effectiveness of CDW reduction.
Scenario 2 (S2): this scenario aims at studying the extent of effectiveness of CDW reduction concerning the measure of governmental subsidies.
Scenario 3 (S3): this scenario is designed to examine the effect of changing both enterprises' economic incentives and governmental subsidies on CDW reduction.

It is obvious that S1 and S2 are sing-policy scenarios, but S3 is multi- policy scenario.

Simulation results of three designed scenarios are shown in Table 95.5. In S1, the model is simulated under three situations, among which one is base run and the other two are designed to test what differences of increasing the enterprises' economic incentive would contribute to the CDW generation,and enterprises' economic incentive is 0.78 and added to 0.88, 0.98 respectively. It is obvious seen from Table 95.5 (a), the amount of CDW decreases from the initial 50.3 million tons to 37.8 million tons in the end, with a decline of 24.85 %. S2 is to test how different levels of effectiveness of governmental subsidies on final CDW generation, and also there are three situations under this scenario. The initial value is 0.73, and added to 0.83 and 0.93 in the other two situations. The results can be told from Table 95.5 (b), the CDW generation has decreased to 38.4 million tons (decreased by 23.66 %). The results of S1 and S2 indicates that adaption of such measures can lead to improvement of CDW reduction so that decrease CDW generation. In scenarios 3, two measures are taken to test in our model, namely enterprises' economic incentive and governmental subsidies. As shown in Table 95.5 (c), the results demonstrate that the generation amount of CDW is reduced significantly from 50.3 million tons to 25.9 million tons (dropped by 48.51 %), indicating the

Table 95.5 Simulation results of scenario analysis (unit: tons)

(a) Scenario 1				(b) Scenario 2			(c) Scenario 3		
Years	1. CDW generation	2. CDW generation	3. CDW generation	1. CDW generation	2. CDW generation	3. CDW generation	1. CDW generation	2. CDW generation	3. CDW generation
1	1.02×10^8	1.02×10^8	1.02×10^8	1.02×10^8	1.02×10^8	1.02×10^8	1.02×10^8	1.02×10^8	1.02×10^8
2	2.05×10^8	2.05×10^8	2.05×10^8	2.05×10^8	2.05×10^8	2.05×10^8	2.05×10^8	2.05×10^8	2.05×10^8
3	3.14×10^8	3.14×10^8	3.14×10^8	3.14×10^8	3.14×10^8	3.14×10^8	3.14×10^8	3.14×10^8	3.14×10^8
4	4.27×10^8	4.26×10^8	4.26×10^8	4.27×10^8	4.26×10^8	4.26×10^8	4.27×10^8	4.26×10^8	4.25×10^8
5	5.40×10^8	5.39×10^8	5.37×10^8	5.40×10^8	5.39×10^8	5.38×10^8	5.40×10^8	5.38×10^8	5.35×10^8
6	6.44×10^8	6.41×10^8	6.38×10^8	6.44×10^8	6.41×10^8	6.38×108	6.44×10^8	6.38×10^8	6.31×10^8
7	7.24×10^8	7.16×10^8	7.07×10^8	7.24×10^8	7.16×10^8	7.08×10^8	7.24×10^8	7.07×10^8	6.88×10^8
8	7.52×10^8	7.32×10^8	7.12×10^8	7.52×10^8	7.33×10^8	7.14×10^8	7.52×10^8	7.13×10^8	6.68×10^8
9	6.89×10^8	6.50×10^8	6.08×10^8	6.89×10^8	6.51×10^8	6.12×10^8	6.89×10^8	6.10×10^8	5.24×10^8
10	5.03×10^8	4.41×10^8	3.78×10^8	5.03×10^8	4.44×10^8	3.84×10^8	5.03×10^8	3.81×10^8	2.59×10^8

adaption of combined measures can contribute to decrease CDW generation to a larger extent.

By comparing the simulation results of S1 and S2, it is obvious that enterprises' economic incentive is more effective to reduce CDW generation than that of governmental subsidies. In addition, combined measures S3 reduce CDW than that in the single one in a more efficient way, which should be highlighted in decision-making in the future.

95.6 Conclusion

Although experts began to focus on the problem of CDW even from 1908s, there always lack an effective way to evaluate the effect of CDW reduction. The main contributions of this paper are in three points. Firstly, a system dynamics model that can be used as a useful platform to assess the effectiveness of different reduction measures on CDW reduction. Secondly, the measure of enterprises' economic incentive to employee is the most significant to CDW reduction, which should be highlighted in the future. Thirdly, three different policy scenarios are designed to provide decision-makers with more scientific information concerning CDW reduction.

Acknowledgments The authors wish to express our sincere gratitude to the Special Fund of the Central Finance for the Development of Local Universities (No. 000022070151) for their financial support on this research.

Appendix1

No.	Abbreviations	Variable name	Variable type
1	CDWR	CDW Reduction	Auxiliary Variable
2	CDWRR	CDW Reduction Rate	Auxiliary Variable
3	CFA	Constructed Floor Area	Auxiliary Variable
4	CS	Classification and Sorting	Parameter
5	CT	Construction Technologies	Stock
6	CW	Construction Waste	Auxiliary Variable
7	CWPFA	Construction Waste Per Floor area	Parameter
8	DFA	Demolished Floor Area	Auxiliary Variable

(continued)

(continued)

No.	Abbreviations	Variable name	Variable type
9	DW	Demolition Waste	Auxiliary Variable
10	DWPFA	Demolition Waste Per Floor Area	Parameter
11	EIE	Economic Incentives from Enterprise	Parameter
12	ETRA	Education and Training on Reduction for Architects	Parameter
13	ETRCW	Education and Training on Reduction for Construction Workers	Parameter
14	GCDW	Generated CDW	Stock
15	GDP	GDP	Stock
16	GEI	Governmental Economic Incentives	Parameter
17	ICDW	Increased CDW	Flow
18	ICS	Impact of Classification and Sorting	Auxiliary Variable
19	ICT	Impact of Construction Technologies	Auxiliary Variable
20	IEI	Impact of Economic Incentives	Auxiliary Variable
21	IEIE	Impact of Economic Incentives from Enterprise	Auxiliary Variable
22	IETR	Impact of Education and Training on Reduction	Auxiliary Variable
23	IETRA	Impact of Education and Training on Reduction for Architects	Auxiliary Variable
24	IETRCW	Impact of Education and Training on Reduction for Construction Workers	Auxiliary Variable
25	IGDP	Increasing GDP	Flow
26	IGEI	Impact of Governmental Economic Incentives	Auxiliary Variable
27	II	Increasing Initiative	Flow
28	IICDWR	Impact of Initiative on CDW Reduction	Auxiliary Variable
29	IIR	Initiative Increasing Rate	Auxiliary Variable
30	ILCF	Impact of Landfill Charging Fee	Auxiliary Variable
31	IMD	Impact of Modular Design	Auxiliary Variable
32	IP	Impact of Prefabrications	Auxiliary Variable
33	IRDC	Impact of Reducing Design Changes	Auxiliary Variable

(continued)

(continued)

No.	Abbreviations	Variable name	Variable type
34	IRGDP	Increasing Rate of GDP	Auxiliary Variable
35	IRLR	Impacts of Related Laws and Regulations	Auxiliary Variable
36	IRMPR	Impact of Related Management Policies on Reduction	Auxiliary Variable
37	IRTR	Increasing Rate of Technologies on Reduction	Auxiliary Variable
38	ISCDWP	Impact of Supervision of CDW Policy	Auxiliary Variable
39	ITR	Improvement of Technologies on Reduction	Flow
40	IUMT	Impact of Using Metal Template	Auxiliary Variable
41	LCF	Landfill Charging Fee	Parameter
42	MCDWM	Measures of CDW Management	Stock
43	MD	Modular Design	Parameter
44	MPM	Mandatory Policies and Measures	Flow
45	P	Prefabrications	Parameter
46	RCDW	Reduced CDW	Flow
47	RDC	Reducing Design Changes	Parameter
48	SCDWP	Supervision of CDW Policy	Parameter
49	SICDWR	Stakeholders' Initiative on CDW Reduction	Stock
50	UMT	Using Metal Template	Parameter
51	W1	Weight 1	Parameter
52	W2	Weight 2	Parameter
53	WCS	Weight of Classification and Sorting	Parameter
54	WCT	Weight of Construction Technologies	Parameter
55	WICDWR	Weight of Initiative on CDW Reduction	Parameter
56	WLCF	Weight of Landfill Charging Fee	Parameter
57	WMD	Weight of Modular Design	Parameter
58	WP	Weight of Prefabrications	Parameter
59	WP1	Weight of Policy1	Parameter
60	WP2	Weight of Policy2	Parameter
61	WRDC	Weight of Reducing Design Changes	Parameter
62	WRMP	Weight of Related Management Policies	Parameter
63	WSCDWP	Weight of Supervision of CDW Policy	Parameter
64	WUMT	Weight of Using Metal Template	Parameter

References

1. Li P (2007) Utilization of construction waste, vigorously develop the circular economy. DC Pract Theor 6:84–87
2. Yuan HP et al (2012) A dynamic model for assessing the effects of management strategies on the reduction of construction and demolition waste. Waste Manag 32:521–531
3. Kulatunga U et al (2006) Attitudes and perceptions of construction workforce on construction waste in Sri Lanka. Manag Environ Qual 17(1):57–72
4. Yuan HP, Shen LY (2011) Trend of the research on construction and demolition waste management. Wastes Manag 31(12):670–679
5. Hao JL et al (2003) Integrated construction waste reduction measures. Chongqing Environ Sci 25(11):10–12
6. Li J et al (2010) Construction site construction waste reduction measures Investigation. J Proj Manag 24(3):332–335
7. Zhu J, Li J (2011) Construction workers construction waste reduction behavior. Influencing Factors Anal Case Shenzhen City Proj Manag Technol 25(6):633–637
8. Jaillon L, Poon CS, Chiang YH (2009) Quantifying the waste reduction potential of using prefabrication in building construction in Hong Kong. Waste Manag 29(1):309–320
9. Ekanayake LL, Ofori G (2004) Building waste assessment score: design-based tool. Build Environ 39(7):851–861
10. Faniran OO, Caban G (1998) Minimizing waste on construction project sites. Eng Constr Archit Manag 5(2):182–188
11. Kautto P, Melanen M (2004) How does industry responds to waste policy instruments—Finnish experiences. J Clean Prod 12(1):1–11
12. Zhou J (2010) PRD construction waste investigation and countermeasures. South China University of Technology, Guangzhou
13. Tam VWY (2008) On the effectiveness in implementing a waste-management-plan method in construction. Waste Manag 28(6):1072–1080
14. Wang Q (1994) System dynamics (Revised Edition). Tsinghua University Press, Beijing
15. Wang J, Yuan H (2009) Construction waste management model based on system dynamics. Syst Eng Theor Pract 29(7):173–180
16. Marzouk M, Azab S (2014) Environmental and economic impact assessment of construction and demolition waste disposal using system dynamics. Resour Conserv Recycl 82:41–49
17. Zhao WH, Ren VS (2011) A system dynamics model for evaluating the alternative of type in construction and demolition waste recycling center—The case of Chongqing. Resour Conserv Recycl 55:604–612
18. Shenzhen Urban Statistic Bureau. http://www.sztj.gov.cn/
19. Shenzhen Urban Planning and Land Resources Committee http://www.sz.gov.cn/

Chapter 96
Characterization of the Generation Rate of Demolition Waste in Shenzhen, a Mega City of China

Huanyu Wu, Huabo Duan, Jiayuan Wang and Guomin Zhang

Abstract The projection of generation of C&D waste is the first process in conducting management. As the basis of the projection, however, there is a lack of data of the demolition waste generation rates (DWGRs) in previous studies and opening sources. This study is aim at providing detailed information of DWGRs by conducting an on-site investigation in 78 demolition projects in Shenzhen city of China, which covers four specific building types and three structure types. The results reveal that the DWGRs range from 877.5 to 2337.5 kg/m^2, and the average amount is 1360.2 kg/m^2, with the standard deviation of 310.5. By analyzing the mean values, concrete is undoubtedly the dominant part among various waste materials, which is 660.1 kg/m^2, followed by brick/block (287.2 kg/m^2), ceramic (139.1 kg/m^2) and mortar (105.1 kg/m^2). Those four inert wastes contribute 88 % to the total weight of demolition waste generated in 1 m2 demolition work. Based on the DWGRs derived from the research, the differences of DWGRs among several building types and structure types are also discussed. Specifically for residential, industrial, commercial and public building, the DWGRs vary from 1304.0 to 1919.5 kg/m^2, from 1065.4 to 1448.5 kg/m^2, from 1310.2 to 1627.4 kg/m^2 and from 1131.7 to 1605.3 kg/m^2, respectively. For different structures, the DWGR of brick-concrete structure lies between 1314.9 and 1553.5 kg/m^2, frame structure lies between 1079.8 and 1490.3 kg/m^2 and frame-shear system lies between 1293.0 and 1696.9 kg/m^2. What's more, a comparative analysis of present DWGRs with findings in previous studies is also conducted. These results could work as the basis

H. Wu · H. Duan (✉) · J. Wang (✉)
College of Civil Engineering, Shenzhen University, Shenzhen, China
e-mail: huabo@szu.edu.cn

J. Wang
e-mail: wangjy@szu.edu.cn

G. Zhang (✉)
School of Civil, Environmental and Chemical Engineering, RMIT University, Melbourne, Australia
e-mail: kevin.zhang@rmit.edu.au

Y. Wu et al. (eds.), *Proceedings of the 20th International Symposium on Advancement of Construction Management and Real Estate*,
DOI 10.1007/978-981-10-0855-9_96

for the future studies regarding the generation of C&D waste, and they do add to the body of knowledge that is currently available for understanding the generation of demolition waste in China.

Keywords Demolition waste · Waste management · Waste generation rate

96.1 Introduction

Since the ongoing urbanization prompted a massive urban development and renovation activities in China, a huge number of construction and demolition (C&D) waste were generated during the process of building demolition and reconstruction [1, 2]. Although the majority of the compositions of C&D waste are materials could be reused and recycled, they were often disposed by landfill and dumps, which trigs serious environmental impacts for sustainable development [3, 4]. In order to reduce these environmental impacts, proper management strategies and plans targeting C&D waste must be programmed, and the first step towards that is to determine the generation and composition [5, 6].

There are a number of studies that report the methods of estimating the generation of C&D waste, ranging from the region level to project level [1, 7–10]. Focusing the attention on China alone, there is a broad variation in C&D waste generation study [1, 8, 11, 12]. For example, according to the previous study, the amount of construction waste generated per m2 is 40.7 kg/m^2, and concrete waste is the largest contributor to the index [1]. Among them, a method based on the principle that projecting the total weight of C&D waste (kg or t), by times the area of buildings constructed or demolished (m2) and the WGRs during those activities (kg/m^2 or m^3/m^2), was widely employed in several studies [6, 13–16]. As these studies, one of the key points to estimate the generation of C&D waste is the determination of waste generation rates (WGRs). However, most of these studies are based on data from previous studies data (Cross-reference), statistical regional data or relatively small sites survey, due to the difficulties involved in conducting survey on large-scale projects or regions. As a consequence, these data can hardly meet the needs of accuracy of information [17, 18].

Accordingly, the data reporting on WGRs and compositions of C&D waste are also heterogeneous [7]. For example, the proportion of ceramic in demolition waste in Spain is 42.3 % [6] while it is only 1.2 % in Portugal [19]. These wide-ranging results may be attributed to a number of different reasons such as the lack of reliable data sources, differences in economic power, city size, construction practices, etc. [6, 10]. Although there are some studies expressed the demolition waste generation rates (DWGRs) in particular regions, studies based on the full information that obtained by first hand are still rare.

Moreover, in most cases, there is a lack of regular statistics on the quantities of demolition waste despite there are several researchers made efforts on the projection of the generation of C&D waste [20]. Given these shortcomings, the objective of this study is to promote the chances for conducting proper management of C&D waste through the contribution of basic data in the original management process. Therefore, this study conducts a survey for the DWGRs in Shenzhen, a mega city in south of China. The compositions of demolition waste and differences in generation rates among different building types and structure types are analyzed. And finally, a comparative analysis of present DWGRs with findings in previous studies is also conducted.

96.2 Research Method

96.2.1 Investigation

The investigation was conducted between 1 November 2014 and 30 January 2015 in Shenzhen, which is a coastal city in South China adjacent to Hong Kong. The reasons to choose Shenzhen as the study region are threefold. Firstly, Shenzhen is the special economic zone and has been regarded as a model for sustainable urban development in China. However, there are a number of urban villages need to be demolished and reconstructed, and this would results a huge number of C&D waste. Secondly, although the Shenzhen government has developed a series of policies of to managing C&D waste treatment and disposal, the recycling rate of entire Shenzhen city is not optimum. Ninety percent of the metal waste could be collected and recycled, but the recycling rates of other waste materials are under 40 %. As a result, the majority of recyclable waste materials are usually disposed to landfill. Thirdly, the space of landfills for C&D waste in Shenzhen is incapable to meeting the needs for disposing C&D waste, and the planning of management of C&D waste is urgent. Thus, a better understanding of the generation and composition of demolition waste could promotes the chances of better reuse and recycle the waste materials and the chances of better planning of landfills.

The interviewees chosen in this study are direct managers who are in charge of a demolition project or have in charge of at least one demolition project in the study region, since their knowledge could reflect the real situations regarding the generation and composition of demolition waste. The survey interviewed 85 respondents, where the effective number of samples is 78, and these projects involving a total demolition area of 464,600 m^2. In this study, the DWGR (namely the total quantity of waste generated from per unit area of demolition work) and the DWGRs for each material were both investigated. The unit of DWGRs is presented by kg/m^2.

96.2.2 Data Processing and Calculation

The study conducted a statistical analysis for data obtained from the survey with the help of the software package SPSS 19.0©. Firstly, analyzing the description statistics of DWGRs and the main calculation parameters include sample mean, variance, standard deviation, skewness, kurtosis. Then, displaying a box diagram based on the statistical analysis results. Secondly, conducting a comparative analysis on building type and structure type with the help of the exploration and analysis function of SPSS software and comparing DWGRs among different building types. Finally, a comparison analysis was conducted on the comparing of the present DWGRs in China with other countries.

96.3 Results and Discussion

96.3.1 General Situation of DWGRs

The results show the value of total DWGRs range from 877.50 to 2337.50 kg/m^2, and the mean value in statistics is 1360.21 kg/m^2 with the standard deviation of 310.54. However, the skewness value is above zero and kurtosis value is less than 3, so it is not subject to normal distribution.

Among the compositions, concrete is the dominant part of demolition waste sine it widely used in all the structure/building types, which bears the largest DWGRs from 470.00 to 1025.00 kg/m^2, with the mean value of 660.13 kg/m^2. The second contributor is brick/block waste (Min: 150.00 kg/m^2; Max: 550.00 kg/m^2; Mean: 287.18 kg/m^2), followed by ceramic and mortar waste, with the mean value of 139.10 and 105.13 kg/m^{2,} respectively. The weight of these four kinds of inert waste accounts for 88 % of total weight with the mean value. As the largest contributor of landfill, these inert wastes usually be expected to be reuse or recycled by making pavement base or aggregates for recycling cement productions (i.e. recycling concrete, recycling mortar, recycling brick/block etc.).

As the major metal materials used in buildings, the DWGRs for Steel are valued from 25–145 kg/m^2, with the mean value of 73.85 kg/m^2, which is the largest part among metal waste. Due to in some demolition buildings, wires and water pipes were still embedded in walls or grounds, there are aluminum and copper waste remain the demolition waste, with the mean value of 10 and 4.94 kg/m^2 respectively. Since the window and door frames of old building were usually made by timber in Shenzhen, there is considerable value of DWGRs of timber generated during the demolition works. The max value of DWGR of timber is up to 100 kg/m^2 and the mean value of that is 31.03 kg/m^2. Normally, the large size timber could be recycled as composite panels or composite firewood, while the small pieces are usually treated along with other non-inert waste by sanitary landfill. Similarly, the demolition works also generated some plastic waste (Max: 60 kg/m^2; Mean: 8.97 kg/m^2) and glass waste (Max: 17.5 kg/m^2; Mean: 6.66 kg/m^2), which normally bought by particular collecting enterprise and gathered together with other same sort materials form industry waste or household waste before recycled.

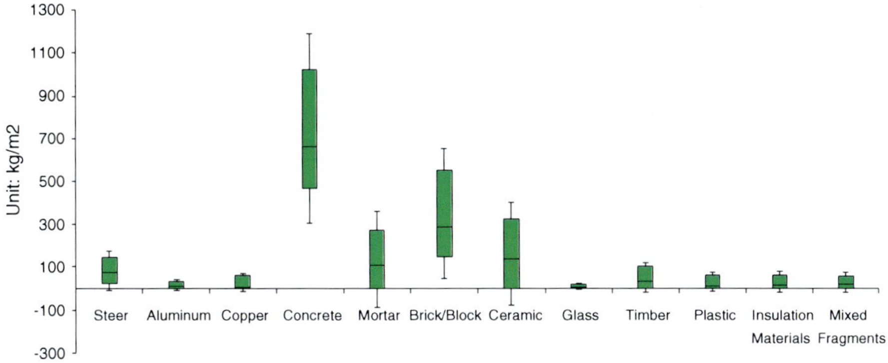

Fig. 96.1 Distribution of demolition waste generation rates in general

Although, the insulation materials are not common in buildings in Shenzhen since the warm weather, the survey results still reveal that there are average 12.56 kg/m^2 insulation materials might generated in the demolition sites of old buildings. This is probably because of the architects came from north area of China in the beginning of developing Shenzhen, and they adopted the design standards in the north. Besides, there are some mixed fragments within the demolition waste, with the quantity up to 55.00 kg/m^2 (Mean: 20.67 kg/m^2). The compositions of mixed fragments are probably composited by many kinds of material presented above or even soil, thus they usually be transport to sanitary landfill. The details of general DWGRs in Shenzhen are shown in Fig. 96.1.

96.3.2 Comparative Analysis on Demolition Waste Generation Rates Among Different Building/Structure Types

Table 96.1 shows the DWGRs of different building types. The mean value of DWGRs of residential building is the largest (1611.74 kg/m^2) among four kinds of types of buildings, and the distribution of that range from 1304.01 to 1919.48 kg/m^2, with the 95 % confidence. The followers are commercial building (Mean: 1468.80 kg/m^2; Min: 1310.23 kg/m^2; Max: 1627.37 kg/m^2) and public building (Mean: 1368.47 kg/m^2; Min: 1131.69 kg/m^2; Max: 1605.25 kg/m^2). However, the standard deviation value of commercial building is the smallest in all building types (79.46), which means the concentration of DWGRs of commercial building is higher than others, sine the structures type of this kind of buildings are similar within samples. While industrial building has the least mean value of DWGRs (Mean: 1256.93 kg/m^2), which is far behind the residential building. This is mainly because that residential building consist more partition walls due to the

Table 96.1 The DWGRs of different building types

Building type	Mean	Standard deviation	95% Confidence intervals	
			Min	Max
Residential building	1611.74	154.22	1304.01	1919.48
Industrial building	1256.93	96.00	1065.36	1448.49
Commercial building	1468.80	79.46	1310.23	1627.37
Public building	1368.47	118.66	1131.69	1605.25

Table 96.2 The DWGRs of different structure types

Structure type	Mean	Standard deviation	95% Confidence intervals	
			Min	Max
Brick-concrete structure	1434.21	59.77	1314.94	1553.49
Frame structure	1285.07	102.87	1079.80	1490.33
Frame-shear structure	1494.94	101.19	1293.02	1696.87

needs of function division, and those walls are normally made by sintered brick/brock in old buildings. On the contrary, due to the larger space span, fewer partition walls, larger doors and windows, the total weight of industrial building is much lighter than other types of building if they have the same gross flow area.

The results also show that the DWGRs were affected by the structure types. As shown in Table 96.2, there is 95 % confidence to say that the DWGRs of frame-shear structure range from 1293.02 to 1690.87 kg/m^2. According to the mean value, the DWGR of frame-shear structure is the largest among three structure types (Mean: 1495.94 kg/m^2), which followed by brick-concrete structure and frame structure, with the mean value of 1434.21 and 1285.07 kg/m^2, respectively. This is probably because of the frame-shear structure involves more massive concrete compositions, which are heavier. Besides, the sintered brick/block with larger density are adopted in brick-concrete structure, while the materials for walls in frame structure are smaller-density materials such as hollow brick and aerated concrete block, which are lighter than normal brick/block.

96.3.3 Comparison of Present DWGRs in China with Other Countries

This study conducts a comparison of present work with other studies (see Table 96.3). For demolition waste generated per m^2, the Norway study shows the value of 1103.24 kg [16], while the Portages study shows that up to 1371 kg [19], which is similar to the value of present study (1360.21 kg). For the Metal, the data

Table 96.3 Comparison of present WGRs in China with other countries

Region	Kuwait [13]	US [14]	Norway [16]	Spain [6]	Portugal [19]	China [12] (Shanghai)	China (Shenzhen)
Year	2004	2007	2007	2010	2013	2014	2015
Total	–	–	1103.24 kg	–	805–1371 kg	–	1360.21 kg
Metal	5 %	0.2 %	4.33 %	2.1 %	–	3.1 %	6.5 %
Concrete	30 %	77 %	Concrete/Brick 84.15 %	Concrete/Mortar 49.1 %	59.6 %	22.6 %	48.5 %
Mortar	25 %	–	–	–	–	–	7.7 %
Brick/Block	30 %	–	–	–	34.5 %	63.8 %	21.1 %
Ceramic	–	–	–	42.3 %	1.2 %	–	10.2 %
Glass	–	–	0.12 %	0.1 %	–	–	0.5 %
Timber	8 %	7 %	6.42 %	4.3 %	–	8.4 %	2.3 %
Gypsum	–	6 %	0.15 %	0.8 %	4.6 %	2.1 %	–
Plastic	–	–	0.27 %	–	–	–	0.7 %
Insulation materials	–	–	0.07 %	–	–	–	0.9 %
Bituminous	–	2 %	–	0.1 %	0.1 %	–	–
Mixed fragments	2 %	9 %	4.13 %	0.8 %	–	–	1.5 %
Note	–	Residential building	–	–	Reinforced concrete (max)	–	Mean value

in China (Shenzhen) shows the highest percentage, which is about 6.5 %, while the value in US is only 0.2 % [14]. What's more, in Shanghai (another city in China), the proportion of metal is less half than in Shenzhen, which is only 3.1 %. It shows that the composition and contribution are quite different in different region even in same country.

In Kuwait, concrete and brick/block are both contribute 30% of total weight for demolition waste generated in unit building demolition work (m^2), and the major inert waste of concrete, mortar and brick/block contribute about 85 % for total weight [13], which is similar to Norway (84.15 %) [16] and China (Shanghai, 86.4 %) [12], and for residential building in US, the proportion of them is 77 % [14]. However, if considering the Ceramic together, the proportions of those four materials in Spain, Portugal and China (Shenzhen) would account 91.4 % [6], 95.3 % [19] and 87.5 %, respectively.

As the major building material used in mainland China before 1980s, the brick/block were widely used in all kinds of buildings, especially residential building. In Shanghai (east of China), these sorts of materials contribute near 64 % of total demolition waste [12], which is double that in Kuwait and Portugal, and almost triple that in Shenzhen (another city in south of China but developed after 1980s). Besides, it is interesting to see that the ceramic contribute 42.5 % to the generated demolition waste in Spain [6], which is much higher than other regions.

96.4 Conclusions

Through a critic review of the previous studies relate the quantification of C&D waste generation, it found that there are servals articles published in international journals in recent years. These studies estimate and project the generations of C&D waste generally based on the equation that by timing the C&D activities areas with the WGRs per unit area for each waste, however, there is a lack of detailed data of WGRs in the majority of previous studies and opening data sources. Thus, it is necessary to obtaining the information of WGRs in particular region as rich as possible, since it significantly affects the accuracy of estimation and projection results.

The investigation conducted in this study reveals that DWGRs in Shenzhen range from 877.5 to 2337.5 kg/m^2, and the average amount is 1360.2 kg/m^2, with the standard deviation of 310.5. From the mean value, concrete is undoubtedly the largest part of various components, accounting 660.1 kg/m^2, followed by brick/block (287.2 kg/m^2) and ceramic (139.1 kg/m^2) and mortar (105.1 kg/m^2). The amount of those four inert wastes accounting for 88 % of the total weight of demolition waste generated in 1 m2 demolition work.

Based on the DWGRs derived from the research, this paper also found the DWGRs are affected by building types and structure types. Specifically for residential, industrial, commercial and public building, the DWGRs vary from 1304.0 to 1919.5 kg/m^2, from 1065.4 to 1448.5 kg/m^2, from 1310.2 to 1627.4 kg/m^2 and from 1131.7 to 1605.3 kg/m^2, respectively. Besides, for different structures, the DWGR of brick-concrete structure lies between 1314.9 and 1553.5 kg/m^2, frame structure lies between 1079.8 and 1490.3 kg/m^2 and frame-shear system lies between 1293.0 and 1696.9 kg/m^2.

When comparing the present results with previous studies, it found that the average value of DWGRs in Shenzhen is similar to the Portages result, and is higher than the Norway founding. For the metal, the DWGR in Shenzhen is much higher than other regions including another city in China (Shanghai). It shows that the composition and contribution are quite different in different region even in same country. These results could work as the basis for the future studies regarding the generation of C&D waste, and they do add to the body of knowledge that is currently available for understanding the generation of demolition waste in China.

Acknowledgements This study was supported by the Scientific Research Foundation of Introduced High Talent Financial Subsidies of Shenzhen University (000044); Australian Research Council Project "Re-considering Sustainable Building and Design: A Cultural Change Approach" (LP110100156); Research Center of Urban Resource Recycling Technology of Graduate School at Shenzhen, Tsinghua University and Shenzhen Green Eco-Manufacturer High-Tech, Co. Ltd. (URRT2014002).

References

1. Li J, Ding Z, Mi X, Wang J (2013) A model for estimating construction waste generation index for building project in China. Resour Conserv Recycl 74:20–26
2. Wang J, Li Z, Tam VW (2015) Identifying best design strategies for construction waste minimization. J Clean Prod 92:237–247
3. Poon CS, Yu AT, Jaillon L (2004) Reducing building waste at construction sites in Hong Kong. Constr Manag Econ 22(5):461–470
4. Yuan H, Shen L (2011) Trend of the research on construction and demolition waste management. Waste Manag 31(4):670–679
5. Wu Z, Ann TW, Shen L, Liu G (2014) Quantifying construction and demolition waste: an analytical review. Waste Manag 34(9):1683–1692
6. Lage IM, Abella FM, Herrero CV, Ordóñez JLP (2010) Estimation of the annual production and composition of C&D Debris in Galicia (Spain). Waste Manag 30(4):636–645
7. Llatas C (2011) A model for quantifying construction waste in projects according to the European waste list. Waste Manag 31(6):1261–1276
8. Shi J, Xu Y (2006) Estimation and forecasting of concrete debris amount in China. Resour Conserv Recycl 49(2):147–158
9. Sáez PV, del Río M, Porras-Amores C (2011) Estimation of construction and demolition waste volume generation in new residential buildings in Spain. Waste Manag Res, 0734242X11423955
10. Butera S, Christensen TH, Astrup TF (2014) Composition and leaching of construction and demolition waste: inorganic elements and organic compounds. J Hazard Mater 276:302–311
11. Lu W, Yuan H, Li J, Hao JJ, Mi X, Ding Z (2011) An empirical investigation of construction and demolition waste generation rates in Shenzhen city, South China. Waste Manag 31 (4):680–687
12. Ding T, Xiao J (2014) Estimation of building-related construction and demolition waste in Shanghai. Waste Manag 34(11):2327–2334

13. Kartam N, Al-Mutairi N, Al-Ghusain I, Al-Humoud J (2004) Environmental management of construction and demolition waste in Kuwait. Waste Manag 24(10):1049–1059
14. Cochran K, Townsend T, Reinhart D, Heck H (2007) Estimation of regional building-related C&D debris generation and composition: case study for Florida, US. Waste Manag 27 (7):921–931
15. Cochran KM, Townsend TG (2010) Estimating construction and demolition debris generation using a materials flow analysis approach. Waste Manag 30(11):2247–2254
16. Bergsdal H, Bohne RA, Brattebø H (2007) Projection of construction and demolition waste in Norway. J Ind Ecol 11(3):27–39
17. Katz A, Baum H (2011) A novel methodology to estimate the evolution of construction waste in construction sites. Waste Manag 31(2):353–358
18. Mercader-Moyano P, Ramírez-de-Arellano-Agudo A (2013) Selective classification and quantification model of C&D waste from material resources consumed in residential building construction. Waste Manag Res, 0734242X13477719
19. Mália M, de Brito J, Pinheiro MD, Bravo M (2013) Construction and demolition waste indicators. Waste Manage Res 31(3):241–255
20. Yuan H (2013) A SWOT analysis of successful construction waste management. J Clean Prod 39:1–8

Chapter 97
Environmental Impact Assessment of Construction and Demolition Waste Recycling in Shenzhen

Jingru Li, Ruirui Xia, Jun Li and Guozhou Chen

Abstract Construction and demolition (C&D) waste recycling can bring significant environmental benefits. This paper analyzes the environmental costs and environmental benefits of inert C&D waste recycling using Life Cycle Assessment method. Taking a C&D waste recycling company in Shenzhen as example, the estimation reveals that recycling per ton of inert C&D waste can bring environmental net benefit of 5.14 yuan. The results provide a useful reference for the development of economic policies on C&D waste recycling.

Keywords Construction and demolition (C&D) waste · Recycling · Life cycle assessment (LCA) · Environmental impact · Shenzhen

97.1 Introduction

Recent years have witnessed a significant amount of construction and demolition (C&D) waste as a result of rapid urbanization and large scale of construction activities in China. It is estimated that the volume of C&D waste generated from new construction projects and old building demolition is approximately 0.4–1.7 billion tons every year [1, 2]. The emissions of Beijing, Shanghai and other big cities are at least 30 million tons [1]. Many cities begin or have been facing the pressure of C&D waste landfill. With the increasing of people's environmental awareness, simple landfill of C&D waste cannot meet people's demands, because waste landfill not only takes up valuable land resources, but also brings potential threat to the soil,

J. Li (✉) · R. Xia
School of Civil Engineering, Shenzhen University, 518060, Shenzhen, China
e-mail: lijr2000@szu.edu.cn

J. Li · G. Chen
Zhongcheng Green Building Technology Co. Ltd, 518060, Shenzhen, Guangdong, China

Y. Wu et al. (eds.), *Proceedings of the 20th International Symposium on Advancement of Construction Management and Real Estate*,
DOI 10.1007/978-981-10-0855-9_97

water, air in the surrounding [3]. In this respect, boosted recycling of C&D waste is required by the China's "Twelfth Five-Year Plan".

Compared with the simple landfill, C&D waste recycling can significantly reduce energy consumption and greenhouse gas emissions, bring significantly environmental benefits [4, 5]. Generally, both at home and abroad, the main barriers for recycling of C&D waste are the low cost of landfill and raw materials, which decrease the demand for recycled aggregate and the interest in developing business from recycling [6–8]. Therefore, it is necessary to estimate the external environmental benefits brought from C&D waste recycling so that the government can promote enterprises' enthusiasm towards waste recycling through the reasonable financial subsidies, tax incentives and other economic policy C&D waste.

97.2 Construction and Demolition Waste

97.2.1 The Definition of C&D Waste

Construction and demolition (C&D) waste is defined by The Technical Code for Construction Waste Reduction of Shenzhen as waste which arises from construction, renovation and demolition activities including land excavation or formation, civil and building construction, site clearance, demolition activities, roadwork, and building renovation [9]. The compositions of C&D waste are divided into the three major categories: inert waste, non-inert waste, and contaminated waste [9].The inert waste including concrete, brick, masonry, stone, mortar, ceramics, tiles and construction over soil and etc. These inert wastes can be used to make recycled building materials. This paper will only discuss the environmental impact of recycling inert waste. The excavated soil is not included because it may be directly used for reclamation or backfill.

97.2.2 The Main Components of C&D Waste

A research to extensively survey the components of C&D waste has been conducted by our research team within the scope of construction projects in Shenzhen. The components and proportions of C&D waste from new construction and demolition activities are shown in Table 97.1. As the table shows, the inert waste including concrete, blocks, mortar are dominated, with the share of these over 60 % in new building construction and more than 85 % in demolition activities. These data indicate that it is necessary to recycle the inert C&D waste.

Table 97.1 The components of C&D waste

	New building construction				Old Building demolition			
Building type	Residential (%)	Commercial (%)	Public building (%)	Industrial (%)	Residential (%)	Commercial (%)	Public building (%)	Industrial (%)
Concrete	50.1	52.0	51.3	55.9	60.7	63.8	64.2	73.5
Bricks and blocks	4.8	5.2	6.3	3.9	12.4	10.9	8.4	3.1
Mortar	3.5	3.5	6.0	3.9	13.8	15.9	16.2	13.3
Metal	10.7	13.0	8.5	8.4	4.5	5.1	5.1	4.9
Wood	20.9	16.5	17.9	18.0				
Ceramics and Tiles					5.2	4.3	6.1	5.3
Others	9.9	9.8	10.0	10.0	3.4	5.1	5.1	4.9

97.3 The Recycling of the Inert C&D Waste

The inert C&D waste can be recycled to produce coarse and fine aggregate. Taking the Zhongcheng Green Building Technology Co., Ltd as an example, it has designed an integrated solutions scheme and developed a set of plants to recycle C&D waste into aggregate. By this treatment process, the inert C&D waste can be reduced to aggregate of various size and active powder. The aggregate can be used to produce various construction materials, such as building blocks, wallboard, floor tiles or to produce recycled concrete and mortar. Active powder can be incorporated into the production of concrete, as a substitute of fly ash, slag-powder, and etc. The main machines used in the treatment scheme are listed in Table 97.2.

Table 97.2 The related equipment of treatment process (Capacity: 200 tons/h)

Name	Amount	Capacity (kW)
Vibrator feeder	1	15
Jaw crusher (the first)	1	90
Belt machine	1	15
Separate material belt	1	7.5
Sorting belt	2	10.5
Jaw crusher (the second)	2	75
Belt machine (the second broken lift)	2	15
Sorting belt	2	15
Mix material belt	2	7.5
Conveyor belt (spiral roller to sand crusher)	1	22
Sand making machine	2	220
Conveyor belt (sand crusher to the material department)	1	10
Debris conveying belt	2	30
Trough belt (debris transport)	1	15
Debris discharge belt	1	15
Stone conveying belt (the overall belt to elevator)	1	20
Elevator	1	45
Material output belt	2	15
Belt machine (lift to elevator)	1	7.5
Vibrating screen	1	11
Belt machine (revert material)	1	7.5
Vibration mill (large-scale)	2	22
Vibration mill (small-scale)	2	75
Total capacity		1250.5

97.4 The Environmental Impact Assessment of the Inert C&D Waste Recycling

Life Cycle Assessment (LCA), a method for assessing the environmental impact of products over their whole life cycle [10], is used in this paper. The focus of the LCA in this study is to provide an estimate the environmental impact of C&D waste recycling in Shenzhen. The recycled utilization of the C&D waste will generate "the environmental cost" due to the consumption of energy and the occupation of land during the process of transportation and production. But it also generates "environmental benefit" because of the substitution of raw materials with renewable materials and reducing the landfill of waste.

97.4.1 The Environmental Cost Assessment of C&D Wastes Recycling

Scope of the Environmental Cost Evaluation

Recycling of construction materials requires energy to transport waste materials from project sites to recycling facilities that will generate negative environmental impact. The negative impact and the negative influence of the environment generated by the production process and the landfill of residual waste also constitute the environmental cost of C&D waste recycling.

The Environment Impact Inventory of Recycling

The inventory selects 1 ton as its analysis units. Considering the characteristics of waste utilization, this paper mainly analyzes the environment impact of the C&D waste recycling from two aspects: the consumption of resource and the damage of ecological environment.

1. Transportation

In the process of transporting the C&D waste from the demolition/renovation/construction site to the recycling plant, the transport vehicle will consume energy and emission exhaust gas. According to the Special Plans (2010–2030) for Construction Waste Landfill of Shenzhen, the service radius of C&D waste facilities is within 10 km. Therefore, this paper assumes that the transportation distance is 10 km. According to the Chinese public environmental emission inventory [11] (see Table 97.3), the environment emissions inventory of transportation of 1 ton construction waste is presented in Table 97.4.

Table 97.3 The Chinese public environmental emission inventory

The public system	Energy	CO_2 (kg)	CO (kg)	C_xH_y (kg)	NO_x (kg)	SO_2 (kg)	Waste (kg)
Electricity Power (1 kWh)	Coal 0.424 kg	0.938	0.000004	0.000072	0.0051	0.0011	0.1023
Distance (5 t/km)	Diesel 0.2 L	0.673	0.002	0.0011	0.0047	0.00014	

Table 97.4 The environment emission inventory in the transport phase (1 ton)

Distance (km)	Diesel (L)	CO_2 (kg)	CO (kg)	C_xH_y (kg)	NO_x (kg)	SO_2 (kg)
10	0.4	1.346	0.004	0.0022	0.0094	0.00028

2. Production

The activities which C&D waste are sorted and manufactured in the recycling plant consume electricity and thus generate indirect environmental emission. According to the plants list provided by the Shenzhen Green Building Technology Co. Ltd, the total power of these recycling equipment's is 1250.5 kW and their treatment capacity is 200 tons waste per hour. Namely, the electricity consumption of disposing each ton waste is 6.25 kW/h. According to the Chinese public environmental emission inventory [11], the environmental emission inventory of 1 ton construction waste in production process can be calculated as presented in Table 97.5.

3. Landfill

About 0.0765 t [12] residue need to be disposed in landfill after recycling 1 ton of C&D waste. As the recycling plants are commonly located in landfill in Shenzhen, the environment emission of transportation is neglected in this study. Only the landfill area for dispose of the waste residue is calculated. As the C&D waste's density is 1.5 t/m^3 generally and the average landfill height is 10 m [3], the land area needed for 0.0765 t C&D wastes is 0.005 m^2.

4. The gross environment emission inventory of recycling C&D waste

The gross environment emission inventory of recycling 1 ton C&D waste is summarized and presented in Table 97.6.

Table 97.5 The environment emission inventory in the production phase (1 ton)

Category	Electricity power (kWh)	Coal (kg)	CO_2 (kg)	CO (kg)	C_xH_y (kg)	NO_x (kg)	SO_2 (kg)	Dust (kg)
Amount	6.25	2.65	5.8625	0.000025	0.00045	0.031875	0.006875	0.6394

Table 97.6 The gross environment emission inventory of 1 ton C&D waste recycling

Category	Coal (kg)	Diesel (L)	CO_2 (kg)	CO (kg)	C_xH_y (kg)	NO_x (kg)	SO_2 (kg)	Dust (kg)	Land (m^2)
Amount	2.56	0.4	7.2085	0.004025	0.00265	0.041275	0.007155	0.6394	0.005
WTP value	0.00098[b]	0.0084[b]	0.072[d]	0.6[a]	0.6[a]	1.26[a]	1.26[a]	0.15[a]	30[c]
Unit	yuan/kg	yuan/kg	yuan/kg	yuan/kg	yuan/kg	yuan/kg	yuan/kg	yuan/kg	yuan/m^2

[a]*data source* The pollution charges standard in Shenzhen; [b]*data source* Li et al. [11]; [c]*data source* land tax in Shenzhen; [d]*data source* The first anniversary report of carbon trading system in Shenzhen

Table 97.7 The environment cost of recycling 1 ton C&D waste

Distance (km)	10	20	30	40	50
Cost (yuan)	0.84	0.95	1.07	1.18	1.30

The Environment Cost of Recycling

Environmental taxes for environment polluters and resource exploiters reflect the social willingness of pay for the environment. This paper converts the environmental impact in Table 97.6 into money according to the actual environmental taxes and other environmental WTP values estimated by Li et al. (2007) [13]. Considering the uncertainties of the transport distance from the construction site to C&D waste recycling plant, this paper also calculates the environmental costs of recycling C&D wastes under the distance of 20, 30, 40, and 50 km, respectively, as shown in Table 97.7.

97.4.2 Environmental Benefit Evaluation of C&D Waste Recycling

The Environmental Benefit of Substitution of Natural Aggregate with Recycled Aggregate

According to the estimation by Zhongcheng Green Building Technology Co., Ltd, 1 ton of C&D waste can produce 0.9 ton of recycled aggregate. The life cycle environmental impact of 1 ton natural aggregates estimated by Gong et al. [14] is presented in Table 97.8. Using the same method as the above section, the environmental benefit presented in Table 97.8 is converted into 4.17 yuan. Therefore, the environmental benefit is 3.75 yuan via recycling 1 ton C&D waste into 0.9 ton recycled aggregate.

The Environmental Benefit of Reduction Waste Landfill

1. Research Scope

The environmental impacts of disposing C&D waste in landfill include two aspects: the environmental emission from transportation of C&D waste and the

Table 97.8 Life cycle inventory analysis of natural aggregate (1ton)

Category	Electricity (kWh)	Coal (kg)	Water (t)	CO_2 (kg)	SO_2 (kg)	NO_x (kg)	CO (kg)	Dust (kg)	Land d (m^2)
Amount	25.4	9.78	2.1	29.001	0.0263	0.133	0.0722	0.316	0.02

Table 97.9 The gross environment impact inventory of landfill (1 ton)

Distance (km)	Diesel (L)	CO_2 (kg)	CO (kg)	C_xH_y (kg)	NO_x (kg)	SO_2 (kg)	Land occupied (m^2)
20	0.8	2.692	0.008	0.0044	0.0188	0.00056	0.0667

consumption of land resource. Thus, reducing C&D waste landfill will achieve environmental benefit from these two sides.

2. The environmental impact inventory

According to the Special Plans (2010–2030) for Construction Waste Landfill of Shenzhen, the service radius of the C&D waste landfill is 20 km. Therefore, this paper assumes that the transportation distance is 20 km. The environmental impact inventory of the C&D waste transportation is calculated and presented in Table 97.9, referring to the calculation of the transportation phase in recycling. About 0.0667 m^2 of land is used for landfill of 1 ton C&D waste. Therefore, the total environmental impact inventory of landfill of 1 ton C&D waste is presented in Table 97.9.

3. The environmental benefit

Similarly, the environmental impact in Table 97.9 is concerted into money and the environmental benefit of 2.23 yuan is achieved via the reduction of 1 ton C&D waste landfill.

97.4.3 The Net Environmental Benefit of C&D Waste Recycling

The net environmental benefit and the cost-benefit ratio are obtained from the calculation results, as shown in Table 97.10. In this paper, when the transport distance from C&D waste site to recycling plant is 10 km as assumed, the net environmental benefit of recycling 1 ton C&D waste is 5.14 yuan and the benefit-cost ratio is 7.12. Even if the transport distance reaches 50 km, the net environmental benefit also reaches 4.68 yuan and the benefit-cost ratio is up to 4.60. This results show that it is economical to recycle the C&D waste.

Table 97.10 The net environment benefit and benefit-cost ratio analysis (Yuan)

Distance (km)	Cost	Benefits	The net environment benefit	Benefit-cost ratio
10	0.84	5.98	5.14	7.12
20	0.95	5.98	5.03	6.29
30	1.07	5.98	4.91	5.59
40	1.18	5.98	4.8	5.07
50	1.30	5.98	4.68	4.60

97.5 Conclusions

This study applied life cycle assessment methodology to analyze the environmental impact of C&D waste recycling. As for the case of Shenzhen C&D waste recycling plant, its environmental benefit is estimated. The calculation results showed that when the transport distance is 10 km, the net environmental benefit of recycling 1 ton C&D waste is 5.14 yuan and the benefit-cost ratio is 7.12. Even if the transport distance reaches 50 km, the recovery of C&D waste still can produce 4.68 yuan of environmental profit. The research proves that the recycling can bring good environmental benefit to the whole society. It should be vigorously promoted in order to solve the huge pressure on the C&D waste.

Acknowledgments The present study is funded by the Ministry of Education of China (No. 11YJAZH047) and National Nature Science Foundation of China (No. 71503171).

References

1. Faguang Leng, Gengxin He, Renyu Zhang et al (2009) Status and development trend of construction waste resource at home and abroad. Environ Sanitation Eng 01:33–35
2. Zhao Li, Lu Jixiang, Gu Hongbin (2011) Study on the industrial operation and countermeasures for comprehensive management of construction and demolition waste. Constr Econ 05:16–20
3. Liu G, Fu D (2008) Research on the implementation of PFI financing mode in the construction waste disposal project of China. Project Manage Technol (05):44–48
4. Hu Mingming, He Qiong, Shi Shiying et al (2011) The cost analysis of construction and demolition waste management-Taking Chongqing as an example. Constr Econ 04:93–97
5. Yuan F, Shen L, Li Q (2011) Emergy analysis of the recycling options for construction and demolition waste. Waste Manag 31(12):2503–2511
6. Du T, Li H, Guo T, Zhou Z (2006) The economic analysis on the application of recycled aggregate concrete. New Build Mater 06:30–33
7. Zhao W, Leeftink RB, Rotter VS (2010) Evaluation of the economic feasibility for the recycling of construction and demolition waste in China-The case of Chongqing. Resour Conserv Recycl 54(6):377–389
8. Nunes KRA, Mahler CF, Valle R et al (2007) Evaluation of investments in recycling centres for construction and demolition wastes in Brazilian municipalities. Waste Manag 27(11):1531–1540
9. SJG 21-2011. Technical code for construction waste reduction. Shenzhen Municipal Housing and Construction Bureau 8
10. Li X, Wang S, Kong X et al (2011) Life cycle assessment of environmental impacts of ready mixed concrete. China Civil Eng J 01:132–138
11. Yang J, Xu C, Wang R (2005) The life cycle assessment method and its application. Meteorological press, Beijing
12. Wang B (2012) Research of Shenzhen construction wastes management base on life cycle assessment. Huazhong University of Science and technology, China
13. Li X, Wu X, Zhang Z (2005) Study on social WTP for environmental impacts based on the LCA theory. J Harbin Inst Technol 11:1507–1510
14. Gong Z, Ding R, Chen B et al (2011) Comparison of embodied environmental impact of recycled aggregates and natural aggregates. J Qinghai Univ (Nat Sci) (06):9–12

Chapter 98
A Study of Climate-Responsive Building Technologies in Different Climate Regions of China

Bingqing Jiang, Peng Mao, Yongtao Tan and Xia Yao

Abstract To reduce the impact of buildings on the environment, climate-responsive buildings emerge on the scene. Due to diversified climate in China, those climate-responsive building technologies unsuitable for local climate could easily cause the waste of resource. Based on a comprehensive literature review, a number of climate-responsive building technologies were identified in this paper. And the particular climate-responsive building technologies for each climate region were also identified according to regional analysis. This study provides valuable references for designers to have a better understanding of climate-responsive building technologies in different climate regions, and promotes the climate-responsive building technologies and their applications in China.

Keywords Climate-responsive · Building · Climate-responsive · Building · Technology · Climate region

98.1 Introduction

The built environment is responsible for 40 % of world materials usage, a third of energy consumed by the world economy and 40 % of greenhouse gas emissions [1]. For high-rise buildings, concrete is widely used due to its low cost and high durability. However, concrete generates more CO_2 emissions and consumes more energy than wood [2]. With the deterioration of environment, it becomes necessary to improve designers' environmental consciousness [3]. Therefore, climate-responsive buildings were introduced to meet the environmental requirements [4–6].

B. Jiang · P. Mao (✉) · X. Yao
School of Civil Engineering, Nanjing Forestry University, Nanjing, China
e-mail: 13805171820@163.com

Y. Tan
Department of Building and Real Estate, The Hong Kong Polytechnic University, Hong Kong, China

Y. Wu et al. (eds.), *Proceedings of the 20th International Symposium on Advancement of Construction Management and Real Estate*,
DOI 10.1007/978-981-10-0855-9_98

Climate-responsive buildings can make positive responses to environmental changes based on the principles of resource conservation and environmental protection [7–9].

With the increasing demand on environmental friendly buildings, various climate-responsive building technologies have been adopted in China. The climate in China is diversified. It is important to select appropriate climate-responsive building technologies with considering local climate conditions [10]. Wrong selection of climate-responsive building technologies may cause the waste of resource [11]. Therefore, this paper aims to conduct a comprehensive study on climate-responsive building technologies and their applications in different climate regions, and to promote a sustainable development of the society.

98.2 Identification of Climate-Responsive Building Technologies

Climate-responsive buildings can make people feel more comfortable than traditional buildings by controlling climate elements. There are some main elements influencing the climate of a region, including temperature, humidity, solar radiation and wind. Climate-responsive buildings can provide a comfort internal environment for people by controlling these four elements [12].

To collect climate-responsive building technologies, a large number of papers in databases, such as "ASCE", "EI", "Web of science", "CNKI" and "WANFANG", were retrieved with key words "ecological building", "passive technology", "temperature control", "humidity control", "building sun-shading" and "ventilation design". Finally, 225 articles related to climate-responsive building technologies were reviewed. And 35 climate-responsive building technologies were identified. These climate-responsive building technologies were mainly used to control temperature, humidity, sunlight and wind.

98.2.1 Climate-Responsive Building Technologies for Temperature Control

Based on the literature review, climate-responsive building technologies for temperature control can be grouped into three main categories, building materials, building components and others.

1. Building Materials

Advanced materials can be used for building temperature control, such as reversible normal temperature thermochromic building coatings, transparent surface, light penetration component with temperature-sensing element, trombe walls and ecological skin. These technologies can reduce heat loss and increase heat gain in winter and vice versa in summer. Among those technologies, reversible normal temperature thermochromic building coatings can change its colour by responding to surrounding temperature. When the outside temperature is low in winter, it becomes dark gray to absorb more solar radiation. Then, the indoor temperature increases. When the outside temperature is high in summer, it becomes white to reflect solar radiation. Then, the heat gains from outside can be reduced [13]. Similarly, the surface of light penetration component with temperature-sensing element becomes transparent in winter while it becomes white in summer [14]. Glass with low transmittance can be installed on the exterior concrete wall. The concrete wall with transparent surface can store heat in winter, and the glass can reflect heat in summer [15]. Similarly, trombe wall is also a combination of glass and wall. The air between the wall and glass is heated due to the greenhouse effect and the indoor temperature rises in winter. In summer, hot air rises to the outside through openings at the bottom of the wall and the top of the glass, and take away the heat [16]. With plants on the exterior wall, ecological skin can cool room in summer and keep room warm in winter [17].

2. Building Components

Temperature in buildings can be adjusted by using different building components, such as double walls, grille skin, movable window and movable corridor. All these technologies can be used to control the temperature in a building. Double walls have better thermal performance than single wall with less heat loss in winter and less heat gains in summer [18]. Grille skin has louvers installed on the exterior walls to adapt to the temperature change [15]. Both movable window and movable corridor can adapt to the temperature change by removable windows. The windows can be removed in summer and be installed in winter.

3. Others

Additionally, two climate-responsive building technologies are also commonly used for temperature control, namely roof pond and movable courtyard. Roof pond uses plastic bags filled with water as the heat sink. Movable insulation panels are used to regulate the heating and cooling. During daytime in the winter, the panels can be opened to absorb heat and closed at night. The heat absorbed in the water will release and keep the rooms below warm. It is reverse in the summer [19, 20]. Movable courtyard can be used to control the temperature by the utilization of transitional spaces, such as porch or hall. Doors and windows of the porch or hall are closed to keep room warm in winter and are opened in summer [21].

98.2.2 Climate-Responsive Building Technologies for Humidity Control

Climate-responsive building technologies for humidity control can be grouped into two categories, dehumidification technologies and humidification technologies.

1. Dehumidification Technologies

The common climate-responsive building technologies for dehumidification include ventilated dehumidification, mechanical dehumidification, rock cooling dehumidification and liquid desiccant dehumidification. Ventilated dehumidification means that moisture in indoor air can be removed by natural ventilation. Both mechanical dehumidification and rock cooling dehumidification can reduce the humidity by refrigerating moist air so that the moisture in air condenses to water [22]. Liquid desiccant dehumidification removes moisture in the air by liquid desiccant [23].

2. Humidification Technologies

There are three common technologies for humidification, including direct indoor humidification, outdoor humidity conditioner and ultrasonic humidifier. Direct indoor humidification increases humidity by humidifier put in the room. Outdoor humidity conditioner supplies moist air to the indoor areas after heating and humidifying the air outdoor [24]. Ultrasonic humidifier creates ultrafine water droplets by ultrasonic, and the fan forces the fine droplets into the air [25].

98.2.3 Climate-Responsive Building Technologies for Sunlight Control

Climate-responsive building technologies for sunlight control can also be grouped into three main categories, building materials, building components and intelligent sunlight control.

1. Building materials

Climate-responsive building technologies for sunlight control can be achieved by using advanced building materials. Low-e glass is one of the mostly used materials. There are three types low-e glasses, namely high transmittance low-e glass, sun-shielding low-e glass and double-silver low-e glass [26]. High transmittance low-e glass has high transmittance of visible light and solar energy, and high mid-and-far infrared reflectivity [27]. It can keep room warm in winter with adequate sunshine in. Sun-shielding low-e glass has adequate transmittance of visible light, low transmittance of solar energy and high far infrared reflectivity [27, 28]. So it can keep rooms warm in winter due to less heat loss and keep room cool in summer due to less heat gains from solar radiation. The double-silver low-e glass has high transmittance of visible light and low transmittance of solar energy. So it can reduce heat gains from solar radiation in summer [29].

2. Building components

Louvers and the illuminator and sunbreaker combination system are two typical climate-responsive building technologies for sunlight control. The illuminator and sunbreaker combination system has a rotatable metal plate which acts as a sunbreaker when the light is too strong and as an illuminator when the light is not sufficient [30].

3. Intelligent sunlight control system

Intelligent window is a typical intelligent sunlight control technology. It can be closed automatically when the sunlight is too strong, and opened automatically when the sunlight is not sufficient [31].

98.2.4 Climate-Responsive Building Technologies for Wind Control

Natural ventilation, movable ventilation and section design of ventilation design are three common climate-responsive building technologies used for wind control.

1. Natural ventilation

Wind pressure ventilation and heat pressure ventilation are two common natural ventilation methods. Wind pressure ventilation uses the pressure difference between windward side and lee side to keep the ventilation indoor. Heat pressure ventilation means that the warm air rises and gets out from the upper outlet, and fresh air can get in from the lower inlet [32].

2. Movable design for ventilation

The common climate-responsive building technologies in movable design for ventilation are movable interior wall, movable roof and movable exterior wall. All these movable components can be closed in winter and be opened in summer [33].

3. Section design for ventilation

Building ventilation is not only determined by the design of site layout but also section design. There are three common section designs, namely atrium-type design, light well and split-level design. There are three types of atrium-type design. Type 1 is enclosed atrium which can be closed totally to reduce ventilation. Type 2 is linear atrium which can be closed in winter and be opened in summer. Type 3 is semi-enclosed atrium which can be opened totally to provide enough wind for building [34].

98.3 Climate-Responsive Building Technologies for Diversified Climate

Buildings have great impacts on the environment and society. Climate-responsive building technologies can make buildings more environmental friendly. Therefore, the demanding on climate-responsive buildings is increasing gradually. There are too many climate-responsive building technologies to choose for a particular building. In this paper, the climate-responsive building technologies for each climate region will be identified based on regional analysis.

98.3.1 Climate in China

China is divided into five climate regions based on the "Thermal design codes for domestic building" (GB50176-93), i.e. severe cold region, cold region, hot summer and cold winter region, hot summer and warm winter region, and temperate region (Fig. 98.1). Severe cold region is cold, dry and windy for the whole year with abundant solar radiation. Cold region is dry and windy also with abundant solar radiation and has a cold and long winter. Hot summer and cold winter region is hot in summer and cold in winter, with moist air for the whole year and few windy days. In hot summer and warm winter region, there are almost no cold and windy days. Meanwhile, the air is moist, and the sunlight is strong. In temperate region, the air is warm and moist all the time. The sunlight is strong and there are almost no windy days.

Climate-responsive buildings can respond to the local climate changes to maintain high indoor air quality by utilizing various climate-responsive building

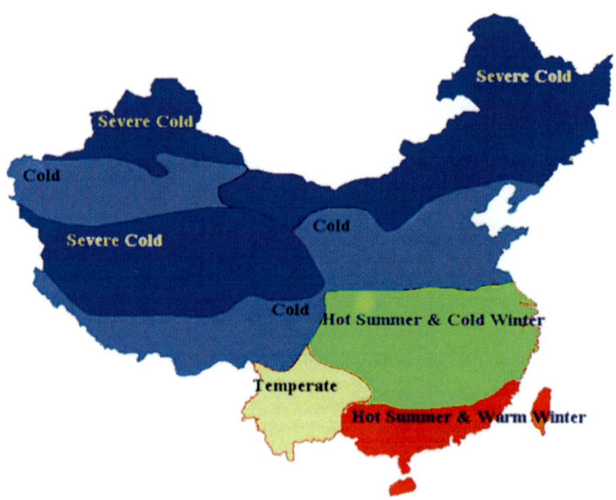

Fig. 1 China regional division map of climate [data from thermal design code for civil building (GB50176-93)]

technologies. Severe cold region and cold region have similar climate so that the climate-responsive building technologies in these two regions are almost the same. Similarly, hot summer and warm winter region and temperate region can adopt the same climate-responsive technologies. Therefore, the five climate regions can be classified to three. Region I includes severe cold region and cold region. Region II represents hot summer and cold winter region. Region III includes hot summer and warm winter region and temperate region. Climate-responsive building technologies for these three regions are identified in the following section.

98.3.2 Regional Analysis of Climate-Responsive Building Technologies

Different climate regions have different climate characteristics. And people in different climate regions have different requirements on indoor temperature, humidity and sunlight. Therefore, there is a need to conduct regional analysis to identify the technologies for each region. Based on the climate characteristics of each region, the design requirements of climate-responsive building technologies for temperature, humidity, solar radiation and wind control in different climate regions were analyzed. And climate-responsive building technologies for each region were identified as shown in Table 98.1.

Climate-Responsive Building Technologies in Region I

The climate-responsive building technologies for temperature control in region I should have good thermal performance to maintain indoor temperature, as shown in Table 98.1. Taking roof pond as an example, it keeps room warm at night due to the heat released from the water bags which adsorb and store heat during the daytime. However, the heat stored on water bags during the daytime is limited, so roof pond cannot meet residents' need of warm in cold winter [35, 36].

For humidity control, the climate-responsive building technologies for dehumidification are not suitable for region I because the air is dry due to low temperature in this region.

For sunlight control, those technologies with low solar energy transmittance are not appropriate in region I, such as the sun-shielding low-e glass and double-silver low-e glass [26], they cannot meet residents' need of warm in cold winter.

For wind control, those technologies causing large heat loss are also not suitable for region I, such as movable roof, movable outer wall and split-level design for ventilation [33].

Table 1. Climate-responsive building technologies for three regions

Function	Type	Climate-responsive technologies	Region I	Region II	Region III
Temperature control	Building materials	Reversible normal temperature thermochromic building coatings	√	√	√
		Transparent surface	√	√	
		The light penetration component with temperature-sensing element	√	√	√
		Trombe walls	√	√	
		Ecological skin	√	√	√
	Building components	Double walls	√	√	√
		Grille skin	√	√	√
		Movable window	√	√	
		Movable corridor	√	√	
	Others	Roof pond			√
		Movable courtyard	√	√	√
Humidity control	Dehumidification technologies	Ventilated dehumidification		√	√
		Mechanical dehumidification		√	√
		Rock cooling dehumidification		√	√
		Liquid desiccant dehumidification		√	√
	Humidification technologies	Direct indoor humidification	√		
		Outdoor humidity conditioner	√		
		Ultrasonic humidifier	√		
Sunlight control	Building materials	High transmittance low-e glass	√		
		Sun-shielding low-e glass		√	
		Double-silver low-e glass			√
		Shading by deciduous plants	√	√	√
	Building components	Louvers	√	√	√
		Illuminator and sunbreaker combination system	√	√	√
	Intelligent sunlight control	Intelligent window	√	√	√
Wind control	Natural ventilation	Wind pressure ventilation	√		
		Heat pressure ventilation		√	√
	Movable design for ventilation	Movable interior wall	√	√	√
		Movable roof		√	√
		Movable outer wall		√	√
	Section design for ventilation	Enclosed atrium	√		
		Linear atrium		√	
		Semi-enclosed atrium			√
		Light well	√	√	√
		Split-level design			√

Climate-Responsive Building Technologies in Region II

The weather is hot in summer and cold in winter in Region II. For temperature control, those technologies which cannot maintain the indoor temperature in winter are not suitable for this region, such as roof pond [36].

The air is humid for the whole year in region II so that those technologies for humidification are not suitable.

With strong sunlight in summer, sun-shading is required in summer which may not be used in winter because solar radiation can help maintain the indoor air warm. Therefore, those climate-responsive building technologies which cannot provide adequate sun shading are not suitable in this region due to the hot summer, such as high transmittance low-e glass [27]. Those technologies which can lower room's temperature in winter are also not suitable in this region, such as double-silver low-e glass [29].

There are not many windy days in region II. Those climate-responsive building technologies with poor ventilation ability are not suitable in this region, such as enclosed atrium. And those technologies with high ventilation efficiency are also not suitable because they may cause high heat loss in winter, such as semi-enclosed atrium. [33, 34].

Climate-Responsive Building Technologies in Region III

The weather is warm to hot in region III all the time. For temperature control, those technologies which can maintain the indoor temperature are not suitable for region III, such as transparent surface and trombe walls [15, 16].

The air is moist in region III. It is necessary to dehumidify the indoor air for the whole year.

This region has abundant solar radiation and the weather is very hot. Those climate-responsive building technologies with high solar energy transmittance are unsuitable in this region, such as high transmittance low-e glass and sun-shielding low-e glass [27, 28].

For wind control, those technologies with poor ventilation ability are not suitable in this region because there are few windy days, such as enclosed atrium and linear atrium. [33, 34].

98.4 Conclusion

Fast economic development has caused many environmental and social problems around the world, especially in developing countries. The building sector is a major contributor to the Greenhouse gas emission. Therefore, there is an increasing demand for environmental friendly buildings. Climate-responsive building is an attempt to face the environmental challenges. Due to diversified climates in China, people in different climate regions have different requirements for climate-responsive building technologies. Based on comprehensive literature review, 35 common climate-responsive building technologies were identified, and particular climate-responsive building technologies for each climate region were also identified. The findings are valuable references for various stakeholders to clearly understand climate-responsive building and relevant technologies. This study applies the theory of climate-responsive building design in China and provides a guideline for designers to choose suitable climate-responsive building technologies in different climate regions. The promotion of climate-responsive buildings in China can meet the requirements of sustainable development.

References

1. Worldwatch Institute (1995) State of the world, 1995. Worldwatch Institute, Washington
2. Park HS, Lee H, Kim Y et al (2014) Evaluation of the influence of design factors on the CO_2 emissions and costs of reinforced concrete columns. Energy Build 82:378–384
3. Surabhi C, ARCH B (2008) Energy efficiency and sustainability buildings. AEI, USA
4. Hyde R (2013) Climate responsive design: a study of buildings in moderate and hot humid climates. Taylor & Francis, UK
5. Laquatra J, Pillai G, Singh A, Syal MM (2008) Green and healthy housing. J Architect Eng 14 (4):94–97
6. Krem M, Hoque ST, Arwade SR, Breña SF (2014) Structural configuration and building energy performance. J Architect Eng (1), 29–40
7. Gut P, Ackerknecht D (1993) Climate responsive buildings: appropriate building construction in tropical and subtropical regions. SKAT Foundation, Switzerland
8. Lv A (2002) Study on building's climate-responsiveness. New Science 4:78–79
9. Upadhyay AK, Yoshida H, Rijal HB (2006) Climate responsive building design in the Kathmandu Valley. J Asian Architect Build Eng 5(1):169–176
10. Li Q (2012) The climate adaptability research on house design. Zhengzhou University, Zhengzhou
11. Mao Y, He J (2010) Architectural design adaptable to local ecoclimate. Indus Constr 40 (8):49–77
12. Ren Y (2009) Research of bioclimatic architecture design strategies. Hebei University of Technology, Hebei
13. Ma Y, Zhu B, Wu K (2001) Research on the reversible normal temperature thermochromic building coatings. J Build Mat 4(3):211–216
14. Ingeborg Flagge (2001) Thomas Herzog architecture + technology. Prestel Verlag Munchen
15. Li Q (2010) Research on the design of passive multistory residence in cold region. Dalian University of Technology, Dalian
16. Saadatian O, Sopian K, Lim CH, Asim N, Sulaiman MY (2012) Trombe walls: a review of opportunities and challenges in research and development. Renew Sustain Energy Rev 16(8):6340–6351
17. Zhang X (2009) Strategy research on ecological design of building skin. Chin Overseas Architect 5:101–103
18. Li B (2007) New idea about elevation design of building. China Build Des Decoration 9:24–26
19. Zeng Y (2007) Theoretical and experimental study of passive solar houses. Xian University of Architecture and Technology, Xian

20. Wongsuwan W, Fongsamootre T, Cole MO (2006) Experimental studies on the roof pond house under tropical climatic conditions. KKU Eng J 33(2):133–139
21. Oktay D (2002) Design with the climate in housing environments: an analysis in Northern Cypru. Build Environ 37(10):1003–1012
22. Wang Y (2013) Humid environment test and dehumidify strategy of residential architecture in HangZhou. Xian University of Architecture and Technology, Xian
23. Yin Y, Qian J, Zhang X (2014) Recent advancements in liquid desiccant dehumidification technology. Renew Sustain Energy Rev 31(2):38–52
24. Xue M (2004) The study of humidity regulation and humidification model. Refrig Air-Conditioning 4(4):39–43
25. Lei Z (2013) Study on ultrasonic humidification test of carbon fiber. New Chem Mat 41(8):148–150
26. Jelle Bjørn Petter, Kalnæs Simen Edsjø, Gao Tao (2015) Low-emissivity materials for building applications: a state-of-the-art review and future research perspectives. Energy Build 96:329–356
27. Ren Y, Guo W, Kong L, Du G (2013) Comparative analysis of energy-saving performance between two low-e glass. Glass 40(6):40–42
28. Zhao QN, Zhao XJ (2001) Effect of substrate temperature on the visible light transmittance and infrared reflectance of low-E glass. In: Proceeings of the international conference on energy conversion and application, pp 646–648
29. Giovannetti F, Foste S, Ehrmann N, Rockendorf G (2012) High transmittance, low emissivity glass covers for flat plate collectors: applications and performance. Sol Energy 104(6):106–115
30. Herzog T (2010) The Oskar Von Miller forum: building for the future. Hirmer, Verlag
31. Meagher M (2010). Dynamic ornament: the design of responsive architectural environments. Epfl, Switzerland
32. Zhong J, Zeng Y (2004) Natural ventilation design in architecture. J Chongqing Jianzhu Univ 26(2):18–21
33. Huang L (2010) Ecological building design strategy of respond to regional natural environment. Chongqing University, Chongqing
34. Yi R (2008) Case study and parametric analysis of daylighting and thermal performance of atriums in subtropical climate. University of Nottingham, UK
35. Kharrufa SN, Adil Y (2008) Roof pond cooling of buildings in hot arid climates. Build Environ 43(1):82–89
36. Zhuang L (2008) The research on building energy-saving design of strain energy conservation. Zhejiang University, Zhejiang

Chapter 99
Study on the Evolution of Building Energy Conservation Standards in China

Liqing Huang, Guiwen Liu, Chengshu Yan and Pengpeng Xu

Abstract High energy consumption in building sector has promoted building energy conservation (BEC) to be of great urgency. A large number of building energy conservation standards have been formulated to guide the development of BEC in China. However, lots of the shortcomings are exposed during the standards implementation, such as lack of integrity, poor suitability etc. This paper aims to explore the evolution process of the BEC standards in China based on the time sequence. Four stages are divided, with the discussion of features and key problems in each stage. On the basis of institutional change theory, two main influence factors and three evolution paths of the BEC standards are investigated at last. It is expected that the findings of this paper can provide references and suggestions on compilation and revision of the standards of building energy conservation in future.

Keywords Building energy conservation · Institutional change · Evolutional paths · Influence factors

99.1 Introduction

Under the background of global energy crisis, energy development and utilization have aroused extensive attention. Construction as a resources-driven industry would expend massive energy during constructing and operating. The study shows that building energy consumption accounts about 30 % of total consumption in China [1], with the speed of 2 billion square each year and less than 5 % of the newly-added floor space are energy-saving, the others are all high-consumption buildings [2, 3]. The situation of building consumption is so severe that a mass of

L. Huang (✉) · G. Liu · P. Xu
Faculty of Construction Management and Real Estate, Chongqing University, Chongqing, China
e-mail: 1562131192@qq.com

C. Yan
College of Civil Engineering, Chongqing University of Science & Technology, Chongqing, China

Y. Wu et al. (eds.), *Proceedings of the 20th International Symposium on Advancement of Construction Management and Real Estate*,
DOI 10.1007/978-981-10-0855-9_99

standards formulated to promote the implement of BEC. However, the defects of standards widen the gap between pre-estimated results and the actual implementation effect, and in a long term, restrict the development of energy conservation. Based on the institutional change theory, this paper aims to explore the problems of existing BEC standards through analyzing the development process according to the goals of BEC, and evolution dynamics and paths, then the proposals would be given for formulation and renewal of BEC standards.

99.2 Research Methodology

Douglass C. North is the representative of institutional change theory. North believes that the institution can be treated as public goods which stem from social demand. However, the external reality that institution is scarce resource which formed for people's limited-rationality. The imbalance between supply and demand of the institution calls for the institutional change and formed the inside power, which eventually generates the evolution the system [4, 5]. American economist Daniel W. Bromley has his own perceptions on the theory of institutional change. Bromley extends the comprehension of institutional change theory on the predecessors' research. He stresses that in the view of the objective causality to understand the process of institutional change. Bromley believes that desired effect of the institution composes the power for institution changing, and current problems of institution promote the formation and adjustment of new institution [6]. Based on institutional change theory, the purpose of the formulation about BEC standards is to improve the effect and efficiency of BEC. The expected efficiency of BEC standards and the gap between actual effect and expected implement consist of the main power which promotes the evolution of BEC standards.

This paper through literature review specification for BEC standards and expert interview method confirms the range of BEC standards which is special for the construction, including that by building materials, construction technology, and management ways to achieve the effect of building energy conservation, the testing of building material also included. Based on the scope of standards, the earliest building energy conservation appeared in 1982 by statistics from this paper. Combining the goal of building energy saving of each phase with the Five Year Forward Planning for national economic social development, the evolution of BEC standard has been divided into 4 stages since 1982. After the analysis of evolution stage, this paper would give the main factors that have greatly impacted the development of BEC standards and the evolution paths, and based on the analysis of above all, the problems of existing standards will be present.

99.3 Evolution Stages of BEC Standards in China

According to the statistics of this paper, the number of BEC codes is 239 totally. Based on different way to implement BEC, the codes are divided into three categories, namely, the comprehensive and auxiliary standards of BEC (CAS), the engineering technical standards of BEC (ETS) and the building material and product standards (BMPS). CAS stresses comprehensive assessment about it. Based on the building life cycle theory, in addition, technology of new energy, ETS is classified into five classes, including design stage, construction stage, operational stage, demolition and renewal stage, and technical on new energy and technology. BMPS includes heat-insulating material codes, exterior protected construction codes, products and instruments codes and performance measuring codes. Various kinds of standards data distribution is shown in Table 99.1.

99.3.1 The Infancy Stage of BEC Standards (1982–1995)

This stage from 1982 to 1995 is the infancy of Chinese BEC standards. The formulation of "The Civil Building Energy-saving Design Code" which applies to the cold area signifies the start of BEC standards formulation. This stage definitely puts forward the building energy-saving goal that is the consumption of new heating residential building is supposed to reduce by 30 % (20 % of construction system and 10 % of heating system) as a threshold of BEC compared to the period of 1980–1981s [7]. The first BEC standard can date back to 1982 when the code named "Plastic Coefficient of Thermal Conductivity Test Method" was issued.

Table 99.1 Statistics of the standards of BEC

Categories	Sub-categories	Numbers
The comprehensive and auxiliary standards of BEC (CAS)	Comprehensive evaluation	3
The engineering technical standards of BEC (ETS)	Design stage	19
	Construction stage	22
	Operational stage	11
	Demolition and renewal stag	2
	Application of new energy and technology	7
The building material and product standards (BMPS)	Heat-insulating material	36
	Exterior protected construction	40
	Products and instruments	35
	Performance measuring	57
	Others	7
Total	/	239

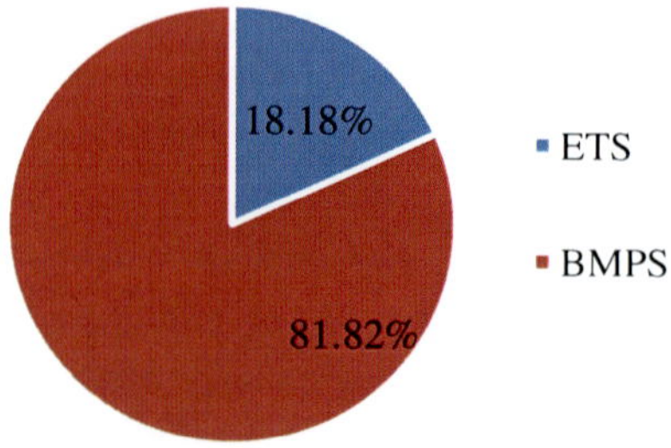

Fig. 99.1 Proportion distribution of standards

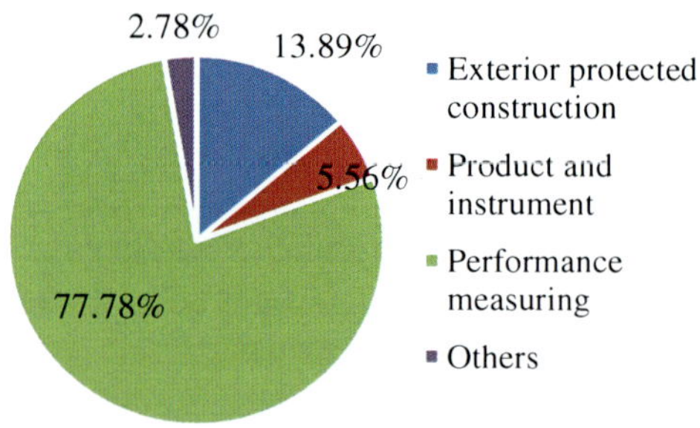

Fig. 99.2 Distribution of sub-category of BMPS

At this stage, the number of related standards is 44 and each category of standards data distribution and proportion is shown in Fig. 99.1.

During this period, the number of BMPS accounts for approximately 81.82 % of all, and performance measuring codes account for 77.78 % of sub-category, mainly including rigid foam and heat-insulating material, shown in Figs. 99.1 and 99.2. The proportion of engineering technical standard of BEC is 18.18 %, and mainly concentrates on design stage, which takes 75 % of this sub-category. In this stage the codes mainly concentrate on building materials and performance testing, as well as the residential building design. Another high light of this stage is that the scope about building design codes from civil buildings extending to public buildings. But the quantity of public construction is few.

99.3.2 Stage of Initial Development of BEC Standards (1986–2000)

The period from 1986 to 2000 is the initial development stage of BEC standards. The target of this stage is that reduce the energy consumption by 30 % of last stage that means reduce energy expense by 50 % compared to the period of 1980–1981s. The distribution of each category of BEC standard is shown in Fig. 99.3.

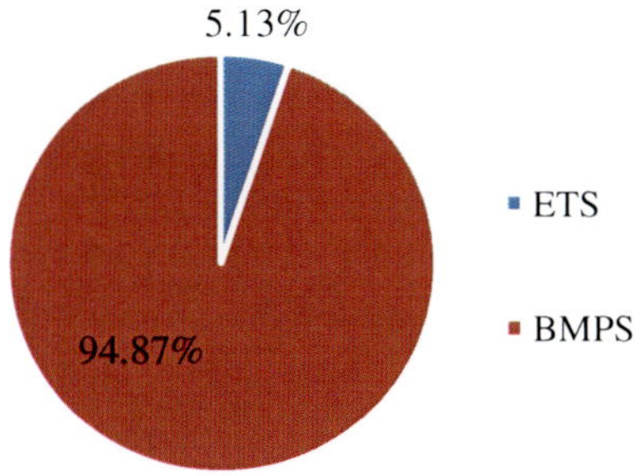

Fig. 99.3 Proportion distribution of standards

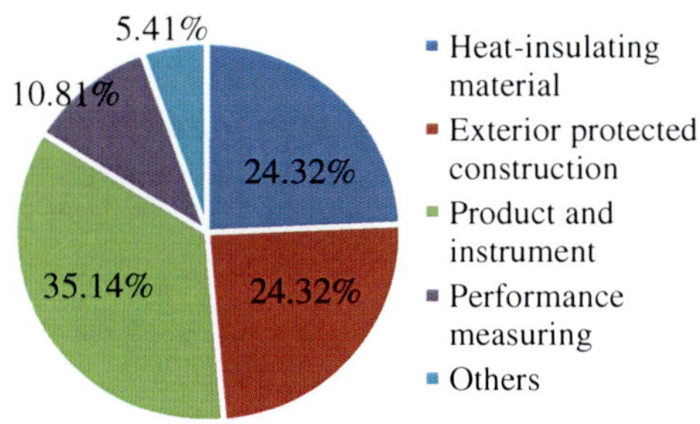

Fig. 99.4 Distribution of sub-category of BMPS

On the whole, the codes of BEC at this period are the expansion of last stage and standards have enriched a lot. The number of ETS accounts for 5.13 % of all, and the quantity of BMPS takes a proportion of 94.87 %, shown in Fig. 99.3. Compared to last period, the proportion of BMPS is larger. The share of heat-insulating material and exterior protected construction all have remarkable increased, accounts for 24.32 and 35.14 % respectively, shows in Fig. 99.4. The development of ETS also has a great progress. The standards gradually pay more attending on demolition and update phase and the range of BEC has expanded to existing buildings from new buildings. However, up to now the distribution of the standards is scattered and the applying range is so limited that still need to get further improved.

99.3.3 Stage of Rapid Development of BEC Standards (2001–2005)

The stage of rapid development of BEC standards is from 2001 to 2005. In this stage, 28 provinces (including autonomous regions and municipalities) according to the development of local BEC, improved the rate of energy saving from 50 up to

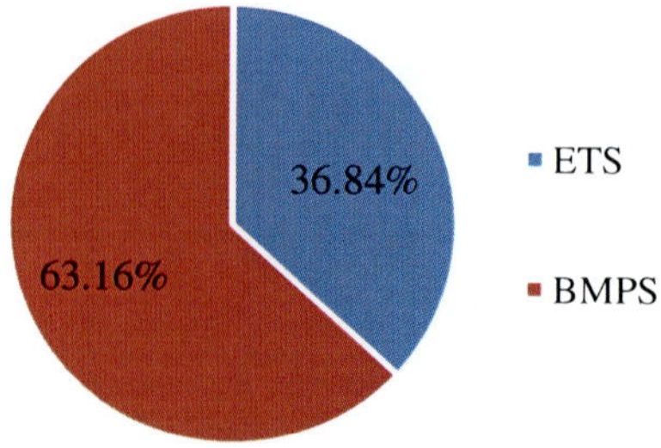

Fig. 99.5 Proportion distribution of standards

57.7 %, and some developed provinces set 65 %as the goal [7]. The number of BEC standards in this period is 76 more than last three stages. Each category of standards distribution and proportion is shown in Fig. 99.5.

In general, this stage is the crucial period in the evolution of standards of BEC. The codes in this period have a great development, especially the design standard compared to last two stages. The number of each type has increased and the number of BET accounts for 36.84 % of total. The codes of construction phase account for 42.86 % of sub-categories; the proportion of operation stage and new energy technology are 7.14 and 17.86 % respectively, shown in Fig. 99.6. Especially design codes of BEC embrace the main climate district of China, which expand from cold region to the zone of hot summer and cold winter, and the region of hot summer and cold warm winter. On the other hand the number of public buildings standard has increased. So far the range of BEC standards has covered the main types of building and the framework of Chinese BEC design has been established. In addition, it is conspicuous that new energy technology standards have developed rapidly during this period which extends from solar energy and biomass expanding to geothermal sources. At the same time, the code of building materials and products keeps developing momentum as last two stages and among which accounts for 63.16 % and each categories has increased. However, the problems of last stage are still not solved.

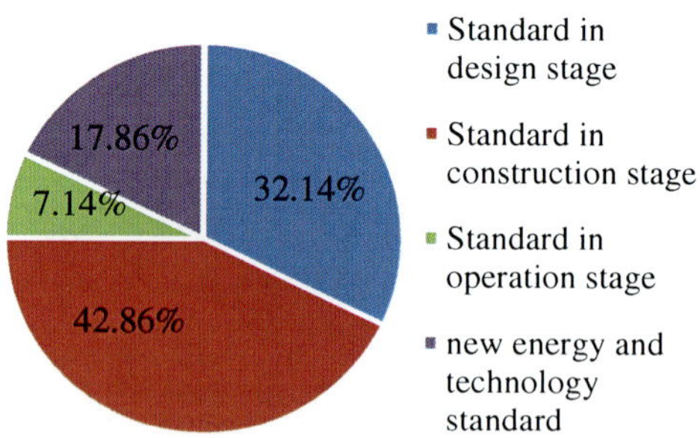

Fig. 99.6 Proportion distribution of ETS

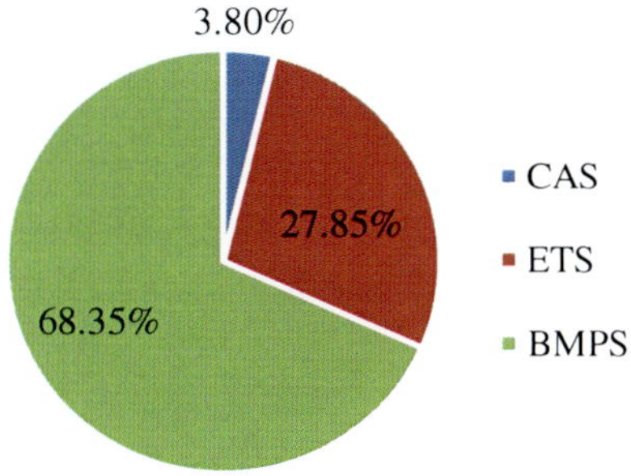

Fig. 99.7 Proportion distribution of standards

99.3.4 Stage of Integrated Development of BEC Standards(2006~)

Since 2006, BEC standards have integrated development. With the development of BEC, more and more provinces set higher goal for BEC. The goal of energy conservation about public buildings is 50 % in Zhejiang province and the target of Chongqing about residential buildings is 65 %. The number of BEC standards in this period is 79 by statistics and the sub-categories items are shown in Fig. 99.7. Compared with first three stages, this phase is more concentrates on amendment and supplement of existing standards that has greatly improved the implementation of BEC. During this stage, half of engineering technical standards has amended version against the originals. The standards of operation phase has enlarged to the aspects about monitoring, testing, and which has increased by 33.19 % compared to last period, the proportion shown in Fig. 99.8. The standards of comprehensive and auxiliary appear firstly during this stage and then bring a new direction of BEC standards from perspective of comprehensive assessment. The number of BMPS also has risen. The conclusion can be draw that with the development of science and technology, building material and product standards would be updated.

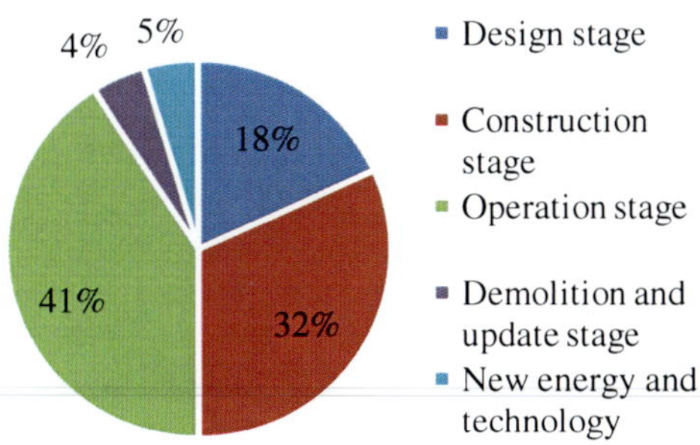

Fig. 99.8 Proportion of BMPS

99.4 Analysis of the Evolution of Chinese BEC Standards

99.4.1 Analysis of Evolutionary Dynamics of Chinese BEC Standards

Subjective Factors

The range of BEC is so wide that the integral framework of BEC standard cannot be provided one-time offer and the standards which cover all aspects of BEC cannot be supported during the early period of BEC standards development. That has already decided the facts which the standards need to be supplemented late. On the other hand, the later usage of code cannot accurately predicted also consist of the reasons why the standards need to be renewed. The inevitable defects of the BEC standards are the inner driving force of the evolution of BEC codes.

Objective Factors

Objective factors also as important factors have impacted the evolution of BEC standards. Firstly, the situation of energy crisis attracts more and more attention for energy development and utilization. The requirements about building energy conservation are rising so that the BEC standards have been revised and renewed correspondingly ultimately. Secondly, standards are suitable for a certain technical level and management requirement of period of time. BEC standards are no exception, with the development of construction technology and the improvement of management requirements, standards will be updated and developed. Thirdly, the influence of government policy cannot be ignored in the evolution of BEC standards. The energy conservation goal of government's five-year plan impacts the development of BEC, thereby impacts the BEC standards. Besides, the government's support and promotion for a certain kind of energy saving technology also generate new standards of BEC.

99.4.2 Evolution Paths of Chinese BEC Standards

From what has been discussed above, based on the institutional change theory, the paths of evolution about BEC standards can be drawn:

(1) Following the path of technicality criteria to management and evaluation criteria.
    In the first three stages, most of the standards belong to technicality criteria and in the last period management of evaluation criteria is spring up. Not until the number of technical standards achieves a certain level that the management or evaluation only can be formed. The conversion from technology to

management and evaluation improves the effect of the implement of BEC standards, which complies with the laws of development of routine.

(2) Following the path of easy control and high consumption region to difficultly control and high consuming region.
The control method behind the evolution of BEC standards is that from easy to difficult and from central heating zones to none central zone but high energy consumption regions. Design standards of BEC have gradually embraced the main climate district, which expand from cold region to the zone of hot summer and cold winter and the region of hot summer and cold warm winter.

(3) Following the path of building design to construction then extend to operation and renovation.
On the whole, the evolution path of BEC standards follows the whole life cycle trace of building that from design phase to construction, then to operation to renewal phase. However, not all the code follows this evolution path and there are some exception standards more concentrated on the testing of material. In addition, the standard of BEC also follows the path that from residential extend to public buildings.

99.4.3 Analysis on Existing Problems of BEC Standards in China

Though the analysis mentioned above, the standard of Chinese BEC has developed rapidly since 1982 and has a crucial guide effect on the implement of BEC. After the analysis of the evolution of BEC standards, this paper summarized the main existing problems. Firstly, the distribution of BEC standards is not uniform. The existing codes of BEC mainly focus on the building materials and products. By statistics, the number of building material and product standard accounts for 73.52 % and the other categories take proportion of 25.12 % and 1.27 % respectively, as shown in Fig. 99.9. Besides, there is the phenomenon that design and construction phase attaches more attentions than operation phase. Furthermore, the

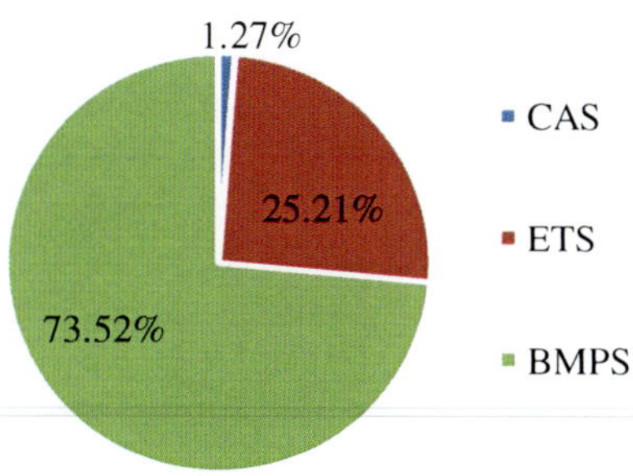

Fig. 99.9 Proportion distribution of standards

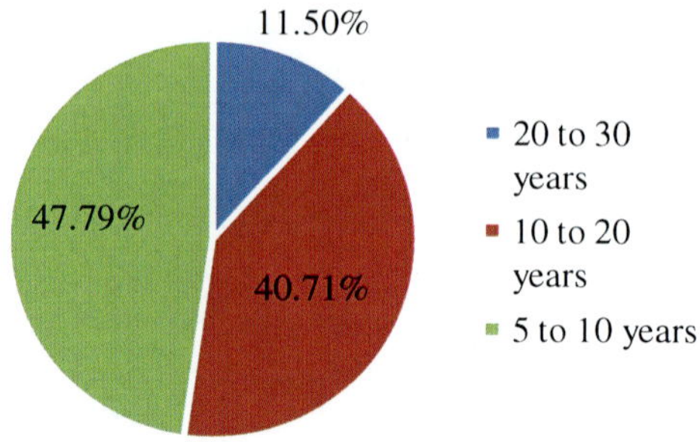

Fig. 99.10 The distribution of standard service time

codes of public buildings still lags behind compared to residential building and the range of which is limited [8]. Secondly, the renewal speed of BEC standards is tardiness. The number of standards which services in 20 to 30 years accounts for 11.50 %. And the codes which service in 10 to 20 years account for 40.71 % and 47.79 % respectively, as shown in Fig. 99.10. With professional development of construction, parts of the existing codes are not suitable for the situation today. And some criterions are out of step with BEC goals. Thirdly, the phenomenon of repetition and omission between standards has usually occurred. For example, "Evaluation standards for green building" and "Standard for energy efficient building assessment" have much same in the range of evaluation and assessment criteria.

99.5 Conclusions and Suggestions

Based on the time sequence, the evolution of the BEC standards divided into four stages. The standards of the infancy stage mainly concentrate on building materials and performance testing, as well as the residential building design; the initial development stage is a transitory period that the range of BEC has expanded to existing buildings from new buildings and the number of all kinds of codes has increased. In the stage of rapid development, the standards developed roundly. On the one hand until now the residential design codes of BEC have covered the main climate zones of China and public building standards have risen frequently. On the other hand the range of codes has extended to the whole life cycle of building. The high light of integrated development stage is that the amendment or supplement for standards improves the integral applicability of the standards.

On the basis of institutional change theory, this paper presents two main influence factors which affect the evolution of BEC standards. The subject factor is that the inevitable defects of the BEC standards and objective factors including energy crisis, technical progress and government policy. The paths of BEC standards are follows: the standards following the evolution route of technicality criteria gradually expands to management and evaluation criteria. The second path is that the application range of the standards from easy control and high energy consumption region to difficultly control high energy consumption region. The third path is that standards from design to construction then extend to operation and renovation phase which follows trace of building life cycle. The three main paths constitute the framework of the development of BEC standards and also the main evolution lines of Chinese BEC. And the main problems of BEC standards are that the distribution of BEC standards is not uniform and the renewal speed of BEC standards is tardiness and there are repetition and omission between standards.

Based on the deficiency and the evolution of the BEC standards, this paper provides references and suggestions on compilation and revision of the code of BEC. Firstly, expediting the speed and increasing efficiency of the renewal and formulation of BEC standards to ensure which are suitable for situation. Secondly, it is BEC standards system that is supposed to be established to guide the formulation of BEC standards. Although the quantity and quality of Chinese BEC standards have reached a certain level, the distribution of standards is dispersed and there have no a system which can guide the compilation and revision of codes. Thirdly, government would reasonably use the policy tools to guide development of BEC. Government can through economic incentives or other policy to promote BEC technologies and energy saving materials, such as the establishment of special fund, tax breaks and so on, also can make a lot of preferential policy to promote energy saving.

Acknowledgements This research is jointly supported by the National Social Science Foundation of China (Project No.: 12AZD064). The authors would love to express the deepest gratitude to the anonymous reviewers.

References

1. Chen G (2008) Overview study on building energy saving of China. Inf China Constr 06:28–31
2. Wang Q, Fan Y, Wan R (2011) Present situation of building energy-saving and energy-saving technologies. Block-Brick-Tile 3:37
3. Su Dequan LJ (2011) The research and practice of building energy efficiency. China Sci Technol Inf 2:45–46
4. North DC (2002) The structure change in economic history (Chen Y, Luo H). Shanghai People's publishing House, Shanghai, p 251
5. North DC (2008) Institutions, institutional change and economic performance. Shanghai Sanlian Bookstore, Shanghai, p 209
6. Bromley DW (2008) Sufficient reason: volitional pragmatism and the meaning of economic institutions (Lian Jian, Xi Yang, Ninghua Zhong). Shanghai People's publishing House, Shanghai, p 269
7. Tu F, Wang M (1991) Comments on building energy saving standard series. Low Temp Archit Technol 02:13–15
8. Yu S (2000) The standardization of building energy saving. China Stand (01):15–18

Chapter 100
Study on the Governance of Energy Performance Contracting Projects in Building

Saina Zheng, Pengpeng Xu and Hongjuan Wu

Abstract With the booming of economy, resource shortage has been the most urgent problem in China and will be a big setback for sustainable development. There is a paradox between the supply and demand of energy in current construction industry where the energy consumption is large while the demand for new building is continuously increasing. It has revealed that construction field consumes almost 1/3 of energy, so the potential of energy saving is quite considerable in Chinese construction industry. EPC (Energy Performance Contracting) is an efficient method to save energy in building. With the advocacy of the government, it has made great progress through the practice of EPC, but various internal and external factors would be barriers to the diffusion of EPC. This paper aims to build a structure of project governance and give some suggestions for solving such problems. This paper first introduces the basic theory of Energy Performance Contracting, then, describes the application of EPC to implement energy-saving issues encountered in practice. By analyzing the problems of energy performance contracting, the paper attempts to solve them from the perspective of project governance by developing a structure of project governance. At last, this paper interprets the structure to find out the governance activities that can promote the mode of EPC.

Keywords Energy performance contracting (EPC) · Building energy efficiency Retrofit (BEER) · Project governance

100.1 Introduction

In any event, human beings, economic development and social progress included, ultimately rely on energy resources. Nowadays, energy consumption in China, the second largest economic entity in the world, has grown sharply due to the rapid

S. Zheng (✉) · P. Xu · H. Wu
Faculty of Construction Management and Real Estate, Chongqing University, Chongqing, China
e-mail: Sena.zheng@yahoo.com

Y. Wu et al. (eds.), *Proceedings of the 20th International Symposium on Advancement of Construction Management and Real Estate*,
DOI 10.1007/978-981-10-0855-9_100

development of industrialization and urbanization. There is a paradox between energy supply and demand in current construction industry where the energy consumption is large while the demand for new building is continuously increasing. According to statistic data, energy consumption in construction industry occupies 1/3 of the total [1]. Underlying all of this is a necessity to arouse people's awareness of construction energy conservation. Besides, as for the proprietor and estate management companies, it is essential to advance their technology and improve risk tolerance capacity, which makes it easier for them to finance. In this regard, the energy performance contracting is needed on effectively curbing problems previously mentioned. There is, however, still a long way to promote EPC.

Despite there is a rapid expansion of the EPC industry and considerable benefits from EPC agreements during recent years, compared with huge potential market, energy conservation and emission reduction goal established by Chinese government are quite insufficient. Many barriers that hinder cost-effective energy efficiency projects and prevent the full development of EPC still exist. Firstly, EPC is a new concept in China without sufficient publicity, As a result, professors in EPC are definitely needed and the training agencies are rare, few of which aim to charge fees from the government. Secondly, EPC projects in China are initially self-financed by Energy Service Companies (ESCOs) because most of them are small and few funded by financial agencies. In Wen and Zhang's [2] opinion, ESCOs in China are undergoing the initial stage without support of policy and fund. Cao [3] argued that some ESCOs get fund from the government according to the relevant policy and then stop the EPC projects just for the supporting money. Thirdly, Sun [4] demonstrated that the main method to judge whether an EPC project is successful is to measure the amount of energy saving. However, in China, the ESCOs industry has no well-rounded industry criteria, such as service standards, energy conservation measurement, verification methods, and contract standards, etc. Fourthly, compared to energy conservation in manufacture and traffic, construction energy conservation is more difficult since construction energy system is composed of many subsystems such as lighting system, air conditioner system, office facilities system and so on. As a result, it needs varieties of ESCOs cooperating with each other, leading to much more stakeholders involved.

Since there are so many problems in EPC, and the traditional project management cannot solve them, it is essential to find a way to solve the problems so that EPC can be applied to save energy. Recently, the concept of "project governance" springs up and projects can be managed at a higher level.

This paper intends to solve problems from the perspective of project governance. The objective of this paper is two-fold. The first one is to a structure of project governance in basis of analyzing the business process of Energy Services Companies (ESCOs). The second one is to summarize measures to solve problems according to the structure and articles.

The paper will be divided as follows. Section 100.2 will present what project governance is and how to apply it in EPC. Section 100.3 will build the structure by analyzing the stakeholders. Section 100.4 will then turn to the governance activities according to the developed structure and the literature. It will interpret the structure to find out which activities can promote the mode of EPC.

100.2 Theory of Project Governance

Although there is an ever-increasing discussion on governance, the concept of project governance remains ambiguous. Crawford and Cooke-Davies [5] viewed project governance as a component of corporate governance. They defined project governance as a set of formal principles, structures and processes. Winch [6] considers that the process of governance may be analyzed at two levels, micro-analytical and macro-analytical, the former referring to how a transaction is carried out between actors and the latter referring to institutions and practices at a broader systemic or societal level. Turner and Simister [7] demonstrated that business challenge contribute to the appropriate governance between client and the contractor. Olsen recognized that a governance structure consists of mechanisms, such as contractual incentives, hierarchical mechanisms, based on authority, and relational or trust-based mechanisms [8]. Aaltonen and Sivonen [9] emphasized the significance of stakeholder management, including the processes of stakeholder identification, classification, and analysis. Artto and Kujala [10] argued that a project is a complex and dynamic system that requires specific governance mechanisms and the adoption of an open systems view. To summarize, project governance is a complex and dynamic system, which focuses on stakeholder management, consisting of mechanisms such as contractual incentives, hierarchical mechanisms based on authority, and relational or trust-based mechanisms.

This paper builds the structure of project governance through the lens of principal–agent theory [11] that is regarded as the most widely recognized theory in the project governance literature [12]. The theory asserts that it is inefficient to assume direct responsibility for the management [13] and highlights the resulting goal conflict that arises when individuals with differing preferences engage in cooperative effort. A core of principal–agent theory is the concept of accountability [14], which is also a critical characteristic of effective governance [15]. To governance the projects, the interests of different stakeholders must be considered while the accountabilities should be allocated.

100.3 Structure of the Governance of Energy Performance Contracting Projects in Building

100.3.1 Systems Thinking

The founder of System Dynamics (SD) Forrester holds the opinion that systems thinking originated from SD. Systems thinking is a holistic analysis approach that focuses on the way that a system's constituent parts interrelate and how systems work over time and within the context of larger systems. Richmond's [16] argued that systems thinking can facilitate developing an increasingly deep understanding of underlying structure. Senge [17] regarded systems thinking as a discipline for seeing wholes and a framework for seeing interrelationships and patterns of change rather than static snapshots. In contrast with traditional analysis, systems thinking seeks to understand how these parts interact instead of studying the parts of a system in isolation.

The governance of EPC projects in building can be regarded as a system. According to the concept of systems thinking, to solve the problems, we should explore the interrelation between factors. This paper intends to find out the operation structure of EPC in building and analyze the stakeholders involved.

100.3.2 Analysis of Stakeholder

Stakeholders involved in the governance of EPC project in building are various, divided as follows:

(1) Government department: The government is responsible for supervising the project.
(2) Relevant community: There is no doubt that the construction and operation of the project will cause pollution to the relevant community. In addition, it may take the residents' housing to extend project, so it is definitely necessary for residents in relevant community to supervise.
(3) Media: With the complement of sound market economy system, the bidding is getting more and more transparent under media control. In addition, every project concerning national economy and the people's livelihood must arouse attention from the society. Thus, media have the responsibility to convey the right message to citizens.
(4) Energy saving customers (Client): The so-called energy saving customer is the object that EPC projects implement on. They can be paid according to the contract and gain not only the equipment but also the precious method to save energy.
(5) ESCOs: ESCOs are the most important stakeholder of EPC projects. They play a leading role and gain the majority benefits of energy saving.
(6) Equipment supplier: Most of the ESCOs do not have energy-saving equipment, so they should buy them from equipment suppliers. ESCOs and equipment suppliers need to cooperate with each other. The suppliers have to install and debug, insure the quality, provide service after sales and so on.
(7) Construction unit: ESCOs provide service to real estate enterprises and they need construction units to install equipment.
(8) Finance agencies: ESCOs in China are usually pint-sized and unable to provide massive funds. As a result, it is essential to get loan from finance agencies.
(9) Third-party certification authorities: Evaluating the energy efficiency is a key work for EPC projects, making third-party certification authorities necessary. The authorities are supposed to check whether the energy efficiency achieves what the contract stipulates in order to decide how much the ESCOs can get.

The EPC projects in building are actually a system of stakeholders, each of whom takes participation in projects seeking the highest profit of their own. If one part gets more, accordingly another part get less and conflict of interest may arise which will definitely affect the ultimate aim. What project governance does is to make a series of systems to arrange rights and liabilities for all stakeholders.

100.3.3 Structure of Energy Performance Contracting in Building

Project governance can be divided into inner governance and external governance. Stakeholders of inner governance are those who take part in the construction, while stakeholders of external governance are those who supervise the social environment. Obviously, In EPC projects, external governance composed with government, media and revenant community who do not take direct participation in. The inner governance involves several stakeholders, to grasp the structure, we can stand from the perspective of ESCOs according to the business process (see Fig. 100.1) since ESCOs take part in most periods and they need to cooperate with the other stakeholders.

(1) Contact signing

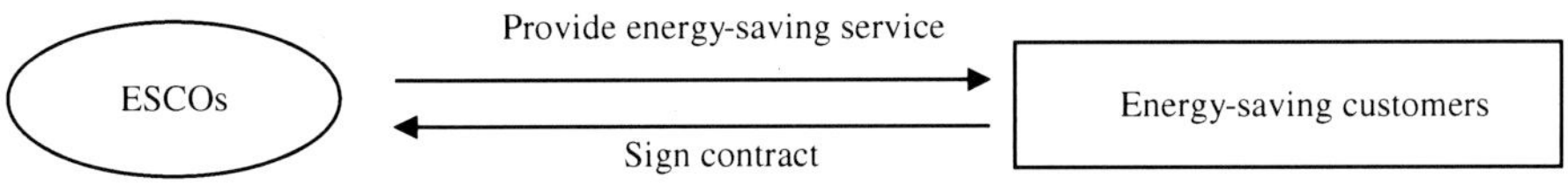

Based on the ESCOs and energy saving customers' agreement, ESCOs get most profit throughout the contract period while getting all profit after contract period.

(2) Financing

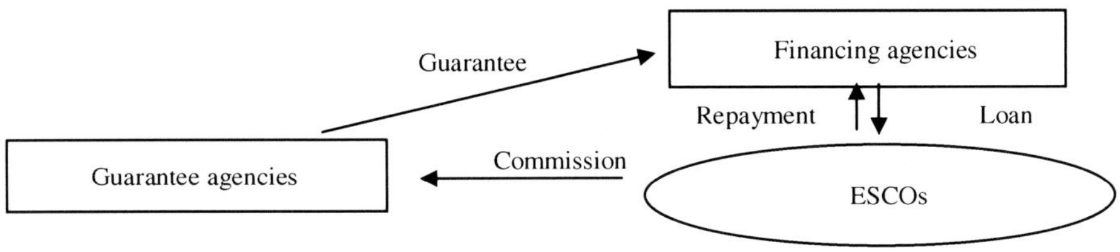

Objects involved in financing are ESCOs, financing agencies, and guarantee agencies.

(3) Equipment Procurement

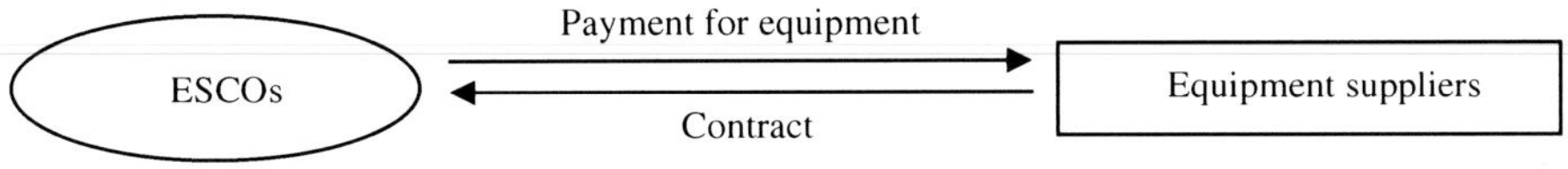

(4) Equipment installation and commissioning

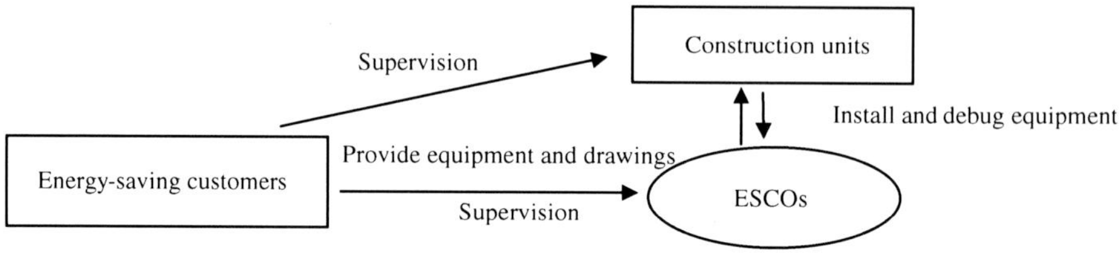

(5) Monitoring energy saving

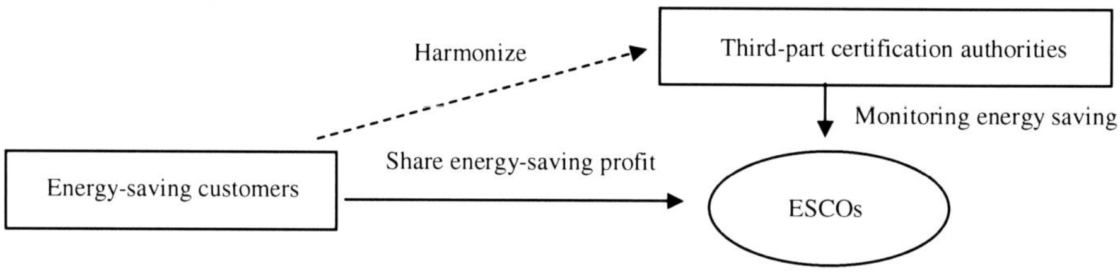

According to the contract, the third-party certification authorities have to complete the monitoring task which can be separated into three kinds-interval one, continuous one and one-off one.

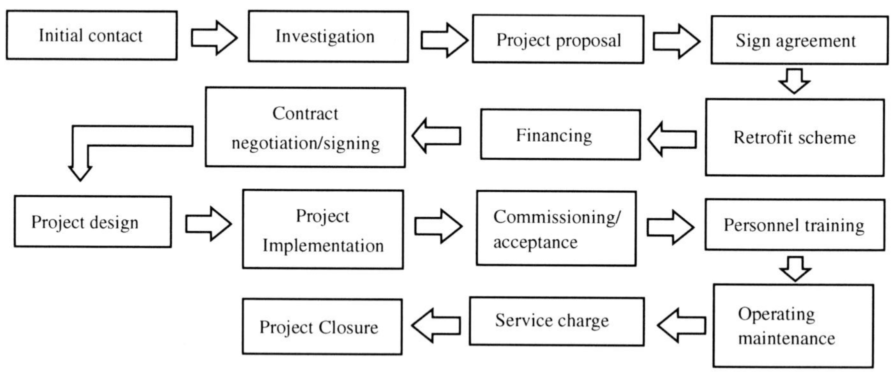

Fig. 100.1 Business process of ESCOs

(6) Project maintenance
According to the contract, ESCOs should provide equipment maintenance service throughout the contract period. What's more, ESCOs have to improve the performance so that profit can become larger.

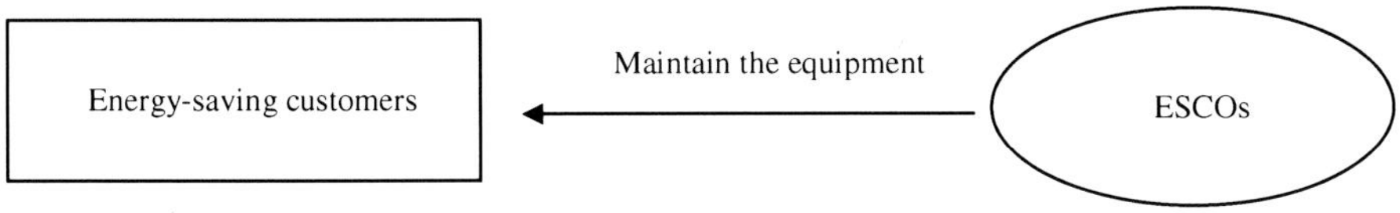

In addition, there are different energy-saving services in construction EPC, mainly transferring maintaining structure, heating installation, nature condition which do not meet the standard. Thus, more than one ESCO is needed and they need to cooperate with each other.

To sum up, it is necessary for ESCOs and customers to believe in and harmonize with each other. When two sides reach an agreement, a principal-agent relationship exists. Based on the analysis above, the governance structure is as Fig. 100.2.

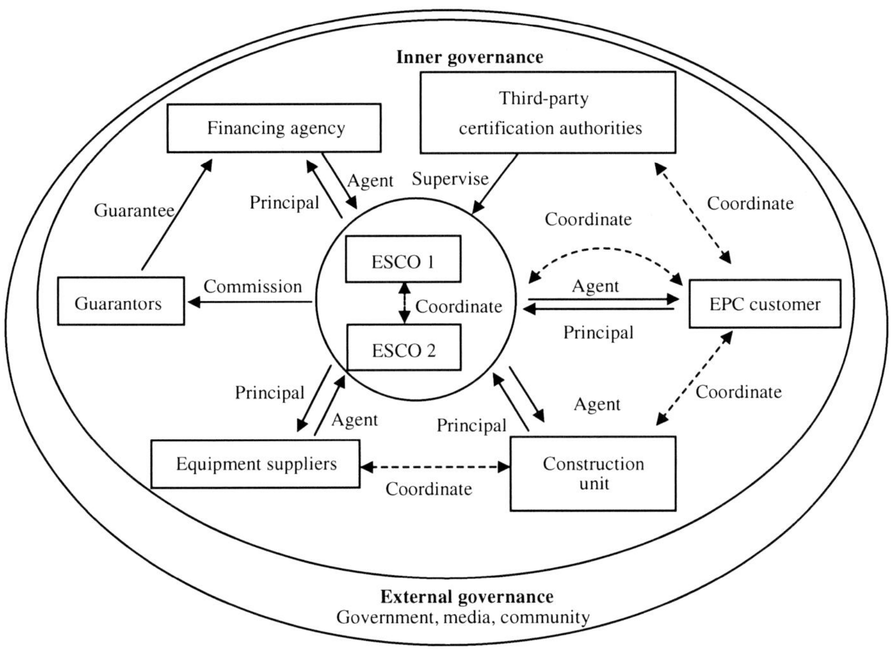

Fig. 100.2 Structure of EPC in building

100.4 Analysis of Governance Activities

100.4.1 External Governance

From the structure of project governance, the external governance is composed of government, media and relevant community, including several governance areas as bellows.

Authority of regulators: Whether a regulator is authoritative greatly affects the structure since EPC in building takes a long time and covers varieties of subjects. All the stakeholders need to be supervised in that they always consider self-interest first.

Economic environment: Because of the difference with other EPC projects, EPC projects develop at a low speed, thus good economic environment can promote its development, advancing the structure.

Awareness of energy conservation and consensus: Only people who realize the importance of energy conservation and disseminate energy conservation in building can promote the development of energy conservation in building.

Policy: Policy is made to promote development of energy conservation in building from the external.

100.4.2 Inner Governance

The aim of inner governance is to manage the principal-agent relationship and coordinate relationship between stakeholders, including some problems existing in stakeholders themselves.

(1) Self-qualification of stakeholders
To ESCOs, the main governance areas include the followings.

Technology: Technology is the foundation of ECSOs and with advanced technology, ESCOs can make more efforts in the industry.
Reputation: To some extent, the reputation can decide whether an ESCO can get fund from the financing, affecting the relationship between guarantors and financing.
Financial condition: Financial condition affects the project throughout the whole period. Once the capital chain breaks, the governance structure will dissolve.
Efficiency of energy saving: Whether an ESCO can make a profit depends on the efficiency of energy saving. In addition, good efficiency can bring more projects. The efficiency can be affected by three stakeholders.

(a) Equipment supplier: The quality of equipment has a direct impact on the efficiency.

(b) Construction unit: The governance activity involved is the quality of installation which affects efficiency.
(c) Third-party certification authorities: The major duty of third-part certification authorities is to check if the energy efficiency achieves what the contract stipulates. Therefore, in this stage the governance activity involved the ability of energy auditing which can decide how much profit an ESCO can get.

(2) The relationship of agent, principal and coordinate
As revealed in the governance structure, there is a contractual relationship between all the stakeholders. In practice, several problems may occur including mistrust to ESCOs, collusion between EPC customers and the third-party authorities, few communication between equipment suppliers and construction units, barriers to get fund for ESCOs and so on.
To summarize, the governance activities include trust, communication and equality. Trust and communication are two necessary preconditions for all the stakeholders working together. When the third-party authorities audit energy, they must comply with the rule of equity. Without equity, the profit of ESCOs will be damaged definitely.

100.4.3 Measures

Since various governance activities are identified including internal ones and external ones, more efforts are put into seek for measures to promote the mode of EPC in building. As for authority of regulators, Zhang and Han [18] regarded that sound legislation and regulation is needed to set standard for the industry. In addition, it is essential to enhance external supervision of media, government and residents. In the aspect of economic environment, it is considerable to place supporting finance policy and set up a special support fund. To raise the awareness of energy conservation and consensus, promoting environmental protection is necessary. Moreover, the government should implement energy efficiency policy [19] and Finance and Taxation policy. From the perspective of ESCOs, it is significant to conduct technological innovation, set up a reputation system, improve the staff's credit, control financial expenditure, and balance the books. To ensure the quality, the equipment suppliers need to reinforce equipment monitoring, select equipment strictly and improve quality of installation. Furthermore, standard protocols for Monitoring and Verification are regarded as one of the most important measures [20]. Last but not least, measures including opening regular meetings, enhancing supervision and punishing unfairness should be taken to ameliorate contractual relationship. Overall, measures can be summarized as Table 100.1.

Table 100.1 Measures to Promote the Mode of EPC

	Areas	Measures		
External governance	Authority of regulators	Establish sound system of legislation and regulation [18]		
		Enhance external supervision		
	Economic environment	Supporting finance policy [18]		
		Set up a special support fund		
	Awareness of energy conservation and consensus	Vigorously promote environmental protection		
		Enhance supervision by public opinion		
	Policy	Vigorously implement energy efficiency policy [19]		
		Supporting finance and taxation policy [18]		
Inner governance	Stakeholders	ESCOs	Technology	Technological innovation
			Reputation	Set up a reputation system
				Train the staff of their credit
			Financial situation	Control financial expenditure
				Make ends meet
			Energy-saving efficiency	*Same as technology*
		Equipment suppliers	Quality of equipment	Reinforce quality monitoring
				Select equipment strictly
				Sampling inspection
		Construction	Quality	Technology training
				Enhance quality of Installation
		Third-party certification authorities	Energy Audit	Develop standard protocols for Monitoring and Verification [20]
	Contractual relationship	Trust	Establish credit system	
		Communication	Open regular meeting	
		Equity	Enhance supervision	
			Punish unfairness	

100.5 Conclusions

At present, China's energy services market is limited, there are still many barriers hampering the successful implementation of EPC projects. Fortunately, China is vigorously promoting energy efficiency and raising people's awareness of energy efficiency. EPC in building is faced with an excellent development opportunity in China.

This article takes an idea of project governance to treat problems meet in practice of EPC. It analyzes main stakeholders including government, relevant community, media, EPC customers (real estate companies), ESCOs, equipment suppliers, construction units, financial agencies and the third-party authorities. Then this paper build a governance structure to manage the practice of EPC. At last, we propose several governance activities to promote the development of the EPC in China. In future studies, the structure will be further improved through verification.

Acknowledgements The work described in the current paper was supported by a grant entitled "Study on Synergetic Mechanism and Evolutionary Model of Building Energy Service Industry Development" from the National Natural Science Foundation of China (Grant No. 71403033). The authors would like to thank anonymous reviewers. All remaining errors are the sole responsibility of the authors.

References

1. Ministry of Housing and Urban–Rural Development of Science and Technology Promotion Centre (2011) Building energy efficiency report in China
2. Wen Y, Zhang YF (2010) Effective solution to the financing problems of energy management contract. Economic Herald, 12–13
3. Cao MD (2011) Analysis of Law and Policy of EPC in China. J East China Univ Polit Sci Law, 63–79
4. Sun H (2012) Energy performance contacting practices, pp 63–70
5. Crawford L, Cooke-Davies T (2009) Project governance: the role and capabilities of the executive sponsor. Project perspectives, vol XXXI, pp 66–74
6. Winch GM (2001) Governing the project process: a conceptual framework. Constr Manag Econ 19(8):799–808
7. Turner JR, Simister S (2001) Project contract management and a theory of organization. Int J Project Manag 19(8):457–464
8. Olsen B, Haugland S, Karlsen E, Husøy G (2005) Governance of complex procurements in the oil and gas industry. J Purchasing Supply Manag 11(1):1–13
9. Aaltonen K, Sivonen R (2009) Response strategies to stakeholder pressures in global projects. Int J Project Manag 27(2):131–141
10. Artto K, Kujala J (2008) Project business as a research field. Int J Manag Project Bus 1 (4):469–497
11. Eisenhardt KM (1989) Agency theory: an assessment and review. Acad Manag Rev 14(1):57–74
12. Turner JR (2009) Governance of project-based management. Handbook of project-based management. McGraw Hill, London
13. Fama EF, Jensen MC (2003) Separation of ownership and control. J Law Econ 1983:301–325

14. Jensen MC (2003) A theory of the firm: governance, residual claims, and organizational forms. Harvard University Press, Cambridge
15. Brand H (2007) Good governance for the public's health. Eur J Public Health 17(6):541
16. Richmond B (1994) Systems dynamics/systems thinking: let's just get on with it. In: International systems dynamics conference, Sterling, Scotland
17. Senge P (1990) The fifth discipline, the art and practice of the learning organization. Doubleday/Currency, New York, NY
18. Zhang Y, Han QM (2008) Analysis for critical success factors of energy performance contracting (EPC) projects in China. In: Proceedings of the 2008 IEEE IEEM, pp 675–679
19. Gao S, Qu S, Geng Z, Yang Q, Guo Y (2004) Review and evaluation of energy strategy and policies. Sub-report 1 of China's National Energy Strategy and Policy. Development Research Center, Beijing
20. Amann JT, Mendelsohn E (2005) Commercial Retrofit Programs: a review of activity and opportunities. April 2005 report number A052

Chapter 101
Public-Private Partnerships for Sustainable Construction Projects: Opportunities and Challenges

Jin Wu

Abstract With the popularity of PPP as an approach for governments providing infrastructure-based service, its potential to promote sustainable construction should be discussed. Based on literature review, this paper aims to identify what attributes in PPP approach could help to make constructions more sustainable, and also discuss whether these attributes can be crystallized in practice. It is suggested that long-term commitment, bundling of finance/construction/operation, as well as payment based on performance against output specifications, are helpful to promote construction more sustainable at first sight. However, all these attributes cannot guarantee sustainable objectives be achieved. It seems that the market opportunity is the root for the public sector to direct the private sector to achieve their objectives, including sustainability. So it is important for the public sector to realize and use this advantage, rather than relying on other fiction depicting the rosy facet of sustainability through PPP.

Keywords Public-Private partnerships · Sustainability · Infrastructure

101.1 Introduction

In recent decades, Public-Private Partnerships (PPP) has been widely used to provide infrastructure-based public services, in both developing and developed countries. Since infrastructures are considered to have a significant impact on sustainability, these two concepts and their relations emerge to be of research interest among academics (see [1–4]. Considering these two concepts are two directions chosen by the practical world, it is not weird to witness research suggesting that sustainable objectives in construction can be better served through PPP, of course with the precondition of proper application.

J. Wu (✉)
School of Investment and Construction Management, Dongbei University of Finance and Economics, Dalian, China
e-mail: wujindufe@yahoo.com

Y. Wu et al. (eds.), *Proceedings of the 20th International Symposium on Advancement of Construction Management and Real Estate*,
DOI 10.1007/978-981-10-0855-9_101

However, whether PPP can really promote sustainability in construction is arguable. In science, even social science, a theory should be grounded on both logic and evidence [5]. Competing evidence are witnessed, since not all PPP projects are successful, let alone their promotion for sustainability. So additional reports on black or white swans cannot demonstrate that all swans are black, neither all in white.

Asking professionals whether PPP can be helpful to achieve sustainable target and key successful factors is another approach to gain support. But opinions are not facts. The results are always decided by who is asked. For example, in UK, government presents their support in PPP, but the parliament preserves a skeptical attitude. Furthermore, this questionnaire approach may be criticized as an abuse to investigate any 'original' area from the combination of sub-topics with fashion concepts, which are not necessarily meaningful [6].

With both success and failure cases on hands, it is not very helpful to generally argue whether PPP can promote sustainability. A meaningful approach is to break the general concept of PPP to its attributes and sustainability to its dimensions, and then discuss what attributes in PPP can influence sustainability on a specific dimension. Some attributes may serve sustainability as an advantage by nature, others may provide a tool but the ultimate effects can be diverse, still others may not be unique trait presented by PPP. There may be even some factors in PPP that impede the sustainable purposes. Open the black-box is helpful to understand why and how elements in PPP can better serve the sustainable target, and then take full advantage of them in practice.

In the following sectors, attributes in PPP and dimensions of sustainability will be reviewed first, and then discussions around whether and how PPP elements can improve sustainability will be launched, specifically on investment and technology commitment in the long term, change in procurement, integration issues, and performance-based payment against output specification.

101.2 PPP and Sustainability

101.2.1 PPP: Definition, Forms and Attributes

Although PPP are now widely used to provide infrastructure-based public services, there still lacks a precise definition for this concept. Since almost all public infrastructures have to involve private sector, in different stages and to different extent of course, it is quite confusing to present a common terminology among business, academia and government to depict the concept of partnership between government and private sectors [7]. Alfen and Jan [8] describe PPP as a continuum ranging from traditional procurement to full privatization. More than 25 terms have been used to describe the models contained within this spectrum [7]. Furthermore, development of PPPs can evolve very differently across national institutional contexts [9] and thus make the definition of this concept quite political (Yescombe 2013).

Definition from international organizations may provide useful ground for further discussion. In the Public-Private Partnerships Reference Guide (version 2), the World Bank defines PPP as 'a long-term contract between a private party and a government entity, for providing a public asset or service, in which the private party bears significant risk and management responsibility, and remuneration is linked to performance.' OECD [10] defines PPP as a model in which the private sector designs and builds new or refurbishes and expands existing infrastructure, provides project financing, manages the asset and operates the service.

From these definitions, some attributes can be extracted, such as long-term contract, performance based payment, and tasks much more than construction are assigned to the private party. The PPP Reference Guide suggests that PPPs can be described in terms of the following three parameters, new assets, task/stage bundling, and payment mechanism. Based on the stages in which private sectors are involved, many models of PPP can be named, such as Build-Operate-Transfer (BOT), Design-Build-Finance-Operate (DBFOM), and so on.

In the World Bank's PPP Reference Guide, it is conclude that PPP may help to improve infrastructure through: (1) additional sources of funding and financing, (2) private sector analysis and innovation which can help to improve the otherwise poor planning and project selection, (3) private sector experience and incentives which can provide a more efficient or effective delivery, and (4) long-term investment.

101.2.2 Sustainability: Triple Dimensions

Since the built environment provides a synthesis of environmental, economic and social issues [11], the idea of sustainability attracts the construction industry. Much research within construction academia has expressed great concerns on this topic (e.g. [12–14]).

However, like PPP, sustainability by itself is also not a quite definite concept. Early promotion of sustainable development was initiated by the World Commission on Environment and Development. The concept was defined to reflect natural resource utilization to attain development objectives by national, regional and individual stakeholders, while fulfilling the need of the present generation will not compromise the ability of future generations to fulfill their need [15]. The original purpose of sustainability was environment based. Koppenjan and Enserink [1] describes this environmental sustainability as the impact of service delivery by public infrastructures on the urban population, urban environments, and the wider surroundings. He further enumerates these impacts may include health, well-being, air quality, water quality, congestion, ecological impacts, depletion or maintenance of resources, impacts on downstream rural communities as a result of water pollution, etc. [1]. These multitudinous impacts imply that even the concept of environment sustainability is not easy to be clearly defined for infrastructures in the real world.

But the idea is in fact much more difficult to elaborate, since social and economic dimensions were suggested to be incorporated. Nair et al. [16] defined financial sustainability as the possibility of authorities to live up to the financial obligations that result from investments in infrastructures, both in the short and in the long run. This concern on financial sustainability has been highly recognized in China, through its regulations. Social sustainability may be more obscure. Although Roseland [17] defined it as the impacts of urban infrastructure on the affordability of and access to public service delivery by poorer groups within urban society, it seems that researchers are still working on how to appraise this concept [18].

In the following discussions, for the sake of simpleness, environment sustainability is used as a default. Social and economic sustainability will be referred to when necessary. Following sections will enumerates advocate for PPP's contribution to sustainability, after each of which arguments will be presented.

101.3 Involvement of Private Sector Commitment

The first support for PPP's contribution to sustainability in the built environment is the involvement of commitment from the private sector in public services. As suggested in World Bank's PPP Reference Guide, additional sources of funding and financing as well as private sector experience and incentives can be introduced. Facts are described, but partially.

What the private sector searches for are business opportunities, no matter sustainable target are met or not. Koppenjan [1] indicates that many PPP projects contribute to infrastructures with negative adaptation and mitigation impacts. Whether the involvement of private sector's funding and knowledge is committed to sustainability depends on the purpose of why involving them. It seems the motive power of this contribution, if any, comes from the public sector's intention to fulfill sustainable targets. This will be discussed in the section of procurement change.

Furthermore, this kind advocate requires a precondition that private sector are more efficient, at least in PPP projects. In the new public management trend, more tasks and responsibilities are assigned to the private sector [18], with the assumption that the private sector is able to initiate innovative solutions, deliver more quickly and at a lower cost, and provide funding and operate facilities more efficiently [19]. However, it may not be the truth. Private funding is always more expensive than public funding, and state-owned enterprises may not be less efficient than their private counterparts. Furthermore, even in the traditional procurement, private sectors are sooner or later involved. The efficiency gains, if any, is not from who are involved, but how they are involved.

101.4 Life Cycle Perspective and Stage Integration

One of the most distinctive and innovative feature of the PPP model is its bundling of different tasks including design, construction, maintenance, operation and financing [20]. Lenferink et al. [21] call this feature as stage integration. Integration of design and construction can improve constructability, while integration of financing responsibility provides a heightened incentive to mitigate cost overruns (Grimsey and Lewis 2004). But the most attractive integration feature in PPP, from sustainability perspective, is including maintenance and operation stages into this package. This allows for a life cycle approach to assess the project by a single party [1]. In PPP arrangements, the private sector has to bear both the construction cost and operation/maintenance expenditure, and can only get rewards in the process of service provision. The private sector is thus incentivized to take the cost of energy use into account of the life cycle cost and optimize the expenditure scheme with a balance between construction stage and operation/maintenance stage. Such arrangements are always with a sustainable infrastructure with low energy consumption [22].

Do the above arguments provide a logical reasoning for the relationship between the improvement of sustainability and the application of PPP? Necessary but not sufficient.

Realization of this stage integration or bundling always takes the form of a consortium that brings together project designers, managers, construction companies and financiers [15]. However, those integrated tasks have to be allocated within this consortium or sub-contracting out, which is called by Roehrich and Caldwell [23] as the unbundling paradox. The question is why the public authority need to assign the integration tasks to a private consortium and let this private consortium to unbundle the integrated tasks, rather than do it by itself. Or more clearly, why this unbundling work can be more effectively fulfilled by the private consortiums than the public authority?

It seems the answer results in incentive motive and mechanisms. With a relatively steady rewards scheme, extra profits results from cost savings. The key point is the impact of energy cost has been successfully internalized in the above case for reasoning. It is the internalization of negative externalities, rather than stage integration that incentivizes the private consortium to consider how to optimize unbundling. Imagine that if operation cost can be substantially reduced through an environment pollution method without any penalty, what would be the choice of the private consortium? That's why I claim that stage integration feature in PPP is just a necessary but not sufficient condition to improve sustainability. Whether sustainability purpose can be served depends on whether negative externalities can be successfully internalized in the contract. This further depends on the public authority's capability to identify future externalities, and the ability to include such concerns in the contract.

101.5 Procurement Change, Out-Put Specification and Performance-Based Payment

In the traditional procurement, the contract is always, although not necessarily, awarded to the bidder with the lowest price, in an open tender procedure. In PPP models, the public authority is mainly responsible for elaborating objectives of the infrastructure-based services, while the private competitors propose plans to fulfill such objectives. Considering that governments are not familiar with which technologies may contribute to sustainable targets, asking the market to present innovative solutions seems not a bad choice. Furthermore, as an incentive mechanism, payment will be made according to service quality, including whether sustainable requirements are met. So the private consortium cannot get full rewards until all the objectives, including sustainable ones, be achieved.

Sounds good? Yes, but procurement change and out-put specification methods are not characters unique in PPP, and payment relating to sustainable targets is based on the contract. So why the laurel should be given to the PPP approach? Even in the traditional procurement approach, sustainable targets could also be incorporated into the selection criteria, and then into the out-put specifications. This inclusiveness of sustainable requirement reflects the target of the public authority, not the purpose of PPP.

Furthermore, once the sustainable requirements are set by the public authority in partner selection or social responsibility regulation, as an economic party with bounded rationality, the private sector is incentivized to meet the sustainable requirements with the minimum cost. Sustainability can only be viewed as a restraint condition, while the minimum cost to meet the restraint is indeed the target of the private sector. Koppenjan and Enserink [1] indicated that governments are bound to be disappointed if they expect private involvement can automatically result in the adoption of innovative environment-friendly technologies.

So, possibility to improve sustainability is not decided by whether a private party is involved, but whether the public authority, as a buyer or regulator, would like and be able to clearly state sustainable requirement and incorporate them into the contract. Considering the obscure dimensions of sustainability, and even the environment sustainability could provide a long list of requirements, such specifications on out-put do not seem easier than those on in-put.

101.6 Long Term Nature

Another feature for the PPP approach is its long time span, in which a concession period of 15, 20 or even 30 years is awarded to the private sector. Although it may lock-in investment on sustainable projects, if have been procured successfully as discussed above, for a long term, it seems that the impact on sustainability are mainly negative.

First worry may come from the long term payment obligation imposed on the government or end user. As a consideration to investment lock-in from the private side, there also exists a lock-in for the public. This long term nature is always with a monopoly status permitted or even protected by the government. In many PPP contracts like toll roads, in attracting private investors, it is not uncommon for the government to admit not allowing a new similar road in the nearby region.

Second, the technology may be environment-friendly when the project was initiated and contracted, but during the long term operation stage, without a technology improvement clause in the PPP contract, the public has to accept the arrangement which is appropriate many years ago. Zhang et al. [29] indicates that technological advancement can impose a substantial influence on project sustainability. And as Koppenjan and Enserink [1] criticized, communities are stuck with old technologies while new ones made old ones outdated.

But the most negative impact of this long-term feature may be from the difficulty to identify and stipulate sustainable requirements in such a long time span. A proper envisage and accurate description of sustainable requirements is essential to incentivize the private sector. All the potential benefits arising from private involvement, stage integration, procurement change, out-put specification and performance-based payment cannot be crystallized.

101.7 Conclusion

As two popular topics in the built environment, PPP and sustainability are always mentioned together by construction academia. The combination of hot sub-topics provides chances to ask opinions from respondent, which can turn out to be good resources for research work. However, whether such combinations are meaningful need support from logical reasoning.

Based on a review of PPP's attributes and dimensions of sustainability, this paper analyzed whether these attributes can help to improve sustainability in infrastructure-based public service. It is argued that involvement of private sector, stage integration, procurement change, out-put specification and performance-based payment are not necessarily helpful to serve for sustainable target. The public authority needs to realize that it is the intention to promote sustainable targets and capability to transfer such purpose to selection criteria and contract clauses by itself that forms the original driving force to purchase sustainability. However, the long term feature in all PPP contracts presents a high level requirement on capability to look-ahead, which makes the contribution to sustainability through PPP more difficult.

References

1. Koppenjan JFM (2015) Public-private partnerships for green infrastructures. Tensions Challenge Curr Opinion Environ Sustain 12:30–34

2. Kumaraswamy M, Zou WW, Zhang JQ (2015) Reinforcing relationships for resilience—by embedding end-user 'people' in public-private partnerships. Civil Eng Environ Syst 32 (1–2):119–129
3. Rathi S (2006) Alternative approaches for better municipal solid waste management in Mumbai, India. Waste Manage 26(10):1192–1200
4. Sha Z, Tiong RLK (2010) First public-private-partnership application in taiwan's wastewater treatment sector: case study of the Nanzih BOT wastewater treatment project. J Constr Eng Manage 136(8):913–922
5. Babbie ER (2013) The practice of social research. Wadsworth Publishing, California
6. Wu J (2010) Securing payment in the mainland China construction industry: the problems of payment arrears and their remedial measures, Ph.D thesis. The University of Hong Kong, Hong Kong
7. Delmon J (2010) Understanding options for public-private partnerships in infrastructure. World Bank policy research working paper 5173. World Bank, Washington, DC
8. Alfen HW, Jan Y-CA (2010) An introduction to PPP concept. In: Wang SQ (ed) Public-private partnership in infrastructure development: case studies from Asia and Europe. North United Press, Shenyang
9. Petersen OH (2011) Public–private partnerships as converging or diverging trends in public management? A comparative analysis of PPP policy and regulation in Denmark and Ireland. Int Public Manag Rev 12(2):1–37
10. Organization for Economic Co-operation and Development (2008) Public–private partnerships: in pursuit of risk sharing and value for money. OECD, Paris
11. Prasad D, Hall M (2004) The construction challenge: sustainability in developing countries. RICS, London
12. Crawley D, Aho I (1999) Building environmental assessment methods: application and development trends. Build Res Inform 27(4/5):300–308
13. Ding GKC (2008) Sustainable construction: the role of environmental assessment tools. J Environ Manage 86(3):451–464
14. Shen LY, Tam VWY, Tam L, Ji YB (2010) Project feasibility study: the key to successful implementation of sustainable and socially responsible construction management practice. J Clean Prod 18(3):254–259
15. WCED (1987) Our common future. Oxford University Press, Oxford
16. Nair DG, Bert EG, Gopikuttan PV, Alex F, Rene D (2005) A conceptual framework for sustainable-affordable housing for the rural poor in less developed economies. In: Paper presented at the 2005 world sustainable building conference SB05. Tokyo, Japan
17. Roseland Mark (1998) Toward sustainable communities: resources for citizens and their communities. New Society Publishers, Gabriola Island, BC
18. England K, Ward K (2007) Neoliberalization: States, networks, peoples. Blackwell, Malden
19. Savas ES (2000) Privatization and public-private partnerships. Chatham House, New York
20. Daniels RJ, Trebilcock MJ (1996) Private provision of public infrastructure: an organizational analysis of the next privatization frontier. Univ Toronto Law J 46:375–426. doi:10.2307/825772
21. Lenferink S, Tillema T, Arts J (2013) Towards sustainable infrastructure development through integrated contracts: experiences with inclusiveness in Dutch infrastructure projects. Int J Project Manage 31(4):615–627
22. Akintoye A, Matthias B, Cliff H (eds) (2003) Public-private partnerships: managing risks and opportunities. Blackwell, Oxford, UK
23. Roehrich JK, Caldwell ND (2012) Delivering integrated solutions in the public sector: The unbundling paradox. Ind Mark Manage 41(6):995–1007
24. Edum-Fotwe FT, Price ADF (2009) A social ontology for appraising sustainability of construction projects and developments. Int J Project Manage 27(4):313–322
25. Koppenjan JFM, Enserink B (2009) Public-private partnerships in urban infrastructures: reconciling private sector participation and sustainability. Public Adm Rev 69(2):284–296

26. Trebilcock M, Rosenstock M (2015) Infrastructure public-private partnerships in the developing world: lessons from recent experience. J Develop Stud 51(4):335–354
27. World Bank, Public-Private Partnerships Reference Guide (version 2)
28. Yescombe ER (2007) Public-private partnerships principles of policy and finance, Amsterdam. Elsevier; Butterworth-Heinemann, Burlington, Mass
29. Zhang XL, Wu YZ, Shen LY, Skitmore M (2014) A prototype system dynamic model for assessing the sustainability of construction projects. Int J Project Manage 32(1):66–76

Chapter 102
An Investigation of Demolition Waste Management: Case of Shenzhen in China

Huanyu Wu, Hongping Yuan, Jiayuan Wang, Lei Ouyang and Zhengdao Li

Abstract A huge amount of demolition waste was produced during the Chinese urban renewal process. How to promote the material recycling rate and therefore to reduce the environmental impacts has been a significant challenge. However, there is a lack of research revealing the flows and key procedures of demolition waste management. To fill the research gap, this study conducted in-depth interviews with 15 on-site managers involved in demolition projects in Shenzhen of China. As a result, the flows, key management processes and measures have been obtained. Particularly, the key procedures and waste management measures have been identified with a Process Mapping Approach (PMA). The results show that main stakeholders in demolition waste management chain include general contractors, professional demolition companies, transport companies, recycling plants, landfill officers, government departments and scavengers. Illegal dumping and low marketing acceptation of recycling products are two major barriers of demolition waste management. This study forms a concrete base for future studies aiming at optimizing the demolition waste management process.

Keywords Demolition waste · Waste recycling · Waste management · Process mapping approach

H. Wu · J. Wang · L. Ouyang
College of Civil Engineering, Shenzhen University, Shenzhen, China

H. Yuan (✉)
School of Economics and Management, Southwest Jiaotong University, Leshan, China
e-mail: hpyuan2005@gmail.com

Z. Li
Department of Building and Real Estate, The Hong Kong Polytechnic University, Kowloon, Hong Kong

Y. Wu et al. (eds.), *Proceedings of the 20th International Symposium on Advancement of Construction Management and Real Estate*,
DOI 10.1007/978-981-10-0855-9_102

102.1 Introduction

While acknowledging the social and economic development in China over the past decades due to large scale of urban development and renewal activities, it should be envisaged that a huge amount of construction and demolition (C&D) waste was produced [1–3]. According to the annual report of the comprehensive utilization of resources in China (2014), in 2013, one billion tons of C&D waste was generated, which accounts for 40 % of municipal waste of the country, but the waste recycling rate, is less than 5 % [4]. Among the C&D waste, demolition activities contributes to 70–75 %, and the generation amount per unite gross floor area of demolition waste is more than 50 times of that of new construction waste [5–7]. This is because in most building and infrastructure demolition cases, the whole building/block are rapidly flattened, leading to a huge amount of demolition wasted materials. The waste is transported to landfills after simply on-site sorting. The rough way of demolition waste management exacerbates the difficulty of waste recycling and simultaneously increases the overall management cost. Also, it takes up a lot of land resources due to landfilling and causes tremendous pressure on urban environment, which is not conducive to environmental governance [8, 9]. By contrast, the waste recycling rates in most of EU countries have reached 50 % in 2005, even up to 60–70 % in France and Britain [10].

Shenzhen is in a leading position in China in managing C&D waste as it claimed that the disposal capacity of C&D waste has reached four million tons and the recycling rates has reached up to 40 % [11]. However, in fact, more than 95 % of wasted materials could be recycled for new constructions with proper processes. If the recycling rate of C&D waste in China could rise to 90 %, it is estimated that 600 billion yuan of GDP can be gained directly [12]. It is therefore clear that there is still a great potential for improvement in recycling C&D waste in China. In this regard, how to promote the overall recycling rate has become an urgent problem waiting to be solved. Since demolition waste management is a complex system business which involves multiple management processes and stakeholders, identifying paths of waste flows and key waste management procedures is the prerequisite to optimize the waste management process accordingly. However, it is found from existing literature that the research on C&D waste management process is mainly concentrated on new construction sites. There is no research examining the paths of waste flows and key management procedures of demolition waste.

Therefore, this study aims to identify the paths of waste flows and key management processes of demolition waste, based on the particular circumstance of Shenzhen, with an ultimate purpose to optimize demolition waste management measures. The information of paths of waste flows and critical demolition waste management processes were obtained through survey and depicted empirically with the help of Process Mapping Approach (PMA). The findings revealed in the study can be useful for further study attempting to optimize the demolition waste management processes in China.

102.2 Research Methodology

In order to identify the whole process of demolition waste flows and management activities, this study firstly conducted an investigation of demolition waste management in Shenzhen by interviewing on-site managers involved in demolition projects. The survey was carried out during November 2014 to January 2015. As a result, the paths of waste flows, main stakeholders, and key management processes were obtained. Finally, the demolition waste management process model was developed using the PMA, and major barriers to managing demolition waste were identified.

First, to obtain first-hand information and detailed data, 15 managers in charge of waste management in demolition projects were selected for inquiries. Among them, nine respondents come from general contractors and the other six from professional demolition companies. In terms of the construction qualification, two companies hold premium qualification, five possess the qualification of Type Class A, one company has Type Class B and one company is Type Class C. Concerning the position of interviewees, seven respondents are project managers, three are engineers, two are CEOs of demolition companies and the rest are other managers of projects. The detailed profile of the interviewees is shown in Fig. 102.1.

Second, based on the on-site investigation and interview outcomes, the boundary and contents of demolition waste management throughout the waste chain are defined. Since there are various processes and waste disposal flows involved in the demolition waste management, an accounting table was employed to account the key demolition waste management processes, measures and major stakeholders concerned.

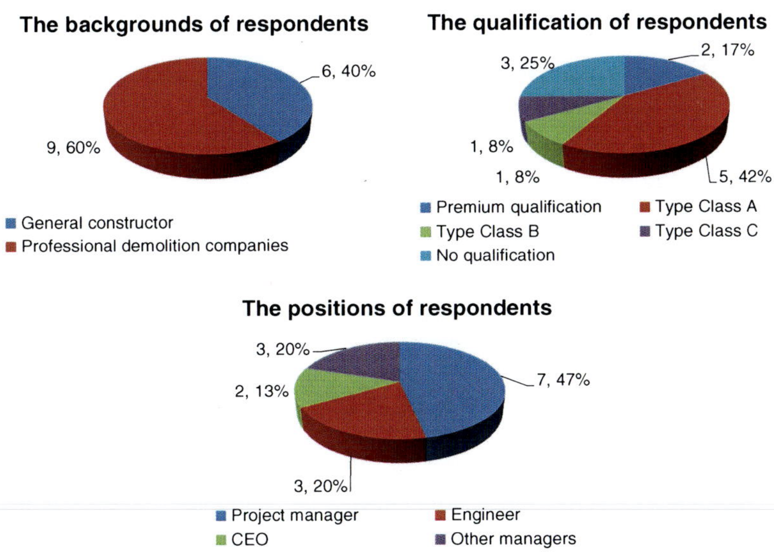

Fig. 102.1 The profile of respondents

Table 102.1 Main element in the PMA

Elements	Definitions
Waste source	Demolition waste
Waste management process	Demolition, waste collection, sorting, classification, loading, transport, and disposal
Waste disposal options	Recycling, landfilling, and illegal dumping
Waste management measures	Measures adopted in each waste management process, which is represented by "M" in the model
Project stakeholders	Participants involved in demolition waste management, such as general contractors, professional demolition enterprises and recycling companies, etc., which is represented by "S" in the model

Third, since the process of demolition waste management involves several stages and different stakeholders, a method capable of systematically characterizing the process flows and disposal options of demolition waste is deserved. Also, the method could logically connect complicated information and their relationships by adopting unified terminologies and graphics to depict the processes visually through an intuitive manner. A similar method called the PMA has been employed by Shen et al. [13] to investigate the on-site construction waste management process in Hong Kong [13]. The PMA consists of five elements, including waste source, waste management process, waste disposal options, waste management measures, and major stakeholders. The demolition waste flows and management activities could be tracked by connecting these elements according to certain logical relationship beset (see Table 102.1).

102.3 Results and Analyses

102.3.1 Processes of Demolition Waste Management

Although the detailed processes of demolition waste management are heterogeneous among different projects, there are some common and also critical procedures within the paths of waste flows, such as waste generation, on-site waste sorting and management, waste transportation, and final disposal. Particularly, waste generation refers to the activity of producing waste by demolition works. On-site sorting and management contain activities such as waste collection, sorting, classification and pre-process if necessary. Waste transportation is responsible for the generated demolition waste placed in appropriate plants or landfills. At present, there are mainly three ways for waste disposal in Shenzhen, namely recycling, landfilling and illegal dumping. Therefore, the entire process of demolition waste management ranges from waste generation to the final disposal.

102.3.2 Stakeholders Involved in Demolition Waste Management

The investigation shows that various stakeholders are involved in demolition waste management, including general contractors, professional demolition companies, transport companies, waste recycling companies, landfills, government departments and scavengers. Among them, general constructors and professional demolition companies are directly in charge of building demolition and site management. Transportation companies delivery part of the sorted waste to either recycling plants or landfills for further processing. Given that the volume of some waste flows such as metal, plastic, timber, glass and other non-inert waste are small, but the economic value of these waste materials are large, they are always collected by special collection companies and processed in recycling plants together with other municipal solid waste. For those inert demolition waste including concrete, mortar, block and other inert waste, they can be recycled by proper recycling companies. Going through some pre-processing procedures, those inert wastes could be recycled to concrete aggregate, recycled blocks and other recycled products. Landfills refer to the landfill sites for waste disposal. In addition, there are several government departments regarding C&D waste management, such as environmental protection department, construction management department, urban planning department, traffic management department. In Shenzhen, there is a special group of people who collect metal and plastic waste in open construction sites and landfills.

The behavior and attitudes of these stakeholders will influence the efficiency and effectiveness of demolition waste management. For example, as the original stakeholders in waste generation, if the general contractors and professional demolition companies hold negative attitudes toward on-site waste sorting and management, the flowing process of demolition waste would be more difficult. In a sense, the attitude is critical to improving their incentives in managing the waste. It is also noted that in waste transportation, some transport companies may dumped the waste illegally due to economical consideration. As a consequence, relevant regulations that forbid the illegal dumping behavior should be enhanced, as well as the awareness of social responsibility of these companies.

At present, the market acceptance to the recycled materials is relatively low in Shenzhen. As a result, most of the constructors are not willing to apply the recycled materials in their projects and only few recycled materials were used in government-appointed projects. Hence, for the waste disposal, the recycling companies need to be guided and encouraged to use advanced technologies, facilities and management methods to improve the performance of recycled materials and products. It will improve public's acceptance for recycled products and consequently promote the rate of waste recycling. In managing demolition waste, relevant government departments should make more reasonable policies and enhance communications among all participants, in order to motivate their incentives for waste management as a whole.

102.3.3 Measures for Demolition Waste Management

Based on an analysis of the measures adopted for demolition waste management in Shenzhen, detailed demolition methods, site management measures, transportation modes and disposal manners are obtained (see Table 102.2). For example, demolition methods mainly include blasting demolition, manual demolition, mechanical demolition and mixed demolition.

Table 102.2 Measures for demolition waste management in Shenzhen

Code	Measures and strategies	Number	Proportion (%)
M1	Demolition methods		
M1.1	Blasting demolition	0	0
M1.2	Manual demolition	0	0
M1.3	Mechanical demolition	12	80
M1.4	Mixed demolition	3	20
M2	On-site management		
M2.1	No sorting	0	0
M2.2	Sorted as metal, plastic, timber, and others	10	67
M2.3	Sorted according to each materials	5	33
M2.4	Classify timber by size	4	27
M2.5	Crush the large volume concrete blocks	6	40
M2.6	Watering	6	40
M2.7	Cover with linoleum	4	27
M2.8	Packaging	2	13
M2.9	Dispose all the waste in field	2	13
M2.10	Special manager arrangement	3	20
M3	Transportation		
M3.1	Transport by the contractors themselves	8	53
M3.2	Entrust professional clearing companies	7	47
M3.3	Watering	8	53
M3.4	Cover with linoleum	7	47
M4	Treatment or disposal		
M4.1	Sell the metal and plastic to collecting companies	15	100
M4.2	Sell timber to timber makers	3	20
M4.3	Backfill inert waste to foundation bitch in field	4	27
M4.4	Sell inert waste to other construction sites to make roadbed	4	27
M4.5	Transport inert waste to landfills	12	80
M4.7	Transport non-inert waste to sanitary landfills	3	20
M.8	Illegal dumping waste except metal	2	13
M4.9	Transport inert waste to recycling plants	4	27

According to the results, 80 % of the projects investigated adopt the mechanical demolition manner. Other 20 % of the projects employ the mixed demolition method combining mechanical and manual methods. Generally, the demolition process is that firstly demolishes the minor parts of buildings such as doors, windows, wires manually, and then demolish the structure parts of buildings by mechanical tools.

In line with the investigation, all the projects adopt the management measures of on-site sorting. 67 % of the respondents sort out the materials having economic value, such as metal, plastic, timber, glass, while dump the inert waste on the sites. The rest of respondents claimed that they would make further efforts to sort out bricks, blocks, concrete in the premise. What's more, 27 % of the respondents mentioned that they would classify timber by size to make it easier to sell. 40 % of the respondents would choose to crush the large volume of concrete blocks for convenience of subsequent transportation.

To the waste piled in the field, 40 % of the respondents adopt watering on the surface to prevent dust. On this basis, 27 % of the respondents would recover linoleum on the surface of waste to further decrease the impacts of dust on the construction site and surrounding environment. In addition, 13 % of the respondents claimed that they would pack the generated waste in bags, where these projects are always in small scale and the amount of the generated waste is not large. For the waste packed in bags, despite some of them would be illegal dumped on the road site, the packages could decrease its environmental impacts.

The investigation shows that special transportation companies are hired to transport waste in half of the projects. In the other projects, general contractors or professional demolition companies transport waste by themselves. In order to minimize environment impacts, they usually make efforts to reduce dust during the transportation through watering or recovering. Since the transportation distance is long and the amount of waste transported is huge, this process contributes a lot to the negative environmental impacts.

There are usual methods like recycling, landfill or illegal dumping for waste treatment and final disposal. The results reveal that all of the respondents would sort out the wasted metal and sell them to collecting companies and 20 % of them would pick up and sell the timber to timber makers. However, 80 % of the respondents claimed that the majority of inert waste like concrete, mortar and brick was disposed of to landfills. Besides, backfill inert waste to foundation bitch in field and sell inert waste to other construction sites to make roadbed are popular ways to dispose of the waste because it can reduce environmental impacts and simultaneously obtain some economic benefits. However, only 27 % of the respondents claimed the waste generated in their projects would be transport to recycling facilities.

102.3.4 Demolition Waste Management Process Model (DWMPM)

Based on the analysis above, a demolition waste management process model (DWMPM) was developed (see Fig. 102.2) by employing the PMA and the paths of waste flows and management processes of demolition waste in Shenzhen. The entire process of demolition waste management is divided into four phases in this model, encompassing waste generation, on-site management, transportation and disposal. The model involves several parts such as building demolition, collection, sorting, pre-treatment, on-site treatment loading and transport, and disposal methods such as recycling, landfill, excavation backfill, road base, and illegally dumping. Meanwhile, the model also includes stakeholders and management measures in each of the activities. In Fig. 102.2, S presents stakeholders and M presents management measures.

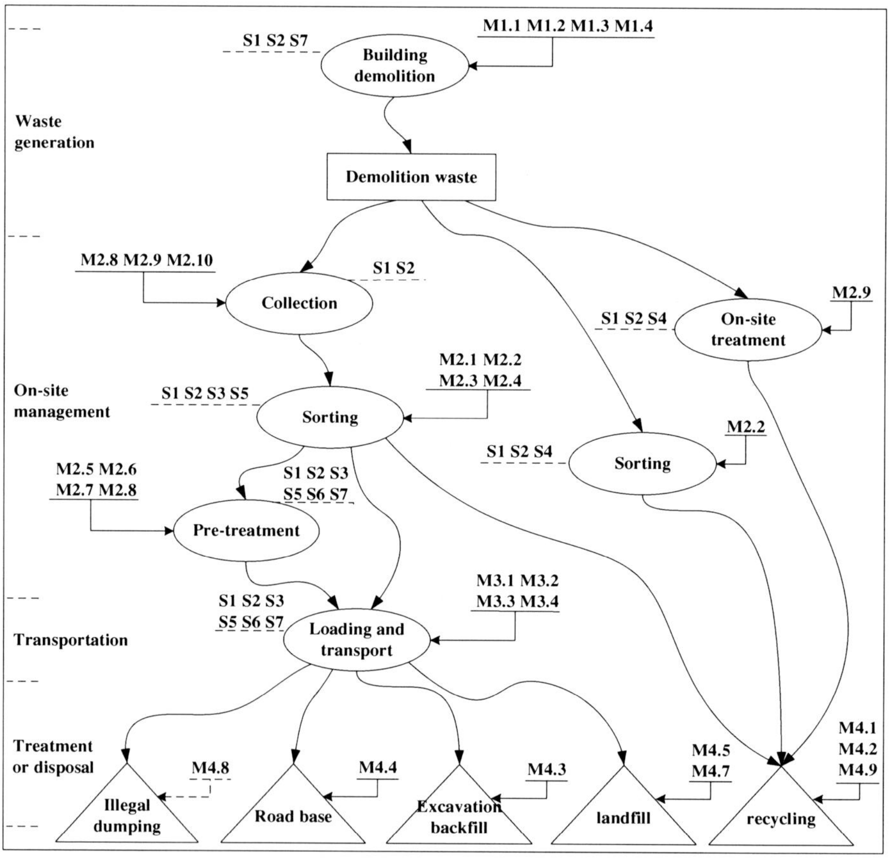

Fig. 102.2 The demolition waste management process model in Shenzhen

As mentioned before, although the paths of waste flows and management processes might different among different projects, there are four common stages namely waste generation, on-site management, transportation, and treatment or disposal. In the beginning of managing demolition waste, the general constructors or professional demolition companies firstly compose a demolition proposal for demolition waste, and then request the government department's approval. For some special projects, the government may take part in the development of demolition proposal.

After the generation of demolition waste, proper on-site management should be conducted. In this stage, particular measures would be proposed according to the scale and demolition methods of the project. For instance, in some small projects, the majority of waste would be dumped as a whole only after the metal was sorted out. But in some large projects, the waste would be sorted according to the particular compositions. If the waste could not be treatment timely, covering measures would be conducted to forbid the dust generated from the waste.

Currently, it is interested to see that the on-site treatment measure is quite popular in Shenzhen. In the on-site treatment cases, the inert demolition waste is crushed by waste mobile treatment equipment after simply sorting. The crushed demolition waste would be classified as large size aggregate (size: 30–50 mm), middle size aggregate (size: 10–30 mm) and small size aggregate (size: 5–10 mm). These aggregates could be used as the materials of recycled concrete, mortal and brick. This model of treatment could decrease the transportation amount of waste, and thus reduce the risk of illegal dumping and increase the recycling rate.

For most cases, however, the generated waste was disposed of to landfill. Although the average fee of processing the waste from the construction site to landfills is about 10 yuan per ton (including collection, loading, transport and landfill), which is much lower than the cost in Hong Kong (about 150 yuan/t), the total cost of disposal waste is quite high due to the huge volume. As a consequence, a large part of waste was unloaded and dumped over the road side illegally since it could reduce the transport cost and hardly be hold accountability.

Although a noticeable part of demolition waste was recycled on site or in recycling facilities, low market acceptation of recycled products is still a major barrier to promoting recycling rate. Since there is a wide concern about the performance of recycled products, developers and constructors are more willing to purchase origin materials rather than recycled materials, even though the recycled products are cheaper. Eventually, the limited profit of producing and selling recycled products reduced the enthusiasm of recycling companies to propagate their recycling products. It is thus suggested that government regulations and economic stimulation policies should be put forward to stimulate the prosperity of C&D waste recycling market in Shenzhen.

102.4 Conclusions

The paths of waste flows and management measures of demolition waste form a complex system involving several processes and stakeholders. In this study, major processes of demolition waste management in Shenzhen were identified, and the stakeholders involved in demolition waste management were analyzed. The stakeholders mainly include general contractors, professional demolition companies, transport companies, recycling plants, landfills, government departments and scavengers. Besides, with the help of the PMA, a demolition waste management process model is developed to display the detailed paths of waste flows and management measures of demolition waste. The study reveals that illegal dumping and low market acceptation of recycled products are two major barriers to demolition waste management in Shenzhen. Therefore, more reasonable government regulations and economic stimulation policies should be put forward to stimulate the prosperity of the waste recycling market. The findings can be useful for further studies on demolition waste management, attempting to quantitatively optimize the path of waste flows and identify the most critical measures for reducing demolition waste.

Acknowledgements This study was supported by the NSFC (71203184), the Australian Research Council Project "Re-considering Sustainable Building and Design: A Cultural Change Approach" (LP110100156) the Sichuan Science and Technology Program (2014ZR0082) and the Fundamental Research Funds for the Central Universities (SWJTU12CX115).

References

1. Li J, Ding Z, Mi X, Wang J (2013) A model for estimating construction waste generation index for building project in China. Resour Conserv Recycl 74:20–26
2. Wang J, Li Z, Tam VW (2015) Identifying best design strategies for construction waste minimization. J Clean Prod 92:237–247
3. Yuan H, Shen L (2011) Trend of the research on construction and demolition waste management. Waste Manag 31(4):670–679
4. NDRC, (2014).Annual report of the comprehensive utilization of resources in China (2014). Available at: http://www.sdpc.gov.cn/xwzx/xwfb/201410/W020141009609573303019.pdf. Ndrc
5. Lu W, Yuan H, Li J, Hao JJ, Mi X, Ding Z (2011) An empirical investigation of construction and demolition waste generation rates in Shenzhen city, South China. Waste Manag 31 (4):680–687
6. Llatas C (2011) A model for quantifying construction waste in projects according to the European waste list. Waste Manag 31(6):1261–1276
7. Butera S, Christensen TH, Astrup TF (2014) Composition and leaching of construction and demolition waste: inorganic elements and organic compounds. J Hazard Mater 276:302–311
8. Poon CS, Yu AT, Jaillon L (2004) Reducing building waste at construction sites in Hong Kong. Constr Manag Econ 22(5):461–470
9. Yuan H (2013) A SWOT analysis of successful construction waste management. J Clean Prod 39:1–8

10. Coronado M, Dosal E, Coz A, Viguri JR, Andrés A (2011) Estimation of construction and demolition waste (C&DW) generation and multicriteria analysis of C&DW management alternatives: a case study in Spain. Waste Biomass Valorization 2(2):209–225
11. Shenzhen Economic Daily (2014) The first regulations of C&D waste recycling in Shenzhen, the recyclingrate up to 40 %. Available at: http://www.concrete365.com/news/content/4503582637295.html. (In Chinese)
12. Dudu (2014) Recycling rate of C&D waste: the awkward number 5 %. J Chin Strat Emer Indus. Available at: http://www.360doc.com/content/14/1128/15/20625606_428757950.shtml. (In Chinese)
13. Shen LY, Tam VW, Tam CM, Drew D (2004) Mapping approach for examining waste management on construction sites. J Constr Eng Manag 130(4):472–481

Chapter 103
A Comparison Study on the Green Building Performance Assessment Tools for Promoting Sustainable Construction

Bei He, Yingying Han, Luyao Zhao and Liyin Shen

Abstract Assessment tools for evaluating green buildings performance have been developed mainly from the perspective of their environmental impacts and sustainable performance since 1990s. These assessment tools have been developed under different aspects for evaluation. The increasing number of application practices of assessment tools presents the growing public interest on this kind of voluntary scheme to promote sustainable construction. This paper presents a comparison study between six different assessment tools for green building performance. These assessment tools are selected with considering their importance, application and influence in the practice. Comprehensive comparisons are analyzed based on the development and application of these assessment tools in promoting sustainable construction practice. The criterions for the comparative analysis include stakeholder, application scope, assessment aspects, performance measurement, assessment process, assessment result, and maturity. Discussion are conducted with regards to identifying potential areas for improvement for the assessment tools.

Keywords Sustainable construction · Green building performance · Assessment tool · Environmental assessment

B. He (✉) · Y. Han · L. Shen
School of Construction Management and Real Estate,
International Research Centre for Sustainable Built Environment,
Chongqing University, Chongqing, China
e-mail: hebei091@163.com

Y. Han
e-mail: hanyy1024@163.com

L. Shen
e-mail: shenliyin@cqu.edu.cn

L. Zhao
School of Construction Management and Real Estate,
Chongqing University, Chongqing, China
e-mail: 576986288@qq.com

Y. Wu et al. (eds.), *Proceedings of the 20th International Symposium on Advancement of Construction Management and Real Estate*,
DOI 10.1007/978-981-10-0855-9_103

103.1 Introduction

The principles of sustainable development define the environmental, economic, and social framework for the activities of communities, enterprises and individual citizens [19]. The construction industry and the built environment are two important areas in pursuing sustainable development of our society as stated by the International Council for Research and Innovation in Building and Construction (CIB) in its CIB Agenda 21 on sustainable construction [5]. Sustainable construction aims to improve the overall building performance and minimize the environmental impacts and costs [29]. Therefore, building performance has been increasingly given consideration in line with the huge amount of ongoing activities in the building sector [19].

Assessment tools for green building performance have been placed as the core instruments for evaluating building performance and achieving sustainability [33]. More attention has been paid to the significance of the development and application of the assessment tools for green building performance. Different assessment tools for evaluating building performance in terms of environmental impacts and sustainability have been developed and already used in the practice. They provide guidelines and checklists which include various environmental-friendly aspects and specific measures for users. These assessment tools are widely used by construction professionals for making design decisions, selecting material and equipment selections, and determining the performance of particular aspects of a building [24]. They have been developed to define comprehensive objectives and provide specific checklists to assess the building design and construction. However, most of them are created by modifying a particular assessment tool developed previously, or integrating several existing tools. In addition, each assessment tool consists of different indicators depending on where they are used and for what purpose. Having recognized the differences between various assessment tools, it is necessary to introduce and develop suitable ones that are suitable to a specific country or region.

Diverse studies relating to assessment tools for green building performance have been carried out previously for various purposes [6, 10, 16, 25, 36]. The categorizations and the relationships between assessment tools for green building performance have been investigated [7, 14]. More specific analysis has been made about the role of the assessment tools on promoting building energy efficiency, the benchmark of their assessment indicators or items, and the discussion on the comprehensiveness of specific assessment tool [17, 21, 22]. The indicators and benchmarks employed in the assessment tools have been also employed for the development of the methods for assessing the environmental performance during the construction stage [26].

Whilst the existing assessment tools for green building performance are widely introduced and effectively developed globally, differences among the typical assessment tools have not been examined. This paper offers a comprehensive analysis on the development and application of the typical assessment tools for green building performance. Parameters for comparison were selected for analyzing their current status of development, potential use for promoting sustainable construction, and contribution for further research and development.

103.2 Research Methods

In the view of the designated research objectives, this paper adopts a comparative approach. Comparative approach is a research methodology that can facilitate knowledge sharing and give insights into future practices [28]. Recently, a number of comparative studies have been conducted in the discipline of environmental researches. For example, Friedrich and Trois [11] compared the existing quantification of greenhouse gas emission from waste in order to identify problematic gaps in Africa. Another research conducted by Geng et al. [12] compared China's green building standards with other countries' with the purpose of identifying benefits and challenges. This approach can be used to derive useful insights.

This research starts with understanding typical assessment tools for green building performance across the world. Six assessment tools for green building performance are selected in this paper by considering their importance, acceptance, and influence in their development and practice. These typical assessment tools include Building Research Establishment Environmental Assessment Method (BREEAM) in the United Kingdom, Comprehensive Assessment System for Building Environmental Efficiency (CASBEE) in Japan, the Green Building Tool (GBTool) for an international project that has involved more than 25 countries, the Leadership in Energy and Environmental Design (LEED) in the United States, Hong Kong Building Environmental Assessment Method (HK-BEAM), and Green Mark in Singapore.

The second stage of the study is to identify the differences and characteristics of these assessment tools for green building performance. For conducting this identification, a list of seven parameters is formulated to guide the comparison on different assessment tools that have been introduced and developed in different countries, as shown in Table 103.1.

Table 103.1 contains seven parameters for comparing assessment tools, including stakeholder (P1), application scope (P2), assessment aspect (P3), performance measurement (P4), assessment process (P5), assessment result (P6), and maturity (P7). The parameter "Stakeholder (P1)" includes specific items of the parties involved and the method for decision. Parameter "Application scope (P2)" includes the type of projects and the type of buildings. Parameter "Assessment Process (P5)" includes the documentation and the verification.

In the further stage of this study, the development and application in the selected six assessment tools are discussed and compared in terms of the seven comparison parameters. For investigating the compliances of these assessment tools with considering different parameters, comparison symbols are used to facilitate the analysis, as shown in Table 103.2.

Table 103.1 Parameters for comparison between assessment tools

Parameter		Specific item		
P1	Stakeholder	P_{1-1}	Parties involved	• Government • Industry • NGO
		P_{1-2}	Method for decision	• Consensus based • Expert opinion
P2	Application scope	P_{2-1}	Type of projects	• Newly or refurbished buildings • Existing buildings • Operation and maintenance
		P_{2-2}	Type of buildings	• Office buildings • Commercial buildings • Residential buildings
P3	Assessment aspect			• Site aspects • Energy use • Water use • Material aspects • Indoor environmental quality • Operational aspects • Others
P4	Performance measurement			• Benchmark • Checklist • Traceability • Quantitative measurements
P5	Assessment process	P_{5-1}	Documentation	• Type of documentation • Documentation scope • Level of detail
		P_{5-2}	Verification	• Verification authority • Third party monitoring • Assessor qualification
P6	Assessment result			• Rates • Product
P7	Maturity	P_{7-1}	Tool development	• Development date • Effective date • Revision date
		P_{7-2}	Applications	• Number of assessments • Number of certified green building

Table 103.2 Comparison symbols

Symbol	Meaning
●	Does meet criterion
○	Does not meet criterion
◊	Meets criterion with exceptions
◐	Under development
n/a	Not applicable
(blank)	Unknown information

The data needed for this study are mainly the contents of these six selected assessment tools for green building performance. They are retrieved from the several resources, including official documentation of assessment tools, application software of assessment tools, and relevant web sites about these tools. The assessment tools include different versions for different type of projects, but for the purpose of this comparative analysis only the versions for the building sector are adopted.

103.3 Principles and Backgrounds of the Selected Assessment Tools for Green Building Performance

The principles and backgrounds of each selected tools will be discussed in this section. This provides the basis for conducting comparisons between these tools.

BREEAM
BREEAM was developed in the United Kingdom in 1990 by the Building Research Establishment Limited (BRE) [3]. It is a reference guide, including a design and procurement checklist for new buildings, and a management and operation checklist for existing buildings. Credits are awarded to each item included in the checklists according to the performance of each item. These credits are added to produce a single overall score. The building under assessment is rated by referring to this overall score on a scale of Pass, Good, Very Good or Excellent.

CASBEE
CASBEE was developed in Japan in 2001 by the Japan Sustainable Building Consortium [20]. It is a weighting and rating tool using Excel 2000. The tool includes a reference manual for self-assessment. The tool includes a checklist, and each category in the checklist is scored from level 1 to level 5, where level 1 represents that the minimum requirements on building sustainability have been met, and level 5 represents the high level of achievement. Results are presented in five assessment levels: Class C, B−, B+, A and S, where C indicates poor, and S for excellent.

Green Building Tool
The Green Building Tool was developed in 1998 by the International Framework Committee for the Green Building Challenge, an international project that has involved more than 25 countries, such as China and Australia who have been applying this assessment tool [18]. It is a weighting and rating tool using Excel 2000. The assessment tool includes four phases in one project: pre-design, design, construction, and operation stages. And it consists of two parts: Module A consisting of the assessment aspects, and their benchmarks and weights; Module B including specific data entry operated by the project team. Qualitative and quantitative assessment benchmarks are applied. The buildings are rated on a scale from

−1 to 5, where −1 represents below typical practice, 0 acceptable performance, 3 good practice, and 5 best practice.

LEED
LEED was developed in the United States in 1998 by the U.S. Green Building Council [32]. It is a reference guide including checklists for design and construction phase. Credits are awarded to each category according with their performance and added to produce a single overall score. The building under assessment is rated according to the overall score on a scale of Certified, Silver, Gold, and Platinum.

HK-BEAM
HK-BEAM was developed in Hong Kong in 1996 by the HK-BEAM Society [15]. It is a reference guide including checklists for planning, design, construction and management, operation, and maintenance phases. The overall assessment grade is based on the credits gained. Awarded classifications are Bronze (above average), Silver (good), Gold (very good), and Platinum (excellent).

Green Mark
Green Mark was developed in Singapore in 2005 by the Building and Construction Authority (BCA) [1]. It is a reference guide including a number of checklists for new buildings, existing buildings, and end verification. The assessment includes documentary reviews and site verifications. The overall assessment grade is awarded in Certified, Gold, Gold Plus and Platinum depending on the points scored.

103.4 Comparative Analysis

By adopting the research methods defined previously, the comparative analysis will be conducted by referring to the seven comparison parameters in this section.

103.4.1 Comparison on Stakeholder (P1)

The development of assessment tools for green building performance enables building developers, operators, and users to accomplish construction performance to their needs and promotes less impact of construction on the environment. The comparison on stakeholder aims to identify the parties involved in the assessment tools and the method for decision in using assessment tools. Table 103.3 provides the comparison on stakeholder between different tools.

It is of great interest to note that non-governmental organizations and industry are the parties with more involvement in the application of these assessment tools.

Table 103.3 P1: stakeholder

	Parties involved			Method for decision	
	Government	Industry	NGO	Consensus based	Expert opinion
BREEAM		●	●		●
CASBEE	◊	●	●	●	●
GBTool	●	○	●	●	●
LEED	●	●	●	●	●
HK-BEAM	○	●	●	●	●
Green mark	●	○	●	○	●

It can be considered that government, public organizations, and construction participants are taking the leading role in pursuing sustainable construction [23]. The designers play an important role in the process of sustainable construction since they produced initial ideas for the built environment. However, as both the source (designer) and the end user (owner) are readily adopting sustainable construction projects, it becomes imperative for the process implementer (contractor) to be more active [30]. Concerning about this fact, Shen et al. [27] proposed a framework that can help contractors to cultivate environmentally friendly culture within organizations.

The method for decision about evaluation result of the assessment tools are of great importance since they define the selection and weighting of the aspects for assessment and the procedures to calculate the overall rates. In general, when there is an absence of scientifically based weights, the weights between different items will be based on consensus between stakeholders. In this approach, groups of experts rank various elements, such as environmental issues in terms of their relative importance or assigning credits to these elements. This ranking or scoring will be used to establish weights [9].

103.4.2 Comparison on Application Scope (P2)

The application scope covered by the assessment tools includes the main type of projects and type of buildings. Table 103.4 shows the application scope of each assessment tool.

All the selected assessment tools clearly define their application scope. The application scope of these assessment tools contains three main types of buildings including office building, commercial and residential building. It was found that Green Mark does not include the section of operation and maintenance, while the section of operation and maintenance is under development.

Table 103.4 P2: application scope

	Type of projects			Type of buildings		
	Newly or refurbished buildings	Existing buildings	Operation and maintenance	Office buildings	Commercial buildings	Residential buildings
BREEAM	●	●	●	●	●	●
CASBEE	●	●	●	●	●	●
GBTool	●	●	○	●	●	●
LEED	●	●	●	●	●	●
HK-BEAM	●	●	●	●	●	●
Green mark	●	●	○	●	●	●

103.4.3 Comparison on Assessment Aspect (P3)

The assessment aspects and the allocation of credits in each assessment tool are compared as shown in Table 103.5.

According to information contained in Table 103.5, the following analysis can be made:

HK-BEAM and GREEN MARK only allocate 13 and 15 % of credits respectively on site aspects. Considering that Hong Kong and Singapore face extreme shortage of land, it would be expected a higher allocation of credits in these aspects for encouraging developers to consider land saving.

It is important to note that the item "energy use" is allocated with the majority of credits in all the assessment tools. This is attributed to the fact that great negative environmental impacts will be encountered when energy resources have been consumed. The effectiveness of energy supply and use is widely acknowledged as the important, and even indispensable components of sustainable development [13].

GREEN MARK is the one that allocates more credits on water use aspects. This is in response to the shortage of water resources that Singapore has been facing. The Singapore government has introduced various policy measures for promoting its sustainable water management in the building sector [34].

HK-BEAM considers greater importance on indoor environmental quality. According to Chau et al. [4], the average person in Hong Kong spends more than 85 % of their time indoors. Besides, Hong Kong is also one of the densely heavy populated cities in the world [31].

LEED and HK-BEAM allocate few credits on operational aspects in the general version. Both of them distribute the operational aspects to the other versions of the tools, for example, LEED provides a specific version for operation and maintenance.

Table 103.5 P3: assessment aspects

	Site aspects (%)	Energy use (%)	Water use	Material aspects	Indoor environmental quality (%)	Operational aspects	Others (%)	Total (%)
BREEAM	15	25	5 %	10 %	15	15 %	15	100
CASBEE	15	20	2 %	13 %	20	15 %	15	100
GBTool	8	22.5	n/a	n/a	18	16 %	35.5	100
LEED	20	25	7 %	19 %	22	n/a	7	100
HK-BEAM	13	39	7 %	12 %	24	n/a	5	100
Green mark	15	35	15 %	7 %	15	3 %	10	100

Other aspects include credits for innovation, social and economic aspects. GBTool is the only tool that includes separate categories for these aspects and allocates more credits on these aspects.

Resource aspect is referred wide range in several assessment tools (e.g., water resources are extended to collection and storage of rain water). This is very important for Hong Kong, which experiences extremely heavy rainfall [8]. Social and economic aspects have yet to be included in most of these selected assessment tools mainly because of the absence for a defined consensus that establishes relevant indicators and weighting. However, all the assessment tools encourage users to attain higher standards of performance and innovation by adding additional credits in the overall results.

103.4.4 Comparison on Performance Measurement (P4)

The specific performance measurement and assessment procedures for the selected assessment tools are compared, as shown in Table 103.6.

It can be seen that most of the selected assessment tools have established their own checklists for performance evaluation according to national and local construction standards and codes. And most of them assess the green building performance in a quantitative term based on the different checklists. When some specific items of performance assessment cannot be quantified, assessment tools usually adopt experts' opinions and grades as alternative.

103.4.5 Comparison on Assessment Process (P5)

Assessment process refers to the type and scope of documentation and the level of detail during the verification process. In addition, it includes the verification authority, the assessor, and the third party which will monitor the assessment process. Table 103.7 provides comparative information for the assessment process.

Table 103.6 P4: performance measurement

	Benchmark	Checklist	Traceability	Quantitative measurements
BREEAM		●		●
CASBEE	●	●	●	●
GBTool	●	○	●	◇
LEED	●	●	●	●
HK-BEAM	●	●	●	●
Green mark	●	●		●

Table 103.7 P5: assessment process

	Documentation			Verification		
	Type of documentation	Documentation scope	Level of detail	Verification authority	Third party monitoring	Assess or qualification
BREEAM	Hardcopy	Design and completion	Detailed assessment of documentary evidence	Trained and licensed by BRE	●	Expert Officer
CASBEE	Excel spreadsheet	Preliminary design, execution design and completion	Review of documentation	Trained and licensed by CASBEE	●	Expert Officer
GBTool	Excel spreadsheet	After design is completed	Review of documentation and site inspection	n/a	n/a	n/a
LEED	Online and/or hardcopy	Design and construction review	Administrative and credit audit	Trained and licensed by USGBC	●	Expert Officer Third party
HK-BEAM	Hardcopy	Initiation at any stage, and at completion	Review of documentation and site inspection	HK-BEAM	◐	Expert Officer Third party
Green mark	Hardcopy	Design, during construction, and completion	Review of documentation and site inspection	BCA	○	Expert Officers

The type of documentation used mostly in the assessment process is hardcopy of evidence documents except for the cases of CASBEE and GBTool that adopt all the information presented on excel spreadsheets. The documentation scope of these assessment tools has contained the whole life cycle of construction except for CASBEE which covers the period after the design. Most of the assessment documentations is in the form of a review of documentation and site inspection, such as GBTool, HK-BEAM, and Green Mark, whilst only review of documentation is required to provide for assessment in CASBEE.

Verification authorities for most of the selected assessment tools are the associations relevant to green building and sustainable construction, such as BRE in the United Kingdom, CASBEE in Japan, USGBC in the United States, and HK-BEAM in Hong Kong. Verification assessors include relevant officers, experts, and third parties. For example, in practicing HK-BEAM, more assessors are recruited to satisfy the demand for assessment of green building. In fact, the third party assessors have been under development in using LEED, which has generated successful experience. It is of vital importance to introduce third-party supervision and monitoring during the application of the assessment tools to ensure the transparency and fairness of assessment, as practiced in BREEAM, CASBEE, and LEED.

103.4.6 Comparison on Assessment Result (P6)

The rates of the overall result from assessment process as well as the assessment product are presented differently in different tools, as shown in Table 103.8.

The results of assessment and verification allow the assessed and certified buildings differentiated in the market thus promote the application of sustainable buildings. Companies are increasingly recognizing that good environmental sense is also good business sense [8]. According to Howard [16] people prefer to work for organizations that are perceived to be ethical and environmentally responsible. A certification plaque on a building is a potent symbol of the organization's environmental ethic for staff, customers and clients.

Table 103.8 P6: assessment result

	Rates	Product
BREEAM	Pass, good, very good, excellent	Certificate
CASBEE	Class C, B, B−, B+, A, S	Certificate
GBTool	−1, 0, 3, 5	Certificate
LEED	Certified, silver, gold, platinum	Award letter, certificate and plaque
HK-BEAM	Bronze, silver, gold, platinum	Certificate
Green mark	Certified, gold, gold plus, platinum	Certificate

Table 103.9 P7: maturity

	Development date	Effective date	Revision date
BREEAM	1990	1990	2007
CASBEE	2001	2002	2006
GBTool	1998	1998	2007
LEED	1998	1998	2005
HK-BEAM	1995	1996	2004
Green mark	2004	2005	2007

Governments will provide financial subsidies for the buildings and projects which get the certifications of assessment tools for green building performance. For example, in the application of "Green Mark Incentive Scheme" in Singapore and the "Special Fund on CASBEE" in Japan, the governments will offer incentives to those who could achieve green building certifications and labels. These assessment tools for green building performance have specified clear subsidization criteria, and this way helps improve the effectiveness in using financial incentives while promoting the development of green buildings.

103.4.7 Comparison on Maturity (P7)

The maturity of a specific assessment tool can examined by the development date, the effective date, and the revision date. These information are presented in Table 103.9.

The concerned governments relating to the selected assessment tools have been revising the tools regularly in order to meet the changes of practices in implementing the assessment and verification for green building. The case number of green building assessments has been increasing in recent years. Assessment tools are experiencing improvements and pursuing opportunities to increase their applicability. For example, the HK-BEAM Society recently changed its name to BEAM to open the opportunity for applications in Mainland China, where they have already installed offices in Shenzhen and Beijing [35]. It is noted that there is no statistics for the case of GBTool. This is because that the tool is still a self-assessment tool.

103.5 Discussion

Assessment tools for green building performance, in comparing to other policy measures for promoting green building, have advantages of flexibility. They can receive quick responses from the practice, thus proper adjustments on these assessment tools can be made in time if necessary. However, the application of assessment tools relies largely on the recognition and willingness between various stakeholders, particularly including building developers and building owners.

Assessment tools for green building performance have become popular internationally for promoting sustainable construction and energy efficiency in the building sector. In line with this development, a number of assessment tools have been adopted globally, such as LEED label program which was established in US and adopted already in many other countries as well. All the governments relating to selected tools have taken efforts in promoting the application of these tools. As various countries and regions have been developing their assessment tools for green building performance, it is necessary to appreciate the differences between these assessment tools. Bemelmans-Videc et al. [2] pointed out that if there are too many similar types of assessment tools, a barrier will present to their application as users need to understand and choose between these alternatives. In other words, understanding on the differentiation between various assessment tools is important to prevent the confusion in application. It is also necessary that assessment tools should provide more holistic approaches and specific checklists for supporting practitioners in use during life cycle phase of construction. It has been demonstrated that the application of the assessment tools for green building performance are essential to the promotion of sustainable construction. Meanwhile, assessment tools should facilitate the dynamic changes which may include more rigorous criterions and indicators in order to meet demand of development.

103.6 Conclusion

The findings from this study suggest that currently the assessment tools for green building performance have represented an important voluntary policy instrument for promoting sustainable construction. Governments throughout the world have realized the importance of applying assessment tools for green building performance in promoting sustainable construction and energy efficiency. In line with this development, various assessment tools have been introduced by governments. Governments have been devoting good efforts in developing and promoting assessment tools, which are evidenced by their wide application scope and regular updating cycle. Specific indicators and criterions have been included in each checklist for these selected assessment tools, while the weighting of them is modified based on the different circumstances. Thus these assessment tools cover comprehensive assessment aspects, and these aspects for assessment. Seven parameters are formulated to comparing the selected tools, including stakeholder, application scope, assessment aspects, performance measurement, assessment process, assessment result, and maturity. The selected tools are different across all seven parameters to certain extent. However, it is considered that assessment tools for green building performance can find more effective application by engaging in effective communication and cooperation between government departments and various building stakeholders, particularly, the public, building developers, and building users.

The findings can help share experiences which are gained in the application of various typical assessment tools for green building performance. And these findings provide references for users of assessment tools to select applicable and effective tools in different situations. Meanwhile, they also provide references for policy makers to develop and introduce more suitable assessment tools.

The research team appreciates, nevertheless, the limitations in this study that the data used are collected only for six typical assessment tools. It is the aim of the further research in this research team to examine the assessment tools with involving much larger sample of countries and regions.

Acknowledgements This research is funded by the Key Project of the National Natural Science Fund of China (Project No: 15AZD025).

References

1. BCA (2007) BCA green mark. Homepage of Building & Construction Authority. Available: http://www.bca.gov.sg/GreenMark/green_mark_buildings.html (Dec 2007)
2. Bemelmans-Videc ML, Rist RC, Vedung EO (eds) (2011) Carrots, sticks, and sermons: policy instruments and their evaluation (vol 1). Transaction Publishers, Piscataway
3. BRE (2007) BREEAM: building and research establishment environmental assessment method. New Offices 2006. Homepage of Building Research Establishment Limited. Available: http://www.breeam.org/ (Nov 2007)
4. Chau CK et al (2002) Estimating the total exposure to air pollutants for different population age groups in Hong Kong. Environ Int 27(8):617–630
5. CIB (1999) CIB Agenda 21 for sustainable construction in developing countries. The International Council for Research and Innovation in Building and Construction, The Netherlands
6. Cole RJ (2005) Building environmental assessment methods: redefining intentions and roles. Build Res Inf 33(5):455–467
7. Cole RJ (2006) Shared markets: coexisting building environmental assessment methods. Build Res Inf 34(4):357–371
8. Davies H (2001) Environmental benchmarking of Hong Kong buildings. Struct Surv 19(1):38–46
9. Dickie I, Howard N (2000) Assessing environmental impacts of construction—industry consensus. BREEAM and UK Ecopoints, BRE, UK
10. Forsberg A, Von Malmborg F (2004) Tools for environmental assessment of the built environment. Build Environ 39(2):223–228
11. Friedrich E, Trois C (2011) Quantification of greenhouse gas emissions from waste management processes for municipalities–a comparative review focusing on Africa. Waste Manag 31(7):1585–1596
12. Geng Y, Dong H, Xue B, Fu J (2012) An overview of Chinese green building standards. Sustain Dev 20(3):211–221
13 Goldemberg J, Johansson TB (1995) Energy as an instrument for socio-economic development.
14. Haapio A, Viitaniemi P (2008) A critical review of building environmental assessment tools. Environ Impact Assess Rev 28(7):469–482

15. HK-Beam Society (2004) HK-BEAM 4/04 'new buildings'. Homepage of HK-BEAM society. Available: http://www.hk-beam.org.hk/general/home.php (Nov 2007)
16. Howard N (2005) Building environmental assessment methods: in practice. In: Proceedings 2005 world sustainable building conference, Tokyo, pp 27–29
17. Hui SC (2003) Energy efficiency and environmental assessment for buildings in Hong Kong. In: MECM LEO seminar–advances on energy efficiency and sustainability in buildings, palace of the Golden Horses, Kuala Lumpur, pp 21–22
18. IISBE (2007) GBTool (SBT07). Homepage of international initiative for a sustainable built environment. Available: http://www.iisbe.org (Nov 2007)
19. ISO (2007) ISO/TC 59/SC 17 N 236. Sustainability in building construction—sustainability indicators. Draft edn. ISO
20. JSBC (2004) CASBEE for new construction tool 1, 2004 ed. V. 1.02. Homepage of Japan sustainable building consortium. Available: http://www.ibec.or.jp/CASBEE/ (Nov 2007)
21. Lee WL, Burnett J (2008) Benchmarking energy use assessment of HK-BEAM, BREEAM and LEED. Build Environ 43(11):1882–1891
22. Mohammad F, Amato A (2006) Public housing and social sustainability indicators: HK-BEAM as a case study. In: Proceedings of the annual research conference of the Royal Institute of Chartered Surveyors, London
23. Riley D, Pexton K, Drilling J (2003) Procurement of sustainable construction services in the United States: the contractor's role in green buildings. Ind Environ 26(2):66–69
24. Seo S, Tucker S, Ambrose M, Mitchell P, Wang CH (2006) Technical evaluation of environmental assessment rating tools. Research and Development Corporation, Project No. PN05, 1019
25. Seo S (2002) International review of environmental assessment tools and databases. 2001-006-B-02
26. Shen LY, Lu WS, Yao H, Wu DH (2005) A computer-based scoring method for measuring the environmental performance of construction activities. Autom Constr 14(3):297–309
27 Shen L, Wu Y, Zhang X (2010) Key assessment indicators for the sustainability of infrastructure projects. J Constr Eng M 137(6):441–451
28. Shen LY, Ochoa JJ, Shah MN, Zhang X (2011) The application of urban sustainability indicators–a comparison between various practices. Habitat Int 35(1):17–29
29. Sjöström C, Holmgren J (2005) From sustainable construction requirements to codes and standards. In: 6th international congress on global construction: ultimate concrete opportunities, Dundee, Scotland pp 455–464, 5–7 July 2005
30. Syal M (2007) Impact of LEED-NC projects on constructors. Michigan State University, USA
31. UNPF (2007) State of world population. United Nations Population Fund, New York
32. US Green Building Council (2005) LEED-NC for new construction: reference guide, version 2.2. US Green Building Council
33. Warnock AC (2007) An overview of integrating instruments to achieve sustainable construction and buildings. Manag Environ Qual Int J 18(4):427–441
34. Wong M (2007) Singapore wins international award for water management. Channel NewsAsia, Singapore News
35. Worldwatch Institute (2007) Vision of sustainable world. Homepage of World Watch Institute. Available: http://www.worldwatch.org/taxonomy/term/53 (Dec 2007)
36. Zimmerman A, Kibert CJ (2007) Informing LEED's next generation with the natural step. Buil Res Inf 35(6):681–689

Chapter 104
A Life Cycle Analysis Approach for Embodied Carbon for a Residential Building

Shengping Li, Hang Yan, Jindao Chen and Liyin Shen

Abstract The rapid urbanization process has led to a dramatic increase in energy consumption and carbon emissions in the world, which has accelerated the trend of global warming significantly. Building sector is one of the greatest contributors to that increase. In this paper, a life cycle inventory analysis model has been formulated to calculate the embodied carbon during the life cycle for a residential building. The life cycle is divided into five stages, including materials production, transportation, construction, maintenance, and demolition & disposal. A case study selected in Chongqing is used to demonstrate the application of the method. The results show that materials production stage contributes the most of the embodied carbon (90.92 %) for a residential building, and those materials contribute mostly to the embodied carbon, including concrete (36.3 %), steel (24.1 %), brick (16.6 %), cement (13.4 %) and others (9.6 %). From the perspective of subsystem to a building, structure is the largest contributor to the total embodied carbon to the building, accounting for 78.4 %. The results reveal the quantity of embodied carbon in different stages, materials and subsystems of buildings, which can help identify the optimal solution to reduce the quantity of embodied carbon in buildings and improve the environment performance.

Keywords Life cycle assessment · Embodied carbon · Residential building · Inventory analysis

S. Li · H. Yan · J. Chen (✉) · L. Shen
School of Construction Management and Real Estate,
Chongqing University, Chongqing, China
e-mail: jindao.chen@qq.com

S. Li · H. Yan · J. Chen · L. Shen
International Research Centre for Sustainable Built Environment,
Chongqing University, Chongqing, China

Y. Wu et al. (eds.), *Proceedings of the 20th International Symposium on Advancement of Construction Management and Real Estate*,
DOI 10.1007/978-981-10-0855-9_104

104.1 Introduction

Energy consumption and carbon emissions experience a noticeable upward trend due to the rapid urbanization process [1]. Global warming has become more and more serious and attracted increasing attention. Building sector is one of the main energy consumers and carbon emitters, which contributes largely to global warming. According to the report by Intergovernmental Panel on Climate Change (IPCC), building sector consumed about 32 % of the global energy sources and emitted about 19 % of the global greenhouse gases [2]. China, the largest carbon emitter in the world, has been predicted that it would build 40 billion square meters of buildings in the next twenty years, equivalent to 85 % of the gross urban and rural residential buildings areas in 2012 [3], which will cause a large increase in carbon emissions.

Life cycle carbon emissions of buildings mainly incorporate embodied carbon and operational carbon. Embodied carbon refers to carbon emissions incurred during the materials manufacture, materials transportation, construction, maintenance, and end-of-life phases. Operational carbon refers to carbon emissions incurred from the use of buildings, including heating, cooling, and lighting. It is acknowledged that the operational carbon of a building is generally higher than embodied carbon [4]. Therefore, many academic efforts have been put into studying solutions for reducing operational carbon. Some renewable energy technologies and low carbon technologies have been applied to promote operational carbon reductions in buildings. Some technologies or measurements such as advanced and effective insulation materials can reduce operational carbon significantly, but on the other hand these measurements may lead to an increase of embodied carbon in the process of materials production. Therefore, the importance of embodied carbon has been recognized in line with the promotion for the reduction of operational carbon. Ayaz and Yang [4] have proved that the embodied carbon could account for a significant part, for example 45 %, if certain environmental friendly measures were taken to reduce carbon emissions in the operational stage. In the future, the implementation of zero carbon target will lead to the consequence that the life cycle carbon are all contributed by embodied carbon. Hence, the embodied carbon is being increasingly concerned by the public. Therefore, how to effectively quantify the embodied carbon of residential building is crucial to achieve low carbon in the whole life cycle of buildings.

Life cycle assessment (LCA) is a widely used method in the analysis of environmental impacts of products, technologies or systems over the entire life cycle of a building. This method is used to identify the whole life cycle environmental impacts of a concerned building based on "cradle to grave" approach. LCA considers a series of stages—from "cradle" (raw material extraction), through material manufacturing, transportation, construction, operation, and demolition to "grave" (disposal). In using LCA approach, there are four steps, namely, goal and scope definition, life cycle inventory analysis, life cycle impact assessment, and interpretation, which are defined in detail in ISO 14040:2006 (E) [5].

This paper aims to conduct a comprehensive study on the embodied carbon of a building located in Chongqing by using LCA method. The results of the study are used to help the designers, contractors and developers understand clearly the embodied carbon of the building and facilitate these stakeholders to take some effective measures to reduce the embodied carbon of buildings.

104.2 Methodology

With referring to embodied carbon, this study considers a building life cycle in five stages, namely, materials production, transportation, construction, maintenance, and demolition and disposal. The carbon emissions occurring in those five stages are all taken into account. The analysis in this study focuses on a residential building, excluding the considerations for infrastructure, such as roads for the transportation of building materials, factories and warehouse for the production of building materials, transportation for on-site workers and temporary works.

A combination of several research methods is used in this study. Literature review is conducted to set theoretical basis of calculation model for embodied carbon. The proposed calculation model by this paper will be presented in different stage. Case study is used to demonstrate the application of this calculation model to quantify the embodied emissions of a building.

104.3 Life Cycle Inventory Analysis

104.3.1 Embodied Carbon in Materials Production Stage

Any building comprises a large range of materials such as steel, concrete, steel, brick, cement, tile, paint and aluminium. The embodied carbon in materials production stage occurs across several phases of the materials production including the extraction of raw materials from the earth, processing and manufacturing of building materials.

A commonly applied method for the embodied carbon of materials and elements production (EC_{mat}) is as follows:

$$EC_{\mathrm{mat}} = \sum_{\mathrm{m}=1}^{\mathrm{M}} Q_{\mathrm{m}} \times CI_{\mathrm{m}}, \tag{104.1}$$

where m stands for the type of building materials and elements; Q_{m} denotes the quantity of the material m; CI_{m} denotes the carbon intensity of material m.

Q_{m} can be found in the Bill of Quantities (BOQ) for a specific building.CI_{m} can be derived from the Inventory of Carbon and Energy (ICE) Version 2.0, which is a

Table 104.1 The carbon intensities of main materials from ICE database [6]

Type of building material	CO_2 intensity (kg CO_2/kg)	Type of building material	CO_2 intensity (kg CO_2/kg)
Aluminum	8.24	Paint	2.42
Asphalt	0.068	Cement	0.73
Brick	0.23	Plywood	0.42
Concrete	0.1	Timber	0.3
Steel	1.37	Tile	0.74
Glass	0.86	Mortar	0.208
Stone	0.073	Sealants and adhesives	5.7
Fly ash	0.008	Sand	0.0048

predominant embodied energy and carbon database for building materials, as shown in Table 104.1.

104.3.2 Embodied Carbon in Transportation Stage

The embodied carbon in transportation stage is the carbon emissions for transporting building materials and components from the factory gate to the site. There are many transport types, such as deep-sea transport, coastal vessel, road freight and railroad. The embodied carbon in this stage (EC_t) is commonly calculated from the following method:

$$EC_t = \sum_{j=1}^{N} \sum_{m=1}^{M} Q_m \times D_{mj} \times F, \qquad (104.2)$$

where j denotes the transport type of building materials and component; Q_m is the quantity of the material m; D_{mj} denotes the total distance of transportation of building materials and components by transport type j; F_j denotes the carbon emission factor by transport type j, which are shown in Table 104.2.

Table 104.2 Carbon emission factors for transportation [7]

Transportation mode	Emission factor (kg CO_2/ton km)
	CO_2
Deep-sea transport	0.01598
Coastal vessel	0.03463
Road Freight	0.16835
Railroad	0.02035

104.3.3 Embodied Carbon in Construction Stage

On-site construction processes require a broad range of machineries, such as loader, crawler crane, and concrete pump car. Energy consumed by these machineries are usually in the forms of electricity and fuel. The energy consumed by on-site accommodation for workers and on-site offices must be considered. Data for energy consumed in construction stage can usually be collected from the on-site record or the BOQ. The embodied carbon related to the energy consumed in construction stage (EC_c) can be calculated through:

$$EC_c = \sum_{r=1}^{R} Q_r \times CI_r, \tag{104.3}$$

where r denotes the type of energy consumed in construction stage; Q_r denotes the quantity of energy r. CI_r denotes the carbon emission intensity of energy r (Carbon emission intensities of different energy are shown in Table 104.3).

104.3.4 Embodied Carbon in Maintenance Stage

This stage involves the maintenance of systems and components of building. In general, the lives of the systems and components in a building will be shorter than the assumed lifespan of the building concerned, thus the maintenance for these components is necessary. Replacement factors are used to describe the ratio of the assumed lifespan of a building to average service lifespan of common building materials. Table 104.4 presents the replacement factors for some common materials

Table 104.3 Carbon emission intensities of energy

Energy	Emission intensity
	CO_2
Diesel (kg CO_2/kg)	0.8686 [8]
Gasoline (kg CO_2/kg)	0.83727 [8]
Electricity (kg CO_2/kWh)	0.685 [9]

Table 104.4 The replacement factors of main common building materials [10]

Building components	Replacement factor	Building components	Replacement factor
Structural elements (columns, beams, etc.)	1.0	Plastic carpeting	2.4
External and interior walls	1.0	Ceiling	2.0
Flooring	1.0	Floor finishes	3.0
Windows and doors	1.3	Painting and wall papering	5.0
Wall and roofing tiles	1.3	Others	1.2

derived from previous studies. Transportation of the materials used for maintaining the building and certain equipment used for replacing the overdue materials both produce some carbon emissions. Whilst there are various methods for calculating embodied carbon in maintenance stage, a typical method is proposed as follows:

$$EC_{man} = \sum_{\mathrm{m}=1}^{\mathrm{M}} (l_m - 1) \times Q_{\mathrm{m}} \times CI_{\mathrm{m}} + \sum_{\mathrm{j}=1}^{\mathrm{N}} \sum_{\mathrm{m}=1}^{\mathrm{M}} (l_m - 1) \times Q_{\mathrm{m}} \times D_{\mathrm{mj}} \times F_{\mathrm{j}} + \frac{\sum_{\mathrm{m}=1}^{\mathrm{M}} (l_m - 1) \times Q_{\mathrm{m}}}{\sum_{\mathrm{m}=1}^{M} Q_{\mathrm{m}}} \times EC_{\mathrm{c}}, \tag{104.4}$$

where l_m denotes the replacement factor of materials or subsystems; Q_{m} denotes the quantity of the material m; CI_{m} denotes the carbon intensity of material m; D_{mj} denotes the total distance of transportation of building materials and components by transport type j; F_{j} denotes the carbon emission factor by transport type j; EC_{c} denotes the embodied carbon in construction stage; EC_{man} denotes the embodied carbon in maintenance stage.

104.3.5 Embodied Carbon in Demolition & Disposal Stage

In general, a building will be demolished when it expires its lifespan. The demolition of the building causes a lot of building wastes. As detailed data for demolition activities usually cannot be acquired directly, coefficient for embodied carbon in demolition is used to estimate the embodied carbon of demolition activities. The coefficient for carbon equivalent embodied in demolition is recommended as 5 kg CO_2e/m^2 gross internal floor area [11]. Coefficient in the form of carbon dioxide is considered to have the same value as that in the form of carbon equivalent because carbon dioxide usually accounts for most of unit carbon equivalent. Since buildings in most of studies are within their service life, the distance of transporting waste from site to landfill is usually an assumed value. In the demolition & disposal stage, embodied carbon comprises the emissions of the fuel used for demolition machineries $(A_b \times F_d)$ and emissions of the fuel used for transportation of the waste to landfill or recovery sties $(Q_{\mathrm{w}} \times D_{\mathrm{w}} \times F_{\mathrm{w}})$. Hence, the embodied carbon in the demolition and disposal stage (EC_{dd}) can be calculated by:

$$EC_{dd} = A_b \times F_d + Q_{\mathrm{w}} \times D_{\mathrm{w}} \times F_{\mathrm{w}}, \tag{104.5}$$

where A_b indicates the area of building to be demolished; F_d indicates the coefficient for carbon embodied in demolition; Q_{w} indicates the quantities of waste in demolition and disposal stage; D_{w} indicates the distance from site to landfill; F_{w} indicates the carbon emission intensity of transportation for waste.

104.3.6 Life Cycle Embodied Carbon in the Building

By integrating the Formulas (104.1)–(104.5), the model for assessing the life cycle embodied carbon of a building can be formulated as follows:

$$\mathrm{LCEC} = EC_{\mathrm{mat}} + EC_{\mathrm{t}} + EC_{\mathrm{c}} + EC_{\mathrm{man}} + EC_{\mathrm{dd}}, \tag{104.6}$$

where LCEC denotes total quantity of embodied carbon during a building's life cycle, including materials production stage, transportation stage, construction stage, maintenance stage, and demolition and disposal stage.

104.4 Case Study

A case study is selected to demonstrate the application of life cycle analysis method introduced in the previous section.

104.4.1 Description of the Case Building

In this study, a typical building serves as a case study for the analysis which represents a typical type of residential buildings in Chongqing. The building, the 3rd Settlement Housing, is located at Cuntan, Chongqing. Detailed data of the case were obtained from the CAD drawings and the Bill of Quantities provided by Chongqing Boma Engineering Consulting Company.

The project is a 28-storey residential building, which has a gross building area of 17,678.49 m^2, including usable commercial floor area of 1064.76 m^2 in the first and second story. The building has typical pile foundation and shear wall structure, and its lifespan is assumed to be 50 years.

104.4.2 Data for Analysis

Data about the quantities of the fifteen most important materials consumed for the erection of the building are acquired from the Bill of Quantities of the building. In order to analyze the embodied carbon of the building across life cycle, the building is divided into six subsystems and other components. The six subsystems are structure, flooring, ceiling, roof, windows and doors, and finishes of walls and columns. The amounts of the most important materials in the each subsystem are shown in Table 104.5. The main mode of transportation for materials is road.

Table 104.5 Inventory of materials in the building

Building subsystem	Materials	Quantity (kg)	Building subsystem	Materials	Quantity (kg)
Structure (walls, columns, beams, foundation etc.)	Steel	903,059.0	Roof	Steel	3056.0
	Cement	197,495.8		Cement	45,135.8
	Brick	3,749,585.6		Concrete	136,414.8
	Concrete	17,156,531.3		Sand	114,538.8
	Stone	145,584.3		Asphalt	3631.5
	Sand	761,439.9		Paint	21,923.9
	Fly ash	147.2		Sealants and adhesives	1666.7
	Timber	749.4			
Flooring	Steel	19.6	Windows and doors	Timber	31,250.5
	Cement	201,345.5		Plywood	14,359.8
	Brick	12,246.8		Paint	577.2
	Concrete	462,335.8	Finishes of walls and columns	Steel	19,206.5
	Stone	88,009.7		Cement	505,391.5
	Sand	268,432.3		Sand	1,401,130.8
	Fly ash	8707.4		Mortar	33,474.6
	Tile	86,849.9		Tile	204,870.4
Ceiling	Cement	15,466.2		Paint	14,685.9
	Paint	8309.7		Timber	2496.4
			Others	Concrete	14,076.0
	Aluminum	7551.0		Stone	26,218.1
	Timber	301.9		Sand	31,589.4
				Paint	305.9

Therefore, this study assumed that all the main materials are transported by road and the average distance of the transportation for materials is 60 km (as most reference literature applied). According to the BOQ, the energy consumption in construction stage is as follows:

- 409.9352 kg for diesel
- 72,255.7329 kwh for electricity
- 10.8881 kg for gasoline.

The assumption of transportation model for materials in maintenance stage is the same as that in transportation stage. According to the discussion with the project members, the quantities of waste in demolition and disposal stage are considered to be equal to the quantities of the fifteen most important materials. It is assumed that the distance from the site to landfill is 100 km and the transport mode for transporting wastes is road transport.

104.4.3 Analysis Results

By applying the data presented in Sect. 104.4.2 to the formula (104.1)–(104.6), the following calculation can be obtained:

- the embodied carbon in building life cycle is 5,789,000 kg CO_2;
- per square metre embodied carbon of the building is 327.46 kg CO_2/m^2.

And the amount of embodied carbon at each life cycle stage for the case building is as follows:

- 5,263,695 kg CO_2 at materials production stage;
- 26,978 kg CO_2 at transportation stage;
- 49,860 kg CO_2 at construction stage;
- 315,178 kg CO_2 at maintenance stage;
- 133,355 kg CO_2 at demolition and disposal stage.

The above data can also be presented graphically in Fig. 104.1.

It can be seen from Fig. 104.1 that the embodied carbon at materials production stage accounts for the largest proportion in building life cycle, assuming 90.92 % of the whole embodied carbon. The stage contributing to the second largest proportion of embodied carbon is maintenance stage, accounting for 5.44 %. The embodied carbon emission in demolition and disposal stage, construction stage, and transportation stage constitute 2.35, 0.86 and 0.47 % of the total embodied carbon, respectively. Therefore, more attention should be paid to the choice of building materials at materials production stage in order to reduce embodied carbon of a building effectively.

In the maintenance stage, the quantity of embodied carbon from the case building is about 315,178 kg CO_2 and accounts for 5.44 % of the total emissions. This stage has produced less embodied carbon than that from materials production

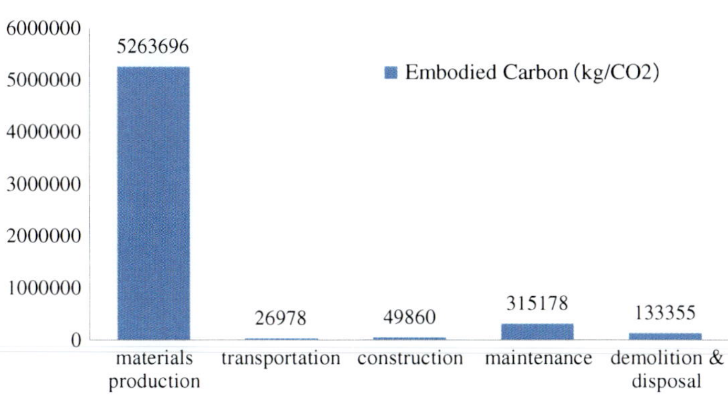

Fig. 104.1 Embodied carbon across life cycle for the case building

stage because only a few type of materials have to be replaced in the maintenance stage.

In the demolition and waste disposal stage, nearly 2.3 % of embodied carbon are generated. The demolition process contributes about two thirds of embodied carbon in this stage and the transportation of waste accounts for the remaining one third of emissions.

Building materials transportation stage contributes about 0.47 % of the life cycle embodied carbon. This indicates that the transportation of building materials does not generate much carbon emissions. Similarly, in construction stage, the proportion of embodied carbon emissions takes a very small value, with less than 1 % of the total amount of emissions.

The distribution of the embodied carbon between the main materials for the case building can also be presented in Fig. 104.2. Most of the embodied carbon is contributed by concrete (36.3 %), steel (24.1 %), brick (16.6 %), cement (13.4 %), tile (4.1 %), paint (3.11 %) and aluminium (1.18 %). The remaining is contributed by stone (0.36 %), sand (0.24 %), sealants and adhesives (0.18 %), mortar (0.13 %), timber (0.2 %), plywood (0.11 %), asphalt (0.005 %) and fly ash (0.001 %). The higher proportion of embodied carbon from steel and concrete compared to other materials is due to the larger amount of steel and concrete used per square meter in the building.

Figure 104.3 shows the distribution of the embodied carbon of materials in each subsystem of the building, from which it can be seen that structure is the greatest contributor to the total embodied carbon of materials (78.41 %), followed by walls and columns finishes (11.78 %), flooring (5.29 %), roof (2.25 %), ceiling (1.85 %), windows and doors (0.33 %) and others (0.08 %).

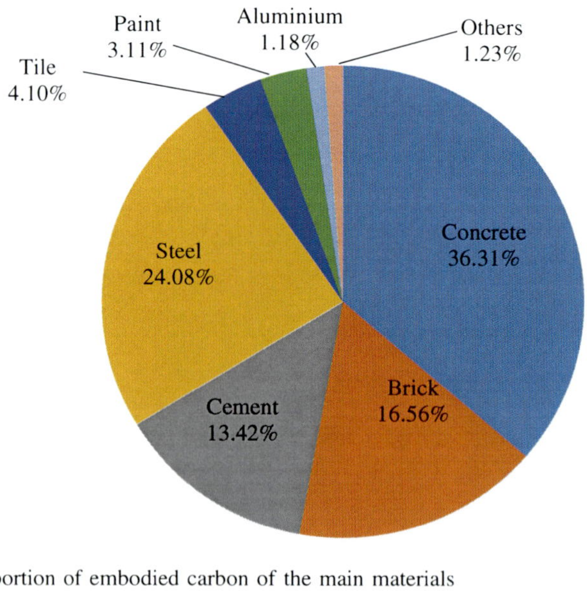

Fig. 104.2 Proportion of embodied carbon of the main materials

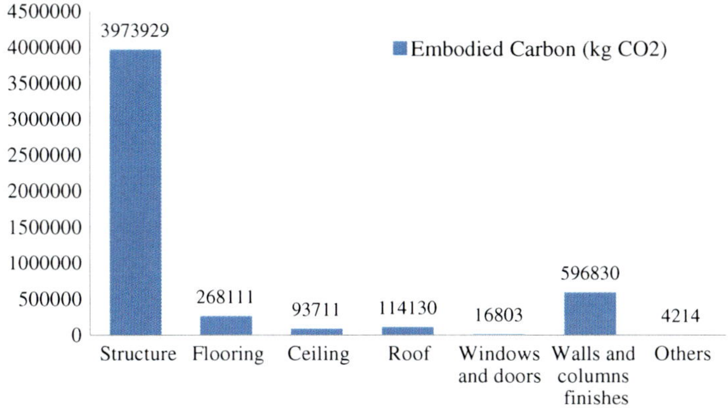

Fig. 104.3 Distribution of embodied carbon between subsystems of a residential building

104.5 Conclusions

Embodied carbon is generated during building life cycle which typically includes building materials stage, materials transportation stage, construction stage, maintenance stage, demolition and disposal stage. Life cycle inventory analysis is used in this paper to quantify the embodied carbon of each stage. Applications of the calculation model for assessing carbon emissions are demonstrated by a case building in Chongqing.

By the calculating of the embodied carbon at each life cycle stage of buildings, it is therefore feasible to assess the quantity of carbon emissions across a building's life cycle. This approach provides an important vehicle for analyzing the embodied carbon of different types of buildings. The selection for different materials and construction methods for lower embodied carbon is possible. This can help identify the optimal solution in choosing materials, construction methods, and modes of transportation towards minimizing the quantity of embodied carbon in buildings. This will in turn improve the environment performance and contribute to the sustainable development. The findings can provide reference for future study on studying optimal design for low carbon building.

Acknowledgements This research was supported by a grant from the Key Project of National Social Science Foundation of China (No. 15AZD025).

References

1. Zhang CX, Zhang BB, Huang YL et al (2010) Study on selection methods of energy carbon emission factors. Constr Econ 10:106–109

2. Edenhofer O, Pichs-Madruga R, Sokona Y et al (2014) Climate change 2014: mitigation of climate change. Cambridge University Press, NewYork
3. Wang WG, Zheng GG (2013) Green book of climate change: annual report on actions to address climate change. Social Sciences Academic Press, Beijing
4. Ayaz E, Yang F (2009) Zero carbon isn't really zero: why embodied carbon in materials can't be ignored (2009-09-01) [2014-11-30]. http://www.di.net/articles/zero_carbon/
5. International Organization for Standardization (2006) Environmental management: life cycle assessment. Requirements and guidelines. ISO
6. Hammond G, Jones C, Lowrie F et al. (2011) Embodied carbon: the inventory of carbon and energy (ICE). BSRIA
7. Zhang XL, Shen LY, Zhang L (2013) Life cycle assessment of the air emissions during building construction process: a case study in Hong Kong. Renew Sustain Energy Rev 17:160–169
8. Intergovernmental Panel on Climate Change (2006) IPCC guidelines for national greenhouse gas inventories. Intergovernmental Panel on Climate Change, 2006
9. Ma CM, Li SC, Ge QS (2014) Greenhouse gas emission factors for grid electricity for chinese provinces. Resour Sci 36(5)
10. Chen TY, Burnett B, Chau CK (2001) Analysis of embodied energy use in the residential building of Hong Kong. Energy 26:323–340
11. Moncaster AM, Symons KE (2013) A method and tool for 'cradle to grave' embodied carbon and energy impacts of UK buildings in compliance with the new TC350 standards. Energy Build 66:514–523

Chapter 105
An Analysis on the Carbon Emission Contributors in the Chinese Construction Industry

Jindao Chen, Ya Wu, Hang Yan and Liyin Shen

Abstract With the rapid development of Chinese economy and increasing improved living standards, the amount of carbon emissions in China has been increasing consistently in a high speed, which consists of the largest percentage of the world's total carbon emissions in recent years. The construction industry, playing an important role in the Chinese economy, accounts for a large proportion of the total carbon emissions in China. In this paper, the carbon emissions from construction industry in China in 2009 are analyzed by adopting Multi Regional Input-output (MRIO) Model and the World Input-Output Database (WIOD). Results show that, according to the data in 2009, the construction industry is the largest carbon emitter among all industries in China, responsible for the emissions of 2,121,649.31 kt CO_2, accounting for 66.54 % of Chinese total carbon emissions. This emission value is contributed by other economic sectors and activities, and it has been found that the industrial sector "Electricity, Gas and Water Supply" is the largest contributor to the carbon emissions of Chinese construction industry, with an amount of 984,830.85 kt CO_2, accounting for 46.42 % of the total carbon emissions of Chinese construction industry. Furthermore the carbon emissions in the construction industry comprise 71,418.19 kt CO_2 (3.37 %) of direct carbon emissions and 2,050,231.12 kt CO_2 (96.63 %) of indirect carbon emissions. The carbon emissions of domestic goods, exports and imports within construction industry are 2,129,974.07, 8663.33 and 338.58 kt CO_2, respectively, consisting of 100.39, 0.41 and 0.02 % of the total carbon emissions of Chinese construction industry. The results can help identify critical areas where policymakers can formulate effective policy measures for carbon emissions reduction in Chinese construction industry.

Keywords Carbon emissions · Construction industry · MRIO · WIOD · China

J. Chen (✉) · Y. Wu · H. Yan · L. Shen
School of Construction Management and Real Estate,
Chongqing University, Chongqing, China
e-mail: jindao.chen@qq.com

J. Chen · Y. Wu · H. Yan · L. Shen
International Research Centre for Sustainable Built Environment,
Chongqing University, Chongqing, China

Y. Wu et al. (eds.), *Proceedings of the 20th International Symposium on Advancement of Construction Management and Real Estate*,
DOI 10.1007/978-981-10-0855-9_105

105.1 Introduction

With rapid economic development and increasingly improved living standards, China is now the largest emitter of carbon emissions in the world with estimated 8.3 billion tons of carbon dioxide produced in 2011 [1]. Historically, the construction industry has been playing a key role in the Chinese economy [2], for example, it contributed to 6.9 % of GDP in 2013. Among secondary industries, the construction industry is one of the largest emitters as it involves manipulation and use of large quantities of natural and artificial materials with high embodied carbon emission, such as cement and steel bars. Analysis on the carbon emissions in the construction industry becomes an important research topic worldwide in recent year. The research results are used to help make proper measures for carbon emissions reduction.

The carbon emissions of construction industry comprise direct carbon emissions and indirect carbon emissions [3]. Chang et al. [4] conducted a study on thc embodied energy and environmental emissions of construction project, and found that the direct and indirect carbon emissions are 49.61 and 1346.55 mt CO_2 respectively, accounting for 3.3 and 96.7 % respectively of the total carbon emissions in the Chinese construction industry. Ji et al. [5] investigated greenhouse gas emission by the Chinese economy, and concluded that construction industry accounted for the highest embodied emissions, 29.79 % of the total emissions, of which 93.91 % are indirect carbon emissions. However, most of researches on carbon emissions in construction industry focus on domestic carbon emissions, excluding the carbon emissions embodied in imported goods, which may largely influence the adequacy of the total carbon emissions of construction industry. Acquaye and Duffy [6] analyzed Irish construction industry greenhouse gas emissions by considering both domestic and imported goods, and its results showed that in 2005 the emissions of Irish construction industry were 13.81 mt CO_2eq, comprising 2.37 mt CO_2eq (17 %) of direct on-site emissions, 5.69 mt CO_2eq (41 %) indirect domestic emissions and 5.75 mt CO_2eq (42 %) of indirect emissions of imported goods sharing a largest proportion of the total emissions.

The existing methods on analyzing carbon emissions of construction industry still suffer some shortcomings. Since the Economic input-output Life-cycle assessment (EIO-LCA) model considered the economic flows and the associated carbon emission flows completely, it has been a widely used in environmental assessment, such as the analysis on carbon emissions and energy. Liu et al. [7] employed EIO-LCA model to analyze China's energy consumption at sectoral level, and concluded that the embodied energy shared a considerable proportions of the total energy use. However, EIO-LCA model in most studies about carbon emissions of construction industry suffered a disadvantage that it can not quantify the actual carbon emissions of imported goods in construction industry. With rapid development of interregional and international trade, more and more researchers put their interests in quantifying the environmental impacts of international trade. And multi-regional Input-output (MRIO) model is the dominating method to analyze the environmental impacts of international trade in recent years. Su and Ang [8] conducted a comprehensive study of China's regional emissions embodied in trade based on MRIO model. Wiedmann et al. [9] studied the carbon emissions embedded in UK trade applying UK-MRIO mode. However, based on MRIO model, most of these studies focused on the carbon emissions of international trade. It appears that there is little existing study of using MRIO model to analyze the total carbon emissions of construction industry.

This paper adopts the MRIO model to calculate and assess the carbon emissions of China's construction industry. To apply the MRIO model, a large amount of data originated from MRIO database are needed. According to the study by Wiedmann et al. [10], there are many types of MRIO databases including OECD, Global Trade Analysis Project database (GTAP), World Input–Output Database (WIOD), Eora, EXIOPOL and Asian International IO Tables. However, the GTAP and the WIOD are the most commonly used databases in policy-related studies in recent years. As illustrated in the study by Arto et al. [11], the composition of carbon emissions recorded in WIOD are more comprehensive, thus the total carbon emissions listed in the database are higher than that in GTAP. In fact, the carbon emissions in WIOD consist of not only the carbon emissions related to the combustion of fossil fuels but also the carbon emissions from industrial processes, while the carbon emissions in GTAP only considered the former part of carbon emissions. Moreover, WIOD offers free-access for acquiring information. Therefore, this paper use the data in WIOD when applying the MRIO model.

This study aims to provide a systematic profile of carbon emissions of Chinese construction industry in order to identify critical areas where policymakers can formulate effective policy measures for carbon emissions reduction. The paper is structured as follows. The next section presents a detailed introduction to the research method, MRIO model. In Sect. 105.3 we illustrates the data collection process. Section 105.4 analyzes the calculation results. Section 105.5 closes with a summary.

105.2 Multi-regional Input-Output (MRIO) Model

The basic input-output model was put forward by Leontief and Ford for dealing with the industrial interrelationships in a national economic system. It can be demonstrated using the following Eqs. (105.1) and (105.2).

$$x = Ax + y \tag{105.1}$$

where vector x means the total output of one specific region. $A = [a_{ij}]$, called the technical coefficient or technology matrix, describes all the product requirements i needed by industry j to produce a unit monetary output. Vector y represents the final demand of a country.

The MRIO model has extended the basic input-output model to describe inter-regional trade flows and environmental flows. The framework of MRIO model can be shown as follows:

$$X^* = A^* X^* + Y^* \tag{105.2}$$

where X^* denotes total output of regions. A^* represents technical coefficient between regions. Y^* represents the final total demand of regions. Y^* consists of demand on domestic production, import and exports.

If there are m regions and region 1 is the region we study on, then Eq. (105.2) becomes

$$\begin{pmatrix} x_1 \\ x_2 \\ x_3 \\ \vdots \\ x_m \end{pmatrix} = \begin{pmatrix} A_{11} & A_{12} & A_{13} & \cdots & A_{1m} \\ A_{21} & A_{22} & A_{23} & \cdots & A_{2m} \\ A_{31} & A_{32} & A_{33} & \cdots & A_{3m} \\ \vdots & \vdots & \vdots & \ddots & \vdots \\ A_{m1} & A_{m2} & A_{m3} & \cdots & A_{mm} \end{pmatrix} \begin{pmatrix} x_1 \\ x_2 \\ x_3 \\ \vdots \\ x_m \end{pmatrix} + \begin{pmatrix} y_{11} + y_1^{ex} \\ y_{21} \\ y_{31} \\ \vdots \\ y_{m1} \end{pmatrix} \tag{105.3}$$

where x_i indicates the output of region i. Each diagonal matrix A_{ii} in A^* is a matrix of technical coefficient on domestic production in region i and A_{ij} denotes the inter-industry technical coefficient from region i to region j. y_{11} and y_1^{ex} represent the final demand on domestic production consumed domestically in region 1 and exports to other regions, respectively. y_{i1} indicates the final demand of goods from region i to region 1.

From above, the output of region 1 is

$$x_1 = A_{11}x_1 + y_{11} + \sum_{j=2}^{m} \left(A_{1j}x_j + y_{1j}\right) \tag{105.4}$$

where $\sum_{j=2}^{m} \left(A_{1j}y_{1j} + y_{1j}\right)$ is a vector representing the total exports from region 1.

$$x_i = A_{ii}x_i + \sum_{j=1, i\neq j}^{m} A_{ij}x_j + y_{i1} \quad \text{for all } \; i \neq 1 \tag{105.5}$$

where $\sum_{j=1, i\neq j}^{m} \left(A_{ij}x_j + y_{i1}\right)$ is a vector representing the final demand on goods imported to region 1.

Based on the linearity assumption of the IO model, it is possible to figure out the output of each region for an arbitrary demand. Therefore, Eqs. (105.4) and (105.5) can be formulated as follows:

$$x_1 = (I - A_{11})^{-1} \left[y_{11} + \sum_{j=2}^{m} \left(A_{1j}x_j + y_{1j}\right) \right] \tag{105.6}$$

$$x_i = (I - A_{ii})^{-1}\left(\sum_{j=1,i\neq j}^{m} A_{ij}x_j + y_{i1}\right) \quad \text{for all } i \neq 1 \tag{105.7}$$

where I represents the identity matrix.

Then, we transform y_{11}, $\sum_{j=2}^{m}\left(A_{1j}x_j + y_{1j}\right)$ and $\sum_{j=1,i\neq j}^{m} A_{ij}x_j + y_{i1}$ these three vectors into three corresponding positive diagonal matrixes D, E, and M_i, in which the diagonal value of each row is equal to the value of each row in y_{11}, $\sum_{j=2}^{m}\left(A_{1j}x_j + y_{1j}\right)$ and $\sum_{j=1,i\neq j}^{m} A_{ij}x_j + y_{i1}$ respectively.

Finally, the carbon emissions matrices can be set up as follows:

$$CE_D = F_1(I - A_{11})^{-1}D \tag{105.8}$$

$$CE_E = F_1(I - A_{11})^{-1}E \tag{105.9}$$

$$CE_M = \sum_{i=2}^{m} F_i(I - A_{ii})^{-1}M_i \tag{105.10}$$

where CE_D denotes the carbon emissions of the goods produced and consumed in region 1. CE_E denotes the carbon emissions of goods exported from region 1. CE_M denotes the carbon emissions of goods imported to region 1. F_i are the direct carbon emissions intensities in region i (i = 1, 2, …, m).

Based on the consumption-based accounting principle, the total carbon emissions are calculated as follows:

$$CE = CE_D - CE_E + CE_M \tag{105.11}$$

105.3 Research Data

In order to apply the MRIO model presented in previous section, the data were collected from the World Input-output database (WIOD). WIOD, a project funded by the European Commission as part of the 7th Framework, comprises a series of databases and covers 35 industries, 59 products and 40 countries, including 27 EU countries and 13 other major countries worldwide, and the rest of the World (Row) for the period from 1995 to 2011. Table 105.1 presents the industries classification in the WIOD. In this paper, we use the World Input-output Tables to acquire the technical coefficients and use the Environmental Accounts to derive the direct carbon emissions intensities. We only consider CO_2 for that it is the most significant part in carbon emissions. Since the latest Environmental Accounts is the year 2009, we adopt the data of 2009.

Table 105.1 Industries classification in the WIOD [13]

Code	Industry name	Code	Industry name
1	Agriculture, Hunting, Forestry and Fishing	21	Retail Trade, Except of Motor Vehicles and Motorcycles; Repair of Household Goods
2	Mining and Quarrying	22	Hotels and Restaurants
3	Food, Beverages and Tobacco	23	Inland Transport
4	Textiles and Textile Products	24	Water Transport
5	Leather, Leather and Footwear	25	Air Transport
6	Wood and Products of Wood and Cork	26	Other Supporting and Auxiliary Transport Activities; Activities of Travel Agencies
7	Pulp, Paper, Paper, Printing and Publishing	27	Post and Telecommunications
8	Coke, Refined Petroleum and Nuclear Fuel	28	Financial Intermediation
9	Chemicals and Chemical Products	29	Real Estate Activities
10	Rubber and Plastics	30	Renting of M&Eq and Other Business Activities
11	Other Non-metallic Mineral	31	Public Admin and Defence; Compulsory Social Security
12	Basic Metals and Fabricated Metal	32	Education
13	Machinery, Nec	33	Health and Social Work
14	Electrical and Optical Equipment	34	Other Community, Social and Personal Services
15	Transport Equipment	35	Private Households with Employed Persons
16	Manufacturing, Nec; Recycling		
17	Electricity, Gas and Water Supply		
18	Construction		
19	Sale, Maintenance and Repair of Motor Vehicles and Motorcycles; Retail Sale of Fuel		
20	Wholesale Trade and Commission Trade, Except of Motor Vehicles and Motorcycles		

According to the China Statistical Yearbook 2010, United States, Japan, Hong Kong, South Korea, Taiwan, Germany and Australia were the biggest China's trade partners in 2009 [12]. To simplify the calculation procedures, China, United States, Japan, South Korea, Taiwan, Germany, Australia were chosen as the representative countries, while the other countries in the 40 countries of WIOD were merged into the ROW with an assumption that the ROW has the same direct carbon emissions intensities with Germany.

105.4 Analysis Results

105.4.1 Total Carbon Emissions of Chinese Construction Industry

Figure 105.1 displays the result for the total carbon emissions of Chinese construction industry and the total carbon emissions of Chinese other industries. From Fig. 105.1, we can see that the total carbon emissions of construction industry (No. 18) are 2,121,649.31 kt CO_2 and construction industry is the largest contributor to Chinese total carbon emission (66.54 %).

105.4.2 Structure of Total Carbon Emissions of Chinese Construction Industry

Figure 105.2 shows the structure of the total carbon emissions of Chinese construction industry, from which we can see "Electricity, Gas and Water Supply" industry is the greatest contributor to the total carbon emissions of Chinese construction industry (46.42 %), followed by "Other Non-Metallic Mineral" (25.32 %), "Basic Metals and Fabricated Metal" (10.97 %), "Mining and Quarrying" (3.43 %), "Construction" (3.28 %) and "Chemicals and Chemical Products" (2.57 %). Each of these industries' carbon emissions are 984,830.85, 537,105.79, 232,725.90, 72,701.92, 69,584.92, and 54,442.85 kt CO_2, respectively.

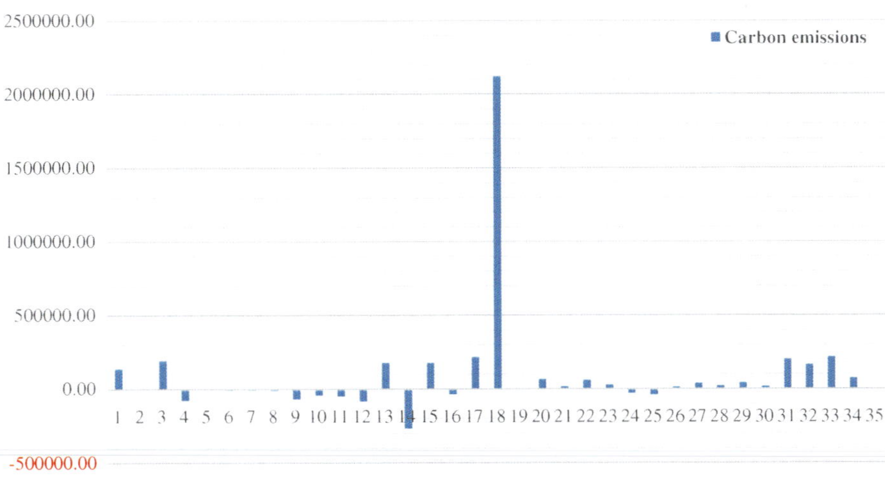

Fig. 105.1 Total carbon emissions of Chinese industries (units: kt CO_2)

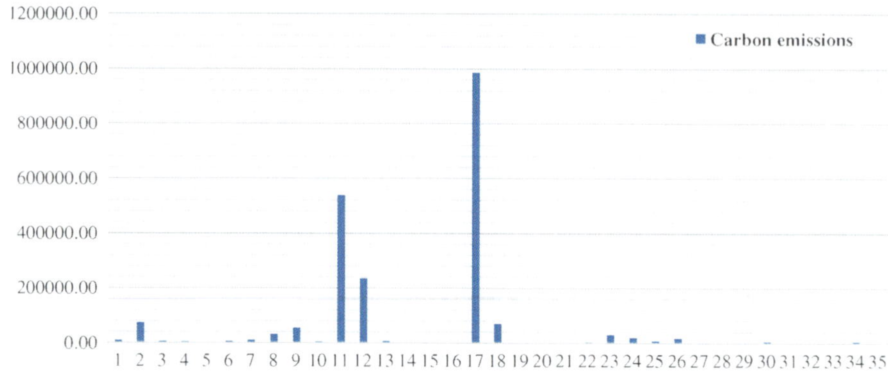

Fig. 105.2 Structure of total carbon emissions of Chinese construction industry (units: kt CO_2)

105.4.3 Direct Carbon Emissions Versus Indirect Carbon Emissions of Chinese Construction Industry

Table 105.2 shows the comparison results between the direct carbon emissions and indirect carbon emissions of Chinese construction industry. The results indicates that the indirect carbon emissions have constituted most proportion of the total carbon emissions of Chinese construction industry. The indirect carbon emissions are 2,050,231.12 kt CO_2, accounting for 96.63 % of the total carbon emissions of construction industry while the direct carbon emissions are 71,418.19 kt CO_2, accounting for 3.37 % of the total carbon emissions of construction industry.

105.4.4 Carbon Emissions of Domestic Goods Versus Carbon Emissions of Trades of Chinese Construction Industry

Table 105.3 shows the comparison results between carbon emissions of domestic goods and trades of Chinese construction industry. The results denote that the carbon emissions of domestic goods are 2,129,974.07 kt CO_2, accounting for 100.39 % of the total carbon emissions of China' construction industry, while the

Table 105.2 Comparison of direct and total carbon emissions of Chinese construction industry

Direct carbon emissions	Indirect carbon emissions	Total carbon emissions
71,418.19[a]	2,050,231.12[a]	2,121,649.31[a]
3.37 %	96.63 %	100 %

[a]Units: kt CO_2

Table 105.3 Comparison between carbon emissions of domestic goods and that of trades

Carbon emissions of domestic goods (CE_D)	Carbon emissions of exports (CE_E)	Carbon emissions of imports (CE_M)	Net carbon emissions of trades ($CE_M - CE_E$)
2,129,974.07[a]	8663.33[a]	338.58[a]	−8324.75[a]
100.39 %	0.41 %	0.02 %	−0.39 %

[a]Units: kt CO_2

net carbon emissions of trades are −8324.75 kt CO_2, accounting for −0.39 % of the total carbon emissions of China' construction industry. For net carbon emissions of trades, the carbon emissions of exports are 8663.33 kt CO_2, accounting for 0.41 % and the carbon emissions of imports are 338.58 kt CO_2, accounting for 0.02 % of the total carbon emissions of Chinese construction industry.

105.5 Conclusions

By using the relevant data for 2009, this paper found that the construction industry is the largest carbon emitter among all industries in China, responsible for the emissions of 2,121,649.31 kt CO_2, accounting for 66.54 % of Chinese total carbon emissions. "Electricity, Gas and Water Supply" industry is the greatest contributor to the carbon emissions of Chinese construction industry, with an amount of 984,830.85 kt CO_2, accounting for 46.42 % of the total carbon emissions of Chinese construction industry. The carbon emissions of Chinese construction industry comprise 71,418.19 kt CO_2 (3.37 %) of direct carbon emissions and 2,050,231.12 kt CO_2 (96.63 %) of indirect carbon emissions. The carbon emissions of domestic goods, exports and imports within construction industry are 2,129,974.07, 8663.33 and 338.58 kt CO_2, respectively, consisting 100.39, 0.41 and 0.02 % of the total carbon emissions of Chinese construction industry. The net carbon emissions of trades are −8324.75 kt CO_2, accounting for −0.39 % of the total carbon emissions of Chinese construction industry.

Based on the above analysis, it is suggested that policymakers should pay more attention to the indirect carbon emissions of Chinese construction industry when they formulate policy measures on carbon emissions reduction. On the other hand, all domestic goods have positive contributions to the total carbon emissions of Chinese construction industry, except for international trades, which have negative contributions. However, international trades consist of very little proportion of the total carbon emissions in the Chinese construction industry. Therefore, international trades may have great potential to reduce carbon emissions in Chinese construction industry.

Acknowledgements This research was supported by a grant from the Key Project of National Social Science Foundation of China (No. 15AZD025).

References

1. World Bank (2014) CO_2 emissions [EB/OL], 1 Aug. http://data.worldbank.org/indicator/EN.ATM.CO2E.KT/countries
2. Zhai X, Reed R, Mills A (2014) Embracing off-site innovation in construction in China to enhance a sustainable built environment in urban housing. Int J Constr Manage 14(3):123–133
3. Qi SJ, Zhang YB, Wang XX (2012) Analysis on structure characteristic of direct energy consumption and carbon emissions in Chinese building sector. Constr Econ 12:015
4. Chang Y, Ries RJ, Wang Y (2010) The embodied energy and environmental emissions of construction projects in China: an economic input–output LCA model. Energy Policy 38 (11):6597–6603
5. Ji J, Liu L, Ma X (2011) Greenhouse gas emissions by Chinese economy: an assessment based on EIO-LCA model. Acta Scientiarum Naturalium Universitatis Pekinensis 47(4):741–749
6. Acquaye AA, Duffy AP (2010) Input–output analysis of Irish construction sector greenhouse gas emissions. Build Environ 45(3):784–791
7. Liu Z, Geng Y, Lindner S et al (2012) Embodied energy use in China's industrial sectors. Energy Policy 49:751–758
8. Su B, Ang BW (2014) Input–output analysis of CO_2 emissions embodied in trade: a multi-region model for China. Appl Energy 114:377–384
9. Wiedmann T, Wood R, Minx J et al (2008) Emissions embedded in UK trade–UK-MRIO model results and error estimates. In: International input–output meeting on managing the environment, pp 9–11
10. Wiedmann T, Wilting HC, Lenzen M et al (2011) Quo Vadis MRIO? Methodological, data and institutional requirements for multi-region input–output analysis. Ecol Econ 70(11):1937–1945
11. Arto I, Rueda-Cantuche JM, Peters GP (2014) Comparing the GTAP-MRIO and WIOD databases for carbon footprint analysis. Econ Syst Res 26(3):327–353
12. National Bureau of Statistics (2010) China statistical yearbook. China Statistics Press, Beijing (in Chinese)
13. Timmer MP, Dietzenbacher E, Los B et al (2015) An illustrated user guide to the world input-output database: the case of global automotive production. Rev Int Econ 23:575–605

Chapter 106
Decoupling Analysis on the Relationship Between Economic Development and Environment Degradation in China

Chenyang Shuai, Liudan Jiao, Xiangnan Song and Liyin Shen

Abstract In recent 20 years, the dramatic economic development in China has also triggered a number of environmental problems such as fast-growing of resource consumption, air pollution, soil erosion and water pollution. This unsustainable economic development pattern will gradually become the bottleneck and impede the further development of the Chinese economy. Therefore, the relationship between economic development and environment degradation in China has been studied by various researchers. This paper adopts the decoupling factor introduced by the Organization for Economic Co-operation and Development (OECD) to investigate the decoupling level between economic growth and environment degradation in China during the period of 1990–2010. This paper concludes that the environment degradation has been gradually improved whilst the good level of economic development has been sustained.

Keywords Economic growth · Environment degradation · Decoupling analysis

106.1 Introduction

Since the strategy of the Reform and opening up was adopted by the Chinese government in 1978, the economic development in China has markedly increased with a high annual rate in these 36 years [8]. The statistics are clear to support this.

C. Shuai (✉) · L. Jiao · X. Song · L. Shen
School of Construction Management and Real Estate, International Research Centre for Sustainable Built Environment, Chongqing University, Chongqing, China
e-mail: 1391843874@qq.com

L. Jiao
e-mail: jld0617@126.com

X. Song
e-mail: 1018500021@qq.com

L. Shen
e-mail: shenliyin@cqu.edu.cn

Y. Wu et al. (eds.), *Proceedings of the 20th International Symposium on Advancement of Construction Management and Real Estate*,
DOI 10.1007/978-981-10-0855-9_106

The GDP per capita in China has increased from RMB 382 in 1978 to RMB 46652 in 2014, accounting for nearly 121 times with an annual average rate of 9.5 %. The whole GDP in China has increased from RMB 365 billion in 1978 to RMB 63646 billion in 2014, giving nearly 173 times with an annual average rate of 10.6 % [14]. In comparison with the world's average GDP of 3.3 % during the same period, the rapid economic growth in China has greatly raised its economic status in the world. As the giant economy, China has overtaken Japan as the second-largest economy in 2010. It is predicted in Goldman report that China could move ahead of the United States in terms of GDP by year 2028.

However, it is commonly appreciated that the traditional Chinese economic development is a kind of extensive development pattern with the conception of "grow (pollute) first, clean up later". This has triggered the problems of fast-growing of resource consumption, air pollution, soil erosion and climate change. According to the Netherlands Environmental Assessment Agency [16], China is one of the largest contributor to CO_2. Liu [9] pointed out that Two-thirds of China's 656 cities (with 390 million people) suffer shortages of water for domestic and industrial use, and pollution is the main cause to the water scarcity. Various existing studies have confirmed that the extensive development pattern is the primarily cause to the environment degradation in China. McGuire [12] opined that the Chinese government put economic growth ahead of the environment protection. Wang et al. [25] proposed that the surface water has been seriously deteriorated almost everywhere in China due to its rapid yet imprudent economic development in the past three decades. Zhang and Anadon [28] suggested that China is facing serious water shortages as a result of the ever-increasing water demand driven by its rapid industrialization and urbanization. Lu et al. [11] identified that increasingly land desertification in Qinghai–Tibetan Plateau resulting from human economic activities in this area. Therefore, it is considered important to change the current practice of extensive development pattern. In order to change this extensive development pattern, it is necessary to understand the relationship between economic development and environment degradation.

Currently, there are various methods used for examining the relationship between economic development and environment degradation. Chen et al. [4] developed an integrated modeling framework to examine the connections between economic driving forces and environment risk. Aşıcı [1] used panel data and regression analysis to explore the relationship between the economic growth and the pressure on nature from the environmental sustainability perspective. Baek and Kim [2] employed a dynamic cointegration framework to analyze the effects of economic growth on the environment in Korea, and concluded that economic growth in Korea plays a role in improving environmental performance. Their findings support the hypothesis of the environmental Kuznets curve (EKC). Xue and Li [26] adopted principal component analysis and fuzzy mathematics to analyze the coordinated development degree of the economic and environment system. Others have adopted decoupling analysis to examine relationship between environment degradation and economic growth. De Freitas and Kaneko [5] examined the relationship between the growth rates in economic activity and CO_2 emissions from energy consumption in Brazil from 2004 and 2009 by using decoupling analysis. Sjöström and Östblom [19] used the decoupling analysis and Computable General Equilibrium model to analyze the relationship between solid waste increase and the economic growth in Sweden. Van Caneghem et al. [22] investigated the environment impact from economic growth in Flemish by adopting the decoupling analysis. Fodha and Zaghdoud [6] studied the relationship between economic growth and pollutant emissions in Tunisia during the period 1961–2004 by using decoupling analysis. Brown-Santirso and Thornly [3] adopted the decoupling analysis to discuss the relationship between consumer energy demand and economic growth in New Zealand.

The above discussion suggested that many researchers have adopted decoupling analysis to investigate the relationship between economic growth and environment degradation. However, There are few research works concluded in analyzing the relationship between economic development and environment degradation by using decoupling analysis method in the context of China. Peng et al. [18] and discussed the relationship between economic growth and carbon emission in China using decoupling method. Wang [24] constructed decoupling indicators and analyzed the dynamic relationship between economic development and energy consumption. Nevertheless, these works only focus on one specific aspect in referring to the environment degradation. In fact, environmental degradation includes many aspects such as air pollution, water pollution, soil erosion and resource pressure etc. it is important to incorporate these aspects in order to analyze effectively the relationship between economic growth and environment degradation. This paper adopts decoupling analysis to investigate the relationship between economic growth and environment degradation with considering multiple environmental aspects. The data used for analysis are collected from Chinese Statistic Bureau and World Bank Database for the period 1990–2010.

106.2 The Principle of the Decoupling Analysis Technique

Decoupling is a concept of physics which represents the transition process from close interactions between particles to their effective independence. The OECD [17] introduced the decoupling factor to analyze the relationship between agriculture policy and trade market, The World Bank adopted this technique for analyzing the relationship between economy and resource and environment.

There are two types of decoupling: relative decoupling and absolute decoupling. Relative decoupling occurs when the growth rate of economy (GDP) is faster than environmental pressure. In other words, the pressure on the environment is increasing but at a slower rate than the growth rate of GDP. On the other hand, absolute decoupling occurs when the growth rate of the environmental pressure is zero or negative and GDP growth is positive.

In using decoupling technique, the parameter "decoupling factor" needs to be measured. There are many different methods to measure the decoupling factor [17, 21, 23], but these methods are based on a same principle proposed by the OECD [17], which measures decoupling factor as follows:

$$\text{decoupling factor} = 1 - \frac{(\mathrm{EP}/\mathrm{DF})_{t+n}}{(\mathrm{EP}/\mathrm{DF})_{t}} \tag{106.1}$$

where EP is the environmental pressure, which is divided into CO_2 [27] SO_2 [27] soot emissions [27] waste water emissions [27] energy consumptions [7] and household waste [20]. DF is the driving force namely, GDP [13, 27]. t indicates the study year and n indicates the study period. There three kinds of parameter n, namely n = 1, n = 10, n = 20.

When decoupling factor is zero or negative indicating the absence of decoupling, On the contrary, the decoupling occurrs when the decoupling factor is positive. Li and Wei [10] classified the decoupling level as follows in Table 106.1.

106.3 Data and Analysis

The data used for the decoupling analysis are about the performance of all individual indicators of EP and DF. The resources for retrieving these performance data are from two database websites, including World Bank Database "http://data.worldbank.org/indicator" and Chinese Economic and Social Development Database "http://tongji.cnki.net/kns55/Dig/dig.aspx". The data are presented in Table 106.2.

106.3.1 Annual Decoupling Factor Analysis

The data in Table 106.3 can be presented as shown in Fig. 106.1, in general, most of the environment indicators were at feeble or even none decoupling level, only few indicators were at weak or normal decoupling level. For example, the energy decoupling factor (as shown in Fig. 106.1a) was always at the feeble decoupling level during the study period. However, there did not exist decoupling of waste water decoupling factor (as shown in Fig. 106.1b) in 91/90. Moreover, the soot decoupling factor (as shown in Fig. 106.1d) in 96/95 was at the strong decoupling

Table 106.1 Classification on the level of decoupling

DF value	[−1 to 0)	[0 to 0.2)	[0.2 to 0.4)	[0.4 to 0.6)	[0.6 to 0.8]	[0.8 to 1]
Level	None	Feeble	Weak	Normal	Obvious	Strong

Table 106.2 Research data of 7 selected individual indicators

	GDP (10^9 \$)	Energy (10^4 tons coals)	CO_2 (10^4 tons)	Waste water (10^4 tons)	SO_2 (10^4 tons)	Soot (10^4 tons)	Household waste (10^4 tons)
1990	3589.7323	98,703	2,460,744.017	3,537,991	1494	1324	6767
1991	3814.547191	103,783	2,584,538.27	3,362,054	1622	1314	7636
1992	4249.340733	109,170	2,695,982.067	3,587,835	1685	1414	8262
1993	4428.745787	115,993	2,878,694.009	3,555,888	1795	1416	8791
1994	5622.611279	122,737	3,058,241.33	3,652,546	1843	1414	9952
1995	7320.320119	131,176	3,320,285.15	3,728,508	1891	1478	10,671
1996	8608.440566	135,192	3,463,089.131	3,943,289	1991	758	10,825.4
1997	9581.594607	135,909	3,469,510.048	4,158,070	1924	1273	10,982
1998	10,252.7685	136,184	3,324,344.519	4,155,035	2091	1455	11,302
1999	10,894.47071	140,569	3,318,055.614	4,153,517.5	1857	1159	11,415.19
2000	12,052.60617	145,531	3,405,179.867	4,152,000	1995.1	1165.4	11,819
2001	13,322.39848	150,406	3,487,566.356	4,330,000	1948	1070	13,470.4
2002	14,619.13993	159431	3694242.143	4,395,000	1927	1013	13,650
2003	16,499.21408	183,792	4,525,177.009	4,593,000	2159	1049	14,857
2004	19,417.45602	213,456	5,288,166.032	4,824,000	2255	1100	15,509.3
2005	22,685.94289	235,997	5,790,016.984	5,250,000	2549.4	1183	15,577
2006	27,297.84032	258,676	6,414,463.08	5,370,000	2589	1089	14,875.5
2007	35,230.94315	280,508	6,791,804.714	5,570,000	2468.1	987	15,215
2008	45,584.31073	291,448	7,035,443.861	5,720,000	2321.3	902	15,438
2009	50,594.19738	306,647	7,692,210.895	5,891,000	2214.4018	848	15,734
2010	60,396.58508	324,939	8,256,969.232	6,173,000	2185.1494	829.1321	15,805

Table 106.3 Decoupling factor results of annual decoupling analysis

	91/90	92/91	93/92	94/93	95/94	96/95	97/96	98/97	99/98	00/99
Energy	0.096	0.083	−0.01	0.184	0.184	0.135	0.111	0.068	0.049	0.054
Soot	0.066	0.034	0.039	0.213	0.197	0.564	−0.51	−0.07	0.250	0.091
SO_2	−0.02	0.067	−0.02	0.191	0.212	0.105	0.132	−0.02	0.164	0.029
CO_2	0.012	0.064	−0.03	0.163	0.166	0.113	0.100	0.105	0.061	0.072
Waste water	0.106	0.042	0.049	0.191	0.216	0.101	0.053	0.066	0.059	0.096
Household waste	−0.06	0.029	−0.02	0.108	0.176	0.137	0.089	0.038	0.049	0.064
	01/00	02/01	03/02	04/03	05/04	06/05	07/06	08/07	09/08	10/09
Energy	0.082	0.044	−0.01	0.030	0.079	0.098	0.187	0.202	0.034	0.079
Soot	0.169	0.137	0.082	0.109	0.079	0.235	0.298	0.294	0.153	0.181
SO_2	0.117	0.099	0.007	0.113	0.032	0.156	0.261	0.273	0.141	0.173
CO_2	0.073	0.035	−0.09	0.007	0.063	0.079	0.180	0.199	0.015	0.101
Waste water	0.057	0.075	0.074	0.108	0.068	0.150	0.196	0.206	0.072	0.122
Household waste	−0.03	0.077	0.036	0.113	0.140	0.206	0.207	0.216	0.082	0.159

level, indicating the soot pollution was improved a lot during that year, while this situation only occurred one year.

106.3.2 Decoupling Factor Analysis for a Period of Ten Years

As shown in Fig. 106.2, from the year 1990 to 2000, all the indicators occur decoupling, however the decoupling level is different, the indicator household waste and CO_2 are at only the normal decoupling level, while the waste water, SO_2, soot, energy are all at the obvious decoupling level. Besides, the CO_2 and soot are the worst and best indicators over the period 1990–2000 at the decoupling factor 0.588 and 0.738 respectively. From the year 2000–2010, all the indicators also occur decoupling, however the some indicators decoupling level is different when compared with 1990–2000. For example, the level of indicator energy decreases from obvious level to normal level, indicating the energy consumption is higher than the last ten years when compared with GDP increment. Besides, CO_2 emission is also aggravated in these ten years, when the decoupling factor decreases from 0.588 to 0.516. While, the decoupling factors of other four indicators, namely, household waste, waste water, soot and SO_2 increase. Among these four indicators, household waste increases from normal level to obvious level, and soot increased from obvious level to strong level, which indicating soot and household waste emission in China are improved in these ten years when compared with GDP increment. Moreover, decoupling factor of waste water and SO_2 increases from 0.65 and 0.602 to 0.703 and 0.78 respectively.

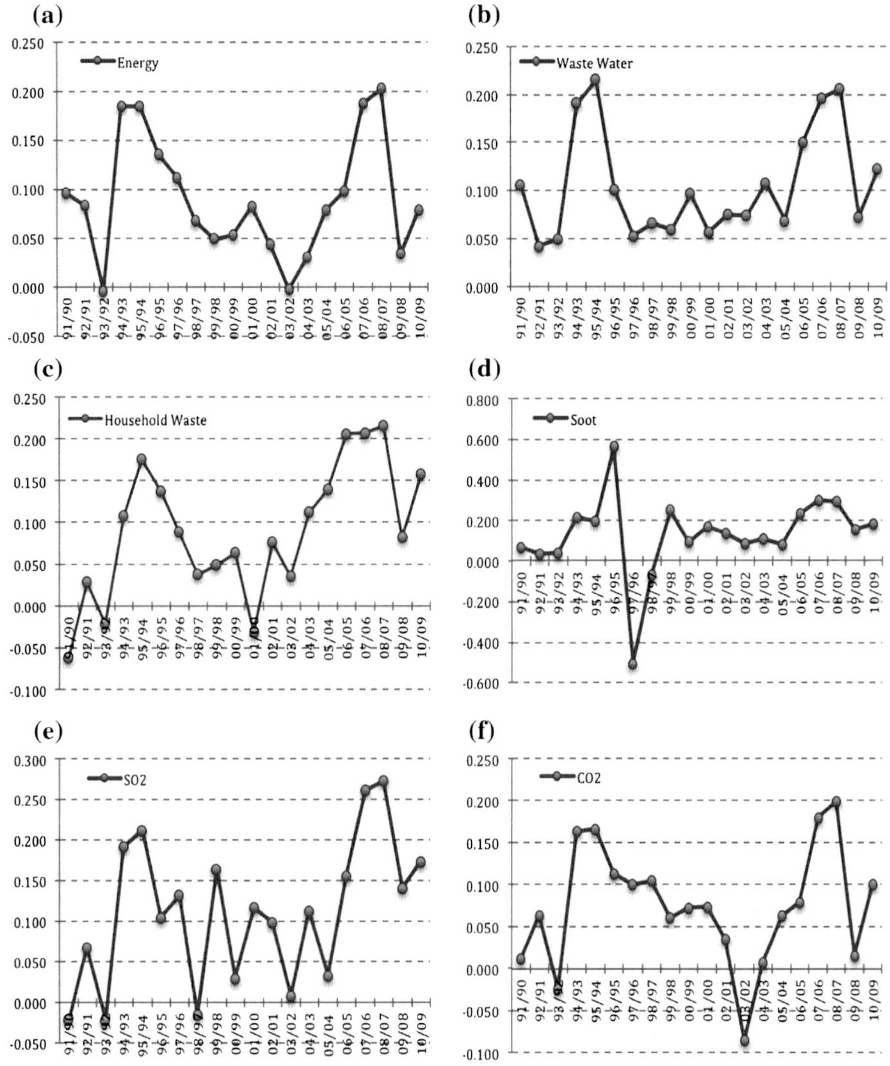

Fig. 106.1 The trends of decoupling factor based on annual analysis

From 1990 to 2010, the decoupling factor of all the indicators increases when compared with other two ten years, meanwhile, almost all the decoupling level are improved at strong level except CO_2, which is still at the obvious level. It is worth mentioning that one example of an environment pressure for which absolute decoupling from economic growth in China occurred is soot, as the volume of soot emissions declines in the period 1990–2010. Besides, the volume of SO_2 emissions is also declined in the period 2006–2010 although absolute decoupling did not occur in the period 1990–2010.

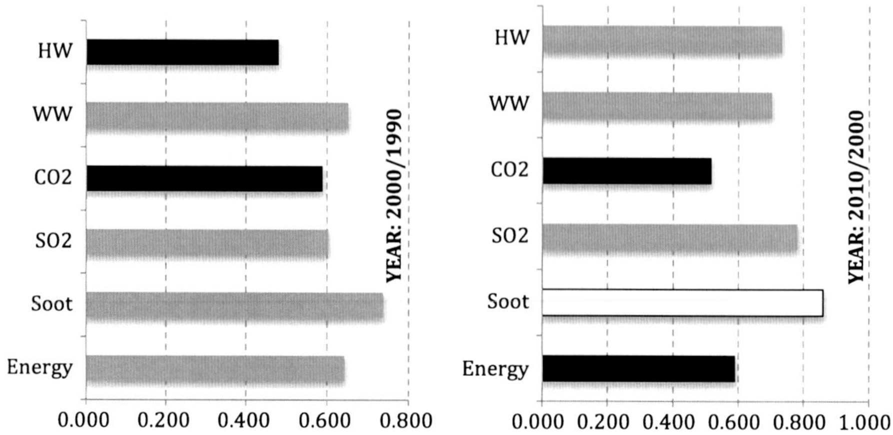

Fig. 106.2 The decoupling factor between ten years

106.3.3 Decoupling Factor Analysis for the Whole Survey Period

As shown in Fig. 106.3, in general, the environment pressure from economic growth has been improved during the period 1990–2010, which is thanks to the fast increase of investment in the treatment of environment pollution from 4.54 billion in 1990 to 761.22 billion in 2010 [15]. Pipe-end treatment technology is applied increasingly widely in industrial sectors, which is prominent to the air pollution emissions except CO_2 for it is not cost-effective and practical [27]. Meanwhile, the processing ratio of household waste and increases from 72.5 % in 2006 to 90.7 % in 2010, the treatment rate of domestic sewage increased from 34.3 % in 2000 to 82.3 % in 2010.

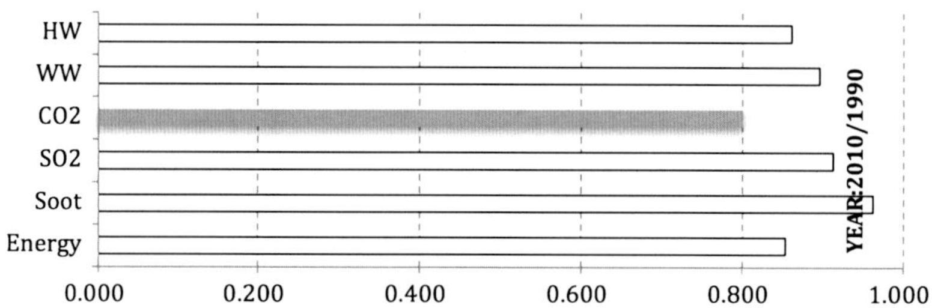

Fig. 106.3 The decoupling factor between the whole surveyed period

106.4 Conclusion

The conclusion from the study can be obtained that the environment pressure from economic growth in China has been improved during the surveyed 21 years, especially for the environmental indicators soot and SO_2. This improvement is mainly contributed by the fast increase of investment in the treatment of environment pollution. However, the decoupling analysis for CO_2, relatively speaking, CO_2 emission is still a tough problem in China.

The research findings demonstrate that the policy for mitigating SO_2, Soot is effective, but that for other pollutants especially for CO_2, the existing policies appear not very effective. It presents the importance of engaging the application of advanced technologies and management strategies in improving these environmental problems. The Chinese government in fact has been implementing the mission of low carbon city. This is considered important strategies. The findings from this paper provide important references for further studies on the methods of building low carbon cities or towns.

Acknowledgements This research is funded by the Key Project of the National Natural Science Fund of China (Project No.: 15AZD025).

References

1. Aşıcı AA (2013) Economic growth and its impact on environment: a panel data analysis. Ecol Ind 24:324–333
2. Baek J, Kim HS (2013) Is economic growth good or bad for the environment? Empirical evidence from Korea. Energy Econ 36:744–749
3. Brown-Santirso M, Thornly A (2006) Decoupling economic growth and energy use in New Zealand. Available from www.stats.govt.nzS
4. Chen PC, Crawford-Brown D, Chang CH, Ma HW (2014) Identifying the drivers of environmental risk through a model integrating substance flow and input–output analysis. Ecol Econ 107:94–103
5. De Freitas LC, Kaneko S (2011) Decomposing the decoupling of CO_2 emissions and economic growth in Brazil. Ecol Econ 70(8):1459–1469
6. Fodha M, Zaghdoud O (2010) Economic growth and pollutant emissions in Tunisia: an empirical analysis of the environmental Kuznets curve. Energy Policy 38(2):1150–1156
7. Hacatoglu K, Dincer I, Rosen MA (2014) Exergy analysis of a hybrid solar–wind–biomass system with thermal and electrical energy storage for a community. In: Progress in exergy, energy, and the environment. Springer International Publishing, Switzerland pp 3–14
8. Kubiszewski I, Costanza R, Franco C, Lawn P, Talberth J, Jackson T, Aylmer C (2013) Beyond GDP: measuring and achieving global genuine progress. Ecol Econ 93:57–68
9. Liu J, Diamond J (2008) Revolutionizing China's environmental protection. Science (New York then Washington) 319(5859):37
10. Li XK, Wei J (2010) Decoupling between economic growth and resource-environmental Pressures: a case study of the Chongqing metropolitan area. J Chongqing Normal Univ (Nat Sci) 27(1):28–35 (in Chinese with English abstract)
11. Lu CX, Yu G, Xiao Y (2013) Wind tunnel simulation and evaluation of soil conservation function of alpine grassland in Qinghai-Tibet Plateau. Ecol Econ 86:16–20

12. McGuire W (2014) The effect of ISO 14001 on environmental regulatory compliance in China. Ecol Econ 105:254–264
13. Menegaki AN, Tsagarakis KP (2015) Rich enough to go renewable, but too early to leave fossil energy? Renew Sustain Energy Rev 41:1465–1477
14. National Bureau of Statistics of China (2014) China statistical yearbook 2014. China Statistical Press, Beijing
15. National Bureau of Statistics of China (2011) China statistical yearbook 2011. China Statistical Press, Beijing
16. Netherlands Environmental Assessment Agency (2007) China now no. 1 in CO_2 emissions; USA in second position. Bilthoven, The Netherlands
17. OECD (2002) Indicators to measure decoupling of environmental pressure from economic growth. Sustainable development. SG/SD (2002) 1/Final. http://www.oecd.org/officialdocuments/displaydocumentpdf/?cote=sg/sd%282002%291/final&doclanguage=en
18. Peng J, Huang X, Zhong T, Zhao Y (2011) Decoupling analysis of economic growth and energy carbon emissions in China. Resour Sci 4:626–633
19. Sjöström M, Östblom G (2010) Decoupling waste generation from economic growth—a CGE analysis of the Swedish case. Ecol Econ 69(7):1545–1552
20. Swami V, Chamorro-Premuzic T, Snelgar R, Furnham A (2011) Personality, individual differences, and demographic antecedents of self-reported household waste management behaviours. J Environ Psychol 31(1):21–26
21. Tapio P (2005) Towards a theory of decoupling: degrees of decoupling in the EU and the case of road traffic in Finland between 1970 and 2001. Transp Policy 12(2):137–151
22. Van Caneghem J, Block C, Van Hooste H, Vandecasteele C (2010) Eco-efficiency trends of the Flemish industry: decoupling of environmental impact from economic growth. J Clean Prod 18(14):1349–1357
23. Wang C (2010) Decoupling analysis of china economic growth and energy consumption. China Popul Res Environ 20(3):35–37 (in Chinese with English abstract)
24. Wang H (2011) Decoupling measure between economic growth and energy consumption of China. Energy Proc 5:2363–2367
25. Wang W, Liu R, Zhang M, Li H (2013) Decomposing the decoupling of energy-related CO_2 emissions and economic growth in Jiangsu Province. Energy Sustain Dev 17(1):62–71
26. Xue XG, Li L (2014). Research on coordination degree of economic development and human settlements environment based on PCA for Guizhou. In: Advanced materials research, vol 962, pp 2055–2060
27. Yu Y, Chen D, Zhu B, Hu S (2013) Eco-efficiency trends in China, 1978–2010: decoupling environmental pressure from economic growth. Ecol Ind 24:177–184
28. Zhang C, Anadon LD (2014) A multi-regional input–output analysis of domestic virtual water trade and provincial water footprint in China. Ecol Econ 100:159–172

Chapter 107
Green Procurement Management in Building Industry: An Alternative Environmental Strategy

Zhenyu Zhang, Jingwei Hu and Liyin Shen

Abstract As environmental issues are attracting ever-greater attention and debate from the government and general public, an increasing number of business companies in the building industry have started to embrace environmental initiatives. As playing a critical role in boundary-spanning function for connecting a firm with its supply chain actors, green procurement (GP) is considered as an alternative environmental strategy to integrate environmental aspects into the whole building supply chain and in turn improve environmental performance of the building industry. Nevertheless, GP management has not garnered considerable attention from both professionals and researchers in the building industry. In this context, this study, by conducting literature review, examined the concept and current understanding of GP in the context of green supply chain management. Furthermore, this research classified GP management into two clusters: project-level and corporate-level, and accordingly identified a variety of GP practices in both of GP management clusters. The benefits by engaging GP practices in improving environmental performance of a building firm are also explicitly identified. The research findings fill the void of researches incorporating GP into environmental management in the building industry, and form an important basis for further development of GP. This study provides an alternative perspective for building industry professionals in adopting environmental initiatives, and the identified GP practices can serve as theoretical guidelines for developers and contractors to design, implement and evaluate their GP strategies.

Keywords Green procurement (GP) · Green building materials (GBM) · Building industry · Green supply chain management

---

Z. Zhang (✉) · J. Hu · L. Shen
School of Construction Management and Real Estate, International Research Centre for Sustainable Built Environment, Chongqing University, Chongqing, China
e-mail: zhangzhenyu@cqu.edu.cn

J. Hu
e-mail: 470332301@qq.com

L. Shen
e-mail: shenliyin@cqu.edu.cn

Y. Wu et al. (eds.), *Proceedings of the 20th International Symposium on Advancement of Construction Management and Real Estate*,
DOI 10.1007/978-981-10-0855-9_107

107.1 Introduction

Given detrimental influence the activities of building industry have wrought on natural environment, the importance of improving building products' environmental performance as a business issue has steadily grown over the past few years [1, 2]. Various environmental practices contributing to integrate environmental concerns into business issues have emerged in the building sector. These practices include green design of building products [3–5], green construction management [6, 7], green marketing of building products [8, 9], operation management of green building [10, 11], and construction and demolition waste management [12, 13]. It is found that a great contribution has been made in reducing environmental adverse effect caused by activities of building industry by promoting these environmental practices [2, 14].

With intense desire for greening the building industry, previous studies contended that greater improvement can be achieved by applying theories from other disciplines [15–17]. Among these disciplines, green supply chain management and GP management have received attention from the academic of the building industry in recent years [18, 36]. There is a consensus in these fields of research that, as the environmental issues are becoming increasingly systemic and complex, a company should work collaboratively with its entities within supply chain in order to achieve the objective of environmental performance improvement [1, 19].

However, a large number of extant environmental practices in the building industry are inter-organizational measures, which are designed and implemented from a perspective of an individual firm rather than that of the supply chain [16]. In other words, the objective of these environmental practices is to incorporate environmental concerns into individual firm instead of the entire supply chain [18]. In the light of theory of green supply chain management, merely adopting environmental practices in isolation is considered not sufficient for a building firm to green the whole building supply chain [15, 20]. Therefore, opening communication and developing collaboration between supply chain actors are considered necessary routes for a company in building sector to make further progress toward greenness of building products.

Green procurement, playing a critical role in boundary-spanning function for connecting a firm with its suppliers, provides an important opportunity for building industry to integrate environmental aspects into the whole building supply chain [21]. The GP management is defined as the set of purchasing policies held, actions taken, and relationships formed in response to concerns associated with the natural environment. These concerns relate to the acquisition of raw materials, including supplier selection, evaluation and development; suppliers' operations; in-bound distribution; packaging; recycling; reuse; resource reduction; and final disposal of the firm's products [22].

There is a recent evidence that firms' response to environmental pressures is resulting in a much more important position for purchasing function [23]. Consequently, a large number of major companies have ranked GP as a strategic priority in their business strategies [24]. The function of GP is no longer narrowed down to the operational practices of buying products and services, but is extended to the strategic practices, such as developing long-term partnerships with suppliers and integrating the whole supply chain towards achieving environmental improvement [22]. In this respect, suppliers are deemed as the environmental value generators who provide a buying company with valuable resources and advanced technologies and continually help them improve environmental performance [25, 26]. Effective implementation of GP function enables a company to work collectively with suppliers and in turn gets a more thorough understanding of environmental benefits and damages of the products and services it procures from its suppliers [25]. This insightful understanding is conducive to effective efforts for improving environmental performance of products within the entire supply chain.

The above research developments enlighten that the adoption of GP in the building industry can make a great contribution to reduction of environmental impact from building products. Nevertheless, as opined by Adetunji et al. [15] that GP management has not garnered adequate attention from professionals in the building sector. Developers and contractors typically procured services and products based on the lowest price bid and rarely considered environmental performance of procured products [27]. The relationships between a building development company or a contractor and its building materials suppliers are short-term; as a result, suppliers are not involved in the design, construction and operation of building products [16, 28]. In this respect, traditional procurement management is subject to a lot of criticism such as generating regular conflicts between the various project stakeholders, leading to divorce between design and construction, and providing little or no guarantee to improvement of environmental performance [27–29]. Extant procurement practices adopted by developers and contractors have considerably impeded the improvement of environmental performance of building industry, and it is therefore considered important for practitioners and researchers to have an insightful understanding of GP and in turn promote development of GP within building industry.

The published academic research associated with GP in the building industry is less developed and only covers a fraction of the GP experience and practices, which make the understanding and further implementation of GP practice a challenging task [16]. For example, Ofori [30] identified a number of GP strategies that can be adopted in the building sector by reviewing the current practices of GP in other sectors. However, these identified practices mainly focused on developing long-term collaboration with building materials suppliers, whilst overlooked GP practices involved in the life cycle of a building project. In contrast, the researches conducted by Uttam et al. [28] and Vrijhoef and Koskela [17] predominantly examined the GP practices adopted within the short-period of process of construction project management. Moreover, Sanchez et al. [31] conducted a case study in the Australia construction industry to examine the current practices of GP associated with greenhouse gas emission rather than overall environmental performance of building products. There is, therefore, a real need to supplement all the existing research works on GP practices in the building sector.

In this respect, the purposes of this paper are to systematically review current literatures about GP in building and other industry in the context of green supply chain management; identify the scope of GP practices and various GP practices that can be applied in building industry; and examine the benefits of adopting these GP practices in the building industry.

107.2 Green Procurement Management Practices in Building Industry

From a synthesis of prior researches, green procurement management practices in the building industry can be divided into two clusters: project-level and corporation-level GP management practices.

107.2.1 Identification of Project-Level Green Procurement Management Practices

The project-level GP management practice, stemming from a project-wide perspective, refers to the process and activity of applying GP initiatives into the process of implementing a specific building project in order to improve environmental performance of the project. Developers and contractors, as the powerful actors in the supply chain of building process, are supposed to strategically influence other supply chain actors' attitudes and actions in a way that these actors will be environmentally responsible [22]. Such influence should be extended over life cycle of a building project. In this light, various strategies of project-level GP management are identified from the perspective of life cycle of a building project, which can be divided into five stages, namely, "Concept Definition", "Design", "Construction", "In use" and "End of life" [20].

Concept Definition: where the initial idea, scope and brief for a new building project is established and agreed. By taking into account of environmental missions and strategies of the company, as well as capital budget available for the project, specific guidelines for GP strategy implemented in this project will be defined [31]. Relevant public mandatory and incentive policies, which place requirements for environmental performance of building materials used in the project, will also be considered in the process of forming GP guidelines [30, 36]. Furthermore, an increase in communication with relevant NGOs will also help establish GP guidelines [1, 24].

Design: where the initial idea for the project is translated into a detailed design embedded into environmental function. A requisition with detail specifications on GBM products in terms of types, qualities and environmental performance will be raised. The specifications of GBM should focus on increasing the efficiency of materials use, reducing the use of hazardous materials, and maximizing construction waste utilization [19, 22]. A bill of quantities of GBM products for specifying the quantities of each type of GBM will also be compiled. Evaluation standards for green building materials need to be established in this stage, which will be used in tendering process subsequently. Environmental impact assessment, life cycle assessment, Green BIM and green rating systems for green building can be used as tools for establishing evaluation standards and guiding procurement of GBM [20, 28]. GP practices in this stage also involve designing strategies to facilitate information transferring and coordination in terms of environmental issues within building supply chain in the life-cycle of a project, such as the involvement of GBM suppliers into design of a building product [16].

Construction: where detailed design of the project is translated into a building product. Typically, the building materials will be procured prior to construction activities commencing. Based on the requisition and bill of quantities, developers will invite for quotations of GBM products. The tendering process often includes GBM supplier prequalification [31], green assessment of the tenders submitted by GBM suppliers [28], and arrangement of contractual agreements with selected suppliers [15]. GP practices in this stage also involve the management of delivery process of GBM on construction site [22]. After the completion of construction, the construction waste is supposed to be disposed, and environmental approaches for disposal often include reuse and recycling [20]. Purchasing professionals are best qualified to account for this task as they are responsible for buying the products and know GBM products better [22].

In Use: covering the operation, maintenance, and renovation of the building project. One of the primary GP practices in this stage is monitoring real environmental effects of GBM and adopting environmental management toward achieving the desired effect of GP [28, 32]. The instrument such as life cycle assessment, green auditing and key performance indicators can be applied to evaluate and monitor environmental performance. Furthermore, post-project feedback is considered as an important project-level GP practice, because it can provide essential experiences for later projects in adopting GP strategies [31]. In this respect, best practices, and case studies are effective instruments for adopting post-project feedback [30].

End of life: where a building is demolished and the demolition waste of building materials is produced. Minimizing demolition waste and maximizing reuse and recycle are considered as vital GP practices in this phase [20]. Achieving a "reserve distribution" in construction supply chain is considered as a crucial and systematic initiative that can effectively eliminate waste at source, reuse and recycle waste [22]. Reserve distribution refers to planning, implementing and controlling materials flow opposite to the traditional supply chain direction for the purpose of recovering value or proper disposal [33].

Based on the above literature analysis, the typical project-level GP practices can be summarized as in Table 107.1.

Table 107.1 The project-level green procurement practices

Life-cycle of project	Green procurement practice	Key reference
Concept Definition	Setting a clear objective of green procurement management to build consensus for the development of green buildings within the project team	Ofori [30]
	Establishing a clear understanding of governmental policies associated with building materials	Appolloni et al. [21]
	Facilitating the communication with Non-business actors, such as green building materials certification association	Crespin-Mazet and Dontenwill [24], Kiron et al. [1]
	Incorporating environmental business policies into project management	United Nations Environmental Programme [20]
	Forming specific guidelines for green procurement strategies	Appolloni et al. [21]
Design	Compiling a requisition with detail environmental specifications and a bill of quantities of GBM products	Benton and McHenry [40]
	Establishing evaluation standards of GBM products	United Nations Environmental Programme [20]
	Integrating building materials suppliers into design process	Dadhich et al. [32]
	Establishing information channel for transmission of environmental information within supply chain	Plambeck et al. [19]
Construction	Implementing suppliers prequalification	Sanchez et al. [31]
	Assessing life-cycle environmental performance of green building materials products	Uttam et al. [28]
	Implementing contract management with GBM suppliers	Adetunji et al. [15]
	Implementing on-site delivery management	Zsidisin and Siferd [22]
	Disposing construction waste	Zsidisin and Siferd [22]
In use	Monitoring real effects of building material products	Sanchez et al. [31]
	Compiling project environment report	Adetunji et al. [15]
	Increasing coordination with downstream members, such as buyers and tenants	Zsidisin and Siferd [22]
	Auditing suppliers to evaluate their environmental performance in life cycle of a project	Kiron et al. [1]
	Implementing post-project feedback and accumulating best practice case studies to ensure continuous improvement	Ofori [30], Sanchez et al. [31]
End of life	Achieving "reverse distribution" and thus enable reuse and recycle of GBM to occur	Zsidisin and Siferd [22]

107.2.2 Identification of Corporation-Level Green Procurement Management Practices

As discussed above, project-level GP management focuses on improving environmental issues of a project. However, there are GP management practices at corporation level. The corporation-level GP management represents activities of making long-term partnering relationships with GBM suppliers for achieving better holistic environmental performance of corporations.

There are various studies on the examination of corporation-level GP management practices. For example, Ofori [30], by reviewing the current practices of GP in the public and private sectors, identified a number of corporation-level GP practices that can be adopted in the building sector. These GP practices include encouraging suppliers to certify their products, work with suppliers to help them reduce environmental impact, inform suppliers of technological developments relating to their operations, and training suppliers on environmental issues. Adetunji et al. [15] found that there is a high level of understanding of GP management among practitioners in the UK construction industry; advanced corporation-level GP strategies, such as requiring contractors to work toward ISO 14001 certification for their depots, making continuous improvement agreement, applying in-housing training and communication tools, and developing strategic alliance and partnering with suppliers, are widely adopted in the UK building industry. The studies conducted by Touboulic and Walker [34] and Bygballe et al. [16] put a high value in GP practices to foster common organizational culture, vision and strategy with suppliers. Based on the literature analysis above, a number of corporate-level GP practices are identified and listed in Table 107.2.

Table 107.2 The corporation-level green procurement practices

Corporation-level green procurement practice	Key Reference
Encouraging green building material suppliers to implement and maintain environmental management systems	Adetunji et al. [15]
Encouraging green building material suppliers to obtain environmental certification of their products	Ofori [30]
Leveraging resources and funding to green building material suppliers to improve their environmental performance	Touboulic and Walker [34]
Encouraging suppliers to establish partnership with each other and thereby forming alliance within building supply chain	Adetunji et al. [15], Bygballe et al. [16]
Fostering common organizational culture, vision and strategy with suppliers	Bygballe et al. [16]
Instituting training programmes for suppliers to increase their knowledge of environmental issues	Ofori [30]
Advising suppliers to share environmental data and plans	United Nations Environmental Programme [20]
Informing suppliers of technological developments relating to their operation	Ofori [30]

107.3 Discussions on the Benefits of Green Procurement

107.3.1 Benefits of Project-Level Green Procurement Management Practices

There are a number of literatures concerned with the benefits of adopting project-level GP management. For example, by conducting empirical studies in Netherlands and Finland construction industry, Vrijhoef and Koskela [17] found that some environmental problems associated with building materials were encountered at interferes between various stages of construction supply chain, such as the interfere between delivery of building materials and site installation. And other environmental problems are found stem from earlier activities performed by prior actors within the supply chain, rather than where the problems surfaced. It is therefore considered that the efficiency of GP practices in the construction project cannot be achieved by business organizations' actions in isolation, and a company needs to increase coordination with its suppliers in the life-cycle of a project [35]. In this respect, setting a clear objective of GP practices and establishing GP guidelines in the beginning of a building project is conducive to building consensus and facility cooperation with supply chain actors.

Dadhich et al. [32] suggested that integrating building suppliers into the design process, and establishing information channel for transmission of environmental issues within supply chain through, are conducive to coming up with innovative ideas of reducing adverse environmental impact of a building product.

Furthermore, establishing cooperation with NGOs has been attached increasingly importance by researchers in GP field [1, 24] due to a large number of benefits NGOs can provide. Expected benefits include developing networks of relationships with relevant stakeholders, accessing relevant expertise, and acquiring the ability to influence public opinion.

The technique of Environmental Impact Assessment and Life-cycle assessment are advised to be incorporated into design, tender, and monitor process for reducing detrimental environmental impacts caused by the purchase and application of building materials [20, 28, 36]. It is considered that the adoption of these tools can not only provide valuable and accurate information about environmental performance of GBM products in the design stage of a construction project, but also help monitor the real environmental effects of GBM and in turn adopt environmental management in the construction and operation stage.

Zsidisin and Siferd [22] highlighted the benefits of establishing reverse distribution, a critical project-level GP practices adopted in the process of construction and demolition waste management, to facilitate the activities of resource reduction, reuse, and recycle. The establishment of reverse distribution will close the loop in the building supply chain, and in turn improve efficiency and effective of applying building materials products.

107.3.2 *Benefits of Corporation-Level Green Procurement Management Practices*

In comparison with the researches on project-level GP management practices, the benefits of adopting corporation-level GP practices are less developed in the building industry. Nevertheless, extant studies dealing with long-term partnering relationships within building supply chain have shed light on the benefits of adopting corporation-level GP practices in achieving better corporate environmental performance. For example, a building firm can save the cost of assessing and auditing suppliers' environmental performance by encouraging its suppliers to obtain third-party environmental certifications for their products, and to implement environmental management system [15, 37]. The cooperation with green suppliers, who proactively engage in environmental activities, can also increasing a buyer firm's public image.

Love et al. [38] argued for the importance of maintaining trust, knowledge and environmental objectives among supply chain actors. Thus developers and contractors are advisable to institute training on environmental issues and cultivate common organizational culture and vision within supply chain entities. In addition, a shared environmental culture and objectives is beneficial to developing long-term cooperation with suppliers, and in turn help a buyer firm fully make use of the advantages of partners' special competence in addressing environmental issues [39].

Plambeck et al. [19] stated that leveraging resources and funding to green building material suppliers can help a building firm to derive environmental performance information more systematically and comprehensibly and in turn increasing opportunity for environmental innovation.

107.4 Conclusion

Green procurement management has a great potential for integrating environmental aspects into and thereby greening the whole building supply chain. It is therefore considered essential to promote GP management in the building industry. The research identified a variety of GP practices that can be applied in the building industry. These practices are divided into two clusters: project-level and corporate-level GP management. The former contains GP practices applied in the life cycle of a building project to eliminate its adverse effect on the natural environment, including the specific practices of incorporation of GBM suppliers into green design of a building product, green assessment of building materials, as well as purchase of building material with environmentally friendly attributes. On the other hand, corporation-level GP practices focus on long-term environmental strategies for developing partnership with suppliers and achieving holistic improvement of environmental performance, including forming strategic alliance with GBM suppliers, developing GBM product collaboratively with suppliers, and training suppliers on environmental issues.

The research findings fill the void of the lack of research on GP management in the building industry and form an important basis for further development of GP management in the building industry. Given adverse effect on environmental performance yielded by traditional procurement management and insufficient implementation of GP management in the building industry, this study can provide an alternative perspective for professionals in adopting environmental initiatives, and enhancing the understanding that there are effective methods for improving our building industry's environmental performance at both project and corporation levels. In particular, the collaboration between different actions on the supply chain can improve to large extent the project-level environmental performance. GP practices identified in this study can serve as theoretical guidelines for developers and contractors to design, implement and evaluate GP strategies.

Future research will be conducted to examine the level of effectiveness of these identified GP practices to be adopted by developers and contractors in the building industry. That way can further promote the benefits by engaging GP practices. The research can also be extended to identifying problems encountered by developers and contractors in implementing GP practices, which will provide valuable information for relevant government departments, business organizations, as well as relevant non-governmental organizations, to take measures for promoting GP practice, and ultimately promote development of green building.

References

1. Kiron D, Kruschwitz N, Haanaes K, Reeves M, Fuisz-Kehrbach SK, Kell G (2015) Joining forces: collaboration and leadership for sustainability. MIT Sloan Manage Rev. Retrieved 1 June 2015
2. Zuo J, Zhao Z-Y (2014) Green building research–current status and future agenda: a review. Renew Sustain Energy Rev 30:271–281
3. Castro-Lacouture D, Sefair JA, Flórez L, Medaglia AL (2009) Optimization model for the selection of materials using a LEED-based green building rating system in Colombia. Build Environ 44:1162–1170
4. Kibert CJ (2012) Sustainable construction: green building design and delivery. Wiley, New York
5. Krygiel E, Nies B (2008) Green BIM: successful sustainable design with building information modeling. Wiley, New York
6. Ahn YH, Pearce AR (2007) Green construction: contractor experiences, expectations, and perceptions. J Green Build 2:106–122
7. Hwang B-G, Ng WJ (2013) Project management knowledge and skills for green construction: Overcoming challenges. Int J Project Manage 31:272–284
8. Eichholtz P, Kok N, Quigley JM (2013) The economics of green building. Rev Econ Stat 95:50–63
9. Kats G, Capital E (2003) Green building costs and financial benefits. Massachusetts Technology Collaborative, Boston, MA

10. Ye L, Cheng Z, Wang Q, Lin W, Ren F (2013) Overview on green building label in China. Renewable Energy 53:220–229
11. Zhang X, Platten A, Shen L (2011) Green property development practice in China: costs and barriers. Build Environ 46:2153–2160
12. Button A, Lee W, Marshall D, Dawood Z, MacLellan S, Umali H, Pagsuyoin S (2014) Management of construction and demolition waste in the Region of Waterloo, systems and information engineering design symposium (SIEDS), 2014. IEEE, pp 53–56
13. Yuan H, Shen L (2011) Trend of the research on construction and demolition waste management. Waste Manag 31:670–679
14. Shi Q, Lai X, Xie X, Zuo J (2014) Assessment of green building policies—a fuzzy impact matrix approach. Renew Sustain Energy Rev 36:203–211
15. Adetunji IO, Price ADF, Fleming PR (2008) Achieving sustainability in the construction supply chain
16. Bygballe LE, Jahre M, Swärd A (2010) Partnering relationships in construction: a literature review. J Purchasing Supply Manage 16:239–253
17. Vrijhoef R, Koskela L (2000) The four roles of supply chain management in construction. Eur J Purchasing Supply Manage 6:169–178
18. Ellegaard C, Bygballe LE, Dubois A, Bankvall L (2010) Purchasing and supply management in the construction industry. J Purchasing Supply Manage 16:219–220
19. Plambeck E, Lee HL, Yatsko P (2012) Improving environmental performance in your Chinese supply chain. MIT Sloan Manage Rev 53(2):43
20. United Nations Environmental Programme (2014) Greening the building supply chain
21. Appolloni A et al (2014) Green procurement in the private sector: a state of the art review between 1996 and 2013. J Cleaner Prod 85:122–133
22. Zsidisin GA, Siferd SP (2001) Environmental purchasing: a framework for theory development. Eur J Purchasing Supply Manage 7:61–73
23. Hoejmose SU, Adrien-Kirby AJ (2012) Socially and environmentally responsible procurement: A literature review and future research agenda of a managerial issue in the 21st century. J Purchasing Supply Manage 18:232–242
24. Crespin-Mazet F, Dontenwill E (2012) Sustainable procurement: building legitimacy in the supply network. J Purchasing Supply Manage 18:207–217
25. Green K, Morton B, New S (1998) Green purchasing and supply policies: do they improve companies' environmental performance? Supply Chain Manage: Int J 3:89–95
26. Wilding R, Wagner B, Miemczyk J, Johnsen TE, Macquet M (2012) Sustainable purchasing and supply management: a structured literature review of definitions and measures at the dyad, chain and network levels. Supply Chain Manage: Int J 17:478–496
27. Crespin-Mazet F, Portier P (2010) The reluctance of construction purchasers towards project partnering. J Purchasing Supply Manage 16:230–238
28. Uttam K, Faith-Ell C, Balfors B (2012) EIA and green procurement: opportunities for strengthening their coordination. Environ Impact Assess Rev 33:73–79
29. Dubois A, Gadde L-E (2000) Supply strategy and network effects—purchasing behaviour in the construction industry. Eur J Purchasing Supply Manage 6:207–215
30. Ofori G (2000) Greening the construction supply chain in Singapore. Eur J Purchasing Supply Manage 6:195–206
31. Sanchez AX, Lehtiranta L, Hampson KD, Kenley R (2014) Evaluation framework for green procurement in road construction. Smart Sustainable Built Environ 3:153–169
32. Dadhich P, Genovese A, Kumar N, Acquaye A (2015) Developing sustainable supply chains in the UK construction industry: a case study. Int J Prod Econ 164:271–284
33. Fleishmann M (2000) Quantitative models for reverse logistics. Ph.D. thesis, Erasmus University Rotterdam, The Netherlands
34. Touboulic A, Walker H (2015) Love me, love me not: a nuanced view on collaboration in sustainable supply chains. J Purchasing Supply Manage 21(3):178–191
35. Edum-Fotwe FT, Thorpe A, McCaffer R (2001) Information procurement practices of key actors in construction supply chains. Eur J Purchasing Supply Manage 7:155–164

36. Shen L, Zhang Z, Zhang X (2016) Key factors affecting green procurement in real estate development: a China study. J Cleaner Production
37. Construction Industry Institute (1991) In search of partnering excellence. Bureau of Engineering Research, Construction Industry Institute, University of Texas Austin, TX
38. Love PED, Tse RYC, Holt GD, Proverbs DG (2002) Transaction costs, learning, and alliances. J Constr Res 3:193–207
39. Beach R, Webster M, Campbell KM (2005) An evaluation of partnership development in the construction industry. Int J Project Manage 23:611–621
40. Benton WC, McHenry LF (2010) Construction purchasing and Supply chain management. McGraw-Hill, New York

Chapter 108
Relationship Between the Energy Consumption for Urban Residential Buildings and Residents' Living Standards—A Case Study of the Four Municipalities in China

Ya Wu, Jindao Chen, Xiangnan Song and Liyin Shen

Abstract The living standards particularly in urban areas in China have been improving over last few decades and will continue to grow. This development will affect the amount of energy consumption on urban residential buildings. This paper presents the findings about the effects of improving living standards on the energy consumption for urban residential buildings. Data used for the analysis in this study are collected from official statistics. Research results show that the energy consumption of urban residential buildings are associated closely with the living standards, which are characterized with the parameters such as per capita disposable income and per capita consumer-spending. This demonstrates that analyzing and controlling these two parameters from the perspective of policy and technology has positive effect on reducing energy consumption for residential buildings.

Keywords Case studies · Urban residential buildings · Energy consumption · Living standards · Correlation · Regression analysis

108.1 Introduction

In its Twelfth Five-Year-Plan for energy conservation and emissions reduction, the Chinese government defined "Strengthening the new building energy conservation and intensifying the energy-saving renovation for existing building" as one of the

Y. Wu (✉) · J. Chen · X. Song · L. Shen
School of Construction Management & Real Estate, Chongqing University, Chongqing, China
e-mail: mswuya@126.com

Y. Wu · J. Chen · X. Song · L. Shen
International Research Centre for Sustainable Built Environment, Chongqing University, Chongqing, China

Y. Wu et al. (eds.), *Proceedings of the 20th International Symposium on Advancement of Construction Management and Real Estate*,
DOI 10.1007/978-981-10-0855-9_108

main strategies for the national development [1]. In line with this, the building energy conservation is the key task in China. According to statistics, the energy consumed by residential buildings in the world accounts for 10–35 % of the total energy consumption [2]. As the world's largest developing country, China's energy consumption on residential buildings accounts for more than 11 % of its total energy consumption [3], and this proportion will grow with the acceleration of urbanization process, the increase of per capita income, and the improvement of living standards [4]. Therefore, studying the relationships between the energy consumption on urban residential buildings and residents' living standards is very important to promote energy conservation for buildings effectively. Thc study results are used to help formulate effective energy saving measures.

The effects of residents' living standards on the energy consumption have been studied by many scholars in recent years. For example, Sun et al. [5], selected five factors, including residents' consumption level, building area, urbanization level, development of third industry, population, to study the degree that these factors influence energy consumption of urban buildings, and found that the biggest influential factor was the residents' consumption level. By using prediction model on the energy consumption of residential buildings, Zeng et al. [6] found that the terminal energy consumption on residential buildings is correlated with the added value of the third industry, the urbanization rate, and the life expenses per capita. Zou et al. [7] built the calculation model on the energy consumption on urban buildings based on IPAT theory and found that the amount of refrigerators, air conditioners per hundred urban households and other indicators of residents' living standards had certain influences on the energy consumption of buildings.

The above researches have shown that the residents' living standards have certain effects on the increase of energy consumption on urban residential buildings. These researches have studied the influences of many synthetic factors such as residents' living standards on energy consumption on urban residential buildings. However, those researches have not considered the residents' living standards as a separate variable to study its relationship with the energy consumption of urban residential buildings. In view of that, this study selects representative parameters for residents' living standards, and establish regression models to study the correlation between these parameters and the energy consumption on urban residential buildings. The data used are from the four municipal cities in Chinese, including Beijing, Tianjin, Shanghai and Chongqing, which are representative cities of China.

108.2 Definitions of Energy Consumption on Urban Residential Buildings and Residents' Living Standards

Energy consumption on residential buildings usually refers to the energy required to maintain the functions of residential building, mainly including air conditioning, electric appliances, lighting, cooking and hot water [8]. China has not yet established a specialized database for building energy consumption, building energy consumption is accounted in various kinds of energy statistics. Many researchers agree that the building energy consumption should be included in the tertiary industry and living energy consumption. Energy consumption on urban residential buildings is part of residents' living energy consumption. This study considers the urban residential energy consumption as the urban residents' consumption on five energy sources, including coal, kerosene, liquefied petroleum gas, natural gas, and electricity energy consumption.

Residents' living standards refer to the level of residents' daily consumption on food, clothing, shelter, transportation, medicine, teaching, entertainment and so on. Evaluation on residents' living standards has been an academic focus. Various evaluation parameters for residents' living standards have been proposed. For example, the quality of life index system, proposed by European Union, involves 13 parameters such as population, housing, leisure, entertainment and culture, family situation, transportation, income and consumption [9]. He et al. [10] proposed that residents' living standards could be evaluated by urban residents annual income, per capita disposable income, per capita consumption expenditure and the Engel's coefficient. Zhu [11] selected to assess the urban residents' living standards in China 13 parameters, such as the urban per capita disposable income, consumer spending, housing construction area.

Considering the data availability and effectiveness, this study chooses 3 variables to represent the residents' living standards, including per capita disposable income of urban households (Yuan), per capita consumption expenditure of urban households (Yuan), and per capita housing area (m^2).

Per capita disposable income of urban households (X1), per capita consumption expenditure of urban households (X2) are the main parameters to reflect residents' income and expenditure. Income and expenditure are the prerequisites to improve residents' living standards [11]. After meeting their basic needs, people will pursue other material and cultural satisfaction, which will bring the increase of residential building energy consumption. Furthermore, Per capita housing area (X3) is the parameter reflecting residents' living conditions. Living is an essential part for people's life, and it is an important indicator to measure living standards. At the same time, per capita housing area impacts directly on the amount of the residential building energy consumption.

108.3 Research Methods

The major research methods adopted in this study are correlation and regression analysis. The correlation between two variables can be analyzed by using SPSS software, which is one of the most commonly used mathematical statistics software, to conduct various types of data process. This study examines the correlation between the energy consumption of urban residential buildings (Y) and the living standards characterized parameters (Xi, i = 1, 2, 3) using SPSS correlation analysis.

The key parameters that have high relevance and significance to the energy consumption of urban residential buildings will be identified, and finally establish regression models to present the relation between the energy consumption of urban residential buildings and the living standards parameters.

108.4 Research Data

108.4.1 Energy Consumption for Urban Residential Buildings

The data of energy consumption for urban residential buildings employed in our analysis are based on the China Energy Statistical Yearbook. The data are about the consumption of coal, kerosene, liquefied petroleum gas, natural gas, and electricity in Beijing, Tianjin, Shanghai and Chongqing for the period of 2003–2012. These data need to be converted into standards coal according to the "General Principles for Calculation of Total Production Energy Consumption". For example, the calculation process of urban residential building energy consumption in Chongqing is recorded in Table 108.1. Accordingly, the data of energy consumption on urban residential building for the four municipalities are shown in Table 108.2.

108.4.2 Residents' Living Standards

In order to establish the relationship between people's living standards (Xi) and urban residential building energy consumption (Y), the data about three living standards parameters [per capita disposable income of urban households (X1), per capita consumption expenditure of urban households (X2), and per capita housing area (X3)] for Beijing, Tianjin, Shanghai and Chongqing are collected, shown in Table 108.3.

Table 108.1 The energy consumption on urban residential building in Chongqing (2003–2012)

Year	Coal (10^4 t)	Kerosene (10^4 t)	Liquefied petroleum gas (10^4 t)	Natural gas (10^9 m^3)	Electricity (10^9 kWh)	Standards coal (10^9 tce)
2003	1.75	0.00	0.00	6.50	31.82	126.81
2004	1.75	0.00	0.00	6.50	36.21	132.20
2005	1.58	0.00	0.00	8.00	41.50	158.53
2006	1.58	0.00	2.52	8.00	51.92	175.66
2007	1.52	1.90	2.58	8.00	51.21	177.64
2008	1.71	2.14	2.62	8.05	57.63	186.75
2009	2.36	2.14	2.87	8.10	62.83	194.70
2010	2.57	2.24	3.37	8.20	64.02	198.65
2011	2.12	2.46	3.94	10.12	80.47	245.38
2012	1.88	2.62	4.12	20.55	81.46	385.69

Table 108.2 Energy consumption (Standard Coal) on urban residential buildings in the four municipalities (2003–2012)

Year	Beijing	Tianjin	Shanghai	Chongqing
2003	367.56	151.33	339.17	207.03
2004	398.47	156.10	373.92	220.08
2005	427.55	183.25	446.19	258.58
2006	467.07	188.65	501.14	298.99
2007	503.86	195.32	542.68	295.09
2008	535.05	198.79	580.00	313.58
2009	534.12	229.89	599.01	328.19
2010	513.52	229.54	631.70	324.62
2011	512.12	222.63	652.61	401.68
2012	643.17	244.70	707.97	540.86

Table 108.3 Residents' living standards in the four municipalities

Cities	Year	Per capita disposable income (1000 yuan) (X_1)	Per capita consumption expenditure (1000 yuan) (X_2)	Per capita housing area (m^2) (X_3)
Beijing	2003	13.88	11.12	20.71
	2004	15.64	12.20	22.07
	2005	17.65	13.24	22.49
	2006	19.98	14.83	24.19
	2007	21.99	15.33	25.12
	2008	24.73	16.46	24.12
	2009	26.74	17.89	23.87
	2010	29.07	19.93	22.80
	2011	32.90	21.98	25.34
	2012	36.47	24.05	25.58
Tianjin	2003	10.31	7.87	16.93
	2004	11.47	8.80	17.72
	2005	12.64	9.65	13.46
	2006	14.28	10.55	18.27
	2007	16.36	12.03	18.57
	2008	19.42	13.42	18.49
	2009	21.43	14.80	18.67
	2010	24.29	16.56	18.29
	2011	26.92	18.42	18.45
	2012	29.63	20.02	18.51

(continued)

Table 108.3 (continued)

Cities	Year	Per capita disposable income (1000 yuan) (X_1)	Per capita consumption expenditure (1000 yuan) (X_2)	Per capita housing area (m^2) (X_3)
Shanghai	2003	14.87	11.04	22.30
	2004	16.88	12.63	23.63
	2005	18.65	13.77	23.79
	2006	20.67	14.76	24.25
	2007	23.62	17.26	22.81
	2008	26.68	19.40	25.19
	2009	28.84	20.99	25.64
	2010	31.84	23.20	25.60
	2011	36.23	25.10	26.40
	2012	40.19	26.25	26.46
Chongqing	2003	8.09	7.12	14.37
	2004	9.22	7.97	17.45
	2005	10.24	8.62	18.26
	2006	11.57	9.40	20.00
	2007	12.59	9.89	20.40
	2008	14.37	11.15	21.24
	2009	15.75	12.14	22.26
	2010	17.53	13.34	24.11
	2011	20.25	14.97	25.95
	2012	22.97	16.57	27.17

108.5 Analysis Results

108.5.1 Correlation Analysis Between the Energy Consumption on Urban Residential Buildings and Residents' Living Standards

Table 108.4 displays the results of correlation analysis between the urban residential building energy consumption (Y) and the residents' living standards (Xi) in four municipalities of China. We can see that the correlation coefficient between Y and X varies between the four municipalities. The urban residential building energy consumption in Beijing, Shanghai and Chongqing (YB, YS, YC) have significant positive correlation with all three living standard parameters [per capita disposable income of urban households (X1), per capita consumption expenditure of urban households (X2), and per capita housing area (X3)]. While the urban residential building energy consumption in Tianjin (YT) have significant positive correlation

Table 108.4 The results of correlation analysis between urban residential building energy consumption and residents' living standards in four Municipalities

Urban residential building energy consumption (Y)	X_1		X_2		X_3	
	Correlation coefficient	Significance level (*P*)	Correlation coefficient	Significance level (*P*)	Correlation coefficient	Significance level (*P*)
Y_B (Beijing)	0.911^{**}	0.000	0.890^{**}	0.001	0.824^{**}	0.003
Y_T (Tianjin)	0.941^{**}	0.000	0.941^{**}	0.000	0.432	0.212
Y_S (Shanghai)	0.964^{**}	0.000	0.972^{**}	0.000	0.874^{**}	0.001
Y_C (Chongqing)	0.944^{**}	0.000	0.937^{**}	0.000	0.911^{**}	0.000

*Note***means the two variables significantly correlate in double sides at 0.01 level, while *means they significantly correlate in double side at 0.05 level with per capita disposable income of urban households (X1), per capita consumption expenditure of urban households (X2), but has no significant positive correlation with per capita housing area (X3).

108.5.2 Regression Analysis Between Energy Consumption on Urban Residential Buildings and Residents' Living Standards

Regression analysis represents the relationship between the energy consumption on urban residential buildings and the most significant living standard parameter. Table 108.5 shows that the regression coefficients and the significance test between the energy consumption on urban residential building and the most correlated parameter of the residents' living standards in the four municipalities. Table 108.5, the larger regression coefficient indicates the larger influence of one independent variable on the dependent variable (Xi) when other independent variables (Y) are constant. Besides, the multiple correlation coefficients in Table 108.5 are all higher than 0.9, indicating that the regression models are effective. The "*p*" values are all less than 0.05 significant level, which show that the regression analysis have statistical significance. Therefore, the regression model for the urban residential building energy consumption in the four municipalities is derived as follows.

Table 108.5 The regression coefficients and the significance test

City	Variables for defining regression	Regression coefficient	Multiple correlation coefficient (R)	Significance level (*p*)
Y_B (Beijing)	Constant	259.146	0.911	0.000
	Most correlated parameter [per capita disposable income of urban households (X_1)]	9.668		0.000
Y_T (Tianjin)	Constant	105.641	0.941	0.000
	Most correlated parameter [Per capita consumption expenditure of urban households (X_2)]	7.143		0.000
Y_S (Shanghai)	Constant	133.107	0.972	0.006
	Most correlated parameter [Per capita consumption expenditure of urban households (X_2)]	21.927		0.000
YC (Chongqing)	Constant	53.755	0.944	0.016
	Most correlated parameter [Per capita disposable income of urban households (X_1)]	18.594		0.000

$$Y_B = 259.146 + 9.668X_1 \quad (108.1)$$

$$Y_T = 105.641 + 7.143X_2 \quad (108.2)$$

$$Y_S = 133.107 + 21.927X_2 \quad (108.3)$$

$$Y_c = 53.755 + 18.594X_1 \quad (108.4)$$

From Table 108.5 and regression models (108.1)–(108.4), we can see that the urban residential building energy consumption in Beijing and Chongqing are primarily correlated with per capita disposable income of urban households, indicating that the urban residential building energy consumption increase with the increase of per capita disposable income of urban households in these two cities. The urban residential building energy consumption in Tianjin and Shanghai are mainly positively correlated with per capita consumption expenditure of urban households, showing that the urban residential building energy consumption increases with the increase of per capita consumption expenditure of urban households in these two municipalities.

108.6 Conclusions

The above regression models show that the energy consumption on urban residential buildings is significantly positively correlated with per capita disposable income of urban households and per capita consumption expenditure of urban households. Therefore, analyzing the changes of the two parameters and controlling these two parameters through policy, technology and other measures are important strategies in controlling the growth of the energy consumption on urban residential buildings.

The effects of the two parameters of living standards, including per capita disposable income of urban households and per capita consumption expenditure of urban households, on urban residential building energy consumption are mainly manifested in two aspects. Firstly, the increase of per capita disposable income is due to the increase of the demands on all kinds of household appliances (such as refrigerator, air conditioner, etc.), this induces the energy consumption increase. Secondly, along with the increase in per capita disposable income of urban households, people have been improving the living environment, especially in Beijing and Chongqing. This causes the increase in the time of using HVAC system or other household electrical appliances and therefore the increase in energy consumption is the consequence. It is considered that the government can implement some policies to guide people's behavior consciousness, and strengthen promotion on residential building energy conservation to reduce the time of using the HVAC or other home appliances. At the same time, the government can also upgrade the technology standards of HVAC system or other household electrical appliances, and strengthen the supervision on the energy application in order to reduce the energy consumption.

The research findings provide valuable reference for further study on relevant subjects, such as the relationship between Y and X in other major cities nationally and internationally. It also provides reference for examining the effects of implementing green building on energy saving in various cities and regions.

Acknowledgements This research was supported by a grant from the Key Project of National Social Science Foundation of China (No. 15AZD025).

References

1. Planning Commission (2013) Twelfth five year plan (2012–2017): faster, more inclusive and sustainable growth. eSocialSciences 5302
2. Lenzen M, Wier M, Cohen C et al (2006) A comparative multivariate analysis of household energy requirements in Australia, Brazil, Denmark, India and Japan. Energy 31(2):181–207
3. Lu HL, Liu GF (2014) Spatial effects of carbon dioxide emissions from residential energy consumption: a county-level study using enhanced nocturnal lighting. Appl Energy 131: 297–306
4. Kelly S, Shipworth M, Shipworth D et al (2013) Predicting the diversity of internal temperatures from the English residential sector using panel methods. Appl Energy 102: 601–621
5. Sun HL, Wang HP (2013) The driving factors of the increase of the building energy consumption based on ridge regression. Urban Develop 5:028
6. Zeng D, Yin S (2012) Research on Prediction to civil building energy consumption based on co-integration theory. Prod Res 1:99–101
7. Zou C, Liu XL (2012) Urban building energy consumption calculation based on the IPAT theory. Chin J Popul Res Environ S2:82–85
8. Signor R, Westphal FS, Lamberts R (2001) Regression analysis of electric energy consumption and architectural variables of conditioned commercial buildings in 14 Brazilian cities. In: Seventh international IBPSA conference, Rio de Janeiro, Brazil. Citeseer, 1373–1379
9. Keng KA, Hooi WS (1995) Assessing quality of life in Singapore: an exploratory study. Soc Indic Res 35(1):71–91
10. He RW, Liu SQ, Liu YW (2012) Evaluation and spatial difference on residents' living standards in representative mountain areas: a case study of Liangshan Yizu autonomous prefecture, Sichuan province. J Moutain Sci 30(3):264–275
11. Zhu FH (2013) Statistical analysis of the interaction relationship between urbanization and urban resident living standard in China. J Hexi Univ 29(4):72–76

Chapter 109
Decomposition of Life Cycle Energy Consumption of Buildings in China

Yujie Lu and Peng Cui

Abstract The building and construction (B&C) sector consumes significant amount of energy in China and such amount has been increased every year during the last decades. However, the influencing factors and their explicit impacts on energy consumption have not been fully investigated. This paper proposed a hybrid LMDI model that can decompose the life cycle energy consumption of the B&C sector to three dimensions attributed by: (1) five driving factors which include area, structure, population intensity, value intensity, and energy intensity; (2) three life cycle phases which include material production, building construction, and building operation; and (3) 30 geographically distributed provinces and municipalities. The model was validated by using the data from China's B&C sector during 2007–2012. The key findings show that building material production phase contributes the most (84.2 %) to energy changes rather than construction and operation phases; while Jiangsu, Sichuan, Shandong, Hebei and Zhejiang rank as top contributors to the national incremental energy consumption.

Keywords Life cycle · Energy consumption · Decomposition · Logarithmic mean divisia index (LMDI) · Buildings

109.1 Introduction

The building and construction (B&C) sector consumes large amount of energy during material production, construction, and operation phase. About 30–40 % of all primary energy is used in buildings all over the world [1], and approximate 49 % of energy produced in the U.S. is consumed by buildings related activities [2]. In European Union (EU), buildings consume roughly 50 % of the total energy use through their life-cycles [3]. In China, energy use by the B&C sector accounted for

Y. Lu (✉) · P. Cui
Department of Building, School of Design and Environment, National University of Singapore, Singapore 117566, Singapore
e-mail: luy@nus.edu.sg

Y. Wu et al. (eds.), *Proceedings of the 20th International Symposium on Advancement of Construction Management and Real Estate*,
DOI 10.1007/978-981-10-0855-9_109

23 % of the total society energy consumption [4]. The share increased to 28 % in 2009 when traditional biomass was included, and it is expected to increase by 40 % from 2009 to 2030, make China the largest building energy consumer in the world [5].

To combat this challenge, China's government has made various efforts to reduce the energy consumption in the B&C sector, including capping the maximal coal power generation, increasing the share of non-fossil fuels in the energy mix and so on [6]. However, current policies focus only on the country level without specifying the energy target for the B&C sector. Meanwhile, the underlying reasons and factors that lead to energy changes in the B&C sector have not been fully understood yet. In particular, in which phase, from which area, and what factor may contribute to the energy change in the B&C sector remain unknown, and therefore the industry-specific strategy to reduce energy is still lacking.

To answer the above question, this paper proposes a hybrid Logarithmic Mean Divisia Index (LMDI) model to decompose energy consumption of the B&C sector into 30 provinces, five key factors, and three life cycle phases. The structure of the study is as follows. Section 109.2 reviews the existing studies for the decomposition method; Sect. 109.3 develops the hybrid LMDI model; Sect. 109.4 collects data for calculations and discusses the results; and Sect. 109.5 concludes the study and suggests future works.

109.2 Literature Review

The generation of "Logarithmic Mean Divisia Index" (LMDI) can be tracked back to 1925 when Divisia [7] came up with the conception of Divisia index which was used to analyze the weight of relative changes (growth rate). In 1978, Boyd, McDonald et al. [8] applied Divisia index approach in the field of decomposition analysis. Then in 1997, Ang and Choi [9] and Choi introduced the logarithmic mean weight function in decomposition analysis. The prototype of LMDI was gradually formed. In 1998, Ang, Zhang [10] and Choi proposed the term of "Logarithmic Mean Divisia Index" (LMDI). Three years later, Ang and Liu [11] formalized the LMDI into LMDI-I and LMDI-II in 2001. In 2000, the term "Index Decomposition Analysis" (IDA) was first used by Ang and Zhang [12] in 2000 to differentiate the technique from "Structural Decomposition Analysis" (SDA).

Nowadays, the LMDI approach is widely used in many fields such as agriculture, manufacturing, industry, commercial and construction and so on in order to decompose the carbon dioxide, energy consumption and other emissions. For instance, Baležentis and Baležentis [13] measured the energy intensity in Lithuania during 1995–2009; Holzmann and Adensam [14] decomposed final energy use for heating in the residential sector in Austria. Later, scholars used improved LMDI methods to analyze complex and multi-scope issues in energy and environmental impacts. Ang and Xu [15] compared the multi-country of energy performance; Fernández and Landajo [16] applied multilevel LMDI to decompose the aggregate energy consumption in the European Union (EU) of its 27 countries; Kang and Zhao [17] conducted a multi-sectoral decomposition analysis of city-level greenhouse gas emissions in Tianjin, China; Choi and Ang [18] introduced multiplicative LMDI approach in the decomposition of energy consumption change in the US manufacturing sector from 1987 to 2004.

The hybrid LMDI model was first used by Xu and Ang [19] to calculate energy consumption of the residential sector in Singapore. They conducted the decomposition analysis from three subsectors, which include environment control, household appliances, and personal devices. The decomposing result of each subsector is then combined to obtain the final effects for the residential sector level.

109.3 Research Methodology and Hybrid Model

To decompose the change of energy consumption during each phase of the life cycle for the B&C sector in China between 2007 and 2012, this study proposes the hybrid LMDI method in 2 steps. Step 1 is to decompose the energy consumption into five effects, and each effect is represented by different indicators; step 2 is to calculate the change of energy consumption results using LMDI method based on the collected data and the hybrid model.

Based on Kaya identity, which was proposed by Japanese energy economist Kaya [20], the energy consumption of the B&C sector in China can be decomposed rewritten as Eq. (109.1) which includes 5 typical effects.

$$E = \sum_{i,j}^{n} A \times \frac{A_i}{A} \times \frac{P_i}{A_i} \times \frac{V_i}{P_i} \times \frac{E_{i,j}}{V_i} \tag{109.1}$$

where, A, P, V, E denotes the area, population, value and energy consumption respectively, and i represents the provinces; j represents the energy type.

The equation above is decomposed by two dimensions, including (1) regions in China indicted by i, and (2) energy sources indicated by j. However, the life cycle of B&C sector is a long process, and is influenced by different factors in various phases. For instance, energy sources could be different in each phase. So the hybrid model is introduced in to consider different factors and characters in each particular phase.

Based on the hybrid LDMI model, a phase disaggregation scheme of the hybrid model is developed in the B&C sector in China. The life cycle of the B&C sector in China is disaggregated into 3 phases which are material production, building construction, and building operation. So the life cycle energy consumption (E^{LC}) of B&C sector equals to the sum of energy consumption in production (E^{P}), construction (E^{C}) and energy in operations (E^{O}).

Table 109.1 The meanings, scopes and sources of the symbols

Symbol	Definition	Scope	Data source
E^P	Embodied energy during building materials production phase	Materials Including steel, timber, cement, glass, and aluminum	[21]
E^{C*}	Energy consumption during construction phase	Including coal, gasoline, diesel, and electricity	
E^O	Energy consumption during the operation phase	Including coal, gasoline, diesel, LPG, heat, and electricity	
A_{con}	Construction area	For housing project	[22]
A_{hou}	Total housing area	Count by the end of year	
P_{lab}**	Number of labor force for housing project	pro-rate by the output value of construction industry	
P_{res}	Population	Refers to the resident population	
V_{out}***	Output value	Housing project	
$V_{\exp}$***	Expenditure	Refers to the expenditure by residential population	
i	Type of provinces	30 kinds, excluding Tibet, Taiwan and Hong Kong	–
j	Type of energy	Different by phases	–

Note *The unit of each type of energy is by physical quantities. So the coal equivalent conversion factor of each type of energy is used which comes from the General principles for calculation of the comprehensive energy consumption [23]. **The number of employees of building construction, the energy consumption of building material production and building construction cannot be obtained in the Yearbooks, so a reduction factor (Building project output value/construction industry output value) is multiplied. ***When calculating the construction output value and residential expenditure, the inflation factor has been considered. The economic values in 2012 have been adjusted to that in the year 2007 according to the inflation rate (17.47 % during this 5-year) from the China's Statistics Yearbook

Each phase consists of different key factors that influence energy consumption. Hence the proposed hybrid model is proposed as Eqs. (109.2)–(109.4). The definition of symbols, scopes, and data sources are shown in Table 109.1.

$$E^P = \sum_{i.j}^{n} A_{con} \times \frac{A_{con,i}}{A_{con}} \times \frac{V_{out,i}}{A_{con,i}} \times \frac{E_{mat,i,j}}{V_{out,i}} \tag{109.2}$$

$$E^C = \sum_{i.j}^{n} A_{con} \times \frac{A_{con,i}}{A_{con}} \times \frac{P_{lab,i}}{A_{con,i}} \times \frac{V_{out,i}}{P_{lab,i}} \times \frac{E_{con,i,j}}{V_{out,i}} \tag{109.3}$$

$$E^O = \sum_{i.j}^{n} A_{hou} \times \frac{A_{hou,i}}{A_{hou}} \times \frac{P_{res,i}}{A_{hou,i}} \times \frac{V_{\exp,i}}{P_{res,i}} \times \frac{E_{ope,i,j}}{V_{\exp,i}} \tag{109.4}$$

where, the effect of area (A) in three phases-material production (P), building construction (C), and building operation (O)-is expressed in A_{con}, A_{con}, A_{hou}, respectively; the effect of structure (S) in these phases is expressed in $\frac{A_{con,i}}{A_{con}}, \frac{A_{con,i}}{A_{con}}, \frac{A_{hou,i}}{A_{hou}}$; the effect of value intensity (VI) is expressed in $\frac{V_{out,i}}{A_{con,i}}, \frac{V_{out,i}}{P_{lab,i}}, \frac{V_{\exp,i}}{P_{res,i}}$; the effect of energy intensity (VI) is expressed in $\frac{E_{mat,i,j}}{V_{out,i}}, \frac{E_{con,i,j}}{V_{out,i}}, \frac{E_{ope,i,j}}{V_{\exp,i}}$; and the effect of population intensity (PI) is expressed in $\frac{P_{lab,i}}{A_{con,i}}$ and $\frac{P_{res,i}}{A_{hou,i}}$ due to no population effect on material production phase.

Based on LMDI approach, the additive form that calculates the change of energy consumption during each phase between the target year (2012) and the basic year (2007) can be developed. Based on five effects described above, the additive form of energy changes during 2002 to 2007 is shown in Eq. (109.5), where the subscripts A, S, PI, VI, and EI represent the five effects, including Area, Structure, Population intensity, Value intensity, and Energy intensity; the superscript m indicates the three different life cycle phases, including production (P), construction (C), and operation (O).

$$\Delta E^{m} = \Delta E_{A}^{m} + \Delta E_{S}^{m} + \Delta E_{PI}^{m} + \Delta E_{VI}^{m} + \Delta E_{EI}^{m} \tag{109.5}$$

Therefore, the aggregate changes of life cyle energy consumption in the B&C sector can be calculated by the sum of changes in all three phases, shown in Eq. (109.6). The proof of the Eq. (109.6) can be referred to the [19]. It is worthy to mention that since there is no population intensity effect on production phase, so the energy changes attributed to PI is $\Delta E_{PI}^{P} = 0$.

$$\Delta E^{LC} = \sum_{m} \Delta E^{m} = \sum_{m} \Delta E_{A}^{m} + \sum_{m} \Delta E_{S}^{m} + \sum_{m} \Delta E_{PI}^{m} + \sum_{m} \Delta E_{VI}^{m} + \sum_{m} \Delta E_{EI}^{m} \tag{109.6}$$

109.4 Data Collection, Result, and Discussion

In this paper, the raw data mentioned in Table 109.1 were collected from the China energy yearbooks, the China construction yearbooks and the China statistics yearbooks from 2007 to 2012. Thirty provinces of China are chosen except Taiwan, Hong Kong, Macao, and Tibet due to the lack of data. The building material embodied energy come from the inventory of carbon and energy (ICE) [24]. After LMDI calculation, the decomposition result is shown in Fig. 109.1 and it presents a 3-dimensional dataset, including 30 provinces, 3 phases, and 5 effects.

It is worth mentioning that the "small value" strategy was used during the LMDI calculation to address the "zero" value in the LMDI data. All zero values were assumed to be replaced by δ, which is an extremely small value (i.e. δ = 10 − 100). Under such a assumption, the result can yield an almost perfect approximation [25].

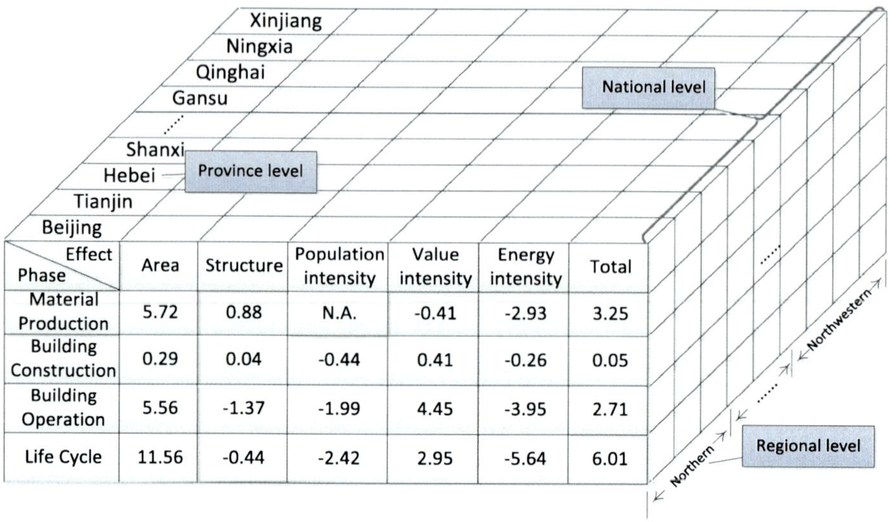

Fig. 109.1 The result of energy consumption in China's B&C sector of 30 provinces (Unit: Mtce)

At the national level, the decomposing result of energy consumption during each phase in the life cycle of the B&C sector in China from 2007 to 2012 can be summarized in Table 109.2. During this 5-year, the change of total energy consumption in China's B&C sector is 766.74 million tons of coal equivalent (Mtce). Among them, the building materials production contributed the most to the energy change up to 84.2 %, followed by the operation phase of 15.1 %, and the construction phase of 0.7 % only.

During five years, 5044.22 million square meters of construction areas were newly built, and such an area effect contributes the most (475.63 Mtce) to the total energy changes. The majority of the increased energy was from the material production phase (327.16 Mtce). The energy intensity effect ranks the second highest

Table 109.2 The addictive form of decomposition result in the B&C section of China (unit: Mtce)

Effect	Production	Construction	Operation	Life cycle (%)
Area effect	327.16	9.83	138.64	475.63 (62.0 %)
Structure effect	2.74	0.33	−18.48	−15.41 (−2.0 %)
Population intensity effect	N.A.	−8.13	−108.68	−116.81 (−15.2 %)
Value intensity effect	77.54	10.35	117.65	205.54 (26.8 %)
Energy intensity effect	238.17	−6.77	−13.61	217.79 (28.4 %)
Total effect	645.62 (84.2 %)	5.60 (0.7 %)	115.52 (15.1 %)	766.74 (100.0 %)

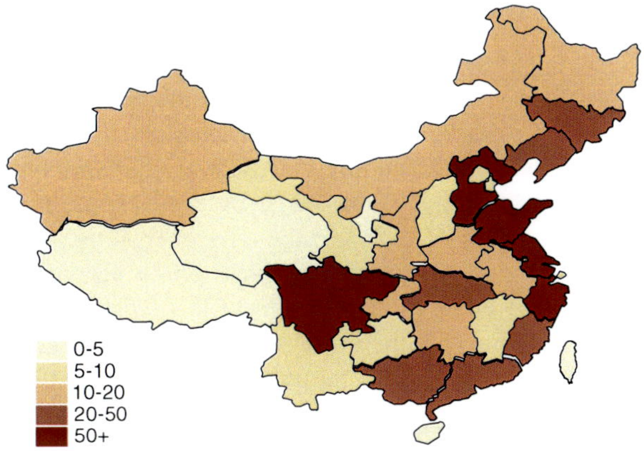

Fig. 109.2 The energy change of each province from 2007 to 2012 in the B&C sector of China (Unit: Mtce)

factor for the energy change with additional energy of 217.79 Mtce. Comparatively, the effect on the construction phase generated the minimum impact (5.60 Mtce and about 0.7 %) to the life cycle energy changes.

It is worth noticing that the growth of the resident population had a negative impact on energy consumption, which reduced 116.81 Mtce of energy consumption change, with 108.68 happening during the operational phase. Structure effect also reduced the energy consumption, indicating that the national distribution of construction projects could change the total energy consumption.

At the province level, Sichuan and four east coast provinces, including Shandong, Jiangsu, Zhejiang, and Hebei, generated the largest amount of energy changes in China's B&C sector. In addition, Hubei, Northeast region (Liaoning and Jilin), the Pearl River Delta area around (Guangdong, Guangxi, and Fujian) also shared a remarkable energy contribution. By contrast, energy consumption changes in the southwest, northwest and most other cities in central China were not significant, as shown in Fig. 109.2. For most provinces, the area effect and the value intensity effect were two key driving factors that increased energy consumption; the change of population normally decreased the energy consumption; while the change caused by structure effect and the energy intensity effect were depended on different provinces.

According to LMDI results and the raw data, the incremental energy consumption per unit of new construction area can be obtained. Take Beijing as an example, the construction area increased from 182.25 million m^2 in 2007 to 416.60 million m^2 in 2012. While the energy changes attributed to the area effect is 6.0 Mtce, so the incremental energy consumption per unit of new construction area in Beijing is about 25.62 kgce. Similarly, the incremental energy consumption caused by the changes of new housing areas, the amount of labor force per unit construction area, residential expenditures, and the population intensity can be also obtained.

109.5 Conclusion

This study is the first attempt to combine the life cycle conception and LMDI method with a hybrid model and to apply it in the B&C sector of China. At the national level, the building materials production contributed the most to the energy change up to 84.2 %, followed by the operation phase of 15.1 %, and the construction phase of 0.7 % only. The energy consumption change due to the area effect is the most during life cycle phase accounting for 475.63 Mtce (62.0 %), followed by energy intensity effect (28.4 %) and value intensity effect (26.8 %). It is worth noticing that the population intensity effect and structure effect had negative impacts of energy consumption, which reduced 116.81 and 15.41 Mtce of energy consumption change respectively.

The result of this study can help both academics and practitioners better understanding the key driving factors of energy consumption in the B&C sector within different provinces and different building phases. It can also facilitate the policy-making process by providing effective and industry-specific evaluation for the B&C sector toward an energy efficiency industry. Further study can elaborate on several aspects, such as to enrich the life cycle phase by including the demolition phase into consideration.

References

1. Kelso JD (2007) Building energy data book. Energy Efficiency and Renewable Energy, USA
2. Energy Information Administration (2011) Annual energy review. US Department of Energy, Washington, DC
3. Pataki D, Emmi P, Forster C, Mills J, Pardyjak E, Peterson T et al (2009) An integrated approach to improving fossil fuel emissions scenarios with urban ecosystem studies. Ecol Complex 6:1–14
4. Hong J, Shen GQ, Feng Y, Lau WS-t, Mao C (2014) Greenhouse gas emissions during the construction phase of a building: a case study in China. J Clean Prod 103, 249–259
5. International Energy Agency (2011) Key world energy statistics. International Energy Agency, Paris, France
6. The Tree (2014) China aims to cap coal as soon as 2020. The tree: content for climate communicators. http://treealerts.org/type/alerts/2014/11/china-aims-to-cap-coal-as-soon-as-2020/2014
7. Divisia F. L'indice monetaire et la theorie de la monnaie: Société anonyme du recueil sirey (1926)
8. Boyd G, McDonald J, Ross M, Hanson DA (1987) Separating the changing composition of US manufacturing production from energy efficiency improvements: a Divisia index approach. Energy J, 77–96
9. Ang B, Choi K-H (1997) Decomposition of aggregate energy and gas emission intensities for industry: a refined Divisia index method. Energy J, 59–73
10. Ang BW, Zhang F, Choi K-H (1998) Factorizing changes in energy and environmental indicators through decomposition. Energy 23:489–495
11. Ang BW, Liu F (2001) A new energy decomposition method: perfect in decomposition and consistent in aggregation. Energy 26:537–548

12. Ang BW, Zhang F (2000) A survey of index decomposition analysis in energy and environmental studies. Energy 25:1149–1176
13. Baležentis A, Baležentis T, Streimikiene D (2011) The energy intensity in Lithuania during 1995–2009: a LMDI approach. Energy Policy 39:7322–7334
14. Holzmann A, Adensam H, Kratena K, Schmid E (2013) Decomposing final energy use for heating in the residential sector in Austria. Energy Policy 62:607–616
15. Ang BW, Xu XY, Su B (2015) Multi-country comparisons of energy performance: the index decomposition analysis approach. Energy Econ 47:68–76
16. Fernández González P, Landajo M, Presno MJ (2014) Multilevel LMDI decomposition of changes in aggregate energy consumption. A cross country analysis in the EU-27. Energy Policy, 68:576–584
17. Kang J, Zhao T, Liu N, Zhang X, Xu X, Lin T (2014) A multi-sectoral decomposition analysis of city-level greenhouse gas emissions: case study of Tianjin, China. Energy 68:562–571
18. Choi K-H, Ang BW (2012) Attribution of changes in Divisia real energy intensity index—An extension to index decomposition analysis. Energy Econ 34:171–176
19. Xu X, Ang B (2014) Analysing residential energy consumption using index decomposition analysis. Appl Energy 113:342–351
20. Kaya Y (1990) Impact of carbon dioxide emission control on GNP growth: interpretation of proposed scenarios. IPCC energy and industry subgroup, response strategies working group, Paris, p 76
21. Department of Energy Statistics National Bureau of Statistics (1995–2003) China energy statistical yearbook. Statistic Press, Beijing, China
22. Department of Investment and Construction Statistics National Bureau of Statistics (1995–2013) China statistical yearbook on construction, China Statistic Press, Beijing
23. National Development and Reform Commission (2008) General principles for calculation of the comprehensive energy consumption. Standardization Administration of the People's Republic of China, Beijing, China
24. Hammond G, Jones C, Lowrie F, Tse P (2011) Embodied carbon: the inventory of carbon and energy (ICE). BSRIA, UK
25. Ang BW (2005) The LMDI approach to decomposition analysis: a practical guide. Energy policy 33:867–871

Chapter 110
Building Regulation: Science Versus Values Based Decision Making

Mark Burgess and John Douglas Thomson

Abstract Research into government regulation indicates difficulties in drawing an equitable balance between the role of science and societal values in building regulation decision making. To examine how these difficulties are being balanced in relation to one country's building regulations, an examination of the administrative processes used in the development of the Building Code of Australia has been undertaken. The inductive methodology uses 192 historical cases provided by the Australian Building Codes Board. This topic area is important and has significant political interest. The paper makes a number of major contributions drawn from the Australian context where building codes regulators are strong in science, engineering and technological qualities but lack specific 'societal values' expression. There is also a significant male gender bias and evidence of complexity in the requirements of submissions from the general public. This has caused inadequate public participation, in addition to which there is no industry union representation. Climate change urgently requires review of current building regulations, with these still largely based on 'deemed to satisfy' provisions. Current building product regulations need remedy.

Keywords Building code · Regulation · Values · Science · Stakeholder

110.1 Introduction

Societal values are broad preferences concerning courses of action or outcomes, and such values tend to influence attitudes, behaviour and risk. The application of risk analysis techniques to the building regulation decision process brings a

M. Burgess
CSIRO, Infrastructure Technologies, Commonwealth Scientific and Industrial Research, Organization, Australia, Private Bag 10, Clayton South, VIC 3169, Australia

J.D. Thomson (✉)
Graduate School of Business and Law, RMIT University, Melbourne, Australia
e-mail: doug.thomson@rmit.edu.au

Y. Wu et al. (eds.), *Proceedings of the 20th International Symposium on Advancement of Construction Management and Real Estate*,
DOI 10.1007/978-981-10-0855-9_110

science/values divide, with differences between the 'analytical' and the 'feeling' aspects of risk [9, 34]. This leaves the ultimate building regulation decision making body with a subjective risk assessment [10, 11]. Rayner and Cantor [32] propose that such evaluation is about both social relations and the evaluation of probabilities. Sundlof [35, p. 137] explores the 'problem of making decisions based on science, without taking into consideration societal values', arguing that the value-neutral nature of science, if used as a sole guide to decision-making, would exclude the application of societal values to decision making. Slovic et al. [34] views the science versus values concept from the perspective of risk perception, coining the terms 'risk as feelings' to describe the instinctive and intuitive, but with 'risk as analysis' describing scientific deliberation and reason [12]. There are instances where qualitative societal culture clashes with quantitative analysis, leading to the introduction of a third reality, 'risk as politics' [2, 34]. But 'science needs values' [15, 16].

This paper examines current building regulation processes and decision making science and values balance, with focus on a representative country, Australia. The research includes the qualitative and quantitative analysis of 192 'proposals for change' to building regulation over the period 2011–2014. These cases were made available for this research by the Australian Building Codes Board.

110.2 National Construction Codes

National construction codes and their development vary from country to country. The International Code Council (ICC) is an organization of building code officials and other building safety professionals which produces the most widely used set of model building codes. These codes are written by a broad group of stakeholders, and ICC staff oversee construction code development, coordination, and publication process. Many Chinese national standards are adoptions from international standards developers [3]. Chinese building standards may be either mandatory or voluntary. Mandatory standards have the force of law as do other technical regulations in China. They concern the protection of human health, personal property and safety. All standards that fall outside these characteristics are considered voluntary standards. There are four levels of Chinese building standards. The most widely implemented are the national, professional, local, and enterprise standards. These levels are hierarchical so that national supersede professional, which supersede local, which supersede enterprise standards. For any given product or service, only one type of Chinese standard applies. The US does not have official national building codes developed through a federal or national process and adopted uniformly nationwide. Instead, there are recognized organizations that develop codes and standards that are adopted by state or local (municipal or county) governments, who may modify model codes to meet their specific needs. The codes and standards gain their enforcement authority through the US state and local government adoption processes. Because of this, states and local governments are able to modify the codes they adopt. In Australia, proposals for regulation change processes evolved through a transition from a series of disparate State based systems [25] to the current code. This was the result of nearly 50 years of development. Transition from disparate State regulations to national unification was formally achieved with an Intergovernmental Agreement in 1994 [7]. The Agreement provided for the development of a uniform national building code, creation of the Australian Building Codes Board to administer that code, and a commitment by all States and Territories to enact legislation adopting the resulting code into regulation [6]. Australia's National Construction Code specifies minimum levels of safety, health and amenity. It is reviewed annually, with provision for community participation.

110.3 Methodology

The methodology is inductive and linked to the case data [23], providing connection to empirical reality. Unlike deductive arguments, inductive reasoning allows for the possibility that the conclusion is false, even if all the premises are true. Instead of being valid or invalid, inductive arguments are either strong or weak, which describes how probable it is that the conclusion is true [31, 37] (Wikipedia). Induction seeks to supply strong evidence for the truth of the conclusions being probable, based upon the evidence of the case data [20].

110.4 National Construction Code Development

Because of the many differences in the way countries develop their building regulations, this research has selected one country, Australia, as its focus. Australia's National Construction Code is a 'performance' based code, the result of deliberations by the Australian Building Codes Board. 'Performance' requirements are specified as the minimum level that buildings, building elements, plumbing and drainage systems must meet. It provides four methods to meet these mandated 'performance' regulations. First, evidence of suitability in meeting 'deemed to satisfy' or 'prescriptive' provisions; second, the verification method contained in the National Construction Code; third, the comparison method where the design is compared to other 'deemed to satisfy' or 'prescriptive' provisions; and fourth, the use of expert judgement which requires a qualified and experienced expert to assess and approve the adequacy of a 'performance' solution. In 2015, the Australian Building Codes Board conducted a survey of its subscribers seeking feedback on the use of 'performance' solutions. The response confirmed that the use of 'performance' solutions could be enhanced with benefit. The significant public interest in this document resulted in over 350 comments to some questions. In summary, the major responses to the survey were to develop building capacity; clarify fire authority involvement; quantify 'performance' requirements; and simplify legislation and the National Construction Code. To develop capacity requires increasing the general understanding of the National Construction Code and the benefits of the mandated 'performance' regulations. This is aimed at both current and future industry participants through the tertiary education sector; undertaking investigations of 'performance' issues; clarifying the involvement of fire authorities in the design approval process; quantifying 'performance' regulations requiring the removal of qualitative terms and their replacement with quantified terms; and inserting new verification methods as an option to verify compliance with 'performance' requirements. However, inclusion of 'deemed to satisfy' and 'prescriptive' building regulations seem counter to the use of 'performance' regulations and may inhibit the innovative development of new products on the basis of 'performance' [24, 29, p. 223].

110.5 Proposals for Building Regulation Change

The Intergovernmental Agreement leading to the development of Australia's National Construction Code states an aim to '…facilitate the development of a more efficient, internationally competitive building and construction industry through reforms to building and plumbing regulation nationally' [6, p. 2]. The initialising step for any Code change is the preparation of a Proposal for Change document using a standardised question-answer template. This document is a common template used by all proponents for change, whether they be the general public, industry, government bodies, or the Australian Building Codes Board itself [5, 17–19] (Commonwealth of Australia 1998). The process is consistent with the Council of Australian Governments regulatory principles to ensure appropriate rigour is used in the assessment of proposals. The role of the Building Codes Committee is predominantly technical and consists of representatives of all levels of government as well as industry to provide advice, guidance, and make recommendations relating to technical matters relevant to the National Construction Code. If a proposal has merit, this committee may recommend changes in the next public comment draft of the relevant volume of the National Construction Code. Once submitted, all proposals follow a common review and approval pathway (Fig. 110.1).

Initial review of each proposal is undertaken by the Building Codes Committee [8] whose recommendations are passed to the Australian Building Codes Board for review and approval. Change proposals supported by the Committee take pathways depending on their expected impact [4]. One pathway is for cases with minor effect on the Building Code, such as editorial corrections and minor changes. These progress directly through to public comment stage. Another pathway is for cases likely to effect a significant change. These undergo a Preliminary Impact Analysis. Where this analysis identifies substantial impact, consultation with the Office of Best Practice Regulation is sought regarding the level of Regulation Impact

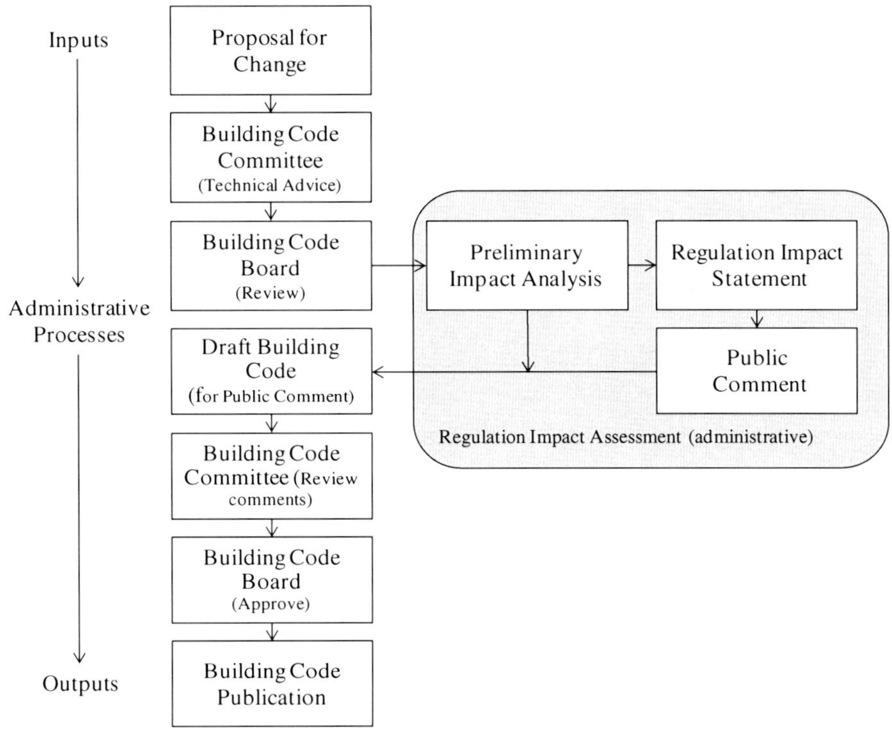

Fig. 110.1 Summary of the proposal for change process. *Source* Burgess and Thomson [14]

Statement required [19]. A Regulation Impact Statement requirement is developed and subjected to Building Codes Committee review and public comment. Australian Government guidelines [21] require proposed changes to the Building Code to be developed into a draft, which is made available to the general public for review and comment.

110.6 No Specific 'Societal Values' Representation

Section 13 of the Intergovernmental Agreement [6] provides the Australian Building Codes Board discretion over the Building Codes Committee composition, size and operation. Table 110.3 is an assessment of the current membership.

Table 110.1 provides a summary of the considerable science, engineering and technical capability that each member organisation brings to the committee. However, there is no specific 'societal values' or union representation on the committee membership, which has a considerable male gender bias.

Table 110.1 Composition of building codes committee and categorisation of technical capability

	Qualified to comment on technical issues							
	General	Building envelope	Fire, bushfire	Access egress	Services	Health	Energy	Ancillary, maintenance
ABCB Office (Chair)	1							1
Australian Government (Commonwealth)	1							
All State and Territory Governments	8							
Australasian Fire and Emergency Service Authorities Council			1					
Australian Institute of Building	1	1		1	1	1		1
Australian Institute of Building Surveyors	1	1		1	1	1		1
Australian Institute of Refrigeration, Air-conditioning and Heating					1		1	
Building Designers Association	1	1		1	1	1	1	
Building Products Innovation Council		1					1	
Commonwealth Scientific and Industrial Research Organisation	1	1	1	1	1	1	1	
Consult Australia	1	1		1	1			1
Engineers Australia	1	1	1	1	1	1	1	1
Fire Protection Association of Australia			1					
Housing Industry Association	1	1		1	1	1	1	1
Master Builders Australia	1	1		1	1	1	1	1
Property Council of Australia	1	1						
Royal Institute of Chartered Surveyors	1	1						
Standards Australia	1	1	1	1	1	1	1	1
Totals	21	12	5	9	10	8	8	8

Source in Burgess [13]

110.7 Royal Commission and 'Societal Values'

Natural disasters involve 'social values' including consideration of settlements and infrastructure, emergency planning and response, insurance, and human health. Recent extreme natural disasters that have frequented Australia recently include the 2009 bushfires in Victoria and New South Wales and the floods in Northern New South Wales and Queensland. In assessing the Building Codes Board achievement of balance between science and societal values, the 2009 bushfire disaster is an example. There was significant societal response to the Royal Commission into the Victorian bushfire in which 173 people died [36, pp. 10–14]. The Royal Commission conducted an extensive investigation into the causes of, the preparation for, the response to and the impact of the fires that burned throughout Victoria in late January and February 2009. The Royal Commission's summary of lessons learnt included how knowledge of the risk and impacts of the fire could be used to minimise fire-related loss of life in future and how bushfires deeply affected people and communities, and that their needs and safety must be at the forefront of government policy. The recommendations reflected the Commission's recognition that individuals, fire agencies and Commonwealth, State and local governments shared responsibility for preparing for bush fires and improving people's safety. The Commission concluded that construction standards for bushfire-prone areas did not adequately cover all the important components of bushfire risk. It recommended improving building standards and clarifying objectives to redress these deficiencies. The high risk to any home built in a designated 'Flame Zone' must be recognised. It considered that the science based 'deemed-to-satisfy' construction standards were not appropriate for such dwellings and that building regulations did not adequately cover the construction of non-residential buildings used by vulnerable groups in bushfire-prone areas, for example, schools, hospitals, child care centres and aged care facilities [36, pp. 10–14]. The recommendations of the Royal Commission indicate a need for greater emphasis on 'societal values' building regulation requirements, such as those represented by the Code's 'performance' regulations.

110.8 Building Product Regulation

In consultation with government, business, industry, community, academia and consumers, product standards but not regulations are developed internationally to help ensure the safety, reliability and performance of a range of products, services and systems. Standards are published documents that set out specifications and procedures designed to ensure products, services and systems are safe, reliable and consistently perform the way they are intended to. They establish a common language that defines quality and safety criteria. Standards specify requirements to achieve minimum objectives of safety, quality or performance of a product or service and are used to specify least-cost solutions to technical requirements expressing characteristics, performance and design criteria compatible with legislative requirements. Standards serve as purchasing specifications or technical conditions of contract between two parties. But on their own, Standards have no legal status and no requirement for compliance by manufacturers, consumers or the public, hence the term 'voluntary Standard'. Australian/New Zealand Standards are used in State and Commonwealth legislation. When this happens, only then do Standards become mandatory and can be subject to the scrutiny of the courts. However, any information that Standards Australia may hold on the legal status of a Standard may be incomplete and not reliable. Standards Australia is not in a position to give legal advice in relation to any implications that may arise if any requirements or specifications in a Standard are disregarded. While the Australian Building Codes Board is committed to ensuring best practice in the building and construction industry, its own building product certification scheme has not provided a nationally consistent quality of certified materials or products. Its voluntary, third party building product certification scheme accredits building product certification agents. These agents evaluate and certify building products to ensure they meet the specified requirements of the National Construction Code and a register of certified products is maintained by the Australian Building Codes Board on its website. Relevant legislation requires building control authorities to accept certified products, but this certification system is not working. Morris [30] reports that 'Australia's construction landscape is riddled with substandard imported materials that are potentially putting property and lives at risk' A survey of the building industry conducted by the Australian Industry Group [1] found nine out of ten companies complained of finding non-compliant products in their industry. Since the survey was conducted the situation had not improved and there hasn't been any regulation or licensing change [30].

110.9 Involvement of General Public

Croley [22] supports the concept of stakeholder involvement in regulatory decision making, considering the extent to which the decision processes encourage participation by those who will most effect and be affected by the regulatory outcomes i.e. the general public. Croley's [22] approach is inclusive of actors beyond stakeholders and interest groups. Slovic [33] suggests promoting public participation in both risk assessment and decision making. This inclusive approach is reflected in Australia's Inter-jurisdictional Regulatory Collaboration Committee's view of building regulation providing 'socially acceptable performance with respect to the building and the welfare of its occupants' [26–28, p. 16]. There were 192 Proposals for Change to the Australian Building Codes Committee over a four year period from 2011 to 2014, of which 29 were from the Board itself. The breakdown is provided in Table 110.2.

All proponents of a proposal for change are required to justify their proposal. The justification required needs to be proportional to the size of the proposed

Table 110.2 Number of proposals for change submissions by proponents

Proponents	Number of submissions
Independent Company	63
Industry/Professional Association	39
Council, Shire or City Government	31
Australian Building Codes Board	29
Other Govt (incl agencies, Government Business Enterprises)	15
General public (Independent Persons)	9
State Government Administration	4
Standards Australia	2
Total	192

Source Adapted from Burgess [13]

change or its potential impacts. It has to include a description of the proposal, an explanation of the problem it is designed to resolve, evidence of the existence of the problem, how the proposal is expected to solve the problem, what alternatives to regulation have been considered and why they are not preferred, who will be affected and how affected, and any consultation that has taken place. For a public person to meet these requirements is onerous, costly and time consuming, as evidenced by only 9 submissions (Table 110.1) being initiated by a private person These few representations over a four year period indicates lack of involvement by the general public. The demands of the regulator in providing pathways for the non-expert to initiate change can influence both the effectiveness and efficiency of the regulatory regime. Review of the Australian Building Codes Board data shows a bias between the number of Board proposals and non Board proposals (Table 110.3). Of the 163 non Board proposals, the non-Board proponents solution was recommended in only 9 instances, with a further 15 recommended after some modification. In 24 non-Board cases an alternative solution was developed and in 21 non-Board cases, a solution developed where the non-Board proponent did not develop one. By comparison, the Australian Building Codes Board, who it could be argued has expertise in the preparation of Proposals for Change, 25 of 29 proposals were recommended without change.

Croley's [22] question of those suited to effect regulatory change is between the non-expert who can request a regulatory change and the expert who participates in the administrative process to develop the change. This indicates a need for a more encouraging interface between the non-Board proposer and the Board regulation implementer. For Australia, this requires reconsideration of the Proposal for Change process to make it more user friendly and transparent, to differentiate in a simpler and better way between problem proposer and solution, and to provide an improved pathway for non-expert proponents to initiate change.

Table 110.3 Building code committee recommendations

Building codes committee recommendation	Frequency		
	Total dataset	Australian building codes board proposals for change	Non Australian building codes board proposals for change
Proposal not supported	96	2	94
Proponent's solution recommended	34	25	9
Proponent's solution with modifications or edits recommended	17	2	15
Problem accepted but alternate solution developed	24	0	24
Proponent provided evidence of a problem, without proposing a solution. Committee prepared a solution	21	0	21
Totals	192	29	163

Source Burgess [13]

110.10 Conclusions

Science versus values in building regulation has received little attention in the literature. This research into government building regulation focusses on Australia and indicates the difficulties in drawing an equitable balance between the role of science and societal values in building regulation decision making. The paper makes a number of major contributions. First, building codes boards and their committees while strong in science, engineering and technology are generally devoid of any specific 'societal values' representation. Second, the greater majority of board and committee members are male, an obvious gender bias. Third, there is evidence of unnecessary complexity causing inadequate participation by the general public and the need for a more encouraging interface between the proposer and the regulation implementer. Fourth, there is no industry union membership. Fifth, building regulations for natural disaster prone areas need to contain specific standards for the construction of buildings in such zones. Sixth, most global and national regulatory building product certification regimes are flawed and need remedy.

References

1. AIG (2014) Burden of government regulation. Australian Industry Group, New South Wales
2. Almedia N, Sousa V, Alves Diaz L, Branco F (2010) A framework for combining risk management and performance based building approaches. In: Building research and information. Routledge, London

3. ANSI (2015) PRC standards system: standards used in China. American National Standards Institute, www.StandardsPortal.org
4. Australian Building Codes Board (2011) Good practice guide for preliminary impact analysis. Australian Building Codes Board, Canberra
5. Australian Building Codes Board (2012) ABCB—proposals for change, Australian government, viewed 10 Oct 2012, http://www.abcb.gov.au/~/link.aspx?_id=A89658AE5694454DB022B691CCAF66F1&_z=z
6. Australian Building Codes Board (2012) An agreement between the governments of the commonwealth of Australia, the states and the territories to continue in existence and provide for the operation of the Australian Building Codes Board
7. Australian Building Codes Board (2013) History of the national construction code, viewed 2 Feb 2013, http://www.abcb.gov.au/en/about-the-national-construction-code/history-of-the-ncc
8. Australian Building Codes Board (2014) NCC 2014 an overview, viewed 29 April 2014, http://www.abcb.gov.au/sitecore/shell/Controls/Rich%20Text%20Editor/~/media/Files/Download%20Documents/About%20the%20NCC/NCC2014%20-%20Overview%20of%20significant%20changes.pdf
9. Baldwin R, Scott C, Hood C (eds) (1998) A reader on regulation. Oxford University Press, Oxford
10. Banks G (2006) Reducing the regulatory burden: the way forward. Paper presented to Inaugural Public Lecture, Monash Centre for Regulatory Studies, Monash University Law Chambers, Melbourne, 17 May 2006
11. Breyer SG (1982) Regulation and its reform, Digital Reprint edn. Harvard University Press, Cambridge
12. Brooks M (2013) Time for science to Sieze political power. New Scientist, vol. Comments Piece
13. Burgess M (2015) Understanding the impact of stakeholder participation in administrative rule making: Decisions on the content of Australia's building code (unpublished)
14. Burgess M, Thomson JD (2015) National construction codes and their inadequacies: Australia's arrangements and difficulties, Chap 84. Springer, Berlin
15. Carolan MS (2006) Scientific knowledge and environmental policy: why science needs values. Environ Sci 31(4):229–237
16. Carolan MS (2006) Science, expertise and the democratisation of the decision making process. Soc Nat Res Int J 19(7):661–668
17. Commonwealth of Australia (2004) Reform of building regulation. Research Report, Productivity Commission, Productivity Commission
18. Commonwealth of Australia (2006) Rethinking regulation. Report on the taskforce on reducing regulatory burdens on business, Regulation Taskforce—Productivity Commission, Australian Government
19. Commonwealth of Australia (2007) Best practice regulation handbook. Office Best Practice Regulation, Australian Government
20. Copi IM, Cohen C, Flage DE (2007) Essentials of logic, 2nd edn. Pearson Education, Upper Saddle River
21. Council of Australian Governments (2007) Best practice regulation—a guide for ministerial councils and national standard setting bodies
22. Croley SP (2008) Regulation and public interests, 1st edn. Princeton University Press, New Jersey
23. Eisenhardt KM (1989) Building theories from case study research. Acad Manag Rev 14 (4):532–550
24. Gann DM, Wang Y, Hawkins R (2010) Do regulations encourage innovation?—The case of energy efficiency in buildings. In: Building research and information. Routledge, London
25. Knox HA (1989) The development of the building code of Australia. Universal Architect Directory, NSW, Feb 1990, pp 15–8
26. Meacham BJ (2010) Performance based building regulatory systems—principles and experiences. In: International Jurisdictional Regulatory Collaboration Committee

27. Meacham BJ (2010) Risk informed performance based approach to building regulation. J Risk Manag 13(7):877–893
28. Meacham BJ (2010) Accommodating innovation in building regulation: lessons and challenges. Buil Res Inf 38(6):686–698
29. Mitusch P (2007) Expert systems for the Norwegian building regulations. In: Building research and practice. Routledge, London
30. Morris M (2015) Imported construction materials putting buildings, lives at risk: industry leaders. Australian Broadcasting Commission, 25 June
31. Popper KR, Miller DW (1983) A proof of the impossibility of inductive probability. Nature 302(5910):687–688
32. Rayner S, Cantor R (1987) How fair is safe enough? The cultural approach to societal technology choice. Risk Anal 7(1):3–9
33. Slovic P (1999) Trust, emotion, sex, politics, and science: surveying the risk-assessment battlefield. Risk Anal 19(4):689–701
34. Slovic P, Finucane M, Peters E, MacGregor D (2004) Risk as analysis and risk as feelings: some thoughts about affect, reason, risk and rationality. Risk Anal 24(2):1–11
35. Sundlof S (2000) The role of science in regulation and decision making. AgBio Forum 3(2&3)
36. Teague B, Mcleod R, Pascoe S (2010) 2009 Victorian bushfires royal commission, Parliament of Victoria, Government Printer for the State of Victoria, Melbourne
37. Vickers J (2010) The problem of induction stanford encyclopedia of philosophy. Stanford, USA

Chapter 111
Chinese Contractors in the International Market: Business Distribution and Competitive Situation

Zhenyu Zhao, Jiahui Yao and Chao Tang

Abstract The past decades have witnessed the dramatic development of Chinese International Contractors (CICs) in the global market. With an increasing number of contractors entering into International Construction Project Contracting (ICPC) market, the CICs has been facing fierce competition environment. The business layout and competitive positioning have significant impacts for the CICs to choose and develop the international market. According to the data collected from the Engineering News Record and other cases, the paper analyzes the structure of ICPC market and the diversified development process of the CICs in overseas different regions. From the perspective of enterprise ecology theory, international contractor populations are classified by countries. The "Market Population Concentration Index" is proposed to calculate the concentration of international construction market; the "Niche Width" is defined to describe the utilization of international market resources; and the "Niche Overlap" is introduced to reflect the competitive level among contractors. Assessment methods are proposed to quantitatively analyze enterprises niche status of international contractors. The results find the dynamic changes of the business market structure of the CICs and their competitiveness characteristics compared with those of other countries. The study supports both Chinese and other countries' contractors to understand the development pattern and competitive structure of the global market, and helps them to select target markets and appropriate strategies for ICPC business.

Keywords International contractors · International market · Competition · China · Enterprise ecology

Z. Zhao (✉) · J. Yao · C. Tang
School of Economics and Management,
North China Electric Power University, Beijing, China
e-mail: zhaozhenyuxm@263.net

Y. Wu et al. (eds.), *Proceedings of the 20th International Symposium on Advancement of Construction Management and Real Estate*,
DOI 10.1007/978-981-10-0855-9_111

111.1 Introduction

Chinese International Contractors (CICs) are one of the largest and fastest expanding groups in the world. According to the statistics from the Ministry of Commerce of China, the annual growth rate of the CICs is more than 20 % over the past 15 years. The official statistics indicates that the amount of overseas contracts was reported to be US$142.4 billion, and the new contracts signed by CICs are worth US $191.76 billion in 2014 [1]. Particularly, the international revenues of CICs, which are included in the top 250 international contractors list of the Engineering News Record 2013, reached US$79 billion. The share of the revenues has exceeded that of the United States' contractors [2]. The CICs are becoming one of the most influential contractor groups next to the European contractors.

With the scale of International construction project contracting (ICPC) business turning to a growth path, an increasing number of contractors pursue the overseas market, adding the competitiveness of the market in the future. The international market environment is complex and volatile, while the performance and strategies of the CICs and their competitors are constantly changing. The CICs and other international contractors benefit from the massive development on the one hand and have to face fierce competition on the other hand. All international contractors, including the CICs, are keen to make development strategies by considering international competitiveness, market selection, and risk management. It is imperative to understand the business distribution and competitive situation of CICs, compared with leading international contractors.

As a new supplement to organization theory, enterprise ecology can be used to explore both the positions of enterprise populations in business ecosystem and the relationships between correlative enterprise populations. This theory is brought into the ICPC research field to analyze the market distribution pattern and the evolution features of international contractors. According to the enterprise ecology theory, ICPC market is regarded as a complex dynamic system and international contractor populations are classified by countries. The aim of this study is to find the dynamic changes of the business market structure of the CICs population and the competitive features compared with other populations. The results will assist Chinese and other countries' international contractors to develop business strategies and to improve their competiveness, also provide a new perspective for them to select a most suitable market when they expanding overseas.

111.2 Literature Review

By an extensive review of the related literatures, the existing researches mainly focus on the following three aspects:

(1) To the development status and trend of ICPC market. Some statistics provided evidence of the development status of international contractors, such

as the data of top Chinese contractors from the "Engineering News Record (ENR)" annually [3]. Du et al. [4] introduced both the "industry structure change rate" and "industry structure change efficiency" to analyze the international contracting market, and found the market structure tended towards tech-intensive development and the efficiency of CICs' resource distribution has been increasing rapidly. As to the green infrastructure construction market, forecasting was performed targeting to the USA, Korea, and global construction market by Du et al., they found the green infrastructure market would rapidly grow in the next decade and consequently lead changes to the industry business practices [5]. By using the statistical data of ENR, Li analyzed the industrial structure and centralizing degree of the international contractors. The results show that international construction market is concentrated on the building, industrial/petroleum and transportation [6].

(2) To the competitiveness of international contractors. Diversified models have been established to discriminate competitiveness of CICs and leading international contractors in existing studies. For example, Guan et al. [7] established a multi-indices comprehensive evaluative model for international competitiveness using weight fuzzy AHP, to appraise the international competitiveness of top contractors. Tan et al. identified the key competitiveness indicators (KCIs) for assessing contractor competitiveness in the Chinese construction market [8]. Zhao et al. [9] also adopted SWOT approach to analyze CICs' strengths, weaknesses, opportunities and threats in the international construction markets.

(3) To the comparative studies of international contractors. For improve the sustainable development situation of Chinese contractors and lead them on the right way from the scale to the capability, the financial capabilities of domestic and foreign large construction contractors in recent five years are compared (ratio analysis, trend analysis, Dumont analysis, etc.) by Liu and Wang, and the gap is quantified from a sustainable financial perspective [10]. Pheng et al. [11] compared the performance of top British and Chinese contractors based on the OLI + S model, which incorporates the ownership, location, internalization and specialty factors, and found the international involvement of top British contractors has declined from a peak in 1996, while that of the top Chinese contractors has grown steadily since the 1980s. In view of the perspective of organizational ecology, Ji et al. [12] compared the developmental trends of the United States and China in the international and Middle East market. The results show that the population density of the two countries has a concave relationship when new firms become listed in the top 225 ENR international contractors.

Overall, the existing literatures mainly concentrated on qualitative studies of ICPC market. There are few quantitative studies to compare the international revenues between CICs and other international contractors, or to build evaluation index system for estimating competitiveness of international contractors. Though some researchers begin to pay attention to the application of enterprise ecology theory, there are also lack of research about international contractors by using corporate ecological indicators. Therefore, the enterprise ecology theory will be introduced in this paper to study the business distribution and competitive situation of CICs by comparing with other international contractors.

111.3 Research Methodology

The ICPC market is complicated and changeable with the characteristics of different regions varies greatly from each other. In this study, according to the statistical date from the "Engineering News Record" and related cases, the methodology of employing typical indicators are used to analyze the performance of CICs and other international contractors. The research methodology and process of this study are shown in Fig. 111.1.

Firstly, regional market revenue curves are set to overview the structure of ICPC market. The data mainly come from the ENR from 2000 to 2013. And then the proportions of international contractors' revenue by regional market are calculated, while CICs' proportions are highlighted. The business layout and the dynamic development process of CICs are shown by using four radar charts.

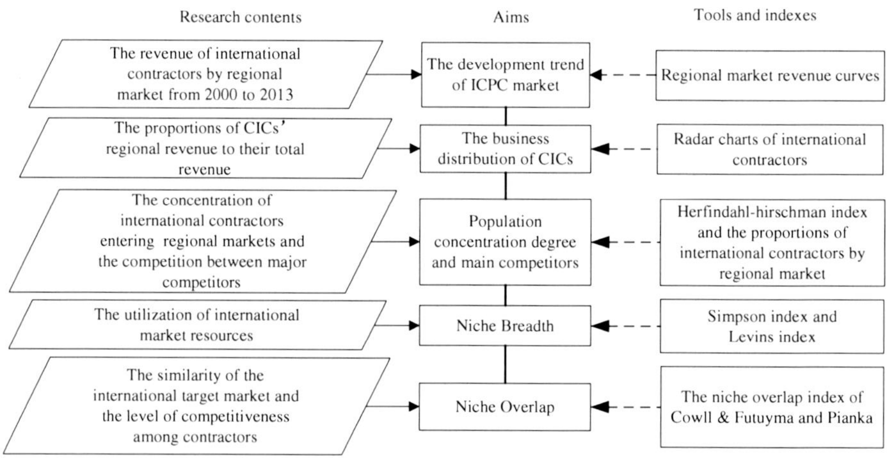

Fig. 111.1 Research methodology

Then according to the enterprise ecology theory, international contractor populations are classified by countries. Herfindahl-hirschman index is utilized, which can reflect the market structure, to analyze the population concentration degree of the seven regional markets. Subsequently, four major international populations are chosen respectively from three mature regional markets which CICs have entered into sufficiently, to understand the competitive status in each regional market.

Furthermore, the niche width degree and niche overlap degree indicators of international contractors are defined to describe the diversification and utilization of international market resources, as well as the similarity of the international target market and the level of competitiveness among contractors. Simpson index, Levins index are used to measure the niche width degree, while the niche overlap index of Cowll, Futuyma and Pianka are used to measure the niche overlap degree. Then enterprises niche statuses of international contractors from different countries in the complex international market environment are analyzed quantitatively.

111.4 The Development Trend of ICPC Market

The past decades have witnessed the development progress of ICPC market. In recent years, the ICPC market structure are becoming complicated and diversified. The seven regions, which are classified by the Engineering News Record, including the Middle East, Asia, Africa, Europe, U.S., Canada and Latin American Caribbean, are different in both environment complexity and market maturity. The revenue curves of the total top 225/250 international contractors from 2000 to 2013 show the evolution of regional markets [13], as shown in Fig. 111.2.

According to the Fig. 111.2, we can find there are two development periods taking 2008 by the global financial crisis as the boundary. From 2000 to 2008, the revenues of most regional markets were at a stable growth stage, of which the revenue of the Middle East and the European grew more rapidly. The European market was the pioneer during the development progress of the international contract market, and drove the evolution of the whole ICPC industry and promoted the progress of project contracting mode and construction technology. Meanwhile, with a large income from oil resources, Middle East market was active to push infrastructure projects which contributed to the increase of the regional contracting volume. In this period, ICPC industry developed rapidly with the integration of regional economy and global investment. However, most regional markets declined since 2008. Both European and American firms suffered a decreasing performance in terms of total revenue after reaching a peak in 2008. While since 2010, the trend of regional market revenue amount changed again, such as the Asia jumped to the first place along a rapid growth curve; the Europe, the Middle East, and the U.S. also revived after a few years of falling. The Latin America and Caribbean market remained rising fortunately.

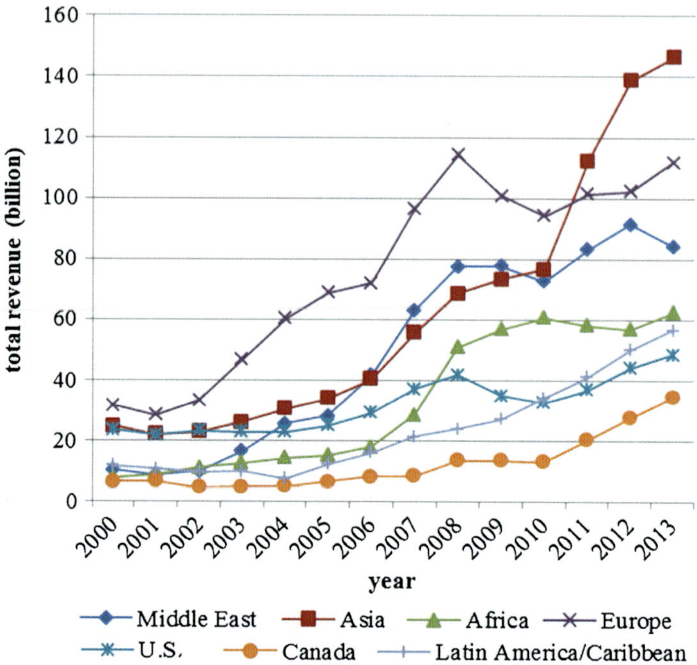

Fig. 111.2 Total revenue of top international contractors in seven regions (2000–2013)

111.5 Business Distributions of CICs

To study the business layout of CICs and compare it with the structure of whole ICPC market, data-processing is used as follows: Firstly, regional market shares are calculated by dividing each regional market's revenue by the total revenue from ENR [13]. Secondly, CICs regional market shares are calculated according to their revenue amount percentage during 2000 to 2013. Finally, four radar charts are concluded as shown in Fig. 111.3, which involves the business layout of both total top 225/250 international contractors and CICs among them in the year 2000, 2004, 2008 and 2013. The charts reveal the dynamic development processes of CICs.

From the Fig. 111.3, we can find the market shares' change of the top 225/250 international contractors in different regions. Compared with the share of 2000, Europe market accounted for 40 % in 2004, significantly higher than other regions. It matured ahead of other regional market, but shrunk after the financial crisis in 2008. Latin American market slightly declined in 2004, while significantly increased in 2013, which is expected to be emerging market in the future. Africa market achieved rapid investment growth, however the unstable social and political environment prevent international contractors entering into the region. Of all the regional markets, Asia market's share beyond Europe firstly in 2013, becoming the only regional market counts for more than 20 %.

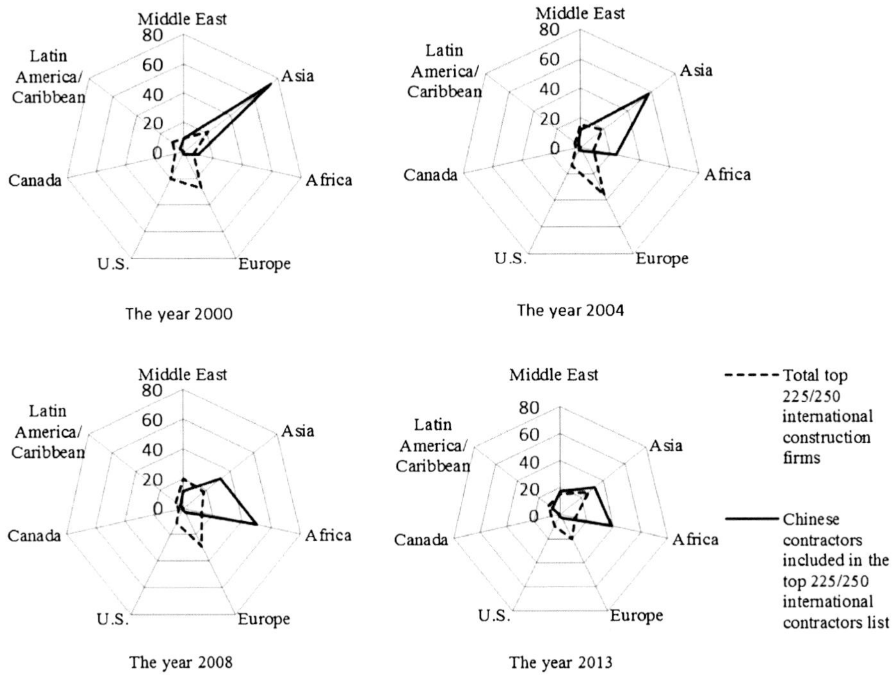

Fig. 111.3 Market percentage shares of top 225/250 international contractors by regional

Compared with the market structure of the whole ICPC market, the business layout of CICs changed more obviously as their regional markets with a large share turns out to diversify.

In 2000, CICs gained nearly 80 % of their total international revenue in Asia. It is far ahead the Middle East and Africa markets. The share of other four regions is low, which accounts for less than 7 %. In 2004, the Middle East and Africa market shares of CICs rose rapidly, up to about 20 %. By contrast, Asia market share was on a decline. In those years, some CICs seized market opportunities, for instant, China State Construction Engineering Corporation and China Railway Group Company signed lots of new contracts in Pakistan, Sudan, Algeria, Hong Kong and other Asian and African regions. More than 80 % of the contracts are worth hundreds of millions of US$.

In 2008, the business distribution of CICs was significantly different from the previous. The Chinese contractors had 50 % of their total international revenue coming from Africa as the largest regional market, while that from Asia reduced to 31.8 %. In particularly, CICs also opened up the U.S., Latin American and Caribbean market. For example, China State Construction Engineering Corporation won the bidding of the Hamilton Alexander bridge renovation project worth 40.7 million from U.S. Department of transportation in New York.

In 2013, the data from ENR show that CICs' market shares in Asia and Africa are about 35 and 40 % respectively attributed to a large number of electric power, housing and transportation construction projects contracts are signed by CICs. In addition, there was an obviously rise in the Latin American Caribbean market as the total value of the contracts signed by CICs up to US$3.2 billion of the year. The hydropower, petrochemical, infrastructure, port construction of Argentina, Brazil, Chile, Mexico and other countries in the Latin American Caribbean market have great potential to develop for CICs [14]. The Middle East market also showed a significant growth for CICs with the infrastructure construction investment increased in Saudi Arabia, UAE, Kuwait and other African countries. Moreover, European governments enhanced the willingness to attract foreign investment, making CICs' entry into 16 countries in Central and Eastern Europe easier. These bring opportunities for CICs to maintain a rapid growth and enter into more regional markets.

In terms of the dynamic development processes of CICs, their business distribution showed a trend of diversification. Thus, the CICs shall pay close attention to the regions which have good performance and experiences, and seize opportunities to open up new markets.

111.6 Population Concentration Degrees and Main Competitors Analysis

Herfindahl-Hirschman Index (HHI) refers to a square sum of percentage calculated from the revenue of market competitors to the industry's total income or total assets [15]. It is a preferable index to measure industry market concentration degree. In this study, the HHI is used in the ICPC industry to measure the impacts of different international contractor populations. It provided a quantitative description of the different structure of ICPC market.

The formula of HHI is

$$HHI = \sum_{i=1}^{n} x_{ij}^2 \tag{111.1}$$

where x_{ij} refers to the market share of population i in the regional market j, $n = 17$.

The bigger the values of HHI are, the higher the population concentration degree and the monopoly degree will be. Generally, when HHI is over 1000, the market structure is oligopoly; and when HHI is less than 1000, the market structure is competitive. Detailed classification is shown in Table 111.1.

Since there are total 17 countries included in the ENR top 225/250 international contractors lists from 2011 to 2013, 17 international contractor populations are classified by the different countries. Market share of each international contractor

Table 111.1 Market structure classification based on HHI value

Market structure	Oligopoly types				Competitive types	
	High oligopoly type I	High oligopoly type II	Low oligopoly type I	Low oligopoly type II	Competitive type I	Competitive type II
HHI scores 0–10,000	HHI ≥ 3000	3000 > HHI ≥ 1800	1800 > HHI ≥ 1400	1400 > HHI ≥ 1000	1000 > HHI ≥ 500	HHI < 500

Table 111.2 Calculation results of HHI in the seven regions from 2011 to 2013

Year	Middle East	Asia	Africa	Europe	U.S.	Canada	Latin America/Caribbean
2011	2055.7	2821.9	3407.3	8378.7	6953.9	6249.1	5520.3
2012	2069.8	3263.6	3308.1	8827.8	9209.8	6454	5016.5
2013	2266.1	3003.5	3379.3	8131.7	9308.5	5582.5	4040.3

population can be regarded as a x_{ij}. Then they are brought into the formula (111.1) to calculate. The results of HHI are shown in Table 111.2.

Table 111.2 shows Middle East market belongs to high oligopoly type II, and the Asia market has gradually translated from high oligopoly type IIto high oligopoly type I. Other regional markets are attributed to high oligopoly type I. Among these regions, the HHI of Europe is 2.8 times higher than the basic standard level; others are about 2 times higher. Overall, the concentration degrees of regional markets are high. But the HHI values of Africa, Canada, Latin America and the Caribbean decreased gradually, which is in favor of CICs to compete in these regions.

Among the seven areas, Africa, Asia and the Middle East are the three major regional markets, where Chinese contractor population occupies larger market shares. In different regions, Chinese contractor population faced different competition populations. Africa was the largest regional market of CICs from the business distribution structure in 2008 (see Fig. 111.3). In this area, Italy, France, South Korea contractor populations are three main competitors of CICs. However, in 2013, CICs accounted for almost 50 % of the African market, significantly larger than the three main competitive populations [16]. In Asia, the main competitive populations to CICs are United States, Germany and Spain contractors. The fierce competition between these countries' contractors resulted in an alternate translocation of the top position. Statistics show American contractor population occupied the largest market share in 2000, and then it decreased rapidly. After 2003, German contractors population's market share grew rapidly once overtook the United States and China, ranking in the first place. However, in 2009, CICs' market share ranked first by slightly higher than that of German contractors [17]. In recent five years, the market shares of the four major competitive populations in Asia are closer, at around 15 %. It is expected that the Asian market competition will grow fiercer. In Middle East, the market share of the United States contractor population has shrunk dramatically since 2006. Italy contractor population's market shares are relatively stable with a slight decline trend. After 2008, the market share of South Korean contractors climbed, followed by Chinese contractors. Although South Korea once held a significant ahead position to China, the gap has been narrowing since 2012. In the future, South Korea' contractors will become the main rival of CICs with their growth potential in Middle East.

Table 111.3 Definitions of variables of niche width

Variables	Definitions
$NW_{(s)i}$	Simpson geography/product niche width of population i
$NW_{(1)i}$	Levins geography/product niche width of population i
p_{ij}	Market share of population i in region/industry j
r	Number of regions/industries j

111.7 Calculations of Niche Breadth and Niche Overlap

111.7.1 Calculation Method of Indexes

According to the proposed concept and measure model of niche width and niche overlap, two formulas and definition of variables are described as following [18, 19] (Table 111.3).

$$NW_{(s)i} = 1 - \sum_{j=1}^{r} p_{ij}^2 \tag{111.2}$$

$$NW_{(1)i} = \frac{1}{r\sum_{j=1}^{r} p_{ij}^2} \tag{111.3}$$

$$C_{ik} = 1 - \frac{\sum \left|N_{ij}/N_i - N_{kj}/N_k\right|}{2} = 1 - \frac{\sum \left|p_{ij} - p_{kj}\right|}{2} \tag{111.4}$$

$$C_p = \frac{\sum p_{ij}p_{kj}}{\left(\sum_{j=1}^{r} p_{ij}^2 \sum_{j=1}^{r} p_{kj}^2\right)^{1/2}} \tag{111.5}$$

Niche overlap is in the range of 0–1. The value “0” indicates completely separate niche while the value “1” represents completely overlapped niche. Different values ranging between 0 and 1 represent varying degrees of overlap (Table 111.4).

111.7.2 Data Analyses

According to the data of Top 225 International Contractors listed in ENR, this paper selected the international contracting revenue of 13 countries from 2010 to 2013, as the contractors in these countries have been involved in the Top 225 International Contractors (ENR) lists over the past decade. Contractors’ average and total international revenue of the 13 countries from 2010 to 2013 in different regions are calculated as shown in Table 111.5.

Table 111.4 Definitions of variables of niche overlap

Variables	Definitions
C_{ik}	Cowll and Futuyma niche overlap between population *i* and population *k*
C_p	Pianka niche overlap between population *i* and population *k*
N_{ij}	Revenue of population *i* in region *j*
N_i	Revenue of population *i* in all regions
N_{kj}	Revenue of population *k* in region *j*
N_k	Revenue of population *k* in all regions
p_{ij}	The portion of revenue in region *j* shared by population *i*
p_{kj}	The portion of revenue in region *j* shared by population *i*
r	Number of regions

Table 111.5 Annual average and total overseas revenue of international contractors in different regions from 2010 to 2013 (billion US$)

Country/Region	Middle East	Asia	Africa	Europe	United States	Canada	Latin Amer. Carib.	Total Revenue
U.S.	11.032	17.515	2.999	6.682	0.00	16.816	6.298	61.342
Canada	0.165	0.389	0.526	0.247	0.729	0.00	0.643	2.698
Britain	2.530	2.119	0.890	2.123	2.732	0.096	0.128	10.616
Germany	1.670	20.904	0.730	6.934	10.237	0.802	0.314	41.590
France	2.274	5.474	5.683	21.795	3.678	2.432	2.380	43.716
Italy	6.986	3.972	8.370	5.573	0.839	0.522	5.170	31.432
Netherlands	0.727	0.814	0.304	5.891	0.198	0.013	0.174	8.121
Spain	3.652	14.201	1.895	16.545	10.602	1.487	13.764	62.145
Australia	1.219	4.164	0.040	1.172	2.404	0.706	0.148	9.853
Japan	3.020	11.921	0.872	0.488	2.274	0.265	0.581	19.420
China	11.092	22.333	25.655	2.003	0.580	0.073	4.726	66.461
Korea	19.192	8.176	2.534	0.194	0.392	0.161	1.322	31.971
Turkey	5.273	3.327	2.102	6.105	0.020	0.00	0.095	16.921

111.7.3 Calculation Results of Niche Width

Using Simpson and Levins niche width as two measure models as formulas (111.2) and (111.3), the niche widths of geography dimension for 13 countries' contractors are shown in Table 111.6. The calculated graph results are basically consistent despite the differences between the values by using the two models (see Fig. 111.4).

Table 111.6 and Fig. 111.4 showed all 13 countries' international contractors' niche width was lower than the niche width (0.821, 0.798) of the total 250/225 International contractors in ENR significantly, indicating there are a vast market development potential for these countries' international contractors (especially for the Asian).

Table 111.6 Geography niche width of contractors from 13 countries listed in ENR (2010–2013)

Geography niche width	Total 250/225 international contractors in ENR	Italy	Canada	Britain	Spain	U.S.	Australia
$NW_{s(i)}$	0.821	0.804	0.799	0.794	0.79	0.786	0.727
$NW_{l(i)}$	0.798	0.73	0.712	0.693	0.68	0.668	0.523
Geography niche width	Turkey	China	France	Germany	Japan	Korea	Netherlands
$NW_{s(i)}$	0.719	0.704	0.703	0.657	0.582	0.566	0.453
$NW_{l(i)}$	0.508	0.483	0.481	0.416	0.341	0.329	0.261

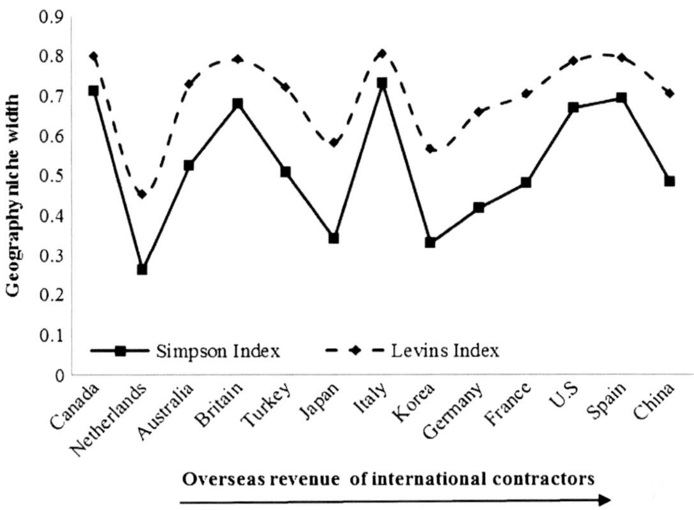

Fig. 111.4 Geography niche width graph

The western countries' contractors have been a leading force in terms of their international business distributions, market share and revenues by their capacity, influence and experience. Tables 111.5 and 111.6 proved that the European and the U.S. contractors had a higher geography niche width. Top five of geography niche widths are Italy, Canada, Britain, Spain and the U.S.:

Italy has the widest geography niche width, calculated by two formulas are 0.804 and 0.730 respectively. Seen from the layout, the Italian contractors are distributed in seven regions, and with the amount of US$8.37 billion in Africa, followed by the Middle East (US$6.986 billion), Europe (US$5.573 billion), and with the minimum amount in Canada (US$5.22 billion). Market development and resource utilization are diversified and balanced for Italian contractors in the seven regions. Canadian and British contractors have a more balanced geography distribution though their international revenue is not high, while the business of Spanish contractors is mainly in Europe, Asia and America. The U.S. contractors are distributed in the six regions by varying revenue degrees. They have much higher revenue in the Asia, Canada and Middle East. Geography niche width of the Netherlands contractors is the narrowest with their small distribution in the international construction market. The revenue generated by the Netherlands companies working in Europe is about 70 % while it is very low in other regions throughout recent years.

By comparison, contractors from Asia, such as China, Korea and Japan, show relatively narrow niche width in geography dimension, suggesting their main business only in fewer areas such as Asia and the Middle East. And the market share of Chinese contractors in Asia, Africa and Latin America keep growing, especially in Latin America with great potential though the revenue is not high yet. The results also showed most ICPC business located in developing regions, where there is a large demand of infrastructure construction projects and having relatively loose restrictions, such as in Asia, Africa and the Middle East.

111.7.4 Calculation Results of Niche Overlap

The results calculated by two niche overlap index formulas (111.4) and (111.5) of Pianka, Cowll and Futuyma are also basically consistent. Two indexes are verified each other and the results of Pianka niche overlap index are shown in Table 111.7.

Table 111.7 indicates that the contractors in the 13 countries have certain niche overlaps. There are 60 pairs of species have the Niche overlap index greater than 0.5, accounting for 77 % of the total number of pairs. There are 4 pairs of contractors have high niche overlap: Australia and Germany, France and Netherlands, Australia and Japan, Germany and Japan, with the niche overlap indexes of 0.976, 0.961, 0.923, and 0.921. The top four countries (Italy, Canada, Britain, and Spain) in spatial niche width ranking also have higher niche overlap. Except the overlap index between Italy and Spain is 0.69, the others were higher than 0.7. Since the Netherlands contractors' business basically located in Europe, the overlap between its contractors and Asian contractors is relatively small. The Overlap indexes between contractors from Netherlands and contractors from South Korea, Japan, and China was 0.181, 0.207, and 0.216. The Niche overlap reflects that those international contractors were in competition with each other as well as sharing market resources.

Table 111.7 Niche overlap of contractors from 13 countries listed in ENR (2010–2013)

Country	U.S.	Canada	Britain	Germany	France	Italy	Netherlands	Spain	Australia	Japan	China	Korea	Turkey
U.S.	1.000												
Canada	0.464	1.000											
Britain	0.611	0.733	1.000										
Germany	0.649	0.617	0.776	1.000									
France	0.502	0.524	0.684	0.541	1.000								
Italy	0.635	0.738	0.715	0.443	0.674	1.000							
Netherlands	0.375	0.318	0.579	0.420	0.961	0.536	1.000						
Spain	0.646	0.815	0.779	0.775	0.792	0.690	0.689	1.000					
Australia	0.734	0.640	0.842	0.976	0.507	0.490	0.379	0.768	1.000				
Japan	0.719	0.510	0.671	0.921	0.328	0.478	0.207	0.628	0.923	1.000			
China	0.619	0.640	0.595	0.596	0.407	0.830	0.216	0.501	0.602	0.720	1.000		
Korea	0.628	0.348	0.681	0.411	0.223	0.670	0.181	0.365	0.544	0.603	0.614	1.000	
Turkey	0.645	0.447	0.808	0.560	0.821	0.819	0.801	0.683	0.597	0.537	0.615	0.714	1.000

111.8 Conclusion

This paper introduced enterprise ecology theory into the study of the ICPC industry development. Based on data collected from the Engineering News Record and related cases, an in-depth analysis of the business distribution and competitive situation of CICs and other countries' international contractors are conducted. The results show that the evolution of Chinese international project contracting industry is rapid. The business layout of CICs is a constant state of change with the diversification development of the international market. At present, in African market, CICs are in an advantageous position temporarily. While in Asian market, CICs face fierce competition with the United States, Germany and Spain contractor. In the Middle East market, competitive pressures of CICs mainly come from South Korea, the United States and Italy contractors. Chinese international contractors shall not only to expand their shares in the Asia, Africa, and Middle East as familiar markets, but also to develop Latin America and the Caribbean as emerging markets.

The ICPC business of Asian contractors represented by the CICs, whose geography niche width are narrow compare with those of the European and the U.S. contractors, is mainly located in Asia, Africa and the Middle East. These areas are China's traditional markets, which need large capital construction and have low restrictions for foreign contractors. The niche width of Chinese contractors is expected to further growth since they have been actively pursuing new business in Latin America, and also strive to open up the European market. The study found the highest niche overlaps in spatial niche width were between Australia and Germany, France and Netherlands, Australia and Japan, Germany and Japan contractors. Low overlaps are between those European contractors with narrow geography niche width and Asian contractors, such as China and Netherlands. However, the international contractors from some countries with narrow geography niche width still have high niche overlap with other contractors, such as Netherlands and France, Australia and Japan. These findings should provide a valuable reference for both China and other countries' contractors who are interested in developing business in the international construction market.

Acknowledgments This study was funded by the project "Co-evolution and dynamic capabilities growth models of the Chinese international contracting business in complex environment" (No. 71371072) supported by the National Natural Science Foundation of China.

References

1. Xing HY (2015) Changing ideas and innovative ways to lead a new era of win-win situation—a review of the development of foreign engineering contracting in 2014. Int Econ Cooperation 3:4–10
2. Du Q, Yin CY, Chen YX (2015) Market analysis of 2013 annual top 250 international contractors. Constr Econ 1:28–32

3. ENR (2015) The top 250 international contractors. Eng News-Rec. http://enr.construction.com/toplists/Top-International-Contractors/001-100.asp July 2015
4. Du Q, Guo JL, Lu N (2013) International market analysis of Chinese contractors in 2002–2012. J Eng Manag 6:30–34
5. Kim SB, Cho JH (2014) A study on forecasting green infrastructure construction market. KSCE J Civil Eng 18(2):430–443
6. Li P (2008) Comparative analysis on the structure and development trend of international engineering contracting market. Constr Econ 2:12–15
7. Guan XD, Guan K, Lai X (2004) Comprehensive evaluation and simulation analysis of international competitiveness of top contractors. J Harbin Inst Technol 36(10):1354–1357
8. Tan YT, Shen LY, Yam MCH et al (2007) Contractor key competitiveness indicators (KCIs): a Hong Kong study. Surveying Built Environ 18(2):33–46
9. Zhao ZY, Shen LY, Zuo J (2009) Performance and strategy of Chinese contractors in the international market. J Constr Eng Manag 135(2):108–118
10. Liu GW, Wang L (2013) Study of the development difference between domestic and foreign top construction contractors: a sustainable financial capability perspective. In: 2013 international conference on construction and real estate management, pp 20–32
11. Pheng LS, Jiang H, Leong CHY (2004) A comparative study of top British and Chinese international contractors in the global market. Constr Manag Econ 22(7):717–731
12. Ji K, Lin YX, Li QM (2012) The Analysis of populations' developmental trend of top international contractors: case of U.S. and China. J Eng Manag 26(1):104–108
13. Peter R, Gary JT (2001–2014) The top 250 international contractors, enr.com, 08
14. China International Contractors Association (2013) An overview of China's foreign contracted projects in 2013. China Explor Des 5:41–45
15. Jin F (2011) Methods for analyzing the competitive environment of international contracting market. J Eng Manag 25(3):260–265
16. Zhang Y (2014) Analysis of the top 250 international contractors in 2014 ENR. J Eng Manag 28(5):131–137
17. Peter R, Gary JT (2010) The top 225 international contractors. Eng News-Record 44–56
18. Guo Y, Xu XY (2009) Overview of research on enterprise niche: concept, measurement and strategy. Ind Econ Rev 6:105–118
19. Levins R (1968) Evolution in changing environments: some theoretical explorations. Princeton University Press, Princeton

Chapter 112
PPP Concession Contract/Guidelines: A Comparative Analysis

Jing Zhang, Jie Li, Sha Xu and Jian Zuo

Abstract As an approach of providing public infrastructure and services, Public-Private-Partnership (PPP) has drawn growing attention all over the world. However, there are a number of problems associated with the implementation of the PPP contracts. PPP concession contract/guidelines from World Bank, UK, Australia and China are selected and compared in this study in terms of principles, structures, features of content and the completeness. This is followed by the deep analysis of failure cases of PPP projects. This paper aims to explore the relationship between the effectiveness and completeness of PPP concession contract/guidelines. This is achieved by means of comparing the effectiveness of solving real problems based on the most frequent market risks. Results show that the contract from UK which is the most completed pays more attention on market demand and market revenue. The guidelines of World Bank emphasize more on responding to the price variation risk. On contrary, guidelines of Australia and China both are guiding only which seems not sufficient to solve real problems. Therefore, the higher the completeness, the higher the effectiveness of PPP concession contract/guidelines is. These findings provide practical advice for practitioners to improve the PPP concession contract terms.

Keywords PPP concession contract/guidelines · Project risks · Core contract terms · Comparative analysis

J. Zhang (✉) · J. Li · S. Xu
School of Civil Engineering, Nanjing Forestry University, Nanjing, China
e-mail: fighting91916@163.com

J. Zuo
School of Architecture and Built Environment, University of Adelaide, Adelaide, Australia

Y. Wu et al. (eds.), *Proceedings of the 20th International Symposium on Advancement of Construction Management and Real Estate*,
DOI 10.1007/978-981-10-0855-9_112

112.1 Introduction

With the development of project financing, Public-Private-Partnership (PPP) has gained a growing public awareness. PPP was introduced by the UK government in 1992. Since then, PPP has been favored by countries around the world. Currently, PPP has become a critical approach in providing public services and improving capital efficiency for public sectors worldwide.

Lu and Cheng [1] pointed out that the content of the concession agreement reflects basic principle or stand of government to the authorized content of franchising projects. Iyer and Sagheer [2] identified 17 risks during the development phase of PPP projects in the Indian road sector which found that delay in financial closure, cost overrun risk, and time overrun risk are most dependent on other risks. Zhang et al. [3] analyzed the relationship chain of PPP contract network and the interaction between each contract. Khanzadi et al. [4] put forward a BOT concession term decision method which combined system dynamics (SD) and the fuzzy theory. Cruz and Marques [5] proposed to treat the uncertainty of the PPP projects as an opportunity where the method of flexible management should be adopted to deal with the uncertainty. Ashuri et al. [6] suggested a risk-neutral pricing method with minimum income guarantee based on the real option theory.

However, there is lack of systematic and comprehensive comparative analysis of various PPP concession contract/guidelines. Therefore, this study aims to fill this gap by studying the effectiveness of core contract terms of PPP concession contract through the case analysis. Findings will provide useful inputs for the design of PPP concession contract.

112.2 The Related Concepts of the PPP Contract

112.2.1 The Definition of the PPP Contract and the Parties

Public-Private-Partnership (PPP) is a long-term cooperative partnership built up in the construction of infrastructure based on the formal agreement between the public and private sector [7]. There are many stakeholders during the implementation of PPP projects, including government, private companies, SPV, financial institutions, contractors, design institutes, supervision firms, etc. The operation flow of PPP projects is: the project sponsors launch project intention; concessions are granted to SPV; SPV is in charge of the project financing, construction, operation, etc.; SPV transfers the project to the government after concession period expires [8].

112.2.2 The Characteristics and Content Framework of the PPP Concession Contract

The PPP concession contract is long-term, complex, diverse and full of risks. The main clauses of PPP concession contracts are largely similar. Generally the following contents are included: definitions, rights and obligations of each party, project financing, construction, operation, maintenance, concession, fees, pricing, transformation and the general terms.

112.3 A Comparative Analysis of PPP Concession Contract/Guidelines

PPP models at UK and Australia are widely recognized as the best over the world. Therefore, PFI Contract Standard (UK) and PPP Victoria (Australia) are selected in this study to compare with General Contract Guidelines of PPP Projects (China). In addition, Concessions of Infrastructure Construction: the Design and Granting of Concessions which was issued by World Bank provides comprehensive guidance. They were chosen to conduct comparative analysis.

112.3.1 The Comparative Analysis of Principles

The results of the comparative analysis of their principles are shown in Table 112.1.

Table 112.1 The comparative analysis of principles

Name	Principles
World Bank	(1) Seek the balance of contractual relationship. (2) Combine with international experience. (3) Broad applicability. (4) Emphasize project performance and treatment of key problems of contract
UK	(1) Make people realize main risks associated with the PFI project. (2) Keep consistency of similar projects in the implementation and pricing. (3) Abandon the lengthy negotiation process
Australia	(1) Government is responsible for the delivery of core services. (2) Emphasize VFM of projects and public interests. (3) Reasonable risk allocation. (4) Broad applicability. (5) The principle of "best practices"
China	(1) Emphasize equal contractual subjects. (2) Emphasize on improving the quality and efficiency of public service. (3) Emphasize that social capital can obtain a reasonable return. (4) Emphasize transparency and openness. (5) Emphasize legality and effective implementation. (6) Emphasize combining international experience with domestic practices

Table 112.2 The comparative analysis of structures

Name	Structures
World Bank	It consists of six parts namely introduction, the environment, design and granting of franchise, regulators, government support. Each section lists the issues and their related items to further elaborate
UK	It consists of 37 chapters and 1 attachment. It lists some likely issues of PPP projects according to the development of projects in each chapter. Possible terms, related items, processing methods, etc. in these issues are described in detail in each clause
Australia	The establishment and development of partnership, project risk management and processing of key problems, overall project management and contract management are explained separately in three guidelines. In each guideline, the related issues in the terms are combined into a part. Related model texts are attached at the end of each file to conduct normalized instruction
China	It uses modular framework to set up 15 modules and 86 clauses. In principle, the body of project contracts in all modes should contain 10 general modules while five modules namely investment and financing, construction, operation, service and transformation can be used flexibly case by case

112.3.2 The Comparative Analysis of Structures

The results of the comparative analysis of their structures are shown in Table 112.2.

112.3.3 The Comparative Analysis of Main Contents

The results of the comparative analysis of their main contents are shown in Table 112.3.

Table 112.3 The comparative analysis of main contents

Name	Main contents
World Bank	(1) Emphasize government's rights and responsibilities. (2) Emphasize the importance of regulators. (3) Regulate the granting of concession
UK	(1) Detailed content. (2) Emphasize the rights and obligations of main credits. (3) Compared with the design and construction, it pays more attention to operational service. (4) Describe changes of the contract elements more fully
Australia	(1) Systematic. (2) Seek for the standardized project management processes. (3) The model instruments are more practical
China	(1) Strong applicability. (2) The coherent content. (3) List the considerations of contract design comprehensively

Table 112.4 The comparative analysis of the completeness

Name	The completeness
UK	It pays attention to the analysis of specific risk factors. It explains the individual issues one by one and describes terms in detail. The contents are the most detailed. It belongs to the demonstrative text and is the most completed
World Bank	It pays attention to the treatment of specific contractual issues. It explains the individual issues one by one. It helps related decision makers to understand the main issues associated with the design, granting of concession, etc. It belongs to the transitional text between demonstrative text and frame-type text and is comparatively completed
Australia	They focus on the macro guidance. They provide a theoretical framework to conduct standardized instructions however consider little about the details. With systematic contents, however, they fail to provide full instructions or guidance for the efficient execution. They belong to the guidance document and are the most incomplete

112.3.4 The Comparative Analysis of the Completeness

Based on above analysis, the results of the comparative analysis of their completeness are shown in Table 112.4.

112.4 The Risks Analysis of PPP Projects Based on Cases

112.4.1 The Cases Analysis

Some failure cases of PPP projects from the UK, Australia and China are analyzed to identify possible causes of failure. These cases are from a variety of sectors such as sewage treatment, water supply, power plants, bridges, railways, roads and tunnels. The details are shown in Table 112.5.

112.4.2 Risks Analysis

Through the analysis of the above cases, the corresponding risk factors are identified and summarized in accordance with their frequency. The results are shown in Table 112.6. From Table 112.6, market risk was found to be the most frequent.

Table 112.5 The failure cases of PPP projects

No.	Area	Project name	Case analysis
1	Sewage treatment plants	A Sewage Treatment Plant of Jiangsu	The project company was forced to negotiate ROI with the government. During this process, the project was interrupted by SARS. Failure causes: policy changes, government credibility, force majeure
2		Changchun Rate Sewage Treatment Plant	In 2000, the government formulated the corresponding management regulations. In February 2003, the regulations were abolished. Since March, the drainage company had no longer paid any sewage treatment fee. In 2005, government repurchased it. Failure causes: the government fails to fulfill commitments
3		Qingdao Veolia Sewage Treatment Plant	The frequent change of government's attitude led to a lengthy negotiation. In case of incomprehension of market price, the higher sewage treatment price was determined. Later, government wanted to negotiate to cut the price. Failure causes: negotiation delay, government credibility
4		Tangxun Lake Sewage Treatment Plant	After the completion of the first phase of the project, supporting pipeline network construction, the collection of sewage charges and so on remained unsettled. As a result, the factory was not operating and was transferred to Wuhan Water Group finally. Failure causes: supporting facilities can't be supplied in time
5	Water supply plants	Lianjiang China-France Water Supply Plant	Originally, the daily amount of water to be purchased from the water company was agreed not less than 60,000 m^3 while the daily consumption was only 20,000 m^3 in fact. As the contract was separated from practice, the plant was forced to be idle until now. Failure causes: government credibility, insufficient market demand, price variation
6		Shenyang Ninth Water Supply Plant	Because of the corruption and excessive ROI promised, the water price Shenyang Water Corporation paid Ninth Water Supply Plant was too high. By 2000, total losses amounted to 200 million yuan and the corporation requested to change the contract. Failure causes: official corruption, government credibility
7		Beijing Tenth Water Supply Plant	In 2003, the reform of water price across the country had suffered public resistance which delayed the project approval. As a result, Anglia withdrew from the project. Failure causes: delay of project approval

(continued)

Table 112.5 (continued)

No.	Area	Project name	Case analysis
8		Desalination Plant of Victoria	Investment costs continued to increase, rising from $2.9 billion to $4 billion. Due to lack of demand, the entire factory immediately entered the idle mode. Cause: increasing costs, insufficient market demand
9		Shanghai Big Field Water Supply Plant	The reform measures of water price could not be carried out which led to delay of project approval. Moreover, the new policy in 2002 led that the project company was forced to negotiate ROI with government. In 2004, government repurchased it. Failure causes: delay of project approval, policy changes
10	Power plants	A Power Plant of Hunan	Choppy political situation made the bidder fail to finance. The government withdrew the project, confiscated bid guarantee and hadn't started the tender anew. The project failed. Failure causes: force majeure, financing risk
11		Shuanggang Waste-to-energy Plant of Tianjin	The contract stated that government would provide subsidies if insufficient market revenue was caused by agreed causes. But the subsidy amount wasn't clearly defined that caused project company assumed the risk of insufficient revenue. Failure causes: government credibility, insufficient market revenue, unclear government subsidies
12		Chinese Power Project of Shandong	The change and reform of Shandong electric power market had a significant impact on the project. The price fell to 0.32 yuan/kwh from 0.41 yuan/kwh. The minimum purchasing power was threatened which caused market demand changes. Failure causes: government credibility, market demand changes, price variation
13	Bridges	Hangzhou Bay Sea-crossing Bridge	In 2010, the vehicle flow was 30 % fewer than expected. Since 2013, competitive projects appeared in succession. It mightn't take back the principal. Failure causes: less-than-expected market demand, competitive projects
14		Xinyuan Min River Fourth Bridge	The government promised if fees couldn't be guaranteed, they would repay foreign investors. After the bridge opened, many vehicles bypassed toll station. The government failed to fulfill promises. Failure causes: competitive projects, government credibility
15		Fujian Quanzhou Erythrina Bridge	Exclusive terms weren't considered. When the revenue was in a better condition, government invested to construct the parallel new bridge which led traffic distribution and loss of interests. Failure causes: competitive projects

(continued)

Table 112.5 (continued)

No.	Area	Project name	Case analysis
16	Railways	London Subway Project	London government signed three PPP contracts with three companies controlled by two big unions. The government provided the guarantee for debt. The disagreements and ineffective management led that unions successively bankrupted. Failure causes: government regulatory inaction, chaotic management of unions
17		Beijing Subway Line 4	The main problem was about rail transportation subsidies mechanism. The approval procedures were also too cumbersome. Failure causes: unreasonable government subsidy mechanism, delay of project approval
18	Tunnels	Yanan East Road Tunnel	In 2002, new policy was enacted. The project company was forced to negotiation ROI with government and was repurchased finally by the government. Failure causes: policy changes
19		Anglo-French Channel Tunnel	The price competition couldn't guarantee a stable ROI due to "uniqueness of the project". Their governments caused cost overruns and schedule delay. The issue of operating license was delayed. They filed for bankruptcy eventually. Failure causes: competitive projects, insufficient market revenue, cost overruns, schedule delay, operation delay
20		New South Wales Intercity Toll Tunnel	Predicted capacity was 90,000 cars per day while real capacity was 34,000 cars per day. Failure causes: Less-than- expected market demand
21		New South Wales Langu Toll Tunnel	Predicted capacity was 115,000 cars per day while real capacity was 58,000 cars per day. It entered into bankruptcy procedures within 3 years. Failure causes: less-than-expected market demand
22		Queensland Cramer Jones Toll Tunnel	Predicted capacity was 100,000 cars per day while real capacity was 21,178 cars per day. A year later, it went bankrupt because of failing to pay the loan. Failure causes: less-than-expected market demand
23		Sydney City-crossing Tunnel	The real capacity was 35,000 cars per day much less than expected 90,000 cars. In 2007, it was sold to private company because of debts. Failure causes: less-than-expected market demand
24	Roads	Beijing-Tongzhou Road	At the beginning, the neighboring roads are free which led to the lack of traffic for a long time. Failure causes: competitive projects

(continued)

Table 112.5 (continued)

No.	Area	Project name	Case analysis
25		Brisbane Airport Line Toll Roads	The predicted capacity was more than 100 % than the real capacity. The market value reached $1.2 billion. Before long, the share price fell 1 %. Failure causes: less-than-expected market demand
26	Sports stadiums	Beijing National Stadium "Bird's Nest"	There was interests conflict among shareholders. The government unilaterally dominated the changes of provisions. The project faced strong market competition and a conflict between commercial and public interest and the project company lacked operating experience. Finally, the government took over the project. Failure causes: interests conflict, government dominated changes of provisions, public opposition, market operation problems

112.5 The Effectiveness of PPP Contract/Guidelines and Relationship Between the Effectiveness and Completeness

112.5.1 The Comparative Analysis of the Effectiveness

According to Table 112.6, risk factors can be summarized into three categories, i.e. price variation, less-than-expected market demand and market demand changes. Moreover, the insufficient market revenue is also considered. The effectiveness of the relevant provisions of texts from World Bank and UK was analyzed (Tables 112.7, 112.8, 112.9 and 112.10.

1. Price variation
2. Less-than-expected market demand
3. Market demand changes
4. Insufficient market revenue

In addition, the guidelines from Australia belong to the guidance document which don't offer specific solutions for specific problems. But its risk management guidelines give lots of guidance for market risks which is worth referring. The guidelines from China also belong to the guidance document and need further perfection as little thought is given to market risks.

Table 112.6 The analysis of main risks in cases

Risks										
No.	Project approval	Protracted negotiations	Financing risk	Completion risk	Production risk	Market risk	Law and policy changes	Government credibility	Corruption risk	Force majeure
Case 1							√	√		√
Case 2								√		
Case 3		√						√		
Case 4					√					
Case 5						√		√		
Case 6								√	√	
Case 7	√									
Case 8				√		√				
Case 9	√						√			
Case 10			√					√		√
Case 11						√		√		
Case 12						√		√		
Case 13						√				
Case 14						√		√		
Case 15						√				
Case 16						√				
Case 17	√									
Case 18							√			
Case 19				√		√		√		
Case 20						√				
Case 21						√				
Case 22						√				

(continued)

Chapter 113
Developing an Ontology-Based Knowledge Base for Residual Value Risks in PPP Projects

Kaiwen Chen, Xin Xu, Faxi Yuan, Jingfeng Yuan and Qiming Li

Abstract Aiming at improving the knowledge management in PPP projects, the development of an ontology-based knowledge base for Residual Value Risk (RVR) in PPPs was conducted. Through literature review and typical case study of knowledge management in PPP projects, the RVR in PPP projects was analyzed from the perspective of risk structure, risk process and vulnerability, based on which the formation mechanism of RVR was explored. The RVR was deconstructed into five risk consequences. Furthermore, an improved RVR framework was established as a platform for organizing all kinds of knowledge for RVR in PPPs. Meanwhile, in order to express knowledge more effectively and promote their sharedness and transferability, the knowledge base was built by means of ontology. Finally, the ontology editing software Protégé was used to build ontology-based RVR knowledge base, to achieve a more in-depth data processing function. The study can be used to analyze risk probability and vulnerability for both of public and private sectors in PPPs.

Keywords Ontology · PPP projects · Residual value risk · Knowledge base

113.1 Introduction

With the increasing demands for infrastructure facilities and lack of funds from the government, public private relationships (PPPs) has been popularly used in the infrastructure facilities including highway, railway, hydropower station, medical

K. Chen · X. Xu · Q. Li
School of Civil, Southeast University, Nanjing, China

F. Yuan
School of Civil, Politecnico di Milano, Milan, Italy

J. Yuan (✉)
Department of Construction and Real Estate, Southeast University, Nanjing, China
e-mail: jingfeng-yuan@seu.edu.cn

Y. Wu et al. (eds.), *Proceedings of the 20th International Symposium on Advancement of Construction Management and Real Estate*,
DOI 10.1007/978-981-10-0855-9_113

facilities and schools all around the world. PPP projects, however, possess characteristics like long periods, high complexity, multi-objective and multi-participant. Consequently, PPP projects will confront various risks in the construction and operation phase, such as residual value risk (RVR), a kind of key risk in PPP project, which may result in the losses of public sectors due to residual value (RV) lower than estimated or anticipated value when the private party agreed to transfer the project to them. Then, losses caused by RVR may influence the project's success. Therefore, how to reduce or prevent project's losses due to such a risk is becoming a managers' urgent issue.

While previous research have been conducted in PPPs, research focusing on RVR in PPPs almost remains blank except a few papers have mentioned concept and distribution methods of RVR [1–4]. Therefore, the significance of RVR in PPPs has not been fully recognized by either the public or private sectors. Furthermore, to share the knowledge of RVR in PPPs with the public, identification of the RVR framework and establishment of the knowledge management system for RVR become necessary.

Moreover, knowledge base can contribute mainly to the presentation, transfer, reasoning and acquisition of knowledge so as to make knowledge retrieval possible to meet the users' demands. Therefore, introducing ontology theory into engineering field can contribute to the collection of related knowledge so that it can provide the common understanding in the corresponding field. Furthermore, ontology's introduction can also help to determine the commonly recognized concept, present the clear definition of the concepts (terminology) and interrelation between these concepts through formalized modes with different levels, and then describe concept meaning with relationships between concepts. Consequently, ontology theory increases the reusability of knowledge by providing a formalized method for structured presentation of field knowledge and also increases the shareability of knowledge by supporting the detachment of static knowledge and operational knowledge.

This paper aims at developing an ontology-based knowledge base for RVR, making such base present a better description of RVR's characteristics in PPPs and also providing a better convenience for the shareability and transferability of RVR knowledge. Furthermore, the adoption of such a base can help to conduct a more effective dynamic management and prediction of RVR, which helps to reduce RVR in all stages before transfer and improve performance in a project. Lastly, application of such a base can provide protection for the project transfer and reasonable decision advices for government's effective monitoring in PPP projects.

113.2 Research Background

113.2.1 Residual Value (RV) and Residual Value Risk (RVR) in PPPs

Residual value, in automotive lease contracts, is defined as the market price or value of the leased vehicle at maturity of the lease contract [5]. While generally, RV is one of the constituents of a leasing operation, which describes the future value of goods in terms of the percentage of depreciation from its initial value [1]. As proposed by Lucko and Vorster [6], terminology that is used to describe the concept of RV varies widely in the literature, including market value, salvage value, resale value, and trade-in value.

As for PPPs, RV is defined as the values transferred to the government at the end of the concession period [2]. Therefore, the research adopted the definition of RV as mentioned by Ke et al. [2], which includes two key points: (a) the time for RV means the end of concession rather than asset retirement; and (b) the meaning of RV includes not only asset but organization and people, intellectual property as patented technology and brand.

Previous studies have listed out the main risk factors by literature review [2] to conduct the comparative analysis of risk allocation preferences from Lam et al. [7], Ng and Loosemore [8], Bing et al. [3], Arndt [9], Wang and Tiong [10], NTSA [11], VDTF [4]. Consequently, residual assets (value) risk and residual risk have been identified as 'operation' and 'other' risk factors respectively. While residual risk has been defined in capital asset pricing model (CAPM) [12], the research adopted residual value risk (RVR) to avoid misunderstanding.

RVR has been defined in many previous research: Yuan et al. [1], VDTF [4], Bing et al. [3] and Ke et al. [2]. Based on their study, the research defined RVR as uncertainty on expiry that RV may be lower than estimated or anticipated value when the private party agreed to transfer it to the public.

113.2.2 Previous Studies on RVR in PPPs

RVR has been mentioned in previous studies while few attention has focused on RVR in PPPs. More exactly, these work have proposed the concept of RVR as mentioned above, perception, causes, losses due to RVR, mitigation measures and distribution methods [1–4].

In terms of perception of RVR in PPPs, Yuan et al. [1] has conducted an open-end questionnaire to discuss with 46 PPP experts and professionals about the significance, the initial definition proposed by the authors, the contents, and the treatment methods of RVR. Consequently, 95 % of the respondents considered RVR as an important or extremely important concern in PPPs and the contents of RVR in PPPs includes facilities, technical documents, project organization, equipment, intellectual property, instruments, project reputation and goodwill, and market shares. Treatment methods for RVR, however, were various among respondents. Moreover, a conceptual model was established to explain how the RVR occurs, how the RV changes, and which factors influence the RVR and to identify the cumulative impacts of the interactions between different risk factors on the RV changes.

As for the causes of RVR, previous study has been done to identify the influencing factors for RVR. Xiong [13] has tidied up all the related factors for RVR in PPPs by literature review, where the project has been divided into 4 stages as preparation, design, construction and operation. Then, based on these collection of influencing factors for RVR in PPPs, Xiong [13] has adopted structure equation model (SEM) to assess RVR in PPPs.

While for mitigation measures, based on the allocation to the government, Partnerships Victoria [4] has proposed: (a) government will impose on the private party maintenance and refurbishment obligations, ensure an acceptable maintenance contractor is responsible for the work, commission regular surveys and inspections; and (b) government may also direct funds from the project into dedicated controlled sinking fund accounts to accumulate funds sufficient to bring the asset to agreed condition and/or (if required) obtain performance bonds to ensure the liability is satisfied.

Lastly, in terms of the distribution of RVR in PPPs, many previous work has been done to solve this issue. The most distinctive studies have been done by Bing et al. [3] and Ke et al. [2] respectively. More precisely, by postal questionnaire method, Bing et al. [3] defined residual (value) risk in meso group and concluded that 22 % of survey respondents considered that RVR should be allocated to public sector and 22 % thought it should be shared for PPP/PFI construction projects in UK. In terms of PPPs in China, Ke et al. [2] conducted a two-round Delphi survey to identify the preference of risk allocation in China's PPP projects. With the help from SPSS software, mean scores for residual assets (value) risk is 3.52 in the range 3.5–4.5, which means RVR should be allocated to the private sector.

In a nutshell, a systemic study which focuses on RVR management has not been conducted so that the current knowledge is not available for sharing in the public, while RVR is becoming a more and more important factor which needs to be considered in the contract phase. Therefore, establishment of a knowledge management system for RVR is necessary and the paper will focus on knowledge management for RVR in PPPs by introducing ontology.

113.2.3 Previous Studies on Ontology

Ontology as a philosophical concept can be traced back to the ancient Greek philosopher Aristotle BC (BC 384–322 BC). It is defined as "systematic description of objective existence in the world", is an abstract and simplified view in certain area. Then, in the 1990s, there was a substantial increase in definitions of ontology.

In 1993 [14], Gruber first put forward the definition of ontology, which is then quoted the most, as "an explicit specification of conceptualization, which refers to a description of the concept in a specific domain and the relationship among them". Later, Studer et al. [15] further defined ontology as "a formal, explicit specification of a shared conceptualization", and considered it mainly contains four aspects: the conceptualization, explication, formalization and share.

The purpose of utilizing ontology is to access relevant domain knowledge, provide common understanding of domain knowledge, confirm common recognized terms of a domain and provide clear definition of the terms and relationship between these terms from different levels of the formal model [16]. It can be applied to relevant research field of artificial intelligence, representation of knowledge, semantic web, system integration and problem solving techniques. In the construction management academia, ontology had been applied to knowledge representation, decision making, and information integration. Nowadays, seeing the importance of ontology in knowledge management, research related to ontology and semantic development increases rapidly, and a batch of researchers in the field of project management has been trying to apply ontology into risk management in construction projects.

Tserng et al. [17] proposed an ontology-based risk management framework (ORM) of construction projects through project life cycle in order to enhance the RM performance. Compared with traditional tools, the ORM framework enabled contractors to transfer organization knowledge more effectively and reduced the complexity of common RM workflow. Mostafa and El-Gohary [18] developed a semantic ontology model for representing and reasoning about the stakeholders, benefits and social equity analysis of transportation projects. Its structured, extendable and flexible format facilitated future evolution and extension of the ontology and its analysis contexts. Jiang and Zhang [19] introduced an ontology-based semantic retrieval method for construction project risk management in order to facilitate project participants to query required information among numerous project documents efficiently. Fidan et al. [20] presented a formal ontology for relating risk and vulnerability to cost overrun in the international projects, which shaped a basic structure and provided a common vocabulary of risk database.

On the whole, ontology has obvious advantages in project description, reasoning, shareability, etc. However, related application practice focus on small projects, many large-scale applications are still in development, and the literature relating to ontology application are not detailedly described. There is a gap in function between the existing ontology editing tools and the more overall application in knowledge database system, so the combination of them should be taken into consideration. As for the study of residual value risks in PPP projects, this paper provides the RVR framework and discusses the detailed concepts as well as their relationships in order to apply ontology into knowledge database of RVR in PPPs.

113.2.4 Development of RVR Framework

In order to establish a knowledge framework of residual value risk in PPP projects, which further analyzes the composition of knowledge system and the relationship between RVRs in PPP projects, we need to firstly study the knowledge framework of risk concepts in construction projects and the relationship between different concepts. Through literature review, we learned that the related concepts mainly contains 2 types, namely risk and vulnerability, and their relationship can be structured with the risk path, as depicted in Fig. 113.1 [20, 21].

As it can be seen from Fig. 113.1, the risk path is composed of different risk sources (RS), risk events (RE) and risk consequences (RC) factors in a causal relationship. While vulnerability (V) is defined to reflect the exposure to risk in one project, or the ability to resist risks when facing an internal characteristic, including robustness (V1), resilience (V2) and sensitivity (V3). Vulnerability enriched the attributes of relationship, so that relationship is no longer simply a link between two risk factors, the relationship itself is adjusted to a third factor (the vulnerability). In addition, it should be mentioned that the structure given in Fig. 113.1 is believed to be applicable for all contractors among the same PPP project, regardless of where or what other specific situation of the PPPs happened.

113.3 Ontology Development

Despite there were a great deal of scholars focus on establishing a system of risk impact in risk in PPP projects, they did not use the same terminology. Therefore, it is difficult for the comparison and communication between the outcomes of various

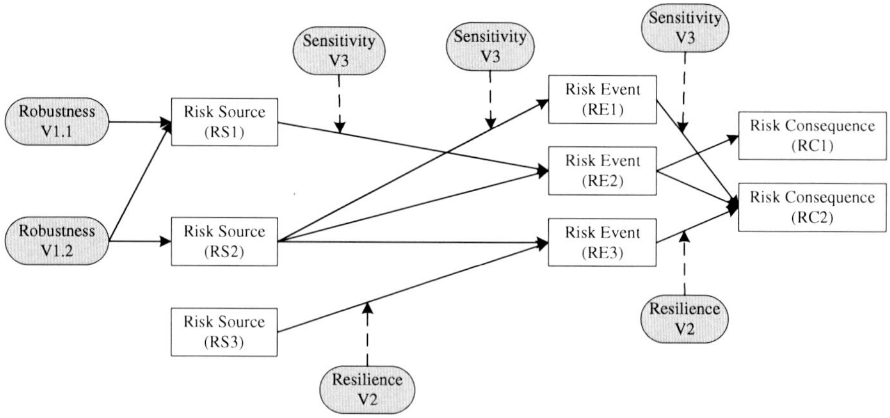

Fig. 113.1 Risk-vulnerability paths

systems. It greatly increase the difficulty of knowledge sharing and reuse. Therefore, building the ontology for RVR in PPP projects is a practical solution.

Ontology knowledge base system refers to the theory of structure. Currently, there is not a widely accepted ontology modeling method for engineering construction. Many scholars built their own methods and procedures according to the guidelines and their experiences of ontology development, but they are essentially the same. According to the process of developing ontology in the School of Medicine in Stanford University [22], we sum up the following five steps for building an ontology system:

- Scope and Specification—identification of basic limitations and necessities of current RVR literature.
- Class and Taxonomy—determination of basic concepts in the RVR framework and description of their hierarchy structure.
- Relationship Expression—development of important interrelations between these concepts through literature review and case studies.
- Property Identification—determination of formal property of object and datatype with regards to the concepts and their interrelations.
- Individual and Validation—evaluation of the established ontology by analyzing one specific individual.

The scope and specification of the ontology to be established in this paper is to define RVR in PPP projects. It mainly studies the concepts of all classes and their relationships based on the risk-vulnerability paths. Then, the following two parts separately discuss the taxonomy of concepts of all classes and relationship knowledge expression in the RVR ontology. The definitions of object property and datatype property will then be demonstrated in the relationship model previously established to express the taxonomy of classes and relationships.

113.3.1 Taxonomy of Concepts in the Ontology

Protégé represents a collection of individuals by class, which is presented by the hierarchy diagram of parent and child classes—Taxonomy. The basic classes of RVR in PPP projects include risk sources (RS), risk events (RE), risk consequences (RC) and vulnerability (V). The risk sources means the risk of actual loss that has not been emerged in one project, while risk events means the risk of actual loss that has been emerged. Thus, they are classed in the same concept system with different codes. Using hierarchical risk-breakdown structures (HRBS) ([12, 23]; Lam et al. 1999), we subdivide the RVR factors in PPP projects into a hierarchical system, shown in Fig. 113.2.

While the framework shown in Fig. 113.2 demonstrates the overall risk in PPP projects, RVR is only one aspect of the overall risk. In order to highlight the RVR

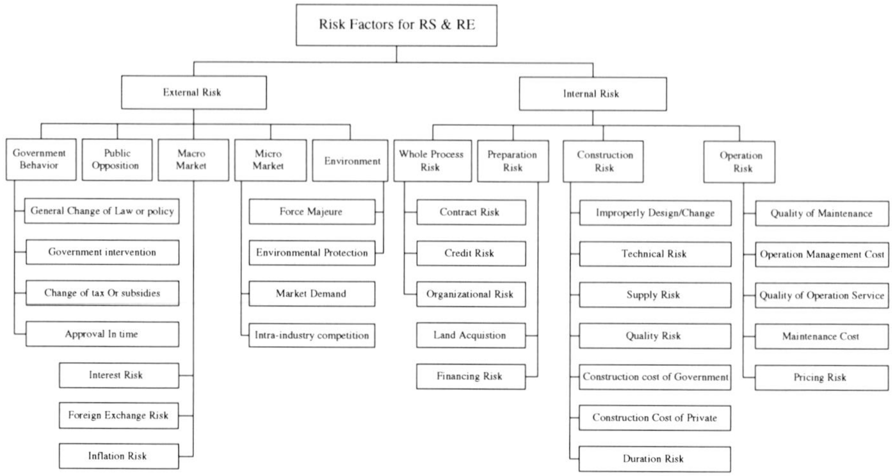

Fig. 113.2 Taxonomy of risk factors for RS and RE

perspective, this paper introduces the risk consequences (RC) into connecting the RVR with other factors in the framework. However, PPP projects are especially complex construction projects, whose value composition contains a variety of factors from many aspects. As discussed by Ji et al. [24], RVR in PPP projects can be divided into two parts, according to the concrete manifestation of each risk, the economic value risk (RVE) and the social value risk (RVS). Among them, RVE refers to risks whose economic value of the project at the handover point cannot meet the government's expectations, including yield rate risk and yield years risk, while RVS represents risks whose social value cannot meet the government's expectations, including governmental credibility risk, ecological and resource risk and social beneficial risk [25]. Then, we consider these five residual value risk concepts as risk consequences. The taxonomy of the RVR consequences related concepts included in the ontology are given in Fig. 113.3.

In spite of the fact that all companies and projects are exposed to risk, some characteristics of firms and projects will influence the impact of risk in the event of its occurrence [25]. The term "vulnerability (V)" is used to explain inborn characteristics of a system that exist within systems independently of external hazards and depend on an organization's capability to manage risks. The identified vulnerability within a PPP project system are categorized considering their places in a risk path as follows: robustness (V1), resilience (V2) and sensitivity (V3) [20]. Robustness (V1) refers to the degree of how one project exposed to risk and the possibility of the existence of one risk in the project. It is influenced by the capacity of the government and the private sector, the cooperation experience, the

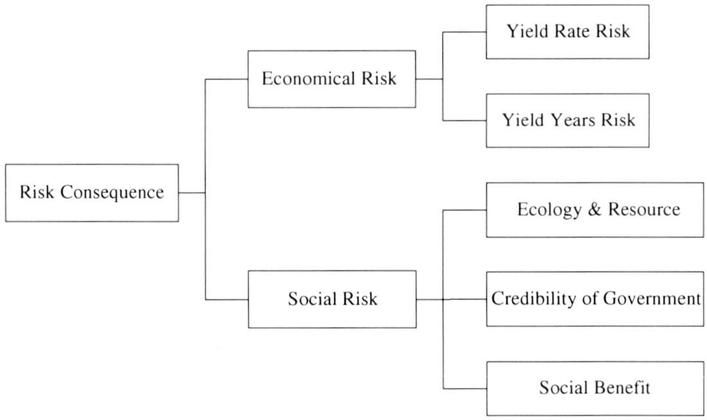

Fig. 113.3 Taxonomy of risk consequences

construction difficulty and financing situation. Resilience (V2) refers to the restorability of the project after the occurrence of one risk source or risk event. It is a more complex indicator, primarily concerned with the government and private sectors' management capability and experience, which exists only in inherited relationships. Sensitivity (V3) is the effect degree of one risk factor effected by non-inherited risk factors in the risk framework. It is concerned with not only the government and private sectors' management capability and experience, but also the sharing mechanism determined by contract terms. V3 only exists in non-inherited relationships. The taxonomy of vulnerability included in the ontology are given in Fig. 113.4.

113.3.2 Relationship Expression and Model

In this paper, we use semantic network, the most commonly used knowledge representation method, to establish the domain ontology. There are mainly four kinds of semantic relations as the following descriptions.

Generic relationship: generic relationship is the most common kind of relationship, referring to the classification relationship between different concepts with the same properties. This kind of relationship can be transferred. Which means, knowing that "RS" belongs to "Risk" and "Risk" belongs to "Thing", we can deduce that "RS" belongs to "Thing", according to the transitivity rule.

Affiliation relationship: this relation is a general property, which means the project has risks or has vulnerabilities. "Has risk" can be broken down to "Has RS",

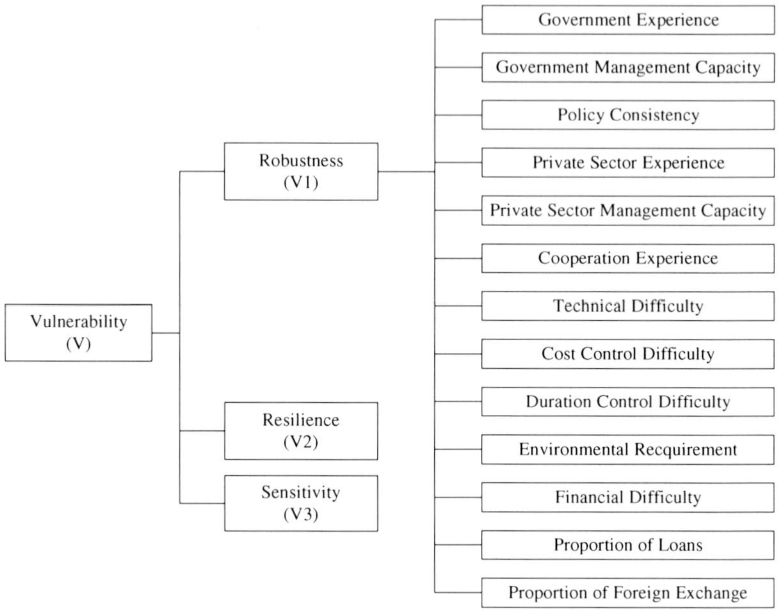

Fig. 113.4 Taxonomy of vulnerability

"Has RE" and "Has RC", while "Has vulnerability" can be subdivided to "Has V1", "Has V2" and "Has V3".

Inherited/non-inherited relationship: inheritance relationship is the relationship between RS and the RE/RC caused by the same source/event during adjacent stage. It is regulated by resilience (V2). Non-inheritance relationship is the relationship between risk and vulnerability except for inheritance relationships. It is regulated by the sensitivity (V3).

Numerical relationship: Numerical relationships are data-related relations. The definition of numerical relationship include name, domain and value limits.

Despite these four main relationships, there still exists some other common relationships as we can see in the RVR framework, such as RS is influenced by V1, RE is generated from RS, RC may be generated from RE (since it possesses conductivity, RC may also be generated from RS, as well as be influenced by V1). Figure 113.5 shows the classes and their relationships diagram of the RVR ontology. Apart from the body structure of ontology it has described, Fig. 113.5 also shows the defined object and datatype properties of each concept.

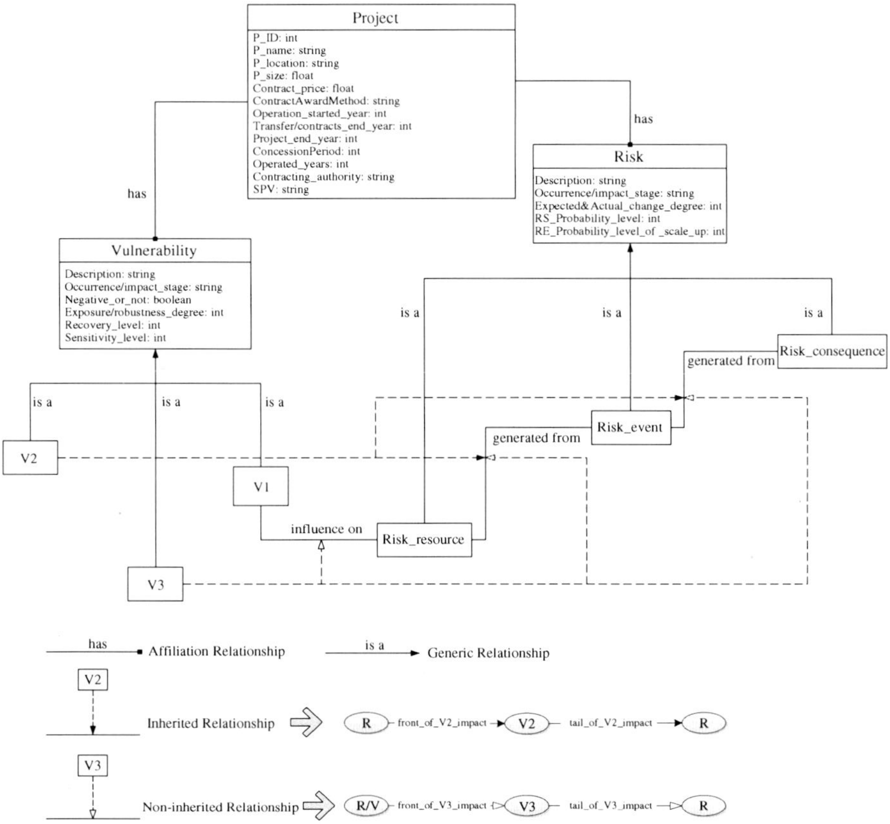

Fig. 113.5 Data model for RVR ontology in PPP projects

113.4 Conclusion

In this paper, we firstly discuss the significance and research status of developing a RVR ontology for PPP projects. Then, on the basis of reviews over risk knowledge framework in PPP projects, combined with the residual risk consequences and vulnerabilities of PPP projects, we build the taxonomy for RVR ontology. On the basis of residual risk framework, to build a residual risk ontologies. Finally, through analyzing the risk path's inherent relationships between different classes and their datatype properties, we build the final inner-connected ontology model. Through this method, this established ontology is fairly complete, general and effective in capturing residual value risk sources, events and consequences in PPP projects, which can be used for analyzing and predicting the RVR situation and related parameters in similar PPPs. In future research, we will focus on the application of ontology editing tools such as Protégé, to further study how to manage the editing and analysis function for better predicting RVR in PPPs.

Acknowledgments The authors' special thanks go to all and reviewers of the paper, and to the National Natural Science Foundation of China (NSFC-71001027, 71472037), Social Foundation of Jiangsu Province, China (13GLB005), Program for Outstanding Young Teachers of Southeast University (2242015R30009), and the Fundamental Research Funds for the Central Universities for financially supporting this research.

References

1. Yuan J, Chan AP, Xiong W, Skibniewski MJ, Li Q (2013) Perception of residual value risk in public private partnership projects: Critical review. J Manage Eng 31(3):04014041
2. Ke Y, Wang S, Chan AP, Lam PT (2010) Preferred risk allocation in China's public–private partnership (PPP) projects. Int J Project Manage 28(5):482–492
3. Bing L, Akintoye A, Edwards PJ, Hardcastle C (2005) The allocation of risk in PPP/PFI construction projects in the UK. Int J Project Manage 23(1):25–35
4. Victorian Department of Treasury and Finance (2001) Partnerships Victoria: risk allocation and contractual issues, Australia, June 2001, pp 178–191
5. Nau K (2012) An empirical analysis of residual value risk in automotive lease contracts. Dissertation, University of Hohenheim, Stuttgart, Germany
6. Lucko G, Vorster MC (2003) Predicting the residual value of heavy construction equipment. In: Proceedings of the 4th joint international symposium on information technology in civil engineering, pp 15–16
7. Lam KC, Wang D, Lee PT, Tsang YT (2007) Modelling risk allocation decision in construction contracts. Int J Project Manage 25(5):485–493
8. Ng A, Loosemore M (2007) Risk allocation in the private provision of public infrastructure. Int J Project Manage 25(1):66–76
9. Arndt RH (1998) Risk allocation in the Melbourne city link project. J Struct Finance 4(3):11–24
10. Wang SQ, Tiong LK (2000) Case study of government initiatives for PRC's BOT power plant project. Int J Project Manage 18(1):69–78
11. National Treasury of South Africa (2004) Public-private partnership manual-Module 4: PPP feasibility study, South Africa, pp 63–66
12. Lam PTI (1999) A sectoral review of risks associated with major infrastructure projects. Int J Project Manage 17(2):77–87
13. Xiong W (2011) Research on residual value risk assessment in PPP project. Dissertation, Southeast University, Nanjing, China
14. Gruber TR (1993) A translation approach to portable ontology specifications. Knowl Acquisition 5(2):199–220
15. Studer R, Benjamins VR, Fensel D (1998) Knowledge engineering: principles and methods. Data Knowl Eng 25(1):161–197
16. Nepal MP, Staub-French S, Pottinger R, Zhang J (2012) Ontology-based feature modeling for construction information extraction from a building information model. J Comput Civil Eng 27(5):555–569
17. Tserng HP, Yin SY, Dzeng RJ, Wou B, Tsai MD, Chen WY (2009) A study of ontology-based risk management framework of construction projects through project life cycle. Autom Constr 18(7):994–1008
18. Mostafa MA, El-Gohary NM (2014) Semantic system for stakeholder-conscious infrastructure project planning and design. J Constr Eng Manage
19. Jiang S, Zhang J (2013) Development of an ontology-based semantic retrieval method for construction project risk management. In: ICCREM 2013's construction and operation in the context of sustainability. ASCE, pp 750–760
20. Fidan G, Dikmen I, Tanyer AM, Birgonul MT (2011) Ontology for relating risk and vulnerability to cost overrun in international projects. J Comput Civil Eng 25:302–315

21. Dikmen I, Birgonul MT, Anac C, Tah JHM, Aouad G (2008) Learning from risks: A tool for post-project risk assessment. Automat Constr 18(1):42–50
22. Noy NF, McGuinness DL (2001) Ontology development 101: A guide to creating your first ontology
23. Hall J (1998) Private opportunity, public benefit? Fiscal Stud 19(2):121–140
24. Ji C, Yuan J, Han R, Li Q (2013) A Case-Based Reasoning System for Residual Value Risk in Public-Private Partnership Projects. In: ICCREM 2013's construction and operation in the context of sustainability. ASCE, Reston, pp 680–692
25. Al Khattab A, Anchor J, Davies E (2007) Managerial perceptions of political risk in international projects. Int J Project Manage 25(7):734–743

Chapter 114
An Analysis of the Adaptive Re-use of Heritage Buildings in South Australia

Tony Ma and Minmei Yu

Abstract The conservation of heritage buildings plays an important role to enhance the sustainable built environment. Adaptive re-use of heritage buildings has become a wider revitalization way to promote sustainability and protect the heritage buildings' significance. However, many building owners and developers still perceive the re-use of heritage buildings as being an unviable option as planning and building regulations may restrict their uses. Therefore the viability of adaptive re-use of heritage buildings is yet to be fully evaluated. The aim of this research is to investigate the perceived benefits and barriers of adaptive re-use of heritage buildings and to suggest recommendations to promote its re-use. Questionnaire survey and case studies collected in South Australia are used to illustrate the research objectives.This research discovers that the adaptive re-use of heritage buildings provides environmental, social and community benefits based on the conservation experts point of views. From the building owners, the adaptive re-use did provide some economic benefits. It also identifies the barriers for the conservation work. The major problem is the conservation cost. Moreover, the compliance of Building Code requirements and earthquake review are also the barriers for the conservation work. The long development approval can be another obstacle for the conservation work.

Keywords Heritage building · Adaptive re-use · Conservation · South australia

114.1 Introduction

Now a days, all Australian governments regard sustainable development as a goal to balance the environment with the health of the economy. There is no double that conservation of heritage buildings plays an important role to enhance the sustain-

---

T. Ma (✉) · M. Yu
School of Natural & Built Environments, University of South Australia,
Adelaide, SA, Australia
e-mail: tony.ma@unisa.edu.au

Y. Wu et al. (eds.), *Proceedings of the 20th International Symposium on Advancement of Construction Management and Real Estate*,
DOI 10.1007/978-981-10-0855-9_114

able built environments. Heritage buildings not only provide a footnote to Australia histories, but also help to identify the places as significant to Australia [1]. Conserving heritage buildings will provide significant economic benefits, cultural and social benefits, and is gaining acceptance within Australia [2]. Adaptive re-use of heritage buildings has become a wider revitalization way to promote sustainability rather than being subjected to demolition [3, 4]. In fact, adaptive re-use of heritage building is not a new strategy in South Australia; instead it has been used for many years. For example, the Belmont House in North Adelaide had been converted from residence to an office house in 19th Century [5].

Cooper [6] suggested that the adaptive re-use of heritage building improves the material and resource efficiency (environmental sustainability), reduces cost (economic sustainability) and improves retention (social sustainability). The Department of Environment and Heritage (DEH) also regards that the adaptive re-use of heritage buildings provides environmental benefits (the retention of the original building's embodied energy), social benefits (protection of heritage buildings will provide benefits to future generations) and economic benefits (embodied energy saving from not demolishing buildings [1]. Besides, the DEH considers that the most successful built heritage adaptive re-use projects are those that best respect and retain the buildings' heritage significance and add a contemporary layer that provides value for the future. For example, Balhnnah, remarkable as the most intact group of 19th Century mining structures in South Australia, was uninhabitable and derelict in 1986; now it has been adapted to fit the needs of 21st century families with its heritage significance.

However, many building owners and developers still regard the re-use of heritage buildings as being an unviable option as planning and building regulations may restrict their functioning [2]. Australia's heritage management system has been well-developed, but its sustainable practices in built heritage is found to be slow [7]. There are still plenty of historic buildings waiting for conservation in Australia, especially in South Australia. Hunter [8] mentioned there are 1500 buildings under protected in the South Australian Heritage Register (Register), and it is believed that the number of heritage buildings in the Register will be increased continually. Some owners may prefer to adapt to a new use with its heritage significance. Heritage buildings that lose relevance or purpose may become vulnerable to lack of care, decay and possibly demolition [1]. In order to protect the invaluable heritage buildings, users, developer, planner and government are all encouraged to participate in the conservation process. A number of literatures and researchers in the past have developed to contribute a more sustainable built environment and a more feasible adaptive re-use strategy for heritage buildings around the world. But not many researches are related to the South Australia. Since some heritage buildings have been adapted to a new function to fit the market demand and provide benefits to the environment, community and society, the experience of those cases is worth as a reference point to new heritage owners and advisors. It would be useful to study the perceived benefits of adaptive reuse of heritage buildings in order to provide critical information for those potential developers, planners and investors who are interested in adaptive re-use and to encourage them to take action to protect the history.

The aim of this research is to investigate the perceived benefits and barriers of adaptive re-use of built heritage and to suggest strategies for the conservation of heritages buildings. Questionnaire surveys were collected based on the opinions of the conservation experts which will focus on the technical side and issues with the regulations. On the other hand, owners' and consultants' point of views is also collected through case studies. It is anticipated that the insights of how well the buildings have been adapted will be investigated.

114.2 Adaptive Re-use of Heritage Buildings in South Australia

114.2.1 Adaptive Re-use Principles

The Department of Environment, Water and Natural Resources (DEWNR) identifies the adaptive re-use as the adaptation of historic buildings from one use to another while preserving their heritage value. In the Year of Built Environment [1], the Commonwealth Department of Environment and Heritage claimed:

"The adaptive reuse of a historic building should have minimal impact on the heritage significance of the building and its setting. Developers should gain an understanding of why the buildings has heritage status, and then pursue development that is sympathetic to the building to give it a new purpose. Adaptive reuse is self-defeating if it fails to protect the building's heritage values".

The DEH also stated at the Year of Built Environment [1]:

"Sometimes, adaptive re-use is the only way that the buildings fabric will be properly cared for, revealed or interpreted, while making better use of the building itself. Where a building can no longer function with its original use, a new use through adaption may be the only way to preserve its heritage significance".

It is suggested that the adaptive re-use of a heritage building should have minimal impact on its heritage significance of the building. However, it is easier said than done. Many heritage buildings are required modifications to fit their new functions. Their heritage significance should be properly cared and considered well before commencement. Therefore, the adaptive re-use is an important process in retaining heritage buildings and are useful for future conservation activities. The adaptive reuse of heritage building is regarded as a vital part of Australian environment planning in order to create a more sustainable future [9].

114.2.2 Adaptive Re-use Development Process

The DEWNR [10] compiles a guide to developing State Heritage Places. According to this guide, the change of use has been defined as 'development' which requires the application for development approval lodged with the relevant planning authority—usually the local council but in some areas the Development Assessment Commission (DAC). In order to get the development approval, the sufficient information must be included. The sufficient information includes clear documentation and as much descriptive materials as is necessary, which can help to speed up the assessment and prevent unnecessary delay. The information required may include details of the Place, the Drawings, a written Specification, Photographs, the Heritage Impact Statement and the Conservation Studies. For a State Heritage Place, its cultural heritage value must be identified in a Statement of Heritage Significance; DEWNR Heritage Advisors will then assess the impact of proposed development based on that Heritage Significance. The Advisers will also assist applicants with design options, building materials and techniques sympathetic to the heritage value of the State Heritage Place. Recommendations will be provided by the Heritage Adviser for the Minister's delegate. The overall Development proposal will be assessed against the principles of the Burra Charter and the requirements of the Heritage Places Act based on the place's significance. The Burra Charter defines the basic principles and procedures to be followed when heritage places are undergoing conservation. Once the planning consent has been granted, the applicant shall arrange building rules consent through private certificate, if applicable.

114.2.3 Examples of Adaptive Re-use of Heritage Building in South Australia

Table 114.1 indicates the examples of adaptive re-use of heritage buildings in South Australia:

From the examples, the adaptive re-use of heritage buildings supports the realization of environmental, social, economic and cultural benefits to wider communities. Despite the many positive outcomes, they are not without planning, financial and commercial risks. The redevelopment of a heritage building must meet the requirement of heritage conservation guidelines and Acts, and proceed in accordance with the authority's recommendations. It is difficult for some heritage buildings to satisfy the present-day's building, health, safety and accessibility requirement of current regulations [11]. If the redevelopment proposal has been disapproved by the planning authority, developers and owners need to re-address the design and plan, which may take a lot of time and additional cost. The risks may also include unknown work, scope changes, compatibility of materials, quality of information and the operation environment, health and safety, design constrains and decanting of occupants [12]. Moreover, the potential latent problems, defects or

Table 114.1 Examples of adaptive re-use of heritage buildings in south australia

No.	Property name	Heritage protection	Original use	New use	The benefits
1	Railway goods shed [14]	State heritage place	Brewery	Goods shed	Visitors to the goolwa wharf can enjoy a unique blending of the old with the new: paddle steamers and steam engines, boatbuilding and brewing
2	Norwood baptist church [14]	State heritage place	Baptist church	Café and bookshop	The commercial stability can provide the incentive and funds for restoration and conservation work
3	The former megaw and hogg building [14]	State heritage place	A warehouse	New office building	The new function and conservation can make the building meets the needs of the 21st century tenants
4	Old treasury building [15]	State heritage listed	Office building	Medina grand hotel	The national trust runs tours of the old cabinet room and tunnels under the building used for the secure transport of gold
5	Adelaide town hall [16]	State heritage listed	Hall	Community centre	The town hall remains a major city landmark and a popular venue for concerts, public meetings and special events
6	Balhannah mine [1]	State heritage place	Private slaughterhouse and implement shed	Home for south australian based family	The mine was uninhabitable and derelict before adaptively reused, now it is the home for a south australian based family
7	Mount gambier city hall [17]	State heritage place	Institute building	Civic club	Inserting a new contemporary multipurpose space to meet the community's ever changing needs
8	Strut street primary school [9]	Local heritage place	Primary school	Integrated early learning center	The building had been disused and dilapidated for more than 10 years

(continued)

Table 114.1 (continued)

No.	Property name	Heritage protection	Original use	New use	The benefits
9	Bray house [18]	State heritage listed	Residential	Office	The building changes its use to fit the market demand
10	Former produce markets [10]	State heritage place	Food market	Café, shops and offices	The building changes its use to fit the market demand

dimensional and material inconsistencies may affect the success of adaptive re-use [4]. In the past, the local government did provide some sort of heritage incentive schemes. The most persuasive incentives includes "relaxation of building requirements for heritage listed buildings", "monetary contributions to construction works" and "flexibility in meeting current construction regulations". In South Australia, the fund is administered by the DEWNR. The funding is from $5,000 up to $25,000 for the conserving the State Heritage Places [13]. Unfortunately, the State Heritage Fund has been canceled. Adelaide City Council has some incentive schemes for city property's conservation work but not for adaptive re-use.

114.3 Research Analysis

114.3.1 Questionnaire Survey

In order to obtain a better understanding about the benefits and barriers of the adaptive re-use of heritage buildings, a questionnaire survey was developed from targeting the heritage and building conservation advisers and consultants, architects and planners. Over 40 invitations were sent out; and after repeated reminders, 20 responses were finally collected. 85 % participants have been involved with the adaptive re-use of heritage buildings. There are four major questions being asked and the answers from respondents were summarized in Table 114.2.

114.3.2 Case Studies Through Interviews

Something would have been missed out without the actual analysis of real life cases. It is anticipated that through the following four case studies, there will be insights of how well these heritages buildings have been adaptively re-used. Table 114.3 summarized the profile of cases, interviewees, questions and answers:

Table 114.2 Questionnaire survey analysis

Q1. *What are the perceived benefits of adaptive re-use of heritage buildings (from 5-strongly agree to 1-strongly disagree)?*		
Benefits	Ranking	Mean
Ability to aesthetically fit streetscape	2	4.75
Benefits of re-use versus redevelopment	2	4.75
Community value of existing buildings	1	4.85
Demand for buildings for economic use	6	3.40
Opportunity for technical innovation	5	3.95
Public awareness of adaptive re-use	4	4.25
Viability of recycling existing building materials	3	4.65
Q2. *What are the perceived barriers of adaptive re-use of heritage buildings (from 5-strongly agree to 1-strongly disagree)?*		
Barriers	Ranking	Mean
Difficult to design to fit the new use of the heritage building	6	2.90
Difficult to choose a new function to fit the market demand	4	3.10
Inadequate information of structural plan	5	3.05
Availability of materials to match existing	3	3.10
Availability of high skilled workers	2	3.70
High conservation cost	1	3.75
Communication problems with owner/planners and constructors	7	2.95
Q3. *What are the factors impacting on your conservation works regarding the approval process (from 5-strongly agree to 1-strongly disagree)?*		
Factors	Ranking	Mean
Long planning approval process	1	3.55
Communication problems with Heritage Advisors' guidelines	3	2.85
Difficult to fit the BCA requirements	2	3.9
Difficult to provide required information for planning consent	4	2.75
Difficult to retain the significance	4	2.75
Q4. *What are the factors that may encourage the owners to go for adaptive re-use of heritage buildings (from 5-strongly agree to 1-strongly disagree)?*		
Factors	Ranking	Mean
Incentive scheme and/or financial support	1	4.75
Faster the development approval process	4	3.95
Heritage building promoting /let people be aware the built heritage	2	4.30
Provide more conservation advice	3	4.10
Reduce the BCA requirement	5	3.70

114.4 Discussions and Conclusion

The aim of this research is to determine the perceived benefits and barriers of adaptive re-use of heritage buildings, and to recommend innovative strategies to advocate its use. Throughout these data collection, it is discovered that the adaptive

Table 114.3 Summary of four case studies

Cases	Torrens parade ground (public)	Tandanya (public)	Westpac bank building (jamies' italian restaurant) (private)	Brecknock hotel (citi zen restaurant) (private)
Interviewees	Heritage architect	Heritage architect	Building surveyor	Manager (owner)
Main concern during conservation	Complying with the burra charter requirements	–Access for disable –Compliance with BCA	–BCA requirements –Earthquake protection	If location suitable for the purpose
Benefits identified	Visitors understand the history of building	–Increase the character of the city –Embodied energy	–Provides for a more vibrant city –Economic boost – Character of the city while maintaining the history	Provide good business opportunity
Barriers identified	–Conservation cost –allow disable access –Lift installation –Additional building services	–Conservation cost –Design challenges	–Conservation cost –Compliance with BCA requirements	–Compliance with BCA requirements (e.g. fire services and disable access) –Harsh limitation of the building facade
After the adaption, would you still advocate the adaptive re-use	Yes	Yes	No	Not sure
Supports from government	Not enough, only from council	Not enough, only from council	No	No
Recommendations	–Adapt the most flexible heritage building –Getting early heritage architect's input –Pre-application agreement or heritage agreement	–Suitable in very good economic condition	–Getting early building surveyor's input –Involvement of builder early	Reduce BCA requirements

re-use of heritage buildings did provide some sort of environmental, social and community benefits based on the conservation experts' point of views. In the owner's opinion, the adaptive re-use also provides some economic benefits. The literature reviews also highlight the environmental, social and economic benefits of the adaptation of heritage buildings.

This research has also identified the barriers for the work. The major problem is the conservation cost. The conservation work involves a significant conservation amount of conservation cost. Moreover, the compliance of BCA requirements and earthquake review are also the barriers for the conservation work. The adaptive re-use of heritage buildings involves the change of building use, which requires complying with the BCA requirements and the heritage building is required to be strengthened for the earthquake review. The long development approval can be another barrier for the conservation work.

Some recommendations have been suggested in order to solve those barriers mentioned above. Firstly, the Government is recommended to provide incentive scheme and financial support in order to encourage the conservation of built heritage, and to reduce the conservation cost spent on upgrading of building structure, building services and fire services. Secondly, the Government is also recommended to lower the BCA requirements for the conservation of heritage buildings because the compliance of BCA requirements is one major cause for the significant conservation cost and long development approval process. Thirdly, the heritage experts, such as heritage advisors, building surveyors and builders, are recommended to be involved at the early stage. Their early involvement would help to solve the conservation problems and providing the innovative solutions for the conservation work.

It can be concluded that the adaptive re-use of heritage buildings do provide some benefits under the constraints of lengthy development process and conservation cost. These barriers may prevent the developer, planners and building owners to take initiatives to protect the heritage buildings by the way of adaptation. The barriers may be mitigated if the Government can provide financial and technical support. One advisor recommended the use of Heritage Agreement for major buildings. That means the range of change of use of buildings has been pre-agreed prior to the sale. If the use is within the range, the new owner does not need to apply for lengthy approval process for adaptive re-use.

References

1. Department of Environment and Heritage (DEH) (2004) Adaptive Reuse-Preserving our past, building our future. Commonwealth of Australia, Canberra
2. Bullen PA, Love PED (2010) The rhetoric of adaptive reuse or reality of demolition: views from the field. Cities 27(4):215–224
3. Ball R (1999) Developers, regeneration and sustainability issues in the reuse of vacant buildings. Build. Res. and Inf. 27(3):140–148

4. Bullen PA, Love PED (2009) Residential regeneration and adaptive reuse: learning from the experience of Los Angeles. Struct. Surv. 27(5):351–360
5. Kent W, Smith (1992) Development adaptation and alternation of heritage buildings. University of South Australia
6. Cooper I (2001) Post-occupancy evaluation-where are you? Build. Res. and Inform. 29 (2):158–163
7. Lynne A, Janine I (2013) The values of built heritage. Property Manag. 31(3):246–259
8. Hunter I (2003) Labour's plan to protect South Australia heritage. National Trust Heritage Management Discussion Paper. Jan 2003, pp. 1–2
9. Government of South Australia (GSA) (2008) State of the Environment Report 2008: built heritage. Government of South Australia
10. Department of Environment, Water and Natural Resources (DEWNR) (2012) SA Guide to Developing State Heritage Places. Government of South Australia
11. Australia ICOMOS (1999) The Burra Charter, Australia, pp.7
12. Reyers J, Mansfield J (2001) The assessment of risk in conservation refurbishment projects. Struct. Surv. 19(5):238–244
13. Department of Environment, Water and Natural Resources (DEWNR) (2014b) Fact Sheet-South Australian heritage fund. Government of South Australia
14. Department of Environment, Water and Natural Resources (DEWNR) (2014a) Adaptive reuse examples-fact sheet. Government of South Australia
15. Adelaide City Heritage (ACH) (2014a) Treasury building. Adelaide City Heritage. viewed 2 Apr 2014. http://www.adelaideheritage.net.au/all-site-profiles/treasury-building
16. Adelaide Town Hall (ATH) (2014) "History". Adelaide Town Hall. viewed 30 Mar 2014. http://www.adelaidetownhall.com.au/visit-discover/history
17. Government of South Australia (GSA) (2012a) Heritage Direction 2012. A future for heritage in South Australia. Government of South Australia
18. Adelaide City Heritage (ACH) (2014b) Bray house. Adelaide City Heritage. viewed 2 Apr 2014. http://www.adelaideheritage.net.au/all-site-profiles/bray-house
19. Bullen PA, Love PED (2011) Adaptive reuse of heritage buildings. Struct. Surv. 29(5):411–421

Chapter 115
Study on Quasi-Public Goods Supply Based on Transaction Cost Theory

Zewen Zhang, Tingting Zhou, Xiao Li, Bo Xu and Shirong Li

Abstract According to the characteristics of quasi-public goods, this paper analyzes the adaptability of supplying quasi-public goods by PPP moderns. Transaction costs of three different pattern, government's supply, enterprise's supply and PPP patterns are made qualitative comparisons, while this paper analyzes the ways of innovation in supply entity in the case of market failure. At last approach of decreasing transaction cost has been discussed and both enterprise and government need to revolute for pushing the development of PPP modern had been proved.

Keywords PPP · Quasi-public good · Transaction cost · Supply innovation

115.1 Introduction

Scholars have made series of research about PPP mode in 10 years, including the concept, critical elements and risks of PPP. Jia Kang point out that roles of PPP are not only including finance but also the role of mechanism innovation according the situation of China. And mechanism innovation has two meanings. The one is public sector transferring from traditional planning to marketing economic, the other is private sector transferring from market economy to planning. This round-tripped can develop into a new incentive system, which can advance and deepen the reform in finance field and enhance the effective disposition of sources.

However, the previous research discussed PPP from a practical point of view, lacking of analysis of PPP from theoretical overview. This paper is base on it and analysis the impact of transaction cost under difference of supply model by the theory of transaction cost theory,and furthermore discussed the way to decrease the transaction cost of PPP mode.

Z. Zhang (✉) · T. Zhou · X. Li · B. Xu · S. Li
School of School of Construction Management and Real Estate,
Chongqing University, Chongqing, China
e-mail: 358484048@qq.com

Y. Wu et al. (eds.), *Proceedings of the 20th International Symposium on Advancement of Construction Management and Real Estate*,
DOI 10.1007/978-981-10-0855-9_115

115.1.1 Definition and Function of PPP Mode

PPP mode comes hot in China and has applied in the whole world. It is defined by ministry of finance as the whole process of cooperation between government and social capital to provide public goods or services. The features include benefit-sharing and risk-sharing,aiming to improve the quality and efficiency of publics or services by taking the advantages of both sides.

Investment in public product construction field is decreased under PPP mode, which can reduce the government's financial pressure effectively. For small and medium-sized private enterprises, PPP mode can not only make more private capital to participate in the project construction, but also guarantee return of private capital in a certain extent, so as to form a government, private enterprise "mutual win" cooperation mode. Secondly, Lian Hongjun put that the more investment and bigger enterprise-scale, can help these small and medium-sized private enterprise establish a better mutual trust partnership and create more communication opportunities. In addition, public and private sectors can help each other and learn from each other, provide as high a quality of service as possible at the lowest possible cost.

115.1.2 The Influence of PPP Mode on Different Subjects

For the government, PPP mode can reduce the financial burden by introducing social capital, thus playing a guiding role of financial funds and mobilizing social capital to play their respective advantages. What's more, PPP can promote the transformation of government functions effectively, so the government can focus more on planning and supervision. In terms of the enterprise, they were mainly in the manufacturing sector before, now PPP mode makes them enter the field of public utilities, which can reflect the decisive role of market. Therefore, PPP mode can enhance the development space of the private sector and promote the non-public ownership economy development. To the society, the supply efficiency of public goods could be improved through the rational allocation of resources. The public can have the opportunity to enjoy a more adequate price of public goods or public services, thus increase the public's confidence and trust of the government directly and indirectly.

115.1.3 The Evolution of Public Product Supply Subject Based on PPP Mode

There are three types of traditional public goods supply: government, market and the third party voluntary supply main body. There is little third party voluntary supply main body in China, so we don't discuss it in this article. Hobbes believes that the supply of public goods has experienced three stages: the "market failure" stage. The second stage, government intervene the market to make up the "market failure", but due to government intervention, financial expansion and information which is not complete, the government has a "government failure"; the third stage, due to "market failure" and "government failure", because of various reasons, the voluntary supply of the main body is little. There is a "voluntary failure" stage.

In the case of the failure of government, market and voluntary supply, whether the new department or the new model will appearing to guide the government and promote the construction and supply of public goods become a breakthrough point.

115.2 Quasi-Public Goods and Transaction Cost Theory

115.2.1 Quasi-Public Goods Theory

According to Samuelson's definition of public goods, public goods are non-competitive and non-exclusive, the marginal cost is zero, and it will not affect other people's use. However, this definition of public goods is an extreme situation. In fact, public goods is treat as between public goods and private goods, and has the following characteristics:

Incomplete Competitive and Incomplete

The club product, which is called by American economist Buchanan, is between private and public products. And one of its characteristics is the non competition, that is within the critical point of the number of users, the individual user's consumption of the club will not affect the use of other people, but once beyond the critical point will crowd, then the non competition will disappear.

The second feature is the local exclusive, that is, the club's products are not exclusive to the members of the club, but it is not for the nonmembers. Therefore, quasi-public goods are not totally noncompetitive or non–exclusive, its feature is between private and public goods.

Positive Externality

The positive externality is refer to the spill over of the income, that is, impacting on other people's benefits in the process of production or consumption behavior, but did not get the appropriate remuneration. And negative externality is actually the cost spillover, which is not paid in the above process. Positive externalities are shown as the use of the product's social gains greater than the private benefits, and therefore will make the product demand decline. Economists believe that the existence of externalities can lead to market failure, and the internalization of external effects is a good way to improve the efficiency of the market.

115.2.2 Elements of Transaction Cost

According to the transaction cost theory, this paper is supposed that the transaction costs incurred in the infrastructure construction include: R for searching information, E for negotiation and signing contract, D is the cost for fulfill the contract, M is the cost of supplying public goods. P is the cost of the contract for avoiding existence of suppliers. Mike Kahn under the framework of Thompson's system of transaction costs proposed that transaction cost could be divided as:

$$\mathrm{TC} = \sum_{t=0}^{T} \beta(\mathrm{Rt} + \mathrm{Et} + \mathrm{Dt} + \mathrm{St} + \mathrm{Pt} + \mathrm{Mt})$$

Among them, the beta is for the discount factor, total transaction cost TC can be obtained by estimating the average time cost of the type required according to the Delphy technology forecasting method and investigation of accidental factors evaluation.

115.2.3 Main Influencing Factors of Transaction Cost

Williamson believes that the transaction costs are generally reflected in the following three ways: (1) uncertainty, (2) transaction frequency, (3) asset specificity. The transaction cost in trade is mainly affected by these three factors.

Asset specificity means that it is hard to change its function when it is used for a specific purpose. If change it, the value will reduce or even become worthless assets. Therefore, the stronger the asset specificity of the industry, the higher the conversion cost of the industry, and the higher the entry barriers. On the contrary, the stronger the homogeneity of assets, the lower the cost of the change the property of assets. Therefore, in the definition, the supply of public goods is essentially a kind of highly idiosyncratic transaction.

The uncertainty of the transaction costs mainly includes: (1) the traders have incomplete or asymmetric information; (2) the limited rationality of the traders. First of all, uncertain trading environment is mainly because traders' information is incomplete or asymmetric. Such as traders cannot accurately get the price and quality of products, if the partners were potential opportunists are uncertain. Secondly, the limited rationality of the traders also result in the uncertainty of the transaction. Due to the limitation of the rationality of the traders, the contracts are inevitably incomplete, and the opportunistic behavior is inevitable.

The transaction frequency is the times of transactions. It does not affect the absolute value of the transaction costs, it only affects the relative cost of various transactions. The relationship between transaction cost and transaction frequency is, the transaction cost is decreasing while the transaction frequency is increasing in the repeated transaction, but the transaction costs can't decrease infinitely, in other words, the transaction costs will not be close to zero.

115.3 Research of Quasi-Public Goods Supply Base on Transaction Cost

115.3.1 Research Hypothesis

(1) All of the public goods of this paper are quasi-public goods.
(2) The public goods supply stakeholders are divided into three modes: government unilateral supply, enterprise unilateral supply and government-enterprise cooperative supply under the PPP mode.
(3) Transactions under the PPP mode is long-term and repeated.

115.3.2 Supply Stakeholder Modes

Government Unilateral Supply

Chinese scholar Yicheng Zhou holds that the government unilateral supply mode of public goods can be described from three points of view:

(1) From the perspective of main structure, the mode of government unilateral supply means it has the unique supply stakeholder, the government. In the paper "Comparative Research of Public Economy" in 1997, Elinoros Trom called the theory of pure government unilateral supply mode as single center theory. The single center theory means that: the amount of supply products determined by the governments, which is a single power center.
(2) From the perspective of process, in the government unilateral supply mode, all the links of provision and production of public is goods are monopolized by governments, the governments is not only responsible for funds input, but also production. The production form of governments includes two types: one is the public goods produced by the governments and their employees, another is public goods produced by the government's enterprises and their employees.
(3) From the perspective of the concrete production way of the governments, it's mainly includes government service, government selling and administrative agreements. Government service means all the public goods are provided by

governments, the governments play roles in both arranger and producer. Government selling means people purchase public goods from government organizations. Administrative agreements emphasizes that one government dab down another government the right to provide public goods, and pay for it.

Enterprise Unilateral Supply

The enterprises in this mode is non-state-owned enterprises. The paper interpreted this mode as following: the government throw a quasi-public goods project to the market, than the enterprise obtains the right of public goods supply by market competition, and undertake investment, construction and operation of the project. During the concession period defined by protocol, the individual enterprise collects proper expense from users, in order to retrieve the investment and financing of the project, and get reasonable returns. The government has the right of supervision and regulation of the infrastructure, when it comes to an end, the individual enterprise should hand over the infrastructure to government free or paid.

Government-Enterprise Cooperative Supply Under the PPP Mode

PPP is a cooperating way to provide public goods used by public sections and private sections. Different from the enterprise unilateral supply mode, public sections and private sections undertake financing and responsibility of risks together, and emphasize the both sides is highly cooperative, jointly owned the control rights. For the point of participation level, the both sides participate together in every link of decision, designing, construction, financing and operation, there is no substitute construction and transfer process.

115.3.3 Comparison of Transaction Cost in Different Supply Stakeholder Mode

For the general assets, due to the assets have no special character, in this case, only if the both contract sides sign the market contract one by one and dispersedly, will the market contract be effective. Since the market can make full competition, the opportunistic motivation is difficult to be transformed into an opportunistic behavior. But for the assets that are highly appropriative, because of the interdependence of each other, opportunistic behavior may emerge from one party in the transaction, so that the transaction cost is greatly increased. At this point, the right way to trade is especially important. Therefore, Due to the lack of effective supervision of the third party in the mode of government unilateral supply of public goods, high assets specificity will lead to the immoral obligation appear a rent-seeking behavior.

For example, immoral managers damage the interests of the state and the workers, they falsify accounts, expenses, take kickbacks or other measures to maximize the revenue, the consequence of these behaviors is resulting that transaction costs increased. In the mode of Enterprise unilateral supply, also for their own benefit maximization, they take speculation such as jerry works, inferior as superior and raising the price of the production factors, in order to seek for a higher profit, this also resulted that transaction costs increased. In the PPP mode of the cooperation between government and the social capitals, it's equivalent to a community of interest formation of governments and enterprises, and formed a two-way supervision so as to reduce the opportunism behavior, and then reduce transaction costs.

In the mode of government unilateral supply, it's dear to plan and lack of market, government often can not get effective market information; The enterprise unilateral public goods supply, which dear to plan and lack of market, is often know or acquire insufficient information about government system. While the PPP mode put the governments and the enterprises together, this model greatly reduces the incomplete acquisition of the market information and government information for traders, and improves the asymmetry of information, so as to reduce the transaction costs of information search to some extent. The incompleteness of information will lead to the emergence of bounded rationality of traders, which is also common in the three models. However, the diffusion and exchange of knowledge and information is an effective way to improve the rationality of the individual. In PPP mode, the traders not only acquire knowledge and accumulate experience in their own practice, and in the process of cooperation with the other party will be a greater probability to promote the transaction rationality.

Government or enterprise unilateral supply, as well as the supply of quasi-public goods in the PPP mode, the transaction frequency is similar. But in terms of single transaction, as the supply stakeholder of PPP mode combines the advantage of government credibility and enterprise's good investment returns, so it is more advantages in the transaction and negotiations. The repeatable transactions based on trust in the PPP mode make less transaction cost than the other two modes. (Table 115.1).

Table 115.1 Qualitative comparison of transaction costs under the three modes (1 = relatively higher, 0 = relatively lower)

The relative amount of the transaction costs caused by the transaction			
Supply mode influencing factors	Assets specificity	Transaction uncertainty	Trading frequency
Government unilateral supply	1	1	1
Enterprise unilateral supply	1	1	1
PPP mode	0	0	0

115.3.4 Quasi-Public Product Supply Model Based on Transaction Cost

The cost of quasi-public product supply stakeholder in the supply process is divided into production cost and transaction cost, and the production cost is pursued by the lowest no matter what kind of supply mode, so it can be assumed that the supply cost is the lowest production cost and equal.

According to Williamson's description (1979), the basic aspects of the transaction is based on the characteristics of assets, transaction frequency and uncertainty, so that this paper compare the transaction cost amount by comparing the asset specificity, transaction frequency and uncertainty of the transaction. That is, the transaction costs will show a growth trend with the enhancement of asset specificity, which is similar to the trend of the exponential function. The extent to which the organization deviates from the optimal supply scale is enhanced with the increase of asset specificity, so as to increase the transaction cost of the organization unit, showing a "decrease first and then increase" trend like the "U"-shaped change (Fig. 115.1). Therefore, the change of transaction costs can play a decisive role in defined organization boundary incase of the direct proportion of the production cost and the supply of public goods. This paper discusses the multiple supply main body of quasi-public goods in the former part, introduce the innovative supply main body into the traditional market unilateral supply, government unilateral supply and third party voluntary supply model, and combined with the main body of the quasi-public goods supply in our country, based on the comparison of the transaction costs of various supply main body, discuss the transformation and innovation of the PPP mode in the supply main body of quasi-public goods.

Figure 115.1 is a model of the optimal supply main body selection of public goods, which is proposed by Wang Lei in his paper "The main body selection of the supply of public goods—Theoretical analysis framework based on the theory of transaction cost economics and its application in China", that is, the paradigm

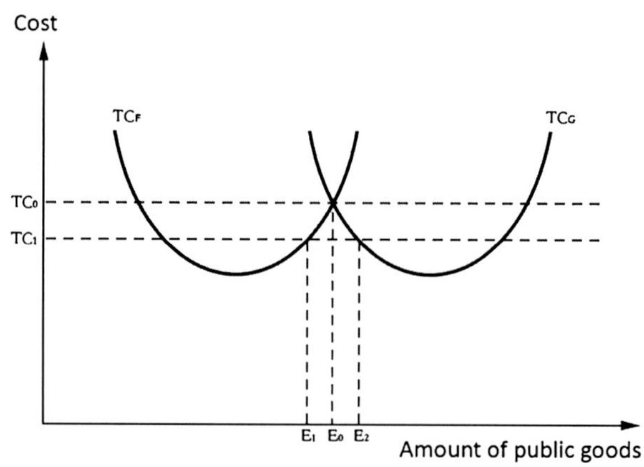

Fig. 115.1 Public goods supply failure

Fig. 115.2 Innovation of supply main body of public goods

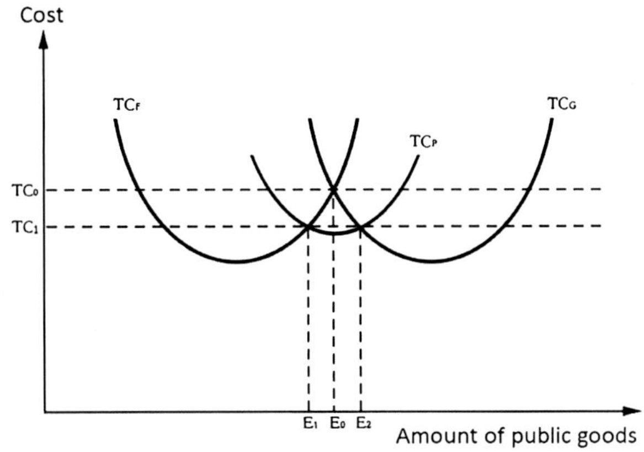

expansion of the Williamson. He pointed out that when the transaction costs of social and economic activities is reduced generally by the effect of social progress, technological innovation and other factors (such as the development of information technology and convenient transportation, etc. to make the social transaction costs reduce from TC1 to TC0), the original form of organization will appear some "empty" (as shown in Fig. 115.2, E1, E2) and can not cover these areas, In these areas, the government, the market and the voluntary supply main body are no longer the best choice to realize the internalization of external effect, and creating the demand space for the innovation of supply main body of public goods(TCf and TCg in the figure respectively represent the transaction costs of the quasi-public goods supplied by the enterprises or the government).

Figure 115.2 is the addition of Williamson paradigm re-expansion of PPP supply mode of quasi-public goods based on the transaction costs. It can be seen from the comparison between Figs. 115.1 and 115.2, when the average transaction cost get down generally (from TC0 to TC1), the supply of quasi-public goods will appear (E1, E2) interval of supply hole in the traditional supply mode, that is based on the transaction costs in this interval, the traditional supply main body are reluctant or unable to provide quasi-public products effectively (Because when the transaction costs are reduced, the transaction costs of traditional supply main body are higher than the average transaction costs, they are not willing to trade); when introduced the PPP mode transaction costs TCP of Fig. 115.2, this paper made a concrete analysis and obtained the transaction costs of government unilateral supply, enterprise unilateral supply and PPP mode supply in the former, TCP is lower than TCf and TCg, standing on the point of view of transaction cost, PPP will become a kind of innovative supply main body, and make up the public goods supply hole of the traditional supply main body.

Table 115.2 analyzes the way to reduce transaction costs in PPP mode based on the transaction cost, the main path of these are: Build internal capital markets, set up comprehensive independent regulatory agencies, and promote the exchange of market information and program information, construct learning organization,

Table 115.2 Way to reduce transaction cost under PPP mode

Mode	Influence factors of transaction cost	Method to reduce transaction cost	Ways to reduce transaction cost
PPP	Asset specificity	Internal transaction	Build internal capital market, cooperation with financial department
		Improve supervision efficiency	Set up comprehensive regulatory agencies at all stages of the project
	Uncertainty	Realize information exchange economy	Strengthen internal communication and achieve communication economy
			Promote the exchange of market information and plan information
		Improve learning ability	Construct learning organization, improve the knowledge and technical ability
			Technological innovation of enterprises
	Trading frequency	Promote the realization of repeated transactions	The government should attract private enterprises through the promotion of public trust, taking competitive mechanism, making reasonable fee price policy. Enterprises should improve the level of profitability, and enhance social responsibility, so as to ensure the long-term effective cooperation between the two sides of the PPP model

innovate enterprise technology, and strength social responsibility. From the point of view of transaction cost, the enterprise and government must make change of the corresponding function to the successful implementation of PPP model.

For the government, the government should establish a comprehensive regulatory institution to improve the regulatory efficiency in cooperation to avoid the occurrence of opportunism, and to avoid extra transaction costs caused by the not in place of supervision and regulation. In addition, the government can strengthen the communication and exchange information with the market in order to reduce the uncertainty in the transaction, and update information in time in the process of learning together with the enterprise,so as to integrate with the market. In addition, government can make use of the effective team and technology of the enterprise to realize the function of the government changed from the plan to the market.

Enterprises can expand the construction capital by the establishment of the internal capital market. Because the project becomes more complexity, the enterprise needs to strengthen technological innovation and update the knowledge to reduce uncertainty of transaction. In the process of cooperation, enterprises have been involved in the development of policies and regulations due to the frequent communication and exchange information with the government. In addition, in order to establish long-term effective cooperation with the government, enterprises need to improve their social responsibility, thus cooperate with the government with the same interest to complete the PPP cooperation, and finally realize the transformation from the market oriented to the planning.

115.4 Conclusions

This paper is basing on the characteristics of quasi-public goods, and analysis the innovation of the supply main of public goods in terms of the transaction cost under the PPP mode. This paper mainly discusses the supply of public goods in the field of quasi-public goods, and make the qualitative comparison of the transaction costs between three modes: the government's supply, the supply of enterprises and the cooperation of government and social capital (PPP). In addition, this paper analyzes how the PPP model can promote the innovation of quasi-public products in the case of market failure. At the end of this paper, it discuss the way to reduce the transaction cost in PPP mode and in order to promote the development of PPP better, what should enterprises and government do to achieve the corresponding functions.

Acknowledgement We gratefully acknowledge the financial support from Chongqing Graduate Student Research Innovation Project (CYS14036).

Bibliography

1. Kang J (2014) Play a positive effect in the process of reform and innovation of the PPP mode. Sub National Fiscal Research (9)
2. Shouqing W (2014) PPP is not from heaven. Financial benchmarking
3. Hongjun L (2011) Empirical research on the relationship between small and medium sized enterprises participating in PPP mode and technological innovation. Reformation & Strategy (5)
4. Huali Z (2013). Investment and pricing of quasi public goods in the model of multi objective programming-Taking hangzhou metro as an example. Zhejiang Gongshang University
5. Fengyuan D (2010) The meaning and measurement of transaction costs: a review and prospect. Research of Institutional economics (5)
6. Jifang Xu (2009) The three failure of public service supply and the innovation of public service supply mode in China. J. of Nanjing Agric. University
7. Williamson OE (1979) Transaction cost economics-regulation of contractual relationship. Economics of Law (10)
8. Dezhang X (2010) Review and Reflection on the theory of asset specificity. Mod. Manage. Science
9. Qingming Y.(2004) Theoretical analysis of Williamson's transaction cost. Financ. Rev. and Practice
10. Qingsong He, Xuming Z (2007) Cluster governance of opportunistic behavior in trading activities. Contemporary finance
11. Lei W (2007) Supply of public goods in the main choice-Based on the theory analysis framework of transaction cost economics and in the application of finance. Trade Economy

Chapter 116
Advantages of Agent Construction in Building University Talent Apartments in China

Zhengrong Liu, Wendan Jiang, Yi Peng and Yuzhe Wu

Abstract Since the termination of national welfare housing allocation in 1998, the housing price has been soaring, which brings certain burdens for the talents to settle down in urban areas. In order to attract more outstanding scholars around the world, as well as mitigate the housing worries of teachers, colleges and universities still have to build housing for the staff personnel. However, colleges and universities lack management experience in building houses. This problem has been partially mitigated by agent construction. Agent construction system is a kind of turnkey engineer construction, based on the principal-agent model, project bidding mode, engineering contracting projects and project supervision system. This paper summarizes the common agent construction modes in China and further presents the stakeholder relationship in corresponding mode. Based on the summarized modes, this study further analyze the advantages of agent construction in building talent apartments through a case study.

Keywords Agent construction · Talent apartments · Company agent construction mode · Institutional agent construction mode

---

Z. Liu
Real Estate Management Division, Zhejiang University, Zhejiang, China
e-mail: liuzr@zju.edu.cn

W. Jiang · Y. Wu
Department of Land Management, Zhejiang University, Zhejiang, China
e-mail: wendyjng@outlook.com

Y. Wu
e-mail: wuyuzhe@zju.edu.cn

Y. Peng (✉)
School of Urban-Rural Planning and Management,
Zhejiang University of Finance and Economics, Zhejiang, China
e-mail: pengyihz@gmail.com

Y. Wu et al. (eds.), *Proceedings of the 20th International Symposium on Advancement of Construction Management and Real Estate*,
DOI 10.1007/978-981-10-0855-9_116

116.1 Introduction

Since the termination of national welfare housing allocation in 1998, the housing price has been soaring, which brings certain burdens for the talents to settle down in urban areas. This is resulted by various reasons including accelerated urbanization, farmland protection, and over investment [1–3]. As shown in Fig. 116.1, except for the year 2008 when global financial crisis occurred, the average housing price has grown year after year. The high housing price also presents barriers for the university staff to work well, which calls for the demand of university talent apartments. University talent apartments are useful to attract outstanding scholars around the world. In addition, the university talent apartments can facilitate to mitigate the housing difficulties of young teachers. Moreover, such apartments usually locates adjacent to campus, which promotes the communications between teachers and students. Therefore, it is necessary to build university talent apartments at current condition of China, although this practice is against the rule of commodity housing market to some extent.

However, colleges and universities lack management experience in building houses. This problem has been partially mitigated by agent construction. Agent construction system is a kind of turnkey engineer construction, based on the principal-agent model, project bidding mode, engineering contracting projects and project supervision system [4–7]. Agent construction system was first launched by the Xiamen in 1993, which aimed to select qualified professional construction companies to do project management through biding. With the acceleration of urbanization process in the 21st century, the construction market as well as agent system is booming. In 2004, the central government promulgated the "Decision of the State Council on Reform of Investment System". This document proposed "to accelerate the implementation of agent system for the non-profit government investment projects". So far, the agent construction system has been widely promoted at the national level [8, 9].

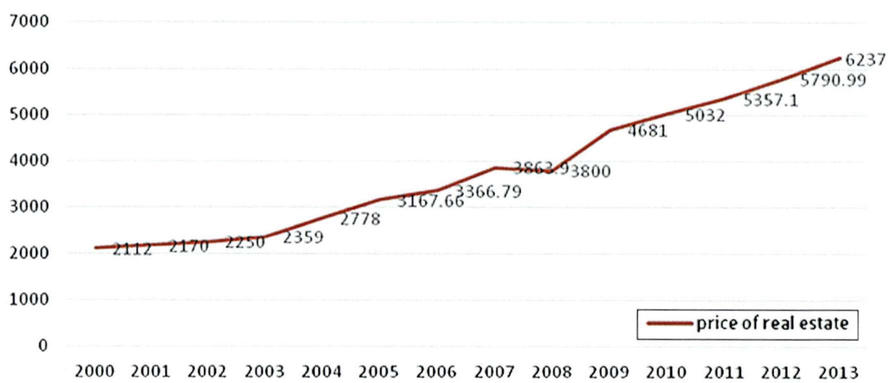

Fig. 116.1 The average price of commercial housing (*Data source* National Bureau of Statistics)

This paper aims to identify the advantages and risks of agent construction in building university talent apartments in China. Section 2 summarizes the common agent construction modes in China and further presents the stakeholder relationship in corresponding mode. Based on the summarized modes, Sect. 3 further analyze the advantages of agent construction in building talent apartments through a case study. Section 4 concludes this study with presenting the future studies.

116.2 Chinese Agent Construction Mode

The agent construction gradually evolved into two types including institutional agent construction mode and company agent construction mode.

116.2.1 Institutional Agent Construction Mode

Institutional agent construction mode means commissioning selects a specific construction agent that is embranchment of commissioning or institutions set up by the government at all levels. The stakeholder relationship in this mode can be found in Fig. 116.2.

Shenzhen adopts the institutional agent construction mode. This model was borrowed from Hong Kong, who established Construction Affairs Bureau to take charge of projects invested by government fiscal funds (in addition to water supply and highway), as well as affordable housing and public building [10]. The institution even plays the role of construction agent if compatible. The project planning under bureau is mainly responsible for agent construction project management. Compared to traditional construction project management, these construction agents are non-profit government agencies or institutions, which can represent commissioning to better realize the controlling of investment, quality and duration. While the agent attaches to the political system, it's easy to communicate and coordinate the operation in the approval process of various sectors. At the same time, this mode achieves the separation of construction agent and use of the unit, with effectively guaranteeing the rational use and distribution of project investment.

However, there are still some drawbacks. The non-profit mode, easily driven by political interests, leads to the neglect of quality engineering in the management process and one-sided pursuit of duration. In addition, the agents directly assigned by the investor or management department will form a market monopoly, easily leading to rent-seeking power.

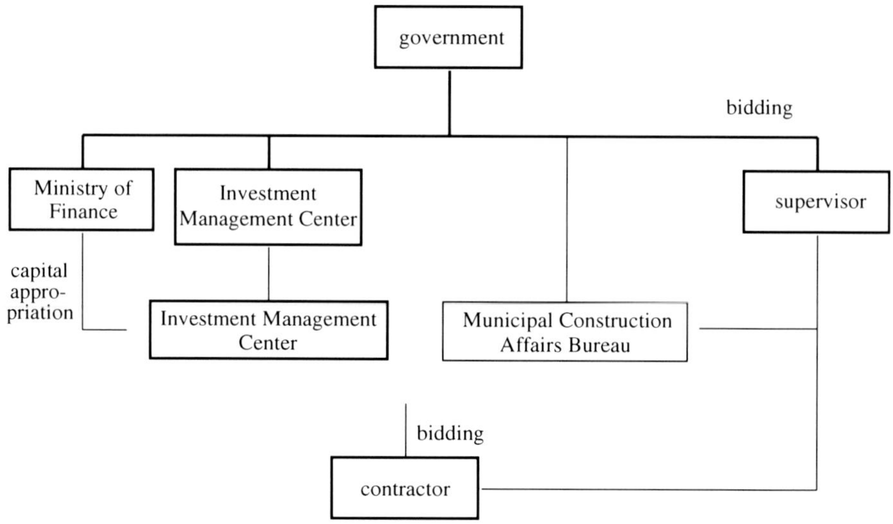

Fig. 116.2 Stakeholders' relationship in institutional agent construction mode

116.2.2 Company Agent Construction Mode

Company agent construction mode refers that the commissioning implements opening tendering, qualified bidders may participate in the bidding, then someone who offers the optimal bidding plan will win the bidding. The stakeholder relationship in this mode can be found in Fig. 116.3.

Shanghai adopts the company agent construction mode. The Shanghai Municipal Engineering Administration, performs the functions of project approval, market regulation, supervision and management. The superior, Shanghai Municipal Engineering Administration Department and the Shanghai Highway Administration are responsible for specific management. According to the "decision on implementing pilot agent construction of Shanghai Municipal Engineering Construction Management", project entity should choose the agent from the market, where participating engineering project management companies are no less than three. Small-scale projects with low-level technology can directly select other project management company, after the consent of the City Council. Project management companies, must possess the project management qualification certificate issued by the Construction and Management Commission or be pilot municipal construction management units.

What differs greatly between two modes is that company agent construction doesn't limit chosen range of construction agents. The agent has to build a contractual relationship with the commissioning party, and be responsible for the whole life cycle of the project to effectively guarantee the quality. Specified by bidding, the contract is more rigorous, with the best completion time and not breakthrough

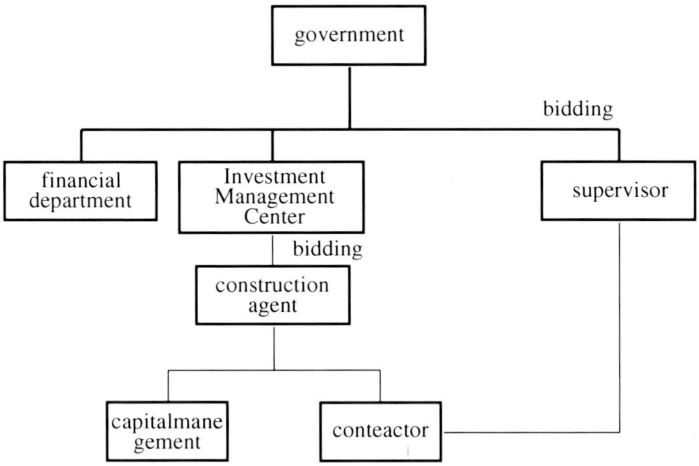

Fig. 116.3 Stakeholders of company agent construction mode

the total investment, realizing "management, control and coordination." This model fits to market competition, effectively curb corruption, eliminate low competitiveness enterprises while enhancing competitive edge of enterprises, help to optimize the allocation of social resources, which integrates social resources to achieve the project objectives by maximizing cost savings, making two sides to achieve "win-win".

116.3 Case Study

116.3.1 Case Background

Due to privacy protection, some information of this case was concealed. Project A lies in Hangzhou Peng Fu district. The project covers an area of 32,000 m^2, with a total construction area of 98,000 m^2 and nearly 700 households. Community mainly consists of slab-type apartment building and multi-layer building, belonging to the affordable housing project. The project provides water, electricity and gas pipeline supply, as well as the elevator service. The community is equipped with leisure fitness sites, elderly activity center, library, reading room, supermarkets. Around the community, there are hospital, kindergarten, elementary school, middle school, service station and shopping malls, etc. The company B, mainly develops the real estate projects or serves as construction agent with the owners of government, project partners or capital market in the form of contracts. The company B insists of "Agent Construction of Land Resources", "Agent Construction of Capital" and "Agent Construction of security housing".

116.3.2 Advantages Discussions

This project has realized a "win-win" for the owners and agents. Company B has various benefits. For example, by agent construction, company B promotes their own brand and generates brand influence on the larger scale. By statistical data, among 150,000 Membership, the 30–40 % will again purchase or recommend friends and relatives to buy the house of Company B. Meanwhile construction for government leads to combined effect, bringing more projects, which in reverse forms the combined effect, such as lower cost of material, scale dividend, and then further forms their own competitiveness. Just in 2014, eight government agent construction projects with a total construction area of 1.58 million square meters was signed with company B, with an increasing rate over 10 %. Government housing projects bring about stable demand, low-risk, less affected by the price fluctuation of the housing market. It is possible to guarantee the overall operation, improve capital efficiency and reduce asset-liability ratio.

The owner C also enjoys the benefits of company agent construction mode. For example, the owner C can effectively compensate for the lack of institutional agent construction. In addition, open bidding through market competition can effectively reduce the rent-seeking. As well, company B signed a contract with the owners on behalf of a legal person, which effectively reduces the risk of the owner. Company B can provide the exactly satisfactory products required by the owner through standardized operation way, fine design and product management, production processes. Such characteristics of the company B can effectively guarantee production quality but not break the total cost.

116.4 Conclusion

University talent apartments are critical for attracting outstanding scholars and settling down the staff personnel. However, the universities usually lack required experiences and abilities to build such apartments by themselves. Agent construction system partially mitigates such problem. This paper summarizes the common agent construction modes in China and further presents the stakeholder relationship in corresponding mode. Based on the summarized modes, this study further analyze the advantages of agent construction in building talent apartments through a case study.

It was found that company agent construction forms the scale effect and chain effect to achieve cost savings through technical innovation, design optimization and efficient management. Under the survival pressure, competition among enterprises effectively promotes the development of agent construction. Future studies can be conducted to identify effective measures to reduce the risk of agent construction in building university talent apartments in China.

Acknowledgements The work described in this paper was supported by the National Natural Science Foundation of China (71373231).

References

1. Shen LY, Peng Y, Zhang XL, Wu YZ (2012) An alternative model for evaluating sustainable urbanization. Cities 29(1):32–39
2. Wu YZ, Peng Y, Zhang XL, Skitmore M, Song Y (2012) Development priority zoning (DPZ)-led scenario simulation for regional land use change: the case of Suichang County. China, Habitat Int 36(2):268–277
3. Hsieh SC (2014) Analyzing urbanization data using rural–urban interaction model and logistic growth model. Comput Environ Urban Syst 45:89–100
4. Al-Harbi KMAS (1998) Sharing fractions in cost-plus-incentive fee contracts. Int J Project Manage 16(2):73–80
5. Chapman C, Ward S (2003) Project risk management: processes techniques and insights. Wiley, Chichester, UK
6. Chang CY (2013) Principal-agent model of risk allocation in construction contracts and its critique. J Constr Eng Manage 140(1):04013032
7. Murray-Webster R (2010) Management of risk: guidance for practitioners. Stationery Office, London, UK
8. Xiang Q, Du L (2005) Agent construction risks and preventive measures. Build Econ 4:18–21 (in Chinese)
9. Su SG, Peng H (2009) Agent system of risk control and prevention on road investment project. Highway Transp Res (Appl Technol Ed) S1:34–37 (in Chinese)
10. Shenzhen Construction Bureau (2002) The reform of public investment project management. Constr Econ 2:5–8 (in Chinese)